Coulson & Richardson's

CHEMICAL ENGINEERING

VOLUME 3

THIRD EDITION

Chemical & Biochemical Reactors & Process Control

Related titles in the Chemical Engineering Series by

J M COULSON & J F RICHARDSON

Chemical Engineering, Volume 1, Fourth edition
Fluid Flow, Heat Transfer and Mass Transfer
(with J R Backhurst and J H Harker)

Chemical Engineering, Volume 3, Third edition
Chemical and Biochemical Reaction Engineering, and Control
(edited by J F Richardson and D G Peacock)

Chemical Engineering, combined Volume 4/5
Solutions to the Problems in Volumes 1, 2 & 3
(J R Backhurst and J H Harker)

Chemical Engineering, Volume 6, Second edition
Chemical Engineering Design
(R K Sinnott)

Coulson & Richardson's

CHEMICAL ENGINEERING

VOLUME 3

THIRD EDITION

Chemical & Biochemical Reactors & Process Control

EDITORS OF VOLUME THREE

J. F. RICHARDSON

Department of Chemical Engineering
University of Wales Swansea

and

D. G. PEACOCK

The School of Pharmacy, London

Butterworth-Heinemann
Linacre House, Jordan Hill, Oxford OX2 8DP
225 Wildwood Avenue, Woburn, MA 01801-2041
A division of Reed Educational and Professional Publishing Ltd

 A member of the Reed Elsevier plc group

OXFORD AUCKLAND BOSTON
JOHANNESBURG MELBOURNE NEW DELHI

First published 1971
Reprinted 1975
Second edition 1979
Reprinted with corrections 1982, 1987, 1991
Third edition 1994

ISBN 0 08 041003 0

Printed and bound in Great Britain by Antony Rowe Ltd, Chippenham, Wiltshire

Contents

Preface to Third Edition

The publication of the Third Edition of *Chemical Engineering* Volume 3 marks the completion of the re-orientation of the basic material contained in the first three volumes of the series. Volume 1 now covers the fundamentals of Momentum, Heat and Mass Transfer, Volume 2 deals with Particle Technology and Separation Processes, and Volume 3 is devoted to Reaction Engineering (both chemical and biochemical), together with Measurement and Process Control.

Volume 3 has now lost both Non-Newtonian Technology, which appears in abridged form in Volume 1, and the Chapter on Sorption Processes, which is now more logically located with the other Separation Processes in Volume 2. The Chapter on Computation has been removed. When Volume 3 was first published in 1972 computers were, by today's standards, little more than in their infancy and students entering chemical engineering courses were not well versed in computational techniques. This situation has now completely changed and there is no longer a strong case for the inclusion of this topic in an engineering text book. With some reluctance the material on numerical solution of equations has also been dropped as it is more appropriate to a mathematics text.

In the new edition, the material on Chemical Reactor Design has been re-arranged into four chapters. The first covers General Principles (as in the earlier editions) and the second deals with Flow Characteristics and Modelling in Reactors. Chapter 3 now includes material on Catalytic Reactions (from the former Chapter 2) together with non-catalytic gas–solids reactions, and Chapter 4 covers other multiphase reactor systems. Dr J. C. Lee has contributed the material in Chapters 1, 2 and 4 and that on non-catalytic reactions in Chapter 3, and Professor W. J. Thomas has covered catalytic reactions in that Chapter.

Chapter 5, on Biochemical Engineering, has been completely rewritten in two sections by Dr R. L. Lovitt and Dr M. G. Jones with guidance from the previous author, Professor B. Atkinson. The earlier part deals with the nature of reaction processes controlled by micro-organisms and enzymes and is prefaced by background material on the relevant microbiology and biochemistry. In the latter part, the process engineering principles of biochemical reactors are discussed, and emphasis is given to those features which differentiate them from the chemical reactors described previously.

The concluding two chapters by Dr A. P. Wardle deal, respectively, with Measurement, and Process Control. The former is a completely new chapter describing the various in-line techniques for measurement of the process variables which constitute the essential inputs to the control system of the plant. The last chapter gives an updated treatment of the principles and applications of process control and concludes with a discussion of computer control of process plant.

January 1994

J F RICHARDSON
Department of Chemical Engineering
University of Wales Swansea
Swansea SA2 8PP
UK

D G PEACOCK
School of Pharmacy
London WC1N 1AX
UK

Preface to Second Edition

Apart from general updating and correction, the main alterations in the second edition of Volume 3 are additions to Chapter 1 on Reactor Design and the inclusion of a Table of Error Functions in the Appendix.

In Chapter 1 two new sections have been added. In the first of these is a discussion of non-ideal flow conditions in reactors and their effect on residence time distribution and reactor performance. In the second section an important class of chemical reactions—that in which a solid and a gas react non-catalytically—is treated. Together, these two additions to the chapter considerably increase the value of the book in this area.

All quantities are expressed in SI units, as in the second impression, and references to earlier volumes of the series take account of the modifications which have recently been made in the presentation of material in the third editions of these volumes.

Preface to the First Edition

Chemical engineering, as we know it today, developed as a major engineering discipline in the United Kingdom in the interwar years and has grown rapidly since that time. The unique contribution of the subject to the industrial scale development of processes in the chemical and allied industries was initially attributable to the improved understanding it gave to the transport processes—fluid flow, heat transfer and mass transfer—and to the development of design principles for the unit operations, nearly all of which are concerned with the physical separation of complex mixtures, both homogeneous and heterogeneous, into their components. In this context the chemical engineer was concerned much more closely with the separation and purification of the products from a chemical reactor than with the design of the reactor itself.

The situation is now completely changed. With a fair degree of success achieved in the physical separation processes, interest has moved very much towards the design of the reactor, and here too the processes of fluid flow, heat transfer and mass transfer can be just as important. Furthermore, many difficult separation problems can be obviated by correct choice of conditions in the reactor. Chemical manufacture has become more demanding with a high proportion of the economic rewards to be obtained in the production of sophisticated chemicals, pharmaceuticals, antibiotics and polymers, to name a few, which only a few years earlier were unknown even in the laboratory. Profit margins have narrowed too, giving a far greater economic incentive to obtain the highest possible yield from raw materials. Reactor design has therefore become a vital ingredient of the work of the chemical engineer.

Volumes 1 and 2, though no less relevant now, reflected the main areas of interest of the chemical engineer in the early 1950s. In Volume 3 the coverage of chemical engineering is brought up to date with an emphasis on the design of systems in which chemical and even biochemical reactions occur. It includes chapters on adsorption, on the general principles of the design of reactors, on the design and operation of reactors employing heterogeneous catalysts, and on the special features of systems exploiting biochemical and microbiological processes. Many of the materials which are processed in chemical and bio-chemical reactors are complex in physical structure and the flow properties of non-Newtonian materials are therefore considered worthy of special treatment. With the widespread use of computers, many of the design problems which are too complex to solve analytically or graphically are now capable of numerical solution, and their application to chemical engineering problems forms the subject of a chapter. Parallel with the growth in complexity of chemical plants has developed the need for much closer control of their operation, and a chapter on process control is therefore included.

Each chapter of Volume 3 is the work of a specialist in the particular field, and the authors are present or past members of the staff of the Chemical Engineering Department of the University College of Swansea. W. J. Thomas is now at the Bath University of Technology and J. M. Smith is at the Technische Hogeschool. Delft.

J. M. C.
J. F. R.
D. G. P.

Acknowledgements

The authors and publishers acknowledge with thanks the assistance given by the following companies and individuals in providing illustrations and data for this volume and giving their permission for reproduction. Everyone was most helpful and some firms went to considerable trouble to provide exactly what was required. We are extremely grateful to them all.

Butterworth-Heinemann for Fig. 3.16
Cambridge University Press for Fig. 3.22.
John Wiley for Fig. 3.24.
McGraw-Hill for Table 3.1.
Endress and Hauser Ltd, Manchester, U.K. for Fig. 6.2*b*, Fig. 6.19*b*.
Foxboro, Great Britain Ltd., Crawley, U.K. for Figs. 6.19*a*, 6.20, 6.52, 6.77.
MTS Systems Corpn, North Carolina, U.S.A. for Fig. 6.33*b*.
Schlumberger Industries, Farnborough, U.K. for Fig. 6.36.
Precision Scientific Inc., Chicago, Illinois, U.S.A. for Fig. 6.39.
Mettler-Toledo Ltd, Leicester, U.K. for Fig. 6.40.
Servomex plc., Crowborough, U.K. for Figs 6.42, 6.49, 6.57, 6.58.
Anacon Corpn, Thame, U.K. for Fig. 6.45.
Nametre Co., Metuchen, New Jersey, U.S.A. for Fig. 6.41.
J. Winter, Gwent Tertiary Campus, Newport, U.K. for Example 6.3.
Fisher Controls Ltd, Rochester, U.K. for Fig. 7.103.
Kent Process Control Ltd, Luton, U.K. for Fig. 7.108.
Samson AG, Frankfurt, Germany for Figs 7.119, 7.121.

List of Contributors

Chapter 1—Reactor Design—General Principles
J. C. LEE (*University of Wales Swansea*)

Chapter 2—Flow Characteristics of Reactors—Flow Modelling
J. C. LEE (*University of Wales Swansea*)

Chapter 3—Gas–Solid Reactions & Reactors
J. C. LEE (*University of Wales Swansea*)
& W. J. THOMAS (*Bath University of Technology*)

Chapter 4—Gas–Liquid & Gas–Liquid–Solid Reactors
J. C. LEE (*University of Wales Swansea*)

Chapter 5—Biochemical Reaction Engineering
R. LOVITT (*University of Wales Swansea*)
& M. JONES (*University of Wales Swansea*)

Chapter 6—Sensors for Measurement & Control
A. P. WARDLE (*University of Wales Swansea*)

Chapter 7—Process Control
A. P. WARDLE (*University of Wales Swansea*)

CHAPTER 1

Reactor Design—General Principles

1.1. BASIC OBJECTIVES IN DESIGN OF A REACTOR

In chemical engineering physical operations such as fluid flow, heat transfer, mass transfer and separation processes play a very large part; these have been discussed in Volumes 1 and 2. In any manufacturing process where there is a chemical change taking place, however, the chemical reactor is at the heart of the plant.

In size and appearance it may often seem to be one of the least impressive items of equipment, but its demands and performance are usually the most important factors in the design of the whole plant.

When a new chemical process is being developed, at least some indication of the performance of the reactor is needed before any economic assessment of the project as a whole can be made. As the project develops and its economic viability becomes established, so further work is carried out on the various chemical engineering operations involved. Thus, when the stage of actually designing the reactor in detail has been reached, the project as a whole will already have acquired a fairly definite form. Among the major decisions which will have been taken is the rate of production of the desired product. This will have been determined from a market forecast of the demand for the product in relation to its estimated selling price. The reactants to be used to make the product and their chemical purity will have been established. The basic chemistry of the process will almost certainly have been investigated, and information about the composition of the products from the reaction, including any byproducts, should be available.

On the other hand, a reactor may have to be designed as part of a modification to an existing process. Because the new reactor has then to tie in with existing units, its duties can be even more clearly specified than when the whole process is new. Naturally, in practice, detailed knowledge about the performance of the existing reactor would be incorporated in the design of the new one.

As a general statement of the basic objectives in designing a reactor, we can say therefore that the aim is to produce a *specified product* at a *given rate* from *known reactants*. In proceeding further however a number of important decisions must be made and there may be scope for considerable ingenuity in order to achieve the best result. At the outset the two most important questions to be settled are:

(a) The type of reactor to be used and its method of operation. Will the reaction be carried out as a batch process, a continuous flow process, or possibly as a hybrid of the two? Will the reactor operate isothermally, adiabatically or in some intermediate manner?

(b) The physical condition of the reactants at the inlet to the reactor. Thus, the basic processing conditions in terms of pressure, temperature and compositions of the reactants on entry to the reactor have to be decided, if not already specified as part of the original process design.

Subsequently, the aim is to reach logical conclusions concerning the following principal features of the reactor:

(a) The overall size of the reactor, its general configuration and the more important dimensions of any internal structures.
(b) The exact composition and physical condition of the products emerging from the reactor. The composition of the products must of course lie within any limits set in the original specification of the process.
(c) The temperatures prevailing within the reactor and any provision which must be made for heat transfer.
(d) The operating pressure within the reactor and any pressure drop associated with the flow of the reaction mixture.

1.1.1. Byproducts and their Economic Importance

Before taking up the design of reactors in detail, let us first consider the very important question of whether any byproducts are formed in the reaction. Obviously, consumption of reactants to give unwanted, and perhaps unsaleable, byproducts is wasteful and will directly affect the operating costs of the process. Apart from this, however, the nature of any byproducts formed and their amounts must be known so that plant for separating and purifying the products from the reaction may be correctly designed. The appearance of unforeseen byproducts on start-up of a full-scale plant can be utterly disastrous. Economically, although the cost of the reactor may sometimes not appear to be great compared with that of the associated separation equipment such as distillation columns, etc., it is the composition of the mixture of products issuing from the reactor which determines the capital and operating costs of the separation processes.

For example, in producing ethylene[1] together with several other valuable hydrocarbons like butadiene from the thermal cracking of naphtha, the design of the whole complex plant is determined by the composition of the mixture formed in a tubular reactor in which the conditions are very carefully controlled. As we shall see later, the design of a reactor itself can affect the amount of byproducts formed and therefore the size of the separation equipment required. The design of a reactor and its mode of operation can thus have profound repercussions on the remainder of the plant.

1.1.2. Preliminary Appraisal of a Reactor Project

In the following pages we shall see that reactor design involves all the basic principles of chemical engineering with the addition of chemical kinetics. Mass transfer, heat transfer and fluid flow are all concerned and complications arise when, as so often is the case, interaction occurs between these transfer processes and the reaction itself. In designing a reactor it is essential to weigh up all the various factors involved and, by an exercise of judgement, to place them in their proper order of importance. Often the basic design of the reactor is determined by what is seen to be the most troublesome step. It may be the chemical kinetics; it may be mass transfer between phases; it may be heat transfer; or it may even be the need to ensure safe operation. For example, in oxidising naphthalene or o-xylene to phthalic anhydride with air, the reactor must be designed so that ignitions, which are not infrequent, may be rendered harmless. The theory of reactor design is being extended rapidly and more precise methods for detailed design and optimisation are being evolved. However, if the final design is to be successful, *the major decisions taken at the outset must be correct*. Initially, a careful appraisal of the basic role and functioning of the reactor is required and at this stage the application of a little chemical engineering common sense may be invaluable.

1.2. CLASSIFICATION OF REACTORS AND CHOICE OF REACTOR TYPE

1.2.1. Homogeneous and Heterogeneous Reactors

Chemical reactors may be divided into two main categories, homogeneous and heterogeneous. In homogeneous reactors only one phase, usually a gas or a liquid, is present. If more than one reactant is involved, provision must of course be made for mixing them together to form a homogenous whole. Often, mixing the reactants is the way of starting off the reaction, although sometimes the reactants are mixed and then brought to the required temperature.

In heterogeneous reactors two, or possibly three, phases are present, common examples being gas–liquid, gas–solid, liquid–solid and liquid–liquid systems. In cases where one of the phases is a solid, it is quite often present as a catalyst; gas–solid catalytic reactors particularly form an important class of heterogeneous chemical reaction systems. It is worth noting that, in a heterogeneous reactor, the chemical reaction itself may be truly heterogeneous, but this is not necessarily so. In a gas–solid catalytic reactor, the reaction takes place on the surface of the solid and is thus heterogeneous. However, bubbling a gas through a liquid may serve just to dissolve the gas in the liquid where it then reacts homogeneously; the reaction is thus homogeneous but the reactor is heterogeneous in that it is required to effect contact between two phases—gas and liquid. Generally, heterogeneous reactors exhibit a greater variety of configuration and contacting pattern than homogeneous reactors. Initially, therefore, we shall be concerned mainly with the simpler homogeneous reactors, although parts of the treatment that follows can be extended to heterogeneous reactors with little modification.

1.2.2. Batch Reactors and Continuous Reactors

Another kind of classification which cuts across the homogeneous–heterogeneous division is the mode of operation—batchwise or continuous. Batchwise operation, shown in Fig. 1.1*a*, is familiar to anybody who has carried out small-scale preparative reactions in the laboratory. There are many situations, however, especially in large-scale operation, where considerable advantages accrue by carrying out a chemical reaction continuously in a flow reactor.

Figure 1.1 illustrates the two basic types of flow reactor which may be employed. In the *tubular-flow reactor* (*b*) the aim is to pass the reactants along a tube so that there is as little intermingling as possible between the reactants entering the tube and the products leaving at the far end. In the *continuous stirred-tank reactor* (C.S.T.R.) (*c*) an agitator is deliberately introduced to disperse the reactants thoroughly into the reaction mixture immediately they enter the tank. The product stream is drawn off continuously and, in the ideal state of perfect mixing, will have the same composition as the contents of the tank. In some ways, using a C.S.T.R., or *backmix reactor* as it is sometimes called, seems a curious method of conducting a reaction because as soon as the reactants enter the tank they are mixed and a portion leaves in the product stream flowing out. To reduce this effect, it is often advantageous to employ a number of stirred tanks connected in series as shown in Fig. 1.1*d*.

The stirred-tank reactor is by its nature well suited to liquid-phase reactions. The tubular reactor, although sometimes used for liquid-phase reactions, is the natural choice for gas-phase reactions, even on a small scale. Usually the temperature or catalyst is chosen so that the rate of reaction is high, in which case a comparatively small tubular reactor is sufficient to handle a high volumetric flowrate of gas. A few gas-phase reactions, examples being partial combustion and certain chlorinations, are carried out in reactors which resemble the stirred-tank reactor; rapid mixing is usually brought about by arranging for the gases to enter with a vigorous swirling motion instead of by mechanical means.

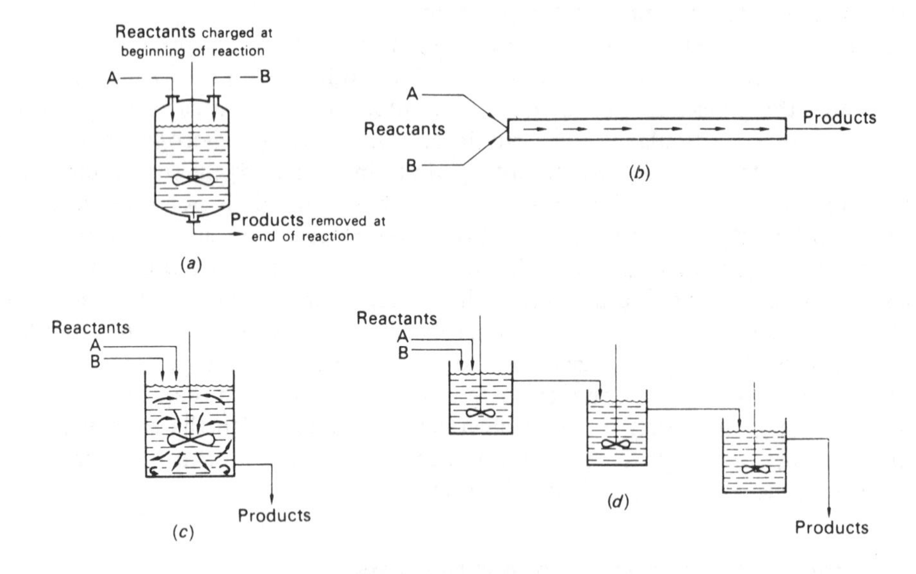

FIG. 1.1. Basic types of chemical reactors

(*a*) Batch reactor
(*b*) Tubular-flow reactor
(*c*) Continuous stirred-tank reactor (C.S.T.R.) or "backmix reactor"
(*d*) C.S.T.R.s in series as frequently used

1.2.3. Variations in Contacting Pattern — Semi-batch Operation

Another question which should be asked in assessing the most suitable type of reactor is whether there is any advantage to be gained by varying the contacting pattern. Figure 1.2*a* illustrates the *semi-batch* mode of operation. The reaction vessel here is essentially a batch reactor, and at the start of a batch it is charged with one of the reactants **A**. However, the second reactant **B** is not all added at once, but continuously over the period of the reaction. This is the natural and obvious way to carry out many reactions. For example, if a liquid has to be treated with a gas, perhaps in a chlorination or hydrogenation reaction, the gas is normally far too voluminous to be charged all at once to the reactor; instead it is fed continuously at the rate at which it is used up in the reaction. Another case is where the reaction is too violent if both reactants are mixed suddenly together. Organic nitration, for example, can be conveniently controlled by regulating the rate of addition of the nitrating acid. The maximum rate of addition of the second reactant in such a case will be determined by the rate of heat transfer.

A characteristic of semi-batch operation is that the concentration C_B of the reactant added slowly, **B** in Fig. 1.2, is low throughout the course of the reaction. This may be an advantage if more than one reaction is possible, and if the desired reaction is favoured by a low value of C_B. Thus, the semi-batch method may be chosen for a further reason, that of improving the yield of the desired product, as shown in Section 1.10.4.

Summarising, a semi-batch reactor may be chosen:

(a) to react a gas with a liquid,
(b) to control a highly exothermic reaction, and
(c) to improve product yield in suitable circumstances.

In semi-batch operation, when the initial charge of **A** has been consumed, the flow of **B** is interrupted, the products discharged, and the cycle begun again with a fresh charge of **A**. If required, however, the advantages of semi-batch operation may be retained but the reactor system designed for continuous flow of both reactants. In

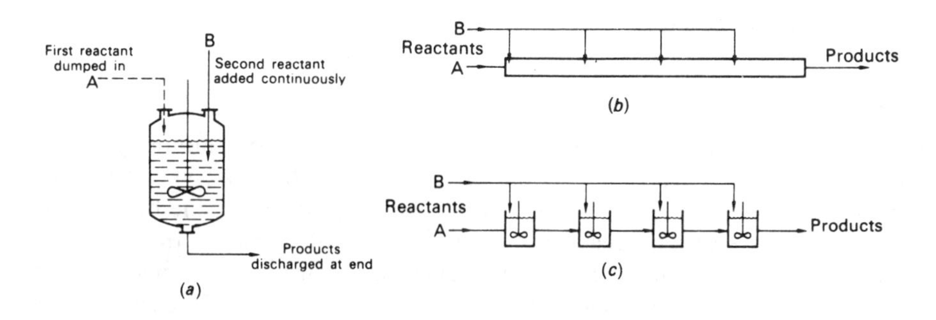

FIG. 1.2. Examples of possible variations in reactant contacting pattern
(*a*) Semi-batch operation
(*b*) Tubular reactor with divided feed
(*c*) Stirred-tank reactors with divided feed
(in each case the concentration of **B**, C_B, is low throughout)

the tubular flow version (Fig. 1.2*b*) and the stirred-tank version (Fig. 1.2*c*), the feed of **B** is divided between several points. These are known as *cross-flow* reactors. In both cases C_B is low throughout.

1.2.4. Influence of Heat of Reaction on Reactor Type

Associated with every chemical change there is a heat of reaction, and only in a few cases is this so small that it can be neglected. The magnitude of the heat of reaction often has a major influence on the design of a reactor. With a strongly exothermic reaction, for example, a substantial rise in temperature of the reaction mixture will take place unless provision is made for heat to be transferred as the reaction proceeds. It is important to try to appreciate clearly the relation between the enthalpy of reaction, the heat transferred, and the temperature change of the reaction mixture; quantitatively this is expressed by an enthalpy balance (Section 1.5). If the temperature of the reaction mixture is to remain constant (isothermal operation), the heat equivalent to the heat of reaction at the operating temperature must be transferred to or from the reactor. If no heat is transferred (adiabatic operation), the temperature of the reaction mixture will rise or fall as the reaction proceeds. In practice, it may be most convenient to adopt a policy intermediate between these two extremes; in the case of a strongly exothermic reaction, some heat-transfer from the reactor may be necessary in order to keep the reaction under control, but a moderate temperature rise may be quite acceptable, especially if strictly isothermal operation would involve an elaborate and costly control scheme.

In setting out to design a reactor, therefore, two very important questions to ask are:

(a) What is the heat of reaction?
(b) What is the acceptable range over which the temperature of the reaction mixture may be permitted to vary?

The answers to these questions may well dominate the whole design. Usually, the temperature range can only be roughly specified; often the lower temperature limit is determined by the slowing down of the reaction, and the upper temperature limit by the onset of undesirable side reactions.

Adiabatic Reactors

If it is feasible, adiabatic operation is to be preferred for simplicity of design. Figure 1.3 shows the reactor section of a plant for the catalytic reforming of petroleum naphtha; this is an important process for improving the octane number of gasoline. The reforming reactions are mostly endothermic so that in adiabatic operation the temperature would fall during the course of the reaction. If the reactor were made as one single unit, this temperature fall would be too large, i.e. either the temperature at the inlet would be too high and undesired reactions would occur, or the reaction would be incomplete because the temperature near the outlet would be too low. The problem is conveniently solved by dividing the reactor into three sections. Heat is supplied externally between the sections, and the intermediate temperatures are raised so that each section of the reactor will operate adiabatically.

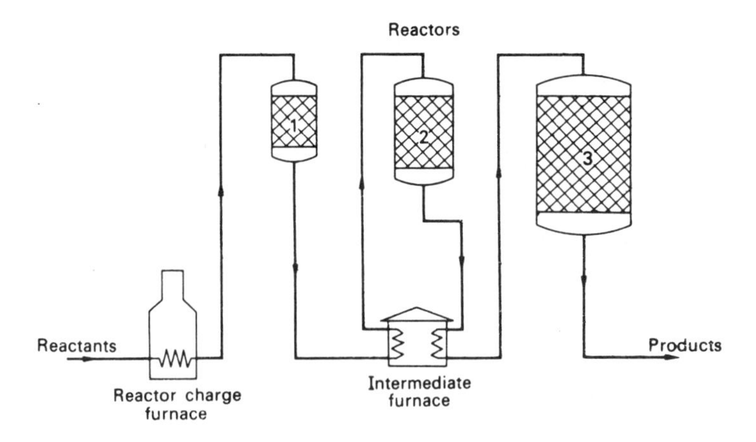

FIG. 1.3. Reactor system of a petroleum naphtha catalytic reforming plant. (The reactor is divided into three units each of which operates *adiabatically*, the heat required being supplied at intermediate stages via an external furnace)

Dividing the reactor into sections also has the advantage that the intermediate temperature can be adjusted independently of the inlet temperature; thus an optimum temperature distribution can be achieved. In this example we can see that the furnaces where heat is transferred and the catalytic reactors are quite separate units, each designed specifically for the one function. This separation of function generally provides ease of control, flexibility of operation and often leads to a good overall engineering design.

Reactors with Heat Transfer

If the reactor does not operate adiabatically, then its design must include provision for heat transfer. Figure 1.4 shows some of the ways in which the contents of a batch reactor may be heated or cooled. In *a* and *b* the jacket and the coils form part of the reactor itself, whereas in *c* an external heat exchanger is used with a recirculating pump. If one of the constituents of the reaction mixture, possibly a

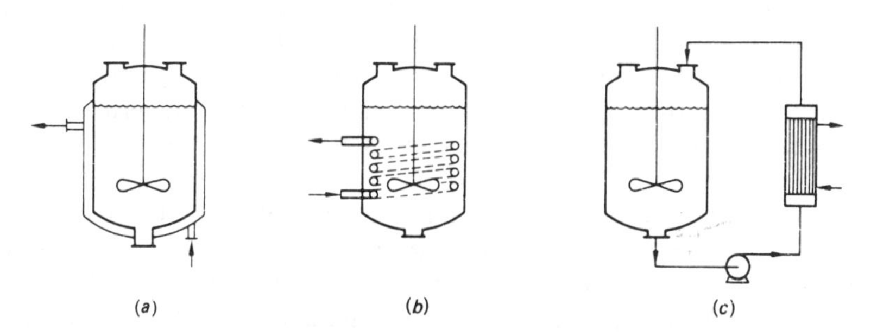

FIG. 1.4. Batch reactors showing different methods of heating or cooling

(*a*) Jacket
(*b*) Internal coils
(*c*) External heat exchangers

solvent, is volatile at the operating temperature, the external heat exchanger may be a reflux condenser, just as in the laboratory.

Figure 1.5 shows ways of designing tubular reactors to include heat transfer. If the amount of heat to be transferred is large, then the ratio of heat transfer surface to reactor volume will be large, and the reactor will look very much like a heat exchanger as in Fig. 1.5*b*. If the reaction has to be carried out at a high temperature and is strongly endothermic (for example, the production of ethylene by the thermal cracking of naphtha or ethane—see also Section 1.7.1, Example 1.4), the reactor will be directly fired by the combustion of oil or gas and will look like a pipe furnace (Fig. 1.5*c*).

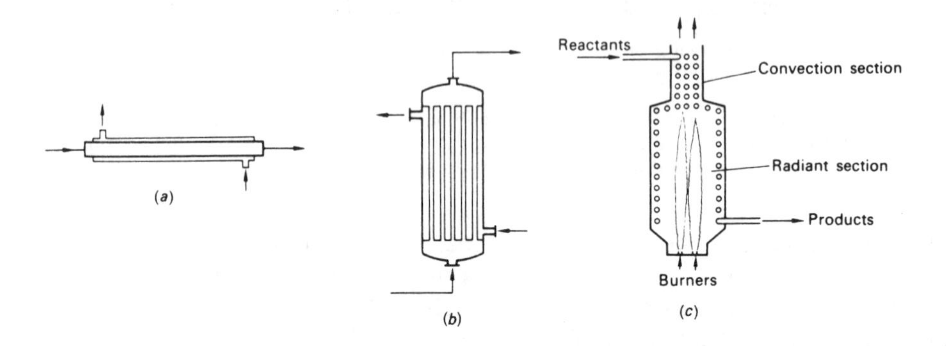

FIG. 1.5. Methods of heat transfer to tubular reactors

(*a*) Jacketed pipe
(*b*) Multitube reactor (tubes in parallel)
(*c*) Pipe furnace (pipes mainly in series although some pipe runs may be in parallel)

Autothermal Reactor Operation

If a reaction requires a relatively high temperature before it will proceed at a reasonable rate, the products of the reaction will leave the reactor at a high temperature and, in the interests of economy, heat will normally be recovered from them. Since heat must be supplied to the reactants to raise them to the reaction temperature, a common arrangement is to use the hot products to heat the incoming feed as shown in Fig. 1.6*a*. If the reaction is sufficiently exothermic, enough heat will be produced in the reaction to overcome any losses in the system and to provide the necessary temperature difference in the heat exchanger. The term *autothermal* is used to describe such a system which is completely self-supporting in its thermal energy requirements.

The essential feature of an autothermal reactor system is the feedback of reaction heat to raise the temperature and hence the reaction rate of the incoming reactant stream. Figure 1.6 shows a number of ways in which this can occur. With a tubular reactor the feedback may be achieved by external heat exchange, as in the reactor shown in Fig. 1.6*a*, or by internal heat exchange as in Fig. 1.6*b*. Both of these are catalytic reactors; their thermal characteristics are discussed in more detail in Chapter 3, Section 3.6.2. Being catalytic the reaction can only take place in that part of the reactor which holds the catalyst, so the temperature profile has the form

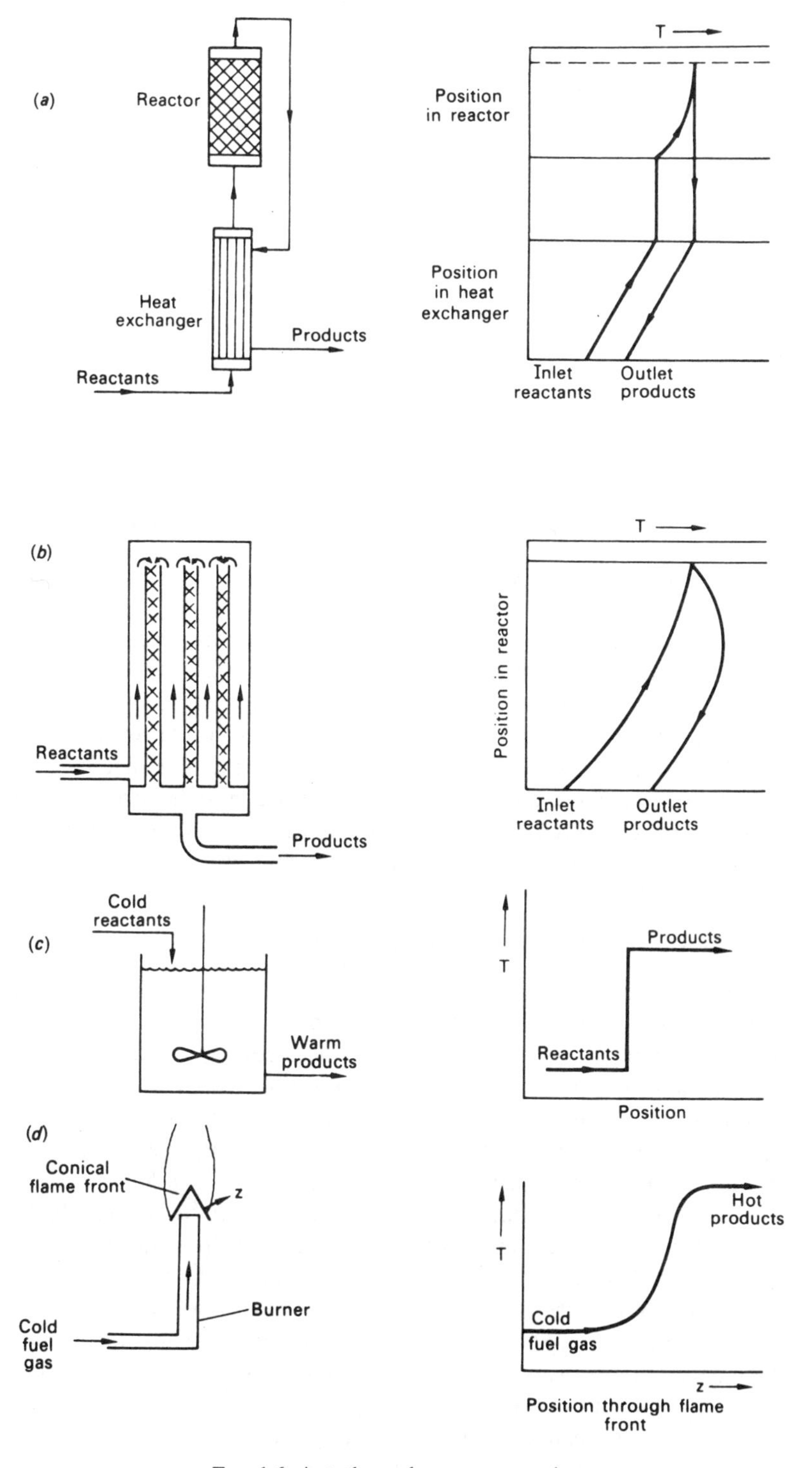

FIG. 1.6. Autothermal reactor operation

indicated alongside the reactor. Figure 1.6*c* shows a continuous stirred-tank reactor in which the entering cold feed immediately mixes with a large volume of hot products and rapid reaction occurs. The combustion chamber of a liquid fuelled rocket motor is a reactor of this type, the products being hot gases which are ejected at high speed. Figure 1.6*d* shows another type of combustion process in which a laminar flame of conical shape is stabilised at the orifice of a simple gas burner. In this case the feedback of combustion heat occurs by transfer upstream in a direction opposite to the flow of the cold reaction mixture.

Another feature of the autothermal system is that, although ultimately it is self-supporting, an external source of heat is required to start it up. The reaction has to be ignited by raising some of the reactants to a temperature sufficiently high for the reaction to commence. Moreover, a stable operating state may be obtainable only over a limited range of operating conditions. This question of stability is discussed further in connection with autothermal operation of a continuous stirred-tank reactor (Section 1.8.4).

1.3. CHOICE OF PROCESS CONDITIONS

The choice of temperature, pressure, reactant feed rates and compositions at the inlet tc the reactor is closely bound up with the basic design of the process as a whole. In arriving at specifications for these quantities, the engineer is guided by knowledge available on the fundamental physical chemistry of the reaction. Usually he will also have results of laboratory experiments giving the fraction of the reactants converted and the products formed under various conditions. Sometimes he may have the benefit of highly detailed information on the performance of the process from a pilot plant, or even a large-scale plant. Although such direct experience of reactor conditions may be invaluable in particular cases, we shall here be concerned primarily with design methods based upon fundamental physico-chemical principles.

1.3.1. Chemical Equilibria and Chemical Kinetics

The two basic principles involved in choosing conditions for carrying out a reaction are thermodynamics, under the heading of chemical equilibrium, and chemical kinetics. Strictly speaking, every chemical reaction is reversible and, no matter how fast a reaction takes place, it cannot proceed beyond the point of chemical equilibrium in the reaction mixture at the particular temperature and pressure concerned. Thus, under any prescribed conditions, the principle of chemical equilibrium, through the equilibrium constant, determines *how far* the reaction can possibly proceed given sufficient time for equilibrium to be reached. On the other hand, the principle of chemical kinetics determines at what *rate* the reaction will proceed towards this maximum extent. If the equilibrium constant is very large, then for all practical purposes the reaction may be said to be *irreversible*. However, even when a reaction is claimed to be *irreversible* an engineer would be very unwise not to calculate the equilibrium constant and check the position of equilibrium, especially if high conversions are required.

In deciding process conditions, the two principles of thermodynamic equilibrium and kinetics need to be considered together; indeed, any complete rate equation for a reversible reaction will include the equilibrium constant or its equivalent (see Section 1.4.4) but complete rate equations are not always available to the engineer. The first question to ask is: in what temperature range will the chemical reaction take place at a reasonable rate (in the presence, of course, of any catalyst which may have been developed for the reaction)? The next step is to calculate values of the equilibrium constant in this temperature range using the principles of chemical thermodynamics. (Such methods are beyond the scope of this chapter and any reader unfamiliar with this subject should consult a standard textbook[2].) The equilibrium constant K_p of a reaction depends only on the temperature as indicated by the relation:

$$\frac{d\ln K_p}{dT} = \frac{\Delta H}{\mathbf{R}T^2} \tag{1.1}$$

where $-\Delta H$ is the heat of reaction. The equilibrium constant is then used to determine the limit to which the reaction can proceed under the conditions of temperature, pressure and reactant compositions which appear to be most suitable.

1.3.2. Calculation of Equilibrium Conversion

Whereas the equilibrium constant itself depends on the temperature only, the conversion at equilibrium depends on the composition of the original reaction mixture and, in general, on the pressure. If the equilibrium constant is very high, the reaction may be treated as being irreversible. If the equilibrium constant is low, however, it may be possible to obtain acceptable conversions only by using high or low pressures. Two important examples are the reactions:

$$C_2H_4 + H_2O \rightleftharpoons C_2H_5OH$$

$$N_2 + 3H_2 \rightleftharpoons 2NH_3$$

both of which involve a decrease in the number of moles as the reaction proceeds, and therefore high pressures are used to obtain satisfactory equilibrium conversions.

Thus, in those cases in which reversibility of the reaction imposes a serious limitation, the equilibrium conversion must be calculated in order that the most advantageous conditions to be employed in the reactors may be chosen; this may be seen in detail in the following example of the styrene process. A study of the design of this process is also very instructive in showing how the basic features of the reaction, namely equilibrium, kinetics, and suppression of byproducts, have all been satisfied in quite a clever way by using steam as a diluent.

Example 1.1

A Process for the Manufacture of Styrene by the Dehydrogenation of Ethylbenzene

Let us suppose that we are setting out from first principles to investigate the dehydrogenation of ethylbenzene which is a well established process for manufacturing styrene:

$$C_6H_5 \cdot CH_2 \cdot CH_3 = C_6H_5 \cdot CH{:}CH_2 + H_2$$

There is available a catalyst which will give a suitable rate of reaction at 560°C. At this temperature the equilibrium constant for the reaction above is:

$$\frac{P_{St} \times P_H}{P_{Et}} = K_p = 100 \text{ mbar} = 10^4 \text{ N/m}^2 \qquad \text{(A)}$$

where P_{Et}, P_{St} and P_H are the partial pressures of ethylbenzene, styrene and hydrogen respectively.

Part (i)

Feed pure ethylbenzene: If a feed of pure ethylbenzene is used at 1 bar pressure, determine the fractional conversion at equilibrium.

Solution

This calculation requires not only the use of the equilibrium constant, but also a material balance over the reactor. To avoid confusion, it is as well to set out this material balance quite clearly even in this comparatively simple case.

First it is necessary to choose a basis; let this be 1 mole of ethylbenzene fed into the reactor: a fraction α_e of this will be converted at equilibrium. Then, from the above stoichiometric equation, α_e mole styrene and α_e mole hydrogen are formed, and $(1 - \alpha_e)$ mole ethylbenzene remains unconverted. Let the total pressure at the outlet of the reactor be P which we shall later set equal to 1 bar.

$C_6H_5 \cdot C_2H_5 \longrightarrow$ REACTOR $\longrightarrow$ $C_6H_5 \cdot C_2H_5$, $C_6H_5 \cdot C_2H_3$, H_2

Temperature 560°C = 833 K
Pressure P (1 bar = 1.0×10^5 N/m^2)

	IN	OUT		
		a	*b*	*c*
	mole	mole	mole fraction	partial pressure
$C_6H_5 \cdot C_2H_5$	1	$1 - \alpha_e$	$\frac{1 - \alpha_e}{1 + \alpha_e}$	$\frac{1 - \alpha_e}{1 + \alpha_e} P$
$C_6H_5 \cdot C_2H_3$	—	α_e	$\frac{\alpha_e}{1 + \alpha_e}$	$\frac{\alpha_e}{1 + \alpha_e} P$
H_2	—	α_e	$\frac{\alpha_e}{1 + \alpha_e}$	$\frac{\alpha_e}{1 + \alpha_e} P$
TOTAL		$1 + \alpha_e$		

Since for 1 mole of ethylbenzene entering, the total number of moles increases to $1 + \alpha_e$, the mole fractions of the various species in the reaction mixture at the reactor outlet are shown in column *b* above. At a total pressure P, the partial pressures are given in column *c* (assuming ideal gas behaviour). If the reaction mixture is at chemical equilibrium, these partial pressures must satisfy equation A above:

$$K_p = \frac{P_{St} \times P_H}{P_{Et}} = \frac{\dfrac{\alpha_e}{(1+\alpha_e)} P \dfrac{\alpha_e}{(1+\alpha_e)} P}{\dfrac{(1-\alpha_e)}{(1+\alpha_e)} P} = \frac{\alpha_e^2}{1 - \alpha_e^2} P$$

i.e.:

$$\frac{\alpha_e^2}{1 - \alpha_e^2} P = 1.0 \times 10^4 \text{ N/m}^2 \qquad \text{(B)}$$

Thus, when $P = 1$ bar, $\alpha_e = 0.30$; i.e. the maximum possible conversion using pure ethylbenzene at 1 bar is only 30 per cent; this is not very satisfactory (although it is possible in some processes to operate at low conversions by separating and recycling reactants). Ways of improving this figure are now sought.

Note that equation B above shows that as P decreases α_e increases; this is the quantitative expression of Le Chatelier's principle that, because the total number of moles increases in the reaction, the decomposition of ethylbenzene is favoured by a reduction in pressure. There are, however, disadvantages in operating such a process at sub-atmospheric pressures. One disadvantage is that any ingress of air through leaks might result in ignition. A better solution in this instance is to reduce the partial pressure by diluting the ethylbenzene with an inert gas, while maintaining the total pressure slightly in excess of atmospheric. The inert gas most suitable for this process is steam: one reason for this is that it can be condensed easily in contrast to a gas such as nitrogen which would introduce greater problems in separation.

Part (ii)

Feed ethylbenzene with steam: If the feed to the process consists of ethylbenzene diluted with steam in the ratio 15 moles steam : 1 mole ethylbenzene, determine the new fractional conversion at equilibrium α_e'.

Solution

Again we set out the material balance in full, the basis being 1 mole ethylbenzene into the reactor.

$C_6H_5 \cdot C_2H_5$, H_2O ⟶ REACTOR ⟶ $C_6H_5 \cdot C_2H_5$, $C_6H_5 \cdot C_2H_3$, H_2, H_2O

Temperature 560°C = 833 K
Pressure P (1 bar = 1.0×10^5 N/m^2)

	IN	OUT		
		a	*b*	*c*
	mole	mole	mole fraction	partial pressure
$C_6H_5 \cdot C_2H_5$	1	$1-\alpha_e'$	$\frac{1-\alpha_e'}{16+\alpha_e'}$	$\frac{1-\alpha_e'}{16+\alpha_e'}P$
$C_6H_5 \cdot C_2H_3$	—	α_e'	$\frac{\alpha_e'}{16+\alpha_e'}$	$\frac{\alpha_e'}{16+\alpha_e'}P$
H_2	—	α_e'	$\frac{\alpha_e'}{16+\alpha_e'}$	$\frac{\alpha_e'}{16+\alpha_e'}P$
H_2O	15	15	$\frac{15}{16+\alpha_e'}$	
TOTAL		$16+\alpha_e'$		

$$K_p = \frac{P_{St} \times P_H}{P_{Et}} = \frac{\dfrac{\alpha_e'}{(16+\alpha_e')}P \dfrac{\alpha_e'}{(16+\alpha_e')}P}{\dfrac{(1-\alpha_e')}{(16-\alpha_e')}P}$$

$$= \frac{\alpha_e'^2}{(16+\alpha_e')(1-\alpha_e')}P$$

i.e.:

$$\frac{\alpha_e'^2}{(16+\alpha_e')(1-\alpha_e')}P = 1.0 \times 10^{-4} \qquad \text{(C)}$$

Thus when $P = 1$ bar, $\alpha_e' = 0.70$; i.e. the maximum possible conversion has now been raised to 70 per cent. Inspection of equation C shows that the equilibrium conversion increases as the ratio of steam to ethylbenzene increases. However, as more steam is used, its cost increases and offsets the value of the increase in ethylbenzene conversion. The optimum steam: ethylbenzene ratio is thus determined by an economic balance.

Part (iii)

Final choice of reaction conditions in the styrene process:

Solution

The use of steam has a number of other advantages in the styrene process. The most important of these is that it acts as a source of internal heat supply so that the reactor can be operated adiabatically. The dehydrogenation reaction is strongly endothermic, the heat of reaction at 560°C being $(-\Delta H) = -125{,}000$ kJ/kmol. It is instructive to look closely at the conditions which were originally worked out for this process (Fig. 1.7). Most of the steam, 90 per cent of the total used, is heated separately from the ethylbenzene stream, and to a higher temperature (710°C) than is required at the inlet to the

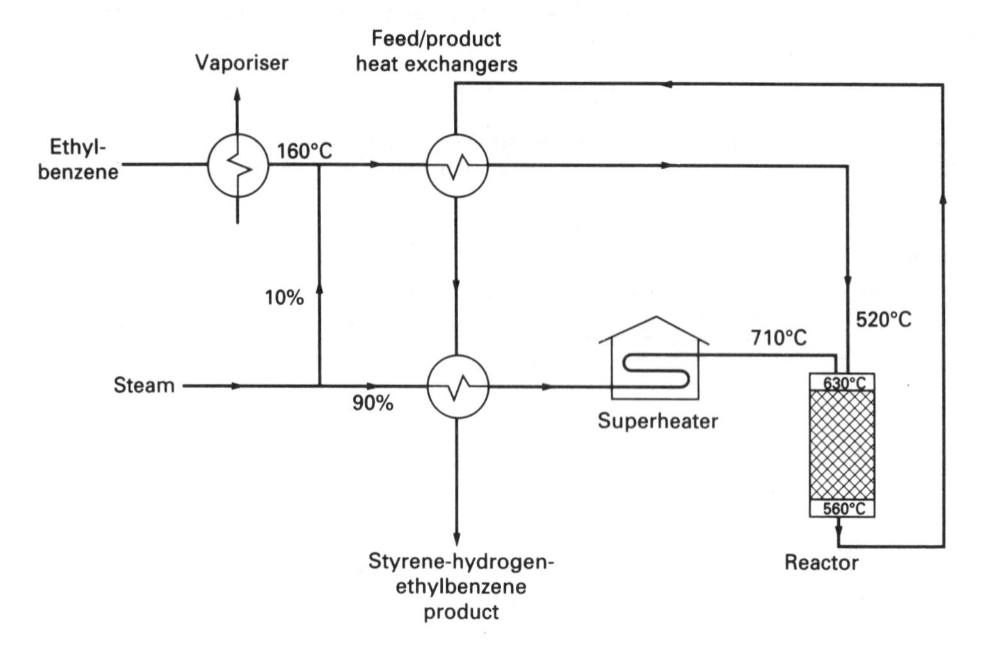

FIG. 1.7. A process for styrene from ethylbenzene using 15 moles steam : 1 mole ethylbenzene. Operating pressure 1 bar. Conversion per pass 0.40. Overall relative yield 0.90

reactor. The ethylbenzene is heated in the heat exchangers to only 520°C and is then rapidly mixed with the hotter steam to give a temperature of 630°C at the inlet to the catalyst bed. If the ethylbenzene were heated to 630°C more slowly by normal heat exchange decomposition and coking of the heat transfer surfaces would tend to occur. Moreover, the tubes of this heat exchanger would have to be made of a more expensive alloy to resist the more severe working conditions. To help avoid coking, 10 per cent of the steam used is passed through the heat exchanger with the ethylbenzene. The presence of a large proportion of steam in the reactor also prevents coke deposition on the catalyst. By examining the equilibrium constant of reactions involving carbon such as:

$$C_6H_5 \cdot CH_2 \cdot CH_3 \rightleftharpoons 8C + 5H_2$$
$$C + H_2O \rightleftharpoons CO + H_2$$

it may be shown that coke formation is not possible at high steam: ethylbenzene ratios.

The styrene process operates with a fractional conversion of ethylbenzene per pass of 0.40 compared with the equilibrium conversion of 0.70. This actual conversion of 0.40 is determined by the *rate* of the reaction over the catalyst at the temperature prevailing in the reactor. (Adiabatic operation means that the temperature falls with increasing conversion and the reaction tends to be quenched at the outlet.) The unreacted ethylbenzene is separated and recycled to the reactor. The overall yield in the process, i.e. moles of ethylbenzene transformed into styrene per mole of ethylbenzene supplied, is 0.90, the remaining 0.10 being consumed in unwanted side reactions. Notice that the conversion per pass could be increased by increasing the temperature at the inlet to the catalyst bed beyond 630°C, but the undesirable side reactions would increase, and the overall yield of the process would fall. The figure of 630°C for the inlet temperature is thus determined by an economic balance between the cost of separating unreacted ethylbenzene (which is high if the inlet temperature and conversion per pass are low), and the cost of ethylbenzene consumed in wasteful side reactions (which is high if the inlet temperature is high).

1.3.3. Ultimate Choice of Reactor Conditions

The use of steam in the styrene process above is an example of how an engineer can exercise a degree of ingenuity in reactor design. The advantages conferred by the steam may be summarised as follows:

(a) it lowers the partial pressure of the ethylbenzene without the need to operate at sub-atmospheric pressures;
(b) it provides an internal heat source for the endothermic heat of reaction, making adiabatic operation possible; and
(c) it prevents coke formation on the catalyst and coking problems in the ethylbenzene heaters.

As the styrene process shows, it is not generally feasible to operate a reactor with a conversion per pass equal to the equilibrium conversion. The rate of a chemical reaction decreases as equilibrium is approached, so that the equilibrium conversion can only be attained if either the reactor is very large or the reaction unusually fast. The size of reactor required to give any particular conversion, which of course cannot exceed the maximum conversion predicted from the equilibrium constant, is calculated from the kinetics of the reaction. For this purpose we need quantitative data on the rate of reaction, and the rate equations which describe the kinetics are considered in the following section.

If there are two or more reactants involved in the reaction, both can be converted completely in a single pass only if they are fed to the reactor in the stoichiometric proportion. In many cases, the stoichiometric ratio of reactants may be the best, but in some instances, where one reactant (especially water or air) is very much cheaper than the other, it may be economically advantageous to use it in excess. For a given size of reactor, the object is to increase the conversion of the more costly reactant, possibly at the expense of a substantial decrease in the fraction of the cheaper reactant converted. Examination of the kinetics of the reaction is required to determine whether this can be achieved, and to calculate quantitatively the effects of varying the reactant ratio. Another and perhaps more common reason for departing from the stoichiometric proportions of reactants is to minimise the amount of byproducts formed. This question is discussed further in Section 1.10.4.

Ultimately, the final choice of the temperature, pressure, reactant ratio and conversion at which the reactor will operate depends on an assessment of the overall economics of the process. This will take into account the cost of the reactants, the cost of separating the products and the costs associated with any recycle streams. It should include all the various operating costs and capital costs of reactor and plant. In the course of making this economic assessment, a whole series of calculations of operating conditions, final conversion and reactor size may be performed with the aid of a computer, provided that the data are available. Each of these sets of conditions may be technically feasible, but the one chosen will be that which gives the maximum profitability for the project as a whole.

1.4. CHEMICAL KINETICS AND RATE EQUATIONS

When a homogeneous mixture of reactants is passed into a reactor, either batch or tubular, the concentrations of the reactants fall as the reaction proceeds. Experimentally it has been found that, in general, the rate of the reaction decreases as the concentrations of the reactants decrease. In order to calculate the size of the reactor required to manufacture a particular product at a desired overall rate of production, the design engineer therefore needs to know how the rate of reaction at any time or at any point in the reactor depends on the concentrations of the reactants. Since the reaction rate varies also with temperature, generally increasing rapidly with increasing temperature, a *rate equation*, expressing the rate of reaction as a function of concentrations and temperature, is required in order to design a reactor.

1.4.1. Definition of Reaction Rate, Order of Reaction and Rate Constant

Let us consider a homogeneous irreversible reaction:

$$\nu_A \mathbf{A} + \nu_B \mathbf{B} + \nu_C \mathbf{C} \rightarrow \text{Products}$$

where **A**, **B**, **C** are the reactants and ν_A, ν_B, ν_C the corresponding coefficients in the stoichiometric equation. The rate of reaction can be measured as the moles of **A** transformed per unit volume and unit time. Thus, if n_A is the number of moles of **A** present in a volume V of reaction mixture, the *rate of reaction* with respect to **A** is defined as:

$$\mathcal{R}_A = -\frac{1}{V}\frac{\mathrm{d}n_A}{\mathrm{d}t} \tag{1.2}$$

However, the rate of reaction can also be measured as the moles of **B** transformed per unit volume and unit time, in which case:

$$\mathcal{R}_B = -\frac{1}{V}\frac{\mathrm{d}n_B}{\mathrm{d}t} \tag{1.3}$$

and $\mathcal{R}_B = (\nu_B/\nu_A)\mathcal{R}_A$; similarly $\mathcal{R}_C = (\nu_C/\nu_A)\mathcal{R}_A$ and so on. Obviously, when quoting a reaction rate, care must be taken to specify which reactant is being considered, otherwise ambiguity may arise. Another common source of confusion is the units in which the rate of reaction is measured. Appropriate units for $\mathcal{R}_A$ can be seen quite clearly from equation 1.2; they are kmol of $\mathbf{A}/\mathrm{m}^3\mathrm{s}$ or lb mol of $\mathbf{A}/\mathrm{ft}^3$ s.

At constant temperature, the rate of reaction $\mathcal{R}_A$ is a function of the concentrations of the reactants. Experimentally, it has been found that often (but not always) the function has the mathematical form:

$$\mathcal{R}_A \left(= -\frac{1}{V}\frac{\mathrm{d}n_A}{\mathrm{d}t}\right) = k\, C_A{}^p C_B{}^q C_C{}^r \tag{1.4}$$

$C_A (= n_A/V)$ being the molar concentration of **A**, etc. The exponents p, q, r in this expression are quite often (but not necessarily) whole numbers. When the functional relationship has the form of equation 1.4, the reaction is said to be of order p with respect to reactant **A**, q with respect to **B** and r with respect to **C**. The order of the reaction overall is $(p + q + r)$.

The coefficient k in equation 1.4 is by definition the *rate constant* of the reaction. Its dimensions depend on the exponents p, q, r (i.e. on the order of the reaction); the units in which it is to be expressed may be inferred from the defining equation 1.4. For example, if a reaction:

$$\mathbf{A} \rightarrow \text{Products}$$

behaves as a simple first-order reaction, it has a rate equation:

$$\mathcal{R}_A = \boldsymbol{k}_1 C_A \tag{1.5}$$

If the rate of reaction $\mathcal{R}_A$ is measured in units of kmol/m^3 s and the concentration C_A in kmol/m^3, then $\boldsymbol{k}_1$ has the units s^{-1}. On the other hand, if the reaction above behaved as a second-order reaction with a rate equation:

$$\mathcal{R}_A = \boldsymbol{k}_2 C_A{}^2 \tag{1.6}$$

the units of this rate constant, with $\mathcal{R}_A$ in kmol/m^3 s and C_A in kmol/m^3, are m^3(kmol)$^{-1}$ s^{-1}. A possible source of confusion is that in some instances in the chemical literature, the rate equation, for say a second order gas phase reaction may be written $\mathcal{R}_A = \boldsymbol{k}_p P_A^2$, where P_A is the partial pressure of **A** and may be measured in N/m^2, bar or even in mm Hg. This form of expression results in rather confusing hybrid units for $\boldsymbol{k}_p$ and is not to be recommended.

If a large excess of one or more of the reactants is used, such that the concentration of that reactant changes hardly at all during the course of the reaction, the effective order of the reaction is reduced. Thus, if in carrying out a reaction which is normally second-order with a rate equation $\mathcal{R}_A = \boldsymbol{k}_2 C_A C_B$ an excess of **B** is used, then C_B remains constant and equal to the initial value C_{B0}. The rate equation may then be written $\mathcal{R}_A = \boldsymbol{k}_1 C_A$ where $\boldsymbol{k}_1 = \boldsymbol{k}_2 C_{B0}$ and the reaction is now said to be *pseudo-first-order*.

1.4.2. Influence of Temperature. Activation Energy

Experimentally, the influence of temperature on the rate constant of a reaction is well represented by the original equation of Arrhenius:

$$\boldsymbol{k} = \mathcal{A} \exp(-\mathbf{E}/\mathbf{R}T) \tag{1.7}$$

where T is the absolute temperature and **R** the gas constant. In this equation **E** is termed the *activation energy*, and $\mathcal{A}$ the *frequency factor*. There are theoretical reasons to suppose that temperature dependence should be more exactly described by an equation of the form $\boldsymbol{k} = \mathcal{A}' T^m \exp(-\mathbf{E}/\mathbf{R}T)$, with m usually in the range 0 to 2. However, the influence of the exponential term in equation 1.7 is in practice so strong as to mask any variation in $\mathcal{A}$ with temperature, and the simple form of the relationship (equation 1.7) is therefore quite adequate. **E** is called the *activation energy* because in the molecular theory of chemical kinetics it is associated with an energy barrier which the reactants must surmount to form an activated complex in the transition state. Similarly, $\mathcal{A}$ is associated with the frequency with which the activated complex breaks down into products; or, in terms of the simple collision theory, it is associated with the frequency of collisions.

Values of the activation energy **E** are in J/kmol in the SI system but are usually quoted in kJ/kmol (or J/mol); using these values **R** must then be expressed as kJ/kmol K. For most reactions the activation energy lies in the range 50,000–250,000 kJ/kmol, which implies a very rapid increase in rate constant with temperature. Thus, for a reaction which is occurring at a temperature in the region of 100°C and has an activation energy of 100,000 kJ/kmol, the reaction rate will be doubled for a temperature rise of only 10°C.

Thus, the complete rate equation for an irreversible reaction normally has the form:

$$\mathcal{R}_A = \mathcal{A} \exp(-\mathbf{E}/\mathbf{R}T)\, C_A{}^p C_B{}^q C_C{}^r \tag{1.8}$$

Unfortunately, the exponential temperature term $\exp(-\mathbf{E}/\mathbf{R}T)$ is rather troublesome to handle mathematically, both by analytical methods and numerical techniques. In reactor design this means that calculations for reactors which are not operated isothermally tend to become complicated. In a few cases, useful results can be obtained by abandoning the exponential term altogether and substituting a linear variation of reaction rate with temperature, but this approach is quite inadequate unless the temperature range is very small.

1.4.3. Rate Equations and Reaction Mechanism

One of the reasons why chemical kinetics is an important branch of physical chemistry is that the rate of a chemical reaction may be a significant guide to its mechanism. The engineer concerned with reactor design and development is not interested in reaction mechanism *per se*, but should be aware that an insight into the mechanism of the reaction can provide a valuable clue to the kind of rate equation to be used in a design problem. In the present chapter, it will be possible to make only a few observations on the subject, and for further information the excellent text of MOORE and PEARSON[3] should be consulted.

The first point which must be made is that the *overall* stoichiometry of a reaction is *no guide whatsoever* to its rate equation or to the mechanism of reaction. A stoichiometric equation is no more than a material balance; thus the reaction:

$$KClO_3 + 6FeSO_4 + 3H_2SO_4 \rightarrow KCl + 3Fe_2(SO_4)_3 + 3H_2O$$

is in fact second order in dilute solution with the rate of reaction proportional to the concentrations of ClO_3^- and Fe^{2+} ions. In the general case the stoichiometric coefficients ν_A, ν_B, ν_C, are not necessarily related to the orders p, q, r for the reaction.

However, if it is known from kinetic or other evidence that a reaction $\mathbf{M} + \mathbf{N} \rightarrow$ Product is a simple *elementary reaction*, i.e., if it is known that its mechanism is simply the interaction between a molecule of **M** and a molecule of **N**, then the molecular theory of reaction rates predicts that the rate of this elementary step is proportional to the concentration of species **M** and the concentration of species **N**, i.e. it is second order overall. The reaction is also said to be *bimolecular* since two molecules are involved in the actual chemical transformation.

Thus, the reaction between H_2 and I_2 is known to occur by an elementary bimolecular reaction:

$$H_2 + I_2 \rightarrow 2HI$$

and the rate of the forward reaction corresponds to the equation:

$$\mathcal{R}_{I_2} = \boldsymbol{k}_f C_{H_2} C_{I_2}$$

For many years the hydrogen–iodine reaction was quoted in textbooks as being virtually the only known example of a simple bimolecular reaction. There is now evidence[4] that in parallel to the main bimolecular transformation, some additional reactions involving iodine atoms do occur.

Whereas in the hydrogen–iodine reaction, atomic iodine plays only a minor part, in the reaction between hydrogen and bromine, bromine and hydrogen atoms are the principal intermediates in the overall transformation.

The kinetics of the reaction are quite different from those of the hydrogen–iodine reaction although the stoichiometric equation:

$$H_2 + Br_2 \rightarrow 2HBr$$

looks similar. The reaction actually has a chain mechanism consisting of the elementary steps:

$$Br_2 \rightleftharpoons 2Br\cdot \quad \textit{chain initiation} \text{ and } \textit{termination}$$

$$\left.\begin{aligned} Br\cdot + H_2 &\rightleftharpoons HBr + H\cdot \\ H\cdot + Br_2 &\rightarrow HBr + Br\cdot \end{aligned}\right\} \textit{chain propagation}$$

The rate of the last reaction, for example, is proportional to the concentration of $H\cdot$ and the concentration of Br_2, i.e. it is second order. When the rates of these elementary steps are combined into an overall rate equation, this becomes:

$$\mathscr{R}_{Br_2} = \frac{k' C_{H_2} C_{Br_2}^{1/2}}{1 + k'' \dfrac{C_{HBr}}{C_{Br_2}}} \tag{1.9}$$

where k' and k'' are constants, which are combinations of the rate constants of the elementary steps. This rate equation has a different form from the usual type given by equation 1.4, and cannot therefore be said to have any order because the definition of order applies only to the usual form.

We shall find that the rate equations of gas–solid heterogeneous catalytic reactions (Chapter 3) also do not, in general, have the same form as equation 1.4.

However, many reactions, although their mechanism may be quite complex, do conform to simple first or second-order rate equations. This is because the rate of the overall reaction is limited by just one of the elementary reactions which is then said to be rate-determining. The kinetics of the overall reaction thus reflect the kinetics of this particular step. An example is the pyrolysis of ethane[4] which is important industrially as a source of ethylene[1] (see also Section 1.7.1; Example 1.4). The main overall reaction is:

$$C_2H_6 \rightarrow C_2H_4 + H_2$$

Although there are complications concerning this reaction, under most circumstances it is first order, the kinetics being largely determined by the first step in a chain mechanism:

$$C_2H_6 \rightarrow 2CH_3\cdot$$

which is followed by the much faster reactions:

$$\left.\begin{aligned} CH_3\cdot + C_2H_6 &\rightarrow C_2H_5\cdot + CH_4 \\ C_2H_5\cdot \qquad &\rightarrow C_2H_4 + H\cdot \\ H\cdot + C_2H_6 &\rightarrow C_2H_5\cdot + H_2 \end{aligned}\right\} \textit{chain propagation}$$

Eventually the reaction chains are broken by termination reactions. Other free radical reactions also take place to a lesser extent leading to the formation of CH_4 and some higher hydrocarbons among the products.

1.4.4. Reversible Reactions

For reactions which do not proceed virtually to completion, it is necessary to include the kinetics of the reverse reaction, or the equilibrium constant, in the rate equation.

The equilibrium state in a chemical reaction can be considered from two distinct points of view. The first is from the standpoint of classical thermodynamics, and leads to relationships between the equilibrium constant and thermodynamic quantities such as free energy and heat of reaction, from which we can very usefully calculate equilibrium conversion. The second is a kinetic viewpoint, in which the state of chemical equilibrium is regarded as a dynamic balance between forward and reverse reactions; at equilibrium the rates of the forward reactions and of the reverse reaction are just equal to each other, making the net rate of transformation zero.

Consider a reversible reaction:

$$\mathbf{A} + \mathbf{B} \underset{\boldsymbol{k}_r}{\overset{\boldsymbol{k}_f}{\rightleftharpoons}} \mathbf{M} + \mathbf{N}$$

which is second order overall in each direction, and first order with respect to each species. The hydrolysis of an ester such as ethyl acetate is an example

$$CH_3COOC_2H_5 + NaOH \rightleftharpoons CH_3COONa + C_2H_5OH$$

The rate of the forward reaction expressed with respect to **A**, $\mathcal{R}_{+A}$ is given by $\mathcal{R}_{+A} = \boldsymbol{k}_f C_A C_B$, and the rate of the reverse reaction (again expressed with respect to **A** and written $\mathcal{R}_{-A}$) is given by $\mathcal{R}_{-A} = \boldsymbol{k}_r C_M C_N$. The net rate of reaction in the direction left to right is thus:

$$\mathcal{R}_A = \mathcal{R}_{+A} - \mathcal{R}_{-A} = \boldsymbol{k}_f C_A C_B - \boldsymbol{k}_r C_M C_N \tag{1.10}$$

At equilibrium, when $C_A = C_{Ae}$ etc., $\mathcal{R}_A$ is zero and we have:

$$\boldsymbol{k}_f C_{Ae} C_{Be} = \boldsymbol{k}_r C_{Me} C_{Ne}$$

or:

$$\frac{C_{Me} C_{Ne}}{C_{Ae} C_{Be}} = \frac{\boldsymbol{k}_f}{\boldsymbol{k}_r} \tag{1.11}$$

But $C_{Me}C_{Ne}/C_{Ae}C_{Be}$ is the equilibrium constant K_c and hence $\boldsymbol{k}_f/\boldsymbol{k}_r = K_c$. Often it is convenient to substitute for $\boldsymbol{k}_r$ in equation 1.10 so that we have as a typical example of a *rate equation for a reversible reaction*:

$$\mathcal{R}_A = \boldsymbol{k}_f \left(C_A C_B - \frac{C_M C_N}{K_c} \right) \tag{1.12}$$

We see from the above example that the forward and reverse rate constants are not completely independent, but are related by the equilibrium constant, which in turn is related to the thermodynamic free energy, etc. More detailed examination of the kinds of kinetic equations which might be used to describe the forward and reverse reactions shows that, to be consistent with the thermodynamic equilibrium constant, the form of the rate equation for the reverse reaction cannot be completely independent of the forward rate equation. A good example is the formation of phosgene:

$$CO + Cl_2 \rightleftharpoons CO\,Cl_2$$

The rate of the forward reaction is given by $\mathcal{R}_{+CO} = \boldsymbol{k}_f C_{CO} C_{Cl_2}^{3/2}$. This rate equation indicates that the chlorine concentration must also appear in the reverse rate equation. Let this be $\mathcal{R}_{-CO} = \boldsymbol{k}_r C_{COCl_2}{}^p C_{Cl_2}{}^q$; then at equilibrium, when $\mathcal{R}_{+CO} = \mathcal{R}_{-CO}$, we must have:

$$\frac{\mathcal{R}_{-CO}}{\mathcal{R}_{+CO}} = 1 = \frac{\boldsymbol{k}_r}{\boldsymbol{k}_f}\left(\frac{C_{COCl_2}{}^p C_{Cl_2}{}^q}{C_{CO} C_{Cl_2}^{3/2}}\right)_{eq} \tag{1.13}$$

But we know from the thermodynamic equilibrium constant that:

$$K_c = \left(\frac{C_{COCl_2}}{C_{CO} C_{Cl_2}}\right)_{eq} \tag{1.14}$$

Therefore it follows that $p = 1$ and $q = 1/2$. The complete rate equation is therefore:

$$\mathcal{R}_{CO} = \boldsymbol{k}_f C_{CO} C_{Cl_2}^{3/2} - \boldsymbol{k}_r C_{COCl_2} C_{Cl_2}^{1/2}$$

or:

$$\mathcal{R}_{CO} = \boldsymbol{k}_f\left(C_{CO} C_{Cl_2}^{3/2} - \frac{C_{COCl_2} C_{Cl_2}^{1/2}}{K_c}\right) \tag{1.15}$$

1.4.5. Rate Equations for Constant-Volume Batch Reactors

In applying a rate equation to a situation where the volume of a given reaction mixture (i.e. the density) remains constant throughout the reaction, the treatment is very much simplified if the equation is expressed in terms of a variable χ, which is defined as the number of moles of a particular reactant transformed per unit volume of reaction mixture (e.g. $C_{A0} - C_A$) at any instant of time t. The quantity χ is very similar to a molar concentration and has the same units. By simple stoichiometry, the moles of the other reactants transformed and products generated can also be expressed in terms of χ, and the rate of the reaction can be expressed as the rate of increase in χ with time. Thus, by definition,

$$\mathcal{R}_A = -\frac{1}{V}\frac{dn_A}{dt} \tag{1.16}$$

and if V is constant this becomes:

$$\mathcal{R}_A = -\frac{d(n_A/V)}{dt} = -\frac{dC_A}{dt} = \frac{d\chi}{dt} \tag{1.17}$$

x being the moles of A which have reacted. The general rate equation 1.4 may then be written:

$$\frac{d\chi}{dt} = k(C_{A0} - \chi)^p \left(C_{B0} - \frac{\nu_B}{\nu_A}\chi\right)^q \left(C_{C0} - \frac{\nu_C}{\nu_A}\chi\right)^r \tag{1.18}$$

where C_{A0} etc. are the initial concentrations. This equation may then in general, at constant temperature conditions, be integrated to give χ as a function of time, so that the reaction time for any particular conversion can be readily calculated.

The equations which result when these integrations are carried out for reactions of various orders are discussed in considerable detail in most texts dealing with the physico-chemical aspects of chemical kinetics[3, 4]. Table 1.1 shows a summary of some of the simpler cases; the integrated forms can be easily verified by the reader if desired.

One particular point of interest is the expression for the *half-life* of a reaction $t_{1/2}$; this is the time required for one half of the reactant in question to disappear. A first order reaction is unique in that the *half-life* is independent of the initial concentration of the reactant. This characteristic is sometimes used as a test of whether a reaction really is first order. Also since $t_{1/2} = \frac{1}{k_1} \ln 2$, a first-order rate constant can be readily converted into a *half-life* which one can easily remember as characteristic of the reaction.

A further point of interest about the equations shown in Table 1.1 is to compare the shapes of graphs of χ (or fractional conversion $\chi/C_{A0} = \alpha_A$) vs. time for reactions of different orders p. Figure 1.8 shows a comparison between first and second-order reactions involving a single reactant only, together with the straight line for a zero-order reaction. The rate constants have been taken so that the curves coincide at 50 per cent conversion. The rate of reaction at any time is given by the slope of the curve (as indicated by equation 1.17). It may be seen that the rate of the second-order reaction is high at first but falls rapidly with increasing time and, compared with first-order reactions, longer reaction times are required for high conversions. The zero-order reaction is the only one where the reaction rate does not decrease with increasing conversion. Many biological systems have apparent reaction orders between 0 and 1 and will have a behaviour intermediate between the curves shown.

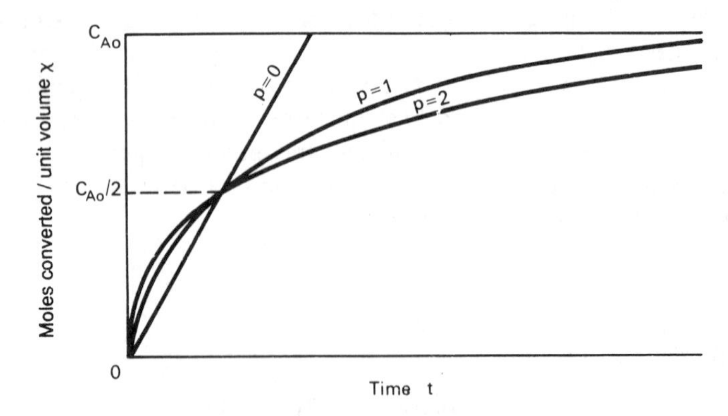

FIG. 1.8. Batch reactions at constant volume: Comparison of curves for zero, first and second-order reactions

TABLE 1.1. *Rate Equations for Constant Volume Batch Reactors*

Reaction type	Rate equation	Integrated form
	Irreversible reactions	
First order $A \rightarrow$ products	$\frac{d\chi}{dt} = k_1(C_{A0} - \chi)$	$t = \frac{1}{k_1} \ln \frac{C_{A0}}{(C_{A0} - \chi)}$ $t_{1/2} = \frac{\ln 2}{k_1}$
Second order $A + B \rightarrow$ products	$\frac{d\chi}{dt} = k_2(C_{A0} - \chi)(C_{B0} - \chi)$	$t = \frac{1}{k_2(C_{B0} - C_{A0})} \ln \frac{C_{A0}(C_{B0} - \chi)}{C_{B0}(C_{A0} - \chi)}$ $C_{A0} \neq C_{B0}$
$2A \rightarrow$ products	$\frac{d\chi}{dt} = k_2(C_{A0} - \chi)^2$	$t = \frac{1}{k_2}\left(\frac{1}{C_{A0} - \chi} - \frac{1}{C_{A0}}\right)$ $t_{1/2} = \frac{1}{k_2 C_{A0}}$
Order p, one reactant $A \rightarrow$ products	$\frac{d\chi}{dt} = k(C_{A0} - \chi)^p$	$t = \frac{1}{k(p-1)}\left[\frac{1}{(C_{A0} - \chi)^{p-1}} - \frac{1}{C_{A0}^{\,p-1}}\right]$ $t_{1/2} = \frac{2^{p-1} - 1}{k(p-1)C_{A0}^{\,p-1}} \quad p \neq 1$
	Reversible reactions	
First order both directions $A \underset{k_r}{\overset{k_f}{\rightleftharpoons}} M$	$\frac{d\chi}{dt} = k_f(C_{A0} - \chi) - k_r(C_{M0} + \chi)$	$t = \frac{1}{(k_f + k_r)} \ln \frac{k_f C_{A0} - k_r C_{M0}}{k_f(C_{A0} - \chi) - k_r(C_{M0} + \chi)}$ or since $K_c = k_f/k_r$ $t = \frac{K_c}{k_f(1 + K_c)} \ln \frac{K_c C_{A0} - C_{M0}}{K_c(C_{A0} - \chi) - (C_{M0} + \chi)}$ If $C_{M0} = 0$ $t = \frac{1}{(k_f + k_r)} \ln \frac{\chi_e}{(\chi_e - \chi)}$
Second order both directions $A + B \underset{k_r}{\overset{k_f}{\rightleftharpoons}} M + N$	If $C_{A0} = C_{B0}$ and $C_{M0} = C_{N0} = 0$ $\frac{d\chi}{dt} = k_f(C_{A0} - \chi)^2 - k_r\chi^2$	$t = \frac{\sqrt{K_c}}{2k_f C_{A0}} \ln \frac{C_{A0} + \chi[1/(\sqrt{K_c}) - 1]}{C_{A0} - \chi[1/(\sqrt{K_c}) + 1]}$

1.4.6. Experimental Determination of Kinetic Constants

The interpretation of laboratory scale experiments to determine order and rate constant is another subject which is considered at length in physical chemistry texts[3, 4]. Essentially, it is a process of fitting a rate equation of the general form given by equation 1.4 to a set of numerical data. The experiments which are carried out to obtain the kinetic constants may be of two kinds, depending on whether the rate equation is to be used in its original (*differential*) form, or in its *integrated* form (see Table 1.1). If the differential form is to be used, the experiments must be designed so that the rate of disappearance of reactant **A**, $\mathcal{R}_A$, can be measured without its concentration changing appreciably. With batch or tubular reactors this has the disadvantage in practice that very accurate measurements of C_A must be made so that, when differences in concentration ΔC_A are taken to evaluate $\mathcal{R}_A$ (e.g. for a batch reactor, equation 1.17 in finite difference form is $\mathcal{R}_A = -\Delta C_A/\Delta t$), the difference may be obtained with sufficient accuracy. Continuous stirred-tank reactors do not suffer from this disadvantage; by operating in the steady state, steady concentrations of the reactants are maintained and the rate of reaction is determined readily.

If the rate equation is to be employed in its integrated form, the problem of determining kinetic constants from experimental data from batch or tubular reactors is in many ways equivalent to taking the design equations and working backwards. Thus, for a batch reactor with constant volume of reaction mixture at constant temperature, the equations listed in Table 1.1 apply. For example, if a reaction is suspected of being second order overall, the experimental results are plotted in the form:

$$\frac{1}{C_{B0} - C_{A0}} \ln \left\{ \frac{C_{A0}(C_{B0} - \chi)}{C_{B0}(C_{A0} - \chi)} \right\} \text{ versus } t$$

If the points lie close to a straight line, this is taken as confirmation that a second-order equation satisfactorily describes the kinetics, and the value of the rate constant k_2 is found by fitting the best straight line to the points by linear regression. Experiments using tubular and continuous stirred-tank reactors to determine kinetic constants are discussed in the sections describing these reactors (Sections 1.7.4 and 1.8.5).

Unfortunately, many of the chemical processes which are important industrially are quite complex. A complete description of the kinetics of a process, including byproduct formation as well as the main chemical reaction, may involve several individual reactions, some occurring simultaneously, some proceeding in a consecutive manner. Often the results of laboratory experiments in such cases are ambiguous and, even if complete elucidation of such a complex reaction pattern is possible, it may take several man-years of experimental effort. Whereas ideally the design engineer would like to have a complete set of rate equations for all the reactions involved in a process, in practice the data available to him often fall far short of this.

1.5. GENERAL MATERIAL AND THERMAL BALANCES

The starting point for the design of any type of reactor is the general material balance. This material balance can be carried out with respect to one of the reactants or to one of the products. However, if we are dealing with a single reaction such as:

$$\nu_A\mathbf{A} + \nu_B\mathbf{B} = \nu_M\mathbf{M} + \nu_N\mathbf{N}$$

then, in the absence of any separation of the various components by diffusion, it is not necessary to write separate material balance equations for each of the reactants and products. The stoichiometric equation shows that if ν_A moles of **A** react, ν_B moles of **B** must also have disappeared, ν_M moles of **M** must have been formed together with ν_N moles of **N**. In such a case the extent to which the reaction has proceeded at any stage can be expressed in terms of the fractional conversion α of any selected reactant, for example **A**. (See also Section 1.4.5 where similarly the rate of the reaction could be expressed by considering one reactant only.) Alternatively, one of the products **M** or **N** could be chosen as the entity for the material balance equation; however, it is usual to use one reactant as a basis because there may be some uncertainty about just what products are present when the procedure is extended to more complex reactions in which several byproducts are formed, whereas usually the chemical nature of the reactants is known for certain.

Basically, the general material balance for a reactor follows the same pattern as all material and energy balances, namely:

$$\text{Input} - \text{Output} = \text{Accumulation}$$

but with the important difference that the reactant in question can disappear through chemical reaction. The material balance must therefore be written:

$$\underset{(1)}{\text{Input}} - \underset{(2)}{\text{Output}} - \underset{(3)}{\text{Reaction}} = \underset{(4)}{\text{Accumulation}}$$

In setting out this equation in an exact form for any particular reactor, the material balance has to be carried out

(a) over a certain element of volume, and
(b) over a certain element of time.

If the compositions vary with position in the reactor, which is the case with a tubular reactor, a differential element of volume δV_t must be used, and the equation integrated at a later stage. Otherwise, if the compositions are uniform, e.g. a well-mixed batch reactor or a continuous stirred-tank reactor, then the size of the volume element is immaterial; it may conveniently be unit volume (1 m^3) or it may be the whole reactor. Similarly, if the compositions are changing with time as in a batch reactor, the material balance must be made over a differential element of time. Otherwise for a tubular or a continuous stirred-tank reactor operating in a steady state, where compositions do not vary with time, the time interval used is immaterial and may conveniently be unit time (1 s). Bearing in mind these considerations the general material balance may be written:

$$\underset{(1)}{\begin{array}{c}\text{Rate of flow of}\\ \text{reactant into}\\ \text{volume element}\end{array}} - \underset{(2)}{\begin{array}{c}\text{Rate of flow of}\\ \text{reactant out of}\\ \text{volume element}\end{array}} - \underset{(3)}{\begin{array}{c}\text{Rate of reactant}\\ \text{removal by reaction}\\ \text{within volume element}\end{array}} = \underset{(4)}{\begin{array}{c}\text{Rate of accumulation}\\ \text{of reactant within}\\ \text{volume element}\end{array}} \quad (1.19)$$

For each of the three basic types of chemical reactor this equation may be reduced to a simplified form. For a *batch reactor* terms (1) and (2) are zero and the *Rate of accumulation*, i.e. the rate of disappearance of the reactant, is equal to the rate of

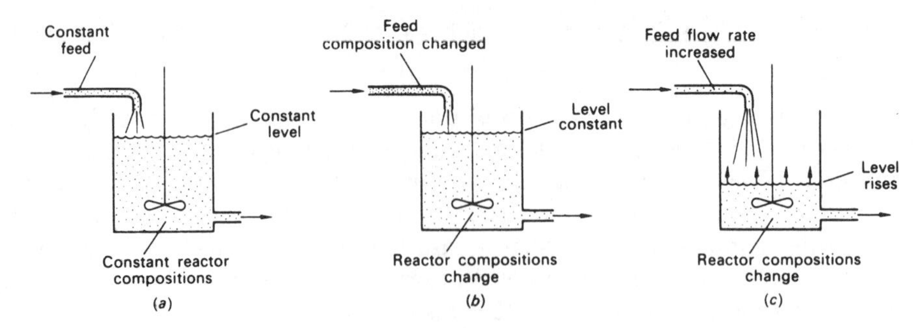

FIG. 1.9. Continuous stirred-tank reactor showing steady state operation (*a*) and two modes of unsteady state operation (*b*) and (*c*)

reaction. For a *tubular reactor* or a *continuous stirred-tank reactor*, if operating in a *steady state*, the *Rate of accumulation* term (4) is by definition zero, and the *Rate of reactant removal by reaction* is just balanced by the difference between inflow and outflow.

For unsteady state operation of a flow reactor, it is important to appreciate the distinction between the *Reaction* term (3) above and the *Accumulation* term (4), which are equal for a batch reactor. Transient operation of a flow reactor occurs during start-up and in response to disturbances in the operating conditions. The nature of transients induced by disturbances and the differences between terms (3) and (4) above can best be visualised for the case of a *continuous stirred-tank reactor* (Fig. 1.9). In Fig. 1.9*a*, the reactor is operating in a steady state. In Fig. 1.9*b* it is subject to an increase in the input of reactant owing to a disturbance in the feed composition. This results in a rise in the concentration of the reactant within the reaction vessel corresponding to the *Accumulation* term (4) which is quite distinct from, and additional to, the *Reactant removal by the reaction* term (3). Figure 1.9*c* shows another kind of transient which will cause compositions in the reactor to change, namely a change in the volume of reaction mixture contained in the reactor. Other variables which must be controlled, apart from feed composition and flowrates in and out, are temperature and, for gas reactions particularly, pressure. Variations of any of these quantities with time will cause a change in the composition levels of the reactant in the reactor, and these will appear in the *Accumulation* term (4) in the material balance.

The heat balance for a reactor has a form very similar to the general material balance, i.e.

Rate of heat inflow into volume element		Rate of heat outflow from volume element		Rate of absorption of heat by chemical reaction in volume element		Rate of heat accumulation in volume element	
(1)	−	(2)	−	(3)	=	(4)	(1.20)

In the *Inflow* and *Outflow* terms (1) and (2), the heat flow may be of two kinds: the first is transfer of sensible heat or enthalpy by the fluid entering and leaving the element; and the second is heat transferred to or from the fluid across heat transfer surfaces, such as cooling coils situated in the reactor. The *Heat absorbed in the chemical reaction*, term (3), depends on the rate of reaction, which in turn depends on the concentration levels in the reactor as determined by the general material balance equation. Since the rate of reaction depends also on the temperature levels in the reactor as determined by the heat balance equation, the material balance and the heat balance interact with each other, and the two equations have to be solved simultaneously. The types of solutions obtained are discussed further under the headings of the various types of reactor—batch, tubular and continuous stirred-tank.

1.6. BATCH REACTORS

There is a tendency in chemical engineering to try to make all processes continuous. Whereas continuous flow reactors are likely to be most economic for large scale production, the very real advantages of batch reactors, especially for smaller scale production, should not be overlooked. Small batch reactors generally require less auxiliary equipment, such as pumps, and their control systems are less elaborate and costly than those for continuous reactors, although manpower needs are greater. However, large batch reactors may sometimes be fitted with highly complex control systems. A big advantage of batch reactors in the dyestuff, fine chemical and pharmaceutical industries is their versatility. A corrosion-resistant batch reactor such as an enamel or rubber-lined jacketed vessel (Fig. 1.4*a*) or a stainless steel vessel with heating and cooling coils (Fig. 1.4*b*) can be used for a wide variety of similar kinds of reaction. Sometimes only a few batches per year are required to meet the demand for an unusual product. In some processes, such as polymerisations and fermentations, batch reactors are traditionally preferred because the interval between batches provides an opportunity to clean the system thoroughly and ensure that no deleterious intermediates such as foreign bacteria build up and spoil the product. Moreover, it must not be forgotten that a squat tank is the most economical shape for holding a given volume of liquid, and for slow reactions a tubular flow reactor with a diameter sufficiently small to prevent backmixing, would be more costly than a simple batch reactor. Although at present we are concerned mainly with homogeneous reactions, we should note that the batch reactor has many advantages for heterogeneous reactions; the agitator can be designed to suspend solids in the liquid, and to disperse a second immiscible liquid or a gas.

In calculating the volume required for a batch reactor, we shall be specifying the volume of liquid which must be processed. In designing the vessel itself the heights should be increased by about 10 per cent to allow freeboard for waves and disturbances on the surface of the liquid; additional freeboard may have to be provided if foaming is anticipated.

1.6.1. Calculation of Reaction Time; Basic Design Equation

Calculation of the time required to reach a particular conversion is the main objective in the design of batch reactors. Knowing the amount of reactant converted, i.e. the amount of the desired product formed per unit volume in this reaction time, the volume of reactor required for a given production rate can be found by simple scale-up as shown in the example on ethyl acetate below.

The reaction time t_r is determined by applying the general material balance equation 1.19. In the most general case, when the volume of the reaction mixture is not constant throughout the reaction, it is convenient to make the material balance over the whole volume of the reactor V_b. For the reactant **A**, if n_{A0} moles are charged initially, the number of moles remaining when the fraction of **A** converted is α_A is $n_{A0}(1 - \alpha_A)$ and, using a differential element of time, the rate at which this is changing, i.e. the *Accumulation* term (4) in equation 1.19 is:

$$\frac{d}{dt}[n_{A0}(1 - \alpha_A)] = -n_{A0}\frac{d\alpha_A}{dt} \tag{1.21}$$

The rate at which **A** is removed by reaction term (3) is $\mathcal{R}_A V_b$ and, since the *Flow* terms (1) and (2) are zero, we have:

$$\underset{\text{Reaction}}{-\mathcal{R}_A V_b} = \underset{\text{Accumulation}}{-n_{A0}\frac{d\alpha_A}{dt}} \tag{1.22}$$

Thus, integrating over the period of the reaction to a final conversion α_{Af}, we obtain the basic design equation:

$$t_r = n_{A0}\int_0^{\alpha_{Af}} \frac{d\alpha_A}{\mathcal{R}_A V_b} \tag{1.23}$$

For many liquid phase reactions it is reasonable to neglect any change in volume of the reaction mixture. Equation 1.23 then becomes:

$$t_r = \frac{n_{A0}}{V_b}\int_0^{\alpha_{Af}} \frac{d\alpha_A}{\mathcal{R}_A} = C_{A0}\int_0^{\alpha_{Af}} \frac{d\alpha_A}{\mathcal{R}_A} \tag{1.24}$$

This form of the equation is convenient if there is only one reactant. For more than one reactant and for reversible reactions, it is more convenient to write the equation in terms of χ, the moles of A converted per unit volume $\chi = C_{A0}\alpha_A$, and obtain:

$$t_r = \int_0^{\chi_f} \frac{d\chi}{\mathcal{R}_A} \tag{1.25}$$

This is the integrated form of equation 1.17 obtained previously; it may be derived formally by applying the general material balance to unit volume under conditions of constant density, when the *Rate of reaction* term (3) is simply $\mathcal{R}_A$ and the *Accumulation* term (4) is:

$$\frac{dC_A}{dt}, \quad \text{i.e.} \quad -\frac{d\chi}{dt}$$

Thus:

$$-\mathcal{R}_A = -\frac{d\chi}{dt}$$

which is equation 1.17.

1.6.2. Reaction Time — Isothermal Operation

If the reactor is to be operated isothermally, the rate of reaction $\mathcal{R}_A$ can be expressed as a function of concentrations only, and the integration in equation 1.24 or 1.25 carried out. The integrated forms of equation 1.25 for a variety of the simple rate equations are shown in Table 1.1 and Fig. 1.8. We now consider an example with a rather more complicated rate equation involving a reversible reaction, and show also how the volume of the batch reactor required to meet a particular production requirement is calculated.

Example 1.2

Production of Ethyl Acetate in a Batch Reactor

Ethyl acetate is to be manufactured by the esterification of acetic acid with ethanol in an isothermal batch reactor. A production rate of 10 tonne/day of ethyl acetate is required.

$$\underset{\mathbf{A}}{CH_3\cdot COOH} + \underset{\mathbf{B}}{C_2H_5OH} \rightleftharpoons \underset{\mathbf{M}}{CH_3\cdot COOC_2H_5} + \underset{\mathbf{N}}{H_2O}$$

The reactor will be charged with a mixture containing 500 kg/m^3 ethanol and 250 kg/m^3 acetic acid, the remainder being water, and a small quantity of hydrochloric acid to act as a catalyst. The density of this mixture is 1045 kg/m^3 which will be assumed constant throughout the reaction. The reaction is reversible with a rate equation which, over the concentration range of interest, can be written:

$$\mathcal{R}_A = k_f C_A C_B - k_r C_M C_N$$

At the operating temperature of 100°C the rate constants have the values:

$$k_f = 8.0 \times 10^{-6}\ \text{m}^3/\text{kmol s}$$

$$k_r = 2.7 \times 10^{-6}\ \text{m}^3/\text{kmol s}$$

The reaction mixture will be discharged when the conversion of the acetic acid is 30 per cent. A time of 30 min is required between batches for discharging, cleaning, and recharging. Determine the volume of the reactor required.

Solution

After a time t, if χ kmol/m^3 of acetic acid (**A**) has reacted, its concentration will be $(C_{A0} - \chi)$ where C_{A0} is the initial concentration. From the stoichiometry of the reaction, if χ kmol/m^3 of acetic acid has reacted, χ kmol/m^3 of ethanol also will have reacted and the same number of moles of ester and of water will have been formed. The rate equation may thus be written:

$$\mathcal{R}_A = k_f(C_{A0} - \chi)(C_{B0} - \chi) - k_r\chi(C_{N0} + \chi).$$

From the original composition of the mixture, its density, and the molecular weights of acetic acid, ethanol and water which are 60, 46 and 18 respectively, $C_{A0} = 4.2$ kmol/m^3; $C_{B0} = 10.9$ kmol/m^3; $C_{N0} = 16.4$ kmol/m^3. Thus, from equation 1.25, t_r in seconds is given by:

$$t_r = \int_0^{\chi_f} \frac{d\chi}{8.0 \times 10^{-6}(4.2 - \chi)(10.9 - \chi) - 2.7 \times 10^{-6}\chi(16.4 + \chi)}$$

This integral may be evaluated either by splitting into partial fractions, or by graphical or numerical means. Using the method of partial fractions, we obtain after some fairly lengthy manipulation:

$$t_r = 16{,}200\left[\log_{10}\frac{29 - \chi_f}{2.4 - \chi_f} - 1.082\right]$$

Since the final conversion of acetic acid is to be 30 per cent, $\chi_f = 0.30 \times C_{A0} = 1.26$ kmol/m^3: whence the reaction time t_r is, from the above equation, 4920 s.

Thus 1 m^3 of reactor volume produces 1.26 kmol of ethyl acetate (molecular weight 88) in a total batch time of 6720 s, i.e. in 4920 s reaction time and 1800 s shut-down time. This is an average production rate of:

$$1.26 \times 88 \times \frac{24 \times 60^2}{6720}$$

i.e. 1420 kg/day per m^3 of reactor volume. Since the required production rate is 10,000 kg/day the required reactor volume is 10,000/1420 = $\underline{\underline{7.1\ \text{m}^3}}$.

Example 1.2 on ethyl acetate is useful also in directing attention to an important point concerning reversible reactions in general. A reversible reaction will not normally go to completion, but will slow down as equilibrium is approached. This progress towards equilibrium can, however, sometimes be disturbed by continuously removing one or more of the products as formed. In the actual manufacture of ethyl acetate, the ester is removed as the reaction proceeds by distilling off a ternary azeotrope of molar composition ethyl acetate 60.1 per cent, ethanol 12.4 per cent and water 27.5 per cent. The net rate of reaction is thereby increased as the rate equation above shows; because C_M is always small the term for the rate of the reverse reaction $\boldsymbol{k}_r C_M C_N$ is always small and the net rate of reaction is virtually equal to the rate of the forward reaction above, i.e. $\boldsymbol{k}_f C_A C_B$.

1.6.3. Maximum Production Rate

For most reactions, the rate decreases as the reaction proceeds (important exceptions being a number of biological reactions which are autocatalytic). For a reaction with no volume change, the rate is represented by the slope of the curve of χ (moles converted per unit volume) versus time (Fig. 1.10), which decreases steadily with increasing time. The maximum reaction rate occurs at zero time, and, if our sole concern were to obtain maximum output from the reactor and the shutdown time were zero, it appears that the best course would be to discharge the reactor after only a short reaction time t_r, and refill with fresh reactants. It would then be necessary, of course, to separate a large amount of reactant from a small amount of product. However, if the shut-down time is appreciable and has a value t_s then as we have seen in the example on ethyl acetate above, the average production rate per unit volume is:

$$\frac{\chi}{t_r + t_s}$$

The maximum production rate is therefore given by the maximum value of:

$$\frac{\chi}{t_r + t_s}$$

This maximum can be most conveniently found graphically (Fig. 1.10). The average production rate is given by the slope of the line ZA; this is obviously a maximum when the line is tangent to the curve of χ versus t, i.e. ZT as shown. The reaction time obtained $t_{r\,\max}$ is not necessarily the optimum for the process as a whole,

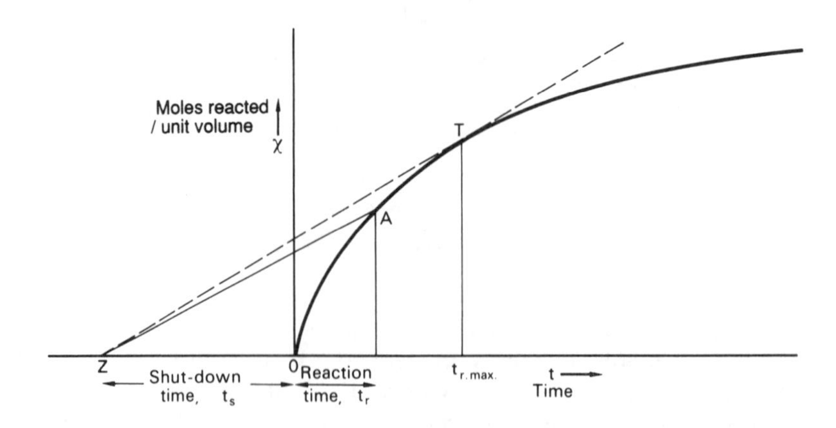

FIG. 1.10. Maximum production rate in a batch reactor with a shut-down time t_S

however. This will depend in addition upon the costs involved in feed preparation, separation of the products, and storage.

1.6.4. Reaction Time — Non-isothermal Operation

If the temperature is not constant but varies during the course of the reaction, then the rate of reaction $\mathfrak{R}_A$ in equation 1.24 or 1.25 will be a function of temperature as well as concentration (equation 1.8). The temperature at any stage is determined by a heat balance, the general form of which is given by equation 1.20. Since there is no material flow into or out of a batch reactor during reaction, the enthalpy changes associated with such flows in continuous reactors are absent. However, there may be a flow of heat to or from the reactor by heat transfer using the type of equipment shown in Fig. 1.4. In the case of a jacketed vessel or one with an internal coil, the heat transfer coefficient will be largely dependent on the agitator speed, which is usually held constant. Thus, assuming that the viscosity of the liquid does not change appreciably, which is reasonable in many cases (except for some polymerisations), the heat transfer coefficient may be taken as constant. If heating is effected by condensing saturated steam at constant pressure, as in Fig. 1.11*b*, the temperature on the coil side T_C is constant. If cooling is carried out with water (Fig. 1.11*c*), the rise in temperature of the water may be small if the flowrate is large, and T_C again taken as constant. Thus, we may write the rate of heat transfer to cooling coils of area A_t as:

$$Q = UA_t(T - T_C) \tag{1.26}$$

where T is the temperature of the reaction mixture. The heat balance taken over the whole reactor thus becomes:

$$-UA_t(T - T_C) + (-\Delta H_A)\, V_b \mathfrak{R}_A = \left(\sum m_j c_j\right) \frac{\mathrm{d}T}{\mathrm{d}t} \tag{1.27}$$

Rate of heat flow out by heat transfer	Rate of heat release by chemical reaction	Rate of heat accumulation

where ΔH_A is the enthalpy change in the reaction per mole of **A** reacting, and $(\Sigma m_j c_j)$ is the sum of the heat capacities (i.e. mass × specific heat) of the reaction

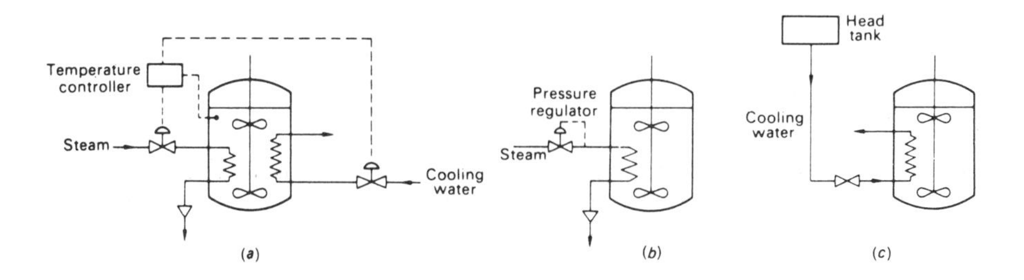

FIG. 1.11. Methods of operating batch reactors
(*a*) Isothermal operation of an exothermic reaction; heating to give required initial temperature, cooling to remove heat of reaction
(*b*) and (*c*) non-isothermal operation; simple schemes

mixture and the reactor itself, including all the various internal components such as the agitator, whose temperatures also change.

Finding the time required for a particular conversion involves the solution of two simultaneous equations, i.e. 1.24 or 1.25 for the material balance and 1.27 for the heat balance. Generally, a solution in analytical form is unobtainable and numerical methods or analogue simulation must be used. Taking, for example, a first-order reaction with constant volume:

$$\mathcal{R}_A = \frac{d\chi}{dt} = k(C_{A0} - \chi) \qquad (1.28)$$

and:

$$k = \mathcal{A} \exp\left(\frac{-\mathbf{E}}{\mathbf{R}T}\right) \qquad (1.29)$$

we have for the material balance:

$$\frac{d\chi}{dt} = \mathcal{A} \exp\left(\frac{-\mathbf{E}}{\mathbf{R}T}\right)(C_{A0} - \chi) \qquad (1.30)$$

and for the heat balance:

$$(-\Delta H_A)\, V_b\, \mathcal{A} \exp\left(\frac{-\mathbf{E}}{\mathbf{R}T}\right)(C_{A0} - \chi) - UA_t(T - T_C) = \left(\sum m_j c_j\right)\frac{dT}{dt} \qquad (1.31)$$

With t as the independent variable, the solution will be:

(a) χ as f(t)
(b) T as F(t).

A typical requirement is that the temperature shall not rise above T_{mx} in order to avoid byproducts or hazardous operation. The forms of the solutions obtained are sketched in Fig. 1.12.

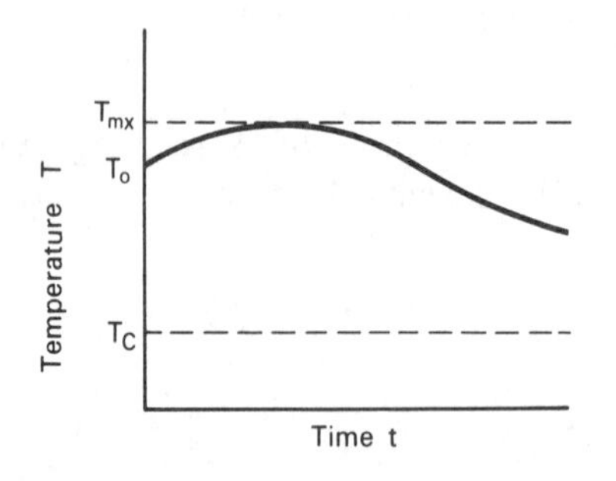

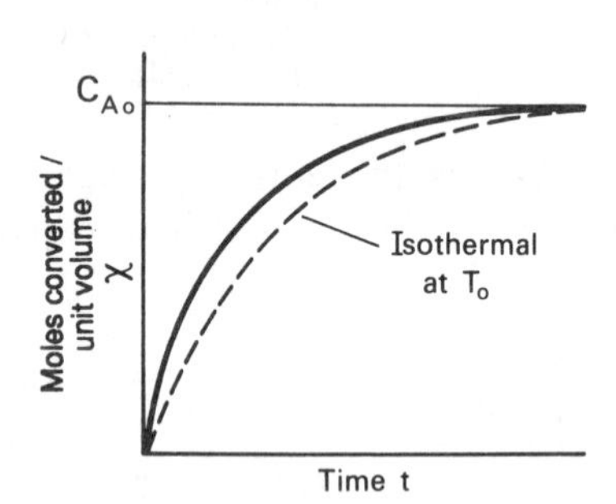

FIG. 1.12. Non-isothermal batch reactor: Typical curves for an exothermic reaction with just sufficient cooling (constant U and T_C) to prevent temperature rising above T_{mx}

1.6.5. Adiabatic Operation

If the reaction is carried out adiabatically (i.e. without heat transfer, so that $Q = 0$), the heat balance shows that the temperature at any stage in the reaction can be expressed in terms of the conversion only. This is because, however fast or slow the reaction, the heat released by the reaction is retained as sensible heat in the reactor. Thus, for reaction at constant volume, putting $Q = 0$ and $\mathfrak{R}_A = \mathrm{d}\chi/\mathrm{d}t$ in equation 1.27:

$$(-\Delta H_A)\, V_b\, \mathrm{d}\chi = \left(\sum m_j c_j\right) \mathrm{d}T \qquad (1.32)$$

Equation 1.32 may be solved to give the temperature as a function of χ. Usually the change in temperature $(T - T_0)$, where T_0 is the initial temperature, is proportional to χ, since $\sum m_j c_j$, the total heat capacity, does not vary appreciably with temperature or conversion. The appropriate values of the rate constant are then used to carry out the integration of equation 1.24 or 1.25 numerically, as shown in the following example.

Example 1.3

Adiabatic Batch Reactor

Acetic anhydride is hydrolysed by water in accordance with the equation:

$$(CH_3 \cdot CO)_2O + H_2O \rightleftharpoons 2CH_3 \cdot COOH$$

In a dilute aqueous solution where a large excess of water is present, the reaction is irreversible and pseudo first-order with respect to the acetic anhydride. The variation of the pseudo first-order rate constant with temperature is as follows:

Temperature	(°C)	15	20	25	30
	(K)	288	293	298	303
Rate Constant	(s^{-1})	0.00134	0.00188	0.00263	0.00351

A batch reactor for carrying out the hydrolysis is charged with an anhydride solution containing 0.30 kmol/m^3 at 15°C. The specific heat and density of the reaction mixture are 3.8 kJ/kgK and 1070 kg/m^3, and may be taken as constant throughout the course of the reaction. The reaction is exothermic, the heat of reaction per kmol of anhydride being 210,000 kJ/kmol. If the reactor is operated adiabatically, estimate the time required for the hydrolysis of 80 per cent of the anhydride.

Solution

For the purposes of this example we shall neglect the heat capacity of the reaction vessel. Since the anticipated temperature rise is small, the heat of reaction will be taken as independent of temperature. Because the heat capacity of the reactor is neglected, we may most conveniently take the adiabatic heat balance (equation 1.32) over the unit volume, i.e. 1 m^3 of reaction mixture. Thus, integrating equation 1.32 with the temperature T_0 when $\chi = 0$:

$$(-\Delta H_A)\chi = mc(T - T_0)$$

i.e.:

$$210{,}000\,\chi = 1070 \times 3.8(T - T_0)$$

$\therefore$

$$(T - T_0) = 52\chi$$

Writing:

$$\chi = C_{A0}\alpha = 0.3\alpha$$

$$(T - T_0) = 15.6\alpha$$

Thus, if the reaction went to completion ($\alpha = 1$), the adiabatic temperature rise would be 15.6°C.

For a pseudo first-order reaction, the rate equation is:

$$\mathfrak{R}_A = k_1(C_{A0} - \chi) = C_{A0}k_1(1 - \alpha)$$

and from equation 1.24:

$$t_r = \int_0^{0.8} \frac{\mathrm{d}\alpha}{k_1(1 - \alpha)}$$

To evaluate this integral graphically, we need to plot:

$$\frac{1}{k_1(1 - \alpha)} \text{ versus } \alpha$$

remembering that k_1 is a function of T (i.e. of α). Interpolating the values of k_1 given, we evaluate:

$$\frac{1}{k_1(1-\alpha)}$$

for various values of α, some of which are shown in Table 1.2.

TABLE 1.2. *Evaluation of Integral to Determine t_r*

Fractional conversion α	Temperature rise $(T - T_0)$ deg K	Temperature K	Rate constant k_1 (s^{-1})	$\frac{1}{k_1(1-\alpha)}$
0	0	288	0.00134	740
0.2	3.1	291.1	0.00167	750
0.4	6.2	294.2	0.00205	810
0.6	9.4	297.4	0.00253	990
0.8	12.5	300.5	0.00305	1630

From the area under the graph up to $\alpha = 0.80$, we find that the required reaction time is approximately 720 s = 12 min.

1.7. TUBULAR-FLOW REACTORS

The tubular-flow reactor (Fig. 1.1*b*) is chosen when it is desired to operate the reactor continuously but without back-mixing of reactants and products. In the case of an *ideal* tubular reactor, the reaction mixture passes through in a state of *plug flow* which, as the name suggests, means that the fluid moves like a solid plug or piston. Furthermore, in the ideal reactor it is assumed that not only the local mass flowrate but also the fluid properties, temperature, pressure, and compositions are uniform across any section normal to the fluid motion. Of course the compositions, and possibly the temperature and pressure also, change between inlet and outlet of the reactor in the longitudinal direction. In the elementary treatment of tubular reactors, *longitudinal dispersion*, i.e. mixing by diffusion and other processes in the direction of flow, is also neglected.

Thus, in the idealised tubular reactor all elements of fluid take the same time to pass through the reactor and experience the same sequence of temperature, pressure and composition changes. In calculating the size of such a reactor, we are concerned with its volume only; its shape does not affect the reaction so long as plug flow occurs.

The flow pattern of the fluid is, however, only one of the criteria which determine the shape eventually chosen for a tubular reactor. The factors which must be taken into account are:

(a) whether plug flow can be attained,
(b) heat transfer requirements,
(c) pressure drop in the reactor,
(d) support of catalyst, if present, and
(e) ease and cheapness of construction.

Figure 1.13 shows various configurations which might be chosen. One of the cheapest ways of enclosing a given volume is to use a cylinder of height approximately equal to its diameter. In Fig. 1.13*a* the reactor is a simple cylinder of this kind. Without packing, however, swirling motions in the fluid would cause serious

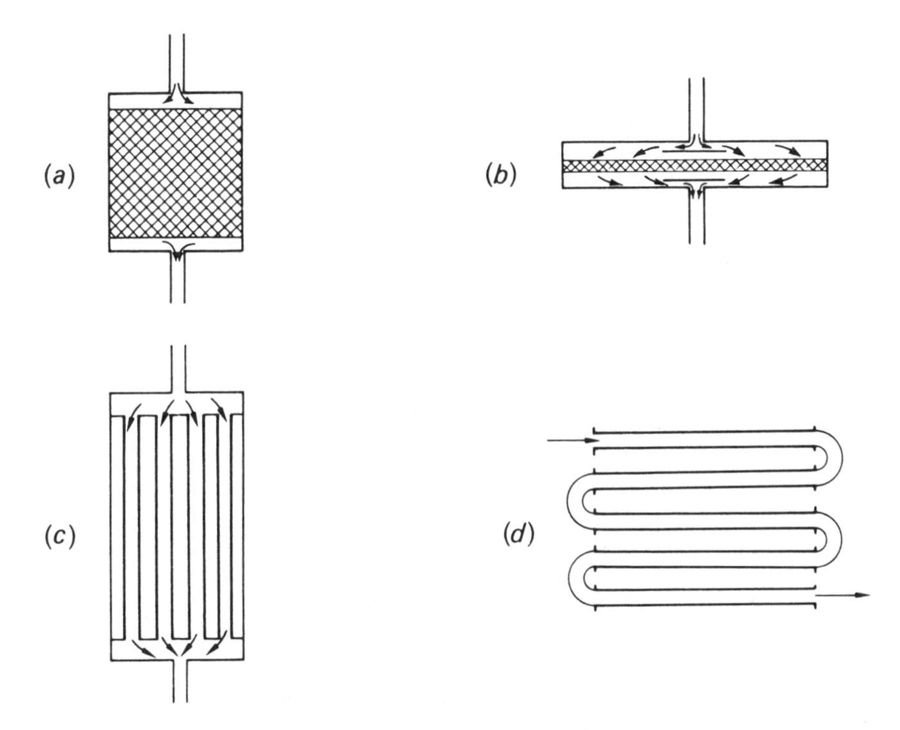

FIG. 1.13. Various configurations for tubular reactors

(*a*) Simple cylindrical shell: suitable only if packed with catalyst
(*b*) Shallow cylinder giving low pressure drop through catalyst bed
(*c*) Tubes in parallel: relatively low tube velocity
(*d*) Tubes in series: high tube velocity

departures from plug flow. With packing in the vessel, such movements are damped out and the simple cylinder is then quite suitable for catalytic reactions where no heat transfer is required. If pressure drop is a problem, the depth of the cylinder may be reduced and its diameter increased as in Fig. 1.13*b*; to avoid serious departures from plug flow in such circumstances, the catalyst must be uniformly distributed and baffles are often used near the inlet and outlet.

When heat transfer to the reactor is required, a configuration with a high surface to volume ratio is employed. In the reactors shown in Fig. 1.13*c* and 1.13*d* the reaction volume is made up of a number of tubes. In *c* they are arranged in parallel, whereas in *d* they are in series. The parallel arrangement gives a lower velocity of the fluid in the tubes, which in turn results in a lower pressure drop, but also a lower heat transfer coefficient (which affects the temperature of the reactant mixture and must be taken into account in calculating the reactor volume). The parallel arrangement is very suitable if a second fluid outside the tubes is used for heat transfer; parallel tubes can be arranged between tube sheets in a compact bundle fitted into a shell, as in a shell and tube heat exchanger. On the other hand, with tubes in series, a high fluid velocity is obtained inside the tubes and a higher heat transfer coefficient results. The series arrangement is therefore often the more suitable if heat transfer is by radiation, when the high heat transfer coefficient helps to prevent overheating of the tubes and coke formation in the case of organic materials.

In practice, there is always some degree of departure from the ideal plug flow condition of uniform velocity, temperature, and composition profiles. If the reactor is not packed and the flow is turbulent, the velocity profile is reasonably flat in the region of the turbulent core (Volume 1, Chapter 3), but in laminar flow, the velocity profile is parabolic. More serious however than departures from a uniform velocity profile are departures from a uniform temperature profile. If there are variations in temperature across the reactor, there will be local variations in reaction rate and therefore in the composition of the reaction mixture. These transverse variations in temperature may be particularly serious in the case of strongly exothermic catalytic reactions which are cooled at the wall (Chapter 3, Section 3.6.1). An excellent discussion on how deviations from plug flow arise is given by DENBIGH and TURNER[5].

1.7.1. Basic Design Equations for a Tubular Reactor

The basic equation for a tubular reactor is obtained by applying the general material balance, equation 1.12, with the plug flow assumptions. In steady state operation, which is usually the aim, the *Rate of accumulation* term (4) is zero. The material balance is taken with respect to a reactant **A** over a differential element of volume δV_t (Fig. 1.14). The fractional conversion of **A** in the mixture entering the element is α_A and leaving it is $(\alpha_A + \delta\alpha_A)$. If F_A is the feed rate of **A** into the reactor (moles per unit time) the material balance over δV_t gives:

$$\underset{\text{Inflow}}{F_A(1-\alpha_A)} - \underset{\text{Outflow}}{F_A(1-\alpha_A-\delta\alpha_A)} - \underset{\text{Reaction}}{\mathcal{R}_A \delta V_t} = 0 \quad \text{(in steady state)} \tag{1.33}$$

$$\therefore \quad F_A \delta\alpha_A = \mathcal{R}_A \delta V_t \tag{1.34}$$

Integrating:

$$\frac{V_t}{F_A} = \int_0^{\alpha_{Af}} \frac{d\alpha_A}{\mathcal{R}_A} \tag{1.35}$$

If the reaction mixture is a fluid whose density remains constant throughout the reaction, equation 1.35 may be written in terms of χ moles of reactant converted per unit volume of fluid. Since $\chi = \alpha_A C_{A0}$, equation 1.35 becomes:

$$\frac{V_t}{F_A} = \frac{1}{C_{A0}} \int_0^{\chi_f} \frac{d\chi}{\mathcal{R}_A} \tag{1.36}$$

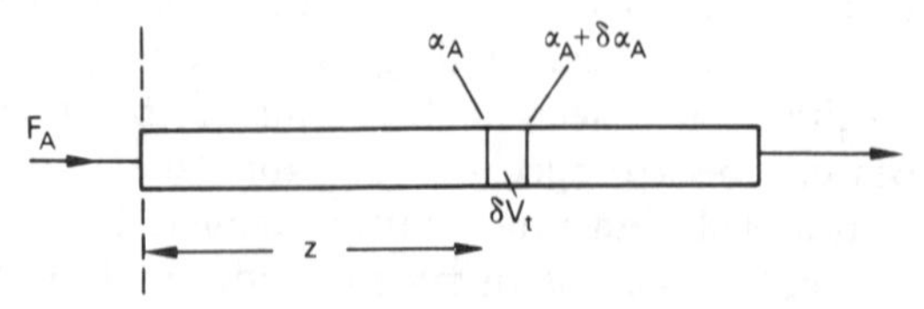

FIG. 1.14. Differential element of a tubular reactor

However, $\dfrac{F_A}{C_{A0}} = v$, the volumetric flowrate of the reaction mixture.

Hence:

$$\frac{V_t}{v} = \int_0^{\chi_f} \frac{d\chi}{\mathcal{R}_A} \tag{1.37}$$

This equation can be derived directly from the general material balance equation above (Fig. 1.14) by expressing the flow of reactant **A** into and out of the reactor element δV_t in terms of the volumetric rate of flow of mixture v, which of course is only valid if v is constant throughout the reactor.

i.e.:

$$\underset{\text{Inflow}}{v(C_{A0} - \chi)} - \underset{\text{Outflow}}{v(C_{A0} - \chi - \delta\chi)} - \underset{\text{Reaction}}{\mathcal{R}_A \delta V_t} = 0 \quad \text{(in steady state)} \tag{1.38}$$

i.e.:

$$v\delta\chi = \mathcal{R}_A \delta V_t \tag{1.39}$$

Hence this leads to equation 1.37

For many tubular reactors, the pressure drop due to flow of the reaction mixture is relatively small, so that the reactor operates at almost constant pressure. An assumption of constant density and the use of equation 1.37 is usually acceptable for liquids and for gas reactions in which there is no change in the total numbers of moles on either side of the stoichiometric equation. If a gas phase reaction does involve a change in the number of moles or if there is a large temperature change, a volume element of reaction mixture will undergo expansion or contraction in passing through the reactor; an assumption of constant v is then unsatisfactory and equation 1.35 must be used. In these circumstances care must be exercised in how the concentrations which appear in the rate equation for $\mathcal{R}_A$ are expressed in terms of α_A, especially when inerts are present; this point is illustrated in the following example. Generally, if the reactor is operated isothermally, $\mathcal{R}_A$ depends on concentrations alone; if it is not operated isothermally, $\mathcal{R}_A$ is a function of temperature also, and the heat balance equation must be introduced. The example is concerned with an isothermal gas-phase reaction in which expansion of a volume element does occur.

Example 1.4

Production of Ethylene by Pyrolysis of Ethane in an Isothermal Tubular Reactor

Ethylene is manufactured on a very large scale[1] by the thermal cracking of ethane in the gas phase:

$$C_2H_6 \rightleftharpoons C_2H_4 + H_2$$

Significant amounts of CH_4 and C_2H_2 are also formed but will be ignored for the purposes of this example. The ethane is diluted with steam and passed through a tubular furnace. Steam is used for reasons very similar to those in the case of ethylbenzene pyrolysis (Section 1.3.2., Example 1.1): in particular it reduces the amounts of undesired byproducts. The economic optimum proportion of steam is, however, rather less than in the case of ethylbenzene. We will suppose that the reaction is to be carried out in an isothermal tubular reactor which will be maintained at 900°C. Ethane will be supplied to the reactor at a rate of 20 tonne/h; it will be diluted with steam in the ratio 0.5 mole steam: 1 mole ethane. The required fractional conversion of ethane is 0.6 (the conversion per pass is relatively low to reduce byproduct formation; unconverted ethane is separated and recycled). The operating pressure is 1.4 bar total, and will be assumed constant, i.e. the pressure drop through the reactor will be neglected.

Laboratory experiments, confirmed by data from large scale operations, have shown that ethane decomposition is a homogeneous first order reaction, the rate constant (s^{-1}) being given by the equation in SI units[6]:

$$k_1 = 1.535 \times 10^{14} \exp(-294{,}000/\mathbf{R}T)$$

Thus at 900°C (1173 K) the value of k_1 is 12.8 s^{-1}

We are required to determine the volume of the reactor.

Solution

The ethane decomposition reaction is in fact reversible, but in the first instance, to avoid undue complication, we shall neglect the reverse reaction; a more complete and satisfactory treatment is given below. For a simple first-order reaction, the rate equation is:

$$\underset{\text{kmol/m}^3\,\text{s}}{\mathcal{R}_A} = \underset{\text{s}^{-1}\,\text{kmol/m}^3}{k_1 C_A}$$

These units for $\mathcal{R}_A$ and C_A will eventually lead to the units m^3 for the volume of the reactor when we come to use equation 1.35. When we substitute for $\mathcal{R}_A$ in equation 1.35, however, to integrate, we must express C_A in terms of α_A, where the reactant A is C_2H_6. To do this we first note that $C_A = y_A C$ where y_A is the mole fraction and C is the molar density of the gas mixture ($kmol/m^3$). Assuming ideal gas behaviour, C is the same for any gas mixture, being dependent only on pressure and temperature in accordance with the ideal gas laws. Thus 1 kmol of gas occupies 22.41 m^3 at 1 bar ($= 1.013 \times 10^5\ N/m^2$) and 273 K. Therefore at 1.4 bar $= 1.4 \times 10^5\ N/m^2$ and 1173 K it will occupy:

$$22.41 \times \frac{1.013 \times 10^5}{1.4 \times 10^5} \times \frac{1173}{273} = 69.8\ \text{m}^3$$

and the molar density C is therefore:

$$1/69.8 = 0.0143\ \text{kmol/m}^3$$

We next determine y_A in terms of α_A by making a subsidiary stoichiometric balance over the reactor between the inlet and the point at which the fractional conversion of the ethane is χ_A. Taking as a basis 1 mole of ethane entering the reactor we have:

C_2H_6, H_2O → REACTOR (α_A) → C_2H_6, C_2H_4, H_2, H_2O

	IN (mole)	AT CONVERSION α_A (a) mole	(b) mole fraction y
C_2H_6	1	$1 - \alpha_A$	$\left(\frac{1-\alpha_A}{1.5+\alpha_A}\right)$
C_2H_4	—	α_A	$\frac{\alpha_A}{1.5+\alpha_A}$
H_2	—	α_A	$\frac{\alpha_A}{1.5+\alpha_A}$
H_2O	0.5	0.5	$\frac{0.5}{1.5+\alpha_A}$
	TOTAL	$1.5 + \alpha_A$	1

Thus:

$$y_A = \frac{1-\alpha_A}{1.5+\alpha_A}$$

Hence, to find the volume of the reactor, using the basic design equation 1.35 and the first-order rate equation:

$$\frac{V_t}{F_A} = \int_0^{\alpha_{Af}} \frac{d\alpha_A}{\mathcal{R}_A} = \int_0^{\alpha_{Af}} \frac{d\alpha_A}{k_1 C_A} = \int_0^{\alpha_{Af}} \frac{d\alpha_A}{k_1 y_A C}$$

$$= \frac{1}{k_1 C}\int_0^{\alpha_{Af}} \frac{(1.5+\alpha_A)}{1-\alpha_A}\, d\alpha_A = \frac{1}{k_1 C}\int_0^{\alpha_{Af}} \left[\frac{2.5}{(1-\alpha_A)} - 1\right] d\alpha_A$$

$$= \frac{1}{k_1 C}\Big[-2.5\ln(1-\alpha_A) - \alpha_A\Big]_0^{\alpha_{Af}} = \frac{1}{k_1 C}\left[2.5\ln\frac{1}{1-\alpha_{Af}} - \alpha_{Af}\right]$$

Introducing the numerical values:

$$\frac{V_t}{F_A} = \frac{1}{12.8 \times 0.0143}\left[2.303 \times 2.5 \log_{10}\frac{1}{(1-0.6)} - 0.6\right]$$

$$= 9.27\ \text{m}^3\ \text{s/kmol}$$

The feed rate of ethane (molecular wt. 30) is to be 20 tonne/h which is equivalent to $F_A = 0.185$ kmol/s. Therefore $V_t = 9.27 \times 0.185 = \underline{\underline{1.72\ \text{m}^3}}$, which is the volume of the reactor required.

The pyrolysis reaction is strongly endothermic so that one of the main problems in designing the reactor is to provide for sufficiently high rates of heat transfer. The volume calculated above would be made up of a series of tubes, probably in the range 50–150 mm diameter, arranged in a furnace similar to that in Fig. 1.5*c*. (For further details of ethylene plants see MILLER[1].)

Calculation with reversible reaction. At 900°C the equilibrium constant K_p for ethane decomposition $P_{C_2H_4}P_{H_2}/P_{C_2H_6}$ is 3.2 bar; using the method described in Example 1.1 the equilibrium conversion of ethane under the conditions above (i.e. 1.4 bar, 0.5 kmol steam added) is 0.86. This shows that the influence of the reverse reaction is appreciable.

The rate equation for a reversible reaction $\mathbf{A} \rightleftharpoons \mathbf{M} + \mathbf{N}$, which is first order in the forward direction and first order with respect to each of **M** and **N** in the reverse direction, may be written in terms of the equilibrium constant K_c:

$$\mathcal{R}_A = k_1\left[C_A - \frac{C_M C_N}{K_c}\right]$$

Expressing this relation in terms of $K_p(= K_c P/C)$ and mole fractions y_A etc. of the various species, ($C_A = Cy_A$ etc):

$$\mathcal{R}_A = k_1 C\left[y_A - \frac{y_M y_N}{K_p}P\right]$$

where P is the total pressure. From the above stoichiometric balance $y_M = y_N = \alpha_A/(1.5 + \alpha_A)$. Hence, substituting, equation 1.35 becomes:

$$\frac{V_t}{F_A} = \int_0^{\alpha_{Af}} \frac{d\alpha_A}{k_1 C\left[\frac{(1-\alpha_A)}{(1.5+\alpha_A)} - \frac{\alpha_A^2}{(1.5+\alpha_A)^2}\frac{P}{K_p}\right]}$$

Substituting numerical values for k_1, C, P and K_p and evaluating this integral for $a_{Af} = 0.6$ gives the result $V_t/F_A = 10.0$ m³ s/kmol: hence $V_t = \underline{\underline{1.86\ \text{m}^3}}$ which may be compared with the previous result above (1.72 m³).

Residence Time and Space Velocity

So far in dealing with tubular reactors we have considered a spatial coordinate as the variable, i.e. an element of volume δV_t situated at a distance z from the reactor inlet (Fig. 1.14), although z has not appeared explicitly in the equations. For a continuous flow reactor operating in a steady state, the spatial coordinate is indeed the most satisfactory variable to describe the situation, because the compositions do not vary with time, but only with position in the reactor.

There is, however, another way of looking at a tubular reactor in which plug flow occurs (Fig. 1.15). If we imagine that a small volume of reaction mixture is encapsulated by a membrane in which it is free to expand or contract at constant pressure, it will behave as a miniature batch reactor, spending a time τ, said to be the *residence time*, in the reactor, and emerging with the conversion α_{Af}. If there is *no expansion or contraction* of the element, i.e. the volumetric rate of flow is constant and equal to v throughout the reactor, the residence time or contact time

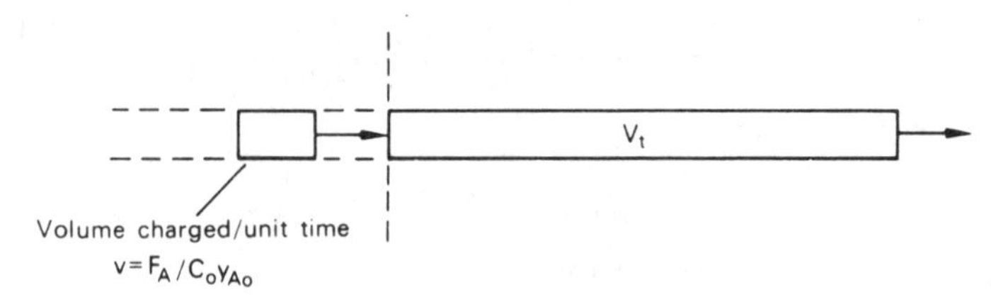

FIG. 1.15. Tubular reactor: residence time $\tau = V_t/v$ if volumetric flowrate constant throughout reactor

is related to the volume of the reactor simply by $\tau = V_t/v$ (e.g. if V_t is 10 m^3 and v is 0.5 m^3/s, i.e. the volume charged to the reactor per second is 0.5 m^3, then the time spent in the reactor is 10/0.5 = 20 s, i.e. V_t/v). To determine the volume of a tubular reactor we may therefore calculate the reaction time t_r for batchwise operation from equation 1.25 and set this equal to the residence time:

$$\frac{V_t}{v} = \tau = t_r = \int_0^{\chi_f} \frac{d\chi}{\mathcal{R}_A} \qquad \text{(equation 1.25)}$$

which is exactly equivalent to equation 1.37. For comparison with equation 1.35 it may be noted that $v = F_A/(C_0 y_{A0})$ where y_{A0} is the mole fraction of **A** at the inlet to the reactor and C_0 is the molar density of the reaction mixture at inlet.

The situation is more complicated if expansion or contraction of a volume element does occur and the volumetric flowrate is not constant throughout the reactor. The ratio V_t/v, where v is the volume flow into the reactor, no longer gives the true residence time or contact time. However, the ratio V_t/v may still be quoted but is called the *space time* and its reciprocal v/V_t the *space velocity*. The space velocity is not in fact a velocity at all; it has dimensions of (time)$^{-1}$ and is therefore really a reactor volume displacement frequency. When a space velocity is quoted in the literature, its definition needs to be examined carefully; sometimes a ratio v_l/V_t is used, where v_l is a liquid volume rate of flow of a reactant which is metered as a liquid but subsequently vaporised before feeding to the reactor.

The space velocity for a given conversion is often used as a ready measure of the performance of a reactor. The use of equation 1.25 to calculate reaction time, as if for a batch reactor, is not to be recommended as normal practice; it can be equated to V_t/v only if there is no change in volume. Further, the method of using reaction time is a blind alley in the sense that it has to be abandoned when the theory of tubular reactors is extended to take into account longitudinal and radial dispersion and other departures from the plug flow hypothesis which are important in the design of catalytic tubular reactors (Chapter 3, Section 3.6.1)

1.7.2. Tubular Reactors — Non-isothermal Operation

In designing and operating a tubular reactor when the heat of reaction is appreciable, strictly isothermal operation is rarely achieved and usually is not economically justifiable, although the aim may be to maintain the local temperatures within fairly narrow limits. On the assumption of plug flow, the rate of temperature rise or fall along the reactor dT/dz is determined by a heat balance (equation 1.20) around an element of reactor volume $\delta V_t = A_c \delta z$, where A_c is the area of cross-section of the reactor:

$$-Gc\frac{\mathrm{d}T}{\mathrm{d}z}\delta z - \frac{\mathrm{d}Q}{\mathrm{d}z}\delta z + (-\Delta H_A)\mathcal{R}_A A_c \delta z = 0 \tag{1.40}$$

Difference between heat inflow and outflow with fluid stream	Heat transferred from element	Heat released by reaction	(in steady state)

where G is the mass flowrate, and $\mathrm{d}Q/\mathrm{d}z$ is the rate of heat transfer per unit length of reactor.

To obtain the temperature and concentration profiles along the reactor, equation 1.40 must be solved by numerical methods simultaneously with equation 1.35 for the material balance.

The design engineer can arrange for the heat transferred per unit length $\mathrm{d}Q/\mathrm{d}z$ to vary with position in the reactor according to the requirements of the reaction. Consider, for example, the pyrolysis of ethane for which a reactor similar to that shown in Fig. 1.5*c* might be used. The cool feed enters the convection section, the duty of which is to heat the reactant stream to the reaction temperature. As the required reaction temperature is approached, the reaction rate increases, and a high rate of heat transfer to the fluid stream is required to offset the large endothermic heat of reaction. This high heat flux at a high temperature is effected in the radiant section of the furnace. Detailed numerical computations are usually made by splitting up the reactor in the furnace into a convenient number of sections.

From a general point of view, three types of expression for the heat transfer term can be distinguished:

(a) *Adiabatic operation*: $\mathrm{d}Q/\mathrm{d}z = 0$. The heat released in the reaction is retained in the reaction mixture so that the temperature rise along the reactor parallels the extent of the conversion α. The material balance and heat balance equations can be solved in a manner similar to that used in the example on an adiabatic batch reactor (Section 1.6.5, Example 1.3). Adiabatic operation is important in heterogeneous tubular reactors and is considered further under that heading in Chapter 3.

(b) *Constant heat transfer coefficient*: This case is again similar to the one for batch reactors and is also considered further in Chapter 3, under heterogeneous reactors.

(c) *Constant heat flux*: If part of the tubular reactor is situated in the radiant section of a furnace, as in Fig. 1.5*c*, and the reaction mixture is at a temperature considerably lower than that of the furnace walls, heat transfer to the reactor occurs mainly by radiation and the rate will be virtually independent of the actual temperature of the reaction mixture. For this part of the reactor $\mathrm{d}Q/\mathrm{d}z$ may be virtually constant.

1.7.3. Pressure Drop in Tubular Reactors

For a homogeneous tubular reactor, the pressure drop corresponding to the desired flowrate is often relatively small and does not usually impose any serious limitation on the conditions of operation. The pressure drop must, of course, be calculated as part of the design so that ancillary equipment may be specified. Only for gases at low pressures or, liquids of high viscosity, e.g. polymers, is the pressure drop likely to have a major influence on the design.

In heterogeneous systems, however, the question of pressure drop may be more serious. If the reaction system is a two-phase mixture of liquid and gas, or if the gas flows through a deep bed of small particles, the pressure drop should be checked at an early stage in the design so that its influence can be assessed. The methods of calculating such pressure drops are much the same as those for flow without reaction (e.g. for packed beds, see Volume 2, Chapter 4).

1.7.4. Kinetic Data from Tubular Reactors

In the laboratory, tubular reactors are very convenient for gas-phase reactions, and for any reaction which is so fast that it is impractical to follow it batchwise. Measurements are usually made when the reactor is operating in a steady state, so that the conversion at the outlet or at any intermediate point does not change with time. For fast reactions particularly, a physical method of determining the conversion, such as ultra-violet or infra-red absorption, is preferred to avoid disturbing the reaction. The conversion obtained at the outlet is regulated by changing either the flowrate or the volume of the reactor.

The reactor may be set up either as a differential reactor, in which case concentrations are measured over a segment of the reactor with only a small change in conversion, or as an integral reactor with an appreciable change in conversion. When the integral method is used for gas-phase reactions in particular, the pressure drop should be small; when there is a change in the number of moles between reactants and products, integrated forms of equation 1.35 which allow for constant pressure expansion or contraction must be used for interpretation of the results. Thus, for an irreversible first-order reaction of the kind $\mathbf{A} \rightarrow \nu_M \mathbf{M}$, using a feed of pure **A** the integrated form of equation 1.35, assuming plug flow, is:

$$\frac{V_t}{F_A} = \frac{1}{k_1 C_{A0}} \left[\nu_M \ln \frac{1}{(1 - \alpha_A)} - (\nu_M - 1)\, \alpha_A \right] \tag{1.41}$$

In order to obtain basic kinetic data, laboratory tubular reactors are usually operated as closely as possible to isothermal conditions. If however a full-scale tubular reactor is to be operated adiabatically, it may be desirable to obtain data on the small scale adiabatically. If the small reactor has the same length as the full-scale version but a reduced cross-section, it may be regarded as a longitudinal element of the large reactor and, assuming plug flow applies to both, scaling up to the large reactor is simply a matter of increasing the cross-sectional area in proportion to the feed rate. One of the problems in operating a small reactor adiabatically, especially at high temperatures, is to prevent heat loss since the surface to volume ratio is large. This difficulty may be overcome by using electrical heating elements wound in several sections along the tube, each controlled by a servo-system with thermocouples which sense the temperature difference between the reaction mixture and the outside of the tube (Fig. 1.16). In this way, the

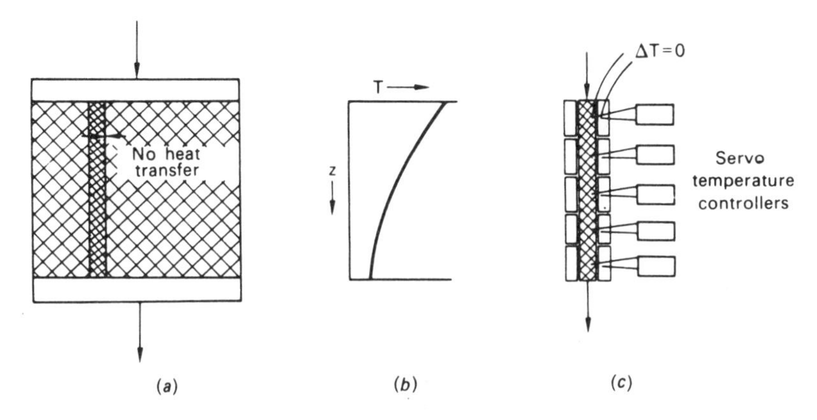

FIG. 1.16. Laboratory-scale reproduction of adiabatic tubular reactor temperature profile
(*a*) Large-scale reactor
(*b*) Adiabatic temperature profile, endothermic reaction
(*c*) Laboratory-scale reactor

temperature of the heating jacket follows exactly the adiabatic temperature path of the reaction, and no heat is lost or gained by the reaction mixture itself. This type of reactor is particularly valuable in developing heterogeneous packed-bed reactors for which it is still reasonable to assume the plug-flow model even for large diameters.

1.8. CONTINUOUS STIRRED-TANK REACTORS

The stirred-tank reactor in the form of either a single tank, or more often a series of tanks (Fig. 1.1*d*), is particularly suitable for liquid-phase reactions, and is widely used in the organic chemicals industry for medium and large scale production. It can form a unit in a continuous process, giving consistent product quality, ease of automatic control, and low manpower requirements. Although, as we shall see below, the volume of a stirred-tank reactor must be larger than that of a plug-flow tubular reactor for the same production rate, this is of little importance because large volume tanks are relatively cheap to construct. If the reactor has to be cleaned periodically, as happens sometimes in polymerisations or in plant used for manufacturing a variety of products, the open structure of a tank is an advantage.

In a stirred-tank reactor, the reactants are diluted immediately on entering the tank; in many cases this favours the desired reaction and suppresses the formation of byproducts. Because fresh reactants are rapidly mixed into a large volume, the temperature of the tank is readily controlled, and hot spots are much less likely to occur than in tubular reactors. Moreover, if a series of stirred tanks is used, it is relatively easy to hold each tank at a different temperature so that an optimum temperature sequence can be attained.

1.8.1. Assumption of Ideal Mixing. Residence Time

In the theory of continuous stirred-tank reactors, an important basic assumption is that the contents of each tank are *well-mixed*. This means that the compositions in the tank are everywhere uniform and that the product stream leaving the tank has the same composition as the mixture within the tank. This assumption is reasonably well borne out in practice unless the tank is exceptionally large, the stirrer inadequate, or the reaction mixture very viscous.

In the treatment which follows, it will be assumed that the mass density of the reaction mixture is constant throughout a series of stirred tanks. Thus, if the volumetric feed rate is v, then in the steady state, the rate of outflow from each tank will also be v. Material balances may then be written on a volume basis, and this considerably simplifies the treatment. In practice, the constancy of the density of the mixture is a reasonable assumption for liquids, and any correction which may need to be applied is likely to be small.

The mean residence time for a continuous stirred-tank reactor of volume V_c may be defined as V_c/v in just the same way as for a tubular reactor. However, in a homogeneous reaction mixture, it is not possible to identify particular elements of fluid as having any particular residence time, because there is complete mixing on a molecular scale. If the feed consists of a suspension of particles, it may be shown that, although there is a distribution of residence times among the individual particles, the mean residence time does correspond to V_c/v if the system is ideally mixed.

1.8.2. Design Equations for Continuous Stirred-Tank Reactors

When a series of stirred-tanks is used as a chemical reactor, and the reactants are fed at a constant rate, eventually the system reaches a steady state such that the concentrations in the individual tanks, although different, do not vary with time. When the general material balance of equation 1.19 is applied, the accumulation term is therefore zero. Considering first of all the most general case in which the mass density of the mixture is not necessarily constant, the material balance on the reactant **A** is made on the basis of F_A moles of **A** per unit time fed to the first tank. Then a material balance for the rth tank of volume V_{cr} (Fig. 1.17) is, in the steady state:

$$\underset{\text{Inflow}}{F_A(1 + \alpha_{Ar-1})} - \underset{\text{Outflow}}{F_A(1 - \alpha_{Ar})} - \underset{\text{Reaction}}{\mathcal{R}_A V_{cr}} = 0 \qquad (1.42)$$

where α_{Ar-1} is the fractional conversion of **A** in the mixture leaving tank $r - 1$ and entering tank r, and

α_{Ar} is the fractional conversion of **A** in the mixture leaving tank r.

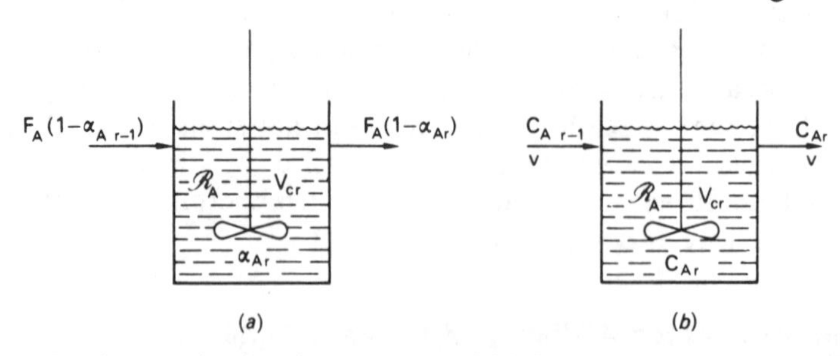

FIG. 1.17. Continuous stirred-tank reactor: material balance over rth tank in steady state

(*a*) General case; feed to first tank F_A
(*b*) Constant volume flowrate v

Assuming that the contents of the tank are well mixed, α_{Ar} is also the fractional conversion of the reactant **A** in tank r.

$$\therefore \qquad \frac{V_{cr}}{F_A} = \frac{\alpha_{Ar} - \alpha_{Ar-1}}{\mathcal{R}_A} \tag{1.43}$$

which is the counterpart of equation 1.35 for a tubular reactor.

Stirred tanks are usually employed for reactions in liquids and, in most cases, the mass density of the reaction mixture may be assumed constant. Material balances may then be taken on the basis of the volume rate of flow v which is constant throughout the system of tanks. The material balance on **A** over tank r may thus be written, in the steady state:

$$\underset{\text{Inflow}}{vC_{Ar-1}} - \underset{\text{Outflow}}{vC_{Ar}} - \underset{\text{Reaction}}{V_{cr}\mathcal{R}_A} = 0 \tag{1.44}$$

where C_{Ar-1} is the concentration of **A** in the liquid entering tank r from tank $r - 1$, and C_{Ar} is the concentration of **A** in the liquid leaving tank r.

$$\therefore \qquad \frac{V_{cr}}{v} = \frac{C_{Ar-1} - C_{Ar}}{\mathcal{R}_A} \tag{1.45}$$

Equation 1.45 may be written in terms of $\chi_r = (C_{A0} - C_{Ar})$; C_{A0} is here the concentration of **A** in the feed to the first tank.

$$\text{Thus:} \qquad \frac{V_{cr}}{v} = \frac{\chi_r - \chi_{r-1}}{\mathcal{R}_A} \tag{1.46}$$

In these equations $V_{cr}/v = \tau_r$, the residence time in tank r. Equation 1.46 may be compared with equation 1.37 for a tubular reactor. The difference between them is that, whereas 1.37 is an integral equation, 1.46 is a simple algebraic equation. If the reactor system consists of only one or two tanks the equations are fairly simple to solve. If a large number of tanks is employed, the equations whose general form is given by 1.46 constitute a set of finite-difference equations and must be solved accordingly. If there is more than one reactant involved, in general a set of material balance equations must be written for each reactant.

In order to proceed with a solution, a rate equation is required for $\mathcal{R}_A$. Allowance must be made for the fact that the rate constant will be a function of temperature, and may therefore be different for each tank. The temperature distribution will depend on the heat balance for each tank (equation 1.20), and this will be affected by the amount of heating or cooling that is carried out. The example which follows concerns an isothermal system of two tanks with two reactants, one of which is in considerable excess.

Example 1.5

A Two-Stage Continuous Stirred-Tank Reactor

A solution of an ester **R**·COOR′ is to be hydrolysed with an excess of caustic soda solution. Two stirred tanks of equal size will be used. The ester and caustic soda solutions flow separately into the first tank at rates of 0.004 and 0.001 m^3/s and with concentrations of 0.02 and 1.0 $kmol/m^3$ respectively. The reaction:

$$R \cdot COOR' + NaOH \rightarrow R \cdot COONa + R'OH$$

is second order with a velocity constant of 0.033 m^3/kmol s at the temperature at which both tanks operate. Determine the volume of the tanks required to effect 95 per cent conversion of the ester ($\alpha_{A_2} = 0.95$).

Solution

Although the solutions are fed separately to the first tank, we may for the purpose of argument consider them to be mixed together just prior to entering the tank as shown in Fig. 1.18. If the ester is denoted by **A** and the caustic soda by **B** and χ is the number of moles of ester (or caustic soda) which have reacted per litre of the combined solutions, a material balance on the ester over the first tank gives, for the steady state:

$$\underset{\text{Inflow}}{vC_{A0}} - \underset{\text{Outflow}}{v(C_{A0} - \chi_1)} - \underset{\text{Reaction}}{V_{c1}k_2(C_{A0} - \chi_1)(C_{B0} - \chi_1)} = 0$$

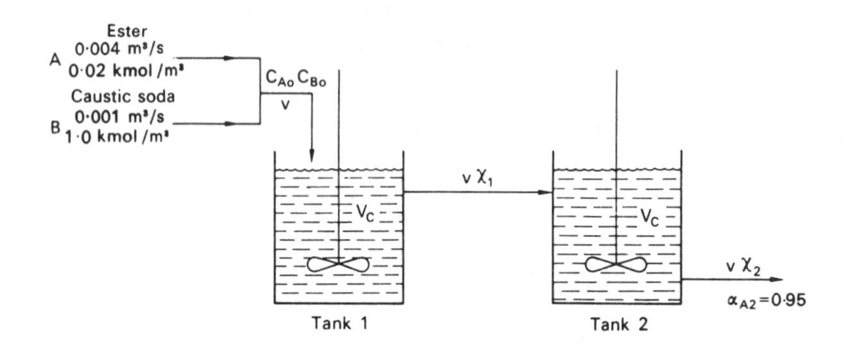

FIG. 1.18. Continuous stirred-tank reactor: worked example

Similarly for tank 2:

$$v(C_{A0} - \chi_1) - v(C_{A0} - \chi_2) - C_{c2}k_2(C_{A0} - \chi_2)(C_{B0} - \chi_2) = 0$$

Because we have used the variable χ, a further set of material balance equations for reactant **B** is not required; they are implicit in the equations for **A** above. Rearranging these, and putting $V_{c1} = V_{c2}$ because the tanks are of equal size:

$$v\chi_1 = V_{c1}k_2(C_{A0} - \chi_1)(C_{B0} - \chi_1) \tag{A}$$

and:

$$v(\chi_2 - \chi_1) = V_{c1}k_2(C_{A0} - \chi_2)(C_{B0} - \chi_2) \tag{B}$$

In these equations we know $\chi_2 = \alpha_{A2}C_{A0}$ and all the other quantities excepting χ_1 and V_{c1}; hence by eliminating χ_1 we may find V_{c1}.
Considering numerical values, the total volume flowrate $v = 0.005$ m^3/s.

$$C_{A0} = 0.02 \times \frac{0.004}{0.005} = 0.016 \text{ kmol/m}^3$$

$$C_{B0} = 1.0 \times \frac{0.001}{0.005} = 0.20 \text{ kmol/m}^3$$

$$\chi_2 = 0.95 \times 0.016 = 0.0152 \text{ kmol/m}^3$$

$$(C_{A0} - \chi_2) = 0.0008 \text{ kmol/m}^3$$

$$(C_{B0} - \chi_2) = 0.1848 \text{ kmol/m}^3$$

At this stage we note that because a substantial excess of caustic soda is used, as a first approximation χ_1 may be neglected in comparison with C_{B0}. We thus avoid the necessity of solving a cubic equation. At a later stage the value of χ_1 obtained using this approximation can be substituted in the term $(C_{B0} - \chi_1)$

and the calculation repeated in an iterative fashion. Setting $(C_{B0} - \chi_1) = 0.20$, equations A and B become:

$$0.005\,\chi_1 = V_{c1} \times 0.033 \times (0.016 - \chi_1) \times 0.20 \tag{C}$$

$$0.005(0.0152 - \chi_1) = V_{c1} \times 0.033 \times 0.0008 \times 0.1848 \tag{D}$$

Eliminating χ_1, and rearranging:

$$V_{c1}^2 + 1.56\,V_{c1} - 11.6 = 0$$

Hence:

$$V_{c1} = 2.71\ \text{m}^3$$

Thus as a first estimate, the two tanks should each be of this capacity. An improved estimate may be obtained by using the value of V_{c1} calculated above and substituting in equation D, giving $\chi_1 = 0.0125\ \text{kmol/m}^3$. Thus the second approximation to the term $(C_{B0} - \chi_1)$ is 0.1875. Repeating the calculation:

$$\underline{\underline{V_{c1} = 2.80\ \text{m}^3}}$$

1.8.3. Graphical Methods

For second order reactions, graphs showing the fractional conversion for various residence times and reactant feed ratios have been drawn up by ELDRIDGE and PIRET[7]. These graphs, which were prepared from numerical calculations based on equation 1.46, provide a convenient method for dealing with sets of equal sized tanks of up to five in number, all at the same temperature.

A wholly graphical method arising from equation 1.45 or 1.46 may be used providing that the rate of reaction $\mathscr{R}_A$ is a function of a single variable only, either C_A or χ. The tanks must therefore all be at the same temperature. Experimental rate data may be used directly in graphical form without the necessity of fitting a rate equation. In order to establish the method, equation 1.45 is firstly rearranged to give:

$$\mathscr{R}_{Ar} = -\frac{v}{V_{cr}}(C_{Ar} - C_{Ar-1}) \tag{1.47}$$

where the subscript r has been added to stress that $\mathscr{R}_{Ar}$ is the value of $\mathscr{R}_A$ for the tank r. Consider now a graph of $\mathscr{R}_A$ versus C_A as shown in Fig. 1.19*a*. From a point on the C_A-axis, $\mathscr{R}_A = 0$, $C_A = C_{Ar-1}$ we construct a line of slope:

$$-\frac{v}{V_{cr}}$$

which therefore has the equation:

$$\mathscr{R}_A = -\frac{v}{V_{cr}}(C_A - C_{Ar-1}) \tag{1.48}$$

Equation 1.47 shows that the point $(\mathscr{R}_{Ar}, C_{Ar})$ lies on this line; however $\mathscr{R}_{Ar}$ must also lie on the rate of reaction versus C_A curve, so that $(\mathscr{R}_{Ar}, C_{Ar})$ is the point of intersection of the two as shown. Thus starting with the first tank, we can draw the first line from the point $O_0(0, C_{A0})$ on Fig. 1.19*b* and locate C_{A1}, the concentration of the reactant leaving the first tank. Then from the point $O_1(0, C_{A1})$ we can draw a second line to locate C_{A2} for the second tank, and so on for the whole series as the following example shows.

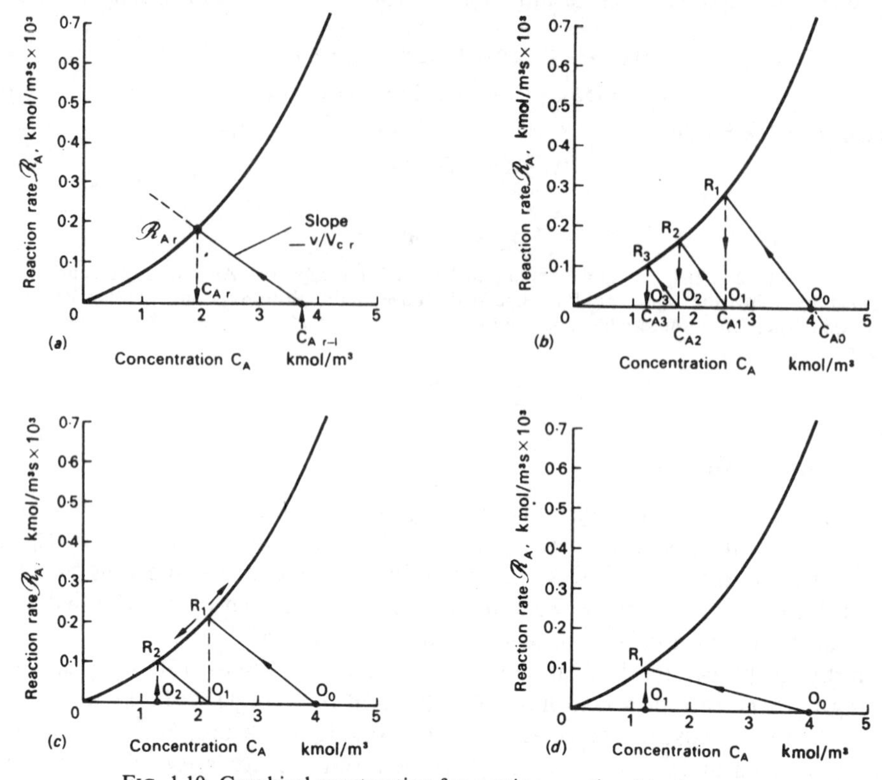

FIG. 1.19. Graphical construction for continuous stirred-tank reactors
(*a*) General method
(*b*) Three equal tanks, outlet concentration C_{A3} unknown
(*c*) Two equal tanks, volume unknown
(*d*) One tank, volume unknown

Example 1.6

Graphical Construction for a Three-Stage Continuous Stirred-Tank Reactor

(a) A system of three stirred tanks is to be designed to treat a solution containing 4.0 kmol/m³ of a reactant A. Experiments with a small reactor in the laboratory gave the kinetic data shown as a graph of rate of reaction versus C_A in Fig. 1.19*b*. If the feed rate to the reactor system is 1.2×10^{-4} m³/s, what fractional conversion will be obtained if each of the three tanks has a volume of 0.60 m³?

(b) Calculate the volumes of the tanks required for the same overall conversion if two equal tanks are used, and if only one tank is used.

Solution

(a) Referring to Fig. 1.19*b*, from the point O_0 representing the feed composition of 4.0 kmol/m³, a line O_0R_1 is drawn of slope $-1.2 \times 10^{-4}/0.60 = -2 \times 10^{-4}\ \text{s}^{-1}$ to intersect the rate curve at R_1. This point of intersection gives the concentration of A in the first tank C_{A1}. A perpendicular R_1O_1 is dropped from R_1 to the C_A axis, and from the point O_1 a second line also of slope $-2 \times 10^{-4}\ \text{s}^{-1}$ is drawn. The construction is continued until O_3 is reached which gives the concentration of A leaving the last tank C_{A3}. Reading from the figure $C_{A3} = 1.23$ kmol/m³. The fractional conversion is therefore $(4.0 - 1.23)/4.0 = \underline{\underline{0.69}}$.

(b) When the volumes of the identical tanks are unknown the graphical construction must be carried out on a trial and error basis. The procedure for the case of two tanks is shown in Fig. 1.19*c*; the points O_0 and O_2 are known, but the position of R_1 has to be adjusted to make O_0R_1 and O_1R_2 parallel because these lines must have the same slope if the tanks are of equal size. From the figure this slope is $1.13 \times 10^{-4}\,s^{-1}$. The volume of each tank must therefore be $1.2 \times 10^{-4}/1.13 \times 10^{-4} = 1.06\,m^3$.

For a single tank the construction is straightforward as shown in Fig. 1.19*d* and the volume obtained is $\underline{\underline{3.16\,m^3}}$.

It is interesting to compare the total volume required for the same duty in the three cases.

Total volume for 3 tanks is $3 \times 0.60 = 1.80\,m^3$
for 2 tanks is $2 \times 1.06 = 2.12\,m^3$
for 1 tank is $1 \times 3.16 = 3.16\,m^3$

These results illustrate the general conclusion that, as the number of tanks is increased, the total volume required diminishes and tends in the limit to the volume of the equivalent plug-flow reactor. The only exception is in the case of a zero order reaction for which the total volume is constant and equal to that of the plug flow reactor for all configurations.

1.8.4. Autothermal Operation

One of the advantages of the continuous stirred-tank reactor is the fact that it is ideally suited to autothermal operation. Feed-back of the reaction heat from products to reactants is indeed a feature inherent in the operation of a continuous stirred-tank reactor consisting of a single tank only, because fresh reactants are mixed directly into the products. An important, but less obvious, point about autothermal operation is the existence of two possible stable operating conditions.

To understand how this can occur, consider a heat balance over a single tank operating in a steady state. The tank is equipped with a cooling coil of area A_t through which flows a cooling medium at a temperature T_C. Because conditions are steady, the accumulation term in the general heat balance, equation 1.20, is zero. The remaining terms in the heat balance may then be arranged as follows (cf. equations 1.31 for a batch reactor and 1.40 for a tubular reactor):

$$(-\Delta H_A)\mathcal{R}_A V_c = Gc(T - T_0) + UA_t(T - T_C) \tag{1.49}$$

Rate of heat generation by reaction	Rate of heat removal by outflow of products	Rate of heat removal by heat transfer

Here, the enthalpy of the products of mass flowrate G and specific heat c is measured relative to T_0, the inlet temperature of the reactants. The term for rate of heat generation on the left-hand side of this equation varies with the temperature of operation T, as shown in diagram (*a*) of Fig. 1.20; as T increases, $\mathcal{R}_A$ increases rapidly at first but then tends to an upper limit as the reactant concentration in the tank approaches zero, corresponding to almost complete conversion. On the other hand, the rate of heat removal by both product outflow and heat transfer is virtually linear, as shown in diagram (*b*). To satisfy the heat balance equation above, the point representing the actual operating temperature must lie on both the rate of heat production curve and the rate of heat removal line, i.e. at the point of intersection as shown in (*c*).

In Fig. 1.20*c*, it may be seen how more than one stable operating temperature can sometimes occur. If the rate of heat removal is high (line 1), due either to rapid outflow or to a high rate of heat transfer, there is only one point of intersection O_1, corresponding to a low operating temperature close to the reactant inlet temperature T_0 or the cooling medium temperature T_C. With a somewhat smaller flowrate or heat transfer rate (line 2) there are three points of intersection

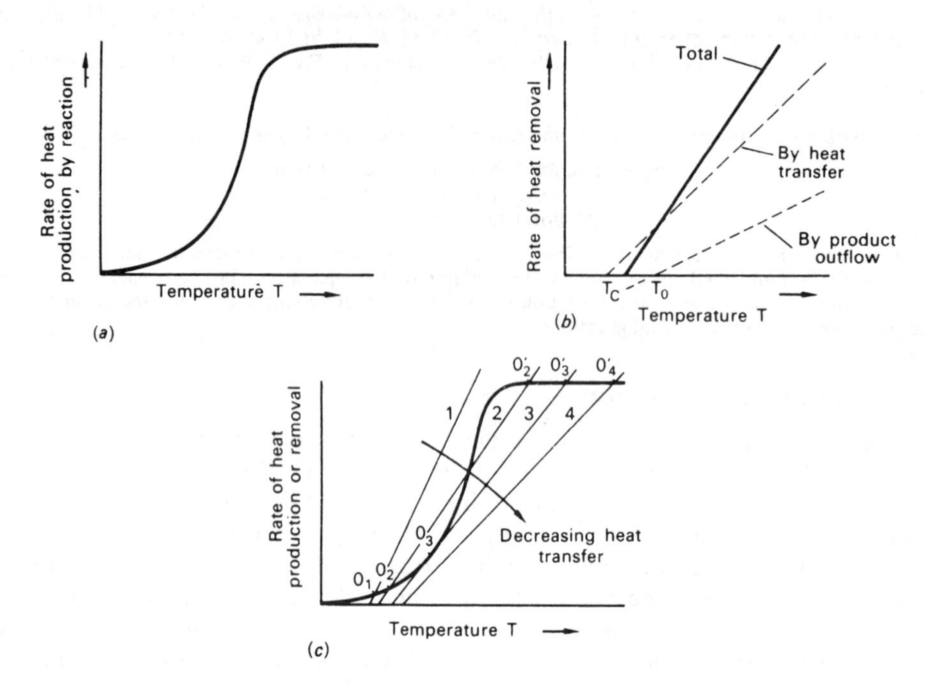

FIG. 1.20. Autothermal operation of a continuous stirred-tank reactor
(*a*) and (*b*) show rates of heat production and removal
(*c*) shows the effects of different amounts of heat transfer on possible stable operating states

corresponding to two stable conditions of operation, O_2 at a low temperature and conversion, and O_2' at a considerably higher temperature and conversion. If the reactor is started up from cold, it will settle down in the lower operating state O_2. However, if a disturbance causes the temperature to rise above the intermediate point of intersection beyond which the rate of heat production exceeds the rate of loss, the system will pass into the upper operating state O_2'. Line 3 represents a heat removal rate for which the lower operating state O_3 can only just be realised, and for line 4, only an upper operating temperature O_4' is possible.

Obviously, in designing and operating a stirred-tank reactor it is necessary to be aware of these different operating conditions. Further discussion of the dynamic response and control of an autothermal continuous stirred-tank reactor is given by WESTERTERP *et al.*[8].

1.8.5. Kinetic Data from Continuous Stirred-Tank Reactors

For fast or moderately fast liquid phase reactions, the stirred-tank reactor can be very useful for establishing kinetic data in the laboratory. When a steady state has been reached, the composition of the reaction mixture may be determined by a physical method using a flow cell attached to the reactor outlet, as in the case of a tubular reactor. The stirred-tank reactor, however, has a number of further advantages in comparison with a tubular reactor. With an appropriate ratio of reactor volume to feed rate and with good mixing, the difference in reactant composition between feed and outflow can be large, without changing the basic situation whereby all the reaction takes place at a single uniform concentration level in the reactor. Comparatively large differences between inlet and outlet compositions are required in order to determine the rate of the reaction with reasonable accuracy. With a tubular reactor, on the other hand, if a large difference in reactant composition is set up across the reactor, the integral method of interpreting the results must be employed and this may give rise to problems when dealing with complex reactions.

There is one further point of comparison. Interpretation of results from a stirred-tank reactor depends on the assumption that the contents of the tank are well mixed. Interpretation of results from a tubular reactor rests on the assumption of plug flow unless the flow is laminar and is treated as such. Which of these two assumptions can be met most satisfactorily in practical experiments? Unless the viscosity of the reaction mixture is high or the reaction extremely fast, a high speed stirrer is very effective in maintaining the contents of a stirred tank uniform. On the other hand, a tubular reactor may have to be very carefully designed if back-mixing is to be completely eliminated, and in most practical situations there is an element of uncertainty about whether the plug flow assumption is valid.

1.9. COMPARISON OF BATCH, TUBULAR AND STIRRED-TANK REACTORS FOR A SINGLE REACTION. REACTOR OUTPUT

There are two criteria which can be used to compare the performances of different types of reactor. The first, which is a measure of reactor productivity, is the output of product in relation to reactor size. The second, which relates to reactor selectivity, is the extent to which formation of unwanted byproducts can be suppressed. When comparing reactions on the basis of output as in the present section, only one reaction need be considered, but when in the next section the question of byproduct formation is taken up, more complex schemes of two or more reactions must necessarily be introduced.

In defining precisely the criterion of reactor output, it is convenient to use one particular reactant as a reference rather than the product formed. The distinction is unimportant if there is only one reaction, but is necessary if more than one reaction is involved. The *unit output* W_A of a reactor system may thus be defined as the moles of reactant **A** converted, per unit time, per unit volume of reaction space; in calculating this quantity it should be understood that the total moles converted by the whole reactor in unit time is to be divided by the total volume of the system. The unit output is therefore an average rate of reaction for the reactor as a whole, and is thus distinct from the specific rate $\mathcal{R}_A$ which is the local rate of reaction.

We shall now proceed to compare the three basic types of reactor—batch, tubular and stirred tank—in terms of their performance in carrying out a single first order irreversible reaction:

$$\mathbf{A} \rightarrow \text{Products}$$

It will be assumed that there is no change in mass density and that the temperature is uniform throughout. However, it has already been shown that the conversion in a tubular reactor with plug flow is identical to that in a batch reactor irrespective of the order of the reaction, if the residence time of the tubular reactor τ is equal to reaction time of the batch reactor t_r. Thus, the comparison rests between batch and plug-flow tubular reactors on the one hand, and stirred-tank reactors consisting of one, two or several tanks on the other.

1.9.1. Batch Reactor and Tubular Plug-Flow Reactor

In terms of χ, moles reacted in unit volume after time t, the material balance for a first order reaction is simply:

$$\frac{d\chi}{dt} = \boldsymbol{k}_1(C_{A0} - \chi) \tag{1.50}$$

where C_{A0} is the initial concentration of the reactant **A**. (The subscript A will from now on be omitted from C, W and α because there is only the one reactant throughout.) Integrating:

$$t_r = \frac{1}{\boldsymbol{k}_1} \ln \frac{C_0}{C_0 - \chi} \tag{1.51}$$

as shown in Table 1.1. Since in unit volume a total of χ moles of **A** have reacted after a time t_r, the unit output W_b from a batch reactor is given by:

$$W_b = \chi / t_r$$

or in terms of the fractional conversion $\alpha(\chi = \alpha C_0)$:

$$W_b = \frac{\alpha C_0}{t_r} = \frac{\alpha C_0 \boldsymbol{k}_1}{\ln [1/(1 - \alpha)]} \tag{1.52}$$

$$\therefore \qquad \frac{W_b}{\boldsymbol{k}_1 C_0} = \frac{\alpha}{\ln [1/(1 - \alpha)]} \tag{1.53}$$

1.9.2. Continuous Stirred-Tank Reactor

One tank

A material balance on the reactant gives (Fig. 1.21*a*), for the steady state:

$$\underset{\text{Inflow}}{vC_0} - \underset{\text{Outflow}}{v(C_0 - \chi)} - \underset{\text{Reaction}}{V_c \boldsymbol{k}_1(C_0 - \chi)} = 0 \tag{1.54}$$

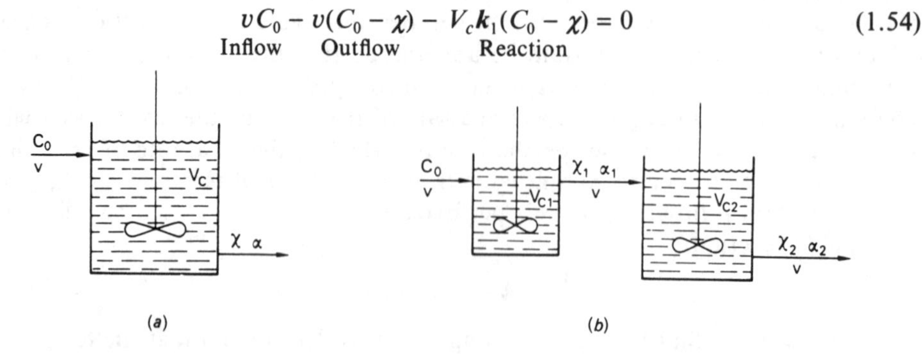

FIG. 1.21. Continuous stirred-tank reactors: calculation of unit output

Considering the outflow, it may bc sccn that in unit time, $v\chi$ moles of reactant are converted in the reactor of volume V_c. Thus:

$$W_{c1} = \frac{v\chi}{V_c}$$

Hence:

$$\frac{v\chi}{V_c} = \boldsymbol{k}_1(C_0 - \chi) = \boldsymbol{k}_1 C_0(1 - \alpha) \tag{1.55}$$

$$\therefore \qquad \frac{W_{c1}}{\boldsymbol{k}_1 C_0} = (1 - \alpha) \tag{1.56}$$

Two tanks

Let us first of all consider the general case in which the tanks are not necessarily of equal size (Fig. 1.21*b*).

Taking a material balance on the reactant **A** over each tank in succession:

Tank 1:

$$vC_0 - v(C_0 - \chi_1) - V_{c1}\boldsymbol{k}_1(C_0 - \chi_1) = 0$$

Tank 2:

$$v(C_0 - \chi_1) - v(C_0 - \chi_2) - V_{c2}\boldsymbol{k}_1(C_0 - \chi_2) = 0$$

i.e.:

$$v\chi_1 = V_{c1}\boldsymbol{k}_1(C_0 - \chi_1)$$

$$v(\chi_2 - \chi_1) = V_{c2}\boldsymbol{k}_1(C_0 - \chi_2)$$

i.e.:

$$\alpha_1 = \frac{V_{c1}}{v}\boldsymbol{k}_1(1 - \alpha_1)$$

$$\alpha_2 - \alpha_1 = \frac{V_{c2}}{v}\boldsymbol{k}_1(1 - \alpha_2)$$

At this point there arises the question of whether the two tanks should be the same size or of different sizes. Mathematically, this question needs to be investigated with some care so that the desired objective function is correctly identified. If we wished to design a two-stage reactor we might be interested in the minimum total volume $V_c = V_{c1} + V_{c2}$ required for a given conversion α_2. Writing the ratio $V_{c2}/V_{c1} = s$, this condition is met by setting:

$$\left(\frac{\partial V_c}{\partial s}\right)_{\alpha_2} = 0$$

After some manipulation this gives $s = 1$, showing that the two reactors should be of equal size. If, however, a fixed total volume is considered and the ratio s is varied to find the effect on α_2, the maximum value of α_2 is found by setting:

$$\left(\frac{\partial \alpha_2}{\partial s}\right)_{V_c} = 0$$

The maximum does not then occur at $s = 1$, but this is not a likely situation in practice. In general, it may be shown that the optimum value of the ratio s depends on the order of reaction and is unity only for a first-order reaction. However, the convenience and reduction in costs associated with having all tanks the same size will in practice always outweigh any small increase in total volume that this may entail.

We will assume henceforth that the tanks are of equal size, i.e.:

$$V_{c1} = V_{c2} = \tfrac{1}{2} V_c$$

Eliminating α_1 from the equations above, we may proceed to determine the unit output W_{c2} of a series of two tanks, where:

$$W_{c2} = \frac{v\chi_2}{V_c} = \frac{vC_0\alpha_2}{V_c}$$

After fairly lengthy manipulation, we obtain:

$$\frac{W_{c2}}{k_1 C_0} = \frac{\alpha_2}{2}\left[\frac{(1-\alpha_2)^{1/2}}{1-(1-\alpha_2)^{1/2}}\right] \tag{1.57}$$

1.9.3. Comparison of Reactors

It may be seen from equations 1.53, 1.56 and 1.57 that unit output is a function of conversion. Some numerical values of the dimensionless quantity $W/k_1 C_0$ representing the unit output are shown in Table 1.3. Shown also in Table 1.3 are values of the following ratios for various values of the conversion:

$$\frac{\text{Unit output batch reactor}}{\text{Unit output stirred-tank reactor}} = \frac{\text{Volume stirred-tank reactor}}{\text{Volume batch reactor}}$$

TABLE 1.3. *Comparison of Continuous Stirred-Tank Reactors and Batch Reactors with Respect to Unit Output $W/k_1 C_0$ and Reactor Volume. First-Order Reaction*

Reactor type		Conversion			
		0.50	0.90	0.95	0.99
Batch or tubular plug-flow	Unit output	0.722	0.391	0.317	0.215
C.S.T.R. one tank	Unit output	0.50	0.10	0.05	0.01
	Vol. ratio C.S.T.R./Batch	1.44	3.91	6.34	21.5
C.S.T.R. two tanks	Unit output	0.604	0.208	0.137	0.055
	Vol. ratio C.S.T.R./Batch	1.19	1.88	2.31	3.91

These show that a single continuous stirred-tank reactor must always be larger than a batch or tubular plug-flow reactor for the same duty, and for high conversions the stirred tank must be very much larger indeed. If two tanks are used, however, the total volume is less than that of a single tank. Although the detailed calculations

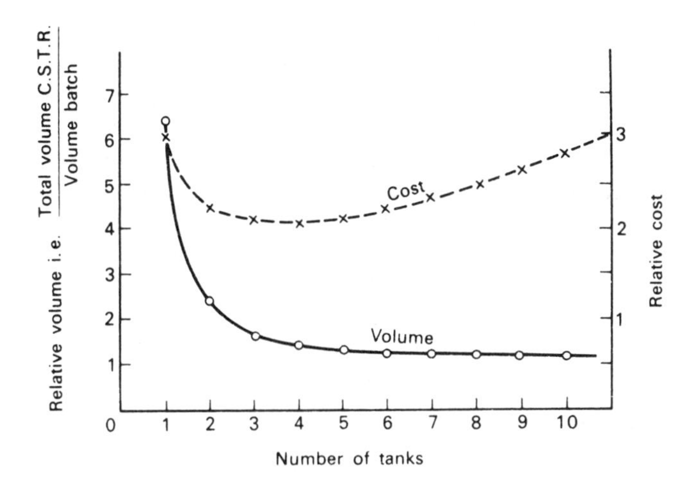

FIG. 1.22. Comparison of size and cost of continuous stirred-tank reactors with a batch or a tubular plug-flow reactor: first-order reaction, conversion 0.95

for systems of three or more tanks are not given here, it can be seen in Fig. 1.22, which is based on charts prepared by LEVENSPIEL[9], that the total volume is progressively reduced as the number of tanks is increased. This principle is evident also in the example on stirred-tank reactors solved by the graphical method used in Example 1.6, which does not refer to a first-order reaction. Calculations such as those in Table 1.3 can be extended to give results for orders of reaction both greater than and less than one. As the order of the reaction increases, so the comparison becomes even less favourable to the stirred-tank reactor.

As the number of stirred tanks in a series is increased, so is the total volume of the system reduced. In the limit with an infinite number of tanks, we can expect the volume to approach that of the equivalent batch or tubular reactor because, in the limiting case, plug flow is obtained. However, although the total volume of a series of tanks progressively decreases with increasing number, this does not mean that the total cost will continue to fall. The cost of a tank and its associated mixing and heat transfer equipment will be proportional to approximately the 0.6 power of its volume. When total cost is plotted against the number of tanks, as in the second curve of Fig. 1.22, the curve passes through a minimum. This usually occurs in the region of 3 to 6 tanks and it is most likely that a number in this range will be employed in practice.

1.10. COMPARISON OF BATCH, TUBULAR AND STIRRED-TANK REACTORS FOR MULTIPLE REACTIONS. REACTOR YIELD

If more than one chemical reaction can take place in a reaction mixture, the type of reactor used may have a quite considerable effect on the products formed. The choice of operating conditions is also important, especially the temperature and the degree of conversion of the reactants. The economic importance of choosing the type of reactor which will suppress any unwanted byproducts to the greatest extent has already been stressed (Section 1.1.1).

In this section, our aim will be to take certain model reaction schemes and work out in detail the product distribution which would be obtained from each of the basic types of reactor. It is fair to say that in practice there are often difficulties in attempting to design reactors from fundamental principles when multiple reactions are involved. Information on the kinetics of the individual reactions is often incomplete, and in many instances an expensive and time-consuming laboratory investigation would be needed to fill in all the gaps. Nevertheless, the model reaction schemes examined below are valuable in indicating firstly how such limited information as may be available can be used to the best advantage, and secondly what key experiments should be undertaken in any research and development programme.

1.10.1. Types of Multiple Reactions

Multiple reactions are of two basic kinds. Taking the case of one reactant only, these are:

(a) Reactions in parallel or competing reactions of the type:

$$\mathbf{A} \rightarrow \mathbf{P} \text{ (desired product)}$$
$$\searrow \mathbf{Q} \text{ (unwanted product)}$$

(b) Reactions in series or consecutive reactions of the type:

$$\mathbf{A} \rightarrow \mathbf{P} \rightarrow \mathbf{Q}$$

where again **P** is the desired product and **Q** the unwanted byproduct.

When a second reactant **B** is involved, the situation is basically unchanged in the case of parallel reactions:

$$\mathbf{A} + \mathbf{B} \rightarrow \mathbf{P}$$
$$\mathbf{A} + \mathbf{B} \rightarrow \mathbf{Q}$$

The reactions are thus in parallel with respect to both **A** and **B**. For reactions in series, however, if the second reactant **B** participates in the reaction with the product **P** as well as with **A**, i.e. if:

$$\mathbf{A} + \mathbf{B} \rightarrow \mathbf{P}$$
$$\mathbf{P} + \mathbf{B} \rightarrow \mathbf{Q}$$

then although the reactions are in series with respect to **A** they are in parallel with respect to **B**. In these circumstances, we have:

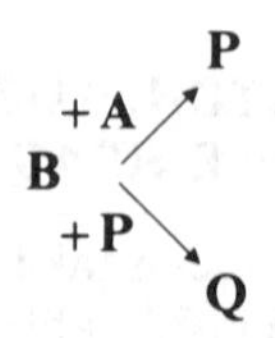

As we shall see, however, the series character of these reactions is the more important, because **B** cannot react to give **Q** until a significant amount of **P** has been formed.

More complex reaction schemes can be regarded as combinations of these basic types of individual reaction steps.

1.10.2. Yield and Selectivity

When a mixture of reactants undergoes treatment in a reactor and more than one product is formed, part of each reactant is converted into the desired product, part is converted into undesired products and the remainder escapes unreacted. The amount of the desired product actually obtained is therefore smaller than the amount expected had all the reactant been transformed into the desired product alone. The reaction is then said to give a certain yield of the desired product. Unfortunately, the term yield has been used by different authors for two somewhat different quantities and care must be taken to avoid confusion. Here these two usages will be distinguished by employing the terms *relative yield* and *operational yield*; in each case the amount of product formed will be expressed in terms of the stoichiometrically equivalent amount of the reactant **A** from which it was produced.

The *relative yield* Φ_A is defined by:

$$\Phi_A = \frac{\text{Moles of } \mathbf{A} \text{ transformed into desired product}}{\text{Total moles of } \mathbf{A} \text{ which have reacted}} \tag{1.58}$$

The relative yield is therefore a net yield based on the amount of **A** actually consumed.

The *operational yield* Θ_A is defined by:

$$\Theta_A = \frac{\text{Moles of } \mathbf{A} \text{ transformed into desired product}}{\text{Total moles of } \mathbf{A} \text{ fed the reactor}} \tag{1.59}$$

It is based on the total amount of reactant **A** entering the reactor, irrespective of whether it is consumed in the reaction or passes through unchanged.

Both these quantities are fractions and it follows from the definitions above that Φ_A always exceeds Θ_A, unless all the reactant is consumed, when they are equal.

If unreacted **A** can be recovered from the product mixture at low cost and then recycled, the relative yield is the more significant, and the reactor can probably be operated economically at quite a low conversion per pass. If it cannot be recovered and no credit can be allotted to it, the operational yield is the more relevant, and the reactor will probably have to operate at a high conversion per pass.

Another way of expressing product distribution is the *selectivity* of the desired reaction. Once more expressing the amount of product formed in terms of the amount of **A** reacted, the selectivity is defined as:

$$\frac{\text{Moles of } \mathbf{A} \text{ transformed into the desired product}}{\text{Moles of } \mathbf{A} \text{ transformed into unwanted products}}$$

It is thus a product ratio and can have any value, the higher the better. It is often used to describe catalyst performance in heterogeneous reactions.

1.10.3. Reactor Type and Backmixing

When more than one reaction can occur, the extent of any backmixing of products with reactants is one of the most important factors in determining the yield of the desired product. In a well-stirred batch reactor, or in an ideal tubular reactor with plug flow, there is no backmixing, whereas in a single continuous stirred-tank reactor there is complete backmixing. Intermediate between these two extremes are systems of two or more continuous stirred-tanks in series, and non-ideal tubular reactors in which some degree of backmixing occurs (often termed longitudinal dispersion).

Backmixing in a reactor affects the yield for two reasons. The first and most obvious reason is that the products are mixed into the reactants; this is undesirable if the required product is capable of reacting further with the reactants to give an unwanted product, as in some series reactions. The second reason is that backmixing affects the level of reactant concentration at which the reaction is carried out. If there is no backmixing, the concentration level is high at the start of the reaction and has a low value only towards the end of the reaction. With backmixing, as in a single continuous stirred-tank reactor, the concentration of reactant is low throughout. As we shall see, for some reactions high reactant concentrations favour high yields, whereas for other reactions low concentrations are more favourable.

If two reactants are involved in a reaction, high concentrations of both, at least initially, may be obtained in a batch or tubular plug-flow reactor. and low concentrations of both in a single continuous stirred-tank reactor. In some circumstances, however, a high concentration of reactant **A** coupled with a low concentration of reactant **B** may be desirable. This may be achieved in a number of ways:

(a) *Without recycle*: For continuous operation a cross-flow type of reactor may be used as illustrated in Fig. 1.2. The reactor can consist of either a tubular reactor with multiple injection of **B** (Fig. 1.2*b*), or a series of several stirred tanks with the feed of **B** divided between them (Fig. 1.2*c*). If a batch type of reactor were preferred, the semi-batch mode of operation would be used (Fig. 1.2*a*). Reaction without recycle is the normal choice where the cost of separating unreacted **A** from the reaction mixture is high.

(b) *With recycle*: If the cost of separating **A** is low, then a large excess of **A** can be maintained in the reactor. A single continuous stirred-tank reactor will provide a low concentration of **B**, while the large excess of **A** ensures a high concentration of **A**. Unreacted **A** is separated and recycled as shown in Fig. 1.23 (Section 1.10.5).

1.10.4. Reactions in Parallel

Let us consider the case of one reactant only but different orders of reaction for the two reaction paths.

i.e.:

$$\mathbf{A} \xrightarrow{k_P} \mathbf{P} \text{ desired product}$$

$$\mathbf{A} \xrightarrow{k_Q} \mathbf{Q} \text{ unwanted product}$$

with the corresponding rate equations:

$$\mathcal{R}_{AP} = \boldsymbol{k}_p C_A{}^p \tag{1.60}$$

$$\mathcal{R}_{AQ} = \boldsymbol{k}_Q C_A^q \tag{1.61}$$

In these equations it is understood that C_A may be (a) the concentration of **A** at a particular time in a batch reactor, (b) the local concentration in a tubular reactor operating in a steady state, or (c) the concentration in a stirred-tank reactor, possibly one of a series, also in a steady state. Let δt be an interval of time which is sufficiently short for the concentration of **A** not to change appreciably in the case of the batch reactor; the length of the time interval is not important for the flow reactors because they are each in a steady state. Per unit volume of reaction mixture, the moles of **A** transformed into **P** is thus $\mathcal{R}_{AP}\delta t$, and the total amount reacted $(\mathcal{R}_{AP} + \mathcal{R}_{AQ})\delta t$. The relative yield under the circumstances may be called the instantaneous or point yield ϕ_A, because C_A will change (a) with time in the batch reactor, or (b) with position in the tubular reactor.

Thus:

$$\phi_A = \frac{\mathcal{R}_{AP}\delta t}{(\mathcal{R}_{AP} + \mathcal{R}_{AQ})\delta t} = \frac{\boldsymbol{k}_P C_A^{\ p}}{\boldsymbol{k}_P C_A^{\ p} + \boldsymbol{k}_Q C_A^{\ q}}$$

or:

$$\phi_A = \left[1 + \frac{\boldsymbol{k}_Q}{\boldsymbol{k}_P} C_A^{(q-p)}\right]^{-1} \tag{1.62}$$

Similarly the local selectivity is given by:

$$\frac{\mathcal{R}_{AP}\delta t}{\mathcal{R}_{AQ}\delta t} = \frac{\boldsymbol{k}_P}{\boldsymbol{k}_Q} C_A^{(p-q)} \tag{1.63}$$

In order to find the overall relative yield Φ_A, i.e. the yield obtained at the end of a batch reaction or at the outlet of a tubular reactor, consider an element of unit volume of the reaction mixture. If the concentration of **A** decreases by δC_A either (a) with time in a batch reactor or (b) as the element progresses downstream in a tubular reactor, the amount of **A** transformed into **P** is $-\phi_A \delta C_A$. The total amount of **A** transformed into **P** during the whole reaction is therefore $\int_{C_{A0}}^{C_{Af}} -\phi_A \mathrm{d}C_A$. The total amount of **A** reacted in the element is $(C_{A0} - C_{Af})$. The overall relative yield is therefore:

$$\Phi_A = -\frac{1}{(C_{A0} - C_{Af})} \int_{C_{A0}}^{C_{Af}} \phi_A \mathrm{d}C_A$$

and thus represents an average of the instantaneous value ϕ_A over the whole concentration range. Thus, from equation 1.62:

$$\Phi_A = -\frac{1}{(C_{A0} - C_{Af})} \int_{C_{A0}}^{C_{Af}} \frac{\mathrm{d}C_A}{1 + \dfrac{\boldsymbol{k}_Q}{\boldsymbol{k}_P} C_A^{(q-p)}} \tag{1.64}$$

For a stirred-tank reactor consisting of a single tank in a steady state, the overall yield is the same as the instantaneous yield given by equation 1.62 because concentrations do not vary with either time or position. If more than one stirred-tank is used, however, an appropriate average must be taken.

Requirements for High Yield

Reactant concentration and reactor type: Although equation 1.64 gives the exact value of the final yield obtainable from a batch or tubular reactor, the nature of the conditions required for a high yield can be seen more readily from equation 1.62. If $p > q$, i.e. the order of the desired reaction is higher than that of the undesired reaction, a high yield ϕ_A will be obtained when C_A is high. A batch reactor, or a tubular reactor, gives a high reactant concentration at least initially and should therefore be chosen in preference to a single stirred-tank reactor in which reactant concentration is low. If the stirred-tank type of reactor is chosen on other grounds, it should consist of several tanks in series. In operating the reactor, any recycle streams which might dilute the reactants should be avoided. Conversely, if $p < q$, a high yield is favoured by a low reactant concentration and a single stirred-tank reactor is the most suitable. If a batch or tubular reactor were nevertheless chosen, dilution of the reactant by a recycle stream would be an advantage. Finally, if $p = q$, the yield will be unaffected by reactant concentration.

Pressure in gas-phase reactions: If a high reactant concentration is required, i.e. $p > q$, the reaction should be carried out at high pressure and the presence of inert gases in the reactant stream should be avoided. Conversely, if $p < q$ and low concentrations are required, low pressures should be used.

Temperature of operation: Adjusting the temperature affords a means of altering the ratio $\mathbf{k}_P/\mathbf{k}_Q$, provided that the activation energies of the two reactions are different.

Thus:

$$\frac{\mathbf{k}_P}{\mathbf{k}_Q} = \frac{\mathcal{A}_P}{\mathcal{A}_Q} \exp\left[-\frac{(\mathbf{E}_P - \mathbf{E}_Q)}{\mathbf{R}T}\right] \tag{1.65}$$

Choice of catalyst: If a catalyst can be found which will enable the desired reaction to proceed at a satisfactory rate at a temperature which is sufficiently low for the rate of the undesired reaction to be negligible, this will usually be the best solution of all to the problem.

Yield and Reactor Output

The concentration at which the reaction is carried out affects not only the yield but also the reactor output. If a high yield is favoured by a high reactant concentration, there is no conflict, because the average rate of reaction and therefore the reactor output will be high also. However, if a high yield requires a low reactant concentration, the reactor output will be low. An economic optimum must be sought, balancing the cost of reactant wasted in undesired byproducts against the initial cost of a larger reactor. In most cases the product distribution is the most important factor, especially (a) when raw material costs are high, and (b) when the cost of equipment for the separation, purification and recycle of the reactor products greatly exceeds the cost of the reactor.

1.10.5. Reactions in Parallel — Two Reactants

If a second reactant **B** is involved in a system of parallel reactions, then the same principles apply to **B** as to **A**. The rate equations are examined to see whether the order of the desired reaction with respect to **B** is higher or lower than that of the undesired reaction, and to decide whether high or low concentrations of **B** favour a high yield of desired product.

There are three possible types of combination between the concentration levels of **A** and **B** that may be required for a high yield:

(a) C_A, C_B both high. In this case a batch or a tubular plug-flow reactor is the most suitable.
(b) C_A, C_B both low. A single continuous stirred-tank reactor is the most suitable.
(c) C_A high, C_B low (or C_A low and C_B high). A cross-flow reactor is the most suitable for continuous operation without recycle, and a semi-batch reactor for batchwise operation. If the reactant required in high concentration can be easily recycled, a single continuous stirred-tank reactor can be used.

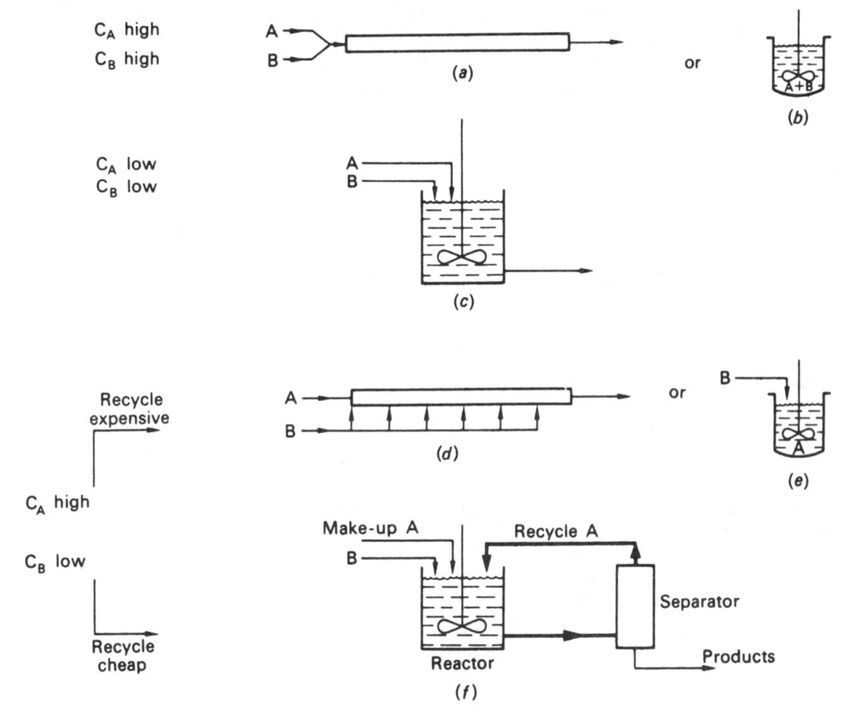

FIG. 1.23. Contacting schemes to match possible concentration levels required for high relative yields in parallel reactions

(*a*) Plug-flow tubular reactor
(*b*) Batch reactor
(*c*) Single continuous stirred-tank reactor
(*d*) Cross-flow tubular reactor
(*e*) Semi-batch (or fed-batch) reactor
(*f*) Single continuous stirred-tank reactor with recycle of A

Alternatively a series of several continuous stirred tanks could be used in place of the tubular reactors in (*a*) and (*d*)

These ways of matching reactor characteristics to the concentration levels required are illustrated in Fig. 1.23.

Example 1.7

Two reactants undergo parallel reactions as follows:

$$\mathbf{A} + \mathbf{B} \rightarrow \mathbf{P} \text{ (desired product)}$$

$$2\mathbf{B} \rightarrow \mathbf{Q} \text{ (unwanted product)}$$

with the corresponding rate equations:

$$\mathcal{R}_{AP} = \mathcal{R}_{BP} = k_P C_A C_B,$$

$$\mathcal{R}_{BQ} = k_Q C_B^2.$$

Suggest suitable continuous contacting schemes which will give high yields of **P** (a) if the cost of separating **A** is high and recycling is not feasible, (b) if the cost of separating **A** is low and recycling can be employed. For the purpose of quantitative treatment set $k_P = k_Q$. The desired conversion of reactant **B** is 0.95.

Solution

Inspection of the rate equations shows that, with respect to **A**, the order of the desired reaction is unity, and the order of the undesired reaction is effectively zero because **A** does not participate in it. The desired reaction is therefore favoured by high values of C_A. With respect to **B**, the order of the desired reaction is unity, and the order of the unwanted reaction is two. The desired reaction is therefore favoured by low values of C_B.

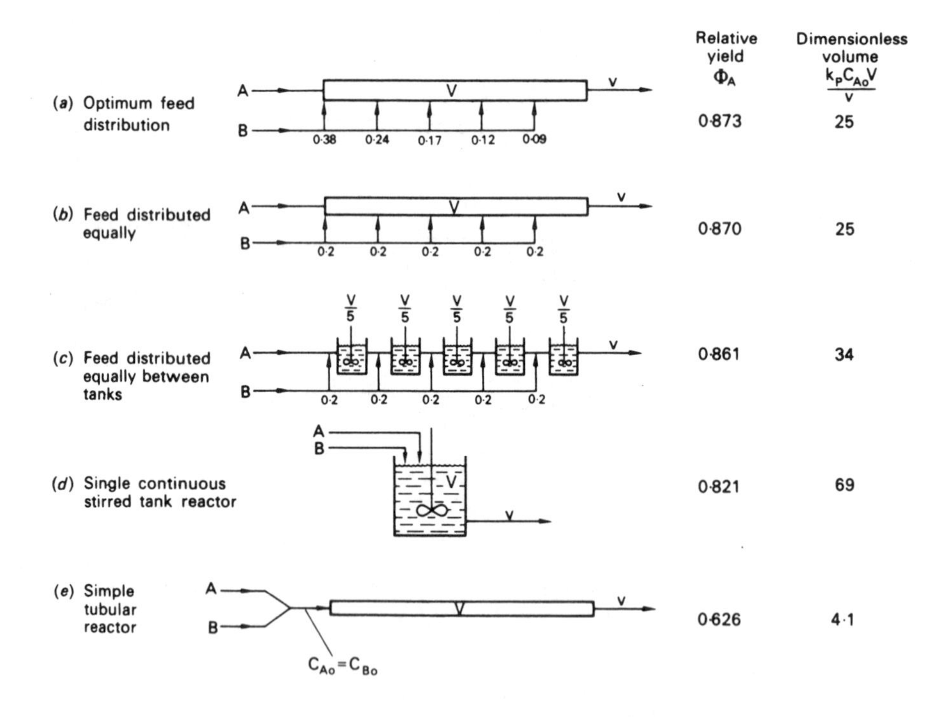

FIG. 1.24. Performance of cross-flow reactors with five equidistant feed points
Parallel reactions: $\mathbf{A} + \mathbf{B} \rightarrow \mathbf{P}$ $\mathcal{R}_{AP} = \mathcal{R}_{BP} = k_P C_A C_B$
$2\mathbf{B} \rightarrow \mathbf{Q}$ $\mathcal{R}_{BQ} = k_Q C_B^2$
with $k_P = k_Q$. Equal molar feed rates of **A** and **B**. Final conversion of **B** = 0.95

(a) If recycling is not feasible, a cross-flow type of reactor will be the most suitable, the feed of **B** being distributed between several points along the reactor. Cross-flow reactors for this particular reaction system have been studied in considerable detail. Figure 1.24 shows some results obtained with reactors employing five equidistant feed positions, together with the performances of a single continuous stirred-tank reactor and of a straight tubular reactor for the purpose of comparison. In case (*d*) the amounts of **B** fed at each point in the tubular reactor have been calculated to give the maximum yield of desired product. It may be seen from (*b*), however, that there is little disadvantage in having the feed of **B** distributed in equal parts. Furthermore, the five sections of the tubular reactor can be replaced by five stirred tanks, as in (*c*), without appreciably diminishing the yield although the total reactor volume is somewhat greater. By way of contrast, a simple tubular reactor (*e*) gives a substantially lower yield.

(b) If **A** can be recycled, a high concentration of **A** together with a low concentration of **B** can be maintained in a single continuous stirred-tank reactor, as shown in Fig. 1.23. By suitable adjustment of the rate of recycle of **A** and the corresponding rate of outflow, the ratio of concentrations in the reactor $C_A/C_B = r'$ may be set to any desired value. The relative yield based on **B**, Φ_B, will then be given by:

$$\Phi_B = \frac{k_P C_A C_B}{k_P C_A C_B + k_Q C_B^2}$$

If $k_P = k_Q$:

$$\Phi_B = \frac{C_A}{C_A + C_B} = \frac{r'}{1 + r'}$$

Even if **A** can be separated from the product mixture relatively easily as the recycle rate and hence r' is increased, so the operating costs will increase and the volumes of the reactor and separator must also be increased. These costs have to be set against the cash value of the increased yield of desired product as r' is increased. Thus, the optimum setting of the recycle rate will be determined by an economic balance.

1.10.6. Reactions in Series

When reactions in series are considered, it is not possible to draw any very satisfactory conclusions without working out the product distribution completely for each of the basic reactor types. The general case in which the reactions are of arbitrary order is more complex than for parallel reactions. Only the case of two first-order reactions will therefore be considered:

$$\mathbf{A} \xrightarrow{k_{11}} \mathbf{R} \xrightarrow{k_{12}} \mathbf{Q}$$

where **P** is the desired product and **Q** is the unwanted product.

Batch Reactor or Tubular Plug-Flow Reactor

Let us consider unit volume of the reaction mixture in which concentrations are changing with time: this unit volume may be situated in a batch reactor or moving in plug flow in a tubular reactor. Material balances on this volume give the following equations:

$$-\frac{dC_A}{dt} = k_{11} C_A; \quad \frac{dC_P}{dt} = k_{11} C_A - k_{12} C_P; \quad \frac{dC_Q}{dt} = k_{12} C_P$$

If $C_P = 0$ when $t = 0$, the concentration of **P** at any time t is given by the solution to these equations which is:

$$C_P = C_{A0} \frac{k_{11}}{k_{12} - k_{11}} \left[\exp(-k_{11} t) - \exp(-k_{12} t)\right] \tag{1.66}$$

Differentiation and setting $dC_P/dt = 0$ shows that C_P passes through a maximum given by:

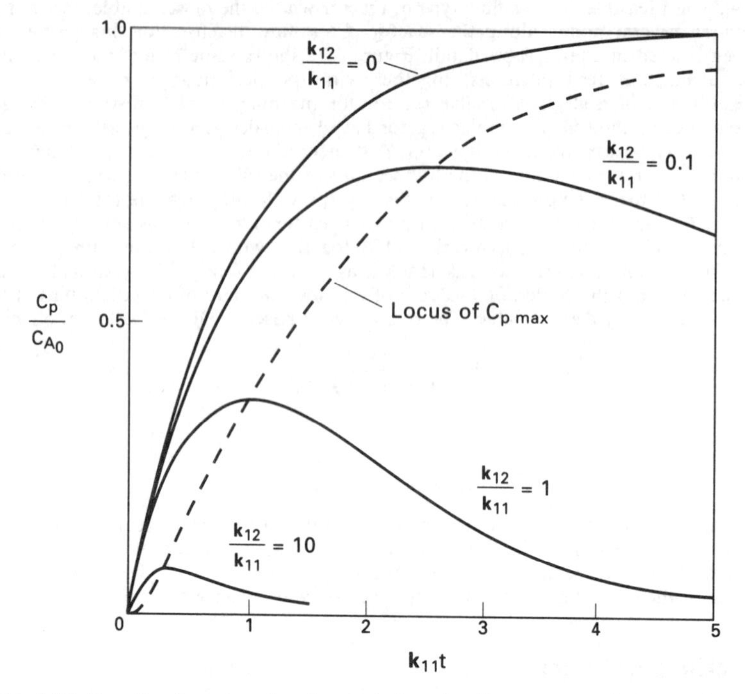

FIG. 1.25. Reaction in series—batch or tubular plug-flow reactor. Concentration C_P of intermediate product **P** for consecutive first order reactions, $\mathbf{A} \rightarrow \mathbf{P} \rightarrow \mathbf{Q}$

$$\frac{C_{P\,max}}{C_{A0}} = \left(\frac{\boldsymbol{k}_{11}}{\boldsymbol{k}_{12}}\right)^{\frac{\boldsymbol{k}_{12}}{\boldsymbol{k}_{12} - \boldsymbol{k}_{11}}} \tag{1.67}$$

which occurs at a time:

$$t_{max} = \frac{\ln(\boldsymbol{k}_{12}/\boldsymbol{k}_{11})}{(\boldsymbol{k}_{12} - \boldsymbol{k}_{11})} \tag{1.68}$$

The relationships 1.66, 1.67 and 1.68 are plotted in Fig. 1.25 for various values of the ratio $\boldsymbol{k}_{12}/\boldsymbol{k}_{11}$.

Continuous Stirred-Tank Reactor — One Tank

Taking material balances in the steady state as shown in Fig. 1.26:

(a) on **A**: $$vC_{A0} - vC_A - V_c\boldsymbol{k}_{11}C_A = 0$$

(b) on **P**: $$0 - vC_P - V_c(\boldsymbol{k}_{12}C_P - \boldsymbol{k}_{11}C_A) = 0$$

whence:

$$\frac{C_P}{C_{A0}} = \frac{\boldsymbol{k}_{11}\tau}{(1 + \boldsymbol{k}_{11}\tau)(1 + \boldsymbol{k}_{12}\tau)} \tag{1.69}$$

$$\frac{C_Q}{C_{A0}} = \frac{\boldsymbol{k}_{11}\boldsymbol{k}_{12}\tau^2}{(1 + \boldsymbol{k}_{11}\tau)(1 + \boldsymbol{k}_{12}\tau)} \tag{1.70}$$

where τ is the residence time ($\tau = V/v$). C_P passes through a maximum in this case also:

$$\frac{C_{P\,\max}}{C_{A0}} = \frac{1}{[(\boldsymbol{k}_{12}/\boldsymbol{k}_{11})^{1/2} + 1]^2} \tag{1.71}$$

at a residence time $t_{\max}$ given by;

$$\tau_{\max} = (\boldsymbol{k}_{11}\boldsymbol{k}_{12})^{-1/2} \tag{1.72}$$

These relationships are plotted in Fig. 1.27.

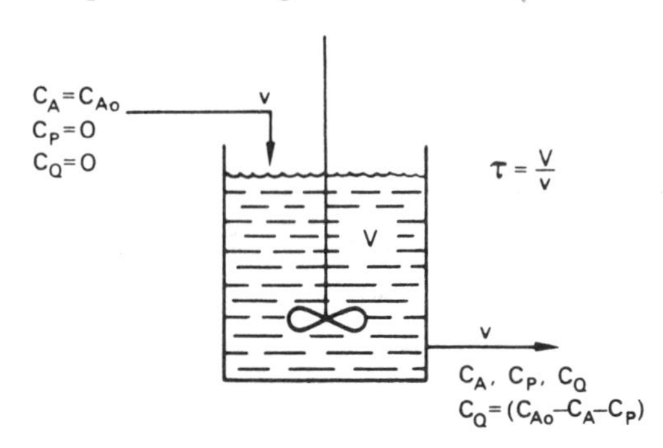

FIG. 1.26. Continuous stirred-tank reactor: single tank, reactions in series, **A** → **P** → **Q**

Reactor Comparison and Conclusions

The curves shown in Figs. 1.25 and 1.27, which are curves of operational yield versus reduced time, can be more easily compared by plotting the relative yield of **P** against conversion of **A** as shown in Fig. 1.28. It is then apparent that the relative yield is always greater for the batch or plug-flow reactor than for the single stirred-tank reactor, and decreases with increasing conversion. We may therefore draw the following conclusions regarding the choice of reactor type and mode of operation:

Reactor type: For the highest relative yield of **P** a batch or tubular plug-flow reactor should be chosen. If a continuous stirred-tank system is adopted on other grounds, several tanks should be used in series so that the behaviour may approach that of a plug-flow tubular reactor.

Conversion in reactor: If $\boldsymbol{k}_{12}/\boldsymbol{k}_{11} >> 1$, Fig. 1.28 shows that the relative yield falls sharply with increasing conversion, i.e. **P** reacts rapidly once it is formed. If possible, therefore, the reactor should be designed for a low conversion of **A** per batch or pass in a tubular reactor, with separation of **P** and recycling of the unused reactant. Product separation and recycle may be quite expensive, however, in which case we look for the conversion corresponding to the economic optimum.

Temperature: We may be able to exercise some control over the ratio $\boldsymbol{k}_{12}/\boldsymbol{k}_{11}$ by sensible choice of the operating temperature. If $E_1 > E_2$ a high temperature should be chosen, and conversely a low temperature if $E_1 < E_2$. When a low temperature is required for the best yield, there arises the problem that reaction rates and reactor output decrease with decreasing temperature, i.e. the size of the reactor required

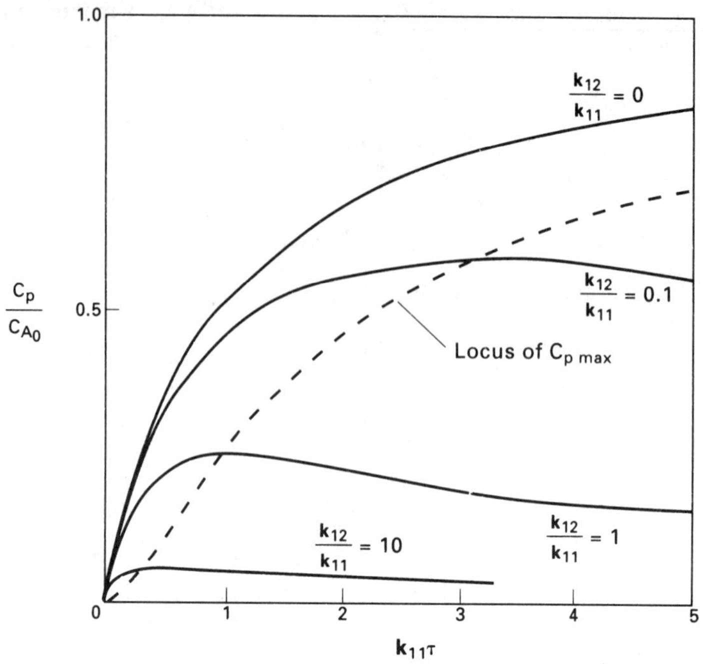

FIG. 1.27. Reactions in series—single continuous stirred-tank reactor. Concentration C_P of intermediate product **P** for consecutive first-order reactions, $\mathbf{A} \rightarrow \mathbf{P} \rightarrow \mathbf{Q}$

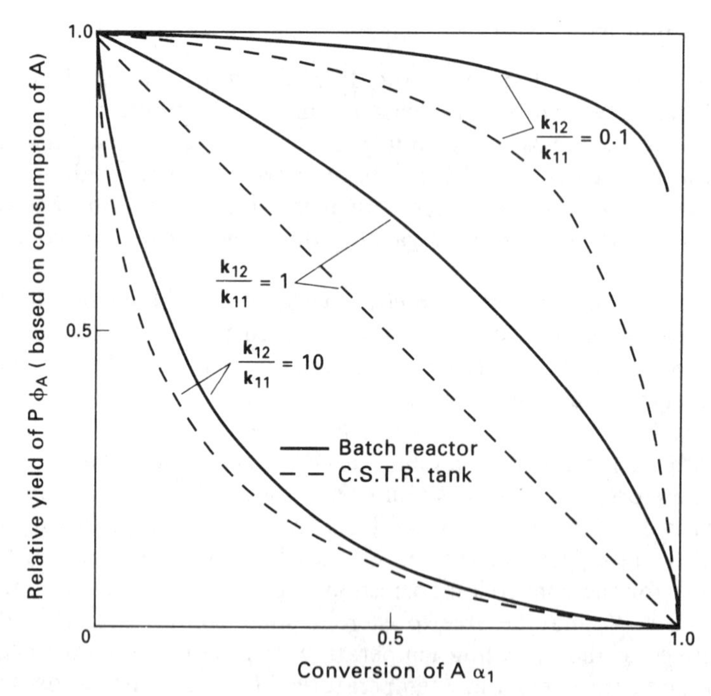

FIG. 1.28. Reactions in series—comparison between batch or tubular plug-flow reactor and a single continuous stirred-tank reactor. Consecutive first-order reactions, $\mathbf{A} \rightarrow \mathbf{P} \rightarrow \mathbf{Q}$

increases. The operating temperature will thus be determined by an economic optimum. There is the further possibility of establishing a temperature variation along the reactor. For example, if for the case of $E_1 < E_2$ two stirred tanks were chosen, the temperature of the first tank could be high to give a high production rate but the second tank, in which the concentration of the product **P** would be relatively large, could be maintained at a lower temperature to avoid excessive degradation of **P** to **Q**.

General conclusions: In series reactions, as the concentration of the desired intermediate **P** builds up, so the rate of degradation to the second product **Q** increases. The best course would be to remove **P** continuously as soon as it was formed by distillation, extraction or a similar operation. If continuous removal is not feasible, the conversion attained in the reactor should be low if a high relative yield is required. As the results for the continuous stirred-tank reactor show, backmixing of a partially reacted mixture with fresh reactants should be avoided.

1.10.7. Reactions in Series — Two Reactants

The series–parallel type of reaction outlined in Section 1.10.1 is quite common among industrial processes. For example, ethylene oxide reacts with water to give monoethylene glycol, which may then react with more ethylene oxide to give diethylene glycol.

$$H_2O + C_2H_4O \rightarrow HO \cdot C_2H_4 \cdot OH$$
$$HO \cdot C_2H_4 \cdot OH + C_2H_4O \rightarrow HO \cdot (C_2H_4O)_2 \cdot H$$

i.e.:

$$\mathbf{A} + \mathbf{B} \rightarrow \mathbf{P}$$
$$\mathbf{P} + \mathbf{B} \rightarrow \mathbf{Q}$$

In such cases the order with respect to **B** is usually the same for both the first and the second reaction. Under these circumstances, the level of concentration of **B** at which the reactions are carried out has no effect on the relative rates of the two reactions, as may be seen by writing these as parallel reactions with respect to **B**:

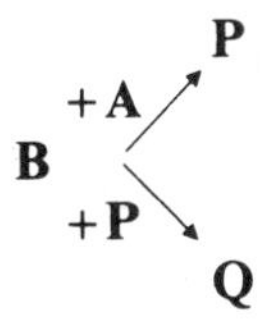

These reactions will therefore behave very similarly to the reactions in series above where only one reactant was involved.

i.e.:

$$\mathbf{A} \xrightarrow{+\mathbf{B}} \mathbf{P} \xrightarrow{+\mathbf{B}} \mathbf{Q}$$

The same general conclusions apply; since backmixing of products with reactants should be avoided, a tubular plug-flow reactor or a batch reactor is preferred. However, there is one respect in which a series reaction involving a second reactant **B** does differ from simple series reaction with one reactant, even when the orders are the same. This is in the stoichiometry of the reaction; the reaction cannot proceed completely to the product **Q**, even in infinite time, if less than two moles of **B** per mole of **A** are supplied. Some control over the maximum extent of the reaction can therefore be achieved by choosing the appropriate ratio of **B** to **A** in the feed. For reactions which are first order in **A, B** and **P**, charts[9] are available showing yields and end points reached for various feed ratios.

1.11. FURTHER READING

ARIS, R.: *Elementary Chemical Reactor Analysis* (Butterworths Reprint Series, 1989).
BARTON, J and ROGERS R.: *Chemical Reaction Hazards* (Institution of Chemical Engineers, Rugby, U.K., 1993).
DENBIGH, K. G. and TURNER, J. C. R.: *Chemical Reactor Theory*, 3rd edn. (Cambridge, 1984).
FOGLER, H. S.: *Elements of Chemical Reaction Engineering*, 2nd edn. (Prentice-Hall, 1992).
FROMENT G. F. and BISCHOFF, K. B.: *Chemical Reactor Analysis and Design*, 2nd edn. (Wiley, 1989).
HOLLAND, C. D. and ANTHONY, R. G.: *Fundamentals of Chemical Reaction Engineering*, 2nd edn. (Prentice-Hall, 1992).
LAPIDUS, L. and AMUNDSON, N. R. (eds.): *Chemical Reactor Theory—A Review*. (Prentice-Hall, 1977).
LEVENSPIEL, O.: *Chemical Reaction Engineering*, 2nd edn. (Wiley, 1972).
LEVENSPIEL, O.: *Chemical Reactor Omnibook* (OSU Book Stores, Corvallis, Oregon, 1989).
NAUMAN, E. B.: *Chemical Reactor Design* (Wiley, 1987).
RASE, H. F.: *Chemical Reactor Design for Process Plants*, Vols. 1 and 2 (Wiley, 1977).
SMITH, J. M.: *Chemical Engineering Kinetics*, 3rd edn. (McGraw-Hill, 1981).
WALAS, S. M.: *Reaction Kinetics for Chemical Engineers* (Butterworths Reprint Series, 1989).
WALAS, S. M.: *Chemical Process Equipment: Selection and Design, Chapter 17—Chemical Reactors* (Butterworths, 1988).
WESTERTERP, K. R., VAN SWAAIJ, W. P. M. and BEENACKERS, A. A. C. M.: *Chemical Reactor Design and Operation*, 2nd edn. (Wiley, 1984).

1.12. REFERENCES

1. MILLER, S. A. (ed.): *Ethylene and its Industrial Derivatives* (Benn, 1969).
2. DENBIGH, K. G.: *Principles of Chemical Equilibria*, 2nd edn. (Cambridge, 1966).
3. MOORE, J. W. and PEARSON, R. G.: *Kinetics and Mechanism*, 3rd edn. (Wiley, 1981).
4. LAIDLER, K. J.: *Chemical Kinetics*, 3rd edn. (Harper and Row, 1987)
5. DENBIGH, K. G. and TURNER, J. C. R.: *Chemical Reactor Theory*, 3rd edn. (Cambridge, 1984).
6. SCHUTT, H. C.: *Chem. Eng. Prog.* **55** (1) (1959) 68. Light hydrocarbon pyrolysis.
7. ELDRIDGE, J. W. and PIRET, E. L.: *Chem. Eng. Prog.* **46** (1950) 290. Continuous flow stirred tank reactor systems.
8. WESTERTERP, K. R., VAN SWAAIJ, W. P. M. and BEENACKERS, A. A. C. M.: *Chemical Reactor Design and Operation*, 2nd edn. (Wiley, 1984).
9. LEVENSPIEL, O.: *Chemical Reaction Engineering*, 2nd edn. (Wiley, 1972).

1.13. NOMENCLATURE

Asterisk (*) indicates that these dimensions are dependent on order of reaction

		Units in SI system	Dimensions in **M, N, L, T, Θ**
$\mathscr{A}$	Frequency factor in rate equation	*	*
A_t	Area of cooling coils	m^2	$\mathbf{L}^2$
A_c	Area of cross-section of a tubular reactor	m^2	$\mathbf{L}^2$
C	Molar density of reaction mixture (at time t)	$kmol/m^3$	$\mathbf{NL}^{-3}$
C_A, C_B	Molar concentration of **A**; molar concentration of **B**, etc.	$kmol/m^3$	$\mathbf{NL}^{-3}$
C_{Ae}	Concentration of **A** at equilibrium, etc	$kmol/m^3$	$\mathbf{NL}^{-3}$
C_{Af}	Concentration of **A** at end of reaction	$kmol/m^3$	$\mathbf{NL}^{-3}$
C_{A0}	Concentration of **A** initially or in feed, etc	$kmol/m^3$	$\mathbf{NL}^{-3}$
C_0	Molar density of reaction mixture at inlet	$kmol/m^3$	$\mathbf{NL}^{-3}$
c	Specific heat at constant pressure	J/kg K	$\mathbf{L}^2\mathbf{T}^{-2}\mathbf{\Theta}^{-1}$
c_j	Specific heat at constant pressure of component j	J/kg K	$\mathbf{L}^2\mathbf{T}^{-2}\mathbf{\Theta}^{-1}$

		Units in SI system	Dimensions in M, N, L, T, Θ
E	Activation energy per mole	J/kmol	$\mathbf{MN^{-1}L^2T^{-2}}$
F_A	Molar feed rate of **A** into reactor, etc	kmol/s	$\mathbf{NT^{-1}}$
G	Mass flowrate	kg/s	$\mathbf{MT^{-1}}$
ΔH	Enthalpy change per mole (Heat of reaction = – ΔH)	J/kmol	$\mathbf{MN^{-1}L^2T^{-2}}$
ΔH_A	Enthalpy change in reaction per mole of **A**	J/kmol	$\mathbf{MN^{-1}L^2T^{-2}}$
j	Component of reaction mixture	—	—
K_c	Equilibrium constant in terms of concentrations	*	*
K_p	Equilibrium constant in terms of partial pressure	*	*
$\boldsymbol{k}$	Rate constant	*	*
$\boldsymbol{k}_f$	Rate constant of forward reaction in a reversible reaction	*	*
$\boldsymbol{k}_P$, $\boldsymbol{k}_Q$	Rate constants for reactions giving products **P**, **Q**, etc.	*	*
$\boldsymbol{k}_p$	Rate constant with concentrations expressed as partial pressures	*	*
$\boldsymbol{k}_r$	Rate constant of reverse reaction in a reversible reaction	*	*
$\boldsymbol{k}_1$	First-order or pseudo-first-order rate constant	s^{-1}	$\mathbf{T^{-1}}$
$\boldsymbol{k}_2$	Second-order rate constant	m^3/kmol s	$\mathbf{N^{-1}L^3T^{-1}}$
$\boldsymbol{k}_{11}$, $\boldsymbol{k}_{12}$	Rate constants of first and second first-order reactions in a series	s^{-1}	$\mathbf{T^{-1}}$
m_j	Mass of component j present in a batch reactor	kg	**M**
N_A	Molar flux across a gas–liquid interface	kmol/m^3 s	$\mathbf{NL^{-2}T^{-1}}$
n_A	Moles of **A** in a given volume of reaction mixture	kmol	**N**
n_{A0}	Moles of **A** in a given volume of reaction mixture initially	kmol	**N**
P	Total pressure	N/m^2	$\mathbf{ML^{-1}T^{-2}}$
P_A	Partial pressure of **A**	N/m^2	$\mathbf{ML^{-1}T^{-2}}$
p	Order of reaction	—	—
Q	Heat flow from batch or tubular reactor	W	$\mathbf{ML^2T^{-3}}$
q	Order of reaction	—	—
R	Gas constant	J/kmol K	$\mathbf{MN^{-1}L^2T^{-2}\Theta^{-1}}$
$\mathcal{R}_A$, $\mathcal{R}_B$	Rate of reaction per unit volume of reactor with respect to reactant **A**, reactant **B**	kmol/m^3 s	$\mathbf{NL^{-3}T^{-1}}$
$\mathcal{R}_{+A}$	Rate of forward reaction with respect to **A** in a reversible reaction	kmol/m^3 s	$\mathbf{NL^{-3}T^{-1}}$
$\mathcal{R}_{-A}$	Rate of reverse reaction with respect to **A** in a reversible reaction	kmol/m^3 s	$\mathbf{NL^{-3}T^{-1}}$
r	Order of reaction	—	—
r'	Concentration ratio C_A/C_B	—	—
s	V_{c2}/V_{c1}	—	—
T	Temperature absolute	K	**Θ**
T_C	Temperature of cooling water	K	**Θ**
T_{mx}	Maximum safe operating temperature	K	**Θ**
T_0	Initial temperature	K	**Θ**
t	Time	s	**T**
t_r	Reaction time, batch reactor	s	**T**
t_s	Shut-down time, batch reactor	s	**T**
$t_{1/2}$	Half-life of reaction, i.e. time for half reactant to be consumed	s	**T**
U	Heat transfer coefficient, overall	W/m^2 K	$\mathbf{MT^{-3}\Theta^{-1}}$
V	Volume of reaction mixture	m^3	$\mathbf{L^3}$
V_b	Volume of batch reactor	m^3	$\mathbf{L^3}$
V_c	Volume of a continuous stirred-tank reactor; total volume if more than one tank	m^3	$\mathbf{L^3}$
V_{cr}	Volume of the rth tank in a series of continuous stirred-tank reactors	m^3	$\mathbf{L^3}$
V_t	Volume of a tubular reactor	m^3	$\mathbf{L^3}$
v	Volume rate of flow into reactor	m^3/s	$\mathbf{L^3T^{-1}}$
W_A	Unit output of a reactor with respect to reactor **A**	kmol/m^3 s	$\mathbf{NL^{-3}T^{-1}}$
W_b	Unit output for batch reactor	kmol/m^3 s	$\mathbf{NL^{-3}T^{-1}}$

		Units in SI system	Dimensions in **M, N, L, T, Θ**
W_{c1}, W_{c2}	Unit output for a continuous stirred-tank reactor comprising one tank; two tanks	kmol/m^3 s	$\mathbf{NL^{-3}T^{-1}}$
x	Distance from interface in direction of transfer	m	**L**
y_A	Mole fraction of **A**	—	—
y_{A0}	Mole fraction of **A** at inlet to reactor	—	—
z	Distance along tubular reactor	m	**L**
α	Fractional conversion	—	—
α_A	Fractional conversion of reactant **A**, i.e. the fraction of **A** which has reacted	—	—
α_{Af}	Fractional conversion of reactant **A**; final value on discharge from reactor	—	—
α_e	Fractional conversion at equilibrium in a reversible reaction	—	—
Θ	Operational yield of desired product	—	—
ν_A, ν_B	Stoichiometric coefficients of **A**, **B**	—	—
τ	Residence time: V_t/v or V_c/v	s	**T**
τ_1	Plug-flow residence time	s	**T**
τ_2	Stirred-tank residence time	s	**T**
Φ	Relative yield of desired product, overall	—	—
ϕ	Relative yield of desired product, instantaneous	—	—
χ	Moles of reactant transformed in unit volume of reaction mixture	kmol/m^3	$\mathbf{NL^{-3}}$
χ_e	Value of χ at equilibrium	kmol/m^3	$\mathbf{NL^{-3}}$
χ_f	Moles of reactant transformed in unit volume of reaction mixture; final value on discharge from reactor	kmol/m^3	$\mathbf{NL^{-3}}$

CHAPTER 2

Flow Characteristics of Reactors—Flow Modelling

2.1. NON-IDEAL FLOW AND MIXING IN CHEMICAL REACTORS

2.1.1. Types of Non-Ideal Flow Patterns

So far we have developed calculation methods only for the ideal cases of plug-flow in tubular reactors (Chapter 1, Section 1.7), and complete mixing in stirred-tank reactors (Chapter 1, Section 1.8). In reality, the flow of fluids in reactors is rarely ideal, and although for some reactors design equations based on the assumption of ideal flow give acceptable results, in other cases the departures from the ideal flow state need to be taken into account. Following the development by DANCKWERTS[1] of the basic ideas, one of the leading contributors to the subject of non-ideal flow has been LEVENSPIEL whose papers and books[2,3,4], especially *Chemical Reaction Engineering*[2] Chapters 9 and 10, should be consulted.

It is possible to distinguish various types of non-ideal flow patterns in reactors (and process vessels generally) the most important being *channelling, internal recirculation* and the presence of *stagnant regions*. These are illustrated in the examples shown in Fig. 2.1. In two-phase (and three-phase) reactors the flow patterns in one phase interact with the flow patterns in the other phase, as in the case of the liquid–liquid reactor of Fig. 2.1*c*. One of the major problems in the scale-up of reactors is that the flow patterns often change with a change of scale, especially in reactors involving two or more phases where flow interactions may occur. A gas–solid fluidised bed is another example of an important class of reactor whose characteristics on scale-up are difficult to predict[4] (see Volume 2, Chapter 6).

2.1.2. Experimental Tracer Methods

When the flow through a reactor or any other type of process vessel is non-ideal, experiments with non-reactive tracers can provide most valuable information on the nature of the flow. The injection of a tracer and the subsequent analysis of the exit stream is an example of the general stimulus–response methods described under Process Control in Chapter 7. In tracer experiments various input signals can be

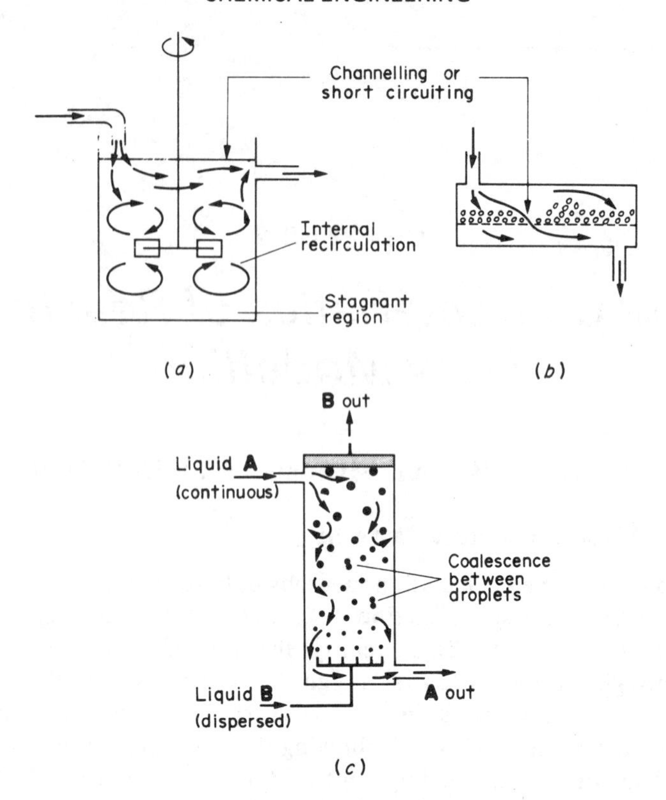

FIG. 2.1. Examples of non-ideal flow in chemical reactors

(*a*) Continuous stirred tank.
(*b*) Gas–solid reaction with maldistribution of solid on a shallow tray.
(*c*) Liquid–liquid reaction: Note the circulation patterns in the continuous phase **A** induced by the rising droplets of B; coalescence may occur between phase **B** droplets giving a range of sizes and upwards velocities.

used, especially the following (Fig. 2.2):

(*a*) *Step input*—**F**-*curves*

The inlet concentration is increased suddenly from zero to C_∞ and maintained thereafter at this value. The concentration–time curve at the outlet, expressed as the fraction C/C_∞ vs. t, is known as an "**F**-curve". The time scale may also be expressed in a dimensionless form as $\theta = t/\tau$, where τ is the *holding time* (or *residence time*), i.e. $\tau = V/v$ where V is the volume of the reactor and v the volumetric rate of flow.

(*b*) *Pulse input*—**C**-*curves*

An instantaneous pulse of tracer is injected into the stream entering the vessel. The outlet response, normalised by dividing the measured concentration C by A_n, the area under the concentration–time curve, is called the "**C**-curve".
Thus:

$$\int_0^\infty \mathbf{C}\,dt = \int_0^\infty \frac{C}{A_n}\,dt = 1 \tag{2.1}$$

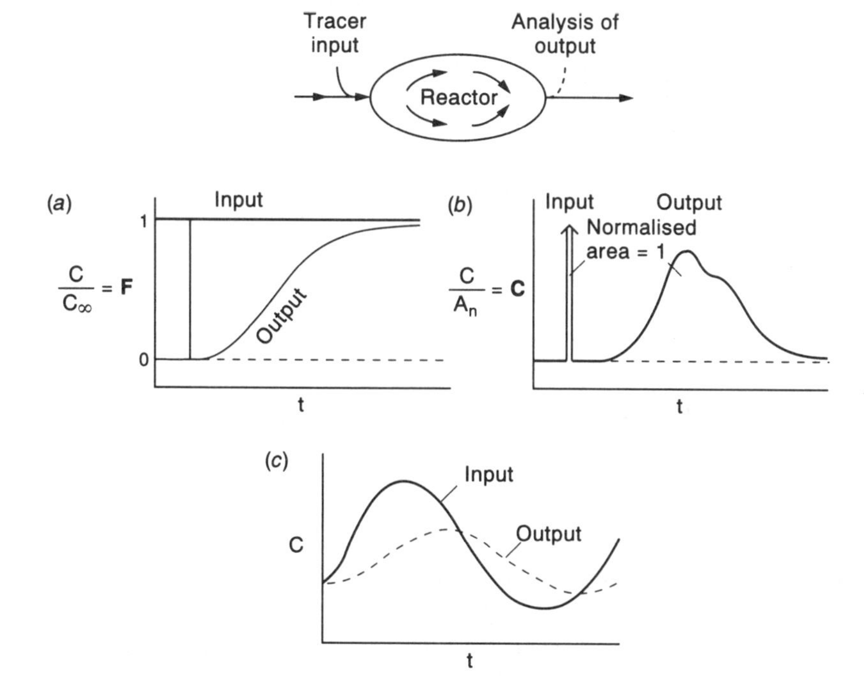

FIG. 2.2. Tracer measurements; types of input signals and output responses
(*a*) Step input—**F**-curve
(*b*) Pulse input—**C**-curve
(*c*) Sinusoidal input

in which:

$$A_n = \int_0^\infty C \, dt \tag{2.2}$$

(Mathematically the instantaneous pulse used here is the unit impulse (Chapter 7, also known as the Dirac delta function.)

A **C**-curve may also be shown against a dimensionless time coordinate $\theta = t/\tau = vt/V$ when it is denoted by $\mathbf{C}_\theta$; from the normalisation of the area under this curve to unity, it follows that $\mathbf{C}_\theta = \tau\mathbf{C}$. The **C**-curves shown in Fig. 2.9*a* are thus $\mathbf{C}_\theta$ curves.

(*c*) *Cyclic, sinusoidal and other inputs*

A cyclic input of tracer will give rise to a cyclic output. This type of input is more troublesome to apply than a pulse or step input but has some advantages for frequency response analysis. A truly instantaneous pulse input is also difficult to apply in practice, but an idealised input pulse is not essential[2,3] for obtaining the desired information (See Section 2.3.5 (*b*)).

2.1.3. Age Distribution of a Stream Leaving a Vessel—E-Curves

The distribution of residence times for a stream of fluid leaving a vessel is called the *exit age distribution function* **E** (synonymous with *residence time distribution* or

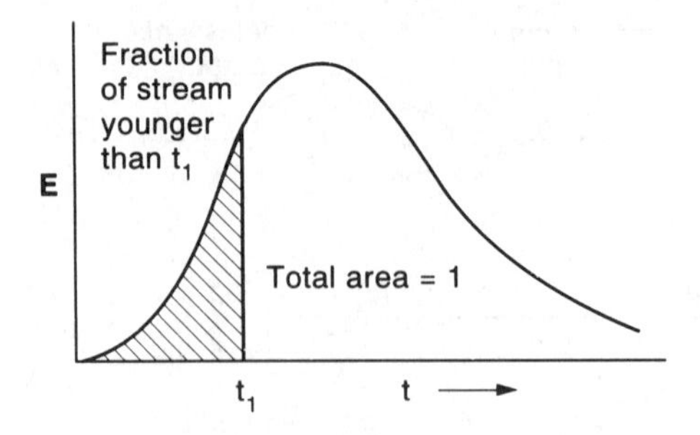

FIG. 2.3. Exit age distribution function or E-curve; also known as the residence time distribution

RTD). By definition, the fraction of the exit stream of age between t and $(t + \delta t)$ is $\mathbf{E}\, \delta t$ (Fig. 2.3.). Integrating over all ages:

$$\int_0^\infty \mathbf{E}\, dt = 1. \tag{2.3}$$

The fraction of material younger than t_1 (shown cross-hatched in Fig. 2.3) is:

$$\int_0^{t_1} \mathbf{E}\, dt$$

Relation between **F**-, **C**- *and* **E**-*Curves*

For steady state flow in a closed vessel (i.e. one in which fluid enters and leaves solely by plug flow) the residence time distribution for any batch of fluid entering must be the same as that leaving (otherwise accumulation would occur).

Hence the **C**-curve generated by injecting a pulse of fluid at the entrance must be identical with the **E**-curve,

i.e.:

$$\mathbf{C} \equiv \mathbf{E} \tag{2.4}$$

Consider an **F**-curve generated by switching, say, to red fluid at $t = 0$. At any time $t > 0$ red fluid, and only red fluid, in the exit stream is younger than age t.

Thus:

(Fraction of red fluid in exit stream) = (Fraction of exit stream younger than age t)

i.e.:

$$\mathbf{F} = \int_0^{t} \mathbf{E}\, dt \tag{2.5}$$

or:

$$\frac{d\mathbf{F}}{dt} \equiv \mathbf{E} \equiv \mathbf{C} \tag{2.6}$$

Note that it may be shown also that the holding time τ is equal to $\bar{t}_E$, the mean of the residence time distribution,

i.e.:

$$\tau = V/v = \bar{t}_E \tag{2.7}$$

2.1.4. Application of Tracer Information to Reactors

Direct Application of Exit Age Distribution

Information on residence time distribution obtained from tracer measurements can be used directly to predict the performance of a reactor in which non-ideal flow occurs. A good example is the conversion obtained when solid particles undergo a reaction in a rotary kiln (Fig. 2.4). Using labelled particles as tracers, a **C**-curve and hence an exit age distribution can be determined; the fraction of particles in the exit stream which have stayed in the reactor for times between t and $(t + \Delta t)$ may thus be expressed as $\mathbf{E}\,\Delta t$. From other experiments α_t the fractional conversion of reactant in those particles staying in the reactor for a time t can be found. Using α_t for the range of times t to $(t + \Delta t)$ providing Δt is small, the average conversion $\bar{\alpha}$ for the mixture of particles which constitute the product is thus $\bar{\alpha} = \sum \alpha_t\,\mathbf{E}\,\Delta t$ where the summation is taken over the whole range of residence times, if necessary 0 to ∞.

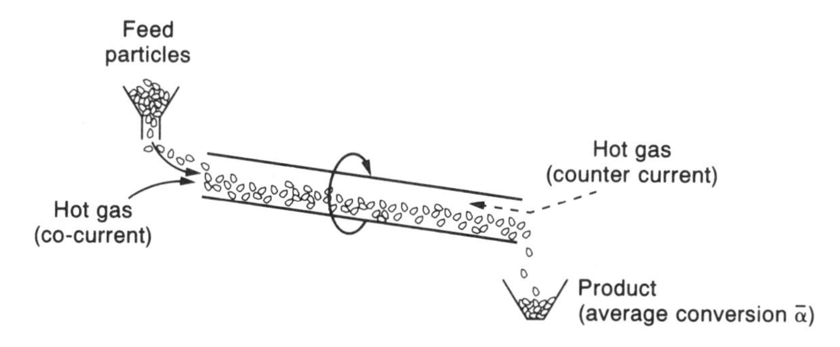

FIG. 2.4. Average conversion from residence time distribution—example of rotary kiln

Macromixing and Micromixing

In the above example of the rotary kiln, although the particles intermingle, each particle remains as a distinct and separate aggregate of molecules which retains its identity as it passes through the process. This type of intermingling of aggregates is called *macromixing*, i.e. mixing but only on a *macroscopic scale*. Perfect macromixing means that the composition averaged over a volume sufficiently great to contain a relatively large number of aggregates does not vary with location in the vessel.

In contrast to this, when a solution of a substance is diluted, the molecules of the solute mix on a *molecular scale* with fresh solvent and essentially lose their identity in the process. This is called *micromixing*. In the same way solutions of two different solutes can undergo micromixing.

If, however, the two solutions to be mixed are very viscous, and diffusion is slow, complete micromixing may not occur; instead we could have a mixture which was intermediate in its *degree of segregation* between macromixing at one extreme and micromixing at the other. As we shall see below, it transpires that in attempting to use tracer measurements directly to predict, say, the average conversion from a reactor, further information on the degree of segregation is required, unless the process (here the reaction) occurring is a *first-order* or *linear* process, in this case, a first-order reaction.

Significance of Linear and Non-linear Processes

As an example, consider the various cases of mixing and reaction shown in Fig. 2.5. In Case I, we have a solution of a reactant concentration C_0 and pure solvent in separate compartments each of volume V. In Case II, the partition is removed and there is dilution of the reactant solution with micromixing so that the reactant concentration is halved. In Case III the partition is removed but there is macromixing only, and the reactant solution remains segregated. It is seen that the number of moles reacting per unit time in all three cases is the same for a first-order reaction which is a linear process. However, for a second-order reaction, which is an example of a non-linear process, the rate of reaction depends on the degree of segregation resulting from the mixing: if micromixing occurs (Case II of Fig. 2.2), the rate is only half the rate with segregation (Cases I and III). If only macromixing occurs, dilution has no effect on the rate of a second-order reaction.

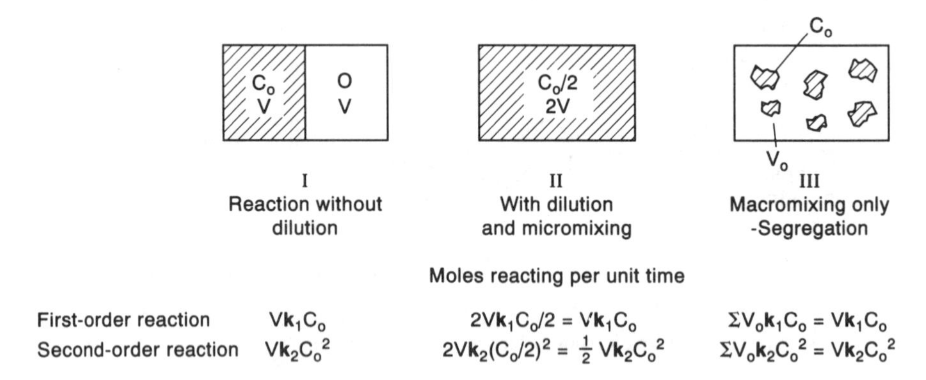

FIG. 2.5. Influence of state of mixing on first and second-order reaction rates

In general, it may be concluded:

Reaction order > 1, reaction rate is greatest with complete segregation.
Reaction order = 1, reaction rate is unaffected by segregation.
Reaction order < 1, reaction rate is greatest with complete micromixing.

Occurrence of Micromixing in Flow Reactors

In flow reactors two extremes of micromixing can be visualised:

(a) Completely segregated flow until mixing occurs at the outlet or as the fluid is sampled for analysis, i.e. segregation throughout the entire system to the outlet, or *late* mixing.
(b) A condition of *maximum mixedness* meaning intimate micromixing at the earliest possible stage, i.e. *early* mixing.

The significance of early or late mixing may be seen from the example given by KRAMERS[5] (Fig. 2.6) of a plug-flow reactor in series with an ideally mixed stirred

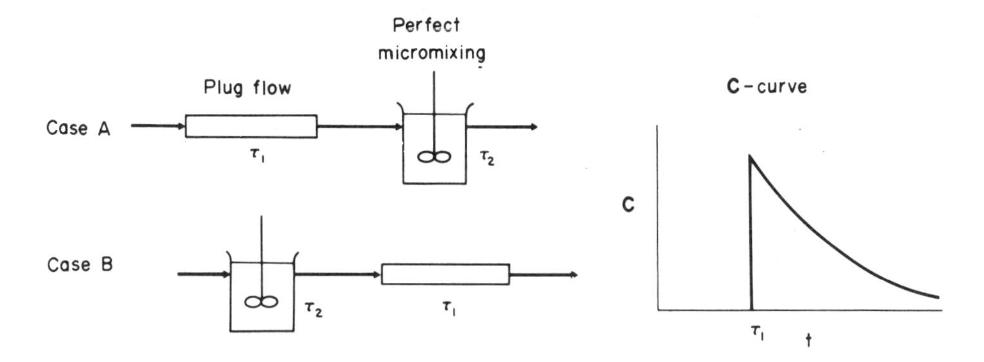

FIG. 2.6. Kramers' example of a plug-flow tubular reactor (residence time τ_1) and an ideal stirred tank (residence time τ_2) in series. Note that the C-curve is the same for both configurations

tank in which macromixing occurs. The two units may be connected either with the tubular reactor first (Case A) or with the stirred tank first (Case B). As far as the response to a pulse input of tracer is concerned, the plug-flow section merely delays the tracer by a time τ_1, the plug flow residence time, irrespective of the configuration, and the **C**-curve is the same for both cases (for the equation of the part due to the stirred tank, see Section 2.2). If we suppose that a *first-order* reaction with rate constant k_1 occurs in this system, the reader may like to show as an exercise that again in both Case A and Case B:

$$\frac{C_e}{C_0} = \frac{e^{-k_1 \tau_1}}{1 + k_1 \tau_2} \tag{2.8}$$

where C_0 is the inlet concentration of reactant, C_e the outlet concentration, and τ_2 is the residence time in the stirred tank.

However, for a *second-order* reaction the conversions in the two cases A and B are *different*. With late mixing as in Case A, the conversion is higher than with early mixing as in Case B. Again as a general rule it is concluded that if the reaction order is > 1 late mixing, i.e. segregation, gives the highest conversion: if the reaction order is < 1 early mixing gives the highest conversion; and for a first-order reaction the conversion is independent of the type of mixing.

A further important conclusion is that for a given **C**-curve or residence time distribution obtained from tracer studies, a *unique* value of the conversion in a chemical reaction is not necessarily obtainable unless the reaction is first order. Tracer measurements can certainly tell us about departures from good macromixing. However, *tracer measurements cannot give any further information about the extent of micromixing* because the tracer stimulus-response is a first-order (linear) process as is a first-order reaction.

Just as knowledge about micromixing can affect predictions about the rates of reactions other than first order, so the reverse is true; measurements of rates of reaction, apart from first order ones, can give valuable information about the extent to which micromixing is occurring in a complex fluid flow field, such as that in a stirred tank[6].

2.2. TANKS-IN-SERIES MODEL

Following the example of Fig. 2.6, if a reactor on which tracer measurements were made gave a **C**-curve similar to that shown for an ideal plug-flow tubular section in series with an ideal stirred-tank section, then this sequence of elements could be used as a model of the reactor (within the limitations about order of reactions outlined above). These ideal reactor sections can be regarded as building blocks for the model. However, before proceeding to more complicated combinations of elements, we need to investigate the flow behaviour that results from connecting a number of similar stirred tanks in series.

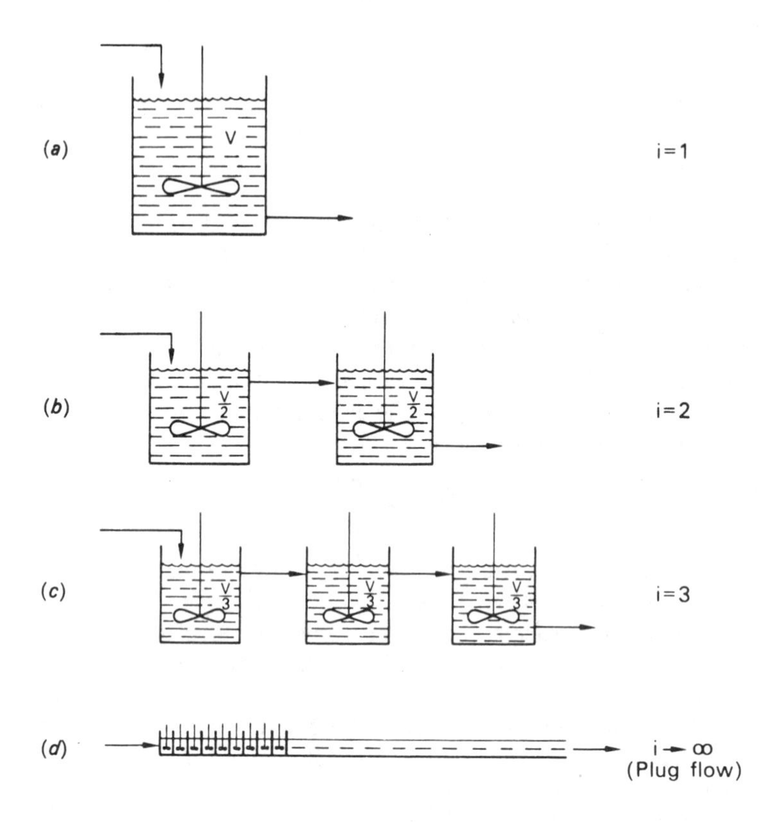

FIG. 2.7. Sets of equal stirred tanks in series, each set having the same total volume V

Several useful ideas and conclusions can be obtained by studying the flow of an unreactive tracer, such as a coloured dye, through a series of tanks each of which is assumed to be ideally mixed. So that we may readily compare the effects of increasing the number of tanks, let us consider a volume V which can be made up of one tank only; or two tanks each of volume $V/2$; or three tanks each of volume $V/3$; or, in general, i equally sized tanks each of volume V/i (Fig. 2.7). Through this series of tanks, a pure liquid flows at a volumetric flowrate v. Suppose that at time $t = 0$ when all the tanks are full and the whole system is in a state of steady flow, a shot of n_0 moles of tracer is injected into the feed. The tracer is assumed to be completely miscible with the liquid in the tanks and to be in the form of a concentrated solution whose volume is negligible compared with V. The aim is to calculate the number of moles which have left the system completely, as a function of time, and the total number of tanks to be employed, i.

At any time t, let tank 1 contain n_1 moles of tracer, whence the concentration in the tank $C_1 = n_1/(V/i)$.

∴ Rate of outflow of tracer from tank 1 $= C_1 v = (n_1 i/V)v$.

Applying a material balance to the tracer over tank 1, and noting that as far as tracer concentration is concerned, the system is not in a steady state:

$$\text{Inflow} - \text{Outflow} - \text{Reaction} = \text{Accumulation}$$

$$0 - \frac{n_1 iv}{V} - 0 = \frac{dn_1}{dt}$$

∴

$$n_1 = n_0 \exp\left(-\frac{vit}{V}\right) \tag{2.9}$$

Similarly for tank 2:

$$\frac{n_1 iv}{V} - \frac{n_2 iv}{V} - 0 = \frac{dn_2}{dt}$$

Substituting for n_1 and integrating:

$$n_2 = n_0\left(\frac{vit}{V}\right)\exp\left(-\frac{vit}{V}\right)$$

Similarly for tank 3:

$$n_3 = \frac{n_0}{2!}\left(\frac{vit}{V}\right)^2 \exp\left(-\frac{vit}{V}\right)$$

and, in general, for the rth tank:

$$n_r = \frac{n_0}{(r-1)!}\left(\frac{vit}{V}\right)^{r-1} \exp\left(-\frac{vit}{V}\right) \tag{2.10}$$

The progress of a tracer through a series of tanks can be followed using these equations. Taking the case of three tanks as an example, Fig. 2.7*c*, the curves shown

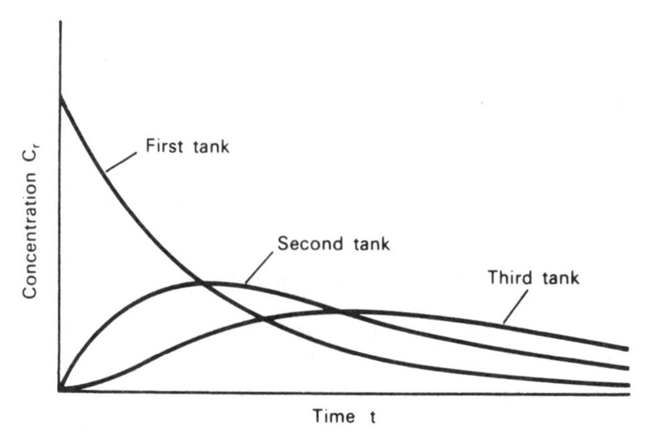

FIG. 2.8. Progress of a pulse of tracer through a series of three stirred tanks

in Fig. 2.8 are obtained. The concentration of tracer in tank 1 after its injection ($t = 0$) is given by $3n_0/V$ and thereafter it diminishes steadily. On the other hand, the concentrations in tanks 2 and 3 rise to a peak as the wave of tracer passes through, the peak of the wave becoming attenuated and the wave more drawn out as it proceeds from one tank to the next.

As the basis for further comparison, let us return to the case of a fixed volume V which may be a single tank (Fig. 2.7*a*) or two tanks (Fig. 2.7*b*) or three tanks and so on. In the general case there will be i tanks and the concentration of the tracer leaving the last tank will be C_i. If we now plot C_i/C_0', where $C_0' = n_0/V$, against the reduced time (vt/V), the family of curves shown in Fig. 2.9*a* is obtained. The curves are "reduced" **C** (i.e. outlet concentration) curves, as already indicated in Section 2.1.2.

Alternatively, we may plot **F**, the fraction of tracer which has escaped completely after a time t, against (vt/V). **F** may be obtained by noting that at any time t the number of moles remaining in the series of i tanks is:

$$n_t = n_1 + n_2 + n_3 + \ldots + n_i$$

therefore:

$$\mathbf{F} = \frac{n_0 - n_t}{n_0} = 1 - \frac{1}{n_0}(n_1 + n_2 + n_3 + \ldots + n_i)$$

or:

$$\mathbf{F} = 1 - \exp\left(-\frac{vit}{V}\right)\left\{1 + \frac{vit}{V} + \frac{1}{2!}\left(\frac{vit}{V}\right)^2 + \ldots + \frac{1}{(i-1)!}\left(\frac{vit}{V}\right)^{i-1}\right\} \tag{2.11}$$

As $i \to \infty$ the term in brackets above tends to $\exp(vit/V)$ for $t < (V/v)$, and **F** becomes zero[7]. Thus, for an infinite number of tanks the fraction of tracer that has escaped is zero for all times less than the residence time V/v. This is exactly the same as for the case of an ideal tubular reactor with plug flow.

We see therefore from the curves in Fig. 2.9 that for a single stirred tank a high proportion of the tracer, 0.632, has escaped from the tank within a time equal to the residence time V/v. This constitutes a bypassing effect which is an inherent disadvantage of a single stirred tank. However, as the number of tanks in the system is increased, the proportion of tracer which escapes within the residence time is diminished, showing that the bypassing effect is reduced by having a larger number of tanks. In the limit as $i \to \infty$, the system becomes identical to an ideal tubular reactor with plug flow.

2.3. DISPERSED PLUG-FLOW MODEL

2.3.1. Axial Dispersion and Model Development

The dispersed plug-flow model can be regarded as the first stage of development from the simple idea of plug flow along a pipe. The fluid velocity and the concentrations of any dissolved species are assumed to be uniform across any section of the pipe, but here mixing or 'dispersion' in the direction of flow (i.e. in the axial z-direction) is taken into account (Fig. 2.10). The axial mixing is described by

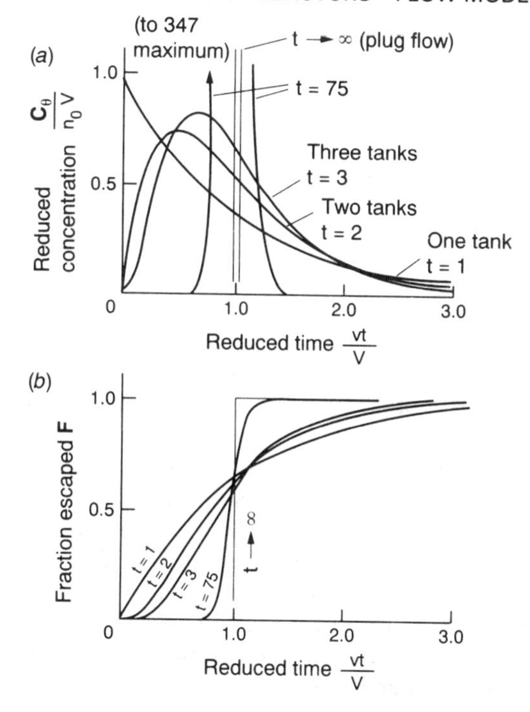

FIG. 2.9. Curves for tracer leaving different sets of stirred tanks in series, each set having the same total volume V

(a) $\mathbf{C}_\theta$-curves: reduced concentration versus reduced time
(b) **F**-curves: fraction of tracer escaped versus reduced time

a flux N_L which is related to the axial concentration gradient $\partial C/\partial z$ through a parameter D_L in exactly the same way as in molecular diffusion.

i.e.:

$$N_L = - D_L \frac{\partial C}{\partial z} \tag{2.12}$$

where D_L is the *dispersion coefficient* in the longitudinal direction. Although equation 2.12 resembles the equation for molecular diffusion, the physical processes involved in flow dispersion are distinctly different.

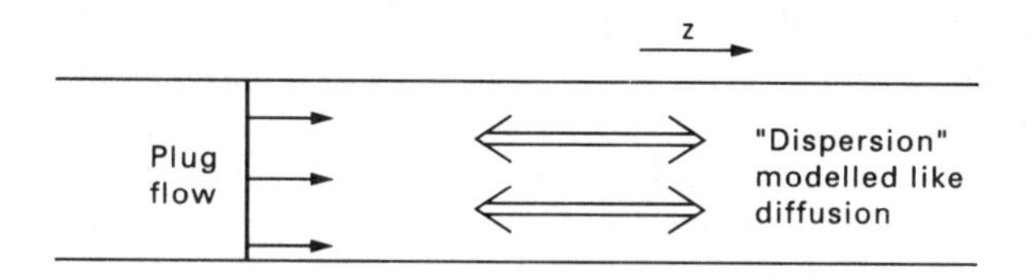

FIG. 2.10. The dispersed plug-flow model with mixing (dispersion) in axial z-direction; dispersion coefficient D_L

Investigations into the underlying flow mechanisms that actually cause axial mixing in a pipe have shown that, in both laminar and turbulent flow, the non-uniform velocity profiles (see Fig. 2.11 and Volume 1, Fig. 3.11.) are primarily

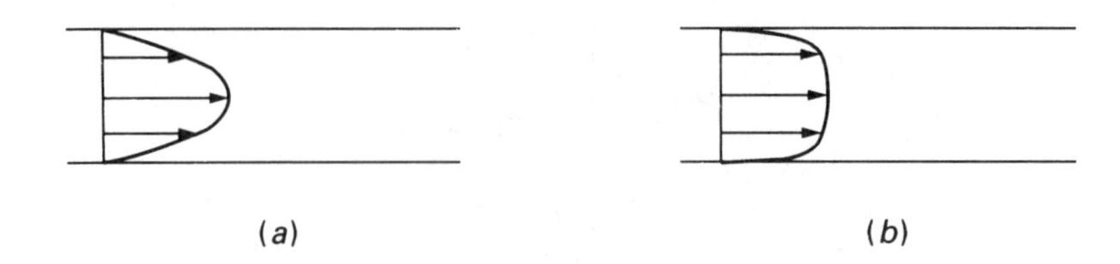

FIG. 2.11. Actual velocity distribution in a pipe. (*a*) Laminar flow; (*b*) Turbulent flow

responsible. Fluid near the centre of the pipe travels more quickly than that near the wall, the overall result being mixing in the axial direction. This effect is most easily visualised for laminar flow in a pipe, although in reactor design the turbulent regime is generally the more important in pipes, vessels and packed beds.

Consider a liquid, initially colourless, passing in laminar flow along a pipe, and then imagine the incoming stream to be suddenly switched to a red liquid of the same viscosity etc. The boundary between the two liquids is initially a plane disc at the inlet of the pipe, (Fig. 2.12*a*). Because the liquid velocity is a maximum at the centre of the pipe and zero at the walls, as time proceeds this boundary becomes stretched out into a cone-like shape with a parabolic profile (Fig. 2.12*b*). After the tip of this cone has passed a particular point such as **A** (Fig. 2.12*c*), a section **AA′** across the pipe will show a central red core whilst the surrounding liquid will still be colourless. The fraction of red liquid averaged across the section will increase with time of flow, giving the impression of mixing in the axial direction. In laminar flow in tubes of small length to diameter ratio, the sharpness of the boundary between the red and colourless liquid is only slightly blurred by the occurrence of molecular diffusion because, in liquids, diffusion is slow. With gases, however, the effect of diffusion will be greater (see, for example, Section 2.3.6, Fig. 2.20) for the effect of Schmidt Number on dispersion coefficient). In turbulent flow, the effect of radial eddy diffusion is to make this boundary so diffuse that concentrations become effectively uniform across the pipe, as required by the *dispersed plug-flow* model. Even in laminar flow, for long small-diameter tubes, there exists a regime known as Taylor–Aris [8,9,10] dispersion in which radial diffusion is sufficient to produce an effectively uniform concentration over any cross-section of the tube. The general conclusion that, even for turbulent flow, axial dispersion in a pipe is caused primarily by differences in velocity at different radial positions rather than by turbulent eddy diffusion in the axial direction may seem strange at first sight to the reader.

If the radial diffusion or radial eddy transport mechanisms considered above are insufficient to smear out any radial concentration differences, then the simple *dispersed plug-flow* model becomes inadequate to describe the system. It is then necessary to develop a mathematical model for simultaneous radial and axial dispersion incorporating both radial and axial dispersion coefficients. This is especially important for fixed bed catalytic reactors and packed beds generally (see Volume 2, Chapter 4).

The model developed below is for dispersion in the axial direction only. Because the underlying mechanisms producing axial dispersion are complex as the discussion above shows, equation 2.12 is best regarded as essentially a mathematical definition

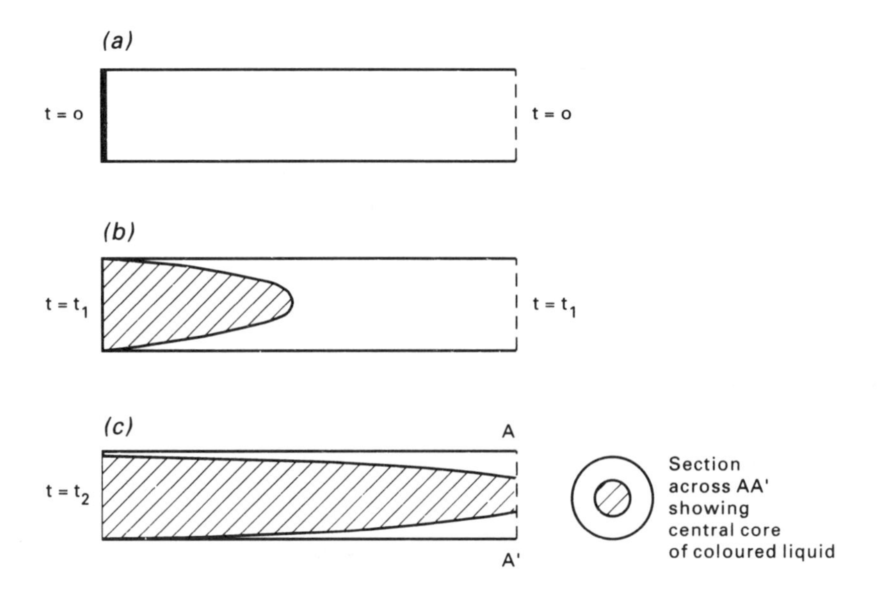

FIG. 2.12. Sudden switch from clear to coloured liquid in laminar flow in a pipe

of the dispersion process and the dispersion coefficient D_L. Note that the dispersed plug-flow model uses just D_L to describe axial mixing and for that reason is classed as a 'one parameter model'. Another such model is that of tanks in series, the parameter being the number of tanks.

2.3.2. Basic Differential Equation

Consider a fluid flowing steadily along a uniform pipe as depicted in Fig. 2.13; the fluid will be assumed to have a constant density so that the mean velocity u is constant. Let the fluid be carrying along the pipe a small amount of a tracer which has been injected at some point upstream as a pulse distributed uniformly over the cross-section; the concentration C of the tracer is sufficiently small not to affect the density. Because the system is not in a steady state with respect to the tracer distribution, the concentration will vary with both z the position in the pipe and, at any fixed position, with time; i.e. C is a function of both z and t but, at any given value of z and t, C is assumed to be uniform across that section of pipe. Consider a material balance on the tracer over an element of the pipe between z and $(z + \delta z)$, as shown in Fig. 2.13, in a time interval δt. For convenience the pipe will be considered to have unit area of cross-section. The flux of tracer into and out of the element will be written in terms of the dispersion coefficient D_L in accordance with equation 2.12. For completeness and for later application to reactors (see Section 2.3.7) the possibility of disappearance of the tracer by chemical reaction is also taken into account through a rate of reaction term $\mathcal{R}$.

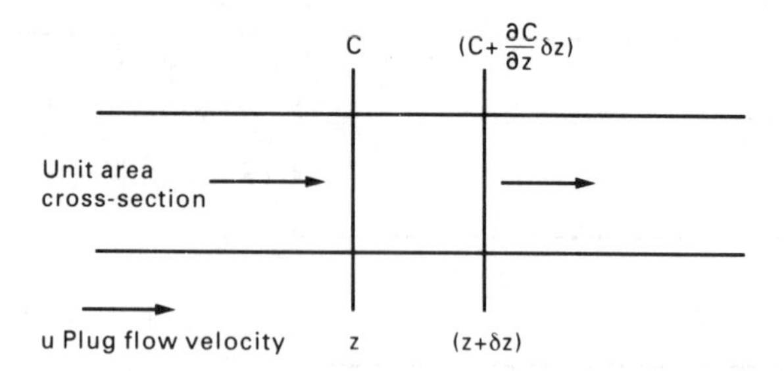

FIG. 2.13. Material balance over an element of pipe

Then, for unit cross-sectional area of pipe:

Inflow − Outflow − Reaction

$$\left(uC - D_L \frac{\partial C}{\partial z}\right)\delta t - \left[u\left(C + \frac{\partial C}{\partial z}\delta z\right) - D_L\left(\frac{\partial C}{\partial z} + \frac{\partial}{\partial z}\left(\frac{\partial C}{\partial z}\right)\delta z\right)\right]\delta t - \mathfrak{R}\,(1 \times \delta z)\delta t$$

$$= \text{Accumulation}$$

$$= \frac{\partial C}{\partial t}(1 \times \delta z)\delta t$$

where $(1 \times \delta z)$ is the volume of the element.

i.e.:

$$-u\frac{\partial C}{\partial z}\delta z + D_L\frac{\partial^2 C}{\partial z^2}\delta z - \mathfrak{R}\,\delta z = \frac{\partial C}{\partial t}\delta z$$

i.e.:

$$D_L\frac{\partial^2 C}{\partial z^2} - u\frac{\partial C}{\partial z} - \mathfrak{R} = \frac{\partial C}{\partial t} \tag{2.13}$$

When there is no reaction:

$$D_L\frac{\partial^2 C}{\partial z^2} - u\frac{\partial C}{\partial z} = \frac{\partial C}{\partial t} \tag{2.14}$$

This is the basic differential equation governing the transport of a dilute tracer substance along a pipe. Being a partial differential equation, its solution, which gives the concentration C as a function of z and t, will be very much dependent on the boundary conditions that apply to any particular case.

2.3.3. Response to an Ideal Pulse Input of Tracer

Figure 2.14 shows, in principle, the result of injecting an instantaneous pulse of tracer into fluid in a pipe in which the differential equation above applies. Following the injection of the pulse at time $t = 0$, profiles of concentration versus position z in the pipe are shown at three successive 'snapshots' in time, t_1, t_2, t_3. The inset graph shows how the concentration of tracer would change with time for measurements taken at a fixed position $z = L$. Although this graph may look roughly similar in shape to the 'snapshot' profiles in the pipe, it is distinctly different in principle. When the results of such measurements are 'normalised' (see Section 2.1.2), the 'C-curve' for the tracer response is obtained.

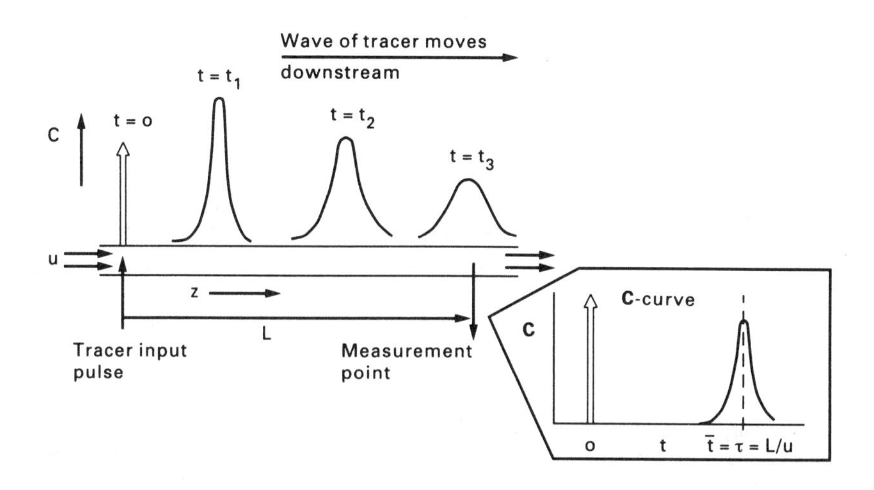

FIG. 2.14. Response of the dispersed plug-flow model to a pulse input of tracer. Three 'snapshot' concentration profiles at different times t_1, t_2, t_3 are shown. The inset shows the 'C-curve' derived from measurements at a fixed position, $z = L$. If axial dispersion occurs to only a small extent (small D_L/uL) the C-curve is almost symmetrical

Mathematically, the solution to the partial differential equation 2.13 for a pulse input of M moles of tracer into a pipe of cross-sectional area A is[2]:

$$C = \frac{M}{2A\sqrt{\pi D_L t}} \mathrm{e}^{-\frac{(z-ut)^2}{4D_L t}} \tag{2.15}$$

That this equation is indeed a solution of equation 2.13. may be verified by partial differentation with respect to t and z and substituting. It also satisfies the initial condition corresponding to an ideal pulse, i.e. when $t \to 0+$, $z = 0$, $C \to \infty$; but, at $z \neq 0$, $C = 0$. In addition, at any time $t > 0$, the total moles of tracer anywhere in the pipe must be equal to M.

i.e.:

$$M = \int_{-\infty}^{\infty} CA \, \mathrm{d}z$$

At any particular time $t = t_1$, equation 2.15 becomes:

$$C_{t_1} = \frac{M}{2A\sqrt{\pi D_L t_1}} \mathrm{e}^{-\frac{(z-ut_1)^2}{4D_L t_1}} \tag{2.16}$$

which has the same form as the general equation for a Gaussian distribution (equation 2.20 below). The 'snapshot' concentration profiles of C versus z shown in Fig. 2.14 are therefore *symmetrical* Gaussian-shaped curves.

Now let us see what happens to the tracer concentration at a fixed position $z = L$; equation 2.15 then becomes:

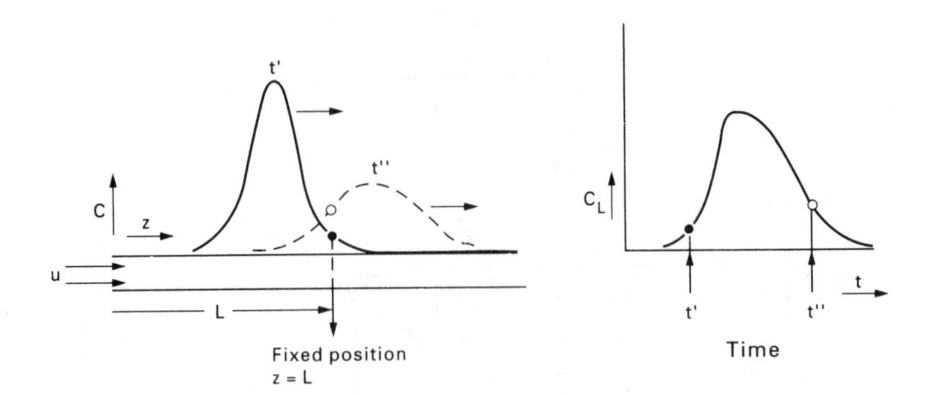

FIG. 2.15. Dispersion of an ideal pulse of tracer injected into a pipe. Formation of an asymmetric curve of C versus t at a fixed point due to change in shape of a symmetrical concentration wave as it broadens while passing the point. Large D_L/uL value

$$C_L = \frac{M}{2A\sqrt{\pi D_L t}} \mathrm{e}^{-\frac{(L-ut)^2}{4D_L t}} \tag{2.17}$$

As an equation for C_L as a function of t, this does not in general give a symmetrical Gaussian curve. The reason why it is not symmetrical can be seen from Fig. 2.15. If the concentration wave of tracer moving along the pipe changes its shape significantly whilst passing a fixed observation point, then the graph of the observed concentration C_L versus time will be skewed.

Mathematically, the skewness of the C_L versus t curve can be identified with the presence of t in the equation 2.17 in the positions arrowed:

$$C_L = \frac{M}{2A\sqrt{\pi D_L t_{\nwarrow}}} \mathrm{e}^{-\frac{(L-ut)^2}{4D_L t_{\nwarrow}}} \tag{2.18}$$

In many instances, however, the change in shape of the concentration wave in passing the observation point is negligibly small. Mathematically, this corresponds to a value $D_L/uL < 0.01$, approximately. In this case the arrowed t in the equation above may be replaced by $\bar{t} = L/u$, the mean residence time of the pulse in the section of pipe between $z = 0$ and $z = L$. (Since the shape of the concentration wave is now considered not to change in passing the observation point, this is also the time at which the peak of the wave passes $z = L$)

Then the term $\dfrac{M}{2A\sqrt{\pi D_L t}}$ becomes $\dfrac{M}{2A\sqrt{\pi D_L \bar{t}}}$ or $\dfrac{M}{2A\sqrt{\dfrac{\pi D_L}{uL} uL\bar{t}}}$

or $\dfrac{M}{A\sqrt{2\pi}\sqrt{\dfrac{2D_L}{uL} u^2\bar{t}^2}}$ or $\dfrac{M}{Au\sqrt{2\pi}\sqrt{\dfrac{2D_L}{uL} \bar{t}^2}}$

Also $e^{-\frac{(z-ut)^2}{4D_L t}}$ becomes $e^{-\frac{(z-ut)^2}{4D_L \bar{t}}}$ and, since $z = L$, this is $e^{-\frac{(L-ut)^2}{(4D_L L/u)}}$ or $e^{-\frac{(\bar{t}-t)^2}{(4D_L)(L/u^3)}}$ or $e^{-\frac{(\bar{t}-t)^2}{4(D_L/uL)(L/u)^2}}$ or $e^{-\frac{(\bar{t}-t)^2}{2(2D_L/uL)\bar{t}^2}}$

Thus:

$$C = \frac{M}{Au} \frac{1}{\sqrt{2\pi}\sqrt{(2D_L/uL)\bar{t}^2}} e^{-\frac{(t-\bar{t})^2}{2(2D_L/uL)\bar{t}^2}} \tag{2.19}$$

We may compare this equation with the general form of equation for a Gaussian distribution:

$$f = \frac{1}{\sqrt{2\pi}\sigma} e^{-\frac{(x-\bar{x})^2}{2\sigma^2}} \tag{2.20}$$

where f is the frequency function for an observation of magnitude x, $\bar{x}$ is the mean and σ^2 is the variance. We see that they correspond, apart from the M/Au factor, if:

$$\sigma^2 = \frac{2D_L}{uL}\bar{t}^2 \tag{2.21}$$

The group D_L/uL which is *dimensionless* is called the *dispersion number* for the flow.

Equations 2.19 and 2.20 can be matched exactly if the C versus t curve is converted into a **C**-curve (Section 2.1.2) by means of a normalising factor.

i.e.:

$$\mathbf{C} = \frac{C}{\int_{-\infty}^{\infty} C\,dt}$$

Here the normalising quantity $\int_{-\infty}^{\infty} C\,dt$ is:

$$\int_{-\infty}^{\infty} \frac{M}{Au} \frac{1}{\sqrt{2\pi}\,\sigma} e^{-\frac{(t-\bar{t})^2}{2\sigma^2}} dt$$

i.e.:

$$\frac{M}{Au} \int_{-\infty}^{\infty} \frac{1}{\sqrt{2\pi}\,\sigma} e^{-\frac{(t-\bar{t})^2}{2\sigma^2}} dt$$

which simply is equal to M/Au, because a fundamental property of the function f in equation 2.20 which follows from its definition as a frequency function in the Gaussian distribution ($f\,dx$ is the probability of observing values of x in the range x to $x + dx$ and the sum of all the probabilities must be unity).

i.e.:

$$\int_{-\infty}^{\infty} f\,dx = 1$$

We conclude therefore that, for small values of the dispersion number ($D/uL < 0.01$), the **C**-curve for a pulse input of tracer into a pipe is symmetrical and corresponds exactly to equation 2.20 for a Gaussian distribution function:

$$\mathbf{C} = \frac{1}{\sqrt{2\pi}\,\sqrt{(2D_L/uL)\bar{t}^2}}\, e^{-\frac{(t-\bar{t})^2}{2(2D_L/uL)\bar{t}^2}} \tag{2.22}$$

The above equations can also be expressed in a fully dimensionless form. Using a dimensionless time $\theta = ut/L$, which may also be written $\theta = t/\bar{t}$ since $\bar{t} = L/u$ (note that in dimensionless form the mean time $\bar{\theta} = 1$), then equation 2.22 becomes:

$$C = \frac{1}{\bar{t}}\,\frac{1}{\sqrt{2\pi}\sqrt{(2D_L/uL)}}\, e^{-\frac{(1-\theta)^2}{2(2D_L/uL)}} \tag{2.23}$$

Carrying out a further stage of normalisation on this equation, i.e. introducing $C_\theta = \bar{t}C$ so that the area under the C_θ versus θ curve is unity, then:

$$C_\theta = \frac{1}{\sqrt{2\pi}\sqrt{(2D_L/uL)}}\, e^{-\frac{(1-\theta)^2}{2(2D_L/uL)}} \tag{2.24}$$

which has a mean $\bar{\theta} = 1$ and a variance $\sigma^2 = 2D_L/uL$.
Essentially this is the solution to equation 2.14, modified by using the dimensionless variables $z' = z/L$ and $\theta = ut/L$

i.e.:

$$\frac{\partial C}{\partial \theta} + \frac{\partial C}{\partial z'} = \frac{D_L}{uL}\frac{\partial^2 C}{\partial z'^2} \tag{2.25}$$

From the appearance of the dispersion number D_L/uL in this dimensionless form of the basic differential equation of the plug-flow dispersion model it can be inferred that the dispersion number must be a significant characteristic parameter in any solution to the equation, as we have seen.

2.3.4. Experimental Determination of Dispersion Coefficient from a Pulse Input

Provided that the introduction of an ideal pulse input of tracer can be achieved experimentally, equation 2.21 allows the determination of dispersion coefficients from measurements of tracer concentration taken at a sampling point distance L downstream. Again, in an ideal situation, a large number of measurements of concentration would be taken so that a continuous graph of C versus t could be plotted (Fig. 2.16). The two most important parameters that characterise such a curve are (*a*) the mean time $\bar{t}$ which indicates when the wave of tracer passes the measuring point, and (*b*) the variance σ^2 which indicates how much the tracer has spread out during the measurement time, i.e. the width of the curve. Thus, given a smooth continous pulse response curve (in general it could be symmetrical or not):

$$\text{Area under the curve} = \int_0^\infty C\,\mathrm{d}t \tag{2.26}$$

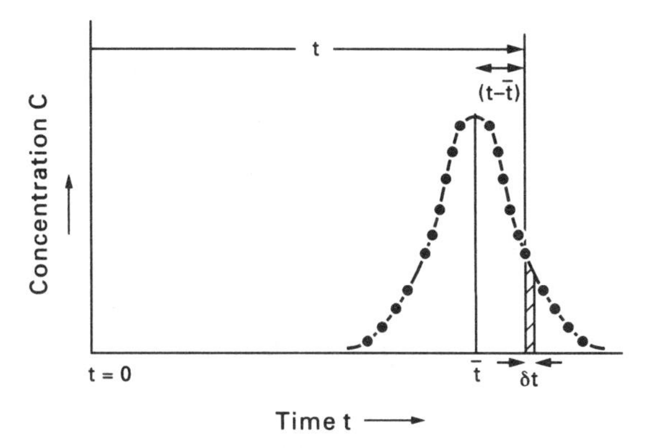

FIG. 2.16. Experimental determination of dispersion coefficient–ideal-pulse injection. Expected symmetrical distribution of concentration measurements. Small D_L/uL value

$$\text{Mean (or first moment)} = \frac{\int_0^\infty Ct\,dt}{\int_0^\infty C\,dt} \tag{2.27}$$

$$\text{Variance (or second moment about the mean)} = \frac{\int_0^\infty C(t-\bar{t})^2\,dt}{\int_0^\infty C\,dt} \tag{2.28}$$

Note:
$$\sigma^2 = \frac{\int_0^\infty C(t-\bar{t})^2\,dt}{\int_0^\infty C\,dt} = \frac{\int_0^\infty C\,t^2\,dt}{\int_0^\infty C\,dt} - \frac{2\int_0^\infty C\,t\bar{t}\,dt}{\int_0^\infty C\,dt} + \frac{\int_0^\infty C\,\bar{t}^2\,dt}{\int_0^\infty C\,dt}$$

Within the integrals of the second and third terms on the right-hand side of this equation, $\bar{t}$ is a constant. The second term may therefore be written:

$$-\frac{2\int_0^\infty C\,t\bar{t}\,dt}{\int_0^\infty C\,dt} = -\frac{2\bar{t}\int_0^\infty C\,t\,dt}{\int_0^\infty C\,dt} = -2\bar{t}^2$$

where equation 2.27 has been used to simplify the expression, while the third term becomes:

$$\frac{\int_0^\infty C\,\bar{t}^2\,\mathrm{d}t}{\int_0^\infty C\,\mathrm{d}t} = \frac{\bar{t}^2\int_0^\infty C\,\mathrm{d}t}{\int_0^\infty C\,\mathrm{d}t} = \bar{t}^2$$

Thus:

$$\sigma^2 = \frac{\int_0^\infty C\,t^2\,\mathrm{d}t}{\int_0^\infty C\,\mathrm{d}t} - \bar{t}^2 \tag{2.29}$$

These relationships can be applied either to the C versus t curve directly or to the already normalised **C**-curve, in which case the area under the curve $\int_0^\infty C\,\mathrm{d}t$ is then unity. There are several different types of experimental data that might be obtained.

A. *Many equally spaced points*

If we have a large number of data points equally spaced at a time interval Δt, the above relationships for mean and variance may be written in finite difference form:

$$\bar{t} = \frac{\sum C_i\,t_i\,(\Delta t)}{\sum C_i\,(\Delta t)} \qquad \sigma^2 = \frac{\sum C_i\,t_i^2\,(\Delta t)}{\sum C_i(\Delta t)} - \bar{t}^2$$

i.e.:

$$\bar{t} = \frac{\sum C_i\,t_i}{\sum C_i} \tag{2.30}$$

$$\sigma^2 = \frac{\sum C_i\,t_i^2}{\sum C_i} - \bar{t}^2 \tag{2.31}$$

B. *Relatively few data points but each concentration C_i measured instantaneously at time t_i* (Fig. 2.17*a*)

Linear interpolation may be used; this is equivalent to joining the data points with straight lines. Thus, the area under the $C - t$ curve is approximated by the sum of trapezium-shaped increments.

$$\bar{t} = \frac{\sum_{i=1}^{n-1}\left\{\left(\frac{C_{i+1}+C_i}{2}\right)\left(\frac{t_{i+1}+t_i}{2}\right)(t_{i+1}-t_i)\right\}}{\sum_{i=1}^{n-1}\left\{\left(\frac{C_{i+1}+C_i}{2}\right)(t_{i+1}-t_i)\right\}} \tag{2.32}$$

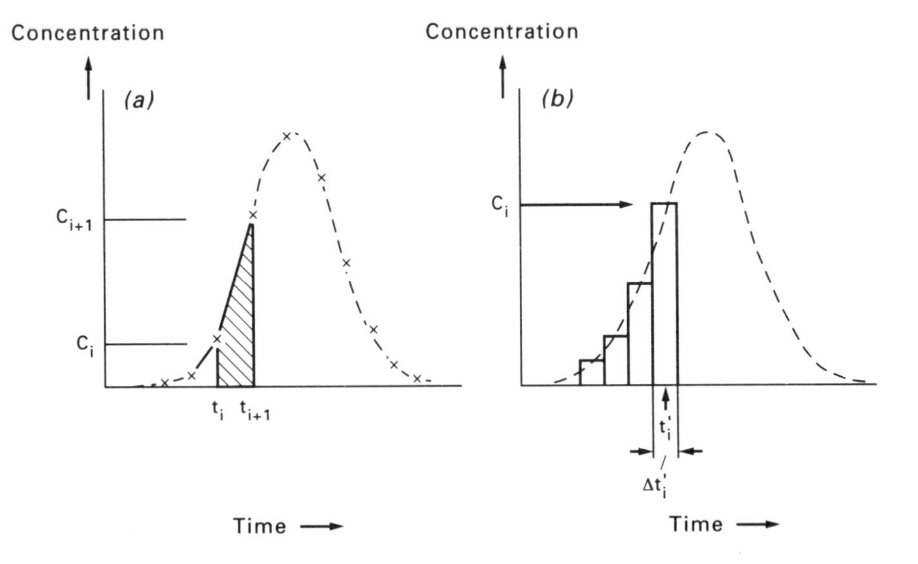

FIG. 2.17. Experimental determination of dispersion coefficient. (*a*) Treatment of data by linear interpolation; (*b*) Treatment of mixing-cup data

$$\sigma^2 = \frac{\sum_{i=1}^{n-1} \left\{ \left(\frac{C_{i+1} + C_i}{2} \right) \left(\frac{t_{i+1} + t_i}{2} \right)^2 (t_{i+1} - t_i) \right\}}{\sum_{i=1}^{n-1} \left\{ \left(\frac{C_{i+1} + C_i}{2} \right) (t_{i+1} - t_i) \right\}} - \bar{t}^2 \qquad (2.33)$$

Note that the terms $(t_{i+1} - t_i)$ in the numerator and denominator of these expressions will cancel out if the concentration data are for equal time intervals.

C. *Data collected by a "Mixing Cup"*

Over a series of time intervals Δt_i each sample is collected by allowing a small fraction of the fluid to flow steadily into a little cup; these samples are stirred to make them uniform and then analysed. Each sample thus represents the concentration in some way averaged over the interval Δt_i. When these concentrations are plotted as the stepwise graph shown in Fig. 2.17*b*, the area under the graph is equal to the summation $\sum C_i \Delta t_i$ but, in order to approximate equations 2.27 and 2.28, the appropriate times at which to take moments (first and second) are not immediately apparent because of the unknown nature of the concentration averaging process. Arbitrarily, we take the mid-increment value t_i' as shown.

$$\bar{t} = \frac{\sum_i C_i \, t_i' \, \Delta t_i}{\sum_i C_i \, \Delta t_i} \qquad (2.34)$$

Similarly:

$$\sigma^2 = \frac{\sum_i C_i t_i'^2 \Delta t_i}{\sum_i C_i \Delta t_i} - \bar{t}^2 \qquad (2.35)$$

Here also the Δt_i in both numerator and denominator will cancel for equal time intervals.

Example 2.1

Numerical Calculation of Residence Time and Dispersion Number.

A pulse of tracer is introduced at the inlet of a vessel through which fluid flows at a steady rate. Samples, each taken virtually instantaneously from the outlet stream at time t_i after injection, are analysed with the following results:

Time t_i (s)	0	20	40	60	80	100	120	140	160	180	200
Concentration C_i (kmol/m^3) $\times 10^3$	0	0	0	0	0.4	5.5	16.2	11.1	1.7	0.1	0

Calculate the mean residence time of the fluid in the vessel and the dispersion number.

Solution

Although for a closed vessel the boundary conditions are different from the case of an open pipe (see Section 2.3.5 below), we will assume that the above methods of data treatment apply because, as will subsequently be verified, the dispersion number in this Example is small.

Method A: Finite difference form—equal time intervals

This is the simplest method, although we cannot expect a very high degree of accuacy because the number of data points is relatively small.

time t_i	0	20	40	60	80	100	120	140	160	180	200
$C_i (\times 10^3)$	0	0	0	0	0.4	5.5	16.2	11.1	1.7	0.1	0
$C_i t_i (\times 10^3)$	0	0	0	0	32	550	1944	1554	272	18	0
$C_i t_i^2 (\times 10^3)$	0	0	0	0	2560	55,000	233,280	217,280	43,520	3240	0

Note that because evaluation of $\bar{t}$ and σ^2 involves ratios, multiplication of the concentrations by a common factor (here 10^3) to avoid inconveniently large or small numbers does not affect the result.

Thus from equation 2.30:

$$\bar{t} = \frac{\sum C_i t_i}{\sum C_i} = \frac{4370}{35} = \underline{\underline{124.9 \text{ s}}}$$

and:

$$\sigma^2 = \frac{\sum C_i t_i^2}{\sum C_i} - \bar{t}_i^2 = \frac{555{,}160}{35} - \left(\frac{4370}{35}\right)^2 = 272.4 \text{ s}^2$$

$\therefore$

$$\frac{\sigma^2}{\bar{t}^2} = \frac{272.4}{124.9^2} = 0.0175$$

Hence the Dispersion Number $\left(\frac{D_L}{uL}\right) = \frac{0.0175}{2} = \underline{\underline{0.0087}}$

Note that the dispersion number is thus confirmed as being 'small', i.e. < 0.01 and within the range for which the Gaussian approximation to the **C**-curve is acceptable.

Method B: Treatment of data by linear interpolation

t_i	40	60		80		100		120		140		160		180		200
$C_i(\times 10^3)$	0	0		0.4		5.5		16.2		11.1		1.7		0.1		0
$\dfrac{(t_{i+1}+t_i)}{2}$			70		90		110		130		150		170		190	
$\dfrac{(C_{i+1}+C_i)}{2}$			0.2		2.95		10.85		13.65		6.4		0.9		0.05	
															$\sum = 35$	
$\left(\dfrac{C_{i+1}+C_i}{2}\right)\left(\dfrac{t_{i+1}+t_i}{2}\right)$			14		265.5		1,193.5		1,774.5		960		153		9.5	
															$\sum = 4370$	
$\left(\dfrac{C_{i+1}+C_i}{2}\right)\left(\dfrac{t_{i+1}+t_i}{2}\right)^2$			980		23,895		131,285		230,685		144,000		26,010		1,805	
															$\sum = 558{,}660$	

Using equations 2.32 and 2.33:

$$\bar{t} = \frac{4370}{35} = \underline{\underline{124.9 \text{ s}}}$$

and:

$$\sigma^2 = \frac{558{,}660}{35} - 124.9^2 = 15{,}961.7 - 15{,}598.3 = 372.4 \text{ s}^2$$

Hence:

$$\frac{\sigma^2}{\bar{t}^2} = \frac{372.4}{124.9^2} = 0.0239$$

Hence: Dispersion Number $\left(\dfrac{D_L}{uL}\right) = \dfrac{0.0239}{2} = \underline{\underline{0.0119}}$

The above example demonstrates that treatment of the basic data by different numerical methods can produce distinctly different results. The discrepancy between the results in this case is, in part, due to the inadequacy of the data provided; the data points are too few in number and their precision is poor. A lesson to be drawn from this example is that tracer experiments set up with the intention of measuring dispersion coefficients accurately need to be very carefully designed. As an alternative to the pulse injection method considered here, it is possible to introduce the tracer as a continuous sinusoidal concentration wave (Fig. 2.2*c*), the amplitude and frequency of which can be adjusted. Also there is a variety of different ways of numerically treating the data from either pulse or sinusoidal injection so that more weight is given to the most accurate and reliable of the data points. There has been extensive research to determine the best experimental method to adopt in particular circumstances[7,11].

2.3.5. Further Development of Tracer Injection Theory

Some interesting and useful conclusions can be drawn from further examination of the response of a system to an input of tracer although the mathematical derivation of these conclusions[7] is too involved to be considered here.

(*a*) *Significance of the boundary conditions*

In the early days of the development of the dispersed plug-flow model, there was considerable controversy over the precise formulation of the boundary conditions that should apply to the basic partial differential equation 2.13 in various situations[1,12]. For a simple straight length of pipe as depicted in Fig. 2.14 the flow is undisturbed as it passes the injection and observation points and the dispersion coefficient has the same value both upstream and downstream of each of these boundary locations. Under these circumstances the boundary is described as 'open'. For a reaction vessel, however, it may be more realistic to assume plug flow in the inlet pipe, and dispersed flow in the vessel itself. In this case the inlet boundary is said to be 'closed', i.e. there is no possibility of dispersion of the tracer back upstream into the plug flow region. For an example of boundary conditions applied to the case of a first order reaction, see Section 2.3.7.

(b) Dispersion coefficients from non-ideal pulse data

The exact formulation of the inlet and outlet boundary conditions becomes important only if the dispersion number (D_L/uL) is large (> 0.01). Fortunately, when D_L/uL is small (< 0.01) and the **C**-curve approximates to a normal Gaussian distribution, differences in behaviour between open and closed types of boundary condition are not significant. Also, for small dispersion numbers D_L/uL it has been shown rather surprisingly that we *do not need to have ideal pulse injection* in order to obtain dispersion coefficients from **C**-curves. A tracer pulse of any arbitrary shape is introduced at any convenient point upstream and the concentration measured over a period of time at both inlet and outlet of a reaction vessel whose dispersion characteristics are to be determined, as in Fig. 2.18. The means $\bar{t}_{in}$ and $\bar{t}_{out}$ and the variances σ_{in}^2 and σ_{out}^2 for each of the **C**-curves are found.

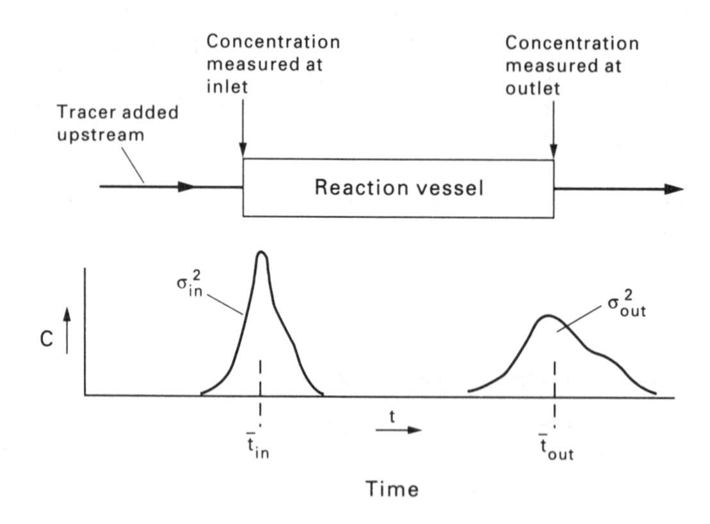

FIG. 2.18. Determination of dispersion coefficient from measurements of **C**-curves at both inlet and outlet of reaction vessel. Tracer added at any convenient point upstream as non-ideal pulse. Small D_L/uL value

Then if:

$$\Delta\bar{t} = \bar{t}_{\text{out}} - \bar{t}_{\text{in}} \quad \text{and} \quad \Delta\sigma^2 = \sigma^2_{\text{out}} - \sigma^2_{\text{in}}$$

$$\Delta L/u = \Delta\bar{t} \quad \text{and} \quad 2(D_L/uL) = \Delta\sigma^2/(\Delta\bar{t})^2$$

where ΔL is the length of the reaction vessel. The mean residence time $\Delta\bar{t}$ may alternatively be expressed in terms of the volume V and the volumetric flowrate, i.e. $\Delta\bar{t} = V/v$.

Example 2.2

Tracer Measurements with Two Sampling Points.

In order to investigate the dispersion characteristics of a reaction vessel, a non-reactive tracer is injected at a convenient location some way upstream from the inlet to the vessel. Subsequently, samples (taken over very short periods of time) are collected from two positions, one being the inlet pipe to the vessel, the other at the vessel outlet, with the following results:

First sampling point:

Time t_i (min)	$\leqslant 1.5$	2.0	2.5	3.0	3.5	4.0	4.5	5.0	5.5	$\geqslant 6$
Concentration C_i (kmol/m^3) $\times 10^3$	< 0.1	0.2	4.7	14.6	23.5	18.5	5.7	1.6	0.4	< 0.1

Second sampling point:-

Time t_i (min)	$\leqslant 5.5$	6.0	6.5	7.0	7.5	8.0	8.5	9.0	9.5	10.0	10.5	$\geqslant 11$
Concentration C_i kmol/m^3) $\times 10^3$	< 0.1	0.2	1.4	5.2	10.9	14.1	12.3	7.7	3.7	1.4	0.4	< 0.1

Estimate (i) the mean residence time in the vessel and (ii) the value of the dispersion number for the vessel.

Solution

Proceeding in the same way as for method A in the preceding example, and for convenience multiplying each concentration by 10^3:

First sampling point:

$$\sum(C_i)_1 = 69.2; \quad \sum(C_it_i)_1 = 248.05; \quad \sum(C_it_i^{\,2})_1 = 912.975$$

$$\bar{t}_1 = \frac{248.05}{69.2} = 3.585 \text{ min}$$

$$\sigma_1^{\,2} = \frac{912.975}{69.2} - 3.585^2 = 0.341 \text{ min}^2$$

Second sampling point:

$$\sum(C_i)_2 = 57.3; \quad \sum(C_it_i)_2 = 468.45; \quad \sum(C_it_i^{\,2})_2 = 3867.075$$

$$\bar{t}_2 = \frac{468.45}{57.3} = 8.175 \text{ min}$$

$$\sigma_2^2 = \frac{3867.075}{57.3} - 8.175^2 = 0.658 \text{ min}^2$$

then: $$\Delta\bar{t} = \bar{t}_2 - \bar{t}_1 = 8.175 - 3.585 = \underline{\underline{4.59 \text{ min}}}$$

Thus, 4.59 min is the mean residence time in the vessel.

$$\Delta\sigma^2 = \sigma_2^2 - \sigma_1^2 = 0.658 - 0.341 = 0.317 \text{ min}^2$$

$$\frac{\Delta\sigma^2}{(\Delta\bar{t})^2} = \frac{0.31}{4.59^2} = 0.0150 = 2\left(\frac{D_L}{u\Delta L}\right)$$

$$\therefore \quad \text{Dispersion number } \left(\frac{D_L}{u\Delta L}\right) = \frac{0.0150}{2} = 0.0075$$

This value thus satisfies the condition that the dispersion number must be small (<0.01) for the above method to apply.

Note that, because the same total amount of tracer must pass the two observation points (assuming that no tracer is lost by adsorption on the vessel walls—which can happen), then we should expect $\int_0^\infty C_i\,dt$ to be the same at both positions; i.e. because the time interval between samples is the same (0.5 min) for each set of data, we would expect $\Sigma(C_i)_1 = \Sigma(C_i)_2$. The fact that they are distinctly different demonstrates the inaccuracies that may enter this type of calculation. Concentrations below 0.1×10^{-3} kmol/m^3 are not taken into account; this means that possibly a long drawn-out 'tail' of concentrations below this value is ignored even though the area under this part of the **C**-curve may be significant in total, especially for the more widely spread curve from the second sampling point.

(c) Pulse of tracer moving through a series of vessels

The principles of additivity of residence times and additivity of variances can be extended to several reaction vessels in series (Fig. 2.19). Thus, for vessels **A**, **B** and **C**, if $\bar{t}_A$, $\bar{t}_B$, $\bar{t}_C$ are the residence times and σ_A^2, σ_B^2, σ_C^2 are the variances that would result from a pulse of tracer passing through each of the vessels in turn, then the overall residence time will be $\bar{t}_A + \bar{t}_B + \bar{t}_C$ and the overall increase in the variance of a pulse will be $\sigma_A^2 + \sigma_B^2 + \sigma_C^2$. The limitation is again that the dispersion number for the vessel must be small.

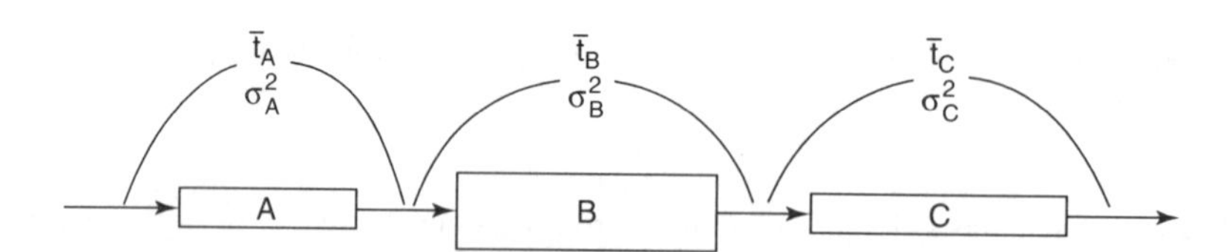

FIG. 2.19. Additivity of mean residence times t_A, t_B, t_C and variances σ_A^2, σ_B^2, σ_C^2 for reaction vessels A, B, C (connected in series by plug-flow sections). Small D_L/uL values for each vessel

2.3.6. Values of Dispersion Coefficients from Theory and Experiment

As might be expected, the dispersion coefficient for flow in a circular pipe is determined mainly by the Reynolds number *Re*. Figure 2.20 shows the dispersion coefficient plotted in the dimensionless form (D_L/ud) versus the Reynolds number $Re = \rho ud/\mu$[2,13]. In the turbulent region, the dispersion coefficient is affected also by the wall roughness while, in the laminar region, where molecular diffusion plays a part, particularly in the radial direction, the dispersion coefficient is dependent on the Schmidt number $Sc(\mu/\rho D)$, where D is the molecular diffusion coefficient. For the laminar flow region where the Taylor-Aris theory[8,9,10] (Section 2.3.1) applies:

$$D_L = D + \frac{1}{192}\frac{u^2d^2}{D} \tag{2.36}$$

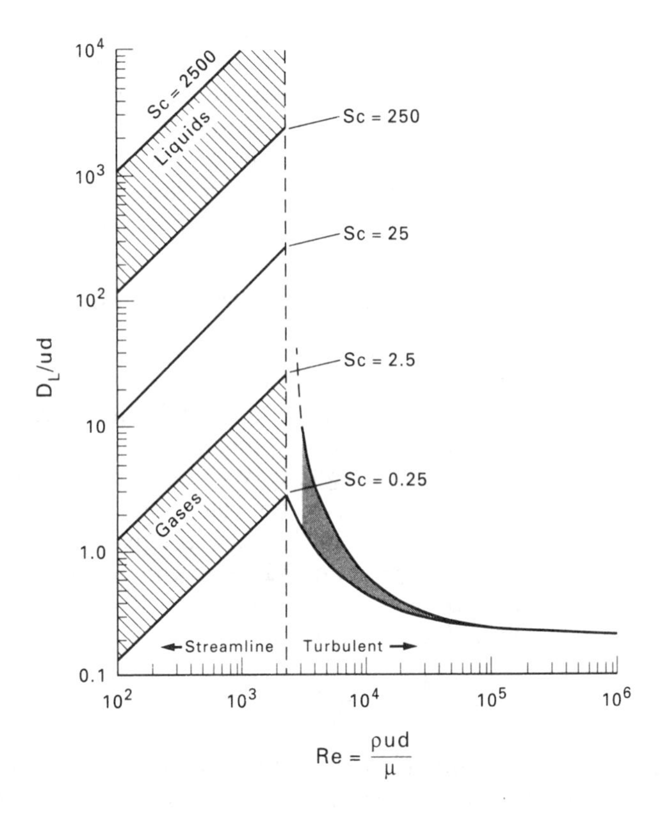

FIG. 2.20. Dimensionless axial-dispersion coefficients for fluids flowing in circular pipes. In the turbulent region, graph shows upper and lower limits of a band of experimentally determined values. In the laminar region the lines are based on the theoretical equation 2.37

where the first term represents the small additional effect on the dispersion coefficient of molecular diffusion in the axial direction. Equation 2.36 may also be written:

$$\frac{D_L}{ud} = \frac{1}{Re\ Sc} + \frac{1}{192} Re\ Sc \tag{2.37}$$

This relation is plotted in Fig. 2.20, the first term having only a small effect in the Reynolds number range shown, except at low Schmidt numbers.

Dispersion coefficients for packed beds are usually plotted in the form of a similar dimensionless group, the Peclet number ud_p/D_L, which uses the diameter of the particles of the bed d_p as the characteristic length rather than the bed diameter. Graphs of experimentally measured values of both axial and radial Peclet numbers plotted against Reynolds number, also based on particle diameter, are shown in Volume 2, Chapter 4.

Another interesting application of the data in Fig. 2.20 for dispersion coefficients in turbulent flow is in calculating the mixing that occurs in long pipelines. Many refined petroleum products are distributed by pipelines which may extend over hundreds of kilometres. The same pipeline is used to convey several different products, each being pumped through the line for several hours before a sudden switch is made to the next product. At the receiving end of the line some mixing will be found to have occurred at the interface between the two products and a slug of the off specification mixture needs to be diverted for reprocessing. Knowing the dispersion coefficient and the product specification required, the volume of mixture to be discarded may be calculated. An example of this calculation can be found in the 'Omnibook' of LEVENSPIEL[14].

2.3.7. Dispersed Plug-Flow Model with First-Order Chemical Reaction

We will consider a dispersed plug-flow reactor in which a homogeneous irreversible first order reaction takes place, the rate equation being $\mathcal{R} = k_1 C$. The reaction is assumed to be confined to the reaction vessel itself, i.e. it does not occur in the feed and outlet pipes. The temperature, pressure and density of the reaction mixture will be considered uniform throughout. We will also assume that the flow is steady and that sufficient time has elapsed for conditions in the reactor to have reached a steady state. This means that in the general equation for the dispersed plug-flow model (equation 2.13) there is no change in concentration with time i.e. $\partial C/\partial t = 0$. The equation then becomes an ordinary rather than a partial differential equation and, for a reaction of the first order:

$$D_L \frac{\mathrm{d}^2 C}{\mathrm{d}z^2} - u\frac{\mathrm{d}C}{\mathrm{d}z} - k_1 C = 0 \tag{2.38}$$

The solution to this equation (unlike the partial differential equation 2.14) has been shown not to depend on the precise formulation of the inlet and outlet conditions, i.e. whether they are 'open' or 'closed'[15]. In the following derivation, however, the reaction vessel is considered to be 'closed', i.e. it is connected at the inlet and outlet by piping in which plug flow occurs and, in general, there is a flow discontinuity at both inlet and outlet. The boundary conditions to be used will be those which properly apply to a closed vessel. (See Section 2.3.5 regarding the significance of the boundary conditions for open and closed systems.)

In order to understand these boundary conditions, let us consider that the inlet pipe in which ideal plug flow occurs has the same diameter (shown by broken lines) as the reactor itself (Fig. 2.21). Inside the reactor, across any section perpendicular to the z-direction, the flux of the reactant, i.e. the rate of transfer is made up of two contributions, the convective flow uC and the diffusion-like dispersive flow $-D_L \frac{\mathrm{d}C}{\mathrm{d}z}$ (equation 2.12).

i.e.: $$\text{Flux within the reactor} = \left(uC - D_L\frac{\mathrm{d}C}{\mathrm{d}z}\right)$$

Because there is no dispersion in the inlet pipe, upstream of the reactor, there is only the convective contribution.

i.e.: $$\text{Flux into the reactor} = uC_{in}$$

Across the plane at the inlet to the reactor, *these two fluxes must be equal*, otherwise there would be a net accumulation at this plane.

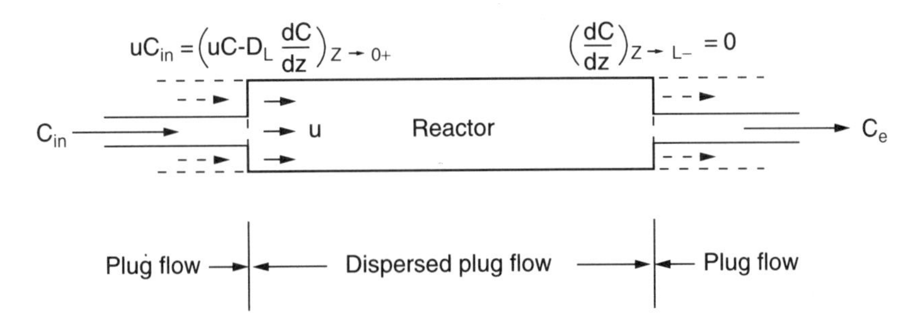

FIG. 2.21. Dispersed plug-flow with first-order homogeneous reaction $\mathscr{R} = \mathbf{k}_1 C$; steady state, illustrating boundary conditions for a 'closed' vessel

i.e.:

$$uC_{\text{in}} = \left(uC - D_L \frac{\mathrm{d}C}{\mathrm{d}z}\right)_{z\to 0+} \tag{2.39}$$

$z \to 0+$ meaning that z tends to zero from the positive direction. However, we do not expect the concentrations on either side of this inlet plane to be the same. On entering, the flow passes from an unmixed plug-flow region to a mixed dispersed flow region. There will be a fall in concentration as the fluid enters the reactor in just the same way as there is a fall in concentration when the feed enters a stirred tank reactor (Section 2.2).

Now consider the outlet pipe from the reactor. The same argument as was used for the inlet plane applies to the flux across the outlet plane (equation 2.39) because again there can be no accumulation.

i.e.:

$$\left(uC - D_L \frac{\mathrm{d}C}{\mathrm{d}z}\right)_{z\to L-} = uC_{\text{ex}} \tag{2.40}$$

However, the fluid is now passing from a mixed region to an unmixed plug-flow region; again as in a stirred-tank reactor, the fluid leaving the reactor must have the same concentration of reactant as the fluid just inside the outlet plane. Thus, if C_L is the concentration at $z = L-$ just inside the reactor and C_{ex} is the concentration in the fluid leaving, then $C_L = C_{\text{ex}}$. But if $C_L = C_{\text{ex}}$, then from the equality of flux (equation 2.40) we must have:

$$-D_L \left(\frac{\mathrm{d}C}{\mathrm{d}z}\right)_{z\to L-} = 0 \tag{2.41}$$

or:

$$\left(\frac{dC}{dz}\right)_{z\to L-} = 0$$

To solve the differential equation 2.38 subject to the boundary conditions 2.39 and 2.41 we adopt the standard method of substituting the trial solution $C = Ae^{mz}$ and obtain the auxiliary equation:

$$D_L m^2 - um - \boldsymbol{k}_1 = 0$$

which has the solutions:

$$m = u(1 + a)/2D_L \quad \text{and} \quad m = u(1 - a)/2D_L$$

where $a = (1 + 4k_1 D_L/u^2)^{1/2}$

i.e.:

$$C = B_1\,\mathrm{e}^{\frac{uz}{2D_L}(1+a)} + B_2\,\mathrm{e}^{\frac{uz}{2D_L}(1-a)} \tag{2.42}$$

where B_1 and B_2 are arbitrary constants to be evaluated from the boundary conditions.

Differentiating equation 2.42:

$$\frac{\mathrm{d}C}{\mathrm{d}z} = \frac{u}{2D_L}\left\{B_1(1 + a)\,\mathrm{e}^{\frac{uz}{2D_L}(1+a)} + B_2(1 - a)\,\mathrm{e}^{\frac{uz}{2D_L}(1-a)}\right\} \tag{2.43}$$

At $z \to 0+$, from equation 2.42: $C_{0+} = B_1 + B_2$

and from equation 2.43: $\left(\frac{\mathrm{d}C}{\mathrm{d}z}\right)_{0+} = \frac{u}{2D_L}\big(B_1(1 + a) + B_2(1 - a)\big)$

and using the inlet boundary condition equation 2.39:

$$u(C_{in} - C_{0+}) = -D_L\left(\frac{\mathrm{d}C}{\mathrm{d}z}\right)_{0+}$$

$$u\big(C_{in} - (B_1 + B_2)\big) = -D_L\frac{u}{2D_L}\big(B_1(1 + a) + B_2(1 - a)\big)$$

or:

$$C_{in} - (B_1 + B_2) = -\tfrac{1}{2}\big(B_1(1 + a) + B_2(1 - a)\big) \tag{2.44}$$

At $z \to L-$, from equation 2.43 and the boundary condition (equation 2.41):

$$\left(\frac{\mathrm{d}C}{\mathrm{d}z}\right)_L = 0 = B_1(1 + a)\,\mathrm{e}^{\frac{uL}{2D_L}(1+a)} + B_2(1 - a)\,\mathrm{e}^{\frac{uL}{2D_L}(1-a)} \tag{2.45}$$

Solving the simultaneous equations 2.44 and 2.45 for B_1 and B_2 and substituting in equation 2.42 gives the concentration C at any position z in the reactor:

$$\frac{C}{C_{in}} = \frac{2\mathrm{e}^{\frac{uz}{2D_L}}\left[(a + 1)\,\mathrm{e}^{\frac{au}{2D_L}(L-z)} + (a - 1)\,\mathrm{e}^{-\frac{au}{2D_L}(L-z)}\right]}{\left[(a + 1)^2\,\mathrm{e}^{\frac{auL}{2D_L}} - (a - 1)^2\,\mathrm{e}^{-\frac{auL}{2D_L}}\right]} \tag{2.46}$$

When $z = L$, $C = C_{ex}$, the concentration of reactant **A** at the outlet of the reactor, so that:

$$\frac{C_{ex}}{C_{in}} = \frac{4a\mathrm{e}^{\frac{uL}{2D_L}}}{(a + 1)^2\,\mathrm{e}^{\frac{auL}{2D_L}} - (a - 1)^2\,\mathrm{e}^{-\frac{auL}{2D_L}}} \tag{2.47}$$

Case of small D_L/uL

Simplification of equation 2.47 for the case of small deviations from ideal plug flow (small D_L/uL) gives some interesting and useful results. From the definition of a which appears in equations 2.42 and 2.47:

$$a = [1 + 4\boldsymbol{k}_1 D_L/u^2]^{1/2} = \left[1 + 4\boldsymbol{k}_1 \tau \left(\frac{D_L}{uL}\right)\right]^{1/2}$$

where τ is the residence time in the reactor, ($\tau = L/u$). This expression is then expanded by the binomial series and, neglecting cubic and higher order terms:

$$a = 1 + 2\boldsymbol{k}_1 \tau \left(\frac{D_L}{uL}\right) - 2(\boldsymbol{k}_1 \tau)^2 \left(\frac{D_L}{uL}\right)^2 \tag{2.48}$$

Dividing numerator and denominator of equation 2.47 by $e^{\frac{uL}{2D_L}}$:

$$\frac{C_{ex}}{C_{in}} = \frac{4a}{(a+1)^2 e^{\frac{uL}{2D_L}(a-1)} - (a-1)^2 e^{-\frac{uL}{2D_L}(a+1)}}$$

The second term in the denominator will be small compared with the first term because:

(i) a is close to 1 and the factor $(a - 1)^2$ will be small,
(ii) the exponent on e is large and negative, since uL/D_L is the reciprocal of D_L/uL which is small.

Thus, neglecting this second term:

$$\frac{C_{ex}}{C_{in}} = \frac{4a}{(a+1)^2 e^{\frac{uL}{2D_L}(a-1)}} \tag{2.49}$$

Substituting from equation 2.48 for a in the exponential term and noting that when a is close to 1, $4a/(a + 1)^2$ can be set approximately equal to 1:

$$\frac{C_{ex}}{C_{in}} = \exp\left[-\boldsymbol{k}_1 \tau + (\boldsymbol{k}_1 \tau)^2 \left(\frac{D_L}{uL}\right)\right] \tag{2.50}$$

Comparison with a simple plug-flow reactor

The case of a simple plug-flow reactor follows very easily from the above result for small values of D_L/uL. When there is no dispersion D_L will be zero and equation 2.50 becomes:

$$\frac{C_{ex}}{C_{in}} = \exp(-\boldsymbol{k}_1 \tau) \tag{2.51}$$

as expected from the basic treatment of plug-flow reactors in Chapter 1.

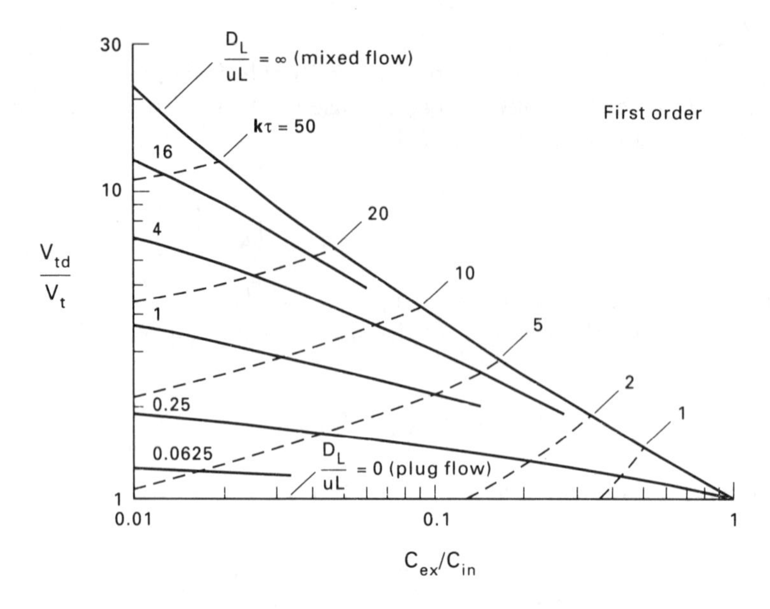

FIG. 2.22. Comparison between plug-flow reactors with and without axial dispersion. V_{td} is the volume with dispersion and V_t is the volume of the simple plug-flow reactor required for the same conversion, i.e. the same value of the ratio C_{ex}/C_{in}

It is interesting to compare a reactor with dispersion (equation 2.47) with a simple plug-flow reactor (equation 2.51). The effect of dispersion at a given desired fractional conversion for the reactant (i.e. at a given value of C_{ex}/C_{in}) is to increase the volume required for reactor V_{td} compared with the volume V_t needed in the case of simple plug-flow. This effect is shown quantitatively in Fig. 2.22, due to LEVENSPIEL[2], where V_{td}/V_t is plotted against C_{ex}/C_{in} for different values of the parameter $k_1\tau$. It can be seen that as D_L/uL increases so does the size of the dispersed plug-flow reactor. The limit is reached when $D_L/uL \rightarrow \infty$, which corresponds to complete mixing such as would be obtained in a single stirred-tank reactor. A similar chart for a second-order reaction may be found in LEVENSPIEL[2].

2.3.8. Applications and Limitations of the Dispersed Plug-Flow Model

For reactors and process vessels where the departure from plug flow is not entirely negligible but not particularly large, the dispersed plug-flow model is widely used. Because of the practical difficulties of introducing tracer substances into process streams feeding large-scale production units, most of the dispersion coefficients reported in the literature are for small-scale laboratory or pilot plant equipment. Use of the plug-flow dispersion model is not necessarily restricted to homogeneous reactors and the model may be applied to one or more of the streams in gas–solid, gas–liquid and multiphase reactors, as discussed later under the heading of the types of reactor concerned.

FIG. 2.23. Types of tracer response curves showing indications that the dispersed plug-flow model is unlikely to be applicable. (*a*) two maxima; (*b*) very skewed; (*c*) long tail

However, if the flow deviates markedly from plug flow (i.e. if D_L/uL is large), it is most likely that the real situation does not conform to the physical processes implicit in the model, namely that the mixing is produced by a large number of independent random fluctations. It is then questionable whether the model should be used. Some forms of tracer response curves following an ideal-pulse type of injection give clear indication that the dispersed plug-flow model is likely to be inadequate. Some examples of these are shown in Fig. 2.23. The occurrence of a double maximum as in (*a*) suggests that some bypassing is taking place. The very skewed curve (*b*) possibly indicates that a well-mixed region is preceded or followed by a nearly plug-flow region similar to the idealised configuration shown in Fig. 2.6. The third curve (*c*) in Fig. 2.23 shows a long tail, suggesting the presence of a dead-water zone with slow transfer of tracer back into the main stream.

When D_L/uL is found to be large and the tracer response curve is skewed, as in Fig. 2.23*b*, but without a significant delay, a continuous stirred-tanks in series model (Section 2.3.2), may be found to be more appropriate. The tracer response curve will then resemble one of those in Fig. 2.8 or Fig. 2.9. The variance σ_c^2 of such a curve with a mean of $\bar{t}_c$ is related to the number of tanks i by the expression $\sigma_c^2 = \bar{t}_c^2/i$ (which can be shown for example by the Laplace transform method[7] from the equations set out in Section 2.3.2). Calculations of the mean and variance of an experimental curve can be used to determine either a dispersion coefficient D_L or a number of tanks i. Thus each of the models can be described as a 'one parameter model', the parameter being D_L in the one case and i in the other. It should be noted that the value of i calculated in this way will not necessarily be integral but this can be accommodated in the more mathematically general form of the tanks-in-series model as described by NAUMAN and BUFFHAM[7].

It can thus be seen that in some ways the dispersed plug-flow model and the tanks-in-series model are complementary to each other. As $D_L/uL \rightarrow \infty$, so the flow corresponds to a single ideally mixed stirred tank. For a series of tanks, as $i \rightarrow \infty$ so the flow corresponds to ideal plug flow (see Section 2.3.2). Generally, small departures from plug flow are better modelled by means of a dispersion coefficient D_L; large deviations are probably better treated by a tanks-in-series model thereby matching more closely the real physical processes. In more complicated situations like those in Fig. 2.23, mixed modes involving more than just a single parameter need to be used as described in the next section.

2.4. MODELS INVOLVING COMBINATIONS OF THE BASIC FLOW ELEMENTS

As well as regions corresponding to ideal plug flow, dispersed plug flow, an ideal stirred tank, and a series of i stirred tanks, as mentioned above a further type of region is a stagnant or dead-water zone (Fig. 2.1*a*). In the construction of mixed models we may also need to introduce by-pass flow, internal recirculation, and cross-flow (which involves interchange but no net flow between regions). Of course, the more complicated the model, the larger is the number of parameters which have to be determined by matching **C**- or **F**-curves. A wide variety of possibilities exists; for example, over twenty somewhat different models have been proposed for representing a gas-solid fluidised bed[4].

In conclusion the following example shows how a real stirred tank might be modelled[16].

Example 2.3

As a model of a certain poorly agitated continuous stirred-tank reactor of total volume V, it is supposed that only a fraction w is well mixed, the remainder being a dead-water zone. Furthermore, a fraction f of the total volume flowrate v fed to the tank by-passes the well-mixed zone completely (Fig. 2.24*a*).

(i) If the model is correct, sketch the **C**-curve expected if a pulse input of tracer is applied to the reactor.

(ii) On the basis of the model, determine the fractional conversion α_f which will be obtained under steady conditions if the feed contains a reactant which undergoes a homogeneous first order reaction with rate constant $\boldsymbol{k}_1$.

Solution

(i) Because of the by-pass which is assumed to have no holding capacity, part of the input pulse appears instantly in the outlet stream. If the input pulse is normalised to unity, the magnitude of this immediate output pulse is f. The remainder of the $\mathbf{C}_\theta$-curve (Fig. 2.24*b*) is a simple exponential decay similar to the

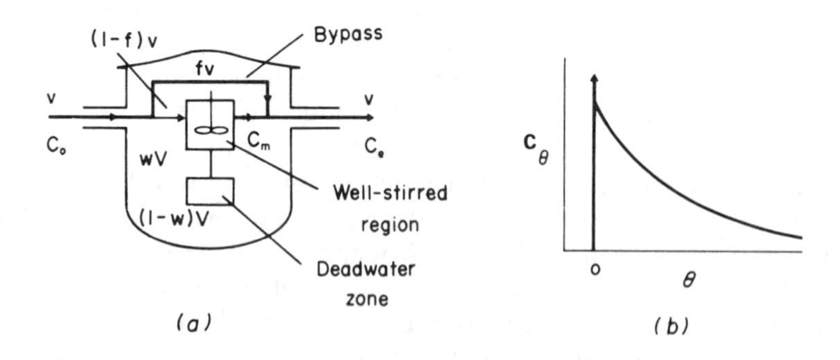

FIG. 2.24. Model of a poorly agitated continuous stirred-tank reactor
(*a*) Flow model: fraction f of flow v by-passes; only a fraction w of tank volume V is well-stirred
(*b*) Equivalent $\mathbf{C}_\theta$ curves for pulse input

curve for one tank shown in Fig. 2.9*a*. However, in the present case, because only a fraction $(1-f)$ of the unit pulse enters the well-mixed region of volume wV, the equation of the curve with $\theta = vt/V$ is:

$$\mathbf{C}_\theta = (1-f)^2 \frac{1}{w} \exp\left\{-\frac{(1-f)}{w}\theta\right\}$$

as may be shown by an analysis similar to that leading to equation 2.10. Note that values for f and w could be determined by comparing experimentally obtained data with this expression.

(ii) Consider a material balance over the well-mixed region only, in the steady state. Let the reactant inlet concentration be C_0 and C_m be the intermediate concentration at the outlet of this region before mixing with the by-pass stream.

$$\underset{\text{Inflow}}{v(1-f)C_0} - \underset{\text{Outflow}}{v(1-f)C_m} - \underset{\text{Reaction}}{\boldsymbol{k}_1 C_m w V} = 0$$

Hence:

$$C_m = \frac{C_0}{1 + \dfrac{\boldsymbol{k}_1 w V}{v(1-f)}}$$

Next consider mixing of this stream with the by-pass giving a final exit concentration C_{ex}.

$$v(1-f)C_m + fvC_0 = vC_{ex}$$

$$\therefore \qquad \frac{C_{ex}}{C_0} = (1-f)\frac{C_m}{C_0} + f$$

The fractional conversion at the exit:

$$\alpha_f = \frac{C_0 - C_{ex}}{C_0} = 1 - \frac{C_{ex}}{C_0}$$

Substituting for C_{ex} and eliminating C_m we find after some manipulation:

$$\alpha_f = \frac{1}{\dfrac{v}{\boldsymbol{k}_1 w V} + \dfrac{1}{(1-f)}}$$

Note that if $f = 0$, $w = 1$, and writing, $V/v = \tau$ this expression reduces to:

$$\alpha_f = \frac{\boldsymbol{k}_1 \tau}{1 + \boldsymbol{k}_1 \tau}$$

as expected for a single ideally mixed stirred-tank reactor.

2.5. FURTHER READING

LEVENSPIEL, O.: *Chemical Reaction Engineering*, 2nd edn. (Wiley, 1972).
LEVENSPIEL, O.: *Chemical Reactor Omnibook* (OSU Book Stores, Corvallis, Oregon, 1989).
NAUMAN, E. B. and BUFFHAM, B. A.: *Mixing in Continuous Flow Systems* (Wiley, 1983).
WEN, C. Y. and FAN, L. T.: *Models for Flow Systems and Chemical Reactors* (Dekker, 1975).

2.6. REFERENCES

1. DANCKWERTS, P. V.: *Chem. Eng. Sci.* **2** (1953) 1. Continuous flow systems. Distribution of residence times.
2. LEVENSPIEL, O.: *Chemical Reaction Engineering*, 2nd edn., (Wiley, 1972).
3. LEVENSPIEL, O. and BISCHOFF, K. B.: In *Advances in Chemical Engineering*, Vol. 4, DREW, T. B., HOOPES, J. W. and VERMEULEN, T. (eds.), (Academic Press, 1963). Patterns of flow in chemical process vessels.
4. KUNII, D. and LEVENSPIEL, O.: *Fluidization Engineering*, 2nd edn., (Butterworth-Heinemann, 1991).
5. KRAMERS, H.: *Chem. Eng. Sci.* **8** (1958) 45. Physical factors in chemical reaction engineering.
6. BOURNE, J. R.: In *Mixing in the Process Industries* by HARNBY, N., EDWARDS, M. F. and NIENOW, A. W. (eds.) (Butterworth-Heinemann, 1992), Chapter 10, Mixing in single phase chemical reactors.
7. NAUMAN, E. B. and BUFFHAM, B. A.: *Mixing in Continuous Flow Systems* (Wiley, 1983).

8. TAYLOR, G. I.: *Proc. Roy. Soc.* **A219** (1953) 186. Dispersion of soluble matter in solvent flowing slowly through a tube.
9. ARIS, R.: *Proc. Roy. Soc.* **A235** (1956) 67. On the dispersion of a solute in a fluid flowing through a tube.
10. PROBSTEIN, R. F.: *Physicochemical Hydrodynamics* (Butterworth-Heinemann, 1989).
11. ABBI, Y. P. and GUNN, D. J.: *Trans. Inst. Chem. Engrs.* **54** (1976) 225. Dispersion characteristics from pulse response.
12. WEHNER, J. F. and WILHELM, R. H.: *Chem. Eng. Sci.* **6** (1956) 89. Boundary conditions of a flow reactor.
13. WEN, C. Y. and FAN, L. T.: *Models for Flow Systems and Chemical Reactors* (Dekker, 1975).
14. LEVENSPIEL, O.: *Chemical Reactor Omnibook*, (OSU Book Stores, Corvallis, Oregon, 1989).
15. FROMENT, G. F. and BISCHOFF, K. B.: *Chemical Reactor Analysis and Design*, 2nd edn. (Wiley, 1989), page 535.
16. CHOLETTE, A. and CLOUTIER, L.: *Can. J. Chem. Eng.* **37** (1959) 105. Mixing efficiency determinations for continuous flow systems.

2.7. NOMENCLATURE

		Units in SI System	Dimensions N, L, T
A	Cross-sectional area	m^2	$\mathbf{L}^2$
A_n	Area under concentration–time curve	kmol s/m^3	$\mathbf{NL}^{-3}\mathbf{T}$
a	Parameter $(1 + 4\boldsymbol{k}_1 D_L/u^2)^{1/2}$	—	—
B_1	Integration constant	kmol/m^3	$\mathbf{NL}^{-3}$
B_2	Integration constant	kmol/m^3	$\mathbf{NL}^{-3}$
C	Concentration	kmol/m^3	$\mathbf{NL}^{-3}$
C_{ex}	Concentration at exit	kmol/m^3	$\mathbf{NL}^{-3}$
C_i	Concentration in sample i	kmol/m^3	$\mathbf{NL}^{-3}$
C_{in}	Concentration in inlet stream	kmol/m^3	$\mathbf{NL}^{-3}$
C_L	Concentration at position L	kmol/m^3	$\mathbf{NL}^{-3}$
C_m	Concentration at intermediate position in stirred-tank model	kmol/m^3	$\mathbf{NL}^{-3}$
C_{0+}	Concentration in reactor at $z = 0+$	kmol/m^3	$\mathbf{NL}^{-3}$
C_0	Concentration in feed	kmol/m^3	$\mathbf{NL}^{-3}$
C_∞	Concentration at infinite time after a step change	kmol/m^3	$\mathbf{NL}^{-3}$
$\mathbf{C}$	Outlet response of concentration divided by area under C–t curve	s^{-1}	$\mathbf{T}^{-1}$
$\mathbf{C}_\theta$	Value of $\mathbf{C}_\delta$ when t is replaced by t/τ	—	—
D	Molecular diffusivity	m^2/s	$\mathbf{L}^2\mathbf{T}^{-1}$
D_L	Dispersion coefficient	m^2/s	$\mathbf{L}^2\mathbf{T}^{-1}$
d	Diameter of pipe	m	$\mathbf{L}$
d_p	Diameter of particle	m	$\mathbf{L}$
$\mathbf{E}$	Exit age distribution function	s^{-1}	$\mathbf{T}^{-1}$
$\mathbf{F}$	C/C_∞	—	—
f	Fraction of feed bypassing reactor	—	—
i	Number of tanks in a series	—	—
$\boldsymbol{k}_1$	First-order rate constant	s^{-1}	$\mathbf{T}^{-1}$
$\boldsymbol{k}_2$	Second-order rate constant	m^3/kmol s	$\mathbf{N}^{-1}\mathbf{L}^3\mathbf{T}^{-1}$
L	Length of pipe or reactor	m	$\mathbf{L}$
M	Moles of tracer injected	kmol	$\mathbf{N}$
m	Parameter in trial solution	m^{-1}	$\mathbf{L}^{-1}$
N_L	Flux due to dispersion	kmol/m^2 s	$\mathbf{NL}^{-2}\mathbf{T}^{-1}$
n	Moles of tracer	kmol	$\mathbf{N}$
n_i	Moles of tracer in last tank i in a series	kmol	$\mathbf{N}$
n_r	Number of moles in rth tank of series	kmol	$\mathbf{N}$
$\mathcal{R}$	Rate of reaction	kmol/m^3 s	$\mathbf{NL}^{-3}\mathbf{T}^{-1}$
t	Time	s	$\mathbf{T}$
$\bar{t}$	Mean time	s	$\mathbf{T}$
t_c	Mean time for a series of continuous stirred-tanks	s	$\mathbf{T}$
t_E	Mean of residence time distribution	s	$\mathbf{T}$
t_i	Time for sample i	s	$\mathbf{T}$

		Units in SI System	Dimensions N, L, T
t_i'	Value of t_i at mid point of Δt_i	s	**T**
u	Velocity in pipe	m/s	$\mathbf{LT^{-1}}$
V	Volume of vessel; total volume if more than one tank	m^3	$\mathbf{L^3}$
V_c	Volume of continuous stirred-tank reactor;	m^3	$\mathbf{L^3}$
V_{cr}	Volume of rth tank in a series of continuous stirred-tank reactors	m^3	$\mathbf{L^3}$
V_t	Volume of plug-flow tubular reactor	m^3	$\mathbf{L^3}$
V_{td}	Volume of dispersed plug-flow tubular reactor	m^3	$\mathbf{L^3}$
v	Volume rate of flow into reactor	m^3/s	$\mathbf{L^3T^{-1}}$
z	Length coordinate in axial direction along pipe or tubular reactor	m	**L**
z'	Dimensionless variable z/L	—	–
α	Fractional conversion	—	—
$\bar{\alpha}$	Average conversion	—	—
α_f	Fractional conversion at exit	—	—
α_t	Fractional conversion of particles in reactor for time t	—	—
θ	Dimensionless time $t/\bar{t}$	—	—
$\bar{\theta}$	Mean value of θ	—	—
σ^2	Variance	s^2	$\mathbf{T^2}$
σ_A^2, σ_B^2	Variance for vessel **A**, vessel **B**	s^2	$\mathbf{T^2}$
σ_C^2	Variance for one or a series of continuous stirred-tanks	s^2	$\mathbf{T^2}$
τ	Residence time in a pipe or reactor L/u or V/v	s	**T**
Re	Reynolds number ($\rho ud/\mu$)	—	—
Sc	Schmidt number ($\mu/\rho D$)	—	—

CHAPTER 3

Gas–Solid Reactions and Reactors

3.1. INTRODUCTION

The design of heterogeneous chemical reactors falls into a special category because an additional complexity enters into the problem. We must now concern ourselves with the transfer of matter between phases, as well as considering the fluid dynamics and chemistry of the system. Thus, in addition to an equation describing the rate at which the chemical reaction proceeds, one must also provide a relationship or algorithm to account for the various physical processes which occur. For this purpose it is convenient to classify the reactions as gas–solid, gas–liquid and gas–liquid–solid processes. The present chapter will be concerned with gas–solid reactions, especially those for which the solid is a catalyst for the reaction.

The very great importance of heterogeneous catalysis in the chemical and petroleum industries can be seen from the examples shown in Table 3.1. Catalytic reactions such as the synthesis of ammonia, which has been practised now for nearly a century, and the cracking of hydrocarbons provide us with fertilisers, fuels and many of the basic intermediates of the chemical industry. As understood from the classic definition of catalysis in chemistry, the key function of a catalyst is to speed up the rate of a desired reaction. In practice this enables the desired reaction to take place under conditions of temperature and pressure such that the rates of competing reactions giving undesirable byproducts are very low. Thus the degree of *selectivity* of a catalyst in promoting the desired reaction, while appearing to suppress unwanted reactions, is a very important consideration in developing or choosing a catalyst.

Some of the reactions listed in Table 3.1, particularly hydrogenations, require the simultaneous contacting of a gas, a liquid, and a solid catalyst and reactors for this type of system are considered in Chapter 4. Otherwise, gas–solid catalytic reactors involve passing a reactant gas through a bed of catalyst particles (an exception being the multichannel 'monolith' reactors now widely used for pollution control of automobile engine exhausts). In the most common *fixed bed* type of reactor, the particles are stationary and form a packed bed through which the gas passes downwards (to avoid lifting the particles from the support plate), or in some designs radially (especially when a low pressure drop is required). For fixed bed reactors, the catalyst particles are typically of a size 1 to 20 mm equivalent diameter and can be of a variety of shapes, as shown in Fig. 3.1. The shape chosen depends partly on how easy that shape of catalyst particle is to manufacture and partly on the need to make the catalyst material readily accessible to the reactant being transferred

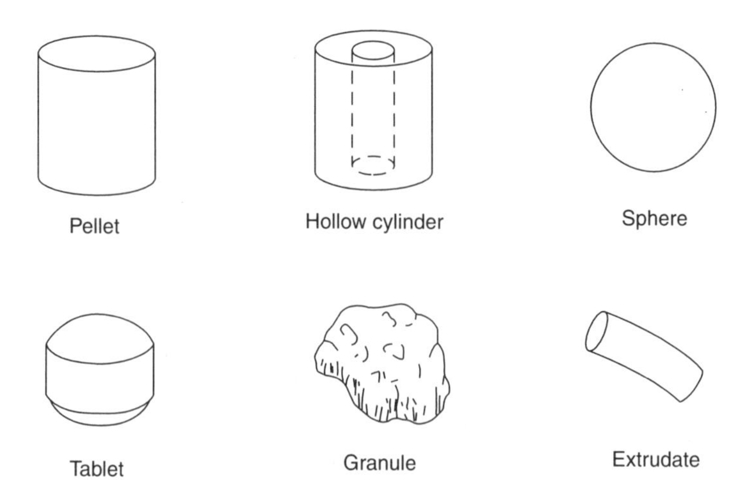

FIG. 3.1. Various forms of catalyst particles

from the gas phase. Thus the hollow cylinder in Fig. 3.1 has the advantage of making the interior of the particle more accessible as well as of providing a low pressure drop for the flow of the reactant mixture through a packed bed of these particles. The other principal type of reactor is a *fluidised bed* (as described in Volume 2), in which the particles are mobile and generally much smaller in size, being typically 50 to 100 μm in diameter.

In a gas–solid catalytic reactor, the reaction itself takes place on the surface of the solid. For this reason, unless the reaction is inherently very fast, it is an advantage to have the particles of catalyst in a porous form giving a large internal surface area available for the reaction. In a reactor using porous catalyst particles, the reactant must first be transported through the gas phase to the surface of the particle and then diffuse along the honeycomb of pores towards the interior. The reactant will then adsorb at all of the accessible active surface in the interior and be chemically transformed into a product which subsequently desorbs, diffuses to the exterior of the particle, and is then transported to the bulk fluid phase. For such reactions the designer usually has to assess the amount of solid material to pack into a reactor to achieve a specified conversion. The rate of transport of gas to and from the surface of the solid and within the pore structure of the particle will affect both the dimensions of the reactor unit required and the selection of particle size. Compared with a reaction uninfluenced by transport processes, the complexity of the design problem is exacerbated since, in the steady state, the rate of each of the transport processes is coupled with the rates of adsorption, of chemical reaction and of desorption.

Whatever the nature of the reaction and whether the vessel chosen for the operation be a packed tubular reactor or a fluidised bed, the essence of the design problem is to estimate the size of reactor required. This is achieved by solving the transport and chemical rate equations appropriate to the system. Prior to this however, the operating conditions, such as initial temperature, pressure and reactant concentrations, must be chosen and a decision made concerning the type of

TABLE 3.1. *Selected Heterogeneous Catalysts of Industrial Importance**

Reaction	Catalyst and reactor type (continuous operation unless otherwise noted)
Dehydrogenation	
C_4H_{10} (butane) $\rightarrow$ butenes	$Cr_2O_3 \cdot Al_2O_3$ (fixed bed, cyclic)
Butenes $\rightarrow$ C_4H_6 (butadiene)	Fe_2O_3 promoted with Cr_2O_3 and K_2CO_3
$C_6H_5C_2H_5 \rightarrow C_6H_5CH{=}CH_2$ (ethyl benzene $\rightarrow$ styrene)	Fe_2O_3 promoted with Cr_2O_3 and K_2CO_3 (fixed bed, in presence of steam)
CH_4 or other hydrocarbons + $H_2O \rightarrow CO + H_2$ (steam reforming)	Supported Ni (fixed bed)
$(CH_3)_2CHOH \rightarrow CH_3COCH_3 + H_2$ (isopropanol $\rightarrow$ acetone + hydrogen)	ZnO
$CH_3CH(OH)C_2H_5 \rightarrow CH_3COC_2H_5 + H_2$	ZnO
Hydrogenation	
Of edible fats and oils	Ni on a support (slurry reactor, batch)
Various hydrogenations of fine organic chemicals	Pd or Pt on carbon (slurry reactor, usually batch)
$C_6H_6 + 3H_2 \rightarrow C_6H_{12}$	Ni or noble metal on support (fixed bed or slurry reactor)
$N_2 + 3H_2 \rightarrow 2NH_3$	Fe promoted with Al_2O_3, K_2O, CaO, and MgO (adiabatic fixed beds)
$C_2H_2 \rightarrow C_2H_6$ (selective hydrogenation of C_2H_2 impurity in C_2H_4 from thermal-cracking plant)	Pd on Al_2O_3 or sulphided Ni on support (adiabatic fixed bed)
Oxidation	
$SO_2 + \frac{1}{2}O_2$(air) $\rightarrow SO_3$	V_2O_5 plus K_2SO_4 on silica (adiabatic, fixed beds)
$2NH_3 + \frac{5}{2}O_2$ (air) $\rightarrow 2NO + 3H_2O$	90% Pt-10% Rh wire gauze, oxidising conditions
$NH_3 + CH_4$ + air $\rightarrow$ HCN (Andrussow process)	90% Pt-10% Rh wire gauze, under net reducing conditions
$C_{10}H_8$ or 1, 2-$C_6H_4(CH_3)_2 + O_2 \rightarrow C_6H_4(CO)_2O$ (naphthalene or *o*-xylene + air $\rightarrow$ phthalic anhydride)	V_2O_5 on titania (multitube fixed bed)
n-$C_4H_{10} + O_2 \rightarrow C_4H_2O_3$ (butane + air $\rightarrow$ maleic anhydride)	Vanadia-phosphate (multitube fixed bed or fluidised bed)
$C_2H_4 + \frac{1}{2}O_2 \rightarrow (CH_2)_2O$ (ethylene oxide)	Ag on α-Al_2O_3, promoted with Cl and Cs (multitube fixed bed)
$CH_3OH + O_2 \rightarrow CH_2O + H_2$ and/or H_2O	Ag (adiabatic reactor) or $Fe_2(MoO_4)_3$ (multitube fixed bed)
$C_3H_6 + O_2 \rightarrow CH_2{=}CHCHO$ (acrolein) and/or $CH_2{=}CHCOOH$ (acrylic acid)	Bismuth molybdate plus other components
$C_3H_6 + NH_3 + \frac{3}{2}O_2 \rightarrow CH_2{=}CHCN + 3H_2O$	Complex metal molybdates (fluidised bed)
Complete oxidation of CO and hydrocarbons, for pollution control	Pt or Pd, or both, on monolith support
Simultaneous control of CO, hydrocarbons, and NO_x in engine exhaust	Same, plus Rh, with careful control of oxidising/reducing conditions
Simultaneous control of NO_x and SO_x in flue gases	Vanadia on titania with addition of NH_3
$C_2H_4 + \frac{1}{2}O_2 + CH_3COOH \rightarrow CH_3COOCH{=}CH_2$ (vinyl acetate)	Pd on acid-resistant support (vapour phase, multitube fixed bed)
$C_4H_8 + \frac{1}{2}O_2 \rightarrow C_4H_6 + H_2O$	Promoted ferrite spinels

*Reproduced by kind permission from SATTERFIELD[1].

Acid-catalysed Reactions	
Catalytic cracking	Zeolite in $SiO_2 \cdot Al_2O_3$ matrix plus other ingredients (transport reactor)
Hydrocracking	Pd on zeolite in an amorphous matrix; NiMo on silica-alumina, various other dual-function catalysts (adiabatic fixed beds)
Paraffin isomerisation	Pt on H-mordenite zeolite in alumina matrix
Catalytic reforming	Pt, Pt-Re or Pt-Sn on acidified Al_2O_3 or on zeolite in matrix (adiabatic, fixed beds, or moving bed, with interstage heating)
Polymerisation	H_3PO_4 on clay (fixed bed)
Hydration, e.g., propylene to isopropyl alcohol	Mineral acid or acid-type ion-exchange resin (fixed bed)
$CH_3OH + isoC_4H_8 \rightarrow$ methyl tert. butyl ether (MTBE)	Acid-type ion-exchange resin
Reactions of Synthesis Gas	
$CO + 2H_2 \rightarrow CH_3OH$	Cu^I-ZnO promoted with Al_2O_3 (adiabatic, fixed beds with interstage cooling or multitube fixed bed)
$CO + 3H_2 \rightarrow CH_4 + H_2O$ (methanation)	Supported Ni (fixed bed)
$CO + H_2 \rightarrow$ paraffins, etc. (Fischer–Tropsch synthesis)	Fe or Co with promoters (multitube fixed bed or transport reactor)
Other	
Oxychlorination (e.g., $C_2H_4 + 2HCl + \frac{1}{2}O_2 \rightarrow C_2H_4Cl_2 + H_2O$)	$CuCl_2/Al_2O_3$ with KCl promoter
Hydrodesulphurisation, hydrodenitrogenation, hydrotreating	$CoMo/Al_2O_3$ or $NiMo/Al_2O_3$, sulphided (adiabatic, fixed beds with interstage cooling)
$SO_2 + 2H_2S \rightarrow 3S + 2H_2O$ (Claus process)	Al_2O_3 (fixed beds)
$H_2O + CO \rightarrow CO_2 + H_2$ (water-gas shift)	Fe_3O_4 promoted with Cr_2O_3 (adiabatic fixed bed); for a second, lower temperature stage, Cu-ZnO on Al_2O_3; CoMo on support

*Reproduced by kind permission from SATTERFIELD[1].

reactor to be used. For example, one might have selected a packed tubular reactor operated adiabatically, or perhaps a fluidised bed reactor fitted with an internal cooling arrangement. Such operating variables constitute the design conditions and must be chosen before a detailed mathematical model is constructed.

3.2. MASS TRANSFER WITHIN POROUS SOLIDS

For gas–solid heterogeneous reactions particle size and average pore diameter will influence the reaction rate per unit mass of solid when internal diffusion is a significant factor in determining the rate. The actual mode of transport within the porous structure will depend largely on the pore diameter and the pressure within the reactor. Before developing equations which will enable us to predict reaction rates in porous solids, a brief consideration of transport in pores is pertinent.

3.2.1. The Effective Diffusivity

The diffusion of gases through the tortuous narrow channels of a porous solid generally occurs by one or more of three mechanisms. When the mean free path of the gas molecules is considerably greater than the pore diameter, collisions between molecules in the gas are much less numerous than those between molecules and pore walls. Under these conditions the mode of transport is Knudsen diffusion. When the mean free path of the gas molecules is much smaller than the pore diameter, gaseous collisions will be more frequent than collisions of molecules with pore walls, and under these circumstances molecular diffusion occurs. A third mechanism of transport which is possible when a gas is adsorbed on the inner surface of a porous solid is surface diffusion. Transport occurs by the movement of molecules over the surface in the direction of decreasing surface concentration. Although there is not much evidence on this point, it is unlikely that surface diffusion is of any importance in catalysis at elevated temperatures. Nevertheless, surface diffusion may contribute to the overall transport process in low temperature reactions of some vapours. Finally, it should be borne in mind that when a total pressure difference is maintained across a pore, as for some catalytic cracking reactions, forced flow in pores is likely to occur, transport being due to a total concentration or pressure gradient.

Both Knudsen and molecular diffusion can be described adequately for homogeneous media. However, a porous mass of solid usually contains pores of non-uniform cross-section which pursue a very tortuous path through the particle and which may intersect with many other pores. Thus the flux predicted by an equation for normal bulk diffusion (or for Knudsen diffusion) should be multiplied by a geometric factor which takes into account the tortuosity and the fact that the flow will be impeded by that fraction of the total pellet volume which is solid. It is therefore expedient to define an effective diffusivity D_e in such a way that the flux of material may be thought of as flowing through an equivalent homogeneous medium. We may then write:

$$D_e = D\frac{\varepsilon}{\tau} \tag{3.1}$$

where D is the molecular diffusion coefficient,
ε is the porosity of the particles, and
τ is a tortuosity factor.

We thus imply that the effective diffusion coefficient is calculated on the basis of a flux resulting from a concentration gradient in a homogeneous medium which has been made equivalent to the heterogeneous porous mass by invoking the geometric factor ε/τ. Experimental techniques for estimating the effective diffusivity include diffusion and flow through pelletised particles. A common procedure is to expose the two faces of a porous disc of the material to the pure components at the same total pressure. An interesting method due to BARRER and GROVE[2] relies on the measurement of the time lag required to reach a steady pressure gradient, while a technique employed by GUNN and PRYCE[3] depends on measuring the dispersion of a sinusoidal pulse of tracer in a bed packed with the porous material. Gas chromatographic methods for evaluating dispersion coefficients have also been employed[4].

Just as one considers two regions of flow for homogeneous media, so one may have molecular or Knudsen transport for heterogeneous media.

The Molecular Flow Region

HIRSCHFELDER, CURTISS and BIRD[5] obtained a theoretical expression for the molecular diffusion coefficient for two interdiffusing gases, modifying the kinetic theory of gases by taking into account the nature of attractive and repulsive forces between gas molecules. Their expression for the diffusion coefficient has been successfully applied to many gaseous binary mixtures and represents one of the best methods for estimating unknown values. On the other hand, Maxwell's formula, modified by GILLILAND[6] and discussed in Volume 1, Chapter 10, also gives satisfactory results. Experimental methods for estimating diffusion coefficients rely on the measurement of flux per unit concentration gradient. An extensive tabulation of experimental diffusion coefficients for binary gas mixtures is to be found in a report by EERKENS and GROSSMAN[7].

To calculate the effective diffusivity in the region of molecular flow, the estimated value of D must be multiplied by the geometric factor ε/τ which is descriptive of the heterogeneous nature of the porous medium through which diffusion occurs.

The porosity ε of the porous mass is included in the geometric factor to account for the fact that the flux per unit total cross-section is ε times the flux which would occur if there were no solid present. The porosity may conveniently be measured by finding the particle density ρ_p in a pyknometer using an inert non-penetrating liquid. The true density ρ_s of the solid should also be found by observing the pressure of a gas (which is not adsorbed) before and after expansion into a vessel containing a known weight of the material. The ratio ρ_p/ρ_s then gives the fraction of solid present in the particles and $(1 - \rho_p/\rho_s)$ is the porosity.

The tortuosity is also included in the geometric factor to account for the tortuous nature of the pores. It is the ratio of the path length which must be traversed by molecules in diffusing between two points within a pellet to the direct linear separation between those points. Theoretical predictions of τ rely on somewhat inadequate models of the porous structure, but experimental values may be obtained from measurements of D_e, D and ε.

The Knudsen Flow Region

The region of flow where collisions of molecules with the container walls are more frequent than intermolecular gaseous collisions was the subject of detailed study by KNUDSEN[8] early in the twentieth century. From geometrical considerations it may be shown[9] that, for the case of a capillary of circular cross-section and radius r, the proportionality factor is $8\pi r^3/3$. This results in a Knudsen diffusion coefficient:

$$D_K = \frac{8r}{3}\sqrt{\frac{\mathbf{R}T}{2\pi M}} \tag{3.2}$$

This equation (which corresponds with equation 17.53 in Volume 2, Chapter 17), however, cannot be directly applied to the majority of porous solids since they are not well represented by a collection of straight cylindrical capillaries. EVERETT[10] showed that pore radius is related to the specific surface area S_g per unit mass and to the specific pore volume V_g per unit mass by the equation:

$$r = \frac{2}{\alpha}\frac{V_g}{S_g} \tag{3.3}$$

where α is a factor characteristic of the particular pore geometry. Values of α depend on the pore structure and for uniform non-intersecting cylindrical capillaries $\alpha = 1$. Although an estimation of precise values of V_g and S_g from experimental data is obtained by the somewhat arbitrary selection of points on an adsorption isotherm representing complete pore filling and the completion of a monolayer, some significance may be given to an average pore dimension derived from pore volume and surface area measurements. Thus if, for the purposes of calculating a Knudsen diffusion coefficient, the pore model adopted consists of non-intersecting cylindrical capillaries and the radius computed from equation 3.3 is a radius r_e equivalent to the radius of a cylinder having the same surface to volume ratio as the pore, then equation 3.2 may be applied. In terms of the porosity ε, specific surface area S_g and particle density ρ_p (mass per unit total particle volume, including the volume occupied by pore space):

$$D_K = \frac{16}{3}\frac{\varepsilon}{\rho_p S_g}\sqrt{\frac{\mathbf{R}T}{2\pi M}} \tag{3.4}$$

In the region of Knudsen flow the effective diffusivity D_{eK} for the porous solid may be computed in a similar way to the effective diffusivity in the region of molecular flow, i.e. D_K is simply multiplied by the geometric factor.

The Transition Region

For conditions where Knudsen or molecular diffusion does not predominate, SCOTT and DULLIEN[11] obtained a relation for the effective diffusivity. The formula they obtained for a binary mixture of gases is:

$$D_e = \frac{1}{\dfrac{1}{D_{eM}} + \dfrac{1}{D_{eK}} - \dfrac{x_A(1 + N'_B/N'_A)}{D_{eM}}} \tag{3.5}$$

where D_{eM} and D_{eK} are the effective diffusivities in the molecular and Knudsen regions,
N'_A and N'_B are the molar fluxes of species **A** and **B**, and
x_A is the mole fraction of **A**.

Equation 3.5 corresponds to equation 17.57 in Volume 2, Chapter 17.

If equimolecular counterdiffusion takes place $N'_A = -N'_B$ (see Volume 1, Chapter 10) and the total pressure is constant, we obtain from equation 3.5 an expression for the effective self-diffusion coefficient in the transition region:

$$\frac{1}{D_e} = \frac{1}{D_{eM}} + \frac{1}{D_{eK}} \tag{3.6}$$

This result has also been obtained independently by other workers[12,13].

Forced Flow in Pores

Many heterogeneous reactions give rise to an increase or decrease in the total number of moles present in the porous solid due to the reaction stoichiometry. In such cases there will be a pressure difference between the interior and exterior of the particle and forced flow occurs. When the mean free path of the reacting molecules is large compared with the pore diameter, forced flow is indistinguishable from Knudsen flow and is not affected by pressure differentials. When, however, the mean free path is small compared with the pore diameter and a pressure difference exists across the pore, forced flow (Poiseuille flow: see Volume 1, Chapter 3) resulting from this pressure difference will be superimposed on molecular flow. The diffusion coefficient D_p for forced flow depends on the square of the pore radius and on the total pressure difference ΔP:

$$D_P = \frac{-\Delta P\, r^2}{8\mu} \tag{3.7}$$

The viscosity of most gases at atmospheric pressure is of the order of 10^{-7} Ns/m^2, so for pores of about 1 μm radius D_P is approximately 10^{-5} m^2/s. Molecular diffusion coefficients are of similar magnitude so that in small pores forced flow will compete with molecular diffusion. For fast reactions accompanied by an increase in the number of moles an excess pressure is developed in the interior recesses of the porous particle which results in the forced flow of excess product and reactant molecules to the particle exterior. Conversely, for pores greater than about 100 μm radius, D_P is as high as 10^{-3} m^2/s and the coefficient of diffusion which will determine the rate of intraparticle transport will be the coefficient of molecular diffusion.

Except in the case of reactions at high pressure, the pressure drop which must be maintained to cause flow through a packed bed of particles is usually insufficient to produce forced flow in the capillaries of the solid, and the gas flow is diverted around the exterior periphery of the pellets. Reactants then reach the interior of the porous solid by Knudsen or molecular diffusion.

3.3. CHEMICAL REACTION IN POROUS CATALYST PELLETS

A porous solid catalyst, whose behaviour is usually specific to a particular system, enhances the approach to equilibrium of a gas phase chemical reaction. Employment of such a material therefore enables thermodynamic equilibrium to be achieved at moderate temperatures in a comparatively short time interval.

When designing a heterogeneous catalytic reactor it is important to know, for the purposes of calculating throughputs, the rate of formation of desired product on the basis of unit reactor volume and under the hydrodynamic conditions obtaining in the reactor. Whether the volume is defined with respect to the total reactor volume, including that occupied by solid, or with respect to void volume is really a matter of convenience. What is important, however, is either that rate data for the reaction be known for physical conditions identical to those prevailing within the reactor (this usually means obtaining rate data *in situ*) or, alternatively, that rate data obtained in the absence of mass transport effects be available. If pilot plant experiments are performed, conditions can usually be arranged to match those within a larger reactor, and then the rate data obtained can be immediately applied to a reactor design problem since the measurements will have taken into account mass transfer effects. If rate data from laboratory experiments are utilised, it is essential to ensure that they are obtained in the absence of mass and heat transfer effects. This being so, the rate should be multiplied by two factors to transpose the experimental rate to the basis of reaction rate per unit volume of a reactor packed with catalyst particles. If we wish to calculate throughput on the basis of total reactor volume the bed voidage e should be taken into account and the rate multiplied by $(1 - e)$ which is the fraction of reactor volume occupied by solid. Account must also be taken of mass transfer effects and so the rate is multiplied by an *effectiveness factor*, η, which is defined as the ratio of the rate of reaction in a pellet to the rate at which reaction would occur if the concentration and temperature within the pellet were the same as the respective values external to the pellet. The factor η therefore accounts for the influence which concentration and temperature gradients (which exist within the porous solid and result in mass and heat transfer effects) have on the chemical reaction rate. The reaction rate per unit volume of reactor is thus written:

$$\mathcal{R}_v = (1 - e)\, \boldsymbol{k} \mathrm{f}(C_A)\, \eta \tag{3.8}$$

where $\mathrm{f}(C_A)$ is that function of concentration of the reactants which describes the specific rate per unit volume in the absence of any mass and heat transfer effects within the particle and $\boldsymbol{k}$ is the specific rate constant per unit volume of reactor. An experimental determination of η merely involves comparing the observed rate of reaction with the rate of reaction on catalyst pellets sufficiently small for diffusion effects to be negligible.

3.3.1. Isothermal Reactions in Porous Catalyst Pellets

THIELE[14], who predicted how in-pore diffusion would influence chemical reaction rates, employed a geometric model with isotropic properties. Both the effective diffusivity and the effective thermal conductivity are independent of position for such a model. Although idealised geometric shapes are used to depict the situation within a particle such models, as we shall see later, are quite good approximations to practical catalyst pellets.

The simplest case we shall consider is that of a first order chemical reaction occurring within a rectangular slab of porous catalyst, the edges of which are sealed so that diffusion occurs in one dimension only. Figure 3.2 illustrates the geometry of the slab. Consider that the first order irreversible reaction:

$$\mathbf{A} \rightarrow \mathbf{B}$$

occurs within the volume of the particle and suppose its specific velocity constant on the basis of unit surface area is $\boldsymbol{k}_a$. For heterogeneous reactions uninfluenced by mass transfer effects experimental values for rate constants are usually based on unit surface area. The corresponding value in terms of unit total volume of particle would be $\rho_p S_g \boldsymbol{k}_a$ where ρ_p is the apparent density of the catalyst pellet and S_g is the specific surface area per unit mass of the solid including the internal pore surface area. We shall designate the specific rate constant based on unit volume of particle as $\boldsymbol{k}$. A component material balance for $\mathbf{A}$ across the element δx gives:

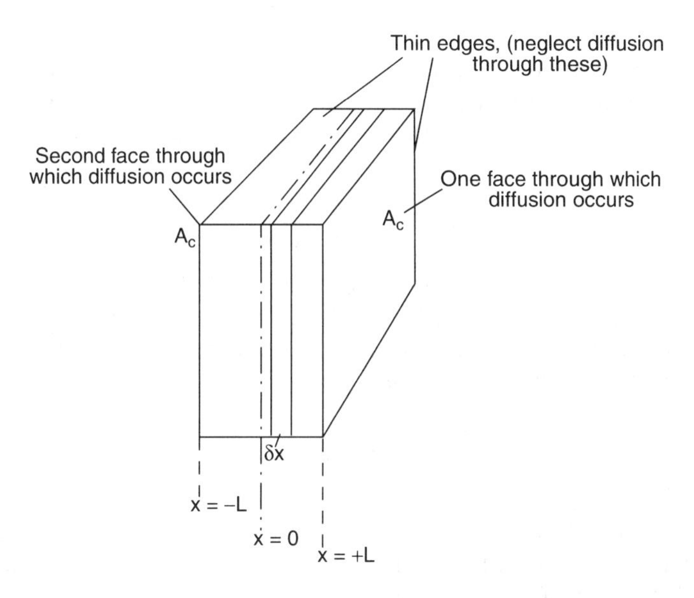

FIG. 3.2. Geometry of slab model for catalyst pellets

$$-A_c D_e \left(\frac{\mathrm{d}C_A}{\mathrm{d}x}\right)_x = -A_c D_e \left(\frac{\mathrm{d}C_A}{\mathrm{d}x}\right)_{x+\delta x} + \mathbf{k} C_A A_c \delta x \tag{3.9}$$

since, in the steady state, the flux of **A** into the element at x must be balanced by the flux out of the element at $(x + \delta x)$ plus the amount lost by reaction within the volume element $A_c \delta x$. If the concentration gradient term at the point $(x + \delta x)$ in equation 3.9 is expanded in a Taylor series about the point x and differential coefficients of order greater than two are ignored, the equation simplifies to:

$$\frac{\mathrm{d}^2 C_A}{\mathrm{d}x^2} - \frac{\mathbf{k} C_A}{D_e} = 0 \tag{3.10}$$

An analogous equation may be written for component **B**. By reference to Fig. 3.2, it will be seen that, because the product **B** diffuses outward, its flux is positive. Reaction produces **B** within the slab of material and hence makes the term depicting the rate of formation of **B** in the material balance equation positive, resulting in an equation similar in form to equation 3.10.

The boundary conditions for the problem may be written by referring to Fig. 3.2. At the exterior surface of the slab the concentration will be that corresponding to the conditions in the bulk gas phase, provided there is no resistance to mass transfer in the gas phase. Hence:

$$C_A = C_{A\infty} \quad \text{at} \quad x = \pm L \tag{3.11}$$

At the centre of the slab considerations of symmetry demand that:

$$\frac{dC_A}{dx} = 0 \quad \text{at} \quad x = 0 \tag{3.12}$$

so that the net flux through the plane at $x = 0$ is zero, diffusion across this boundary being just as likely in the direction of increasing x as in the direction of decreasing x. The solution of equation 3.9 with the boundary conditions given by equations 3.11 and 3.12 is:

$$C_A = C_{A\infty} \frac{\cosh \lambda x}{\cosh \lambda L} \tag{3.13}$$

where λ denotes the quantity $\sqrt{(\boldsymbol{k}/D_e)}$. Equation 3.13 describes the concentration profile of **A** within the catalyst slab. In the steady state the total rate of consumption of **A** must be equal to the total flux of **A** at the external surfaces. From symmetry, the total transfer across the external surfaces will be twice that at $x = L$. Thus:

$$-A_c \boldsymbol{k} \int_{-L}^{+L} C_A(x)\,dx = -2A_c D_e \left(\frac{dC_A}{dx}\right)_{x = \pm L}$$

If the whole of the catalyst surface area were available to the exterior concentration $C_{A\infty}$ there would be no diffusional resistance and the rate would then be $-2A_c L \boldsymbol{k} C_{A\infty}$. The ratio of these two rates is the effectiveness factor:

$$\eta = \frac{-A_c \boldsymbol{k} \int_{-L}^{+L} C_A(x)\,dx}{-2A_c L \boldsymbol{k} C_{A\infty}} = \frac{-2A_c D_e \left(\frac{dC_A}{dx}\right)_{x = L}}{-2A_c L \boldsymbol{k} C_{A\infty}} \tag{3.14}$$

By evaluating either the integral or the differential in the numerator of equation 3.14 the effectiveness factor may be calculated. In either case, by substitution from equation 3.13:

$$\eta = \frac{\tanh \lambda L}{\lambda L} = \frac{\tanh \phi}{\phi} \tag{3.15}$$

where:

$$\phi = \lambda L = L \sqrt{\frac{\boldsymbol{k}}{D_e}} \tag{3.16}$$

where ϕ is a dimensionless quantity known as the Thiele modulus. If the function η is plotted from equation 3.15, corresponding to the case of a first-order irreversible reaction in a slab with sealed edges, it may be seen from Fig. 3.3 that when $\phi < 0.2$, η is close to unity. Under these conditions there would be no diffusional resistance, for the rate of chemical reaction is not limited by diffusion. On the other hand when $\phi > 5.0$, $\eta = 1/\phi$ is a good approximation and for such conditions internal diffusion is the rate determining process. Between these two limiting values of ϕ the effectiveness factor is calculated from equation 3.15 and the rate process is in a region where neither in-pore mass transport nor chemical reaction is overwhelmingly rate determining.

Only a very limited number of manufactured catalysts could be approximately described by the slab model but there appear to be many which conform to the shape of a cylinder or sphere. Utilising the same principles as for the slab, it may

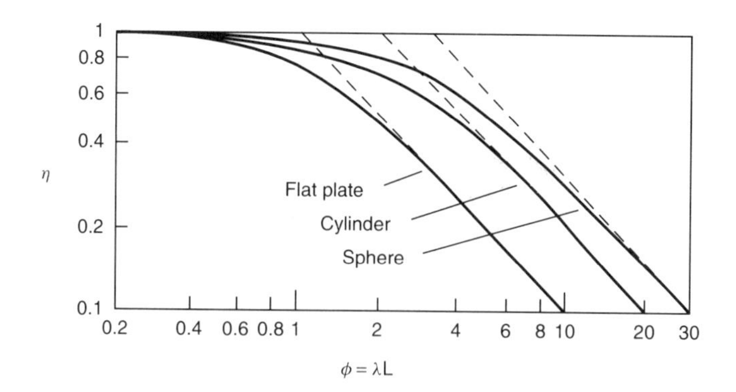

FIG. 3.3. Effectiveness factors for flat plate, cylinder and sphere

be shown (see the next example) that for a cylinder of radius r_0 sealed at the flat ends the effectiveness factor is:

$$\eta = \frac{2}{\lambda r} \frac{\mathbf{I}_1(\lambda r_0)}{\mathbf{I}_0(\lambda r_0)} \tag{3.17}$$

where $\mathbf{I}_0$ and $\mathbf{I}_1$ denote zero and first order modified Bessel functions of the first kind[15].

For a sphere of radius r_0 (see Example 3.1):

$$\eta = \frac{3}{\lambda r_0} \left\{ \coth(\lambda r_0) - \frac{1}{\lambda r_0} \right\} \tag{3.18}$$

GUNN[16] discusses the case of the hollow cylindrical catalyst particle. Such catalyst particles reduce the difficulties caused by excessive pressure drops.

Example 3.1

Derive an expression for the effectiveness factor of a cylindrical catalyst pellet, sealed at both ends, in which a first-order chemical reaction occurs.

Solution

The pellet has cylindrical symmetry about its central axis. Construct an annulus with radii $(r + \delta r)$ and r and consider the diffusive flux of material into and out of the cylindrical annulus, length L.

A material balance for the reactant **A** gives (see Figure 3.4):

Diffusive flux in at r − Diffusive flux out at $(r + \delta r)$ = Amount reacted in volume $2\pi L r \delta r$

i.e.:

$$\left\{ -2\pi D_e L \left(r \frac{\mathrm{d}C_A}{\mathrm{d}r} \right)_r \right\} - \left\{ -2\pi D_e L \left(r \frac{\mathrm{d}C_A}{\mathrm{d}r} \right)_{r+\delta r} \right\} = 2\pi L r \delta r \mathbf{k} C_A$$

Expanding the first term and ignoring terms higher than $(\delta r)^2$:

$$\frac{\mathrm{d}^2 C_A}{\mathrm{d}r^2} + \frac{1}{r}\frac{\mathrm{d}C_A}{\mathrm{d}r} - \lambda^2 C_A = 0 \text{ where } \lambda = \sqrt{\frac{\mathbf{k}}{D_e}}$$

This is a standard modified Bessel equation of zero order whose solution is[15]:

$$C_A = A\mathbf{I}_0(\lambda r) + B\mathbf{K}_0(\lambda r)$$

where $\mathbf{I}_0$ and $\mathbf{K}_0$ represent zero-order modified Bessel functions of the first and second kind respectively.

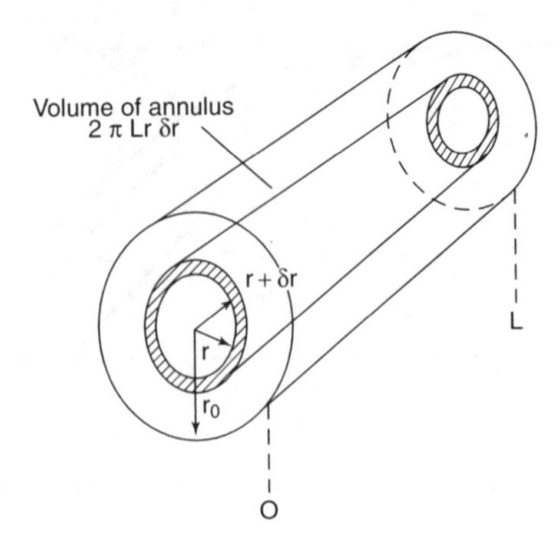

FIG. 3.4. Geometry of cylindrical catalyst pellet

The boundary conditions for the problem are $r = r_0$, $C_A = C_{A0}$; $r = 0$, C_A is finite. Since C_A remains finite at $r = 0$ and $\mathbf{K}_0(0) = \infty$ then we must put $B = 0$ to satisfy the physical conditions. Substituting the boundary conditions therefore gives the solution:

$$\frac{C_A}{C_{A0}} = \frac{\mathbf{I}_0(\lambda r)}{\mathbf{I}_0(\lambda r_0)}$$

For the cylinder:

$$\eta = \frac{2\pi r D_e L(\mathrm{d}C_A/\mathrm{d}r)_{r_0}}{\pi r_0^2 L\boldsymbol{k} C_{A0}}$$

From the relation between C_A and r:

$$\left(\frac{\mathrm{d}C_A}{\mathrm{d}r}\right)_{r_0} = \lambda C_{A0}\frac{\mathbf{I}_1(\lambda r)}{\mathbf{I}_0(\lambda r)}$$

since:

$$\frac{\mathrm{d}}{\mathrm{d}r}\{\mathbf{I}_0(\lambda r)\} = \mathbf{I}_1(\lambda r)$$

so:

$$\eta = \frac{2}{\lambda r_0}\frac{\mathbf{I}_1(\lambda r_0)}{\mathbf{I}_0(\lambda r_0)}$$

Example 3.2

Derive an expression for the effectiveness factor of a spherical catalyst pellet in which a first-order isothermal reaction occurs.

Solution

Take the origin of coordinates at the centre of the pellet, radius r_0, and construct an infinitesimally thin shell of radii $(r + \delta r)$ and r (see Fig. 3.5). A material balance for the reactant **A** across the shell gives:

Diffusive flux in at r – Diffusive flux out at $(r + \delta r)$ = Amount reacted in volume $4\pi r^2 \delta r$

i.e.:

$$\left\{-4\pi D_e\left(r^2\frac{\mathrm{d}C_A}{\mathrm{d}r}\right)_r\right\} - \left\{-4\pi D_e\left(r^2\frac{\mathrm{d}C_A}{\mathrm{d}r}\right)_{r+\delta r}\right\} = 4\pi r^2 \delta r \boldsymbol{k} C_A$$

Expanding the first term and ignoring terms higher than $(\delta r)^2$:

$$\frac{\mathrm{d}^2 C_A}{\mathrm{d}r^2} + \frac{2}{r}\frac{\mathrm{d}C_A}{\mathrm{d}r} = \frac{\boldsymbol{k}}{\boldsymbol{D_e}}C_A$$

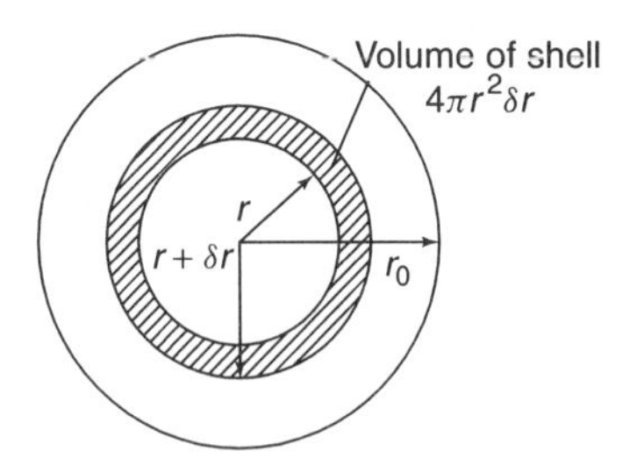

FIG. 3.5. Geometry of spherical catalyst pellet

or:
$$\frac{1}{r^2}\frac{\mathrm{d}}{\mathrm{d}r}\left\{r^2\frac{\mathrm{d}C_A}{\mathrm{d}r}\right\} - \lambda^2 C_A = 0 \text{ where } \lambda = \sqrt{\frac{\boldsymbol{k}}{D_e}}$$

Substituting:
$$C_A = \mathrm{f}(r)/r$$
$$\mathrm{f}''(r) - \lambda^2 \mathrm{f}\,(r) = 0$$

therefore:
$$\mathrm{f}\,(r) = Ae^{\lambda r} + Be^{-\lambda r}$$

The boundary conditions for the problem are $r = r_0$, $C_A = C_{A0}$; $r = 0$, $\dfrac{\mathrm{d}C_A}{\mathrm{d}r} = 0$. At r_0 we have $\mathrm{f}(r_0) = C_{A0}r_0$. Substituting these boundary conditions.

$$\mathrm{f}\,(r) = C_A r = \frac{C_{A0}r_0 \sinh(\lambda r)}{\sinh(\lambda r_0)}$$

Now for a sphere:
$$\eta = \frac{4\pi r_0^2 D_e(\mathrm{d}C_A/\mathrm{d}r)_{r_0}}{4/3\pi r_0^3 \boldsymbol{k} C_{A0}}$$

From the relation between C_A and r:
$$\left(\frac{\mathrm{d}C_A}{\mathrm{d}r}\right)_{r_0} = \frac{C_{A0}}{r_0}\{\lambda r_0 \coth(\lambda r_0) - 1\}$$

Hence:
$$\eta = \frac{3}{\lambda r_0}\left\{\coth(\lambda r_0) - \frac{1}{\lambda r_0}\right\}$$

The Thiele moduli for the cylinder and sphere differ from that for the slab. In the case of the slab we recall that $\phi = \lambda L$, whereas for the cylinder it is conveniently defined as $\phi = \lambda r_0/2$ and for the sphere as $\phi = \lambda r_0/3$. In each case the reciprocal of this corresponds to the respective asymptote for the curve representing the slab, cylinder or sphere. We may note here that the ratio of the geometric volume V_p of each of the models to the external surface area S_x is L for the slab, $r_0/2$ for the cylinder and $r_0/3$ for the sphere. Thus, if the Thiele modulus is defined as:

$$\phi = \lambda \frac{V_p}{S_x} = \frac{V_p}{S_x}\sqrt{\frac{\boldsymbol{k}}{D_e}} \tag{3.19}$$

the asymptotes become coincident. The asymptotes for large ϕ correspond to $\eta = 1/\phi$ for any shape of particle because, as ARIS[17] points out, diffusion is rate determining under these conditions and reaction occurs, therefore, in only a very thin region of the particle adjacent to the exterior surface. The curvature of the surface is thus unimportant.

The effectiveness factor for the slab model may also be calculated for reactions other than first order. It turns out that when the Thiele modulus is large the asymptotic value of η for all reactions is inversely proportional to the Thiele modulus, and when the latter approaches zero the effectiveness factor tends to unity. However, just as we found that the asymptotes for a first-order reaction in particles of different geometry do not coincide unless we choose a definition for the Thiele modulus which forces them to become superimposed, so we find that the asymptotes for reaction orders $n = 0$, 1 and 2 do not coincide unless we define a generalised Thiele modulus:

$$\bar{\phi} = \frac{V_p}{S_x}\sqrt{\frac{n+1}{2}\frac{kC_{A\infty}^{(n-1)}}{D_e}} \tag{3.20}$$

The modulus $\bar{\phi}$ defined by equation 3.20 has the advantage that the asymptotes to η are approximately coincident for all particle shapes and for all reaction orders except $n = 0$; for this latter case[18] $\eta = 1$ for $\bar{\phi} < 2$ and $\eta = 1/\bar{\phi}$ for $\bar{\phi} > 2$. Thus η may be calculated from the simple slab model, using equation 3.20 to define the Thiele modulus. The curve of η as a function of $\bar{\phi}$ is therefore quite general for practical catalyst pellets. For $\bar{\phi} > 3$ it is found that $\eta = 1/\bar{\phi}$ to an accuracy within 0.5 per cent, while the approximation is within 3.5 per cent for $\bar{\phi} > 2$. It is best to use this generalised curve (i.e. η as a function of $\bar{\phi}$) because the asymptotes for different cases can then be made almost to coincide. The errors involved in using the generalised curve are probably no greater than errors perpetrated by estimating values of parameters in the Thiele modulus.

3.3.2. Effect of Intraparticle Diffusion on Experimental Parameters

When intraparticle diffusion occurs, the kinetic behaviour of the system is different from that which prevails when chemical reaction is rate determining. For conditions of diffusion control ϕ will be large, and then the effectiveness factor $\eta(= 1/\phi \tanh \phi$, from equation 3.15) becomes $\bar{\phi}^{-1}$.From equation 3.19, it is seen therefore that η is proportional to $\boldsymbol{k}^{-1/2}$. The chemical reaction rate on the other hand is directly proportional to $\boldsymbol{k}$ so that, from equation 3.8 at the beginning of this section, the overall reaction rate is proportional to $\boldsymbol{k}^{1/2}$. Since the specific rate constant is directly proportional to $e^{-\mathbf{E}/\mathbf{R}T}$, where $\mathbf{E}$ is the activation energy for the chemical reaction in the absence of diffusion effects, we are led to the important result that for a diffusion limited reaction the rate is proportional to $e^{-\mathbf{E}/2\mathbf{R}T}$. Hence the apparent activation energy $\mathbf{E}_D$, measured when reaction occurs in the diffusion controlled region, is only half the true value:

$$\mathbf{E}_D = \mathbf{E}/2 \tag{3.21}$$

A further important result which arises because of the functional form of $\bar{\phi}$ is that the apparent order of reaction in the diffusion controlled region differs from that which is observed when chemical reaction is rate determining. Recalling that the reaction order is defined as the exponent n to which the concentration $C_{A\infty}$ is raised in the equation for the chemical reaction rate, we replace $f(C_A)$ in equation 3.8 by $C_{A\infty}^n$. Hence the overall reaction rate per unit volume is $(1 - e)\eta \boldsymbol{k} C_{A\infty}^n$. When diffusion is rate determining, η is (as already mentioned) equal to ϕ^{-1}; from equation

3.20 it is therefore proportional to $C_{A\infty}^{-(n-1)/2}$. Thus the overall reaction rate depends on $C_{A\infty}^{n} C_{A\infty}^{-(n-1)/2} = C_{A\infty}^{(n+1)/2}$. The apparent order of reaction n_D as measured when reaction is dominated by intraparticle diffusion effects is thus related to the true order as follows:

$$n_D = (n + 1)/2 \tag{3.22}$$

A zero-order reaction thus becomes a half-order reaction, a first-order reaction remains first order, whereas a second-order reaction has an apparent order of 3/2 when strongly influenced by diffusional effects. Because $\boldsymbol{k}$ and n are modified in the diffusion controlled region then, if the rate of the overall process is estimated by multiplying the chemical reaction rate by the effectiveness factor (as in equation 3.8), it is imperative to know the true rate of chemical reaction uninfluenced by diffusion effects.

The functional dependence of other parameters on the reaction rate also becomes modified when diffusion determines the overall rate. If we write the rate of reaction for an nth order reaction in terms of equation 3.8 and substitute the general expression obtained for the effectiveness factor at high values of $\bar{\phi}$, where η is proportional to $1/\bar{\phi}$ and $\bar{\phi}$ is defined by equation 3.20, we obtain:

$$\mathcal{R}_V = (1 - e)\, \boldsymbol{k} C_{A\infty}^{n} \eta = (1 - e) \boldsymbol{k} C_{A\infty}^{n} \frac{S_x}{V_p} \sqrt{\frac{2D_e}{(n+1)\boldsymbol{k} C_{A\infty}^{n-1}}} \tag{3.23}$$

The way in which experimental parameters are affected when intraparticle diffusion is important may be deduced by inspection of equation 3.23. Referring the specific rate constant to unit surface area, rather than unit reactor volume, the term $(1 - e)\, \boldsymbol{k}$ is equivalent to $\rho_b S_g \boldsymbol{k}_a$ where ρ_b is the bulk density of the catalyst and S_g is its surface area per unit mass. On the other hand, the rate constant $\boldsymbol{k}$ appearing in the denominator under the square root sign in equation 3.23 is based on unit particle volume and is therefore equal to $\rho_p S_g \boldsymbol{k}_a$ where ρ_p is the particle density. Thus, if bulk diffusion controls the reaction, the rate becomes dependent on the square root of the specific surface area, rather than being directly proportional to surface area, in the absence of transport effects. We do not include the external surface area S_x in this reckoning since the ratio V_p/S_x is, for a given particle shape, an independent parameter characteristic of the particle size. On the other hand, if Knudsen diffusion determines the rate, then because the effective diffusivity for Knudsen flow is inversely proportional to the specific surface area (equation 3.4) the reaction rate becomes independent of surface area.

The pore volume per unit mass V_g (a measure of the porosity) is also a parameter which is important and is implicitly contained in equation 3.23. Since the product of the particle density ρ_p and specific pore volume V_g represents the porosity, then ρ_p is inversely proportional to V_g. Therefore, when the rate is controlled by bulk diffusion, it is proportional not simply to the square root of the specific surface area but to the product of $\sqrt{S_g}$ and $\sqrt{V_g}$. If Knudsen diffusion controls the reaction then the overall rate is directly proportional to V_g since the effective Knudsen diffusivity contained in the quantity $\sqrt{(D_e/\rho_p)}$ is, from equation 3.4, proportional to the ratio of the porosity ψ and the particle density ρ_p.

TABLE 3.2. *Effect of Intraparticle Diffusion on Various Parameters*

Rate determining step	Order	Activation energy	Surface area	Pore volume
Chemical reaction	n	E	S_g	independent
Bulk diffusion	$(n+1)/2$	$E/2$	$\sqrt{S_g}$	$\sqrt{V_g}$
Knudsen diffusion	$(n+1)/2$	$E/2$	independent	V_g

Table 3.2 summarises the effect which intraparticle mass transfer effects have on parameters involved explicitly or implicitly in the expression for the overall rate of reaction.

3.3.3. Non-isothermal Reactions in Porous Catalyst Pellets

So far the effect of temperature gradients within the particle has been ignored. Strongly exothermic reactions generate a considerable amount of heat which, if conditions are to remain stable, must be transported through the particle to the exterior surface where it may then be dissipated. Similarly an endothermic reaction requires a source of heat and in this case the heat must permeate the particle from the exterior to the interior. In any event a temperature gradient within the particle is established and the chemical reaction rate will vary with position.

We may consider the problem by writing a material and heat balance for the slab of catalyst depicted in Fig. 3.2 For an irreversible first-order exothermic reaction the material balance on reactant **A** is:

$$\frac{d^2C_A}{dx^2} - \frac{\boldsymbol{k}C_A}{D_e} = 0 \qquad \text{(equation 3.10)}$$

A heat balance over the element δx gives:

$$\frac{d^2T}{dx^2} + \frac{(-\Delta H)\boldsymbol{k}C_A}{k_e} = 0 \qquad (3.24)$$

where ΔH is the enthalpy change resulting from reaction, and k_e is the effective thermal conductivity of the particle defined by analogy with the discussion on effective diffusivity leading to equation 3.1. In writing these two equations it should be remembered that the specific rate constant $\boldsymbol{k}$ is a function of temperature, usually of the Arrhenius form ($\boldsymbol{k} = \mathscr{A}e^{-E/RT}$, where $\mathscr{A}$ is the frequency factor for reaction). These two, simultaneous differential equations are to be solved together with the boundary conditions:

$$C_A = C_{A\infty} \quad \text{at} \quad x = \pm L \qquad (3.25)$$

$$T = T_\infty \quad \text{at} \quad x = \pm L \qquad (3.26)$$

$$\frac{dC_A}{dx} = \frac{dT}{dx} = 0 \quad \text{at} \quad x = 0 \qquad (3.27)$$

Because of the non-linearity of the equations the problem can only be solved in this form by numerical techniques[18, 19]. However, an approximation may be made which gives an asymptotically exact solution[20], or, alternatively, the exponential function of temperature may be expanded to give equations which can be solved analytically[21, 22]. A convenient solution to the problem may be presented in the form of families of curves for the effectiveness factor as a function of the Thiele modulus. Figure 3.6 shows these curves for the case of a first-order irreversible reaction occurring in spherical catalyst particles. Two additional independent dimensionless parameters are introduced into the problem and these are defined as:

$$\beta = \frac{(-\Delta H) D_e C_{A\infty}}{k_e T_\infty} \tag{3.28}$$

$$\varepsilon' = \mathbf{E}/\mathbf{R}T \tag{3.29}$$

The parameter β represents the maximum temperature difference that could exist in the particle relative to the temperature at the exterior surface, for if we recognise that in the steady state the heat flux within an elementary thickness of the particle is balanced by the heat generated by chemical reaction then:

$$k_e \frac{\mathrm{d}T}{\mathrm{d}x} = \Delta H \, D_e \frac{\mathrm{d}C_A}{\mathrm{d}x} \tag{3.30}$$

If equation 3.30 is then integrated from the exterior surface where $T = T_\infty$ and $C_A = C_{A\infty}$ to the centre of the particle where (say) $T = T_M$ and $C_A = C_{AM}$ we obtain:

$$\frac{T_M - T_\infty}{T_\infty} = \frac{(-\Delta H) D}{k_e T_\infty} (C_{A\infty} - C_{AM}) \tag{3.31}$$

When the Thiele modulus is large C_{AM} is effectively zero and the maximum difference in temperature between the centre and exterior of the particle is $(-\Delta H) D_e C_{A\infty}/k_e$. Relative to the temperature outside the particle this maximum temperature difference is therefore β. For exothermic reactions β is positive while for endothermic reactions it is negative. The curve in Fig. 3.6 for $\beta = 0$ represents isothermal conditions within the pellet. It is interesting to note that for a reaction in which $-\Delta H = 10^5$ kJ/kmol, $k_e = 1$ W/mK, $D_e = 10^{-5}$ m^2/s and $C_{A\infty} = 10^{-1}$ kmol/m^3, the value of $T_M - T_\infty$ is 100°C. In practice much lower values than this are observed but it does serve to show that serious errors may be introduced into calculations if conditions within the pellet are arbitrarily assumed to be isothermal.

On the other hand, it has been argued that the resistance to heat transfer is effectively within a thin gas film enveloping the catalyst particle[23]. Thus, for the whole practical range of heat transfer coefficients and thermal conductivities, the catalyst particle may be considered to be at a uniform temperature. Any temperature increase arising from the exothermic nature of a reaction would therefore be across the fluid film rather than in the pellet interior.

Figure 3.6 shows that, for exothermic reactions ($\beta > 0$), the effectiveness factor may exceed unity. This is because the increase in rate caused by the temperature rise inside the particle more than compensates for the decrease in rate caused by the negative concentration gradient which effects a decrease in concentration towards the centre of the particle. A further point of interest is that, for reactions which are highly exothermic and at low value of the Thiele modulus, the value of η is not uniquely defined by the Thiele modulus and the parameters β and ε'. The shape of

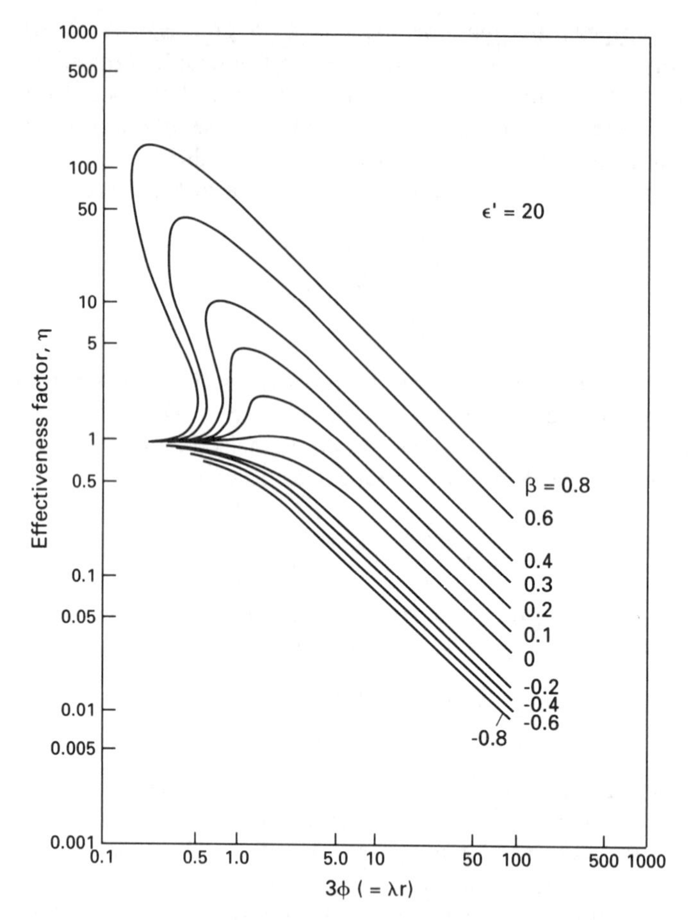

FIG. 3.6. Effectiveness factor as a function of the Thiele modulus for non-isothermal conditions for a sphere[21]

the curves in this region indicates that the effectiveness factor may correspond to any one of three values for a given value of the Thiele modulus. In effect there are three different conditions for which the rate of heat generation within the particle is equal to the rate of heat removal. One condition represents a metastable state and the remaining two conditions correspond to a region in which the rate is limited by chemical reaction (relatively low temperatures) and a region where there is diffusion limitation (relatively high temperatures). The region of multiple solutions in Fig. 3.6, however, corresponds to large values of β and ε' seldom encountered in practice.

McGreavy and Thornton[23] have developed an alternative approach to the problem of identifying such regions of unique and multiple solutions in packed bed reactors. Recognising that the resistance to heat transfer is probably due to a thin gas film surrounding the particle, but that the resistance to mass transfer is within the porous solid, they solved the mass and heat balance equations for a pellet with modified boundary conditions. Thus the heat balance for the pellet represented by equation 3.24 was replaced by:

$$h(T - T_\infty) = h_D(-\Delta H)(C_{A\infty} - C_A) \tag{3.32}$$

and solved simultaneously with the mass balance represented by equation 3.10. Boundary conditions represented by equations 3.25 and 3.26 were replaced by:

$$D_e \frac{dC_A}{dx} = h_D(C_{A\infty} - C_A) \quad \text{at} \quad x = L \tag{3.33}$$

and:

$$k_e \frac{dT}{dx} = h(T - T_\infty) \quad \text{at} \quad x = L \tag{3.34}$$

respectively.

A modified Thiele modulus may be defined by rewriting $\phi = L\sqrt{k/D_e}$ (see equation 3.16) in the form:

$$\phi' = L\sqrt{\frac{\mathscr{A}}{D_e}} \tag{3.35}$$

where $\mathscr{A}$ is the frequency factor in the classical Arrhenius equation. The numerical solution is then depicted in Fig. 3.7[23], which resembles a plot of effectiveness factor as a function of Thiele modulus (cf. Fig. 3.6). Whereas an effectiveness factor chart describes the situation for a given single particle in a packed bed (and is therefore of limited value in reactor design), Fig. 3.7 may be used to identify the region of multiple solutions for the whole reactor. If local extrema are calculated from Fig. 3.7 by finding conditions for which $dT_\infty/dT = 0$, the bounds of T_∞ may be located. It is then possible to predict the region of multiple solutions corresponding to unstable operating conditions. Figure 3.8, for example, shows two reactor trajectories which would intersect the region of metastable conditions within the reactor. Such a

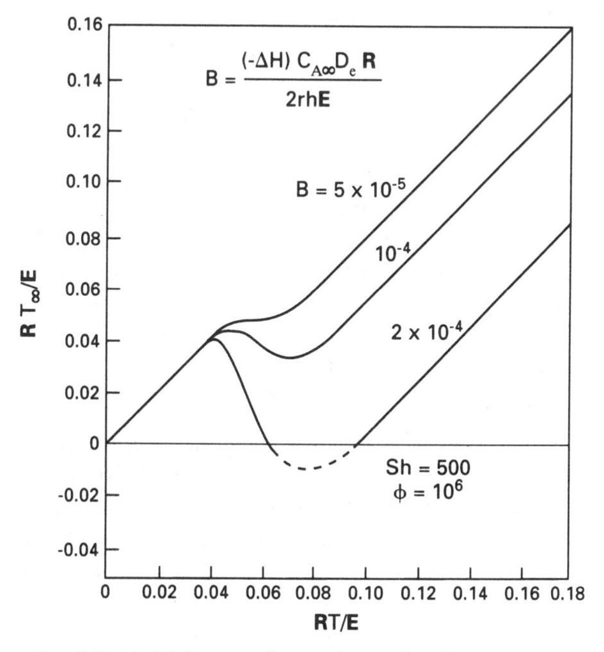

FIG. 3.7. Multiple states for catalyst pellets in reactor

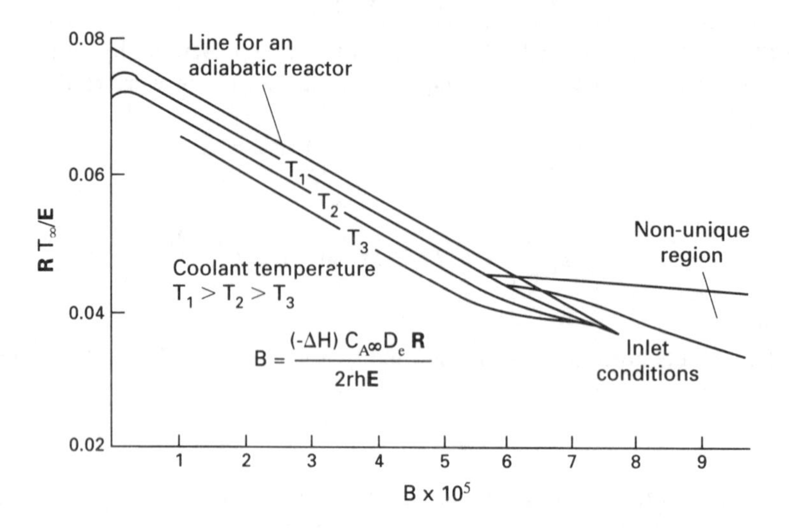

FIG. 3.8. Reactor trajectories in adiabatic and cooled catalyst beds

method can predict regions of instability for the packed reactor rather than for a single particle, because use has been made of the modified Thiele modulus employing the kinetic Arrhenius factor $\mathscr{A}$ which is independent of position along the bed.

3.3.4. Criteria for Diffusion Control

In assessing whether a reactor is influenced by intraparticle mass transfer effects WEISZ and PRATER[24] developed a criterion for isothermal reactions based upon the observation that the effectiveness factor approaches unity when the generalised Thiele modulus is of the order of unity. It has been shown[17] that the effectiveness factor for all catalyst geometries and reaction orders (except zero order) tends to unity when the generalised Thiele modulus falls below a value of one. Since η is about unity when $\phi < \sqrt{2}$ for zero-order reactions, a quite general criterion for diffusion control of simple isothermal reactions not affected by product inhibition is $\bar{\phi} < 1$. Since the Thiele modulus (see equation 3.19) contains the specific rate constant for chemical reaction, which is often unknown, a more useful criterion is obtained by substituting $\mathscr{R}'_V / C_{A\infty}$ (for a first-order reaction) for $\boldsymbol{k}$ to give:

$$\left\{\frac{V_p}{S_x}\right\}^2 \frac{\mathscr{R}'_V}{D_e C_{A\infty}} < 1 \tag{3.36}$$

where $\mathscr{R}'_V$ is the measured rate of reaction per unit volume of catalyst particle.

PETERSEN[25] points out that this criterion is invalid for more complex chemical reactions whose rate is retarded by products. In such cases the observed kinetic rate expression should be substituted into the material balance equation for the particular geometry of particle concerned. An asymptotic solution to the material balance equation then gives the correct form of the effectiveness factor. The results indicate that the inequality 3.36 is applicable only at high partial pressures of product. For low partial pressures of product (often the condition in an experimental differential tubular reactor) the criterion will depend on the magnitude of the constants in the kinetic rate equation.

The usual experimental criterion for diffusion control involves an evaluation of the rate of reaction as a function of particle size. At a sufficiently small particle size the measured rate of reaction will become independent of particle size and the rate of reaction can be safely assumed to be independent of intraparticle mass transfer effects. At the other extreme, if the observed rate is inversely proportional to particle size the reaction is strongly influenced by intraparticle diffusion. For a reaction whose rate is inhibited by the presence of products, there is an attendant danger of misinterpreting experimental results obtained for different particle sizes when a differential reactor is used for, under these conditions, the effectiveness factor is sensitive to changes in the partial pressure of product.

WEISZ and HICKS[21] showed that when reaction conditions within the particle are non-isothermal a suitable criterion defining conditions under which a reaction is not controlled by mass and heat transfer effects in the solid is:

$$\left\{\frac{V_p}{S_x}\right\}^2 \frac{\mathcal{R}'_V}{D_e C_{A\infty}} \exp\left\{\frac{\varepsilon'\beta}{1+\beta}\right\} < 1 \qquad (3.37)$$

3.3.5. Selectivity in Catalytic Reactions Influenced by Mass and Heat Transfer Effects

It is rare that a catalyst can be chosen for a reaction such that it is entirely specific or unique in its behaviour. More often than not products additional to the main desired product are generated concomitantly. The ratio of the specific chemical rate constant of a desired reaction to that for an undesired reaction is termed the kinetic selectivity factor (which we shall designate by S) and is of central importance in catalysis. Its magnitude is determined by the relative rates at which adsorption, surface reaction and desorption occur in the overall process and, for consecutive reactions, whether or not the intermediate product forms a localised or mobile adsorbed complex with the surface. In the case of two parallel competing catalytic reactions a second factor, the thermodynamic factor, is also of importance. This latter factor depends exponentially on the difference in free energy changes associated with the adsorption–desorption equilibria of the two competing reactants. The thermodynamic factor also influences the course of a consecutive reaction where it is enhanced by the ability of the intermediate product to desorb rapidly and also the reluctance of the catalyst to re-adsorb the intermediate product after it has vacated the surface.

The kinetic and thermodynamic selectivity factors are quantities which are functions of the chemistry of the system. When an active catalyst has been selected for a particular reaction (often by a judicious combination of theory and experiment) we ensure that the kinetic and thermodynamic factors are such that they favour the formation of desired product. Many commercial processes, however, employ porous catalysts since this is the best means of increasing the extent of surface at which the reaction occurs. Chemical engineers are therefore interested in the effect which the porous nature of the catalyst has on the selectivity of the chemical process.

WHEELER[26] considered the problem of chemical selectivity in porous catalysts. Although he employed a cylindrical pore model and restricted his conclusions to the effect of pore size on selectivity, the following discussion will be based on the simple geometrical model of the catalyst pellet introduced earlier (see Fig. 3.2 and Section 3.3.1).

Isothermal Conditions

Sometimes it may be necessary to convert into a desired product only one component in a mixture. For example, it may be required to dehydrogenate a six-membered cycloparaffin in the presence of a five-membered cycloparaffin without affecting the latter. In this case it is desirable to select a catalyst which favours the reaction:

$$\mathbf{A} \xrightarrow{k_1} \mathbf{B}$$

when it might be possible for the reaction:

$$\mathbf{X} \xrightarrow{k_2} \mathbf{Y}$$

to occur simultaneously.

Suppose the desired product is **B** and also suppose that the reactions occur in a packed tubular reactor in which we may neglect both longitudinal and radical dispersion effects. If both reactions are first-order, the ratio of the rates of the respective reactions is:

$$\frac{\mathcal{R}'_{VA}}{\mathcal{R}'_{VX}} = \frac{k_1 C_A}{k_2 C_X} \tag{3.38}$$

If the reactions were not influenced by in-pore diffusion effects, the intrinsic kinetic selectivity would be $k_1/k_2(= S)$. When mass transfer is important, the rate of reaction of both **A** and **X** must be calculated with this in mind. From equation 3.9, the rate of reaction for the slab model is:

$$-A_c D_e \left(\frac{\mathrm{d}C_A}{\mathrm{d}x}\right)_{x=\pm L}$$

The concentration profile of **A** through the slab is given by:

$$C_A = C_{A\infty} \frac{\cosh(\lambda x)}{\cosh(\lambda L)} \qquad \text{(equation 3.13)}$$

so that by differentiation of C_A we obtain the rate of decomposition of **A**:

$$\mathcal{R}'_{VA} = \frac{-A_c D_e}{V_p}\left(\frac{\mathrm{d}C_A}{\mathrm{d}x}\right)_{x=L} = \frac{-A_c D_e C_{A\infty}}{L V_p}\phi_1 \tanh\phi_1 \tag{3.39}$$

where $\phi_1 = L\sqrt{k_1/D_e}$ is the Thiele modulus pertaining to the decomposition of A. A similar equation may be written for the decomposition of **X**:

$$\mathcal{R}'_{VX} = \frac{-A_c D_e C_{X\infty}}{L V_p}\phi_2 \tanh\phi_2 \tag{3.40}$$

where ϕ_2 corresponds to the Thiele modulus for the decomposition of **X**. If we were dealing with a general type of catalyst pellet, then because of the properties of the general Thiele parameter $\bar{\phi}$ we need only replace ϕ_1 and ϕ_2 by $\bar{\phi}_1$ and $\bar{\phi}_2$ respectively and substitute (V_p/S_x) for the characteristic dimension L. The ratio of the rates of decomposition of **A** and **X** then becomes:

$$\frac{\mathcal{R}'_{VA}}{\mathcal{R}'_{VX}} = \frac{C_{A\infty}}{C_{X\infty}} \frac{\bar{\phi}_1 \tanh \bar{\phi}_1}{\bar{\phi}_2 \tanh \bar{\phi}_2} \tag{3.41}$$

Although equation 3.41 is only applicable to competing first-order reactions in catalyst particles, at large values of $\bar{\phi}$ where diffusion is rate controlling the equation is equivalent at the asymptotes to equations obtained for reaction orders other than one.

Since $\bar{\phi}$ is proportional to $\sqrt{\boldsymbol{k}}$ we may conclude that for competing simultaneous reactions strongly influenced by diffusion effects (where $\bar{\phi}$ is large and $\tanh\bar{\phi} \simeq 1$) the selectivity depends on $\sqrt{S}$ (where $S = \boldsymbol{k}_1/\boldsymbol{k}_2$), the square root of the ratio of the respective rate constants. The corollary is that, for such reactions, maximum selectivity is displayed by small sized particles and, in the limit, if the particle size is sufficiently small (small $\bar{\phi}$ so that $\tanh\bar{\phi} \simeq \bar{\phi}$) the selectivity is the same as for a non-porous particle, i.e. S itself.

We should add a note of caution here, however, for in the Knudsen flow region D_e is proportional to the pore radius. When the pores are sufficiently small for Knudsen diffusion to occur then the selectivity will also be influenced by pore size. Maximum selectivity would be obtained for small particles which contain large diameter pores.

When two simultaneous reaction paths are involved in a process the routes may be represented by:

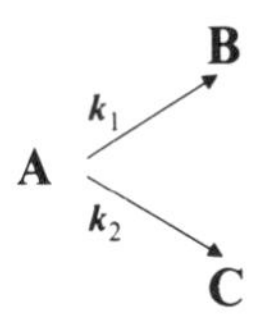

A classical example of this type of competitive reaction is the conversion of ethanol by a copper catalyst at about 300°C. The principal product is acetaldehyde but ethylene is also evolved in smaller quantities. If, however, an alumina catalyst is used, ethylene is the preferred product. If, in the above reaction scheme, **B** is the desired product then the selectivity may be found by comparing the respective rates of formation of **B** and **C**. Adopting the slab model for simplicity and remembering that, in the steady state, the rates of formation of **B** and **C** must be equal to the flux of **B** and **C** at the exterior surface of the particle, assuming that the effective diffusivities of **B** and **C** are equal:

$$\frac{\mathcal{R}'_{VB}}{\mathcal{R}'_{VC}} = \left(\frac{\mathrm{d}C_B}{\mathrm{d}x}\right)_{x=L} \Big/ \left(\frac{\mathrm{d}C_C}{\mathrm{d}x}\right)_{x=L} \tag{3.42}$$

The respective fluxes may be evaluated by writing the material balance equations for each component and solving the resulting simultaneous equations. If the two reactions are of the same kinetic order, then it is obvious from the form of equation 3.42 that the selectivity is unaffected by mass transfer in pores. If, however, the kinetic orders of the reactions differ then, as the example below shows, the reaction of the lower kinetic order is favoured. The rate of formation of **B** with respect to **C** would therefore be impeded and the selectivity reduced if the order of the reaction producing **B** were less than that for the reaction producing **C**. In such cases the highest selectivity would be obtained by the use of small diameter particles or particles in which the effective diffusivity is high.

Example 3.3

Two gas-phase concurrent irreversible reactions

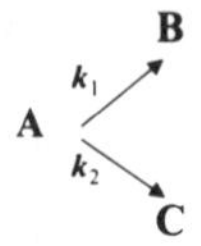

occur isothermally in a flat slab-shaped porous catalyst pellet. The desirable product **B** is formed by a first-order chemical reaction and the wasteful product **C** is formed by a zero-order reaction. Deduce an expression for the catalyst selectivity.

Solution

Taking a material balance across the element δx in Fig. 3.2, the flux in at $(x + \delta x)$ minus the flux out at x is equal to the amount reacted in volume $2A_C\delta x$ where A_C represents the area of each of the faces. If the slab is thin, diffusion through the edges may be neglected. For the three components the material balance equations therefore become:

$$\frac{d^2 C_A}{dx^2} - f^2 C_A = g^2$$

$$\frac{d^2 C_B}{dx^2} + f^2 C_A = 0$$

$$\frac{d^2 C_C}{dx^2} = -g^2$$

where $f^2 = k_1/D_e$ and $g^2 = k_2/D_e$.
The boundary conditions are:

$$x = \pm L, \quad C_A = C_{A0}, \quad C_B = C_C = 0$$

and:

$$x = 0, \quad \frac{dC_A}{dx} = \frac{dC_B}{dx} = \frac{dC_C}{dx} = 0$$

The solution satisfying the above set of differential equations is $C_A = Ae^{fx} - g^2/f^2$ where the term g^2/f^2 represents the particular integral. On inserting the boundary conditions the complete solution is:

$$C_A = \left(C_{A0} + \frac{g^2}{f^2}\right)\frac{\cosh(fx)}{\cosh(fL)} - \frac{g^2}{f^2}$$

The concentration C_B may now be found, but since the selectivity will be given by:

$$S = \frac{(dC_B/dx)_{x=L}}{(dC_C/dx)_{x=L}}$$

the material balance equation need only be integrated once. We obtain:

$$\left(\frac{dC_B}{dx}\right)_{x=L} = -f^2 \int_0^L \left\{\left(C_{A0} + \frac{g^2}{f^2}\right)\frac{\cosh(fx)}{\cosh(fL)} - \frac{g^2}{f^2}\right\} dx = -f\left(C_{A0} + \frac{g^2}{f^2}\right)\tanh(fL) + g^2 L$$

and:

$$\left(\frac{\mathrm{d}C_C}{\mathrm{d}x}\right)_{x=L} = -g^2 L$$

Then writing:

$$\phi = L\sqrt{\frac{k_1}{D_e}}$$

$$S = \left(\frac{k_1}{k_2}C_{A0} + 1\right)\frac{\tanh\phi}{\phi} - 1$$

Another important class of reactions, which is common in petroleum reforming reactions, may be represented by the scheme:

$$\mathbf{A} \xrightarrow{k_1} \mathbf{B} \xrightarrow{k_2} \mathbf{C}$$

and exemplified by the dehydrogenation of six-membered cycloparaffins to aromatics (e.g. cyclohexane converted to cyclohex-1-ene and ultimately benzene) catalysed by transition metals and metal oxides. Again we suppose **B** to be the desired product while **C** is a waste product. (If **C** were the desired product we would require a low selectivity for the formation of **B**). On the basis of first order kinetics and using the flat plate model, the material balance equation for component **B** is:

$$D_e \frac{\mathrm{d}^2 C_B}{\mathrm{d}x^2} = k_2 C_B - k_1 C_A \tag{3.43}$$

For component **A** the material balance equation is, as previously obtained:

$$D_e \frac{\mathrm{d}^2 C_A}{\mathrm{d}x^2} = k_1 C_A \qquad \text{(equation 3.10)}$$

The boundary conditions for the problem are:

$$C_A = C_{A\infty} \quad \text{and} \quad C_B = C_{B\infty} \quad \text{at} \quad x = L \tag{3.44}$$

$$\mathrm{d}C_A/\mathrm{d}x = \mathrm{d}C_B/\mathrm{d}x = 0 \qquad \text{at} \quad x = 0 \tag{3.45}$$

Solving the two simultaneous linear differential equations with the above boundary conditions leads to:

$$C_A = C_{A\infty}\frac{\cosh(\lambda_1 x)}{\cosh(\lambda_1 L)} \qquad \text{(equation 3.13)}$$

and:

$$C_B = C_{A\infty}\left\{\frac{k_1}{k_1 - k_2}\right\}\left\{\frac{\cosh(\lambda_2 x)}{\cosh(\lambda_2 L)} - \frac{\cosh(\lambda_1 x)}{\cosh(\lambda_1 L)}\right\} + C_{B\infty}\frac{\cosh(\lambda_2 x)}{\cosh(\lambda_2 L)} \tag{3.46}$$

where λ_1 and λ_2 are $\sqrt{k_1/D_e}$ and $\sqrt{k_2/D_e}$ respectively. The selectivity of the reaction will be the rate of formation of **B** with respect to **A** which, in the steady state, will be equal to the ratio of the fluxes of **B** and **A** at the exterior surface of the particle. Thus:

$$-\frac{(\mathrm{d}C_B/\mathrm{d}x)_{x=L}}{(\mathrm{d}C_A/\mathrm{d}x)_{x=L}} = \left\{\frac{k_1}{k_1 - k_2}\right\}\left\{1 - \frac{\phi_2 \tanh\phi_2}{\phi_1 \tanh\phi_1}\right\} - \frac{C_{B\infty}}{C_{A\infty}}\frac{\phi_2 \tanh\phi_2}{\phi_1 \tanh\phi_1} \tag{3.47}$$

Although the ratio of the fluxes of **B** and **A** at the exterior surface of the particle is really a point value, we may conveniently regard it as representing the rate of change of the concentration of **B** with respect to **A** at a position in the reactor corresponding to the location of the particle. The left side of equation 3.47 may thus be replaced by $-\mathrm{d}C_B/\mathrm{d}C_A$ where C_B and C_A are now gas phase concentrations. With this substitution and for large values of ϕ_1 and ϕ_2 one of the limiting forms of equation 3.47 is obtained:

$$-\frac{\mathrm{d}C_B}{\mathrm{d}C_A} = \frac{\sqrt{S}}{1+\sqrt{S}} - \frac{C_B}{C_A}\frac{1}{\sqrt{S}} \tag{3.48}$$

where S is the kinetic selectivity ($= \boldsymbol{k}_1/\boldsymbol{k}_2$). Integrating equation 3.48 from the reactor inlet (where the concentration of **A** is, say, C_{A0} and that of **B** is taken as zero) to any point along the reactor:

$$\frac{C_B}{C_A} = \left\{\frac{S}{S-1}\right\}\left\{\left(\frac{C_A}{C_{A0}}\right)^{\frac{1-\sqrt{S}}{\sqrt{S}}} - 1\right\} \tag{3.49}$$

The other limiting form of equation 3.47 is obtained when ϕ_1 and ϕ_2 are small:

$$-\frac{\mathrm{d}C_B}{\mathrm{d}C_A} = 1 - \frac{1}{S}\frac{C_B}{C_A} \tag{3.50}$$

which, on integration, gives:

$$\frac{C_B}{C_A} = \left(\frac{S}{S-1}\right)\left\{\left(\frac{C_A}{C_{A0}}\right)^{\frac{1-S}{S}} - 1\right\} \tag{3.51}$$

Comparison of equations 3.48 and 3.50 shows that at low effectiveness factors (when ϕ_1 and ϕ_2 are large) the selectivity is less than it would be if the effectiveness factor for each reaction were near unity (small ϕ_1 and ϕ_2). The consequence of this is seen by comparing equations 3.49 and 3.51, the respective integrated forms of equations 3.48 and 3.50. The yield of **B** is comparatively low when in-pore diffusion is a rate limiting process and, for a given fraction of **A** reacted, the conversion to **B** is impeded. A corollary to this is that it should be possible to increase the yield of a desired intermediate by using smaller catalyst particles or, alternatively, by altering the pore structure in such a way as to increase the effective diffusivity. If, however, the effectiveness factor is below about 0.3 (at which value the yield of **B** becomes independent of diffusion effects) a large reduction in pellet size (or large increase in D_e) is required to achieve any significant improvement in selectivity. WHEELER[26] suggests that when such a drastic reduction in pellet size is necessary, a fluidised bed reactor may be used to improve the yield of an intermediate.

Non-isothermal Conditions

The influence which the simultaneous transfer of heat and mass in porous catalysts has on the selectivity of first-order concurrent catalytic reactions has recently been investigated by ØSTERGAARD[27]. As shown previously, selectivity is not affected by any limitations due to mass transfer when the process corresponds to two concurrent first-order reactions:

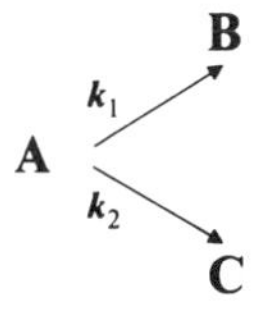

However, with heat transfer between the interior and exterior of the pellet (made possible by temperature gradients resulting from an exothermic diffusion limited reaction) the selectivity may be substantially altered. For the flat plate model the material and heat balance equations to solve are:

$$\frac{d^2 C_A}{dx^2} - \left(\frac{k_1 + k_2}{D_e}\right) C_A = 0 \tag{3.52}$$

$$\frac{d^2 C_B}{dx^2} + \frac{k_1}{D_e} C_A = 0 \tag{3.53}$$

and:

$$\frac{d^2 T}{dx^2} + \frac{(-\Delta H_1)k_1 + (-\Delta H_2)k_2}{k_e} C_A = 0 \tag{3.54}$$

where:

$$k_i = \mathscr{A}_i \exp(-\mathbf{E}_i/\mathbf{R}T), \quad i = 1,2 \tag{3.55}$$

and ΔH_1 and ΔH_2 correspond to the respective enthalpy changes. The boundary conditions are:

$$\frac{dC_A}{dx} = \frac{dC_B}{dx} = \frac{dT}{dx} = 0 \quad \text{at} \quad x = 0 \tag{3.56}$$

$$C_A = C_{A\infty}, \quad C_B = C_{B\infty}, \quad T = T_\infty \quad \text{at} \quad x = L \tag{3.57}$$

This is a two point boundary value problem and, because of the non-linearity of the equations, cannot be solved analytically. If, however, $\mathbf{E}_1 = \mathbf{E}_2$ the selectivity is

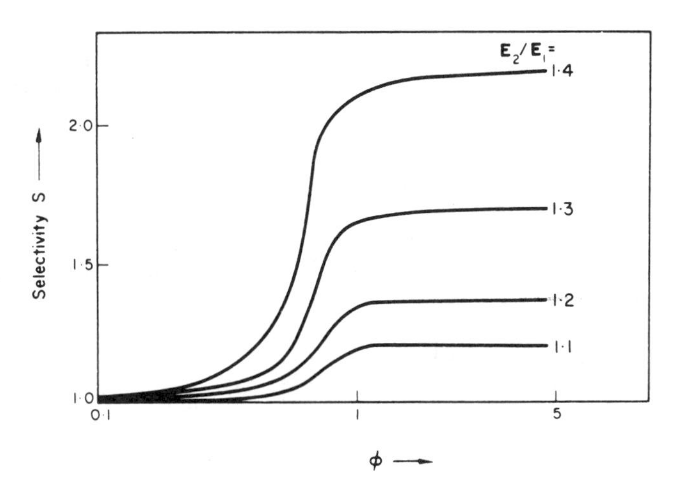

FIG. 3.9. Selectivity as a function of the Thiele modulus for non-isothermal conditions

the same as if there were no resistance to either heat or mass transfer (see Example 3.4). For the case where $\Delta H_1 = \Delta H_2$ but $\mathbf{E}_1 \neq \mathbf{E}_2$, the selectivity is determined by the effects of simultaneous heat and mass transfer. Figure 3.9 shows that if the activation energy of the desired reaction is lower than that of the reaction leading to the wasteful product ($\mathbf{E}_2/\mathbf{E}_1 > 1$) the best selectivity is obtained for high values of the Thiele modulus. When $\mathbf{E}_2/\mathbf{E}_1 < 1$ a decrease in selectivity results. In the former case the selectivity approaches an upper limit asymptotically, and it is not worth while increasing the Thiele modulus beyond a value where there would be a significant decrease in the efficiency of conversion.

Example 3.4

Show that the selectivity of two concurrent first-order reactions occurring in flat-shaped porous catalyst pellets is independent of the effect of either heat or mass transfer if the activation energies of both reactions are equal.

Solution

For the concurrent first-order reactions:

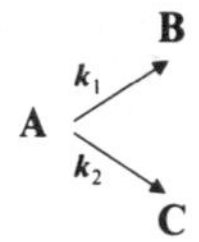

the mass and heat transfer equations are:

$$\frac{d^2 C_A}{dx^2} - \frac{k_1 + k_2}{D_e} C_A = 0$$

$$\frac{d^2 C_B}{dx^2} + \frac{k_1}{D_e} C_A = 0$$

$$\frac{d^2 T}{dx^2} + \frac{(-\Delta H_1)k_1 + (-\Delta H_2)k_2}{k_e} C_A = 0$$

where $k_i = \mathcal{A}_i \exp(-\mathbf{E}_i/\mathbf{R}T)$, $i = 1, 2$.
The boundary conditions are given by:

$$\frac{dC_A}{dx} = \frac{dC_B}{dx} = \frac{dT}{dx} = 0 \quad \text{at} \quad x = 0$$

$$C_A = C_{A\infty}, \quad C_B = C_{B\infty}, \quad T = T_\infty \quad \text{at} \quad x = L$$

When $\mathbf{E}_1 = \mathbf{E}_2$ we see that $k_1/k_2 = \mathcal{A}_1/\mathcal{A}_2$ — a ratio independent of temperature. Hence, if the mass-transfer equations are divided they may, for the case $\mathbf{E}_1 = \mathbf{E}_2$, be integrated directly and the gradients evaluated at the slab surface, $x = L$. Thus:

$$\left(\frac{dC_A}{dx}\right)_{x=L} = \frac{\left(1 + \dfrac{\mathcal{A}_2}{\mathcal{A}_1}\right)}{\left(1 + \dfrac{(-\Delta H_2)\mathcal{A}_2}{(-\Delta H_1)\mathcal{A}_1}\right)} \frac{k_e}{D_e(-\Delta H_1)} \left(\frac{dT}{dx}\right)_{x=L}$$

$(dC_B/dx)_{x=L}$ may be found similarly. The selectivity is determined by the ratio of the reaction rates at the surface. In the steady state, this is equal to the ratio of the fluxes of **C** and **B** at the slab surface. Hence we obtain for the selectivity:

$$\frac{\left(\dfrac{dC_C}{dx}\right)_{x=L}}{\left(\dfrac{dC_B}{dx}\right)_{x=L}} = \frac{\left(\dfrac{dC_A}{dx}\right)_{x=L}}{\left(\dfrac{dC_B}{dx}\right)_{x=L}} - 1 = 1 + \frac{\mathcal{A}_2}{\mathcal{A}_1} - 1 = \frac{\mathcal{A}_2}{\mathcal{A}_1}$$

and this is the same result as would have been obtained if there were no resistance to either mass or heat transfer within the pellets.

Selectivity of Bifunctional Catalysts

Certain heterogeneous catalytic processes require the presence of more than one catalyst to achieve a significant yield of desired product. The conversion of *n*-heptane to *i*-heptane, for example, requires the presence of a dehydrogenation catalyst, such as platinum, together with an isomerisation catalyst such as silica-alumina. In this particular case the *n*-heptane would be dehydrogenated by the platinum catalyst and the product isomerised to *i*-heptene which, in turn, would be hydrogenated to give finally *i*-heptane. When the hydrogenation and the isomerisation are carried out simultaneously the catalyst is said to act as a bifunctional catalyst. The recent trend is towards the production of catalyst in which both functions are built into the individual catalyst particles. This has the great advantage that it is not necessary to ensure that the component particles are mixed together well.

Many organic reactions involving the upgrading of petroleum feedstocks are enhanced if a bifunctional catalyst is used. Some of the reactions that take place may be typified by one or more of the following:

1: $$\mathbf{A} \xrightarrow{X} \mathbf{B} \xrightarrow{Y} \mathbf{C}$$

2: $$\mathbf{A} \overset{X}{\rightleftharpoons} \mathbf{B} \overset{Y}{\rightleftharpoons} \mathbf{C}$$

3: $$\mathbf{A} \overset{X}{\rightleftharpoons} \mathbf{B} \begin{cases} \xrightarrow{Y} \mathbf{C} \\ \xrightarrow{X} \mathbf{D} \end{cases}$$

in which **A** represents the initial reactant, **B** and **D** unwanted products and **C** the desired product. **X** and **Y** represent hydrogenation and isomerisation catalysts respectively. These reaction schemes implicitly assume the participation of hydrogen and pseudo first-order reaction kinetics. To a first approximation, the assumption of first-order chemical kinetics is not unrealistic, for SINFELT[28] has shown that under some conditions many reactions involving upgrading of petroleum may be represented by first-order kinetics. GUNN and THOMAS[29] examined mass transfer effects accompanying such reactions occurring isothermally in spherical catalyst particles containing the catalyst components **X** and **Y**. They demonstrated that it is possible to choose the volume fraction of **X** in such a way that the formation of **C** may be maximised. Curves 1, 2 and 3 in Fig. 3.10 indicate that a tubular reactor may be packed with discrete spherical particles of **X** and **Y** in such a way that the throughput of **C** is maximised. For given values of the kinetic constants of each step, the effective diffusivity, and the particle size, the amount of **C** formed is maximised by choosing the ratio of **X** to **Y** correctly. Curve 1 corresponds to the irreversible reaction 1 and requires more of component **X** than either reactions 2 or 3 for the same value of chosen parameters. On the other hand, the mere fact that reaction 2 has a reversible step means that more of **Y** is required to produce the maximum throughput of **C**. When a second end-product **D** results, as in reaction 3, even more of catalyst component **Y** is needed if the formation of **D** is catalysed by **X**.

Even better yields of **C** result if components **X** and **Y** are incorporated in the same catalyst particle, rather than if they exist as separate particles. In effect, the intermediate product **B** no longer has to be desorbed from particles of the **X** type catalyst, transported through the gas phase and thence readsorbed on **Y** type particles prior to reaction. Resistance to intraparticle mass transfer is therefore reduced or eliminated by bringing **X** type catalyst sites into close proximity to **Y** type catalyst sites. Curve 4 in Fig. 3.10 illustrates this point and shows that for such a composite catalyst, containing both **X** and **Y** in the same particle, the yield of **C** for reaction 3 is higher than it would have been had discrete particles of **X** and **Y** been used (curve 3).

The yields of **C** found for reactions 1, 2 and 3 were evaluated with the restriction that the catalyst composition should be uniform along the reactor length. However, there seems to be no point in **Y** being present at the reactor inlet where there is no **B** to convert. Similarly, there is little point in much **X** being present at the reactor outlet where there may be only small amounts of **A** remaining unconverted to **B**. THOMAS and WOOD[30] examined this question for reaction schemes represented by reactions 1 and 2. For an irreversible consecutive reaction, such as 1, they showed analytically that the optimal profile consists of two catalyst zones, one containing pure **X** and the other containing pure **Y**. When one of the steps is reversible, as in 2, numerical optimisation techniques showed that a catalyst composition changing along the reactor is superior to either a constant catalyst composition or a two-zone reactor. However, it was also shown that, in most cases, a bifunctional catalyst of constant composition will give a yield which is close to the optimum, thus obviating the practical difficulty of packing a reactor with a catalyst having a variable composition along the reactor length. The exception noted was for reactions in which there may be a possibility of the desired product undergoing further unwanted side reactions. Under such circumstances

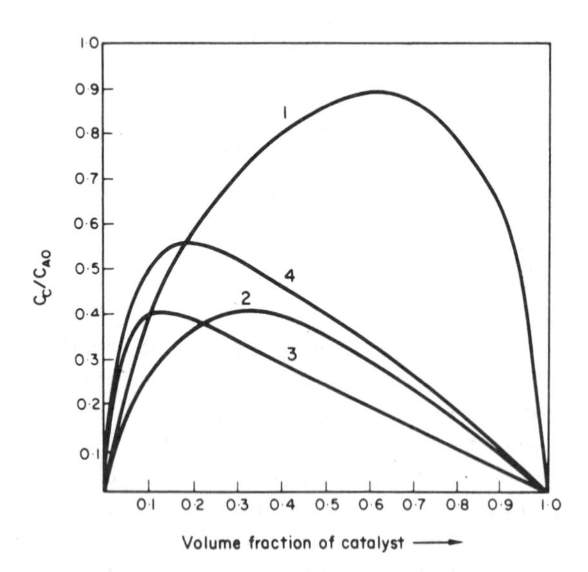

FIG. 3.10. Optimum yields of desired product in bifunctional catalyst systems

the holding time should be restricted and a two-zone reactor, containing zones of pure **X** and pure **Y** is to be preferred. Similar results were obtained independently by GUNN[31] who calculated, using a numerical method, the maximum yield of desired product for discrete numbers of reactor stages containing different proportions of the two catalyst components. JACKSON[32] has produced an analytical solution to the problem for the case of reaction 2. JENKINS and THOMAS[33] have demonstrated that, for the reforming of methylcyclopentane at 500°C (773 K) and 1 bar pressure, theoretical predictions of optimum catalyst compositions based on a kinetic model are close to the results obtained employing an experimental hill-climbing procedure.

3.3.6. Catalyst De-activation and Poisoning

Catalysts can become de-activated in a number of ways during the course of the operation. A common cause of de-activation is the sintering of the particles to form a continuous matte as a result of the development of local hot spots within the body of the reactor. Another is the deposition of carbon on the surface of the particles, as in the case of the fluidised cracking of hydrocarbons in the oil industry, a process described in Volume 2, Chapter 6. In this case the spent catalyst is transported to a second fluidised bed in which the carbon coating is burned off and the regenerated catalyst is then recycled.

Catalysts may also become poisoned when the feed stream to a reactor contains impurities which are deleterious to the activity of the catalyst. Particularly strong poisons are substances whose molecular structure contains lone electron pairs capable of forming covalent bonds with catalyst surfaces. Examples are ammonia, phosphine, arsine, hydrogen sulphide, sulphur dioxide and carbon monoxide. Other poisons include hydrogen, oxygen, halogens and mercury. The surface of a catalyst becomes poisoned by virtue of the foreign impurity adsorbed within the porous structure of the catalyst and covering a fraction of its surface, thus reducing the overall activity. The reactants participating in the desired reaction must now be transported to the unpoisoned part of the surface before any further reaction ensues, and so poisoning increases the average distance over which the reactants must diffuse prior to reaction at the surface. We may distinguish between two types of poisoning:

(i) homogeneous poisoning, in which the impurity is distributed evenly over the active surface;
(ii) selective poisoning in which an extremely active surface first becomes poisoned at the exterior surface of the particle, and then progressively becomes poisoned towards the centre of the particle.

When homogeneous poisoning occurs, since no reaction will be possible on the poisoned fraction (ζ, say, as shown in Fig. 3.11) of active surface it is reasonable to suppose that the intrinsic activity of the catalyst is in proportion to the fraction of active surface remaining unpoisoned. To find the ratio of activity of the poisoned catalyst to the activity of an unpoisoned catalyst one would compare the stationary flux of reactant to the particle surface in each case. For a first-order reaction

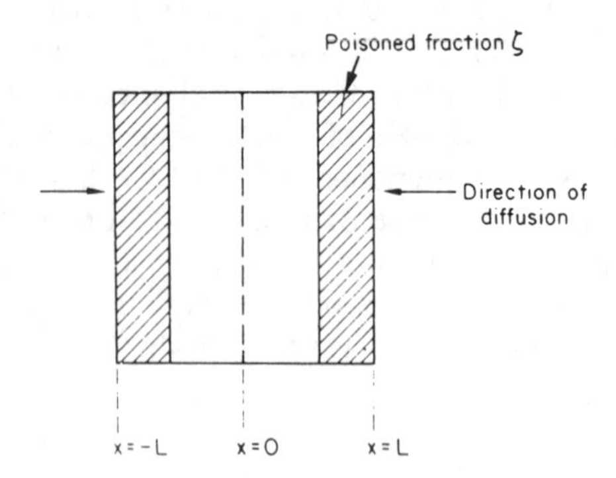

FIG. 3.11. Geometry of partially poisoned (selective) slab

occurring in a flat plate (slab) of catalyst one finds (see Example 3.5) this ratio to be:

$$F = \frac{\sqrt{1-\zeta}\tanh\{\phi\sqrt{1-\zeta}\}}{\tanh\phi} \tag{3.58}$$

where ϕ is the Thiele modulus for a first-order reaction occurring in a flat plate of catalyst and ζ is the fraction of active surface poisoned. The two limiting cases of equation 3.58 correspond to extreme values of ϕ. Where ϕ is small F becomes equal to $(1 - \zeta)$ and the activity decreases linearly with the amount of poison added, as shown by curve 1 in Fig. 3.12. On the other hand, when ϕ is large $F = \sqrt{1-\zeta}$ and the activity decreases less rapidly than linearly due to the reactants penetrating the interior of the particle to a lesser extent for large values of the Thiele modulus (curve 2, Fig. 3.12).

Example 3.5

A fraction ζ of the active surface of some porous slab-shaped catalyst pellets becomes poisoned. The pellets are used to catalyse a first-order isothermal chemical reaction. Find an expression for the ratio of the activity of the poisoned catalyst to the original activity of the unpoisoned catalyst when (a) homogeneous poisoning occurs, (b) selective poisoning occurs.

Solution

(a) If homogeneous poisoning occurs the activity decreases in proportion to the fraction $(1 - \zeta)$ of surface remaining unpoisoned. In the steady state the rate of reaction is equal to the flux of reactant to the surface. The ratio of activity F of the poisoned slab to the unpoisoned slab will be equal to the ratio of the reactant fluxes under the respective conditions. Hence:

$$F = \left(\frac{\mathrm{d}C_A}{\mathrm{d}x}\right)'_{x=L} \Bigg/ \left(\frac{\mathrm{d}C_A}{\mathrm{d}x}\right)_{x=L}$$

where the prime denotes conditions in the poisoned slab.

Now from equation 3.13 the concentration of reactant is a function of the distance the reactant has penetrated the slab. Thus:

$$C_{AL} = C_{A\infty}\frac{\cosh(\lambda x)}{\cosh(\lambda L)} \quad \text{where} \quad \lambda = \sqrt{\frac{k}{D_e}}.$$

If the slab were poisoned the activity would be $k(1-\zeta)$ rather than k and then:

$$C_{AL} = C_{A\infty}\frac{\cosh(\lambda' x)}{\cosh(\lambda' L)} \quad \text{where} \quad \lambda' = \sqrt{\frac{k(1-\zeta)}{D_e}}$$

Evaluating the respective fluxes at $x = L$:

$$F = \frac{\sqrt{1-\zeta}\,\tanh\left\{\phi\sqrt{(1-\zeta)}\right\}}{\tanh\phi} \quad \text{where} \quad \phi = L\sqrt{\frac{k}{D_e}}$$

(b) When selective poisoning occurs the exterior surface of the porous pellet becomes poisoned initially and the reactants must then be transported to the unaffected interior of the catalyst before reaction may ensue. When the reaction rate in the unpoisoned portion is chemically controlled the activity merely falls off in proportion to the fraction of surface poisoned. However if the reaction is diffusion limited, in the steady state the flux of reactant past the boundary between poisoned and unpoisoned surfaces is equal to the chemical reaction rate (see Fig. 3.11). Thus:

$$\text{Flux of reactant at the boundary between poisoned and unpoisoned portion of slab} = D_e\frac{C_{A\infty} - C_{AL}}{\zeta L}$$

$$\text{Reaction rate in unpoisoned length } (1-\zeta)L = D_e\left(\frac{\mathrm{d}C_A}{\mathrm{d}x}\right)_{x=(1-\zeta)L}$$

The concentration profile in the unpoisoned length is, by analogy with equation 3.13:

$$C_A = C_{AL}\frac{\cosh(\lambda x)}{\cosh\{\lambda(1-\zeta)L\}}$$

therefore:

$$D_e\left(\frac{\mathrm{d}C_A}{\mathrm{d}x}\right)_{x=(1-\zeta)L} = \frac{D_e}{L}C_{AL}\phi\tanh\{\phi(1-\zeta)\}$$

where:

$$\phi = \lambda L = L\sqrt{\frac{k}{D_e}}.$$

In the steady state then:

$$\frac{D_e(C_{A\infty} - C_{AL})}{\zeta L} = \frac{D_e}{L}C_{AL}\phi\tanh\{\phi(1-\zeta)\}$$

Solving the above equation explicitly for C_{AL}:

$$C_{AL} = \frac{C_{A\infty}}{1 + \phi\zeta\tanh\{\phi(1-\zeta)\}}$$

Hence the rate of reaction in the partially poisoned slab is:

$$\frac{D_e(C_{A\infty} - C_{AL})}{L} = \frac{C_{A\infty}D_e}{L}\,\frac{\phi\tanh\{\phi(1-\zeta)\}}{1 + \phi\zeta\tanh\{\phi(1-\zeta)\}}$$

In an unpoisoned slab the reaction rate is $\left(\frac{C_{A\infty}D_e}{L}\right)\phi\tanh\phi$ and so:

$$F = \frac{\tanh\{\phi(1-\zeta)\}}{1 + \phi\zeta\tanh\{\phi(1-\zeta)\}}\,\frac{1}{\tanh\phi}$$

For an active catalyst, $\tanh\{\phi(1-\zeta)\} \to 1$ and then $F \to (1+\phi\zeta)^{-1}$.

Selective poisoning occurs with very active catalysts. Initially, the exterior surface is poisoned and then, as more poison is added, an increasing depth of the interior surface becomes poisoned and inaccessible to reactant. If the reaction rate in the unpoisoned portion of catalyst happens to be chemically controlled, the reaction rate will fall off directly in proportion to the fraction of surface poisoned and the activity decreases linearly with the amount of poison added (curve 1, Fig. 3.12). When the reaction is influenced by diffusion, in the steady state the flux of reactant past the boundary between the poisoned and unpoisoned parts of the surface will equal the reaction rate in the unpoisoned portion. For the slab model, as the foregoing example shows, the ratio of activity in a poisoned catalyst to that in an unpoisoned catalyst is

$$F = \frac{\tanh\{\phi(1-\zeta)\}}{1+\phi\zeta\tanh\{\phi(1-\zeta)\}}\,\frac{1}{\tanh\phi} \tag{3.59}$$

For large values of the Thiele modulus the fraction $\phi(1-\zeta)$ will usually be sufficiently large that $F = (1+\phi\zeta)^{-1}$. Curve 3 in Fig. 3.12 depicts selective poisoning of active catalysts near the particle exterior and is the function represented by equation 3.59. Curve 4 describes the effect of selective poisoning for large values of the Thiele modulus. For the latter case the activity decreases drastically, after only a small amount of poison has been added.

It is apparent from the above discussion that ζ is time dependent, for an increasing amount of poison is being added quite involuntarily as the impure feed continually flows to the catalyst contained in the reactor. The general problem of obtaining $\zeta(t)$ is complex but, in principle, can be treated by solving the unsteady state conservation equation for the poison. When the poison is strongly and rapidly adsorbed, the fraction of surface poisoned is dependent on the square root of the time for which the feed has been flowing. In general, if $\zeta(t)$ is known, a judgement can be made concerning the optimum time for which the catalyst may be used before its activity falls to a value which produces an uneconomical throughput or yield of product[34-37].

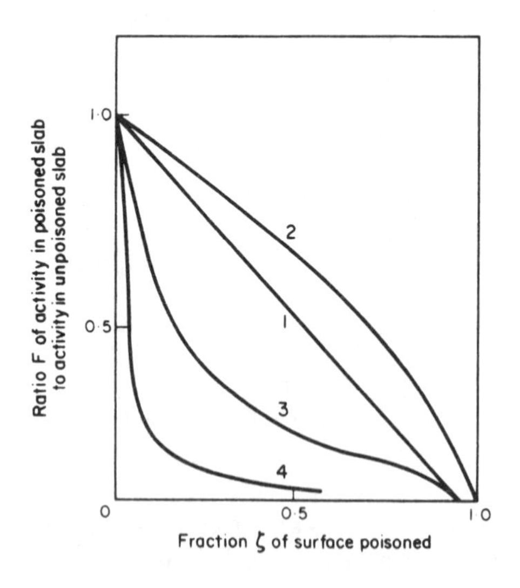

FIG. 3.12. Catalyst poisoning

3.4. MASS TRANSFER FROM A FLUID STREAM TO A SOLID SURFACE

Under some circumstances there will be a resistance to the transport of material from the bulk fluid stream to the exterior surface of a catalyst particle. When such a resistancc to mass transfer exists, the concentration C_A of a reactant in the bulk fluid will differ from its concentration C_{Ai} at the solid–gas interface. Because C_{Ai} is usually unknown it is necessary to eliminate it from the rate equation describing the external mass transfer process. Since, in the steady state, the rates of all of the steps in the process are equal, it is possible to obtain an overall rate expression in which C_{Ai} does not appear explicitly.

Based on such analyses, which of course do imply a film model in which the resistance to mass transfer is supposed to be confined to a film of finite thickness (see Volume 1, Chapter 10), it is possible to estimate the effect which mass transport external to the solid surface has on the overall reaction rate. For equimolar counterdiffusion of a component **A** in the gas phase, the rate of transfer of **A** from the bulk gas to the interface can be expressed as:

$$N_A = h_D(C_{AG} - C_{Ai}) \tag{3.60}$$

where h_D is the gas film mass transfer coefficient per unit external surface area, and C_{AG} and C_{Ai} are the molar concentration of the component **A** in the bulk gas and at the interface respectively.

The right-hand side of equation 3.60 contains the mass transfer coefficient h_D which is used if the driving force is expressed in terms of gas concentrations. Because of the stoichiometric demands imposed by chemical reaction, equimolar counterdiffusion of components may not necessarily occur and the effects of bulk flow must be taken into account (see Volume 1, Chapter 10).

The rate $\mathcal{R}_V'$ of chemical reaction per unit volume of particle for a first-order reaction is given by:

$$\mathcal{R}_V' = \boldsymbol{k} C_{Ai}\eta \tag{3.61}$$

where η is the efficiency factor, defined by equation 3.8, which allows for mass transfer effects in the pores of the catalyst articles. The rate is proportional to the concentration C_{Ai} of reactant at the interface. In the steady state, the flux of reactant will balance the rate of chemical reaction. From equations 3.60 and 3.61 we find the unknown concentration, since:

$$N_A a = \mathcal{R}_V' \tag{3.62}$$

where a is the interfacial area per unit volume.

Hence:
$$C_{Ai} = \frac{h_D a C_{AG}}{h_D a + \boldsymbol{k}\eta} \tag{3.63}$$

The rate at which the overall process of mass transfer and chemical reaction occurs may be found by substituting for C_{Ai} in equation 3.61 to give:

$$\mathcal{R}_V' = \left\{\frac{1}{1/\boldsymbol{k}\eta + 1/h_D a}\right\} C_{AG} \tag{3.64}$$

If, under the operating conditions, $h_D a >> k\eta$ the overall rate approaches $k\eta C_{AG}$. In this case the chemical reaction is said to be rate determining. If, on the other hand, $h_D a << k$ the rate approaches $h_D a C_{AG}$ and the transport process is rate determining.

The transfer coefficient can be correlated in the form of a dimensionless Sherwood number $Sh(= h_D d_p/D)$. The particle diameter d_p is often taken to be the diameter of the sphere having the same area as the (irregular shaped) pellet. THALLER and THODOS[38] correlated the mass transfer coefficient in terms of the gas velocity u and the Schmidt number $Sc(= \mu/\rho D)$:

$$j_d = \frac{Sh}{ReSc^{1/3}} = \frac{h_D}{u} Sc^{2/3} \tag{3.65}$$

where j_d is the mass transfer factor (see Volume 1, Chapter 12, and Volume 2, Chapter 4), and Re is the Reynolds number $u\rho d_p/\mu$ based on the particle diameter.

The mass transfer factor has also been correlated as a function of the Reynolds number only and thus taking account only of hydrodynamic conditions. If e is the voidage of the packed bed and the total volume occupied by all of the catalyst pellets is V_p, then the total reactor volume is $V_p/(1 - e)$. Hence the rate of mass transfer of component **A** per unit volume of reactor is $N_A S_x(1 - e)/V_p$. If we now consider a case in which only external mass transfer controls the overall reaction rate we have:

$$\mathcal{R}_V = \frac{N_A S_x(1 - e)}{V_p} = (1 - e)\frac{S_x}{V_p} h_D (C_A - C_{Ai}) \tag{3.66}$$

Alternatively, equation 3.66 may be written in terms of the j-factor. The unknown interface concentration C_{Ai} can now be eliminated in the usual way, by equating the rate of mass transfer to the rate of chemical reaction.

We see that, in principle, the overall reaction rate can be expressed in terms of coefficients such as the reaction rate constant and the mass transfer coefficient. To be of any use for design purposes, however, we must have knowledge of these parameters. By measuring the kinetic constant in the absence of mass transfer effects and using correlations to estimate the mass transfer coefficient we are really implying that these estimated parameters are independent of one another. This would only be true if each element of external surface behaved kinetically as all other surface elements. Such conditions are only fulfilled if the surface is uniformly accessible. It is fortuitous, however, that predictions of overall rates based on such assumptions are often within the accuracy of the kinetic information, and for this reason values of k and h_D obtained independently are frequently employed for substitution into overall rate expressions.

3.5. CHEMICAL KINETICS OF HETEROGENEOUS CATALYTIC REACTIONS

To complete the discussion of factors involved in the design of gas–solid heterogeneous catalytic reactors we will examine several aspects of the kinetics of chemical reactions occurring in the presence of a catalyst surface. We consider, for heuristic purposes, the equilibrium reaction:

$$\mathbf{A} + \mathbf{B} \underset{\mathbf{k}'}{\overset{\mathbf{k}}{\rightleftharpoons}} \mathbf{P}$$

occurring at an active catalyst surface. The net rate* of surface chemical reaction will be the difference between the rates of the forward and reverse reactions. If the forward rate is determined by the simultaneous presence of chemisorbed **A** and **B** (i.e. a Langmuir–Hinshelwood mechanism applies; see Volume 2, Chapter 17, Section 17.73) then it will be proportional to the number of pairs of adjacent catalyst sites occupied by **A** and **B**. This is given by the product of the number $\boldsymbol{n}_A$ of adsorbed reactant molecules and the ratio of occupied to unoccupied sites. This latter quantity is expressed as $\theta_B/(1 - \theta_A - \theta_B)$ where θ_A and θ_B represent the fractions of active surface covered by **A** and **B** respectively. The reverse rate, on the other hand, is proportional to the product of the number $\boldsymbol{n}_P$ of adsorbed product molecules and the fraction of active sites left vacant $(1 - \sum_j \theta_j)$ where $\sum_j \theta_j$ is the total fraction of sites occupied by any species. The net rate of the surface reaction per unit surface is thus:

$$\mathcal{R}_S'' = \mathbf{k}\boldsymbol{n}_A\left(\frac{\theta_B}{1 - \theta_A - \theta_B}\right) - \mathbf{k}'\boldsymbol{n}_P(1 - \sum_j \theta_j) \tag{3.67}$$

We may conveniently define $\boldsymbol{n}_A/S'$, where S' is the surface area of the catalyst, as the surface concentration C_A. The corresponding surface concentration, C_B, of **B** is defined as $\boldsymbol{n}_B/S'$ and the total concentration, C_S, of catalyst sites as $\boldsymbol{n}_S/S'$ where $\boldsymbol{n}_S$ is the number of available active sites. The fraction θ_A of surface covered by **A** will be $\boldsymbol{n}_A/\boldsymbol{n}_S$, which, in terms of surface concentrations becomes C_A/C_S. Similar substitutions may be made for any other species so the term $(1 - \sum_j \theta_j)$ representing the fraction of vacant sites can be written C_V/C_S where C_V is $(C_S - \sum_j C_j)$. Substituting these quantities into equation 3.67† gives:

$$\mathcal{R}_S'' = \mathbf{k}S'\frac{C_A C_B}{C_S} - \mathbf{k}'S'\frac{C_P C_V}{C_S} = \mathbf{k}_S\frac{C_A C_B}{C_S} - \mathbf{k}_S'\frac{C_P C_V}{C_S} \tag{3.68}$$

provided the surface coverage is sufficiently small that $1 >> (\theta_A + \theta_B)$, which condition is usually fulfilled during catalysis. If the forward rate of reaction is determined by gaseous or physically adsorbed **B** interacting with chemisorbed **A** (an Eley–Rideal mechanism) the expression corresponding to equation 3.68 would be:

$$\mathcal{R}_S'' = \mathbf{k}S'C_A a_B - \mathbf{k}'S'\frac{C_P C_V}{C_S} = \mathbf{k}_S C_A a_B - \mathbf{k}_S'\frac{C_P C_V}{C_S} \tag{3.69}$$

where a_B is the (thermodynamic) activity of gaseous or physically adsorbed **B**.

As well as the net rate of surface reaction we must also consider the net rate of each reactant at the surface. We may think of this as the difference between the rates of adsorption and desorption of both **A** and **B**. Since the rate of adsorption is usually expressed in terms of the product of the prevailing partial pressure of

*If the rate is measured as the number of molecules transformed per unit area per unit time, **k** and **k′** have units $\mathbf{L}^{-2}\mathbf{T}^{-1}$.

†The rate constants $\mathbf{k}_S$ and $\mathbf{k}_S'$ in equations 3.68 and 3.69 have been substituted in place of $\mathbf{k}S'$ and $\mathbf{k}'S'$ respectively and, since the rate of surface reaction $\mathcal{R}_S''$ has units $\mathbf{L}^{-2}\mathbf{T}^{-1}$, $\mathbf{k}_S$ and $\mathbf{k}_S'$ have units $\mathbf{T}^{-1}$.

adsorbate and the extent of free surface, while the rate of desorption is proportional to the fraction of surface covered, the net rate of adsorption of **A** in molecules per unit area will be:

$$\mathcal{R}''_{Aa} = \mathbf{k}_{Aa} P_A (1 - \sum_j \theta_j) - \mathbf{k}_{Ad} \theta_A \tag{3.70}$$

which in terms of surface concentrations becomes:

$$\mathcal{R}''_{Aa} = \mathbf{k}_{Aa} P_A \frac{C_V}{C_S} - \mathbf{k}_{Ad} \frac{C_A}{C_S} \tag{3.71}$$

A similar equation can be written for the reactant **B**.

Since **P** is the final product of reaction it is also necessary to consider the net rate of desorption. In an analogous way to the net rate of adsorption of **A**, the net rate of desorption of **P** may be written:

$$\mathcal{R}''_{Pd} = \mathbf{k}_{Pd} \frac{C_P}{C_S} - \mathbf{k}_{Pa} P_P \frac{C_V}{C_S} \tag{3.72}$$

It should be noted that the partial pressures in equations 3.71 and 3.72 are those corresponding to the values at the interface between solid and gas. If there were no resistance to transport through the gas phase then, as discussed in an earlier section, the partial pressures will correspond to those in the bulk gas.

The overall chemical rate may be written in terms of the partial pressures of **A**, **B** and **P** by equating the rates $\mathcal{R}''_S$, $\mathcal{R}''_{Aa}$ and $\mathcal{R}''_{Pd}$ and eliminating the surface concentrations C_A, C_B and C_P from equations 3.68 (or 3.69 as the case may be), 3.71 and 3.72. The final equation so obtained is cumbersome and unwieldy and contains several constants which, for practical reasons, cannot be determined independently. For this reason it is convenient to consider limiting cases in which one parameter, adsorption surface reaction or desorption is rate determining.

3.5.1. Adsorption of a Reactant as the Rate Determining Step

If the adsorption of **A** is the rate determining step in the sequence of adsorption, surface reaction and desorption processes, then equation 3.71 will be the appropriate equation to use for expressing the overall chemical rate. To be of use, however, it is first necessary to express C_A, C_V and C_S in terms of the partial pressures of reactants and products. To do this an approximation is made; it is assumed that all processes except the adsorption of **A** are at equilibrium. Thus the processes involving **B** and **P** are in a state of pseudo-equilibrium. The surface concentration of **B** can therefore be expressed in terms of an equilibrium constant K_B for the adsorption–desorption equilibrium of **B**:

$$K_B = \frac{C_B}{P_B C_V} \tag{3.73}$$

and similarly for P:

$$K_P = \frac{C_P}{P_P C_V} \tag{3.74}$$

The equation for the surface reaction between **A** and **B** may be written in terms of the equilibrium constant for the surface reaction:

$$K_S = \frac{C_P C_V}{C_A C_B} \tag{3.75}$$

Substituting equations 3.73, 3.74 and 3.75 into equation 3.71:

$$\mathcal{R}''_{Aa} = \frac{\mathbf{k}_{Aa} C_V}{C_S}\left\{P_A - \frac{K_P}{K_A K_B K_S}\frac{P_P}{P_B}\right\} \tag{3.76}$$

where K_A is the ratio $\mathbf{k}_{Aa}/\mathbf{k}_{Pd}$. The ratio C_V/C_S can be written in terms of partial pressures and equilibrium constants since the concentration of vacant sites is the difference between the total concentration of sites and the sum of the surface concentrations of **A, B** and **P**:

$$C_V = C_S - (C_A + C_B + C_P) \tag{3.77}$$

and so, from equations 3.73, 3.74 and 3.75:

$$\frac{C_V}{C_S} = \frac{1}{1 + K_B P_B + K_P P_P + \dfrac{K_P}{K_B K_S}\dfrac{P_P}{P_B}} \tag{3.78}$$

If we define the equilibrium constant for the adsorption of **A** as:

$$K_A = \frac{C_A}{P_A C_V} \tag{3.79}$$

and recognise that the thermodynamic equilibrium constant for the overall equilibrium is:

$$K = \frac{P_P}{P_A P_B} \tag{3.80}$$

equation 3.76 gives, for the rate of chemical reaction:

$$\mathcal{R}''_A = \frac{k_{Aa}\left\{P_A - \dfrac{1}{K}\dfrac{P_P}{P_B}\right\}}{1 + K_B P_B + K_P P_P + \dfrac{K_A}{K}\dfrac{P_P}{P_B}} \tag{3.81}$$

The rate $\mathcal{R}''_A$ expressed in units $\mathbf{L}^{-2}\mathbf{T}^{-1}$ must be transposed to units of moles per unit time per unit catalyst mass for reactor design purposes. This is accomplished by multiplying $\mathcal{R}''_A$ by $S_g/\mathbf{N}$, where **N** is Avogadro's number.

It should be noted that equation 3.81 contains a driving force term in the numerator. This is the driving force tending to drive the chemical reaction towards the equilibrium state. The collection of terms in the denominator is usually referred to as the adsorption term, since terms such as $K_B P_B$ represent the retarding effect of the adsorption of species **B** on the rate of disappearance of **A**. New experimental techniques enable the constants K_B etc. to be determined separately during the course of a chemical reaction[39] and hence, if it were found that the adsorption of

A controls the overall rate of conversion, equation 3.81 could be used directly as the rate equation for design purposes. If, however, external mass transfer were important the partial pressures in equation 3.81 would be values at the interface and an equation (such as equation 3.66) for each component would be required to express interfacial partial pressures in terms of bulk partial pressures. If internal diffusion were also important, the overall rate equation would be multiplied by an effectiveness factor either estimated experimentally, or alternatively obtained by theoretical considerations similar to those discussed earlier.

3.5.2. Surface Reaction as the Rate Determining Step

If, on the other hand, surface reaction determined the overall chemical rate, equation 3.68 (or 3.69 if an Eley–Rideal mechanism operates) would represent the rate. If it is assumed that a pseudo-equilibrium state is reached for each of the adsorption–desorption processes then, by a similar method to that already discussed for reactions where adsorption is rate determining, it can be shown that the rate of chemical reaction is (for a Langmuir–Hinshelwood mechanism):

$$\mathcal{R}''_{AS} = \frac{C_S \mathbf{k}_S K_A K_B \left\{ P_A P_B - \dfrac{1}{K} P_P \right\}}{(1 + K_A P_A + K_B P_B + K_P P_P)^2} \tag{3.82}$$

This equation also contains a driving force term and an adsorption term. A similar equation may be derived for the case of an Eley–Rideal mechanism and its form is interpreted in Table 3.3.

3.5.3. Desorption of a Product as the Rate Determining Step

In this case pseudo-equilibrium is assumed for the surface reaction and for the adsorption–desorption processes involving **A** and **B**. By similar methods to those employed in deriving equation 3.81, it can be shown that the rate of chemical reaction is:

$$\mathcal{R}''_{Pd} = \frac{\boldsymbol{k}_{Pa} K \left\{ P_A P_B - \dfrac{1}{K} P_P \right\}}{1 + K_A P_A + K_B P_B + K K_P P_A P_B} \tag{3.83}$$

when the desorption of **P** controls the rate. This equation, it should be noted, also contains a driving force term and an adsorption term.

3.5.4. Rate Determining Steps for Other Mechanisms

In principle it is possible to write down the rate equation for any rate determining chemical step assuming any particular mechanism. To take a specific example, the overall rate may be controlled by the adsorption of **A** and the reaction may involve the dissociative adsorption of **A**, only half of which then reacts with adsorbed **B** by a Langmuir–Hinshelwood mechanism. The basic rate equation which represents such a process can be transposed into an equivalent expression in terms of partial pressures and equilibrium constants by methods similar to those employed to obtain the rate equations 3.81, 3.82 and 3.83. Table 3.3 contains a number of selected mechanisms for each of which the basic rate equation,, the driving force term and the adsorption term are given.

TABLE 3.3. *Structure of Reactor Design Equations*

Reaction		Mechanism	Driving force	Adsorption term
1. **A** ⇌ **P**	(i)	Adsorption of **A** controls rate	$P_A - \frac{P_P}{K}$	$1 + \frac{K_A}{K} P_P + K_P P_P$
(Equilibrium constant K is dimensionless)	(ii)	Surface reaction controls rate, single site mechanism	$P_A - \frac{P_P}{K}$	$1 + K_A P_A + K_P P_P$
	(iii)	Surface reaction controls rate, adsorbed **A** reacts with adjacent vacant site	$P_A - \frac{P_P}{K}$	$(1 + K_A P_A + K_P P_P)^2$
	(iv)	Desorption of **P** controls rate	$P_A - \frac{P_P}{K}$	$1 + K K_P P_A + K_A P_A$
	(v)	**A** dissociates when adsorbed and adsorption controls rate	$P_A - \frac{P_P}{K}$	$\left\{1 + \left(\frac{K_A}{K} - P_P\right)^{1/2} + K_P P_P\right\}^2$
	(vi)	**A** dissociates when adsorbed and surface reaction controls rate	$P_A - \frac{P_P}{K}$	$\{1 + (K_A P_A)^{1/2} + K_P P_P\}^2$
2. **A** + **B** ⇌ **P**		Langmuir–Hinshelwood mechanism (adsorbed **A** reacts with adsorbed **B**)		
(Equilibrium constant K has dimensions $\mathbf{M^{-1}LT^2}$)	(i)	Adsorption of **A** controls rate	$P_A - \frac{P_P}{K P_B}$	$1 + \frac{K_A P_P}{K P_B} + K_B P_B + K_P P_P$
	(ii)	Surface reaction controls rate	$P_A P_B - \frac{P_P}{K}$	$\{1 + K_A P_A + K_B P_B + K_P P_P\}^2$
	(iii)	Desorption of **P** controls rate	$P_A P_B - \frac{P_P}{K}$	$1 + K_A P_A + K_B P_B + K K_P P_A P_B$

Footnote:

1. The expression for the rate of reaction in terms of partial pressures is proportional to the driving force divided by the adsorption term.
2. To derive the corresponding kinetic expressions for a bimolecular–unimolecular reversible reaction proceeding via an Eley–Rideal mechanism (adsorbed **A** reacts with gaseous or physically adsorbed **B**), the term $K_B P_B$ should be omitted from the adsorption term. When the surface reaction controls the rate the adsorption term is not squared and the term $K_B K_B$ is omitted.
3. To derive the kinetic expression for irreversible reactions simply omit the second term in the driving force.
4. If two products are formed (**P** and **Q**) the equations are modified by (a) multiplying the second term of the driving force by P_Q, and (b) adding $K_Q P_Q$ within the bracket of the adsorption term.
5. If **A** or **B** dissociates during the process of chemisorption, both the driving force and the adsorption term should be modified (see for example cases 1(v) and 1(vi) above). For a full discussion of such situations see HOUGEN and WATSON[40].

A useful empirical approach to the design of heterogeneous chemical reactors often consists of selecting a suitable equation, such as one in Table 3.3 which, with numerical values substituted for the kinetic and equilibrium constants, represents the chemical reaction in the absence of mass transfer effects. Graphical methods are often employed to aid the selection of an appropriate equation[41] and the constants determined by a least squares approach[40]. It is important to stress, however, that while the equation selected may well represent the experimental data, it does not necessarily represent the true mechanism of reaction; as is evident from Table 3.3, many of the equations are similar in form. Nevertheless the equation will adequately represent the behaviour of the reaction for the conditions investigated and can be used for design purposes. A fuller and more detailed account of a wide selection of mechanisms has been given by HOUGEN and WATSON(40).

3.5.5. Examples of Rate Equations for Industrially Important Reactions

For most gas–solid catalytic reactions, usually a rate equation corresponding to one form or another of the Hougen and Watson type described above can be found to fit the experimental data by a suitable choice of the constants that appear in the adsorption and driving force terms. The following examples have been chosen to illustrate this type of rate equation. However, there are some industrially important reactions for which rate equations of other forms have been found to be more appropriate, of particular importance being ammonia synthesis and sulphur dioxide oxidation(42).

(a) The hydration of ethylene to give ethanol was studied by MACE and BONILLA(43) using a catalyst of tungsten trioxide WO_3 on silica gel at a total pressure of 136 bar:

$$\underset{\mathbf{A}}{C_2H_4} + \underset{\mathbf{B}}{H_2O} \rightleftharpoons \underset{\mathbf{M}}{C_2H_5OH}$$

They concluded that the applicable rate equation was:

$$\mathcal{R} = \frac{\boldsymbol{k} K_A K_B (P_A P_B - P_M / K_P)}{(1 + K_A P_A + K_B P_B)^2} \tag{3.84}$$

Equation 3.84 was consistent with a mechanism whereby the reaction between ethylene and water adsorbed on the surface was rate-controlling but without strong adsorption of the product ethanol. Commercially at the present time, phosphoric acid on kieselguhr is the preferred catalyst.

(b) In the process outlined in Chapter 1 (Example 1.1), for the dehydrogenation of ethylbenzene to produce styrene in the presence of steam, the reaction

$$\underset{\mathbf{E}}{C_6H_5 \cdot CH_2 \cdot CH_3} = \underset{\mathbf{S}}{C_6H_5 \cdot CH{:}CH_3} + \underset{\mathbf{H}}{H_2}$$

is catalysed by iron oxide. Potassium carbonate also is present in the catalyst to promote the reaction of carbon with steam and so to keep the catalyst free from carbon deposits. Present day catalysts contain in addition other promoters such as molybdenum and cerium oxides Mo_2O_3 and Ce_2O_3(44). Kinetic studies using an older type of catalyst resulted in the rate equation(45):

$$\mathcal{R} = \boldsymbol{k}\left[P_E - \frac{P_S P_H}{K_P}\right] \tag{3.85}$$

This equation can be interpreted as indicating a surface reaction being rate-controlling (although the controlling step could be adsorption), but all the species involved being only weakly adsorbed; i.e. the adsorption terms $K_E P_E$, $K_S P_S$, $K_H P_H$, which would otherwise appear in the denominator, are all much less than unity.

Rate equations for the promoted catalysts currently used have the form(44):

$$\mathcal{R} = \frac{kK_E\left(P_E - \dfrac{P_H P_S}{K_P}\right)}{(1 + K_E P_E + K_S P_S)} \tag{3.86}$$

This indicates that, for these catalysts, the adsorption terms $K_E P_E$ and $K_S P_S$ are significant compared with unity.

3.6. DESIGN CALCULATIONS

The problems which a chemical engineer has to solve when contemplating the design of a chemical reactor packed with a catalyst or reacting solid are, in principle, similar to those encountered during the design of an empty reactor, except that the presence of the solid somewhat complicates the material and heat balance equations. The situation is further exacerbated by the designer having to predict and avoid those conditions which might lead to instability within the reactor.

We shall consider, in turn, the various problems which have to be faced when designing isothermal, adiabatic and other non-isothermal tubular reactors, and we shall also briefly discuss fluidised bed reactors. Problems of instability arise when inappropriate operating conditions are chosen and when reactors are started up. A detailed discussion of this latter topic is outside the scope of this chapter but, since reactor instability is undesirable, we shall briefly inspect the problems involved.

3.6.1. Packed Tubular Reactors

In a substantial majority of the cases where packed tubular reactors are employed the flow conditions can be regarded as those of plug flow. However, dispersion (already discussed in Chapter 2 in relation to homogeneous reactors) may result in lower conversions than those obtained under truly plug flow conditions.

In the first instance, the effects of dispersion will be disregarded. Then the effects of dispersion in the axial (longitudinal) direction only will be taken into account and this discussion will then be followed by considerations of the combined contributions of axial and radial (transverse) dispersion.

Behaviour of Reactors in the Absence of Dispersion

Isothermal conditions. The design equation for the isothermal fixed bed tubular reactor with no dispersion effects represents the simplest form of reactor to analyse. No net exchange of mass or energy occurs in the radial direction, so transverse dispersion effects can also be neglected. If we also suppose that the ratio of the tube length to particle size is large (> *ca.* 50) then we can safely ignore longitudinal dispersion effects compared with the effect of bulk flow. Hence, in writing the conservation equation over an element δz of the length of the reactor (Fig. 3.13), we may consider that the fluid velocity u is independent of radial position; this implies a flat velocity profile (plug flow conditions) and ignores dispersion effects in the direction of flow.

Suppose that a mass of catalyst, whose bulk density is ρ_b, is contained within an elementary length δz of a reactor of uniform cross-section A_c. The mass of catalyst

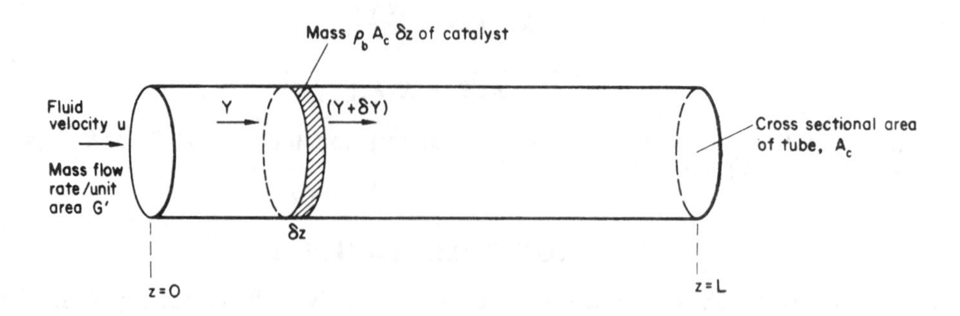

FIG. 3.13. One-dimensional tubular reactor

occupying the elementary volume will therefore by $\rho_b A_c \delta z$. Let G' be the steady state mass flowrate per unit area to the reactor. If the number of moles of product per unit mass of fluid entering the elementary section is Y and the amount emerging is $(Y + \delta Y)$, the net difference in mass flow per unit area across the element will be $G'\delta Y$. In the steady state this will be balanced by the amount of product formed by chemical reaction within the element so:

$$G'\delta Y = \mathcal{R}_Y^M \rho_b \delta z \tag{3.87}$$

where the isothermal reaction rate $\mathcal{R}_Y^M$ is expressed in units of moles per unit time per unit mass of catalyst*. To calculate the length of reactor required to achieve a conversion corresponding to an exit concentration Y_L, equation 3.87 is integrated from Y_0 (say) to Y_L, where the subscripts 0 and L refer to inlet and exit conditions respectively. Thus:

$$L = \frac{G'}{\rho_b} \int_{Y_0}^{Y_L} \frac{\mathrm{d}Y}{\mathcal{R}_Y^M} \tag{3.88}$$

If the reaction rate is a function of total pressure as well as concentration, and there is a pressure drop along the reactor due to the solid packing, the conversion within the reactor will be affected by the drop in total pressure along the tube. In most cases it is perfectly reasonable to neglect the change in kinetic energy due to the expansion of the fluid in comparison with the pressure force and the component of the drag force in the direction of flow created by the surface of the solid particles. The pressure drop $-\mathrm{d}P$ along an elementary length $\mathrm{d}z$ of the packed bed may thus be written in terms of R_1, the component of the drag force per unit surface area of particles in the direction of flow:

$$-\frac{\mathrm{d}P}{\mathrm{d}z} = \frac{R_1}{d_m'} = \left(\frac{R_1}{\rho u_1^2}\right)\frac{\rho u_1^2}{d_m'} \tag{3.89}$$

* If intraparticle diffusion effects are important and an effectiveness factor η is employed (as in equation 3.8) to correct the chemical kinetics observed in the absence of transport effects, then it is necessary to adopt a stepwise procedure for solution. First the pellet equations (such as 3.10) are solved in order to calculate η for the entrance to the reactor and then the reactor equation (3.87) may be solved in finite difference form, thus providing a new value of Y at the next increment along the reactor. The whole procedure may then be repeated at successive increments along the reactor.

in which u_1 is the mean velocity in the voids ($= u/e$ where u is the average velocity as measured over the whole cross-sectional area of the bed and e is the bed voidage) and d'_m (= volume of voids/total surface area) is one quarter of the hydraulic mean diameter. ERGUN[46] has correlated the friction factor ($R_1/\rho u_1^2$) in terms of the modified Reynolds number (see equation 4.21 in Volume 2, Chapter 4). In principle, therefore, a relation between P and z may be obtained by integration of equation 3.89 and it is therefore possible to allow for the effect of total pressure on the reaction rate by substituting $P = f(z)$ in the expression for the reaction rate. Equation 3.88 may then be integrated directly.

When $\mathcal{R}_Y$ is not a known function of Y an experimental programme to determine the chemical reaction rate is necessary. If the experimental reactor is operated in such a way that the conversion is sufficiently small—small enough that the flow of product can be considered to be an insignificant fraction of the total mass flow—the reactor is said to be operating differentially. Provided the small change in composition can be detected quantitatively, the reaction rate may be directly determined as a function of the mole fraction of reactants. On this basis a differential reactor provides only initial rate data. It is therefore important to carry out experiments in which products are added at the inlet, thereby determining the effect of any retardation by products. An investigation of this kind over a sufficiently wide range of conditions will yield the functional form of $\mathcal{R}_Y$ and, by substitution in equation 3.88, the reactor size and catalyst mass can be estimated. If it is not possible to operate the reactor differentially, conditions are chosen so that relatively high conversions are obtained and the reactor is now said to be an integral reactor. By operating the reactor in this way Y may be found as a function of W/G' and $\mathcal{R}_Y$ determined, for any given Y, by evaluating the slope of the curve at various points.

If axial dispersion were an important effect, the reactor performance would tend to fall below that of a plug flow reactor. In this event an extra term must be added to equation 3.87 to account for the net dispersion occurring within the element of bed. This subject is considered later.

Example 3.6

Carbon disulphide is produced in an isothermally operated packed tubular reactor at a pressure of 1 bar by the catalytic reaction between methane and sulphur.

$$CH_4 + 2S_2 = CS_2 + 2H_2S$$

At 600°C (=873 K) the reaction is chemically controlled, the rate being given by:

$$\mathcal{R}_Y^M = \frac{8 \times 10^{-2} \exp(-115{,}000/\mathbf{R}T) P_{CH_4} P_{S_2}}{1 + 0.6P_{S_2} + 2.0P_{CS_2} + 1.7P_{H_2S}} \text{ kmol/s(kg catalyst)}$$

where the partial pressures are expressed in bar, the temperature in K, and $\mathbf{R}$ as kJ/kmol K.

Estimate the mass of catalyst required to produce 1 tonne per day of CS_2 at a conversion level of 90 per cent.

Solution

Assuming plug flow conditions within the reactor, the following equation applies:

$$L = \frac{G'}{\rho_b} \int_0^{Y_L} \frac{dY}{\mathcal{R}_Y^M}$$

where Y is the concentration of CS_2 in moles per unit mass, G' is the mass flow per unit area and ρ_b the catalyst bulk density. In terms of catalyst mass this becomes:

$$W = G\int_0^{Y_L}\frac{\mathrm{d}Y}{\mathscr{R}_Y^M}$$

where G is now the molar flowrate through the tube.

The rate expression is given in terms of partial pressures and this is now rewritten in terms of the number of moles χ of methane converted to CS_2. Consider 1 mole of gas entering the reactor, then at any cross-section (distance z along the tube from the inlet) the number of moles of reactant and product may be written in terms of χ as follows:

	Moles at inlet	Moles at position z
CH_4	1/3	$1/3 - \chi$
S_2	2/3	$2/3 - 2\chi$
CS_2	—	χ
H_2S	—	2χ
Total	1.00	1.00

For a pressure of 1 bar it follows that the partial pressures in the rate expression become:

$$P_{CH_4} = (1/3 - \chi); \quad P_{S_2} = (2/3 - 2\chi); \quad P_{CS_2} = \chi \quad \text{and} \quad P_{H_2S} = 2\chi$$

Now if M is the mean molecular weight of the gas at the reactor inlet then χ kmol of CS_2 are produced for every M kg of gas entering the reactor. Hence:

$$Y = \frac{\chi}{M}$$

and the required catalyst mass is now:

$$W = \frac{G}{M}\int_0^{\chi_L}\frac{\mathrm{d}\chi}{\mathscr{R}_Y^M}$$

For an expected conversion level of 0.9, the upper limit of the integral will be given by $\dfrac{\chi_L}{\frac{1}{3} - \chi_L} = 0.9$, i.e. by $\chi_L = 0.158$. Thus 1 tonne per day (= 1.48 kmol/s CS_2) is produced by using an inlet flow of $G = \dfrac{1.48 \times 10^{-4} \times M}{0.158}$ kg/s.

Hence:

$$W = \frac{533}{0.158}\int_0^{0.158}\frac{\mathrm{d}\chi}{\mathscr{R}_\chi^M}$$

Substituting the expressions for partial pressures into the rate equation the mass of catalyst required may be determined by direct numerical (or graphical) integration. Thus:

$$\underline{\underline{W = 23 \text{ kg}}}$$

Adiabatic conditions. Adiabatic reactors are more frequently encountered in practice than isothermal reactors. Because there is no exchange of heat with the surroundings, radial temperature gradients are absent. All of the heat generated or absorbed by the chemical reaction manifests itself by a change in enthalpy of the fluid stream. It is therefore necessary to write a heat balance equation for the reactor in addition to the material balance equation 3.87. Generally heat transfer between solid and fluid is sufficiently rapid for it to be justifiable to assume that all the heat generated or absorbed at any point in the reactor is transmitted instantaneously to or from the solid. It is therefore only necessary to take a heat balance for the fluid entering and emerging from an elementary section δz. Referring to Fig. 3.13 and neglecting the effect of longitudinal heat conduction:

$$G'\bar{c}_p\delta T = \rho_b(-\Delta H)\mathscr{R}_{YT}^M\,\delta z \tag{3.90}$$

where $\bar{c}_p$ is the mean heat capacity of the fluid,

ΔH is the enthalpy change on reaction, and

$\mathcal{R}^M_{YT}$ is the reaction rate (moles per unit time and unit mass of catalyst)—now a function of Y and T.

Simultaneous solution of the mass balance equation 3.87 and the heat balance equation 3.90 with the appropriate boundary conditions gives z as a function of Y.

A simplified procedure is to assume $-\Delta H/\bar{c}_p$ as constant. If equation 3.90 (the heat balance equation) is divided by equation 3.87 (the mass balance equation) and integrated, we immediately obtain:

$$T = T_0 + \frac{(-\Delta H)\,Y}{\bar{c}_p} \tag{3.91}$$

where T_0 is the inlet temperature. This relation implies that the adiabatic reaction path is linear. If equation 3.91 is substituted into the mass balance equation 3.87:

$$\frac{\mathrm{d}Y}{\mathrm{d}z} = \frac{\rho_b}{G'}\,\mathcal{R}^M_{YT_0} \tag{3.92}$$

where the reaction rate $\mathcal{R}^M_{YT_0}$ along the adiabatic reaction path is now expressed as a function of Y only. Integration then gives z as a function of Y directly.

Because the adiabatic reaction path is linear a graphical solution, also applicable to multi-bed reactors, is particularly apposite. (See Example 3.7 as an illustration.) If the design data are available in the form of rate data $\mathcal{R}^M_{YT}$ for various temperatures and conversions they may be displayed as contours of equal reaction rate in the (T, Y) plane. Figure 3.14 shows such contours upon which is superimposed an adiabatic reaction path of slope $\bar{c}_p/(-\Delta H)$ and intercept T_0 on the abscissa. The reactor size may be evaluated by computing:

$$\frac{\rho_b L}{G'} = \int_{Y_0}^{Y_L} \frac{\mathrm{d}Y}{\mathcal{R}^M_{YT_0}} \tag{3.93}$$

from a plot of $1/\mathcal{R}^M_{YT_0}$ as a function of Y. The various values of $\mathcal{R}^M_{YT_0}$ correspond to those points at which the adiabatic reaction path intersects the contours.

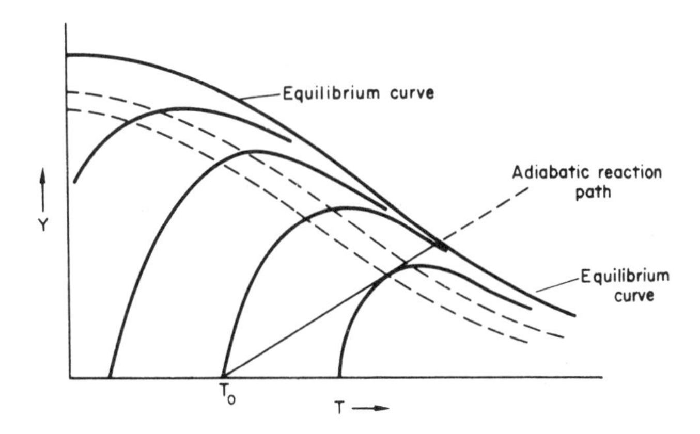

FIG. 3.14. Graphical solution for an adiabatic reactor

It is often necessary to employ more than one adiabatic reactor to achieve a desired conversion. In the first place chemical equilibrium may have been established in the first reactor and it is then necessary to cool and/or remove the product before entering the second reactor. This, of course, is one good reason for choosing a catalyst which will function at the lowest possible temperature. Secondly, for an exothermic reaction, the temperature may rise to a point at which it is deleterious to the catalyst activity. At this point the products from the first reactor are cooled prior to entering a second adiabatic reactor. To design such a system it is only necessary to superimpose on the rate contours the adiabatic temperature paths for each of the reactors. The volume requirements for each reactor can then be computed from the rate contours in the same way as for a single reactor. It is necessary, however, to consider carefully how many reactors in series it is economic to operate.

Should we wish to minimise the size of the system it would be important to ensure that, for all conversions along the reactor length, the rate is at its maximum[47, 48]. Since the rate is a function of conversion and temperature, setting the partial differential $\partial \mathcal{R}_{YT}/\partial T$ equal to zero will yield, for an exothermic reaction, a relation $T_{MX}(Y_{MX})$ which is the locus of temperatures at which the reaction rate is a maximum for a given conversion. The locus T_{MX} of these points passes through the maxima of curves of Y as a function of T shown in Fig. 3.15 as contours of constant rate. Thus, to operate a series of adiabatic reactors along an optimum temperature path, hence minimising the reactor size, the feed is heated to some point A (Fig. 3.15) and the reaction allowed to continue along an adiabatic reaction path until a point such as B, in the vicinity of the optimum temperature curve, is reached. The products are then cooled to C before entering a second adiabatic reactor in which reaction proceeds to an extent indicated by point D, again in the vicinity of the curve T_{MX}. The greater the number of adiabatic reactors in the series the closer the

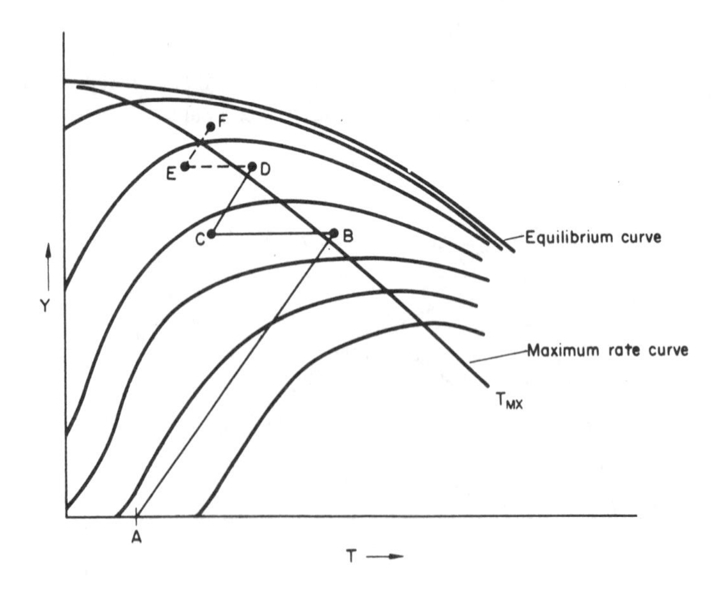

FIG. 3.15. Optimum design of a two-stage and three-stage adiabatic reactor

optimum path is followed. For a given number of reactors, three say, there will be six design decisions to be made corresponding to the points A to F inclusive. These six decisions may be made in such a way as to minimise the capital and running costs of the system of reactors and heat exchangers. However, such an optimisation problem is outside the scope of this chapter and the interested reader is referred to ARIS[17]. It should be pointed out, nevertheless, that the high cost of installing and operating heat transfer and control equipment so as to maintain the optimum temperature profile militates against its use. If the reaction is not highly exothermic an optimal isothermal reactor system may be a more economic proposition and its size may not be much larger than the adiabatic system of reactors. Each case has to be examined on its own merits and compared with other alternatives.

A single-stage adiabatic reactor is illustrated in Fig 3.16 which shows the graded spherical packing used as a catalyst support at the bottom of the bed, and as a

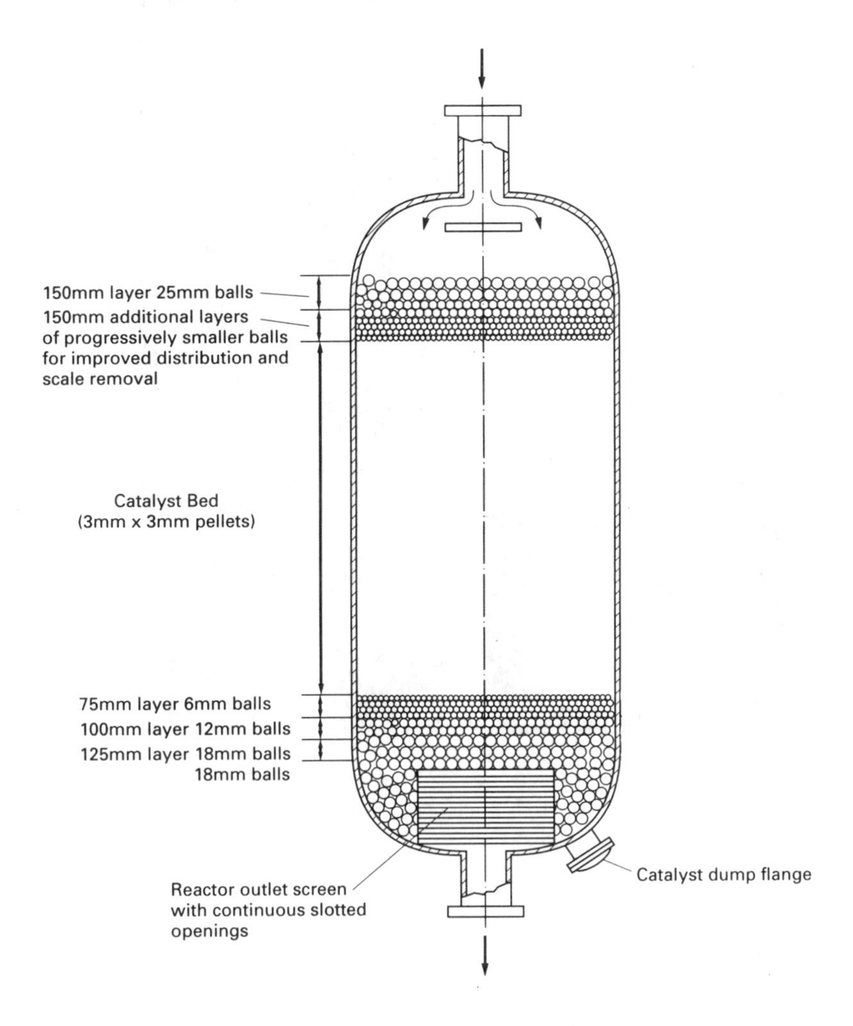

FIG. 3.16. Single-stage adiabatic reactor

mechanical stabiliser for the bed at the top. For conditions where the gas flowrate is high and where the pressure drop allowable across the bed is low, much higher diameter:height ratios will be used.

Figure 3.17 shows a radial-flow reactor with the inlet gas flowing downwards at the walls of the vessel and then flowing radially inwards through the bed before passing through a distributor into the central outlet. Uniform flow along the length of the reactor is ensured by the use of a distributor with appropriate resistance to the flow of gas.

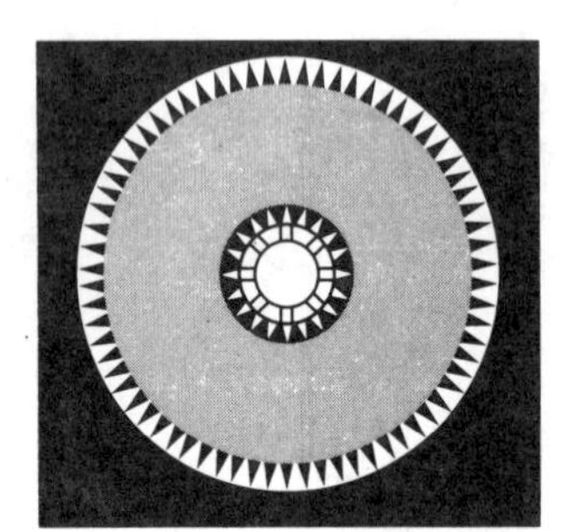

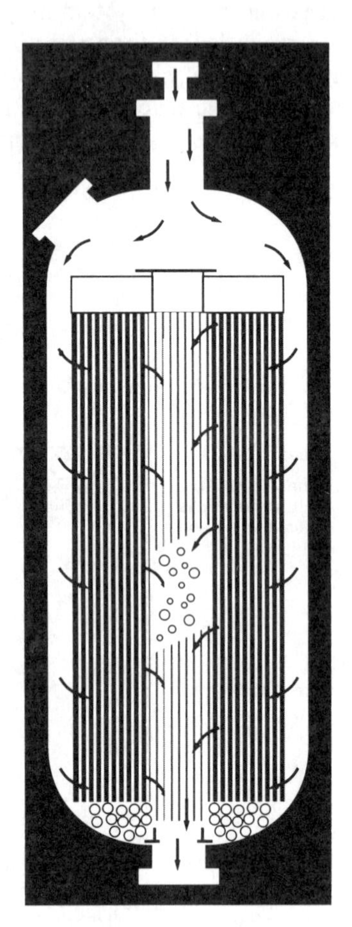

FIG. 3.17. Radial-flow reactor

Example 3.7

SO_3 is produced by the catalytic oxidation of SO_2 in two packed adiabatic tubular reactors arranged in series with intercooling between stages. The molar composition of the mixture entering the first reactor is 7 per cent SO_2, 11 per cent O_2 and 82 per cent N_2. The inlet temperature of the reactor is controlled at 688 K and the inlet flow is 0.17 kmol/s. Calculate the mass of catalyst required for each stage so that 78 per cent of the SO_2 is converted in the first stage and a further 20 per cent in the second stage.

Thermodynamic data for the system are as follows:

Component	Mean specific heat in range 415°C to 600°C (688 to 873 K) (kJ/kmol K)
SO_2	51.0
SO_3	75.5
O_2	33.0
N_2	30.5

For the reaction:

$$SO_2 + \tfrac{1}{2}O_2 \rightleftharpoons SO_3$$

the standard enthalpy change at 415°C (= 688 K) is $\Delta H^0_{600} = -97{,}500$ kJ/kmol

Solution

The chemical reaction which ensues is:

$$SO_2 + \tfrac{1}{2}O_2 \rightleftharpoons SO_3$$

and if 0.17×0.07 kmol of SO_2 are converted to χ kmol of SO_3 a material balance for the first reactor may be drawn up:

Component	kmol at inlet	kmol at outlet
SO_2	0.17×0.07	$0.17 \times 0.07 - \chi$
O_2	0.17×0.11	$0.17 \times 0.11 - \chi/2$
SO_3	—	χ
N_2	0.17×0.82	0.17×0.82

From the last column of the above table and the specific heat data, the difference in sensible heat between products and reactants is:

$$H_p - H_R = (T - 688)\{[(0.17 \times 0.07) - \chi] \times 51.0 + [(0.17 \times 0.11) - \chi/2] \times 33.0 + [75.5\chi] + [0.17 \times 0.82 \times 30.5]\}$$

where T (K) is the temperature at the exit from the first reactor. The conversion x may be written:

$$x = \frac{\chi}{0.17 \times 0.07}$$

and on substitution and simplification:

$$H_P - H_R = (T - 688)(5.476 + 0.095x)$$

A heat balance (in kJ) over the first reactor gives:

$$H_P - H_R = \Delta H^0_{688} = 97{,}500\chi = 1160x$$

where $\Delta H^0_{688}(= -97{,}500$ kJ/kmol) is the standard enthalpy of reaction at 688K.

From the two expressions obtained:

$$T = 688 + \frac{1160x}{5.476 + 0.095x} \approx 688 + 212x$$

This provides a linear relation between temperature and conversion in the first reactor.

Figure 3.18 is a conversion chart for the reactant mixture and shows rate curves in the conversion–temperature plane. The line AA′ is plotted on the chart from the linear T—x relation. The rate of reaction at any conversion level within the reactor may therefore be obtained by reading the rate corresponding to the intersection of the line with various conversion ordinates.

Now for the first reactor:

$$W_1 = 0.17 \times 0.07 \int_0^{0.78} \frac{dx}{\mathcal{R}^M}$$

and this is easily solved by graphical integration as depicted in Fig. 3.19, by plotting $1/\mathcal{R}^M$ versus conversion x and finding the area underneath the curve corresponding to the first reactor. The area corresponds to 0.336×10^6 kg s (kmol)$^{-1}$ and $W_1 = 4000$ kg = 4 tonne.

A similar calculation is now undertaken for the second reactor but, in this particular case, a useful approximation may be made. Because the reactant mixture is highly diluted with N_2, the same heat

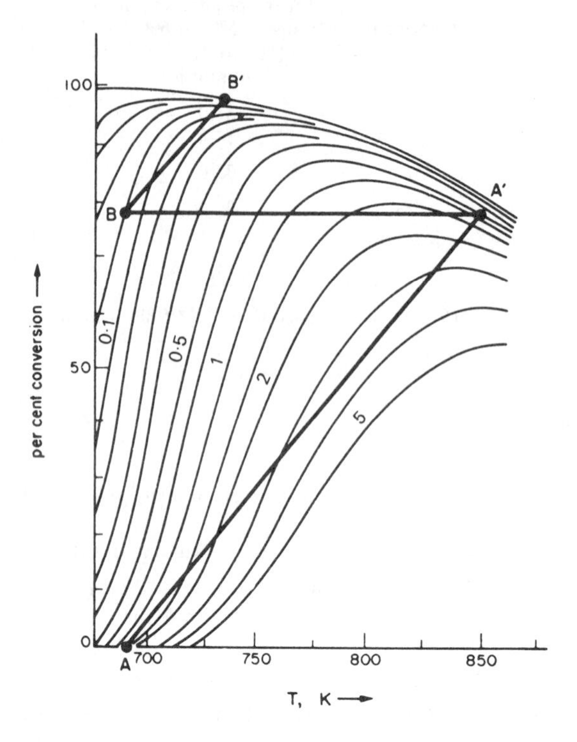

FIG. 3.18. Graphical solution for design of SO_2 reactor. Figures on curves represent the reaction rate in units of ($\text{kmol s}^{-1}\ \text{kg}^{-1} \times 10^6$)

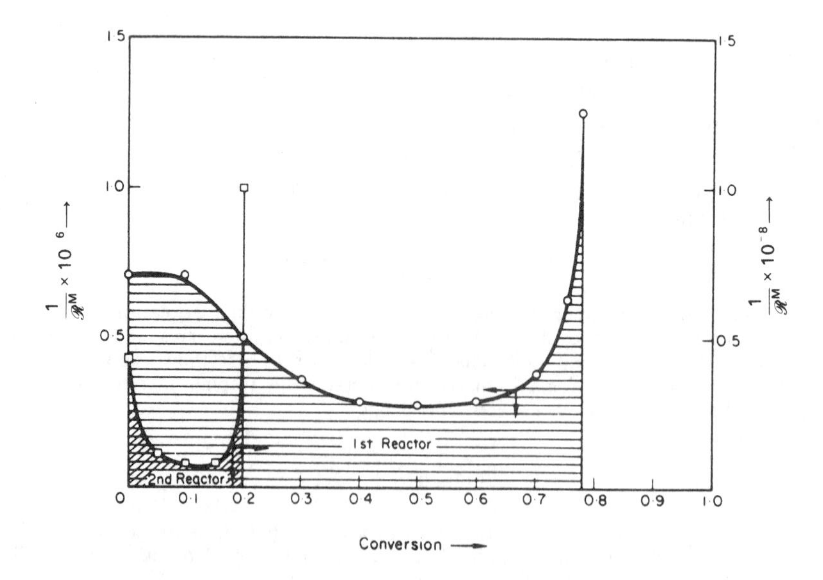

FIG. 3.19. Graphical integration to determine catalyst masses

conservation equation may be used for the second reactor as for the first reactor. The line BB′ starting at the point (688, 0.78) and parallel to the first line AA′ may therefore be drawn. For the second reactor:

$$W_2 = 0.17 \times 0.07 \int_0^{0.2} \frac{\mathrm{d}x}{\mathscr{R}^M}$$

and this is also solved by the same procedure as beforc. $1/\mathscr{R}^M$ is plotted as a function of x by noting the rate corresponding to the intersection of the line BB′ with the selected conversion ordinate. The area under the curve $1/\mathscr{R}^M$ versus x for the second reactor is 4.9×10^6 kg s/kmol.
Hence $W_2 = 6 \times 10^4$ kg = 60 tonne.

It will be noted that the catalyst requirement in the second reactor is very much greater because, being near to equilibrium, the reaction rate is between one and two orders of magnitude lower.

Non-isothermal and non-adiabatic conditions. A useful approach to the preliminary design of a non-isothermal fixed bed reactor is to assume that all the resistance to heat transfer is in a thin layer near the tube wall. This is a fair approximation because radial temperature profiles in packed beds are parabolic with most of the resistance to heat transfer near the tube wall. With this assumption a one-dimensional model, which becomes quite accurate for small diameter tubes, is satisfactory for the approximate design of reactors. Neglecting diffusion and conduction in the direction of flow, the mass and energy balances for a single component of the reacting mixture are:

$$G' \frac{\mathrm{d}Y}{\mathrm{d}z} = \rho_b \mathscr{R}_{YT}^M \qquad \text{(equation 3.87)}$$

$$G' \bar{c}_p \frac{\mathrm{d}T}{\mathrm{d}z} = \rho_b(-\Delta H)\, \mathscr{R}_{YT}^M - \frac{h}{a'}(T - T_W) \qquad (3.94)$$

where h is the heat transfer coefficient (dimensions $\mathbf{MT^{-3}\theta^{-1}}$) expressing the resistance to the transfer of heat between the reactor wall and the reactor contents,
a' is the surface area for heat transfer per unit hydraulic radius (equal to the area of cross-section divided by the perimeter), and
T_W is the wall temperature.

Inspection of the above equations shows that, if the wall temperature is constant, for a given inlet temperature, a maximum temperature is attained somewhere along the reactor length if the reaction is exothermic. It is desirable that this should not exceed the temperature at which the catalyst activity declines. In Fig. 3.20 the curve ABC shows a non-isothermal reaction path for an inlet temperature T_0 corresponding to A. Provided $T_0 > T_W$, it is obvious that $\frac{\mathrm{d}T}{\mathrm{d}Y} < \frac{-\Delta H}{\bar{c}_p}$ and the rate of temperature increase will be less than in the adiabatic case. The point B, in fact, corresponds to the temperature at which the reaction rate is at a maximum and the locus of such points is the curve T_{MX} described previously. The maximum temperature attained from any given inlet temperature may be calculated by solving, using an iterative method[49], the pair of simultaneous equations 3.87 and 3.94 and finding the temperature at which $\frac{\mathrm{d}\mathscr{R}_{YT}^M}{\mathrm{d}T}$ or, equivalently, $\frac{\mathrm{d}Y}{\mathrm{d}T} = 0$. We will see later that a packed tubular reactor is very sensitive to change in wall temperature. It is therefore important to estimate the maximum attainable temperature for a given inlet

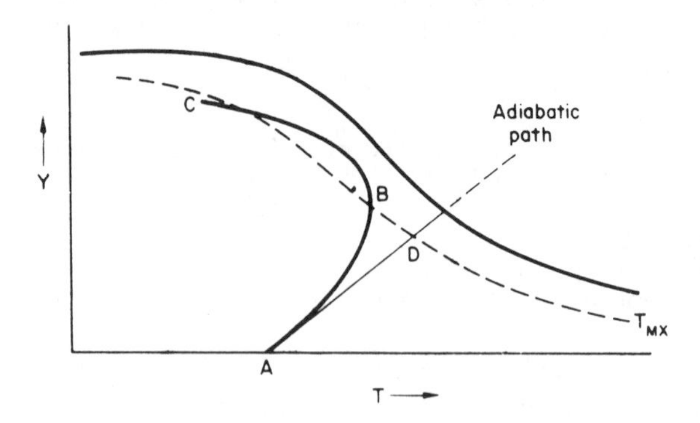

FIG. 3.20. Reaction path of a cooled tubular reactor

temperature, from the point of view of maintaining both catalyst activity and reactor stability.

An important class of reactors is that for which the wall temperature is not constant but varies along the reactor length. Such is the case when the cooling tubes and reactor tubes form an integral part of a composite heat exchanger. Figures 3.21*a* and 3.21*b* show, respectively, cocurrent and countercurrent flow of coolant and reactant mixture, the coolant fluid being entirely independent and separate from the reactants and products. However, the reactant feed itself may be used as coolant prior to entering the reactor tubes and again may flow cocurrent or countercurrent to the reactant mixture (Fig. 3.21*c* and 3.21*d* respectively). In each case heat is exchanged between the reactant mixture and the cooling fluid. A heat balance for a component of the reactant mixture leads to:

$$G'\bar{c}_p \frac{\mathrm{d}T}{\mathrm{d}z} = \rho_b(-\Delta H)\,\mathcal{R}^M_{YT} - \frac{U}{a'}(T - T_C) \qquad (3.95)$$

an equation analogous to 3.94 but in which T_C is a function of z and U is an overall heat transfer coefficient for the transfer of heat between the fluid streams. The variation in T_C may be described by taking a heat balance for an infinitesimal section of the cooling tube:

$$G'_C\,\bar{c}_{pC}\frac{\mathrm{d}T_C}{\mathrm{d}z} \pm \frac{U}{a'}(T - T_C) = 0 \qquad (3.96)$$

where G'_C is the mass flowrate of coolant per unit cross-section of the reactor tube, and
$\bar{c}_{pC}$ is the mean heat capacity of the coolant.

If flow is cocurrent the lower sign is used; if countercurrent the upper sign is used. Since the mass flowrate of the cooling fluid is based upon the cross-sectional area of the reactor tube the ratio $G'\bar{c}_p/G'_C\,\bar{c}_{pC}(=\Gamma)$ is a measure of the capacities of the two streams to exchange heat. In terms of the limitations imposed by the one-dimensional model, the system is fully described by equations 3.95 and 3.96 together with the mass balance equation:

$$G'\frac{\mathrm{d}Y}{\mathrm{d}z} = \rho_b\mathcal{R}^M_{YT} \qquad \text{(equation 3.87)}$$

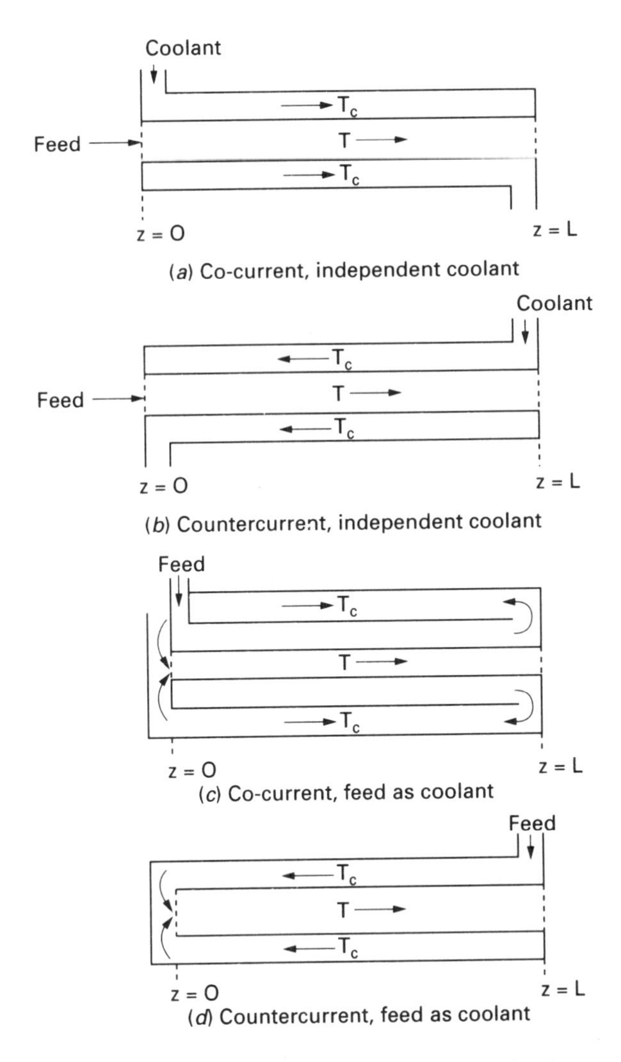

FIG. 3.21. Cocurrent and countercurrent cooled tubular reactors

The boundary conditions will depend on whether the flow is cocurrent or countercurrent, and whether or not the reactant mixture is used in part of the cooling process.

The reaction path in the *T*, *Y* plane could be plotted by solving the above set of equations with the appropriate boundary conditions. A reaction path similar to the curve ABC in Fig. 3.20 would be obtained. The size of reactor necessary to achieve a specified conversion could be assessed by tabulating points at which the reaction path crosses the constant rate contours, hence giving values of $\mathcal{R}_{YT}$ which could be used to integrate the mass balance equation 3.87. The reaction path would be suitable provided the maximum temperature attained was not deleterious to the catalyst activity.

In the case of a reactor cooled by incoming feed flowing countercurrent to the reaction mixture (Fig. 3.21*d*), and typified by the ammonia converter sketched in

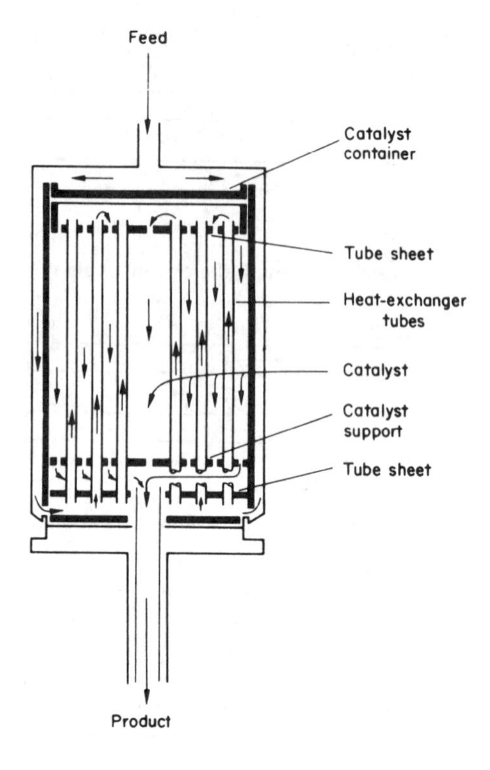

FIG. 3.22. Ammonia converter

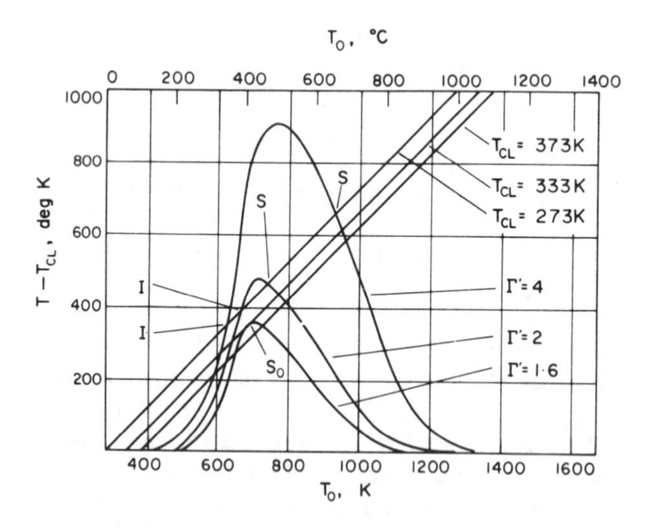

FIG. 3.23. Heat transfer characteristics of countercurrent cooled tubular reactor

Fig. 3.22, the boundary conditions complicate the integration of the equations. At the point $z = L$ where the feed enters the cooling tubes, $T_C = T_{CL}$, whereas at the reactor entrance ($z = 0$) the feed temperature is, of course, by definition the reactor inlet temperature, so here $T = T_0 = T_C$ and $Y = Y_0$. VAN HEERDEN[50] solved this two-point boundary value problem by assuming various values for T_0 and then integrating the equations to find the reactor exit temperature T_L and the incoming feed temperature T_{CL} at $z = L$. The quantity $(T - T_{CL})$ is a measure of the heat transferred between reactor and cooling tubes so a plot of $(T - T_{CL})$ versus T_0 will describe the heat exchanging capacity of the system for various reactor inlet conditions. Such curves are displayed in Fig. 3.23 with $\Gamma'(= UaL/G_C' \bar{c}_{pC})$ as a parameter. Superimposed on the diagram are parallel straight lines of unit slope, each line corresponding to the value of T_{CL}, the boundary condition at $z = L$. The solution to the three simultaneous equations and boundary conditions will be given by the intersections of the line T_{CL} with the heat transfer curves. In general an unstable solution I and a stable solution S are obtained. That S corresponds to a stable condition may be seen by the fact that if the system were disturbed from the semi-stable condition at I by a sudden small increase in T_0, more heat would be transferred between reactor and cooling tubes, since $(T - T_{CL})$ would now be larger and the heat generated by reaction would be sufficiently great to cause the inlet temperature to rise further until the point S was reached, whereupon the heat exchanging capacity would again match the heat generated by reaction. The region in which it is best to operate is in the vicinity of S_0, the point at which I and S are coincident. As the catalyst activity declines the curves of $(T - T_{CL})$ versus T_0 are displaced downwards and so the system will immediately become quenched. For continuing operation, therefore, more heat exchange capacity must be added to the system. This may be achieved by decreasing the mass flowrate. The temperature level along the reactor length therefore increases and the system now operates along a reaction path (such as ABC in Fig 3.20), displaced to regions of higher temperature but lower conversion. Production is thus maintained but at the cost of reduced conversion concomitant with the decaying catalyst activity.

In industrial practice, a multiple-bed reactor (Fig. 3.24) is normally used for the synthesis of ammonia, rather than the single-stage reactor illustrated in Fig. 3.22. Because the reaction takes place at high pressures, the whole series of reactions is contained within a single pressure vessel, the diameter of which is minimised for reasons of mechanical design.

Dispersion in Packed Bed Reactors

The importance of dispersion and its influence on flow pattern and conversion in homogeneous reactors has already been studied in Chapter 2. The role of dispersion, both axial and radial, in packed bed reactors will now be considered. A general account of the nature of dispersion in packed beds, together with details of experimental results and their correlation, has already been given in Volume 2, Chapter 4. Those features which have a significant effect on the behaviour of packed bed reactors will now be summarised. The equation for the material balance in a reactor will then be obtained for the case where plug flow conditions are modified by the effects of axial dispersion. Following this, the effect of simultaneous axial and radial dispersion on the non-isothermal operation of a packed bed reactor will be discussed.

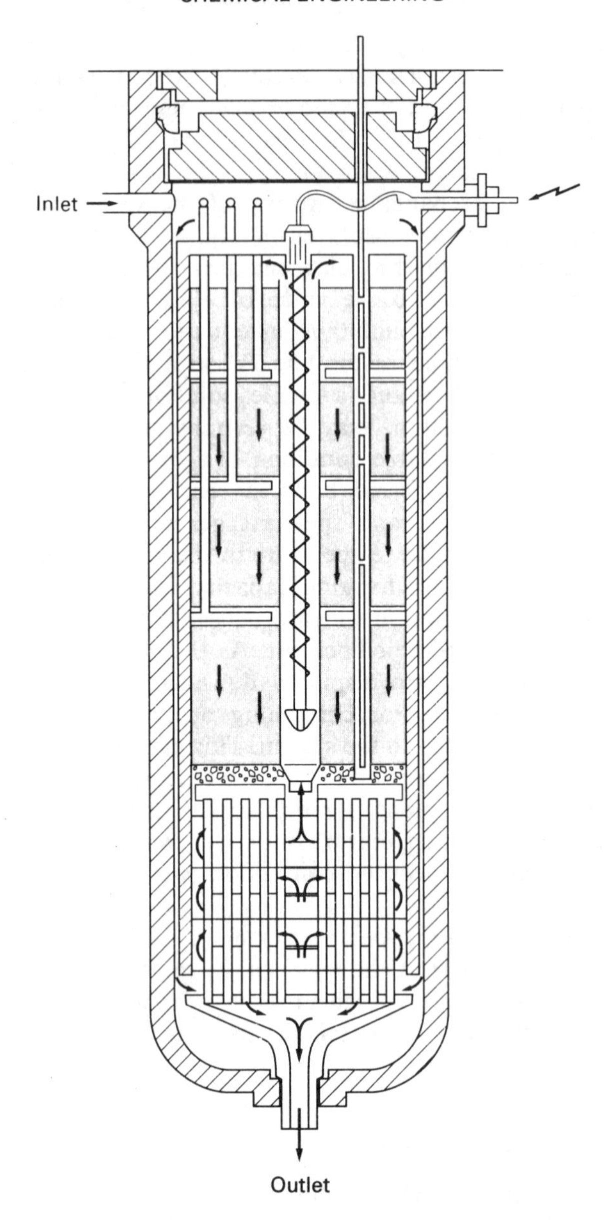

FIG. 3.24. Multiple-bed reactor for ammonia synthesis

The nature of dispersion. The effect which the solid packing has on the flow pattern within a tubular reactor can sometimes be of sufficient magnitude to cause significant departures from plug flow conditions. The presence of solid particles in a tube causes elements of flowing gas to become displaced randomly and therefore produces a mixing effect. An eddy diffusion coefficient can be ascribed to this mixing effect and becomes superimposed on the transport processes which normally occur in unpacked tubes—either a molecular diffusion process at fairly low Reynolds numbers, or eddy motion due to turbulence at high Reynolds numbers. Both transverse and longitudinal components of the flux attributed to this dispersion effect are of importance but operate in opposite ways. Transverse dispersion tends to bring the performance of the reactor closer to that which would be predicted by the simple design equation based upon plug flow. On the other hand, longitudinal dispersion is inclined to invalidate the plug flow assumptions so that the conversion would be less than would be expected if plug flow conditions obtained. The reason for this is that transverse mixing of the fluid elements helps to smooth out the parabolic velocity profile which normally develops in laminar flow in an unpacked tube, whereas longitudinal dispersion in the direction of flow causes some fluid elements to spend less time in the reactor than they would if this additional component of flux due to eddy motion were not superimposed.

The magnitude of the dispersion effect due to transverse or radial mixing can be assessed by relying on theoretical predictions[51] and experimental observations[52] which confirm that the value of the Peclet number $Pe(= ud_p/D$, where d_p is the particle diameter) for transverse dispersion in packed tubes is approximately 10. At bed Reynolds numbers of around 100 the diffusion coefficient to be ascribed to radial dispersion effects is about four times greater than the value for molecular diffusion. At higher Reynolds numbers the radial dispersion effect is correspondingly larger.

Longitudinal dispersion in packed reactors is thought to be caused by interstices between particles acting as mixing chambers. Theoretical analysis of a model[53] based on this assumption shows that the Peclet number Pe_l for longitudinal dispersion is about 2 and this has been confirmed by experiment[52, 54]. Thus the diffusion coefficient for longitudinal dispersion is approximately five times that for transverse dispersion for the same flow conditions. The flux which results from the longitudinal dispersion effect is, however, usually much smaller than the flux resulting from transverse dispersion because axial concentration gradients are very much less steep than radial concentration gradients if the ratio of the tube length to diameter is large. Whether or not longitudinal dispersion is important depends on the ratio of the reactor length to particle size. If the ratio is less than about 100 then the flux resulting from longitudinal dispersion should be considered in addition to the flux due to bulk flow when designing the reactor.

Axial dispersion. An axial (longitudinal) dispersion coefficient may be defined by analogy with BOUSSINESQ's concept of eddy viscosity[55]. Thus both molecular diffusion and eddy diffusion due to local turbulence contribute to the overall dispersion coefficient or effective diffusivity in the direction of flow for the bed of solid. The moles of fluid per unit area and unit time an element of length δz entering by longitudinal diffusion will be $-D_L'\,(\mathrm{d}Y/\mathrm{d}z)_z$, where D_L' is now the dispersion coefficient in the axial direction and has units $\mathbf{ML^{-1}T^{-1}}$ (since the concentration gradient has units $\mathbf{NM^{-1}L^{-1}}$). The amount leaving the element will be $-D_L'\,(\mathrm{d}Y/\mathrm{d}z)_{z+\delta z}$. The material balance equation will therefore be:

$$-D_L'\left(\frac{\mathrm{d}Y}{\mathrm{d}z}\right)_z + G'Y_z = -D_L'\left(\frac{\mathrm{d}Y}{\mathrm{d}z}\right)_{z+\delta z} + G'Y_{z+\delta z} + \rho_b \mathcal{R}_Y^M \delta z \tag{3.97}$$

which on expansion becomes:

$$D'_L \frac{d^2 Y}{dz^2} - G' \frac{dY}{dz} = \rho_b \mathscr{R}_Y^M \tag{3.98}$$

The boundary condition at $z = 0$ may be written down by invoking the conservation of mass for an element bounded by the plane $z = 0$ and any plane upstream of the inlet where $Y = Y_0$, say:

$$G' Y_0 = G'(Y)_{z=0} - D'_L \left(\frac{dY}{dz}\right)_{z=0} \tag{3.99}$$

from which we obtain the condition:

$$z = 0, \quad G' Y = G' Y_0 + D'_L \frac{dY}{dz} \tag{3.100}$$

Similarly, at the reactor exit we obtain:

$$z = L, \quad \frac{dY}{dz} = 0 \tag{3.101}$$

since there is no further possibility of Y changing from Y_L after the reactor exit. Equation 3.97 may be solved analytically for zero or first-order reactions, but for other cases resort to numerical methods is generally necessary. In either event Y is obtained as a function of z and so the length L of catalyst bed required to achieve a conversion corresponding to an exit concentration Y_L may be calculated. The mass of catalyst corresponding to this length L is $\rho_b A_c L$.

Having pointed out the modifications to be made to a design based upon the plug flow approach, it is salutary to note that axial dispersion is seldom of importance in fixed bed tubular reactors. This point is illustrated in Example 3.8.

Example 3.8

Show that the effect of axial dispersion on the conversion obtained in a typical packed bed gas–solid catalytic reactor is small. As the starting point consider the following relationship (see Chapter 2, equation 2.50).

$$\frac{C_A}{C_{Ain}} = \exp\left[-k\tau + (k\tau)^2 \frac{D_L}{uL}\right] \tag{A}$$

which applies to a first-order reaction, rate constant k, carried out in a *homogeneous* tubular reactor in which the Dispersion Number D_L/uL is small. In equation A:

- C_A is the concentration at a distance L along the reactor,
- C_{Ain} is the concentration at the inlet,
- D_L is the longitudinal dispersion coefficient for an empty tube homogeneous system,
- u is the linear velocity, and
- τ is the residence time (L/u).

For a typical packed bed, take the value of the Peclet number (ud_p/eD_L^*) as 2. D_L^* is the dispersion coefficient in a packed tube. Then show that for a fractional conversion of 0.99 and a ratio of d_p/L of 0.02, the length L of the reactor with axial dispersion exceeds the length of the simple ideal plug-flow reactor by only 4.6 per cent where d_p is the diameter of the particles.

Solution

Let L_p be the length of the simple ideal plug-flow homogeneous reactor with no dispersion for the same value of C_A/C_{Ain}, and τ_p be the required residence time in the plug-flow reactor i.e. in equation A $D_L = 0$, so; $$\frac{C_A}{C_{A\text{in}}} = \exp(-\,k\,\tau_p) \qquad \text{(B)}$$

∴ $$\exp(-\,k\,\tau_p) = \exp\left[-\,k\,\tau + (k\,\tau)^2\,\frac{D_L}{uL}\right]$$

i.e., dividing by $k\tau_p$: $$\frac{\tau}{\tau_p} = 1 + k\,\frac{\tau^2}{\tau_p}\,\frac{D_L}{uL}$$

The second term on the RHS may be regarded as a "correction" term which will be small cf 1. Therefore, setting $\tau = \tau_p$ in this term:

$$L/L_p = \frac{\tau}{\tau_p} = 1 + k\,\tau_p\,\frac{D_L}{uL} = 1 + k\,\tau_p\,\frac{D_L}{u}\,\frac{1}{L}$$

For a packed bed: Peclet number $Pe = \dfrac{u\,d_p}{e\,D_L^*} = 2$ or: $\dfrac{e\,D_L^*}{u} = \dfrac{d_p}{2}$

Furthermore, D_L for the empty tube may be replaced by $e\,D_L^*$.

i.e.: $$L/L_p = 1 + k\,\tau_p\,\frac{d_p}{2}\,\frac{1}{L}$$

From equation B:

$$k\,\tau_p = \ln\frac{C_{A\text{in}}}{C_A}$$

∴ $$L/L_p = 1 + \left(\ln\frac{C_{A\text{in}}}{C_A}\right)\frac{1}{2}\,\frac{d_p}{L}$$

Thus, for 99 per cent conversion $\dfrac{C_{A\text{in}}}{C_A} = \dfrac{1}{0.01}$ and for $\dfrac{d_p}{L} = 0.02$:

∴ $$L/L_p = 1 + \ln(100 \times \tfrac{1}{2} \times 0.02) = 1 + 0.046$$

i.e.: Effect of axial dispersion would be to increase length of plug-flow reactor by 4.6 per cent.

Axial and radial dispersion—Non-isothermal conditions. When the reactor exchanges heat with the surroundings, radial temperature gradients exist and this promotes transverse diffusion of the reactant. For an exothermic reaction, the reaction rate will be highest along the tube axis since the temperature there will be greater than at any other radial position. Reactants, therefore, will be rapidly consumed at the tube centre resulting in a steep transverse concentration gradient causing an inward flux of reactant and a corresponding outward flux of products. The existence of radial temperature and concentration gradients, of course, renders the simple plug flow approach to design inadequate.

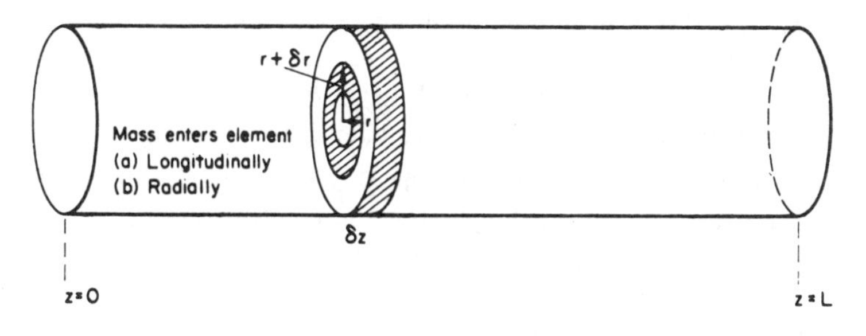

FIG. 3.25. Two-dimensional tubular reactor

It is now essential to write the mass and energy balance equations for the two dimensions z and r. For the sake of completeness, we will include the effect of longitudinal dispersion and heat conduction and deduce the material and energy balances for one component in an elementary annulus of radius δr and length δz. We assume that equimolar counterdiffusion occurs and by reference to Fig. 3.25 write down, in turn, the components of mass which are entering the element in unit time longitudinally and radially:

Moles entering by longitudinal bulk flow $= 2\pi r \delta r G'(Y)_z$

Moles entering by transverse diffusion $= -D_r' 2\pi r \delta z \left(\dfrac{\partial Y}{\partial r}\right)_r$

Moles entering by longitudinal diffusion $= -D_L' 2\pi r \delta r \left(\dfrac{\partial Y}{\partial z}\right)_z$

In general, the longitudinal and radial dispersion coefficients D_r' and D_L' will differ. The moles leaving the element in unit time can be written similarly as a series of components:

Moles leaving by longitudinal bulk flow $= 2\pi r \delta r G'(Y)_{z+\delta z}$

Moles leaving by transverse diffusion $= -D_r' 2\pi(r+\delta r)\delta z \left(\dfrac{\partial Y}{\partial r}\right)_{r+\delta r}$

Moles leaving by longitudinal diffusion $= -D_L' 2\pi r \delta r \left(\dfrac{\partial Y}{\partial z}\right)_{z+\delta z}$

Moles of component produced by chemical reaction $= 2\pi r \delta r \delta z \rho_b \mathcal{R}_{YT}^M$

In the steady state the algebraic sum of the moles entering and leaving the element will be zero. By expanding terms evaluated at $(z + \delta z)$ and $(r + \delta r)$ in a Taylor series about the points z and r respectively and neglecting second order differences, the material balance equation becomes:

$$D_r'\left\{\frac{\partial^2 Y}{\partial r^2} + \frac{1}{r}\frac{\partial Y}{\partial r}\right\} + D_L'\frac{\partial^2 Y}{\partial z^2} - G'\frac{\partial Y}{\partial z} = \rho_b \mathcal{R}_{YT}^M \tag{3.102}$$

A heat balance equation may be deduced analogously:

$$k_r\left\{\frac{\partial^2 T}{\partial r^2} + \frac{1}{r}\frac{\partial T}{\partial r}\right\} + k_l\frac{\partial^2 T}{\partial z^2} - G'\bar{c}_p\frac{\partial T}{\partial z} = \rho_b(-\Delta H)\mathcal{R}_{YT}^M \tag{3.103}$$

where k_r and k_l are the thermal conductivities in the radial and longitudinal directions respectively,

$\bar{c}_p$ is the mean heat capacity of the fluid, and

$(-\Delta H)$ is the heat evolved per mole due to chemical reaction.

When writing the boundary conditions for the above pair of simultaneous equations the heat transferred to the surroundings from the reactor may be accounted for by ensuring that the tube wall temperature correctly reflects the total heat flux through the reactor wall. If the reaction rate is a function of pressure then the momentum balance equation must also be invoked, but if the rate is insensitive or independent of total pressure then it may be neglected.

It is useful at this stage to note the assumptions which are implicit in the derivation of equations 3.102 and 3.103. These are as follows: dispersion coefficients (both axial and radial) and thermal conductivity all constant, instantaneous heat transfer between the solid catalyst and the reacting ideal-gas mixture, and changes in potential energy negligible. It should also be noted that the two simultaneous equations are coupled and highly non-linear because of the effect of temperature on the reaction rate. Numerical methods of solution are therefore generally adopted and those employed are based upon the use of finite differences. The neglect of longitudinal diffusion and conduction simplifies the equations considerably. If we also suppose that the system is isotropic we can write a single effective diffusivity D_e' for the bed in place of D_r' and D_L' and a single effective thermal conductivity k_e.

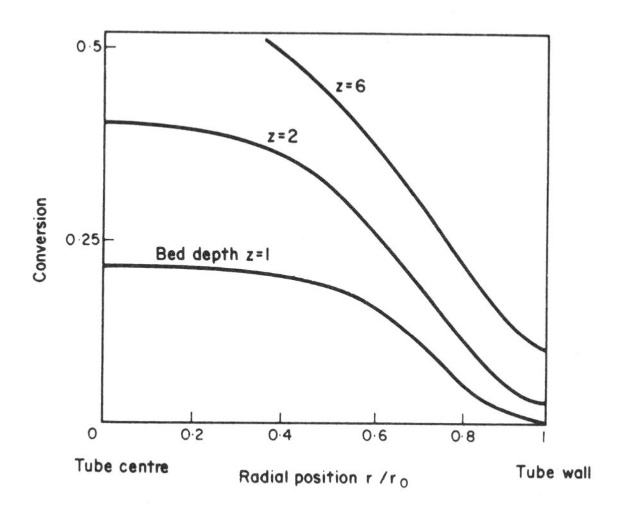

FIG. 3.26. Conversion profiles as a function of tube length and radius

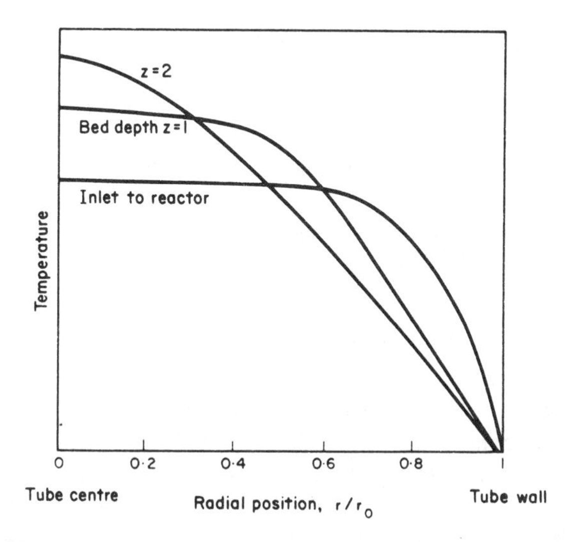

FIG. 3.27. Temperature profiles as a function of tube length and radius

The results of calculations typical of an exothermic catalytic reaction are shown in Figs. 3.26 and 3.27. It is clear that the conversion is higher along the tube axis than at other radial positions and that the temperature first increases to a maximum before decreasing at points further along the reactor length. The exothermic nature of the reaction, of course, leads to an initial increase in temperature but, because in the later stages of reaction the radial heat transfer to the wall and surroundings becomes larger than the heat evolved by chemical reaction, the temperature steadily decreases. The mean temperature and conversion over the tube radius for any given position along the reactor length may then be computed from such results and consequently the bed depth required for a given specified conversion readily found.

Further advancements in the theory of fixed bed reactor design have been made(56,57) but it is unusual for experimental data to be of sufficient precision and extent to justify the application of sophisticated methods of calculation. Uncertainties in the knowledge of effective thermal conductivities and heat transfer between gas and solid make the calculation of temperature distribution in the bed susceptible to inaccuracies, particularly in view of the pronounced effect of temperature on the reaction rate.

3.6.2. Thermal Characteristics of Packed Reactors

There are several aspects of thermal sensitivity and instability which are important to consider in relation to reactor design. When an exothermic catalytic reaction occurs in a non-isothermal reactor, for example, a small change in coolant temperature may, under certain circumstances, produce undesirable hotspots or regions of high temperature within the reactor. Similarly, it is of central importance to determine whether or not there is likely to be any set of operating conditions which may cause thermal instability in the sense that the reaction may either become extinguished or continue at a higher temperature level as a result of fluctuations in the feed condition. We will briefly examine these problems.

Sensitivity of Countercurrent Cooled Reactors

To illustrate the problem of thermal sensitivity we will analyse the simple one-dimensional model of the countercurrent cooled packed tubular reactor described earlier and illustrated in Fig. 3.25. We have already seen that the mass and heat balance equations for the system may be written:

$$G'\frac{\mathrm{d}Y}{\mathrm{d}z} = \rho_b.\mathscr{R}^M \qquad \text{(equation 3.87)}$$

$$G'\bar{c}_p\frac{\mathrm{d}T}{\mathrm{d}z} = -\Delta H\,\rho_b.\mathscr{R}^M - \frac{h}{a'}(T - T_W) \qquad \text{(equation 3.94)}$$

where T_W represents a constant wall temperature. Solution of these simultaneous equations with initial conditions $z = 0$, $T = T_0$ and $Y = Y_0$ enables us to plot the reaction path in the (T, Y) plane. The curve ABC, portrayed in Fig. 3.20, is typical of the reaction path that might be obtained. There will be one such curve for any given value T_0 of the reactor inlet temperature. By locating the points at which

$dT/dY = 0$ the loci of maximum temperature may be plotted, each locus corresponding to a chosen value T_W for the constant coolant temperature along the reactor length. Thus, a family of curves such as that sketched in Fig. 3.28 would be obtained, representing loci of maximum temperatures for constant wall temperatures T_W, increasing sequentially from curve 1 to curve 5. Since the maximum temperature along the non-isothermal reaction path (point B in Fig. 3.20) must be less than the temperature given by the intersection of the adiabatic line with the locus of possible maxima (point D in Fig. 3.20), the highest temperature achieved in the reactor must be bounded by this latter point. If the adiabatic line—and therefore the non-isothermal reaction path—lies entirely below the locus of possible maxima, then $dT/dY > 0$, and the temperature will increase along the reactor length because $dT/dz > 0$ for all points along the reactor; if, however, the adiabatic line lies above the locus, $dT/dY < 0$ and the temperature will therefore decrease along the reactor length. For a particular T_W corresponding to the loci 1 or 2 in Fig. 3.28, T_0, the inlet temperature to the reactor, must be the maximum temperature in the reactor because the adiabatic line lines entirely above these two curves making dT/dz always negative. If T_W corresponds to curve 3, however, we might expect the temperature to increase from T_0 to a maximum and thence to decrease along the remaining reactor length since the adiabatic line now intersects this particular locus at point A. In the case of curve 4 the adiabatic line is tangent to the maximum temperature locus at B, and then intersects it at point C where the temperature would be much higher. The adiabatic line intersects curve 5 at point D. From this analysis, therefore, we are led to expect that, if the steady wall temperature of the reactor is increased through a sequence corresponding to curves 1 to 4, the maximum temperature in the reactor gradually increases from T_0 to B as T_W increases from T_{W1} to T_{W4}. If the wall temperature were to be increased to T_{W5} (the maximum temperature locus corresponding to curve 5), there would be a discontinuity and the maximum temperature in the reactor would suddenly jump to point D. This type of sensitivity was predicted by BILOUS and AMUNDSON[58] who calculated temperature maxima within a non-isothermal reactor for various constant wall temperatures. The results of their computations are shown in Fig. 3.29 and it is seen that, if the wall temperature increases from 300 K to 335 K, a

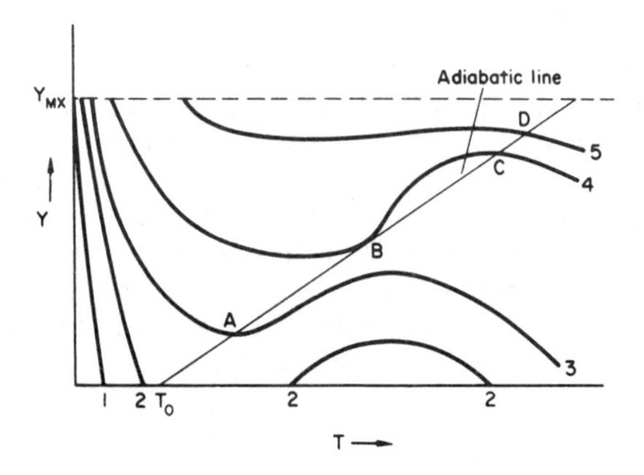

FIG. 3.28. Loci of maximum wall temperatures

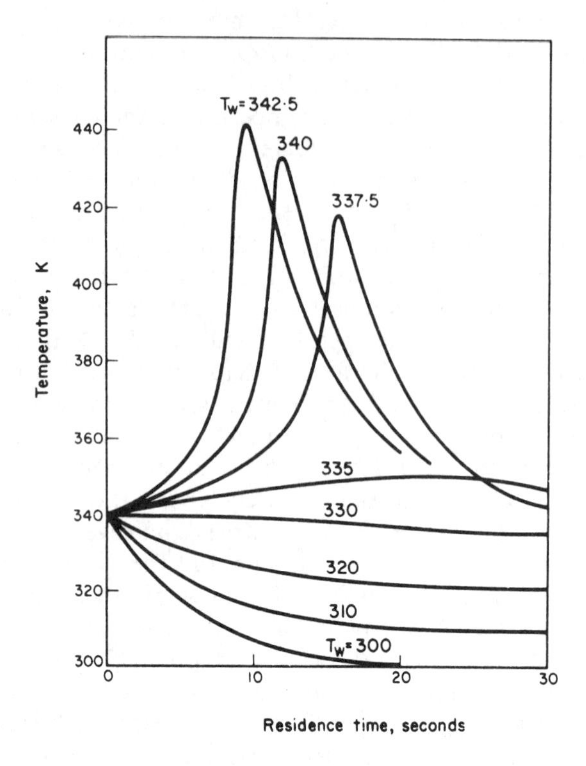

FIG. 3.29. Sensitivity of reactor to change in wall temperature

temperature maximum appears in the reactor. Further increase in wall temperature causes the maximum to increase sharply to higher temperatures. Bilous and Amundson also found that sensitivity to heat transfer can produce temperature maxima in much the same way as sensitivity to change in wall temperature. In view of the nature of equations 3.87 and 3.94 this is to be expected.

The Autothermal Region

Clearly it is desirable to utilise economically the heat generated by an exothermic reaction. If the heat dissipated by reaction can be used in such a way that the cold incoming gases are heated to a temperature sufficient to initiate a fast reaction, then by judicious choice of operating conditions the process may be rendered thermally self-sustaining. This may be accomplished by transferring heat from the hot exit gases to the incoming feed. The differential equations already discussed for the countercurrent self-cooled reactor are applicable in this instance. Combining the three simultaneous equations 3.95, 3.96 and 3.87 (utilising the upper sign in equation 3.96 for the countercurrent case and remembering that for a self-cooled—or in this particular case self-heated—reactor $G_c' \bar{c}_{pc} = G' \bar{c}_p$) we obtain:

$$G' \frac{\mathrm{d}}{\mathrm{d}z}\left\{\frac{(-\Delta H)}{\bar{c}_p} Y - T + T_C\right\} = 0 \tag{3.104}$$

Integration of this equation yields:

$$T_C = T - \frac{(-\Delta H)}{\bar{c}_p}(Y - Y_0) \qquad (3.105)$$

Substituting equation 3.105 into the heat balance equation 3.95 for the reactor:

$$G'\frac{\mathrm{d}T}{\mathrm{d}z} = \frac{(-\Delta H)}{\bar{c}_p}\rho_p\mathcal{R}^M - \frac{G'(-\Delta H)}{\bar{c}_p}\frac{\Gamma'}{L}(Y - Y_0) \qquad (3.106)$$

where the heat exchanging capacity of the system has been written in terms of $\Gamma'(= UaL/G'\bar{c}_{pc})$. The mass balance equation:

$$G'\frac{\mathrm{d}Y}{\mathrm{d}z} = \rho_b\mathcal{R}^M \qquad \text{(equation 3.87)}$$

must be integrated simultaneously with equation 3.106 from $z = 0$ to $z = L$ with the initial conditions (at $z = 0$), $Y = Y_0$ and $T = T_0$ to obtain T and Y as functions of z. In this event the exit temperature T_L and concentration Y_L can, in principle, be computed. Numerical integration of the equations is generally necessary. To estimate whether the feed stream has been warmed sufficiently for the reaction to give a high conversion and be thermally self-supporting, we seek a condition for which the heat generated by reaction equals the heat gained by the cold feed stream. The heat generated per unit mass of feed by chemical reaction within the reactor may be written in terms of the concentration of product:

$$Q_1^M = (-\Delta H)(Y_L - Y_0) \qquad (3.107)$$

whereas the net heat gained by the feed would be:

$$Q_2^M = \bar{c}_p(T_L - T_{CL}) \qquad (3.108)$$

The system is thermally self-supporting if $Q_1^M = Q_2^M$, i.e. if:

$$T_L - T_{CL} = \frac{(-\Delta H)}{\bar{c}_p}(Y_L - Y_0) \qquad (3.109)$$

The temperature of the non-reacting heating fluid at the reactor inlet is T_0 and at the exit T_{CL} so the difference, say ΔT, is

$$\Delta T = T_0 - T_{CL} = T_0 - T_L + \frac{(-\Delta H)}{\bar{c}_p}(Y_L - Y_0) \qquad (3.110)$$

Since the numerical values of T_L and Y_L are dependent on the heat exchanging capacity (as shown by equation 3.106), the quantity on the right-hand side of equation 3.110 may be displayed as a function of the inlet temperature to the bed, T_0, with Γ' as a parameter. The three bell-shaped curves in Fig. 3.30 are for different values of Γ' and each represents the locus of values given by the right-hand side of this equation. The left-hand side of the equation may be represented by a straight line of unit slope through the point (T_{CL}, 0). The points at which the line intersects the curve represent solutions to equation 3.110. However, we seek only a stable solution which coincides with a high yield. Such a solution would be represented by

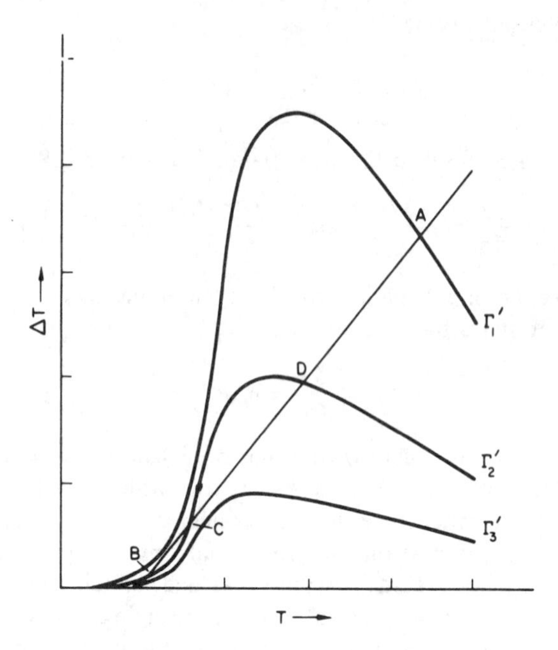

FIG. 3.30. Temperature increase of feed as function of inlet temperature

the point A in Fig. 3.30, for the straight line intersects the curve only once and at a bed temperature corresponding to a high yield of product. The condition for autothermal reaction would therefore be represented by the operating condition (T_{CL}, Γ_3'), i.e. the cold feed temperature should be $T_0 = T_{CL}$ and the heat exchange capacity (which determines the ratio L/G, the length of solid packing divided by the mass flow rate) should correspond to Γ_3'. The corresponding temperature profiles for the reactor and exchanger are sketched in Fig. 3.31.

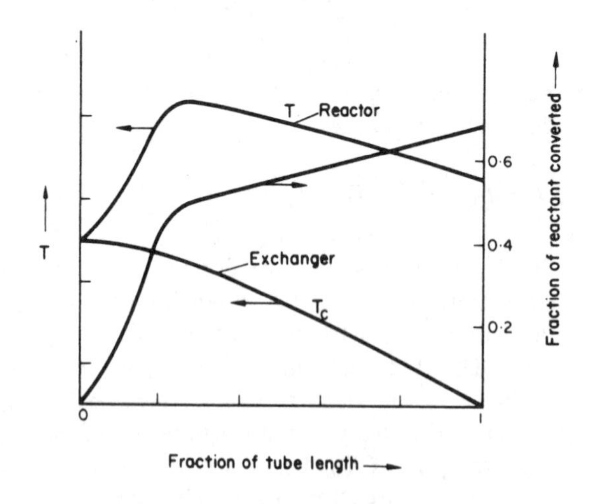

FIG. 3.31. Countercurrent self-sustained reactor

Stability of Packed-bed Tubular Reactors

Referring to Fig. 3.30, which in effect is a diagram to solve a heat balance equation, it may be seen that the line representing the heat gained by the feed stream intersects two of the heat generation curves at a single point and one of them at three points. As already indicated, intersection at one point corresponds to a solution which is stable, for if the feed state is perturbed there is no other state of thermal equilibrium at which the system could continue to operate. The intersection at point A corresponds to a state of thermal equilibrium at a high reaction temperature and therefore high conversion. On the other hand, although the system is in a state of thermal equilibrium at point B, the temperature and yield are low and the reaction is almost extinguished—certainly not a condition one would choose in practice. A condition of instability is represented by the intersection at C. Here any small but sudden decrease in feed temperature would cause the system to become quenched, for the equilibrium state would revert to B. A sudden increase in feed temperature would displace the system to D. Although D represents a relatively stable condition, such a choice should be avoided for a large perturbation might result in the system restabilising at B where the reaction is quenched. A reactor should therefore be designed such that thermal equilibrium can be established at only one point (such as A in Fig. 3.30) where the conversion is high.

It is interesting to note from Fig. 3.32 that if the feed to an independently cooled reactor (in which an exothermic reaction is occurring) is gradually varied through a sequence of increasing temperatures such as 1 to 5, the system will have undergone some fairly violent changes in thermal equilibrium before finally settling at point I. Corresponding to a sequential increase in feed temperature from 1 to 4 the reactor will pass through states A, B, and C to D, at which last point any slight increase in feed temperature causes the thermal equilibrium condition to jump to a point near H. If the feed temperature is then raised to 5 the system gradually changes along the smooth curve from H to I. On reducing the feed temperature from 5 to 2 on the other hand, the sequence of equilibrium states will alter along the path IHGF. Any

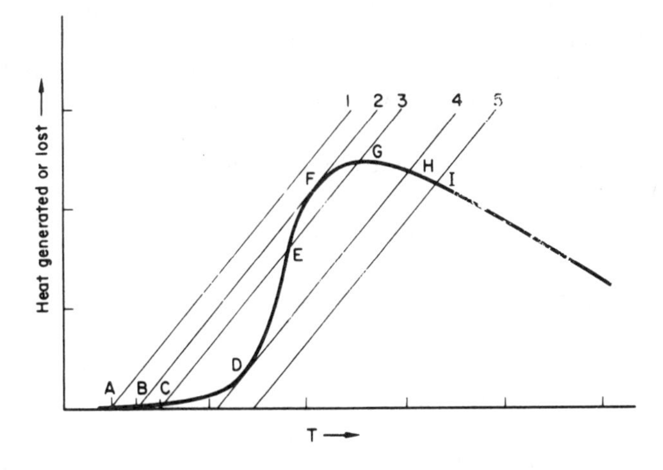

FIG. 3.32. Stability states in reactors

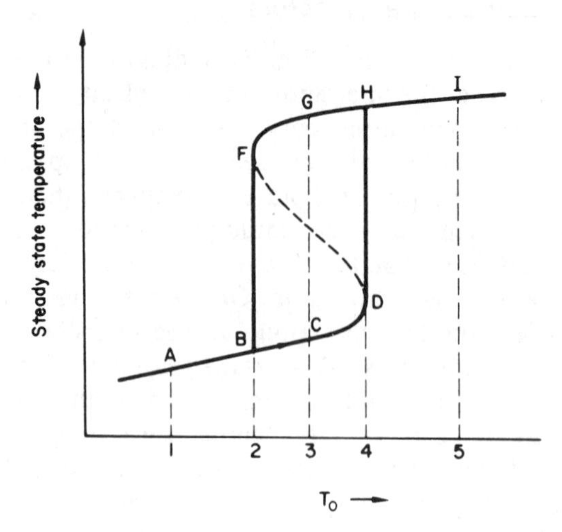

FIG. 3.33. Hysteresis effect in packed-bed reactor

further small decrease in feed temperature would displace the system to a point near B and subsequently along the curve BA as the feed temperature gradually changes from 2 to 1. Thus, there is a hysteresis effect when the system is taken through such a thermal cycle and this is clearly illustrated in Fig. 3.33, which traces the path of equilibrium states as the feed temperature is first increased from 1 to 5 and subsequently returned to 1. These effects are important to consider when start-up or shut-down times are estimated.

Another type of stability problem arises in reactors containing reactive solid or catalyst particles. During chemical reaction the particles themselves pass through various states of thermal equilibrium, and regions of instability will exist along the reactor bed. Consider, for example, a first-order catalytic reaction in an adiabatic tubular reactor and further suppose that the reactor operates in a region where there is no diffusion limitation within the particles. The steady state condition for reaction in the particle may then be expressed by equating the rate of chemical reaction to the rate of mass transfer. The rate of chemical reaction per unit reactor volume will be $(1 - e)\boldsymbol{k}C_{Ai}$ since the effectiveness factor η is considered to be unity. From equation 3.66 the rate of mass transfer per unit volume is $(1 - e)(S_x/V_p)h_D(C_{AG} - C_{Ai})$ so the steady state condition is:

$$\boldsymbol{k}C_{Ai} = h_D \frac{S_x}{V_p}(C_{AG} - C_{Ai}) \tag{3.111}$$

Similarly, equating the heat generated by reaction to the heat flux from the particle:

$$(-\Delta H)\boldsymbol{k}C_{Ai} = h \frac{S_x}{V_p}(T_i - T) \tag{3.112}$$

where h is the heat transfer coefficient between the particle and the gas, and
T_i is the temperature at the particle surface.

Eliminating C_{Ai} from equations 3.111 and 3.112:

$$T_i - T = (-\Delta H)\frac{V_p}{S_x}\frac{1}{h}\frac{\boldsymbol{k}C_{AG}}{1+\dfrac{V_p\boldsymbol{k}}{S_x h_D}} \tag{3.113}$$

Now the maximum temperature T_{MX} achieved for a given conversion in an adiabatic reactor may be adduced from the heat balance:

$$T_{MX} = T + \frac{(-\Delta H)C_{AG}}{\rho \bar{c}_p} \tag{3.114}$$

which is analogous to equation 3.91, except that concentration has been expressed in terms of moles per unit volume, $C_{AG}(=\rho Y)$, with the gas density ρ regarded as virtually constant. Combining equations 3.113 and 3.114, the temperature difference between the particle surface and the bulk gas may be expressed in dimensionless form:

$$\frac{T_i - T}{T_{MX} - T} = \frac{\mathrm{a}\boldsymbol{k}}{1 + \mathrm{b}\boldsymbol{k}} \tag{3.115}$$

where $\boldsymbol{k}$ is the usual Arrhenius function of temperature,

a is the constant $\dfrac{\rho \bar{c}_p V_p}{S_x h}$, and

b is the constant $\dfrac{V_p}{S_x h_D}$.

The condition of thermal equilibrium for a particle at some point along the reactor is thus established by equation 3.115. The function on the right-hand side of this equation is represented by a sigmoidal curve as shown in Fig. 3.34, and is as a function of bed temperature $T(z)$ for some position z along the reactor. The left-hand side of equation 3.115 (as a function of T) is a family of straight lines all

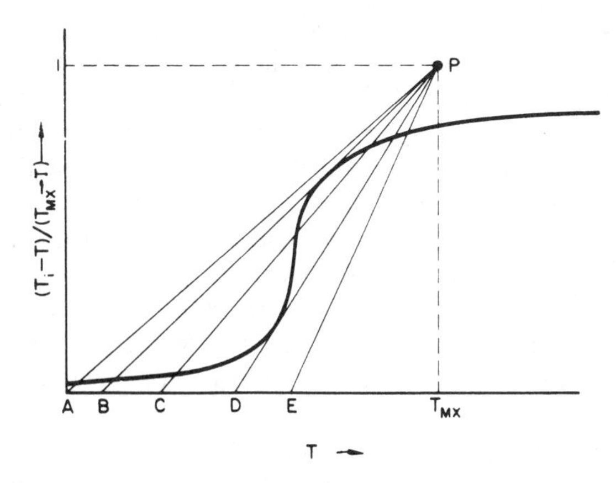

FIG. 3.34. Stability states of particles in adiabatic bed

of which pass through the common pole point (T_{MX}, 1) designated P. It is apparent from Fig. 3.34 that as the temperature increases along the bed it is possible to have a sequence of states corresponding to lines such as PA, PB, PC, PD and PE. As soon as a position has been reached along the reactor where the temperature is B there is the possibility of two stable temperatures for a particle. Such a situation would persist until a position along the reactor has been reached where the temperature is D, after which only one stable state is possible. Thus, there will be an infinite family of steady state profiles within this region of the bed. Which profile actually obtains depends on start-up and bed entrance conditions. However, for steady state operating conditions, the work of McGreavy and Thornton[23] offers some hope to designers in that it may be possible to predict and thus avoid regions of instability in cooled packed-bed catalytic reactors.

3.6.3. Fluidised Bed Reactors

The general properties of fluidised beds are discussed in Volume 2, Chapter 6. Their application to gas–solid and catalytic reactions has certain advantages over the use of tubular type reactors. High wall-to-bed heat transfer coefficients enable heat to be abstracted from, or absorbed by, the reactor with considerable efficiency. A mechanical advantage is also gained by the relative ease with which solids may be conveyed and, because of solids mixing, the whole of the gas in the reactor is at substantially the same temperature. Extremely valuable is the large external surface area exposed by the solid to the gas; because of this, reactions limited by intraparticle diffusion will give a higher conversion in a fluidised bed than in a packed bed tubular reactor.

May[59] considered a model for describing reactions in a fluidised bed. He wrote the mass conservation equations for the bubble phase (identified with rising bubbles containing little or no solid particles) and the dense phase (in which the solid particles are thoroughly mixed). He tacitly assumed that mass transferred by cross-flow between the two phases was not a function of bed height, and took an average concentration gradient for the whole height of the bed. Conversions within the bed were calculated by solving the conservation equations for particular cross-flow ratios. May compared the predicted conversion in beds in which there was no mixing of solids with beds in which there was complete back-mixing. It was found that conversion was smaller for complete back-mixing and relatively low cross-flow ratios than for no mixing and high cross-flow ratios. Thus, if back-mixing is appreciable, a larger amount of catalyst has to be used in the fluidised bed to achieve a given conversion than if a plug flow reactor, in which there is no back-mixing, is used. It was also shown that, for two concurrent first-order reactions, the extent of gas–solid contact affects the selectivity. Although the same total conversion may be realised, more catalyst is required when the contact efficiency is poor than when the contact efficiency is good, and proportionately more of the less reactive component in the feed would be converted at the expense of the more reactive component.

The difficulty with an approach in which flow and diffusive parameters are assigned to a model is that the assumptions do not conform strictly to the pattern of behaviour in the bed. Furthermore, it is doubtful whether either solid dispersion or gas mixing can be looked upon as a diffusive flux. ROWE[60] calculated, from a knowledge of the fluid dynamics in the bed, a mean residence time for particles at the surface of which a slow first-order gas–solid reaction occurred. Unless the particle size of the solid material was chosen correctly most of the reactant gas passed through the bed as bubbles and had insufficient time to react. The most effective way of increasing the contact efficiency was to increase the particle size, for this caused more of the reactant gas to pass through the dense phase. Doubling the particle size almost halved the contact time required for the reactants to be completely converted. Provided that the gas flow is sufficiently fast to cause bubble formation in the bed, the heat transfer characteristics are good.

3.7. GAS–SOLID NON-CATALYTIC REACTORS

There is a large class of industrially important heterogeneous reactions in which a gas or a liquid is brought into contact with a solid and reacts with the solid transforming it into a product. Among the most important are: the reduction of iron oxide to metallic iron in a blast furnace; the combustion of coal particles in a pulverised fuel boiler; and the incineration of solid wastes. These examples also happen to be some of the most complex chemically. Further simple examples are the roasting of sulphide ores such as zinc blende:

$$ZnS + \tfrac{3}{2}O_2 = ZnO + SO_2$$

and two reactions which are used in the carbonyl route for the extraction and purification of nickel:

$$NiO + H_2 = Ni + H_2O$$

$$Ni + 4CO = Ni(CO)_4$$

In the first of these reactions the product, impure metallic nickel, is in the form of a porous solid, whereas in the second, the nickel carbonyl is volatile and only the impurities remain as a solid residue.

It can be seen from even these few examples that a variety of circumstances can exist. As an initial approach to the subject, however, it is useful to distinguish two extreme ways in which reaction within a particle can develop.

(a) *Shrinking core reaction mode.* If the reactant solid is non-porous, the reaction begins on the outside of the particle and the reaction zone moves towards the centre. The core of unreacted material shrinks until it is entirely consumed. The product formed may be a coherent but porous solid like the zinc oxide from the oxidation of zinc sulphide above, in which case the size of the particle may be virtually unchanged (Fig. 3.35*a*(i)). On the other hand, if the product is a gas or if the solid product formed is a friable ash, as in the combustion of some types of coal, the particle decreases in size during the reaction until it disappears (Fig. 3.35*a*(ii)). When the reactant solid is porous a fast reaction may nevertheless still proceed via a shrinking core reaction mode because diffusion of the gaseous reactant into the interior of the solid will be a relatively slow process.

(b) *Progressive conversion reaction mode.* In a porous reactant solid an inherently slower reaction can, however, proceed differently. If the rate of diffusion of the

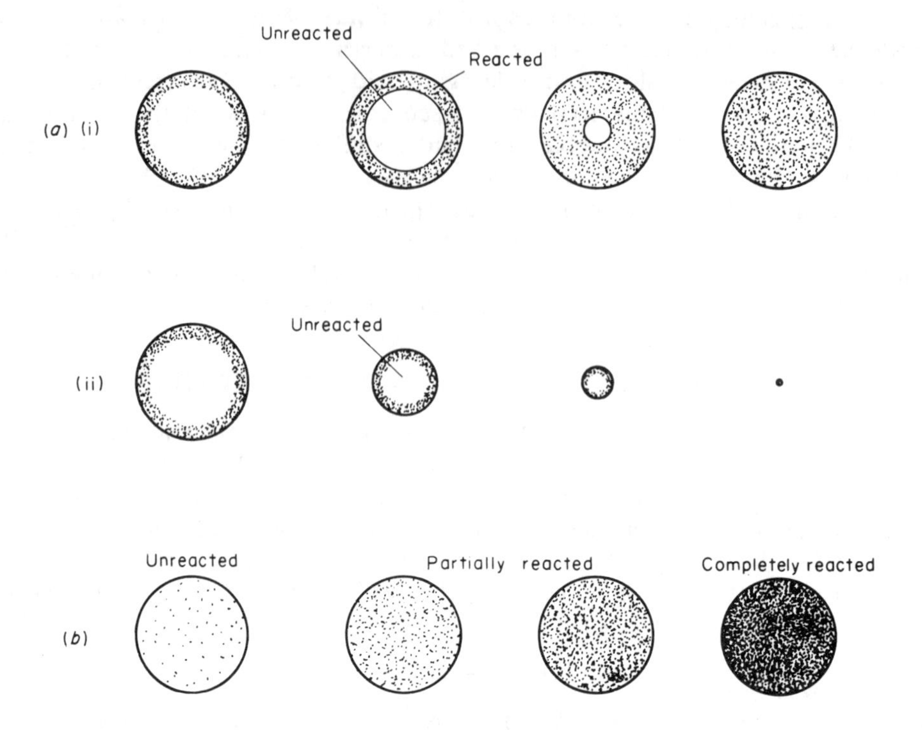

FIG. 3.35. Stages in the reaction of a single particle by
(*a*) Shrinking core reaction mode with the formation of either (i) a coherent porous solid or (ii) friable ash or gaseous product
(*b*) Progressive conversion reaction mode

gaseous reactant into the interior of the particle can effectively keep pace with the reaction, the whole particle may be progressively and uniformly converted into the product (Fig. 3.35*b*). If the product is itself a coherent solid, the particle size may be virtually unchanged, but if the product is volatile or forms only a weak solid structure, the particle will collapse or disintegrate towards the end of the reaction.

3.7.1. Modelling and Design of Gas–Solid Reactors

The basic approach is to consider the problem in two parts. Firstly, the reaction of a single particle with a plentiful excess of the gaseous reactant is studied. A common technique is to suspend the particle from the arm of a thermobalance in a stream of gas at a carefully controlled temperature; the course of the reaction is followed through the change in weight with time. From the results a suitable kinetic model may be developed for the progress of the reaction within a single particle. Included in this model will be a description of any mass transfer resistances associated with the reaction and of how the reaction is affected by concentration of the reactant present in the gas phase.

The second part of the problem is concerned with the contacting pattern between gas and solid in the equipment to be designed. Material and thermal balances have to be considered, together with the effects of mixing in the solid and gas phases.

These will influence the local conditions of temperature, gas composition and fractional conversion applicable to any particular particle. The ultimate aim, which is difficult to achieve because of the complexity of the problem, is to estimate the overall conversion which will be obtained for any particular set of operating conditions.

3.7.2. Single Particle Unreacted Core Models

Most of the gas–solid reactions that have been studied appear to proceed by the shrinking core reaction mode. In the simplest type of unreacted core model it is assumed that there is a non-porous unreacted solid with the reaction taking place in an infinitely thin zone separating the core from a completely reacted product as shown in Fig. 3.36 for a spherical particle. Considering a reaction between a gaseous reactant **A** and a solid **B** and assuming that a coherent porous solid product is formed, five consecutive steps may be distinguished in the overall process:

(1) Mass transfer of the gaseous reactant **A** through the gas film surrounding the particle to the particle surface.
(2) Penetration of **A** by diffusion through pores and cracks in the layer of product to the surface of the unreacted core.
(3) Chemical reaction at the surface of the core.
(4) Diffusion of any gaseous products back through the product layer.
(5) Mass transfer of gaseous products through the gas film.

If no gaseous products are formed, steps (4) and (5) cannot contribute to the resistance of the overall process. Similarly, when gaseous products are formed, their counterdiffusion away from the reaction zone may influence the effective diffusivity

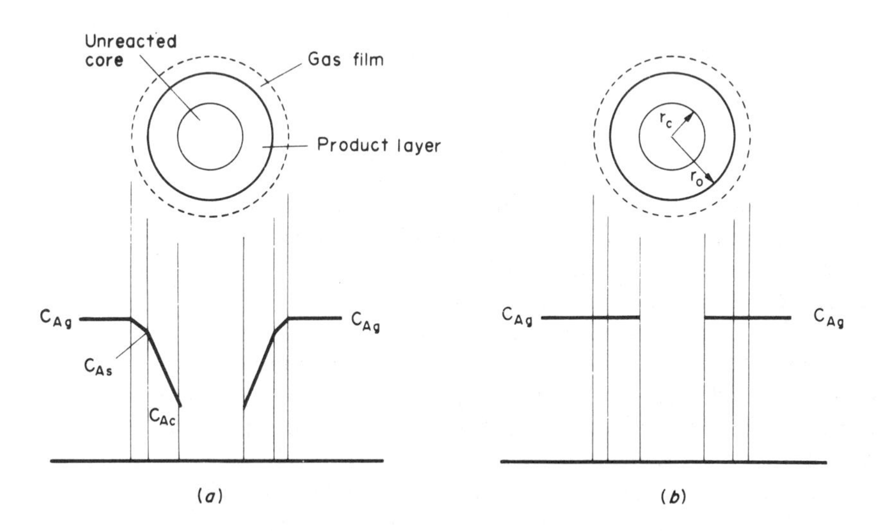

FIG. 3.36. Unreacted core model, impermeable solid, showing gas phase reactant concentration
(*a*) With significant gas film and product layer resistances
(*b*) Negligible gas film and product layer resistances

of the reactants, but gaseous products will not otherwise affect the course of the reaction unless it is a reversible one. Of the three remaining steps (1) to (3), if the resistance associated with one of these steps is much greater than the other resistances then, as with catalytic reactors, that step becomes rate-determining for the overall reaction process.

The shrinking core reaction mode is not necessarily limited to non-porous unreacted solids. With a fast reaction in a porous solid, diffusion into the core and chemical reaction occur in parallel. The mechanism of the process is very similar to the mechanism of the catalysed gas–solid reactions where the Thiele modulus is large, the effectiveness factor is small and the reaction is confined to a thin zone (Section 3.3.1). This combination of reaction with core diffusion gives rise to a reaction zone which, although not infinitely thin but diffuse, nevertheless advances into the core at a steady rate.

Before a simple mathematical analysis is possible a further restriction needs to be applied; it is assumed that the rate of advance of the reaction zone is small compared with the diffusional velocity of **A** through the product layer, i.e. that a pseudo steady-state exists.

Although general models in which external film mass transfer, diffusion through the product layer, diffusion into the core and chemical reaction are all taken into account have been developed[(61)], it is convenient to consider the special cases which arise when one of these stages is rate determining. Some of these correspond to the shrinking core mode of reaction, while others lead to the progressive conversion mode. Owing to limitations of space, only the shrinking core reaction mode in which chemical reaction or a combination of chemical reaction and core diffusion is rate determining will be considered further. A more extended treatment may be found in the book by SZEKELY *et al.*[(62)].

Unreacted Core Model—Fast Chemical Reaction

Consider a reaction of stoichiometry

$$\mathbf{A}(g) + b\mathbf{B}(s) = \text{Products}$$

which is first order with respect to the gaseous reactant **A**. If neither the gas-film nor the solid-product layer presents any significant resistance to mass transfer, the concentration of **A** at the reaction surface at the radial position r_c will be C_{AG}, the same as in the bulk of the gas (Fig. 3.36*b*).

If the reacting core is impermeable, reaction will take place at the surface of the core, whereas if the core has some degree of porosity the combination of chemical reaction and limited core diffusivity will give rise to a more extended reaction zone. In either case, the overall rate of reaction will be proportional to the area of the reaction front.

Taking therefore unit area of the core surface as the basis for the reaction rate, and writing the first order rate constant as $\boldsymbol{k}_s$, then the rate at which moles of **A** are consumed in the reaction is given by:

$$-\frac{\mathrm{d}n_A}{\mathrm{d}t} = 4\pi r_c^2 \boldsymbol{k}_s C_{AG} \tag{3.116}$$

If the core is porous, $\boldsymbol{k}_s$ is not a simple rate constant but incorporates the core diffusivity as well.

From the stoichiometry of the reaction:

$$-\frac{dn_A}{dt} = -\frac{1}{b}\frac{dn_B}{dt}$$

where n_B is the moles of **B** in the core. $n_B = C_B \frac{4}{3}\pi r_c^3$, C_B being the molar density of the solid **B**. Thus:

$$-\frac{1}{b}\frac{dn_B}{dt} = -\frac{1}{b}C_B 4\pi r_c^2 \frac{dr_c}{dt} \tag{3.117}$$

Hence, from equations (3.116) and (3.117):

$$-C_B\frac{dr_c}{dt} = b\boldsymbol{k}_s C_{AG}$$

Integrating:

$$-C_B\int_{r_0}^{r_c} dr_c = b\boldsymbol{k}_s C_{AG}\int_0^t dt$$

i.e.:

$$t = \frac{C_B}{b\boldsymbol{k}_s C_{AG}}(r_0 - r_c) \tag{3.118}$$

The time t_f for complete conversion corresponds to $r_c = 0$,

i.e.:

$$t_f = \frac{C_B r_0}{b\boldsymbol{k}_s C_{AG}} \tag{3.119}$$

When the core has a radius r_c and the fractional conversion for the particle as a whole is α_B the ratio t/t_f is given by:

$$t/t_f = (1 - r_c/r_0) = 1 - (1 - \alpha_B)^{1/3} \tag{3.120}$$

Example 3.9

Spherical particles of a sulphide ore 2 mm in diameter are roasted in an air stream at a steady temperature. Periodically small samples of the ore are removed, crushed and analysed with the following results:

Time (min)	15	30	60
Fractional conversion	0.334	0.584	0.880

Are these measurements consistent with a shrinking core and chemical reaction rate proportional to the area of the reaction zone? If so, estimate the time for complete reaction of the 2 mm particles, and the time for complete reaction of similar 0.5 mm particles.

Solution

Using the above data, time t can be plotted against $[1 - (1 - \alpha_B)^{1/3}]$ according to equation 3.120; a straight line confirms the assumed model and the slope gives t_f, the time to complete conversion.

Alternatively, calculate $t_f = t/[1 - (1 - \alpha_B)^{1/3}]$ from each data point:

Time (min)	15	30	60
Fractional conversion	0.334	0.584	0.880
Calculated t_f (min)	118	118	118

The constancy of t_f confirms the model; estimated time to complete conversion of 2 mm particles is thus 118 min.

Because $t_f \propto r_0$, the estimated time to complete conversion of 0.5 mm diam. particles is $118.4 \times (0.5/2) = 29.6$ min.

Limitations of Simple Models—Solids Structure

In practice, reaction in a particle may occur within a diffuse front instead of a thin reaction zone, indicating a mechanism intermediate between the shrinking core and progressive conversion modes of reaction. Often it is not realistic to single out a particular step as being rate determining because there may be several factors, each of which affects the reaction to a similar extent. As a further complication there may be significant temperature gradients around and within the particle with fast chemical reactions; for example, in the oxidation of zinc sulphide temperature differences of up to 55 K between the reaction zone and the surrounding furnace atmosphere have been measured[(63)].

Recent progress in understanding and modelling uncatalysed gas–solid reactions has been based on the now well-established theory of catalysed gas–solid reactions. The importance of characterising the structure of the solids, i.e. porosity, pore size and shape, internal surface area and adsorption behaviour, is now recognised. The problem with uncatalysed reactions is that structural changes necessarily take place during the course of the reaction; pores in the reacting solid are enlarged as reaction proceeds; if a solid product is formed there is a process of nucleation of the second solid phase. Furthermore, at the reaction temperature, some sintering of the product or reactant phases may occur and, if a highly exothermic reaction takes place in a thin reaction zone, drastic alterations in structure may occur near the reaction front.

Some reactions are brought about by the action of heat alone, for example the thermal decomposition of carbonates, and baking bread and other materials. These constitute a special class of solid reactions somewhat akin to the progressive conversion type of reaction models but with the rate limited by the rate of heat penetration from the exterior.

3.7.3. Types of Equipment and Contacting Patterns

There is a wide choice of contacting methods and equipment for gas–solid reactions. As with other solids-handling problems, the solution finally adopted may depend very much on the physical condition of the reactants and products, i.e. particle size, any tendency of the particles to fuse together, ease of gas–solid separation, etc. One type of equipment, the rotary kiln, has already been mentioned (Chapter 2, Fig. 2.4) and some further types of equipment suitable for continuous operation are shown in Fig. 3.37. The concepts of macromixing in the solid phase and dispersion in the gas phase as discussed in the previous section will be involved in the quantitative treatment of such equipment.

The principle of the moving bed of Fig. 3.37*a* in cross-flow to the air supply is used for roasting zinc blende and for the combustion of large coal on chain grate stokers. Another kind of moving bed is found in hopper-type reactors in which particles are fed at the top and continuously move downwards to be discharged at

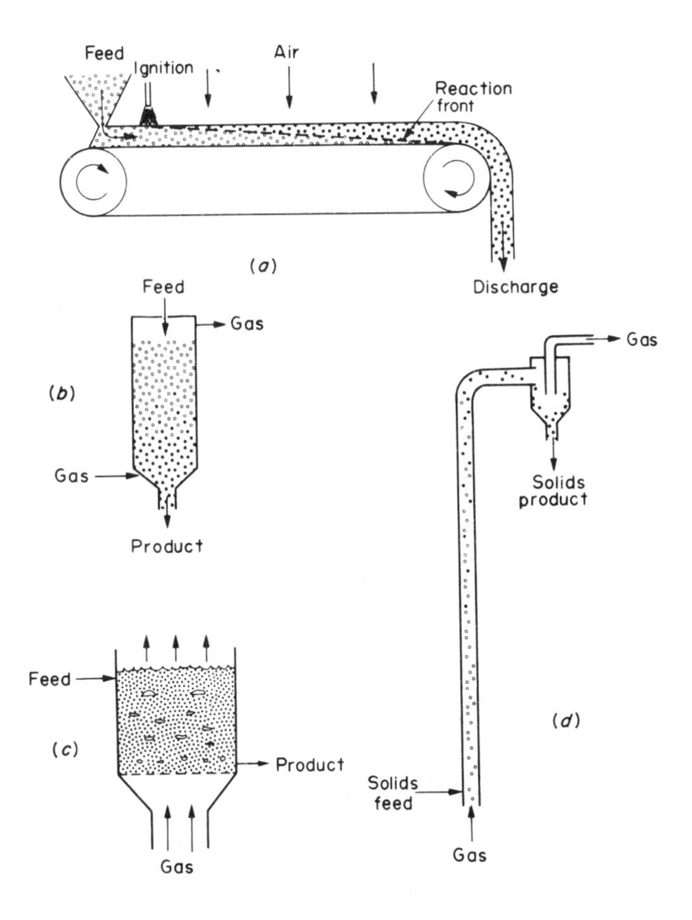

FIG. 3.37. Types of reactor and contacting patterns for gas–solid reactions
(*a*) Moving bed in cross flow
(*b*) Hopper type of reactor, particles moving downwards in countercurrent flow
(*c*) Fluidised bed reactor, particles well mixed
(*d*) Transfer line reactor, solids transported as a dilute phase in co-current flow

the bottom; an example is the decomposer used for nickel carbonyl where pure nickel is deposited on metallic nickel balls which grow in size as they pass down the reactor (Fig. 3.37*b*).

Rotary kilns (Chapter 2, Fig. 2.4) have advantages where the solid particles tend to stick together as in cement manufacture, and in the reduction and carbonylation steps in the purification of nickel. In rotary kilns, the flow of gas may be co-current or countercurrent to the solids.

If the solid particles can be maintained in the fluidised state without problems of agglomeration or attrition, the fluidised bed reactor Fig. 1.45*c* is likely to be preferred. For short contact times at high temperatures the dilute phase transfer line reactor (Fig. 3.37*d*) has advantages.

Raked hearth reactors were once extensively used in the metals extraction industries but are now being superseded.

Semi-batch reactors, where the gas passes through a fixed bed of solids which are charged and removed batchwise, are common for small-scale operations.

Fluidised Bed Reactor

As the following example shows, the fluidised bed reactor is one of the few types that can be analysed in a relatively simple manner, providing (a) complete mixing is assumed in the solid phase, (b) a sufficient excess of gas is used so that the gas-phase composition may be taken as uniform throughout.

Example 3.10

Particles of a sulphide ore are to be roasted in a fluidised bed using excess air: the particles may be assumed spherical and uniform in size. Laboratory experiments indicate that the oxidation proceeds by the unreacted core mechanism with the reaction rate proportional to the core area, the time for complete reaction of a single particle being 16 min at the temperature at which the bed will operate. The particles will be fed and withdrawn continuously from the bed at a steady rate of 6 tonnes of product per hour (1.67 kg/s). The solids hold-up in the bed at any time is 10 tonnes.

(a) Estimate what proportion of the particles leaving the bed still contain some unreacted material in the core.
(b) Calculate the corresponding average conversion for the product as a whole.
(c) If the amount of unreacted material so determined is unacceptably large, what plant modifications would you suggest to reduce it?

Solution

The fluidised bed will be considered as a continuous stirred tank reactor in which ideal macromixing of the particles occurs. As shown in the section on mixing (Chapter 2, Section 2.1.3), in the steady state the required *exit age distribution* is the same as the **C**-curve obtained using a single shot of tracer. In fact the desired **C**-curve is identical with that derived in Chapter 2, Fig. 2.3, for a tank containing a liquid with ideal micromixing, but now the argument is applied to particles as follows:

In Fig. 3.38 let W_F be the mass hold-up of particles in the reactor, G_0 the mass rate of outflow in the steady state (i.e. the holding time or residence time $\tau = W_F/G_0$), and c' the number of tracer particles per unit mass of all particles. Consider a shot of n_{m0} tracer particles input at $t = 0$, giving an initial value $c'_0 = n_{m0}/W_F$ in the bed. Applying a material balance:

$$\text{Inflow} - \text{Outflow} = \text{Accumulation}$$

$$0 \quad - \quad c'G_0 \quad = \quad W_F\frac{\mathrm{d}c'}{\mathrm{d}t}$$

i.e.:

$$\frac{\mathrm{d}c'}{\mathrm{d}t} = -\frac{G_0}{W_F}c' = -\frac{c'}{\tau}; \quad \text{whence} \quad c' = c'_0\,\mathrm{e}^{-t/\tau}$$

Feed
W_F
G_0
Air

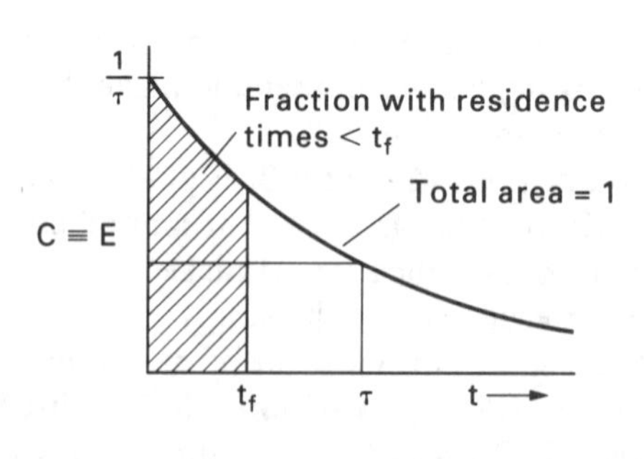

FIG. 3.38. Flow of particles through a fluidised bed showing a normalised **C**-curve which is identical with the exit age distribution function **E**

For a normalised **C**-curve we require:

$$\mathbf{C} = \frac{c'}{\int_0^\infty c'\,\mathrm{d}t} = \frac{c'_0\,\mathrm{e}^{-t/\tau}}{c'_0\int_0^\infty \mathrm{e}^{-t/\tau}\,\mathrm{d}t} = \frac{1}{\tau}\,\mathrm{e}^{-t/\tau} = \mathbf{E}$$

By definition the exit age distribution function **E** is such that the fraction of the exit stream with residence times between t and $t + \delta t$ is given by $\mathbf{E}\,\delta t$.

(a) To determine the proportion of the particles which leave the bed still containing some unreacted material in the core we calculate **F**, that fraction of particles which leave the bed with residence times less than t_f, the time for complete reaction of a particle:

$$\mathbf{F} = \int_0^{t_f} \frac{1}{\tau}\,\mathrm{e}^{-t/\tau}\,\mathrm{d}t = 1 - \mathrm{e}^{-t_f/\tau}$$

For the fluidised bed $\tau = \dfrac{10}{6/60} = 100$ min.

For a particle $t_f = 16$ min.

$\therefore$ Fraction with some unreacted material $= 1 - \mathrm{e}^{-16/100} = \underline{\underline{0.148}}.$

(b) The conversion α of reactant in a single particle depends on its length of stay in the bed. Working in terms of the fraction unconverted $(1 - \alpha)$ which, from equation 3.120, is given by:

$$(1 - \alpha) = \left(1 - \frac{t}{t_f}\right)^3$$

the mean value of the fraction unconverted:

$$(1 - \bar{\alpha}) = \sum (1 - \alpha)\,\mathbf{E}\,\delta t$$

where $\mathbf{E}\,\delta t$ is the fraction of the exit stream which has stayed in the reactor for times between t and $t + \delta t$ and in which $(1 - \alpha)$ is still unconverted,

i.e.:
$$(1 - \bar{\alpha}) = \int_0^{t_f} (1 - \alpha)\,\frac{1}{\tau}\,\mathrm{e}^{-t/\tau}\,\mathrm{d}t$$

where t_f is taken as the upper limit rather than ∞ because even if the particle stays in the reactor longer than the time for complete reaction, the conversion cannot exceed 100 per cent and for such particles $(1 - \alpha)$ is zero. Substituting for $(1 - \alpha)$:

$$(1 - \bar{\alpha}) = \int_0^{t_f} \left(1 - \frac{t}{t_f}\right)^3 \frac{1}{\tau}\,\mathrm{e}^{-t/\tau}\,\mathrm{d}t$$

Using repeated integration by parts:

$$(1 - \bar{\alpha}) = 1 - 3\left(\frac{\tau}{t_f}\right) + 6\left(\frac{\tau}{t_f}\right)^2 - 6\left(\frac{\tau}{t_f}\right)^3 (1 - \mathrm{e}^{-t_f/\tau})$$

If t_f/τ is small, it is useful to expand the exponential when several of the above terms cancel,

$$(1 - \bar{\alpha}) = \frac{1}{4}\left(\frac{t_f}{\tau}\right) - \frac{1}{20}\left(\frac{t_f}{\tau}\right)^2 + \frac{1}{120}\left(\frac{t_f}{\tau}\right)^3 - \ldots$$

In the present problem $t_f/\tau = 16/100$

i.e.:
$$(1 - \bar{\alpha}) = 0.04 - 0.0013 + \ldots = 0.039$$

Hence the average conversion for the product $\bar{\alpha} = 0.961$.

(c) To reduce the amount of unreacted material the following modifications could be suggested:

(i) reduce the solids feed rate,
(ii) increase the solids hold-up in the bed,
(iii) construct a second fluidised bed in series,
(iv) partition the proposed fluidised bed.

3.8. FURTHER READING

ARIS, R.: *Elementary Chemical Reactor Analysis* (Prentice-Hall, 1969).
ARIS, R.: *The Mathematical Theory of Diffusion and Reaction in Permeable Catalysts*, Vols. I and II (Oxford University Press, 1975).
BOND, J. C.: *Heterogeneous Catalysis: Principles and Applications*, 2nd edn. (Oxford Science Publications, 1987).
CAMPBELL, I. M.: *Catalysis at Surfaces* (Chapman and Hall, 1988).
CARBERRY, J. J.: *Chemical and Catalytic Reaction Engineering* (McGraw-Hill, 1976).
DORAISWAMY, L. K. and SHARMA, M. M.: *Heterogeneous Reactions*, Vol. 1, Gas–Solid and Solid–Solid Reactions (Wiley, 1984).
FROMENT, G. F. and BISCHOFF, K. B.: *Chemical Reactor Analysis and Design*, 2nd edn. (Wiley, 1990).
GATES, B. C.: *Catalytic Chemistry* (Wiley, 1992).
LAPIDUS, L. and AMUNDSON, N. R. (eds.): *Chemical Reactor Theory—A Review* (Prentice-Hall, 1977).
PETERSEN, E. E.: *Chemical Reaction Analysis* (Prentice-Hall, 1965).
RASE, H. F.: *Chemical Reactor Design for Process Plants*, Vols. 1 and 2 (Wiley, 1977).
RICHARDSON, J. T.: *Principles of Catalyst Development* (Plenum Press, 1989).
SATTERFIELD, C. N.: *Mass Transfer in Heterogeneous Catalysis* (M.I.T. Press, 1970).
SATTERFIELD, C. N.: *Heterogeneous Catalysis in Industrial Practice*, 2nd edn. (McGraw-Hill, 1991).
SMITH, J. M.: *Chemical Engineering Kinetics*, 3rd edn. (McGraw-Hill, 1981).
SZEKELY, J., EVANS, J. W. and SOHN, H. Y.: *Gas-Solid Reactions* (Academic Press, 1976).
THOMAS, J. M. and THOMAS, W. J.: *Introduction to the Principles of Heterogeneous Catalysis* (Academic Press, 1967).
TWIGG, M. V. (ed.): *Catalyst Handbook*, 2nd edn. (Wolfe Publishing, 1979).

3.9. REFERENCES

1. SATTERFIELD, C. N.: *Heterogeneous Catalysis in Industrial Practice*, 2nd edn. (McGraw-Hill, 1991).
2. BARRER, R. M. and GROVE, D. M.: *Trans. Faraday Soc.* **47** (1951) 826, 837. Flow of gases and vapours in a porous medium and its bearing on adsorption problems: I. Steady state of flow, II. Transient flow.
3. GUNN, D. J. and PRYCE, C.: *Trans. Inst. Chem. Eng.* **47** (1969) T341. Dispersion in packed beds.
4. MACDONALD, W. R. and HABGOOD, H. W.: *19th Canadian Chem. Eng. Conf., Edmonton* (1969) Paper 45. Measurement of diffusivities in zeolites by gas chromatographic methods.
5. HIRSHFELDER, J. O., CURTISS, C. F. and BIRD, R. B.: *Molecular Theory of Gases and Liquids* (Wiley, 1954).
6. GILLILAND, E. R.: *Ind. Eng. Chem.* **26** (1934) 681. Diffusion coefficients in gaseous systems.
7. EERKENS, J. W. and GROSSMAN, L. M.: *Tech. Report* HE/150/150 (Univ. Calif. Inst. Eng. Res., Dec. 5th 1957). Evaluation of the diffusion equation and tabulation of experimental diffusion coefficients.
8. KNUDSEN, M.: *Ann. Phys.* **28** (1909) 75. Die Gesetze der Molekularströmung und der inneren Reibungströmung der Gase durch Röhren. *Ann. Phys.* **35** (1911) 389. Molekularströmung des Wasserstoffs durch Röhren und das Hitzdrahtmanometer.
9. HERZFELD, K. and SMALLWOOD, M.: in *Treatise on Physical Chemistry* (edited by H. S. Taylor) **1**, 169 (Princeton Univ. Press, 1931).
10. EVERETT, D. H.: *Colston Res. Symp.* (Bristol, 1958). The structure and properties of porous materials: Some problems in the investigation of porosity by adsorption methods.
11. SCOTT, D. S. and DULLIEN, F. A. L.: *A.I.Ch.E.Jl.* **8** (1962) 113. Diffusion of ideal gases in capillaries and porous solids.
12. BOSANQUET, C. H.: *British TA Report* BR/507 (Sept. 27th 1944). The optimum pressure for a diffusion separation plant.
13. POLLARD, W. G. and PRESENT, R. D.: *Phys. Rev.* **73** (1948) 762. Gaseous self-diffusion in long capillary tubes.
14. THIELE, E. W.: *Ind. Eng. Chem.* **31** (1939) 916. Relation between catalytic activity and size of particle.
15. MICKLEY, H. S., SHERWOOD, T. K. and REED, C. E.: *Applied Mathematics in Chemical Engineering*, 2nd edn. (McGraw-Hill, 1957).
16. GUNN, D. J.: *Chem. Eng. Sci.* **22** (1967) 1439. Diffusion and chemical reaction in catalysis.
17. ARIS, R.: *Introduction to the Analysis of Chemical Reactors*, 236 (Prentice-Hall, 1965).
18. CARBERRY, J. J.: *A.I.Ch.E.Jl.* **7** (1961) 350. The catalytic effectiveness factor under nonisothermal conditions.
19. TINKLER, J. D. and METZNER, A. B.: *Ind. Eng. Chem.* **53** (1961) 663. Reaction rate in nonisothermal catalysts.
20. PETERSEN, E. E.: *Chem. Eng. Sci.* **17** (1962) 987. Nonisothermal chemical reaction in porous catalysts.

21. WEISZ, P. B. and HICKS, J. S.: *Chem. Eng. Sci.* **17** (1962) 265. The behaviour of porous catalyst particles in view of internal mass and heat diffusion effects.
22. GUNN, D. J.: *Chem. Eng. Sci.* **21** (1966) 383. Nonisothermal reaction in catalyst particles.
23. MCGREAVY, C. and THORNTON, J. M.: *Can. J. Chem. Eng.* **48** (1970) 187. Generalized criteria for the stability of catalytic reactors.
24. WEISZ, P. B. and PRATER, C. D.: *Adv. Catalysis* **6** (1954) 143. Interpretation of measurements in experimental catalysis.
25. PETERSEN, E. E.: *Chem. Eng. Sci.* **20** (1965) 587. A general criterion for diffusion influenced chemical reactions in porous solids.
26. WHEELER, A.: *Adv. Catalysis* **3** (1951) 249. Reaction rates and selectivity in catalyst pores.
27. ØSTERGAARD, K.: *Proc. 3rd Int. Cong. Catalysis, Amsterdam* **2** (1964) 1348. The influence of intraparticle heat and mass diffusion on the selectivity of parallel heterogeneous catalytic reactions.
28. SINFELT, J. H.: In *Advances in Chemical Engineering* Vol. 5, DREW, T. B., HOOPES, J. W. and VERMEULEN, T., eds. (Academic Press, 1964) p. 37. Bifunctional catalysis.
29. GUNN, D. J. and THOMAS, W. J.: *Chem. Eng. Sci.* **20** (1965) 89. Mass transport and chemical reaction in multifunctional catalyst systems.
30. THOMAS, W. J. and WOOD, R. M.: *Chem. Eng. Sci.* **22** (1967) 1607. Use of maximum principle to calculate optimum catalyst composition profile for bifunctional catalyst systems contained in tubular reactors.
31. GUNN, D. J.: *Chem. Eng. Sci.* **22** (1967) 963. The optimisation of bifunctional catalyst systems.
32. JACKSON, R.: *J. Optim. Theory Applic.* **2** (1968) 1. Optimal use of mixed catalysts.
33. JENKINS, B. and THOMAS, W. J.: *Can. J. Chem. Eng.* **48** (1970) 179. Optimum catalyst formulation for the aromatization of methylcyclopentane.
34. HORN, F.: *Z. Electrochem.* **65** (1961) 209. Optimum temperature regulation for continuous chemical processes.
35. CHOU, A., RAY, W. H. and ARIS, R.: *Trans. Inst. Chem. Eng.* **45** (1967) T153. Simple control policies for reactors with catalyst decay.
36. JACKSON, R.: *Trans. Inst. Chem. Eng.* **45** (1967) T160. An approach to the numerical solution of time-dependent optimisation problems in two-phase contacting devices.
37. OGUNYE, A. F. and RAY, W. H.: *Trans. Inst. Chem. Eng.* **46** (1968) T225. Non-simple control policies for reactors with catalyst decay.
38. THALLER, L. and THODOS, G.: *A.I.Ch.E.Jl.* **6** (1960) 369. The dual nature of a catalytic reaction: the dehydration of *sec*-butyl alcohol to methyl ethyl ketone at elevated pressures.
39. TAMARU, K.: *Trans. Faraday Soc.* **55** (1959) 824. Adsorption during the catalytic decomposition of formic acid on silver and nickel catalysts.
40. HOUGEN, O. A. and WATSON, K. M.: *Chemical Process Principles*, Part 3, 938 (Wiley, 1947).
41. THOMAS, W. J. and JOHN, B.: *Trans. Inst. Chem. Eng.* **45** (1967) T119. Kinetics and catalysis of the reactions between sulphur and hydrocarbons.
42. FROMENT, G. F. and BISCHOFF, K. B.: *Chemical Reactor Analysis and Design*, 2nd edn. (Wiley, 1990).
43. MACE, C. V. and BONILLA, C. F.: *Chem. Eng. Prog.* **50** (1954) 385, The conversion of ethylene to ethanol.
44. RASE, H. F.: *Fixed-Bed Reactor Design and Diagnostics* (Butterworths, 1990).
45. WENNER, R. R. and DYBDAL, E. C.: *Chem. Eng. Prog.* **44** (1948) 275. Catalytic dehydrogenation of ethylbenzene
46. ERGUN, S.: *Chem. Eng. Prog.* **48** No. 2 (Feb. 1952) 89. Fluid flow through packed columns.
47. DENBIGH, K. G.: *Trans. Faraday Soc.* **40** (1944) 352. Velocity and yield in continuous reaction systems.
48. DENBIGH, K. G.: *Chem. Eng. Sci.* **8** (1958) 125. Optimum temperature sequences in reactors.
49. PETERSEN, E. E.: *Chemical Reaction Analysis*, 186 (Prentice-Hall, 1965).
50. VAN HEERDEN, C.: *Ind. Eng. Chem.* **45** (1953) 1242. Autothermic processes—properties and reactor design.
51. WEHNER, J. F. and WILHELM, R. H.: *Chem. Eng. Sci.* **6** (1956) 89. Boundary conditions of flow reactor.
52. MCHENRY, K. W. and WILHELM, R. H.: *A.I.Ch.E.Jl.* **3** (1957) 83. Axial mixing of binary gas mixtures flowing in a random bed of spheres.
53. ARIS, R. and AMUNDSON, N. R.: *A.I.Ch.E.Jl.* **3** (1957) 280. Longitudinal mixing or diffusion in fixed beds.
54. KRAMERS, H. and ALBERDA, G.: *Chem. Eng. Sci.* **2** (1953) 173. Frequency-response analysis of continuous-flow systems.
55. BOUSSINESQ, M. J.: *Mém. prés. div. Sav. Acad. Sci. Inst. Fr.* **23** (1877) 1–680 (see 31). Essai sur la théorie des eaux courantes.
56. DEANS, H. A. and LAPIDUS, L.: *A.I.Ch.E.Jl.* **6** (1960) 656, 663. A computational model for predicting and correlating the behavior of fixed-bed reactors: I. Derivation of model for nonreactive systems, II. Extension to chemically reactive systems.

57. BEEK, J.: In *Advances in Chemical Engineering* Vol. **3**, DREW, T. B., HOOPES, J. W. and VERMEULEN, T., eds. (Academic Press, 1962) p. 203. Design of packed catalytic reactors.
58. BILOUS, O. and AMUNDSON, N. R.: *A.I.Ch.E.Jl.* **2** (1956) 117. Chemical reactor stability and sensitivity II. Effect of parameters on sensitivity of empty tubular reactors.
59. MAY W. G.: *Chem. Eng. Prog.* **55** No. 12 (Dec. 1959) 49. Fluidized-bed reactor studies.
60. ROWE, P. N.: *Chem. Eng. Prog.* **60** No. 3 (March 1964) 75. Gas-solid reaction in a fluidized bed.
61. WEN, C. Y.: *Ind. Eng. Chem.* **60** (9) (1968) 34. Noncatalytic heterogeneous solid fluid reaction models.
62. SZEKELY, J., EVANS, J. W. and SOHN, H. Y.: *Gas–Solid Reactions* (Academic Press, 1976).
63. DENBIGH, K. G. and BEVERIDGE, G. S. G.: *Trans. Inst. Chem. Eng.* **40** (1962) 23. The oxidation of zinc sulphide spheres in an air stream.

3.10. NOMENCLATURE

		Units in SI system	Dimension in M, N, L, T, θ
$\mathscr{A}$	Frequency factor in Arrhenius equation	s^{-1*}	$\mathbf{T}^{-1*}$
A_c	Cross-sectional area	m^2	$\mathbf{L}^2$
a	Constant in equation 3.115	—	—
a	Interfacial area per unit volume	m^{-1}	$\mathbf{L}^{-1}$
a_B	Activity of component **B**	—	—
a'	Surface area for heat transfer per unit hydraulic radius	m	**L**
B	$\left(=\dfrac{-\Delta H C_{A\infty} D_e \mathbf{R}}{2rh\mathbf{E}}\right)$ Dimensionless parameter introduced by McGREAVY and THORNTON (see Figs. 3.7 and 3.8)	—	—
b	Constant in equation 3.115	—	—
b	Stoichiometric ratio	—	—
C	Outlet response of concentration divided by area under C–t curve	s^{-1}	$\mathbf{T}^{-1}$
C	Total molar concentration	kmol/m³	$\mathbf{NL}^{-3}$
C_A	Molar concentration of component A	kmol/m³	$\mathbf{NL}^{-3}$
C_A	Surface concentration of A		$\mathbf{L}^{-2}$
C_{Ai}	Molar concentration at interface	kmol/m³	$\mathbf{NL}^{-3}$
C_M	Molar concentration at centre of pellet	kmol/m³	$\mathbf{NL}^{-3}$
C_S	Concentration of available sites		$\mathbf{L}^{-2}$
C_V	Concentration of vacant sites		$\mathbf{L}^{-2}$
$C_{A\infty}$	Molar concentration at infinity	kmol/m³	$\mathbf{NL}^{-3}$
$\bar{c}_p$	Mean heat capacity of fluid per unit mass	J/kgK	$\mathbf{L}^2\mathbf{T}^{-2}\boldsymbol{\theta}^{-1}$
$\bar{c}_{pc}$	Mean heat capacity of coolant per unit mass	J/kgK	$\mathbf{L}^2\mathbf{T}^{-2}\boldsymbol{\theta}^{-1}$
c'	Number of tracer particles per unit mass of all particles	kg^{-1}	$\mathbf{M}^{-1}$
c'_0	Initial value of c'	kg^{-1}	$\mathbf{M}^{-1}$
D	Molecular bulk diffusion coefficient	m²/s	$\mathbf{L}^2\mathbf{T}^{-1}$
D_e	Effective diffusivity	m²/s	$\mathbf{L}^2\mathbf{T}^{-1}$
D_{eK}	Effective diffusivity in Knudsen regime	m²/s	$\mathbf{L}^2\mathbf{T}^{-1}$
D_{eM}	Effective diffusivity in molecular regime	m²/s	$\mathbf{L}^2\mathbf{T}^{-1}$
D_K	Knudsen diffusion coefficient	m²/s	$\mathbf{L}^2\mathbf{T}^{-1}$
D_p	Diffusion coefficient for forced flow	m²/s	$\mathbf{L}^2\mathbf{T}^{-1}$
D'_e	Effective diffusivity based on concentration expressed as Y	kg/ms	$\mathbf{ML}^{-1}\mathbf{T}^{-1}$
D'_L	Dispersion coefficient in longitudinal direction based on concentration expressed as Y	kg/ms	$\mathbf{ML}^{-1}\mathbf{T}^{-1}$
D'_r	Radial dispersion coefficient based on concentration expressed as Y	kg/ms	$\mathbf{ML}^{-1}\mathbf{T}^{-1}$
d	Tube diameter	m	**L**
d_p	Particle diameter	m	**L**
d'_M	Volume of voids per total surface area	m	**L**
E	Exit age distribution	s^{-1}	$\mathbf{T}^{-1}$
E	Activation energy per mole for chemical reaction	J/kmol	$\mathbf{MN}^{-1}\mathbf{L}^2\mathbf{T}^{-2}$
$\mathbf{E}_D$	Apparent activation energy per mole in diffusion controlled region	J/kmol	$\mathbf{MN}^{-1}\mathbf{L}^2\mathbf{T}^{-2}$

* For first-order reaction. Dimensions are a function of order of reaction.

		Units in SI system	Dimension in **M, N, L, T, θ**
e	Void fraction	—	—
F	C/C_∞	—	—
F	Ratio of activity in poisoned catalyst pellet to activity in unpoisoned pellet	—	—
G	Mass flowrate	kg/s	$\mathbf{MT^{-1}}$
G_0	Mass outflow ratio in steady state	kg/s	$\mathbf{MT^{-1}}$
G'	Mass flow per unit area	kg/m²s	$\mathbf{ML^{-2}T^{-1}}$
G'_C	Mass flowrate of coolant per unit cross-section of reactor tube	kg/m²s	$\mathbf{ML^{-2}T^{-1}}$
ΔH	Difference of enthalpy between products and reactants (heat of reaction = $-\Delta H$)	J/kmol	$\mathbf{MN^{-1}L^{2}T^{-2}}$
h	Heat transfer coefficient	W/m²K	$\mathbf{MT^{-3}\theta^{-1}}$
h_D	Mass transfer coefficient	m/s	$\mathbf{LT^{-1}}$
j	Mass transfer factor	—	—
K	Thermodynamic equilibrium constant—for **A** + **B** ? **P**	ms²/kg	$\mathbf{M^{-1}LT^{2}}$
	—for **A** ? **P**	—	—
K_A	Adsorption–desorption equilibrium constant for **A**	ms²/kg	$\mathbf{M^{-1}LT^{2}}$
K_B	Adsorption–desorption equilibrium constant for **B**	ms²/kg	$\mathbf{M^{-1}LT^{2}}$
K_P	Adsorption–desorption equilibrium constant for **P**	ms²/kg	$\mathbf{M^{-1}LT^{2}}$
K_S	Equilibrium constant for surface reaction	—	—
k	Chemical rate constant for forward reaction	s^{-1}*	$\mathbf{T^{-1}}$*
k'	Chemical rate constant for reverse reaction, per unit volume of particle or reactors	s^{-1}*	$\mathbf{T^{-1}}$*
$\bar{k}$	Overall rate *or* reaction constant per unit volume	s^{-1}*	$\mathbf{T^{-1}}$*
k .	Chemical rate constant for forward surface reaction (molecules per unit area and unit time)	$m^{-2}s^{-1}$	$\mathbf{L^{-2}T^{-1}}$
k′	Chemical rate constant for reverse surface reaction (molecules per unit area and unit time)	$m^{-2}s^{-1}$	$\mathbf{L^{-2}T^{-1}}$
$\mathbf{k}_{Aa}$	Rate constant for adsorption of **A** (see equation 3.70)	s/kg m	$\mathbf{M^{-1}L^{-1}T}$
$\mathbf{k}_{Ad}$	Rate constant for desorption of **A** (see equation 3.70)	$m^{-2}s^{-1}$	$\mathbf{L^{-2}T^{-1}}$
k_e	Effective thermal conductivity	W/mK	$\mathbf{MLT^{-3}\theta^{-1}}$
k_G	Gas film mass transfer coefficient	s/m	$\mathbf{L^{-1}T}$
k_l	Thermal conductivity in longitudinal direction	W/mK	$\mathbf{MLT^{-3}\theta^{-1}}$
$\mathbf{k}_{Pa}$	Rate constant for adsorption of **P** (see equation 3.72) (molecules per unit area, unit time and unit partial pressure of P)	s/kgm	$\mathbf{M^{-1}L^{-1}T}$
$\mathbf{k}_{Pd}$	Rate constant for desorption of **P** (see equation 3.72) (molecules per unit area and unit time)	$m^{-2}s^{-1}$	$\mathbf{L^{-2}T^{-1}}$
k_r	Thermal conductivity in radial direction	W/mK	$\mathbf{MLT^{-3}\theta^{-1}}$
k_s	Chemical rate constant for forward surface reaction (molecules per unit time)	s^{-1}	$\mathbf{T^{-1}}$
$\mathbf{k}'_s$	Chemical rate constant for reverse surface reaction (molecules per unit time)	s^{-1}	$\mathbf{T^{-1}}$
$\mathbf{k}_s$	Rate constant in equation 3.116	m/s	$\mathbf{LT^{-1}}$
$\mathbf{k}'_e$	Rate constant for reverse surface reaction (molecules per unit time)	s^{-1}	$\mathbf{T^{-1}}$
L	Length of reactor	m	**L**
M	Molecular weight	kg/kmol	$\mathbf{MN^{-1}}$
N	Avogadro's number (6.023×10^{26} molecules per kmol or 6.023×10^{23} molecules per mole)	$kmol^{-1}$	$\mathbf{N^{-1}}$
N'	Molar flux of material	kmol/ms²	$\mathbf{NL^{-2}T^{-1}}$
n	Order of chemical reaction	—	—
n_A	Number of moles of A	kmol	**N**
n_B	Number of moles of B	kmol	**N**
$\mathbf{n}_{M0}$	Number of tracer particles injected	—	—
$\mathbf{n}_A$	Number of adsorbed molecules of component **A**	—	—
$\mathbf{n}_D$	Apparent order of chemical reaction in diffusion controlled regime	—	—
$\mathbf{n}_P$	Number of adsorbed molecules of component **P**	—	—
$\mathbf{n}_S$	Number of available active sites	—	—
P	Pressure	N/m²	$\mathbf{ML^{-1}T^{-2}}$

Symbol	Description	Units in SI system	Dimension in **M, N, L, T, θ**
P_A	Partial pressure of component **A**	N/m^2	$\mathbf{ML^{-1}T^{-2}}$
P_{Ai}	Partial pressure of component **A** at interface	N/m^2	$\mathbf{ML^{-1}T^{-2}}$
Q	Heat transferred per unit volume and unit time to surroundings	W/m^3	$\mathbf{ML^{-1}T^{-3}}$
Q_1^M	Heat generated by chemical reaction per unit mass	J/kg	$\mathbf{L^2T^{-2}}$
Q_2^M	Heat gain by feed to the reactor per unit mass	J/kg	$\mathbf{L^2T^{-2}}$
$\mathbf{R}$	Gas constant	J/kmolK	$\mathbf{MN^{-1}L^2T^{-2}\theta^{-1}}$
R_1	Component of drag force per unit area in direction of flow	N/m^2	$\mathbf{ML^{-1}T^{-2}}$
$\mathcal{R}^M$	Reaction rate (moles per unit mass of catalyst and unit time)	kmol/kg s	$\mathbf{NM^{-1}T^{-1}}$
$\mathcal{R}^M_{YT}$	Reaction rate (moles per unit mass of catalyst and unit time) when a function of Y and T	kmol/kg s	$\mathbf{NM^{-1}T^{-1}}$
$\mathcal{R}_V$	Reaction rate (moles per unit volume of reactor and unit time)	kmol/m^{3}s	$\mathbf{NL^{-3}T^{-1}}$
$\mathcal{R}'_V$	Reaction rate (moles per unit volume of particle and unit time)	kmol/m^{3}s	$\mathbf{NL^{-3}T^{-1}}$
$\mathcal{R}''$	Surface reaction rate (molecules per unit area and unit time)	m^{-2}s^{-1}	$\mathbf{L^{-2}T^{-1}}$
$\mathcal{R}''_{Aa}$	Rate of adsorption of **A** (molecules per unit area and unit time)	m^{-2}s^{-1}	$\mathbf{L^{-2}T^{-1}}$
$\mathcal{R}''_{AS}$	Rate of surface reaction of **A** (molecules per unit area and unit time)	m^{-2}s^{-1}	$\mathbf{L^{-2}T^{-1}}$
$\mathcal{R}_{Pd}$	Rate of desorption of **P** (molecules per unit area and unit time)	m^{-2}s^{-1}	$\mathbf{L^{-2}T^{-1}}$
$\mathcal{R}''_S$	Rate of surface reaction (molecules per unit area and unit time)	m^{-2}s^{-1}	$\mathbf{L^{-2}T^{-1}}$
r	Radius	m	**L**
r_c	Radial position in core at which reaction is taking place	m	**L**
r_0	Initial value of r_c or radius of sphere or cylinder	m	**L**
r_e	Radius of cylinder with same surface to volume ratio as pore	m	**L**
S	Catalyst selectivity	—	—
S_g	Specific surface area (per unit mass)	m^2/kg	$\mathbf{M^{-1}L^2}$
S_x	External surface area	m^2	$\mathbf{L^2}$
S'	Total catalyst surface area	m^2	$\mathbf{L^2}$
T	Temperature	K	$\boldsymbol{\theta}$
T_C	Temperature of coolant stream	K	$\boldsymbol{\theta}$
T_{CL}	Temperature of coolant at reactor outlet	K	$\boldsymbol{\theta}$
T_{C0}	Temperature of coolant at reactor inlet	K	$\boldsymbol{\theta}$
T_i	Temperature at exterior surface of pellet	K	$\boldsymbol{\theta}$
T_L	Temperature at exit of reactor	K	$\boldsymbol{\theta}$
T_M	Temperature at particle centre	K	$\boldsymbol{\theta}$
T_{MX}	Maximum temperature	K	$\boldsymbol{\theta}$
T_W	Wall temperature	K	$\boldsymbol{\theta}$
ΔT	($= T_0 - T_{CL}$) Temperature difference of the non-reacting fluid over length of reactor	K	$\boldsymbol{\theta}$
t	Time	s	**T**
t_f	Time for complete conversion	s	**T**
U	Overall heat transfer coefficient	W/m^{2}K	$\mathbf{MT^{-3}\theta^{-1}}$
u	Gas velocity	m/s	$\mathbf{LT^{-1}}$
u_1	Mean velocity in voids	m/s	$\mathbf{LT^{-1}}$
V_g	Specific pore volume per unit mass	m^3/kg	$\mathbf{M^{-1}L^3}$
V_p	Particle volume	m^3	$\mathbf{L^3}$
v	Volumetric rate of flow	m^3/s	$\mathbf{L^3T^{-1}}$
W	Mass of catalyst	kg	**M**
W_F	Mass of particles in reactor (holdup)	kg	**M**
x	Distance in x-direction	m	**L**
x_A	Mole fraction of A	—	—
Y	Moles of reactant per unit mass of fluid	kmol/kg	$\mathbf{NM^{-1}}$
Y_L	Moles per unit mass of reactant (or product) at reactor exit	kmol/kg	$\mathbf{NM^{-1}}$
Y_{MX}	Moles per unit mass of reactant (or product) for which temperature is maximum	kmol/kg	$\mathbf{NM^{-1}}$
Y_0	Moles per unit mass of reactant (or product) at reactor inlet	kmol/kg	$\mathbf{NM^{-1}}$

		Units in SI system	Dimension in $\mathbf{M, N}$, L, $\mathbf{T}$, $\boldsymbol{\theta}$
z	Length variable for reactor tube	m	$\mathbf{L}$
Pe	$(= ud_p/D)$ Peclet number	—	—
Re	$(= u\rho d_p/\mu)$ Reynolds number	—	—
Sc	$(= \mu/\rho D)$ Schmidt number	—	—
Sh	$(= h_D d_p/D)$ Sherwood number	—	—
α	Fractional conversion	—	—
$\bar{\alpha}$	Average value of α	—	—
β	Parameter representing maximum temperature difference in particle relative to external surface (equation 3.28)	—	—
Γ	$(= G'\bar{c}_P/G'_C\bar{c}_{PC})$ Relative heat capacities of two fluid streams to exchange heat	—	—
Γ'	$(= UaL/G'_C\bar{c}_{PC})$ Heat transfer factor	—	—
ε	Particle porosity	—	—
ε'	Dimensionless activation energy (equation 3.29)	—	—
ζ	Fraction of surface poisoned	—	—
η	Effectiveness factor		
θ_A	Fraction of surface occupied by component $\mathbf{A}$	—	—
λ	Defined as $\sqrt{\dfrac{k}{D_e}}$ for first order process (equation 3.15)	m^{-1}	$\mathbf{L}^{-1}$
μ	Viscosity	Ns/m^2	$\mathbf{ML}^{-1}\mathbf{T}^{-1}$
ρ	Fluid density	kg/m^3	$\mathbf{ML}^{-3}$
ρ_b	Bulk density	kg/m^3	$\mathbf{ML}^{-3}$
ρ_p	Particle density	kg/m^3	$\mathbf{ML}^{-3}$
ρ_s	True density of solid	kg/m^3	$\mathbf{ML}^{-3}$
τ	Residence time	s	$\mathbf{T}$
τ^2	Tortuosity factor	—	—
ϕ	Thiele modulus (equation 3.19)	—	—
$\bar{\phi}$	Generalised Thiele modulus (equation 3.20)	—	—
ϕ'	Modified Thiele modulus (equation 3.35)	—	—

Subscripts

A	Refers to component **A**
B	Refers to component **B**
G	Refers to bulk value in gas phase
P	Refers to product **P**
X	Refers to component **X**
∞	Value as time approaches infinity
in	Inlet value

CHAPTER 4

Gas–Liquid and Gas–Liquid–Solid Reactors

4.1. GAS–LIQUID REACTORS

4.1.1. Gas–Liquid Reactions

In a number of important industrial processes, it is necessary to carry out a reaction between a gas and a liquid. Usually the object is to make a particular product, for example, a chlorinated hydrocarbon such as chlorobenzene by the reaction of gaseous chlorine with liquid benzene. Sometimes the liquid is simply the reaction medium, perhaps containing a catalyst, and all the reactants and products are gaseous. In other cases the main aim is to separate a constituent such as CO_2 from a gas mixture; although pure water could be used to remove CO_2, a solution of caustic soda, potassium carbonate or ethanolamine has the advantages of increasing both the absorption capacity of the liquid and the rate of absorption. The subject of gas–liquid reactor design thus really includes absorption with chemical reaction which is discussed in Volume 2, Chapter 12.

4.1.2. Types of Reactors

The types of equipment used for gas–liquid reactions are shown in Fig. 4.1. That shown in *a* is the conventional packed column which is often used when the purpose is to absorb a constituent from a gas. The liquid is distributed in the form of films over the packing, and the gas forms the continuous phase. The pressure drop for the gas is relatively low and the packed column is therefore very suitable for treating large volume flows of gas mixtures. Figure 4.1*b* shows a spray column; as in the packed column the liquid hold-up is comparatively low, and the gas is the continuous phase. In the sieve tray column shown in *c*, however, and in bubble columns, gas is dispersed in the liquid passing over the tray. Because the tray is relatively shallow, the pressure drop in the gas phase is fairly low, and the liquid hold-up, although a little larger than for a packed column, is still relatively small. The tray column is useful when stagewise operation is required and a relatively large volumetric flow-rate of gas is to be treated.

When a high liquid hold-up is required in the reactor, one of the types shown in Fig. 4.1 *d*, *e* and *f* may be used. The bubble column *d* is simply a vessel filled with liquid, with a sparger ring at the base for dispersing the gas. In some cases a draught

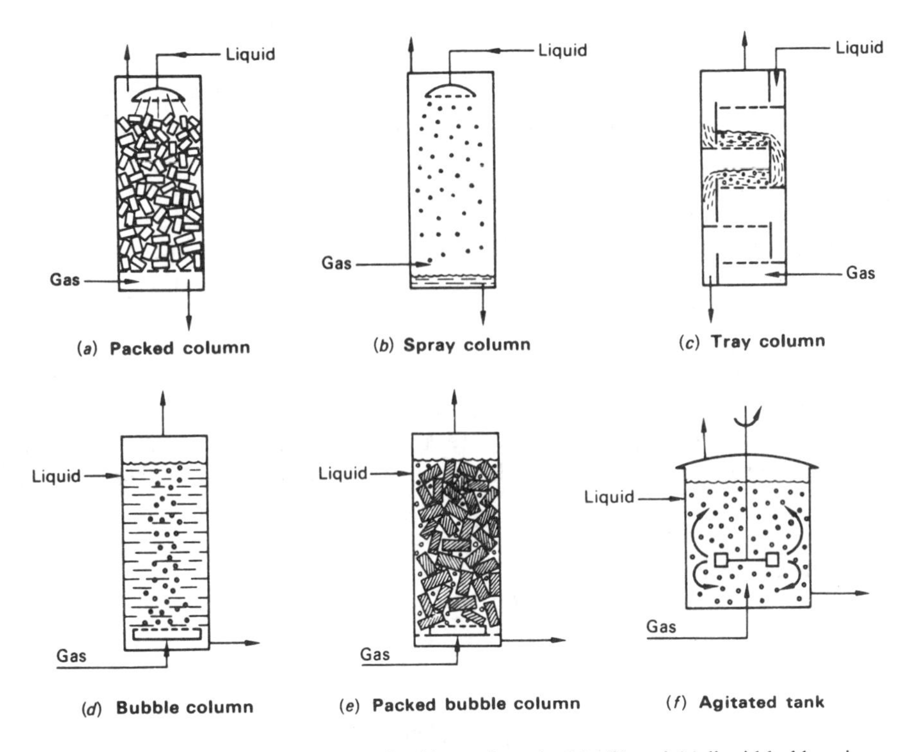

FIG. 4.1. Equipment used for gas–liquid reactions: in (*a*), (*b*) and (*c*) liquid hold-up is low; in (*d*), (*e*) and (*f*) liquid hold-up is high

tube is used to direct recirculation of the liquid and to influence the bubble motion. One of the disadvantages of the simple bubble column is that coalescence of the bubbles tends to occur with the formation of large slugs whose upper surfaces are in the form of spherical caps. By packing the vessel with Raschig rings, for example, as in *e*, the formation of very large bubbles is avoided. The reactor thus becomes an ordinary packed column operated in a flooded condition and with a sparger to disperse the gas; naturally the maximum superficial gas velocity is much less than in an unflooded packed column. Finally in *f* an agitator is used to disperse the gas into the liquid contained in a tank. A vaned-disc type of impeller is normally used (see Volume 2, Chapter 12, Section 12.9.1). An agitated tank provides small bubbles and thus a high interfacial area of contact between gas and liquid phases, but its greater mechanical complication compared with a simple bubble column is a disadvantage with corrosive materials and at high pressures and temperatures. If necessary, stagewise operation can be achieved by having several compartments in a vertical column, with impellers mounted on a common shaft (see Section 4.1.7 below)

4.1.3. Equations for Mass Transfer with Chemical Reaction

In designing a gas–liquid reactor, there is the need not only to provide for the required temperature and pressure for the reaction, but also to ensure adequate interfacial area of contact between the two phases. Although the reactor as such is classed as heterogeneous, the chemical reaction itself is really homogeneous, occurring in either the liquid phase, which is the most common, or in the gas phase, and only rarely in both. Many different varieties of gas–liquid reactions have been considered theoretically, and a large number of analytical or numerical results can be found in the literature[1,2].

The absorption of a gas by a liquid with simultaneous reaction in the liquid phase is the most important case. There are several theories of mass transfer between two fluid phases (see Volume 1, Chapter 10; Volume 2, Chapter 12), but for the purpose of illustration the film theory will be used here. Results from the possibly more realistic penetration theory are similar numerically, although more complicated in their mathematical form[3,4].

Consider a *second order* reaction in the liquid phase between a substance **A** which is transferred from the gas phase and reactant **B** which is in the liquid phase only. The gas will be taken as consisting of pure **A** so that complications arising from gas film resistance are avoided. The stoichiometry of the reaction is represented by:

$$v_A\mathbf{A} + v_B\mathbf{B} \rightarrow \text{Products}$$

with the rate equation:

$$\mathcal{R}_A = k_2 C_A C_B \tag{4.1}$$

Note that:

$$\mathcal{R}_B = \frac{v_B}{v_A}\mathcal{R}_A$$

In the film theory, steady state conditions are assumed in the film such that, in any volume element, the difference between the rate of mass transfer into and out of the element is just balanced by the rate of reaction within the element. Carrying out such a material balance on reactant **A** the following differential equation results:

$$D_A \frac{d^2 C_A}{dx^2} - \mathcal{R}_A = 0 \tag{4.2}$$

Similarly a material balance on **B** gives:

$$D_B \frac{d^2 C_B}{dx^2} - \mathcal{R}_B = 0 \tag{4.3}$$

Introducing the rate equation (4.1) above, the differential equations become:

$$D_A \frac{d^2 C_B}{dx^2} - k_2 C_A C_B = 0 \tag{4.4}$$

$$D_B \frac{d^2 C_B}{dx^2} - k_2 C_A C_B \frac{v_B}{v_A} = 0 \tag{4.5}$$

Typical conditions within the film may be seen in Fig. 4.2*a*. Boundary conditions at the gas–liquid interface are:

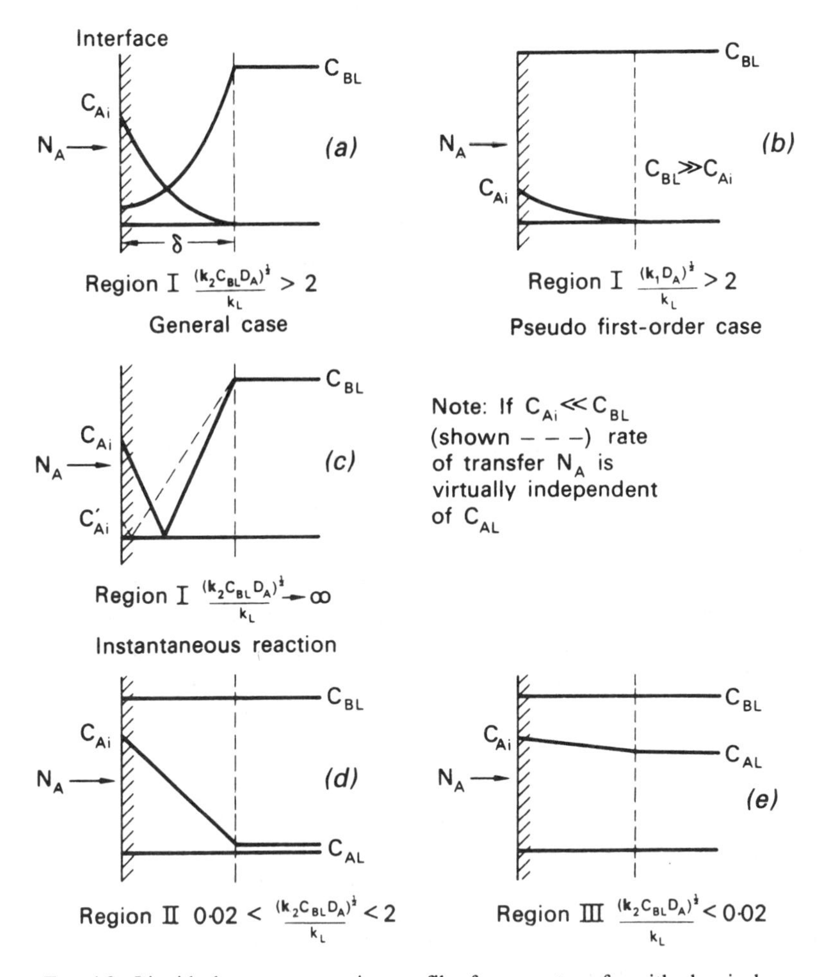

FIG. 4.2. Liquid phase concentration profiles for mass transfer with chemical reaction – film theory

for **A**:

$$C_A = C_{Ai}, \qquad x = 0 \tag{4.6}$$

for **B**:

$$\frac{dC_B}{dx} = 0, \qquad x = 0 \tag{4.7}$$

On the liquid side of the film where $x = \delta$, the boundary condition for **B** is simply:

$$C_B = C_{BL}, \qquad x = \delta \tag{4.8}$$

To obtain the boundary condition for **A** we note that, except for the amount of **A** which reacts in the film, **A** is transferred across this boundary and reacts in the bulk of the liquid. For unit area of interface, the volume of this bulk liquid may be written as $[(\varepsilon_L/a) - \delta]$, where ε_L is the volume of liquid per unit volume of reactor space (i.e. the liquid hold-up), a is the gas–liquid interfacial area per unit volume of reactor space (i.e. specific area), and δ is the thickness of the film. The boundary condition for **A** is thus:

$$-D_A\left(\frac{dC_A}{dx}\right)_{x=\delta} = \boldsymbol{k}_2 C_{AL} C_{BL}\left(\frac{\varepsilon_L}{a} - \delta\right) \tag{4.9}$$

Mass transfer across boundary — Reaction in bulk liquid

Although a complete analytical solution of the set of equations 4.4 to 4.9 is not possible, analytical solutions are obtainable for part of the range of variables[3,5] and a numerical solution has been obtained for the remainder[6].

The results of these solutions will be discussed in terms of a reaction factor f_A which is defined by the expression:

$$N_A = k_L C_{Ai} f_A \tag{4.10}$$

which may be compared with equation 4.11 which applies when there is no reaction:

$$N_A = k_L(C_{Ai} - C_{AL}) \tag{4.11}$$

where N_A is the molar flux of **A** at the interface, and k_L the mass transfer coefficient for physical absorption of **A**. When greater than 1, f_A represents the enhancement of the rate of transfer of **A** caused by the chemical reaction, as compared with pure physical absorption with a zero concentration of **A** in the bulk liquid.

The complete solution of the set of equations 4.4 to 4.9 is shown in graphical form in Fig. 4.3, in which the reaction factor f_A is plotted against a dimensionless parameter:

$$\frac{\sqrt{\boldsymbol{k}_2 C_{BL} D_A}}{k_L}$$

(The right-hand side of this diagram, region I, is the same as Fig. 12.12 of Volume 2.) The physical significance of the various regions covered by this diagram is important and is best appreciated by considering the corresponding concentration profiles of Fig. 4.2 as follows.

Rate of Transformation of A Per Unit Volume of Reactor

The value of the parameter:

$$\frac{\sqrt{\boldsymbol{k}_2 C_{BL} D_A}}{k_L}$$

provides an important indication of whether a large specific interfacial area a or a large liquid hold-up ε_L is required in a reactor to be designed for a particular reaction of rate constant $\boldsymbol{k}_2$. Let the rate of transformation of **A** per unit volume of reactor be J_A. Three regions may be distinguished as shown in Fig. 4.3.

Region I:

$$\frac{\sqrt{\boldsymbol{k}_2 C_{BL} D_A}}{k_L} > 2$$

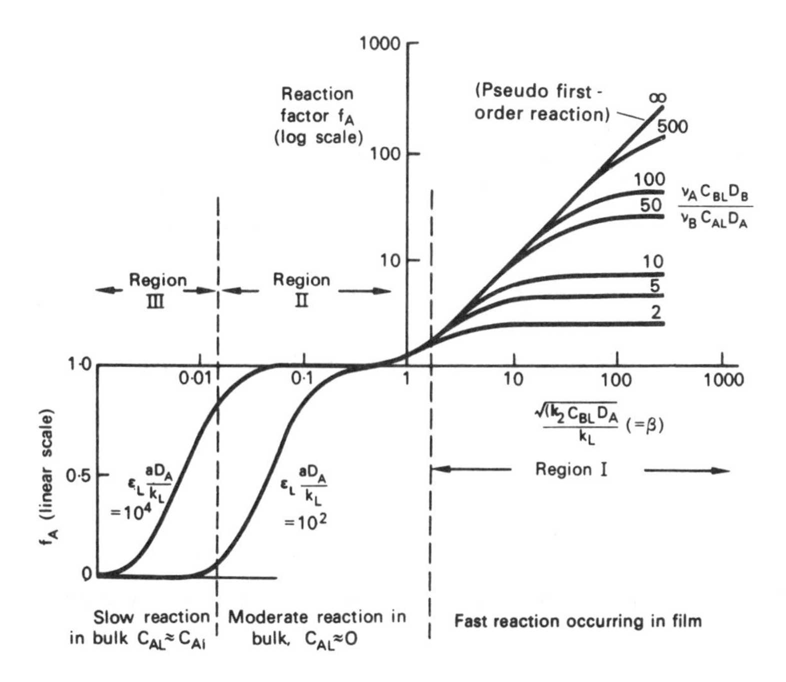

FIG. 4.3. Reaction factor f_A for a second-order reaction (numerical solution) and pseudo first-order reaction (analytical solution)

The reaction is fast and occurs mainly in the film as **A** is being transported (Fig. 4.2*a*). C_{AL}, the concentration of **A** in the bulk of the liquid, is virtually zero. Thus:

$$J_A = k_L C_{Ai} f_A a \tag{4.12}$$

Therefore, in this set of circumstances, J_A will be large if a is large; a large interfacial area a is required in the reactor but the liquid hold-up is not important. A packed column, for example, would be suitable.

If the concentration of **B** C_{BL} in the bulk liquid is much greater than C_{Ai}, a common case being where **A** reacts with a pure liquid **B**, the kinetics of the reaction become pseudo first-order, and the above equations can be solved analytically to give:

$$f_A = \frac{\beta}{\tanh \beta} \quad \text{where } \beta = \frac{\sqrt{\boldsymbol{k}_1 D_A}}{k_L} \tag{4.13}$$

$\boldsymbol{k}_1$ is the pseudo first-order rate constant. The concentration profile is sketched in Fig. 4.2*b*. Note that for large β, $f_A \rightarrow \beta$ and hence:

$$N_A = \sqrt{\boldsymbol{k}_1 D_A}\, C_{Ai} \tag{4.14}$$

N_A is then independent of mass transfer coefficient k_L.

On the other hand, if the reaction between **A** and **B** is virtually instantaneous, as shown in Fig. 4.2*c*, and $C_{Ai} \ll C_{BL}$, the rate of transformation is nearly independent of C_{Ai}, being determined mainly by the rate of mass transfer of **B** to the reaction zone.

Region II:

$$0.02 < \frac{\sqrt{k_2 C_{BL} D_A}}{k_L} < 2$$

This is an intermediate region in which the reaction is sufficiently fast to hold C_{AL} close to zero (Fig. 4.2*d*), but, although C_{AL} is small, nearly all the reaction occurs in the bulk of the liquid. The hold-up ε_L is important because unless:

$$\varepsilon_L \bigg/ \frac{a D_A}{k_L} > 10^2$$

f_A will be substantially less than 1 in the part of the region (see Fig. 4.3) where:

$$\frac{\sqrt{k_2 C_{BL} D_A}}{k_L} < 0.1$$

When $f_A = 1$:

$$J_A = k_L C_{Ai} a \tag{4.15}$$

Thus, both interfacial area and hold-up should be high. For example, an agitated tank will give a high value of J_A.

Region III:

$$\frac{\sqrt{k_2 C_{BL} D_A}}{k_L} < 0.02$$

The reaction is slow and occurs in the bulk of the liquid. Mass transfer serves to keep the bulk concentration C_{AL} of **A**, close to the saturation value C_{Ai} (Fig. 4.2*e*) and sufficient interfacial area should be provided for this purpose. However, a high liquid hold-up is the more important requirement, and a bubble column is likely to be suitable. If $C_{AL} \approx C_{Ai}$, then:

$$J_A = k_2 C_{Ai} C_{BL} \varepsilon_L \tag{4.16}$$

Near the boundary between the regions II and III, where virtually none of the reaction occurs in the film, the concentration C_{AL} of **A** in the bulk of the liquid is determined by the simple relation:

$$J_A = N_A a = \underbrace{k_L a (C_{Ai} - C_{AL})}_{\text{Mass transfer through film}} = \underbrace{k_2 C_{AL} C_{BL} \varepsilon_L}_{\text{Reaction in bulk}} \tag{4.17}$$

i.e.:

$$C_{AL} = C_{Ai} \left(1 + \frac{k_2 C_{BL} \varepsilon_L}{k_L a} \right)^{-1} \tag{4.18}$$

4.1.4. Choice of a Suitable Reactor

Choosing a suitable reactor for a gas–liquid reaction is a question of matching the characteristics of the reaction system, especially the reaction kinetics, with the characteristics of the reactors under consideration. As we have seen above, two of the most important characteristics of gas–liquid reactors are the specific interfacial area a, and the liquid hold-up ε_L. Table 4.1 shows some representative values of these quantities for various gas–liquid reactors.

TABLE 4.1. *Comparison of Specific Interfacial Area a and Liquid Hold-up ε_L for Various Types of Reactor*

Type of contactor	Specific area a m^2/m^3	Liquid hold-up ε_L (fraction)
Spray column	60	0.05
Packed column (25 mm Raschig rings)	220	0.08
Plate column	150	0.15
Bubble contactor	200	0.85
Agitated tank	500	0.80

Example 4.1

You are asked to recommend, with reasons, the most suitable type of equipment for carrying out a gas–liquid reaction between a gas **A** and a solution of a reactant **B**. Particulars of the system are as follows:

Concentration of **B** in solution = 5 $kmol/m^3$

Diffusivity of **A** in the solution = 1.5×10^{-9} m^2/s

Rate constant of the reaction (**A** + **B** → Products; second order overall) = 0.03 $m^3/kmol\ s$.

For bubble dispersions (plate columns, bubble columns, agitated vessels) take k_L as having a range from 2×10^{-4} to 4×10^{-4} m/s. For a packed column, take k_L as having a range 0.5×10^{-4} to 1.0×10^{-4} m/s.

Solution

We first calculate the value of the parameter:

$$\beta = \frac{\sqrt{k_2 C_{BL} D_A}}{k_L}$$

From the data above:

$$\sqrt{k_2 C_{BL} D_A} = \sqrt{0.03 \times 5 \times 1.5 \times 10^{-9}} = 1.5 \times 10^{-5}\ \text{m/s}$$

For the bubble dispersions:

$$\frac{\sqrt{k_2 C_{BL} D_A}}{k_L}$$

will therefore lie in the range:

$$\frac{1.5 \times 10^{-5}}{2 \times 10^{-4}} = 0.075 \quad \text{to} \quad \frac{1.5 \times 10^{-5}}{4 \times 10^{-4}} = 0.038$$

For a packed column, β will have the range:

$$\frac{1.5 \times 10^{-5}}{0.5 \times 10^{-4}} = 0.3 \quad \text{to} \quad \frac{1.5 \times 10^{-5}}{1.0 \times 10^{-4}} = 0.15$$

Referring to Fig. 4.3 these values lie in region II indicating that the reaction is only moderately fast and that a relatively high liquid hold-up is required. A packed column would in any case therefore be unsuitable. We therefore conclude from the above considerations that an agitated tank, a simple bubble column or a packed bubble column should be chosen. The final choice between these will depend on such factors as operating temperature and pressure, corrosiveness of the system, allowable pressure drop in the gas, and the possibility of fouling.

Note that to continue further with the design of the reactor by taking a value of f_A from Fig. 4.3, or using equation 4.15 or 4.17, requires a knowledge of ε_L and a for the bubble dispersion as well as a more exact value of k_L. These will depend on the type and configuration of the reactor chosen, the flowrates at which it is operated and the physical properties of the chemical species involved.

4.1.5. Information Required for Gas–Liquid Reactor Design

Once the type of reactor has been chosen, designing the reactor from basic principles (as opposed to pilot plant construction and development) first of all requires some tentative decisions about the size, shape and mechanical arrangement of the equipment in the light of knowledge of the flowrates to be used. These preliminary decisions will eventually be subject to confirmation or amendment as the design proceeds. Then there is a variety of other information that needs to be either estimated from the literature, or measured experimentally if the opportunity exists. Altogether the kind of information required may be considered under three headings:

(*a*) *Kinetic Constants of the Reaction*:

The kinetics of the reaction need to be known or measured, in particular the rate constant and how it may be affected by temperature. Many gas–liquid reactions, like chemical reactions generally, are accompanied by the evolution or absorption of heat. Even if there are arrangements within the reactor for the removal of heat (e.g. cooling coils in a stirred tank reactor), it is unlikely that the temperature will be maintained constant at all stages in the process. Experimental methods for measuring the kinetics of reactions are considered in a later section.

(*b*) *Physical Properties of the Gas and Liquid*:

The two most important physical properties of the system are the solubility of the gas and its diffusivity in the liquid.

The solubility of the gas is needed so that C_{Ai} the concentration of the reactant **A** at the interface, can be calculated. The solubility is often expressed through the Henry Law constant $\mathscr{H}$; this is defined by $P_{Ae} = \mathscr{H}C_A$, where P_{Ae} is the partial pressure of the gas **A** at equilibrium with liquid in which the concentration of the dissolved gas is C_A. This immediately raises a problem for many systems: how can the solubility, which requires gas and liquid to be at equilibrium, be determined when the gas reacts with the liquid? The answer is by one of several methods.

(i) The Henry law constant can be calculated, in addition to the reaction rate constant, from the results of experiments designed to investigate the kinetics of a gas–liquid reaction (see Section 4.1.9).

(ii) Where the second reactant **B** is dissolved in a solvent (as in the case of sodium hydroxide in water in the example below), the solubility of the reactant **A**, (carbon dioxide in the example), in the solvent (water), can be used and a correction made for any secondary 'salting out' effect of the reactant **B**.

(iii) The solubility can be estimated by semi-empirical correlations based on the theory of solutions.

(iv) The solubility of an unreactive gas with a similar type of molecule (e.g. N_2O for CO_2) can be used with an adjustment based on relative solubilities in other solvents.

A similar problem exists in determining the diffusivity of a gas in a liquid with which it reacts. Diffusivities are not easy to measure accurately, even under the best experimental circumstances. As in the case of solubility, the diffusivity D_{AB} needed in the basic equations can be estimated from a semi-empirical correlation, and usually this is the best method available[7]. As before, where the system involves a second reactant **B** dissolved in a solvent, it may be possible to measure experimentally the diffusivity in the solvent in the absence of **B**. A laminar jet or wetted-wall column (as described in Section 4.1.9 for measuring the kinetics of the reaction) would be suitable for this purpose.

Other physical properties required are viscosities, especially the viscosity of the liquid; densities of the liquid and gas; surface tension of the liquid, including the influence of surfactants (e.g. on bubble coalescence behaviour) and, if the gas is a mixture, the gas-phase diffusivity of the reactant **A**. These physical properties are needed in order to evaluate the equipment characteristics as follows.

(*c*) *Equipment Characteristics*

The remaining quantities that need to be known are the performance characteristics of the particular equipment that has been chosen. These are principally the liquid-phase mass transfer coefficient k_L, the interfacial area of contact per unit volume a, the liquid-phase volume fraction ε_L, and, if the gas is a mixture, the gas-phase mass transfer coefficient k_G or volume coefficient $k_G a$. In addition, if the gas or liquid flow in the reactor is to be modelled using, the dispersed plug flow model, for example, then the dispersion coefficients will also need to be known. These performance characteristics are essentially physical quantities; they depend on the fluid mechanics of liquid films flowing over packing elements in packed columns or on the behaviour of bubbles in bubble columns and agitated tanks. They will generally be the same for purely physical absorption of a gas as in the case where chemical reaction is occurring, unless there are strong local temperature or concentration gradients. For the design of a reactor, values of k_L etc can be estimated from correlations based on gas absorption experiments carried out on the same type of equipment. A number of such correlations for columns with various kinds of packing are discussed in Volume 2. Further correlations can be found for agitated tanks in the book by TATTERSON[8] and in other literature sources, and for bubble columns in DECKWER's book[9].

4.1.6. Examples of Gas–Liquid Reactors

In the following examples, the aim is to illustrate how the chemical reaction is coupled with the mass transfer processes. To this end the reactor operating characteristics such as k_L, a, ε_L, and the gas–liquid physical properties such as the Henry law constant and liquid-phase diffusivity will be used but the details of their calculation will not be considered.

Packed Column Reactors

The design of packed column reactors is very similar to the design of packed columns without reaction (Volume 2, Chapter 12). Usually plug flow is assumed for both gas and liquid phases. Because packed columns are used for fast chemical reactions, often the gas-side mass transfer resistance is significant and needs to be taken into account. The calculation starts on the liquid side of the gas–liquid interface where the chemical reaction rate constant is compounded with the liquid side mass transfer coefficient to give a reaction-enhanced liquid-film mass transfer coefficient k_L' as in the following example. The specific interfacial area a for the irrigated packing is then introduced to give the volumetric coefficient $k_L'a$. This is then combined with the gas side volumetic coefficient k_Ga to give an overall gas-phase coefficient K_Ga on which the performance of the column is finally based.

Example 4.2

A Packed Column Reactor

Carbon dioxide is to be removed from an air stream at 1 bar total pressure by absorption into a 0.5 M solution (0.5 kmol/m^3) of NaOH at 20°C in a column 1 m diameter packed with 25 mm ceramic rings. Air entering at a rate of 0.02 kmol/s will contain 0.1 mole per cent CO_2 which must be reduced to 0.005 mole per cent at the exit. (Such a very low CO_2 concentration might be required for example in the feed stream to an air liquefaction process.) The NaOH solution is supplied to the column at such a rate that its concentration is not appreciably changed in passing through the column. From the data below calculate the height of packing required. Is a packed column the most suitable type of reactor for this purpose?

Data[3]: Second-order rate constant for the reaction

$$CO_2 + OH^- = HCO_3^- \quad : \quad k_2 = 9.5 \times 10^3 \text{ m}^3/\text{kmol s}$$

(In a solution of NaOH this reaction is followed instantaneously by:

$$HCO_3^- + OH^- = CO_3^{2-} + H_2O$$

corresponding to an overall reaction $CO_2 + 2NaOH = Na_2CO_3 + H_2O$)
For 0.5M NaOH at 20°C:

Diffusivity of CO_2: 1.8×10^{-9} m^2/s
Solubility of CO_2: Henry law constant $\mathscr{H} = 25$ bar m^3/kmol (2.5×10^6Nm/kmol)

Equipment performance characteristics (see Volume 2, Chapter 12):

Effective interfacial area of packing: 280 m^2/m^3
Film mass transfer coefficients:

Liquid: $k_L = 1.2 \times 10^{-4}$ m/s
Gas (Volume coefficient): $k_Ga = 0.056$ kmol/m^3 s bar

Solution

Note first of all that the concentration of the OH^- ion, 0.5M assuming complete dissociation, will be very much greater than the concentration of the carbon dioxide in the solution. The reaction can therefore be treated as a pseudo first order reaction with a rate constant:

$$k_1 = k_2 \times C_{OH^-} = 9.5 \times 10^3 \times 0.5 = 4.75 \times 10^3 \text{ s}^{-1}$$

The value of the parameter β (equation 4.13) is thus:

$$\beta = \sqrt{k_1 D_{CO_2}}/k_L = \sqrt{4.75 \times 10^3 \times 1.8 \times 10^{-9}}/1.2 \times 10^{-4} = 24.4$$

Referring to Fig. 4.3, it can be seen that with this value for β the system will lie in the fast reaction regime, Region I, and that the packed column will be a suitable reactor. Also, β in equation 4.13 is sufficiently large for tanh β to be effectively 1, so that equation 4.14 applies:

$$N_{CO_2} = \sqrt{k_1 D_{CO_2}}\, C_{CO_2}$$

This equation can be written $N_{CO_2} = k_L' C_{CO_2}$, where $k_L' = \sqrt{k_1 D_{CO_2}}$ and may be regarded as a liquid-film mass transfer coefficient enhanced by the fast chemical reaction. This is very convenient because it allows us to use the expression in Volume 2, Chapter 12 for combining liquid-film and gas-film coefficients to give an overall gas-film coefficient;

i.e.:
$$\frac{1}{K_Ga} = \frac{1}{k_Ga} + \frac{\mathscr{H}}{k_L'a}$$

where k_L' has been substituted for k_L which would apply in the absence of any chemical reaction. In this way we can take into account that, in this problem, the gas is not pure carbon dioxide but a dilute mixture in air which will give rise to a gas-phase resistance. Substituting the numerical values:
$k_L' = \sqrt{4.75 \times 10^3 \times 1.8 \times 10^{-9}} = 2.9 \times 10^{-3}$ m/s

and: $$\frac{1}{K_G a} = \frac{1}{0.056} + \frac{25}{2.9 \times 10^{-3} \times 280} = \frac{1}{0.056} + \frac{1}{0.032}$$

∴ $$K_G a = 0.021 \text{ kmol/m}^3 \text{ s bar}$$

This value of $K_G a$ will apply to any position in the column.

Height of packing

We can now proceed to the second part of the calculation and find the height of packing required. Plug flow for the gas phase will be assumed; because the composition of the liquid is assumed not to change, the flow pattern in the liquid does not enter into the problem. Note that using $K_G a$ requires that the driving force for mass transfer be expressed in terms of gas-phase partial pressures.

Referring to Fig. 4.4, let flowrate of the carrier gas (air) per unit cross-sectional area be G kmol/m^2 s, and the mole fraction of CO_2 be y (because the gas is very dilute, the mole fraction is virtually the same as the mole ratio which should appear in the following material balance). Taking a balance across the element shown in Fig. 4.4:

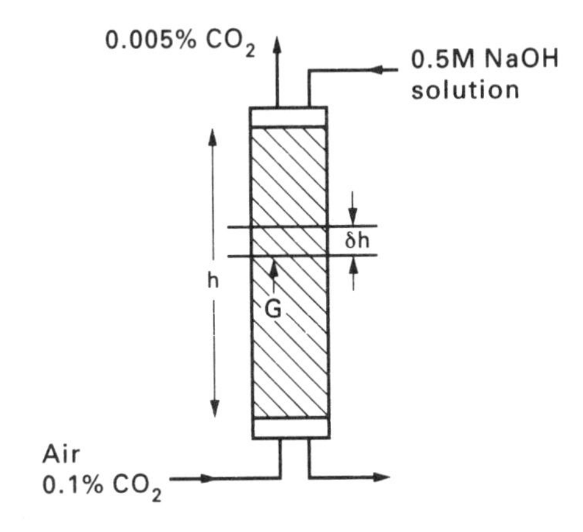

FIG. 4.4. Absorption of CO_2 in 0.5M NaOH solution

$$-G\delta y = K_G a(P_A - P_{Ae})\delta h$$

where P_{Ae} is the partial pressure that would be in equilibrium with the bulk liquid; here, because the liquid is a concentrated solution of NaOH, the partial pressure of CO_2 P_{Ae} in equilibrium with it is virtually zero. Also, $P_A = yP$ where P is the total pressure.

∴ $$-G\,\delta y = K_G a y P\,\delta h$$

∴ $$h = \frac{G}{K_G a P}\int_{y_{out}}^{y_{in}} \frac{dy}{y} = \frac{G}{K_G a P}\ln\frac{y_{in}}{y_{out}}$$

and: $$G = \frac{0.02}{\pi \times 1^2/4} = 0.0255 \text{ kmol/m}^2 \text{ s}$$

∴ *Height of packing required* $$h = \frac{0.0255}{0.021 \times 1}\ln\frac{0.001}{0.00005} = \underline{\underline{3.64 \text{ m}}}$$

Confirmation of pseudo first-order behaviour

When pseudo first-order behaviour is assumed for a reaction of this type, there is always the possibility that the second reactant **B**, here the NaOH, will not diffuse into the reaction zone sufficiently fast to maintain its concentration close to the value in the bulk liquid. A criterion for the pseudo-first order assumption to be applicable has been shown by DANCKWERTS[3] to be:

$$\beta < \frac{1}{2}\left(1 + \frac{\nu_A C_{BL} D_B}{\nu_B C_{Ai} D_A}\right)$$

For the present problem, $\nu_A/\nu_B = 0.5$, $D_B/D_A = 1.7$, and $C_{BL} = 0.5$.

The largest value that C_{Ai} could possibly have is that at the base of the column if there were no gas-phase resistance.

Then: $$C_{Ai} = y_{in} P/\mathscr{H} = (0.001 \times 1)/25 = 0.00004\ \text{kmol/m}^3$$

The right-hand side of the above inequality is then 1.1×10^4, which is much greater than the value for β, which has already been calculated as 24.4.

Further comments

It should be appreciated that this example is one of the few very simple practical cases of a packed column reactor. The removal of carbon dioxide by reaction is a step in many important industrial processes, e.g. the purification of natural gas and of hydrogen in the manufacture of ammonia. Most of these processes[10,11] use as the liquid absorbent a solution which can be regenerated, for example solutions of amines or potassium carbonate[12], and the design of these columns is distinctly more complicated.

The value of the superficial gas velocity used in the packed column of the above example may be calculated as follows:

Gas mass flowrate per unit area $G = 0.0255\ \text{kmol/m}^2\text{s}$ at a pressure of 1 bar and 20°C.

i.e.: Volume of gas passing up the column in 1 second $= \dfrac{0.0255 \times 8314 \times 293}{1 \times 10^5}$

$= 0.621\ \text{m}^3$ over $1\ \text{m}^2$ cross-sectional area.

Thus, the superficial gas velocity is 0.621 m/s. Note that this is a much higher value than would typically be used in a bubble column.

Agitated Tank Reactors: Flow Patterns of Gas and Liquid

The most important assumptions in the design of agitated tank reactors are that *both* the liquid phase and the gas phase may be regarded as *well mixed.* That the liquid phase should be well mixed is the more obvious of these assumptions because one of the main functions of the impeller is to induce a rapid circulatory motion in the liquid. If the liquid circulation velocities are sufficiently intense, gas bubbles are drawn down with the liquid and recirculated to the impeller zone. Behind the blades of most types of gas dispersing impellers, gas-filled 'cavities' are formed[8,13]. It has been found that both fresh gas entering the vessel from the sparger and recirculated bubbles merge into these cavities and are then redispersed in the highly turbulent vortices shed from the blade tips. As bubbles of the mixed gas rise towards the surface, some will separate into the headspace above the liquid, and the gas will leave the vessel. The assumption of ideal mixing of the gas phase therefore means that the composition of the gas in the bubbles is the same as that of the gas *leaving* the vessel. (Note that one of the problems in the design or scale-up of agitated tank gas–liquid reactors is that the flow pattern of the gas may tend more towards plug flow as the size of the tank increases.) A further assumption, borne out by many experimental investigations of mass transfer from gas bubbles, is that any gas-film resistance is negligible compared with the liquid-film resistance.

Example 4.3

An Agitated Tank Reactor

o-Xylene is to be oxidised in the liquid-phase to *o*-methylbenzoic acid by means of air dispersed in an agitated vessel[14, 15]:

$C_6H_4(CH_3)CH_3 + 1.5\ O_2 = C_6H_4(CH_3)COOH + H_2O$

The composition of the reaction mixture in the vessel will be maintained at a constant low conversion by continuous withdrawal of the product stream and a steady inflow of fresh *o*-xylene. Under these conditions the reaction rate is approximately independent of the *o*-xylene concentration and is pseudo first-order with respect to the dissolved oxygen concentration C_O.

i.e.
$$\mathcal{R}_O = \boldsymbol{k}_1 C_O$$

where $\mathcal{R}_O$ is the rate of the reaction in terms of oxygen reacting, i.e. kmol O_2/s (m^3 liquid) and C_O is in kmol O_2/m^3.

At the operating temperature of 160°C the rate constant $\boldsymbol{k}_1 = 1.05\ s^{-1}$.

The reactor will operate at a pressure such that the sum of the partial pressures of the oxygen and nitrogen inside the vessel is 14 bar. Under steady conditions the reactor will contain 5 m^3 of liquid into which air will be dispersed at a flowrate of 450 m^3 per hour, measured at reactor conditions.

The quantities to be calculated are: (a) the fraction of the oxygen supplied that reacts, and (b) the corresponding rate of production of *o*-methylbenzoic acid in kmol per hour. There is also the question of whether an agitated tank is the most suitable reactor for this process.

Further data:

Physical properties of system:

Henry Law coefficient for O_2 dissolved in *o*-xylene $\mathscr{H} = 127\ m^3\ bar/kmol$ (1.27×10^7 Nm/kmol).

Diffusivity of O_2 in liquid xylene $D_O = 1.4 \times 10^{-9}\ m^2/s$

Equipment performance characteristics

Gas volume fraction in the dispersion $(1 - \varepsilon_G) = 0.34$

Mean diameter of the bubbles present in the dispersion = 1.0 mm

Liquid-phase mass transfer coefficient $k_L = 4.1 \times 10^{-4}$ m/s

Solution

Assumptions—Flow patterns of gas and liquid:

It will be assumed that, within the tank itself, both gas and liquid phases are well-mixed. Figure 4.5 shows a diagram of the reactor. The inflow of fresh *o*-xylene and the outflow of the products are shown as broken lines since they do not enter into the calculation. Also it is assumed that because the reaction is relatively fast and the solubility of oxygen is low, only a negligible amount of dissolved but unreacted oxygen leaves in the product stream.

General assessment of reaction regime

To assess where the reactor fits into the general scheme of gas–liquid reactions (Fig. 4.3), the value of β will first be calculated.

$$\beta = \sqrt{\boldsymbol{k}_1 D_O}/k_L = \sqrt{(1.05 \times 1.4 \times 10^{-9})} \Big/ (4.1 \times 10^{-4}) = 0.0935$$

This value of β lies in Region II, but fairly close to the boundary with Region III. The value of a, the interfacial area per unit volume of dispersion, is 2040 m^2/m^3 (see below) which is quite large even for an agitated tank, so that the parameter $\varepsilon_L \Big/ \dfrac{aD_O}{k_L}$ on Fig. 4.3 is:

$$(1 - 0.34) \Big/ \left(\frac{2040 \times 1.4 \times 10^{-9}}{4.1 \times 10^{-4}}\right) = 95$$

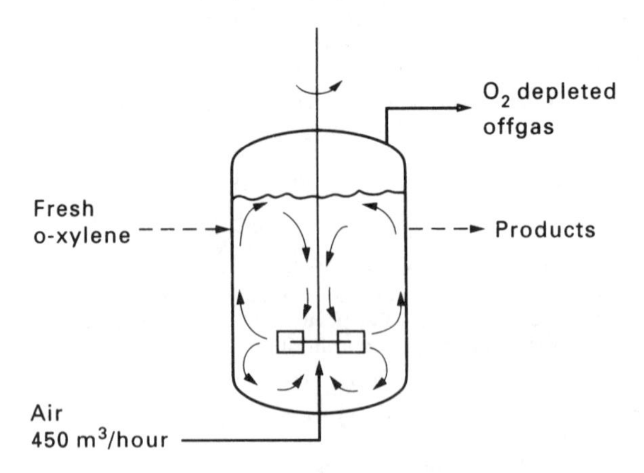

FIG. 4.5. Oxidation of *o*-xylene. Diagram shows vaned-disc impeller to disperse gas, and circulation pattern of liquid and bubbles in tank

Conditions in the tank will therefore be quite close to the curve drawn on Fig. 4.3 for $\varepsilon_L \Big/ \dfrac{aD_A}{k_L} = 100$.

Rather than proceed by trying to read a reaction factor f_A from Fig. 4.3, it is better to set out the basic material balance for mass transfer and reaction as below. Locating the position of β on Fig. 4.3 does however confirm that reaction will be occurring in the main bulk of the liquid and that an agitated tank is a suitable type of reactor.

Feed rate of oxygen

As the next step, the flowrate of the air (molar composition O_2 20.9%; N_2 79.1%) into the reactor will be calculated in kmol/s.

$$450 \text{ m}^3/\text{h is equivalent to } 450/3600 = 0.125 \text{ m}^3/\text{s of air.}$$

From the ideal gas law $PV = n\mathbf{R}T$, in 1 second the volume entering is equivalent to

$$n = \frac{14 \times 1.00 \times 10^5 \times 0.125}{8314 \times 433} = 0.0486 \text{ kmol air.}$$

This corresponds to an oxygen feed rate of $0.0486 \times 0.209 = 0.0102$ kmol/s.

Basis for Material Balance

Let f_u = Fraction of O_2 supplied that is unreacted.

1 kmol of air into reactor; i.e. 0.791 kmol N_2 and 0.209 kmol O_2

$\therefore$ kmol *out* of reactor: 0.791 kmol N_2 and $0.209 f_u$ kmol O_2

$$\therefore \quad \text{Mole fraction } O_2 = \frac{0.209 f_u}{0.791 + 0.209 f_u}$$

$$\therefore \quad \text{Partial pressure } O_2 = \frac{0.209 f_u \times 14}{0.791 + 0.209 f_u} \text{ bar}$$

From Henry's Law:

$$O_2 \text{ concentration at interface } C_{Oi} = \frac{0.209 f_u \times 14}{127(0.791 + 0.209 f_u)} \text{ kmol/m}^3$$

Considering the rate R_t for the whole tank and assuming a steady state:

Rate of mass transfer of O_2 from interface $\equiv$ Rate of reaction

$$R_t = k_L a V_d(C_{Oi} - C_{OL}) = V_l \boldsymbol{k}_i C_{OL}$$

where V_d is the volume of the gas–liquid dispersion in the tank, and
V_l is the volume of the liquid in the tank; note that $V_l = V_d \varepsilon_L$

$$\therefore \qquad C_{Oi} - C_{OL} = \frac{V_l \boldsymbol{k}_1}{k_L a V_d} C_{OL} = \frac{\varepsilon_L \boldsymbol{k}_1}{k_L a} C_{OL}$$

$$\therefore \qquad C_{OL} = \frac{C_{Oi}}{(1 + \varepsilon_L \boldsymbol{k}_1 / k_L a)}$$

Calculation of interfacial area a

In order to substitute numerical values, the interfacial area per unit volume of dispersion a needs to be calculated from the mean bubble diameter $d_b = 1.0$ mm and the gas volume fraction $\varepsilon_G = (1 - \varepsilon_L) = 0.34$. Let there be n_b bubbles per unit volume of the dispersion, all of the same size.

Volume of the bubbles $\varepsilon_G = n_b \pi d_b^3/6$

Surface area of the bubbles $a = n_b \pi d_b^2$

Dividing and rearranging: $a = 6\varepsilon_G/d_b$
Substituting numerical values: $a = (6 \times 0.34)/0.001 = 2040 \text{ m}^2/\text{m}^3$
Note that re-arranging the above relationship to give $d_b = 6\varepsilon_G/a$ shows how a mean bubble size might be calculated from measurements of ε_G and a. A mean bubble diameter defined in this way (Volume 2, Chapter 1) is called a Sauter mean (i.e. a surface volume mean; see also Section 4.3.4).

Numerical calculation of reaction rate

Substituting numerical values in the above equation for C_{OL}:

$$C_{OL} = \frac{\dfrac{0.209 f_u \times 14}{127(0.791 + 0.209 f_u)}}{1 + \dfrac{1.05(1 - 0.34)}{4.1 \times 10^{-4} \times 2040}} = \frac{0.0603 \times 0.209 f_u}{0.791 + 0.209 f_u}$$

Rate at which O_2 reacts in whole vessel $R_t = V_l \boldsymbol{k}_i C_{OL}$

$$= 5 \times 1.05 \times \frac{0.0603 \times 0.209 f_u}{0.791 + 0.209 f_u}$$

This must be equal to the rate of O_2 removal from the air feed.

$$\text{i.e.:} \qquad 0.0102(1 - f_u) = 5 \times 1.05 \times \frac{0.0603 \times 0.209 f_u}{0.791 + 0.209 f_u} = \frac{5 \times 1.05 \times 0.0603 f_u}{\left(\dfrac{0.791}{0.209} + f_u\right)}$$

$$\text{i.e.:} \qquad 31.04 f_u = (3.785 + f_u)(1 - f_u) = 3.785 - 2.785 f_u - f_u^2$$

$$\therefore \qquad f_u^2 + 33.82 f_u - 3.785 = 0$$

$$f_u = \frac{-33.82 \pm \sqrt{33.82 + 4 \times 3.785}}{2} = \frac{-33.82 + 34.04}{2} = 0.112$$

the positive root being taken because f_u cannot be negative.

Thus, the fraction of the O_2 supplied that reacts is $(1 - 0.112) = \underline{\underline{0.888}}$

$\therefore$ kmol of O_2 reacting per second $= 0.0102 \times 0.888 = 0.0091$

From the stoichiometry of the reaction, kmol per second of *o*-xylene reacting, and therefore kmol per second of the product *o*-methylbenzoic acid formed, will be $\dfrac{0.0091}{1.5} = 0.0060$ kmol/s. The hourly rate of production of the *o*-methylbenzoic acid will thus be $0.0060 \times 3600 = \underline{\underline{21.6 \text{ kmol/h}}}$

Further comments

This reactor is typical of those used in a range of processes involving the direct oxidation of hydrocarbons with air or oxygen. Another example is the oxidation of cyclohexane to adipic acid which is an intermediate in the manufacture of polyesters. Chemically, the reactions proceed by free radical chain mechanisms usually initiated by homogeneous catalysts such as cobalt or manganese carboxylic acid salts or naphthenates[16]. From a chemical engineering point of view, some of these processes are extremely hazardous, involving as they do, quantities of volatile hydrocarbons maintained in the liquid phase at high temperatures by using high pressures. Mechanical failure of the reactor could lead to the escape of a highly explosive vapour cloud with disastrous consequences. Understandably, measurements of agitation equipment performance characteristics, such as gas and liquid circulation patterns, bubble diameters and interfacial areas are much more difficult if not well nigh impossible at reaction conditions, compared with experiments on air and water which are very often used in laboratory measurements. Reactors of this type are not generally designed from first principles, but from extensive pilot plant development. Nevertheless a thorough understanding of fundamental principles is an invaluable guide to interpreting and extending the results of pilot plant studies.

Well-Mixed Bubble Column Reactors: Gas–Liquid Flow Patterns and Mass Transfer

In some cases, as in Example 4.4 that follows, a bubble column which is relatively short in relation to its diameter may be used (see Fig. 4.6). The bubbling gas will then generate sufficient circulation and turbulence in the liquid phase for the liquid to be assumed to be well mixed and uniform in composition[9] (except, in principle, in the thin films immediately surrounding the bubbles). The circulating liquid will also drag down smaller bubbles of gas which can then mix with fresh gas. Under these circumstances, the gas phase can also be assumed to be well mixed, just as in the case of the agitated tank in Example 4.3. The question of taller bubble columns will be considered in the following section.

One of the purposes of giving Example 4.4 (on the chlorination of toluene) is to demonstrate the effect of different gas flowrates on the performance of a bubble column. The higher the gas flowrate, the larger the interfacial area a per unit volume of dispersion; gas–liquid mass transfer will take place more readily and the concentration of the dissolved gas in the liquid will rise. Although the rate of reaction will increase, this is offset, as will be seen, by the disadvantage of a lower

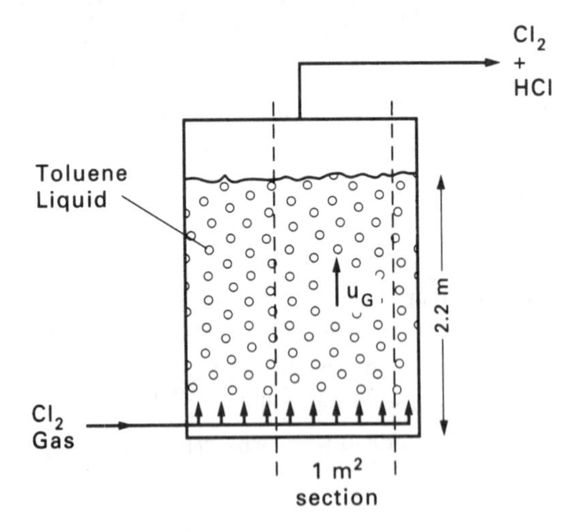

FIG. 4.6. Batch chlorination of toluene in a well-mixed bubble-column reactor (Example 4.4)

fractional conversion for the gas-phase reactant, i.e. in this example, the chlorine in the off-gas from the reactor. To show that this effect of gas flowrate on off-gas composition depends more on the height chosen for the column than on its diameter, the calculations will be carried out for a section of column of unit cross-sectional area (1 m^2) (see Fig. 4.6) and the gas flowrate will be expressed as a superficial gas velocity u_G(m/s).

Chlorination processes in bubble column reactors[9] are unusual in showing a significant gas-phase resistance to mass transfer. It will be seen from the low value of the Henry law constant $\mathscr{H}$ in the list of data for the example below, that the solubility of chlorine in toluene is much greater than the solubility of either the carbon dioxide or oxygen considered in the previous examples. This means that when the gas-phase mass transfer resistance is taken in combination with the liquid-phase resistance according to equation 4.19 which is derived in Volume 2, Chapter 12, then the gas side contribution to the resistance is much greater if $\mathscr{H}$ is small.

$$\frac{1}{K_L a} = \frac{1}{k_L a} + \frac{1}{\mathscr{H} k_G a} \tag{4.19}$$

where K_L is the overall mass transfer coefficient based on liquid side concentrations.

Example 4.4

A Bubble Column Reactor—Batch Chlorination of Toluene[9]

The chlorination of toluene to give a mixture of *ortho*-and *para*-monochlorotoluenes is typical of many similar chlorinations carried out industrially.

$$C_6H_5CH_3 + Cl_2 = o\text{-}ClC_6H_4CH_3 \text{ or } p\text{-}ClC_6H_4CH_3 + HCl$$

As the reaction proceeds and the concentrations of the monochlorotoluenes increase, so further chlorination of the aromatic ring takes place to give dichlorotoluenes. The present example (see Fig. 4.6) is restricted to relatively low toluene conversions so that any subsequent reactions need not be considered. The process is carried out using a homogeneous (i.e. dissolved) catalyst such as ferric chloride $FeCl_3$; in this example, the catalyst used is stannic chloride $SnCl_4$, which, being more active than ferric chloride, is effective even at the relatively low temperature of 20°C. Note that the HCl formed as a byproduct in the reaction is transferred back to the gas phase and leaves the reactor mixed with the unreacted chlorine. Thus, when the reaction is proceeding steadily, 1 mole of chlorine is replaced by 1 mole of HCl and the volume of gas passing through the reactor is unaffected by the reaction.

The example is concerned with a batch chlorination process. At the beginning, the fresh toluene charged to the reactor will contain no dissolved chlorine. After bubbling of the chlorine has commenced, a period of time will need to elapse before the concentration of the dissolved chlorine rises to a level that just matches the rate at which it is being removed from the solution by reaction. To avoid such a complication in this example, calculations are carried out for the stage when, after chlorine bubbling has continued at a steady rate, the fractional conversion of the toluene has reached a value of 0.10. It is then assumed that, at any instant in time, the rate of mass transfer of chlorine from the gas phase is just equal to the rate at which it reacts in the bulk of the liquid, i.e. the rate is given by equation 4.17.

The calculations required in this example may now be summarised as follows:

Toluene is to be chlorinated in a batch-operated bubble column such that, for any particular gas rate chosen, the height of the dispersion in the column is 2.2 m. The reactor will operate at a temperature of 20°C and a pressure of 1 bar; (any effect of hydrostatic pressure differences in the column may be neglected). The catalyst will be stannic chloride at a concentration of 5×10^{-4} kmol/m^3.

(a) Determine whether a bubble column is a suitable reactor for this process.
(b) Consider the stage when 10 per cent of the toluene has been converted. For superficial gas velocities of 0.01, 0.02, 0.03, 0.04, and 0.05 m/s, calculate:

(i) the mole fraction of chlorine in the outlet gas and the fractional conversion of the chlorine,
(ii) the steady-state concentration of dissolved chlorine in the liquid phase, and
(iii) the rate of the reaction in terms of kmol/s toluene reacting per m^3 of dispersion.

(c) For each of the superficial gas velocities in the range above, calculate the reaction time required for the fractional conversion of toluene to increase from 0.1 to 0.6.

Further design data are as follows:

The rate equation proposed for the reaction at 20°C, catalysed by stannic chloride is:

$$\mathcal{R} = k_3 C_{Sn} C_T C_{Cl}$$

where C_T, C_{Cl} and C_{Sn} are the concentrations of toluene, chlorine and stannic chloride, respectively, and $k_3 = 0.134\ (m^3/kmol)^2 s^{-1}$ This equation may be written in the form of equation 4.1 as:

$$\mathcal{R} = k_2 C_T C_{Cl}$$

where $k_2 = k_3 C_{Sn}$ incorporates the catalyst concentration C_{Sn} which remains unchanged during the course of the reaction (any small changes in the molar density of the liquid phase will be neglected).

Molar concentration of pure liquid toluene (calculated from the density of the liquid and the molecular weight of toluene): 9.36 kmol/m^3.

Henry Law constant $\mathscr{H}$ for dissolved chlorine at 20°C: 0.45 bar m^3/kmol.
Gas volume fraction in dispersion $\varepsilon_G = 4.0u_G$ where u_G is the superficial gas velocity (m/s).
Liquid-side volumetric mass transfer coefficient $k_L a = 0.078u_G\ (s^{-1})$.
Gas-side volumetric mass transfer coefficient $k_G a = 0.104\, u_G$ (kmol/m^3s bar) (= 1.04×10^{-6} kmol/N s m).
Diffusion coefficient of chlorine in toluene at 20°C $D_{Cl} = 3.5 \times 10^{-9}\ m^2/s$
Liquid-side mass transfer coefficient $k_L = 4.0 \times 10^{-6}$ m/s.

(Note: this experimental value appears to be remarkably low; the reason for the anomaly has yet to be fully resolved.)

Solution

(a) Suitability of a bubble column reactor

To assess the suitability of a bubble column using Fig. 4.3, the parameter $\beta = \sqrt{k_2 C_T D_{Cl}}/k_L$ will be evaluated. At the beginning of the reaction the toluene concentration C_T is 9.36 kmol/m^3 (neglecting any dilution effect of the stannic chloride catalyst and the dissolved chlorine). The value of $k_2 = k_3 C_{Sn}$ is therefore $k_2 = 0.134 \times 5 \times 10^{-4} = 0.67 \times 10^{-4}\ m^3$/kmol s.

Thus:
$$\beta = \sqrt{0.67 \times 10^{-4} \times 9.36 \times 3.5 \times 10^{-9}}\Big/4.0 \times 10^{-6} = 0.37$$

This value lies in the region of Fig. 4.3 corresponding to a moderately fast reaction occurring in the bulk of the liquid and not in the film. A bubble column, and a stirred tank reactor as sometimes used industrially, are therefore suitable types of reactor for the process.

(b) Calculation of the chlorine conversion and the toluene reaction rate

As in Example 4.3 with an agitated tank, let the fraction of chlorine passing through the reactor unreacted be f_u and, because Cl_2 is replaced by HCl, f_u is also the mole fraction of chlorine in the off-gas. Since the total pressure is 1 bar, the partial pressure of the chlorine will be f_u bar. Because the gas phase in the reactor is assumed to be well mixed, the equivalent interfacial chlorine concentration C_{Cli} is $f_u/\mathscr{H}$, i.e. $f_u/0.45 = 2.22 f_u$ kmol/m^3. Considering unit volume, i.e. 1 m^3 of dispersion, and following equation 4.17, the rate of mass transfer across the interface is now equated to the rate of the reaction in the bulk of the liquid where the concentration of the chlorine is C_{ClL}:

$$J_{Cl} = K_L a(C_{Cli} - C_{ClL}) = k_2 C_{ClL} C_T (1 - \varepsilon_G) \qquad \text{(A)}$$

where K_L is the overall mass transfer coefficient (there may be significant gas-side resistance) and $(1 - \varepsilon_G)$ has been substituted for ε_L. The value of C_T when a fraction α of the toluene has been converted, i.e. a fraction $(1 - \alpha)$ remains unconverted, will be $9.36(1 - \alpha)$ kmol/m^3. Using numerical values in equation 4.19:

$$\frac{1}{K_L a} = \frac{1}{0.078u_G} + \frac{1}{0.45 \times 0.104u_G} = \frac{1}{0.029u_G} \quad \text{or } K_L a = 0.029u_G.$$

Then:

$$J_{Cl} = 0.029u_G(2.22f_u - C_{ClL}) = 0.67 \times 10^{-4} C_{ClL} 9.36(1 - \alpha)(1 - 4.0u_G) \qquad \text{(B)}$$

Solving this equation for C_{ClL} gives:

$$C_{ClL} = \frac{0.0644u_G f_u}{0.029u_G + 0.627 \times 10^{-3}(1 - \alpha)(1 - 4.0u_G)} \qquad \text{(C)}$$

Now, using as a basis 1 m^2 cross-sectional area of the reactor and 1 second in time, the amount of chlorine *fed* to the reactor (from the ideal gas law $PV = n\mathbf{R}T$) is $n = \dfrac{1.0 \times 10^5 u_G}{8314 \times 293} = 0.0411u_G$ kmol; of this, a quantity $0.0411u_G(1 - f_u)$ reacts. The rate at which chlorine is removed by reaction in a column 1 m^2 cross-sectional area and 2.2 m high, i.e. dispersion volume 2.2 m^3, is 2.2 J_{Cl}; thus in 1 second, using equation B for J_{Cl} and substituting for C_{ClL} from equation C:

$$\text{Rate of removal of chlorine} = \frac{0.89 \times 10^{-4} u_G f_u(1 - \alpha)(1 - 4.0u_G)}{0.029u_G + 0.627 \times 10^{-3}(1 - \alpha)(1 - 4.0u_G)} = 0.0411u_G(1 - f_u).$$

Equating these two quantities and rearranging for f_u gives:

$$f_u = \left(1 + \frac{0.00216(1 - \alpha)(1 - 4.0u_G)}{0.029u_G + 0.627 \times 10^{-3}(1 - \alpha)(1 - 4.0u_G)}\right)^{-1} \qquad \text{(D)}$$

Substituting this expression in equation C gives:

$$C_{ClL} = \frac{0.0644u_G}{0.029u_G + 0.00279(1 - \alpha)(1 - 4.0u_G)} \qquad \text{(E)}$$

For a fractional conversion α of toluene of 0.1, Table 4.2 shows how f_u (which is the same as the mole fraction of chlorine in the outlet gas) and $(1 - f_u)$ (the fractional conversion of the chlorine) vary with the superficial gas velocity u_G over the range 0.01 to 0.05 m/s, as required in the problem. Table 4.2 also shows the concentration of the dissolved chlorine in the bulk liquid C_{ClL} calculated from equation E. Shown also is the relative saturation of the liquid C_{ClL}/C^*_{ClL}, where C^*_{ClL} is the saturation chlorine concentration in the bulk liquid at the operating pressure of the reactor, i.e. 1.0 bar.

Thus, $C^*_{ClL} = P/\mathscr{H} = 1.0/0.45 = 2.22\ \text{kmol/m}^3$.

The rate of the reaction in terms of kmol of toluene reacting per unit volume of dispersion, i.e. J_T, which is equal to the rate at which chlorine reacts J_{Cl}, is given by equation A. Substituing for C_T, this equation then becomes:

$$J_T = J_{Cl} = k_2 C_{ClL} 9.36(1 - \alpha)(1 - \varepsilon_G) \qquad \text{(F)}$$

Values of J_T over the required range of superficial gas flowrates with $\alpha = 0.1$, $\varepsilon_G = 4.0u_G$, and C_{ClL} given by equation E, are shown in Table 4.2.

(c) Time required for conversion of the toluene from a mole fraction $\alpha = 0.1$ to $\alpha = 0.6$

The reactor is operated batchwise with respect to the toluene. Considering now the change in the fractional conversion of the toluene with time, the rate of reaction J_T in terms of the rate of change in the number of moles of toluene per unit volume of dispersion is:

$$-\frac{d[(1 - \varepsilon_G)9.36(1 - \alpha)]}{dt}.$$

Substituting this value of J_T into equation F and cancelling the $9.36(1 - \varepsilon_G)$ term on each side:

$$\frac{d\alpha}{dt} = k_2 C_{ClL}(1 - \alpha) \qquad \text{(G)}$$

Substituting for C_{ClL} from equation E:

$$\frac{d\alpha}{dt} = k_2 \frac{0.0644u_G(1 - \alpha)}{0.029u_G + 0.00279(1 - \alpha)(1 - 4.0u_G)}$$

Integrating:

$$k_2 t_r = \int_{0.1}^{0.6} \frac{0.029u_G + 0.00279(1 - \alpha)(1 - 4.0u_G)}{0.0644u_G(1 - \alpha)} d\alpha$$

where t_r is the time required for the reaction to proceed from a toluene conversion of 0.1 to a conversion of 0.6.

Simplifying:

$$k_2 t_r = \int_{0.1}^{0.6} \left[\frac{0.45}{1-\alpha} + 0.0433\left(\frac{1}{u_G} - 4.0\right)\right] d\alpha$$

i.e.:

$$k_2 t_r = \left[-0.45 \ln(1-\alpha) + 0.0433\left(\frac{1}{u_G} - 4.0\right)\alpha\right]_{0.1}^{0.6}$$

Substituting the value $k_2 = 0.67 \times 10^{-4}$ m^3/kmol s and evaluating the integral, this equation for t_r becomes:

$$t_r = \frac{1}{0.67 \times 10^{-4}}\left[0.365 + 0.0217\left(\frac{1}{u_G} - 4.0\right)\right] \text{ seconds.}$$

TABLE 4.2. *Results of Calculations for Batchwise Chlorination of Toluene*

At toluene fractional conversion $\alpha = 0.1$	Superficial gas velocity u_G(m/s)				
	0.01	0.02	0.03	0.04	0.05
Fraction chlorine unreacted f_u	0.308	0.381	0.444	0.500	0.550
Chlorine fractional conversion $(1 - f_u)$	0.692	0.619	0.556	0.500	0.450
Chlorine conc. in liquid (kmol/m^3)	0.238	0.446	0.627	0.788	0.931
% Chlorine saturation in liquid	10.7	20.1	28.3	35.5	41.9
Rate of reaction of toluene (kmol/m^{3}s $\times 10^4$)	1.29	2.32	3.11	3.74	4.20
Time for reaction of toluene from $\alpha = 0.1$ to $\alpha = 0.6$ (hour)	10.1	5.6	4.2	3.4	2.9

Further comments

The above results demonstrate how a bubble column reactor responds to increases in the gas feed rate. At low superficial gas velocities, the gas hold-up and hence the gas–liquid interfacial area are small; the degree of saturation of the liquid is low and therefore the reaction rate is low. This is especially noticeable in the long period of time, 12.5 hours, for the conversion of the toluene from 10 per cent to 60 per cent. On the other hand, the fractional conversion of the gaseous reactant, the chlorine, is greatest at the lowest superficial gas velocity. Increasing the gas rate markedly increases the rate of conversion of the liquid reactant, toluene, but also increases the fraction of the gaseous reactant, chlorine, that passes through the column unreacted. As discussed in the following section, this has important consequences in the design of a bubble column reactor and in the design of the chemical process of which it forms a part.

The results given in Table 4.2 are based on the assumption that steady-state conditions have been reached and that, as indicated earlier, the gas flowrate through the reactor is unaffected by the reaction. More detailed calculations show that, in the time taken for the toluene conversion to reach a value of 0.1, the concentration of chlorine reaches more than 99 per cent of its steady-state value, thus justifying the assumption made in the solution.

4.1.7. High Aspect-Ratio Bubble Columns and Multiple-Impeller Agitated Tanks

The relatively low conversions of the gaseous reactant in many processes, like the chlorination of toluene above, pose problems for the process design of such operations. One approach for the chlorine–toluene process would be simply to pass the Cl_2–HCl mixture through a downstream section of plant to separate the unreacted chlorine which would then be recycled to the reactor. In an alternative design chlorine might be passed from the first reactor into a second reactor in series, and then if necessary into a third reactor, and so on, as shown in Fig. 4.7. However, the problem arises because a bubble column with a low aspect ratio or a single-impeller agitated tank behaves essentially as a well-mixed reactor (aspect ratio is

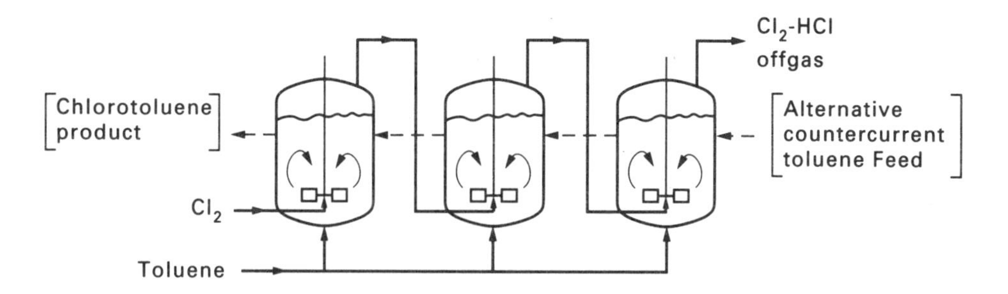

FIG. 4.7. Three well-mixed stirred tank reactors (or low aspect-ratio bubble columns) for the chlorination of toluene showing the chlorine gas supplied in series to the reactors to improve the fractional conversion of chlorine

defined as L/d_C where L is the height of the gas–liquid dispersion in the column and d_C the column diameter). A much more desirable situation would be to arrange for the gas to be more nearly in a state of plug flow. One of the reasons why the carbon dioxide concentration of the gas in the packed column of Example 4.2 could easily be reduced by a factor of 20 was that the gas could be assumed to be in plug flow.

Continuing this idea of trying to approximate to plug flow, the logical development is to use a bubble column which is tall in relation to its diameter, i.e. is of high aspect ratio. As Fig. 4.8 shows, if instead of having a given volume of liquid in a shallow, large diameter bubble column, the same volume of liquid is contained in a tall, smaller diameter column, there are two advantages. Firstly, there is the probability that the gas (and possibly the liquid) will be more nearly in a state of plug flow; and secondly, assuming that gas is supplied at the same volumetric flowrate, the superficial gas velocity through the column will be increased. Because the specific interfacial area a and the gas hold-up ε_G increase with superficial gas velocity (as demonstrated by the equations used in Example 4.4), this will give an increase in the rate of reaction per unit volume of dispersion. (Note however that bubble columns generally operate at lower superficial gas velocities than packed columns; obviously, if the superficial gas velocity in the short column is already close to the limit above which the liquid would be entrained, no further increase is practicable.) One disadvantage of a tall column is the cost of compressing the gas to overcome the additional hydrostatic head.

For agitated tanks, a similar train of logic can be followed. Instead of the three separate stirred vessels shown in Fig. 4.7, all three impellers can be mounted on the same shaft[17] as shown in Fig. 4.9. In Chapter 2 Section 2.2, it was seen that as the number of stirred tanks in a series is increased, so does the overall flow behaviour tend more nearly to plug flow. In biological reaction engineering (Chapter 5), aerobic stirred tank fermenters are often of the three impeller type of Fig. 4.9, in order to increase the fraction of oxygen usefully transferred from the compressed air supply to an economic level. Figure 4.9 also illustrates that the similarity between a multiple-impeller stirred tank system and a bubble column extends further than might at first be imagined. The gas bubbles rising in a bubble column create circulation cells in which the mixing effect, at least for the liquid phase, is similar

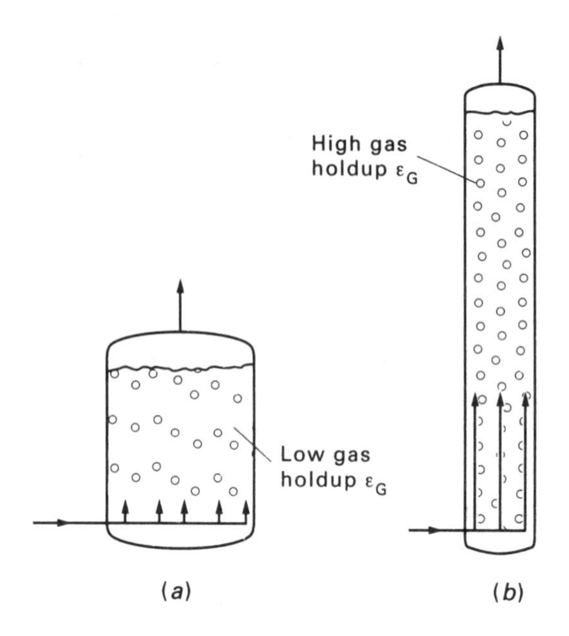

FIG. 4.8. Alternative bubble columns each using the same volume of liquid, and the same total gas flowrate. (*a*) Low aspect-ratio column: low superficial gas velocity u_G; small interfacial area a; low rate of reaction. (*b*) High aspect-ratio column: high superficial gas velocity u_G; large interfacial area a; higher rate of reaction. Note: For case (*b*) it must not be assumed that both gas and liquid phases are well-mixed

to that in the stirred compartments of the impeller-driven column. Both experimental data and theoretical considerations have indicated that the height of these mixing cells is approximately equal to the column diameter.

4.1.8. Axial Dispersion in Bubble Columns

Dispersion Coefficients for Gas and Liquid Phases

Although the mixing patterns in bubble columns do not obviously correspond to simple axial dispersion, the dispersed plug flow model has been found to hold reasonably well in practice. For a two-phase gas–liquid system, the equation for gas-phase convection and dispersion (Chapter 2, equation 2.14) becomes:

$$\frac{\partial C_G}{\partial t} = \varepsilon_G \mathscr{D}_G \frac{\partial^2 C_G}{\partial z^2} - u_G \frac{\partial C_G}{\partial z} \tag{4.20}$$

where C_G is the concentration of dispersed species in the gas phase and $\mathscr{D}_G$ is the gas-phase dispersion coefficient. The gas hold-up ε_G is introduced because it is equal to the fraction of the tube cross-sectional area occupied by the gas, i.e. the region in which gas dispersion occurs. The liquid-phase dispersion coefficient $\mathscr{D}_L$ is defined similarly.

Experimental measurements of dispersion coefficients[9] have shown that, unless the liquid velocity is unusually high, both gas and liquid phase dispersion coefficients

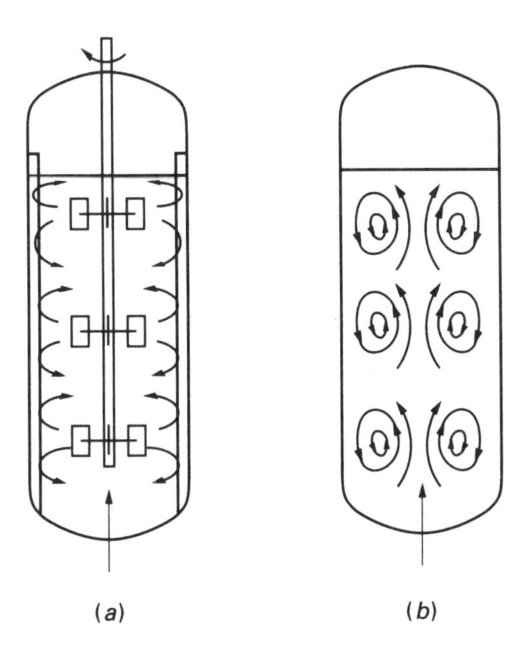

FIG. 4.9. Multiple-impeller stirred vessel (*a*) and tall bubble column (*b*) showing similarity of liquid circulation patterns

depend primarily on the superficial gas velocity u_G and the column diameter d_c. An empirical equation for the gas-phase dispersion coefficient is:

$$\mathscr{D}_G = 50\left(\frac{u_G}{\varepsilon_G}\right)^3 d_C^{1.5} \tag{4.21}$$

where the units are $\mathscr{D}_G(\mathrm{m^2/s})$, $u_G(\mathrm{m/s})$ and $d_C(\mathrm{m})$. Note that 50 is a dimensional constant with units $\mathrm{s^2/m^{2.5}}$. u_G/ε_G in this expression is essentially the rising velocity of the bubbles relative to the liquid. This equation is recommended for use over a wide range of column diameters (from 0.20 to 3.2 m). An equation for the liquid-phase dispersion coefficient is:

$$\mathscr{D}_L = 0.35(u_G g)^{1/3} d_C^{4/3} \tag{4.22}$$

where g is the gravitational acceleration. The constant 0.35 is here dimensionless. BAIRD and RICE[18] have shown that the form of this equation can be derived theoretically using dimensional analysis and the Kolmogoroff theory of isotropic turbulence.

Modelling the Flow in Bubble Columns

Although the most realistic model for a bubble column reactor is that of dispersed plug-flow in both phases, this is also the most complicated model; in view of the uncertainty of some of the quantities involved, such a degree of complication may not be warranted. Because the residence time of the liquid phase in the column is usually much longer than that of the gas, it is likely that the liquid phase will be well-mixed even when the gas phase is not. Therefore, in modelling bubble column reactors, it is often acceptable to assume that the liquid phase is well-mixed while the gas may be either well-mixed, or in plug flow, or somewhere between these two extremes and described by the dispersed plug-flow model.

One particular case of modelling a bubble column reactor will now be considered in more detail. This is the case where the reaction is very fast and occurs wholly within the liquid film surrounding the bubbles, i.e. in Region I of Figs 4.2 and 4.3. In this case, the concentration in the bulk of the liquid of the species **A** being transferred from the gas phase is zero, and the mixing pattern in the liquid has therefore no influence on the rate of transfer. The advantage of considering this case in the present context is that it will allow attention to be focussed on the influence of gas-phase mixing on the performance of a bubble column.

To incorporate mixing by the dispersed plug flow mechanism into the model for the bubble column, we can make use of the equations developed in Chapter 2 for dispersed plug flow accompanied by a first-order chemical reaction. In the case of the very fast gas–liquid reaction, the reactant **A** is transferred and thus removed from the gas phase at a rate which is proportional to the concentration of **A** in the gas, i.e. as in a homogeneous first-order reaction. Applied to the two-phase bubble column for steady-state conditions, equation 2.38 becomes:

$$\varepsilon_G \mathcal{D}_G \frac{\mathrm{d}^2 C_{AG}}{\mathrm{d}z^2} - u_G \frac{\mathrm{d}C_{AG}}{\mathrm{d}z} - \boldsymbol{k}_{1G} C_{AG} = 0 \tag{4.23}$$

where C_{AG} is the concentration of **A** in the gas phase, $\varepsilon_G \mathcal{D}_G$ replaces D_L as in equation 4.20, and $\boldsymbol{k}_{1G}$ is now the first-order rate constant for the transfer of **A** from the gas-phase. The solution to this equation gives the ratio of the concentrations of the reactant **A** at the exit ($C_{A\mathrm{ex}}$) and inlet of the column ($C_{A\mathrm{in}}$): From equation 2.47:

$$\frac{C_{A\mathrm{ex}}}{C_{A\mathrm{in}}} = \frac{4\mathcal{A}\,\mathrm{e}^{\frac{u_G L}{2\varepsilon_G \mathcal{D}_G}}}{(\mathcal{A}+1)^2 \mathrm{e}^{\frac{\mathcal{A} u_G L}{2\varepsilon_G \mathcal{D}_G}} - (\mathcal{A}-1)^2 \mathrm{e}^{-\frac{\mathcal{A} u_G L}{2\varepsilon_G \mathcal{D}_G}}} \tag{4.24}$$

where:

$$\mathcal{A} = [1 + 4\boldsymbol{k}_{1G}\varepsilon_G \mathcal{D}_G / u_G^2]^{1/2}$$

The first-order constant $\boldsymbol{k}_{1G}$ will now be expressed in terms of the rate constant $\boldsymbol{k}_1$ of the reaction in the liquid phase. From equation 4.14, the rate of transfer of **A** per unit area of gas–liquid interface is $\sqrt{(\boldsymbol{k}_1 D_A)}\, C_{Ai}$; i.e. in terms of an enhanced mass transfer coefficient $k_L' = \sqrt{(\boldsymbol{k}_1 D_A)}$ this rate of transfer is $k_L' C_{Ai}$. The rate of transfer per unit volume of dispersion J_A is thus:

$$J_A = k_L' a C_{Ai} \tag{4.25}$$

where a is the interfacial area per unit volume of dispersion. Now C_{Ai} is the liquid-phase concentration of **A** at the interface. To convert this to a gas-phase concentration C_{AG}, first the Henry law constant is introduced so that, assuming no gas-side mass transfer resistance, $C_{Ai} = P_A/\mathscr{H}$ where P_A is the partial pressure of **A** in the gas. From the ideal gas law, $PV = n\mathbf{R}T$; i.e. the gas-phase concentration $C_{AG} = n_A/V = P_A/\mathbf{R}T$. Substituting in equation 4.25:

$$J_A = k_L' a \frac{\mathbf{R}T}{\mathscr{H}} C_{AG} = \boldsymbol{k}_{1G} C_{AG}$$

i.e.:

$$\boldsymbol{k}_{1G} = k_L' a \frac{\mathbf{R}T}{\mathscr{H}} = \sqrt{\boldsymbol{k}_1 D_A}\, a \frac{\mathbf{R}T}{\mathscr{H}} \tag{4.26}$$

The influence of gas-phase mixing on the performance of a bubble column will now be demonstrated in Example 4.5.

Example 4.5

Effect of Gas-Phase Dispersion on the Gas Outlet Concentration from a Bubble Column: The Absorption of CO_2 *from an Air Stream*

Carbon dioxide at an inlet concentration of 0.1 mole per cent is to be removed from an air stream at a total pressure of 1 bar by bubbling through a 0.05M (0.05 kmol/m^3) solution of NaOH at 20°C in a bubble column. The caustic soda solution passes through the column at such a rate that its composition is not significantly affected by the absorption. The superficial velocity of the gas will be 0.06 m/s. The ratio of outlet concentration of CO_2 in the air to inlet concentration is to be calculated for the following cases; the height of the gas–liquid dispersion will be 1.5 m in each case:

(a) The diameter of the column is 0.20 m and the gas is in plug flow.
(b) The diameter of the column is 0.20 m and the dispersed plug flow model applies to the gas.
(c) The diameter of the column is 1.5 m and the dispersed plug flow model applies to the gas.
(d) The diameter of the column is 1.5 m and the gas phase is assumed to be well mixed.

Data: With respect to the chemical reaction, the conditions in this problem are very similar to those in Example 4.2 except that the concentration of the caustic soda solution here is 0.05M (0.05 kmol/m^3); the same physico-chemical data will therefore be assumed, i.e. a second-order rate constant for the reaction

$$CO_2 + OH^- = HCO_3^- \qquad \boldsymbol{k}_2 = 9.5 \times 10^3\ \text{m}^3/\text{kmol s}$$

Diffusivity of CO_2 in the liquid: 1.8×10^{-9} m^2/s

Solubility of CO_2: Henry law constant $\mathscr{H} = 25$ bar m^3/kmol (2.5×10^6 Nm/kmol)

Bubble column mass transfer characteristics: The following estimates for a superficial gas velocity of 0.06 m/s will be used for both columns:

Gas hold-up: $\varepsilon_G = 0.23$

Gas–liquid interfacial area per unit volume of dispersion: $a = 195$ m^2/m^3

Liquid-side mass transfer coefficient: $k_L = 2.5 \times 10^{-4}$ m/s

Note: Bubble column mass transfer parameters are difficult to estimate reliably.[9] The above figures are based on results from the sulphite oxidation method at a higher electrolyte concentration than the 0.05M solution to be used in the present example, and therefore the values for ε_G and a may be overestimates.

Solution

First of all the value of $\beta = \sqrt{\boldsymbol{k}_2 C_{\overline{OH}} D_{CO_2}}/k_L$ will be calculated in order to check that the fast-reaction regime will apply.

i.e.:

$$\beta = \sqrt{9.5 \times 10^3 \times 0.05 \times 1.8 \times 10^{-9}}/2.5 \times 10^{-4} = 3.7$$

From Fig. 4.3 it may be seen that this value, although smaller than that in Example 4.2, is still sufficiently large for all the reaction to occur in the film. Also the concentration of dissolved CO_2 at the interface in contact with the incoming gas containing CO_2 at a partial pressure of 0.001 bar will be only $0.001/\mathscr{H}$, i.e. $0.001/25 = 4 \times 10^{-5}$ kmol/m^3 which is much less than 0.05 kmol/m^3, the concentration of the OH^- ion, so that the reaction will still behave as pseudo first-order. The enhanced liquid-side mass transfer coefficient k_L' will thus be $k_L' = \sqrt{(9.5 \times 10^3 \times 0.05 \times 1.8 \times 10^{-9})} = 0.925 \times 10^{-3}$ m/s.

The first-order rate constant $\boldsymbol{k}_{1G}$ is given by equation 4.26:

$$\boldsymbol{k}_{1G} = 0.925 \times 10^{-3} \times 195 \times \frac{8314 \times 293}{25 \times 10^5} = 0.176\ \text{s}^{-1}$$

(the factor 10^5 occurs in order to convert bar, which appears in the units of the Henry law constant, to N/m^2).

(a) Gas in plug flow in the small diameter column

(Note that the actual diameter of the column does not appear in the calculation because the gas rate is expressed as a superficial velocity; however, plug flow for the large column would be extremely improbable.)

Very conveniently, equation 2.51 of Chapter 2 (which was derived as the limiting case of dispersed plug flow as the dispersion coefficient tends to zero) exactly describes this case if $\boldsymbol{k}_1$ in that equation is replaced by $\boldsymbol{k}_{1G}$.

i.e.:

$$\frac{C_{A\text{ex}}}{C_{A\text{in}}} = \exp(-\boldsymbol{k}_{1G}\tau)$$

where in the present case, $\tau(= L/u_G)$ is the residence time of the gas in the column.

Thus:

$$\frac{C_{CO_2\text{ex}}}{C_{CO_2\text{in}}} = \exp\left(-0.176 \times \frac{1.5}{0.06}\right) = \underline{\underline{0.0123.}}$$

This means that in a column with plug flow, the ratio of inlet to outlet concentration is $\dfrac{1}{0.0123} = \underline{\underline{81}}$.

(b) Gas in dispersed plug flow in the small diameter column

For this case the gas-phase dispersion coefficient needs to be calculated from equation 4.21 and the result will now depend on d_C the diameter of the column. Inserting numerical values:

$$\mathcal{D}_G = 50\left(\frac{0.06}{0.23}\right)^3 0.2^{1.5} = 0.0794\ \text{m}^2/\text{s}$$

Evaluating the parameters that appear in equation 4.24:

$$\frac{u_G L}{2\varepsilon_G \mathcal{D}_G} = \frac{0.06 \times 1.5}{2 \times 0.23 \times 0.0794} = 2.464$$

$$\mathcal{A} = \left[1 + \frac{4 \times 0.176 \times 0.23 \times 0.0794}{0.06^2}\right]^{1/2} = 2.138$$

Substituting in equation 4.24:

$$\frac{C_{CO_2\text{ex}}}{C_{CO_2\text{in}}} = \frac{4 \times 2.138 \times e^{2.464}}{3.138^2 e^{2.138 \times 2.464} - 1.138^2 e^{-2.138 \times 2.464}} = \underline{\underline{0.0526}}$$

This means that in the 0.20 m column with dispersed plug flow, the ratio of inlet to outlet concentration is $1/0.0526 = \underline{\underline{19}}$.

(c) Gas in dispersed plug flow in the large diameter column

Using again equation 4.21 but now with $d_C = 1.5$ m:

$$\mathcal{D}_G = 50\left(\frac{0.06}{0.23}\right)^3 1.5^{1.5} = 1.63\ \text{m}^2/\text{s}$$

Again evaluating the parameters that appear in equation 4.24:

$$\frac{u_G L}{2\varepsilon_G \mathcal{D}_G} = \frac{0.06 \times 1.5}{2 \times 0.23 \times 1.63} = 0.120$$

$$\mathcal{A} = \left[1 + \frac{4 \times 0.176 \times 0.23 \times 1.63}{0.06^2}\right]^{1/2} = 8.62$$

Substituting in equation 4.24:

$$\frac{C_{CO_2ex}}{C_{CO_2in}} = \frac{4 \times 8.62 \times e^{0.120}}{9.62^2 e^{8.62 \times 0.120} - 7.62^2 e^{-8.62 \times 0.120}} = \underline{\underline{0.162}}$$

This means that in the 1.5 m column with dispersed plug flow, the ratio of inlet to outlet concentration is $1/0.162 = \underline{\underline{6.2}}$.

(d) Gas phase in the larger column is well-mixed

(In this case also, the actual diameter of the column does not appear in the calculation.)

This is equivalent to a single stirred tank reactor with a first-order reaction and equation 1.25 of Chapter 1 can be used.

i.e.:

$$\tau = \frac{(C_{in} - C_{ex})}{\boldsymbol{k}_1 C_{ex}} = \frac{1}{\boldsymbol{k}_1}\left(\frac{C_{in}}{C_{ex}} - 1\right)$$

$\therefore$

$$\frac{C_{ex}}{C_{in}} = \frac{1}{(1 + \boldsymbol{k}_1 \tau)}$$

In the present context, $\boldsymbol{k}_1$ is replaced by $\boldsymbol{k}_{1G} = 0.176$, and:

$$\tau = \frac{L}{u_G} = \frac{1.5}{0.06} = 25 \text{ s}$$

Thus:

$$\frac{C_{CO_2ex}}{C_{CO_2in}} = \frac{1}{(1 + 0.176 \times 25)} = \underline{\underline{0.185}}.$$

This means that in the 1.5 m column with the gas phase well-mixed, the ratio of inlet to outlet concentration is $1/0.185 = \underline{\underline{5.4}}$.

Further comments

The above example demonstrates that, due to increased gas-phase mixing, the performance of a bubble column can worsen dramatically as the column diameter is increased at the same superficial gas rate. This is an important point to remember when scaling up from a small-diameter pilot-plant reactor to a full-scale production unit. One of the reasons why this effect is particularly marked in this example is that the plug flow reactor should, in theory, give a very low outlet concentration. Whenever the predicted ratio of inlet to outlet concentration for a reactor is very high (81 in this example) then the performance will be very susceptible to small departures from ideal plug flow. In cases where such a high ratio is definitely required, it is advisable to carry out the process in two separate stages.

4.1.9. Laboratory Reactors for Investigating the Kinetics of Gas–Liquid Reactions

To design a gas–liquid reactor on a fundamental basis, a knowledge of the kinetics is essential. Some apparently straightforward, industrially important gas–liquid reactions, such as the absorption of nitrogen oxides NO_x in water to form nitric acid, are in fact extremely complex[19]. Although laboratory reactors can be very useful in investigating the mechanisms of complex reactions, discussion here will be limited to measuring the rate constants of simple first or second-order reactions.

Types of laboratory gas–liquid reactors[3,9]

Choosing a laboratory reactor for the purpose of investigating a particular reaction is rather like choosing a reactor for an industrial scale operation, in that the choice depends mainly on the intrinsic speed of the reaction—*fast*, *moderately fast*, or *slow*. As with large scale reactors, the value of $\beta = \sqrt{\boldsymbol{k}_2 C_{BL} D_A}/k_L$ is a useful guide, although β can only be calculated with hindsight *after* the rate constant has been measured. The aim for most types of laboratory reactor is to control the hydrodynamics so that each fresh element of liquid is exposed to the gas for a known contact time t_e (or a controlled spread of contact times). The different types of laboratory reactor can be broadly classified according to the range of contact times that each of them offers. The idea is to match approximately the contact time with the reaction time (which, for a first-order reaction, can be identified as the half-life of the reaction $(\ln 2)/k_1$ see Chapter 1). The contact times for the main types of laboratory reactor described below are shown in Table 4.3[20].

TABLE 4.3. *Range of Contact Times for Gas–Liquid Laboratory Reactors*

Reactor type	Contact Time (s)	Reaction characteristic
(a) Laminar jet	0.001–0.1	Very fast to fast
(b) Laminar film	0.1–2	Fast to moderately fast
(c) Stirred Cell	1–10*	Moderately fast

* Average values calculated from the mass transfer coefficient k_L.

In these types of laboratory reactor, the flow of the liquid is very carefully controlled so that, although the mass transfer step is coupled with the chemical reaction, the mass transfer characteristics can be disentangled from the reaction kinetics. For some reaction systems, absorption of the gas concerned may be studied as a purely physical mass transfer process in circumstances such that no reaction occurs. Thus, the rate of absorption of CO_2 in water, or in non-reactive electrolyte solutions, can be measured in the same laboratory contactor as that used when the absorption is accompanied by the reaction between CO_2 and OH^- ions from an NaOH solution. The experiments with purely physical absorption enable the diffusivity of the gas in the liquid phase D_L to be calculated from the average rate of absorption per unit area of gas–liquid interface $\bar{N}_A$ and the contact time t_e. As shown in Volume 1, Chapter 10, for the case where the incoming liquid contains none of the dissolved gas, the relationship is:

$$\bar{N}_A = 2C_{Ai}\sqrt{\frac{D_A}{\pi t_e}} \tag{4.27}$$

When the reaction is accompanied by a fast first-order or pseudo first-order reaction (as for example, CO_2 being absorbed into a solution of high OH^- ion concentration) with a rate constant k_1, then the average rate of absorption per unit area is given by*:

$$\bar{N}_A = C_{Ai}\left(\frac{D_A}{k_1}\right)^{1/2}\left[\left(k_1 + \frac{1}{2t_e}\right)\operatorname{erf}\sqrt{k_1 t_e} + \sqrt{\frac{k_1}{\pi t_e}}\,e^{-k_1 t_e}\right] \tag{4.28}$$

from which, k_1 may be calculated if D_A is known.

* Derivation in Volume 1, Chapter 10 (4th edition, 1994 revision)

As an alternative to investigating the kinetics of a gas–liquid reaction on a laboratory scale, the mass transfer resistance may be minimised or eliminated so that the measured rate corresponds to the rate of the homogeneous liquid-phase reaction. This method of approach will be considered after first describing those reactors giving rise to controlled surface exposure times.

Laboratory Reactors with Controlled Gas–Liquid Mass Transfer Characteristics

(a) Laminar-jet contactor. The laminar-jet apparatus[3,21] shown in Fig. 4.10 is simple in principle, but quite difficult to operate successfully in practice. The aim is to produce a vertical jet having a uniform velocity profile across any section. Various shapes of entry nozzles have been tried, but a simple orifice in a thin plate appears to provide the best way of avoiding a surface layer of more slowly moving liquid at the beginning of the jet. The jet tapers slightly towards the lower part as the liquid is accelerated under the influence of gravity. The receiving tube at the bottom of the jet must have a diameter that matches fairly closely the diameter of the jet; if the receiving tube is too small, liquid overflows, whereas if the receiver is too large, gas is entrained as bubbles and there is uncertainty about the amount of gas absorbed in the jet itself. A further complication is that ripples, which can be difficult to eliminate, may sometimes appear on the surface of the jet. If the jet is perfectly smooth and regular, then all elements of liquid moving down the jet have the same time of exposure t_e (i.e. the contact time) which can be calculated from the length and diameter of the jet, and the flowrate of the liquid. These same parameters can be varied to produce the range of contact times shown in Table 4.3. An advantage of the laminar jet is that the contact time is independent of the viscosity and density of the liquid. The rate of absorption of the gas is found by

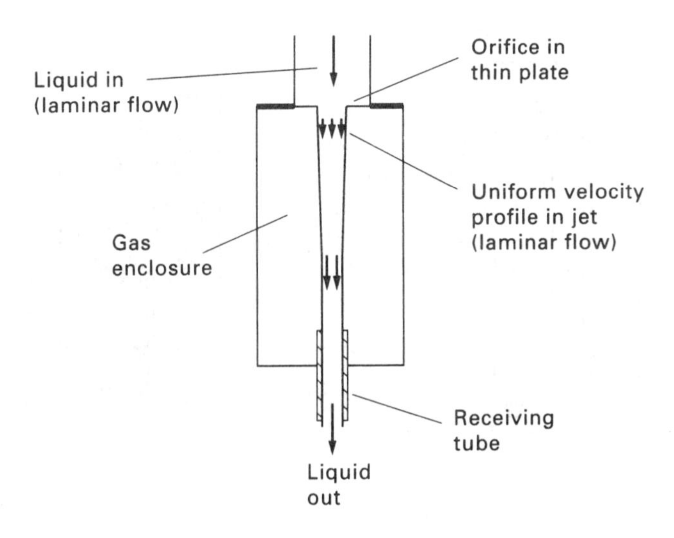

FIG. 4.10. Laminar-jet contactor

measuring the difference between the rates of inflow and outflow of the gas supplied to the enclosure.

(b) Laminar-film contactors. Two forms of laminar-film contactor[3] are shown in Fig. 4.11. Both use a solid surface of regular geometry to support and guide the thin film of liquid which is exposed to the gas. The wetted-tube contactor illustrated in Fig. 4.11*a* is very similar to the wetted-wall type of column described in Volume 2, Chapter 12, where the liquid flows down the inside wall of a tube. The advantage of using the outside wall of a tube, as shown in Fig. 4.11*a*, is that the distribution of liquid at the inlet can be more easily adjusted to give a film of uniform thickness around the whole circumference of the tube. A uniform thickness is important because the contact times between gas and liquid for different parts of the film will vary if the thickness varies. Also a satisfactory liquid take-off arrangement, such as the collar indicated in Fig. 4.11*a*, is more easily designed when the liquid flows on the outer wall of the tube. The hydrodynamic theory of laminar flow in a vertical film is treated in Volume 1, Chapter 3, where it is shown that the velocity at the free surface, and therefore the contact time t_e, in the present application, may be calculated from the flowrate of the liquid together with its viscosity and density. The thickness of the film can also be calculated from these quantities. When using a laminar-film contactor, both extremes of unduly thick and unduly thin films are to be avoided. In thick films, any ripples which tend to develop on the surface of the film are less likely to be suppressed by the dampening effect of the solid wall. For thin films, there may be a tendency for imperfectly wetted areas to appear on the tube surface. Also, for a very thin film, the theory of gas absorption into a semi-infinite medium, which can be applied satisfactorily to thicker films,

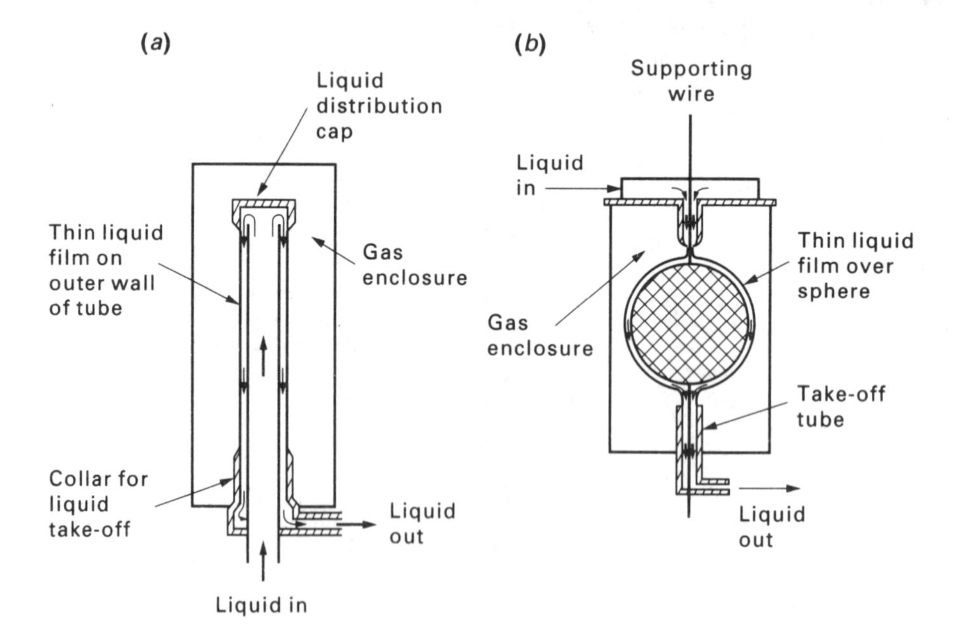

FIG. 4.11. Laminar-film contactors (*a*) wetted-wall and (*b*) wetted-sphere

may need modification. For a film of medium thickness, which can be treated as semi-infinite in extent, equations 4.27 and 4.28 above can be used to determine the diffusivity in the liquid D_A or the rate constant k_1, just as in the case of the laminar jet.

In the laminar-film contactor shown in Fig. 4.11*b*, the supporting surface has the form of a sphere held in place by a wire[3]. This arrangement has the advantage that, in the regions where the liquid runs on to the sphere, and where it leaves the sphere, the surface areas exposed to the gas are relatively small so that, even if the hydrodynamics of the liquid flow in these regions is not ideal, the effect on the rate of absorption of the gas will be small. As in the case of the cylindrical tube, the contact time for each element of liquid as it flows around the sphere can be calculated from the liquid flowrate, although the mathematics of the analytical treatment is somewhat more complicated than in the case of the tube.

Instead of a single sphere, some authors[22] have used a vertical string of several adjoining spheres with the purpose of increasing the range of achievable contact times without the necessity of having to use a large diameter single sphere; a disadvantage, however, is that it is necessary to make some assumptions about the mixing process that occurs in the liquid as it passes from one sphere to the next.

(c) Stirred cell reactor. In this reactor the flow regime in the liquid phase is turbulent so that there is no unique exposure time that applies to all elements of liquid, but a spectrum of exposure times exists. Therefore, instead of attempting to specify contact times for the reactor, the mass transfer characteristics are described in terms of the liquid-phase mass transfer coefficient k_L.

The main features of the stirred cell reactor[3,23,24] shown in Fig. 4.12 are:

(i) the area of the surface between gas and liquid is clearly defined (the stirrer speed is so low that the surface remains flat), and

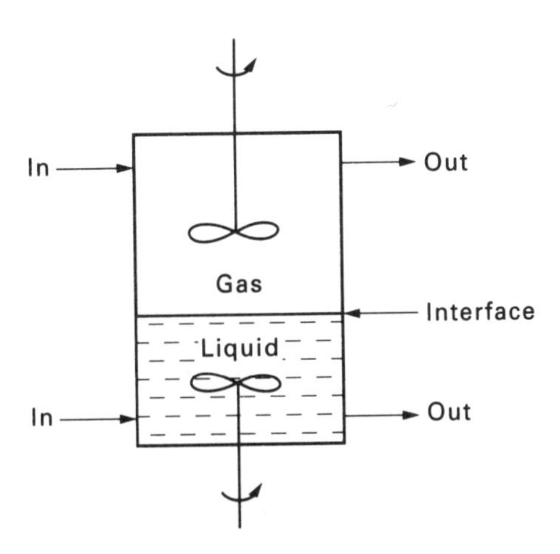

FIG. 4.12. Stirred cell laboratory reactor. Note: In some designs a perforated plate is positioned in the interface to prevent excessive disturbance and to maintain a known area of contact between gas and liquid

(ii) the stirrer in the liquid phase provides a means of maintaining steady and reproducible levels of turbulence and therefore consistent mass transfer characteristics.

The cell contains a relatively large volume of liquid so that, when operated as a stirred tank reactor with steady inflow and outflow of the liquid reagent, the composition of the liquid phase can be held constant. In fact, one of the original aims in developing this reactor was to simulate and maintain conditions steady at any particular point in a packed column[3]. The stirrer in the gas phase serves to keep the gas composition uniform and, in many cases, to eliminate any gas-phase mass transfer resistance.

In order to investigate the kinetics of a reaction with the stirred cell, firstly experiments are carried out using a suitably related system in which gas absorption takes place by a purely physical mass transfer process (i.e. no reaction occurs). This establishes values of the physical mass transfer coeffient k_L for the range of stirrer speeds employed. Then the rate of gas absorption into the liquid with the reaction occurring is measured. Finding the rate constant for a fast first-order reaction, for example, is then a matter of working back through equation 4.13 to find the value of β and hence of $\boldsymbol{k}_1$.

Reactors Eliminating Gas–Liquid Mass Transfer Resistance

For *slow* reactions, a rather different approach may be adopted whereby the mass transfer resistance is virtually eliminated altogether. This is achieved by dispersing the gas as small bubbles in a laboratory-scale stirred tank reactor as shown in Fig. 4.13, with the aim of saturating the liquid with the gas. If the bubbles are sufficiently small, i.e. if the interfacial area a is sufficiently large, then the concentration of the gas-phase reactant **A** in the bulk of the liquid C_{AL} becomes almost equal to the interface value, as shown in Fig. 1.2*e*. The rate of transformation is then the rate of the homogeneous reaction occurring in the liquid phase. As a test of whether the gas–liquid mass transfer resistance has been wholly eliminated, the agitator speed is increased, so increasing the gas–liquid surface area, until the overall rate of transformation levels off to an almost constant value. This type of reactor would be suitable for measuring the kinetics of the chlorine–toluene reaction which is involved in Example 4.4.

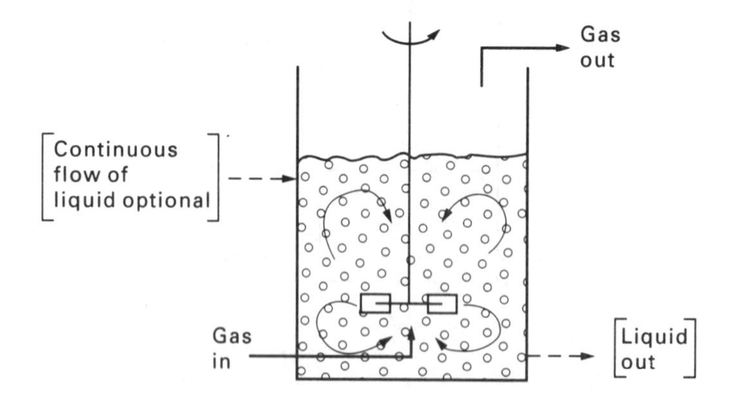

FIG. 4.13. Laboratory stirred-tank reactor for investigating the kinetics of slow reactions. Agitator disperses the gas with the aim of saturating the liquid with dissolved gas

For some *fast* reactions, there is another method of eliminating the gas–liquid mass transfer step and thereby of studying only the homogeneous liquid-phase reaction. A classic example is the measurement of the rate constant for the reaction between CO_2 and OH^- ions using a fast flow reactor[25,26] (Fig. 4.14). The CO_2 is dissolved in water and a flowing stream of the solution is mixed very rapidly with a stream of the alkaline solution. Downstream from the mixing junction is situated a series of sensors that register the change in a property such as temperature, which is the property that has been used for monitoring the exothermic CO_2–OH^- reaction. Assuming that the reaction takes place adiabatically and under steady flow conditions, the temperature rise is a measure of the progress of the reaction as the mixed solutions proceed downstream. Other properties such as conductivity, pH or light transmission can also be used to register the course of the reaction. A further development of the fast flow reactor is the 'stopped-flow' technique[27] in which the flows are arrested soon after the start of the reaction. The progress of the reaction in the tube downstream of the mixing chamber is then followed as a function of time. In this method much less of the liquid reagents is used than in the continuous steady-flow method; it is therefore suitable for reactions involving enzymes or other biological fluids of limited availability.

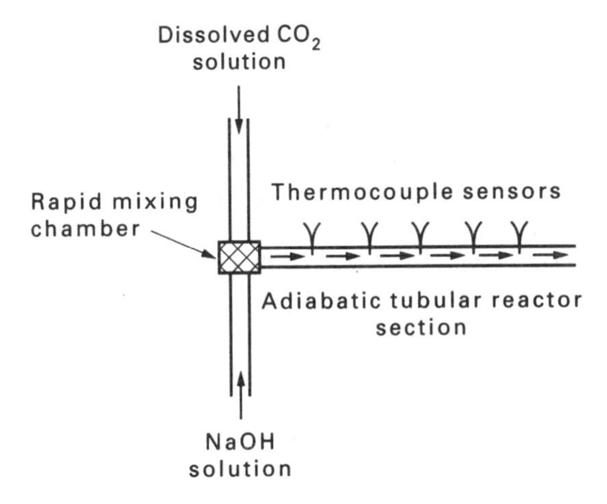

FIG. 4.14. Rapid thermal flow reactor as used to determine the kinetics of the fast first-order reaction between CO_2 and OH^- ions: $CO_2 + OH^- = HCO_3^-$

4.2. GAS–LIQUID–SOLID REACTORS

4.2.1. Gas–Liquid–Solid Reactions

In a surprisingly large number of industrially important processes reactions are involved that require the simultaneous contacting of a gas, a liquid and solid particles[28]. Very often the solid is a catalyst and it is on the surface of the solid that the chemical reaction occurs. The need for three-phase contacting can be appreciated by considering, as an example, the hydro-desulphurisation of a residual petroleum fraction, i.e. of the liquid taken from the base of a crude oil distillation column.

$$H_2 + R \cdot SH = R \cdot H + H_2S$$

where R · SH is representative of a wide range of sulphur compounds present in the oil. The catalyst is of necessity a solid, typically molybdenum disulphide (MoS_2) deposited on a highly porous alumina support[29]. The liquid, being a residue, cannot under any practical conditions be vaporised; and hydrogen, sometimes referred to as a 'permanent' gas, cannot be liquefied except at extremely low temperatures. The reaction is normally carried out at high temperatures (300–350°C) and at high pressures (50–100 bar) to increase the solubility of the hydrogen in the oil[30]. Many industrial three-phase reactors involve hydrogenations or oxidations. Another reaction which has been extensively investigated, because of its importance if crude oil becomes scarce, is the conversion of carbon monoxide–hydrogen mixtures, derived from the gasification of coal, to hydrocarbon liquids in the Fischer–Tropsch process[9]. A finely divided solid catalyst based on iron is commonly used, the catalyst being suspended in a hydrocarbon liquid in a slurry-type reactor.

4.2.2. Mass Transfer and Reaction Steps

The individual mass transfer and reaction steps occurring in a gas–liquid–solid reactor may be distinguished as shown in Fig. 4.15. As in the case of gas–liquid reactors, the description will be based on the film theory of mass transfer. For simplicity, the gas phase will be considered to consist of just the pure reactant **A**, with a second reactant **B** present in the liquid phase only. The case of hydro-desulphurisation by hydrogen (reactant **A**) reacting with an involatile sulphur compound (reactant **B**) can be taken as an illustration, applicable up to the stage where the product H_2S starts to build up in the gas phase. (If the gas phase were not pure reactant, an additional gas-film resistance would need to be introduced, but for most three-phase reactors gas-film resistance, if not negligible, is likely to be small compared with the other resistances involved.) The reaction proceeds as follows:

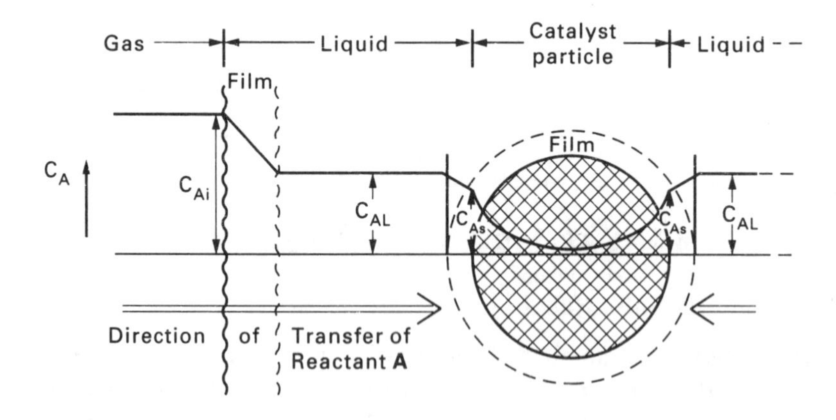

FIG. 4.15. Mass transfer and reaction steps in gas–liquid–solids reactors (solids completely wetted by liquid), showing concentration gradients for reactant **A** being transferred from the gas phase

(a) Starting in the gas phase, the first step is that **A** (hydrogen) dissolves at the gas–liquid interface where its concentration in the liquid will be C_{Ai} (equilibrium with the gas is assumed), and is transferred across the liquid film into the bulk of the liquid where its concentration is C_{AL}.

(b) (i) Reactant **A** is transferred, across the stagnant liquid film surrounding the particle, to the external surface of the solids where its concentration is C_{AS}.
(ii) The second reactant **B** (mercaptan or similar sulphur compound) is also transferred from the bulk liquid to the surface of the solid.

(c) Both reactants then diffuse through the porous structure of the catalyst and react at active sites on the surface of the solid. As in the case of gas–solid catalytic reactions, the driving force required for diffusion will reduce the concentrations of the reactants towards the interior of the particle, and so lead to a reduction in the efficiency of the catalyst. To allow for this, an *effectiveness factor* η is introduced in just the same way as for gas–solid catalytic reactors (Chapter 3). An important difference, however, with three-phase reactors is that the pores of the catalyst will be filled with liquid, and diffusivities in liquids are much lower than in gases. This means that effectiveness factors are likely to be much below unity unless the high resistance to internal diffusion is offset by a very small particle size for the catalyst.

The final stage in the complete reaction process is the diffusion of the products of the reaction out of the particle and their eventual transfer back into the liquid or gas phase. However, only if the reaction is reversible will any build-up of products affect the rate of the reaction itself. For an irreversible reaction, generally the fate of the products is not so important and need not be taken into account in determining the rate of the forward transfer and reaction steps.

4.2.3. Gas–Liquid–Solid Reactor Types: Choosing a Reactor

Significance of Particle Size

The design of a gas–liquid–solid reactor is very much dependent upon the size of the solid particles chosen for the reaction and the anticipated value of the effectiveness factor is one of the most important considerations. Generally, the smaller the particle size the closer the effectiveness factor will be to unity. Particles smaller than about 1 mm in diameter cannot, however, be used in the form of a fixed bed. There would be problems in supporting a bed of smaller particles; the pressure drop would be too great; and perhaps, above all, the possibility of the interstices between the particles becoming blocked too troublesome. There may, however, be other good reasons for choosing a fixed-bed type of reactor.

When small particles are chosen, they have to be used in the form of a suspension in the liquid. The lower limit of particle size may be set by the nature of the catalyst, but is most likely to depend on how easily the particles can be separated from the liquid products. To avoid filtration, the most convenient way of separating the particles is to allow them to settle under gravity, and the minimum particle size is set accordingly.

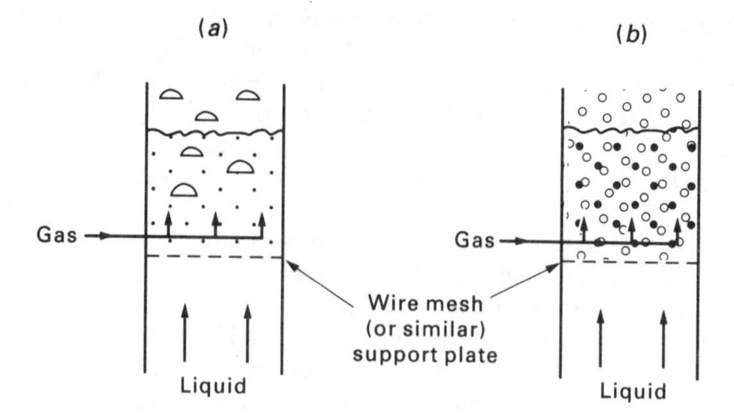

FIG. 4.16. Effect of interaction between particles and bubbles in three-phase fluidised beds. (*a*) Fluidised bed of small particles giving large spherical-cap bubbles. (*b*) Fluidised bed of large particles giving small, more nearly spherical bubbles Note: Bubble sizes and breakup/coalescence behaviour is very dependent on gas and liquid flowrates

Types of Reactor

Three-phase reactors[28] can thus be divided into two main classes: A. *Suspended-bed reactors*, and B. *Fixed-bed reactors*.

A. *Suspended-bed reactors*. These may be further subdivided according to how the particles are maintained in suspension in the reactor.

(i). Bubble columns: If the particles are very small, i.e. have relatively low sedimentation velocities, then they can quite easily be kept in suspension just by bubbling the gas through the liquid in a bubble column as shown in Fig. 4.1*d*. Special attention needs to be given to the design of the gas sparger to ensure that there are no stagnant regions where solids might be deposited.

(ii). Agitated tanks: A more positive way of suspending the particles is to use an agitated tank like that shown in Fig. 4.1*f*. For three-phase operation the impeller needs to be properly designed and positioned nearer to the bottom of the tank so that it will keep the solids in suspension as well as dispersing the incoming gas (see also Fig. 4.20). Information on the performance of such impellers can be found in the book *Mixing in the Process Industries*[17]. In some types of tall and narrow agitated tanks, two or three impellers are mounted on a single shaft (see Fig. 4.9*a*). Because of the higher local velocities produced by the impeller in an agitated tank compared with those in a bubble column, an agitated tank can be used for larger sizes of particles.

Note that three-phase bubble columns and agitated tank reactors are sometimes referred to as *slurry reactors*.

(iii). Three-phase fluidised beds: A further method of suspending even larger particles is to fluidise the bed using an upwards flow of liquid. If gas is then introduced at the bottom of the bed, three-phase contacting is achieved. As in bubble columns and agitated tank reactors, hydrodynamic interactions between the bubbles and the particles occur (see Fig. 4.16). Large particles can cause break up of any large bubbles, so increasing the interfacial area for gas–liquid transfer.

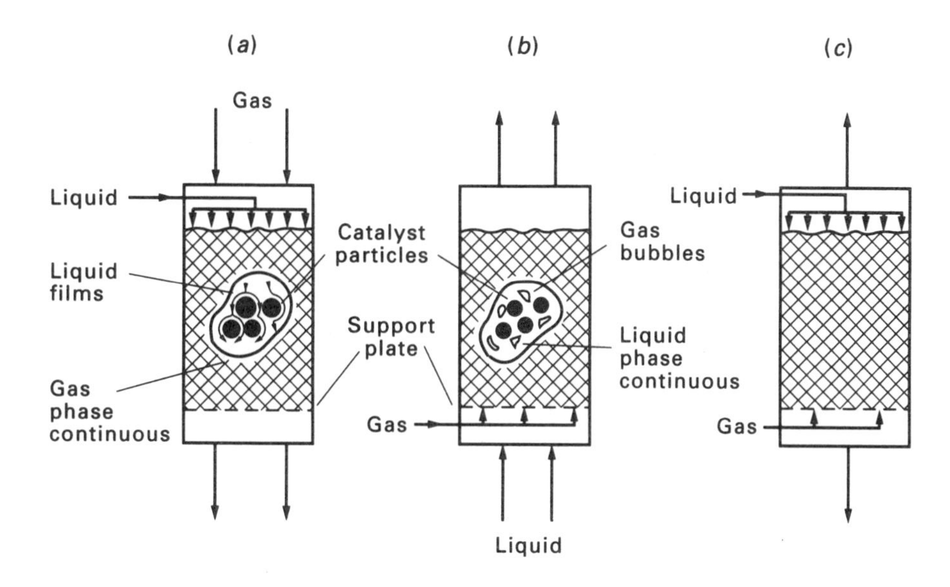

FIG. 4.17. Flow regimes in three-phase fixed-bed reactors. (*a*) Gas and liquid in co-current downwards flow (trickle-bed operation). (*b*) Gas and liquid in co-current upwards flow (liquid floods bed). (*c*) Gas and liquid in countercurrent flow (not often used for catalytic reactors)

B. *Fixed-bed reactors.* Apart from particle size, the main choice to be made with fixed-bed reactors is the direction of flow, i.e. upwards or downwards for the gas and liquid phases. As shown in Fig. 4.17, there are three possibilities, any of which may be chosen for reasons particular to a given process.

(i). Liquid and gas in co-current downflow:—This configuration is sometimes called a *trickle-bed* reactor[31] because, at low to moderate gas and liquid flows, the gas phase is continuous and the liquid flows as a thin film over the surface of the catalyst. In this regime, the hydrodynamic behaviour of the liquid is influenced only to a small extent by changes in the gas flowrate. At higher gas flowrates there is more interaction between the liquid and gas flow regimes[32], especially for liquids which have a tendency to foam. Figure 4.18 is a flow map[33] showing the conditions under which trickling flow changes to a pulsing type of flow and, at very high gas rates, to spray flow.

Trickle-bed reactors are widely used in the oil industry because of reliability of their operation and for the predictability of their large-scale performance from tests on a pilot-plant scale. Further *advantages* of trickle-bed reactors are as follows:

The flow pattern is close to plug flow and relatively high reaction conversions may be achieved in a single reactor. If warranted, departures from ideal plug flow can be treated by a dispersed plug-flow model with a dispersion coefficient for each of the liquid and gas phases.

Pressure drop through the reactor is smaller than with upflow, and there is no problem with flooding which might otherwise impose a limit on the gas and liquid flowrates used.

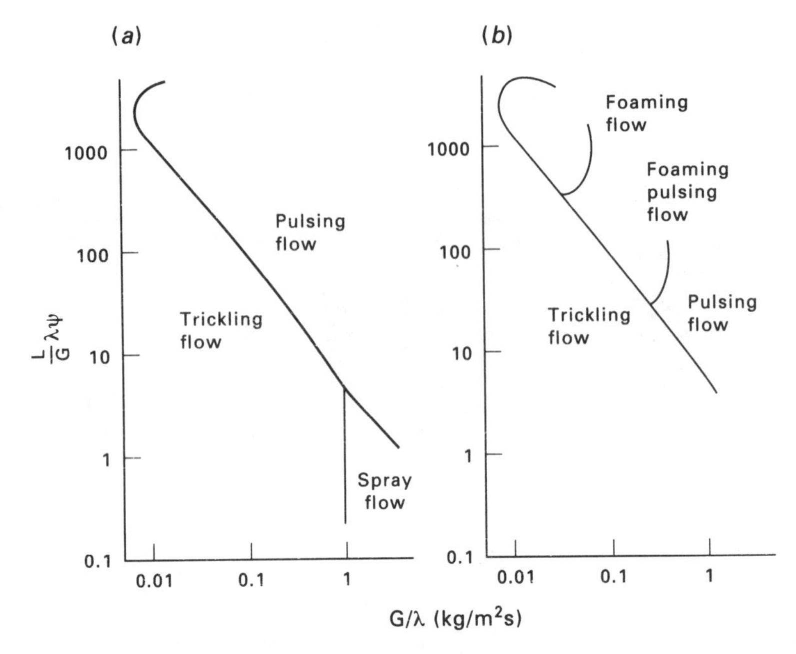

FIG. 4.18. Flow regimes for co-current downwards flow of gas and liquid through a fixed bed of particles. (*a*) Non-foaming liquids. (*b*) Foaming liquids. L, G are liquid, gas flowrates per unit area (kg/m^{2}s); $\lambda = [(\rho_G/\rho_{air})(\rho_L/\rho_{water})]^{1/2}$;
$\psi = [\sigma_{water}/\sigma_L][\mu_L/\mu_{water})/(\rho_{water}/\rho_L)^2]^{1/3}$
where ρ_L, ρ_G, ρ_{air}, ρ_{water} are densities of liquid, gas, air, water;
μ_L, μ_{water} are viscosities of liquid and water;
σ_L, σ_{water} are surface tensions of liquid and water

The particles of the bed are held firmly in place against the bottom support plate as a result of the combined effect of the forces attributable to gravity and fluid drag.

On the other hand some *disadvantages* of trickle-bed reactors are encountered.

Heat transfer in the radial direction is poor, so large reactors are best suited to adiabatic operation. If the temperature rise is significant, it may be controlled in some processes by recycling the liquid product or introducing a quench stream at an intermediate point.

At low liquid flowrates incomplete wetting of the catalyst by the liquid may occur, as illustrated in Fig. 4.19, leading to channelling and a deterioration in reactor performance[34].

(ii). Liquid and gas in co-current upflow: At low flowrates the packed-bed reactor (see Fig. 4.17*b*), behaves rather like the packed bubble column shown in Fig. 4.1*e* with the liquid phase continuous and the gas dispersed as bubbles. The higher liquid hold-up compared with a trickle-bed reactor is likely to be the main reason for choosing the upflow mode of operation; it has the added advantage that higher gas–liquid mass transfer coefficients are also obtained. However, at high flow rates, there is the possibility that the catalyst particles will be lifted and fluidised unless constrained from above by a perforated plate pressing down on the bed. In any case the pressure drop will be higher than for a downflow reactor.

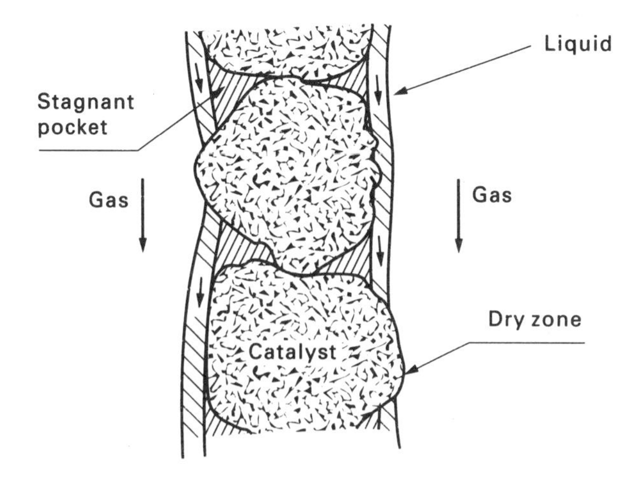

FIG. 4.19. Trickle-flow of liquid over a bed of particles, showing possible complications of stagnant pockets of liquid and dry zones due to poor wetting of the solid surface

(iii). Liquid in downflow—gas in upflow: This situation is similar to that in the packed column shown in Fig. 4.1*a* with countercurrent flow of gas and liquid as in a conventional gas absorption process. In a three-phase reaction, consideration of the concentration driving forces for mass transfer shows that countercurrent operation is usually of little advantage unless the reaction is reversible. Because with small catalyst particles countercurrent operation can easily lead to overloading and flooding of the column, it is rarely used in reactors.

4.2.4. Combination of Mass Transfer and Reaction Steps

Suspended-Bed Reactor

The individual mass transfer and reaction steps outlined in Fig. 4.15 will now be described quantitatively. The aim will be firstly to obtain an expression for the overall rate of transformation of the reactant, and then to examine each term in this expression to see whether any one step contributes a disproportionate resistance to the overall rate. For simplicity we shall consider the gas to consist of just a pure reactant **A**, typically hydrogen, and assume the reaction which takes place on the interior surface of the catalyst particles to be first order with respect to this reactant only, i.e. the reaction is pseudo first-order with rate constant $\boldsymbol{k}_1$. In an agitated tank suspended-bed reactor, as shown in Fig. 4.20, the gas is dispersed as bubbles, and it will be assumed that the liquid phase is "well-mixed", i.e. the concentration C_{AL} of dissolved **A** is uniform throughout, except in the liquid films immediately surrounding the bubbles and the particles. (It will be assumed also that the particles are not so extremely small that some are present just beneath the surface of the liquid within the diffusion film and are thus able to catalyse the reaction before **A** reaches the bulk of the liquid.)

Choosing a clearly defined basis is an all important first stage in the treatment of three-phase reactors. In this case the overall reaction rate $\mathcal{R}_t$ will be based on unit volume of the whole dispersion, i.e. gas + liquid + solid. On this basis let:

a = gas–liquid interfacial area per unit volume of *dispersion*,

ε_G = volume fraction of gas, i.e. bubbles,

ε_L = volume fraction of liquid, and

ε_P = volume fraction of particles.

$$\varepsilon_G + \varepsilon_L + \varepsilon_P = 1 \tag{4.29}$$

For the particles, the external surface area is most conveniently defined as:

a_p = external surface area per unit volume of *particles*.

For both bubbles and particles, there will be a distribution of sizes in the dispersion. The above quantities can be related to the volume/surface or Sauter mean bubble and particle diameters d_b and d_p (see Volume 2, Chapter 1). Thus, for bubbles (as in Example 4.3):

$$d_b = \frac{6\varepsilon_G}{a} \tag{4.30}$$

and for the particles:

$$d_P = \frac{6}{a_P} \tag{4.31}$$

Note that, per unit volume of dispersion, the surface area of the particles is $a_P\varepsilon_P$.

As Fig. 4.15 demonstrates, the two mass transfer steps, gas to liquid and liquid to solid, and then the chemical reaction, take place in series. This means that in the steady state each must proceed at the same rate as the overall process. Continuing then on the basis of unit volume of dispersion, and using the reactant concentrations shown in Fig. 4.15, with the gas–liquid and liquid–solid film mass transfer coefficients k_L and k_s shown in Fig. 4.20, the overall rate $\mathcal{R}_t$ may be written as:

$$\mathcal{R}_t = k_L a(C_{Ai} - C_{AL}) \text{ corresponding to gas–liquid mass transfer} \tag{4.32}$$

or: $$\mathcal{R}_t = k_s a_P \varepsilon_P (C_{AL} - C_{AS}) \text{ corresponding to liquid–solid mass transfer} \tag{4.33}$$

or: $$\mathcal{R}_t = \mathbf{k}_1 C_{As} \eta \varepsilon_P \text{ from the rate of reaction within the particle} \tag{4.34}$$

The reaction term follows from the assumption of a first-order reaction with a rate constant $\mathbf{k}_1$ defined on the basis of unit volume of *particle* (see Chapter 3); ε_P allows for the change in basis to unit volume of dispersion. The effectiveness factor η is also included to take into account any diffusional resistance within the pores of the particle.

Equations 4.32, 4.33 and 4.34 are then rearranged to give:

$$\frac{\mathcal{R}_t}{k_L a} = C_{Ai} - C_{AL}$$

$$\frac{\mathscr{R}_t}{k_S a_P \varepsilon_P} = C_{AL} - C_{AS}$$

$$\frac{\mathscr{R}_t}{\boldsymbol{k}_1 \eta \varepsilon_P} = C_{AS}$$

On adding these equations, C_{AL} and C_{AS}, which are unknown, cancel out so that:

$$\mathscr{R}_t = \left(\frac{1}{k_L a} + \frac{1}{k_S a_P \varepsilon_P} + \frac{1}{\boldsymbol{k}_1 \eta \varepsilon_P}\right)^{-1} C_{Ai} \tag{4.35}$$

Equation 4.35 expresses the overall rate $\mathscr{R}_t$ in terms of the overall driving force C_{Ai}, the concentration existing at the gas–liquid interface, which is known from the solubility of the gas in the liquid. Thus, if the pressure of the reactant **A** in the gas-phase is P_A then C_{Ai} is given by:

$$C_{Ai} = P_A/\mathscr{H} \tag{4.36}$$

where $\mathscr{H}$ is the Henry law constant. The three terms in parentheses on the right hand side of equation 4.35 can be regarded as the effective resistances associated with each of the three individual steps, $1/k_L a$ for the gas–liquid mass transfer step, $1/k_S a_P \varepsilon_P$ for the liquid–solid mass transfer step and $1/\boldsymbol{k}_1 \eta \varepsilon_P$ for the pore diffusion with reaction step. Equation 4.35 thus has the form, familiar in other cases of mass transfer and heat transfer where resistances are in series, and the overall resistance is obtained by adding terms involving the reciprocals of transfer coefficients. In the numerical example that follows, we will evaluate each of these resistances and discuss how the process might be changed in order to minimise the contribution of the step offering the largest proportion of the resistance.

Example 4.6

The Hydrogenation of α-Methyl Styrene in an Agitated Tank Slurry Reactor

Note: The hydrogenation of α-methyl styrene, although not itself a commercially important reaction, has been studied as a model for other more complex catalytic hydrogenations carried out on an industrial scale. The kinetics of the hydrogenation reaction catalysed by palladium deposited within porous alumina particles were investigated by SATTERFIELD *et al.*[35]. The particles were crushed to a very small size to eliminate diffusional resistance within the pores; i.e. the effectiveness factor in the experiments was effectively unity. Also, very intense agitation was used to disperse the hydrogen and so eliminate interphase mass transfer resistances. The result of these experiments was that the reaction appeared to be pseudo first-order with respect to hydrogen with a rate constant $\boldsymbol{k}_1$ equal to 16.8 s^{-1} at 50°C. Also from measurements of the porosity of the larger sized catalyst particles, the effective diffusivity D_e of hydrogen in the particles, with α-methyl styrene filling the pores, was estimated to be 0.11×10^{-8} m^2/s.

We will now consider the design of an agitated tank slurry reactor which might be used industrially for this reaction. We will choose a particle size of 100 μm rather than the very small size particles used in the laboratory experiments, the reason being that industrially we should want to be able easily to separate the catalyst particles from the liquid products of the reaction. If we use spherical particles (radius r_0, diameter d_p), the Thiele modulus ϕ (see Chapter 3) is given by:

$$\phi = \lambda\frac{r_0}{3} = \lambda\frac{d_p}{6} \qquad \text{(from equation 3.19)}$$

where $\lambda = \sqrt{\dfrac{\boldsymbol{k}_l}{D_e}}$

and the effectiveness factor η by:

$$\eta = \frac{3}{\lambda r_0}\left\{\coth(\lambda r_0) - \frac{1}{\lambda r_0}\right\} \qquad \text{(equation 3.18)}$$

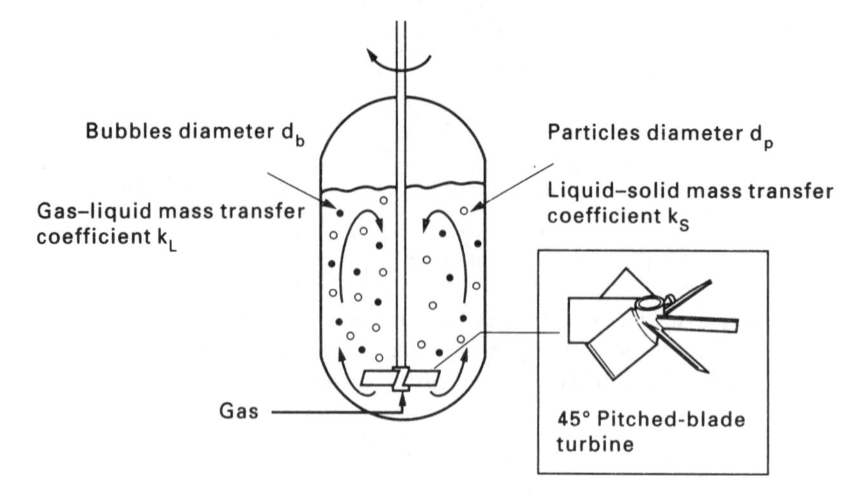

FIG. 4.20. Suspended-bed agitated-tank reactor: Combination of mass transfer and reaction steps. Impeller used would typically be a pitched-blade turbine, pumping downwards as shown, serving both to suspend particles and to disperse gas

Substituting for λr_0:

$$\eta = \frac{1}{\phi}\left(\coth 3\phi - \frac{1}{3\phi}\right)$$

Using the above values for d_p, k_1 and D_e:

$$\phi = 2.06 \text{ and } \eta = 0.407$$

The fact that η, which can be regarded as a catalyst efficiency, has a value of 0.407, rather than unity, is the price we pay for choosing catalyst particles of such a size that they can be separated easily.

The size of the bubbles produced in the reactor and the gas volume fraction will depend on the agitation conditions, and the rate at which fresh hydrogen is fed to the impeller, as shown in Fig. 4.20. (Some hydrogenation reactors use gas-inducing impellers to recirculate gas from the head-space above the liquid while others use an external compressor.) For the purposes of the present example, the typical values, $d_b = 0.8$ mm and $\varepsilon_G = 0.20$, will be taken. Also dependent to some extent on the agitation conditions are the values of the transfer coefficients k_L and k_s, although these will depend mainly on the physical properties of the system such as the viscosity of the liquid and the diffusivity of the dissolved gas. The values taken here will be: $k_L = 1.23 \times 10^{-3}$ m/s and $k_s = 0.54 \times 10^{-3}$ m/s.
(These have been estimated for typical reactor conditions from the correlations described in textbooks such as those by TATTERSON[8] and by HARNBY *et al.*[17]).

The last important design parameter to be specified is the catalyst loading to be used in the process, i.e. the ratio of catalyst to liquid charged to the reactor. There is little point in using more catalyst than is really necessary so, as a first trial, a solid/liquid ratio of 0.10 by volume will be used. The pressure of the hydrogen gas in the reactor also needs to be decided. Hydrogen is one of the least soluble gases so that industrial hydrogenations are usually carried out at high pressures to increase the solubility under reactor conditions, i.e. to increase C_{Ai} in equation 4.36. In this example, a pressure of 30 bar will be used compared with which the vapour pressure of the α-methyl styrene is negligible. At 50°C the value of the Henry law constant for hydrogen dissolved in α-methyl styrene is 285 m^3 bar/kmol.

Using the above values, from equations 4.30 and 4.31, the gas–liquid interfacial area a is 1500 m^2/m^3 and the particle surface area 6×10^4 m^2/m^3. In the dispersion, where the gas volume fraction ε_G is 0.20, the solid to liquid ratio of 0.10 for the feed charged to the reactor corresponds to the ratio $\varepsilon_P/\varepsilon_L$ so that, from equation 4.29, ε_P is 0.073 and ε_L 0.727. The resistance terms in equation 4.35 can now be examined in turn:

$$\frac{1}{k_L a} = \frac{1}{1.23 \times 10^{-3} \times 1500} = \underline{\underline{0.54 \text{ s}}}$$

$$\frac{1}{k_s a_P \varepsilon_P} = \frac{1}{0.54 \times 10^{-3} \times 6 \times 10^4 \times 0.073} = \underline{\underline{0.42 \text{ s}}}$$

$$\frac{1}{\boldsymbol{k}_1 \eta \varepsilon_P} = \frac{1}{16.8 \times 0.407 \times 0.073} = \underline{\underline{2.00 \text{ s}}}$$

It is clear from these calculations that the term $1/\boldsymbol{k}_1 \eta \varepsilon_P$ arising from pore diffusion with reaction is considerably larger than either of the others. Unless a more active catalyst can be found, or possibly the operating temperature varied, little can be done to increase $\boldsymbol{k}_1$. The effectiveness factor η is considerably less than unity and could be increased by choosing a smaller particle size, with the disadvantage of a more difficult separation. However, a distinct improvement can be made by using a higher catalyst loading, i.e. a higher ratio of solid to liquid in the charge to the reactor. There will be a limit to the extent that the amount of catalyst can be increased because, as the solids content rises, the slurry will become non-Newtonian, dispersion of the gas will become more difficult, and the bubble size d_b will increase to the detriment of the gas–liquid interfacial area a in the $1/k_L a$ resistance term. This example demonstrates the importance of achieving an optimised balance in the choice of conditions for a gas–liquid–solid reactor.

Finally a value for the overall rate $\mathscr{R}_t$ of the hydrogenation process can be calculated from equation 4.35. For a pressure of 30 bar, equation 4.36 gives a value for C_{Ai} of 0.105 kmol/m^3. Using the above values for the individual resistances, the overall rate becomes:

$$\mathscr{R}_t = (0.54 + 0.42 + 2.00)^{-1} \times 0.105 = \underline{\underline{0.036 \text{ kmol/m}^3 \text{ s}}}.$$

Three-Phase Fluidised Suspended-Bed Reactor—Combination of Mass Transfer and Reaction Steps

At the present time, three-phase fluidised-beds are not often chosen for gas–liquid–solid reactions despite their advantages of good heat and mass transfer and, in principle, freedom from the blockages that can occur with fixed-beds[30]. The reason may be that, because of the pronounced hydrodynamic interactions between the phases as indicated in Fig. 4.16, development of a three-phase fluidised-bed

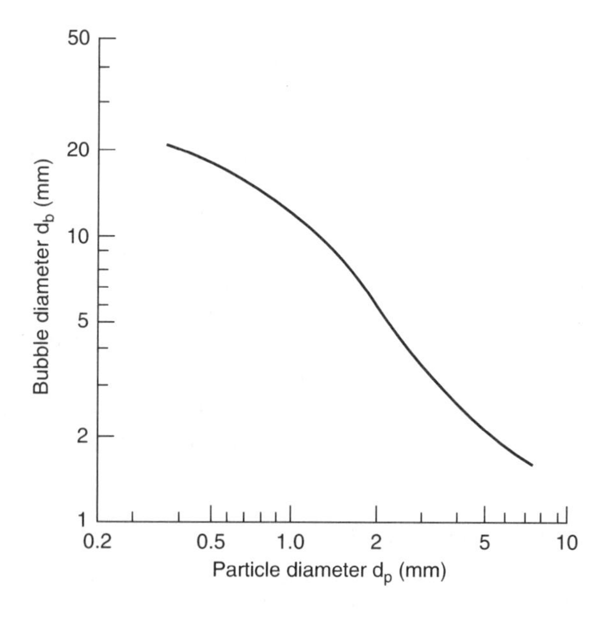

FIG. 4.21. Estimated bubble sizes in α-methyl styrene with catalyst particles in a three-phase fluidised bed; particle volume fraction $\varepsilon_P = 0.5$, gas volume fraction $\varepsilon_G = 0.2$

reactor for a new process would necessarily require extensive experimental work which would be both time-consuming and costly. Also in the present state of knowledge, scale-up from pilot plant operation to a full-scale design would still be uncertain. By contrast, scale-up of trickle-bed reactors is relatively well understood so that a trickle-bed reactor might be chosen for a process despite any of the advantages that, in theory, a three-phase fluidised-bed reactor might have to offer.

The three-phase fluidised-bed reactors so far developed commercially for processes such as the hydrogenation of petroleum residues use beds of various particle sizes depending on the type of feedstock and the particular experience of the company that developed the process. Reactors with small particles will tend to behave in the way shown in Fig. 4.16*a*, producing comparatively large bubbles. Apart from having an upwards flow of liquid, these are somewhat similar to the bubble column type of slurry reactor. A three-phase fluidised-bed reactor using relatively large particles, as shown in Fig. 4.16*b*, requires higher liquid flowrates in order to maintain the particles in the fluidised state, but has the advantage of producing smaller bubbles and therefore a larger gas–liquid surface area for mass transfer. To compare this type of three-phase fluidised-bed reactor with the agitated-slurry type of reactor using small particles, as previously considered in Example 4.6, the hydrogenation of α-methyl styrene will again be taken as a model reaction in the example that follows.

One of the areas of uncertainty in three-phase fluidised-bed design is the mixing that occurs within each of the liquid and gas phases. As the particle size increases, so the flow patterns approximate more nearly to plug flow. For the hydrogenation of α-methyl styrene using pure hydrogen, gas phase mixing will not affect the process and, because the reaction rate is independent of the α-methyl styrene concentration, mixing in the liquid phase will have an effect only through the dissolved hydrogen concentration. In the example, it will be assumed that a steady state has been established between the rates of mass transfer and the rate of the reaction (which is the case for the agitated tank suspended-bed reactor in Example 4.6). Unless the liquid is already saturated with dissolved hydrogen before entering, this steady state may not apply near the liquid inlet. (If the liquid is in plug flow, the variation of dissolved hydrogen concentration near the liquid inlet will be similar to that shown in Fig. 4.22 where, in Example 4.8, the effect is considered in more detail for a trickle-bed reactor.)

Although bubble sizes for large particles with the α-methyl styrene system have not been measured directly and any prediction must be regarded as highly speculative, Fig. 4.21 shows an estimate of the variation of mean bubble diameter with size of catalyst particles[36]. This estimate is based on measurements with glass beads in water, with subsequent adjustments to allow for the different densities of the particles and differences in the viscosity and surface tension of the liquids.

Example 4.7

The Hydrogenation of α-Methyl Styrene in a Three-Phase Fluidised-Bed Reactor

The performance of a three-phase fluidised-bed is to be assessed in relation to a model reaction, the hydrogenation of α-methyl styrene at 50°C. The same catalyst will be used as in Example 4.6, excepting that the diameter of the spherical particles will be 2.0 mm. The liquid and gas flowrates will be such that the volume fraction of particles in the bed will be 0.50 and the volume fraction of gas 0.20; i.e. the relationship shown in Fig. 4.21 will apply, from which the estimated bubble size is 5.0 mm.

Solution

Because the conditions for the reaction are the same as in Example 4.6, the pseudo first-order rate constant $\boldsymbol{k}$ will again be 16.8 s^{-1}, and the effective diffusivity of hydrogen in the liquid filled pores of the catalyst D_e will be 0.11×10^{-8} m^2/s. Also because the transfer coefficients k_L and k_S depend mainly on the physical properties of the system, the same values, namely, $k_L = 1.23 \times 10^{-3}$ m/s and $k_S = 0.54 \times 10^{-3}$ m/s, will be used even though the hydrodynamics of the three-phase fluidised-bed will be different and the particle size is larger.

As in Example 4.6 the aim is again to evaluate the individual resistance terms in equation 4.35 and to examine them in turn. Also the overall rate of hydrogenation at the same hydrogen pressure of 30 bar will be calculated.

(a) Gas–liquid mass transfer resistance $1/k_L a$:

Since $\varepsilon_G = 0.20$ and $d_b = 0.005$ m, from equation 4.30 the gas–liquid interfacial area $a = 6\varepsilon_G/d_b = 6 \times 0.20/0.005 = 240$ m^2/m^3.

$$\therefore \quad \frac{1}{k_L a} = \frac{1}{1.23 \times 10^{-3} \times 240} = \underline{\underline{3.39 \text{ s}}}$$

(b) Liquid–solid mass transfer resistance $1/k_S a_P \varepsilon_P$:

From equation 4.31, $a_P = 6/d_P = 6/0.002 = 3000$ m^2/m^3.

$$\therefore \quad \frac{1}{k_S a_P \varepsilon_P} = \frac{1}{0.54 \times 10^{-3} \times 3000 \times 0.50} = \underline{\underline{1.23 \text{ s}}}$$

(c) Reaction within the particle $1/\boldsymbol{k}_1 \eta \varepsilon_P$:

As in Example 4.6, the effectiveness factor η for spherical particles is given by: $\eta = \frac{1}{\phi}\left(\coth 3\phi - \frac{1}{3\phi}\right)$

where:

$$\phi = \frac{d_P}{6}\sqrt{\frac{\boldsymbol{k}_l}{D_e}}$$

Thus, in the present case, $\phi = 41.2$ and $\eta = 0.0241$.

$$\therefore \quad \frac{1}{\boldsymbol{k}_1 \eta \varepsilon_P} = \frac{1}{16.8 \times 0.0241 \times 0.5} = \underline{\underline{4.94 \text{ s}}}$$

The overall rate of reaction $\mathscr{R}_t$ is therefore obtained from equation 4.35, using the value for C_{Ai} of 0.105 $kmol/m^3$, from Example 4.6:

$$\mathscr{R}_t = (3.39 + 1.23 + 4.94)^{-1} \times 0.105 = \underline{\underline{0.0110 \text{ kmol/m}^3\text{s.}}}$$

If even larger particles are considered, repeating the above calculation shows that, although the gas–liquid resistance (*a*) decreases, both the liquid–solid resistance (*b*) and the reaction term (*c*) increase. As a result of these opposing tendencies, there must exist an optimum particle size for which the overall resistance is a minimum and therefore the overall reaction rate $\mathscr{R}_t$ is a maximum. More detailed calculations(36) show that for the value of the rate constant $\boldsymbol{k}_1 = 16.8$ s^{-1} used above, the optimum particle size is indeed approximately 2 mm which is the size taken for the example. However, for lower values of the rate constant, the optimum shifts towards smaller particle sizes, the reason being that the gas–liquid interfacial area for mass transfer becomes less important.

The overall rate of reaction calculated for the three-phase fluidised-bed reactor above is approximately one tenth of the rate calculated for the agitated tank slurry reactor in Example 4.6. The main reasons are the very poor effectiveness factor and the relatively smaller external surface area for mass transfer caused by using the larger particles. Even the gas–liquid transfer resistance is greater for the three-phase fluidised-bed, in spite of the larger particles being able to produce relatively small bubbles; these bubbles are not however as small as can be produced by the mechanical agitator. Although the three-phase fluidised bed does not appear to perform very favourably compared with the agitated tank slurry reactor, it should be remembered that the fluidised-bed does not have the sealing and other mechanical complications associated with an agitator. Also the gas and liquid will be more nearly in plug flow in the fluidised bed, although this advantage may to some extent be lost if the liquid has to be recycled in order to provide a high liquid velocity for fluidisation.

Trickle-Bed Reactor—Combination of Mass Transfer and Reaction Steps

One of the advantages of the co-current downward flow, trickle-bed type of three-phase reactor is that both gas and liquid move nearly in plug flow. For the gas, any departure from ideal plug flow is usually unimportant, but for the liquid-phase, the departure from plug flow may be sufficient in some cases to warrant using the dispersed plug-flow model. In the example that follows, it will be assumed for simplicity that the reactor operates under isothermal conditions, that the distribution of both gas and liquid is uniform, and that there are no radial concentration gradients. The reaction involved is the hydrogenation of thiophene to remove the sulphur as hydrogen sulphide. This reaction has been studied as a model for industrial hydro-desulphurisation processes involving feedstocks consisting of complex mixtures of sulphur-containing hydrocarbons. The catalyst for the reaction is sulphided cobalt–molybdenum oxide on an alumina support.

In general, for a trickle-bed reactor, a material balance is required for each of the components present, taken over each of the gas and liquid phases. In the example, however, pure hydrogen will be used and the volatility of the other components will be assumed to be sufficiently low that they do not enter the vapour phase. This means that the material balance on the gas phase can be omitted. It also means that there will be no gas-film resistance to gas–liquid mass transfer. In the liquid phase, material balances are required for (i) the hydrogen (reactant **A**), and (ii) the thiophene (reactant **B**). The amounts of each of the reactants consumed will be linked by the stoichiometric equation:

$$C_4H_4S \text{ (thiophene)} + 4\,H_2 = C_4H_{10} + H_2S$$

In deriving the material balance equations, the dispersed plug flow model will first be used to obtain the general form but, in the numerical calculations, the dispersion term will be omitted for simplicity. As used previously throughout, the basis for the material balances will be unit volume of the whole reactor space, i.e. gas plus liquid plus solids. Thus in the equations below, for the transfer of reactant **A**:
k_La is the volumetric mass transfer coefficient for gas–liquid transfer, and
k_sa_s is the volumetric mass transfer coefficient for liquid–solid transfer.

Consider an element height δz of a reactor of unit cross-sectional area, as shown in Fig. 4.22*a*. Let the gas phase consist of a pure reactant **A** (hydrogen in the example) at a pressure P_A. The concentration of **A** dissolved in the liquid at the gas–liquid interface will therefore be $C_{Ai} = P_A/\mathscr{H}$, where $\mathscr{H}$ is the Henry law constant. We will now consider each of the terms in the mass balance equation for **A**, in the liquid phase, over the element δz:

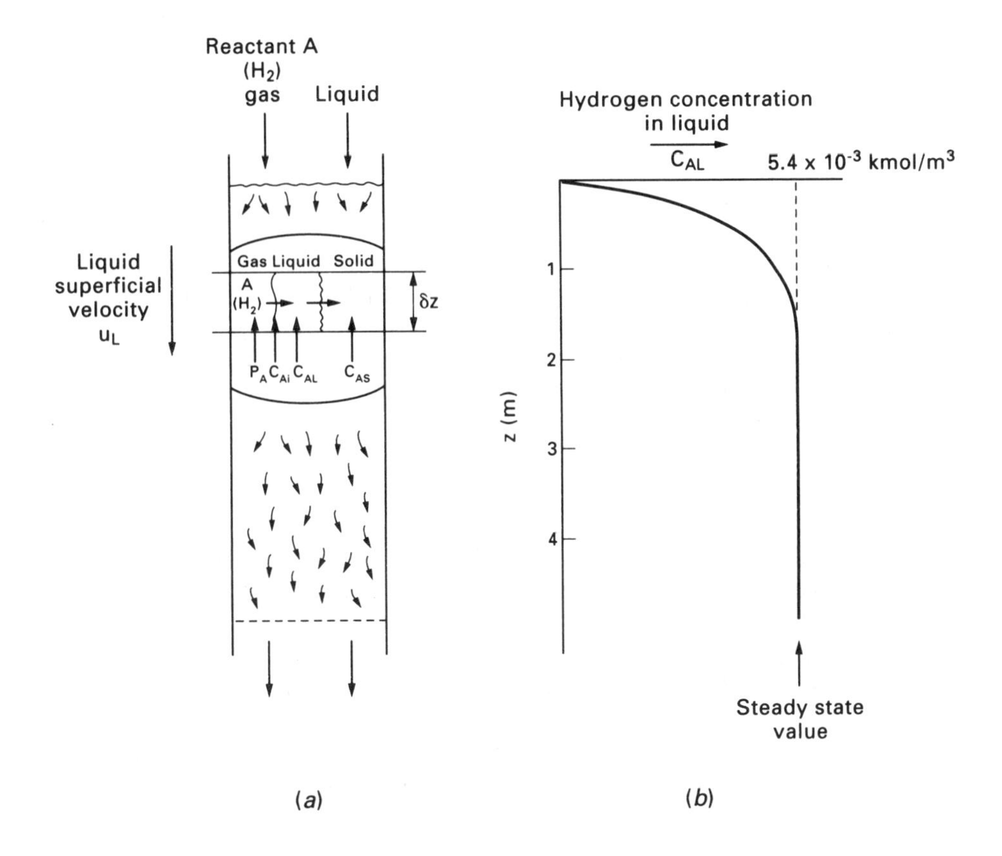

FIG. 4.22. Trickle-bed reactor. (*a*) Schematic diagram showing individual mass transfer steps for a hydrogenation process; (*b*) profile of dissolved hydrogen in the liquid flowing down the column, calculated in Example 4.8

Rate of transfer of **A** into the element from the gas phase: $k_L a(C_{Ai} - C_{AL})\delta z$

Rate of transfer of **A** out of the element to the solid: $k_S a_S(C_{AL} - C_{AS})\delta z$

Rate of convection of A into the element with the liquid flow: $u_L C_{AL}$

Rate of convection of **A** out of the element by liquid flow: $u_L\left(C_{AL} + \dfrac{\mathrm{d}C_{AL}}{\mathrm{d}z}\,\delta z\right)$

Rate of transport of **A** into element by axial dispersion: $-\mathscr{D}_L\dfrac{\mathrm{d}C_{AL}}{\mathrm{d}z}$

Rate of transport of **A** out of element by axial dispersion: $\mathscr{D}_L\left(\dfrac{\mathrm{d}C_{AL}}{\mathrm{d}z} + \dfrac{\mathrm{d}^2C_{AL}}{\mathrm{d}z^2}\,\delta z\right)$

Because conditions in the reactor do not change with time, taken over one second, the net sum of all the transfer terms above, on cancelling out δz, gives the following equation:

$$\mathscr{D}_L \frac{d^2 C_{AL}}{dz^2} - u_L \frac{dC_{AL}}{dz} + k_L a(C_{Ai} - C_{AL}) - k_s a_s(C_{AL} - C_{As}) = 0 \qquad (4.37)$$

Dispersion Convection Gas to liquid Liquid to solid

The concentration C_{As} is determined by the rate of reaction within the solid which must be equal to the rate of transfer from the liquid:

$$k_s a_s(C_{AL} - C_{As}) = \boldsymbol{k}_1 C_{As} \eta \varepsilon_P \qquad (4.38)$$

In this equation the reaction is taken to be pseudo first-order with respect to **A** (as is the case for hydrogen in the reaction with thiophene) with an effectiveness factor η. The factor ε_P appearing in this equation converts the basis of the reaction rate constant from unit volume of particles to unit volume of bed in order to be consistent with the basis used for the volumetric mass transfer coefficients.

From equation 4.38, solving for C_{As}:

$$C_{As} = \frac{k_s a_s}{(k_s a_s + \boldsymbol{k}_1 \eta \varepsilon_P)} C_{AL} \qquad (4.39)$$

Also since C_{Ai} is given by $C_{Ai} = P_A/\mathscr{H}$, C_{Ai} and C_{As} from equation 4.39 may be substituted into equation 4.37 to give a differential equation in C_{AL}:

$$\mathscr{D}_L \frac{d^2 C_{AL}}{dz^2} - u_L \frac{dC_{AL}}{dz} + k_L a\left(\frac{P_A}{\mathscr{H}} - C_{AL}\right) - \left(\frac{1}{k_s a_s} + \frac{1}{\boldsymbol{k}_1 \eta \varepsilon_P}\right)^{-1} C_{AL} = 0 \qquad (4.40)$$

The equation corresponding to 4.37 for the material balance on the reactant **B** (thiophene in the example), present in the liquid phase only, can be derived in the same way giving:

$$\mathscr{D}_L \frac{d^2 C_{BL}}{dz^2} - u_L \frac{dC_{BL}}{dz} - k_{sB} a_s(C_{BL} - C_{Bs}) = 0 \qquad (4.41)$$

where k_{sB} is the liquid–solid mass transfer coefficient for reactant **B** which will differ from k_s owing to the different physical properties of **B**. The concentration C_{BL} is determined similarly to C_{AL} by the rate of reaction within the solid, but because the reaction is pseudo first-order and the reactant **B** (thiophene) is the one in considerable excess, the rate of reaction term involves just the concentration of the reactant **A**, C_{As}.

i.e. the equation for transfer and reaction of reactant **B** is:

$$k_{sB} a_s(C_{BL} - C_{Bs}) = \tfrac{1}{4} \boldsymbol{k}_1 C_{As} \eta \varepsilon_P \qquad (4.42)$$

so that equation 4.41 becomes:

$$\mathscr{D}_L \frac{d^2 C_{BL}}{dz^2} - u_L \frac{dC_{BL}}{dz} - \tfrac{1}{4} \boldsymbol{k}_1 C_{As} \eta \varepsilon_P = 0$$

where the factor of $\frac{1}{4}$ arises because 4 moles of hydrogen are needed to react with 1 mole of thiophene.

Substituting for C_{As} from equation 4.39, the concentration C_{BL} is given by the solution of the differential equation:

$$\mathscr{D}_L \frac{d^2 C_{BL}}{dz^2} - u_L \frac{dC_{BL}}{dz} - \frac{1}{4}\left(\frac{1}{k_s a_s} + \frac{1}{\boldsymbol{k}_1 \eta \varepsilon_P}\right)^{-1} C_{AL} = 0 \tag{4.43}$$

which requires equation 4.40 to be solved first so that C_{AL} can be expressed as a function of z.

When dispersion is neglected, i.e. ideal plug flow is assumed, the two equations, 4.40 for C_{AL}, and 4.43 for C_{BL}, are simplified and become:

$$-u_L \frac{dC_{AL}}{dz} + k_L a \left(\frac{P_A}{\mathscr{H}} - C_{AL}\right) - \boldsymbol{k}^0 C_{AL} = 0$$

or:

$$-u_L \frac{dC_{AL}}{dz} - (k_L a + \boldsymbol{k}^0) C_{AL} + \frac{P_A}{\mathscr{H}} k_L a = 0 \tag{4.44}$$

and:

$$-u_L \frac{dC_{BL}}{dz} - \frac{1}{4} \boldsymbol{k}^0 C_{AL} = 0 \tag{4.45}$$

where $\boldsymbol{k}^0$ is given by $\dfrac{1}{\boldsymbol{k}^0} = \dfrac{1}{k_s a_s} + \dfrac{1}{\boldsymbol{k}_1 \eta \varepsilon_P}$ (4.46)

Trickle-Bed Reactor—Simplified Steady State Treatment

A further simplification of the case of the hydrogenation reaction above can be obtained if it is assumed that, across any section through the reactor, the rate of hydrogen transfer from gas to liquid, is equal to the rate of transfer from liquid to solid, which in turn is just equal to the rate of reaction of hydrogen within the solid catalyst. This assumption implies that as soon as the liquid enters the reactor, the steady state concentration of hydrogen in the liquid C_{AL} is instantly established. Because the rate of the reaction is independent of the thiophene concentration and pure hydrogen is used, this concentration of hydrogen in the liquid C_{AL} is the same throughout the reactor. In this case the overall rate of the reaction $\mathscr{R}_t$ can be expressed as a combination of the individual mass transfer and reaction steps in exactly the same way as for the suspended-bed reactors previously considered. Thus, in terms of $k_s a_s$ the volumetric coefficient for mass transfer to the catalyst particles (a_s is equivalent to $a_P \varepsilon_P$), equation 4.35 becomes:

$$\mathscr{R}_t = \left(\frac{1}{k_L a} + \frac{1}{k_s a_s} + \frac{1}{\boldsymbol{k}_1 \eta \varepsilon_P}\right)^{-1} C_{Ai} \tag{4.47}$$

This is the rate at which the hydrogen reacts. From the stoichiometry of the reaction, the rate at which the thiophene reacts is $\frac{1}{4}\mathscr{R}_t$. Thus this term replaces the $\frac{1}{4}\boldsymbol{k}^0 C_{AL}$ term in equation 4.45 for the thiophene material balance which then becomes:

$$-u_L \frac{dC_{BL}}{dz} - \frac{1}{4} \mathscr{R}_t = 0 \tag{4.48}$$

Although the steady state concentration of hydrogen in the liquid-phase does not appear in equation 4.47 for the overall rate, if the value of this concentration denoted $C_{\overline{AL}}$, is required, it is most easily obtained from equation 4.44 by setting dC_{AL}/dz equal to zero because there will then be no variation of C_{AL} with z.

Thus:

$$C_{\overline{AL}} = \frac{P_A}{\mathscr{H}} \frac{k_L a}{(k_L a + k^0)} \tag{4.49}$$

Alternatively, this equation can be obtained by eliminating $\mathscr{R}_t$ and C_{AS} from equations 4.32, 4.33 and 4.34, which are the basic equations for the trickle-bed reactor with the steady-state assumption.

The numerical results from these two treatments will now be compared in the following example[37].

Example 4.8

The Hydro-Desulphurisation of Thiophene in a Trickle-Bed Reactor

A trickle-bed reactor is to be designed to remove 75 per cent of the thiophene present in a hydrocarbon feed which contains 0.012 kmol/m^3 thiophene. The bed of catalyst particles will operate at a temperature of 200°C and a pressure of 40 bar. Under these conditions the rate of the reaction may be assumed to be first order with respect to dissolved hydrogen and independent of the thiophene concentration. The superficial velocity of the liquid in the reactor will be 0.05 m/s.

The *objective* of the calculation is to determine the depth of bed required to give a fractional conversion of thiophene at the outlet of 0.75. As part of the calculation, a graph of the concentration of hydrogen in the liquid phase C_{AL} versus the distance z from the top of the catalyst bed is to be drawn. Further data are as follows:

First-order rate constant with respect to hydrogen (here expressed as k_1^m, the rate constant per unit mass of catalyst, which is often the practice for heterogeneous reactions catalysed by solids, as in Chapter 3):

$$k_1^m = 0.11 \times 10^{-3}\ \text{m}^3/(\text{kg catalyst})\ \text{s}.$$

Bulk density of catalyst: $\rho_b = 960$ kg/m^3. Effectiveness factor $\eta = 1$.

Volumetric mass transfer coefficients for hydrogen:

Gas to liquid: $k_L a = 0.030$ s^{-1}. Liquid to solid: $k_s a_s = 0.50$ s^{-1}.
Henry law constant for dissolved hydrogen under the above conditions: $\mathscr{H} = 1940$ bar m^3/kmol.

Solution

The rate constant given, k_1^m has, as its basis, unit mass of catalyst, whereas the term $k_1 \varepsilon_P$ in equation 4.46 is the rate constant k_1 per unit volume of particles converted to a basis of *unit volume of bed.* To convert k_1^m similarly it has to be multiplied by ρ_b the bulk density of the bed.

i.e.:

$$k_1 \varepsilon_P = \rho_b k_1^m = 960 \times 0.11 \times 10^{-3} = 0.106\ \text{s}^{-1}$$

Then from equation 4.46 with $\eta = 1$:

$$\frac{1}{k^0} = \frac{1}{0.05} + \frac{1}{0.106 \times 1}. \qquad \therefore\ k^0 = 0.085\ \text{s}^{-1}$$

Proceeding now to the differential equation whose solution will give C_{AL}, the concentration of hydrogen in the liquid phase as a function of z, i.e. equation 4.44, this becomes:

$$-0.05\frac{dC_{AL}}{dz} - (0.030 + 0.085)C_{AL} + \frac{40}{1940}0.030 = 0$$

i.e.:

$$\frac{dC_{AL}}{dz} = 0.0124 - 2.30C_{AL}$$

At the inlet to the bed, when $z = 0$, $C_{AL} = 0$, so integrating with this condition:

$$\int_0^{C_{AL}} \frac{1}{(0.0124 - 2.30C_{AL})}\, dC_{AL} = z$$

i.e.: $$-\frac{1}{2.30}\ln(0.0124 - 2.30C_{AL}) + \frac{1}{2.30}\ln 0.0124 = z$$

or: $$\frac{0.0124 - 2.30C_{AL}}{0.0124} = e^{-2.30z}$$

or $$(1 - 185C_{AL}) = e^{-2.30z}$$

$\therefore$ $$C_{AL} = 5.4 \times 10^{-3}(1 - e^{-2.30z}) \quad \text{(A)}$$

Equation A gives the required dependence of the hydrogen concentration in the liquid phase upon the distance from the inlet to the reactor and is plotted in Fig. 4.22*b*.

The corresponding differential equation representing the material balance on thiophene is equation 4.45. Substituting numerical values and C_{AL} from equation A, this becomes:

$$-0.05\frac{dC_{LB}}{dz} - \frac{0.085}{4}5.4 \times 10^{-3}(1 - e^{-2.30z})$$

i.e.: $$-\frac{dC_{LB}}{dz} = 0.0023(1 - e^{-2.30z})$$

Integrating: $$(C_{LB})_{in} - (C_{LB})_{out} = 0.0023\left(z + \frac{1}{2.30}e^{-2.30z} - \frac{1}{2.30}\right)$$

Thus the fractional conversion of the thiophene α_T is:

$$\alpha_T = \frac{(C_{LB})_{in} - (C_{LB})_{out}}{(C_{LB})_{out}} = \frac{0.0023}{0.012}\left(z + \frac{1}{2.30}e^{-2.30z} - 0.435\right) \quad \text{(B)}$$

It can be seen from this expression that if $z > 1$, $\frac{1}{2.30}e^{-2.30z} < 0.044$ and if $z > 4$, $\frac{1}{2.30}e^{-2.30z} < 0.00004$ so that neglecting this term in equation B corresponds to an error of less than 0.0012 per cent when $z > 4$. When this term is neglected, equation B becomes explicit for z, so that with $\alpha_t = 0.75$, the desired fractional conversion for the thiophene, equation B becomes:

$$0.75 = \frac{0.0023}{0.012}(z - 0.435)$$

Thus: $z = \underline{\underline{4.35 \text{ m}}}$ which is the answer required.

Calculation using simplified steady state treatment

In this case the rate of reaction with respect to hydrogen is given by equation 4.47 with C_{Ai} equal to $P_A/\mathscr{H}$.

i.e.: $$\mathscr{R}_t = \left(\frac{1}{0.030} + \frac{1}{0.50} + \frac{1}{0.106}\right)^{-1}\frac{40}{1940} = 0.00046 \text{ kmol/m}^3\text{ s}$$

Equation 4.48 for the thiophene material balance therefore becomes:

$$-0.05\frac{dC_{LB}}{dz} - \frac{0.00046}{4} = 0$$

Integrating: $$(C_{LB})_{in} - (C_{LB})_{out} = \frac{0.00046}{0.05 \times 4}z = 0.0023z$$

and: $$\alpha_T = \frac{(C_{LB})_{in} - (C_{LB})_{out}}{(C_{LB})_{out}} = \frac{0.0023}{0.012}z = 0.192z$$

Thus, when $\alpha_t = 0.75$, $z = \frac{0.75}{0.192} = \underline{\underline{3.9 \text{ m}}}$ which is the answer required. The steady state concentration of hydrogen $C_{\bar{A}L}$ is given by equation 4.49, so that:

$$C_{\bar{A}L} = \frac{40}{1940}\frac{0.030}{(0.030 + 0.085)} = 5.4 \times 10^{-3} \text{ kmol/m}^3.$$

This value, which is independent of z, is plotted in Fig. 4.22*b* for comparison with the more exact variation with z obtained in the first calculation.

4.3. FURTHER READING

ASTARITA, G.: *Mass Transfer with Chemical Reaction* (Elsevier, 1967).

DANCKWERTS, P. V.: *Gas–Liquid Reactions* (McGraw-Hill, 1970).

DECKWER, W.-D.: *Bubble Column Reactors* (Wiley, 1992).

DORAISWAMY, L. K. and SHARMA, M. M.: *Heterogeneous Reactions*, Vol. 2, *Fluid–Fluid–Solid Reactions* (Wiley, 1984).

FAN, L.-S.: *Gas–Liquid–Solid Fluidization Engineering* (Butterworth, 1989).

KASTANEK, F., ZAHRADNIK, J., KRATOCHVIL, J. and CERMAK, J.: *Chemical Reactors for Gas–Liquid Systems* (Ellis Horwood, 1991).

RAMACHANDRAN, P. A. and CHAUDHARI, R. V.: *Three-Phase Catalytic Reactors* (Gordon and Breach, 1983).

SHAH, Y. T.: *Gas–Liquid–Solid Reactor Design* (McGraw-Hill, 1979).

SMITH, J. M.: *Chemical Engineering Kinetics*, 3rd edn. (McGraw-Hill, 1981).

4.4. REFERENCES

1. DORAISWAMY, L. K. and SHARMA, M. M.: *Heterogeneous Reactions*, Vol. 2, *Fluid–Fluid–Solid Reactions* (Wiley, 1984).
2. VAN SWAAIJ, W. P. M. and VERSTEEG, G. F.: *Chem. Eng. Sci.* **47** (1992) 3181. Mass transfer accompanied with complex reversible chemical reactions in gas–liquid systems: An overview.
3. DANCKWERTS, P. V.: *Gas–Liquid Reactions* (McGraw-Hill, 1970).
4. RESNICK, W. and GAL-OR, B.: *Advances in Chemical Engineering*, Vol. 7, DREW, T. B., COKELET, G. R., HOOPES, J. W. and VERMEULEN, T. eds. (Academic Press, 1968) p. 295. Gas–liquid dispersions.
5. LIGHTFOOT, E. N.: *A.I.Ch.E.J.* **4** (1958) 499. Steady state absorption of a sparingly soluble gas in an agitated tank with simultaneous first order reaction.
6. VAN KREVELEN, D. W. and HOFTIJZER, P. J.: *Rec. Trav. Chim. Pays-Bas* **67** (1948) 563. Kinetics of gas–liquid reactions. Part 1. General theory.
7. REID, R. C., PRAUSNITZ, J. M. and POLING, B. E.: *The Properties of Gases and Liquids*, 4th edn. (McGraw-Hill, 1987).
8. TATTERSON, G. B.: *Fluid Mixing and Gas Dispersion in Agitated Tanks* (McGraw-Hill, 1991).
9. DECKWER, W.-D.: *Bubble Column Reactors* (Wiley, 1992).
10. KOHL, A. L. and RIESENFELD, F. C.: *Gas Purification* (Gulf Publishing, 1979).
11. ASTARITA, G., SAVAGE, D. W. and BISIO, A.: *Gas Treating with Chemical Solvents* (Wiley-Interscience, 1983).
12. DANCKWERTS, P. V. and SHARMA, M. M.: *The Chemical Engineer* No. 202 (1966) CE 244. The absorption of carbon dioxide into solutions of alkalis and amines (with some notes on hydrogen sulphide and carbonyl sulphide).
13. BRUIJN, W., VAN'T RIET, K. and SMITH, J. M.: *Trans. Inst. Chem. Eng.* **52** (1974) 88. Power consumption with aerated Rushton turbines.
14. SITTIG, M.: *Organic Chemical Process Encyclopedia* (Noyes Development Corporation, 1969).
15. FROMENT, G. F. and BISCHOFF, K. B.: *Chemical Reactor Analysis and Design*, 2nd edn. (Wiley, 1989), p. 642.
16. GATES, B. C.: *Catalytic Chemistry* (Wiley, 1992).
17. HARNBY, N., EDWARDS, M. F. and NIENOW, A. W. eds.: *Mixing in the Process Industries*, 2nd edn. (Butterworth–Heinemann, 1992).
18. BAIRD, M. H. I. and RICE, R. G.: *Chem. Eng. J.* **9** (1975) 171. Axial dispersion in large unbaffled columns.
19. SHERWOOD, T. K., PIGFORD, R. L. and WILKE, C. R.: *Mass Transfer* (McGraw-Hill, 1975).
20. WESTERTERP, K. R., VAN SWAAIJ, W. P. M. and BEENACKERS, A. A. C. M.: *Chemical Reactor Design and Operation* (Wiley, 1984).
21. DUDA, J. L. and VENTRAS, J. S.: *A.I.Ch.E.J.* **14** (1968) 286. Laminar liquid jet diffusion studies.
22. DAVIDSON, J. F., CULLEN, E. J., HANSON, D. and ROBERTS, D.: *Trans. Inst. Chem. Eng.* **37** (1959) 122. The hold-up and liquid film coefficient of packed towers. Part 1: Behaviour of a string of spheres.
23. LEVENSPIEL, O. and GODFREY, J. H.: *Chem. Eng. Sci.* **29** (1974) 1723. A gradientless contactor for experimental study of interphase mass transfer with/without reaction.
24. SHARMA, M. M.: *Trans. Inst. Chem. Eng. Part A* **71** (1993) 595. Some novel aspects of multiphase reactions and reactors.

25. PEARSON, L., PINSENT, B. R. W. and ROUGHTON, F. J. W.: *Faraday Soc. Discussions* 'Study of Fast Reactions' **17** (1954) 141. The measurement of the rate of rapid reactions by a thermal method.
26. PINSENT, B. R. W., PEARSON, L. and ROUGHTON, F. J. W.: *Trans. Faraday Soc.* **52** (1956) 1512. The kinetics of combination of carbon dioxide with hydroxide ions.
27. HAGUE, D. N.: *Fast Reactions* (Wiley-Interscience, 1971).
28. SHAH, Y. T.: *Gas–Liquid–Solid Reactor Design* (McGraw-Hill, 1979).
29. TOPSØE, H., CLAUSEN, B. S., TOPSØE, N.-Y. and PEDERSEN, E.: *Ind. Eng. Chem. Fundam.* **25** (1986) 25. Recent basic research in hydro-desulfurization catalysts.
30. FAN, L.-S.: *Gas–Liquid–Solid Fluidization Engineering* (Butterworth, 1989).
31. HERSKOWITZ, M. and SMITH, J. M.: *A.I.Ch.E.J.* **29** (1983) 1. Trickle-bed reactors: A review.
32. NG, K. M.: *A.I.Ch.E.J.* **32** (1986) 115. A model for flow regime transitions in cocurrent down-flow trickle-bed reactors.
33. CHARPENTIER, J. C. and FAVIER, M.: *A.I.Ch.E.J.* **21** (1975) 1213. Some liquid hold-up experimental data in trickle-bed reactors for foaming and non-foaming hydrocarbons.
34. GIANETTO, A. and SPECCHIA, V.: *Chem. Eng. Sci.* **47** (1992) 3197. Trickle-bed reactors: State of art and perspectives.
35. SATTERFIELD, C. N., PELOSSOF, A. A. and SHERWOOD, T. K.: *A.I.Ch.E.J.* **15** (1969) 226. Mass transfer limitations in a trickle-bed reactor.
36. LEE, J. C., SHERRARD, A. J. and BUCKLEY, P. S.: *Fluidization and Its Applications*, ANGELINO, H., COUDERC, J. P., GIBERT, H. and LAGUERIE, C. eds. (Société de Chimie Industrielle, Toulouse, 1974) p. 407. Optimum particle size in three phase fluidized bed reactors.
37. SMITH, J. M.: *Chemical Engineering Kinetics*, 3rd edn. (McGraw-Hill, 1981).

4.5. NOMENCLATURE

		Units in SI System	Dimensions **M, N, L, T, θ**
$\mathcal{A}$	$\left[1 + 4k_{1G}\varepsilon_G\mathcal{D}_G/u_G^2\right]^{\frac{1}{2}}$	—	—
a	Interfacial area per unit volume of dispersion	m^{-1}	$\mathbf{L^{-1}}$
a_P	Surface area per unit volume of solid particle	m^{-1}	$\mathbf{L^{-1}}$
a_s	Surface area per unit volume of fixed bed	m^{-1}	$\mathbf{L^{-1}}$
C_A, C_B	Molar concentrations of **A**, **B**	$kmol/m^3$	$\mathbf{NL^{-3}}$
C_{AG}	Value of C_A in bulk gas	$kmol/m^3$	$\mathbf{NL^{-3}}$
C_{AL}, C_{BL}	Value of C_A, C_B in bulk liquid	$kmol/m^3$	$\mathbf{NL^{-3}}$
$\overline{C_{AL}}$	Concentration of **A** in liquid phase with steady state assumption	$kmol/m^3$	$\mathbf{NL^{-3}}$
C_G	Molar concentration of dispersed species in gas phase	$kmol/m^3$	$\mathbf{NL^{-3}}$
C_L^*	Saturation concentration in bulk liquid	$kmol/m^3$	$\mathbf{NL^{-3}}$
D_A, D_B	Diffusivity of **A**, **B** in mixture	m^2/s	$\mathbf{L^2T^{-1}}$
D_{AB}	Diffusivity of **A** in **B**	m^2/s	$\mathbf{L^2T^{-1}}$
D_e	Effective diffusivity	m^2/s	$\mathbf{L^2T^{-1}}$
D_L	Liquid-phase diffusivity	m^2/s	$\mathbf{L^2T^{-1}}$
$\mathcal{D}_G$	Gas-phase dispersion coefficient	m^2/s	$\mathbf{L^2T^{-1}}$
$\mathcal{D}_L$	Liquid-phase dispersion coefficient	m^2/s	$\mathbf{L^2T^{-1}}$
d_b	Bubble diameter	m	**L**
d_C	Column diameter	m	**L**
d_p	Particle diameter	m	**L**
f_A	Reaction factor for **A** (equation 4.10)	—	—
f_u	Fraction of gas unreacted	—	—
$\mathcal{H}$	Henry law constant	Nm/kmol	$\mathbf{MN^{-1}L^2T^{-2}}$
G	Molar gas flowrate per unit area	$kmol/m^2s$	$\mathbf{NL^{-2}T^{-1}}$
h	Height of packing	m	**L**
J_A	Transformation rate of **A** per unit volume of reactor	$kmol/m^3s$	$\mathbf{NL^{-3}T^{-1}}$

		Units in SI System	Dimensions **M, N, L, T, θ**
K_G	Overall mass transfer coefficient based on gas phase (cf k_G)	kmol/Ns	$\mathbf{NM^{-1}L^{-1}T}$
K_L	Overall mass transfer coefficient based on liquid phase (cf k_L)	m/s	$\mathbf{LT^{-1}}$
k_G	Gas-film mass transfer coefficient, driving force as partial pressure difference	kmol/Ns	$\mathbf{NM^{-1}L^{-1}T}$
k_L	Liquid-film mass transfer coefficient, driving force as molar concentration difference	m/s	$\mathbf{LT^{-1}}$
k'_L	Reaction-enhanced liquid-film mass transfer coefficient	m/s	$\mathbf{LT^{-1}}$
k_S	Mass transfer coefficient from liquid to solid	m/s	$\mathbf{LT^{-1}}$
$\boldsymbol{k}_1$	Reaction constant for first-order reaction	s^{-1}	$\mathbf{T^{-1}}$
$\boldsymbol{k}_2$	Reaction constant for second-order reaction	m^3/kmol s	$\mathbf{N^{-1}L^3T^{-1}}$
$\boldsymbol{k}_3$	Reaction constant for third-order reaction	$(m^3/kmol)^2/s$	$\mathbf{N^{-2}L^6T^{-1}}$
$\boldsymbol{k}_{1G}$	First-order rate constant for transfer from gas phase	s^{-1}	$\mathbf{T^{-1}}$
$\boldsymbol{k}^0$	Combination of liquid–solid transfer coefficient and rate constant, defined by equation 4.46	s^{-1}	$\mathbf{T^{-1}}$
$\boldsymbol{k}_1^m$	First-order rate constant defined on basis of unit mass of catalyst	m^3/kg s	$\mathbf{M^{-1}L^3T^{-1}}$
L	Height of column	m	$\mathbf{L}$
N_A	Molar rate of transfer of **A** per unit area	kmol/m^2s	$\mathbf{NL^{-2}T^{-1}}$
$\bar{N}_A$	Mean value of N_A over exposure time	kmol/m^2s	$\mathbf{NL^{-2}T^{-1}}$
n	Number of moles	kmol	$\mathbf{N}$
n_b	Number of bubbles per unit volume dispersion	m^{-3}	$\mathbf{L^{-3}}$
P	Total pressure	N/m^2	$\mathbf{ML^{-1}T^{-2}}$
P_A	Partial pressure of **A**	N/m^2	$\mathbf{ML^{-1}T^{-2}}$
P_{Ae}	Partial pressure of **A** in equilibrium with liquid concentration C_A	N/m^2	$\mathbf{ML^{-1}T^{-2}}$
R_t	Reaction rate in whole volume of tank	kmol/s	$\mathbf{NT^{-1}}$
R	Universal gas constant	J/kmol K	$\mathbf{MN^{-1}L^2T^{-2}\theta^{-1}}$
$\mathscr{R}_A, \mathscr{R}_B$	Reaction rate of **A, B** per unit volume	kmol/m^3s	$\mathbf{NL^{-3}T^{-1}}$
$\mathscr{R}_t$	Overall reaction rate per unit volume	kmol/m^3s	$\mathbf{NL^{-3}T^{-1}}$
r_0	Radius of sphere	m	$\mathbf{L}$
T	Absolute temperature	K	θ
t	Time	s	$\mathbf{T}$
t_e	Exposure time	s	$\mathbf{T}$
t_r	Reaction time	s	$\mathbf{T}$
u_G	Superficial gas velocity	m/s	$\mathbf{LT^{-1}}$
V	Volume	m^3	$\mathbf{L^3}$
V_d	Volume of dispersion	m^3	$\mathbf{L^3}$
V_L	Volume of liquid	m^3	$\mathbf{L^3}$
x	Distance in direction of transfer	m	$\mathbf{L}$
y	Mole fraction	—	—
α	Fractional conversion	—	—
α_T	Fractional conversion of thiophene	—	—
β	$\sqrt{\boldsymbol{k}_1 D_A}/k_L$ or $\sqrt{(\boldsymbol{k}_2 C_{BL} D_A)}/k_L$ (equation 4.13)	—	—
δ	Film thickness	m	$\mathbf{L}$
ε_G	Volume fraction of gas	—	—
ε_L	Volume fraction of liquid	—	—
ε_P	Volume fraction of particles	—	—
ν_A, ν_B	Stoichiometric coefficients for **A, B**	—	—
η	Effectiveness factor	—	—

		Units in SI System	Dimensions M, N, L, T, θ
ρ	Density	kg/m^3	$\mathbf{ML^{-3}}$
σ	Surface tension	N/m	$\mathbf{MT^{-2}}$
γ	Residence time	s	**T**
μ	Viscosity	Ns/m^2	$\mathbf{ML^{-1}T^{-1}}$
ϕ	Thiele modulus	—	—

Suffixes

A	Component A
B	Component B
e	Equilibrium value
i	Value at interface
L	Liquid
s	Solid
ex	Value at exit
in	Value at inlet
air	Value for air
water	Value for water

CHAPTER 5

Biochemical Reaction Engineering

5.1. INTRODUCTION

Previous chapters in this volume have been concerned with chemical reaction engineering and refer to reactions typical of those commonplace in the chemical process industries. There is another class of reactions, often not thought of as being widely employed in industrial processes, but which are finding increasing application, particularly in the production of *fine chemicals*. These are biochemical reactions, which are characterised by their use of enzymes or whole cells (mainly micro-organisms) to carry out specific conversions. The exploitation of such reactions by man is by no means a recent development—the fermentation of fruit juices to make alcohol and its subsequent oxidation to vinegar are both examples of biochemical reactions which have been used since antiquity.

Living organisms are complex chemically and are capable of carrying out a wide range of transformations which can often be manipulated by controlling their environment or by changing their genetic constitution. Whilst the primary interest of organisms is to replicate themselves and the formation of other products is incidental to this, the chemical engineer can often modify the environment in order to harness the biological transformations to suit his own ends. It may be that it is in the engineer's interests to minimise the proportion of the nutrient supply which goes towards replication so that a greater yield of his desired product is obtained. On the other hand, the *biomass* itself might be the desired product so that the engineer would be concerned with enhancing the natural inclination of the organism to replicate itself.

The active biochemical constituents of cells are a particular group of proteins which have catalytic properties. These catalytic proteins, or *enzymes*, are in some ways similar to inorganic catalysts but are distinctive in other, quite important respects. Enzymes are very powerful catalysts, capable of enhancing the overall rates of reactions much more markedly; they are much more specific than the average inorganic catalyst.

Furthermore, the conditions under which the reactions proceed are typically mild, temperatures, for example, being generally under 100°C and frequently below 50°C. Enzymes are water-soluble but are frequently bound to membranes within the cells or retained in the microbe by the cell walls. The structure of those cell walls is such that they permit the ingress of nutrients, or *substrates*, and the egress of the by-products of the cell's growth. Enzymes usually retain their catalytic activity when isolated from the cell and are often used as such, thus removing the need for providing for the maintenance of the cell itself whilst retaining the desirable specificity and activity of the protein. However, because the enzymes are soluble in water, it is often necessary to immobilise them, perhaps by attaching them to a suitable support, so that they are not carried away with the product.

There is an increasing number of areas where bioreactors are serious alternatives to conventional chemical reactors, particularly when their mild conditions and high selectivity can be exploited. In the pharmaceutical industry micro-organisms and enzymes can be used to produce specific stereo-isomers selectively, a very desirable ability since it can be that only the one isomer (possibly an optical isomer) may possess the required properties. In such applications the limitations of bioreactors are clearly outweighed by the advantages in their use. The fact that the products are formed in rather dilute aqueous solution and at relatively low rates may be of secondary importance and it may then be economically feasible to employ multiple separation stages in their purification.

Many of the problems encountered in the processing of biological materials are similar to those found in other areas of chemical engineering, and the separation processes used are frequently developments from counterparts in the chemical industry. However, biological materials frequently have rheological properties which make then difficult to handle, and the fact that their density differs little from that of water and the interfacial tensions are low can give rise to difficulties in *physical* separation of product.

Table 5.1 shows the major areas of the biological sciences that are of significance to the process engineer. Systematics, genetics, biochemistry and physiology are all

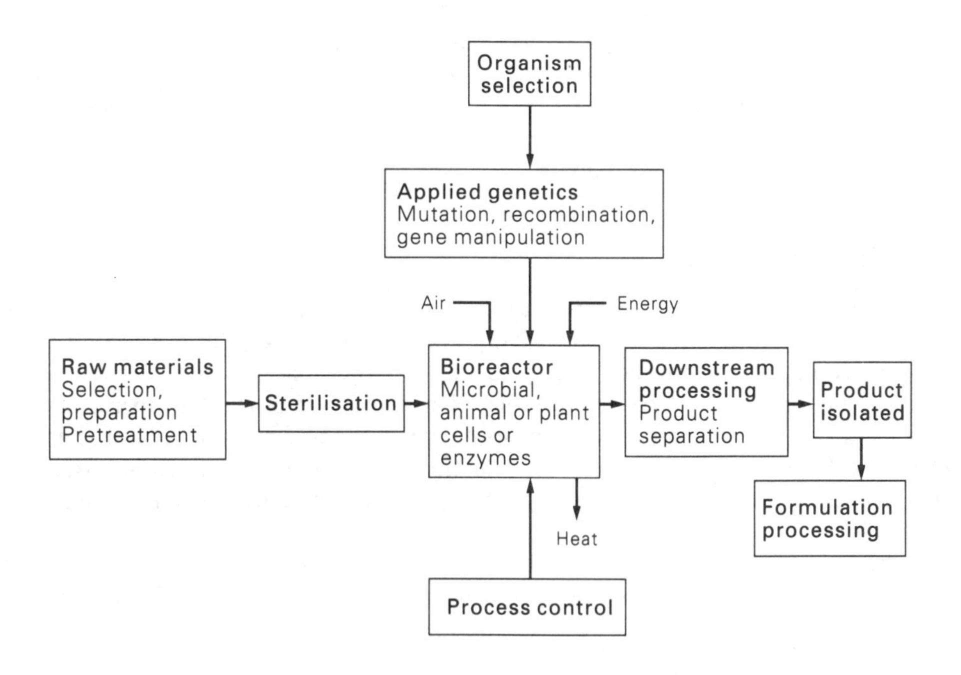

FIG. 5.1. Schematic overview of a biotechnological process[1]

important when considering the applications of biological processes. These areas all impinge on the choice of microbe, the type of culture/reactor used, how the culture will react to such an environment and how the process is operated. The utility of these systems is legion and it is the ingenuity of man that is the limiting factor in the exploitation of this resource. It is the challenge to the biochemical engineer to exploit these materials in a variety of processes, central to which are biological reactions. However, it should be noted that the underlying principles and disciplines associated with the industrialisation of biological systems are those of chemical engineering.

Figure 5.1 shows an outline of a biological process and at its centre is the reaction. Clearly, there are three important aspects to reactors; the nature and processing of the raw materials, the choice and manipulation of the catalysts and the control of the reaction process from which the products must be recovered. Consideration has to be given to the behaviour and properties of the biological materials employed if the process is to be successfully designed.

This chapter is divided into two sections which are aimed at:

(a) giving a general background of the nature of living systems for the benefit of those without a background of biological science, and
(b) discussing the characteristics of biochemical reactions and the design features of reactors.

TABLE 5.1. *Process Engineering and its Relations to Biological Sciences and Biochemical Engineering*

Biological Sciences	Process	Chemical Engineering Science	
Systematics	Culture choice		
Genetics	Mass culture	Reactor design	
Biochemistry	Cell responses	Control	Biochemical engineering
Physiology			
	Process operation	Unit operations	
Chemistry		Energy & material utilisation	
	Product recovery		

5.1.1. Cells as Reactors

In order to survive, an organism must grow and reproduce itself using resources from the surrounding environment. Living systems capture and utilise energy from their environment to produce highly ordered structures so as to give rise to autocatalytic processes. Another notable, if not more important property, is the fundamental variation within living systems on which the principles of natural selection (selection of the fittest) may act. This selection allows the natural optimisation of living processes for the evolution of new biological information structures and even new life forms.

The basis of life lies in chemistry, albeit of a rather specialised form, which creates a variety of novel solutions to a set of problems associated with life. These include:

(1) the acquisition of food, which will provide the energy and other nutrients that make up the structure of the cell,
(2) the conversion of the food into the structure of the living system,

(3) the retention of information such that the chemical structures may be reproduced when required, and
(4) the introduction of variation in the biological information so as to encourage the change or adaptation of the organisms to the environment.

The complexity of the bacterial cell (for example) is thus a match for any chemical plant in the power and sophistication of its chemistry. Inside every cell there is a controlled environment in which several thousand catalysed reactions occur and where their products and reactants may be found. Cells also contain the control systems to match, and a store of information to co-ordinate and reproduce the whole system. All of these activities take place at relatively mild temperatures and pressures and, overall, these reaction processes make up the *metabolism* of the cells.

5.1.2. The Biological World and Ecology

Living systems, as far as is known, are limited to this planet. The power of living systems is immense if their activities are considered on the global scale. The biosphere, or the environment in which life exists, is a rather limited part of the planet, the most important of which is that near or on surface, or in the oceans.

Evidence for activity of living processes appears to expand steadily. There is good evidence that life can successfully exist in pressurised water at 120°C, in ice at − 10°C, and at extremes of pH (1–11), in saturated salt solutions and even when subjected to high levels of radiation.

Living systems are important in the control of rates of flow of many elemental materials in the environment. Indeed, much of environmental chemistry is dependent upon the activities of living systems. The nutrient cycles of carbon, oxygen, nitrogen, sulphur and phosphorus (the major constituents of living cells) are good examples of processes mediated by living systems. In all of these cycles, cells obtain materials from the environment and transform them by oxidation and reduction processes. There are two basic types of nutrition which are associated with the carbon cycle. The first involves the fixation and reduction of carbon dioxide to carbohydrate and the oxidation of water to oxygen in the presence of an external inorganic energy source; when it involves light it is called *photosynthesis*. This form of nutrition, called *autotrophy*, is the ability to use materials derived from inorganic sources. Plants, for example, are *autotrophs*. The second basic mode of nutrition is the reverse process and the most important reaction involved is the oxidation of organic materials to carbon dioxide and water in the presence of oxygen (*respiration*). This form of nutrition is called *heterotrophy*, being able to live on other organic materials in the environment. Animals, for example, are *heterotrophs*.

The carbon cycle is illustrated in Fig. 5.2. All living systems require an external source of energy, either in the form of chemical bond energy, as chemical (redox) potential or as some form of electromagnetic radiation usually in or near the visible light region.

Energy is initially captured from the sun's radiation and is then used to fix carbon dioxide and convert it into the organic molecules that constitute living tissue. The energy is stored in organic molecules which may be oxidised to release part of that energy to drive other processes. It is this flow of materials and energy that drives all living systems. By interacting with one another, various cell types can become

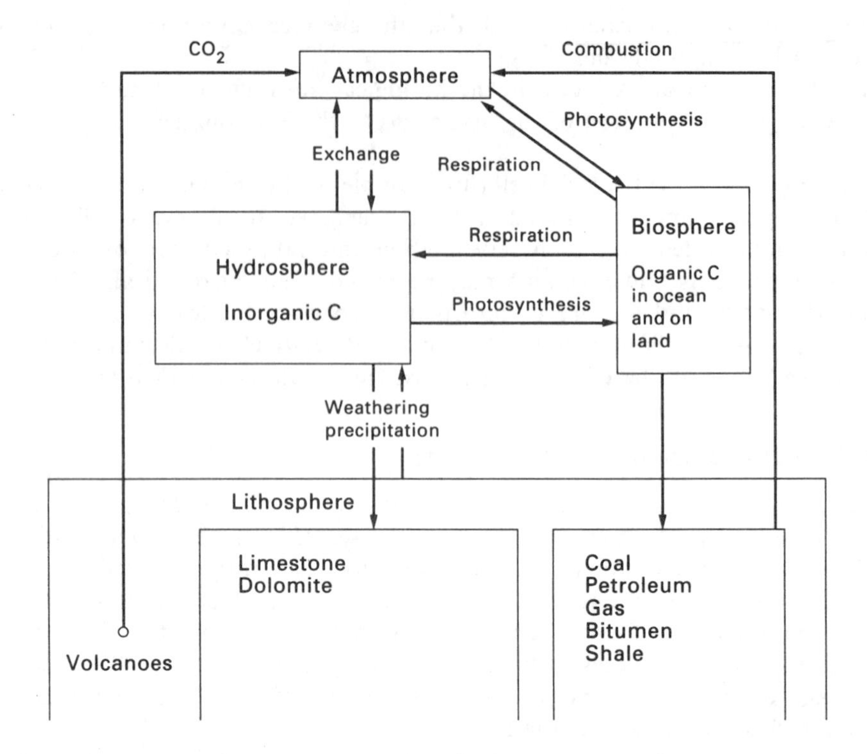

FIG. 5.2. The circulation of carbon in nature

associated and heavily interdependent. The net result is that the energy absorbed from the sun is rapidly dissipated while the materials associated with the biosphere are rapidly circulated[(2)].

The biosphere has a complex chemistry and many aspects of the processes and their consequences are not fully understood. However, pollution arises from the cyclic processes being out of balance, or from the accumulation of other undesirable materials. Indeed, a major perception of pollution is the disruption of these cycles. The activities of man are now on such a scale that they significantly affect the environment and its chemistry.

5.1.3. Biological Products and Production Systems

Table 5.2[(3)] shows the markets that are associated with the products of biotechnology. Not shown here is the impact of biological systems on waste treatment and environmental systems. The products illustrated in the table are a mixture of traditional products such as alcoholic beverages and food materials, of commodity chemicals which are produced by large-scale fermentation (e.g. citrates), in addition to specialised compounds (e.g. enzymes and vaccines). More recently the products of so-called genetic engineering, such as insulin and other therapeutic proteins, have begun to make an impact. The advent of genetic engineering (manipulation of cells' genetic material) is now providing a powerful tool, not only for the fundamental understanding of biological systems, but also in the manipulation and exploitation of biological materials and processes in medicine and for industry. It should also be noted that the production systems are primarily based on micro-organisms. This is due mainly to their rapid growth rates and the ease with which they can be manipulated in the laboratory and on the plant.

TABLE 5.2. *World Sales of Major Products of Biotechnology*[3]

Product	Quantity (1000 tonne)	Price ($ per tonne)	Value ($ millions)
Fuel/industrial ethanol			
U.S.A.	1000	576	576
Brazil	4500	576	2600
India	400	576	230
Others	400	576	230
Total			3636
High fructose syrups			
U.S.A.	3150	400	1260
Japan	600	400	240
Europe and others	200	400	80
Total			1580
Antibiotics			
Penicillin and synthetic penicillins			3000
Cephalosporins			2000
Tetracyclines			1500
Others			1500–2000
Total			8000–8500
Other products			
Citric acid	300	1600	480
Monosodium glutamate	220	2500	550
Yeast biomass	450	1000	450
Enzymes			400
Lysine	40	4000	160
Total			2040
Total of all products listed			15,256–15,756

5.1.4. Scales of Operation

The scale of bioreactions is listed in Table 5.3. In today's economic environment there are three distinct scales of operation. On the large scale the emphasis of research is generally on fermentation technology and process engineering, while this emphasis changes as the scale of operation is reduced and ultimately the research is dominated by work in genetic manipulation and catalyst development.

Large Scale Processes

Typically, large scale processes are dominated by the cost of raw materials and the downstream processes are usually well understood. However, the treatment of effluent is a significant problem and can make an appreciable contribution to the cost of production. Surplus biomass, spent slurry, and either utilisation or disposal of feed residues represent problems that must be solved economically. This is the major problem associated with the production of biofuels.

TABLE 5.3. *Classification of Industrial Microbial Processes by Scale of Operation*[2]

Process	Phase of product in slurry	Typical broth volume	Typical scale or form	Typical volume or weight reduction ratio from broth
Large-scale				
Ethanol and soluble products (conventional process)	Liquid	Widely varying scale	5–12% in slurry	5–10
Single-cell protein from methanol	Solid	> 4000 m^3/d	170 t/d	
Single-cell protein + lipids from methanol	Solid	Unknown	3–5% in slurry	5–7
Single-cell protein from gas oil	Solid	> 2000 m^3/d		
Medium-scale				
Polysaccharides	Liquid + solid	40–200 m^3 per batch	3% in slurry	Unknown*
Stable organic acids (e.g. citric acid)	Liquid	50–250 m^3 per batch	10% in slurry	4
Small-scale				
Antibiotics				
Cephalosporin	Liquid	50–200 m^3 per batch	3% in slurry	5–10
Penicillin G (K salt)	Liquid	40–200 m^3 per batch	3% in slurry	7–10
Streptomycin	Liquid	40–200 m^3 per batch	1.5% in slurry	5–10
Vitamins				
Cyanocobalamin (Vitamin B_{12})	Solid	40–100 m^3 per batch	3% in slurry	7
Enzymes				
Extracellular	Liquid	10–200 m^3 per batch	< 10.5% pure enzyme	
Intracellular				
Conventional route product	Solid	Not widely established		8
Glucose isomerase +	Solid	40–200 m^3 per batch	3% in slurry	5–10
Two-phase aqueous process	Solid	Not widely established		
Vaccines and Antibodies				
Poliomyelitis vaccine	Liquid + solid	> 1m^3 per batch	250,000 doses in 1 m^3	5–10

* Viscosity effects limit the concentration that can be achieved.
+ Product is dried whole cells that exhibit enzymatic activity.

Medium Scale Processes

Medium scale process are usually associated with commodity organic chemicals and with the substitution of natural products in place of materials from other sources. These processes generally produce stable products and utilise conventional process operations for the recovery of the product. The economic viability of such processes is sensitive to the concentration of the product in the fermenter broth.

Small Scale Processes

Small scale processes are concerned with the production of fine chemicals, pharmaceuticals and enzymes. Generally, products from these fermentations are not available from other sources. A significant cost of production is the development of the process and catalyst, together with the safety testing of the products. The materials produced are chemically and physically diverse in form, and consequently a wide variety of separation processes is employed. In many cases the capacity of the production equipment may be less than 1 m^3, yet the potency of the product may be such that 300,000 doses can be produced, and only a few batches a week will satisfy the world market for such a product. As a consequence, various recovery operations operate which are economically acceptable on a small scale.

The Role of Biochemical Engineering

Whilst it may be said that biochemical engineering involves the application of chemical engineering principles to biological systems and the manufacture of biologically derived products in general, there is, however, a considerable emphasis on processes involving the growth of micro-organisms because either the organisms themselves represent the product, or the formation of product is in some way related to the growth process. The typical rates of growth of microbes are such that they double their total mass in a few hours and, in some cases, in a matter of minutes. This, when compared with the growth rates of more complex life forms with doubling times of weeks or months, makes them attractive systems on which to base commercial process, even though their care and manipulation present their own problems and difficulties.

Batch reactions, or semi-continuous operations, have in the past tended to dominate processes involving microbial growth. This has largely been due to problems in maintaining sterility in the case of enzyme reactors and the maintenance of a pure culture in the case of microbial reactors. Apart from those cases where the desired product is formed in a manner not simply related to the growth of the organism, there can be a clear economic attraction in making a process continuous rather than batch. The one large scale process not requiring pure culture conditions is waste water treatment, where the objective is to consume feed substrates as rapidly as possible. It was the first to be developed as a widespread, successful continuous operation.

Within the constraints imposed, particularly by the requirements of sterility, for example, when mono-cultures are used, a major role of biochemical engineering is to provide appropriate mathematical descriptions of biochemical reaction processes and from these to devise suitable design criteria for economically viable processes.

This Chapter is essentially in two parts. In Sections 5.2 to 5.7 is included the basic microbiology and biochemistry of cells and enzymes and the reactions which they can undergo or catalyse. This is followed by a disccussion of the utilisation of biological reactions under industrial process conditions—Sections 5.8 to 5.14.

5.2. CELLULAR DIVERSITY AND CLASSIFICATION OF LIVING SYSTEMS

The concept of the cell, or cell theory, was developed by the middle of the nineteenth century when living structures had been observed with the microscopes which were then available. The theory stated that all living systems are composed of cells which, with their products, form the basic building blocks of living systems. It was later found that the cells were made up of various chemicals shared by all cells. This was a very important observation, as the study of simple single-cell systems such as bacteria has led into the understanding of more complex systems. J. Monod once said "What is true for *Escherichia coli* is true for elephants, only more so". Organisms can thus consist of either of a single cell or of a series of specialised cell types which make up multicellular organisms. Cells are between 1 and 50 μm in diameter and cannot normally be seen with the naked eye; free living forms are called micro-organisms.

The cells of most multicellular organisms are differentiated both in structure and function. This differentiation is most highly developed in plants and animals, where there is profusion of cell types, each performing a different function within the collective whole of the organism. Generally, once a cell has become differentiated to perform a specific activity it is rarely able to change its function; it is committed until it dies.

Cells, or systems of cells, may be derived from micro-organisms (protists), animals and plants; the differences between these cell types are principally ones of organisation. For growth and survival, the cell must contain the basic metabolic systems and an information store for the co-ordination of its synthesis. Various types of cells are used in biological processes and these can be derived from almost any living material. Fortunately, there is a common biochemical system found in all living organisms. All that is required is a knowledge of their chemical and physical requirements for growth, from which the growth and reaction conditions may be formulated.

5.2.1. Classification

Organisms are classified according to structure and function. The biological world is divided into three kingdoms: plants, animals and protists. Animals and plants are generally classified according to their visible structure. Protists do not have visible structural differences and they are therefore classified according to chemical differences and biochemical properties. Organisms are named in Latin, or in latinised terms, using a binary nomenclature where the first name represent the group or *genus*, while the second represents the *species*. Usually, a species may be subdivided into strains which nowadays correspond to strain numbers associated with international culture collections, such as the National Collection of Industrial and Marine Bacteria (NCIMB) or the American Type Culture Collection (ATCC). These collections hold many thousands of strains for research and industrial applications; many hold the same strains, e.g. *Escherichia coli* NCIMB 9481 which is equivalent to ATCC 12435 and is a mutant strain derived from K12 (NCIMB 10214, ATCC 23716). Similarly, yeasts and fungi and other micro-organisms are obtainable from culture collections[4].

Micro-organisms, the major form of organisms for biotechnology, can be split into several groups based upon biochemical activity and structure. The basic classification of the protist is shown in Fig. 5.3. There are two major types, the *Prokaryotes* and the *Eukaryotes* and Table 5.4 illustrates the main distinguishing features between the two cell types. The major differences are reviewed in the following sections.

TABLE 5.4. *Features Distinguishing Prokaryotic and Eukaryotic Cells*

Feature	Prokaryotic cells	Eukaryotic cells
Size of organism	< 1 − 2 × 1 − 4 μm	> 5 μm in width or diameter
Genetic system		
Location	Nucleoid, chromatin body or nuclear material	Nucleus, mitochondria, chloroplasts
Structure of nucleus	Not bounded by nuclear membrane	Bounded by nuclear membrane
	One circular chromosome	One or more linear chromosomes
	Chromosome does not contain histones	Chromosomes have histones
	No mitotic division	Mitotic nuclear division
	Nucleolus absent	Nucleolus present
	Functionally related genes may be clustered	Functionally related genes not clustered
Sexuality	Zygote partially diploid (merozygotic)	Zygote diploid
Cytoplasmic nature and structures		
Cytoplasmic streaming	Absent	Present
Pinocytosis	Absent	Present
Gas vacuoles	Can be present	Absent
Mesosome	Present	Absent
Ribosomes	70S distributed in the cytoplasm*	80S* arrayed on membranes (e.g. endoplasmic reticulum), 70S in mitochondria and chloroplasts
Mitochondria	Absent	Present
Chloroplasts	Absent	May be present
Golgi structures	Absent	Present
Endoplasmic reticulum	Absent	Present
Membrane-bound (true) vacuoles	Absent	Present
Outer cell structures		
Cytoplasmic membranes	Generally do not contain sterols	Sterols present
	Contain part of respiratory and, in some, photosynthetic machinery	Do not carry out respiration and photosynthesis
Cell wall	Peptidoglycan (murein or mucopeptide) as component	Absence of peptidoglycan
Locomotor organelles	Simple fibril	Multifibrilled with microtubules
Pseudopodia	Absent	May be present
Metabolic mechanisms	Varied, particularly that of anaerobic energy-yielding reactions, some fix atmospheric N_2, some accumulate poly-β-hydroxybutyrate as reserve material	Glycolysis is pathway for anaerobic energy-yielding mechanism
DNA base ratios (C + G)†	28–73%	About 40%

* S refers to the Svedberg unit, the sedimentation coefficient of a particle in the ultracentrifuge.
† C, cytosine; G, guanosine.

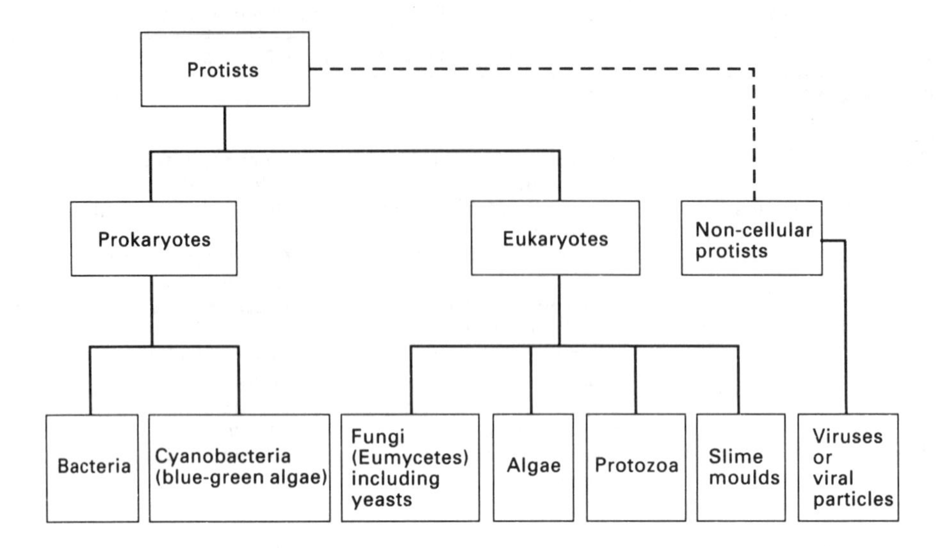

FIG. 5.3. Classication within the kingdom of the Protists[2]

5.2.2. Prokaryotic Organisms

Microbes are relatively simple in structure in that, when viewed under the light and electron microscope, few complex structures are observed. The general structure of prokaryotic cells is shown in Fig. 5.4[5].

Prokaryotic organisms do not include nuclear material bounded by a membrane and therefore have no true nucleus. The membrane is the most important structural component; Fig. 5.5[6] is a diagrammatic representation of a bacterial membrane. This not only contains lipid which partitions the interior of the cell from the environment, but also is packed with proteins and carrier molecules. It is these molecules that make it an energised selective barrier for importing and exporting materials. The membrane also provides an essential component in the mechanism of energy generation, by electron transport processes (Zone A, Fig. 5.5) in which differences of chemical potential (usually in the form of pH gradient) over the membrane are exploited to generate ATP (Zone D, Fig. 5.5), to cause the movement of *flagella* (Zone B, Fig. 5.5) and to drive *active transport* processes (Zone C, Fig. 5.5).

The most important cell types of the prokaryotes are bacteria. Bacterial morphology is rather limited and, only under the light microscope, cocci (spherical forms), bacilli (rod or cylindrical forms) and spirilla (spiral forms), and aggregations thereof, may be observed. The structures of these systems can be further subdivided on the basis of capsules, flagella and other sub-cellular materials which may be present. Table 5.5 summarises the structures found in various bacteria. An important division between prokaryotic types is made on the basis of cell wall structure. The *Gram stain* (named after its inventor) is capable of showing this distinction and the bacteria are divided into two main groups on the basis of this stain i.e. Gram positive (no Lipopolysaccharide or LPS layer) and Gram negative (extensive LPS layer), Fig 5.4[7, 8].

Prokaryotic organisms are metabolically the most diverse of all living systems and are responsible for most degradative processes in the biosphere. They can be grown

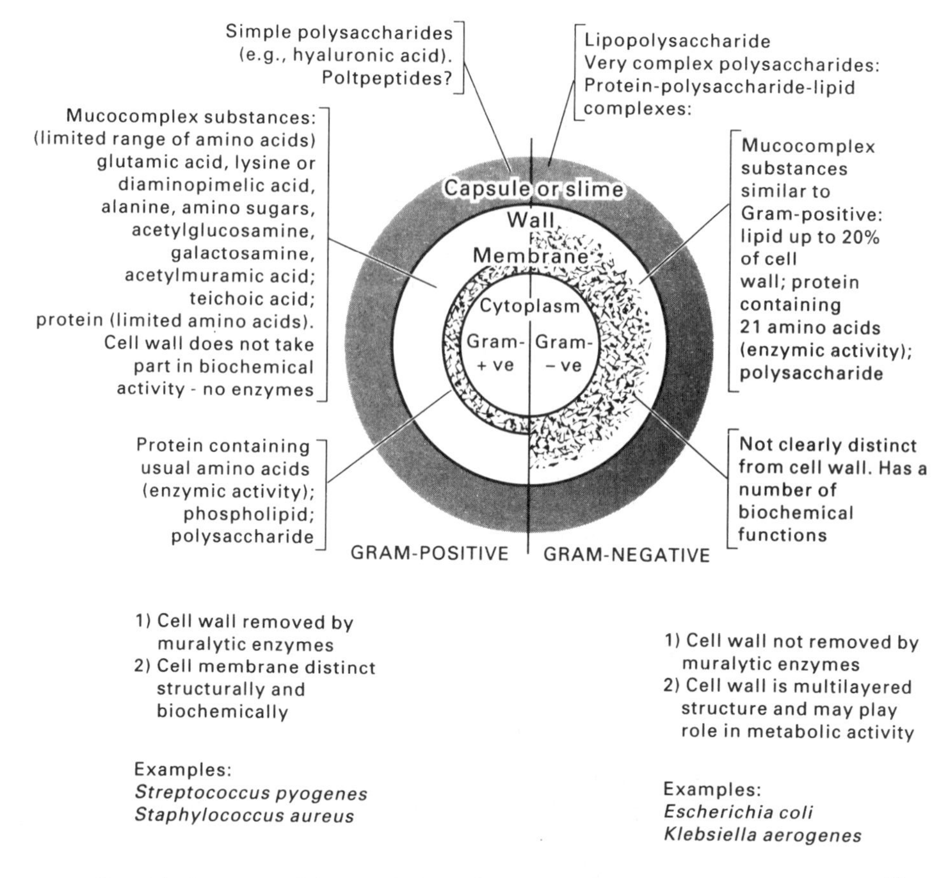

FIG. 5.4. Summary of the morphology of Gram-positive and Gram-negative bacteria[5]

in the presence or absence of oxygen and form a wide range of organic products. This property has both positive and negative impacts on society. On the positive side, they represent a massive resource of biocatalysis for the biotransformation of organic materials and the degradation of herbicides, insecticides and other man-made chemicals. On the negative side, they represent the principal agents causing the deterioration of biomaterials, e.g. food and wood, and are major hazards to public health (food poisoning and other diseases).

The ability of bacteria to grow rapidly in a wide range of environments is a very important property which is exploited in many processes both in mono and mixed cultures.

Life Cycles of Prokaryotes

Most bacteria divide by *binary fission*: the cell cycle involves a mature cell in which the nuclear material of the cell first separates, and then an ordered and usually symmetric division of the remaining material occurs to produce two daughter cells.

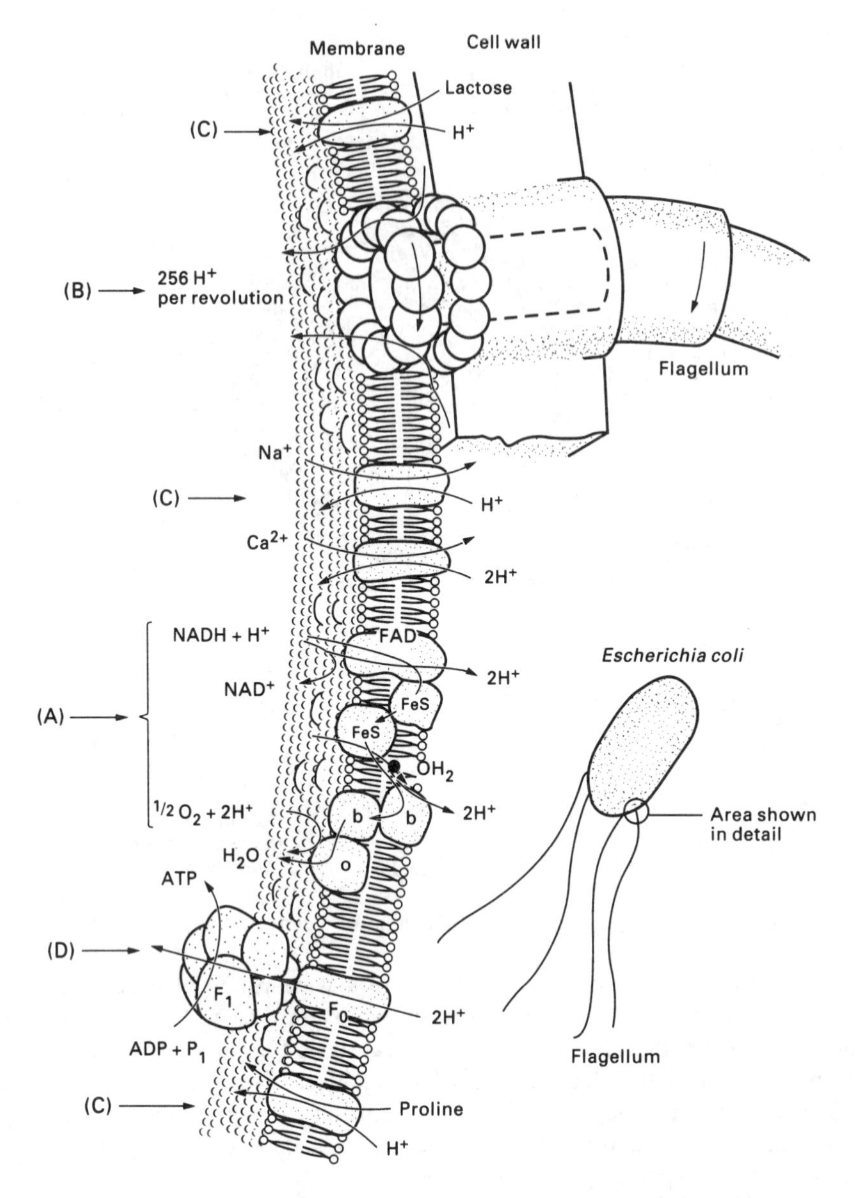

FIG. 5.5. The structure and activities of the cytoplasmic membrane involving proton transfer[(6)]

One form of asymmetric division caused by unfavourable conditions is sporulation, which is a highly coordinated process ultimately producing spores which are resistant to adverse environmental conditions. When conditions are favourable, out-growth to vegetative cells then occurs. Streptomycetes and the blue-green bacteria are exceptions to this relatively simple cellular morphology and each of these types of bacteria shows various forms of cellular differentiation, both biochemical and morphological[(7, 8)].

TABLE 5.5. *Size and Composition of Various Parts of Bacteria, adapted from* ATKINSON *and* MAVITUNA[2]

Part	Size	Comments
Slime Layer	5–500 nm	
Microcapsule Capsule Slime		Complex materials that vary in composition mainly polysaccharide but may contain signicant proteins. Responsible for antigenic properties of cells.
Cell Wall	10–20 nm	20% of cell dry wt
Gram-positive organisms		Mainly a mixed polymer of muramic acid and peptide techoic acids and polysaccharide
Gram-negative organisms		Have outer semi-permeable structure of lipopolysacharide with inner structure of muramic polymer, between these structures there is a space containing protein.
Cell Membrane	10–20 nm	Doubled-layered membrane; main semi-permeable barrier of cell; 5–10% of cell dry wt: 50% protein, 30% lipid and 20% carbohydrate
Flagellum	0.1 × 12,000 nm	Protein structure arises from membrane. Responsible for motility
Inclusions		
Spores	1–2 μm dia	Specialised resistant intracellular structures
Storage granules	0.05–2 μm	Consist of polysaccharide, lipid, polyhydroxybutyrate and sulphur.
Chromatophores	50–100 nm	Specialised structures containing photosynthetic apparatus
Ribosomes	10–30 nm	Organelles for protein synthesis; consist of RNA and protein and make up to 20% dry wt of the cell and is a function of growth rate.
Nuclear material		Poorly aggregated materials but can occupy up to 50% cell volume. Consist of DNA usually as a single molecule and makes up to 3% cell dry wt.
Cytoplasm		Free proteins (enzymes); about 50% of dry wt

5.2.3. Eukaryotic Organisms

There also exist structurally more complex organisms in the microbial world; these, exemplified by the yeast and the fungi, are called eukaryotes. Animals and plants are even more complex in that the cells are organised in complex interacting multicellular structures, e.g. leaves, stems and roots in plants, or muscles, nerves, organs and the cardiovascular system in animals. It is possible to grow individual cellular components of multicellular organisms or organs by the use of specialised media and culture conditions in reactors. The most important eukaryotic cells in biotechnology are the yeasts and fungi, although animal cells are becoming increasingly used. Most eukaryotic cells undergo a form of asymmetric division whereby the parent and daughter cells are distinguishable and the cultures have population profiles. With the exception of immortalised cancer cells, and cell fusions thereof, it is thought that most cells have a finite life.

The cell cycles of eukaryotic organisms are complex and not only involve changes in morphology but also variation in the genetic complement of the cell. Typically the simplest organisms contain a single set of chromosomes for most of their life cycle, and in this condition cells are termed *haploid*. Cell division in this state is termed *mitosis* and involves only replication of DNA. However, in some situations cells can fuse to contain two or more sets of chromosomes. When two sets of chromosomes are present, the cells are termed *diploid*. In the diploid state, a specialised form of cell division called *meiosis* can take place, where the chromosomes are segregated between two cells to return the cell to the haploid state. Generally, this is associated with a sexual process where specialised cell types fuse to form the diploid state before returning to the more usual haploid state on meiotic cell division. The time spent in one or other of these two states varies considerably from organism to organism; however, in most organisms of industrial importance, cells multiply by mitotic processes.

Yeast

Yeasts reproduce by budding, by fission and by sporulation, the most common process being budding. In budding, the nucleus which is present in the parent cell, enlarges and then extends into a bud forming a dumbbell shaped structure that later divides giving rise to a nucleus in both the parent and daughter cell. On division, an inactive surface scar remains on the mother cell surface, and ultimately after several more divisions the surface becomes covered with the scar material; it is thought that this eventually brings about the death of the cell. Sporulation can occur in yeast and usually involves the formation of spores via either asexual or sexual processes; the spores are called conidia and ascospores, respectively.

The overall life cycle of a particular yeast *Saccharomyces* (*S.*) *cerevisiae*, is summarised in Fig. 5.6 which shows how it is possible for the cells to fuse to form various cell and spore types. The figure shows the possible types of reproduction in yeast. Generally, industrial strains of *S. cerevisiae*, brewers' yeast, reproduce by budding/fission processes and only sporulate under specialised conditions. However, many strains of yeast are capable of cell fusion to form spores or cells with increased genetic complements. Such strains have many sets of chromosomes and are termed *polyploid*. Active fermentation of industrial strains involves growth by mitotic division and nutrient depletion which results in stationary cells with little or no spore formation.

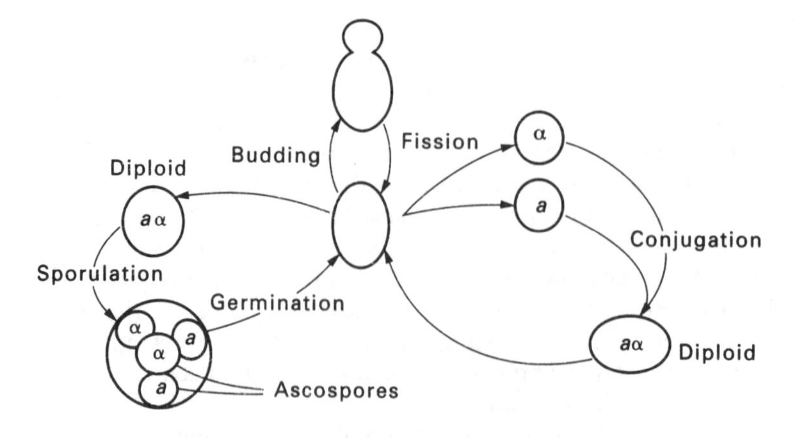

FIG. 5.6. The Yeast cell cycle[2]

Fungi

The name, fungi, is derived from their most obvious representatives, the mushrooms (Greek mykes and Latin fungus). They share many properties with plants, possessing a cell wall and liquid-filled intracellular vacuoles; microscopically they exhibit visible streaming of the cytoplasm and they lack motility. However, they do not contain photosynthetic pigments and are heterotrophic. Nearly all grow aerobically and obtain their energy by oxidation of organic substances. Table 5.6. shows the properties and parts of fungi which are typical of eukaryotic cells, the most interesting component is the cell wall with its great diversity of structure and composition. Fungi do show a limited morphological differentiation which is associated with spore formation or fruiting bodies. Some species have forms of sexual reproduction.

TABLE 5.6. *Size and Properties of Parts of Fungi* [2]

Part	Size (μm)	Comments
Outer fibrous layer	0.1–0.5	Very electron-dense material
Cell wall	0.1–0.25	Zygomycetes, Ascomycetes and Basidiomycetes contain chitin (2–26% dry wt); Oomycetes contain cellulose, not chitin; yeast cells contain glucan (29%), mannan (31%), protein (13%) and lipid (8.5%)
Cell membrane	0.007–0.01	Much-folded, double-layered membrane; semipermeable to nutrients
Endoplasmic reticulum	0.007–0.01	Highly invaginated membrane or set of tubules, probably connected with both the cell membrane and the nuclear membrane and concerned in protein synthesis and probably other metabolic functions
Nucleus	0.7–3	Surrounded by a double membrane (10 nm), containing pores (40–70 nm wide); flexible and contains cytologically distinguishable chromosomes. Nucleolus about 3 nm
Mitochondria	0.5–1.2 × 0.7–2	Analogous to those in animal and plant cells containing electron transport enzymes and bounded by an outer membrane and an inner membrane forming cristae; probably develop by division of existing mitochondria
Inclusions		Lipid and glycogen-like granules are found in some fungi. Ribosomes in all fungi

The vegetative body is a *thallus*. It consists of filaments about 5 μm in diameter which are multi-branched or spread over or into the nutrient medium. The filaments or *hyphae*, can be present without cross walls as in lower fungi; or divided into cells by *septa* in higher fungi. The total hyphal mass of the fungal thallus is called the *mycelium*. In certain situations during transition between asexual and sexual reproduction, various other tissue structures are formed, e.g. *plectrenchyma* (mushroom flesh).

Growth usually involves elongation of hyphae at their tips or apices. In most fungi every part of the mycelium is capable of inoculation and a small piece of mycelium is sufficient to produce a new thallus (Fig. 5.7).

There are many forms of structure and mechanism involved in the reproductive processes of fungi. Most fungi are capable of reproduction in two ways, asexual

FIG. 5.7. The mycelium structure of moulds

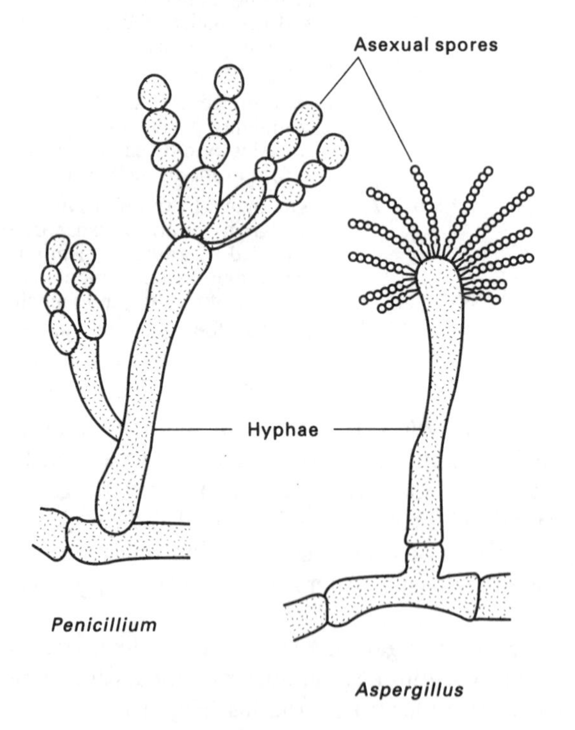

FIG. 5.8. The hyphae of Aspergillus and Penicillium, two industrially important moulds

and sexual. Asexual methods are by budding, fragmentation and formation of spores, while the sexual process involves cell fusion. Spore formation is the most widely distributed and highly differentiated method. Further details of the biology and industrial uses of fungi are given by BERRY[9] and ONIONS *et al.*[10] The structural diversity of sporulating bodies is used as a basis for classification. Figure 5.8 shows the structural differences in fruiting bodies between penicillium and aspergillus.

Algae and Protozoa

Algae and protozoa are now being used in industrial fermentations; generally they are used for the production of glycerol, pigments and their derivatives. Algae are autotrophic and generally require light for growth. Some of the most useful forms are however heterotrophic which means that in the presence of an organic carbon source they will grow in traditional fermenters in the absence of light. Usually, algal fermentations are not enclosed and take place in ponds having a significant impact on waste treatment processes[11].

Protozoa, again, are found in waste treatment processes where these generally motile forms graze on bacterial populations. The most important and interesting application concerns the use of these organisms as biological control agents.

Animal and Plant Cells

Animal and plant cells are also used in fermentations but rarely in fully differentiated forms. Most differentiated cells have a finite life and so, to grow cells successfully from plants, undifferentiated cells or callus are used. Animal cells for fermentations are derived from the fusion products where the cell is immortalised by fusing the differentiated cell with a cancer cell. The most interesting types of cells are those which produce antibodies; these are highly specific reagents and of high value. The most important considerations with these specialised fermentations are the expense of the medium, the relatively slow growth rates, the low productivity and shear sensitivity. These problems are being overcome: shear sensitivity can be obviated by the use of correct cell selection methods, and media are being reformulated to reduce the content of expensive components. Specialised reactor systems, such air-lift reactors and immobilised cell reactors, have now been developed[2, 12] for this purpose.

5.2.4. General Physical Properties of Cells

There are many important physical properties of cells of significance in biological processing but there is a lack of detailed information concerning the physical properties of cells and their components. For example, the colloidal and surface properties of cells, cell walls and proteins which are of considerable importance in separation and purification processes are not fully understood.

The size and shape of cells, as shown in Table 5.7, can be used as the basis of methods to separate cells from a liquid medium. Cells are only slightly denser than water and separation from liquids by centrifugation requires the use of intense fields. The shape of cells is important because of its effect on the rheology of the fermentation broth. Generally bacteria and yeast present few problems until very high concentrations are encountered, e.g. 50 g dry wt/l. However, filamentous organisms at relatively low concentrations can cause significant changes in rheological properties and many broths quickly become non-Newtonian in character[(2)].

Another important parameter is the surface charge associated with microbial particles. This reflects the composition of the cell wall surrounding the cell membranes; it is complex and dependent on environmental conditions and on the stages of the life cycle[(7)].

5.2.5. Tolerance to Environmental Conditions

Although individual microbial cell systems usually grow under only a narrow range of physical conditions, there are many cell types which are adapted to extreme but specific environmental conditions. Cells have a series of cardinal conditions, i.e. a maximum, a minimum and an optimum with respect to physical conditions. Environmental conditions such as temperature, pressure, pH and ion concentration can all affect the growth of cells. For the more detailed explanation of the physical conditions of growth the reader is referred to specific texts on the subject[(13, 14)].

TABLE 5.7. *Shape and Size of Biological Particles*[(2)]

Microorganisms	Shape	Size
Bacteria		
Bacillus megaterium	Rod-shaped	2.8–1.2–1.5 μm
Serratia marcescens	Rod-shaped	0.7–1.0 × 0.7 μm
Staphylococcus albus	Spherical	1.0 μm
Streptomyces scabies	Filamentous	0.5–1.2 μm dia
Fungi		
Botrytis cinerea conidiophores	Filamentous	11–23 μm dia
B. cinerea, conidia	Ellipsoidal	9–15 × 5–10 μm
Rhizopus nigricans sporangiophores	Filamentous	24–42 μm dia
R. nigricans sporangiospores	Almost spherical	11–14 μm dia
R. nigricans zygospores	Spherical	150 – 200 μm dia
Saccharomyces uvarum (*S. carlsbergensis*) vegetative cells	Ellipsoidal	5–10.5 × 4–8 μm
Algae		
Chlamydomonas kleinii	Ovoid	28–32 × 8–12 μm
Ulothrix subtilis	Filamentous	4–8 μm dia
Protozoa		
Amoeba proteus	Amorphous	< 600 μm dia
Euglena viridis	Spindle-shaped	40–65 × 14–20 μm
Paramecium caudatum	Slipper-shaped	180–300 μm length
Slime moulds		
Badhamia utricularis plasmodia	Amorphous	Indefinite, extensive
B. utricularis spores	Spherical	9–12 μm dia
Protein molecules		
Egg albumin		4.0 nm
Serum albumin		5.6 nm
Serum globulin		6.3 nm
Haemocyanin		22.0 nm

5.3. CHEMICAL COMPOSITION OF CELLS

5.3.1. Elemental Composition

The elemental composition of various micro-organisms is shown in Table 5.8. Six major elements—carbon, hydrogen, nitrogen, oxygen, phosphorus and sulphur—are associated with cells and make up over 90 per cent of their weight. The remaining elemental material consists mainly of the alkali metals and transition metals. Whilst the compositions shown in Table 5.8 are typical of living cells, the composition of the cell can vary considerably during the growth cycle and with the type of cell. Materials associated with energy storage functions, such as carbohydrates (starch) and fats (polyhydroxbutyrate), are noted in this respect and are dependent on environmental conditions. Similarly, materials related to growth rate, such as ribonucleic acid (RNA), can undergo significant changes in their concentration during the growth cycle.

TABLE 5.8. *Elemental Composition of Microorganisms*[12]

Micro organism	Limiting nutrient	Dilution (h^{-1}) rate*	Composition (% by wt)							Empirical chemical formula	Formula 'molecular' weight
			C	H	N	O	P	S	Ash		
Bacteria			53.0	7.3	12.0	19.0			8	$CH_{1.666}N_{0.20}O_{0.27}$	20.7
			47.0	4.9	13.7	31.3				$CH_2N_{0.25}O_{0.5}$	25.5
Enterobacter (Aerobacter) aerogenes			48.7	7.3	13.9	21.1			8.9	$CH_{1.78}N_{0.24}O_{0.33}$	22.5
Klebsiella aerogenes	Glycerol	0.1	50.6	7.3	13.0	29.0				$CH_{1.74}N_{0.22}O_{0.43}$	23.7
Yeast	Glycerol	0.85	50.1	7.3	14.0	28.7				$CH_{1.73}N_{0.24}O_{0.43}$	24.0
			47.0	6.5	7.5	31.0			8	$CH_{1.66}N_{0.13}O_{0.49}$	23.5
			50.3	7.4	8.8	33.5				$CH_{1.75}N_{0.15}O_{0.5}$	23.9
			44.7	6.2	8.5	31.2	1.08	0.6		$CH_{1.64}N_{0.16}O_{0.52}P_{0.01}S_{0.005}$	26.9
Candida utilis	Glucose	0.08	50.0	7.6	11.1	31.3				$CH_{1.82}N_{0.19}O_{0.47}$	24.0
	Glucose	0.45	46.9	7.2	10.9	35.0				$CH_{1.84}N_{0.2}O_{0.56}$	25.6
	Ethanol	0.06	50.3	7.7	11.0	30.8				$CH_{1.82}N_{0.19}O_{0.46}$	23.9
	Ethanol	0.43	47.2	7.3	11.0	34.6				$CH_{1.84}N_{0.2}O_{0.55}$	25.5

*Dilution rate in continuous fermentation, generally equal to the specific growth rate μ and is given by the ratio of volumetric throughput to liquid volume in the fermenter (See Section 5.11.3).

There is a molecular hierarchy within cells. It is possible for some organisms, *autotrophs*, to synthesise all their structure from simple chemicals, such as carbon dioxide, water and ammonia, using light energy. On the other hand, *heterotrophs* use organic materials to form the building blocks of cells, carbohydrates, nucleotides, amino acids and fatty acids. These basic building blocks are then polymerised into polysaccharides, nucleic acids, proteins and fats. These in turn form macromolecular structures, *organelles* (small membrane-bound structures) and ultimately cells and multicellular structures, as illustrated in Fig. 5.9.

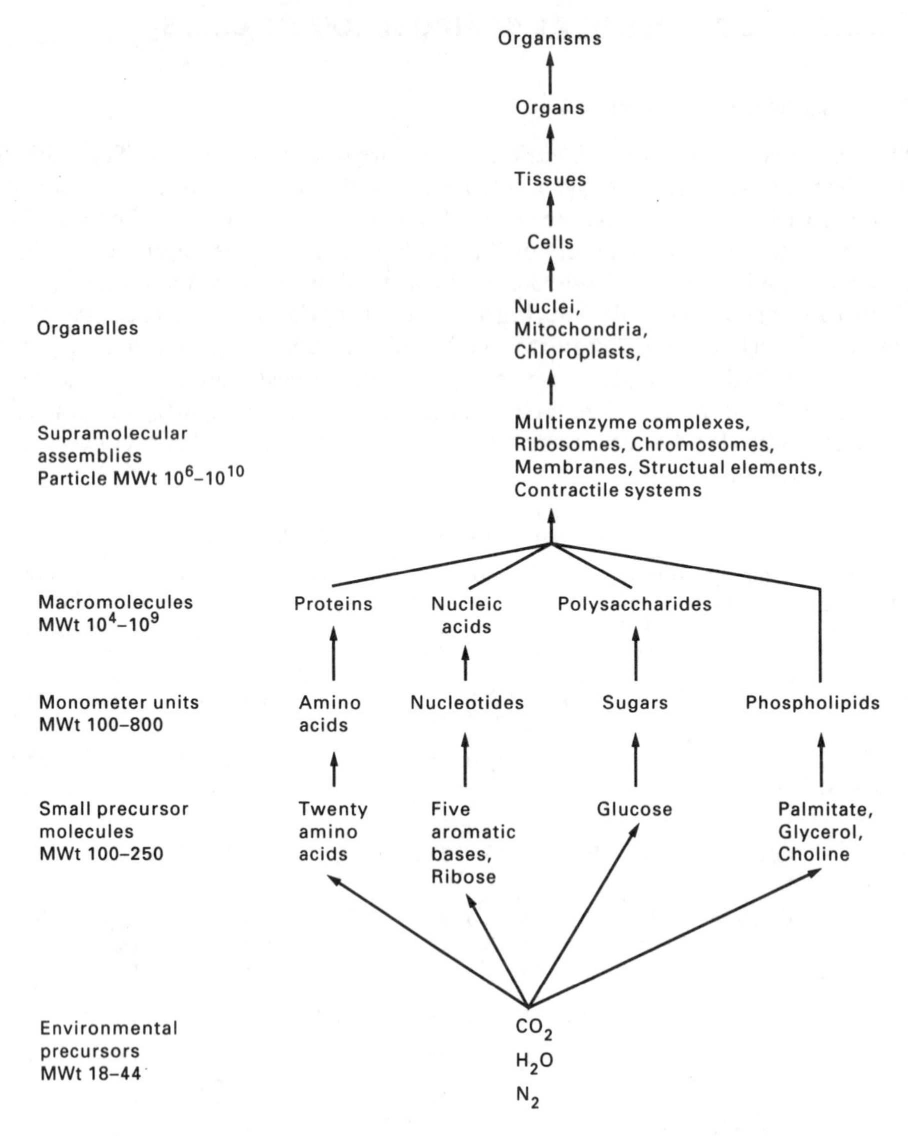

FIG. 5.9. The hierarchy of biological structure. The approximate relative sizes as molecular weights (MWt) or equivalent[15]

A typical molecular analysis of various micro-organisms is shown in Table 5.9[16]. Most of the elemental composition of cells is found in three basic types of materials—proteins, nucleic acids and lipids. In Table 5.10, the molecular composition of a bacterium is shown in more detail. Water is the major component of the cell and accounts for 80–90 per cent of the total weight, whilst proteins form the next most abundant group of materials and these have both structural and functional properties. Most of the protein present will be in the form of enzymes. Nucleic acids are found in various forms—ribonucleic acid (RNA) and deoxyribonucleic acid (DNA). Their primary function is the storage, transmission and utilisation of genetic information. Polysaccharides and lipids are associated with wall and membrane structures and they can also act as energy-storage materials within the cell. Small molecules, monomers of the polymers described above together with metal salts make up 0.5 per cent weight of the cell; however, within the cell, this pool of small molecules is constantly and rapidly turning over.

TABLE 5.9. *Chemical Analyses, Dry Weights and the Populations of Different Microorganisms Obtained in Culture*[16]

Microorganism	Protein	Composition (% dry wt) Nucleic acid	Lipid	Population in culture (numbers ml^{-1})	Dry weight in culture (g $(100\ ml)^{-1}$)	Comments
Algae (small unicellular)	10–60 (50)	1–5 (3)	4–80 (10)	4×10^7–8×10^7	0.4–0.9	Figure in parentheses is a commonly found value; composition varies with the growth conditions
Bacteria	40–50	13–25	10–15	2×10^8–2×10^{11}	0.02–2.9	Mycobacteriam may contain 30% lipid
Fungi (filamentous)	10–25	1–3	2–7		3–5	Some *Aspergillus* and *Penicillium* species contain 50% lipid
Viruses	50–90	5–50	< 1	10^8–10^9	0.0005*	Viruses with a lipoprotein sheath may contain 25% lipid
Yeasts	40–50	4–10	1–6	10^8–10^9	1–5	Some *Rhodotorula* and *Candida* species contain 50% lipid

* For a virus 200 nm in diameter.

Much of the chemistry of the cell is common to all living systems and is directed towards ensuring growth and cell multiplication, or at least the survival of the cell. Organisms also share various structural characteristics. They all contain genetic material (DNA), membranes (the boundary material between the cell and the environment), cytoplasm (small particulate materials, ribosomes and enzyme complexes), and cell walls or surfaces (complex structures external to the membrane). In addition, there are various distinct membrane-bound organelles in eukaryotic organisms which have specialised functions within the cell (Tables 5.4, 5.5 and 5.6)[8, 17].

5.3.2. Proteins

Of all the macromolecules present in living systems proteins and nucleic acids are the most significant. Nucleic acid in the form of DNA forms the stored structural and regulatory material for the organism, while proteins and their expression represent the functional form of this information.

TABLE 5.10. *Molecular Composition of a Typical Bacterium*[2]

Component	Mass ($\times 10^{13}$g)	Percentage of total mass	Molecular weight, M_r	Molecules per cell*
Entire cell	15	100		
Water	12	80	18	4×10^{10}
Dry weight	3	20		
Protein				
Ribosomal	0.22	1.5	4×10^4	3.3×10^5
Nonribosomal	1.5	10	5×10^4	1.8×10^6
RNA				
Ribosomal, 16S	0.15	1	6×10^5	1.5×10^4
Ribosomal, 23S	0.30	2	1.2×10^6	1.5×10^4
tRNA	0.15	1	2.5×10^4	3.5×10^5
mRNA	0.15	1	10^6	9×10^5
DNA	0.15	1	4.5×10^9	2
Polysaccharides	0.15	1	1.8×10^2	5×10^7
Lipids	0.15	1	10^3	9×10^6
Small molecules	0.08	0.5	4×10^2	1.2×10^7

* Calculated by dividing the weight of a component in grams by its molecular weight to give the number of gram molecular weights (moles) of the component. Since there are approximately 6×10^{23} molecules in one mole of any compound, the number of moles multiplied by 6×10^{23} gives the number of molecules. For example, the number of moles of water in 12×10^{-13}g of water is $12 \times 10^{-13}/18$ or about 7×10^{-14}; the number of molecules of water per cell is then $7 \times 10^{-4} \times 6 \times 10^{23}$ or about 4×10^{10}.

Protein is formed mainly of polymerised amino-acids. The primary structure, unlike that of synthetic polymers, is non-repetitive and, for its production, requires a chemical template stored in the structure of the DNA molecule. The sizes of proteins vary considerably (in a range of molecular weights from 6000 to 1,000,000). Proteins fulfill many roles within the cell, the most important of which is that of catalysis. Proteins which have catalytic activity are called enzymes whilst other proteins have important roles in storage, transport, protection (antibodies), as chemical messengers (hormones) and in structure[17, 18].

Chemical Properties of Proteins

The chemical and physical behaviour of proteins is very complex and the prediction of these characteristics has met with limited success as studies on these materials are based upon empirical measurements. The difficulties lie in the lack of understanding of protein structures; only a few three dimensional structures are known in detail, and the systems in which have been studied are very simple. Studies of complex systems of mixtures or of inter-molecular interactions are limited and the appropriate techniques have yet to be developed. However, with advances in the molecular modelling of proteins, it is becoming possible to predict more of their physical and chemical behaviour and modelling is now a major area of research activity.[18]

The chemical properties of a protein are dictated by the chemical properties of the amino acids of which it is composed. There are many functional groups which may participate in reaction processes; these include alkyl, hydroxy, carboxy, amino and thiol groups. Some of these groups can be used in protein immobilisation, for example, where the protein is covalently fixed to other materials using such functional groups.

Some important reactions are those which are used to estimate the concentration of proteins. These assays all assume that the reacting groups of a protein have an average distribution through the protein molecule.

In the *Biuret reaction*, a purple colour develops when the protein is treated with alkaline copper sulphate. This reaction is dependent on peptide bonds and not on the side chains of individual amino-acids present. In the *Folin-Ciocalteu reaction*, the protein is treated with tungstate and molybdate under alkaline conditions and the formation of a complex such phenylalanine and tyrosine gives rise to a blue colour. LOWRY developed one of the most widely used protein assays in which a combination of the above reactions is involved[17, 18].

5.3.3. Physical Properties of Proteins

A rational understanding of the physical properties of proteins can be based on a sound knowledge of protein structure[18], and this is discussed in Appendix 5.1. The more important properties are outlined below.

(a) Acid—Base properties

By definition, a base is a proton acceptor while an acid is a proton donor and the general equation for any acid–base reaction can be written:

$$HA \rightleftharpoons H^+ + A^-$$

The strength of an acid is determined by its ability to give up protons while the strength of a base is determined by its ability to take up protons. This strength is indicated by the dissociation or equilibrium constant, (pK_a), for the acid or base; strong acids have a low affinity for protons, while weak acids have a higher affinity and only partially dissociate (e.g. HCl (strong) and acetic acid (weak)).

Thus, for weakly acidic groups, as found in amino-acids:

$$R\text{–}COOH \rightleftharpoons H^+ + RCOO^-$$

and for weakly basic groups, as found in amino-acids:

$$R\text{–}NH_3 \rightleftharpoons H^+ + R\text{–}NH_2$$

The side groups of amino-acids consist of carboxylic acid, amino groups, thiols, alcohols and alkyl groups. Proteins will thus have a charge which will also be dependent on pH (Table 5.11). For example, at higher pH values a protein will be negatively charged and will move towards the anode when exposed to an electric field. When the charges on a molecule are balanced, then there is no movement in an electric field. The pH at which this occurs is known as the *isoelectric point* of the protein. This an important physical property which is used to characterise a protein and it depends on the amino-acid content and the three-dimensional molecular structure of protein. Generally, proteins with a high isoelectric point have a high concentration of basic amino-acids, while those with a low isoelectric point will be high in acidic amino-acids.

TABLE 5.11. *The Ionisable Groups which Contribute to the Acid–Base Properties of Proteins, Shown with their Approximate pKa Values*

Ionisable group	Dissociation reaction	Approximate pK_a
α-Carboxyl	$—COOH \rightleftharpoons H^+ + —COO^-$	3.0
Aspartyl carboxyl	$—CH_2COOH \rightleftharpoons H^+ + —CH_2COO^-$	3.9
Glutamyl carboxyl	$—CH_2CH_2COOH \rightleftharpoons H^+ + —CH_2CH_2COO^-$	4.1
Histidine imidazole	$—CH_2 \rightleftharpoons H^+ + —CH_2$	6.0
α-amino	$—NH_3^+ \rightleftharpoons H^+ + —NH_2$	8.0
Cyteine sulphydryl	$—CH_2SH \rightleftharpoons H^+ + —CH_2S^-$	8.4
Tyrosyl hydroxyl	$—OH \rightleftharpoons H^+ + —O^-$	10.1
Lysyl amino	$—(CH_2)_4NH_3^+ \rightleftharpoons H^+ + —(CH_2)_4NH_2$	10.8
Arginine guanidine	$—(CH_2)_3NH.C(=\!{}^+NH_2).NH_2 \rightleftharpoons H^+ + —(CH_2)_3NH.C(=NH).NH_2$	12.5

pK_a is equilibrium constant.

The conformation of a protein is dependent on weak intramolecular interactions and is also likely to change with varying pH. Therefore its function and its catalytic activity will be highly dependent on pH[18].

(b) Solubility of globular proteins

Another property of proteins which is important in the understanding of the limits of their catalytic activity, as well as being useful in their recovery, is solubility. The solubility of globular proteins in aqueous solution is enhanced by weak ionic interactions, including hydrogen bonding between solute molecules and water. Therefore, any factor which interferes with this process must influence solubility. Electrostatic interactions between protein molecules will also affect solubility, since repulsive forces will hinder the formation of insoluble aggregates[18].

Salt concentration. The addition of a small amount of neutral salt usually increases the solubility of a protein, and changes the interaction between the molecules as well as changing some amino-acid charges. The overall effect is to increase the solubility. This phenomenon is known as *salting in*. However, at high concentrations of salts the solvating interactions between protein and water are reduced, and the protein may be precipitated from solution—a process termed *salting out*.

Heavy metals (such as Hg and Zn) combine irreversibly with protein to form salts. Similarly, acid-insoluble salts can also be formed; this effect is often used to deproteinise solutions.

pH. At extremes of pH, the charge on the side chains of amino-acids will be different from that at the physiological pH (pH 5 to 8). The change in charge on the side chains will cause disruption of the tertiary structure, usually irreversibly. The solubility of the proteins also changes considerably around the isoelectric point; this is due to the fact that protein molecules will interact causing aggregation, in contrast to the repulsion which occurs when molecules have an overall charge.

Organic molecules in the solvent. The introduction of a water-miscible organic component lowers the dielectric constant of water. This has the net effect of increasing the attractive forces between groups of opposite charge and therefore of reducing the interactions between the water and the protein molecules; consequently, the solubility of the protein will decrease.

Temperature. The range of temperature over which proteins are soluble depends on their source. Typically the solubility of globular proteins increases with temperature up to a maximum. Above this critical temperature, thermal agitation within the molecule disrupts the tertiary structure and gives rise to denaturation. In enzymes this behaviour is paralleled to some extent by changes in activity.

5.3.4. Protein Purification and Separation

Protein separation processes generally utilise differences in one or more of the physical properties of the materials which can be purified by the use of three or four basic operations, including the following:

Gel filtration (size),
Ultrafiltration (size),
Ultracentrifugation (density and size),
Electrophoresis (charge and size),
Ion exchange (charge),
Precipitation by salts or solvent (hydrophobicity and charge).

Generally speaking, these methods result in very mild treatments which will maintain the biological function of proteins which are usually associated with their tertiary structure. If the configuration of the protein is altered, the process is generally reversible. A discussion of many of these unit operations can be found in Volume 2, Chapters 17–20.

5.3.5. Stability of Proteins

One of the major problems that a biochemical engineer will encounter is that of the stability of protein materials. The biological function of the molecule is determined by its secondary and tertiary structures and if these are upset irreversibly then the protein becomes denatured. Denaturation can occur under relatively mild conditions as proteins are usually stable over only very narrow ranges of pH (e.g. 5–8) and of temperature (e.g. 10–40°C). The boiling of an egg illustrates this point. In its uncooked form the white is a slimy clear protein solution, but under acidic conditions, or when put in boiling water, the solution gels to white solid.

The stability of tertiary structure can be appreciated by the relative strengths of these associations as compared with that of a covalent bond, as discussed in Appendix 5.1.

5.3.6. Nucleic Acids

The basic monomers of nucleic acids are nucleotides which are made up of heterocyclic nitrogen-containing compounds, purines and pyrimidines, linked to pentose sugars. There are two types of nucleic acids and these can be distinguished on the basis of the sugar moiety of the molecule, Ribonucleic acids (RNA) contain ribose, while deoxyribonucleic acid (DNA) contains deoxyribose. The bases cytosine (C) adenine (A) and guanine (G) are common in both RNA and DNA. However, RNA molecules contain a unique base, uracil (U), while the unique DNA base is thymidine (T). These differences in the base structure markedly affect the secondary structures of these polymers. The structures of DNA and RNA are outlined in Appendix 5.2.

5.3.7. Lipids and Membranes

Lipids are the major components of membranes; they have complex structures comprising fatty acids esterified with alcohols to form glycerides, and other lipids based upon esters of phosphatidylethanolamine. Other important lipid components are based on sterols. Within this hydrophobic structure, proteins provide ports of entry and exit from the interior of the cell and distinguish the inside from the outside of the cell. Figure 5.5 illustrates the complexity of this structure.

5.3.8. Carbohydrates

Carbohydrates form the major structural components of the cell walls. The most common form is cellulose which makes up over 30 per cent of the dry weight of wood. Other structural forms are hemicellulose (a mixed polymer of hexose and pentose sugars), pectins and chitin. Apart from contributing to the structure, some polymers also act as energy storage materials in living systems. Glycogen and starch form the major carbohydrate stores of animals and plants, respectively. Carbohydrate structure, like that of nucleic acids and proteins, is complex, and various levels of structure can be identified.

5.3.9. Cell Walls

Cell walls are derived from carbohydrates and additionally contain a number of polymeric materials, ranging from peptidoglycan and other complex polymers in bacteria, to glycans/mannans in yeast, and to chitin and cellulose in fungi. The most important function of these walls is to provide the cell with mechanical strength, in the same way that a tyre reinforces the inner tube and the leather of the rugby ball supports the bladder. It therefore provides a rigid structure against which the membrane sits and, in consequence, the cells are sufficiently robust to cope with osmotic stress and other environmental shocks(7,8).

Cell walls have a variable structure which changes according to environmental conditions and the surface properties of the cell are largely determined by the cell wall materials. Generally, the wall is negatively charged and can act as an ion exchanger.

5.4. ENZYMES

5.4.1. Biological Versus Chemical Reaction Processes

Biological reactions are like any other chemical processes, except that they are catalysed by a variety of sophisticated catalysts whose structure is based upon proteins and ancillary associated materials, such as metal ions, carbohydrates and nucleotides. Biological catalysts, or enzymes, have a very precise architecture and this is reflected in their reactivity with both their products and substrates (e.g. reactants, food materials for growth); usually these are stereo-specific. Enzymes provide the means to generate and reproduce the complex structure of living systems. Whole cells may be used to provide catalytic materials, or if the cells are ruptured, the catalyst may be extracted and purified. Table 5.12 compares biological and chemical catalysts and it may be seen that most enzymes compare well with their chemical counterparts in terms of *turnover number* (See Section 5.4.2) and operational temperature range. With all proteins, there is, however, a major problem in relation to their stability, although strategies for obtaining more stable catalysts are now being developed through the study of protein structure and its manipulation (now called *protein engineering*), and the isolation of organisms from extreme environments.

TABLE 5.12. *Some Turnover Numbers for Enzyme and Inorganic-Catalysed Reactions*[19]

Catalyst	EC number	Reaction	Turnover number (s^{-1})	Temperature (°C)
Enzymes				
Bromelain	3.4.22.4	Hydrolysis of peptides	4×10^{-3}–5×10^{-1}	0–37
Carbonic anhydrase	4.2.1.1	Hydration of carbonyl compounds	8×10^{-1}–6×10^{5}	0–37
Fumarase	4.2.1.2	L-Malate ⇌ fumarate + H_2O	10^3 (forward) 3×10^3 (backward)	0–37
Papain	3.4.22.2	Hydrolysis of peptides	8×10^{-2}–10	0–37
Ribonuclease	3.1.4.22	Transfer phosphate of polynucleotide	2–2×10^3	0–37
Trypsin	3.4.21.4	Hydrolysis of peptides	3×10^{-3}–10^2	0–37
Inorganic catalysts				
Aluminium trichloride/alumina		*n*-Hexane isomerisation	10^{-2}	25
			1.5×10^{-2}	60
Copper/silver		Formic acid dehydrogenation	2×10^{-7}	25
Cu_3Au			3×10^{10}	327
Silica-alumina		Cumene cracking	3×10^{-8}	25
			2×10^{4}	420
Vanadium trioxide		Cyclohexene dehydrogenation	7×10^{-11}	25
			10^2	350

5.4.2. Properties of Enzymes

As with normal chemical reactions, there are many reasons for studying the kinetics of enzyme-catalysed reactions: to be able to predict how the rate of reaction will be affected by changes in reaction conditions; to aid in the determination of the mechanism of the reaction—that is to identify the sequence of reaction events that intervene between reactants and terminal products; to identify and characterise the enzyme molecule in functional terms, as protein molecules are structurally too complex for the application of coventional structural-analysis; to explain aspects of metabolic regulation and cell differentiation in terms of mechanisms that control either the variety or the quantity of enzymes in the cell; to understand the kinetic aspects of reactions which are critical in the design and operation of bioreactors.

An understanding of the influence of environmental conditions on the kinetics of enzyme reactions is essential for the design of processes based upon the use of these materials as catalysts. Their growth kinetics are also governed by similar kinetic equations (c.f. Monod growth equations, Section 5.9).

In five respects, the catalytic activity of enzymes differs from that of other catalysts.

(*a*) *Efficiency*. Turnover number (molecules reacted per catalytic site per unit time) at room temperature is usually much higher than for industrial chemical catalysts (see Table 5.12).

(*b*) *Specificity*. A characteristic feature of enzymes is that they are specific in action, some showing complete (or absolute) specificity (or discrimination) for only one type of molecule. If a substance exists in two stereochemical forms, L and D isomers, enzymes may recognise only one of the two forms. For example, *glucose oxidase* will oxidise D(+) glucose only and no other hexose isomer. Other enzymes have group specificity, showing activity towards a series of closely related compounds (e.g. branched-chained amino-acids, keto-acids, alcohols, etc.).

The stereochemical specificity of enzymes depends on the existence of at least three different points of interaction, each of which must have a binding or catalytic function. A catalytic site on the molecule is known as an *active site* or *active centre* of the enzyme. Such sites constitute only a small proportion of the total volume of the enzyme and are located on or near the surface. The active site is usually a very complex physico-chemical space, creating micro-environments in which the binding and catalytic areas can be found. The forces operating at the active site can involve charge, hydrophobicity, hydrogen-bonding and redox processes. The determinants of specificity are thus very complex but are founded on the primary, secondary and tertiary structures of proteins (see Appendix 5.1).

Several models have been put forward to explain specificity. The Fischer 'lock and key' hypothesis was the earliest hypothesis proposed to explain the interaction between the substrate and the active site which is considered rigid throughout the binding process. Here the active site of the protein is matched and fitted by a specific substrate and the hypothesis explains in simple terms the matching of the two complex architectures. Later, as protein structure became better understood, the Koshland 'induced fit' hypothesis was put forward to explain the possible conformational change associated with the formation of the enzyme–substrate complex in structural enzymes. Here, both the enzyme and the substrate change their structure slightly to accommodate each other. Finally, the Haldane and Pauling 'transition state' hypothesis represents yet a further development. It takes into account the way the active site might actually catalyse the reaction by forcing the formation and stabilisation of a transitional state between substrate and enzymes. This assists the reaction by which the desired product is formed[17].

(c) *Versatility*. The versatility of enzyme catalysis is shown by the types of reactions that can be catalysed. There are six major groups of enzymes arranged according to their reactivity:

1. *Oxidoreductase*—Oxidation–reduction reactions,
2. *Transferases*—Transference of an atom or group between two molecules,
3. *Hydrolases*—Hydrolysis reactions,
4. *Lyases*—Removal of a group from a substrate (not by hydrolysis),
5. *Isomerases*—Isomerisation reactions,
6. *Ligases*—The synthetic joining of two molecules, coupled with the breakdown of a pyrophosphate bond in a nucleoside triphosphate.

The International Union of Biochemistry has drawn up a numeric system of classification based on the type of reaction catalysed. Enzyme numbers can be generated using the first digit to represent the enzyme group indicated above, then three other numbers added to identify the enzyme uniquely. The additional numbers are based on properties such as reaction mechanism, substrate and *cofactor* type, (i.e. secondary substrate, suppling reducing power, energy, or catalytic intermediate). The number conventions are unique within each major reaction group:

e.g. Lactate dehydrogenase, L-lactate: NAD^+ oxidoreductase (E.C. 1.1.1.27),
Lactate dehydrogenase, D-lactate: NAD^+ oxidoreductase (E.C. 1.1.1.28),
Alcohol dehydrogenase, ethanol: NAD^+ oxidoreductase (E.C. 1.1.1.1.).

(d) *Controlled expression*. Enzyme expression and activity can be controlled in a number of ways at the metabolic level by specialised enzymes (e.g. Allosteric enzymes), whose activity can be modulated by environmental conditions, and at the genetic level by the environmentally controlled synthesis of the enzymes. These processes are discussed further in Sections 5.6 and 5.7.

(e) *Stability*. Enzymes are relatively unstable as the functional tertiary structure is held together by only weak molecular forces. The kinetic aspects of stability are discussed below, and the structure of protein is discussed in more detail in Appendix 5.1.

5.4.3. Enzyme Kinetics

Factors Responsible for Enzyme Catalysis

Enzymes, like all catalysts, enhance the rate of reaction but do not alter the thermodynamic equilibrium. The reactant molecules (substrates) must collide for a reaction to take place and the collision must both be in the correct orientation and have sufficient energy (the activation energy) for the reaction to take place.

The Arrhenius equation expresses the relationship between the rate constant $\boldsymbol{k}$ and the activation energy $\mathbf{E}_a$:

$$\boldsymbol{k} = \mathcal{A} e^{-\mathbf{E}_a/\mathbf{R}T} \tag{5.1}$$

where **R** is the universal gas constant, T is the absolute temperature and $\mathscr{A}$ is the Arrhenius constant. In enzyme-catalysed reactions, the rate constant k may be many orders of magnitude larger than that of the uncatalysed reaction. To accommodate the activity of an enzyme into the model, the concept of the formation of enzyme–substrate and enzyme–product complexes is introduced. Thus, the first step of an enzyme-catalysed reaction is the formation of an enzyme–substrate complex (ES) which is generally a more stable entity than the enzyme alone. The stability is attributed to the hydrogen bonding, van der Waals forces and ionic interactions between the enzyme and the substrate. The enhancement of the rate at which the reaction takes place depends on the lowering of the activation energy occurring as a result of the intra- and inter-molecular interactions between the enzyme and the substrate. The products, once formed, are released when the enzyme and product dissociate. A more detailed and extensive explanation of enzyme catalysts is given by FERSHT[20].

5.4.4. Derivation of the Michaelis–Menten Equation

For most enzymes, the rate of reaction can be described by the Michaelis–Menten equation which was originally derived in 1913 by MICHAELIS and MENTEN[21]. Its derivation can be achieved by making one of two assumptions, one of which is a special case of the more general Briggs–Haldane scheme, whilst the alternative is the 'rapid-equilibrium' method given in Appendix 5.3[20].

The Steady-State Assumption (Briggs–Haldane)

Consider the simplest reaction scheme between an enzyme and a substrate, Scheme 1. It is assumed that the decomposition of the enzyme–substrate complex to yield product is not reversible, while its decomposition to substrate **S** and free enzyme **E** is reversible. The decomposition enzyme–substrate complex (**ES**) is assumed to be rate-limiting. The rate constants for the formation and breakdown of the enzyme substrate (**ES**) complex are given by:

$$\mathrm{E + S} \underset{\mathbf{k}_{f2}}{\overset{\mathbf{k}_{f1}}{\rightleftharpoons}} \mathrm{ES} \xrightarrow{\mathbf{k}_{f2}} \mathrm{E + P} \qquad \text{(Scheme 1)}$$

The steady-state assumption is valid when the concentration ES of the enzyme–substrate complex **ES** is constant, and when the total enzyme concentration, E_{tot}, is small relative to that of the substrate, i.e. $E_{tot} << S$. In most cases this assumption holds over a long period of the reaction, as illustrated in Fig. 5.10, which shows the significance of this assumption during reaction processes.

Since the concentration of the enzyme–substrate complex is constant, its rate of change is zero. That is:

$$\frac{dES}{dt} = 0 \qquad (5.2)$$

As the enzyme–substrate is formed by one pathway but disappears by two (reversal of formation and product formation), then:

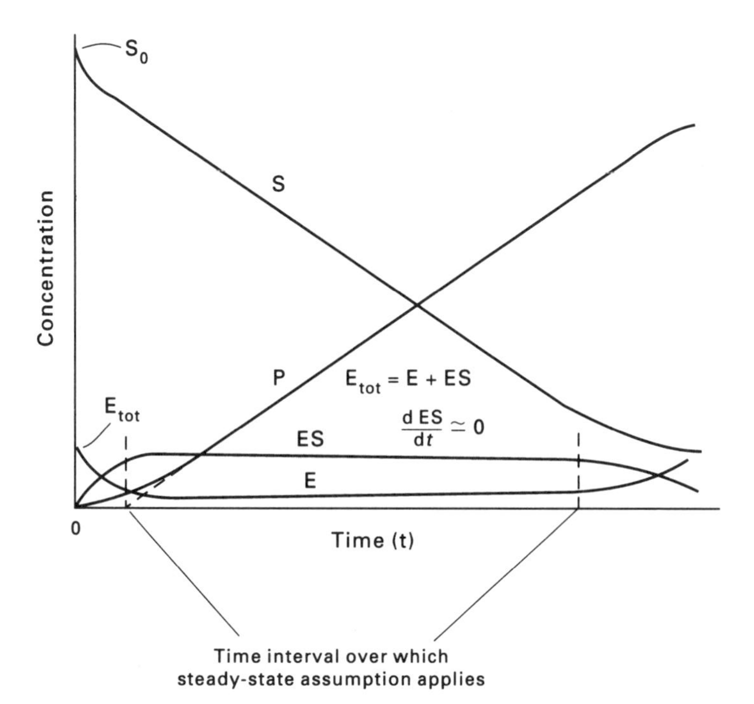

FIG. 5.10. The steady-state assumption. Concentration of substrate, **S**, free enzyme, **E**, product **P**, and enzyme–substrate complex, **ES**, over time t. After an initial burst, the concentration of enzyme–substrate complex is almost constant and the steady-state assumption is applicable

$$\boldsymbol{k}_{f1} E\, S - \boldsymbol{k}_{f2} ES - \boldsymbol{k}_{r1} E\, S = \frac{\mathrm{d}ES}{\mathrm{d}t} = 0 \tag{5.3}$$

and since:

$$ES = E_{\text{tot}} - E \tag{5.4}$$

substituting equation 5.3 into equation 5.4 then gives:

$$ES = \frac{E_{\text{tot}} S}{\dfrac{(\boldsymbol{k}_{r1} + \boldsymbol{k}_{f2})}{\boldsymbol{k}_{f1}} + S} \tag{5.5}$$

and, since:

$$\mathcal{R}_0 = \boldsymbol{k}_{f2} E\, S \tag{5.6}$$

where $\mathcal{R}_0$ is the initial rate of reaction, then substituting from equation 5.5 into equation 5.6 yields the rate law for the reaction:

$$\mathcal{R}_0 = \frac{\boldsymbol{k}_{f2} E_{\text{tot}} S}{\dfrac{(\boldsymbol{k}_{r1} + \boldsymbol{k}_{f2})}{\boldsymbol{k}_{f1}} + S} \tag{5.7}$$

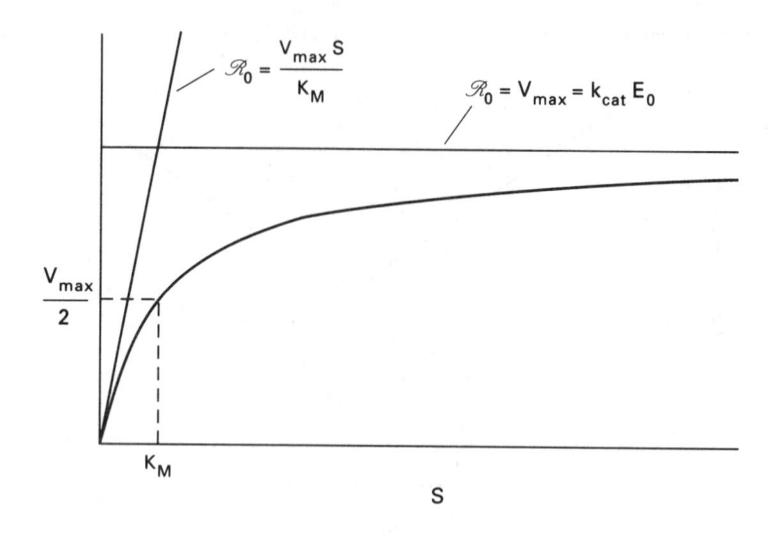

FIG. 5.11. Initial reaction rate $\mathcal{R}_0$, plotted against substrate concentration S for a reaction obeying Michaelis–Menten kinetics

The composite term $((k_{r1} + k_{f2})/k_{f1})$ in the above equation is the Michaelis constant K_m.
Writing:

$$K_m = \frac{(k_{r1} + k_{f2})}{k_{f1}} \tag{5.8}$$

$$\mathcal{R}_0 = \frac{k_{f2} E_{tot} S}{S + K_m} \tag{5.9}$$

where $k_{f2} E_{tot} = V_{max}$, the maximum rate of reaction, and therefore:

$$\mathcal{R}_0 = \frac{V_{max} S}{S + K_m} \tag{5.10}$$

From the derivation shown here and in Appendix 5.3 (detailing the various forms of the Michaelis–Menten rate law) it should be noted that nothing is implied concerning the mechanism of action of the enzyme, and different enzyme reaction mechanisms cannot be distinguished kinetically. Figure 5.11 shows the relation between rate of the reaction and substrate concentration. At low substrate concentrations it is a first-order with respect to the substrate, while at high (saturating) concentration the reaction is zero-order and limited by the enzyme concentration and the turnover number k_{cat}. (In the simple situation above, $k_{f2} = k_{cat}$, but in more complex situations this may not be so and k_{cat} will be a sum of decomposition terms[20].) The K_m value for the reaction is the concentration of substrate required to give half the maximum rate. Typically values for K_m are in the range of 0.01 to 20 mM (10^{-5} – 0.02 kmol/m^3).

5.4.5. The Significance of Kinetic Constants

K_m and K'_S

The derivation using the steady-state assumptions shows that:

$$K_m = \frac{k_{r1} + k_{f2}}{k_{f1}} \qquad \text{(equation 5.8)}$$

while the rapid equilibrium hypothesis of Michaelis–Menten (Appendix 5.3) shows that the relationship between K_m and K'_S is:

$$K_m = \frac{k_{f2}}{k_{f1}} + K'_S$$

(as the dissociation of **ES**, $K'_S = k_{r1}/k_{f1}$).
It would appear, therefore, that the view of the Michaelis–Menten situation envisaged by Briggs and Haldane is a special case in which k_{f2} is so much smaller than k_{r1} that the value of the ratio is negligible compared with K'_S, and:

$$K_m = K'_S \text{ when } k_{f2} << k_{r1}$$

The apparent dissociation constant may be treated as an overall dissociation constant for much more complex situations involving several linked rate constants[(21)].

The maximum rate V_{max}

According to the interpretation of the two-step mechanism offered above, the definition of V_{max} reflects the fact that the decomposition of the **ES** complex is rate limiting so that when all the enzyme is in the form of **ES** a maximum rate will be observed, i.e. $ES = E_{tot}$.

Thus:
$$V_{max} = k_{f2} E_{tot}$$

It should also be noted that k_{f2} is equivalent to k_{cat} in the Michaelis–Menten rapid-equilibrium hypothesis when the decomposition rate of the enzyme–substrate complex is fast, as described above in Scheme 1. However, the value of k_{cat} may be attributable to more complex situations involving several decomposition terms.

It is also feasible that, following changes in the value of V_{max} under different reaction conditions, it might be possible to obtain information concerning the kinetics of the rate-limiting step in the decomposition of **ES**. The catalytic constant or turnover number (k_{cat}) is a first-order rate constant that refers to the properties and reactions of the enzyme–substrate, enzyme–intermediate, and enzyme–product complexes. The units of k_{cat} are *time*$^{-1}$, and $1/k_{cat}$ is the time required to turn over a molecule of substrate on an active site.

The Ratio V_{cat}/K_m

This ratio is of fundamental importance in the relationship between enzyme kinetics and catalysis. In the analysis of the Michaelis–Menten rate law (equation 5.8), the ratio k_{cat}/K_m is an apparent second-order rate constant and, at low substrate concentrations, only a small fraction of the total enzyme is bound to the substrate and the rate of reaction is proportional to the free enzyme concentration:

$$\mathcal{R}_0 = \frac{k_{cat} ES}{K_m} \tag{5.11}$$

Enzyme specificity is also related to the ratio k_{cat}/K_m. Suppose that two substrates are competing for the same enzyme.

Consider the general case:

$$E + A \rightleftharpoons EA \longrightarrow E + P$$

$$E + B \rightleftharpoons EB \longrightarrow E + Q$$

The rate for these two reactions will be:

$$\mathcal{R}_0^A = \frac{k_{cat}^A EA}{K_m^A} \qquad \text{where } A << K_m^A \tag{5.12}$$

$$\mathcal{R}_0^B = \frac{k_{cat}^B EB}{K_m^B} \qquad \text{where } B << K_m^B \tag{5.13}$$

The ratio of the two rates $\mathcal{R}_0^A/\mathcal{R}_0^B$ expresses the relative ability of the enzyme to catalyse the conversion of **A** and **B** to products **P** and **Q**. In other words, $\mathcal{R}_0^A/\mathcal{R}_0^B$ is a measure of enzyme specificity. When the concentrations of **A** and **B** are equal, the ratio of the rates is:

$$\frac{\mathcal{R}_0^B}{\mathcal{R}_0^B} = \frac{k_{cat}^A/K_m^A}{k_{cat}^B/K_m^B} \tag{5.14}$$

This equation indicates that enzyme specificity depends upon both k_{f2} and K_m, not upon K_m alone. This interpretation of specificity also provides a theoretical basis for the optical resolution of racemic mixture[22].

5.4.6. The Haldane Relationship

All reactions are to some degree reversible, and many enzyme-catalysed reactions can take place in either direction inside a cell. It is therefore interesting to compare the forward and back reactions, especially when the reaction approaches equilibrium, as in an enzyme reactor. Haldane derived a relationship between the kinetic and equilibrium constants. The derivation of the relationship is shown in Appendix 5.4.

Consider the situation in Scheme 2 in which the conversion of substrate to product proceeds via the formation of a single intermediate complex; in the forward direction it would be regarded as an **ES** complex, while in the back reaction it would be regarded as an **EP** complex

$$\mathrm{S + E} \underset{\mathbf{k}_{r2}}{\overset{\mathbf{k}_{f1}}{\rightleftharpoons}} \mathrm{ES/EP} \underset{\mathbf{k}_{r2}}{\overset{\mathbf{k}_{f2}}{\rightleftharpoons}} \mathrm{P + E} \qquad \text{(Scheme 2)}$$

Therefore:

$$K_{eq} = \frac{\boldsymbol{k}_{f1}\boldsymbol{k}_{f2}}{\boldsymbol{k}_{r1}\boldsymbol{k}_{r2}} = \frac{V_{max}^{\ \mathrm{S}} K_m^{\mathrm{S}}}{V_{max}^{\ \mathrm{P}} K_m^{\mathrm{P}}} \ \text{(The Haldane relationship)} \tag{5.15}$$

If the equilibrium constant is known, this relationship can be used to check the validity of the kinetic constants which have been determined. In general, the equilibrium will favour the metabolically important direction. However, it should be noted that the direction of metabolic flow in cell systems will also be dependent of the concentrations of substrate and products.

5.4.7. Transformations of the Michaelis–Menten Equation

(i) The Lineweaver–Burk Plot—The Double Reciprocal Plot

Graphical transformation of the representation of enzyme kinetics is useful as the value of V_{max} is impossible to obtain directly from practical measurements. A series of graphical transformations/linearisations may be used to overcome this problem. Lineweaver and Burk (see reference[17]) simply inverted the Michaelis-Menten equation (equation 5.10). Thus:

$$\frac{1}{\mathscr{R}_0} = \frac{S}{V_{max}S} + \frac{K_m}{V_{max}S} \tag{5.16}$$

therefore:

$$\frac{1}{\mathscr{R}_0} = \frac{K_m}{V_{max}}\frac{1}{S} + \frac{1}{V_{max}} \tag{5.17}$$

A plot $\dfrac{1}{\mathscr{R}_0}$ against $\dfrac{1}{S}$ gives a straight line graph with an intercept of $\dfrac{1}{V_{max}}$ and a slope of $\dfrac{K_m}{V_{max}}$ as shown in Fig. 5.12.

This transformation suffers from a number of disadvantages. The data are reciprocals of measurements, and small experimental errors can lead to large errors in the graphically determined values of K_m, especially at low substrate concentrations. Departures from linearity are also less obvious than on other kinetic plots such as the Eadie–Hofstee and Hanes plots (see reference[17]).

(ii) Eadie-Hofstee Plot and Hanes Plot

If the Lineweaver-Burk double–reciprocal equation (equation 5.17) is multiplied by V_{max} throughout then, after rearrangement, the following equation is obtained:

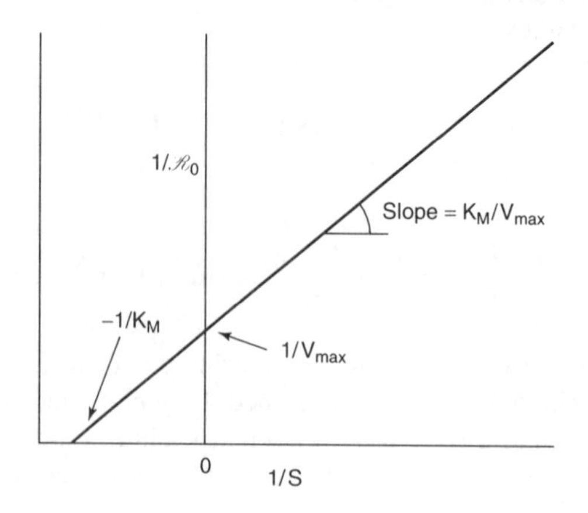

FIG. 5.12. The Lineweaver-Burk plot

$$\mathscr{R}_0 = -K_m \frac{\mathscr{R}_0}{S} + V_{\max} \tag{5.18}$$

The Eadie-Hofstee plot of $\mathscr{R}_0$ against $\mathscr{R}_0/S$ has an intercept of $V_{\max}$ on the $\mathscr{R}_0$ axis, while on the other axis the intercept is $V_{\max}/K_m$ (Fig. 5.13).

The Hanes plot also starts with the Lineweaver-Burk transformation (equation 5.15) of the Michaelis-Menten equation which in this instance is multiplied by S throughout; on simplification this yields:

$$\frac{S}{\mathscr{R}_0} = \frac{1}{V_{\max}} S + \frac{K_m}{V_{\max}} \tag{5.19}$$

A plot of $S/\mathscr{R}_0$ versus S is linear with a slope of $1/V_{\max}$. The intercept on the $S/\mathscr{R}_0$ axis gives $-K_m$.

These equations compress data at high substrate concentrations, but the relationship between rate and concentration is less obvious than in the Lineweaver-Burk

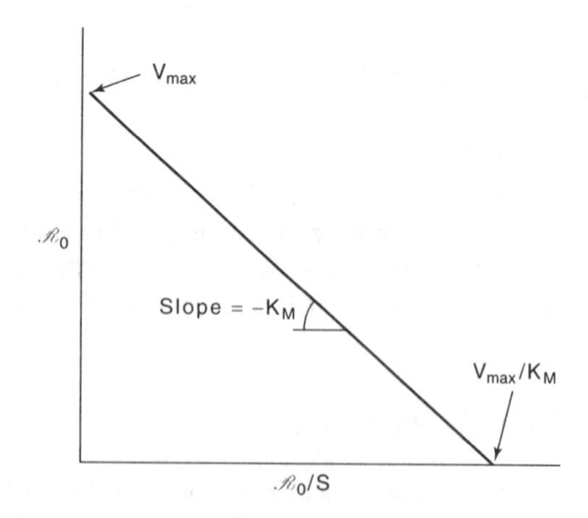

FIG. 5.13. The Eadie-Hofstee plot

plot. As they only involve one reciprocal term, they tend to give more accurate figures than those obtained in a Lineweaver–Burk plot.

5.4.8. Enzyme Inhibition

Inhibitors are substances that tend to decrease the rate of an enzyme-catalysed reaction. Although some act on the substrate, the discussion here will be restricted to those inhibitors which combine directly with the enzyme. Inhibitors have many uses, not only in the determination of the characteristics of enzymes, but also in aiding research into metabolic pathways where an inhibited enzyme will allow metabolites to build up so that they are present in detectable levels. Another important use is in the control of infection where drugs such as sulphanilamides competitively inhibit the synthesis of tetrahydrofolates which are vitamins essential to the growth of some bacteria. Many antibiotics are inhibitors of bacterial protein synthesis (e.g. tetracyclin) and cell-wall synthesis (e.g. penicillin).

There are two types of inhibitors. *Reversible inhibitors* bind to an enzyme in a reversible fashion and can be removed by dialysis (or dilution) to restore full enzyme activity. *Irreversible inhibitors* cannot be removed by dialysis and, in effect, permanently deactivate or *denature* the enzyme.

Reversible Inhibition

Reversible inhibition occurs rapidly in a system which is near its equilibrium point and its extent is dependent on the concentration of enzyme, inhibitor and substrate. It remains constant over the period when the initial reaction velocity studies are performed. In contrast, irreversible inhibition may increase with time. In simple single-substrate enzyme-catalysed reactions there are three main types of inhibition patterns involving reactions following the Michaelis–Menten equation: *competitive, uncompetitive* and *non-competitive* inhibition. Competitive inhibition occurs when the inhibitor directly competes with the substrate in forming the enzyme complex. Uncompetitive inhibition involves the interaction of the inhibitor with only the enzyme–substrate complex, while non-competitive inhibition occurs when the inhibitor binds to either the enzyme or the enzyme–substrate complex without affecting the binding of the substrate. The kinetic modifications of the Michaelis–Menten equation associated with the various types of inhibition are shown below. The derivation of these equations is shown in Appendix 5.5.

$$\mathcal{R}_0 = \frac{V_{max}\,S}{S + K_m\left(1 + \dfrac{I}{K_I}\right)} \quad \text{Competitive} \tag{5.20}$$

$$\mathcal{R}_0 = \frac{V_{max}\,S}{\left(1 + \dfrac{I}{K_I}\right)} \Bigg/ \left\{S + \frac{K_m}{\left(1 + \dfrac{I}{K_I}\right)}\right\} \quad \text{Uncompetitive} \tag{5.21}$$

$$\mathcal{R}_0 = \frac{V_{max}}{\left(1 + \frac{I}{K_I}\right)} \frac{S}{(S + K_m)} \quad \text{Non-Competitive} \tag{5.22}$$

Here, I is the inhibitor concentration and K_I is the inhibitor constant.

The effects of different types of inhibition are shown in the Eadie–Hofstee and Lineweaver–Burk plots indicated in Fig. 5.14.

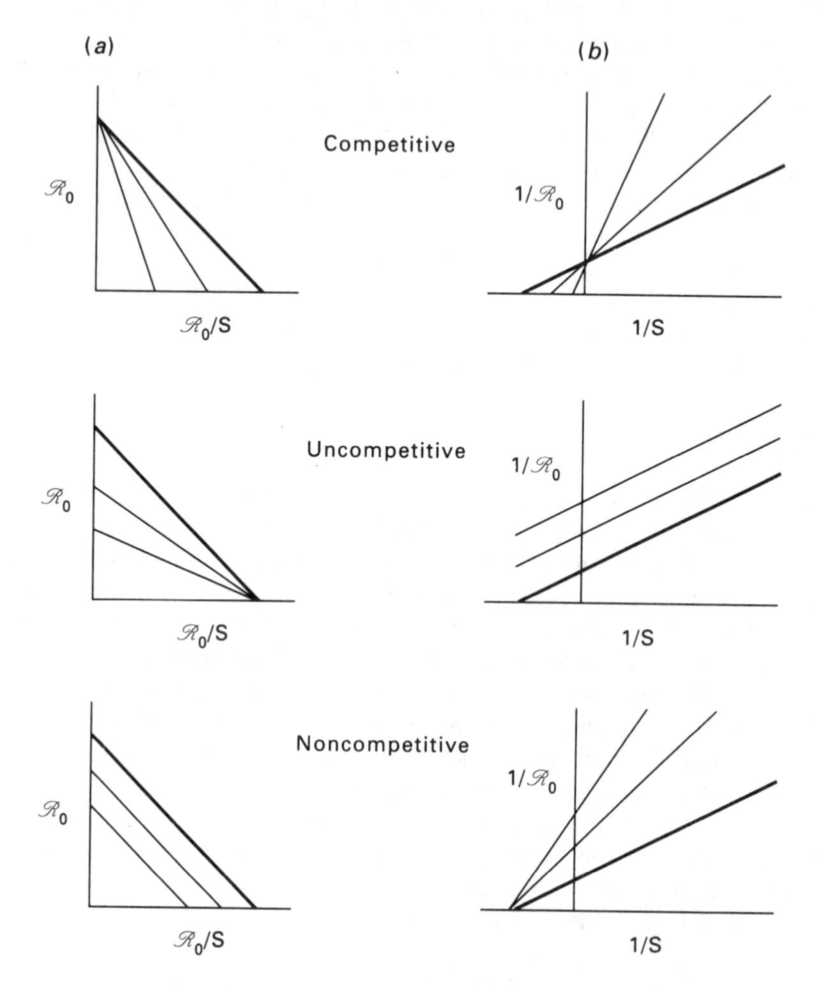

FIG. 5.14. Graphical representation of enzyme inhibition. (*a*) Eadie–Hofstee and (*b*) Lineweaver–Burk plots of different types of inhibition. The bold line indicates initial reaction rate in the absence of the inhibitor; the lighter lines show initial rates in the presence of inhibitors

Mixed Inhibition

In mixed inhibition, a second assumption in the derivation of non-competitive inhibition must be made—that the binding of the substrate and the inhibitor to the enzyme are no longer independent of one another. With Lineweaver–Burk plots of such inhibition phenomena the slopes and intercepts, and hence K_M and V_{max}, are all affected by the inhibitor. As a result, neither the consistent intercepts nor the constant slopes which are characteristic of the other forms of inhibition are obtained.

Substrate Inhibition

At very high substrate concentrations deviations from the classical Michaelis–Menten rate law are observed. In this situation, the initial rate of a reaction increases with increasing substrate concentration until a limit is reached, after which the rate declines with increasing concentration. Substrate inhibition can cause such deviations when two molecules of substrate bind immediately, giving a catalytically inactive form. For example, with succinate dehydrogenase at very high concentrations of the succinate substrate, it is possible for two molecules of substrate to bind to the active site and this results in non-functional complexes. Equation 5.19 gives one form of modification of the Michaelis–Menten equation.

$$\mathcal{R}_0 = \frac{V_{max} S}{K_m + S + \dfrac{S^2}{K_I}} \tag{5.23}$$

where K_I is the equilibrium constant for substrate in the inhibitory form.

5.4.9. The Kinetics of Two-Substrate Reactions

Most enzymes catalyse reactions involving two interacting substrates and such reactions show much more complicated kinetics than the simple one-substrate kinetics discussed previously. Enzymes catalysing bi-substrate reactions include tranferase reactions (transferring functional groups from one of the substrates to the other), e.g. aminotransferase and phosphorylases. Another important group consists of the dehydrogenases where NAD(P) or NAD(P)H are also co-substrates with the carbon substrates, e.g. alcohol dehydrogenase (ADH)[17].

$$CH_3CH_2OH + NAD \underset{}{\overset{ADH}{\rightleftharpoons}} CH_3\ CHO + NADH + H^+$$

The kinetic analysis of two-substrate reactions is more complex due to the increased number of possible enzyme–substrate complexes. Consider the following reaction:

$$A + B \overset{Enzyme}{\rightleftharpoons} P + Q$$

A large number of possible enzyme–substrate complexes may form, e.g. the binary complexes EA, EB, EP and EQ, ternary complexes EAB, EPQ, EAQ, and EBQ. Most two-substrate reactions can be grouped into two major classes, based upon the reaction sequence in the two-substrate reactions, single displacement reactions and double displacement reactions.

Single Displacement Reactions

In single displacement reactions both substrates **A** and **B** simultaneously must be present on the active site of the enzyme to yield a ternary complex EAB in order that the reaction may proceed. Single displacement reactions take place in two forms, random and ordered, and they are distinguished by the way the two substrates bind to the enzyme.

In *random* bi-substrate reactions, either substrate may bind to the enzyme first, indicating that the ternary complex (sometimes called the central complex) EAB can be formed equally well in two different ways.

e.g.:

$$E + A \rightleftharpoons EA$$

$$EA + B \rightleftharpoons EAB$$

or:

$$E + B \rightleftharpoons EB$$

$$EB + A \rightleftharpoons EAB$$

In *ordered* single displacements there is a compulsory sequence for the reaction which dictates that a specific substrate, the leading substrate, must be bound first, before the second, or following substrate, can be bound, as shown below:

$$E + A \rightleftharpoons EA \quad \textit{and then} \quad EA + B \rightleftharpoons EAB$$

In this case **A** is the leading substrate. Many dehydrogenase reactions which utilise NAD as a co-substrate are good examples of ordered single displacements, e.g. malate dehydrogenase (MDH):

$$\text{Malate} + NAD^+ \overset{\text{MDH}}{\rightleftharpoons} \text{Oxaloacetate} + NADH + H^+$$

In this reaction, NAD must bind first to yield the E–NAD^+ complex to which malate then combines to form a ternary complex, E–NAD–malate; the main reaction then takes place to yield NADH and oxaloacetate.

Random bi-substrate reactions can be distinguished from ordered reactions experimentally. The final reaction product can inhibit the overall reaction by competing with only the first (leading) substrate of the reaction. The reaction involving malate dehydrogenase outlined above is ordered and is inhibited by excess NADH, which competes with a normal leading-substrate NAD^+ for binding to the enzyme. NADH does not, however, compete with the malate.

Such considerations are important when optimising two-substrate reaction processes; for example, in the increasingly important area of biotransformation reactions.

Double Displacement Reactions

In bi-substrate reactions of the double displacement type, one substrate must be bound and one product released before the entry of the second substrate. In such reactions, the first substrate reacts with the enzyme to yield a chemically modified form of the enzyme (usually a functional group is changed) and the first product. In the second step, the functional group of the modified enzyme is transferred from the enzyme to the second substrate to form the second product. A good example is the aminotransferase class of enzymes, where an amino group is transferred from an amino-acid to the enzyme, from which it is transfered to a keto-acid.

Kinetic Constant Determination

Determination of the kinetic constant for a bi-substrate reaction is carried out in a similar manner to that for single substrate reactions. This is achieved by investigating only one substrate at a time, while the other is kept at a set concentration which is usually its saturation concentration. Thus, to determine the K_m and V_{max} of substrate **A**, **B** is kept constant at a saturating level while the reaction of **A** is investigated at different concentrations. The experimental conditions are then reversed to determine the kinetic constants of **B**. Thus, the kinetic constants for a bi-substrate reaction are determined using two separate kinetic plots, as discussed previously for the conditions where concentrations of **A** or **B** limit the rate of the reaction. Clearly, the conditions under which the rates are determined must be quoted for any determination.

The Alberty Equation

As stated above, many two-substrate reactions obey the Michaelis–Menten equation with respect to one substrate while the concentration of the other substrate remains constant. This is true of reactions involving only one site, or those involving several sites provided that there is no interaction between sites. Alberty derived the following general equation for the reaction[23]:

$$AX + B \rightleftharpoons BX + A$$

$$\mathcal{R}_0 = \frac{V_{max} A B}{K_m^B A + K_m^A B + A B + K_S^A K_m^B} \tag{5.24}$$

where V_{max} is the maximum possible value of $\mathcal{R}_0$ when **A** and **B** are both saturating, K_m^A is the concentration of **A** which gives $1/2\, V_{max}$ when **B** is saturating, K_m^B is the concentration of **B** which gives $1/2\, V_{max}$ when **A** is saturating and K_S^A is the dissociation constant for $E + A \rightleftharpoons EA$.

At very large concentrations of **B** (10–20 times K_m^B) the general equation simplifies to:

$$\mathcal{R}_0 = \frac{V_{max} A}{A + K_m^A} \tag{5.25a}$$

or for very large concentrations of **A**:

$$\mathcal{R}_0 = \frac{V_{max} B}{B + K_m^B} \tag{5.25b}$$

A good example of this type of simplification occurs when one considers water to be a substrate in hydrolytic reactions where the concentration of water is in vast excess (55 M = 55 $kmol/m^3$), while K_m for reactions (although not determined) would be in the region of 10^{-3}–10^{-5} $kmol/m^3$.

5.4.10. The Effects of Temperature and pH on Enzyme Kinetics and Enzyme De-activation

The two major environmental factors that significantly affect the kinetics of enzymes are pH and temperature.

The Effect of the Ionisation State on Catalytic Activity

The amino-acids that make up the primary structure of proteins will change their charge when the pH of the solution is altered due to their acid–base properties (Section 5.3 and Appendix 5.1). The effects of pH on enzyme-catalysed reactions can be complex since both K_m and V_{max} may be affected. Here, only the effects on V_{max} are considered, as this usually reflects a single constant rather than several that may be associated within the constant K_m (see Section 5.4.4.). It is assumed that pH does not change the limiting step in a multi-step process and that the substrate is saturating at all times.

The simplest case, therefore, involves a single ionising side chain in the enzyme where – E^- is the active form while EH^+ and E^{2-} are inactive forms:

$$\underset{\text{inactive}}{-\,EH^+} \overset{K_{a1}}{\rightleftharpoons} \underset{\text{active}}{-\,E^- + H^+} \overset{K_{a2}}{\rightleftharpoons} \underset{\text{inactive}}{E^{2-} + H^+}$$

The acid dissociation constant K_a is given by:

$$K_a = \frac{EH^+}{EH^+} \qquad (5.26)$$

and therefore:

$$EH^+ = \frac{EH^+}{K_a} \qquad (5.27)$$

The fraction f of enzyme in the active unprotonated form is given by:

$$f = \frac{E}{E + EH^+} = \frac{K_a}{K_a + H^+} \qquad (5.28)$$

If $(V_{max})_m$ is equal to the maximum rate which can occur for the enzyme in the unprotonated form, then at any pH, the observed V_{max} is given by:

$$V_{max} = (V_{max})_m f \qquad (5.29)$$

$$V_{max} = (V_{max})_m \frac{K_a}{K_a + H^+} \qquad (5.30)$$

Similarly, for a second ionising group which de-activates the active site, the effect on V_{max} at any pH can be shown to be:

$$V_{max} = \frac{(V_{max})_m}{1 + \frac{H^+}{K_{a1}} + \frac{K_{a2}}{H^+}} \tag{5.31}$$

The Effects of Temperature on the Kinetics of the Reactions

In general, the effects of changes in temperature on the rate of an enzyme-catalysed reaction do not provide much useful information about the mechanism of catalysis. However, as with most chemical reactions, the rate of an enzyme-catalysed reaction increases with increasing temperature and it is therefore useful to be able to predict activity at various temperatures.

Unlike most uncatalysed reactions as the temperature is raised the rate of an enzyme-catalysed reaction rises to a maximum and then decreases as the protein is denatured by the heat. Temperature affects not only activity but also the stability of the enzyme since the tertiary structure is particularly susceptible to thermal damage.

For most simple reactions the activation energy is 300 MJ/kmol or less. The activation energy is very low for very fast reactions and the rate of reaction is then affected less by temperature. However, typically the activation energy is around 50 MJ/kmol, and an increase in temperature from say 25 to 35°C approximately doubles the rate of reaction, whereas with higher activation energies (around 85 MJ/kmol) the rate can triple over the same temperature range.

The ratio of rate constants of a reaction at two temperatures 10°C apart is called the Q_{10} value of the reaction. This ranges from 1.7 to 2.5 in enzyme-catalysed reactions.

In the Arrhenius equation (equation 5.1) $\boldsymbol{k}$ can be replaced by $\boldsymbol{k}_{cat}$ and, at constant enzyme concentration, E_{tot} will be directly related to V_{max}

$$V_{max} = \mathcal{A}e^{-\mathbf{E}_a/\mathbf{R}T} \tag{5.32}$$

Thus, a standard plot of V_{max} versus $1/T$ gives a straight line with a slope of $-\mathbf{E}_a/\mathbf{R}$, allowing the determination of activation energy.

5.4.11. Enzyme De-activation

Most literature on enzyme kinetics is devoted to initial rate data and the analysis of reversible effects on enzyme activity. In many applications and process settings, however, the rate at which the enzyme activity declines is of critical importance. This is especially true when considering its long-term use in continuous reactors. In such situations the economic feasibility of the process may hinge on the useful lifetime of the enzyme biocatalyst. The focus of this section is on the mechanisms and kinetics of loss of enzyme activity. It should also be recognised that the alteration of protein structure is central to the practical manipulation of proteins (e.g. precipitation, affinity and other forms of protein chromatography, and purification in general).

Protein Denaturation

As noted earlier, protein structure is stabilised by a series of weak forces which often give rise to the properties which are functionally important (models of active sites and substrate binding are discussed above). On the other hand, because active sites involve a set of subtle molecular interactions involving weak forces, they are vulnerable and can be transformed into less active configurations by small perturbations in environmental conditions. It is therefore not surprising that a multitude of physical and chemical parameters may cause perturbations in native protein-geometry and structure. Thus, enzyme deactivation rates are usually multi-factorial, e.g. enzyme sensitivity to temperature varies with pH and/or ionic strength of the medium.

In most cases the de-activation caused by temperature or other single environmental factors, is a first-order decay process.

Consider the deactivation of active enzyme E_{act} to inactive form E_I which may be described by a first-order rate constant ($\boldsymbol{k}_{de}$).

$$E_{\text{act}} \xrightarrow{k_{de}} E_I$$

$$-\frac{dE_{\text{act}}}{dt} = \boldsymbol{k}_{de} E_{\text{act}} \tag{5.33}$$

Integrating between time $t = 0$ and time t:

$$\int_{E_{\text{act}0}}^{E_{\text{act}t}} \frac{dE_{\text{act}}}{E_{\text{act}}} = \int_0^t -\boldsymbol{k}_{de}\, dt \tag{5.34}$$

Therefore:

$$\ln \frac{E_{\text{act}t}}{E_{\text{act}0}} = -\boldsymbol{k}_{de} t \tag{5.35}$$

or:

$$E_{\text{act}t} = E_{\text{act}0}\, e^{-\boldsymbol{k}_{de} t} \tag{5.36}$$

With the reaction equation in this form, the half-life may be determined by setting the enzyme activity at half the initial value;

i.e.:

$$E_{\text{act}t} = \frac{E_{\text{act}0}}{2} \tag{5.37}$$

Substituting from equation 5.37 into equation 5.36, then:

$$\frac{E_{\text{act}0}}{2} = E_{\text{act}0}\, e^{-\boldsymbol{k}_{de} t_{1/2}} \tag{5.38}$$

and:

$$1/2 = e^{-\boldsymbol{k}_{de} t_{1/2}} \tag{5.39}$$

$\therefore$

$$t_{1/2} = \frac{\ln 2}{\boldsymbol{k}_{de}} \tag{5.40}$$

Generally, deactivation rates are determined in the absence of substrate, but enzyme deactivation rates can be considerably modified by the presence of substrate and other materials.

The effect of combining the deactivation model with the simple catalytic sequence of the Michaelis–Menten relation is shown below:

$$\begin{array}{l} E_{act} + S \rightleftharpoons E_{act}S \longrightarrow E_{act} + P \\ \Big\Updownarrow \\ E_{act} \longrightarrow E_I \end{array}$$

Assuming that the deactivation process is much slower than the reaction represented by Section 5.4.4, scheme 1, and that enzyme E will deactivate faster than the enzyme in the bound state (i.e. ES complex), then equation 5.32 may be written as:

$$\frac{dE_{act}}{dt} = -E\,k_{de} \tag{5.41}$$

but:

$$ES = E_{tot} - E \qquad \text{(equation 5.4)}$$

and replacing E_{tot} with E_{act} in equation 5.4 and elimination of E from equation 5.41 gives:

$$\frac{dE_{act}}{dt} = -k_{de}\,(E_{act} - ES) \tag{5.42}$$

Substitution for ES from equation 5.5 and using the definition of K_m given in equation 5.8 after rearranging, equation 5.42 becomes:

$$\frac{dE_{act}}{dt} = -\frac{k_{de}\,E_{act}}{1 + S/K_m} \tag{5.43}$$

This implies that the rates of enzyme deactivation and of substrate conversion can be linked. If E and ES are deactivated at the same rate, then this rate will be the same as that for the substrate/enzyme preparation. Extending these notions, an enzyme will deactivate at different rates depending on which of the complex forms is present, and the overall deactivation rate will vary according to the proportions of the different forms of the enzyme that are present.

Mechanical Forces acting on Enzymes

Mechanical forces can disturb the elaborate structure of the enzyme molecules to such a degree that de-activation can occur. The forces associated with flowing fluids, liquid films and interfaces can all cause de-activation. The rate of denaturation is a function both of intensity and of exposure time to the flow regime. Some enzymes show an ability to recover from such treatment. It should be noted that other enzymes are sensitive to shear stress and not to shear rate. This characteristic mechanical fragility of enzymes may impose limits on the fluid forces which can be tolerated in enzyme reactors. This applies when stirring is used to increase mass transfer rates of substrate, or in membrane filtration systems where increasing flux through a membrane can be accompanied by increased fluid shear at the surface of the membrane and within membrane pores. Another mechanical force, surface tension, often causes denaturation of proteins and consequent de-activation of enzymes. Thus, foaming or frothing in protein solutions commonly results in denaturation of protein at the air–water interface.

In the processing context, a combination of mechanical forces and chemical reactions (oxidation, etc.) deactivates enzymes. It should be noted that there is at present no systematic way of improving the stability of an enzyme. Each system has unique properties and stabilising agents must be selected in an empirical fashion.

5.5. METABOLISM

5.5.1. The Roles of Metabolism

In Section 5.1 it is stated that the main purpose of a living system is to ensure its survival and reproduction. This is achieved through the chemistry of the cell and there are two primary activities.

(a) The Synthesis of Materials for Cell Structure

Synthesis within the cell is a complex process. Figure 5.15 illustrates how the general metabolic system of the cell is used to acquire and create the basic building blocks of the cell. Once assembled, the building blocks are then used to produce polymeric materials, most importantly proteins and nucleic acids (see also Fig. 5.9).

(b) The Generation of Energy for Growth, and for Chemical and Mechanical Work

To maintain a highly ordered structure in relation to the surrounding environment, the cell must consume energy. The major source of energy for living systems is ultimately traceable to light (electromagnetic radiation) which is converted into various forms of chemical energy. This captured energy can then drive synthetic processes, such as the reduction of carbon dioxide to glucose (i.e. photosynthesis). Glucose and other organic compounds can also be considered as sources of energy as they can be oxidised to carbon dioxide and water with the release of energy.

Enzymes are organised into metabolic pathways which collectively constitute metabolism. Two types of metabolism are found in cells, *catabolism* (breakdown pathways) and *anabolism* (synthetic pathways). Linking these two types of metabolic reactions are the intermediary reactions of central metabolism. Cells, which contain many complex polymers, thus have the means to generate and convert monomeric materials into the complex biological structure. The sources of these materials are the simpler components from the cell's environment, such as inorganic salts and glucose (Fig. 5.9).

5.5.2. Types of Reactions in Metabolism

Catabolic Metabolism

The breakdown of materials from the environment and their conversion into usable forms take place through a series of catabolic pathways and the process is one of

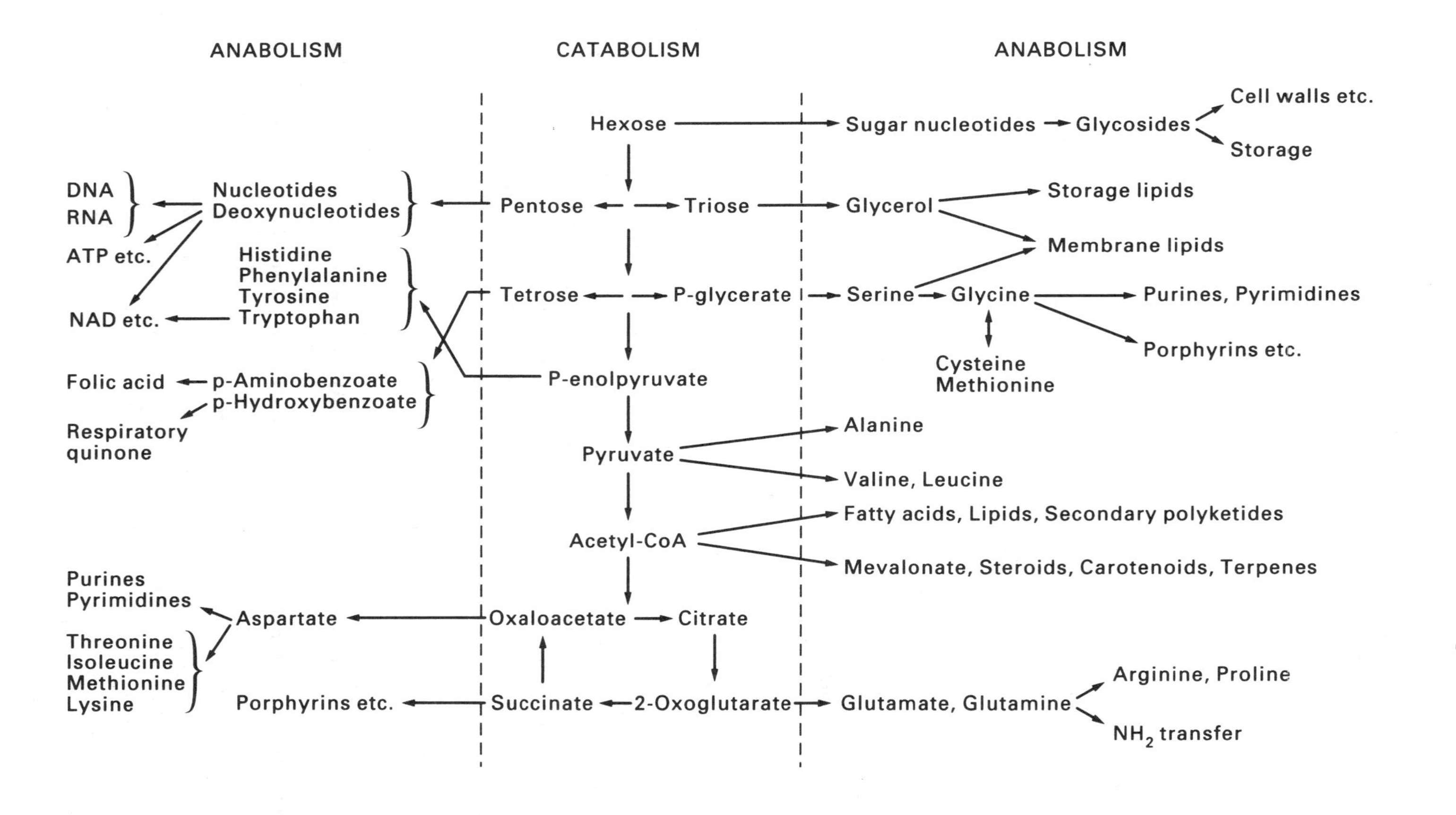

FIG. 5.15. Anabolic pathways (synthetic) and central catabolic pathways[24]. Only the main biosynthetic routes, and their principal connections to catabolic routes, are shown, in simplified version. The relationships between energy (ATP), and redox (NAD^+, $NADP^+$) metabolism, nitrogen metabolism, etc. are omitted

catabolism. The general pattern of metabolism is one of converging pathways in that many materials are broken down to the common materials of central metabolism. A good analogy is that of the flow of materials and resources to a central point for conversion. For example, a steel works receives coal and iron ore which is smelted to form molten iron, converted to steel, and then formed into rod, sheet and bar. The metabolic breakdown processes of herbicides, insecticides and natural polymeric materials follow an analogous pattern. The breakdown of glucose along the glycolytic pathway is another example (Section 5.5.6). These pathways provide useful routes for the isolation of new enzymes for biotransformation processes[25].

Anabolic Metabolism

Anabolism concerns the synthetic pathways of metabolism which are involved in the generation of the building blocks and the subsequent synthesis of macromolecules. The pattern of these pathways is generally divergent. They start from the components of central metabolism and then diverge into the many pathways associated with synthesis, the polymer building blocks and then the polymers themselves. An analogue of this would be the distribution of materials, e.g. steel to a component manufacturer, nuts and bolts to an automotive manufacturing plant and the final distribution of the car to the end user. Figure 5.15 shows the synthesis of amino-acids, sugars and nucleotides to polymeric materials such as proteins, carbohydrates and nucleic acids (see also Fig. 5.9).

Intermediary Metabolism

Anabolism and catabolism are complementary to one another, and may be compared with the biologically mediated cycling of carbon (and other materials) in the environment (Fig. 5.2). However, there is an area of metabolism where the two types of metabolism coincide and become indistinguishable. These pathways are known as intermediary metabolism and are the central part of cellular metabolism where the basic requirements for growth are met. At this metabolic crossroads, materials are either rearranged into synthetic precursors or are oxidised to generate energy.

Primary and Secondary Metabolism

The metabolism of an organism is in most cases related to growth and reproduction and it is then termed *primary* metabolism. In many cases, however, metabolism can be uncoupled from growth by limiting some nutrient or by environmental control. This non-growth associated metabolism of an organism is called *secondary metabolism.*

The most obvious product of primary metabolism is the biomass itself or the fermentation products of anaerobic organisms. Table 5.13 shows primary *metabolites* (biochemical intermediates and products) of industrial interest.

Secondary metabolism is commonly achieved by uncoupling the anabolic from the growth pathways. The subsequent overflow metabolites are then channelled towards *secondary products* which may include such antibiotics as penicillin, tetracyclin and streptomycin. Figure 5.16 shows the relationship between secondary products and central anabolic pathways.

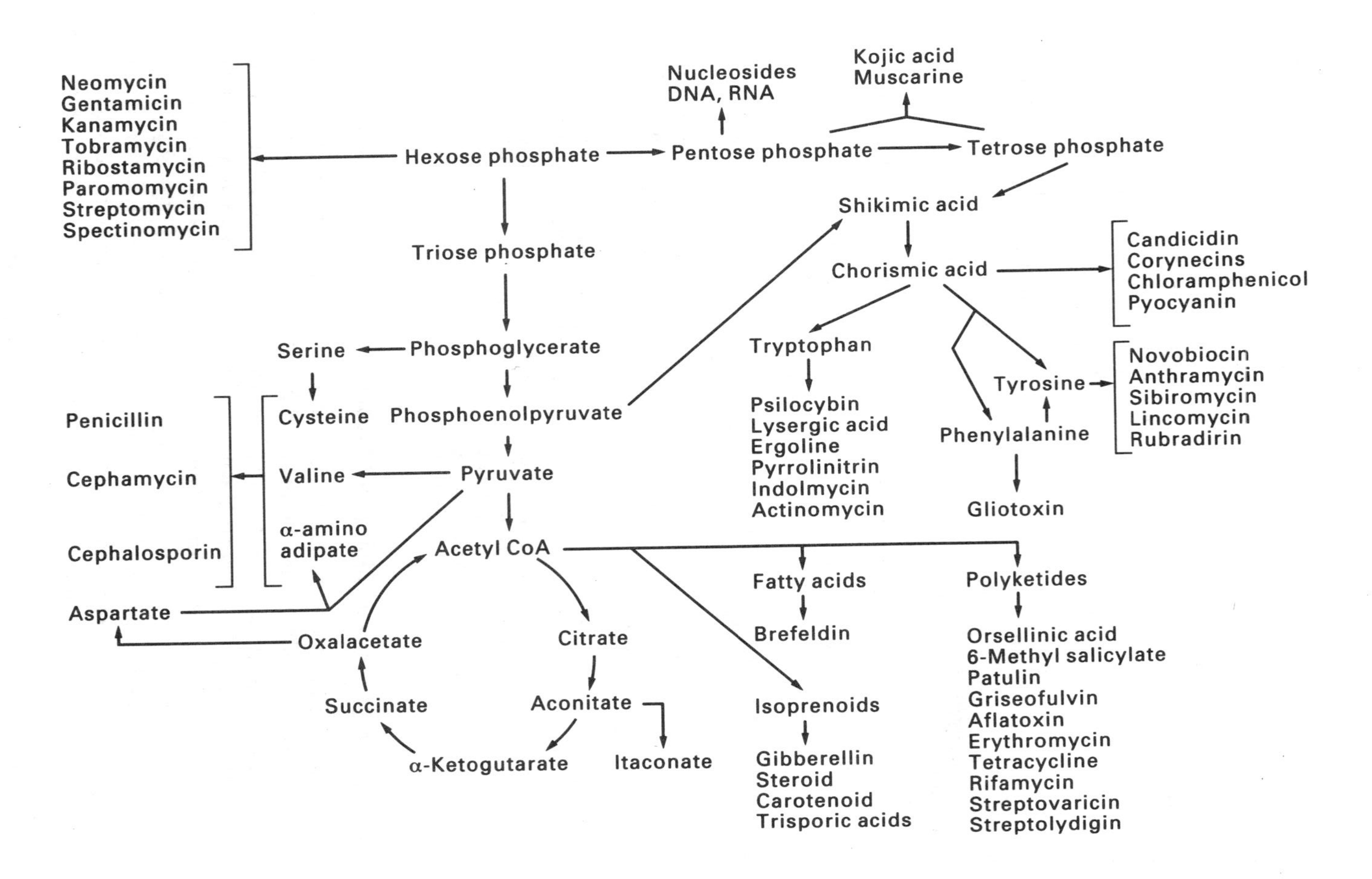

FIG. 5.16. Summary of different metabolic routes leading to secondary metabolites[26]

TABLE 5.13. *Some Primary Metabolites of Industrial Interest and Some Overproducing Species*[27]

Metabolite	Organism	Uses (present and future)
Glutamate	*Corynebacterium glutamicum*, *Brevibacterium* spp.	Food industry as flavour-accentuating agent
Lysine, threonine	*Corynebacterium glutamicum*, *Brevibacterium flavum*, *Escherichia coli*	Essential amino-acids, added to supplement low grade protein
Guanylic, inosinic and xanthylic acids	*Bacillus subtilis*, *Corynebacterium glutamicum*	Food industry as flavour-accentuating agents
Citric acid	*Aspergillus niger*, *Candida* spp.	Acidulant in food industry; pharmaceuticals (effervescent powders); esters used as plasticizers
Itaconic acid	*Aspergillus terreus*, *Aspergillus itaconicus*	Synthetic fibre and resin manufacture; copolymers (e.g. styrene-butadiene)
Fumaric acid	*Rhizopus arrhizus*, *Rhizopus nigricans*	Plastics and food industries
Cyanocobalamin (vitamin B_{12})	*Propionibacterium shermanii*, *Pseudomonas dentrificans*	Food and animal feed supplement
Riboflavin (vitamin B_2)	*Eremothecium ashbyii*, *Ashbya gossypii*	Food and animal feed supplement
β-Carotene	*Blakeslea* spp., *Choanephora* spp.	Precursor of vitamin A: colouring agent in food industry; food supplement
Xanthan gum, dextran	*Xanthomonas campestris*, *Leuconostoc mesenteroides*	Food, pharmaceutical, textile industries, useful for thickening, stiffening and setting properties
Acetic acid	*Saccharomyces ellipsoideus* or *Saccharomyces cerevisiae*, plus *Acetobacter* spp.; *Clastridium* spp.	Vinegar, preservative in food industry; chemical feedstocks; polymer industry
Acetone, butanol	*Clostridium acetobutylicum*	Solvents in chemical industry; thinners; synthetic polymers
Ethanol	*Saccharomyces cerevisiae*, *Zymomonas mobilis*, *Clostridium thermocellum* and other *Clostridium* spp.	Alcoholic beverages; solvent in chemical industry; fuel extender

5.5.3. Energetic Aspects of Biological Processes

In most engineering processes thermal energy balances form one of the most important criteria in design, and the heat balance frequently approximates to the energy balance of the system. However, in biological systems the total energy balance must be considered and heat transfer within living cells is relatively unimportant compared with the transport, storage and utilisation of chemical energy. This is because enzymes operate over a narrow range of temperature, typically 15–40°C.

Utilisable forms of free energy in biological systems occur at several levels—as stored energy, e.g. polysaccharides; as energy from the environment, e.g. food; as energy at the intermediate level, e.g. glucose in the cell; as utilisable chemical energy, e.g. adenosine triphosphate.

In all organisms the free energy released in redox reactions is conserved in the energy-carrier molecule adenosine triphosphate (ATP) which is the universal carrier

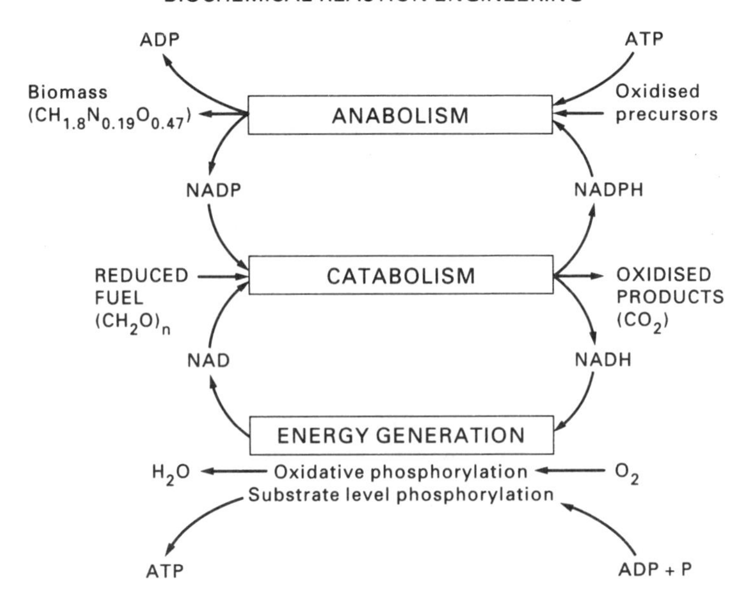

FIG. 5.17. The inter-relationships between metabolism, energy and redox processes

of free energy in biological systems. ATP functions as an energy carrier by virtue of its two 'high energy' phosphate anhydride bonds. When these bonds are hydrolysed:

$$ATP + H_2O \rightleftharpoons ADP + P_i + H^+ \quad \Delta G^{0\prime} = -\ 30.55\ MJ/kmol$$

$$ADP + H_2O \rightleftharpoons AMP + P_i + H^+ \quad \Delta G^{0\prime} = -\ 30.55\ MJ/kmol$$

The high, negative $\Delta G^{0\prime}$ values indicate that hydrolysis of the phosphate bond is strongly favoured thermodynamically.

ATP formation from adenosine diphosphate (ADP) and inorganic phosphate (P_i) is driven by the energy generated when the fuel molecules are oxidised in heterotrophs, or when light energy is trapped in *phototrophs*. The energy in ATP is utilised, via breakdown to ADP + P_i, to drive thermodynamically unfavourable processes such as biosynthesis, active transport and motility. ATP is not used in energy storage, but rather as a carrier between storage compounds such as fats and carbohydrate and the energy-consuming processes including biosynthesis and transport. Thus ATP and ADP cycle between energy requiring and energy generating reactions as shown in Fig. 5.17.

Energy transfer via ATP is accomplished by the transfer of phosphate groups. In the breakdown of fuel molecules (catabolism), energetically favourable reactions produce the high energy phosphorylated compounds for which the $\Delta G^{0\prime}$ value for hydrolysis is greater than that of ATP. Such compounds can spontaneously transfer phosphate groups to ADP, producing ATP, with a net negative $\Delta G^{0\prime}$. In biosynthesis, ATP is used to phosphorylate intermediates, thereby producing activated derivatives capable of further reaction. The detailed mechanisms by which ATP is generated are discussed in Section 5.5.6.

Another important set of reactions in living systems comprise those that involve redox processes. Reduced pyridine nucleotides, nicotinamide adenine dinucleotide

(NAD) and nicotinamide adenine dinucleotide phosphate (NADP), are the universal carriers of hydrogen and electrons, or metabolic reducing power. Many redox processes may be linked together in electron transport chains.

All organisms use the same pair of pyridine nucleotides as carrier molecules for hydrogen and electrons. Both of these molecules accept hydrogen and electrons in the redox reactions of catabolism and become reduced. The oxidative half-reactions of catabolism generally produce two H^+ and two electrons. The nicotinamide ring can accept two electrons and one H^+ and, since the second H^+ is released into the solution, most redox reactions in biological systems take the form:

$$AH_2 + NAD(P) \rightleftharpoons A + NAD(P)H + H^+$$

NADH and NADPH have different metabolic functions. NADPH generally transfers H^+ and $2e^-$ to oxidised precursors in the reduction reactions of biosynthesis. Therefore, NADPH cycles between catabolic and biosynthetic reactions and serves as a carrier in the same way that ATP serves as the energy carrier. NADH is used almost exclusively in redox processes associated with energy metabolism.

The roles of NADH and NADPH in the overall strategy of metabolism are shown in Fig. 5.17. Fuel molecules, such as glucose, are oxidised in catabolism; they lose electrons and these reducing equivalents are transfered to an environmental acceptor such as oxygen, with concomitant ATP production (see oxidative phosphorylation, Section 5.5.6). However, some reducing equivalents are conserved and re-utilised in the synthesis of cellular components, with the consumption of ATP, as oxidised intermediates are reduced to synthetic precursors with subsequent polymerisation. The pyridine nucleotides thus have roles in both synthetic and energy generation process.

5.5.4. Energy Generation

Energy evolving reactions (exogonic) are coupled as described above to a reaction requiring energy (endogonic). It was noted that energy in the cell may be stored in various forms (storage chemicals, e.g. fats and carbohydrates; and glucose) and converted rapidly to the universal energy chemical ATP, which is then employed to do osmotic, chemical and mechanical work. There are two basic systems by which the synthesis of ATP can be achieved—*substrate level phosphorylation* and *electron transport linked phosphorylation* of which *oxidative phosphorylation* is the more important process.

5.5.5. Substrate Level Phosphorylation

Substrate level phosphorylation refers to those reactions associated with the generation of energy by the transfer of phosphate groups in metabolism, and is exemplified by fermentative metabolism where it is the sole source of energy (e.g. yeast and bacteria growing in anaerobic conditions).

Glycolysis

The sequence of biochemical steps by which glucose is degraded to pyruvic acid is called glycolysis or the Embden–Meyerhof–Parnas (EMP) Pathway, and is shown

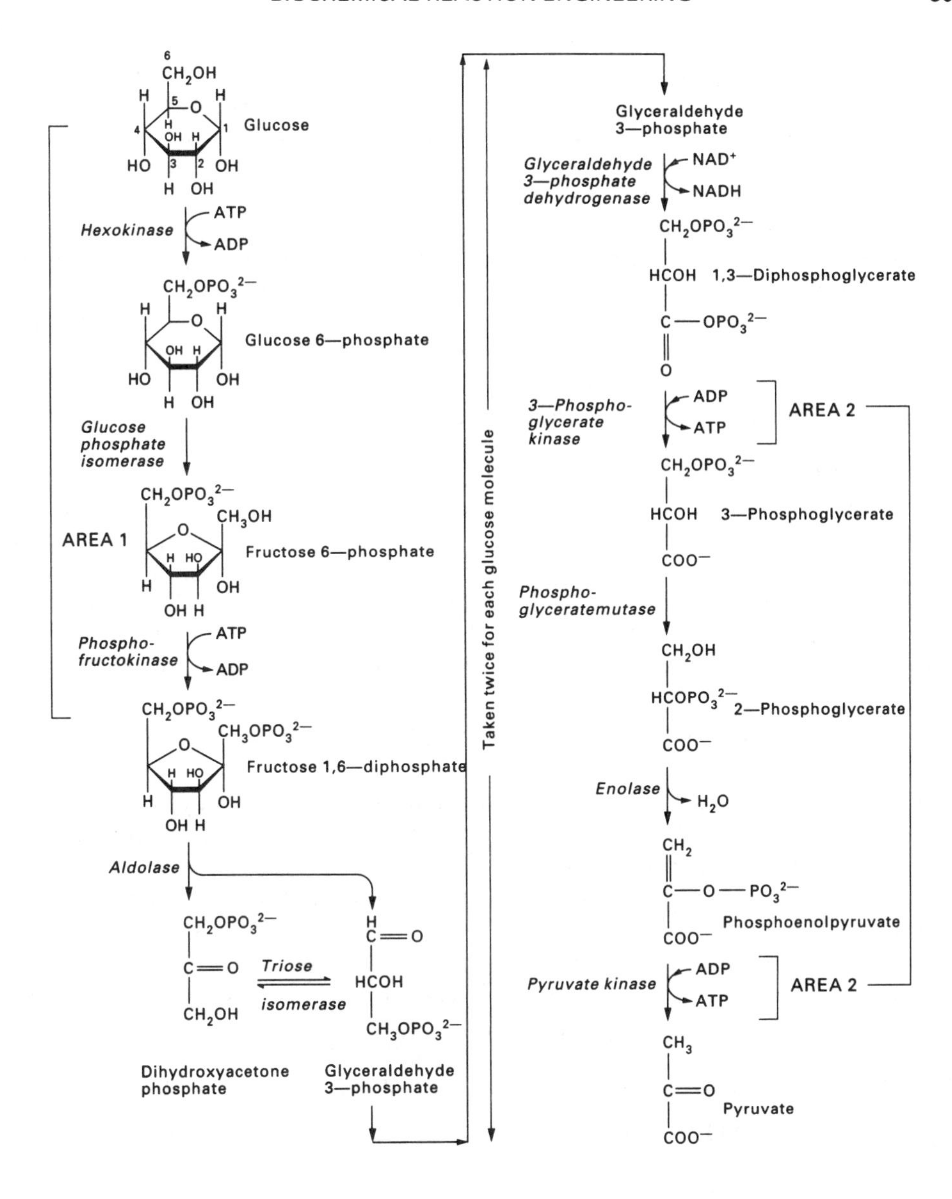

FIG. 5.18. The Embden–Meyerhof–Parnas pathway. The six carbon substrate, (glucose) yields two three carbon intermediates to produce two moles of pyruvate

in Fig. 5.18. This is the best known metabolic pathway and is central to the metabolism of carbohydrates, having roles in energy generation and in providing the carbon skeletons used in biosynthetic reactions. This pathway occurs in both anaerobic and aerobic systems, but the fate of pyruvate produced by this pathway is determined by environmental conditions and the type of organism. For example, pyruvate is completely oxidised to carbon dioxide and water in most aerobic systems, but under anaerobic conditions a variety of organic acids and solvents can be produced.

The net products of the oxidation of glucose, two molecules each of NADH, ATP and pyruvate, are formed by a series of simple reactions, either ring splitting or the transfer of a small group such as phosphate or hydrogen. Two molecules of ATP are hydrolysed to ADP when used to activate glucose to fructose 1, 6 di-phosphate via hexokinase and phosphofructokinase enzymes (area 1 of Fig. 5.20). Once activated, the six-carbon-atom molecule is split into two three-carbon compounds, following which two reactions are effective in transferring phosphate to ADP to form ATP, for each three-carbon moiety catalysed by 3-phosphoglycerate kinase and pyruvate kinase (area 2 of Fig. 5.18). Overall, the pathway results in the formation of a net 2 molecules ATP, as two molecules of ATP are consumed in the activation stage. The other remaining feature of the pathway is the net production of reducing equivalents in the form of NADH from NAD.

The overall stoichiometry of the EMP pathway is thus:

$$C_6H_{12}O_6 + 2\,P_i + 2\,ADP + 2\,NAD^+ \rightarrow 2\,C_3H_4O_3 + 2\,ATP + 2\,NADH + H^+$$

Apart from the energy and reducing power derived from these reactions, the pathway also provides carbon skeletons for the synthesis of cellular structures. The energy provided by this pathway is used to drive the synthetic processes of the cell, and the EMP pathway is the major route for glucose metabolism. However, there are specialised alternative routes by which glucose may be metabolised and the reader is referred to specialised biochemical texts for an account of these[17, 28].

The Metabolic Fates of Pyruvate in Anaerobic Conditions

(a) Lactic acid fermentation. Another feature of the EMP pathway is that there is only a small pool of NAD/NADH within the cell and NADH must be oxidised back to NAD before it can participate in another set of glycolytic reactions. There is a whole series of routes by which this can be achieved. The simplest way is by the reduction of pyruvate to lactate (Fig. 5.19). This type of metabolism is widespread and is encountered in both lactic acid bacteria and human muscle. Thus, the pyruvate and the reducing equivalents in NADH + H⁺ react to produce lactic acid:

$$C_3H_4O_3 + NADH + H^+ \rightarrow C_3H_6O_3 + NAD^+$$

The overall glycolytic sequence from glucose to lactic acid can be written as follows:

$$C_6H_{12}O_6 + 2\,Pi + 2\,ADP \rightarrow 2\,C_3H_6O_3 + 2\,ATP + 2\,H_2O \quad (\Delta G^{0\prime} = -\,135.7\ \text{MJ/kmol})$$

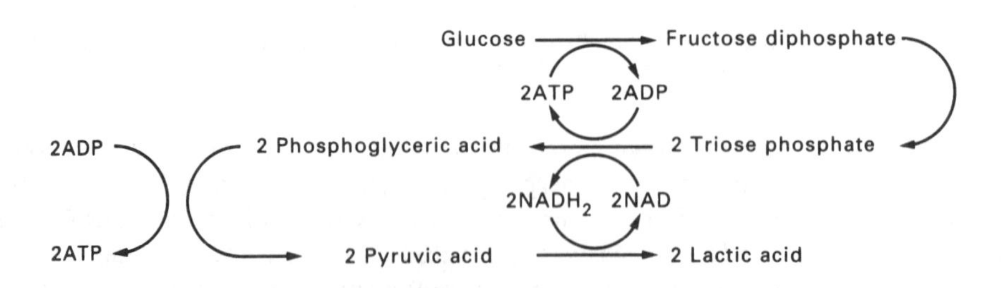

FIG. 5.19. The lactic acid fermentation. The reaction sequence results in two molecules of lactic acid, two molecules of ATP and a balanced redox process

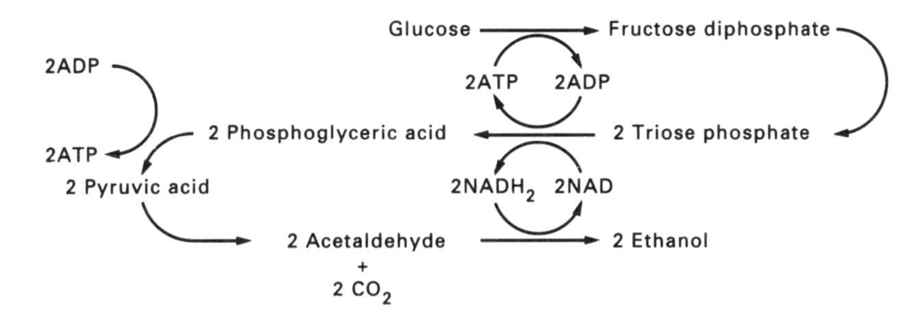

FIG. 5.20. The alcoholic fermentation sequence results in 2 molecules of ethanol from glucose, 2 molecules of ATP and a balanced redox process

The free energy change for glucose breakdown alone is:

$$\text{Glucose} \rightarrow 2\,C_3H_6O_3 \qquad \Delta G^{0\prime} = -\ 196.8\ \text{MJ/kmol}$$

The difference between the free energy changes of the two reactions shows how much of the energy is retained by the molecules in the pathway (i.e. $-\ 196.8 + 135.7 = -61.1$ MJ/kmol) as 2 moles of ATP ($30.55 \times 2 = 61.1$ kJ/mol). The efficiency of free energy transfer from the glucose molecule to the phosphate bonds of ATP is thus $(61.1/196.8) \times 100 = 31$ per cent. Correction of the standard free energy data for the concentrations and the pH found in living cells suggests that this estimate is low and should more realistically be above 50 per cent[17].

(b) Other fermentations. The lactic acid fermentation represents the simplest fermentative metabolism but, economically, the most important anaerobic fermentation is the alcohol fermentation of yeast. The major difference between alcoholic and the homolactic fermentation is that pyruvate is decarboxylated to acetaldehyde in the alcohol fermentation and this is then reduced to ethanol, as illustrated in Fig. 5.20.

Other products can be produced in fermentative bacteria but the central feature of all these pathways is the strict maintenance of the oxidation–reduction balance within the fermentation system. This gives rise to another important tool in assessing fermentation pathways—a mass balance of the substrate and products. The amount of carbon, hydrogen and oxygen in the fermentation products (including cells) must correspond to the quantities in the substrate utilised.

e.g. Homolactic fermentation: $C_6H_{12}O_6 \rightarrow 1.9\,C_3H_6O_3 + \text{cells}$

Alcohol fermentation: $C_6H_{12}O_6 \rightarrow 1.8\,C_2H_5OH + 1.8\,CO_2 + \text{cells}$

The diversity of fermentative metabolism, summarised in Fig. 5.21, results in the production of a variety of products such as neutral compounds (ethanol, isopropanol acetone, butanediol and butanol), organic acids (formic, acetic, propionic, butyric, succinic), and gases (hydrogen and carbon dioxide)[28]. There are various physiological advantages in producing this diverse collection of products but the most important is the generation of additional energy via the acid–phosphate intermediates as shown:

$$\text{Acetyl CoA} + P_i \rightarrow \text{Acetyl} - P + \text{CoA}$$

$$\text{Acetyl} - P + \text{ADP} \rightarrow \text{Acetate} + \text{ATP}$$

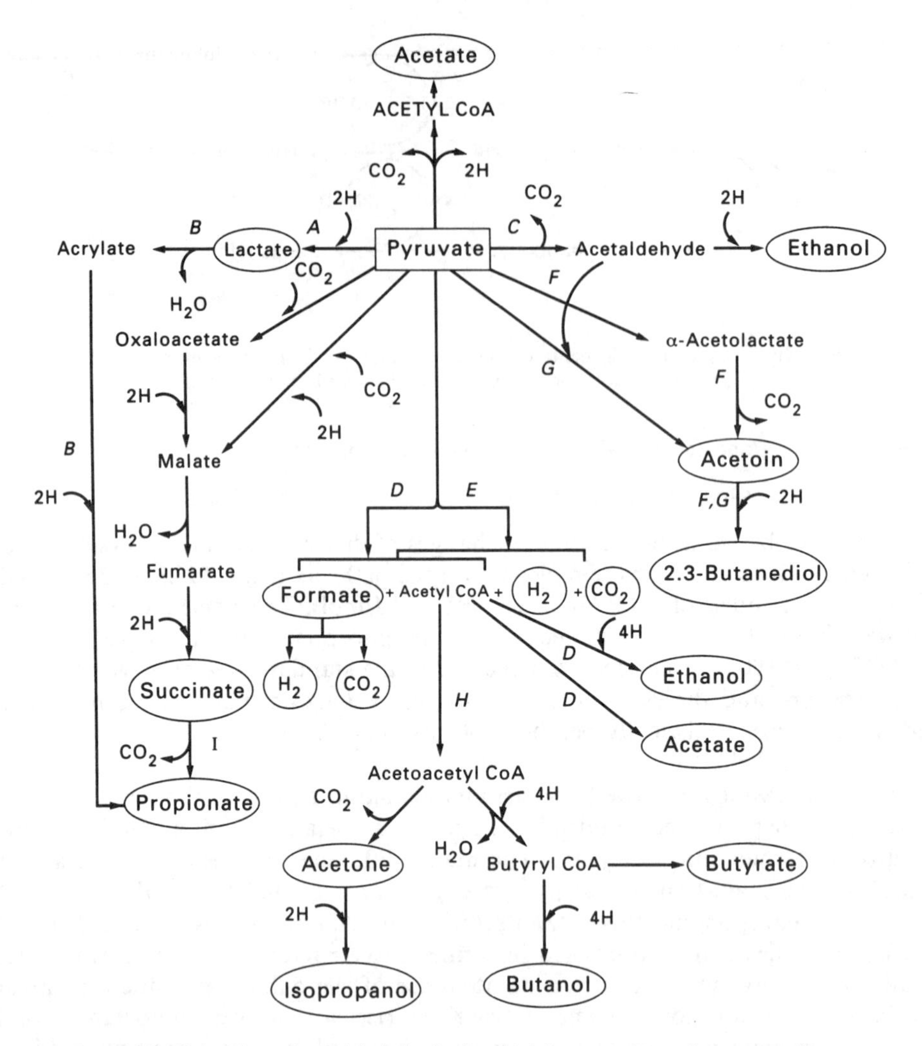

FIG. 5.21. The end-products (circled) of microbial fermentations of pyruvate. Letters indicate the organisms able to perform these reactions. (*A*) Lactic acid bacteria (*Streptococcus, Lactobacillus*); (*B*) *Clostridium propionicum*; (*C*) Yeast, *Zymomonas mobilis, Sarcina ventriculi*; (*D*) *Enterobacteriaceae* (*Coli-aerogenes*); (*E*) *Clostridia*; (*F*) *Aerobacter, Bacillus polymyxa* (*G*) Yeast; (*H*) *Clostridia*; (*I*) Propionic acid bacteria[29]

Although acids are generally more highly oxidised than the original substrates, if the fermentation can at the same time reduce another compound or release its reducing equivalents in the form of hydrogen, then energy-generating acids can be produced. For example, a mixed-acid fermentation of the enteric bacteria produces lactate, acetate, ethanol, formate, hydrogen and carbon dioxide.

(c) Fermentation of other materials. Many other carbohydrates can be fermented, including most sugars (hexoses and pentoses), disaccharides (maltose) and polysaccharides (starch and cellulose). Extracellular enzymes are used to hydrolyse the polymeric materials down to sugar monomers. Once in monomeric forms, hexose

and pentose sugars can be taken up and metabolised in the cell where they eventually feed into the glycolytic pathway. Some of these sugars may be more highly reduced than glucose and accordingly the product spectrum reflects this and the redox balance of the system is maintained.

Anaerobic fermentations are thus capable of producing a wide variety of products from carbohydrates. Energy generation and chemicals production by fermentation are becoming attractive possibilities. At present these fermentations are being investigated for the possible conversion of wastes into useful chemicals, particularly in the western world. In developing countries, especially in the tropics and where there are no fossil fuel reserves, this type of process is being used for the production of fuels and chemicals, e.g. the 'gasohol' programme of Brazil where ethanol is produced from sugar cane. Ethanol can either be used as a motor fuel or it can be dehydrated to ethylene by chemical means for the production of other bulk chemicals. In Europe and America the production of bulk chemicals from biomass is more expensive than by petrochemical routes, but the situation will change as oil becomes more difficult to find and recover. By utilising the anaerobic fermentation pathways, it is possible to produce many of the basic precursor molecules of the modern petrochemical industry and thereby to maintain a sustainable technology for chemicals production.

5.5.6. Aerobic Respiration and Oxidative Phosphorylation

Phosphorylation linked to electron transport is the other major mechanism of generation of ATP in cells. Carbon substrates are completely oxidised in the presence of oxygen to carbon dioxide and reducing equivalents, the latter usually in the form of NADH. NADH is a source of electrons which are then used to generate via electron transport processes before they ultimately reduce oxygen to form water. This process is commonly called *aerobic respiration*.

Respiration is a broad term for the oxidation of organic materials to carbon dioxide. When the electron acceptor is not oxygen they are called *anaerobic respirations*. Table 5.14 illustrates possible types of respiration. Nitrate, sulphate and carbon dioxide are significant electron acceptors in anaerobic respiration and many are of environmental significance.

Aerobic respiration can be subdivided into a number of distinct but coupled processes, such as the carbon flow pathways resulting in the production of carbon dioxide and the oxidation of NADH + H^+ and $FADH_2$ (flavin adenine dinucleotide) to water via the electron transport systems or the respiratory chain.

TABLE 5.14. *Reductants and Oxidants in Bacterial Respirations*

Reductant	Oxidant	Products	Organism
H_2	O_2	H_2O	Hydrogen bacteria
H_2	SO_4^{2-}	$H_2O + S^{2-}$	*Desulfovibrio*
Organic compounds	O_2	$CO_2 + H_2O$	Many bacteria, all plants and animals
NH_3	O_2	$NO_2^- + H_2O$	Nitrifying bacteria
NO_2^-	O_2	$NO_3^- + H_2O$	Nitrifying bacteria
Organic compounds	NO_3^-	$N_2 + CO_2$	Denitrifying bacteria
Fe^{2+}	O_2	Fe^{3+}	*Ferrobacillus* (iron bacteria)
S^{2-}	O_2	$SO_4^{2-} + H_2O$	*Thiobacillus* (sulphur bacteria)

(a) Carbon Flow and the Generation of Reducing Power in Oxidative Phosphorylation

Glucose is converted to pyruvate by the glycolytic pathway (Fig. 5.20). Pyruvate is then oxidised to acetyl–CoA and CO_2 by the pyruvate decarboxylase enzyme:

$$CH_3COCOOH + NAD^+ + CoA\text{–}SH \rightarrow CH_3CO\text{–}S\text{–}CoA + CO_2 + NAD + H^+$$

The transformation of pyruvate to carbon dioxide is achieved by the several steps in a cyclical series of reactions known as the tricarboxylic acid (TCA) cycle. The name of the cycle comes from the first step where acetyl–CoA is condensed with oxaloacetic acid to form citric acid, a tricarboxylic acid. Once citrate is formed the material is converted back to oxaloacetate through a series of 10 reactions, as illustrated in Fig. 5.22, with the net production of 2 molecules of carbon dioxide and reducing equivalents in the form of 4 molecules of $NADH + H^+$ and 1 molecule of $FADH_2$, together with 1 mole of ATP. The overall stoichiometry of the TCA cycle from pyruvate is:

$$CH_3COCOOH + ADP + P_i + 2H_2O + FAD + 4NAD \rightarrow 3CO_2 + ATP + FADH_2 + 4(NADH + H^+)$$

The Respiratory Chain

Carbon from many sources can be successfully oxidised to carbon dioxide, and the reduced electron carriers produced are ultimately oxidised by oxygen to form water. It is possible to pass electrons through a series of proteins which participate in a series of interlinked redox reactions, known collectively as the *electron transport chain* or *system*. Thus, if redox proteins are ordered correctly, electrons will shuttle down the series, as illustrated in Fig. 5.23, giving rise to chain-like transport behaviour. Figure 5.23 shows quite clearly how the term electron transport chain originated. As an electron is passed down through the redox carriers, the first is reduced by electrons and then reoxidised as the electron is passed from it to the next carrier. Several carriers may be linked together, each one more highly oxidised than the previous one, so that the electron is passed along a gradient of redox potential to a terminal electron acceptor which, in the case of aerobic respiration, is oxygen. (Fig. 5.24.)

Electron Transport Processes Linked to Phosphorylation

Energy can be generated by electron transport processes, the operation of which can be understood by looking at the orientation and the location of the electron transport systems. The electron transport systems are sited in the cell membranes of bacteria or in the membranes of *mitochondria*, specialised organelles, in yeast and in animal cells (eukaryotes). The redox carriers of the electron transport chain straddle the membrane and, as electrons are passed down the chain, they also transport protons across the membrane, as shown in Fig. 5.25. The result is the passage of protons from the inside to the outside of the membrane and the creation of a proton or pH gradient. The pH gradient represents a source of potential energy (a concentration gradient and a gradient of charge) which can now be utilised to generate ATP. This generation of ATP is achieved by channelling the transport of the protons

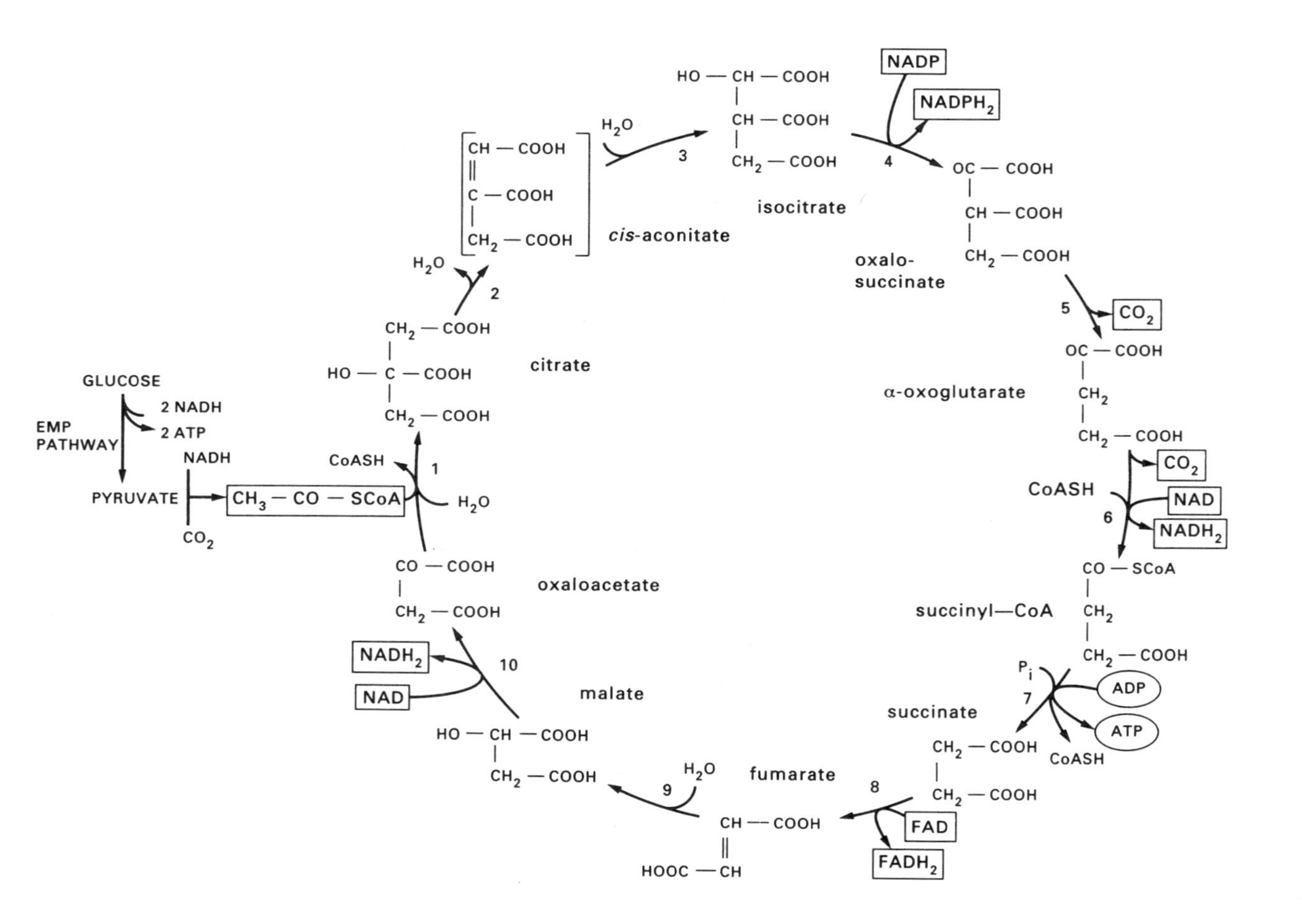

FIG. 5.22. Oxidation of acetyl–CoA via the tricarboxylic acid (TCA) cycle. Individual enzymes of the pathway are marked. 1, citrate synthase; 2 and 3, cis-aconitate hydratase; 4 and 5, isocitrate dehydrogenase; 6, α-oxo glutarate dehydrogenas; 7, succinate thiokinase; 8, succinate dehydrogenase; 9, fumarase; 10, malate dehydroganase[28]

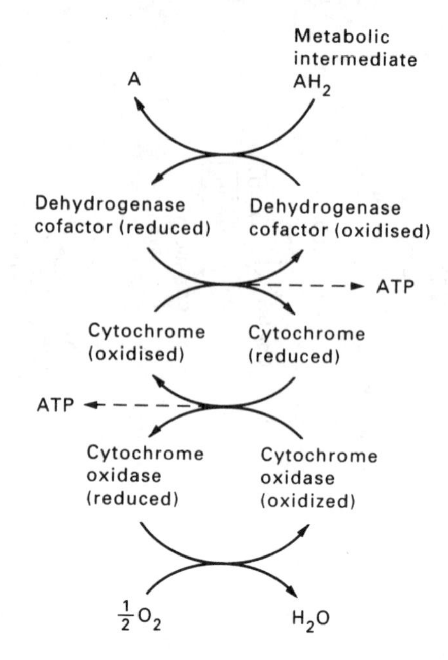

FIG. 5.23. Simplified scheme showing route taken by electrons. At each step the reduced form of one carrier reduces (or passes electrons to) the oxidised form of the next

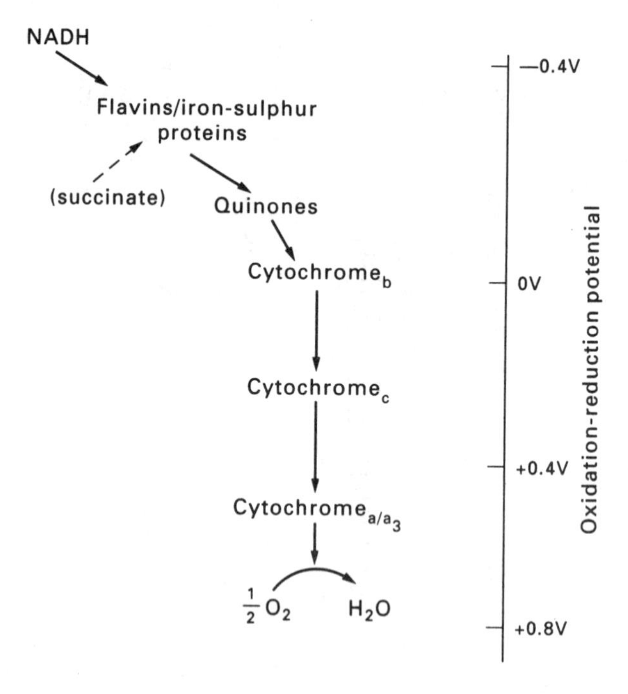

FIG. 5.24. The passage of electrons through an electron transport chain involving the passing of electrons *down* a potential gradient

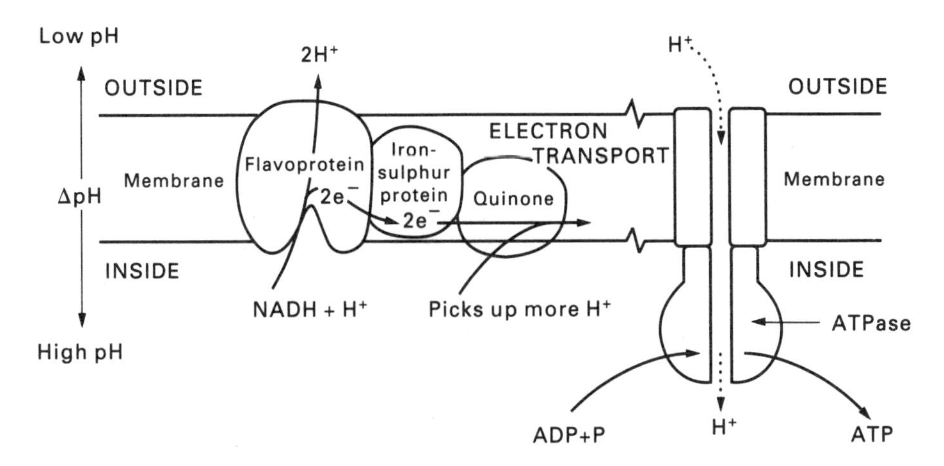

FIG. 5.25. Energy generation linked to electron transport process

across the membrane from the outside back into the inside of the cell through a specific protein pore called ATPase. The energy released by this process is coupled to phosphorylation of ADP + P_i to ATP. The high energy electron released from oxidation of carbohydrate can thus drive ADP phosphorylation, as illustrated in Fig. 5.25, but the precise biochemical coupling of proton transport and ATP generation is yet to be elucidated.

The nature of this process means that the stoichiometry of the these reactions is variable, although it is thought that the process will produce a maximum of 3 moles of ATP from the oxidation of 1 mole NADH + H^+ and 2 moles ATP from 1 mole $FADH_2$.
i.e.

$$NADH + H^+ + 1/2\ O_2\ 3\ ADP + 3\ P_i \rightarrow NAD + 4\ H_2O + 3\ ATP$$

$$FADH_2 + 1/2\ O_2\ 2\,ADP + 2\,P_i \rightarrow FAD + 4\,H_2O + 2\,ATP$$

Thus from one mole of glucose the following numbers of moles of product are generated:

Process	*Product* NADH + H^+	$FADH_2$	ATP
Glycolysis	2		2
Oxidation of two pyruvates	8	2	2
Oxidation of NADH + H^+	− 10	− 2	34
	0	0	38

i.e., overall 38 moles of ATP may be generated from the metabolism of 1 mole of glucose. The overall process is summarised in the Fig. 5.26.

Energy Efficiency of Aerobic Respiration (Oxidative Phosphorylation)

Aerobic respiration potentially makes available much more energy for use by the

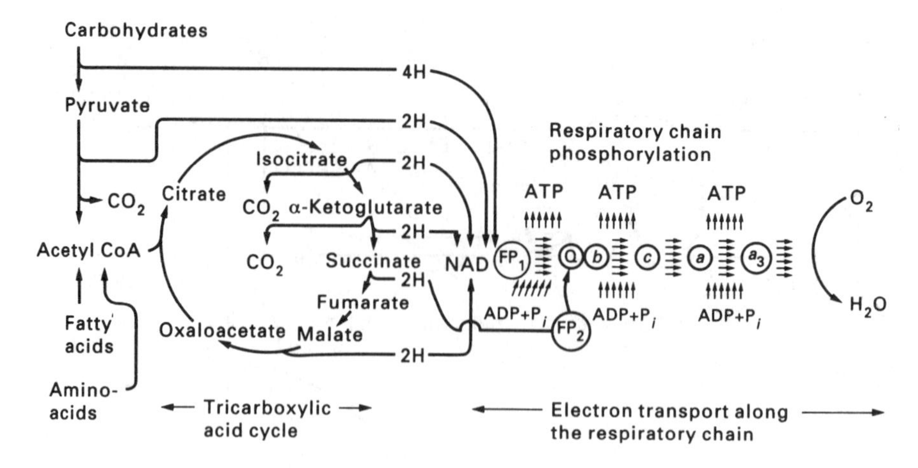

FIG. 5.26. Summary of the potential energy released from oxidation of carbon materials to carbon dioxide and water via the tricarboxylic acid cycle and electronc transport in the respiratory chain[17]

cell than glycolysis (19 times). Now, for the oxidation of glucose:

$$C_6H_{12}O_6 + 6\,O_2 \rightarrow 6\,CO_2 + 6\,H_2O \qquad \Delta G^{0\prime} = -\,2871\ \text{MJ/kmol}$$

whereas in glycolysis and respiration:

$$C_6H_{12}O_6 + 38\,\text{ADP} + 38\,P_i \rightarrow 6\,CO_2 + 38\,\text{ATP} + 44\,H_2O$$

Since ATP hydrolysis has a standard free energy change of – 30.55 MJ/kmol the free energy change of the reaction is approximately:

$$\Delta G^{0\prime} = 38 \times 30.55 = -1161\ \text{MJ/kmol}$$

Therefore, the energy capture efficiency $= \dfrac{1161}{2871} \times 100 = 40$ per cent.

Again, corrections which take pH and the non-standard concentration of reactants into account suggest that the energy capture is in fact greater than 70 per cent, with most of the remaining energy being dissipated as heat.

Energy Capture and Electron Transport Processes

One of the most significant problems associated with electron-transport mediated processes is that they are not directly linked to the energy generation process. The amount of energy generated by these processes is variable, due primarily to leakage of the protons against a pH gradient over the membrane. Alternatively, some electron transport processes may avoid or nullify the transport of protons cross the membrane and thus reduce the number of protons transported and reduce the membrane potential. Energised membranes also do work connected with the biosynthetic process by maintaining *homeostasis* (meaning self-maintenance of a constant environment) within cells. As a consequence, the yield of cell material from a substrate is variable and depends on a number of factors, including growth rate, environmental conditions and the organism employed.

If energy is generated solely by substrate-level phosphorylation, as with anaerobic fermentative metabolism of bacteria and yeast, then the yield is more tightly linked to the amount of energy generated. Generally 10 to 12 kg of cell dry matter can be synthesised per kmol of ATP generated in metabolism.

5.5.7. Photosynthesis

Glucose is a major source of energy for heterotrophs; however, the main source of energy on the planet is sunlight. This ultimate source of energy can be captured by organisms (e.g. plants, algae and bacteria) in the process called photosynthesis. Electromagnetic radiation in the wavelength range 300–800 nm (visible light) drives the fixation of carbon dioxide from the atmosphere into glucose. As is common with synthetic processes, the fixation requires reducing power, NADPH (CO_2 is very highly oxidised compared with glucose $C_6H_{12}O_6$) and chemical energy (ATP). It can therefore be considered as an energy capture processes and is, in effect, the reverse of respiration. It also completes the cycling of carbon material in the environment; carbon dioxide, water and light are utilised to reduce organic materials and oxygen, while glucose and oxygen are then oxidised back to carbon dioxide and water.

An Overview of the Photosynthetic Process

The radiant energy of the sun is captured and converted to chemical energy by photosynthetic organisms by the overall stoichiometry:

$$6\,CO_2 + 6\,H_2O \xrightarrow{\text{light}} C_6H_{12}O_6 + O_2$$

The photosynthetic reactions can be split into two major groups, the light and the dark reactions. The *light* reactions are those involved with light energy captured in a variety of pigments including chlorophyll. This light energy is used to split water and to create reducing power in the form of NADPH and energy (ATP), with the simultaneous formation of molecular oxygen.

$$H_2O + NADP + HPO_4^{=} + ADP \rightarrow O_2 + ATP + NADPH + H^+$$

The resultant NADPH and ATP provide the reducing power and free energy to drive the reduction of carbon dioxide (*the dark reactions*) via the pentose phosphate pathway (or Calvin cycle[17]) and lead ultimately to the synthesis of glucose according to the overall stoichiometry below.

$$6\,CO_2 + 12(NADPH + H^+) + 18\,ATP \rightarrow C_6H_{12}O_6 + 12\,NADP + 18\,ADP + P_i + 6\,H_2O$$

Both the 'light' and 'dark' reactions occur in the *chloroplast*, a specialised membrane-bound organelle. For further details of this process the reader is referred to a general biochemistry textbook such as that by LENINGER[17].

5.6. STRAIN IMPROVEMENT METHODS

The improvement of the biocatalyst, in the form of either whole cell or isolated enzyme, can be achieved by a change in the structure of the DNA of the cell. This can be effected either by random selection of mutants or by a concerted development of a strain by careful genetic manipulation. There is a good analogy between modern information technology using computers and data storage and the storage of biological information, its utilisation and manipulation. Indeed some of the terms used in computing are derived from biological processes; for example, computer viruses and their ability to infect machines and programs.

The manipulation of the stored information in the DNA can result in the creation of new strains of organisms with new or enhanced capabilities, and this may be achieved by a variety of methods including mutation and *in vivo* and *in vitro* recombination (recombinant DNA technology). 'Genetic engineering' covers the manipulation of DNA in the laboratory and can be utilised for the improvement of traditional processes and provide the enabling technology for the generation of new ones; for example, the production of human insulin by bacteria.

The structure of nucleic acids is discussed in Appendix 5.2, while Appendix 5.6 gives details of how information can be stored in the cell in the form of DNA, genes, plasmids, and of the mechanisms of protein synthesis.

5.6.1. Mutation and Mutagenesis

Chromosomes are by no means inert, stable structures, holding dead information in storage. They are constantly undergoing changes of various kinds, some of which are involved in genetic recombination. However, other changes are accidental and random in nature and, if not repaired, result in a permanent genetic mutation. In Appendices 5.2 and 5.6, some of these repair mechanisms are discussed (e.g. DNA repair), but the results of these phenomena will now be discussed further.

DNA is subject to damage by electromagnetic radiation or reactive chemicals, many of which have been introduced into the environment as a by-product of industrial activity. Some chemicals may not be dangerous *per se*, but may be metabolised to products which are dangerous. There are three major classes of such chemicals[17].

(1) De-aminating agents, such as nitrous acid or compounds that can be metabolised to nitrous acid or nitrites. These reagents are capable of removing amino groups from cytosine, adenine and guanine.
(2) Alkylating agents that can cause methylation and ethylation of bases. Guanine, for example, reacts with dimethylsulphate to form o-methylguanine.
(3) Analogues of bases which mimic the normal bases present in DNA. Thus, many aromatic and polyaromatic compounds produced by industry are potential mutagens.

Although DNA proof-reading and repair mechanisms within cells are very effective, inevitably some errors in replication remain uncorrected, and as a result become perpetuated in the DNA of the organism. Such permanent changes are called *mutations*.

Mutations caused by the replacement of a single base with an incorrect one are called *substitution mutations*. Such mutations will have the effect of changing one codon-triplet and, depending on the code, this may alter an amino-acid in the polypeptide chain. Examples of hypothetical mutations are shown in Table 5.15. *Silent mutations* are changes in the base sequence which cause no change to the functionality product, e.g. an enzyme activity. Other mutations are so catastrophic that the organism can no longer function. If, for example, the enzyme is on a central pathway, and becomes critically mutated, this process is termed a *lethal mutation*. *Leaky mutants*, are those in which the alteration in amino-acids changes the kinetic characteristics of the enzyme so that, for example, it does not function very well or, conversely (but rarely), it may be an improvement on the original enzyme.

TABLE 5.15. *Effects of Some Hypothetical Single-Base Mutations on the Biological Activity of the Resulting Protein Products*

Mutation		Wild type (unmutated DNA triplet)	Mutated triplet
A single-base substitution causing no change in the amino-acid sequence; a silent mutation	DNA template	(3′)-GGT-(5′)	-GGA-
	RNA codon	(5′)-CCA-(3′)	-CCU-
	Amino acid	-Pro-	-Pro-
A single-base mutation resulting in an amino-acid change that may not alter the biological activity of the protein because the amino-acid replacement is in a noncritical position and also resembles the normal amino-acid; also a silent mutation		(3′)-TAA-(5′)	-GGA-
		(5′)-AUU-(3′)	-CUU-
		-Ile-	-Leu-
A lethal single-base mutation in which a serine residue essential for enzyme activity is replaced by phenylalanine to give an enzymatically inactive product		(3′)-AGA-(5′)	-AAA-
		(5′)-UCU-(3′)	-UUU-
		-Ser-	-Phe-
A leaky mutation in which the amino-acid change results in a protein that retains at least some of its normal activity		(3′)-CGT-(5′)	-CCT-
		(5′)-GCA-(3′)	-GGA-
		-Ala-	-Gly-
A hypothetical beneficial mutation, in which the amino-acid replacement yields a protein with improved biological activity, giving the mutated organism an advantage: it is not possible to predict advantageous amino-acid replacements		(3′)-TTC-(5′)	-TCC-
		(5′)-AAG-(3′)	-AGG-
		-Lys-	-Arg-

Substitution mutation is only one type; *insertion* and *deletion mutations* are much more numerous and more lethal. Insertion or deletion of nucleotides causes *frame shift* mutations, in which a base pair is inserted in or deleted from a gene, giving rise to a more extensive type of mutation damage. The consequence of such a mutation is the disruption of the linear order of the codons in the DNA and thence of the amino-acid sequence for which it codes. The disruption begins at the site where the base has been gained or lost, and the result of such a mutation is a garbled protein sequence. For example, consider the analogy of words in a sentence. Using following sentence of three letter words:

THE CAT WAS RED AND BAD

On deletion of the W, it would read:

THE CAT ASR EDA NDB ADX

and the meaning is lost. Sometimes, one frame-shift mutation can be cancelled out by a second one; e.g. when the first adds a base while the second takes another base out. Such a mutation is called a *suppressor mutation* as it suppresses the first mutation. Frame-shift mutations are caused by molecules that intercalate between two adjacent base pairs and, in effect, add an additional base to the sequence. Thus, on replication, these molecules cause an additional base to be added to the sequence; for example, acridine is a mutagen that intercalates with DNA and causes such a phenomenon.

Mutation is a rare event as far as an individual human being is concerned; the probability that a mutation will occur in any one protein is about 1 in 10^9. For a human cell it is probably 1 in 10^5 (calculated from the incidence of haemophilia), but mutagenic agents in effect increase the mutation frequency. Furthermore, a specific mutation in the enzymes involved in DNA repair will also increase the level of mutation.

Statistical evidence strongly suggests that continued exposure of human beings to certain chemical agents, especially in the work place, results in an increased incidence of cancer. For example, workers involved in the production of naphylamines have a much higher incidence of bladder cancer than the general population and, interestingly, many cancers are the result of loss of the growth control mechanisms of the cell. Dyes and many other aromatic compounds, exhaust gases, medicines and cosmetics all contain chemicals that are possibly carcinogenic. To protect humans from exposure to such chemicals, extensive animal tests have had to be developed. This is now a significant part of the cost (£100,000 to millions) of developing a drug and takes a considerable time (2–3 years).

On the positive side, the use of mutagens to increase the degree of variation in a culture, followed by selecting and screening individuals with enhanced capabilities, has been one of the most successful methods in improving microbial strains. A classic case is the development of penicillin G fermentation where the productivity of culture strains was improved by a factor of more than a thousand over a period of about 20 years, using mutant screening programmes.

5.6.2. Genetic Recombination in Bacteria

There are several mechanisms by which genetic recombination occurs in nature. Genes or sets of genes can also be recombined in the test tube to produce new combinations that do not normally occur in nature. It is possible to isolate the genes necessary for the formation of a specific protein and to splice them with other forms of DNA to yield these new combinations. Such artificial recombinant DNA's are extremely useful as tools in genetic research and in biotechnology. The development of methods for isolating and splicing genes into recombinations has been a major scientific and technological advance.

Biological exchange of genes to form a modified chromosome is called *genetic recombination* and can occur between DNA molecules that are similar. Their similarity allows the molecules to associate closely with one another and to exchange genetic material. Recombination occurs in a many situations and has been studied in great depth for bacteria. A summary of the basic types of recombination in bacteria is shown in Fig. 5.27. All require the new DNA to enter the recipient cell before the recombination event can take place. The simplest is called *transformation* and involves the uptake by the cell of a small piece of DNA from the environment. Once absorbed it may then undergo recombination with the chromosome. *Transduction*

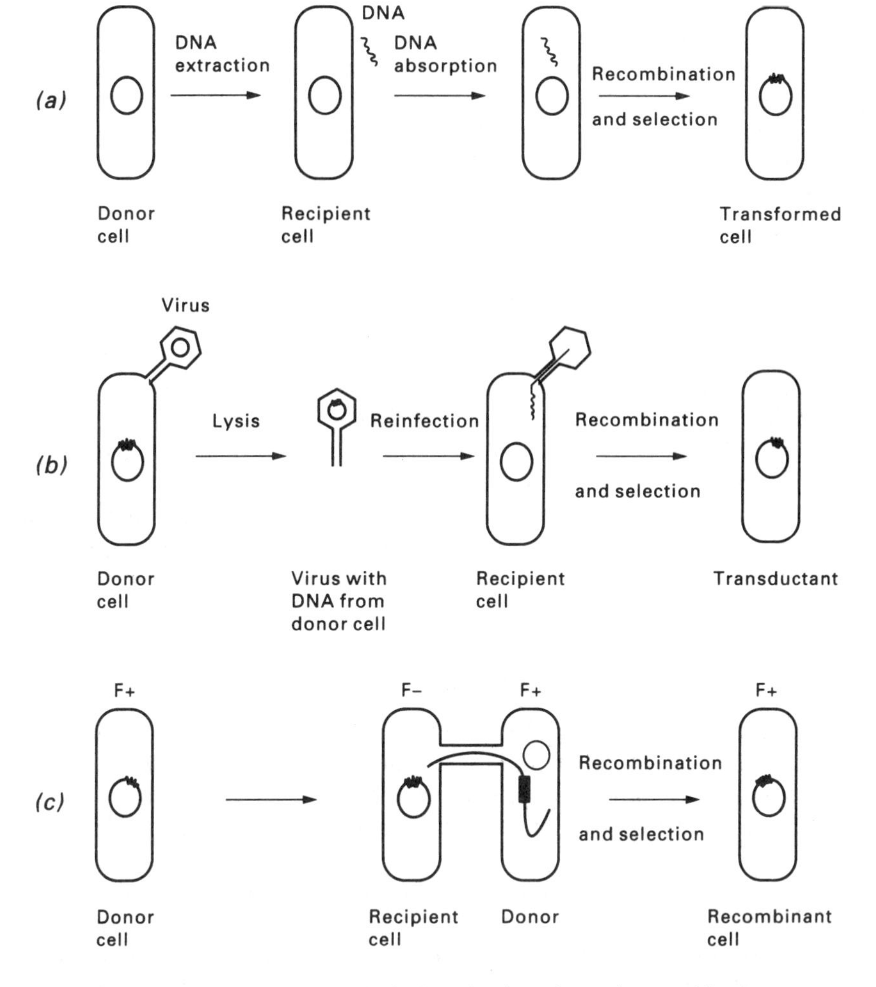

FIG. 5.27. Summary of the principal mechanism of genetic recombination.

(*a*) Transformation (the absorption of DNA from the environment);
(*b*) Transduction (the viral-mediated transfer of DNA from one cell to another);
(*c*) Conjugation (the transfer of DNA between touching cells and involving specialised plasmid)

involves the virus-mediated transfer of DNA from one cell to the other. The bacterial virus is able to take up small fragments of donor cell DNA during its reproductive cycle. If the modified virus is then introduced to another culture, then DNA may be transferred into the recipient culture. The final form of bacterial recombination is the concerted passage of material between touching cells. DNA from one organism (designated F^+) containing a sex plasmid is able to pass the replicated chromosome from one organism to another which does not contain the plasmid (F^- strain). This process occurs in a linear fashion and, if it is allowed to proceed for long enough, a complete transfer of material will take place. This process, called *conjugation*, has been a very useful tool for the mapping of genes on the chromosome.

5.6.3. Genetic Engineering

The increase in knowledge about the structure and properties of DNA, the discovery of the genetic systems of bacteria and the development of practical methods in the manipulation of DNA, have led to the development of gene cloning (clone is derived from the greek *klon* = *cutting*, as in plant propagation). This concept has been extended to the cellular and molecular levels so that material is either derived from a single cell or from a piece of DNA. Another term used to describe these methods is recombinant DNA (rDNA) technology.

In pure research, gene cloning has very important applications in the understanding of gene function (e.g. control of gene expression or protein synthesis); control of DNA replication and repair; obtaining DNA sequence information; determining structure–function relationships in proteins (protein engineering).

In applied research and industry, gene cloning offers a mechanism for producing many valuable products from animals, employing some of the more appropriate techniques used for the production of bacteria. Typical of the pre-molecular biology products are antibiotics, organic acids, amino-acids and solvents. The introduction of the techniques of molecular biology to industrial biology has led to the development of processes for animal and plant proteins using bacteria and yeast which are more easily used on a large scale than the original animal and plant cells. Examples of such products are hormones and specialised proteins such as vaccines and antibodies.

Physical and Biochemical Techniques used in the Manipulation of DNA

Following the discovery of DNA structure and the genetic code, various associated techniques involving the manipulation and separation of DNA fragments have been developed. Several of the physical methods are useful in nucleic-acid biochemistry; these include, centrifugation, electrophoresis and chromatographic separation of DNA fragments.

The production of DNA fragments can be achieved by specific enzymatic hydrolysis. Enzymes that hydrolyse DNA, usually at the phosphodiester bonds, are collectively known as nucleases. Those specific to RNA are ribonucleases (RNAases) and those specific to DNA are deoxyribonucleases (DNAases). Nucleases are also classified according to how they attack the polymer chain. Exonucleases attack the polymer chains from the end; some are specific to the ends of the polymer (See Appendix 5.2), while others, endonucleases, attack the polymer from within the chain. There is a very important sub-group of the endonucleases, the *restriction endonucleases* which are very useful reagents in the study of DNA. Although their normal physiological functions involve DNA replication, DNA repair and DNA recombination, they can be used in the laboratory to cut DNA selectively, yielding several products. A single cut in a double strand called a *NICK*; cutting several bases from a strand is called a *GAP*; if both strands are cut, then the result is called a *BREAK*.

5.6.4. Recombinant DNA Technology

The development of the means to isolate and splice genes into new combinations was a major scientific and technological advance[30, 31]. Figure 5.28 outlines how an animal gene may be transfered to a bacterium. Recombinant DNA (rDNA)

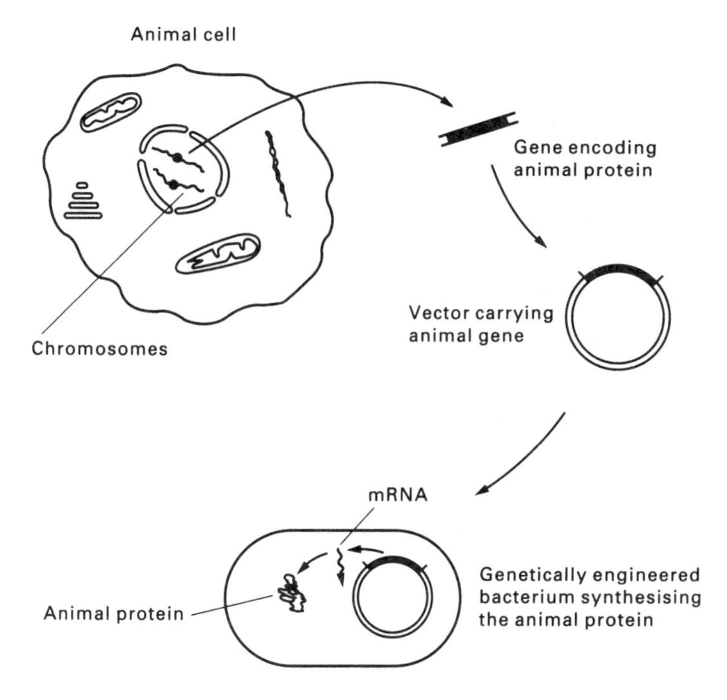

FIG. 5.28. A possible scheme for the production of an animal protein by a bacterium

technology follows a basic outline (Fig. 5.29). A fragment of DNA containing the gene to be cloned is isolated, and then inserted into a circular DNA molecule called a *vector* to form a recombinant DNA molecule or a *chimaera* (an organism made up of two distinct pieces of genetic material; in Greek mythology a chimaera was a fire-spouting monster with a lion's head, a goat's body and a serpent's tail). The vector acts as a vehicle that transports genes into the host cell, which is usually a plasmid or a virus. Within the cell the vector multiplies, producing numerous identical copies of itself including the cloned gene. As the organism multiplies, further vector replication takes place. After a large number of cell divisions, a colony or clone of identical host cells, each containing one or more copies of the cloned DNA, is produced.

There are several specific techniques for:

(a) *Cutting and splicing DNA*. The discovery of restriction endonucleases gave the first clue as to how genes can be recombined in the laboratory. The analogy of film or tape editing can be used to illustrate the principle of these techniques in which a story can be retold in a number different ways. A piece of DNA is cut at a specific site with a restriction endonuclease to leave a staggered two-strand break to give 'sticky' or cohesive ends, which enables DNA from two sources to be mixed and joined together. Other enzymes may be used to produce a sticky end from a blunt end, as shown for example in Fig. 5.29.

(b) *Introduction of DNA into living cells*. The next step in developing recombinant DNA technology was finding a means of introducing foreign DNA into the host cell. This is now done by using either plasmids (via transformation) or viruses (via transduction). Plasmids (see Appendix 5.6) are small circular, autonomous DNA

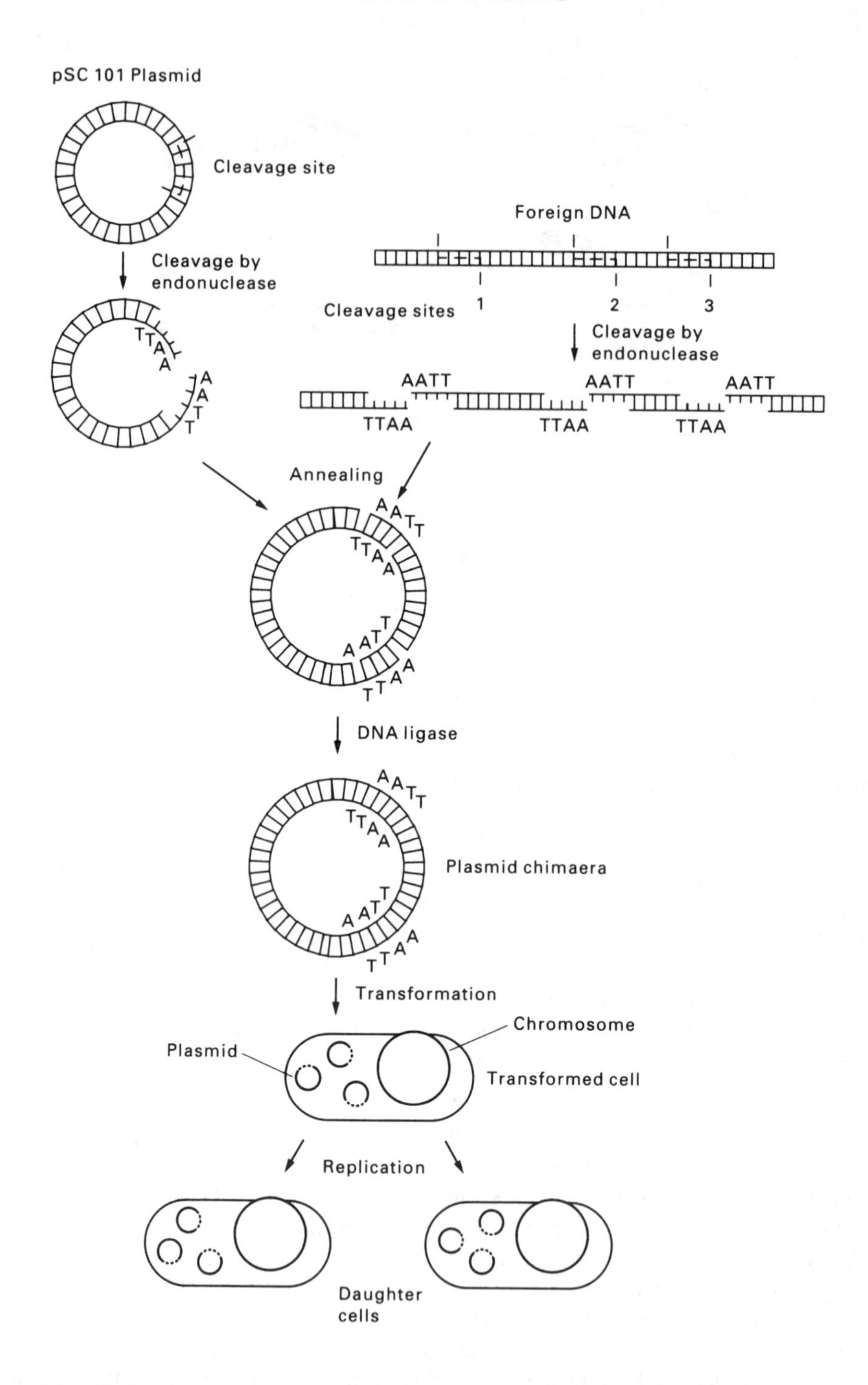

FIG. 5.29. The basic steps of recombinant DNA technology. Using this sequence of steps, a fragment of foreign DNA can be annealed and sealed into a plasmid which in turn can be introduced into a living cell where the foreign DNA is replicated and expressed

units which can contain 2000–1,000,000 bases; the smaller plasmids may have 20 or so copies per cell. In each plasmid several genes are replicated, transcribed and translated independently, but simultaneously, with chromosomal genes. They are readily isolated from chromosomal DNA as they are smaller and less dense. They have two important properties:

(i) They are able to pass from one cell to another, indeed from one species to another. For example, *Salmonella typhimurium* can acquire permanent resistance to certain antibiotics, such as penicillin, when it is mixed with a strain of *E. coli* cells resistant to penicillin;

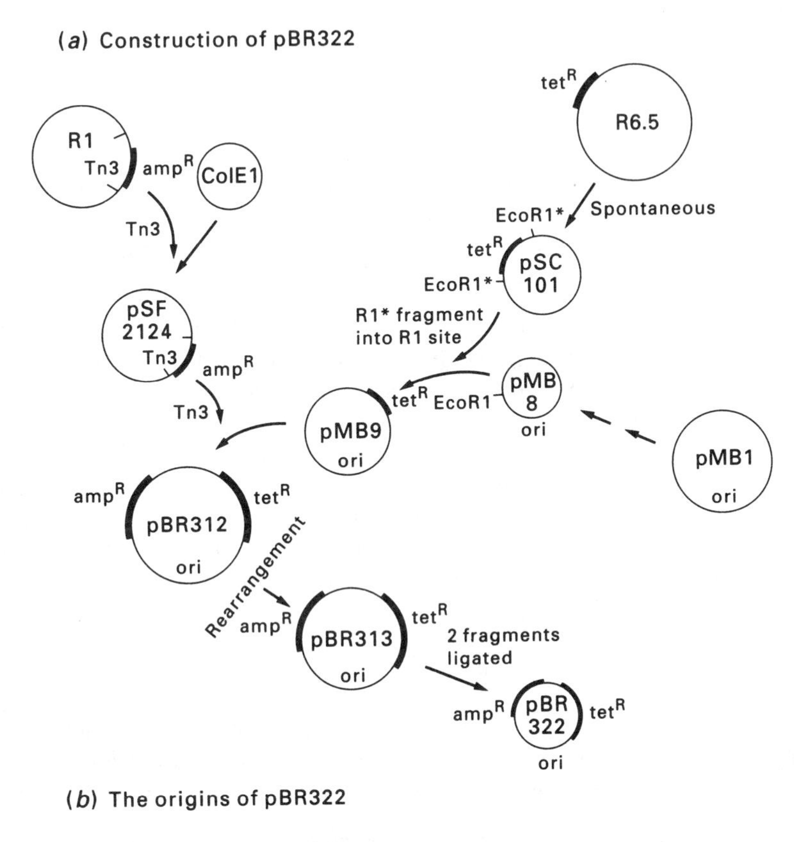

(b) The origins of pBR322

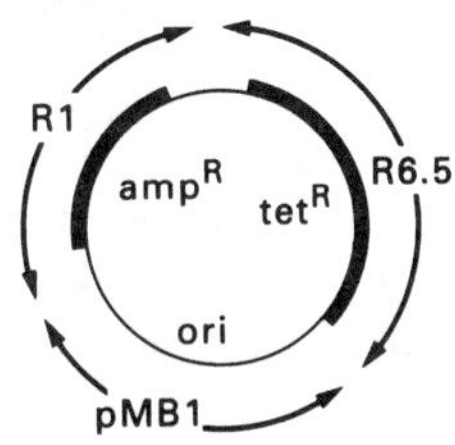

FIG. 5.30. The pedigree of pBR322

(*a*) Manipulation involved in the construction of pBR322

(*b*) Summary of the origins of pBR322[30]

(ii) Foreign genes can be spliced into the plasmids which can then be carried as passengers into *E. coli* cells and become part of the host cell *genome* (meaning genetic complement).

A good cloning vehicle must generally have a number of features:

(i) It must be able to replicate in the foreign organism;
(ii) It must be small, so as to ease purification;
(iii) It must occur at high copy numbers (e.g. 10–50 copies of the plasmid per cell).

Considerable effort has been spent on developing the vectors (for example pBR322) so that they incorporate desirable features. These include:

(i) Ease of purification;
(ii) Enzyme manipulation (i.e. specific manipulation of DNA by restriction endonucleases);
(iii) Antibiotic resistances for ease of selection (i.e. if pieces of DNA are added to the sites within a resistance gene, they become sensitive to the antibiotic—this is called *insertional inactivation*);
(iv) High copy number of the plasmid is possible and can reach a 1000 copies under the right conditions.

The plasmid pBR322 has such desirable features (Fig. 5.30) and is a good example of how such plasmids are developed. It has 4363 base-pairs and is thus small and, when pieces of cloned DNA have been added, the molecule is still small enough to handle. It has antibiotic resistances to two materials, ampicillin (amp^r) and tetracyclin (tet^r), and has many sites where restriction enzymes can cut the molecule. Furthermore, it is produced in high copy numbers. Such plasmid cloning vectors do not arise by chance but are themselves products of gene manipulation. The construction of pBR322 is shown in Fig. 5.30. It is constructed from three natural plasmids isolated from drug-resistant organisms and then manipulated to the appropriate size and configuration.

An Example of Cloning using a Plasmid

The plasmid can be used, not only to transfer and express foreign genes in *E. coli* as with pBR322, but also to study the control of gene expression etc. Figures 5.28 and 5.29 show some of the procedures, each of which involves a number of steps:

(1) *Isolation of genes and the preparation of cloned DNA*. There are two general methods for obtaining cloned DNA:

(i) The 'shotgun' approach where the complete cell DNA is treated with endonucleases which generate staggered ends. These fragments are then combined with plasmids which have been 'opened' by the same exonuclease. The product is an exceedingly complex mixture of recombinant plasmid-containing organisms which are then screened for an organism with the correct plasmid composition.
(ii) Specific methods used to select the required pieces of DNA by isolating its complementary mRNA. In rapidly growing cells, mRNA can make up a large proportion of cell mass, and it is possible to isolate the mRNA and

use it as a template for selection of DNA fragments. Several other methods are used, but for the sake of simplicity these will not be considered here.

(2) *Construction of the gene-bearing vector*. Once the clone of DNA has been isolated, this is put into a vector (in this case a plasmid) which will transfer it from the host to the recipient organism. Again, by the use of restriction enzymes to cut the plasmid vector and the cloned DNA, the cloned DNA can now be inserted into the vector.
(3) *Insertion of the loaded plasmid*. The plasmid is then introduced into the host in one of a number of ways, either in conjugation, or more commonly by transformation using the purified plasmid and the desired host.
(4) *Expression of cloned genes*. Once inside the microbe, the genes are then expressed by the normal protein synthesis equipment of the cell. The expression or production of the cloned material can be a significant problem as the promoters and operators may not be compatible with the control systems of these cells. Therefore, careful manipulation of these control sequences can enhance the production of the protein required. Similarly, the subsequent folding of the protein and/or excretion of the protein may well not be accomplished, and this is a symptom of a basic incompatibility between the host cell and the production of the foreign protein.

Enhanced gene expression can be achieved by the addition of a promoter to the cloned gene. For example, the *lac promoter* may be added upstream of the protein DNA sequence, giving rise to the expression of the protein in the presence of lactose and in the absence of glucose. (cf. operon hypothesis Section 5.7).

5.6.5. Genetically Engineered Products

An example of the use of genetic engineering is the production of somatostatin. This is a growth hormone which regulates the secretion of insulin from the pancreas and is a polypeptide with 14 amino-acid residues. As the gene is so small, it can be synthesised chemically and then joined to the end of the β-galactosidase gene. Thus, when the β-galactosidase gene is expressed, the protein chimaera (or hybrid) is produced and excreted into the medium and, after selective enzymatic hydrolysis, it yields a biologically active somatostatin molecule.

The engineering of eukaryotic genes in eukaryotic organisms (yeast) is still in its infancy and its application is not as well developed as that of the bacterial systems. This is due to the increased complexity encountered in the structure and function of eukaryotic chromosomes. However, many advances have been made in the development of this system, particularly for the production of materials that require post-translational modification of the protein, and where other additional materials must be added before a fully functional molecule is produced (e.g. glycoproteins).

Many chemicals important to genetic research have been produced at high levels, e.g. DNA ligase production in *E. coli* can be enhanced several hundred fold. Other enzymes can also be produced for industrial use and enhanced degradative ability has been generated; for instance, enhanced petroleum degraders can be added to oil spills to accelerate clean-up operations. However, it is in the production of medically related compounds that this technology has been most successfully applied as, for example, in the production of insulin.

Historically, insulin has been produced by extracting it from the pancreas of pigs and oxen, but the protein is not precisely the same as that found in humans and, although it functions in the same manner, its use can produce unwelcome side effects. Insulin produced from cloned DNA is identical to human insulin and is consequently considered safer. Another major advantage is that the production of insulin is not then limited by the number of pigs slaughtered. Using *E. coli* or yeasts, the process can be far more easily controlled and matched to demand so that this, and many other hormones, are now produced by this method.

Other proteins which are being made using genetically engineered organisms include vaccines for animals and humans. Table 5.16 shows some examples of products that have used rDNA technology.

TABLE 5.16. *Products of Recombinant DNA Technology*

Substance	Application
Blood coagulation factors VIII IX	Haemophilia
Human gonadotropin	Sterility
Human insulin	Diabetes
Human serum albumin	Blood substitute
Interferon alpha 2	Antiviral/Antitumor
Interferon alpha	Antiviral/Antitumor
Interferon beta	Antiviral/Antitumor
Hepatitis B vaccine	Vaccine
Human growth hormone	Dwarfism
Lymphokinines (interleukin 2)	Stimulation of immune system
Tissue plasminogen activator	Myocardial infarction

5.7. CELLULAR CONTROL MECHANISMS AND THEIR MANIPULATION

5.7.1. The Control of Enzyme Activity

In previous sections the basic aspects of cell structure and cell biochemistry have been discussed. The control and integration of cellular processes to produce appropriate structures and functions is vital for the efficient reproduction and use of resources. The activity of the cell may be controlled in a number of ways. Figure 5.31 illustrates the interactions that have been shown to take place within cells. Two major systems of control exist.

The first involves various feedback and feed forward control mechanisms associated with metabolic pathways. Here a chemical present in the cell, usually an end product of a metabolic sequence, will influence the activity of an enzyme at the beginning of the pathway. This is usually achieved by the presence of an *allosteric enzyme* whose properties are significantly changed by effector molecules (Fig. 5.31).

A second major set of controls is associated with the expression or synthesis of the enzyme itself, so that cells adapt to environmental conditions with an optimised spectrum of enzymes. In bacteria, this control is usually achieved by changing the rate of transcription of DNA to mRNA. Studies of such systems have shown that the genome (or complete genetic material) is partitioned into sections relating to groups of enzymes (*structural genes*) and their control systems (*regulatory sequences*) which are collectively known as *operons*. Knowledge of these control systems may

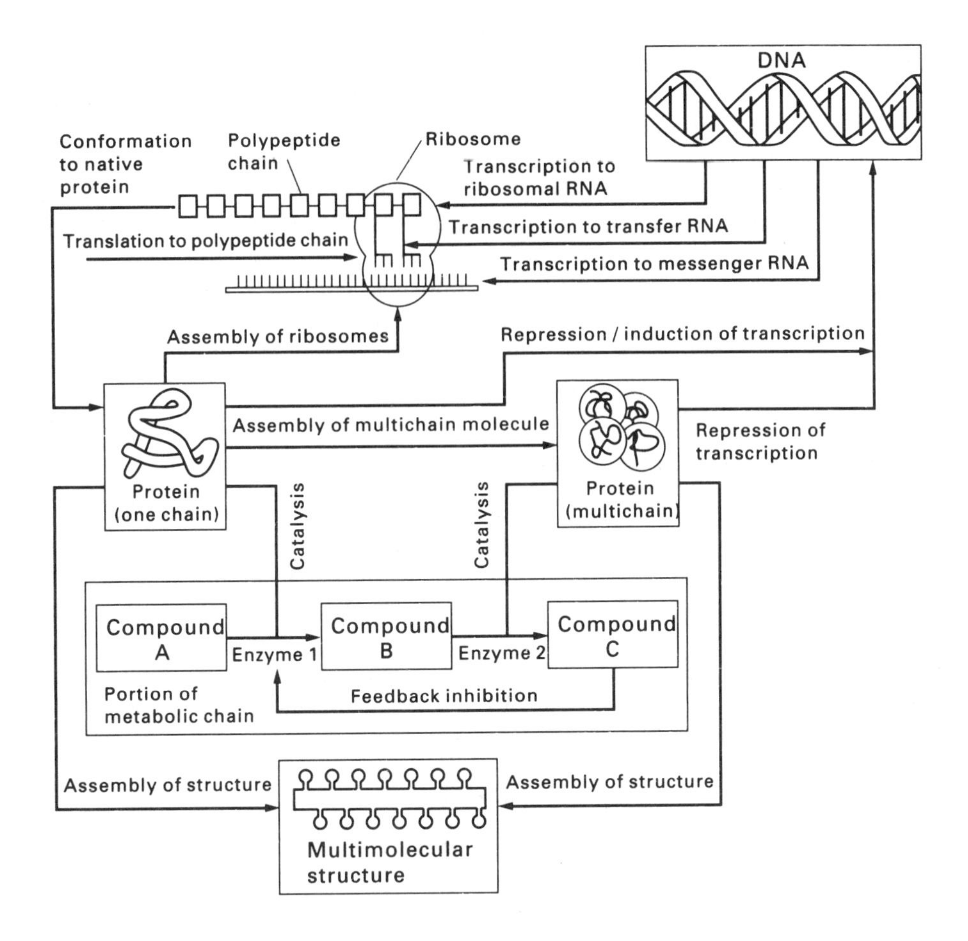

FIG. 5.31. A hierarchy of control and information transmission in the cell[19]

be exploited to produce systems capable of enhanced production of metabolic products and enzymes.

5.7.2. The Control of Metabolic Pathways

Regulatory Enzymes

All enzymes exhibit various features that could conceivably be elements in the regulation of their activity in cells. All have a characteristic pH optimum which makes it possible for their catalytic rates to be altered by changes in intracellular pH (e.g. in ribulose diphosphate carboxylase[17]). The activities also depend on the concentration of substrates, which may vary according to intracellular conditions. Moreover, many require metal ions or vitamins, and the activity of enzymes may be a function of the concentrations in which such materials are present (e.g. the effect of iron limitation on the citric acid fermentation of *Aspergillus niger*[2]). However, over and above these factors, some enzymes have other properties that can endow them with specific regulatory roles in metabolism. Such specialised forms are called *regulatory* enzymes, of which there are three major types.

In *allosteric enzymes*, the activity of the enzyme is modulated by a non-covalently bound metabolite at a site on a protein other than the catalytic site. Normally, this results in a conformational change, which makes the catalytic site inactive or less active. *Covalent modulated enzymes* are interconverted between active and inactive forms by the action of other enzymes, some of which are modulated by allosteric-type control. Both of these control mechanisms are responsive to changes in cell conditions and typically the response time in allosteric control is a matter of seconds as compared with minutes in covalent modulation. A third type of control, the control of enzyme synthesis at the transcription stage of protein synthesis (see Appendix 5.6), can take several hours to take effect.

Allosteric Enzymes and the Regulation of Biosynthetic Pathways

The concept of control of metabolic activity by allosteric enzymes or the control of enzyme activity by ligand-induced conformational changes arose from the study of metabolic pathways and their regulatory enzymes. A good example is the multi-enzymatic sequence catalysing the conversion of L-threonine to L-isoleucine shown in Fig. 5.32.

The first enzyme in the sequence, L-threonine dehydratase, is strongly inhibited by L-isoleucine, the end product, but not by any other intermediates in the sequence.

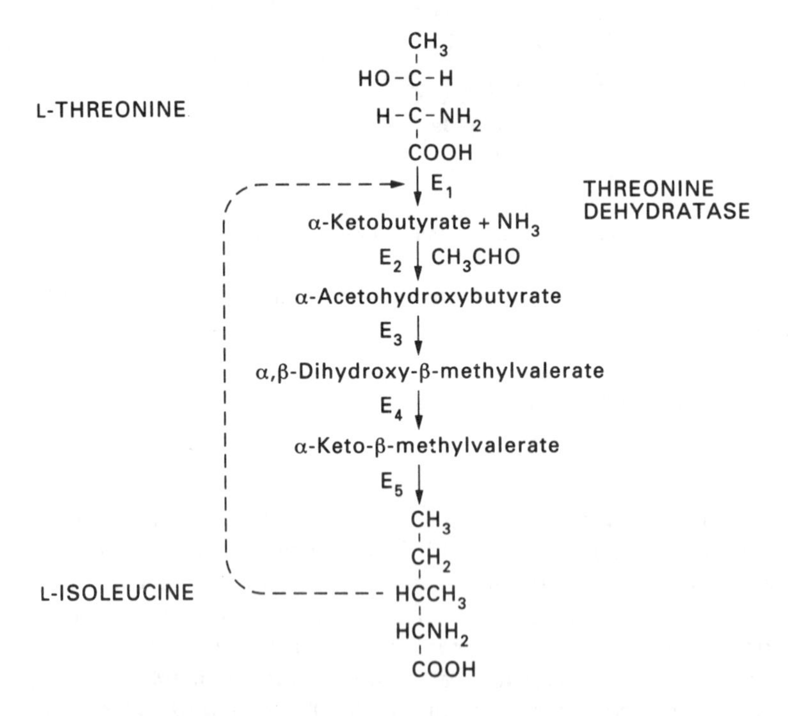

FIG. 5.32. Feedback inhibition of the formation of isoleucine from threonine. This enzymatic pathway (E_1 to E_5) is inhibited by its product isoleucine at the first step E_1 (threonine dehydratase)

The kinetic inhibition is not competitive with the substrate, nor is it non-competitive or uncompetitive. Isoleucine is quite specific in this characteristic and other amino-acids or other analogues do not inhibit. This type of inhibition is called end-product inhibition or feedback inhibition.

This first enzyme, whose activity is modulated by an end-product, is an allosteric enzyme where, in addition to the active site, it has another space specific for binding the ligand which modulates the active site. Some negative modulators inhibit, as shown above with isoleucine on threonine hydratase, while others may stimulate or positively modulate the enzyme. Some enzymes have only one modulator and are called *monovalent*, while others have have several and are called *polyvalent* modulators. Moreover, some allosteric enzymes have both negative and positive modulators. Figure 5.33 illustrates some patterns of allosteric modulation. The advantage of these control systems is that cellular materials are economically used.

Allosteric enzymes have unique structural characteristics; they are much larger in molecular weight than average enzymes and are structurally more complex. This leads to a difficulty in purification because they are oligomeric (have more than one peptide chain) and are sensitive to low as well as high temperatures. They also

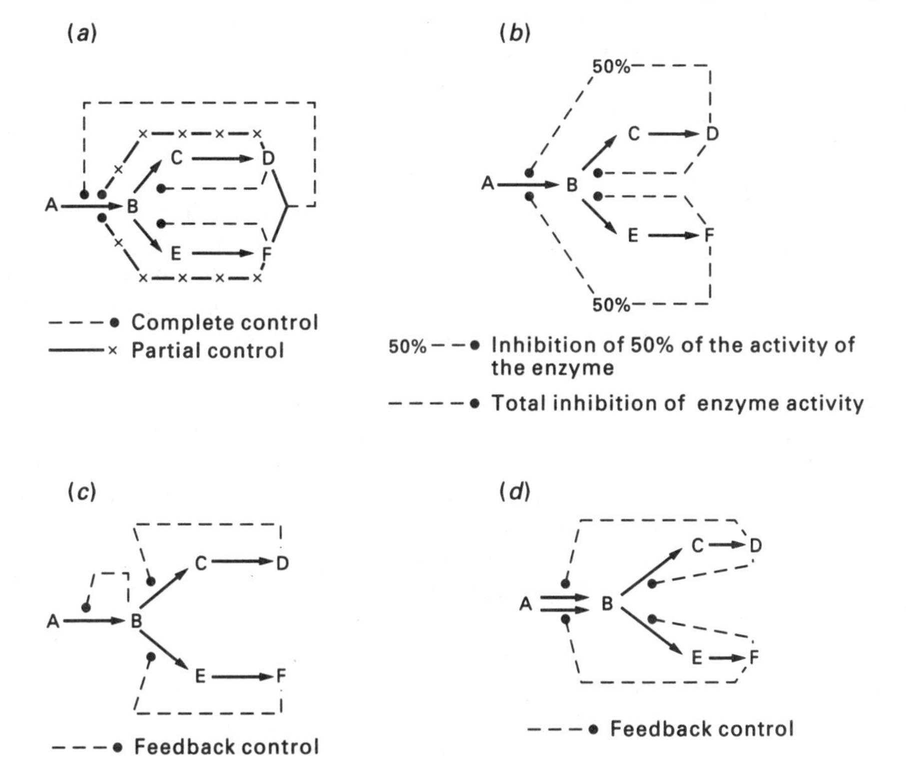

FIG. 5.33. Patterns of allosteric modulation in anabolic pathways

(*a*) The control of a biosynthetic pathway by the co-operative control by end products **D** and **F**
(*b*) The control of a biosynthetic pathway by sequential feedback control
(*c*) The control of a biosynthetic pathway by the cumulative control of products **D** and **F**
(*d*) The control of two isoenzymes (catalysing the conversion of **A** to **B**) by end products **D** and **F**

show atypical kinetic characteristics which fail to conform to the classic Michaelis–Menten relationship.

There are two types of control shown by allosteric enzymes. They can be modulated by a molecule other than a substrate of the enzyme (termed *heterotrophic enzymes*, e.g. threonine dehydratase), or by the substrate itself (termed *homotrophic enzymes*, e.g. oxygen binding to haemoglobin). They contain two or more binding sites for the substrate, and activity is modulated by the number of binding sites which are filled.

Kinetics of Allosteric Enzymes

Allosteric enzymes do not follow the Michaelis–Menten kinetic relationships between substrate concentration $V_{\max}$ and K_m because their kinetic behaviour is greatly altered by variations in the concentration of the allosteric modulator. Generally, homotrophic enzymes show sigmoidal behaviour with reference to the substrate concentration, rather than the rectangular hyperbolae shown in classical Michaelis–Menten kinetics. Thus, to increase the rate of reaction from 10 per cent to 90 per cent of maximum requires an 81-fold increase in substrate concentration, as shown in Fig. 5.34*a*. *Positive cooperativity* is the term used to describe the substrate concentration–activity curve which is sigmoidal; an increase in the rate from 10 to 90 per cent requires only a nine-fold increase in substrate concentration (Fig. 5.34*b*). *Negative cooperativity* is used to describe the flattening of the plot (Fig. 5.34*c*) and requires requires over 6000-fold increase to increase the rate from 10 to 90 per cent of maximum rate.

Some allosteric enzymes are also classified by the way in which they are affected by the binding of a modulator; some affect the value of K_m without affecting that of $V_{\max}$. They are classed as K-series enzymes while others, which affect $V_{\max}$ without affecting K_m, are called M-series enzymes. Figure 5.35 shows the characteristic kinetic patterns observed for K-series and M-series enzymes. There are, of course, exceptions to these two extremes of kinetic behaviour.

Altering the control of metabolic pathways can also be achieved by genetic manipulation. As proteins are generated using the template stored as a fragment of DNA, the structure of the allosteric enzyme may be altered so that there is little or no regulatory control. It is therefore possible to generate mutants that over-produce metabolites, and techniques based on this principle have been most widely exploited in amino-acid and nucleotide production[(2)].

Amino-acid fermentations are usually performed with mutants that remove the control by feedback inhibition. This is achieved by chemical mutation and screening for the appropriately modified organisms. The result of such a mutation is the channelling of the feedstock carbon to the desired product. Consider the system illustrated in Fig. 5.36. The production of lysine, threonine and methionine are controlled by a number of mechanisms, the most notable of which is the feedback control of aspartate kinase by threonine (Thr) and lysine (Lys). Two basic strategies are employed. The first uses an *auxotrophic mutant* that requires threonine (designated Thr^-), for growth. In this situation, only limiting amounts of threonine are fed to the organism and this results in very low levels of threonine in the cell, and thus a reduction of feedback inhibition by threonine. The organism will now produce high levels of lysine. The second strategy uses regulatory mutants of aspartate kinase. If the structure of aspartate kinase is altered so that it becomes insensitive to the molecules that cause feedback inhibition, then the organism will overproduce lysine. This is achieved by feeding the organism with an analogue of lysine, S-2-aminoethyl-L-cysteine (AEC), a powerful feedback inhibitor of the aspartate kinase. In the presence of AEC, complete feedback inhibition occurs and no growth is possible as neither lysine nor threonine will be synthesised. If the enzyme structure is altered so that they become resistant to AEC or AEC^r, the AEC no longer binds and then the synthesis of threonine and lysine will take place and growth will occur. As a consequence of the enzyme alteration, there is a good chance that lysine will also no longer act as a feedback inhibitor and so an enhanced production of lysine is achieved in a fermentation involving such a mutant[32].

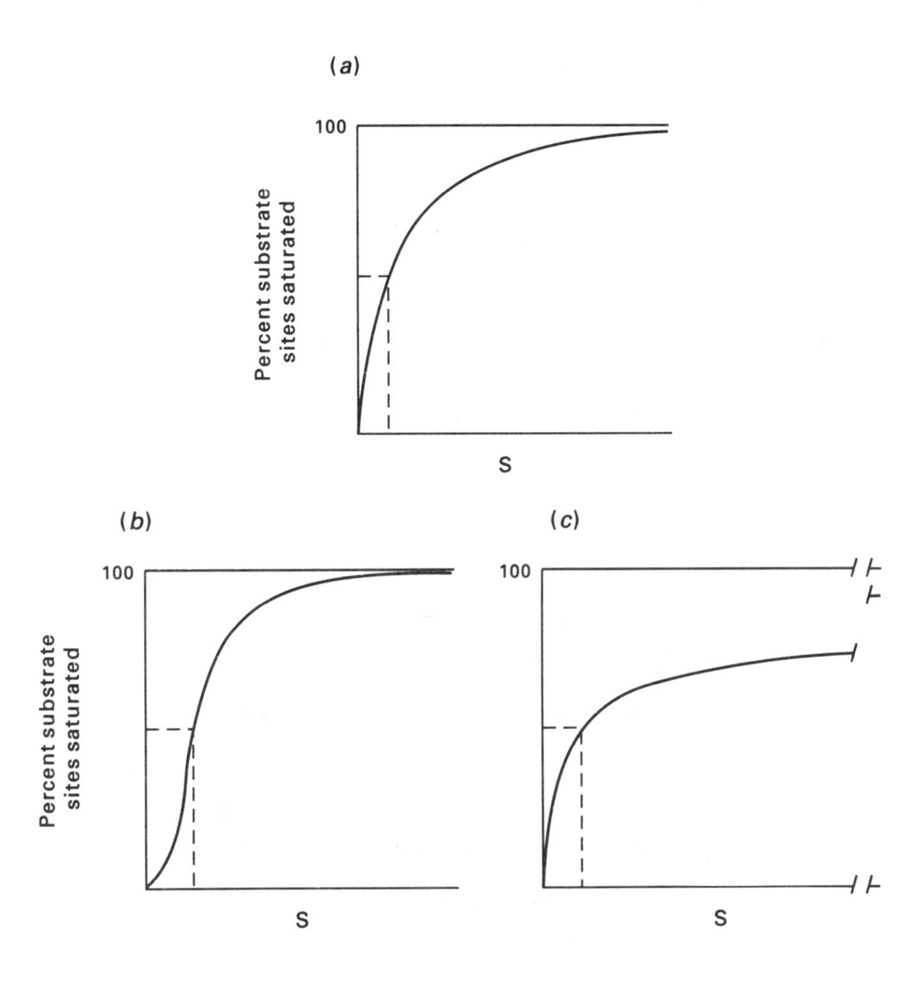

FIG. 5.34. Comparison of idealised plots of per cent maximum reaction rate as a function of substrate concentration for:

(*a*) A non-regulatory enzyme obeying Michaelis–Menten kinetics;
(*b*) Regulatory enzyme following positive cooperativity;
(c) Regulatory enzyme showing negative cooperativity

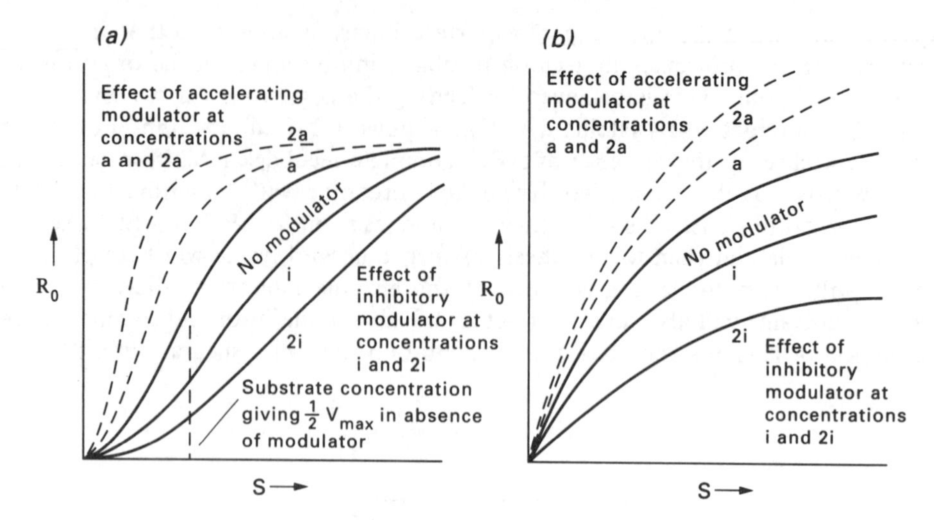

FIG. 5.35. Effect of positive and negative modulators of reaction rate. Substrate concentration – rate curves for

(*a*) K enzymes which modulate activity by altering K_m;
(*b*) M enzymes which modulate activity by altering the maximum rate

Covalent Modulated Enzymes

In a second class of regulatory enzymes the active and inactive forms are interconverted by covalent modifications of their structures by enzymes. The classic example of this type of control is the use of glycogen phosphorylase from animal tissues to catalyse the breakdown of the polysaccharide glycogen yielding glucose-1-phosphate, as illustrated in Fig. 5.37.

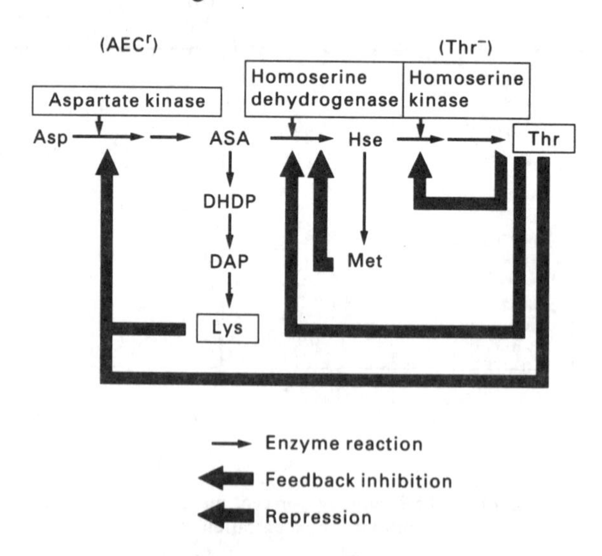

FIG. 5.36. Regulation of lysine (Lys), threonine (Thr) and methionine (Met) biosynthesis in *Brevibacterium flavum*[32]

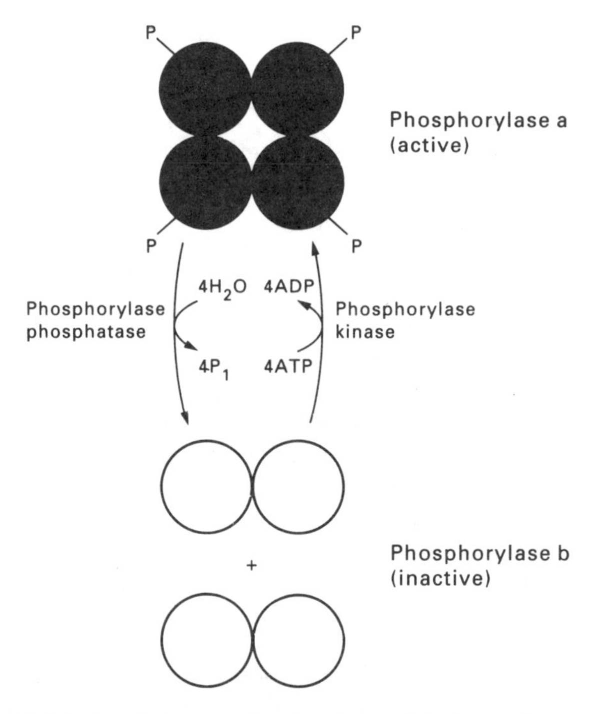

FIG. 5.37. Modulation of glycogen phosphorylase activity by covalent modulation

There are two forms of the enzyme, phosphorylase *a* and the less active form phosphorylase *b*. Phosphorylase *a* is an oligomeric protein with four major subunits. Phosphorylation of a serine hydroxy-group produces an active form of the enzyme. Removal of the phosphate causes a breakdown of the tetramer to a dimeric form which is the less active phosphorylase *b*. Re-activation is achieved by the enzyme phosphorylase kinase which catalyses the phosphorylation at the expense of ATP.

Another important characteristic of these enzymes is the amplification cascade involving phosphorylase kinase and phosphorylase. The chemical signal, here phosphorylase kinase, can react with many molecules of phosphorylase *b* which, in turn, produce phosphorylase *a* which converts glycogen into glucose 1-phosphate.

A different type of covalent regulation of enzyme activity is the enzyme-catalysed activation of inactive precursors of enzymes (zymogens) to give catalytically active forms. The best examples are the digestive enzymes, e.g. trypsin. Proteolytic enzymes would digest the inside of the cells that produce the enzyme, so they are produced in an inactive form which is activated to the true enzyme once they have entered the digestive system of the animal.

Iso-enzymes

The final form of metabolic regulation is effected by the use of iso-enzymes, which are multiple forms of an enzyme. For example, lactate dehydrogenase exists in five forms in a rat. They differ in primary structure and have different isoelectric points, but they all catalyse the reversible reduction of pyruvate to lactate.

$$CH_3COCOO^- + NADH + H^+ \rightleftharpoons CH_3CHOHCOO^- + NAD$$

A careful kinetic study has shown that, although all catalyse the same reaction, both the kinetic constants K_m and V_{max} differ. The kinetic characteristics match the requirements of the tissues, e.g. V_{max} is high in skeletal muscle but low in heart muscle.

5.7.3. The Control of Protein Synthesis

Living cells have many control mechanisms for regulating the synthesis of proteins so that each cell has the correct amount of each protein to carry out its metabolic activities smoothly and efficiently. For example, although *E. coli* contains 3000 proteins, it can have different amounts of each protein under different conditions. For example, when growing on glucose, *E. coli* has very high copy numbers of the glycolytic enzymes while it will only have five or so copies of β-galactosidase. However, when grown on lactose, the copy number of this enzyme increases rapidly. Regulation of enzyme synthesis provides each cell with the proper ensemble of enzymes for balanced growth. It also allows the organism to economise on the highly energy-expensive process of protein synthesis.

The control and expression of protein in eukaryotic cells is more complex than in bacteria. Research into this field is ongoing and, as such, is beyond the scope of this text. Many systems of control are known in bacteria and can be used to illustrate the type of control mechanisms and the importance of the environmental control of protein synthesis. The first and best documented example is that of the lactose (lac) operon is *Escherichia coli.*

It has been established that some enzymes are produced only in the presence of their substrates and they are then said to be *inducible*. The quantity of such enzymes will vary considerably with changing environmental conditions. Other enzymes which are always detectable at constant levels, irrespective of nutrition and environmental conditions, are called *constitutive* enzymes.

In the induction of enzymes of galactose metabolism in *E. coli*, three enzymes are involved: β-galactosidase (which catalyses the hydrolysis of the β-glycosidic bonds of lactose), galactose permease (which is responsible for transport of lactose across the cell membrane); and a third enzyme, A-protein, apparently not directly involved in galactose metabolism. The system has an environmental *inducer*, galactose, and in its presence the number of β-galactosidase molecules rises from 5–10 to 10,000 within the cell. The addition of the inducer can increase the protein production in less than five minutes after its addition. Protein synthesis of these enzymes stops almost immediately in the absence of lactose.

Another important change in the concentration of an enzyme in bacteria, seemingly the opposite to that observed in enzyme induction, is *enzyme repression*. An example of enzyme repression is demonstrated when *E. coli* is grown in the presence of ammonium salts which act as the sole source of nitrogen for growth. Such systems require all the biosynthetic apparatus to be present to make all the amino-acids required by the cell. If, for example, histidine is added to the medium, a whole set of enzymes is no longer synthesised. Repression, like induction, is a reflection of the principle that the cell uses energy as economically as possible. In general, repression operates in anabolic pathways (biosynthetic), while in catabolic pathways (degradation for energy) induction usually operates.

The Operon Hypothesis

The molecular and genetic relationship between enzyme induction and repression was clarified by the genetic research of Jacob and Monod at the Pasteur Institute, Paris (see reference[17]). Their classic work led them to develop the *operon hypothesis* for the control of protein synthesis in prokaryotes, which has since been verified by direct biochemical experiments.

Jacob and Monod proposed that the operon consists of a sequence of genes which are partitioned into *structural genes* and *regulatory genes*. In the case of the lactose, or *lac*, operon there are z, y and α coding for the three enzymes galactosidase, galactoside permease and A-protein. It was proposed that an area refered to as the i-region was responsible the for *regulatory gene*, coding the amino-acid sequence for a regulatory protein called the *repressor* protein. The repressor protein in the absence of the inducer will bind to another regulatory site called the *operator*, or the *o*-site. If the repressor protein is bound to the *o*-site, then the RNA polymerase which is bound to the promoter site (p), is unable to transcribe the structural genes. However, in the presence of the inducer the repressor protein is inactivated and no longer binds to the operator. The RNA polymerase can now transcribe the structural genes. Figure 5.38 illustrates the control mechanisms associated with an operon; the mechanisms of both induction and repression are illustrated.

An operon thus consists of a series of functionally related structural genes which are turned on and off together, plus their regulatory gene, the operator. This overall

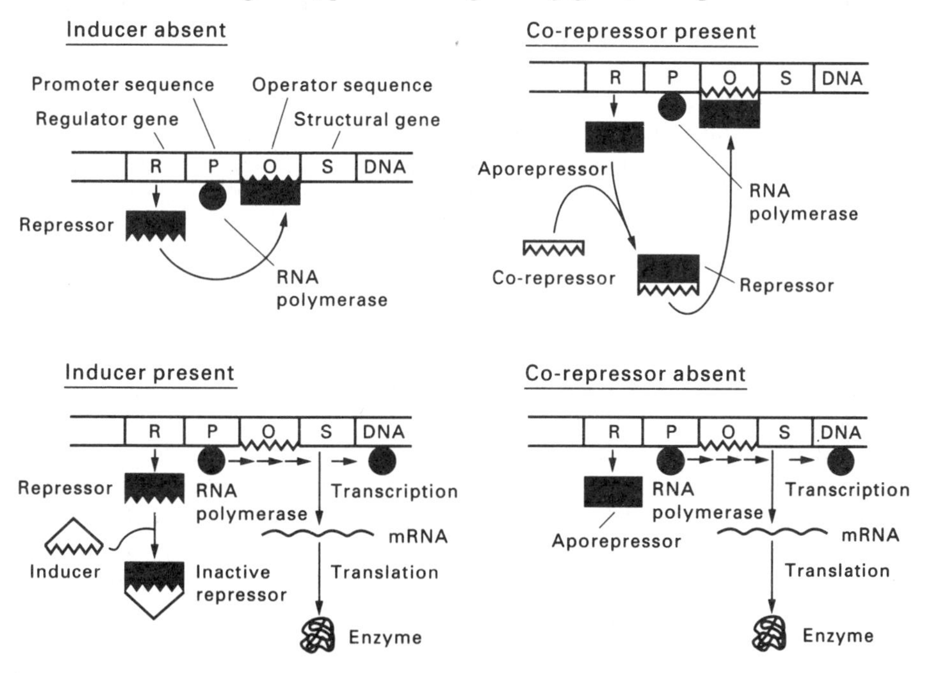

FIG. 5.38. The control of protein (enzyme) synthesis at the level of DNA transcription. Two types of control are illustrated. On the left, the inducer acts by inactiviating a repressor protein allowing transcription of the structural genes. On the right, a repressor combines with an aporepressor protein to form an active repressor complex which blocks transcription of the structural genes

process is called *transcriptional control* since control is primarily directed at the rate of transcription of genes, then at their corresponding mRNAs and ultimately at the amount of protein synthesised (Appendix 5.6).

Another way in which gene expression is regulated is by translational control, where the rate of protein synthesis is controlled at the point of transcription of mRNA into polypeptides (Appendix 5.6). Generally, the majority of the control mechanisms in bacteria is at the transcriptional level. Translational control is less well understood and appears to be a secondary mechanism in bacteria, but it is thought to be very important in eukaryotic organisms.

Catabolic Repression

The operon theory has to be further elaborated when considering another common phenomenon, catabolite repression. Suppose glucose and lactose are present in the growth medium. Under these conditions *E. coli* will utilise only the glucose and ignore the lactose until the glucose is exhausted. In the presence of glucose the cells no longer make the proteins required for lactose metabolism. This repression of the lactose, or lac, proteins is called *catabolite repression.*

The organism senses whether glucose is available by another regulatory mechanism which cooperates with the lac repressor and the lac operator. The promoter is therefore sub-divided into two specific regions, each of distinctive function. One is the RNA polymerase entry site, where RNA polymerase first becomes bound to DNA (cf. DNA transcription, Appendix 5.6), and the other is the protein binding site for the catabolite activator protein (CAP) (Fig. 5.39). The CAP protein binding site controls the polymerase site which, when bound to the DNA, allows successful transcription provided that the repressor is not bound. When the CAP protein is not bound, then RNA polymerase cannot bind and transcription cannot take place.

The binding of the CAP protein to the promoter site on DNA can be modulated by the chemical, cyclic AMP (cAMP). When this is bound to the CAP protein, the CAP protein will bind to the promoter site on the DNA which allows the RNA polymerase to bind to DNA at the promoter site. Now, if glucose is absent but lactose is present, then the operator site will be open and, where these two conditions are met, the RNA polymerase will function to produce the lac proteins required. On the other hand, when ample glucose is available, then the concentration of cyclic AMP is low and the CAP–cAMP complex cannot form. The net result of this is that the RNA polymerase cannot enter the promoter site for the lac genes; thus the lac genes are only synthesised when glucose is unavailable. The lac operon thus has both positive and negative control characteristics. The CAP 'senses' whether glucose is available by having two binding sites, one for the binding of DNA at the promoter site, and the other for cAMP. In *E. coli*, cAMP is a chemical messenger (or hunger signal) which signals the presence or absence of glucose for use as cell fuel. Figure 5.39 illustrates this type of control mechanism.

As research has progressed in this area, many systems have been shown to follow a similar behaviour. Such phenomena are important in the production of enzymes and metabolites. In industrial applications, these very important control mechanisms usually have to be bypassed for the reliable generation of high concentrations of enzymes. Good examples may be seen in carbohydrase production systems,

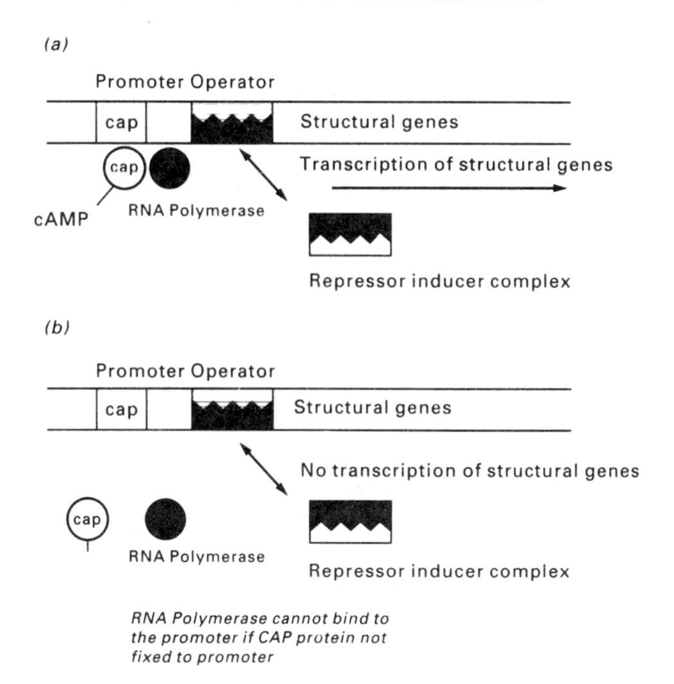

FIG. 5.39. The control of enzyme synthesis by catabolite repression. A control region of the lac operon contains the CAP binding site within the promoter region

(*a*) At low energy levels, the CAP protein binds enhancing protein synthesis
(*b*) When glucose is present and energy levels are high the CAP protein does not bind, so reducing the rate of protein synthesis

such as those for amylases and cellulases.[2] Similarly, inducible enzymes may be made constitutive by altering the structure of the control site of the operon. Alternatively, the structure of the repressor/inducer proteins may be altered so as to change the binding characteristics of sites which can then no longer bind to DNA, a similar principle to that outlined when using control mutants of allosteric enzymes (Section 5.7.2).

5.8. STOICHIOMETRIC ASPECTS OF BIOLOGICAL PROCESSES

As indicated in the introduction and in earlier sections of this chapter, micro-organisms will, under appropriate conditions, proliferate and, in many cases, yield metabolic products which are of economic value. The necessary conditions are that the microbes are provided with a source of energy, a supply of essential elements for their growth, that they are at a suitable temperature and pH and that there are no inhibitory substances present. Different microbes will have differing requirements according to their individual needs and these may vary considerably between species. Whilst the elements principally associated with living organisms are carbon, hydrogen, oxygen, nitrogen, sulphur and phosphorus, other elements such as sodium, calcium, potassium, iron, cobalt and molybdenum are also required, although in lesser quantities. An exhaustive list would be very difficult to compile, especially since some microbes require certain combinations in specific forms; for example, there are species which will accept their nitrogen as urea whereas others require it as ammonium ions.

Industrial manufacturing processes involving micro-organisms generally use organic compounds to supply the microbes with their source of energy and carbon. The commonest form in which the carbon is supplied is as a carbohydrate, although, again, it should be noted that this is by no means always the case. During the course of growth the organic material is assimilated into the cellular material of the micro-organism to produce more cellular material, i.e. the organism grows, and providing that there are no complicating factors it will replicate itself. This process can be represented as:

$$\text{Organic compounds} \xrightarrow{\text{micro-organisms}} \text{Micro-organisms}$$

This process frequently involves the synthesis of more complex compounds than those in the feed and they tend to have a higher free energy of formation than the starting materials. Consequently, the micro-organisms must consume some portion of the feed to meet this energy requirement. In the case where the feed is a carbohydrate, the waste materials from the metabolism are carbon dioxide and water, so that the reaction might be written:

$$\text{Organic compounds} \xrightarrow{\text{micro-organisms}} \text{Carbon dioxide} + \text{Water} + \text{Energy}$$

This biochemical reaction proceeds in parallel with the synthetic reaction, and the overall process would be observed to be that:

$$\text{Organic compounds} \xrightarrow{\text{micro-organisms}} \text{Micro-organisms} + \text{Carbon dioxide} + \text{Water}$$

The energy released would be observed as heat, causing the culture to increase in temperature.

HERBERT[33] has reviewed the methods for elemental analysis of microbial cells. Typically, analyses of carbon, hydrogen, oxygen and nitrogen content are quoted [e.g. BATTLEY[34], ABBOTT and CLAMEN[35]], these being utilised to create an empirical formula for the cells. Batley analysed the growth of the yeast *Saccharomyces cerevisiae* on glucose in the absence of oxygen and quoted the formula of the yeast as:

$$CH_{1.737}N_{0.200}O_{0.451}$$

Whilst such a formula may be meaningless in terms of the biochemistry of the cell, and its accuracy should be viewed with reference to Table 5.9, it does serve a useful purpose when considering the stoichiometry of the microbial activity.

Battley reports that 23 per cent of the glucose consumed during the fermentation is utilised to produce biomass whilst 77 per cent is utilised to produce the energy required by the cell. He gives the stoichiometry for the anabolic reaction as:

$$\begin{aligned} 0.23\,C_6H_{12}O_6\,(aq) + 0.240\,C_2H_5OH\,(aq) + 0.118\,NH_3 \rightarrow\ & 0.590\,CH_{1.737}N_{0.200}O_{0.451}\ \text{(cells)} \\ & + 0.432\,C_3H_8O_3\,(aq) \\ & + 0.036\,H_2O\,(1) \end{aligned}$$

$$\Delta G^{0\prime} = 0.0\ \text{MJ/kmol}$$

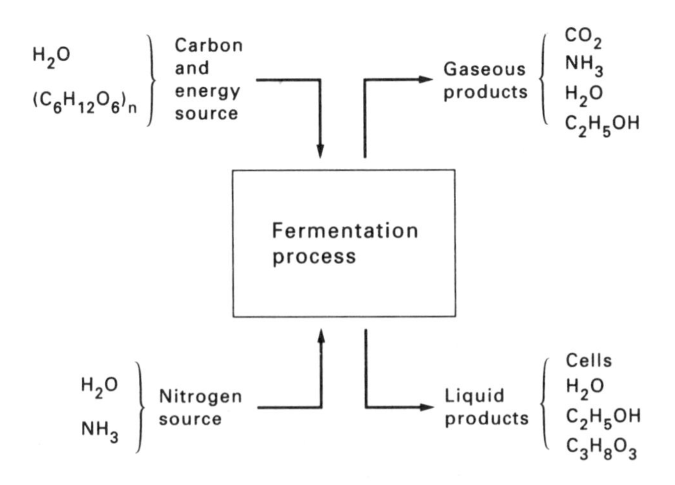

FIG. 5.40. Schematic representation of a fermentation process

and the catabolic reaction:

$$0.770\,C_6H_{12}O_6\,(aq) \rightarrow 1.540\,CO_2\,(aq) + 1.540\,C_2H_5OH \qquad \Delta G^{0\prime} = -167.9\ \text{MJ/kmol}$$

and these are summed to give the overall stoichiometry as:

$$\begin{aligned} C_6H_{12}O_6\ (aq) + 0.188\ NH_3 \rightarrow\ & 0.590\ CH_{1.737}N_{0.200}O_{0.451} \\ & + 0.432\ C_3H_8O_3 \\ & + 1.540\ CO_2\,(aq) \\ & + 1.300\ C_2H_5OH\,(aq) \\ & + 0.036\ H_2O\,(1) \end{aligned}$$

$$\Delta G^{0\prime} = -167.9\,\text{MJ/kmol}$$

Such equations allow calculations to be carried out to quantify the materials used and produced during the course of a fermentation in the same manner as for a chemical reaction process. If the fermentation scheme is simplified to the situation shown in Fig. 5.40, then an input–output table can be drawn up for the streams shown, given the composition of, say, the carbon and energy feed stream and the gaseous product stream.

Quantification in this manner is an important stage in the design of a biochemical reaction process and provides a means of specifying the performance requirements of the biochemical reactor. Furthermore, since the change in entropy is small in this case, the change in free energy approximates to the change in enthalpy associated with the process so that an estimate of the heat balance may be made. This, in turn, allows consideration of the temperature effects to be anticipated and, of course, estimates to be made of any cooling requirement.

5.8.1. Yield

The growth of a microbial culture, consuming substrate for energy purposes, for incorporation into its own cellular material, or for the synthesis of a product as discussed above, gives rise to the concept of yield. This is a familiar notion in chemical process technology and is used to relate the quantities of materials consumed and produced in a particular reaction; it is an expression of the law of conservation of matter. A yield coefficient may be defined as being the ratio of the mass of product obtained to that of reactant consumed and, for given reaction conditions, temperature and pressure, this is normally expected to be a constant quantity. The same concept has been applied in biochemical reaction engineering but, since the system is generally more complex, the specification of the yield coefficient must be more closely defined. Ultimately, in the more sensitive experiments, the yield coefficient appears not to be a constant quantity, but may be a function of time as well as the physico-chemical environment. This is the result of the changing composition of the microbial cell, and the fact that the cell organises itself in a different manner—the phenomenon of adaptation (see Section 5.5.4). There are occasions when adaptation may be difficult to distinguish from mutation but the net result in either case is that the yield coefficient may vary. It must be emphasised at this stage that the conservation laws are not by-passed or violated in any way and that the change in yield coefficient only reflects different processes dominating the reactions occurring in the cell.

The yield coefficient can, however, be advantageously considered to be a constant for a given microbial/substrate combination under given physical conditions. Mathematically, this may be written in terms of concentration* as:

$$Y = \frac{\Delta X}{-\Delta S} \tag{5.44}$$

where ΔX represents the change in the biomass concentration X which results from the change ΔS in the substrate concentration. The negative sign is included to indicate that an increase in biomass concentration is accompanied by a decrease in substrate, or nutrient concentration. In the case of, say, a yeast culture which consumed carbohydrate as its sole carbon and energy source to produce more yeast cells and alcohol, then clearly two yield coefficients may be defined; that where the biomass is related to the feed, and that where the quantity of alcohol produced is related to the quantity of substrate consumed. These may be represented as:

for biomass:

$$Y_{X/S} = \frac{\Delta X}{-\Delta S} \tag{5.45}$$

and for alcohol:

$$Y_{P/S} = \frac{\Delta P}{-\Delta S} \tag{5.46}$$

where X relates to biomass, S to the nutrient feed or substrate consumed and P to the product (alcohol in this case). It is worth noting at this point that for products $1, 2, \ldots, i, \ldots, N$ resulting from the microbial action, each has a yield coefficient associated with it:

$$Y_{i/S} = \frac{\Delta P_i}{-\Delta S} \tag{5.47}$$

* Because molecular weights are ill-defined in biological systems, concentrations and reaction rates are expressed in mass rather than molar units employed in Chapters 1 to 4 dealing with chemical reactions.

and, by the law of conservation of matter, and expressing the yield on, say, the carbon balance, then:

$$\sum_{i=1}^{N} Y_{i/S} = 1 \tag{5.48}$$

When the yield is based upon the mass of material produced and consumed, then the balance may appear to be contravened, and yield coefficients over unity are obtainable. The reason for this is that the reaction may well incorporate other material into the product, as well as the substrate on which the yield coefficient is based, thus increasing the relative mass of product.

PIRT[36] has pointed out that the yield coefficient relating to biomass production, as measured experimentally, is affected by the maintenance energy requirement of the microbes. The concept of maintenance energy requirement results from recognition of the fact that the microbes consume a certain amount of energy to carry out the basic functions such as maintaining pH and osmality within the cell, functions which are in themselves independent of growth rate. If the overall or 'observed' yield coefficient for microbe growth is:

$$Y_{X/S} = \frac{\Delta X}{-\Delta S} \tag{5.49}$$

and the substrate consumed, $-\Delta S$, is notionally divided into two components, ΔS_G being that part being converted into cellular material and the remainder being that used for maintenance.

Thus we may define:

$$Y_G = \frac{\Delta X}{-\Delta S_G} \tag{5.50}$$

to be the yield coefficient for the conversion of feed or substrate into the mass of the cells; this is referred to as the 'true growth yield'.

A material balance for the consumption of substrate gives:

$$\begin{Bmatrix}\text{Total consumption}\\ \text{of substrate}\end{Bmatrix} = \begin{Bmatrix}\text{Substrate used}\\ \text{for growth}\end{Bmatrix} + \begin{Bmatrix}\text{Substrate used}\\ \text{for maintenance}\end{Bmatrix}$$

or:

$$\frac{dS}{dt} = \frac{dX/dt}{Y_G} + mX \tag{5.51}$$

where m is the specific requirement for maintenance. The above can be rewritten as:

$$\frac{1}{X}\frac{dS}{dt} = \frac{1}{X}\frac{dX}{dt}\frac{1}{Y_G} + m \tag{5.52}$$

but $\frac{1}{X}\frac{dX}{dt}$ is the specific growth rate of the microbe, usually denoted by the symbol μ. Thus, with this substitution:

$$\frac{dS}{dt}\bigg/\frac{dX}{dt} = \frac{1}{Y_G} + \frac{m}{\mu} \tag{5.53}$$

or:

$$\frac{dS}{dX} = \frac{1}{Y_{X/S}} = \frac{1}{Y_G} + \frac{m}{\mu} \tag{5.54}$$

The quantity $Y_{X/S}$ is the observed yield, and equation 5.51 indicates that this is dependent on the actual growth rate, irrespective of whether or not the processes occurring within the cell are constant. At high specific growth rates μ, the specific requirement for maintenance is insignificant and $Y_{X/S}$ approaches the value of Y_G, but when μ is comparable with m then cognisance has to be made of the above relationship.

5.9. MICROBIAL GROWTH

5.9.1. Phases of Growth of a Microbial Culture

The mathematical description of the rate of growth of a microbial culture frequently makes use of the concept of doubling time and, by implication, an exponential growth pattern. This arises from the premise that the growth rate is directly proportional to the existing population and the proportionality constant is a function of the organism type. In most analyses the cell number, representative of population, is replaced by cell mass. This is both more easily measured and avoids problems in material balances which might arise when cell size varies with growth rate. If then, the mass of cells per unit volume of a batch culture is represented by X, exponential growth of the microbes is given by Malthus' law which may be stated as:

$$\frac{\mathrm{d}X}{\mathrm{d}t} = \mu X \tag{5.55}$$

where t represents time and μ is the specific growth rate of the culture. The quantity X is frequently referred to as the 'microbial density' or the 'biomass concentration'.

Integration of this equation between the limits X_0 at time $t = 0$ and X at some time t gives:

$$\ln\left(\frac{X}{X_0}\right) = \mu t \tag{5.56}$$

or:

$$X = X_0 e^{\mu t} \tag{5.57}$$

that is, the growth is exponential. If the time for the biomass concentration to double is t_d then substitution in equation 5.56 gives:

$$\ln\left(\frac{2X_0}{X_0}\right) = \mu t_d \tag{5.58}$$

or:

$$t_d = \frac{\ln 2}{\mu} \tag{5.59}$$

Exponential growth, however, occurs only for a limited time during the course of the development of a microbial culture with a fixed supply of nutrients. The classic time course of the growth of such a culture is shown in Fig. 5.41 where the logarithm of the microbial density (or alternatively microbial number) is plotted against time.

The culture passes through a series of phases characterised by the rate of growth, starting with a lag phase during which little or no increase in the microbial density occurs. This gives way during the acceleration phase to a period of exponential

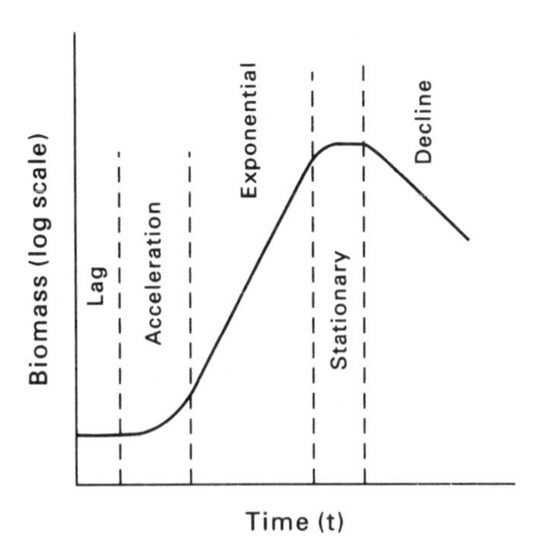

FIG. 5.41. Phases of growth of a microbial culture

growth which continues until the food supply becomes exhausted. It then enters a stationary phase during which the organisms cease to grow and then the final decline phase in which the active mass diminishes as the cell population dies away.

The initial phase, the so-called lag phase, represents a period of time in which subtle but complicated changes occur in the internal organisation of the individual cells. The cause of the delay in the development of exponential growth can be traced to a variety of factors, such as change in food type or its concentration, change in pH or the presence of an inhibitor. Much of the theory relating to microbial growth is restricted in that it deals with cells which are fully adapted to their environment and are growing exponentially. In the case of continuous cultures this requirement is normally fulfilled, but in batch cultures the lag phase assumes more importance since it represents a significant proportion of the time over which the fermentation takes place. The mathematical modelling of the lag phase is a difficult subject, as is the collection of experimental data against which the model can be tested.

It is commonplace to account for the lag by a method attributed to LODGE and HINSHELWOOD[37]. This involves extrapolating the straight sections of the growth

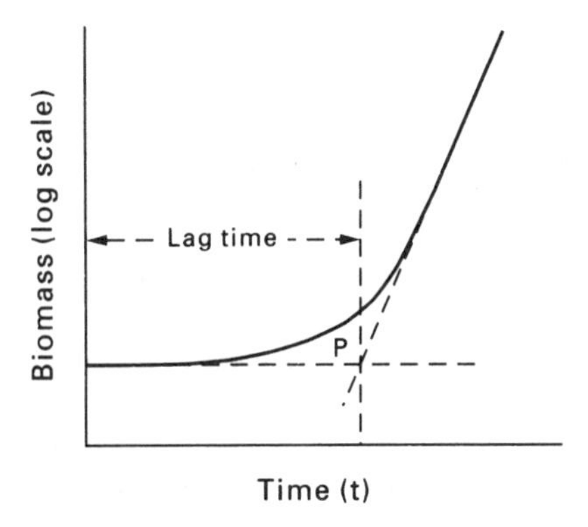

FIG. 5.42. Estimation of lag time

curve when plotted on a logarithmic scale as shown in Fig. 5.42. The vertical dropped from the point of intersection P is then used to read off the lag time on the abscissa. Clearly, this is a crude device but it gives a numerical value to the combined lag and so-called 'acceleration' phases which can be used for scheduling calculations.

Exponential growth is the most extensively quantified of the phases of microbial growth and, in the case of growth associated products, is also the most important from an industrial point of view. The vast bulk of the growth of a microbial culture occurs in this period and, as referred to earlier, is characterised by the periodic doubling of the biomass. This growth, however, cannot be sustained indefinitely and for one reason or another will lead to the stationary phase. PEARL and REED[38] modified equation 5.55 by adding a further term to account for 'inhibition' at high biomass concentration.

$$\frac{dX}{dt} = kX - k\gamma X^2 \tag{5.60}$$

The second term assumes that the 'inhibition' is proportional to the square of the biomass concentration and this equation, on integration, yields the 'logistic' equation:

$$X = \frac{X_0 e^{kt}}{1 - \gamma X_0(1 - e^{kt})} \tag{5.61}$$

As may be seen from Fig. 5.43, a plot of X against t for this equation gives rise to a sigmoidal shaped curve with the section at smaller values of t approximating to an exponential curve.

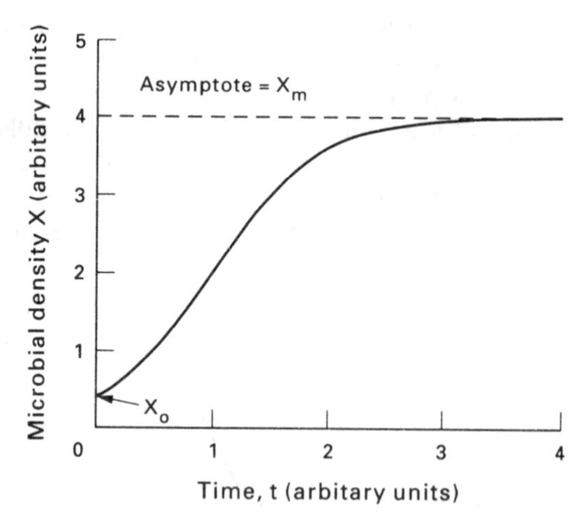

FIG. 5.43. Microbial density as predicted by equation 5.61

However, whilst equation 5.61 is not based on any theory which relates to biological observation other than that the growth curve is sigmoidal, it does serve to present data in a compact form. It can be used to describe the lag, exponential and stationary phases of microbial growth and the constants involved can be related to some physical features of the fermentation. The constant $\boldsymbol{k}$ is, in fact, the maximal specific growth rate of the culture and γ is the reciprocal of the final biomass concentration (X_m). The main problem with its use is its inflexibility and its lack of cognisance of the substrate concentration. Modifications of the equation (by EDWARDS and WILKE[39]) have been suggested which involve substituting for the exponential term $\boldsymbol{kt}$ a polynomial function of the form $a_0 + a_1 t + a_2 t^2 + \ldots$, which, although successful in terms of accuracy, increases the number of constants required and hides the physical significance of $\boldsymbol{k}$.

5.9.2. Microbial Growth Kinetics

Monod Kinetics

MONOD[40] proposed the use of a saturation-isotherm type of equation to relate the growth rate of a micro-organism culture to the prevailing feed concentration. This has become known as the 'Monod' equation and is usually expressed as:

$$\mu = \frac{\mu_m S}{K_S + S} \tag{5.62}$$

where, as before, μ is the specific growth rate, S is the feed or substrate concentration, μ_m is a constant sometimes known as the maximum specific growth rate and K_S is a constant referred to as the 'Monod constant'. Figure 5.44 shows the general form of the relation. Note that the microbe may well require several substrates for its growth to proceed, but it is assumed that all but one are in excess of requirements and the substance to which S relates is the *limiting* substrate component. The value of K_S will be linked to the particular limiting substrate component in any given mixture, and for different substrate mixtures it would also be expected that μ_m would change. It is usual to quote μ_m and K_S for the substance which is the prime supplier of carbon (and energy) for the organism being grown.

The graphical significance of the constants in the Monod equation are identical to the corresponding constants in the Michaelis–Menten relationship for enzyme kinetics (see Section 5.4.4). The specific growth rate initially increases with increas-

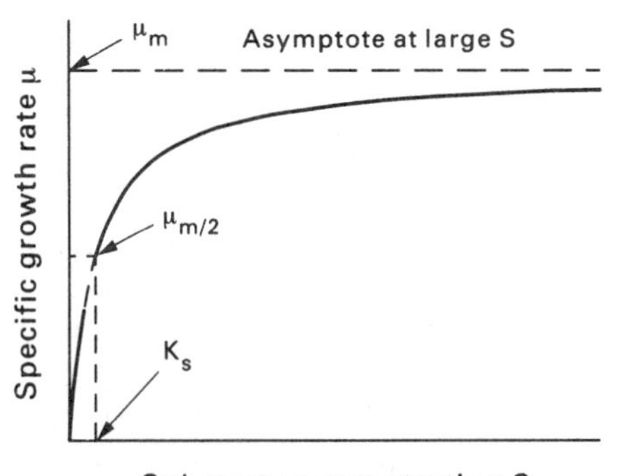

FIG. 5.44. The general form of the Monod equation

ing substrate concentration and reaches a limiting value of μ_m at high substrate levels. In equation 5.62 when $S >> K_S$:

$$\mu = \frac{\mu_m S}{K_S + S} \Rightarrow \frac{\mu_m S}{S}$$

or:

$$\mu = \mu_m \tag{5.63}$$

Also, when $\mu = \mu_m/2$ the substitution in 5.62 gives:

$$\frac{\mu_m}{2} = \frac{\mu_m S}{K_S + S}$$

or, after division by μ_m and re-arranging:

$$K_S = S \tag{5.64}$$

i.e. when the specific growth rate is half its maximum value the limiting substrate concentration is numerically equal to K_S.

Two possible explanations can be readily put forward as to why this form of equation should be suitable for describing the dependence of microbial growth rate on feed concentration. The first of these is that the equation has the same form as the theoretically based Michaelis–Menten equation used to describe enzyme kinetics. The chemical reactions occurring inside a microbial cell are generally mediated by enzymes, and it would be reasonable to suppose that one of these reactions is for some reason slower than the others. As a result the growth kinetics of the micro-organism would be expected to reflect the kinetics of this enzyme reaction, probably modified in some way, but in essence having the form of the Michaelis–Menten equation.

The other possible cause of the saturation-isotherm type of response of specific growth rate to feed concentration is that the rate-determining step may be controlled by the transport of the feed through the boundary layers to the micro-organism or through the membranes forming the cell wall itself. In these cases, even if the transport is 'active', the rate equation will be of the same general form and result in the Monod type equation given above.

Inhibition

The important point to note, however, is that the Monod equation is, like that of PEARL and REED[38], essentially empirical, and there is no reason to limit equations describing microbial growth rates to this form alone. Indeed, the use of other kinetic equations which were originally derived for enzyme kinetics has met with considerable success. EDWARDS[41] used an equation originally derived for substrate inhibited enzyme reactions (see equation 5.23) to describe the growth rates of micro-organisms on feed material which, at high concentrations, was inhibitory to growth. The growth of microbes on, for example, acetate is an example of the use of the

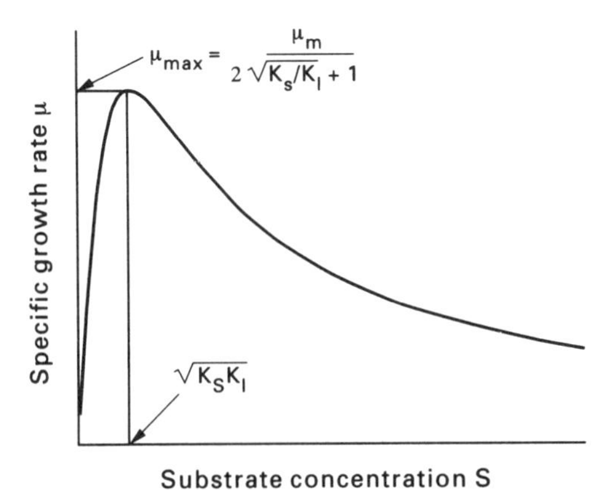

FIG. 5.45. Influence of substrate inhibition on kinetics

rate equation:

$$\mu = \mu_m \frac{S}{K_S + S + \left(\dfrac{S^2}{K_I}\right)} \tag{5.65}$$

where K_I is the 'inhibition' constant. In this case it should be observed that whilst the notation used is similar to that in the Monod equation the significance of the constants has changed. The form of this relation is shown in Fig. 5.45.

The maximum growth rate in this instance does not occur at the highest substrate concentration, but rather at some intermediate value. Differentiation of equation 5.65 with respect to S and setting the result equal to zero enables the maximum growth rate to be determined. The maximum rate, μ_{max} is shown to occur when:

$$S = \sqrt{K_S K_I} \tag{5.66}$$

and substitution of this result into 5.65 leads to:

$$\mu_{max} = \frac{\mu_m}{2\sqrt{K_S/K_I} + 1} \tag{5.67}$$

The success with which this equation predicts the specific growth rate of a substrate-inhibited fermentation is somewhat varied. Typically, the constants may be adjusted to represent the experimental data with a fair degree of accuracy at low substrate values, but at substrate concentrations above $\sqrt{K_S K_I}$ the predicted growth rate tends to be slightly greater than that observed in practice.

Another important case of inhibition of microbial growth is that of product inhibition. Examples of this are widespread, the effect of ethanol on the fermentation of sugars by yeast being one, and here again the Monod equation does not give an adequate estimate of the specific growth rate. The classic approach to the mathematical description is to consider the product as depressing the value of μ_m and introducing a further pair of parameters into what otherwise would be a

Monod-type equation. One expression used (AIBA *et al.*[42]) is:

$$\mu = \frac{\mu_m S}{K_S + S} e^{-k_1 P} \tag{5.68}$$

where P is the concentration of the product and k_1 is a constant. An alternative, also proposed by AIBA *et al.*[16], is:

$$\mu = \frac{\mu_m S}{K_S + S} \frac{K_P}{K_P + P} \tag{5.69}$$

K_P being the constant of the inhibitory system. In both cases the expressions reduce to the Monod equation when the product concentration is zero.

Table 5.17 shows some of the different expressions which have been used to describe microbial growth kinetics. (Note that the constants used in the table may not fulfill the same function and their units may be different, the table being intended to give some indication of the breadth of forms which have been employed.)

TABLE 5.17. *Expressions used for Microbial Growth Rates*

$\mu = \mu_m \frac{S}{K_S + S}$	MONOD[40]
$\mu = \mu_m \frac{(K_C S)^a}{1 + (K_C S)^a}$	MOSER[43]
$\mu = \mu_m \frac{1}{1 + \frac{K_k X}{S}}$	CONTOIS[44]
$\mu = \mu_m \frac{S}{K_S + S + \left(\frac{S^2}{K_I}\right)}$	EDWARDS[41]
$\mu = \mu_m \frac{S}{K_S + S} \frac{K_P}{K_P + P}$	AIBA *et al.*[16]

Maintenance

The most successful and widely used of the equations in Table 5.17 is that due to Monod and, although it may not be universally applicable, it gives a reasonable description of the variation of growth rate with substrate concentration in a surprisingly large number of cases. Whilst it does not allow for the lag phase at the beginning of a batch process, it may be modified by the addition of one extra term to allow for the consumption of cellular material to produce maintenance energy.

$$\mu = \mu_m \frac{S}{K_S + S} - k_d \tag{5.70}$$

The constant k_d is referred to as the endogenous respiration coefficient or the specific maintenance rate. PIRT[45] points out that k_d is proportional to m, the maintenance coefficient defined in equation 5.51:

$$k_d = m Y_G \tag{5.71}$$

At high growth rates the endogenous respiration rate has little net effect on the formation of biomass but in continuous culture with low nutrient levels then endogenous respiration becomes significant and has to be included in the expression for growth rate.

pH

Whilst the kinetic equations considered are concerned with substrate and product concentrations in relation to the specific growth rate of microbial cultures, other factors may also have a profound effect on the rate. In general, micro-organisms have an optimum pH for growth, usually near to neutrality, although a number of species can tolerate rather hostile environments and even grow in them. The effect of pH on the specific growth rate is conveniently represented in a manner similar to product inhibition, that is, by considering it as a modification of μ_m. Figure 5.46 shows a typical plot of μ_m against pH (HARWOOD[46]).

SINCLAIR and KRISTIANSEN[47] describe the use of another function, previously used to describe enzyme kinetics, to quantify this effect:

$$\mu_m(\mathrm{pH}) = \frac{\mu_m}{1 + \dfrac{[\mathrm{H^+}]}{K_{a1}} + \dfrac{K_{a2}}{[\mathrm{H^+}]}} \tag{5.72}$$

where K_{a1} and K_{a2} are constants and the hydrogen ion concentration is represented by $[\mathrm{H^+}]$. The specific growth rate μ_m is not necessarily a maximum at the optimum pH and the expression suffers from the fact that it produces a symmetrical curve as shown in Fig. 5.47, a pattern not always observed in practice. WASUNGU and SIMARD[48] show an example of this in the variation of the growth rate of *Saccharomyces cerevisiae* at 30°C (Fig. 5.48).

Much of the work carried out to investigate the effect of pH on microbial growth has been done using buffer solutions to stabilise, rather than to control, its value

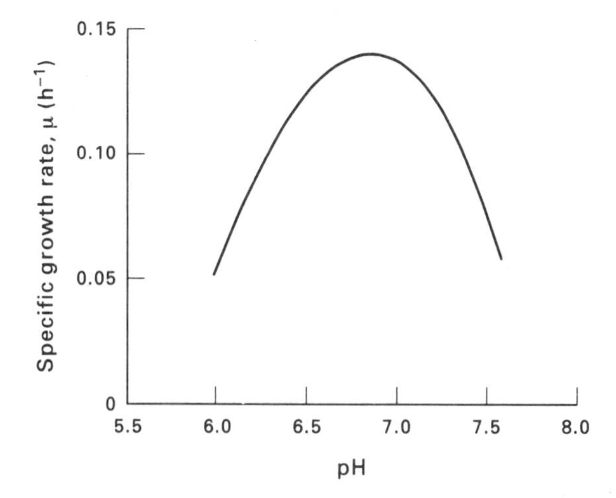

FIG. 5.46. Maximal growth rate of *Methylococcus capsulatus* as a function of pH (HARWOOD[46])

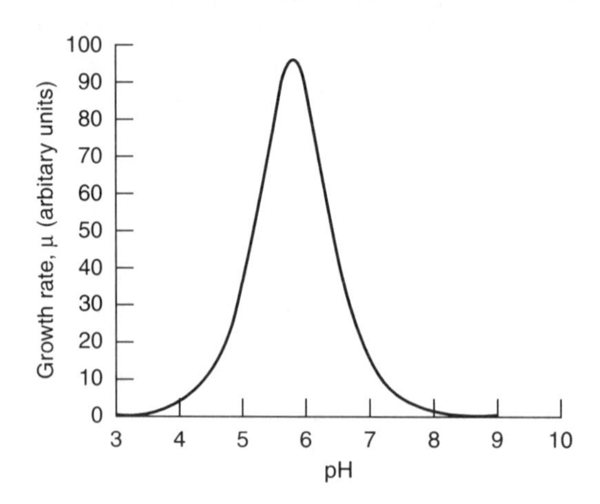

FIG. 5.47. Plot of growth rate against pH as predicted by equation 5.72

accurately. The internal components of a cell are protected to some extent by active transport mechanisms, as discussed in Section 5.5.6, which are able to generate large pH gradients across the cell wall boundary. This, clearly, represents an expenditure of energy and results in a change in the overall activity of the cell. It is to be expected then that the pH of the growth medium may have a profound effect on secondary metabolism; e.g. the rate of melanin formation in *Aspergillus nidulans* undergoes a 20-fold change when the external pH is changed from 7 to 7.9 (ROWLEY and PIRT[49]).

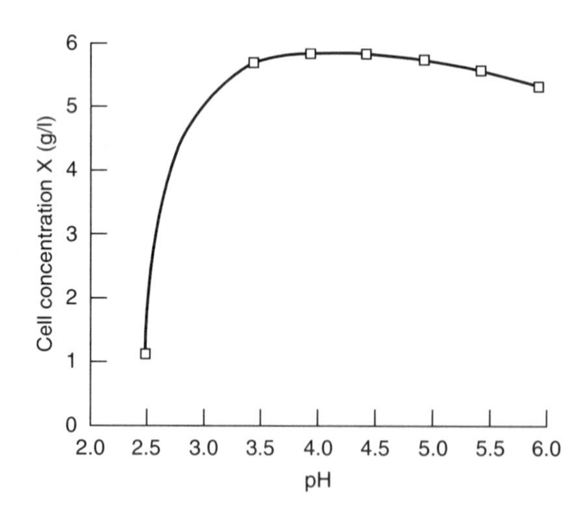

FIG. 5.48. Effect of pH on the growth rate of *S. cerevisae* (WASUNGU and SIMARD[48])

Temperature

There is considerable variation in temperature tolerance and optima between species and three broad categories of micro-organisms have been distinguished in this respect. *Psychrophiles* have temperature optima below 20°C, *mesophiles* between 20 and 45°C and *thermophiles* have their optimum growth temperatures above 45°C (ATKINSON and MAVITUNA[2]). RYU and MATELES[50] showed that the growth rate of *E. coli* on a nitrogen-limited glycerol medium varied according to an Arrhenius function:

$$\mu(T) = \mu_0 \exp\left\{-\frac{\mathbf{E}_a}{\mathbf{R}}\left(\frac{1}{T} - \frac{1}{T_0}\right)\right\} \tag{5.73}$$

where $\mu(T)$ is the specific growth rate at temperature T, μ_0 is that at temperature T_0, $\mathbf{R}$ is the gas constant and $\mathbf{E}_a$ the activation energy, in their particular case having a value of 3.2×10^3 kJ/kmol.

A simplified form of this equation, recommended for use in waste water treatment (Metcalf and Eddy Inc.[51]), has the form:

$$\mu(\theta) = \mu(20)\,\alpha^{(\theta - 20)} \tag{5.74}$$

where $\mu(\theta)$ is the specific growth rate at θ°C, $\mu(20)$ is that at 20°C and α is a constant in the range 1.00–1.08 with a typical value of 1.04. It should be noted that the exponent in this equation is not dimensionless and temperature must be expressed in °C.

Temperature is recognised as having an effect on the growth yield, the endogenous respiration rate and the Monod kinetic parameters K_S and μ_m. Within the temperature range of 25 to 40°C these have been shown to have dependencies which could be accounted for by Arrhenius-type exponential equations (TOPIWALA and SINCLAIR[52]). If the temperature-dependent nature of the constants has to be taken into account, equation 5.70 must be written as:

$$\mu(T) = \mu_m(T)\frac{S}{K_S(T) + S} - k_d(T) \tag{5.75}$$

the relationships proposed were of the form:

$$\mu_m(T) = \mathscr{A}_1 \exp\left(\frac{-\mathbf{E}_{a1}}{\mathbf{R}T}\right) - \mathscr{A}_2 \exp\left(\frac{-\mathbf{E}_{a2}}{\mathbf{R}T}\right) \tag{5.76}$$

$$\frac{1}{K_S(T)} = \mathscr{A}_3 \exp\left(\frac{-\mathbf{E}_{a3}}{\mathbf{R}T}\right) \tag{5.77}$$

and

$$k_d(T) = \mathscr{A}_4 \exp\left(\frac{-\mathbf{E}_{a4}}{\mathbf{R}T}\right) \tag{5.78}$$

It must be emphasised that these expressions are valid over only a relatively narrow range of temperatures, reflecting the extremely complex nature of the interacting biochemical reactions taking place within the cell. Plots of $\ln(\mu)$ *v* $(1/T)$ typically give several regions of only local linearity indicating the danger of extrapolating outside the ranges investigated experimentally.

Other Factors

Other factors which may affect the specific growth rate include the ionic strength of the medium. In the case of the mixed culture used in waste water treatment, an increase in salinity to that of sea water is accompanied by a sharp drop in the growth rate (KINCANNON and GAUDY[53]). This, however, is a transient response and the culture recovers a large proportion of its activity when allowed to acclimatise. The effect is difficult to quantify, as it applies to a mixed culture and is a transient, but it does underline the complexity of environmental influences on growing micro-organisms. For the most part therefore the mathematical treatment of biological reactors is restricted to considering the effect of substrate concentration on the growth rate, it being assumed that temperature, pH, ionic strength and other parameters are constant.

5.9.3. Product Formation

Microbial products may be formed as a result of a variety of processes which occur within the cells of a microbial culture. In some cases the cell itself may be the desired product, or it may be that the product is formed as the direct consequence of its growth. There is in the latter circumstances a direct link between growth and the accumulation of product. However, there are many important microbial products which are not growth-associated, where there may exist mechanisms within the cells, as outlined in Section 5.5.2, which operate to produce a particular material only under certain special circumstances. As a result, the yield coefficient $Y_{P/S}$ is of little value in predicting the course of product accumulation, except when the overall productivity of a complete batch operation is considered.

GADEN[54] proposed a scheme which grouped fermentations according to the manner in which the microbial product appeared in the broth (Table 5.18). Whilst this is simpler than that put forward later by DEINDOERFER[55], it does form an useful basis from which to develop a quantitative description of the process.

TABLE 5.18. *Gaden's Classification Scheme for Fermentations*

Fermentation type	Characteristic	Example
1	Product formation is directly related to carbohydrate utilisation	Ethanol
2	Product formation indirectly related to carbohydrate utilisation	Citric acid
3	Product formation apparently not associated with carbohydrate utilisation	Penicillin

LUEDEKING and PIRET[56] proposed that the formation of microbial products could be described by a two parameter model. This was based on their studies of lactic acid fermentation using *Lactobacillus delbruekii* in which the accumulation of the acid is not directly linked to the growth rate. The function used was:

$$\frac{dP}{dt} = \delta\frac{dX}{dt} + \varepsilon X \tag{5.79}$$

and the quantities δ and ε were found to be pH-dependent. The first term on the right-hand side of the expression relates to that part of the product which *is* growth associated, whilst the second refers to that portion of the product which results from cellular activity associated with maintenance functions. Dividing equation 5.79 by the biomass concentration X gives the relationship in terms of the specific rates, i.e.:

$$\frac{1}{X}\frac{dP}{dt} = \delta\frac{1}{X}\frac{dX}{dt} + \varepsilon \tag{5.80}$$

and substituting for the specific growth rate:

$$\frac{1}{X}\frac{dP}{dt} = \delta\mu + \varepsilon \tag{5.81}$$

The values of δ and ε may be estimated by calculating the slope and intercept of the plot of the specific rate of product formation $(1/X)(dP/dt)$ against μ. Clearly, for a purely growth associated product ε will tend to zero but will dominate the expression for a growth independent product.

SHU[57] devised a product formation model which took into account the fact that the individual cells making up the growing culture are not identical and that their metabolic functions vary accordingly. The differentiation of the population in this manner gives rise to a *segregated* model of microbial culture. It is arguably a more realistic approach than that where the biomass is considered as being homogeneous, but suffers from the drawback that it is mathematically complex. In view of the apparent dependence of product formation on the age of the individual cells, Shu's model makes the age distribution the central theme.

If τ is the cell age then the instantaneous rate of product formation under constant environmental conditions is assumed to take the form:

$$\frac{dP}{dt} = \sum_i A_i e^{-\mathbf{k}_i \tau} \tag{5.82}$$

where A_i and $\boldsymbol{k}_i$ are positive constants. The concentration of product accumulated in the broth over the lifetime of the cell per unit mass of the cell will be:

$$P_t = \int_0^\tau \sum_i A_i e^{-\mathbf{k}_i \tau} d\tau \tag{5.83}$$

If the culture contains a range of cells of all possible ages, represented by $\mathscr{C}(\tau)$ and the culture time is t then the overall cell concentration will be:

$$X_t = \int_0^t \mathscr{C}(\tau)\, d\tau \tag{5.84}$$

and the overall product concentration in the fermenter becomes:

$$P = \int_0^t \mathscr{C}(\tau) \left\{ \int_0^\tau \sum_i A_i e^{-\mathbf{k}_i \tau} d\tau \right\} d\tau \tag{5.85}$$

Implementation of the model requires the choice of a suitable form for $\mathscr{C}(\tau)$ (for example, $X_t = X_0\, e^{k(t-\tau)}$ would be appropriate for exponential growth), as well as the values of A_i and k_i. Shu developed from equation 5.85 a double exponential expression to describe penicillin production. The formula is open to criticism for its extreme flexibility and generality but is noteworthy as being comparable with the Ludeking and Piret model in its form and performance.

5.10. IMMOBILISED BIOCATALYSTS

The remarkable catalytic properties of enzymes make them very attractive for use in processes where mild chemical conditions and high specificity are required. Cheese manufacture has traditionally used rennet, an enzyme preparation from calf stomach, as a specific protease which leads to the precipitation of protein from milk. 'Mashing' in the malting of grain for the brewing of beer makes use of β-amylase from germinating grain to hydrolyse starch to produce sugars for the fermentation stage. In both of these examples the enzymes are not recovered from the reaction mixture and a fresh preparation is used for each batch. Similarly, in more modern enzyme reaction applications, such as in biological washing detergents, the enzyme is discarded after single use but there are, however, situations where it may be desirable to recover the enzyme. This may be because the product is required in a pure state or that the cost of the enzyme preparation is such that single use would be uneconomic. To this end, immobilised biocatalysts have been developed where the original soluble enzyme has been modified to produce an insoluble material which can be easily recovered from the reaction mixture.

Many industrially important micro-organisms tend to agglomerate during their growth and form flocs suspended in the culture medium or films which adhere to the internal surfaces of the fermenter. This tendency may or may not be advantageous to the process and is dependent on a variety of parameters such as the pH and ionic strength of the medium and the shear rate experienced in the growth vessel. In some cases the formation of substantial flocs is essential to the proper operation of the process. In the case of the activated sludge waste water treatment the settling properties of the flocculated micro-organisms are utilised in order to produce a concentrated stream of biomass for the recycle. The so-called 'trickling filter', also in widespread use in waste-water treatment, is reliant on the formation of a film of organisms on the surfaces of its packing material. The operation is not that of a filter, in which material would be removed on the basis of its particle size, but that of a biological reactor in which the waste material forms the substrate for the growth of the microbes in the zoogleal film. The presence of the film provides a means of retaining a higher microbial concentration in the reactor than would be retained in a comparable stirred-tank fermenter. The formation of flocs and films for the retention of high microbial densities or to facilitate separation of microbes from the growth medium may be desirable in other instances as well. However, in some cases the microbe used may neither be amenable to the natural formation of large flocs nor adhere as surface films, and recourse may be made to the artificial immobilisation of microbes.

Immobilisation Techniques

There are various methods which have been developed for enzyme and micro-organism immobilisation and some of these have found commercial application.

The two largest scale industrial processes utilising immobilised enzymes are the hydrolysis of benzyl penicillin by penicillin acylase and the isomerisation of glucose to a glucose–fructose mixture by immobilised glucose isomerase. The immobilisation techniques used in general may be broadly categorised as:

(*a*) *Physical adsorption on to an inert carrier*. The first of these methods has the advantage of requiring only mild chemical conditions so that enzyme deactivation during the immobilisation stage is minimised. The natural formation of microbial flocs and films may be considered to be in this category, although the subsequent adhesion of the microbes to the surface may not be a simple phenomenon. Special materials may be used as supports which provide the microbes with environments which are particularly amenable to their adhesion; such materials include foam plastics which provide conditions of low shear in their pores. The process may also be relatively cheap but it does tend to have the drawback that desorption of the enzyme may also occur readily or that the microbial film may slough and be carried into the bulk of the growth medium. The process is dependent on the nature of the specific enzyme or microbe used and its interaction with the carrier and, whilst it is common in the case of immobilised microbes, it has found only limited application in the case of immobilised enzymes.

(*b*) *Inclusion in the lattices of a polymer gel or in micro-capsules*. This method attempts to overcome the problem of leakage by enclosing the relatively large enzyme molecules or microbes in a tangle of polymer gel (one analogy made is that of a football trapped in a heap of brushwood), or to enclose them in a membrane which is porous to the substrate. It is theoretically possible to immobilise any enzyme or micro-organism using these methods but they too have their problems. Some leakage of the entrapped species may still occur, although this tends to be minimal, particularly in the case of micro-encapsulation of enzymes or respiring but non-growing cells. The main problem is due to mass transfer limitations to the introduction of the necessarily small substrate molecules into the immobilised structure, and to the slow outward diffusion of the product of the reaction. If the substrate is itself a macro-molecule, such as a protein or a polysaccharide, then it will be effectively screened from the enzyme or microbes and little or no reaction will take place.

(*c*) *Covalent binding*. Biological catalysts may be made insoluble and hence immobilised by effectively increasing their size. This can be done either by chemically attaching them to otherwise inert carrier materials or by crosslinking the individuals to form large agglomerations of enzyme molecules or micro-organisms. The chemical reagents used for the linking process are usually bifunctional, such as the carbo-di-imides, and many have been developed from those used in the chemical synthesis of peptides and proteins. The use of a carrier is the most economical in terms of enzyme usage since the local enzyme activity in the crosslinked enzyme will be less of a limitation than the rate of transfer of substrate to the active centres. This has the result that in many cases only about 10 per cent of the original enzyme activity can be realised. The inert carriers used tend to be hydrophilic materials, such as cellulose and its derivatives, but in some cases the debris of the original cells has been used, the cells having been broken and then crosslinked with the enzyme and each other to form large particles. The latter technique has the advantage of missing out some of the purification steps (with their loss of total activity) which are normally associated with enzyme recovery and also avoids the need to satisfy the maintenance requirement of the living cells.

Loss of Activity

In general, when enzymes (or microbes) are immobilised for use in engineering systems a significant decrease in overall activity is observed. The decrease may be ascribed to three effects:

(a) *Loss due to deactivation of the catalytic activity by the immobilising procedure itself*. This includes destruction of the active sites of the enzyme by the reagents used and the obstruction of the active sites by the support material.

(b) *Loss of overall activity by diffusional limitation external to the immobilised system*. This refers to the apparent loss in activity when the rate of reaction is controlled by transport of the substrate from the bulk of the solution to the surface of the carrier of the immobilised biocatalyst. This is particularly important when an enzyme is attached to the surface of a carrier or the microbes form a very thin film, with negligible activity within the support.

(c) *Loss of overall activity due to diffusional limitation within the immobilised catalyst matrix*. This can clearly arise when gel entrapment is being considered but it can also occur when enzymes or microbes are covalently attached within pores in the inert carrier.

The consumption or biotransformation of substrate by immobilised micro-organisms results in most cases in the growth of the micro-organisms. The growth which gives rise to a significant increase of thickness in an established biofilm, occurs at a rate which is essentially slow in comparison with the rates of the diffusion processes. Simultaneously, the attrition of biofilms or flocs arising from the effects of fluid flow tends to maintain their thickness or size, and, overall, the immobilised system can be considered to be in a quasi-steady state when short time intervals are involved. The mathematical similarity of enzyme and microbial kinetics then means that a common set of equations can be used to describe the behaviour of both immobilised enzymes and microbial cells. The following discussion is therefore valid for both kinds of immobilised biosystems which are, in many respects, comparable with the conventional chemical catalysts discussed in Chapter 3.

5.10.1. Effect of External Diffusion Limitation

If the activity of the immobilised catalyst is sufficiently high, the reaction which it mediates occurs essentially at the interface between the catalyst and the substrate solution. In the case of the surface immobilised enzyme or a thin microbial film this will, of course, occur irrespective of the level of activity. Under these conditions the limiting process for transporting substrate from the bulk of the solution to the immobilised enzyme is molecular or convective diffusion through the layer of solution immediate to the carrier. Under steady-state conditions, the rate of reaction at the active sites is equal to the rate at which substrate arrives at the site. This material balance may be written as:

$$h_D(S_b - S_s) = \mathcal{R}' \tag{5.86}$$

where:

h_D is the mass transfer coefficient
S_b is the bulk substrate concentration
S_s is the substrate concentration at the surface, and
$\mathcal{R}'$ is the rate of reaction per unit surface area of catalyst

Assuming that the local rate of enzyme reaction follows Michaelis–Menten kinetics, or that the microbe film follows Monod kinetics regardless of immobilisation, then equation 5.86 becomes:

$$h_D(S_b - S_s) = \frac{\mathcal{R}'_m S_s}{K_x + S_s} \tag{5.87}$$

where $\mathcal{R}'_m$ is the maximal rate of reaction per unit surface area, and
K_x is the Michaelis *or* the Monod constant.

The dimensionless substrate concentration may be defined by $\beta = \dfrac{S}{K_x}$ so that substitution in equation 5.87 and rearranging gives:

$$(\beta_b - \beta_s) = \frac{\beta_s}{(1 + \beta_s)} \frac{\mathcal{R}'_m}{h_D K_x} \tag{5.88}$$

where the subscripts *b* and *s* refer to bulk and surface parameters as before. The dimensionless group $\dfrac{\mathcal{R}'_m}{h_D K_x}$ is the Damköhler number (*Da*) and equation 5.88 may be written:

$$(\beta_b - \beta_s) = \frac{\beta_s}{(1 + \beta_s)} Da \tag{5.89}$$

The Damköhler number represents the ratio of the maximum rate of reaction to the maximum transport rate of substrate to the surface. A large value for *Da* therefore indicates that the transport is the limiting step in the consumption of the substrate, whereas a small value would show that the rate of reaction is more important. Figure 5.49 shows the variation of the rate of reaction with substrate concentration for various values of *Da*. The rate has been normalised by expressing it as a fraction of $\mathcal{R}'_m$ and it may be seen that, where there is no diffusional resistance ($Da = 0$), the form of the plot is that which would be obtained if the enzyme were used in free solution. With increase in *Da* the observed rate of reaction decreases and, in particular, the initial slope of the curve (Fig. 5.49) is less. This would be observed as an increase in the apparent value of the affinity constant (K_S or K_m) for the reaction if such a determination were to be carried out disregarding the immobilised nature of the biocatalyst. On the other hand, $\mathcal{R}'_m$ would be found to be unchanged, although difficult to determine, because the plot of rate against substrate concentration would be linear for a much greater range of concentrations. The straight line is a reflection of the dominance of diffusional effects in controlling the overall rate, since under these conditions $S_b - S_s \Rightarrow S_b$ and the rate of reaction is then given by $h_D S_b$.

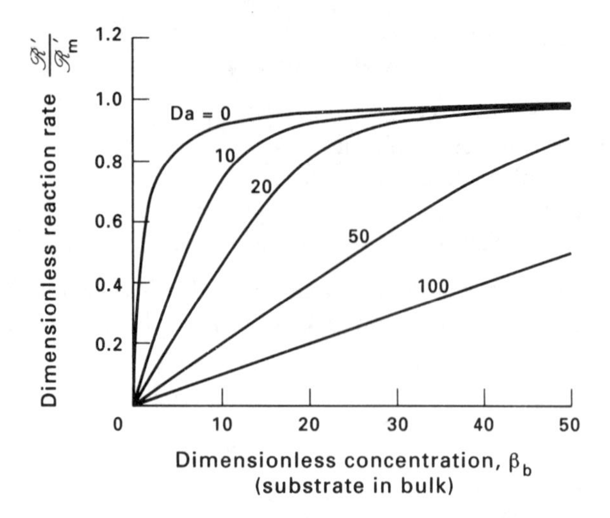

FIG. 5.49. Dimensionless plot of overall reaction rate against bulk substrate concentration for a surface immobilised biocatalyst

If the rate of reaction obtained with no diffusional restrictions is denoted by the symbol $\mathcal{R}_k'$ then an effectiveness factor may be defined as:

$$\eta_e = \frac{\mathcal{R}'}{\mathcal{R}_k'} \tag{5.90}$$

A plot of η_e against the Damköhler number is shown in Fig. 5.50 with the bulk concentration as parameter. On the graph it may be seen that η_e is dependent on both β_b and Da, but that three identifiable regions exist. At low values of Da, kinetic control of the reaction is observed and the curves show that η_e approaches unity for most substrate concentrations, whilst at high bulk substrate concentrations the

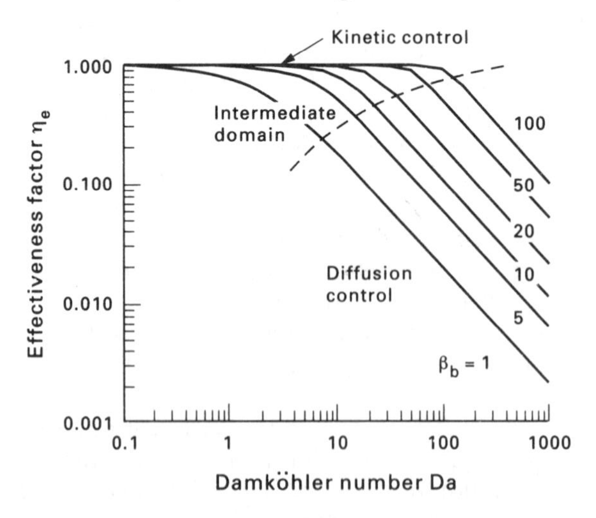

FIG. 5.50. Variation of effectiveness factor η_e for a surface immobilised enzyme with Damköhler number

effectiveness factor still approaches unity, even for Damköhler numbers of a hundred. The diffusion-controlled domain is indicated on the diagram as the region below the broken line.

The deviation of the reaction rate $\mathcal{R}'$, from the rectangular hyperbola which would be shown by a true Michaelis–Menten reaction law, is best illustrated by considering the data as represented by an Eadie–Hofstee plot. The original equation for the Michaelis–Menten or Monod kinetics:

$$\mathcal{R}' = \frac{\mathcal{R}'_m S}{K_x + S} \tag{5.91}$$

is converted to its dimensionless form again by dividing by $\mathcal{R}'_m$ and replacing S and K_x by $\beta = \dfrac{S}{K_x}$ to give:

$$\frac{\mathcal{R}'}{\mathcal{R}'_m} = \frac{\beta}{1 + \beta} \tag{5.92}$$

Then, as previously, inverting the equation and multiplying through by $\dfrac{\mathcal{R}'}{\mathcal{R}'_m}$ gives, after re-arranging:

$$\frac{\mathcal{R}'}{\mathcal{R}'_m} = -\frac{\mathcal{R}'}{\mathcal{R}'_m}\frac{1}{\beta} + 1 \tag{5.93}$$

Now, for Michaelis–Menten or Monod kinetics a plot of $\dfrac{\mathcal{R}'}{\mathcal{R}'_m}$ against $\dfrac{\mathcal{R}'}{\mathcal{R}'_m}\dfrac{1}{\beta}$ as shown in Fig. 5.12, would result in a straight line with a slope of -1.

The plot at larger values of *Da* shows a marked departure from this pattern with the line tending to become vertical and near to the ordinate in the extreme case.

The rate of reaction of a surface-immobilised biocatalyst may be determined graphically, as shown in Fig. 5.52. The curve marked $\mathcal{R}'_k$ is that relating the substrate concentration to the rate of reaction when there is no diffusional

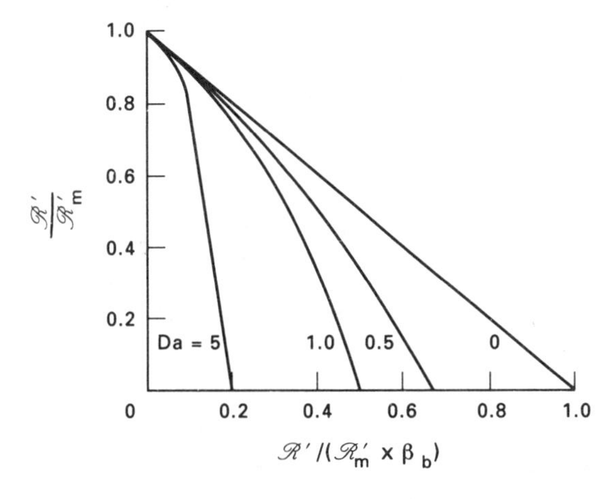

FIG. 5.51. Eadie–Hofstee type plot showing departure from Michaelis–Menten kinetics due to external diffusion limitation

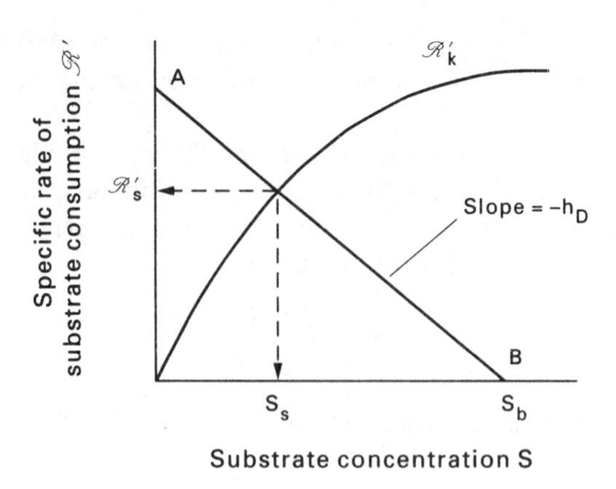

FIG. 5.52. Graphical solution of equation 5.86

limitation (i.e. when the reaction is kinetically controlled). The intersecting line AB is drawn such that B is on the axis where $S_s = S_b$, which represents the condition when the rate of reaction is zero, and the surface substrate concentration is, as a result, the same as that in the bulk. The slope of this line is $-h_D$ and the line cuts the $\mathcal{R}'$ axis at the point where $\mathcal{R}' = h_D S_b$. The point at which this line intersects the kinetic curve will correspond to the effective rate of reaction and the substrate concentration at the surface.

5.10.2. Effect of Internal Diffusion Limitation

In the case of gel entrapped biocatalysts, or where the biocatalyst has been immobilised in the pores of the carrier, then the reaction is unlikely to occur solely at the surface. Similarly, the consumption of substrate by a microbial film or floc would be expected to occur at some depth into the microbial mass. The situation is more complex than in the case of surface immobilisation since, in this case, transport and reaction occur in parallel. By analogy with the case of heterogeneous catalysis, which is discussed in Chapter 3, the flux of substrate is related to the rate of reaction by the use of an effectiveness factor η. The rate of reaction is itself expressed in terms of the surface substrate concentration which in many instances will be very close to the bulk substrate concentration. In general, the flux of substrate will be given by:

$$\mathcal{R}'' = \eta\, \mathcal{R}''_m \frac{S_b}{K_x + S_b} \tag{5.94}$$

where $\mathcal{R}''_m$ is $V''_{\max}$ (the rate of reaction per unit volume of immobilised catalyst) for an enzyme and $\dfrac{\mu_m X}{Y_{X/S}}$ for the case of microbes.

The simplest case to consider is that of an uniform microbial film or of an enzyme which is immobilised uniformly through a slab of supporting material which has infinite area but finite depth. As in the previous discussion the local rate of reaction is assumed to be described by Michaelis–Menten or Monod kinetics, so that at steady state a material balance for any point in the slab gives, using the same nomenclature as before:

$$D_e \frac{d^2 S}{dx^2} = \frac{\mathcal{R}''_m S}{K_x + S} \tag{5.95}$$

where:

D_e is the effective diffusivity of the substrate in the slab, and
x is the distance measured from the surface of the slab.

This equation may be made dimensionless by putting $\beta = \frac{S}{K_x}$ as before and letting $z = \frac{x}{L}$ where L is the total thickness of the slab. Equation 5.95 then becomes:

$$\frac{d^2\beta}{dz^2} = \phi^2 \frac{\beta}{1 + \beta} \tag{5.96}$$

where ϕ is the Thiele modulus defined by:

$$\phi = L \sqrt{\frac{\mathcal{R}''_m}{D_e K_m}} \tag{5.97}$$

for an immobilised enzyme system, and by:

$$\phi = L \sqrt{\frac{\mu_m X}{Y_{X/S} D_e K_S}} \tag{5.98}$$

for a microbial film.

Equation 5.96 can be solved by numerical integration using the boundary values $\beta = \beta_s$ when $z = 0$ and $\frac{d\beta}{dz} = 0$ when $z = 1$ to yield the set of curves shown in Fig. 5.53. The graph shows the overall rate of reaction normalised as $\frac{\mathcal{R}''}{\mathcal{R}''_m}$, in the same

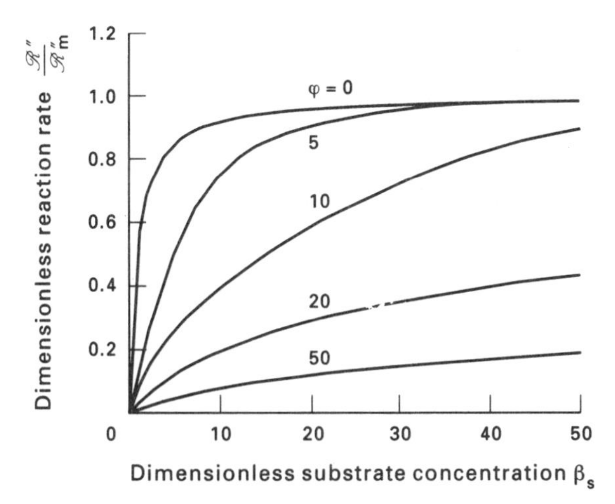

FIG. 5.53. The overall rate or reaction for uniformly distributed immobilised biocatalyst in an infinite slab of carrier

manner as in Fig. 5.49 for the surface-immobilised enzyme, but with ϕ as the parameter.

As with the external diffusion limitation, a family of curves is obtained which shows that the overall rate of reaction decreases with increase in the Thiele modulus (as compared with the Damköhler number for the external diffusion limitation). The rate is essentially under kinetic control at low values of ϕ ($\phi < 1$); that is, there is negligible diffusion limitation. In contrast to the case of external diffusion, it may be seen from the curvature of the lines in Fig. 5.53 that the rate of reaction is always a function of the kinetic parameters, even at the higher values of ϕ.

HORVATH and ENGASSER[58] point out that, whilst the curves are superficially similar to those produced by the Michaelis–Menten reaction scheme, or Monod kinetics in the case of micro-organisms, the functional relationship with the surface concentration is different. This difference may be most marked when the Eadie–Hofstee type plot is considered, as shown in Fig. 5.54. This time the curves obtained for the cases where ϕ is not zero have a sigmoidal shape, even for low values of ϕ.

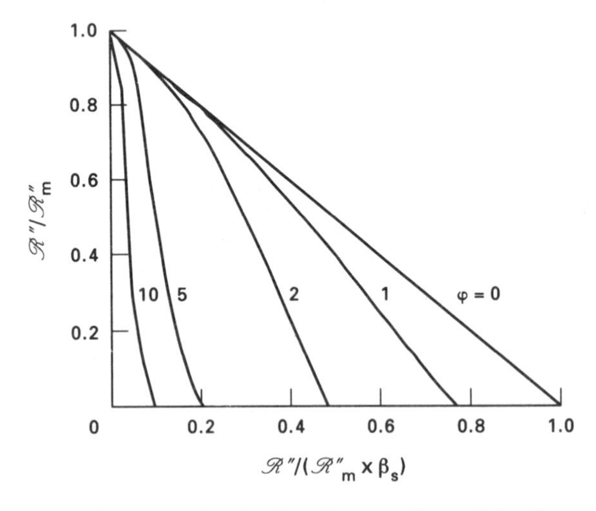

FIG. 5.54. Eadie–Hofstee type plot showing departure from Michaelis–Menten kinetics due to internal diffusion limitation

The effectiveness factor η may be given in terms of the modified Thiele modulus ϕ' which is defined by ATKINSON[59] and ATKINSON and MAVITUNA[2] as:

$$\phi' = \frac{\beta_s}{(1+\beta_s)} \frac{\phi}{\sqrt{2(\beta_s - \ln(1+\beta_s))}} \tag{5.99}$$

so that:

$$\eta = 1 - \frac{\tanh \phi}{\phi}\left(\frac{\phi'}{\tanh \phi'} - 1\right) \quad \text{for } \phi' \leqslant 1 \tag{5.100}$$

and:

$$\eta = \frac{1}{\phi'} - \frac{\tanh \phi}{\phi}\left(\frac{\phi'}{\tanh \phi'} - 1\right) \quad \text{for } \phi' \geqslant 1 \tag{5.101}$$

For spherical particle geometry, as in the case of a microbial floc, a pellet of mould or a bead of gel-entrapped enzyme, the expression for the effectiveness factor can again be derived by a procedure similar to that used in Chapter 3 for a spherical pellet of conventional catalyst. A material balance for the substrate across an elementary shell of radius r and thickness $\mathrm{d}r$ within the pellet will yield:

$$D_e\left(\frac{\mathrm{d}^2S}{\mathrm{d}r^2}+\frac{2}{r}\frac{\mathrm{d}S}{\mathrm{d}r}\right)=\mathcal{R}''_s \tag{5.102}$$

where:

S is the local substrate concentration, and
$\mathcal{R}''_s$ is the local rate of reaction of the substrate per unit volume of the sphere.

Equation 5.102 can be converted into a dimensionless form by letting:

$$\overline{S}=\frac{S}{S_b} \text{ and } \overline{r}=\frac{r}{r_p}$$

where r_p is the radius of the particle. The equation then becomes:

$$\frac{\mathrm{d}^2\overline{S}}{\mathrm{d}\overline{r}^2}+\frac{2}{\overline{r}}\frac{\mathrm{d}\overline{S}}{\mathrm{d}\overline{r}}=\frac{\mathcal{R}''_s\, r_p}{D_e S_b} \tag{5.103}$$

or:

$$\frac{\mathrm{d}^2\overline{S}}{\mathrm{d}\overline{r}^2}+\frac{2}{\overline{r}}\frac{\mathrm{d}\overline{S}}{\mathrm{d}\overline{r}}=\phi^2\frac{\overline{S}}{1+\beta_s\overline{S}} \tag{5.104}$$

where $\beta_s=\dfrac{S_s}{K_X}$. The Thiele modulus ϕ is defined as in equations 5.97 and 5.98 but with the thickness L of the film replaced by the radius r_p of the particle.

At low values of S_s the reaction will be pseudo first-order and the solution given in Chapter 3 will apply. The effectiveness factor will in that case be given by:

$$\eta=\frac{1}{\phi}\left(\frac{1}{\tanh 3\phi}-\frac{1}{3\phi}\right) \tag{5.105}$$

Sufficiently large values of S_s will give rise to zero-order kinetics and in that case it may be shown that:

$$\eta=1-\left(1-\frac{6\beta}{\phi^2}\right) \tag{5.106}$$

In the case where neither of these asymptotic conditions applies, that is the system shows intermediate order, then equation 5.104 has to be solved numerically. The boundary conditions for that solution are that $\overline{S}=1$ when $\overline{r}=1$ and $\dfrac{\mathrm{d}\overline{S}}{\mathrm{d}\overline{r}}=0$ when $\overline{r}=0$.

5.11. REACTOR CONFIGURATIONS

5.11.1. Enzyme Reactors

Enzymes are frequently used as catalysts to promote specific reactions in free solution. They are typically required in small amounts and are attractive in that they obviate both the need to provide the nutritional support which would be required for micro-organisms to perform the same conversion, and the possible subsequent removal of those microbes. Furthermore, the enzyme need not necessarily be of microbial origin so that a wider choice of operating conditions and characteristics may be available.

In a batch reactor where substrate is being converted to product by the enzyme, a material balance for the substrate gives:

$$-\left\{\begin{array}{l}\text{Rate of}\\\text{formation of}\\\text{substrate}\end{array}\right\} = \left\{\begin{array}{l}\text{Rate of}\\\text{enzyme}\\\text{reaction}\end{array}\right\}$$

If the substrate concentration is S, the volume of the reactor is V and t is the time then this becomes:

$$-V\frac{\mathrm{d}S}{\mathrm{d}t} = V\mathcal{R}_0 \tag{5.107}$$

where $\mathcal{R}_0$ is the activity of the enzyme expressed in quantity of substrate used per unit volume in unit time. Thus, if the initial concentration of substrate is S_0 (i.e. at time $t = 0$) and S is the concentration at time t, then these are related by the integral:

$$-\int_{S_0}^{S} \mathrm{d}S = \int_0^S \mathcal{R}_0\,\mathrm{d}t \tag{5.108}$$

As discussed before, the rate of an enzyme-catalysed reaction is dependent, not only on the substrate concentration at any instant, but also on the temperature, pH and the degree of decay of that enzyme. For given values of pH and temperature, the specific rate of reaction v is given, in the simplest case, by the Michaelis–Menten equation:

$$\mathcal{R}_0 = \frac{V_{\max,t}\,S}{K_m + S} \qquad \text{(Equation 5.10)}$$

and the constant $V_{\max,t}$ at any time t by:

$$V_{\max,t} = V_{\max,0}\exp(-\boldsymbol{k}_{de}t) \tag{5.109}$$

where $V_{\max,0}$ is the specific activity at the beginning of the batch reaction and $\boldsymbol{k}_{de}$ is the first-order kinetic constant for the decay of the enzyme. Combining these equations gives:

$$\mathcal{R}_0 = \frac{V_{\max,0}S}{K_m + S}\exp(-\boldsymbol{k}_{de}t) \tag{5.110}$$

which may be substituted into the expression derived from the material balance above to give:

$$-\,\mathrm{d}S = \frac{V_{\max,0}\,S}{K_m + S}\exp(-\,k_{de}t)\,\mathrm{d}t \tag{5.111}$$

Separating the variables:

$$-\int_{S_0}^{S}\frac{K_m + S}{S}\,\mathrm{d}S = V_{\max,0}\int_0^t \exp(-\,k_{de}t)\mathrm{d}t \tag{5.112}$$

and integrating between the limits shown:

$$K_m \ln\left(\frac{S_0}{S}\right) + (S_0 - S) = V_{\max,0}\,\frac{1 - \exp(-\,k_{de}t)}{k_{de}} \tag{5.113}$$

In the case where the decay of the enzyme is negligible, then $k_{de} \Rightarrow 0$ and $\dfrac{1 - \exp(-\,k_{de}t)}{k_{de}} \Rightarrow t$ which results in the expression:

$$K_m \ln\left(\frac{S_0}{S}\right) + (S_0 - S) = V_{\max}t \tag{5.114}$$

These design equations for a free-enzyme batch reaction may be formulated in a similar manner for kinetic schemes other than the case used above. If, for example, the back reaction is significant then the procedure would make use of equation A5.4.13 to define $\mathcal{R}_0$, and at infinite time an equilibrium mixture of substrate and product would be obtained.

5.11.2. Batch Growth of Micro-Organisms

The batch growth of micro-organisms involves adding a small quantity of the micro-organisms or their spores (the seed culture or inoculum) to a quantity of nutrient material in a suitable vessel. In the case of an aerobic fermentation (i.e. a growth process requiring the presence of molecular oxygen) the contents of the vessel (or fermenter) are aerated and the growth of the micro-organisms allowed to proceed. For convenience, the case where the feed material is present in aqueous solution is considered and, furthermore, it is assumed that in the feed there is contained a carbon and energy source which is the limiting substrate for the growth of the culture. Whilst for an aerobic culture aeration is of prime importance, the fact that air enters the vessel and leaves enriched in carbon dioxide will be ignored in this discussion and the analysis focused on the changes occurring in the liquid phase.

After inoculation, assuming no lag phase, the resultant growth can be analysed by considering the unsteady-state material balances for the substrate and biomass. The general form of this balance for a fermenter is:

$$\left\{\begin{matrix}\text{Flow of}\\ \text{material}\\ \text{in}\end{matrix}\right\} + \left\{\begin{matrix}\text{Formation by}\\ \text{biochemical}\\ \text{reaction}\end{matrix}\right\} - \left\{\begin{matrix}\text{Flow of}\\ \text{material}\\ \text{out}\end{matrix}\right\} = \text{Accumulation}$$

Since a batch process is being considered, the flow in and out of the fermenter are both zero and the expression reduces to:

$$\begin{Bmatrix}\text{Formation by} \\ \text{biochemical} \\ \text{reaction}\end{Bmatrix} = \text{Accumulation}$$

So, for the case of the biomass:

$$\mathcal{R}_X V = \mu V X = \frac{\mathrm{d}X}{\mathrm{d}t} V \tag{5.115}$$

where μ is the specific growth rate, V is the volume of the vessel and X is the instantaneous concentration of the biomass. If Y is the overall yield coefficient for the formation of biomass and the limiting substrate concentration is S, then the equivalent expression for substrate is:

$$\mathcal{R}_S V = \frac{\mathrm{d}S}{\mathrm{d}t} V \tag{5.116}$$

where $\mathcal{R}_S$ is the rate of conversion of substrate per unit volume of the reactor. Equation 5.116 makes no assumptions regarding the uniformity of the yield coefficient $Y_{X/S}$, but if that can be taken to be constant then equation 5.44 may be used to relate equations 5.115 and 5.116. This condition is met when μ is large in comparison with m (see equation 5.54) so that, dispensing with the subscript, the differential form of equation 5.44 can be written:

$$Y\frac{\mathrm{d}S}{\mathrm{d}t} = -\frac{\mathrm{d}X}{\mathrm{d}t} \tag{5.117}$$

which gives:

$$Y\mathcal{R}_S = -\mu X \tag{5.118}$$

Equation 5.116 thus becomes:

$$-\frac{1}{Y}\mu X = \frac{\mathrm{d}S}{\mathrm{d}t} \tag{5.119}$$

The yield coefficient may also be expressed in its integral form as:

$$Y = \frac{X - X_0}{S_0 - S} \tag{5.120}$$

which can be re-arranged:

$$S = S_0 - \frac{X - X_0}{Y} \tag{5.121}$$

If the growth follows the Monod kinetic model, then equation 5.62 may be substituted into equation 5.115 to give:

$$\frac{\mathrm{d}X}{\mathrm{d}t} = \frac{\mu_m S X}{K_S + S} \tag{5.122}$$

The condition of the fermentation after any time t would then be given by:

$$\int_{X_0}^{X} \frac{K_S + S}{\mu_m S} \frac{dX}{X} = \int_0^t dt \tag{5.123}$$

However, S is a function of X and substitution using equation 5.121 must be made before carrying out the integration. The result is:

$$\frac{(K_S Y + S_0 Y + X_0)}{\mu_m (Y S_0 + X_0)} \ln\left(\frac{X}{X_0}\right) + \frac{K_S Y}{\mu_m (Y S_0 + X_0)} \ln\left(\frac{Y S_0}{Y S_0 + X_0 - X}\right) = t \tag{5.124}$$

A similar expression can be obtained for the substrate concentration:

$$\frac{(K_S Y + S_0 Y + X_0)}{\mu_m (Y S_0 + X_0)} \ln\left(1 + \frac{Y(S_0 - S)}{X_0}\right) - \frac{K_S Y}{\mu_m (X_0 + Y S_0)} \ln\left(\frac{S}{S_0}\right) = t \tag{5.125}$$

These rather unwieldy equations can be used to generate a graph showing the changes in biomass and substrate concentrations during the course of a batch fermentation (see Fig. 5.55). Their main disadvantage is that they are not explicit in X and S so that a trial and error technique has to be used to determine their values at a particular value of t.

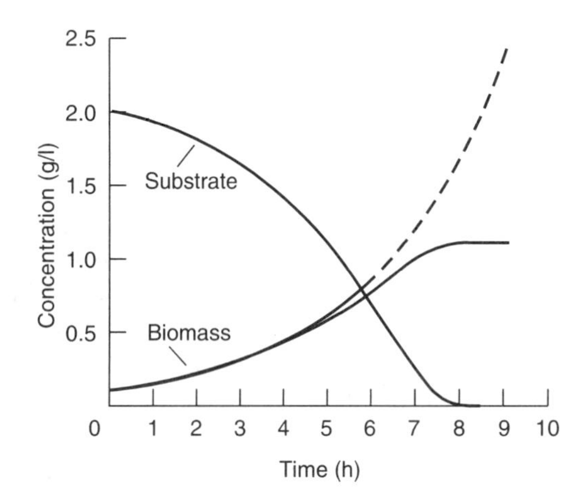

FIG. 5.55. The time course of a batch fermentation as predicted by equations 5.124 and 5.125. The parameters used in the calculation were: $K_S = 0.015$ g/l, $\mu_m = 0.35\ h^{-1}$, $Y = 0.5$, $X_0 = 0.1$ g/l, $S_0 = 2$ g/l

It is worth noting that the curves obtained in Fig. 5.55 show an inflexion towards the final stages of the fermentation, whereas the broken line showing the values of X generated by equation 5.57 shows no such characteristic and predicts that the growth would proceed to give an infinite value of X.

5.11.3. Continuous Culture of Micro-Organisms

The continuous growth of micro-organisms, as with continuous chemical reactions, may be carried out either in tubular fermenters (plug flow) or in well-mixed tank (back-mix) fermenters. Such fermenters are idealised forms and a practical fermenter exhibits, to a greater or lesser extent, some of the features of both forms. The continuous stirred-tank fermenter (CSTF) is characterised by containing a homogeneous liquid phase into which nutrients are continuously fed and from which the suspension of micro-organisms and depleted feed are continuously removed. Since the fermenter is well-mixed, samples taken from any location in the fermenter will be identical and, in particular, the composition of the exit stream will be identical to that of the liquid in the fermenter. Whilst the CSTF can be operated as a *turbidostat*—where the feed is metered in such a way as to maintain a constant biomass concentration (as measured by its turbidity) in the fermenter—it is more commonly used as a *chemostat*. In that case constant feed conditions are maintained and the system is allowed to attain a steady state.

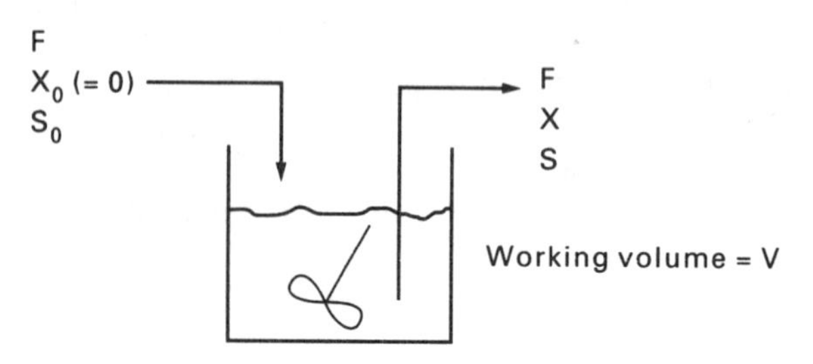

FIG. 5.56. The continuous stirred-tank fermenter

If a CSTF is considered (Fig. 5.56), which has a volume V, volumetric feed flow rate F, with influent substrate and biomass concentrations S_0 and X_0 respectively, then suppose that the substrate and biomass concentrations in the fermenter are S and X. A material balance can be established over the fermenter in the same manner as for the batch fermenter. This is:

$$\left\{\begin{matrix}\text{Flow of}\\ \text{material}\\ \text{in}\end{matrix}\right\} + \left\{\begin{matrix}\text{Formation by}\\ \text{biochemical}\\ \text{reaction}\end{matrix}\right\} - \left\{\begin{matrix}\text{Flow of}\\ \text{material}\\ \text{out}\end{matrix}\right\} = \text{Accumulation}$$

The balance may be carried out for both biomass and substrate, noting that the formation term for the substrate becomes negative since biomass is formed by the consumption of substrate. For the biomass this becomes:

$$FX_0 + \mathcal{R}_X V - FX = \frac{dX}{dt} V \tag{5.126}$$

where $\mathcal{R}_X$ is the rate of formation of biomass per unit volume of the fermenter. For the substrate the balance gives:

$$FS_0 + \mathcal{R}_S V - FS = \frac{dS}{dt} V \tag{5.127}$$

$\mathcal{R}_S$ being the corresponding rate of formation of substrate (negative).

Equation 5.126 may be developed by introducing the dilution rate D, defined by:

$$D = \frac{F}{V}$$

so that at steady state when $\frac{dX}{dt} = 0$, then:

$$D(X_0 - X) = - \mathcal{R}_X \tag{5.128}$$

Now, the formation rate of biomass $\mathcal{R}_X$ is μX so that:

$$D(X_0 - X) = - \mu X \tag{5.129}$$

or:

$$D = \mu \frac{X}{X - X_0} \tag{5.130}$$

If the feed contains no micro-organisms (i.e. it is sterile) then $X_0 = 0$ and the expression, for steady state, becomes:

$$D = \mu \tag{5.131}$$

It should be noted that this result is independent of the type of growth kinetics which the micro-organisms follow. If the dependence of specific growth rate on substrate concentration is described by the Monod equation, then substitution from equation 5.62 will give for the steady-state condition:

$$D = \frac{\mu_m S}{K_S + S} \tag{5.132}$$

and re-arranging this gives the steady-state substrate concentration as:

$$S = \frac{DK_S}{\mu_m - D} \tag{5.133}$$

It is interesting to note that this expression implies that the steady-state substrate concentration is independent of the feed substrate concentration S_0. The biomass concentration under this condition does, however, depend on the value of S_0. The behaviour of the biomass may be deduced by similar development of equation 5.127 and relating $\mathcal{R}_S$ to μ using equation 5.118. More conveniently, equation 5.120 may be rearranged to give:

$$X = X_0 + Y(S_0 - S) \tag{5.134}$$

Then, substituting for S using equation 5.133 so that for a CSTF at steady state utilising sterile feed, the biomass concentration is:

$$X = Y\left(S_0 - \left[\frac{DK_S}{\mu_m - D}\right]\right) \tag{5.135}$$

Example 5.1

Calculate the steady-state substrate and biomass concentrations in a continuous fermenter which has an operating volume of 25 l when the sterile feed stream contains limiting substrate at 2000 mg/l and enters the vessel at 8 l/h. The values of K_S and μ_m are 10.5 mg/l and 0.45 h^{-1}, respectively, and the yield coefficient may be taken to be 0.48.

Solution

Dilution rate for the fermenter $= \frac{F}{V} = \frac{8.1}{25} = 0.32\ h^{-1}$

From equation 5.133: $$S = \frac{0.32 \times 10.5}{(0.45 - 0.32)} = 25.8 \text{ mg/l}$$

and substituting in equation 5.134 (remembering that $X_0 = 0$ for sterile feed) gives:

$$X = 0 + 0.48 \times (2000 - 25.8) = \underline{\underline{948 \text{ mg/l}}}$$

Equations 5.133 and 5.135 indicate that the substrate and biomass concentrations will vary in an inverse relationship to each other. At low dilution rates, such that $D << K_S$, then S will be small and X will be large. Increasing the dilution rate will result in larger values of the steady-state substrate concentration and lower values of biomass concentration. The graph shown in Fig. 5.57 has been generated using these two equations and illustrates this behaviour. In the graph it may also be seen that, as D approaches the value of μ_m, X becomes infinitesimally small. This condition is referred to as *wash-out* and the value of D at which it occurs is D_{crit}.

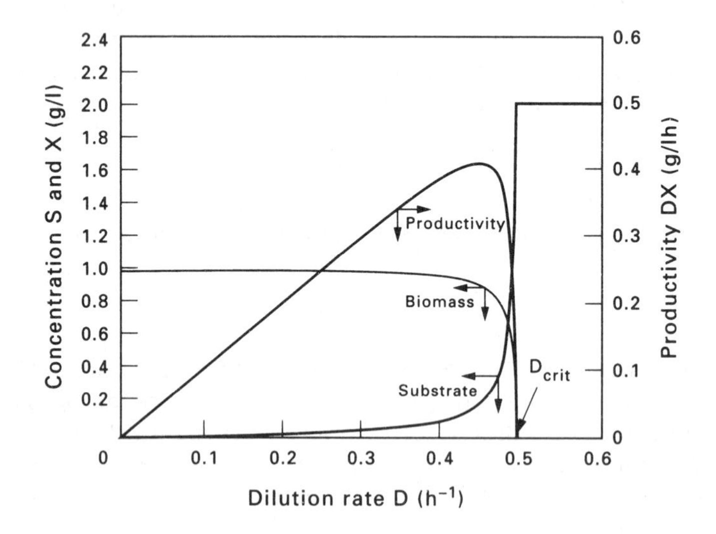

FIG. 5.57. Performance curves for CSTF at steady state

The specific production rate of biomass by the CSTF is the product of the biomass concentration and the volumetric flowrate of feed divided by the volume of the fermenter.

But: $$\frac{F}{V} = D$$

Thus:

$$\text{Cell productivity } \frac{FX}{V} = DX \tag{5.136}$$

The behaviour of the cell productivity DX with varying dilution rate and Monod constant K_S is shown in Fig. 5.58. It may be seen that it becomes zero when the dilution rate is D_{crit} and also that it passes through a maximum when plotted against

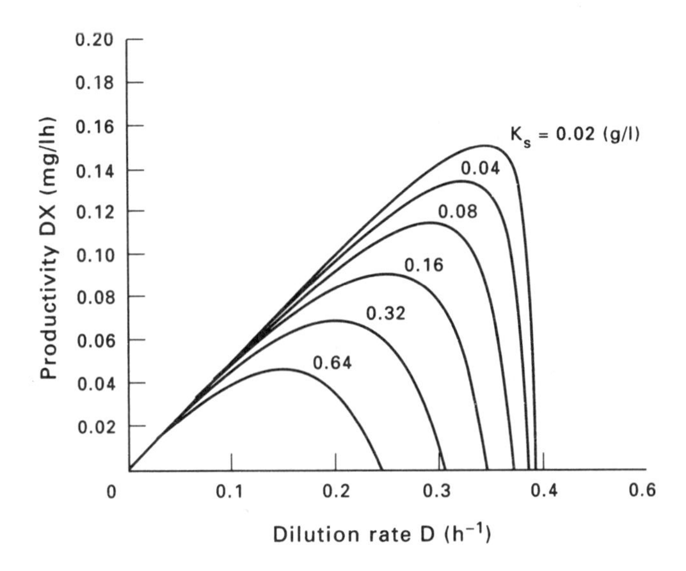

FIG. 5.58. Variation of productivity DX with K_S

the dilution rate. The location of this maximum may be obtained by first multiplying equation 5.135 by D to give the expression for productivity:

$$DX = DY\left(S_0 - \left[\frac{DK_S}{\mu_m - D}\right]\right) \tag{5.137}$$

Now when DX is a maximum:

$$\frac{d}{dD}\left\{DY\left(S_0 - \left[\frac{DK_S}{\mu_m - D}\right]\right)\right\} = 0 \tag{5.138}$$

so that on differentiating:

$$S_0 - \frac{(\mu_m - D_{op})2D_{op}K_S + D_{op}^2 K_S}{(\mu_m - D_{op})^2} = 0 \tag{5.139}$$

where D_{op} is the value of the dilution when the productivity is a maximum. The equation may be solved for D_{op} to give:

$$D_{op} = \mu_m\left(1 - \sqrt{\frac{K_S}{K_S + S_0}}\right) \tag{5.140}$$

Equation 5.140 represents an important design criterion for the continuous production of microbial cells, but in arriving at the equation an assumption has been made regarding the value of the yield coefficient Y. As shown in equation 5.54, the overall yield is a function of the growth rate (and hence the dilution rate). Furthermore, there may be biological factors intervening, possibly triggered by a change in dilution rate, which might alter Y, so some caution needs to be exercised when using equation 5.140.

The biomass concentration prevailing in the CSTF at the condition for maximum productivity may be obtained by substitution of the expression for D_{op} into equation 5.135. On simplification this gives:

$$X_{op} = Y(S_0 - \sqrt{K_S(K_S + S_0)} + K_S) \tag{5.141}$$

and so the productivity itself is given by:

$$D_{op}X_{op} = \mu_m Y\left(1 - \sqrt{\frac{K_S}{K_S + S_0}}\right)(S_0 - \sqrt{K_S\,(K_S + S_0)} + K_S) \tag{5.142}$$

It is reasonable to assume that, since a high productivity is being considered or sought, then the feed substrate concentration would be high, and in particular $S_0 >> K_S$ so that the expression for maximum productivity of a CSTF reduces to:

$$D_{op}X_{op} = \mu_m Y S_0 \tag{5.143}$$

In the case of a batch fermenter the cell productivity is given by the ratio of the biomass produced to the batch cycle time. The latter comprises the actual growth time period t_g and the 'unproductive' time t_u, which includes the time required for vessel cleaning, sterilisation, lag time, and harvesting time. The growth time may be calculated using equation 5.56, assuming again that $S_0 >> K_S$ so that $\mu \Rightarrow \mu_m$. Thus:

$$t_g = \frac{1}{\mu_m}\ln\left(\frac{X_m}{X_0}\right) \tag{5.144}$$

where X_m is the final biomass concentration which is assumed to occur when all the substrate has been consumed, i.e. $X_m - X_0 = Y_{X/S}S_0$. The productivity may then be obtained by dividing, as outlined above, to give:

$$\text{Productivity of batch fermenter} = \frac{Y_{X/S}S_0}{\dfrac{1}{\mu_m}\ln\left(\dfrac{X_m}{X_0}\right) + t_u} \tag{5.145}$$

The ratio of the CSTF productivity to that of the batch fermenter is obtained by dividing equation 5.143 by 5.145 to give:

$$\text{Productivity Ratio} = \frac{\mu_m Y_{X/S}S_0\left(\dfrac{1}{\mu_m}\ln\left(\dfrac{X_m}{X_0}\right) + t_u\right)}{Y_{X/S}S_0}$$

$$= \ln\left(\frac{X_m}{X_0}\right) + t_u\,\mu_m \tag{5.146}$$

The continuous fermenter will be the more productive whenever this expression is greater than unity and, regardless of the turn-round time t_u, this will be the case if $\dfrac{X_m}{X_0} > 2.27$. Typical values for $\dfrac{X_m}{X_0}$ lie in the range 8–10 so that, under normal circumstances, a continuous fermenter will give a higher production rate of biomass than a batch unit of the same volume. It should be noted, however, that this may not be the case for product formation where the relationship between growth and formation may not be linear. Problems associated with the maintenance of a monoculture in the fermenter may also cause a batch fermentation to be favoured in practice.

Another design criterion for a CSTF is the critical dilution rate above which biomass would be removed from the fermenter at a rate faster than it could regenerate itself and ultimately lead to the absence of biomass in the reactor. This condition is referred to as *washout*. Since there is no biomass present, washout is characterised by the substrate concentration in the reactor becoming equal to that in the feed solution. Rewriting equation 5.132 as:

$$\frac{D}{\mu_m} = \frac{S}{K_S + S} \tag{5.147}$$

and then inserting the condition $S \to S_0$ as $D \to D_{crit}$:

$$\frac{D_{crit}}{\mu_m} = \frac{S_0}{K_S + S_0} \tag{5.148}$$

The quantity $\dfrac{S_0}{K_S + S_0}$ will always be less than unity, and D_{crit} for a CSTF will always be less than μ_m by that factor.

The shape of the performance curve for a continuous stirred-tank fermenter is dependent on the kinetic behaviour of the micro-organism used. In the case where the specific growth rate is described by the Monod kinetic equation, then the productivity versus dilution rate curve is given by equation 5.137 and has the general shape shown by the curve in Fig. 5.58. However, if the specific growth rate follows substrate inhibition kinetics and equation 5.65 is applicable then, at steady state, equation 5.131 becomes:

$$D = \frac{\mu_m S}{K_S + S + S^2/K_I} \tag{5.149}$$

This equation is quadratic in S, and may be written in the form:

$$\frac{1}{K_I}S^2 + \left(1 - \frac{\mu_m}{D}\right)S + K_S = 0 \tag{5.150}$$

As a result, there are two possible values for the substrate concentration at a certain dilution rate, and two possible corresponding biomass concentrations. Figure 5.59 shows the performance curves produced by solving equation 5.150 and calculating the biomass concentration using the expression for yield coefficient.

The lower part of the biomass concentration curve in Fig. 5.59 represents a set of solutions for equation 5.150 which, whilst being a physical possibility, nevertheless represents an unstable state since any perturbation will cause the prevailing conditions either to move to a stable steady-state (with a corresponding decrease in S) or to result in washout and failure of the fermenter.

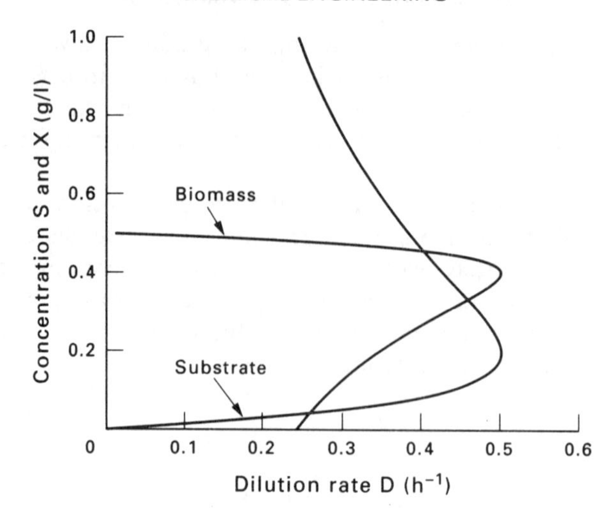

FIG. 5.59. Performance curves for a continuous stirred-tank fermenter with substrate inhibition

Stirred Tank Reactor with Recycle of Biomass

It is frequently desirable, particularly in the field of waste-water treatment, to operate a continuous fermenter at high dilution rates. With a simple stirred-tank this has two effects—one is that the substrate concentration in the effluent will rise, and the other is that such a system in practice tends to be unstable. One solution to this problem is to use a fermenter with a larger working volume, but an alternative strategy is to devise a method to retain the biomass in the fermenter whilst allowing the spent feed to pass out. There are several methods by which this may be achieved (see Fig. 5.60), and the net effect is the same in each case, but the analysis might

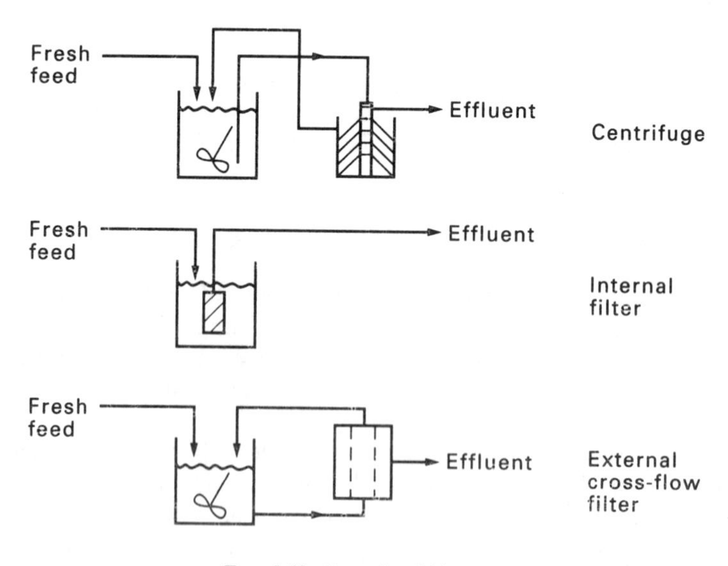

FIG. 5.60. Recycle of biomass

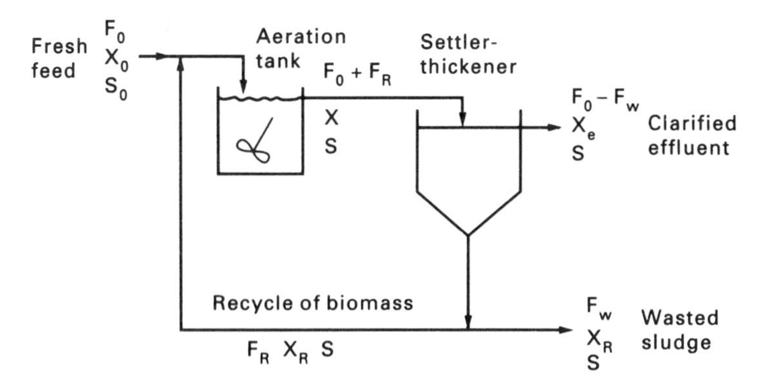

FIG. 5.61. CSTF with settler–thickener and recycle of biomass

vary according to the configuration. One of the most important methods is that where the effluent from the fermenter is passed to a settler–thickener and the concentrated biomass, or a portion of it, is returned to the growth vessel.

The operation of the system outlined in Fig. 5.61 is analysed by taking material balances over the fermenter vessel. It is assumed that in this idealised case, there is no biochemical reaction or growth occurring in the separator, so that the substrate concentration S in the entering stream is the same as that in the clarified liquid effluent stream, in the recycle stream and in the exit biomass rich stream. The material balance then becomes:

$$\left\{\begin{matrix}\text{Material}\\ \text{in with}\\ \text{fresh feed}\end{matrix}\right\} + \left\{\begin{matrix}\text{Material in}\\ \text{with recycle}\\ \text{stream}\end{matrix}\right\} + \left\{\begin{matrix}\text{Formation by}\\ \text{biochemical}\\ \text{reaction}\end{matrix}\right\} - \left\{\begin{matrix}\text{Output}\\ \text{to}\\ \text{separator}\end{matrix}\right\} = \text{Accumulation}$$

so that for the substrate:

$$F_0S_0 + F_RS + \mathcal{R}_SV - (F_0 + F_R)S = V\frac{\mathrm{d}S}{\mathrm{d}t} \tag{5.151}$$

and for the biomass:

$$F_0X_0 + F_RX_R + \mathcal{R}_XV - (F_0 + F_R)X = V\frac{\mathrm{d}X}{\mathrm{d}t} \tag{5.152}$$

For the biomass the rate of formation $\mathcal{R}_X$ is μX and under steady-state conditions $\frac{\mathrm{d}X}{\mathrm{d}t} = 0$ so that, assuming a sterile feed, equation 5.152 may be re-written:

$$\frac{F_R}{F_0}X_R + \mu X\frac{V}{F_0} - X - \frac{F_R}{F_0}X = 0 \tag{5.153}$$

Now $\frac{F_R}{F_0}$ is the recycle ratio R (based on the fresh feed) and the dilution rate D, also based on the fresh feed, is $\frac{F_0}{V}$ so that substitution into equation 5.153 gives:

$$\frac{\mu X}{D} = X(1 + R) - X_R \tag{5.154}$$

At this stage the performance of the separator may be taken into account. If the concentrating ability of the separator is represented by ξ and defined by:

$$\xi = \frac{X_R}{X} \tag{5.155}$$

then division of equation 5.154 by X and substitution gives:

$$\frac{\mu}{D} = 1 + R - \xi R \tag{5.156}$$

and after re-arranging:

$$D = \frac{\mu}{1 - R(\xi - 1)} \tag{5.157}$$

The denominator in equation 5.157 is clearly less than unity, provided the separator has any concentrating ability (i.e. works at all). As in the case for the expression linking D and μ (equation 5.131) for a simple CSTF, it is independent of the microbial kinetics. If $R = 0$ or $\xi = 1$ then equation 5.157 reduces to equation 5.131, corresponding to a CSTF at steady-state with no recycle. The effect of the recycle is to allow the fermenter to be operated at higher dilution rates than would otherwise be possible.

If the growth rate can be described by Monod kinetics, then using equation 5.157 to substitute into equation 5.62, the substrate concentration at steady-state is given by:

$$S = \frac{K_S D(1 - R\xi + R)}{\mu_m - D(1 - R\xi + R)} \tag{5.158}$$

It may also be be shown that under steady-state conditions equation 5.151 reduces to give:

$$D(S_0 - S) = -\mathcal{R}_S \tag{5.159}$$

and, if Monod kinetics are introduced with a constant yield coefficient Y, this becomes:

$$D(S_0 - S) = \frac{\mu_m S X}{Y(K_S + S)} \tag{5.160}$$

which can be re-arranged to give the steady-state biomass concentration as:

$$X = \frac{D Y(S_0 - S)(K_S + S)}{\mu_m S} \tag{5.161}$$

The conditions for washout with the recycle stream of biomass can be determined by putting $S \rightarrow S_0$ as $D \rightarrow D_{crit}$ so that with the incorporation of Monod kinetics into equation 5.157:

$$D = \frac{\mu_m S}{\{1 - R(\xi - 1)\}(K_S + S)} \tag{5.162}$$

and:

$$\frac{D_{\text{crit}}}{\mu_m} = \frac{S_0}{\{1 - R(\xi - 1)\}(K_S + S_0)} \tag{5.163}$$

This confirms the result anticipated from equation 5.157 that the dilution rate for washout will be greater than that for the simple CSTF.

Continuous Stirred Tank Fermenters in Series

There are certain circumstances when it may be desirable to operate a series of stirred-tanks in cascade, with the effluent of one forming the feed to the next in the chain. In the case of two such fermenters the first will behave in the manner of a simple chemostat and the performance equations of the chemostat will apply. The second tank, however, will not have a sterile feed so that some of the simplifications which led to those relationships will not be valid.

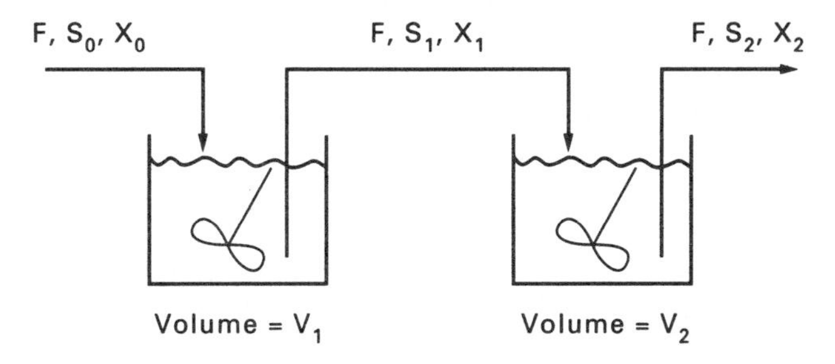

FIG. 5.62. Two continuous stirred-tanks in series

As indicated in Fig. 5.62, the feed for the second fermenter will have biomass concentration X_1 and substrate concentration S_1 which, when the first tank has zero biomass in its feed (i.e. $X_0 = 0$), will be given by:

$$S_1 = \frac{D_1 K_S}{\mu_m - D_1} \tag{5.164}$$

and from the definition of yield:

$$X_1 = Y\left\{S_0 - \frac{D_1 K_S}{\mu_m - D_1}\right\} \tag{5.165}$$

where D_1 is the dilution rate for the first vessel. Now a material balance for the second vessel will give:

$$FX_1 + \mu_2 V_2 X_2 - FX_2 = \frac{dX_2}{dt} V_2 \tag{5.166}$$

At steady state this balance becomes:

$$D_2 = \frac{\mu_2 X_2}{X_2 - X_1} \tag{5.167}$$

where D_2 is the dilution rate and μ_2 is the specific growth rate for the second tank. When Monod kinetics are incorporated into this expression it becomes:

$$D_2 = \frac{\mu_m S_2}{K_S + S_2} \frac{X_2}{X_2 - X_1} \tag{5.168}$$

A material balance over both fermenters gives the biomass concentration in the second vessel as:

$$X_2 = Y(S_0 - S_2)$$

so that substituting for X_2 in equation 5.168 gives:

$$D_2 = \frac{\mu_m S_2}{K_S + S_2} \frac{(S_0 - S_2)}{\left(\dfrac{D_1 K_S}{\mu_m - D_1} - S_2\right)} \tag{5.169}$$

Solving this for S_2 results in:

$$(\mu_m - D_2)S_2^2 + \left\{\frac{D_1 D_2 K_S}{(\mu_m - D_1)} - D_2 K_S - \mu_m S_0\right\} S_2 + \frac{D_1 D_2 K_S^2}{(\mu_m - D_1)} = 0 \tag{5.170}$$

Only one of the roots of this quadratic will give a positive value for the substrate concentration S_2 in the second fermenter in the train.

Example 5.2

Two continuous stirred-tank fermenters are connected in series, the first having an operational volume of 100 l and that of the second being 50 l. The feed to the first fermenter is sterile and contains 5000 mg/l of substrate, being delivered to the fermenter at 18 l/h. If the microbial growth can be described by the Monod kinetic model with $\mu_m = 0.25\ \text{h}^{-1}$ and $K_S = 120$ mg/l, calculate the steady-state substrate concentration in the second vessel. What would happen if the flow were from the 50 l fermenter to the 100 l fermenter?

Solution

Dilution rate for 1st fermenter, $D_1 = \dfrac{18}{100} = 0.18\ \text{h}^{-1}$

Dilution rate for 2nd fermenter, $D_2 = \dfrac{18}{50} = 0.36\ \text{h}^{-1}$

Substitution in equation 5.170 gives:

$$(0.25 - 0.36)S_2^2 + \left\{\frac{0.18 \times 0.36 \times 120}{(0.25 - 0.18)} - 0.36 \times 120 - 0.25 \times 5000\right\} S_2 + \frac{0.18 \times 0.36 \times 120 \times 120}{0.25 - 0.18} = 0$$

which simplifies to;

$$-0.11S_2^2 - 1182S_2 + 13,330 = 0$$

from which:

Substrate concentration in second vessel $\underline{\underline{S_2 = 11.3 \text{ mg/l}}}$

If the order of the fermenters were reversed then the dilution rate for the first vessel would be above D_{crit} and washout would occur in that vessel. The feed would enter the second fermenter unchanged and containing no micro-organisms so that the steady state substrate concentration in it would be given by substitution in equation 5.133.

i.e.:

$$S_2 = \frac{0.18 \times 120}{(0.25 - 0.18)} = \underline{\underline{309 \text{ mg/l}}}$$

The more generalised case of several continuous stirred-tank fermenters in series may be analysed by considering N such fermenter vessels, each of volume V and with fresh feed introduced to the first tank at a volumetric flowrate F (see Fig. 5.63). Because no streams enter or leave intermediately, the flowrates between stages and of the final product will also be F.

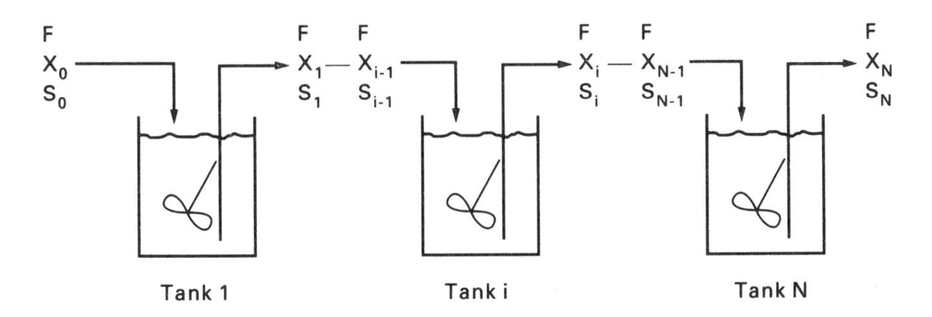

FIG. 5.63. Stirred-tank fermenters in series

The subscript flows on the variables indicated on Fig. 5.63 refers to the number of the tank from which that stream flow. The fresh feed has the subscript zero and is distinctive in that it may be sterile, i.e. $X_0 = 0$. In general, a material balance for biomass over the ith tank will give:

$$FX_{i-1} - FX_i + V\mu_i X_i = V\frac{dX_i}{dt} \tag{5.171}$$

where μ_i is the specific growth rate in the ith fermenter. At steady state $\frac{dX_i}{dt} = 0$, so that the above may be rearranged to give:

$$X_i = \frac{DX_{i-1}}{D - \mu_i} \tag{5.172}$$

where D is the dilution rate based on the volume of each individual tank. For the last tank in the series, where $i = N$, the biomass concentration is given by:

$$X_N = \frac{DX_{N-1}}{D - \mu_N} \tag{5.173}$$

and similarly for the penultimate tank, where $i = (N - 1)$, equation 5.172 becomes:

$$X_{N-1} = \frac{DX_{N-2}}{D - \mu_{N-1}} \tag{5.174}$$

so that, by substitution of equation 5.174 into equation 5.173, X_{N-1} may be eliminated.

Thus:

$$X_N = \frac{D^2 X_{N-2}}{(D - \mu_N)(D - \mu_{N-1})} \tag{5.175}$$

This substitution may be repeated for progressively earlier tanks in the series so that the effluent concentration for N growth tanks may be expressed as:

$$X_N = \frac{D^{N-1}X_1}{\prod_{k=2}^{N}(D - \mu_k)} \tag{5.176}$$

Note that this expression does not involve the feed to the first fermenter, which may well operate under sterile feed conditions. In that case, and provided that the tanks are of equal size, the washout condition is analogous to the case of the single stirred-tank fermenter, that is:

$$\frac{D_{crit}}{\mu_m} = \frac{S_0}{K_S + S_0} \qquad \text{(Equation 5.148)}$$

since washout of the first tank in the series will inevitably result in washout in each successive vessel.

The performance of a set of fermenters in series may be improved by the inclusion of a recycle stream, as in the case of a single stirred-tank fermenter. The situation for N vessels each of volume V may be analysed in a similar manner to the case for no recycle.

For a series of fermenters as shown in Fig. 5.64, the material balance for biomass

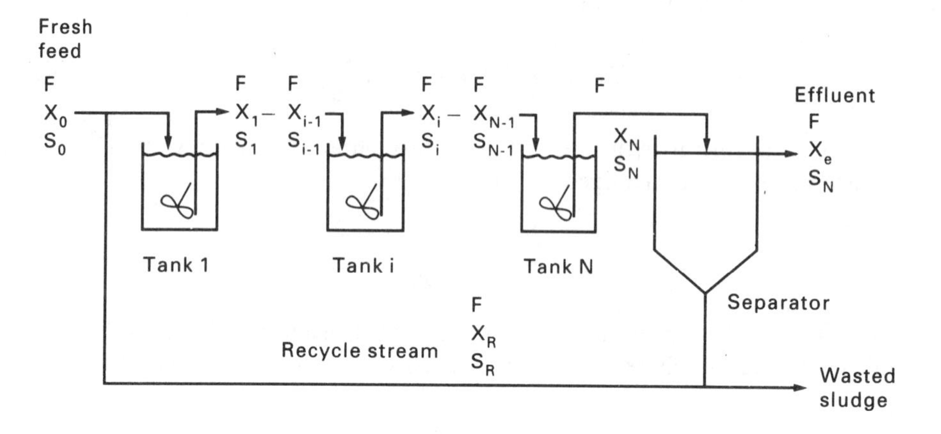

FIG. 5.64. Stirred-tank fermenters in series with recycle of biomass

over the ith fermenter gives:

$$F_0(1 + R)X_{i-1} - F_0(1 + R)X_i + \mu_i X_i V = \frac{dX_i}{dt}V \tag{5.177}$$

At steady state, equation 5.177 reduces to:

$$D(1 + R)\frac{(X_i - X_{i-1})}{X_i} = \mu_i \tag{5.178}$$

Rearranging for X_i gives:

$$X_i = \frac{DX_{i-1}}{D(1 + R) - \mu_i} \tag{5.179}$$

and, by a similar argument to that used above, it may be deduced that the biomass concentration in the Nth tank will be given by:

$$X_N = \frac{D^{N-1}X_1}{\prod_{k=2}^{N}(D(1 + R) - \mu_i)} \tag{5.180}$$

In this case, however, whilst the fresh feed entering the system may be sterile, the biomass concentration entering the first tank is not zero. If this is represented by X'_0, then:

$$X'_0 = X_R \frac{F_R}{F_R + F_0} \tag{5.181}$$

$$= X_R \frac{R}{(R + 1)} \tag{5.182}$$

Equation 5.182 may now be developed a stage further to give:

$$X_N = \frac{D^N X_R \frac{R}{(R+1)}}{\prod_{k=1}^{N}(D(+ R) - \mu_k)} \tag{5.183}$$

For incipient washout, in any tank, then $\mu_k \Rightarrow \mu_{\text{crit}}$ as $D \Rightarrow D_{\text{crit}}$. This condition may be applied to equation 5.183 so that:

$$X_N = \frac{D_{\text{crit}}^N X_R \frac{R}{(R+1)}}{\left[D_{\text{crit}}(1 + R) - \mu_{\text{crit}}\right]^N} \tag{5.184}$$

or using $\xi = \frac{X_R}{X_N}$ and rearranging to give D:

$$D_{\text{crit}} = \frac{\mu_{\text{crit}}}{(1 + R)(1 - \left\{\frac{\xi R}{1 + R}\right\}^{1/N}} \tag{5.185}$$

The washout criterion is typified by $S_k \Rightarrow S_0$ then, if Monod kinetics are assumed, as in the previous considerations:

$$\mu_{\text{crit}} = \frac{\mu_m S_0}{K_S + S_0}$$

Then this can be substituted into equation 5.185 to give:

$$D_{\text{crit}} = \frac{\mu_m S_0/(K_S + S_0)}{(1+R)\left(1 - \left\{\dfrac{\xi R}{1+R}\right\}^{1/N}\right)} \tag{5.186}$$

It may be seen that equation 5.186 reduces to equation 5.163 when $N = 1$; that is, it agrees with the expression derived for the critical dilution rate for a single stirred-tank fermenter with recycle of biomass.

Plug-Flow Fermenters

The plug-flow tubular fermenter (PFTF) is in some respects the opposite limiting form of fermenter to the CSTF. In its idealised form, it is characterised by the fact that the liquid phase passes through the fermenter without back-mixing. The fresh feed and inoculum enter at one end of the fermenter and the mixture of feed and growing cells progress in unison towards the the exit point. A small portion of the feed behaves as in a batch fermenter from its moment of entry up to the time when it leaves the fermenter, with the difference that the time co-ordinate of the batch fermentation must be replaced by the time taken to travel the axial distance along the fermenter tube. In practice, the fermenter may not necessarily be a tube and the idealised requirement of no back-mixing is never achieved. Analysis of the operation of the idealised fermenter may be carried out by considering an elementary volume of the fermenter as shown in Fig. 5.65.

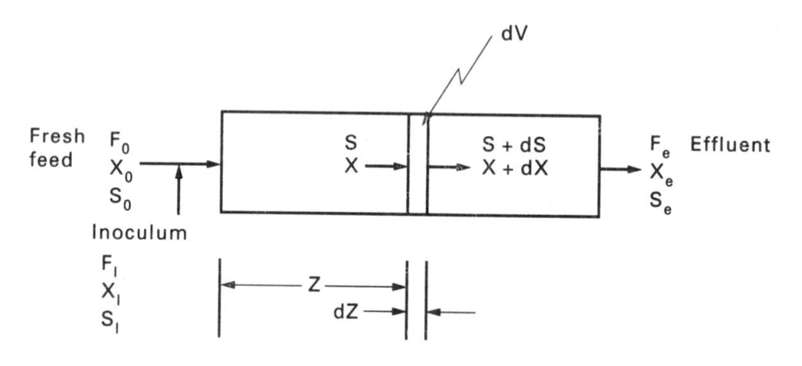

FIG. 5.65. The plug-flow fermenter

As with the batch fermenter, the growth has to be initiated by the addition of an inoculum; in this case it is represented by the stream at a volumetric flowrate of F_I with biomass concentration X_I and substrate concentration S_I. This is mixed with the fresh-feed stream which has a volumetric flowrate of F_0, biomass concentration X_0 and substrate concentration S_0 to produce the entry stream of flowrate F_A, biomass concentration X_A and substrate concentration S_A.

The material balance over the elementary volume gives:

$$\left\{\begin{matrix}\text{Material}\\ \text{entering}\\ \text{by bulk}\\ \text{flow}\end{matrix}\right\} + \left\{\begin{matrix}\text{Formation}\\ \text{by}\\ \text{biochemical}\\ \text{reaction}\end{matrix}\right\} - \left\{\begin{matrix}\text{Material}\\ \text{leaving}\\ \text{by bulk}\\ \text{flow}\end{matrix}\right\} = \text{Accumulation}$$

If the biomass concentration at the point of entry to the volume $\mathrm{d}V$ is X then this results in:

$$(F_0 + F_I)X + \mathcal{R}_X \mathrm{d}V - (F_0 + F_I)(X + \mathrm{d}X) = \frac{\mathrm{d}X}{\mathrm{d}t}\mathrm{d}V \tag{5.187}$$

For a fermenter with cross-sectional area a_c the volume term can be replaced by $\mathrm{d}V = a_c \mathrm{d}Z$ and, at steady state, equation 5.187 becomes:

$$\mathcal{R}_X a_c \mathrm{d}Z - F_A \mathrm{d}X = 0 \tag{5.188}$$

The performance of the fermenter will be given by the integral of:

$$\int_{X_A}^{X_e} \frac{\mathrm{d}X}{\mathcal{R}X} = \frac{a_c}{F_A}\int_0^Z \mathrm{d}Z \tag{5.189}$$

Now, the right-hand side of this equation may be readily integrated to give:

$$\int_{X_A}^{X_e} \frac{\mathrm{d}X}{\mathcal{R}_X} = \frac{a_c Z}{F_A} = \nu \tag{5.190}$$

where ν is the residence time of the liquid in the fermenter. This expression is similar to that for microbial growth in a batch fermenter (see equation 5.123). It would be expected to give a similar integral form, but with ν replacing t the fermentation time, and the boundary conditions being altered. The boundary condition at the entry point is:

$$X_A = \frac{F_0 X_0 + F_I X_I}{F_0 + F_I} \tag{5.191}$$

At the exit point of the reactor $X = X_e$. The incorporation of Monod kinetics for the growth rate results in:

$$\frac{(K_S Y + S_A Y + X_A)}{\mu_m(YS_A + X_A)} \ln\left(\frac{X_e}{X_A}\right) + \frac{K_S Y}{\mu_m(YS_A + X_A)} \ln\left(\frac{YS_A}{YS_A + X_A - X_e}\right) = \nu \tag{5.192}$$

A similar expression can be obtained for the substrate concentration. The material balance for substrate will result in:

$$\mathcal{R}_S a_c \mathrm{d}Z - F_A \mathrm{d}S = 0 \tag{5.193}$$

and integrating:

$$\int_{S_A}^{S_e} \frac{\mathrm{d}X}{\mathcal{R}_S} = \nu \tag{5.194}$$

where S_A is given by:

$$S_A = \frac{F_0 S_0 + F_I S_I}{F_0 + F_I} \tag{5.195}$$

The result for Monod growth kinetics, and assuming a constant yield coefficient is:

$$\frac{(K_S Y + S_A Y + X_A)}{\mu_m(YS_A + X_A)} \ln\left(1 + \frac{Y(S_A - S_e)}{X_A}\right) - \frac{K_S Y}{\mu_m(X_A - YS_A)} \ln\left(\frac{S_e}{S_A}\right) = \nu \tag{5.196}$$

This integration relies on the fact that S_A and X_A are independent of the exit concentrations, there being no recycle stream in this instance. Plug-flow fermenters are, however, operated with recycle of micro-organisms and the condition is not valid for this generalised case.

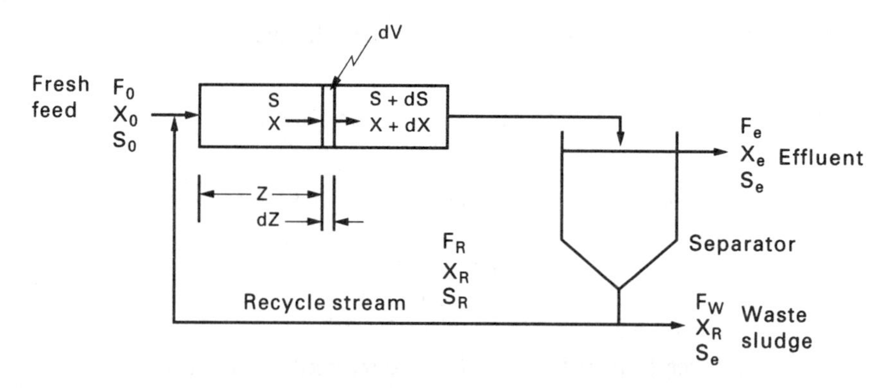

FIG. 5.66. The plug-flow fermenter with recycle of biomass

When recycle of biomass is included, equation 5.187 becomes:

$$(F_0 + F_R)X + \mathcal{R}_X \mathrm{d}V - (F_0 + F_R)(X + \mathrm{d}X) = \frac{\mathrm{d}X}{\mathrm{d}t} V \tag{5.197}$$

and again, since $\mathrm{d}V = a_c \mathrm{d}Z$, at steady state equation 5.197 becomes:

$$\mathcal{R}_X a_c \mathrm{d}Z - (F_0 + F_R)\mathrm{d}X = 0 \tag{5.198}$$

or:

$$\mathcal{R}_X a_c \mathrm{d}Z - F_0(1 + R)\mathrm{d}X = 0 \tag{5.199}$$

where the recycle ratio R is defined by:

$$R = \frac{F_R}{F_0}$$

The performance of the reactor will then be expressed in terms of the integrated form of this equation:

$$\int_{X_A}^{X_e} \frac{\mathrm{d}X}{\mathcal{R}_X} = \int_0^Z \frac{a_c \mathrm{d}Z}{F_0(1 + R)} \tag{5.200}$$

Similarly, for the substrate the form is:

$$\int_{S_A}^{S_e} \frac{\mathrm{d}S}{\mathcal{R}_S} = \int_0^Z \frac{a_c \mathrm{d}Z}{F_0(1 + R)} \tag{5.201}$$

In this case the material balance for the substrate about the point of the mixing of the fresh feed and the recycle stream gives:

$$S_0 F_0 + S_R F_R = S_A(F_0 + F_R) \tag{5.202}$$

Thus, the exit concentration for the substrate is given by:

$$S_A = \frac{S_0 + S_e R}{(1 + R)} \tag{5.203}$$

Similarly, for the biomass:

$$X_0 F_0 + X_R F_R = X_A(F_0 + F_R) \tag{5.204}$$

or, if $X_R = \xi X_e$, then this may be re-arranged to give the other boundary condition, for the biomass as:

$$X_A = X_e \frac{\xi R}{1 + R} \tag{5.205}$$

The resultant equations are non-linear and in this general case numerical solution techniques must be used. However, there exists a special case where an analytical solution may be obtained. If the increase in biomass concentration during flow through the reactor is small then an average value for the biomass concentration, independent of the distance Z along the fermenter, may be used. The material balance for the substrate over the reactor element may then be written:

$$(F_0 + F_R)S + \mathcal{R}_S \mathrm{d}V - (F_0 + F_R)(S + \mathrm{d}S) = \frac{\mathrm{d}S}{\mathrm{d}t} V \tag{5.206}$$

If the recycle ratio R is defined as $R = \frac{F_R}{F_0}$, then at steady state equation 5.206 becomes:

$$- F_0(1 + R)\mathrm{d}S + \mathcal{R}_S \mathrm{d}V = 0 \tag{5.207}$$

and, since $\mathrm{d}V = a_c \mathrm{d}Z$, then:

$$- F_0(1 + R)\,\mathrm{d}S + \mathcal{R}_S a_c \mathrm{d}Z = 0 \tag{5.208}$$

For Monod kinetics and constant yield the reaction rate with respect to the substrate $\mathcal{R}_S$ may be replaced by:

$$\mathcal{R}_S = -\frac{\mu_m S X_A}{Y(K_S + S)} \tag{5.209}$$

to give:

$$F_0(1 + R)\,\mathrm{d}S = \frac{-\mu_m S X_A a_c}{Y(K_S + S)}\,\mathrm{d}Z \tag{5.210}$$

and, since $S = S_I$ when $Z = 0$ and $S = S_e$ at the exit, then for a reactor of length Z:

$$\int_{S_A}^{S}\frac{(K_S + S)\,\mathrm{d}S}{S} = \frac{-\mu_m X_A a_c}{YF_0(1 + R)}\int_0^Z \mathrm{d}Z \tag{5.211}$$

which becomes:

$$(S_I - S_e) + K_S \ln\left[\frac{S_I}{S_e}\right] = \frac{\mu_m X_A a_c Z}{YF_0(1 + R)} \tag{5.212}$$

The definition of the yield coefficient may now be used to derive an expression for the biomass concentration, since by substituting:

$$Y = \frac{X_e - X_I}{S_I - S_e} \tag{5.213}$$

it may be shown that:

$$(X_e - X_I) = \frac{\mu_m X_A a_c Z}{F_0(1 + R)} - YK_S \ln\left[\frac{S_I}{S_e}\right] \tag{5.214}$$

The validity, or otherwise of the assumption regarding the change in X over the length of the reactor may now be checked; that is, the difference between X_e and X_I should be small.

5.12. ESTIMATION OF KINETIC PARAMETERS

The domination of microbial growth kinetics by the Monod equation has led to the development of techniques to determine the constants K_S and μ_m used in that equation. Whilst μ_m is in fact dependent on other parameters, such as temperature and pH, in the usual case these are specified, and a design procedure requires values of the Monod constants under these conditions. The yield coefficient Y will also be required in order to link calculations of microbial growth to substrate concentrations.

The measurement of these constants may be carried out using either batch or continuous fermenter experiments, but it should be noted that there is a possibility that different results might be obtained from the different methods. This is due to the fact that the growth conditions in each case are typically very different, with the substrate concentration in the batch fermenter being much higher than that in a continuous fermenter. Complications can arise when adaptation to a particular substrate occurs, with the batch experiment necessarily spanning a transient phase with the continuous flow experiment being performed at steady-state. For the present, it will be assumed that the values obtained by each method are identical, and that the constants derived using one configuration will be applicable to calculations on the performance of fermenters operating in the other configuration.

5.12.1. Use of Batch Culture Experiments

One simple, rapid approach to the problem is to perform a batch experiment and ensure that the initial substrate concentration S_0, is very much greater than the probable value of K_S. This is not an unreasonable condition to satisfy and, for a normal batch fermentation, the substrate concentration would typically be much larger than the value of K_S for most of the growth period. In that case the Monod equation (5.62) reduces, as has been shown, to:

$$\mu = \mu_m$$

and, by combining this with equation 5.56, then:

$$\ln\left[\frac{X}{X_0}\right] = \mu_m t \tag{5.215}$$

or:

$$\ln X = \mu_m t + \ln X_0 \tag{5.216}$$

This shows that a plot of ln X against t will have a slope equal to μ_m. The experiment, therefore, simply involves making measurements of the biomass concentration X at a series of times during the growth phase.

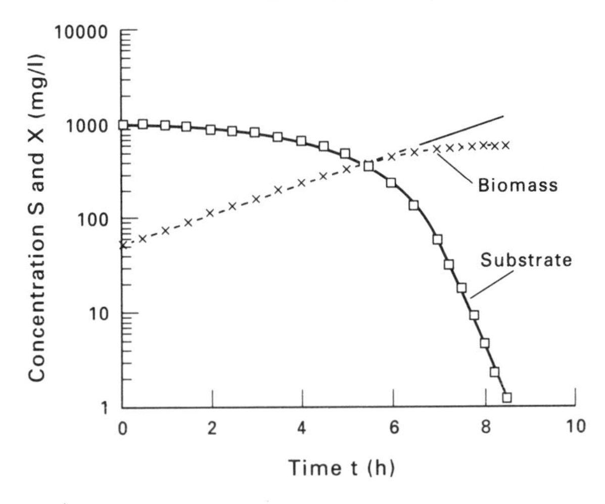

FIG. 5.67. Logarithmic plot of biomass concentration against time for a batch fermentation

Such a plot is shown in Fig. 5.67 for which the data were generated using equation 5.124. The true values of the constants are therefore known and the performance of the technique may be judged by comparing its results with these. In Fig. 5.67 the generated data are represented by crosses and the dotted line shows the result of a linear regression fit. Clearly, the straight line is not a very good representation of the data, there being a marked deviation towards the end of the growth period. This is due to the fact that, by this time, the substrate concentration has declined sufficiently to become comparable with K_S and the assumption that $\mu = \mu_m$, as explained above, is no longer valid. The linear regression line for the data points tends to give an under-estimate of the value of μ_m and a more realistic value would be given by the slope of a line based on the earlier points only. This is not therefore a satisfactory method of determination; not only does a subjective assessment have to be made of the extent of the data to be taken into account, but it also ignores that part of the curve which is influenced by the value of K_S, and consequently does not offer any estimate of its value.

GATES and MARLAR[60] devised a graphical procedure to determine the kinetic constants which took into account the data in the final stages of the growth-phase of a batch fermentation. As a result, their procedure can be used to estimate values of μ_m, K_S and the yield coefficient Y. The method involves rewriting equation 5.125, which is the complementary expression to the integrated form of the Monod equation 5.124, but refers to the substrate:

$$\frac{1}{t}\ln\left[\frac{S}{S_0}\right] = \frac{1}{t}\left[1 + \frac{X_0 + YS_0}{K_S Y}\right]\ln\left[Y\frac{(S_0 - S)}{X_0} + 1\right] - \frac{(X_0 + S_0 Y)}{K_S Y}\mu_m \qquad (5.217)$$

At this stage, the following substitutions may be made:

$$a = \frac{Y}{X_0}$$

and:

$$b = 1 + \frac{X_0 + YS_0}{K_S Y}$$

$$q = S_0 - S$$

and:

$$r = \frac{X_0 + YS_0}{K_S Y}\mu_m$$

The expression becomes:

$$\frac{1}{t}\ln\left[\frac{S}{S_0}\right] = \frac{b}{t}\ln(aq + 1) - r \qquad (5.218)$$

From this equation it may be seen that a plot of $\frac{1}{t}\ln\left[\frac{S}{S_0}\right]$ against $\frac{1}{t}\ln(aq + 1)$, (Fig. 5.68), should produce a straight line graph with slope b and intercept $-r$. It is worthwhile noting at this stage that equation 5.217, and hence equation 5.218, does not explicitly use the value of X, the biomass concentration, at various times during the fermentation. The parameter measured during the course of the experiment is the substrate concentration. Only the initial value of the biomass X_0 and one other value at a later stage in the fermentation (together with the corresponding value of S) are required for an estimate of the yield coefficient to be made. This requirement has the practical advantage that measurement of substrate concentration is normally easier and more accurate than that of biomass. A value of the yield coefficient Y is estimated from the data gathered, and a series of values computed for q and $\frac{1}{t}\ln\left[\frac{S}{S_0}\right]$. At this stage only an approximate value of a is available, but this is sufficient for a first attempt at a solution by calculating the corresponding series of $\frac{1}{t}\ln(aq + 1)$ values and plotting the graph described above. The graph is then inspected for linearity and at this stage considerable curvature would be expected at values derived from the later measurements in the experiment. A trial-and-error technique is then embarked upon, with the value of $\mathscr{a}$ being adjusted to improve progressively the quality of the 'best' straight line through the points generated.

When the optimum value of $\mathscr{a}$ is obtained the kinetic parameters can be calculated from the relationships:

$$\mu_m = \frac{\mathscr{r}}{\mathscr{b} - 1}$$

$$K_S = \frac{1/\mathscr{a} + S_0}{\mathscr{p} - 1}$$

and:

$$Y = \mathscr{a} X_0$$

	□	×	◇	
$\mathscr{a}$	0.0095	0.01	0.00105	1/mg
K_s	19.7	15.0	11.7	mg/l
μ_m	0.412	0.400	0.389	h^{-1}
Y	0.475	0.500	0.525	–

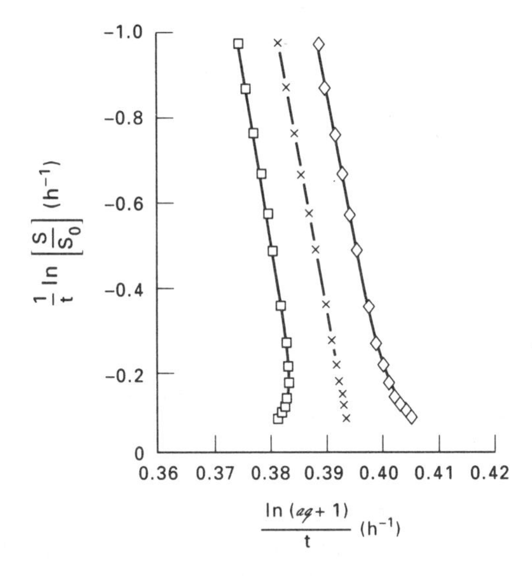

FIG. 5.68. Graphical determination of the Monod kinetic constants using the method of GATES and MARLAR[60]

Figure 5.68 shows the result of three calculations of this nature using the same data as that used in the estimation of μ_m in Fig. 5.67. The optimum value of $\mathscr{a}$ in this case is 0.01 l/mg corresponding to μ_m of 0.4 h^{-1}, K_S of 15.0 mg/l and a yield coefficient of 0.5. Changing $\mathscr{a}$ to 0.0095 or to 0.0105 l/mg may be seen to give distinctly curved lines which, as such, may be rejected. The values of K_S derived from this method are, as expected, much more sensitive to the quality of the line fit than are those of μ_m.

ONG[61] proposed that the best straight line, and the quality of that line, could most readily be determined by its correlation coefficient. In this way the problem would be translated into one of one-dimensional optimisation, that is, maximising the value of the correlation coefficient by adjusting the value of a in equation 5.218. One problem with this approach is that the method itself requires very high quality data for reliable results to be obtained. Figure 5.69 shows the effect of performing the same calculations as above on the same data but with a random error of $\pm$ 1 per cent imposed on the 'measured' value of S. It now becomes quite difficult to distinguish which is the 'best' straight line and relying on the regression coefficient does in fact lead to the wrong choice in this case.

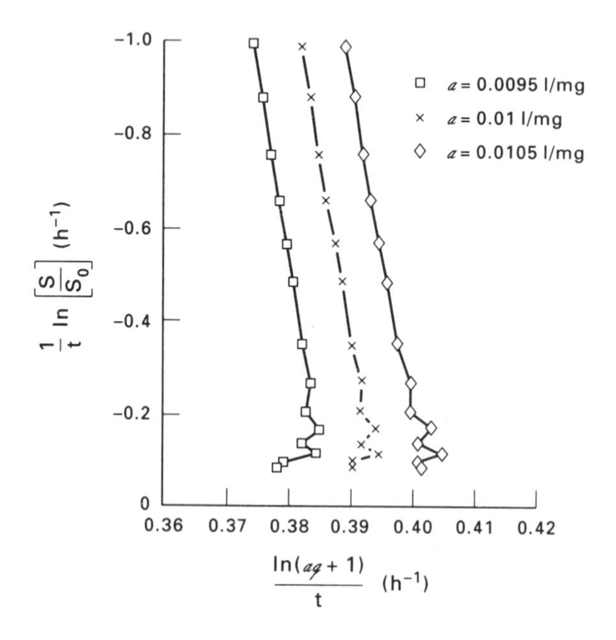

FIG. 5.69. Graphical determination of the Monod kinetic constants using the method of Gates and Marlar but with error imposed on the experimental values

A limitation of the methods described so far is that they have assumed a constant overall yield coefficient and do not allow the endogenous respiration coefficient k_d (or alternatively the maintenance coefficient, m) to be evaluated. Equation 5.54 shows that the overall yield, as measured when monitoring a batch reactor, is affected by the growth rate and has the greatest impact when the growth rate is low. Consequently, it is desirable to be able to estimate the values of k_d or m, so that the yield coefficient reflects the true growth yield. An equivalent method would be one where the specific rates of formation of biomass and consumption of substrate were determined independently, again without the assumption of a constant overall yield-coefficient.

The approach made by Gates and Marlar is unsuitable for use with equation 5.70

since the result of the material balance for the biomass in the batch fermentation is:

$$\frac{1}{X}\frac{\mathrm{d}X}{\mathrm{d}t} = \mu_m \frac{S}{K_S + S} - \boldsymbol{k_d} \qquad \text{(equation 5.70)}$$

Equation 5.70 cannot be integrated analytically. ESENER *et al.*[62], however, propose a method of estimating $\boldsymbol{k_d}$ which makes use of a fed-batch fermentation. In such an experiment the growth is arranged initially to proceed in the same manner as in a batch fermentation but, at a later stage when exponential growth has been well established, more feed is introduced into the growth vessel. This differs from a continuous fermentation in that no material is simultaneously withdrawn from the vessel. As with the simple batch fermentation, the initial substrate concentration is arranged to be much larger than the value of K_S so that the specific growth rate will be equal to μ_m. Thus, measurements of biomass concentration and substrate concentration at this stage will allow the estimation of μ_m and Y.

The experiment is then continued by adding feed to the fermentation broth, at a known flowrate which declines in a preset manner (usually linearly). This will cause the system to come to a net zero growth rate at some stage. Figure 5.70 shows the variation of the total biomass and substrate in the fermenter with time, noting that in this case the feed has been started at the beginning of the fermentation.

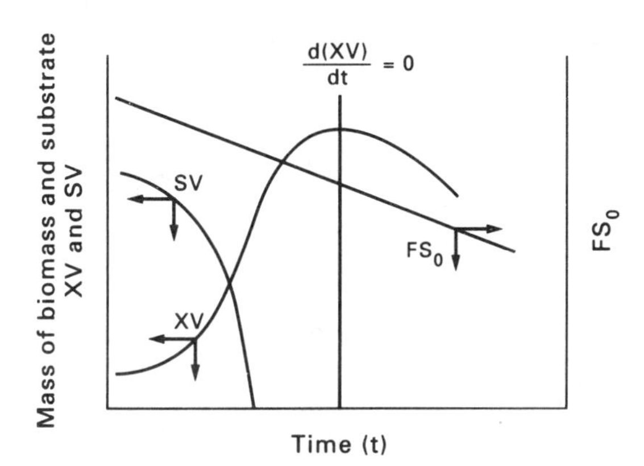

FIG. 5.70. The variation of total substrate and biomass in a fed-batch fermentation

The material balance for the biomass in the fed-batch fermentation gives:

$$\frac{\mathrm{d}(XV)}{\mathrm{d}t} = \frac{\mu_m SXV}{K_S + S} - \boldsymbol{k_d} XV \qquad (5.219)$$

and for the substrate:

$$\frac{\mathrm{d}(SV)}{\mathrm{d}t} = FS_0 - \frac{\mu_m SXV}{Y(K_S + S)} \qquad (5.220)$$

Now, at the time when $\dfrac{\mathrm{d}(XV)}{\mathrm{d}t} = 0$ and $\dfrac{\mathrm{d}(SV)}{\mathrm{d}t} = 0$, then equation 5.220 becomes:

$$YFS_0 = \frac{\mu_m SXV}{K_S + S} \qquad (5.221)$$

and combining this with equation 5.70 gives:

$$\boldsymbol{k}_d = \frac{YFS_0}{XV} \tag{5.222}$$

Hence $\boldsymbol{k}_d$ can be determined knowing the volume V at the time when $\mathrm{d}/(XV)/\mathrm{d}t$ is zero.

KNIGHTS[63] devised a procedure for determining the kinetic constants of a batch fermentation system which involves monitoring both the biomass and substrate concentrations, but without assuming a constant yield coefficient. Exponential growth in a batch fermenter is represented by equation 5.55 written as:

$$\mu = \frac{1}{X}\frac{\mathrm{d}X}{\mathrm{d}t} = \frac{\mu_m S}{K_S + S} \tag{5.223}$$

for the biomass, and the material balance for the substrate gives the specific rate of reaction ψ as:

$$\psi = \frac{-1}{X}\frac{\mathrm{d}S}{\mathrm{d}t} = \frac{\psi_m S}{K_S + S} \tag{5.224}$$

The quantity ψ_m is the constant corresponding to μ_m but relating to the consumption of substrate. The analytic simultaneous solution of these equations is not possible unless the yield is constant, but the integration of equation 5.224 leads to:

$$\mathbf{F} + \frac{K_S}{S_0}\ln\left[\frac{1}{1-F}\right] = \frac{\psi_m}{S_0}\int_0^t X\mathrm{d}t \tag{5.225}$$

where:

$$\mathbf{F} = \frac{S_0 - S}{S_0}$$

The biomass concentration X is a function of time so that the right hand side of equation 5.225 cannot be integrated directly, but the form of the integral will be that of the logistic equation (see equation 5.61). If, therefore, that integral is replaced by the logistic expression then, after rearranging, equation 5.225 becomes:

$$\mathbf{F} + \frac{K_S}{S_0}\ln\left[\frac{1}{1-\mathbf{F}}\right] = \frac{\psi_m}{\mu_m S_0}\frac{X_0}{\chi_m}\ln\left[\chi_m e^{(\mu_m t)} - \chi_m + 1\right] \tag{5.226}$$

Here χ_m is the ratio X/X_m where X_m is a specified maximum biomass concentration. At this stage the substitutions:

$$\mathbf{F}_f = \mathbf{F} + \frac{K_S}{S_0}\ln\left[\frac{1}{1-\mathbf{F}}\right] \tag{5.227}$$

and:

$$\mathbf{T} = \ln\left[\chi_m e^{(\mu_m t)} - \chi_m + 1\right] \tag{5.228}$$

will produce:

$$\mathbf{F}_f = \frac{\psi_m}{S_0}\frac{X_0}{\chi_m}\mathbf{T} \tag{5.229}$$

Thus it may be seen that a plot of $\mathbf{F}_f$ against $\mathbf{T}$ will lie on a straight line passing through the origin.

The experimental data gathered during the exponential growth phase of a batch reaction are used initially to provide an approximate value for μ_m by analysing the $\ln(X)$ v time graph as described before. This will probably give a low estimate of μ_m which will, however, be adequate to serve as a starting value for further calculations. As with the method of Gates and Marlar, Knights' procedure now devolves into a trial-and-error technique. A series of values is chosen for K_S/S_0 and for χ_m, and then the experimental data are used to plot a set of curves computed using equation 5.229. The true values of μ_m, ψ_m and K_S are calculated using the combination of parameters which produces the best agreement with a straight line plot.

5.12.2. Use of Continuous Culture Experiments

In contrast to the batch fermentation based methods of determining kinetic constants, the use of a continuous fermenter (Fig. 5.71) requires more experiments to be performed, but the analysis tends to be more straightforward. In essence, the experimental method involves setting up a continuous stirred-tank fermenter to grow the micro-organisms on a sterile feed of the required substrate. The feed flowrate is adjusted to the desired value which, of course, must produce a dilution rate below the critical value for washout, and the system is allowed to reach steady state. Careful measurements of the microbial density X, the substrate concentration S, and the flowrate F are made when a steady state has been achieved, and the operation is then repeated at a series of suitable dilution rates.

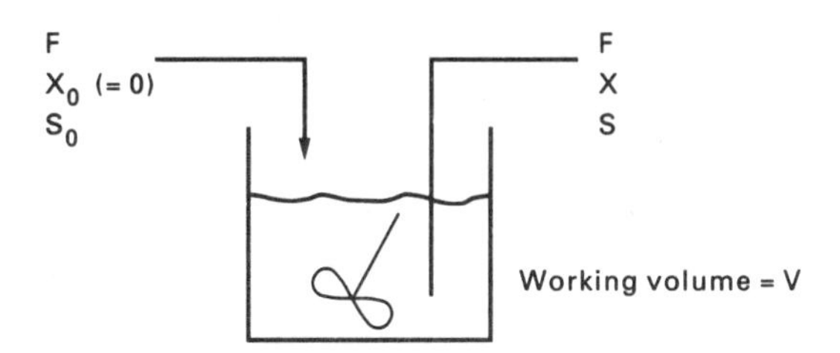

FIG. 5.71. The chemostat used for kinetic parameter estimation

It has been shown (equation 5.127) that the material balance for substrate across a continuous stirred-tank fermenter gives:

$$FS_0 + \mathcal{R}_S V - FS = \frac{dS}{dt} V \qquad \text{(equation 5.127)}$$

which at steady state reduces to:

$$D(S_0 - S) = -\mathcal{R}_S \qquad (5.230)$$

If the yield coefficient Y for the conversion of substrate into microbial cells is assumed to be constant, then when Monod kinetics are applicable the material balance becomes:

$$D(S_0 - S) = \frac{\mu_m SX}{Y(K_S + S)} \tag{5.231}$$

Inversion of this equation gives the linearised form:

$$\frac{X}{D(S_0 - S)} = \frac{K_S Y}{\mu_m} \frac{1}{S} + \frac{Y}{\mu_m} \tag{5.232}$$

It may be seen that a plot of $\frac{X}{D(S_0 - S)}$ against $\frac{1}{S}$ (Fig. 5.72) will give a straight line, and evaluating its slope and intercept on the abscissa will allow the value of K_S to be determined.

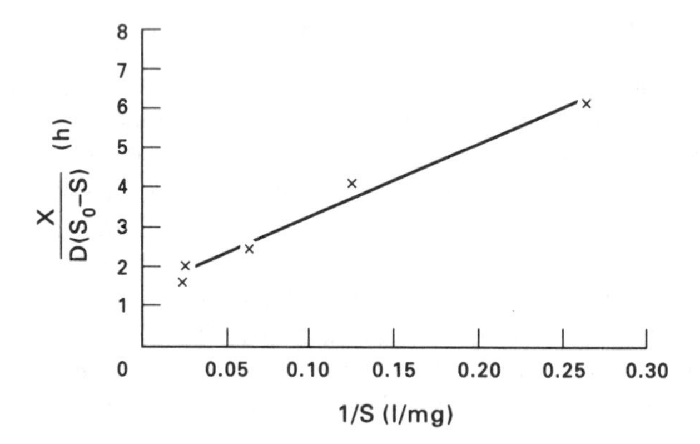

FIG. 5.72. Plot of $\frac{X}{D(S_0 - S)}$ against $\frac{1}{S}$ for chemostat culture

The evaluation of μ_m first requires the value of Y to be determined. Where the maintenance requirements of the culture may be ignored, then the yield coefficient for growth is equal to the overall observed yield coefficient and will be given by:

$$Y = \frac{X}{S_0 - S} \tag{5.233}$$

since the use of sterile feed means that X_0 will be zero. The mean value of Y calculated at each dilution rate used in the experiments will thereafter allow μ_m to be determined from the graph.

However, inspection of Fig. 5.57 suggests that the operation of a chemostat at dilution rates other than those approaching washout conditions results in growth at low values of substrate concentration S. For this condition the endogenous respiration rate becomes significant and $Y_G \neq Y_{X/S}$, as indicated in equation 5.54. It therefore becomes more realistic to express the specific growth rate μ as:

$$\mu = \frac{\mu_m S}{K_S + S} - k_d \qquad \text{(equation 5.70)}$$

and from equation 5.131:

$$D = \frac{\mu_m S}{K_S + S} - \boldsymbol{k}_d \tag{5.234}$$

Equation 5.231 may be rearranged to give:

$$\frac{\mu_m S}{K_S + S} = YD\frac{(S_0 - S)}{X} \tag{5.235}$$

so that substitution of this result in equation 5.234 produces:

$$\frac{S_0 - S}{X} = \frac{\boldsymbol{k}_d}{YD} + \frac{1}{Y} \tag{5.236}$$

Thus, a plot of $\frac{S_0 - S}{X}$ against $1/D$ will produce a straight line graph as shown in Figure 5.73 with slope $\boldsymbol{k}_d/Y$ and intercept $1/Y$. This allows the determination of both $\boldsymbol{k}_d$ and Y and, in conjunction with the calculations associated with Figure 5.72, the complete set of kinetic parameters for the microbial growth.

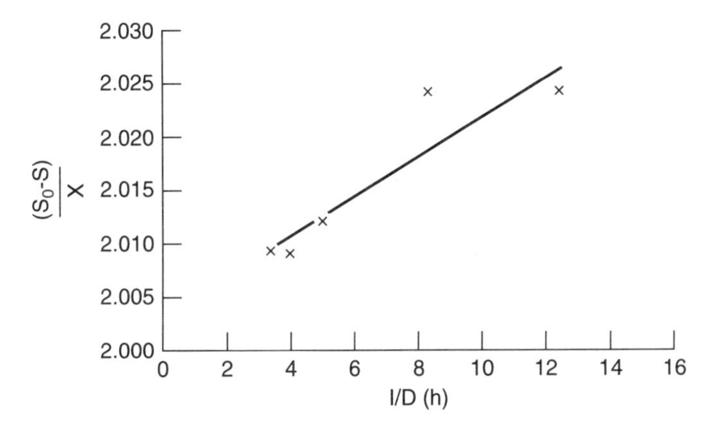

FIG. 5.73. Kinetic parameter determination with chemostat

Example 5.3

The steady-state substrate and biomass concentrations for a continuous stirred-tank fermenter operated at various dilution rates are given below. Given that the fresh feed concentration is 700 mg/l, calculate the values of the Monod constants μ_m and K_S the yield coefficient, Y and the endogenous respiration coefficient $\boldsymbol{k}_d$.

Dilution Rate (h^{-1})	Substrate Concentration (mg/l)	Biomass Concentration (mg/l)
0.30	45	326
0.25	41	328
0.20	16	340
0.12	8.0	342
0.08	3.8	344

Solution

From the data given above the values of $(S_0 - S)/X$ and $X/D(S_0 - S)$ may be calculated:

$\frac{1}{D}$	$\frac{S_0 - S}{X}$	$\frac{1}{S}$	$\frac{X}{D(S_0 - S)}$
3.33	2.0092	0.0222	1.659
4.00	2.0091	0.0244	1.991
5.00	2.0118	0.0625	2.485
8.33	2.0234	0.1250	4.118
12.50	2.0238	0.2632	6.176

Graphs can then be plotted of $(S_0 - S)/X$ against $1/D$ (Fig. 5.73) and $X/D(S_0 - S)$ (Fig. 5.72) against $1/S$ and their slopes and intercepts determined. Now for Fig. 5.73:

$$\text{intercept} = \frac{1}{Y} = 2.0033$$

∴ yield coe $\quad Y = 1/2.0033 = 0.50$

$$\text{slope} = \frac{k_d}{Y} = 0.0018 \text{ (h)}$$

∴ $\quad k_d = 0.0018/2.0033 = 0.0009 \text{ h}^{-1}$

From Fig. 5.72:

$$\text{intercept} = \frac{Y}{\mu} = 1.448 \text{ h}$$

∴ $\quad \mu_m = 0.50/1.448 = 0.34 \text{ h}^{-1}$

$$\text{slope} = \frac{K_S Y}{\mu_m} = 18.48 \text{ mg l}^{-1} \text{ h}^{-1}$$

∴ $\quad K_S = 18.48 \times 0.34/0.5 = \underline{\underline{12.8 \text{ mg/l}}}$

5.13. NON-STEADY STATE MICROBIAL SYSTEMS

5.13.1. Predator–Prey Relationships

Natural food chains frequently start at the level of the microbe, and gradually larger and more sophisticated animals enter into the sequence. The starting point is usually a photosynthetic micro-organism, which grows in the presence of a supply of minerals and carbon but uses light as its energy source. In a defined situation the starting point need not be at the photosynthetic level, but in any case the next species in the chain will regard the prey as its food and energy source. In the case of the activated-sludge waste-treatment process, such interactions can be very important in limiting the growth of unflocculated micro-organisms. These prey microbes consume the suspended waste and are themselves consumed by a variety of predators which also exist in the sludge.

The modelling of real food webs can be an exceedingly complicated task but, to illustrate the basic technique, a situation may be defined where a continuous stirred-tank biological reactor contains two species, one the predator, the other the prey. The food for the prey is assumed to enter as the sterile feed stream to the reactor, so that the predator may only consume the prey which grows in the reactor. Material balances can be drawn up for the process in much the same way as has been done for the earlier cases of microbial growth. Three relationships can be obtained from:

$$\begin{Bmatrix}\text{Material}\\ \text{in with}\\ \text{feed}\end{Bmatrix} + \begin{Bmatrix}\text{Rate of}\\ \text{formation}\\ \text{by reaction}\end{Bmatrix} - \begin{Bmatrix}\text{Rate of}\\ \text{removal}\\ \end{Bmatrix} = \text{Accumulation}$$

which, for the substrate becomes:

$$D(S_0 - S) - \frac{\mu_{mj} S X_j}{Y_j(K_{sj} + S)} = \frac{\mathrm{d}S}{\mathrm{d}t} \tag{5.237}$$

for the prey:

$$- DX_j + \frac{\mu_{mj} S X_j}{K_{sj} + S} - \frac{\mu_{mk} X_j X_k}{Y_k(K_{sk} + X_k)} = \frac{\mathrm{d}X_j}{\mathrm{d}t} \tag{5.238}$$

and for the predator:

$$- DX_k + \frac{\mu_{mk} X_j X_k}{K_{sk} + X_j} = \frac{\mathrm{d}X_k}{\mathrm{d}t} \tag{5.239}$$

where the subscript j refers to the prey and i to the predator.

For steady state operation the right-hand-sides of these equations become zero; however, the more interesting solutions are obtained under unsteady-state conditions. The integration of the differential equations may be carried out numerically after suitable values for the constants have been chosen. Several workers (LOTKA[64], TSUCHIYA *et al.*[65] *inter alia*) have studied this condition, typically using bacteria as prey and amoebae as predators. Figures 5.74–5.76 show the results of solving the equations. Changing the conditions, particularly the feed concentration and the dilution rate, can cause the system to react differently, stable steady-states being possible, as well as washout of the predator or of both species.

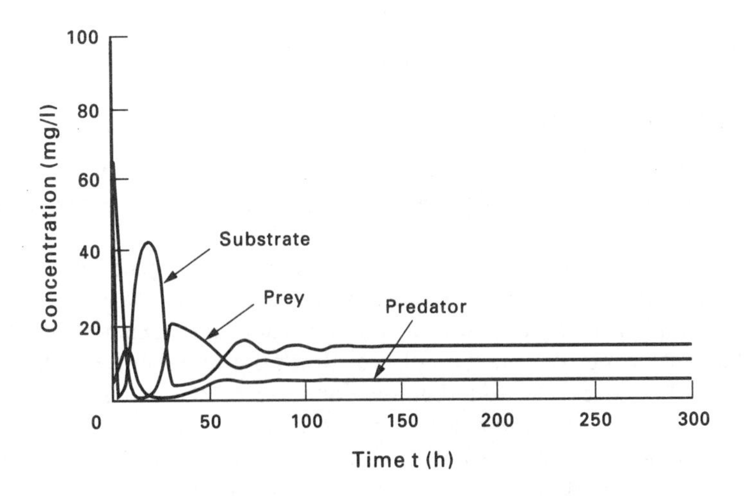

FIG. 5.74. Damped oscillations in population of both species leading to steady state

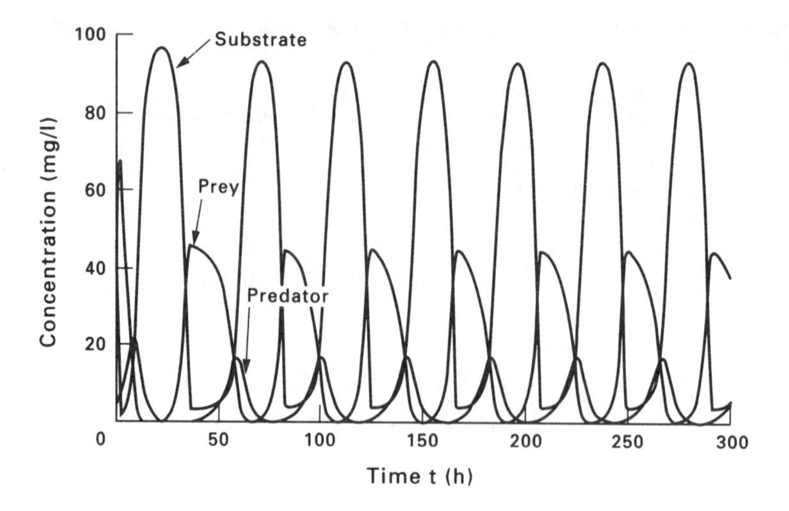

FIG. 5.75. Sustained oscillations in populations

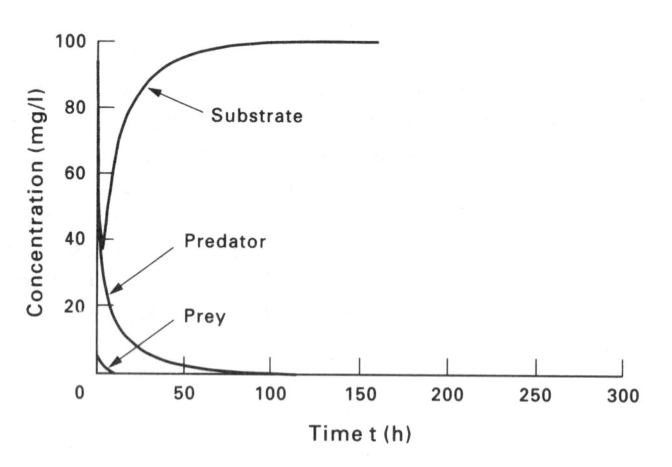

FIG. 5.76. Washout of both species

5.13.2. Structured Models

The preceding consideration of the growth of microbial cultures implies the assumption that the biomass may be taken to be an homogeneous material. Typically, two material balances have been carried out, one for the substrate and the other for the biomass, and these have resulted in a pair of simultaneous differential equations which may be solved to describe the behaviour of the culture. This has the implication that during growth each cell experiences the same conditions as every other cell in the culture, that all components of the microbial cell increase at the same rate and that the function of the cell is invariate. Such growth is referred to as *balanced* growth and the models involved as being *unstructured* and *unsegregated*. Whilst these models are still the mainstay of the design and analysis of biological reactors, they do not make much use of the wealth of biological knowledge referred to in earlier sections of this chapter. The models themselves are relatively simple mathematically and have the advantage of being amenable to algebraic manipulation, justifying their use whenever possible.

When a more detailed analysis of microbial systems is undertaken, the limitations of unstructured models become increasingly apparent. The most common area of failure is that where the growth is not exponential as, for example, during the so-called lag phase of a batch culture. Mathematically, the analysis is similar to that of the interaction of predator and prey, involving a material balance for each component being considered.

One of the earliest structured models is that put forward by WILLIAMS[66] who proposed that the material of a cell could be divided into two categories. One of these is referred to as the active component, the other being the structural component. The model considered that all the cells in the fermentation broth were identical and substrate was incorporated initially into the active component and thence was used to form the structural component. The second, structural, component controlled the observed growth of the culture in that doubling of that component would be a necessary and sufficient condition for the cells to divide.

RAMKRISHNA *et al.*[67] proposed a similar model at about the same time—this too was an unsegregated model which also divided the biomass into two *compartments*. They referred to the material in the two compartments as **G**-*mass* and **D**-*mass*, respectively, and suggested that these materials were formed in parallel. They also proposed that the micro-organism produced a toxic substance which inhibited its growth. They produced a set of differential equations obtained from material-balance considerations, to describe the behaviour of such a system in both batch and continuous culture. For batch culture:

$$\left.\begin{aligned} \frac{dC_G}{dt} &= \mathcal{R}_G - \boldsymbol{k}_{IG} C_G I \\ \frac{dC_D}{dt} &= \mathcal{R}_D - \boldsymbol{k}_{ID} C_D I \\ \frac{dS}{dt} &= -b_1 \mathcal{R}_G - b_2 \mathcal{R}_D \\ \frac{dI}{dt} &= b_3 \mathcal{R}_D + b_4 \boldsymbol{k}_{IG} C_G I \end{aligned}\right\} \qquad (5.240)$$

where:

C_G is the concentration of **G**-mass,
C_D is the concentration of **D**-mass,
I is the concentration of inhibitor,
$\boldsymbol{k}_{IG}$ and $\boldsymbol{k}_{ID}$ are rate constants, and
b_1, b_2, b_3 and b_4 are stoichiometric constants.

The rates $\mathcal{R}_G$ and $\mathcal{R}_D$ are given by:

$$\mathcal{R}_G = \frac{\mu_G S C_D}{(K_{SG} + S)} \frac{C_G}{(K_{GG} + C_G)}$$

$$\mathcal{R}_D = \frac{\mu_D S C_D}{(K_{SD} + S)} \frac{C_G}{(K_{GD} + C_G)} \qquad (5.241)$$

where μ_G and μ_D are rate constants and K_{SD}, K_{SG}, K_{GG} and K_{GD} are affinity constants. Solving these equations numerically using the values for the constants as given in Table 5.19 produced the graph of the batch fermentation shown in Fig. 5.77.

TABLE 5.19. *Constants Used in the Model of* RAMKRISHNA *et al.*[67]

Constant	Customary biological units		S.I. units	
	Value	Units	Value	Units
μ_G	0.5	h^{-1}	0.14×10^{-3}	s^{-1}
μ_D	2.5	h^{-1}	0.69×10^{-3}	s^{-1}
K_{SG}	0.2	g/l	0.2	kg/m^3
K_{SD}	0.1	g/l	0.1	kg/m^3
b_1	8		8	
b_2	2		2	
k_{IG}	150	g/h	42×10^{-6}	kg/s
k_{ID}	70	g/h	19×10^{-6}	kg/s
K_{GG}	30×10^{-6}	g/l	30×10^{-6}	kg/m^3
K_{GD}	5×10^{-6}	g/l	5×10^{-6}	kg/m^3
b_3	20×10^{-6}		20×10^{-6}	
b_4	26.7×10^{-3}		26.7×10^{-3}	

The success of the model may be seen from the shape of the curve shown in Fig. 5.77. In the initial stages of growth a lag phase is predicted which duly gives way to the exponential growth period. After this, the calculations indicate that a stationary phase is attained which is followed by a decline in the active biomass concentration.

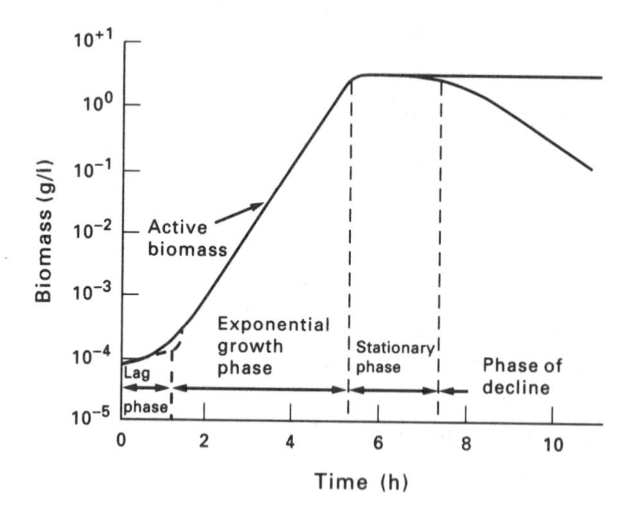

FIG. 5.77. Batch growth curve produced by the model of RAMKRISHNA *et al.*[67]

This model may be criticised in two respects, respectively that the materials referred to are not quantifiable in a precise way (ESENER *et al.*[68]), and that there is no account taken of the dilution effect on the intra-cellular material by the expansion of the biomass (FREDRICKSON[69]). In respect of the latter item, a distinction must be drawn between the concentrations of the components in the culture and those in the biomass itself. For a batch culture, a material balance for an active material which forms a fraction f of the growing biomass is undertaken and its concentration increases from X to $X + \mathrm{d}X$ in a time period $\mathrm{d}t$ then:

$$\begin{Bmatrix}\text{Final}\\ \text{activity}\end{Bmatrix} - \begin{Bmatrix}\text{Initial}\\ \text{activity}\end{Bmatrix} = \begin{Bmatrix}\text{Rate of generation}\\ \text{of active material}\end{Bmatrix} \times \begin{Bmatrix}\text{Time}\\ \text{period}\end{Bmatrix}$$

which may be stated mathematically as:

$$\left\{fX + \frac{\mathrm{d}}{\mathrm{d}t}(fX)\mathrm{d}t\right\} - fX = Xw\,\mathrm{d}t \qquad (5.242)$$

where w is some function which describes the specific rate of generation of f (which may or may not be dependent on f itself). Simplifying equation 5.242:

$$X\frac{\mathrm{d}f}{\mathrm{d}t} + f\frac{\mathrm{d}X}{\mathrm{d}t} = Xw \qquad (5.243)$$

Rearranging gives:

$$\frac{\mathrm{d}f}{\mathrm{d}t} = w - f\frac{1}{X}\frac{\mathrm{d}X}{\mathrm{d}t} \qquad (5.244)$$

and, substituting for the specific growth rate:

$$\frac{\mathrm{d}f}{\mathrm{d}t} = w - f\mu \qquad (5.245)$$

The last term in equation 5.245 represents the dilution of active component f, by the expansion of the biomass. ESENER *et al.*[(68)] also present a two-compartment model which takes this effect into account and they emphasise the need to devise the theory so that it can be tested by experiment. In their model they identify a '**K**' compartment of the biomass which comprised the RNA and other small cellular molecules. The other compartment contained the larger genetic material, enzymes, and structural material. The model assumes that the substrate is absorbed by the cell to produce, in the first instance, **K** material, and thence it is transformed into **G** material. Additionally, the **G** material can be reconverted to **K** material, a feature intended to account for the maintenance requirement of the micro-organism. A series of material balances for the cellular components during growth in a CSTF produced the following differential equations:
For the substrate:

$$\frac{\mathrm{d}S}{\mathrm{d}t} = -\frac{\psi_m SX}{K_S + S} + D(S_0 - S) \qquad (5.246)$$

$$\frac{\mathrm{d}X}{\mathrm{d}t} = Y_{S/K}\frac{\psi_m SX}{K_S + S} + (Y_{G/K} - 1)\boldsymbol{k}_K M_K M_G X - DX \qquad (5.247)$$

$$\frac{\mathrm{d}M_K}{\mathrm{d}t} = (1 - M_K)Y_{S/K}\frac{\psi_S S}{K_S + S} - \boldsymbol{k}_K M_K M_G[1 + (Y_{G/K} - 1)C_K] + m_G M_G \qquad (5.248)$$

and:

$$\frac{\mathrm{d}M_G}{\mathrm{d}t} = -\frac{\mathrm{d}M_K}{\mathrm{d}t} \qquad (5.249)$$

where:

M_G is the mass fraction of **G** compartment in the biomass,
M_K is the mass fraction of **K** compartment in the biomass,
k_K is the rate constant for **K** consumption,
m_G is the maintenance rate for **G** compartment,
$Y_{K/S}$ is the yield coefficient for conversion of substrate to **K** compartment, and
$Y_{G/K}$ is the yield coefficient for conversion of **K** compartment to **G** compartment.

The model was used by Esener *et al.* to calculate the outcome of both fed-batch and continuous culture experiments and the results were compared with the experimental values.

The curves shown in Fig. 5.78 indicate the ability of the model to give a reasonable description of the performance of the fermentations. The authors of the paper, however, draw attention to the fact that that the prediction of the RNA component was nowhere as accurate and concluded that the model had failed. This, they pointed out, was a necessary test to prevent the model from becoming merely a curve fitting exercise as opposed to a mechanistic model.

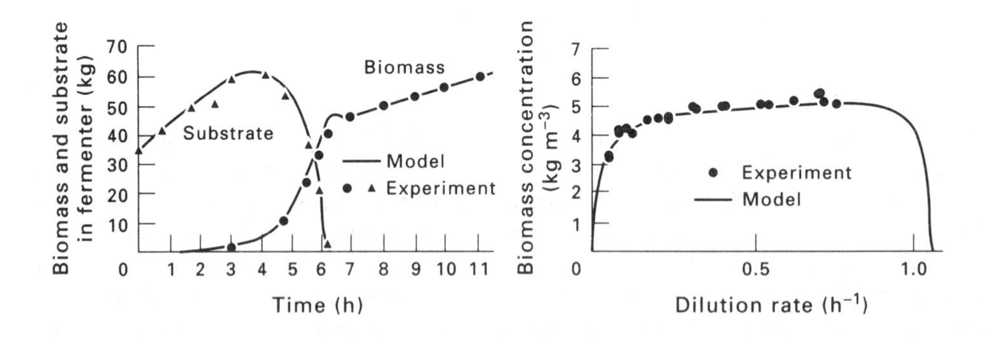

FIG. 5.78. Comparison of theoretical and experimental results using the model of ESENER *et al.*[68]

Other workers (IAMANAKA *et al.*[70], PAPAGEORGAKOPOULU and MAIER[71], *inter alia*) have developed more intricate structured models for the description of microbial growth. These models in particular have been have been based on a mechanistic approach, taking into account a series of cellular functions and products. A set of six simultaneous differential equations arise in the treatment of Iamanaka *et al.* and nine in that of Papageorgakopoulu and Maier, each having a total of over twenty constants associated with them. Whilst both models produced good agreement between calculated and experimental values, their complicated nature makes them rather unwieldy and of limited use for design purposes.

5.14. FURTHER DESIGN CONSIDERATIONS

The primary function of a fermenter is to provide conditions which are conducive to microbial growth. Assuming that the fermenter contains the appropriate nutrients, the most important conditions which must be provided, or controlled, are temperature, pH and dissolved oxygen. The last item is included because the vast majority of industrially important fermentations (with the exception of brewing beer and wine) involve the aerobic growth of micro-organisms. Fermenters, as with other items of process engineering equipment, may vary considerably in their design according to their particular application. The stirred-tank and plug-flow tubular fermenters which have been considered are idealised forms, which in some respects are extreme cases, which nevertheless serve as suitable design models for real installations. However, the ideal forms considered do not give much insight into how the operational requirement of a fermenter sets it apart from any other chemical reactor.

The constructional details of a fermenter vary considerably according to its intended application, particularly with regard to the scale of its operation and the necessity or otherwise of sterilisation before the initiation of microbial growth. In broad terms, small fermenters for laboratory use would be constructed mainly of borosilicate glass with some parts, such as sensor ports and heating surfaces, in stainless steel. Such a fermenter is illustrated in Fig. 5.79, although in other designs the base may also be of glass. A vessel of this size may be easily cleaned when dismantled and is conveniently sterilised by placing the assembled unit (often containing the substrate necessary for the subsequent experiment) in a steam autoclave.

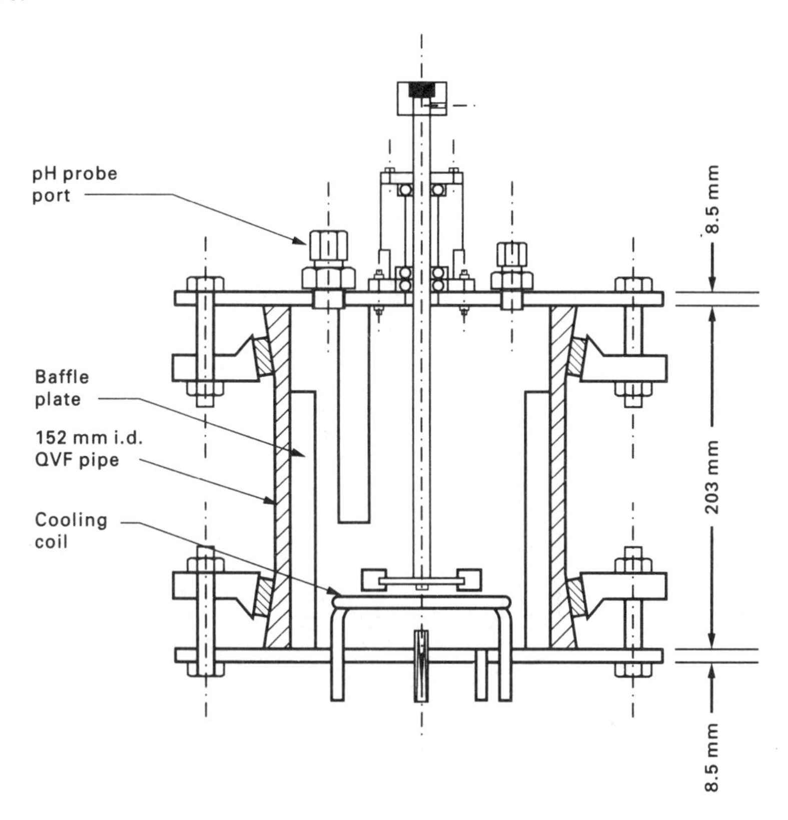

FIG. 5.79. A typical laboratory fermenter

Fermenters with a capacity of over about 10 litres are too heavy to sterilise in autoclaves. Whilst they may still be laboratory-sized, they have to be constructed so that they may be sterilised *in situ*. They become, as a consequence, pressure vessels and the extensive use of glass becomes impractical and the preferred material of construction is a stainless steel. Seals are typically of silicone or other synthetic rubber or fluorinated plastics, with borosilicate glass being retained for sighting windows. This format is retained for vessels which are far larger than the laboratory scale, and Fig. 5.80 outlines the construction of a typical industrial *deep tank* fermenter.

This design is extensively used for the growth of single cultures of microbes, but it is by no means the only form of practical fermenter. Larger installations, particularly those intended for the growth of shear-sensitive organisms, may rely on the aeration system for mixing the contents. In the *air lift* design, the difference in bulk density between aerated and unaerated fermenter broth is used to induce circulation within the vessel. Microbial growth vessels intended for waste water treatment differ even more dramatically, largely due to the fact that the constraint of maintaining a single species whilst excluding all others (*mono culture*) has been removed. They are usually open tanks, frequently constructed of concrete and often employing surface-aeration equipment which also serve to mix the contents.

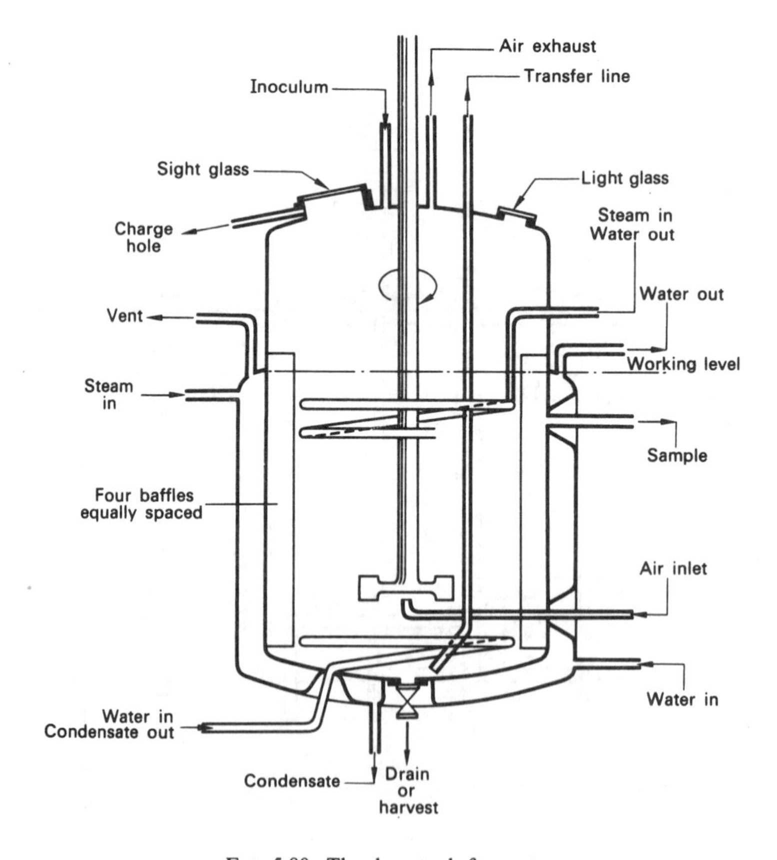

FIG. 5.80. The deep-tank fermenter

5.14.1. Aseptic Operation

A fermenter used for the growth of a specified micro-organism must usually offer a containment facility for the microbial culture, normally such that contamination from outside is avoided, although the retention of the cultured organisms within the fermenter is also a frequent requirement. Mono culture of microbes is the rule rather than the exception and the fermenter vessel, and the feed, must be sterilised before the introduction of the seed micro-organisms. The entry of wild organisms from the outside world generally has an adverse effect on the progress of a fermentation. The contaminating species may compete with the desired organism for the nutrients available, or may consume it or its products as substrate, or may spoil the fermentation by releasing toxic products into the growth medium. For these reasons it is usually necessary not only to destroy all residual microbes in the fermenter (and the medium) before the growth is initiated, but also to ensure that the air entering during the fermentation is free of microbial contamination by passing it through a suitable filtration system beforehand. Similarly, the adventitious introduction of microbes or their spores during sampling or during the transfer of additives to the broth must be avoided, and this alone considerably complicates the design of a fermenter. Figure 5.81 shows a piping arrangement suitable for the aseptic transfer of sterilised material from a holding vessel into a fermenter. The arrangement of valves allows the transfer lines to be flushed with steam and sterilised before the vessels are connected. The condensate formed is removed via the steam traps which are then isolated before the sterile liquid is transferred.

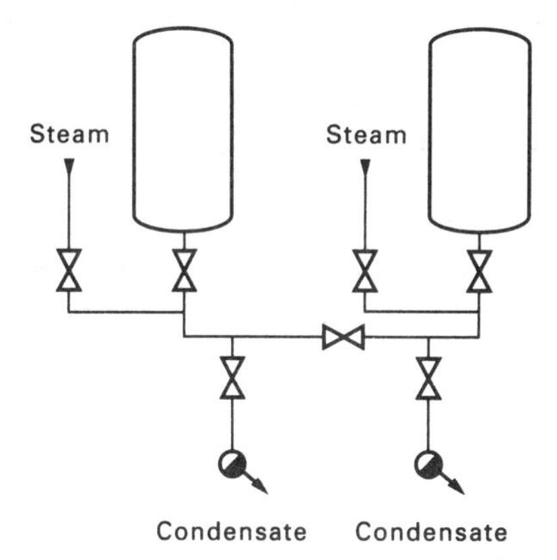

FIG. 5.81. Aseptic transfer of material to a fermenter

5.14.2. Aeration

Many micro-organisms are capable of respiring in the absence of molecular oxygen (anaerobic conditions) and, in some cases, this is a necessary condition for either their well-being, or necessary for them to produce the desired product. On the other hand, some micro-organisms are *obligate aerobes* and cease to function when deprived of oxygen. Fulfilling the oxygen requirement of a microbial fermentation presents another series of problems in the design of a fermenter. The solubility of oxygen in pure water at a typical fermentation temperature of 30°C is 7.7 mg/l and this figure may be depressed by the presence of solutes such as carbohydrates and mineral salts. The respiration rate of the culture is usually such that the residual oxygen in the broth represents less than a minute's supply for the cells, and consequently the oxygen has to be continuously replenished during the fermentation.

The respiration rate of an aerobic microbe is dependent on the dissolved oxygen concentration in the growth medium. The specific volumetric oxygen uptake rate, (Q_{O_2}, the mass of oxygen consumed by unit mass of cells in unit time), when plotted against dissolved oxygen concentration, exhibits an initial rapid rise to a plateau level, as shown in Fig. 5.82. The point at which the rather sharp inflexion in the curve occurs defines the 'critical' dissolved oxygen concentration, above which the dissolved oxygen concentration does not limit the rate of reaction. The specific respiration rate is related to the specific growth rate by the yield coefficient for oxygen utilisation Y_{O_2} as:

$$Q_{O_2} = \frac{\mu}{Y_{O_2}} \tag{5.250}$$

The value for the 'critical' oxygen concentration tends to be rather small and this makes its direct measurement difficult. Operational fermenters are frequently equipped with oxygen sensing probes so that control systems can be used to ensure that the dissolved oxygen concentration is well above the critical value, typically above 20 per cent of the saturation value. The practical problem related to aerobic fermenters is one of mass transfer and, in some cases, an answer has been found by increasing the driving force for the dissolution process. This has been achieved by raising the partial pressure of oxygen in the gas being sparged into the fermenter either by using oxygen enriched air or increasing the overall gas pressure. Generally, the overall mass transfer coefficient for the dissolution process is maximised, by a combination of efficient stirring of the broth and sparging of the gas used.

The dissolution of a sparingly soluble gas in a liquid is discussed in Volumes 1 and 2, and the reader is referred to the relevant sections (Volume 1, Chapter 10 and Volume 2, Chapter 12) for detailed discussion of the mechanisms involved. In the

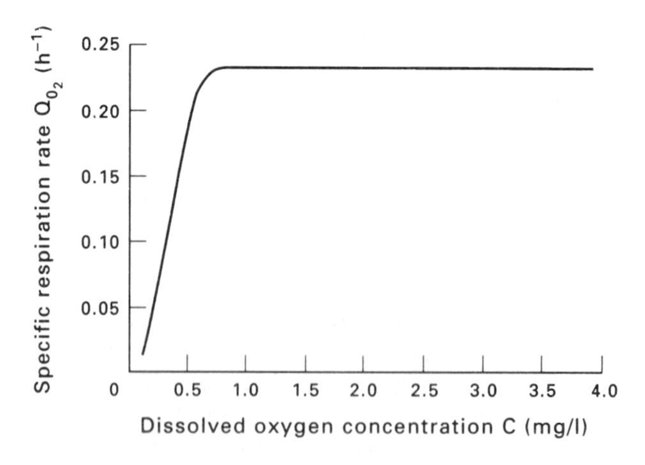

FIG. 5.82. Oxygen consumption rate of yeast at various oxygen concentrations

case of the aeration of a fermenter, the controlling resistance to mass transfer is usually in the liquid phase so that it is appropriate to describe the process mathematically as:

$$\frac{dC}{dt} = K_L a(C^* - C) \tag{5.251}$$

where C is the *mass* concentration of the gas in the liquid phase,
C^* is the equilibrium (saturation) gas *mass* concentration,
K_L is the overall mass transfer coefficient, and
a is the interfacial area per unit volume of liquid.

The interfacial area per unit volume a is, again, a difficult quantity to measure in a fermenter since it depends on the number and size of the air bubbles entrained in the fermentation broth. These, in turn, are dependent on such factors as the aeration rate, the rheology of the broth at that instant and the presence or otherwise of surfactants. As a result, the quantity $K_L a$ tends to be treated as a single entity in experimental measurements of oxygen transfer rates in fermenters.

The simplest method of determining $K_L a$ is to monitor the oxygen concentration during the aeration of a fermenter containing water or growth medium which has initially been rendered oxygen free. The removal of the oxygen can be effected by *gassing out* using a nitrogen supply or by the addition of an oxygen consuming agent, such as sodium sulphite, to the liquid. This is the so-called *static* method.

Equation 5.251 may be integrated to give:

$$\ln\left(\frac{C^*}{C^* - C}\right) = K_L a t \tag{5.252}$$

where C is the dissolved oxygen concentration at any time t. $K_L a$ may be determined from the slope of a plot of this equation, as shown in Fig. 5.83.

The static method of estimating $K_L a$ for a fermenter is not always satisfactory. The use of pure water instead of the fermentation medium can lead to inaccuracies, and the deoxygenation of a complex broth with sodium sulphite can result in undefined changes to the broth and be expensive to carry out on a large scale. An

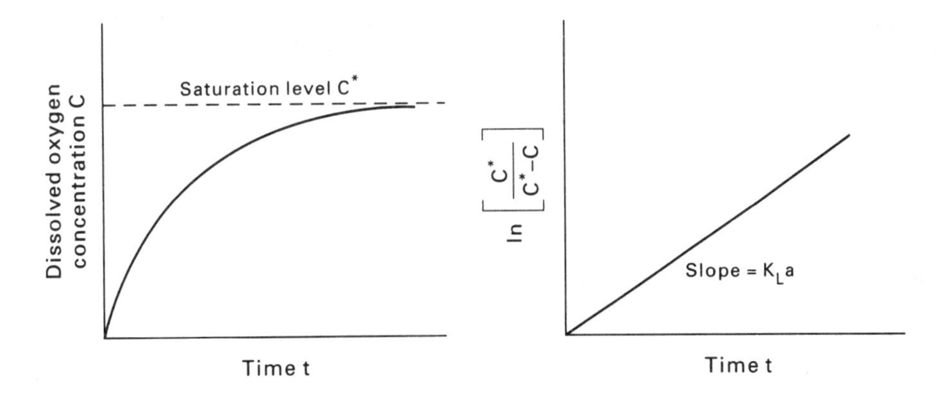

FIG. 5.83. Measurement of $K_L a$ using the static method

alternative dynamic method of determining K_La involves monitoring the course of a perturbation to the dissolved oxygen concentration during a fermentation. This is brought about by temporarily ceasing and then resuming the aeration (Fig. 5.84).

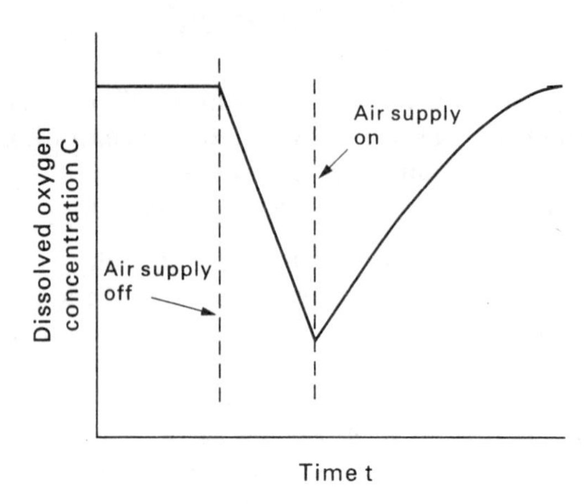

FIG. 5.84. The dynamic method of determining K_La

If during the beginning of the experiment the fermentation is at a quasi-steady state, that is the growth rate is such that the biomass concentration X may be considered to be constant during the experiment, then a material balance for the oxygen gives:

$$-\begin{Bmatrix}\text{Oxygen}\\ \text{uptake}\\ \text{rate by}\\ \text{microbes}\end{Bmatrix} + \begin{Bmatrix}\text{Oxygen}\\ \text{transfer}\\ \text{rate}\end{Bmatrix} = \begin{Bmatrix}\text{Accumulation rate}\\ \text{of oxygen in}\\ \text{fermenter}\end{Bmatrix}$$

This becomes:

$$-Q_{O_2}X + K_La(C^* + C) = \frac{dC}{dt} \tag{5.253}$$

During the period when the aeration is suspended, the slope of the curve will be determined by the oxygen uptake rate (OUR) so that the first term in equation 5.253 may be evaluated. When aeration is resumed the curve will be that described by the whole of the equation rearranged as:

$$C = C^* - \frac{1}{K_La}\left(\frac{dC}{dt} + Q_{O_2}X\right) \tag{5.254}$$

A plot of the concentration C against $\left(\frac{dC}{dt} + Q_{O_2}X\right)$ will give a straight line with slope $-1/K_La$.

Whilst this method has its advantages over the static method it too presents some practical difficulties. One of these is that the dissolved oxygen concentration must not be allowed to fall below the critical value C_{crit} as defined above, since this may be accompanied by a change in the metabolism of the microbe which is reflected in a change in Q_{O_2}.

Many correlations have been proposed for the estimation of $K_L a$ and there is an extensive literature on the subject. Among the most useful are those relating it to the power input to the stirrer and the aeration rate. These generally take the form:

$$K_L a = \text{const} \times (\text{power})^n \times (\text{aeration rate})^w$$

where n and w are usually fractions. A typical example is that given by MIDDLETON[72] for non-coalescing solutions:

$$K_L a = 1.2\left(\frac{\mathbf{P}_g}{V_L}\right)^{0.7} (u_s)^{0.6} \tag{5.255}$$

where:

$\mathbf{P}_g$ is the power input under gassed conditions (W),
V_L is the liquid volume (m^3), and
u_s is the superficial velocity of the air (m/s^{-1}).

This expression is quoted as having a typical error of ± 10 per cent (MIDDLETON[72]). It is not dimensionally consistent and the value of the coefficient therefore depends on the units being used.

5.14.3. Special Aspects of Biological Reactors

As mentioned earlier, the use of biological reactors for waste water treatment is widespread and some very large scale units exist. Since such installations are intended primarily to purify the waste water, and do not, as such, result in a saleable product, they are not expected to show an operational profit in the same way as a conventional process. Some recovery of operational costs may be realised by the use of anaerobic digesters whose off-gas can be used to reduce the requirement for natural gas or to generate electricity on site. Some waste streams, such as whey from cheese making, have been utilised to produce potable alcohol and edible protein, again taking advantage of the potential of biological reactors to make valuable end products. In contrast to waste water treatment, biological reactors can be used to manufacture fine chemicals, particularly pharmaceuticals, and other very valuable materials. These benefits have traditionally been sufficient to offset some of the less desirable aspects of biochemical reactors which might otherwise render the process uneconomic.

Many of the materials produced by biological reactors are chemically labile and may, in conventional terms, be present in aqueous solutions of low concentrations. Mild chemical and physical conditions in the reactor are essential for their formation and survival, and such conditions must usually be maintained in the subsequent stages of the process. As a result, biochemical processes tend to have multiple separation stages, each one operating at near ambient temperature, or below if chillers are used. Centrifugation, micro- and ultra-filtration, solvent extraction, precipitation and species-specific adsorption are unit operations frequently used in the separation stages. Although the high value of the product may make it economically feasible to use multiple stages, there is still an incentive to reduce their number in the overall process. Biological reactors which combine reaction and separation in one stage are currently under development. The use of immobilised enzymes and micro-organisms is an example of such an approach. Another is the inclusion of a micro-filter in a fermenter which allows the removal of the desired product whilst retaining the biomass in the bioreactor.

The near-ambient temperatures used in most enzyme reactors and fermenters are themselves not normally determining factors in the design specification of the vessels concerned. The preferred method of sterilising fermenters is the use of steam at temperatures above 100°C (typically 120–140°C) and hence at pressures above atmospheric. This presents little problem for small laboratory units since they may be conveniently sterilised in a steam autoclave so that there is no net pressure difference between the inside and the outside of the vessel. On the other hand, pilot and production scale fermenters are pressure vessels which must be designed to the same standards as pressure vessels used in conventional chemical manufacturing processes. Heat transfer during microbial growth conditions can present some problems too since the temperature differences involved can then be small. Whilst the high surface area to volume ratio of bench-scale units means that they may actually require heating during operation, the metabolic heat evolved in large fermenters frequently necessitates the use of cooling jackets or coils and the circulation of large quantities of cooling water.

Dealing with clean or sterile materials is not a problem peculiar to processes involving biochemical reactors, nor is the use of material of non-Newtonian rheology, which may give rise to persistent, unwanted foams. Established techniques of chemical engineering are usually applicable, albeit allowing for some of the constraints mentioned above. However, the advantages of the biochemical reactors often outweigh these and other, similar, problems and their increased use in the process industries in future may be expected.

5.15. APPENDICES

Appendix 5.1. Proteins

The monomers from which proteins are synthesised are the α L-amino-acids. The amino-acids are differentiated by the various radical groups attached to the α carbon (or C 2). Since there are generally four different functional groups attached to the C2 atom, it is asymmetrical and thus will have two possible optical isomers. However, with few exceptions, only the L-amino acids are found in proteins.

Amino-acids and their Properties

Amino-acids have at least two groups which can become ionised in aqueous solution.

$$\underset{\text{Anion}}{\text{R—}\underset{\substack{|\\ NH_2}}{\text{CH}}\text{—COO}^-} \underset{-H^+,\ pK_a\ 10}{\overset{+H^+}{\rightleftharpoons}} \underset{\text{Zwitterion}}{\text{R—}\underset{\substack{|\\ NH_3^+}}{\text{CH}}\text{—COO}^-} \underset{-H^+,\ pK_a\ 3}{\overset{+H^+}{\rightleftharpoons}} \underset{\text{Cation}}{\text{R—}\underset{\substack{|\\ NH_3^+}}{\text{CH}}\text{—COOH}}$$

Net charge: Anion −1; Zwitterion 0; Cation +1

An amino-acid therefore can have a number of charged forms, it is an anion at high pH and a cation at low pH; at neutral pH it normally has no net charge, but exists as a dipolar ion or *zwitterion* with both a positive and negative charge.

The 20 common naturally occurring amino-acids are shown in Fig. 5.A1. They are subdivided into three groups depending on the nature of the side chains, i.e. non-polar with alkyl and hydroxy side chains, polar with basic groups and polar with acid groups.

Protein Structure

The simple condensation between two amino-acids to form one molecule yields a dipeptide. Polypeptides are thus large molecules of amino-acids linked together via peptide bonds.

$$\underset{\substack{|\\ R_1}}{NH_2CHCOOH} + H_2N\,\underset{\substack{|\\ R_2}}{CH}\,COOH \longrightarrow H_2N\,\underset{\substack{|\\ R_1}}{CH}\,CONH\,\underset{\substack{|\\ R_2}}{CH}COOH + H_2O$$

Simple proteins are composed only of amino-acids linked together via a *peptide bond* as illustrated above. Note that the molecule as such now has an *N-terminal end* (free amino group) and a *C-terminal end* (free carboxylic acid group). Peptides and proteins are differentiated by size and interaction of the molecule—proteins have a peptide backbone which is sufficient for long distance intermolecular interactions to take place[18]. This phenomenon occurs at about 2000–3000 molecular weight and above.

There is a vast number of proteins in living systems. It has been estimated that at least 1000 different proteins are present at any one time in the bacterium *E. coli.* They form a vital part of living systems and have a wide variety of uses. However, for the function of individual proteins to be clearly understood, they usually have to be purified.

The structure of proteins, as with the structure of carbohydrates, has various levels—primary, secondary, tertiary and quaternary structure. The tertiary and quaternary structures and their subtleties are most important in the biological function of the molecule. Consider an enzyme (a protein-based catalyst)—its structure allows the binding of specific molecules which then react catalytically to give products. Conversely, enzymes are very susceptible to environmental conditions which alter their tertiary structure.

Protein structure may be further complicated by the inclusion of materials such as metal ions and porphyrin rings and by the addition of carbohydrates, lipids and nucleic acids. Such compounds are called *conjugated proteins.*

Primary Structure

A major problem, until recently, was the determination of the protein primary structure, but with the advent of modern analysis of DNA this has become comparatively easy. One of the first structures to be described was that of insulin which contains 60 amino-acids and has a molecular weight of 12,000. Once the primary structure is known, it is possible to predict the secondary and tertiary structures using additional information obtained through X-ray crystallography of the crystallised protein.

Unlike polysaccharides, proteins do not have branched chains, but several chains may be linked together via disulphide bridges rather than peptide bonds. The primary structure of ox insulin is shown in Fig. 5.A2. The protein consists of two peptide chains which are linked via the formation of the disulphide bridges. Disulphide bridges are formed by the condensation of the thiol groups of two cysteine residues.

Secondary Structure

Secondary structure arises from the way in which the primary structure is folded, maximising the number of hydrogen bonds, and in effect lowering the free energy of the molecule and its interaction with water. Generally, these interactions occur over relatively short range between different parts of the molecule.

A major subdivision due to different secondary structures is shown by the differences between those of globular and fibrous proteins. Fibrous proteins are generally structural proteins which are insoluble in water—for example, keratin and collagen which are major components of skin and connective tissues. Globular proteins are usually soluble in water and have functional roles, enzymes being typical examples. Secondary structure also generates a number of structural interactions and conformations. These structures are largely determined by the side groups of the individual amino-acids which will hinder or aid certain conformational structures. There are two forms, the *α helix* and the *β pleated sheet.*

The α helix structure is a right-handed helix, meaning that the chain spirals clockwise as it is viewed down the peptide chain. Because of the limited rotation around either side of the peptide bond, only a

Non-polar side chains		Polar side chains	
-R	**Amino-acid**	**-R**	**Amino-acid**
$-CH_3$	Alanine (Ala)	Negative charge at pH 7	
$-CH.CH_3$ $\vert$ CH_3	Valine (Val)	$-CH_2C(=O)O^-$	Aspartic Acid (Asp) or Asparate
$-CH_2CH.CH_3$ $\vert$ CH_3	Leucine (Leu)	$-CH_2CH_2C(=O)O^-$	Glutamic Acid (Glu) or Glutamate
$-CH.CH_2CH_3$ $\vert$ CH_3	Isoleucine (Ile)	Positive charge at pH 7	
		$-(CH_2)_4\overset{+}{N}H_3$	Lysine (Lys)
		$-(CH_2)_3NHC.NH_2$ $\Vert$ $+NH_2$	Arginine (Arg)
$-CH_2-C_6H_5$ (benzene ring)	Phenylalanine (Phe)	Uncharged at pH 7	
$-CH_2-$ (indole ring, N H)	Tryptophan (Trp)	$-H$	Glycine (Gly)
		$-CH_2OH$	Serine (Ser)
$-CH_2CH_2-S-CH_3$	Methionine (Met)	$-CH.CH_3$ $\vert$ OH	Threonine (Thr)
(ring) $CH.CO_2^-$, $\overset{+}{N}$ H_2 (complete structure)	Proline (Pro)	$-CH_2SH$	Cysteine (Cys)
		$-CH_2-C_6H_4-OH$ (benzene ring)	Tyrosine (Tyr)
		$-CH_2C(=O)NH_2$	Asparagine (Aspn)
		$-CH_2CH_2C(=O)NH_2$	Glutamine (Glun)
		$-CH_2-$ 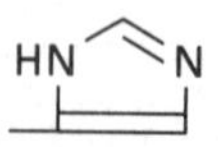 (imidazole ring, HN, N)	Histidine (His)

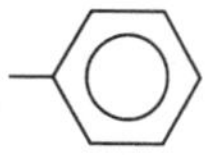

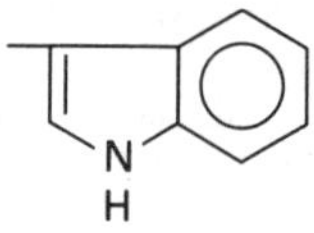

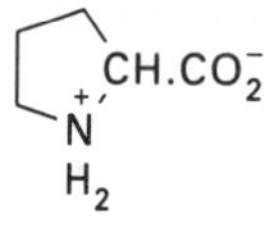

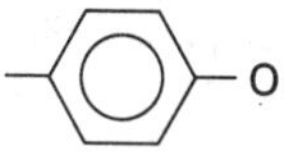

FIG. 5.A1. The side chains of the 20 amino-acids commonly found in proteins

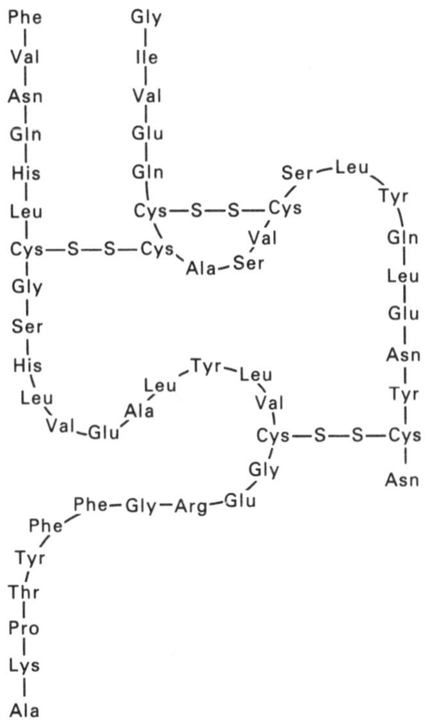

```
Gly—Ile—Val—Glu—Gln—Cys—Cys—Ala—Ser—Val—Cys—Ser—Leu—Tyr—Gln—Leu—Glu—Asn—Tyr—Cys—Asn
Phe—Val—Asn—Gln—His—Leu—Cys—Gly—Ser—His—Leu—Val—Glu—Ala—Leu—Tyr—Leu—Val—Cys—Gly—Glu—Arg—Gly—Phe—Phe—Tyr—Thr—Pro—Lys—Ala
```

FIG. 5.A2. Protein primary structure. The amino-acid sequence of ox insulin

few angles are energetically stable. In the case of the helix, a full turn is achieved by 3.6 amino-acid units. The structure is stabilised by the weak hydrogen bonding as shown in Fig. 5.A3*a*.

The β pleated sheet structure occurs commonly in insoluble structural proteins and only to a limited extent in soluble proteins. It is characterised by hydrogen-bonding between polypeptide chains lying side by side, as illustrated in Fig. 5.A3*b*.

Figure 5.A4. shows the structure of haemoglobin, a conjugate protein where there are extensive regions of α helix around the prosthetic group which is an iron porphyrin.

Tertiary Structure of Proteins and the Forces that Maintain it

Although steric effects and hydrogen-bonding maintain the secondary structure of proteins, these same factors also contribute to the stability of the tertiary structure. These involve long range interactions between the amino-acids of the polymer. Tertiary structure plays a key role in the function of proteins, including enzymes in biological systems. The forces responsible for maintaining the tertiary structure of proteins are the so-called weak forces, consisting of Van der Waals forces, hydrogen-bonding and weaker electrostatic interactions (Fig. 5.A5).

TABLE 5.A1. *Bond Energies of Types of Bond Occurring in Macromolecules*

Type of bond	Energy (MJ/kmol)
Covalent	200–400
Hydrogen	up to 20
Van der Waals	up to about 4
'Hydrophobic'	4–8
Ionic	up to 4

The behaviour of the side chains in an aqueous environment is important in determining the tertiary folding (structure) of the protein. Non-polar groups are *hydrophobic* (showing aversion to water) while polar groups are *hydrophilic* (showing affinity for water). Alteration of these interactions can therefore markedly affect the activity of enzymes. A change in the pH for example will affect the charge on the protein molecule, thence its structure and function. Table 5.A1 compares the bond energies of various

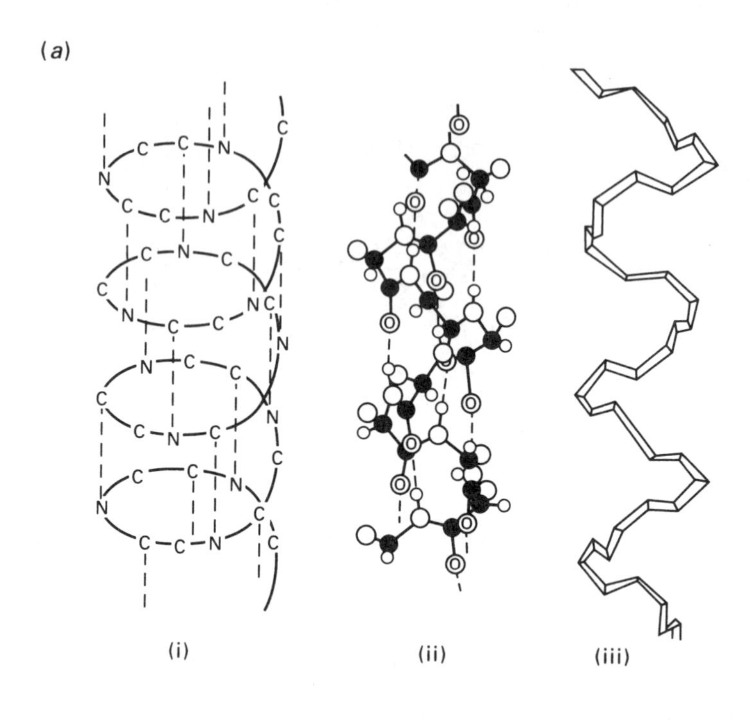

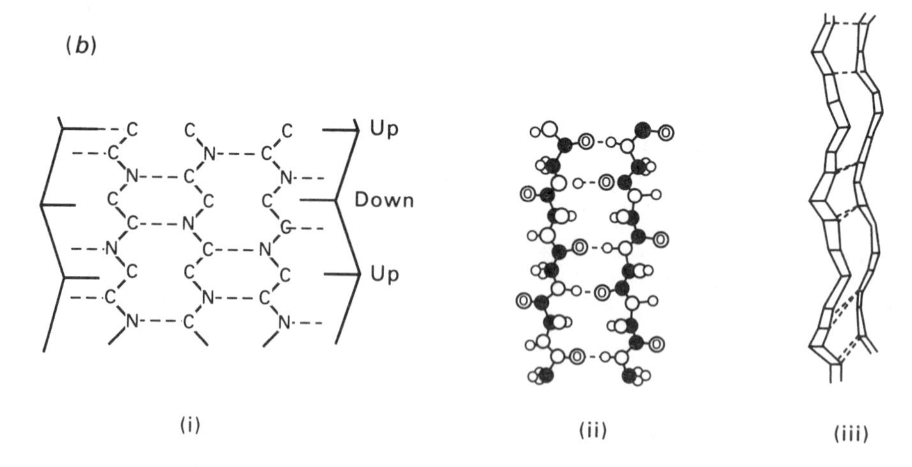

FIG. 5.A3. Protein secondary structure:
(*a*) α helix; (*b*) β pleated sheet, and the interactions that stabilise these structures. Each structure is shown in three ways as (i) hydrogen bonding, (ii) ball and stick, and (iii) ribbon structure

types of interaction in macromolecules, and it should be noted that these are 5–10 per cent of that of a covalent bond.

Changing the environmental conditions can easily provide sufficient energy to alter the tertiary structure of proteins significantly.

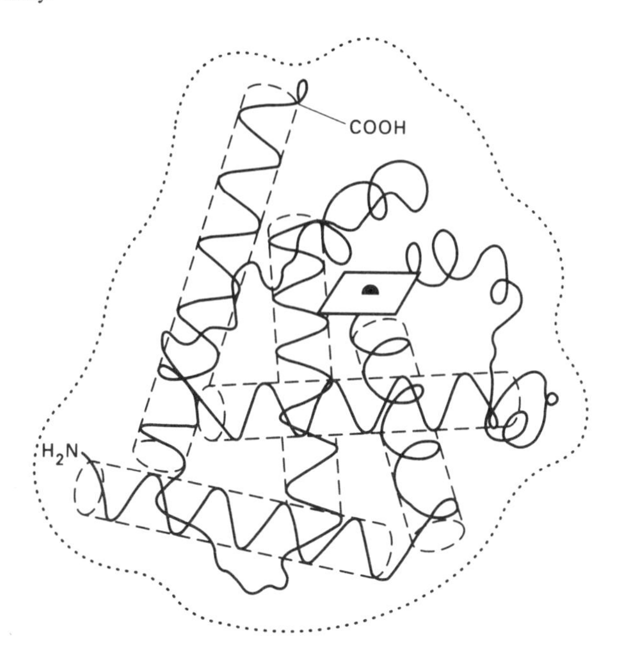

FIG. 5.A4. The structure of haemoglobin and its iron porphyrin prosthetic group. This diagram is highly simplified and the dotted line indicates the overall shape of the folded molecule and the zones of the α helical structure within the molecule are also shown

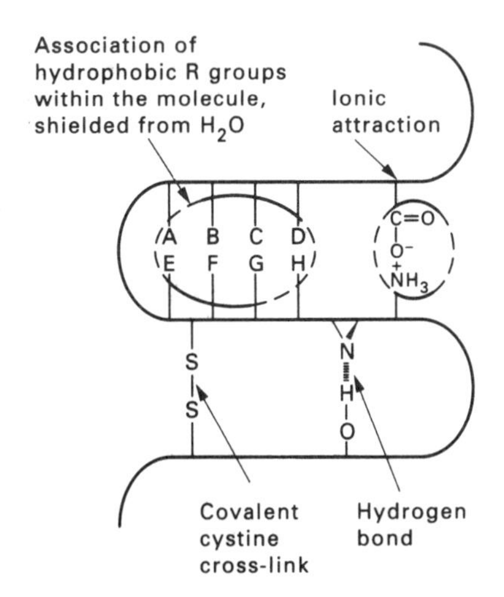

FIG. 5.A5. Forces maintaining the tertiary structure of globular proteins

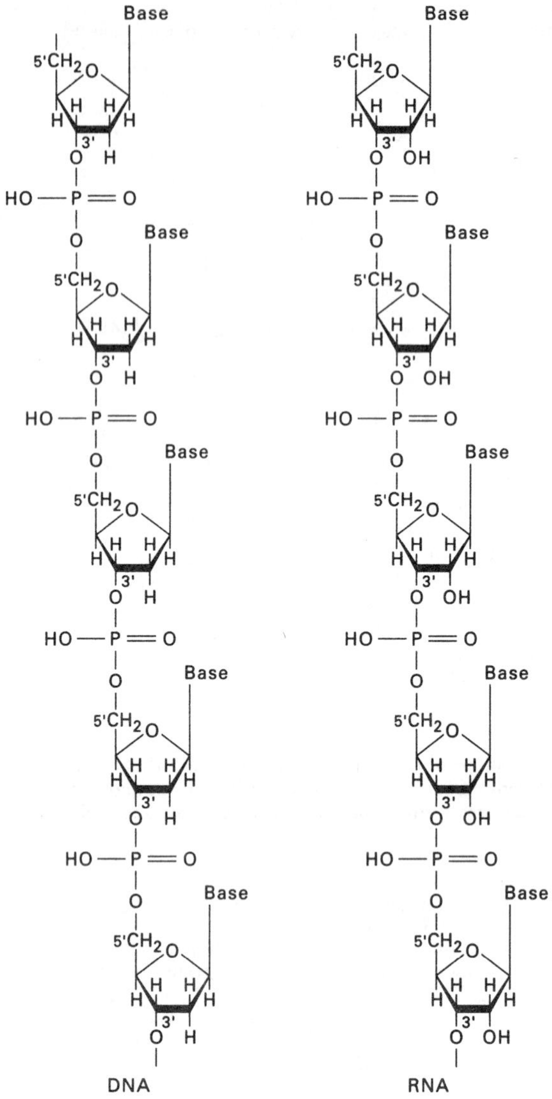

FIG. 5.A6. The covalent backbone structure of nucleic acids

Appendix 5.2. Nucleic Acids

DNA Structure

Deoxyribonucleic acid (DNA) is a linear polymer of nucleotides linked via phosphodiester bonds which are formed between the 3′ and 5′ hydroxy groups of the ribose moiety of the nucleotide. Figure 5.A6 shows the covalent backbone of DNA. These polymers are normally found as an aggregated pair of strands held together by hydrogen bonding (Fig. 5.A7). DNA is composed of four nucleotides, adenine (A), thymine (T), guanine (G) and cytosine (C). The primary structure of DNA is the linear sequence of nucleotide residues comprising the polydeoxyribonucleotide chain. DNA varies in length and in nucleotide composition, the shortest molecules being in the order of 10^3 bases in viruses, 10^6 in bacteria and 10^8 in humans.

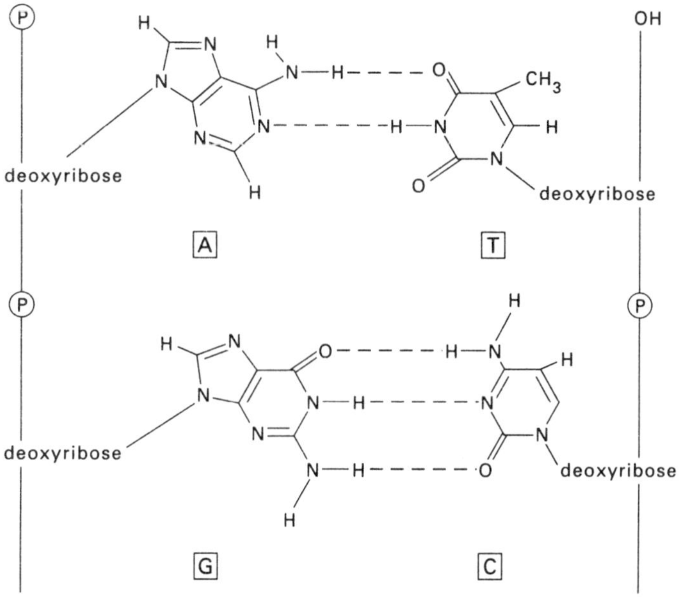

FIG. 5.A7. AT and GC base pairs held together by hydrogen bonding

The secondary structure of DNA was deduced by Watson and Crick in 1953[17]. The key piece of data came from the results of their interpretation of some X-ray diffraction patterns of DNA crystals. This, together with Chargaff's data[17] showing that base ratios A:T and G:C are constant, some model building, and a large amount of intuition and inspiration, enabled Watson and Crick to deduce that native DNA was a double-helix structure. The two strands of DNA are held together by a number of weak forces, the most important being hydrogen bonding. Figure 5.A8 shows the basic double-helix structure and the nature of the hydrogen-bonding between A and T and G and C base pairs.

The tertiary structure of DNA is complex. DNA does not normally exist as a straight linear polymer, but as a supercoiled structure. Supercoiling is associated with special proteins in eukaryotic organisms. Prokaryotic organisms have one continuous molecule while eukaryotes have many (e.g. humans have 46). Viruses also contain nucleic acids and their genetic material can be either DNA or RNA.

DNA stores the complete genetic information required: to specify the structure of all proteins and RNAs of each organism; to programme in space and time the orderly biosynthesis of cells or tissues; to determine the activities of the organism throughout its life cycle; and to determine the individuality of the organism.

Ribonucleic Acid (RNA)

RNA consists of long strings of ribonucleotides, polymerised in a similar way to DNA, but the chains are considerably shorter than those of DNA. RNA contains ribose rather than deoxyribose and also contains uracil instead of thymidine. This has important connotations in the secondary structure of RNA which does not form the long helices found in DNA. RNA is usually much more abundant than DNA in the cell and its concentration varies according to cell activity and growth. This is because RNA has several roles in protein synthesis. There are three major classes: messenger RNA (mRNA); ribosomal RNA (rRNA); and transfer RNA (tRNA).

mRNA is the template used by ribosomes for the translation of genetic material into an amino-acid sequence and it is derived from a specific DNA sequence. The genetic code is made up of trinucleotide sequences, or *codons*, on mRNA. Each mRNA has a unique sequence coding for each protein (polypeptide chain).

tRNA is a single strand of RNA but it is in a highly folded conformation. The molecules are usually 70–95 ribonucleotides long (equivalent 23,000 to 30,000 MW). Each of the 23 amino-acids has one or more tRNAs to which it is able to bind. In this bound form the amino-acids are transported into the ribosome. tRNA molecules therefore serve as adaptors for translating the genetic code or codons of the mRNA into the sequence of amino-acids or proteins. Each tRNA contains a trinucleotide sequence called

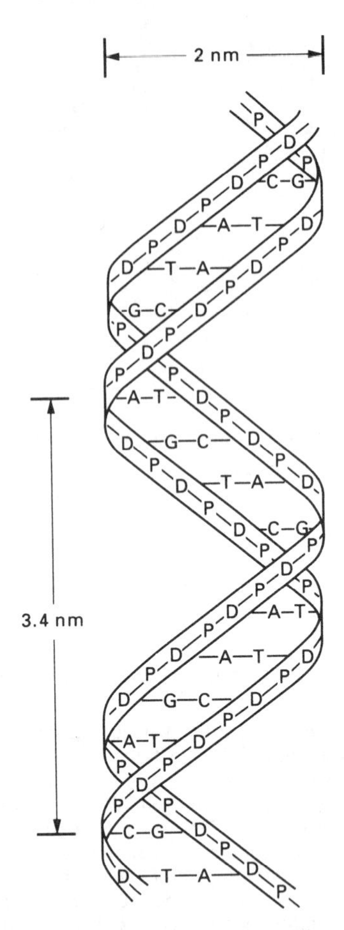

FIG. 5.A8. The double-helix structure of DNA. D, deoxyribose; P, phosphate ester bridge; A, adenine; C, cytosine; G, guanine; and T, thymine

an *anticodon sequence* which is complementary to a codon, the trinucleotide sequence of mRNA that codes for one amino-acid[17].

rRNAs are the major components of ribosomes and make up 65 per cent by weight of the structure. The role of ribosomal RNA is very complex but it is essential for protein synthesis which occurs within the ribosome organelle. Each has a role in the biosynthesis of proteins and this is discussed in Appendix 5.6.

Appendix 5.3. Derivation of the Michaelis–Menten Equation using the Rapid Equilibrium Assumption

The fundamental equation in enzyme kinetics is the Michaelis–Menten equation. These workers worked on the hypothesis that the reaction proceeds through an enzyme–substrate complex (ES) that forms rapidly from the free enzyme (E) and the substrate (S) and may be described by an equilibrium (or Michaelis) constant K_S'. Upon reaction, the ES complex then decomposes and is converted to product by a rate-determining step with a rate constant k_{cat}. The scheme is shown below:

$$E + S \underset{}{\overset{K_S'}{\rightleftharpoons}} ES \xrightarrow{k_{cat}} E + P \qquad \text{(Scheme 5.A3.1)}$$

The major assumption in Scheme 5.A3.1 is that enzymes and substrate remain in thermodynamic equilibrium with the enzyme–substrate complex at all times. K_S' for the enzyme substrate complex is defined as:

$$K_s' = \frac{E\,S}{ES} \tag{5.A3.1}$$

The rate of reaction $\mathcal{R}_0$ depends upon the catalytic rate constant and upon the enzyme–substrate concentration.

$$\mathcal{R}_0 = \boldsymbol{k}_{cat}\, ES \tag{5.A3.2}$$

Note that only for the initial stages of the reaction can the reverse reaction from P to ES in Scheme 1 above be ignored. The total enzyme present is given by the conservation equation:

$$E_{tot} = E + ES \tag{5.A3.3}$$

Thus:

$$E = E_{tot} - ES \tag{5.A3.4}$$

Substituting for E and solving for ES in equation 5.A3.1 gives:

$$ES = \frac{E_{tot}\, S}{K_s' + S} \tag{5.A3.5}$$

Then substituting equation 5.A3.5 in equation 5.A3.2 gives:

$$\mathcal{R}_0 = \frac{E_{tot} S\, \boldsymbol{k}_{cat}}{S + K_s'} \quad \text{where} \quad E_{tot}\, \boldsymbol{k}_{cat} = V_{max} \tag{5.A3.6}$$

or:

$$\mathcal{R}_0 = \frac{V_{max}\, S}{S + K_S'} \tag{5.A3.7}$$

Appendix 5.4. The Haldane Relationship

All reactions are to some degree reversible, and many enzyme-catalysed reactions can function in either direction inside a cell. It is therefore interesting to compare the forward and back reactions, especially when the reaction runs to equilibrium as in an enzyme reactor.
Consider the reaction:

$$\mathrm{S} \rightleftharpoons \mathrm{P}$$

This reaction proceeds via the formation of a single intermediate complex. In the forward reaction it would be regarded as an ES complex while in the back reaction it would be regarded as an EP complex:

i.e.:

$$\mathrm{S + E} \underset{\mathbf{k}_{r1}}{\overset{\mathbf{k}_{f1}}{\rightleftharpoons}} \mathrm{ES/EP} \underset{\mathbf{k}_{r2}}{\overset{\mathbf{k}_{f2}}{\rightleftharpoons}} \mathrm{P + E} \qquad \text{(Scheme 2)}$$

The rate of accumulation of product in this situation is given by:

$$\frac{\mathrm{d}P}{\mathrm{d}t} = \boldsymbol{k}_{f2}\, ES - \boldsymbol{k}_{r2}\, P\, E \tag{5.A4.1}$$

At steady state $\frac{\mathrm{d}ES}{\mathrm{d}t} = 0$, so that:

$$\frac{\mathrm{d}ES}{\mathrm{d}t} = \boldsymbol{k}_{f1} S\, E + \boldsymbol{k}_{r2} P\, E - \boldsymbol{k}_{r1} ES - \boldsymbol{k}_{f2} ES = 0 \tag{5.A4.2}$$

Rearranging to give:

$$\frac{E}{ES}\left(S + \frac{\boldsymbol{k}_{r2}}{\boldsymbol{k}_{f1}}\, P\right) = \frac{\boldsymbol{k}_{r1} + \boldsymbol{k}_{f2}}{\boldsymbol{k}_{f1}} \tag{5.A4.3}$$

Now a material balance for the enzyme gives:

$$E_{tot} = E + ES \quad \text{or} \quad E = E_{tot} - ES \tag{5.A4.4}$$

Thus, from equation 5.A4.3:

$$\frac{E_{tot}}{ES}\left(S + \frac{\boldsymbol{k}_{r2}}{\boldsymbol{k}_{f1}}\, P\right) = \frac{\boldsymbol{k}_{r1} + \boldsymbol{k}_{f2}}{\boldsymbol{k}_{f1}} + S + \frac{\boldsymbol{k}_{r2}}{\boldsymbol{k}_{f1}} \tag{5.A4.5}$$

and rearranging:

$$ES = \frac{E_{\text{tot}}\left(S + \frac{k_{r2}}{k_{f1}}P\right)}{\frac{k_{r1}+k_{f2}}{k_{f1}} + S + \frac{k_{r2}}{k_{r1}}P} \tag{5.A4.6}$$

Similarly, by substituting for ES in equation 5.A4.3 and rearranging:

$$\frac{E_{\text{tot}}}{E} = \left(\frac{k_{r1}+k_{f2}}{k_{f1}}\right) = \frac{k_{r1}+k_{f2}}{k_{f1}} + S + \frac{k_{r2}}{k_{f1}}P \tag{5.A4.7}$$

and:

$$E = \frac{E_{\text{tot}}\left(\frac{k_{r1}+k_{f2}}{k_{f1}}\right)}{\frac{k_{r1}+k_{f2}}{k_{f1}} + S + \frac{k_{r2}}{k_{f1}}P} \tag{5.A4.8}$$

Equations 5.A4.6 and 5.A4.8 may be substituted in equation 5.A4.1 giving:

$$\frac{dP}{dt} = \frac{k_{f2}E_{\text{tot}}\left(S + \frac{k_{r2}}{k_{f1}}P\right) - k_{r2}PE_{\text{tot}}\left(\frac{k_{r1}+k_{f2}}{k_{f1}}\right)}{\left(\frac{k_{r1}+k_{f2}}{k_{f1}}\right) + S + \frac{k_{r2}}{k_{f1}}P} \tag{5.A4.9}$$

On simplification the above equation gives:

$$\frac{dP}{dt} = \frac{k_{f2}E_{\text{tot}}\left(S + \frac{k_{r1}k_{r2}}{k_{f1}k_{f2}}P\right)}{\left(\frac{k_{r1}+k_{f2}}{k_{f1}}\right) + S + \frac{k_{r2}}{k_{f1}}P} \tag{5.A4.10}$$

The equilibrium constants for the forward (K_M^S) and back (K_M^P) reactions are given by:

$$K_M^S = \frac{k_{r1}+k_{f2}}{k_{f1}} \quad \text{and} \quad K_M^P = \frac{k_{r1}+k_{f1}}{k_{r2}} \tag{5.A4.11}$$

and the equilibrium constant K_{eq} is given by:

$$K_{eq} = \frac{k_{r1}k_{f2}}{k_{f1}k_{r2}} = \frac{P}{S} \tag{5.A4.12}$$

Using the relationships in equations 5.A4.11 and 5.A4.12 and noting that $k_{f1}E_{\text{tot}} = V_{\text{max}}^S$, then equation 5.A4.10 for the rate of product accumulation becomes:

$$\frac{dP}{dt} = \frac{V_{\text{max}}^S\left(S - \frac{P}{K_{eq}}\right)}{K_M^S + S + \frac{K_M^S}{K_M^P}P} \tag{5.A4.13}$$

Haldane derived a useful relationship between the kinetic constants and the equilibrium constant of the reaction. At constant enzyme concentration and at equilibrium, the rate of the forward reaction equals the rate of the back reaction. Under these conditions, from Scheme 2:

$$k_{r1}ES = k_{f1}ES \tag{5.A4.14}$$

therefore:

$$\frac{ES}{E} = \frac{k_{f1}S}{k_{r1}} \tag{5.A4.15}$$

Also under these conditions:

$$k_{f1}ES = k_{r2}EP \tag{5.A4.16}$$

$$\frac{ES}{E} = \frac{k_{r2}\,P}{k_{f2}} = \frac{k_{f1}\,S}{k_{r1}} \tag{5.A4.17}$$

Therefore:

$$K_{eq} = \frac{P}{S} = \frac{k_{r1}\,k_{f2}}{k_{f1}\,k_{r2}} \tag{5.A4.18}$$

But as $V^{S}_{max} = k_2 E_{tot}$ and $V^{P}_{max} = k_{r1} E_{tot}$:

$$V^{S}_{max} = k_{f2} \quad \text{and} \quad V^{P}_{max} = k_{r1}$$

Also:

$$\frac{K^{S}_{M}}{K^{P}_{M}} = \frac{(k_{r1} + k_{f2})}{k_{f1}} \; \frac{k_{r2}}{(k_{f1} + k_{f2})} = \frac{k_{r2}}{k_{f1}} \tag{5.A4.19}$$

Therefore:

$$K_{eq} = \frac{k_{f1}\,k_{f2}}{k_{r1}\,k_{r2}} = \frac{V^{s}_{max}\,K^{P}_{m}}{V^{P}_{max}\,K^{S}_{m}} \quad \text{(The Haldane relationship)} \tag{5.A4.20}$$

This relation is useful in confirming the accuracy of kinetic constants which are determined by kinetic measurements.

Appendix 5.5. Enzyme Inhibition

Competitive Inhibition

Competitive inhibitors often closely resemble in some respect the substrate whose reactions they inhibit and, because of this structural similarity, compete for the same binding site on the enzyme. The enzyme–inhibitor complex either lacks the appropriate reactive groups or is held in an unsuitable position with respect to the catalytic site of the enzyme which results in a complex which does not react (i.e. gives a dead-end complex). The inhibitor must first dissociate before the true substrate may enter the enzyme and the reaction can take place. An example is malonate, which is a competitive inhibitor of the reaction catalysed by succinate dehydrogenase. Malonate has two carboxyl groups, like the substrate, and can fill the substrate binding site on the enzyme. The subsequent reaction, however, requires that the molecule be reduced with the formation of a double bond. If malonate is the substrate, this cannot be achieved without the loss of one of the carboxy-groups and therefore no reaction occurs.

The effect of a competitive inhibitor will depend not only upon the inhibitor but also the substrate concentration and the relative affinities of the substrate and inhibitor for the enzyme. Therefore, if one considers the situation where the substrate concentration is low, then the inhibitor will successfully compete but, as the substrate concentration rises, the inhibitor will become less successful in competing with the substrate until, at very high concentrations, one would expect to see very little or no inhibition. Thus, in this situation V_{max} remains unchanged but K_M for the substrate is increased, as illustrated below, and the value of the new K_M is designated K'_M.

The steady-state kinetics of a simple single-substrate, single-binding site, single-intermediate-enzyme catalysed reaction in the presence of competitive inhibitor are shown in Scheme A5.5.1.

$$\begin{array}{l} E + S \underset{k_{r1}}{\overset{k_{f1}}{\rightleftharpoons}} ES \overset{k_{f2}}{\rightleftharpoons} E + P \\ + \\ I \\ \Big\Updownarrow K_I \\ EI \end{array} \qquad \text{(Scheme A5.A5.1)}$$

The equilibrium constant for the reaction between E and I is K_I, where

$$K_I = \frac{E\,I}{EI} \tag{5.A5.1}$$

In this case, K_I is known as the inhibitor constant and the equilibrium between enzyme and inhibitor is almost instantaneous on mixing as $\mathcal{R}_0 = k_2 ES$ (the rate of reaction) is directly dependent on the concentration of the enzyme–substrate complex. The total enzyme present in the present of the inhibitor will be:

$$E_{tot} = E + ES + E\,I \tag{5.A5.2}$$

or, from above, then:

$$E_{tot} = E + ES + \frac{E\,I}{K_I} \tag{5.A5.3}$$

$$= E\frac{(1 + I)}{K_I} + ES \tag{5.A5.4}$$

$$E = E_{tot}\frac{-\,ES}{\dfrac{(1 + I)}{K_I}} \tag{5.A5.5}$$

Now substituting for E in the steady state assumption derivation of the Michaelis–Menten constant (equation 5.A5.5) gives:

$$K_m = \frac{(E_{tot} - ES)S}{\dfrac{(1 + I)}{K_I}\,ES} \tag{5.A5.6}$$

Therefore:

$$K_m\frac{(1 + I)}{K_I} = \frac{(E_{tot} - ES)S}{ES} \tag{5.A5.7}$$

and:

$$ES = \frac{E_{tot}\,S}{S + K_m\dfrac{(1 + I)}{K_I}} \tag{5.A5.8}$$

Since $\mathcal{R}_0 = k_{f2}\,ES$ then:

$$\mathcal{R}_0 = \frac{k_{f2}E_{tot}S}{S + K_m\dfrac{(1 + I)}{K_I}} \tag{5.5A.9}$$

and since $V_{max} = k_{f2}\,E_{tot}$ then

$$\mathcal{R}_0 = \frac{V_{max}\,S}{S + K_m\dfrac{(1 + I)}{K_I}} \tag{5.5A.10}$$

Under conditions where the concentrations of the inhibitor and the substrate are much greater than that of the enzyme, the above equation holds. Therefore, for simple competitive inhibition, V_{max} is unchanged, while the apparent value of K_m is given by:

$$K'_m = K_m\left(1 + \frac{I}{K_I}\right) \tag{5.A5.11}$$

The value of K_I is equal to the concentration of a competitive inhibitor which gives an apparent doubling of the value of K_m. Graphically, a form of the Lineweaver–Burk plot[17] is used (see Section 5.4.6).

$$\frac{1}{\mathcal{R}_0} = \frac{K'_m}{V_{max}}\frac{1}{S} + \frac{1}{V_{max}} \tag{5.A5.12}$$

A plot of $1/\mathcal{R}_0$ versus $1/S$ for the reaction in the presence of inhibitor gives the graph illustrated in Fig. 5.14.

Uncompetitive Inhibition

Uncompetitive inhibitors bind only to the enzyme–substrate complex and not to the free enzyme. For example, the substrate binds to the enzyme causing a conformational change which reveals the inhibitor binding site, or it could bind directly to the enzyme-bound substrate. In neither case does the enzyme compete for the same binding site, so the inhibition cannot be overcome by increasing the substrate concentration. Scheme 5.A5.2 below illustrates this uncompetitive behaviour.

$$\mathrm{E} + \mathrm{S} \underset{k_{r1}}{\overset{k_{f1}}{\rightleftharpoons}} ES \overset{k_{f2}}{\rightleftharpoons} E + P \qquad ES \overset{K_I}{\rightleftharpoons} ESI$$ (Scheme 5.A5.2)

ESI is the dead-end complex; the inhibitor constant $K_I = ES \dfrac{I}{ESI}$
Under steady state conditions:

$$K_m = \frac{k_{f1} + k_{f2}}{k_{f1}} = \frac{E\ S}{ES}$$

For this system:

$$E_{tot} = E + ES + ESI \tag{5.5A.13}$$

$$= E + ES + \frac{ES\ I}{K_I} \tag{5.5A.14}$$

$$= E + ES(1 + \frac{I}{K_I}) \tag{5.5A.15}$$

Therefore, on rearrangement and substitution the outcome is:

$$\mathcal{R}_0 = \frac{V_{max}\ S}{S(1 + \frac{I}{K_I})} + K_m \tag{5.5A.16}$$

Dividing throughout by $(1 + \frac{I}{K_I})$ gives:

$$\mathcal{R}_0 = \frac{V_{max}\ S}{(1 + \frac{I}{K_I})} \Bigg/ \frac{S + K_m}{(1 + \frac{I}{K_I})} \tag{5.5A.17}$$

Equation 5.A5.17 is now in the same form as the Michaelis–Menten equation, the constants K_m and V_{max} both being divided by a factor $(1 + \frac{I}{K_I})$

Thus, for uncompetitive inhibition, the apparent values of both K_m and V_{max} are modified.

$$V'_{max} = \frac{V_{max}}{(1 + \frac{I}{K_I})} \quad \text{and} \quad K'_m = \frac{K_m}{(1 + \frac{I}{K_I})} \tag{5.5A.18}$$

Here V'_{max} is the value of V_{max} in the presence of inhibitor of initial concentration I, of uncompetitive inhibitor, and K'_m is the apparent value of K'_m under the same conditions. An inhibitor concentration equal to K_I will halve the value of both V_{max} and K_m.

The Lineweaver–Burk equation[17] for a reaction in the presence of an uncompetitive inhibitor is:

$$\frac{1}{\mathcal{R}_0} = \frac{K'_m}{V'_{max}} \frac{1}{S} + \frac{1}{V'_{max}} \tag{5.A5.19}$$

and the slope of a Lineweaver–Burk plot is equal to:

$$\frac{K'_m}{V'_{max}} = \frac{K_m}{V_{max}} \frac{(1 + \frac{I}{K_I})}{(1 + \frac{I}{K_m})} = \frac{K_m}{V_{max}} \tag{5.A5.20}$$

The slope of the Lineweaver–Burk plot is not altered by the presence of an uncompetitive inhibitor, but both intercepts change (Fig. 5.14).

Non-Competitive Inhibition

Non-competitive inhibition is more complicated than either competitive or uncompetitive inhibition. A non-competitive inhibitor can combine with an enzyme molecule to produce a dead-end complex regardless of whether a substrate molecule is bound or not. Hence the inhibitor must bind at a different site from the substrate. The only case considered here is when the inhibitor destroys the catalytic activity of the enzyme, either binding to the catalytic site or, as a result of a conformational change affecting the catalytic site, without affecting the substrate binding site. The situation for a simple single-substrate reaction is shown in Scheme 5.A5.3.

$$\begin{array}{ccccc} E + S & \underset{k_{r1}}{\overset{k_{f1}}{\rightleftharpoons}} & ES & \overset{k_{f2}}{\rightleftharpoons} & E \\ + & & + & & \\ I & & I & & \\ \updownarrow K_i & & \updownarrow & & \\ EI + S & \underset{k_{r1}}{\overset{k_{f1}}{\rightleftharpoons}} & ESI & & \end{array} \qquad \text{(Scheme 5.A5.3)}$$

Even this scheme represents a complex situation, for ES can be arrived at by alternative routes, making it impossible for an expression of the same form as the Michaelis–Menten equation to be derived using the general steady-state assumption. However, types of non-competitive inhibition consistent with the Michaelis–Menten type equation and a linear Linweaver–Burk plot can occur if the rapid-equilibrium assumption is valid (Appendix 5.A3). In the simplest possible model, involving simple linear non-competitive inhibition, the substrate does not affect the inhibitor binding. Under these conditions, the reactions $E + I \longrightarrow EI$ and $ES + I \longrightarrow ESI$ are assumed to have an identical dissociation constant K_I which is again called the inhibitor constant. The total enzyme concentration is effectively reduced by the inhibitor, decreasing the value of V_{max} but not altering K_m since neither the inhibitor nor the substrate affects the binding of the other.

In the derivation of an initial velocity equation only the special case where $K_M = K_S'$ is being considered. As before:

$$\frac{E\,S}{ES} = K_m$$

In the presence of the non-competitive inhibitor which will bind equally well to E or to ES:

$$K_I = \frac{E\,I}{EI} = \frac{ES\,I}{SI} \qquad (5.5A.21)$$

But:

$$E_{tot} = E + ES + EI + ESI \qquad (5.5A.22)$$

$$= E + ES + \frac{E\,I}{K_I} + \frac{ES\,I}{K_I} \qquad (5.A5.23)$$

$$= (E + ES)(1 + \frac{I}{K_I}) \qquad (5.A5.24)$$

and so:

$$\mathcal{R}_0 = \frac{V_{max}}{(1 + \frac{I}{K_I})} \frac{S}{(S + K_m)} \qquad (A5.5.25)$$

This is in the form of the Michaelis–Menten equation with V_{max} being divided by the factor $(1 + (I/K_I))$. Thus, for a simple linear non-competitive inhibitor, K_m remains unchanged while V_{max} is altered so that:

$$V'_{max} = \frac{V_{max}}{(1 + \frac{I}{K_I})} \qquad (5.A5.26)$$

or:

$$\frac{1}{V'_{max}} = \frac{1}{V_{max}} \frac{(1 + I)}{K_I} \qquad (5.A5.27)$$

where V'_{max} is the value of V_{max} in the presence of concentration I of an inhibitor concentration.

The Lineweaver–Burk equation for simple linear non-competitive inhibition is:

$$\frac{1}{\mathcal{R}_0} = \frac{K_m}{V'_{max}} \frac{1}{S} + \frac{1}{V'_{max}} \qquad (5.A5.28)$$

Figure 5.14 shows a plot of such an inhibition pattern. There are few clear-cut examples of non-competitive inhibition of a single-substrate reaction, as might be expected from this special case. Normally the inhibitor constants in Scheme 5.A5.3 are different.

Appendix 5.6. Information Storage and Retrieval in the Cell

Introduction

Biological information is stored in the nucleotide base structure of DNA. By manipulating the structure of DNA it is possible to improve and understand the structure and functional relationships of catalysts (termed *protein engineering*). Considerable improvements have been made to products derived from enzyme systems, both in the quality and amount of metabolites, enzymes and proteins produced by organisms (often termed *genetic engineering*).

To-day, knowledge of the molecular aspects of genetics has arisen from the convergence of three disciplines: (i) genetics; (ii) biochemistry; and (iii) molecular physics. This is epitomised by the discovery of the structure of DNA as a double helix by Watson and Crick (Appendix A5.2). This structure was verified using data obtained from these three fields; (i) genetic coding in the form of genes; (ii) X-ray analysis of DNA crystal structure; and (iii) the chemical composition of DNA. The Watson and Crick model not only accounted for the structure, but also showed how replication could be performed with precision.

The central dogma of molecular genetics The 'central dogma' was based upon the findings of Watson and Crick and states that the flow of information is essentially in one direction from DNA to protein. Three major steps can be defined in the process: replication, transcription and translation of genetic material, as shown in Fig. 5.A9.

The control of processes involving biological systems should start from a sound knowledge of their biological properties and characteristics. It has been shown that metabolism consists of many reactions which supply both the energy and the chemical materials for the synthesis and reproduction of the cells. Central to the control of these systems is the utilisation and retrieval of information stored within the cell's DNA.

Two major areas are covered: (i) the transmission and translation of genetic material to produce protein; and (ii) mutation, genetic recombination and manipulation. The reader is referred to Appendix 5.2 for the nature, dimension and conformation of genetic material.

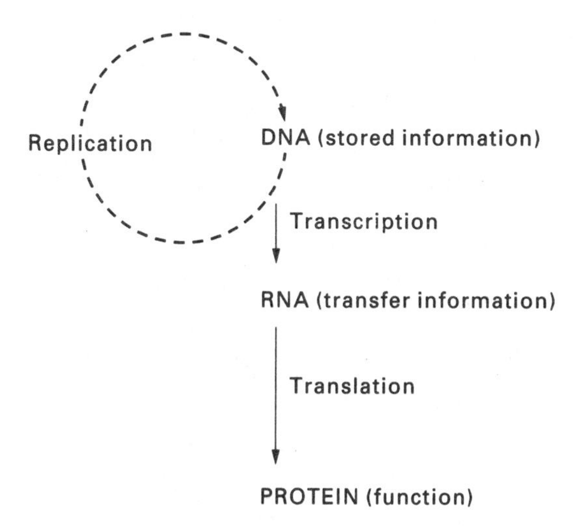

FIG. 5.A9. Information storage and flow in cell systems, showing the major processes involved

DNA and Gene Structure

Two molecules of DNA make up the single chromosome of bacteria. In eukaryotes, chromosome structure is very complex and there can be a large number of chromosome within the cell, e.g. some yeasts have seven while humans have 23 pairs.

The first evidence that DNA is the bearer of genetic information came in 1943, through the work of Avery McCloud and McCarty. They found that DNA extracted from a virulent (disease causing) strain of the bacterium *Streptococcus pneumoniae* permanently transformed a non-virulent strain of this organism into a virulent form.[17]

Genes Genes are segments of DNA, and are defined in the classical biological sense as a portion of a chromosome which determines or specifies a single character, for example, eye colour. This was later redefined to mean that one gene is equivalent to one protein, i.e. a segment of DNA equals a protein. Today the definition has been further refined, to take account of the two function types of DNA sequences; first there are *structural genes*, and secondly these are *regulatory genes* or sequences. The regulatory sequences consist of several components, denoting the stop and start of the structural genes and others that control the rate of transcription of structural genes. There are many genes on a single chromosome; in bacteria such as *E. coli* there are over 1100 which can be identified on the only chromosome. However, taking into account the regulatory sequences, 3000–5000 genes is thought to be a better estimate. Various viral and bacterial chromosomes have been mapped and the task of mapping the human chromosomes systematically (the Human Genome Project) started seriously in 1992.

Because in *E. coli* there are at least 3000 genes and the chromosome has 4×10^6 nucleotide pairs then the average size of a gene is 4,000,000/300 = 1300, or less, nucleotide pairs. The true value is more likely to be 900–1000 pairs per gene.

In prokaryotes there is only one copy of DNA per cell, and in nearly all cases there is only a single copy of a gene. Apart from regulatory DNA and signalling sequences, there are very few *silent*, or non-translated DNA sequences, in prokaryotes. Moreover, each gene is typically collinear with the amino-acid sequence for which it codes. This is in contrast to the organisation of eukaryotic DNA which is structurally and functionally far more complex; the structural parts of genes are split by silent DNA and there are multiple copies of the chromosomes.

Plasmids Apart from chromosomes, prokaryotes and eukaryotes also contain small pieces of autonomous DNA. Autonomous extrachromosomal fragments of DNA in bacteria are called plasmids. Most plasmids are very small and contain only a few genes as compared with the chromosome. Plasmids carry genetic information and, like chromosomes, are passed from each cell into daughter cells at division, or they can be taken up from the environment. They can apparently lead separate lives from the chromosomal DNA for many generations but some plasmids can, however, integrate with the chromosome DNA and leave again in a co-ordinated manner. The best examples of plasmids are those which carry resistance to various antibiotics, such as tetracycline and streptomycin. Bacterial cells that contain such plasmids become resistant to antibiotics, and they can become a significant problem in bacterial infections, especially in hospitals where the use of antibiotics is prevalent. Further, such behaviour is significant as it is possible for plasmids to be transferred to antibiotic sensitive strains. From the biotechnological point of view, plasmid DNA is easily isolated from bacterial cells and is small enough to be easily manipulated. A major strategy in the transfer of foreign DNA into bacteria is putting the foreign DNA into the plasmid and then transferring the plasmid into the host cell. There the plasmid DNA is transcribed together with the foreign DNA and synthesis takes place of the protein coded by the foreign DNA. Eukaryotes also have extrachromosomal DNA which can be found in the cytoplasm, mitochondria and chloroplasts.

Processes Involving the Nucleic Acid Synthesis

Replication and Transcription DNA is replicated to yield daughter molecules. The enzymes and other proteins participating in replication and transcription of DNA are among the most remarkable biological catalysts known. Not only do they form enormously long macromolecules from the mono-nucleotide precursors using guanosine triphosphate (GTP and the energy equivalent ATP), but they also synthesise genetic information from the template strand to an extraordinary fidelity. In addition, these enzymes must solve complex mechanical problems, since the parental duplex DNA must be unwound in advance of the replication enzymes, so that they can gain access to the information stored in the base sequences of the molecule. The transcription enzymes also have extraordinary properties. Not only can they make a large assortment of different RNAs, but they also start and stop the translation at specific parts of the DNA molecule. They also respond to various regulatory signals, so that different proteins are expressed at various specific times in the life cycle. DNA and RNA polymerases and other proteins that help carry out replication and transcription of DNA are thus vitally important materials in the perpetuation of genetic information.

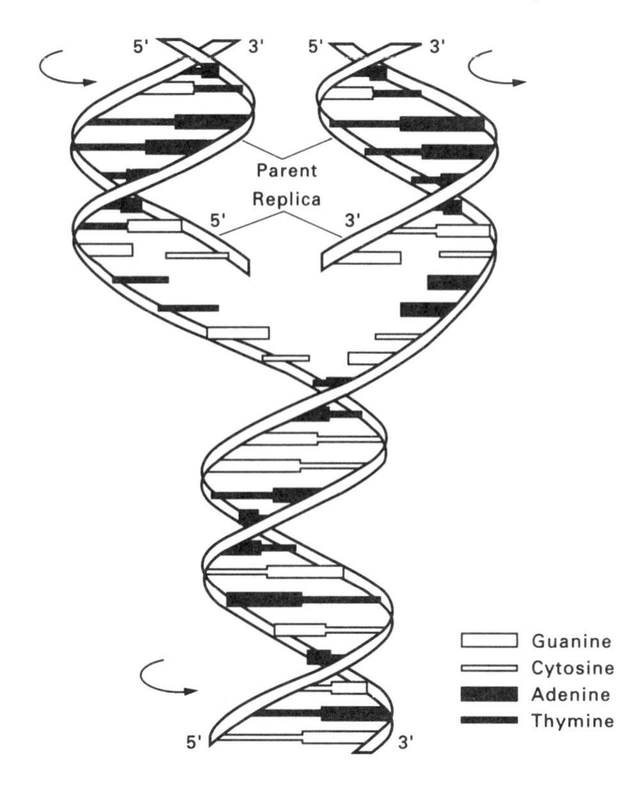

FIG. 5.A10. A simplified diagram of DNA replication

The Watson and Crick hypothesis for DNA replication proposed that each strand of DNA is used as a template for the production of one of the daughter DNA molecules. Thus the result of replication would be that one strand of DNA is present in each daughter molecule of DNA. This is a semi-conservative mechanism of replication. A simplified diagram of replication is shown in Fig. 5.A10; however, the replication patterns are different in bacteria and in eukaryotes.

DNA Transcription

The first step in the utilisation of information held within the DNA structure is the transcription of DNA into RNA. The product of transcription is a complementary strand of RNA, produced with high accuracy. There are three types of RNA products from the transcription of DNA.

Transfer (tRNA) and ribosomal RNA (rRNA) are transcribed and used in protein synthesising processes. Messenger RNA (mRNA) codes the amino-acid sequence of proteins and 95 per cent of the total DNA transcribed is used for this purpose. In prokaryotes a single mRNA molecule may code for a single polypetide or for two or more polypeptide chains. There is a triplet code for each amino-acid; 300 ribonucleotides code for a 100 amino-acid sequence. Fig. 5.A11 shows the relationship between the nucleotide sequence on DNA and RNA and the amino-acid sequence of protein.

The most important difference between replication and transcription is that not all the DNA is used in the transcription. Usually, only small groups of genes are transcribed at any one time. Thus, the transcription of DNA is selective, turned on by specific regulatory sequences which indicate the beginning and ending of the region to be transcribed. The control of protein synthesis is discussed further in Section 5.7.

The discovery of DNA polymerase and its dependence on a DNA template led to the search for enzymes which could make an RNA molecule complementary to the DNA. RNA synthesis does not require a primer strand; it does, however, require a specific initiation signal on the DNA template strand to allow binding and initiation. As the RNA strand is synthesised it forms a temporary helix with the template DNA, but when complete the mRNA breaks off at the stop site on DNA. Once released from DNA, some of the RNA is processed further, for the specific structures of rRNA and tRNA.

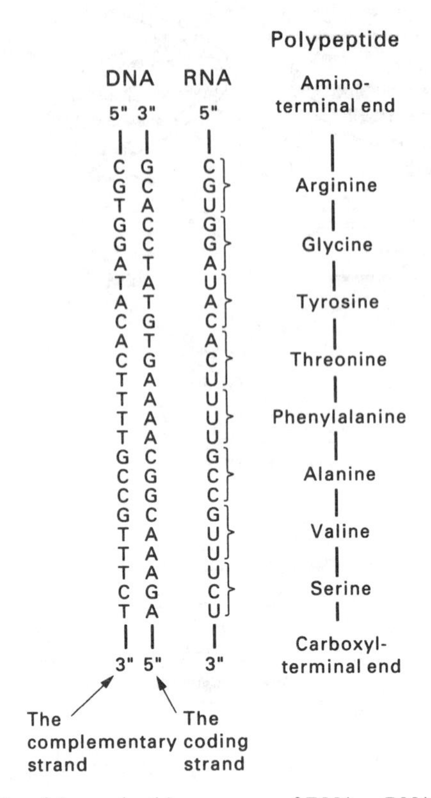

FIG. 5.A11. The collinearity of the nucleotide sequences of DNA, mRNA and amino-acid sequence

Protein Synthesis (Translation)

An overview of protein synthesis is shown in Fig. 5.A12. The linear sequence in mRNA that is translated to protein contains four bases, adenine, uracil, guanine and cytosine. The four 'letters' A,U,G and C constitute the mRNA 'alphabet'. This basic alphabet is used in triplets of bases called codons. The codons on mRNA pair up with anticodon or complementary triplets on the tRNA, thus matching the mRNA code to an amino-acid sequence.

Protein synthesis occurs in five major stages:

(1) *Activation of the amino-acids.* This stage takes place in the cytoplasm. Each of the 20 amino-acids is covalently attached to a specific tRNA at the expense of ATP hydrolysis (i.e. it is an energy-driven process). Each amino-acid has a specific enzyme for this reaction to ensure that the correct amino-acid is linked to the tRNA molecule.

(2) *Initiation of the polypeptide chain.* mRNA bearing the code for the polypeptide is bound to the small sub-unit of RNA, followed by the initiating amino-acid, and is attached to its tRNA to form an initiation complex. The tRNA of the initiating amino-acid–base pairs with a specific nucleotide triplet or codon on the mRNA that signals the beginning of the polypeptide chain. This process requires GTP (ATP equivalent), plus three proteins called initiation factors.

(3) *Elongation.* The polypeptide chain is now lengthened by covalent attachment of successive amino-acid units, each of which has been carried to the ribosome by a tRNA, which is base paired to the corresponding codon. Two molecules of GTP are required for each amino-acid residue added to the peptide

(4) *Termination and release.* The completion of the polypeptide is signalled by a termination codon in the mRNA and is followed by the release of the polypeptide from the ribosome.

(5) *Folding and processing.* In order to achieve its native, biologically-active form, a polypeptide must undergo folding into its proper three-dimensional conformation. Before or after folding, the new polypeptide may undergo processing by enzymes so that prosthetic groups, carbohydrates or nucleotides, may be added to the protein.

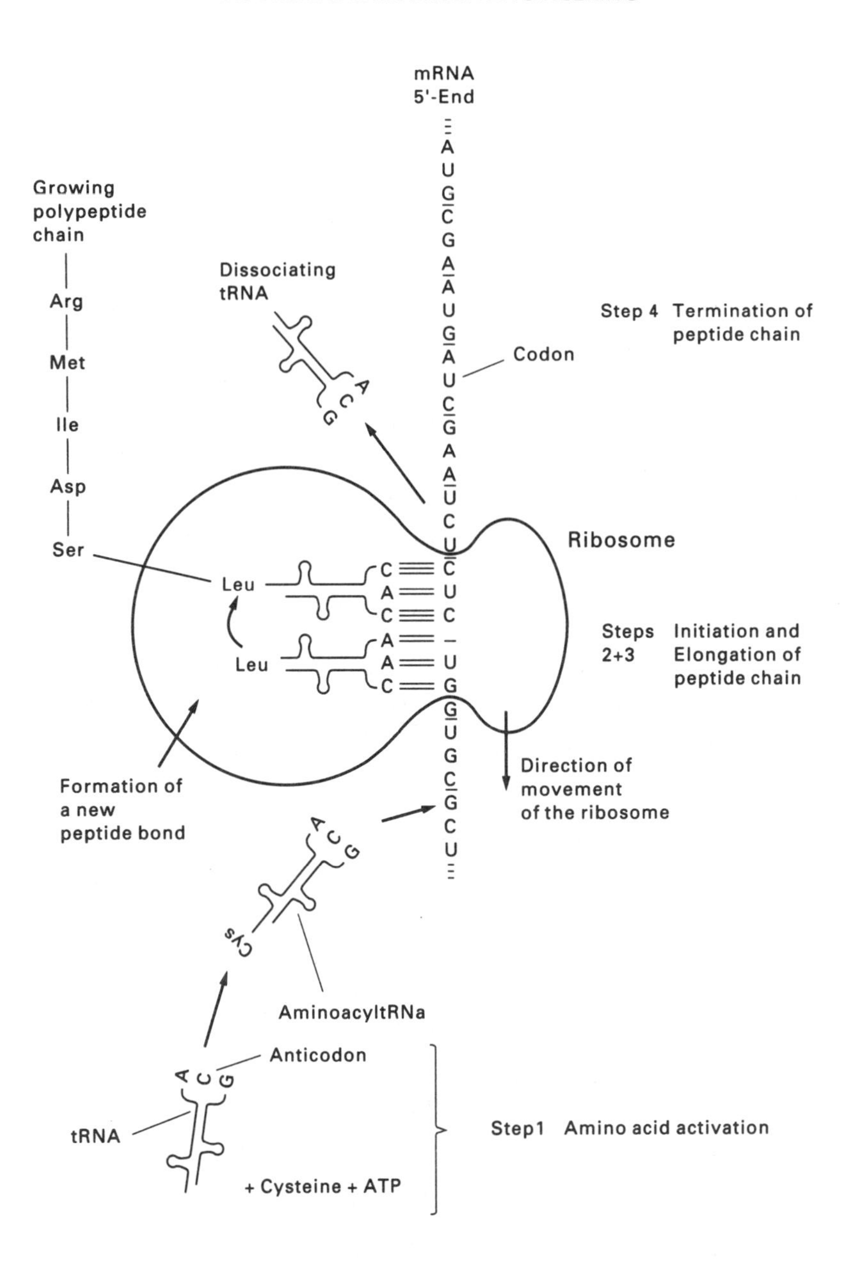

FIG. 5.A12. The major steps involved in protein synthesis

Energy consumption in protein synthesis The energy consumed by the process is the guarantee of fidelity of protein synthesis. Four molecules (or more) of GTP (ATP equivalent) are required for the synthesis of each peptide bond.

Two high energy bonds in the synthesis of aminoacyl tRNA—
One is consumed in the elongation step (GTP),
One is consumed in the translocation step,
plus additions for corrections.

Thus (4 × 30.55 =) 122.2 MJ/kmol. of phosphate group energy is required to generate a peptide bond, a bond energy of −20.6 MJ/kmol. This represents an enormous thermodynamic push as net energy of 101.6 MJ/kmol remains from synthesis. This process may appear wasteful but it is the price that is paid for near perfect fidelity. In bacteria there is a very tight coupling between transcription and translation, as shown in the Fig. A5.12. Another feature is the fact that mRNA is very short-lived and may only last a few minutes, before being degraded by the cell nucleases. The short life of the mRNA makes for a rapid control of synthesis of protein when it is no longer needed.

The Genetic Code

In Appendix 5.2 the structure of DNA was discussed and the basis for storage of information is shown to be in the base sequence of the DNA molecule. There are four types of bases in nucleic acid structure and these represent a four letter alphabet. Combinations of two bases give only 4^2 or 16 unique combinations; while a three base code gives 4^3 or 64 combinations and a four base code will give 4^4 (256) combinations. The code must contain at least three letters to give enough bases to code for all the amino-acid and stop sequences. A problem in a three letter code, however, is that there are too many codes for all the possible amino-acids and stop/start sequences. In fact, different codes have been found to have the same meaning, i.e. some amino-acids have more than one codon and the code is thus said to be degenerate.

The genetic code is shown in Table 5.A2. All the possible 64 combinations are shown and there are codes for either start and stop signals or for amino-acids. There are up to six codes for some amino-acids, e.g. arginine, leucine and serine, while tryptophan and methionine have unique codes. The major features of the genetic code are:

(i) Methionine and tryptophan have only one codon. All the others have at least two, and serine and leucine have six. Many have four codons;
(ii) Those amino-acids which have four codons are coded by the first two bases of the triplet while the third can be variable;
(iii) Similarly coded acids have similar chemistry, e.g. serine and threonine. U (uracil) in the second position will inevitably code for hydrophobic amino-acids; a mutation here will result in the replacement of one hydrophobic acid by another; and
(iv) The code is the same for all organisms.

TABLE 5.A2. *The Genetic Code*

	Second position				
First position	U	C	A	G	Third position
U	*Phe*	*Ser*	*Tyr*	*Cys*	U
	Phe	*Ser*	*Tyr*	*Cys*	C
	Leu	*Ser*	*Stop*	*Trp*	A
	Leu	*Ser*	*Stop*	*Trp*	G
C	*Leu*	*Pro*	*His*	*Arg*	U
	Leu	*Pro*	*His*	*Arg*	C
	Leu	*Pro*	*Gln*	*Arg*	A
	Leu	*Pro*	*Gln*	*Arg*	G
A	*Ile*	*Thr*	*Asn*	*Ser*	U
	Ile	*Thr*	*Asn*	*Ser*	C
	Ile	*Thr*	*Lys*	*Arg*	A
	Met	*Thr*	*Lys*	*Arg*	G
G	*Val*	*Ala*	*Asp*	*Gly*	U
	Val	*Ala*	*Asp*	*Gly*	C
	Val	*Ala*	*Glu*	*Gly*	A
	Val	*Ala*	*Glu*	*Gly*	G

5.16. FURTHER READING

ATKINSON, B. and MAVITUNA, F.: *Biochemical Engineering and Biotechnology Handbook*, 2nd edn. (Macmillan Publishers, Basingstoke, 1991).
BAILEY, J. E. and OLLIS, D. F.: *Biochemical Engineering Fundamentals*, 2nd edn. (McGraw-Hill, New York, 1986).
CANO, R. J. and COLOME, J. S.: *Microbiology* (West Publishing Co., St. Paul, Minnesota, 1986).
PALMER, T.: *Understanding Enzymes*, 3rd edn. (Ellis Horwood Limited, Chichester, 1991).
SEGAL, I. H.: *Biochemical Calculations*, 2nd edn. (John Wiley, New York, 1976).
STANBURY, P. F. and WHITAKER, A.: *Principles of Fermentation Technology* (Pergamon Press, Oxford, 1984).
STRYER, L.: *Biochemistry*, 3rd edn. (Freeman, New York, 1988).

5.17. REFERENCES

1. SMITH, J. E.: *Biotechnology Principles* (Van Nostrand Reinhold (UK) Co Ltd, Wokingham 1985).
2. ATKINSON, B. and MAVITUNA, F.: *Biochemical Engineering and Biotechnology Handbook*, 2nd edn. (Macmillan Publishers, Basingstoke, 1991).
3. HACKING, A. J.: *Economic Aspects of Biotechnology* (Cambridge University Press, Cambridge, 1986).
4. KIRSOP, B. E. and SNELL, J. J. S. (eds): *Maintenance of Microorganisms—A Manual of Laboratory Methods* (Academic Press, Inc. London, 1984)
5. HIBBERT, D. B. and JONES, A. M. *Macmillan Dictionary of Chemistry* (Macmillan Press, London, 1987).
6. HINCKLE, P. C. and MCCARTHY, R.: *Scientific American* **238** (1978) 104. How cells make ATP.
7. SCHELGEL, H. G.: *General Microbiology* (Cambridge University Press, Cambridge, 1986).
8. STANIER, R. Y., ADELBERG, E. A. and INGRAHAM, J. L.: *General Microbiology*, 4th edn. (Macmillan Press, London, 1975).
9. BERRY, R. (ed.): *Physiology of Industrial Fungi* (Blackwell Scientific Publications, Oxford, 1988).
10. ONIONS, A. H. S., ALLSOPP, D. and EGGINS, H. O. W.: *Smith's Introduction to Industrial Mycology*, 7th edn. (Edward Arnold, London, 1981).
11. HORAN, N. J.: *Biological Waste Water Treatment Systems* (John Wiley and Sons, Chichester, 1989).
12. STANBURY, P. F. and WHITAKER, A.: *Principles of Fermentation Technology* (Pergamon Press, Oxford, 1984).
13. INGRAHAM, J. L., MAALOE, O, and NEIDHARDT, F. C.: *Growth of the Bacterial Cell* (Sinauer Associates, Inc., Sunderland Massachusetts, 1983).
14. DEMAIN, A. L. and SOLOMAN, N. A. (eds): *Manual of Industrial Microbiology and Biotechnology* (American Society for Microbiology, Washington, 1986).
15. WOOD, W. B., WILSON, J. H., BENBOW, R. M. and HOOD, L. F.: *Biochemistry—A Problems Approach* (W. A. Benjamin Inc., Menlo Park, California, 1974).
16. AIBA, S., HUMPHREY, A. E., and MILLIS, N. F,: *Biochemical Engineering*, 2nd edn. (Academic Press, New York, 1973).
17. LENINGER, A. L.: *Biochemistry*, 2nd edn. (Worth Publishers Inc., New York, 1975).
18. CREIGHTON, T. E.: *Proteins: Structures and Molecular Properties*, 2nd edn. (W. H. Freeman and Company, New York, 1993).
19. BAILEY, J. E. and OLLIS, D. F.: *Biochemical Engineering Fundamentals*, 2nd edn. (McGraw-Hill, New York, 1986).
20. FERSHT, A.: *Enzyme Structure and Mechanism*, 2nd edn. (W. H. Freeman and Co., New York, 1985).
21. MICHAELIS, L. and MENTEN, M. L.: *Biochem. Z.* **49** (1913) 333. Die Kinetik der Invertinwirkung. (The kinetics of invertin action.)
22. CHEN, C–S., FUJIMOTO, Y., GIRDAUKAS, G. and SIH, C. J.: *J. Amer. Chem. Soc.* **104** (1982), 7294–7299. Qualitative analysis of biochemical resolutions of enantiomers.
23. PALMER, T.: *Understanding Enzymes*, 2nd edn. (Ellis Horwood Publishers, Chichester, 1985).
24. RATLEDGE, C.: In *Basic Biotechnology*, BU'LOCK J. D. and KRISTIANSEN B. eds. (Academic Press, London, 1987) p. 39. Biochemistry of growth and metabolism.
25. KIESLECH, K. ed., Biotechnology Vol 6A, *Biotransformations*, REHM H. J. and REED G. eds. (Verlag Chemie, Weinheim, Germany, 1984).
26. MALIK, V. S.: *Trends in Biochemical Science* **5** (1980)68. Microbial secondary metabolism.
27. BRITZ, J. E. and DEMAIN, A. L.: In *Comprehensive Biotechnology*, Vol. 1, MOO-YOUNG, M. ed. (Pergamon, Oxford, 1985) p. 617. Regulation of metabolite synthesis.
28. GOTTSCHALK, G.: *Bacterial Metabolism* (Springer Verlag, New York, 1979).
29. MANDLESTAM, J., MCQUILLAN, K. and DAWES, I. (eds.).: *Biochemistry of Bacterial Growth*, 3rd edn. (Blackwell Scientific Publications, Oxford, 1985).
30. BROWN, T. A.: *Gene Cloning—An Introduction* (Van Nostrand Reinhold (UK) Co Ltd, Wokingham, 1986).

31. OLD, R. W. and PRIMROSE, S. B.: *Principles of Gene Manipulation*—An *Introduction to Genetic Engineering*, 3rd edn. (Blackwell Scientific Publications, Oxford, 1985).
32. HIROSE, Y., SANO, K., and SHIBIA, H.: In *Ann. Reports on Fermentation Process*, PERLMAN, D. ed. (Academic Press, New York, 1978) Vol. 2, p. 155. Amino acids.
33. HERBERT, D. W.: *Symp. Soc. Gen. Microbiol.* **11** (1961) 391. The composition of micro-organisms as a function of their environment.
34. BATTLEY, E. H.: *Biotechnol. Bioeng.* **21** (1979) 1929. Alternate methods of calculating free energy change accompanying the growth of *S. cerevisiae* (Hansen) on three substrates.
35. ABBOTT, B. J. and CLAMEN, A.: *Biotechnol. Bioeng.* **15** (1973) 117. The relationship of substrate, growth rate, and maintenance coefficient to single cell protein production.
36. PIRT, S. J.: *Proc. R. Soc.* **B163** (1965) 224. The maintenance energy of bacteria in growing cultures.
37. LODGE, R. M. and HINSHELWOOD, C. N.: *J. Chem. Soc.* **213** (1943) 148. Physicochemical aspects of bacterial growth.
38. PEARL, R. and REED, L. J.: *Proc. Natl. Acad. Sci.* **6** (1920) 275. Growth equation with inhibition factor leading to logistic equation.
39. EDWARDS, V. H. and WILKE, C. R.: *Biotechnol. Bioeng.* **10** (1968) 205. Mathematical representation of batch culture data.
40. MONOD, J.: *Recherches sur des Croissances des Cultures Bacteriennes* (Hermann et Cie, Paris, 1942).
41. EDWARDS, V. H.: *Biotechnol. Bioeng.* **12** (1970) 679. The influence of high substrate concentration on microbial kinetics.
42. AIBA, S., SHODA, M. and NAGATANI M.: *Biotechnol. Bioeng.* **10** (1968) 845. Kinetics of product inhibition in alcohol fermentation.
43. MOSER, H.: *Carnegie Institute Publ. No.* 614 (1958). The dynamics of bacterial populations in the chemostat.
44. CONTOIS, D. E.: *J. Gen. Microbiol.* **21** (1959) 40. Kinetics of microbial growth. Relationship between population density and specific growth rate of continuous culture.
45. PIRT, S. J.: *Principles of Microbe and Cell Cultivation* (Blackwell Scientific Publications, Oxford, 1975)
46. HARWOOD, J. H.: University of London, Ph.D. thesis (1970). Studies on the physiology of methylococcus capsulatis growing on methane.
47. SINCLAIR, C. G. and KRISTIANSEN, B.: *Fermentation Kinetics and Modelling*, BU'LOCK J. D., ed. (Open University Press, Milton Keynes, 1987).
48. WASUNGU, K. M. and SIMARD, R. E.: *Biotechnol. Bioeng.* **24** (1982) 1125. Growth characteristics of baker's yeast in ethanol.
49. ROWLEY, B. I. and PIRT, S. J.: *J. Gen. Microbiol.* **72** (1972) 553. Melanin production by *Aspergillus nidulans* in batch and chemostat cultures.
50. RYU, D. D. and MATELES, R. I.: *Biotechnol. Bioeng.* **10** (1968) 385. Transient response of continuous cultures to changes in temperature.
51. METCALF and EDDY INC. *Wastewater Engineering: Treatment, Disposal and Reuse*, 3rd edn. TCHOBANOGLOUS, G. and BURTON, F. L. (McGraw-Hill, New York, 1991).
52. TOPIWALA, H. H. and SINCLAIR, C. G.: *Biotechnol. Bioeng.* **13** (1971) 795. Temperature relationship in continuous culture.
53. KINCANNON, D. F. and GAUDY, A. F.: *Biotechnol. Bioeng.* **10** (1968) 483. Response of biological waste treatment systems to changes in salt concentrations.
54. GADEN, E. L.: *J. Biochem. Microbiol. Tech.* **1** 413 (1959) Fermentation kinetics and productivity.
55. DEINDOERFER, F. H.: *Adv. Appl. Microbiol.* **2** (1955) 321. Fermentation kinetics and model processes.
56. LUEDEKING, R. and PIRET, E. L.: *J. Biochem. Microbiol. Technol. and Eng.* **1** (1959) 393. A kinetic study of the lactic acid fermentation. Batch process at controlled pH.
57. SHU, P.: *J. Biochem. Microbiol. Technol. and Eng.* **3** (1961) 95. Mathematical models for the product accumulation in microbial processes.
58. HORVATH, C. and ENGASSER, J–M.: *Biotechnol. Bioeng.* **16** (1974) 909. External and internal diffusion in heterogeneous enzyme systems.
59. ATKINSON, B.: *Biochemical Reactors* (Pion, 1974).
60. GATES, W. E. and MARLAR, J. T.: *J. Water Pollut. Cont. Fed.* **40** (1968) R469. Graphical analysis of batch culture data using the Monod expressions.
61. ONG, S. L.: *Biotechnol. Bioeng.* **25** (1983) 2347. Least-squares estimation of batch culture kinetic parameters.
62. ESENER, A. A., ROELS, J. A. and KOSSEN, N. W. F.: *Biotechnol. Bioeng.* **25** (1983) 2803. Theory and applications of unstructured growth models: Kinetic and energetic aspects.
63. KNIGHTS, A. J.: University of Wales, Ph.D. thesis (1981). Determination of the biological kinetic parameters of fermenter design from batch culture data.
64. LOTKA, A. J.: *J. Am. Chem. Soc.* **42** (1920) 1595. Undamped oscillations derived from the law of mass action.
65. TSUCHIYA, H. M., DRAKE, J. F., JOST, J. L. and FREDRICKSON, A. G.: *J. Bacteriol.* **110** (1972) 1147. Predator-prey interactions of *Dictyoselium discoidem* and *Eschericha coli* in continuous culture.

66. WILLIAMS, F. M.: *J. Theoret. Biol.* **15** (1967) 190. A model of growth dynamics.
67. RAMKRISHNA, D., FREDRICKSON, A. G., and TSUCHIYA, H. M.: *Biotechnol. Bioeng.* **9** (1967) 129. Dynamics of microbial propagation.
68. ESENER, A. A., VEERMAN, T., ROELS, J. A. and KOSSEN, N. W. F.: *Biotechnol. Bioeng.* **24** (1982) 1749. Modeling of bacterial growth; formulation and evaluation of a structured model.
69. FREDRICKSON, A. G.: *Biotechnol. Bioeng.* **18** (1976) 1481. Structured models.
70. IAMANAKA, T., KAIEDA, T., SATO, K. and TACUGHI, H.: *J. Ferment. Technol.* **50**(9) (1972) 633. Optimisation of α-galactosidase production by mold.
71. PAPAGEORGAKOPOULU, H. and MAIER, W. J.: *Biotechnol. Bioeng.* **26** (1974) 275. A new modeling technique and computer simulation of computer growth.
72. MIDDLETON, J. C.: In *Mixing in the Process Industries*, 2nd edn., HARNEY, N., EDWARDS, M. F. and NIENOW, A. W. eds. (Butterworth Heinemann, Oxford, 1992), p. 322. Gas–liquid dispersion and mixing.

5.18. NOMENCLATURE

		Units in SI System	Dimensions in **M,N,L,T, θ**
A	Concentration of **A**	kg/m^3	$\mathbf{ML^{-3}}$
$\mathcal{A}$, $\mathcal{A}_1$, $\mathcal{A}_2$, $\mathcal{A}_4$	Arrhenius constants	s^{-1}	$\mathbf{T^{-1}}$
$\mathcal{A}_3$	Arrhenius constant	m^3/kg	$\mathbf{M^{-1}L^3}$
A_i	Constant in equation 5.82	kg/m^{3}s^1	$\mathbf{ML^{-3}T^{-1}}$
a	Surface area per unit volume	m^{-1}	$\mathbf{L^{-1}}$
$\mathscr{a}$	Parameter in equation 5.218	m^3/kg	$\mathbf{M^{-1}L^3}$
a_c	Cross-sectional area	m^2	$\mathbf{L^2}$
B	Concentration of **B**	kg/m^3	$\mathbf{ML^{-3}}$
$b_1 \, . \, . b_4$	Stoichiometric constants	—	—
$\mathscr{b}$	Parameter in equation 5.218	—	—
C	Concentration of gas in liquid phase	kg/m^3	$\mathbf{ML^{-3}}$
C^*	Saturation concentration of gas in liquid	kg/m^3	$\mathbf{ML^{-3}}$
C_G	Concentration of **G**-mass	kg/m^3	$\mathbf{ML^{-3}}$
C_D	Concentration of **D**-mass	kg/m^3	$\mathbf{ML^{-3}}$
D	Dilution rate	s^{-1}	$\mathbf{T^{-1}}$
Da	Damköhler number	—	—
D_{crit}	Critical dilution rate for wash-out	s^{-1}	$\mathbf{T^{-1}}$
D_e	Effective diffusion coefficient	m^2/s	$\mathbf{L^2T^{-1}}$
D_{op}	Dilution rate for maximum biomass production	s^{-1}	$\mathbf{T^{-1}}$
D_1	Dilution rate for CSTF 1	s^{-1}	$\mathbf{T^{-1}}$
D_2	Dilution rate for CSTF 2	s^{-1}	$\mathbf{T^{-1}}$
$\mathbf{E}_a$, $\mathbf{E}_{a1}$, . . $\mathbf{E}_{a4}$	Activation energy	J/kmol	$\mathbf{MN^{-1}L^2T^{-2}}$
E	Enzyme concentration	kg/m^3	$\mathbf{ML^{-3}}$
E_{act}	Concentration of active enzyme	kg/m^3	$\mathbf{ML^{-3}}$
E_{actt}	Active enzyme concentration at time t	kg/m^3	$\mathbf{ML^{-3}}$
E_{act0}	Initial active enzyme concentration	kg/m^3	$\mathbf{ML^{-3}}$
E_i	Concentration of inactive enzyme	kg/m^3	$\mathbf{ML^{-3}}$
E_{tot}	Total enzyme concentration	kg/m^3	$\mathbf{ML^{-3}}$
EA	Concentration of enzyme–substrate complex with substance A	kg/m^3	$\mathbf{ML^{-3}}$
EAB	Concentration of ternary complex of enzyme with substances **A** and **B**	kg/m^3	$\mathbf{ML^{-3}}$
EB	Concentration of enzyme–substrate complex with substance **B**	kg/m^3	$\mathbf{ML^{-3}}$
EH^+	Concentration of protonated enzyme	kg/m^3	$\mathbf{ML^{-3}}$
EP	Concentration of enzyme product complex with substance **P**	kg/m^3	$\mathbf{ML^{-3}}$
ES	Enzyme–substrate concentration	kg/m^3	$\mathbf{ML^{-3}}$
ESI	Enzyme–substrate–inhibitor concentration	kg/m^3	$\mathbf{ML^{-3}}$
F	Volumetric flowrate	m^3/s	$\mathbf{L^3T^{-1}}$
F_0	Volumetric flowrate of feed	m^3/s	$\mathbf{L^3T^{-1}}$
F_R	Volumetric flowrate of recycle stream	m^3/s	$\mathbf{L^3T^{-1}}$
$\mathbf{F}$	Dimensionless substrate concentration	—	—

		Units in SI System	Dimensions in M,N,L,T, θ
$\mathbf{F}_f$	Dimensionless function defined in equation 5.211	—	—
f	Fraction of biomass or enzyme	—	—
$\Delta G^{0\prime}$	Gibbs free energy at pH7	J/kmol	$\mathbf{MN^{-1}L^2T^{-2}}$
H^+	Proton concentration	kg/m^3	$\mathbf{ML^{-3}}$
h_D	Mass transfer coefficient	m/s	$\mathbf{LT^{-1}}$
I	Concentration of inhibitor	kg/m^3	$\mathbf{ML^{-3}}$
i	*i*th item in series	—	—
K_{GG}	Affinity constant for **G** compartment	kg/m^3	$\mathbf{ML^{-3}}$
K_{GD}	Affinity constant for **D** compartment	kg/m^3	$\mathbf{ML^{-3}}$
K_I, K_i	Inhibition constants	kg/m^3	$\mathbf{ML^{-3}}$
K_L	Mass transfer coefficient	m/s	$\mathbf{LT^{-1}}$
K_m	Michaelis constant	kg/m^3	$\mathbf{ML^{-3}}$
K_p	Product inhibition constant	kg/m^3	$\mathbf{ML^{-3}}$
K_S	Monod constant	kg/m^3	$\mathbf{ML^{-3}}$
K_{sj}, K_{si}	Monod constants for prey and predator	kg/m^3	$\mathbf{ML^{-3}}$
K_S'	Enzyme–substrate equilibrium constant	m^3/kg	$\mathbf{M^{-1}L^3}$
K_{SG}	Affinity constant for **G** compartment	kg/m^3	$\mathbf{ML^{-3}}$
K_{SD}	Affinity constant for **D** compartment	kg/m^3	$\mathbf{ML^{-3}}$
$K_S(T)$	Monod constant at temperature T	kg/m^3	$\mathbf{ML^{-3}}$
K_x	Michaelis *or* Monod constant	kg/m^3	$\mathbf{ML^{-3}}$
K_a	Equilibrium constants for protonated and unprotonated enzyme	kg/m^3	$\mathbf{ML^{-3}}$
K_m'	Modified Michaelis constant	kg/m^3	$\mathbf{ML^{-3}}$
K_m^A	Michaelis constant for substance **A**	kg/m^3	$\mathbf{ML^{-3}}$
K_m^B	Michaelis constant for substance **B**	kg/m^3	$\mathbf{ML^{-3}}$
K_s^A	The dissociation constant for the reaction $E + A \rightleftharpoons EA$	kg/m^3	$\mathbf{ML^{-3}}$
K_{eq}	Equilibrium constant for enzyme reaction	kg/m^3	$\mathbf{ML^{-3}}$
K_m^P	Michaelis constant for substance **P**	kg/m^3	$\mathbf{ML^{-3}}$
K_m^S	Michaelis constant for substance **S**	kg/m^3	$\mathbf{ML^{-3}}$
k	Rate constant	s^{-1}	$\mathbf{T^{-1}}$
k_{f1}	Rate constant for forward reaction for the formation of ES from E and S	m^3/kg s	$\mathbf{M^{-1}L^3T^{-1}}$
k_{f2}	Rate constant for forward reaction decomposition from ES	s^{-1}	$\mathbf{T^{-1}}$
k_{r1}	Rate constant for back reaction first step	s^{-1}	$\mathbf{T^{-1}}$
k_{r2}	Rate constant for back reaction second step	m^3/kg s	$\mathbf{M^{-1}L^3T^{-1}}$
k_{cat}	Rate constant for decomposition of ES complex	s^{-1}	$\mathbf{T^{-1}}$
k_{cat}^A	Rate constant for decomposition of enzyme substrate complex involving substance **A**	s^{-1}	$\mathbf{T^{-1}}$
k_{cat}^B	Rate constant for decomposition of enzyme substrate complex involving substance **B**	s^{-1}	$\mathbf{T^{-1}}$
k_1	Rate constant in logistic equation	s^{-1}	$\mathbf{T^{-1}}$
k_d	Endogenous respiration rate	s^{-1}	$\mathbf{T^{-1}}$
$k_d(T)$	Endogenous respiration rate at temperature T	s^{-1}	$\mathbf{T^{-1}}$
k_{de}	Enzyme decay coefficient	s^{-1}	$\mathbf{T^{-1}}$
k_K	Rate constant for **K** compartment	s^{-1}	$\mathbf{T^{-1}}$
k_{IG}, k_{ID}	Rate constants for deactivation of **G** and **D**	kg/s	$\mathbf{MT^{-1}}$
L	Biological film thickness	m	$\mathbf{L}$
M_G	Mass fraction of **G** in biomass	—	—
M_K	Mass fraction of **K** in biomass	—	—
m	Specific maintenance requirement	s^{-1}	$\mathbf{T^{-1}}$
m_G	Maintenance rate constant for **G** compartment	s^{-1}	$\mathbf{T^{-1}}$
N	Total number	—	—
n_1, n_2	Indices in expression for K_La	—	—
P	Product concentration	kg/m^3	$\mathbf{ML^{-3}}$
$\mathbf{P}_G$	Power input under gassed condition	W	$\mathbf{ML^2T^{-3}}$
Q	Concentration of substance **Q**	kg/m^3	$\mathbf{ML^{-3}}$
Q_{O_2}	Specific respiration rate	s^{-1}	$\mathbf{T^{-1}}$

		Units in SI System	Dimensions in M,N,L,T, θ
Q_{10}	Ratio of rates at 10 K temperature difference	—	—
$\mathscr{g}$	Parameter in equation 5.218	kg/m³	$\mathbf{ML^{-3}}$
$\mathscr{R}_0$	Initial rate of reaction	kg/m³s	$\mathbf{ML^{-3}T^{-1}}$
$\mathscr{R}_0^A$	Initial rate of reaction of substance **A**	kg/m³s	$\mathbf{ML^{-3}T^{-1}}$
$\mathscr{R}_0^B$	Initial rate of reaction of substance **B**	kg/m³s	$\mathbf{ML^{-3}T^{-1}}$
R	Universal gas constant	8314J/kmolK	$\mathbf{MN^{-1}L^2T^{-2}\theta^{-1}}$
R	Recycle ratio	—	—
$\mathscr{R}$	Rate of reaction per unit volume of reactor	kg/m³s	$\mathbf{ML^{-3}T^{-1}}$
$\mathscr{R}_D$, $\mathscr{R}_G$	Rates of reaction of **D** and **G** materials	kg/m³s	$\mathbf{ML^{-3}T^{-1}}$
$\mathscr{R}_s$	Rate of substrate reaction	kg/m³s	$\mathbf{ML^{-3}T^{-1}}$
$\mathscr{R}_s''$	Local rate of reaction of substrate per unit volume of sphere	kg/m³s	$\mathbf{ML^{-3}T^{-1}}$
$\mathscr{R}_x$	Rate of biomass reaction	kg/m³s	$\mathbf{ML^{-3}T^{-1}}$
$\mathscr{R}'$	Rate of reaction per unit area of biocatalyst	kg/m³s	$\mathbf{ML^{-2}T^{-1}}$
$\mathscr{R}_m'$	Maximal rate of reaction based on unit area of biocatalyst	kg/m²s	$\mathbf{ML^{-2}T^{-1}}$
$\mathscr{R}''$	Rate of reaction per unit volume of biocatalyst	kg/m³s	$\mathbf{ML^{-3}T^{-1}}$
$\mathscr{R}_m''$	Saturation constant based on unit volume of biocatalyst	kg/m³s	$\mathbf{ML^{-3}T^{-1}}$
$\mathscr{R}_k'$	Rate of reaction per unit area of biocatalyst with no diffusional limitation	kg/m²s	$\mathbf{ML^{-2}T^{-1}}$
r	Radius	m	**L**
r_p	Radius of particle	m	**L**
$\bar{r}$	Dimensionless radius	—	—
$\mathscr{r}$	Parameter in equation 5.218	s⁻¹	$\mathbf{T^{-1}}$
S	Substrate concentration	kg/m³	$\mathbf{ML^{-3}}$
$\bar{S}$	Dimensionless substrate concentration	—	—
S_A	Average substrate concentration	kg/m³	$\mathbf{ML^{-3}}$
S_e	Effluent substrate concentration	kg/m³	$\mathbf{ML^{-3}}$
S_G	Concentration of substrate used for growth	kg/m³	$\mathbf{ML^{-3}}$
S_O	Initial or feed substrate concentration	kg/m³	$\mathbf{ML^{-3}}$
S_1, S_2	Substrate concentration in 1st and 2nd vessels	kg/m³	$\mathbf{ML^{-3}}$
S_b	Substrate concentration in bulk of liquid	kg/m³	$\mathbf{ML^{-3}}$
S_s	Substrate concentration at surface	kg/m³	$\mathbf{ML^{-3}}$
T	Absolute temperature	K	**θ**
T	Function defined in equation 5.228	—	—
t	Time	s	**T**
t_d	Doubling time	s	**T**
t_g	Productive fermentation time	s	**T**
t_u	Unproductive fermentation time	s	**T**
u_s	Superficial gas velocity	m/s	$\mathbf{LT^{-1}}$
V	Volume	m³	$\mathbf{L^{-1}}$
V_1, V_2	Volume of first and second vessels	m³	$\mathbf{L^3}$
V_L	Liquid volume	m³	$\mathbf{L^3}$
V_{max}	Enzyme velocity constant	kg/m³s	$\mathbf{ML^{-3}T^{-1}}$
$(V_{max})_m$	Maximum enzyme velocity constant in unprotonated form	kg/m³s	$\mathbf{ML^{-3}T^{-1}}$
$V_{max,0}$	Initial enzyme velocity constant	kg/m³s	$\mathbf{ML^{-3}T^{-1}}$
V''_{max}	Enzyme velocity constant based on unit volume of immobilised biocatalyst	kg/m³s	$\mathbf{ML^{-3}T^{-1}}$
V^S_{max}	Maximum rate of reaction involving substance **S**	kg/m³s	$\mathbf{ML^{-3}T^{-1}}$
V^P_{max}	Maximum rate of reaction involving substance **P**	kg/m³s	$\mathbf{ML^{-3}T^{-1}}$
w	Specific rate of generation of biomass fraction	s⁻¹	$\mathbf{T^{-1}}$
X	Biomass concentration	kg/m³	$\mathbf{ML^{-3}}$
X_0	Initial or feed biomass concentration	kg/m³	$\mathbf{ML^{-3}}$
X_A	Average biomass concentration	kg/m³	$\mathbf{ML^{-3}}$
X_J	Concentration of prey	kg/m³	$\mathbf{ML^{-3}}$
X_L	Concentration of predator	kg/m³	$\mathbf{ML^{-3}}$
X_{op}	Biomass concentration at optimum dilution rate	kg/m³	$\mathbf{ML^{-3}}$

		Units in SI System	Dimensions in M,N,L,T, θ
X_e	Effluent biomass concentration	kg/m^3	$\mathbf{ML^{-3}}$
X_R	Biomass concentration in recycle stream	kg/m^3	$\mathbf{ML^{-3}}$
X_1, X_2	Biomass concentration in vessels 1 and 2	kg/m^3	$\mathbf{ML^{-3}}$
X_i	Biomass concentration in *i*th vessel	kg/m^3	$\mathbf{ML^{-3}}$
X_m	Final biomass concentration	kg/m^3	$\mathbf{ML^{-3}}$
x	Distance	m	**L**
Y	Yield coefficient	—	—
Y_G	Yield coefficient for growth	—	—
Y_J	Yield coefficient for growth of prey	—	—
Y_L	Yield coefficient for growth of predator	—	—
$Y_{P/S}$	Yield coefficient for product formation	—	—
Y_1, Y_2	Yield coefficient for 1st and 2nd vessels	—	—
$Y_{G/S}$	Yield coefficient for formation of **G** from **S**	—	—
$Y_{K/S}$	Yield coefficient for formation of **K** from **S**	—	—
$Y_{G/K}$	Yield coefficient for formation of **G** from **K**	—	—
$Y_{X/S}$	Overall yield coefficient for growth	—	—
$Y_{i/S}$	Yield coefficient for *i*th product from **S**	—	—
Y_{O_2}	Yield coefficient for oxygen utilisation	—	—
Z	Length of reactor	m	**L**
z	Dimensionless distance	—	—
α	Empirical constant in equation 5.71	—	—
β	Dimensionless substrate concentration	—	—
β_b	Bulk dimensionless substrate concentration	—	—
β_s	Surface dimensionless substrate concentration	—	—
χ	Dimensionless biomass concentration	—	—
δ	Coefficient in equation 5.79	—	—
ε	Coefficient in equation 5.79	s^{-1}	$\mathbf{T^{-1}}$
ξ	Biomass concentration ratio	—	—
ϕ	Thiele modulus	—	—
ϕ'	Modified Thiele modulus	—	—
γ	Inverse of maximum biomass concentration	m^3/kg	$\mathbf{M^{-1}L^3}$
η	Effectiveness factor	—	—
η_e	Effectiveness factor for surface reaction	—	—
μ	Specific growth rate	s^{-1}	$\mathbf{T^{-1}}$
μ'	Specific growth rate based on unit area	s^{-1}	$\mathbf{T^{-1}}$
μ''	Specific growth rate based on unit volume	s^{-1}	$\mathbf{T^{-1}}$
μ_0	Specific growth rate at reference temperature	s^{-1}	$\mathbf{T^{-1}}$
μ_m	Maximal specific growth rate	s^{-1}	$\mathbf{T^{-1}}$
μ_J	Maximal specific growth rate of prey	s^{-1}	$\mathbf{T^{-1}}$
μ_L	Maximal specific growth rate of predator	s^{-1}	$\mathbf{T^{-1}}$
$\mu_m(T)$	Maximal specific growth rate at temperature T	s^{-1}	$\mathbf{T^{-1}}$
μ_m'	Maximal specific growth rate based on unit area	s^{-1}	$\mathbf{T^{-1}}$
μ_m''	Maximal specific growth rate based on unit vol.	s^{-1}	$\mathbf{T^{-1}}$
μ_{max}	Maximum observed specific growth rate	s^{-1}	$\mathbf{T^{-1}}$
ψ	Specific rate of substrate consumption	s^{-1}	$\mathbf{T^{-1}}$
ψ_m	Maximum specific rate of substrate consumption	s^{-1}	$\mathbf{T^{-1}}$
τ	Cell age	s	**T**
θ	Celsius temperature	°C	**θ**
ν	Residence time	s	**T**

Subscripts

$_0$	Initial value
$_b$	Bulk
$_i$	*i*th component
$_n$	*n*th component
$_{op}$	Optimum
$_s$	Surface
$_t$	Value at time t

CHAPTER 6

Sensors for Measurement and Control

6.1. INTRODUCTION

No chemical plant can be operated unless it is adequately instrumented. The monitoring of flows, pressures, temperatures and levels is necessary in almost every process in order that the plant operator can see that all parts of the plant are functioning as required. Additionally, it may be necessary to record and display many other quantities which are more specific to the particular process in question, e.g. the composition of a process stream, the heat radiation produced in a crude oil heater or the humidity of a gas stream. In many instances, the sensor forms an essential part of a control system or strategy for a process. This may be quite complex and rely upon the performance and characteristics of a substantial number of different sensors. For example, a large catalytic cracking unit may contain upwards of a thousand instruments of various kinds. Each measurement has to be transmitted to a display unit or indicator or to a recording device which will generally have a visible indicator attached to it. The indicator or recorder may be situated locally with respect to the instrument or will, more commonly, be placed in the plant control room. This often requires the conversion of a measurement signal into a form which is suitable for transmission over what might be quite long distances. Such conversion is frequently called *signal conditioning* (Fig. 6.1) and the term *transducer* is used often in reference to a total sensor package containing both conditioning and sensing elements. Further modification, viz. *signal processing*, is generally necessary in order that the signal be changed to one which is suitable as an input to the next component in the system. This may simply take the form of an amplification or may be a conversion from a continuous (*analog*) to a discontinuous (*digital* or *discrete-time*) signal or vice versa. This can occur after direct transmission to the plant control room as illustrated in Fig. 6.1, or immediately after the signal conditioning stage where the data are to be supplied to a local indicator or to a *data hiway* as part of a *distributed control system* (Section 7.20). Hence, the topics of signal conditioning, signal transmission (*telemetry*) and signal processing are inevitably connected with the study of the sensors themselves.

Two of the most important considerations for a measurement system are accuracy and reliability. Clearly, a measurement is of little use if it is substantially inaccurate. It may be possible to allow for a consistent error in an instrument if it is known, but low frequency, randomly occurring errors will mean that incorrect actions may be taken by the operator as a result of reading that instrument, or by the control system employing it as a sensor. High frequency random variations in the measurement

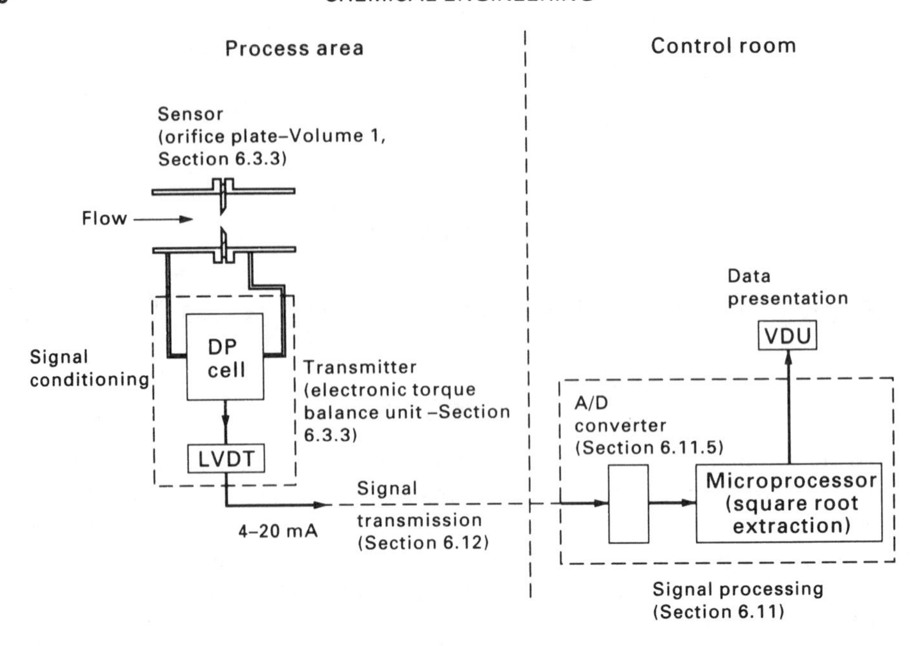

FIG. 6.1. Typical process measurement system showing the stages between sensor and presentation of the data to the process operator (DP—differential pressure cell, LVDT—linear variable differential transformer, A/D—analogue to digital, VDU—visual display unit)

signal (often termed *noise*) can be an equally serious problem. A measurement system should be reliable, i.e. produce consistent readings, in order that the process plant operator can be confident in the value of the particular process variable that it is indicating, and it is well to note that a control system (however sophisticated or complex) will not produce the desired result if the measurements employed are unreliable and/or incorrect.

Different types of sensor, together with the general static characteristics of measurement systems, are described in this chapter with the exception of those previously covered in Volumes 1 and 2. Because of their close connection, signal conditioning and signal processing elements are considered together in Section 6.11. The chapter concludes with a discussion of signal transmission (or telemetry) and a brief description of *smart* instruments and currently developing industry-based protocols. The dynamics of measuring elements are presented in Chapter 7.

6.2. THE MEASUREMENT OF FLOW

6.2.1. Methods Dependent on Relationship Between Pressure Drop and Flowrate

Instruments which are employed to measure flow often use the relation between rate of flow and pressure drop along a section of pipe, around some obstruction or across a restriction within the pipe. The more widely employed flow measuring devices (or *flowmeters*) are discussed in Volume 1, Chapter 6. These are as follows:

(1) The pitot tube (Volume 1, Section 6.3.1).
(2) The orifice meter (Volume 1, Section 6.3.3).
(3) The nozzle (Volume 1, Section 6.3.4).
(4) The venturi meter (Volume 1, Section 6.3.5).
(5) The Dall tube (Volume 1, Section 6.3.6).

Other sensors which are described in Volume 1 (Sections 6.3.7–6.3.9) are the variable area meter, the notch or weir, the hot wire anemometer, the electromagnetic flowmeter and the positive displacement meter. Some of these flowmeters are relatively less suitable for producing signals which can be transmitted to the control room for display (e.g. weir, rotameter) and others are used in more specialist or limited applications (e.g. magnetic flowmeter, hot wire anemometer). The major characteristics of different types of flow sensor are summarised in Table 6.1. Brief descriptions follow of the principles underlying the more important types of flowmeter not described in Volume 1. In many instances such flow sensors are taking the place of those more traditional meters which rely upon pressure drop measurement. This is for reasons of versatility, energy conservation and convenience.

6.2.2. Further Methods of Measuring Volumetric Flow

The Averaging Pitot Tube

Because of its low pressure drop capability, the pitot tube is gaining popularity in the form of the *averaging pitot tube* or *Annubar* which is a variation of the standard pitot tube. This uses multiple sampling points across a pipe or duct in order to provide a representation of the full flow profile.

Vortex Flowmeters

The basic principle of vortex shedding is shown in Fig. 6.2*a* and is discussed in Volume 2, Section 3.2. A barrier (*bluff body*) is placed in the path of the fluid flow such that the fluid splits into two streams. The resulting instability of the boundary layer as it breaks away from the obstruction causes the fluid to *roll up* into a well-defined vortex. After the vortex is formed, it is *shed* from the body and a second vortex begins to form on the opposite side of the body. If the vortex shedding is stable, then the time taken for the formation of each vortex is proportional to the fluid velocity for a given bluff body geometry. This pattern is repeated producing a distinctive series of vortices known as the *von Karman vortex street*. The formation and shedding of these vortices results in regular pressure variations around and downstream of the bluff body. These can be measured by a suitably located pressure, thermal or ultrasonic detector (Fig. 6.2*b*). STROUHAL[1] showed that the free stream velocity u_{free} is related to the frequency ω of the changes in pressure by:

$$u_{free} = \frac{\omega B_f}{S} \tag{6.1}$$

where B_f is the width of the bluff body and S is the *Strouhal number* which is an experimentally determined quantity dependent only upon the geometry of the flowmeter.

TABLE 6.1. *Principal Characteristics of Common Flowmeters*[2–5]

Type of Flowmeter	Reference	Measures Volume or Mass Flow	Gases or Vapours		Steam	Liquids				Slurries		Operating Temperature of Sensor (K)
			Clean	Dirty		Clean	Viscous	Dirty	Corrosive	Fibrous	Abrasive	
Orifice–square edge	Volume 1 Section 6.3.3	Volume	Yes	No	Yes	Yes	No	Usually	Usually	No	No	< 800
Orifice–conic edge	Volume 1 Section 6.3.3	Volume	No	No	No	Yes	Yes	Usually	Usually	No	No	< 800
Venturi	Volume 1 Section 6.3.5	Volume	Yes	Usually	Yes	Yes	Usually	Usually	Usually	Usually	Usually	< 800
Flow Nozzle	Volume 1 Section 6.3.4	Volume	Yes	Usually	Yes	Yes	Usually	Usually	Usually	No	No	< 800
Annubar	Volume 3 Section 6.2.1	Volume	Yes	No	Yes	Yes	No	No	Usually	No	No	< 800
Electromagnetic	Volume 1 Section 6.3.9	Mass	No	No	No	Yes	Yes	Yes	Yes	Yes	Yes	230–450
Positive Displacement	Volume 1 Section 6.3.9	Volume	Yes	No	No	Yes	No	No	Usually	No	No	Gases < 400 Liquids < 580
Turbine	Volume 1 Section 6.3.9	Volume	Yes	No	No	Yes	No	No	Usually	No	No	10–530
Ultrasonic–time of flight	Volume 3 Section 6.2.1	Volume	No	No	No	Yes	Usually	No	Yes	No	No	100–530
Doppler	Volume 3 Section 6.2.1	Volume	No	No	No	No	Usually	Yes	Yes	Yes	Yes	100–400
Variable area	Volume 1 Section 6.3.7	Volume	Yes	No	No	Yes	Yes	No	Usually	No	No	Glass < 400 Metal < 800
Vortex	Volume 3 Section 6.2.1	Volume	Yes	Usually	Yes	Yes	No	Usually	Usually	No	No	80–680
Coriolis	Volume 3 Section 6.2.2	Mass	At high pressure only	At high pressure only	No	Yes	Yes	Yes	Yes	Yes	Yes	See comments
Thermal	Volume 1 Section 6.3.9	Mass	Yes	No	No	No	No	No	No	No	No	< 350

(a) FSD: full scale deflection.
(b) *R*: per cent of flowrate.
(c) See Section 6.10.1.
(d) All flow sensors based on pressure drop measurement become insensitive below about 10 per cent of their full scale deflection.

Maximum Operational Pressure (MPa)	$-\Delta P$ (Comparative)	Pipe Size (mm)	Minimum *Re*	Suitability for Very Low Flow Measurement	Maximum Accuracy (per cent FSD[a])	Maximum Turndown[c]	Comparative Cost (including transmitter)	Comments	Type of Flowmeter
< 40	High	> 40	2000	No	± 1 to ± 2	4:1	Medium	Inaccurate below 10 per cent of FSD[d]	Orifice–square edge
< 40	High	> 40	200	No	± 2	4:1	Medium	Inaccurate below 10 per cent of FSD[d]	Orifice–conic edge
< 40	80–90 per cent of energy recovered	> 50	7.5×10^4	No	± 1 to ± 2	4:1	High	Inaccurate below 10 per cent of FSD[d]	Venturi
< 40	High	> 50	10^4	No	± 1 to ± 2	4:1	Medium	Inaccurate below 10 per cent of FSD[d]	Flow Nozzle
< 40	Low	> 25	10^4	No	± 1.25	4:1	Low	Inaccurate below 10 per cent of FSD[d]	Annubar
< 10	Low	6–2000	(see Table 6.2)	Yes (Table 6.2)	± 0.2*R*[b] to ± 1	10:1	High	Conductive liquids only	Electromagnetic
< 10	Medium	< 300	$\nu < 8000$ (see Table 6.2)	Yes (Table 6.2)	Gases ± 1 Liquids ± 0.25*R*[b]	Better than 10:1	High	Insensitive to viscosity. Filter fluid to 10 μm	Positive Displacement
< 20	Medium	6–600	$\nu < 20$ (see Table 6.2)	Yes (Table 6.2)	± 0.5*R*[b]	20:1	High	Mainly for small pipes. Accurate measurement	Turbine
Pipe rating	Low	> 12	Very small	No	± 1*R*[b] to ± 5	Better than 50:1	High	Also used in partly filled pipes and open channels	Ultrasonic–time of flight
Pipe rating	Low	> 12	Very small	No	± 1*R*[b] to ± 5	Better than 50:1	High	Also used in partly filled pipes and open channels	Doppler
Glass 2.4 Metal 5	Low	≤ 75	Very small	Yes (Table 6.2)	± 0.5*R*[b] to ± 1	10:1	Medium	Can be used to measure a highly viscous fluid	Variable area
< 10	Low	15–400	> 10^4	No	± 0.5*R* to ± 1*R*[b]	20:1	High		Vortex
50	Low	10–100	(See Table 6.2)	Yes (Table 6.2)	± 1	20:1	High	Insensitive to temperature, pressure and viscosity	Coriolis
4	Low	10–100	(See Table 6.2)	Yes (Table 6.2)	± 1	100:1	Medium to high		Thermal

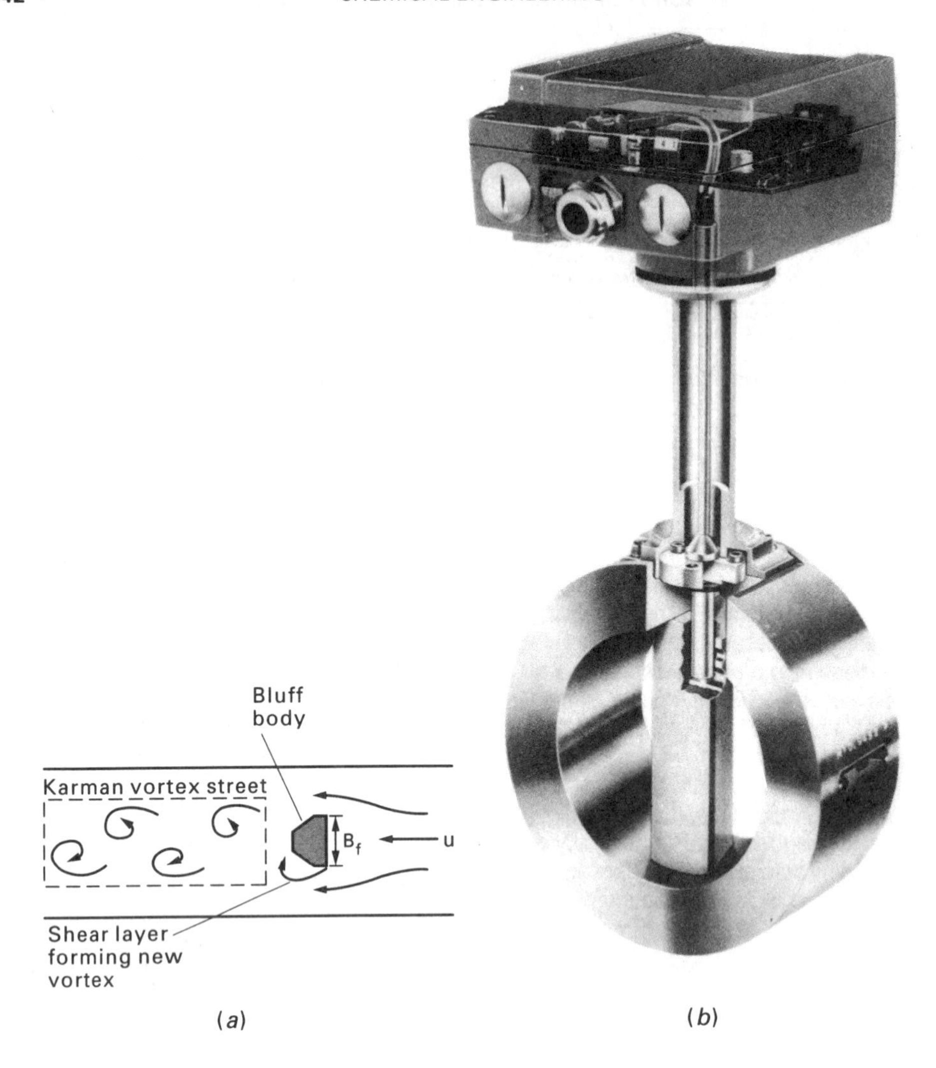

FIG. 6.2. Vortex-shedding flowmeter: (*a*) basic principle of meter; (*b*) general view

Vortex meters are being employed increasingly because of their accuracy, range, precision and relative insensitivity to fluid properties. They can be used for gases, saturated and superheated steam, low viscosity clean and dirty liquids, and cryogenic fluids[6]. They are not recommended for use when $Re < 10^4$.

Ultrasonic Methods[7]

Ultrasonic meters are finding increasing application because of their ability to measure clear and dirty liquids in difficult situations. They are usually non-intrusive and present little or no obstruction to the flow. They are effective also in measuring flow in open channels (Section 6.2.5) and in partially filled pipes. They are, however, highly sensitive to flow conditions and should be calibrated with care.

There are two principal types, viz. the *time-of-flight* instrument and the *Doppler* flowmeter.

The Time-of-Flight Meter

This instrument is used to measure the flow of clean liquids and involves the determination of the time required for an acutely angled, high frequency pressure wave to reach the opposite wall of a pipe. The elapsed time depends upon the velocity of the liquid u_ℓ, whether the pressure wave is moving with, or against the flow and upon the speed of sound in the liquid u_s. The most common time-of-flight meter is the *counter-propagating* type in which two transducers are placed on opposite sides of the liquid stream as shown in Fig. 6.3.

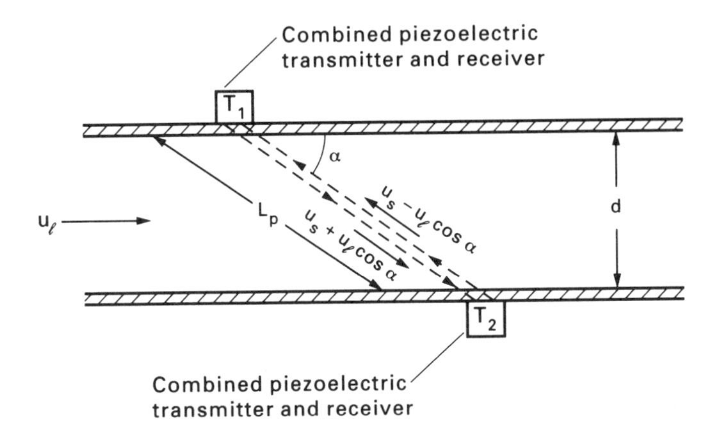

FIG. 6.3. Counter-propagating time-of-flight ultrasonic flowmeter

Both transducers are self-excited oscillating systems (usually piezoelectric crystals—see Section 6.3.1) which act alternately as transmitters and receivers, the receiving pulses being used to trigger the transmitted pulses in a feedback arrangement. A pulse from transmitter T_1 is directed downstream over a path length L_p to transducer T_2 (Fig. 6.3). The downstream elapsed time is:

$$t_{\text{down}} = \frac{L_p}{u_s + u_\ell \cos\alpha}$$

On receipt of the signal from T_1, T_2 is triggered and transmits a pulse upstream to T_1. The upstream elapsed time is:

$$t_{\text{up}} = \frac{L_p}{u_s - u_\ell \cos\alpha}$$

Hence:

$$\frac{1}{t_{\text{down}}} - \frac{1}{t_{\text{up}}} = \frac{2u_\ell \cos\alpha}{L_p}$$

Thus:

$$u_\ell = \frac{L_p}{2\cos\alpha}\left(\frac{t_{up} - t_{down}}{t_{up}t_{down}}\right) \tag{6.2}$$

and the average volumetric rate of flow is given by:

$$Q = K_{vp}\left(\frac{\pi d^3 \tan\alpha}{8}\right)\left(\frac{t_{up} - t_{down}}{t_{up}t_{down}}\right)$$

$$= K_{TF}\left(\frac{t_{up} - t_{down}}{t_{up}t_{down}}\right) \tag{6.3}$$

where K_{vp} is a correction for the velocity profile and K_{TF} is normally a constant for a particular installation.

This type of meter can simply be clamped on the outside surface of the pipe which avoids the possibility of the transducer probes being affected by fouling. In this instance care must be taken to avoid *acoustic short-circuiting*, i.e. transducer signals being transmitted and received via the pipe wall. Both 'clamp-on' and wetted versions of the counter-propagating meter are sensitive to variations in the velocity profile of the measured liquid—but not to the velocity of sound in the medium (equation 6.3).

The Doppler Meter

Doppler ultrasonic flowmeters depend upon the reflection of a continuous ultrasonic wave (frequency 0.5–10 MHz) from particulate matter (*scatterers*) contained in the fluid. Hence they may be used to monitor the rate of flow of dirty liquids. The transducer involved can act both as transmitter and receiver and is generally of the clamp-on type (Fig. 6.4). If the scatterers can be assumed to be moving at the velocity of the liquid, then the volumetric rate of flow Q is related to the Doppler frequency shift $\Delta\omega_D$ by:

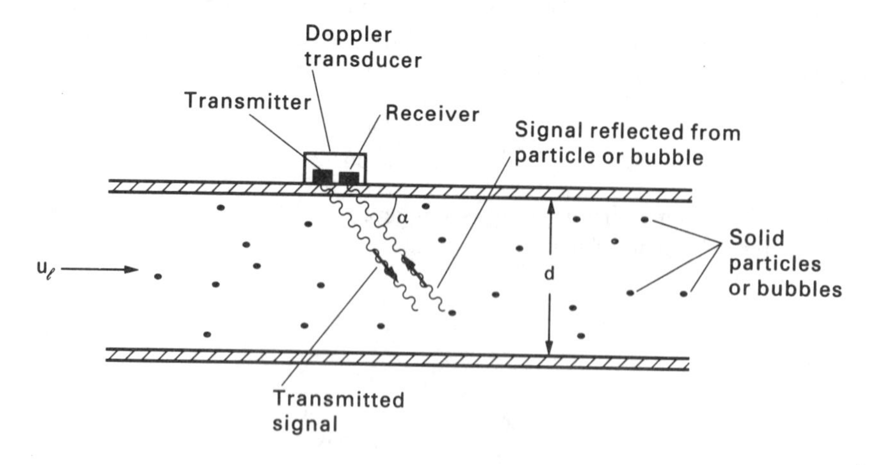

FIG. 6.4. Doppler flowmeter

$$Q = \left(\frac{\pi d^2}{4} \frac{u_s}{\cos \alpha} \right) \frac{\Delta \omega_D}{\omega} = K_D \Delta \omega_D \tag{6.4}$$

where K_D is a constant provided that u_s remains constant and ω is the transmitter frequency.

The Doppler flowmeter can be employed where other methods fail, e.g. for the measurement of two-phase systems such as slurries in the nuclear industry, of food products such as pickles and sauces and of gas–liquid mixtures.

6.2.3. The Measurement of Mass Flow

One class of flow measurement which is becoming of increasing importance (particularly in the form of sensors for control systems) is the monitoring of mass flow. This is rapidly superseding the measurement of volumetric flow—especially where it is required to determine accurately the transfer of large quantities of gas and liquid in the oil, gas and water industries. Two principal approaches are employed to measure mass flow. One is indirect and uses a combination of volumetric flow and density and the other is direct in that it involves the measurement of properties which are sensitive to variations in the mass rate of flow itself.

Indirect (Inferential) Methods

Any combination of flowmeter and suitable specific gravity (density) meter (Section 6.6) can in theory be employed to produce a mass rate of flow. However, these so-called *indirect meters* have a limited range and accuracy and require careful and frequent calibration and maintenance. An instrument based on the turbine flowmeter (Volume 1, Section 6.3.9) and acoustic density meter (Section 6.6.1) has been employed with good results[8] (see Fig. 6.5).

The turbine meter produces a pulsed output of frequency ω_1 such that:

$$Q = K_1 \omega_1 \tag{6.5}$$

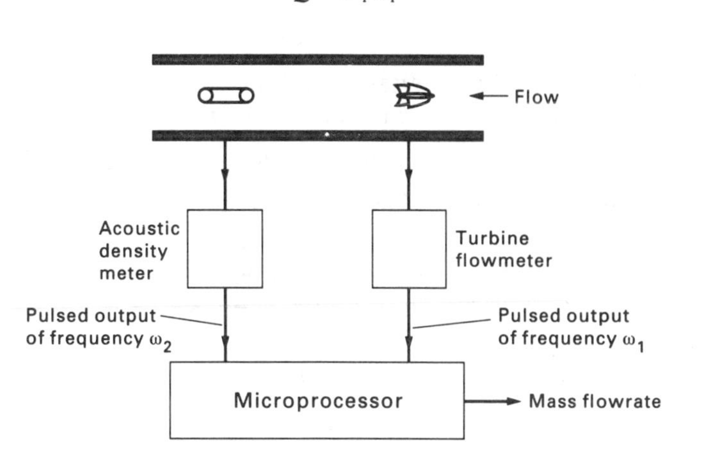

FIG. 6.5. Combination of sensors for determining the mass flow of fluid in a pipe

where Q is the volumetric rate of flow and K_1 is a constant characteristic of the flow measuring device. The density meter also gives a pulsed output of frequency ω_2 which is related to the fluid density ρ by:

$$\rho = \frac{K_2}{\omega_2^2} + \frac{K_3}{\omega_2} \tag{6.6}$$

The microprocessor measures ω_1 and ω_2 by counting pulses over a fixed time period and calculates the mass m of fluid which has passed in time t from:

$$m = \delta t \sum_{j=1}^{N} \rho_j Q_j \tag{6.7}$$

where ρ_j and Q_j are the values of ρ and Q at the jth sampling instant, δt is the sampling interval and $N = t/\delta t$.

Direct Methods

These instruments can be divided into meters which employ thermal effects and those which are energised by variations in the momentum of the fluid passing through the meter. The former rely upon the rate of loss of heat from a heated wire immersed in a fluid and this rate of loss of heat is a function of the mass flowrate. This type of instrument is described in Volume 1, Section 6.3.9. A discussion of two of the more common types of momentum transfer meter follows.

The Axial Flow Transverse Momentum Flowmeter[9]

This consists of an impeller and turbine mounted adjacently within a pipeline. The impeller is rotated at a constant angular velocity ω which imparts the same angular velocity to the fluid as it flows past. On leaving the impeller, the fluid immediately enters the turbine where the angular momentum gained by the fluid is used to drive the turbine. If a mass of fluid dm enters the impeller during time dt, then:

$$dI = r_g^2 dm$$

where r_g and dI are the radius of gyration and the moment of inertia of the mass dm respectively. The transverse angular momentum of the fluid leaving the impeller is:

$$\omega dI = \omega r_g^2 dm$$

The deflecting torque on the turbine $\mathbf{T}$ is equal to the rate of change of angular momentum, i.e.:

$$\mathbf{T} = \omega \frac{dI}{dt} = \omega r_g^2 \frac{dm}{dt}$$

Thus:

$$m = \frac{1}{\omega r_g^2} \int_0^t \mathbf{T} dt \tag{6.8}$$

The turbine is linked to an integrator which computes the mass m passing in a given time t from $\mathbf{T}$ using equation 6.8.

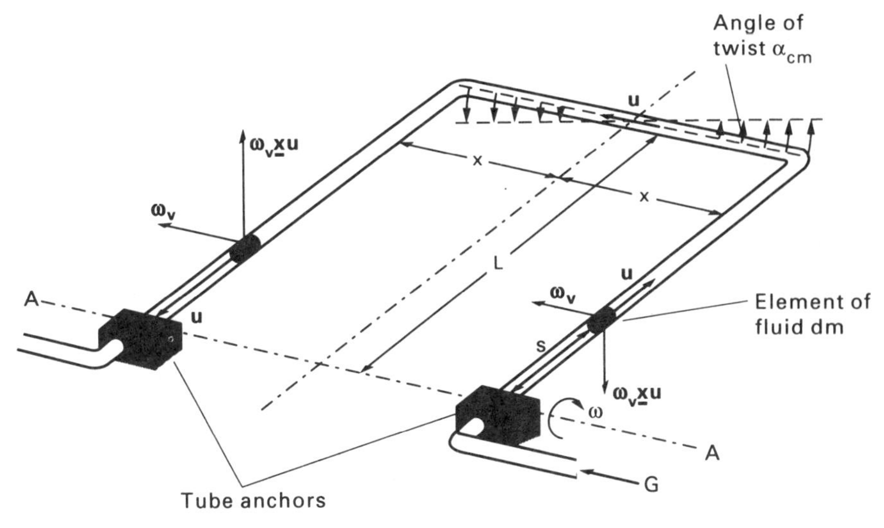

FIG. 6.6. Principle of the Coriolis meter

The Coriolis Meter

This class of meter depends upon the fluid attaining an angular velocity ω whose vector is perpendicular to the fluid velocity. A simplified representation is shown in Fig. 6.6[(10)].

The rectangular pipe section is kept in constant oscillation in a vertical plane about AA at its resonant frequency (50–80 Hz) by an electromagnetic feedback system. As the fluid flows through the pipe, inertial forces cause changes in the shape of the rectangular section. Only those inertial forces which produce a twisting action in the direction opposite to $\boldsymbol{\omega}_\mathbf{v} \, \underline{\mathbf{x}} \, \mathbf{u}$ (the *Coriolis acceleration*) are measured where $\mathbf{u}$ is the fluid velocity vector, $\boldsymbol{\omega}_\mathbf{v}$ is the vector of the angular velocity experienced by the fluid perpendicular to $\mathbf{u}$, and $\underline{\mathbf{x}}$ represents the vector cross-product. An element of fluid of mass dm defined by s (where s is the magnitude of the position vector for dm relative to the origin) flowing in the direction indicated in Fig. 6.6 produces an inertia force of magnitude $\mathrm{d}m(2\boldsymbol{\omega}_\mathbf{v} \, \underline{\mathbf{x}} \, \mathbf{u})$. An equal and opposite force will be impressed upon the return leg of the rectangular section of the pipe. Hence, the torque vector on the element is:

$$\mathrm{d}\mathbf{T} = 2(2\boldsymbol{\omega}_\mathbf{v} \, \underline{\mathbf{x}} \, \mathbf{u})(\mathrm{d}m)x$$

$$= 2(2\boldsymbol{\omega}_\mathbf{v} \, \underline{\mathbf{x}} \, \mathbf{u})\left(\frac{G}{\mathbf{u}}\,\mathrm{d}s\right)x$$

where G is the mass rate of flow and $2x$ is the distance between the limbs of the rectangular section. Thus the total torque $\mathbf{T}$ has a magnitude:

$$K_{CM}\alpha_{CM} = 4\,\omega_v Gx \int_0^L \mathrm{d}s = 4\,\omega_v GxL \tag{6.9}$$

ω varies sinusoidally, hence $\mathbf{T}$ is sinusoidal also and acts as a driving torque which twists the rectangular section through an angle α_{CM}. K_{CM} is a constant for the meter. α_{CM} is obtained by measuring (optically or electrically) the difference in times at which each limb of the beam passes sensing points during each cycle of the oscillation. Thus G can be computed as the output signal from the instrument. The total mass flow over any time interval can be obtained.

This type of meter does not restrict the flow of fluid, can be used for two-phase mixtures (e.g. foams, slurries, liquids containing entrained gases), and is insensitive to variations in temperature, pressure and viscosity. It responds quickly to flow variations and, although primarily a liquid flowmeter, it can be used to monitor gases where pressures are high enough to produce viscosities of sufficient magnitude to actuate the sensing tubes. In alternative versions of the instrument the fluid passes through two straight parallel pipes which are made to resonate. A phase shift occurs between inlet and outlet which is proportional to the mass rate of flow. The resonant frequency of the measuring pipes is a function of the oscillating mass and therefore of the density of the fluid. Most currently manufactured Coriolis meters have a double tube arrangement which produces a nominally balanced system and hence greater stability[11].

6.2.4. The Measurement of Low Flowrates

There are many applications where it is necessary to monitor very low rates of flow (e.g. addition of mercaptans to natural gas in the gas industry or of anti-foaming agents to fermentation processes in the pharmaceutical industry). A *low flow* is considered to be of the order of 100 mm^3/s and an *ultra-low flow* of around 0.1 mm^3/s for both liquids and gases. Table 6.2 includes commonly accepted minimum measurable rates of flow for types of flowmeter normally employed for this purpose[12].

6.2.5. Open Channel Flow

Flow measurement in open unpressurised channels is a requirement generally associated with waste water systems and sewers. The use of a weir or notch to measure the flow of a liquid presenting a free surface is described in Volume 1, Section 6.3.8. The flow through a rectangular notch (Fig. 6.7*a*) is given by (Volume 1, equation 6.42):

$$Q = \tfrac{2}{3} C_D B \sqrt{2g}\, D^{1.5} \tag{6.10}$$

where B is the width of the notch and D is the depth of the liquid at the notch.

For a triangular notch (Volume 1, equation 6.45):

$$Q = \tfrac{8}{15} C_D \tan\alpha \sqrt{2g}\, D^{2.5} \tag{6.11}$$

where the notch is of angle 2α (Fig. 6.7*b*) and C_D is the coefficient of discharge.

Alternatively, a *flume* can be employed where the rate of fall of a stream is slight or where the stream contains a large quantity of solid material (silt or debris). The most widely used is the *venturi flume* (Fig. 6.7*c*) for which the volumetric flow can be determined from[13]:

$$Q = C_D B_2 D_2 \sqrt{\frac{2g(D_1 - D_2)}{1 - (B_2 D_2 / B_1 D_1)^2}} \tag{6.12}$$

TABLE 6.2. *Flowmeters that can be Used to Measure Low Flowrates*

Type of Flowmeter	Suitable for	Minimum Measurable Flowrate (mm^3/s)	Principle of Operation	Typical Applications	Comments
Turbines, Pelton wheels	Liquids and gases	140	(See Volume 1, Section 6.3.9)	Fuel metering in aircraft	High repeatability. Can be used at high temperatures and pressures
Positive displacement	Liquids only	20	(See Volume 1, Section 6.3.9)	Hydrocarbon liquids, glues, food products (jams, chocolate, etc.)	Suitable for high viscosity fluids. High accuracy and wide range
Coriolis	Liquids. Gases at high pressure	20	(See Section 6.2.3)	Can be used for almost any liquid provided that the viscosity is not so great as to impose too high a pressure drop	(See Section 6.2.3)
Electro-magnetic	Liquids only	1.5	(See Volume 1, Section 6.3.9)	Additives in pharmaceuticals, paint, pulp and paper, water treatment using chlorine	Good chemical resistance. Hygenic and intrinsically safe. Wide range. Requires some conductivity
Thermal volumetric	Liquids only	1	Measures time taken for liquid to flow between two detectors by means of a heat pulse	High pressure liquid phase chromatography (Section 6.8.3 and Volume 2, Chapter 19)	No moving parts. Suitable for high pressures and temperatures

where B_1 and B_2 are the width of the channel and the venturi throat respectively, and D_1 and D_2 are the depths immediately upstream of the flume and the minimum depth within the throat of the flume respectively.

In each case, the volumetric rate of flow can be determined by measuring the liquid levels in the appropriate place. This is often achieved using an ultrasonic measuring system (Fig. 6.7*d*) in which the time taken for an ultrasonic wave to be reflected from the liquid surface is measured (see also Section 6.5.5). Accuracies of ±2.5 mm/m distance between sensor and liquid surface are not uncommon. Standard designs of open channel restrictions can be found in BS 3680[14].

6.2.6. Flow Profile Distortion

Pipe fittings such as elbows, tee-pieces, reducers, expanders, valves, etc. can all alter the symmetry of the flow profile. MILLER[2] has listed the following effects:

(a) Pure *swirl* (i.e. rotation of the fluid about the axis of the pipe caused by a succession of sharp elbows).

(b) Secondary flows (e.g. vortices caused by pipe enlargement).

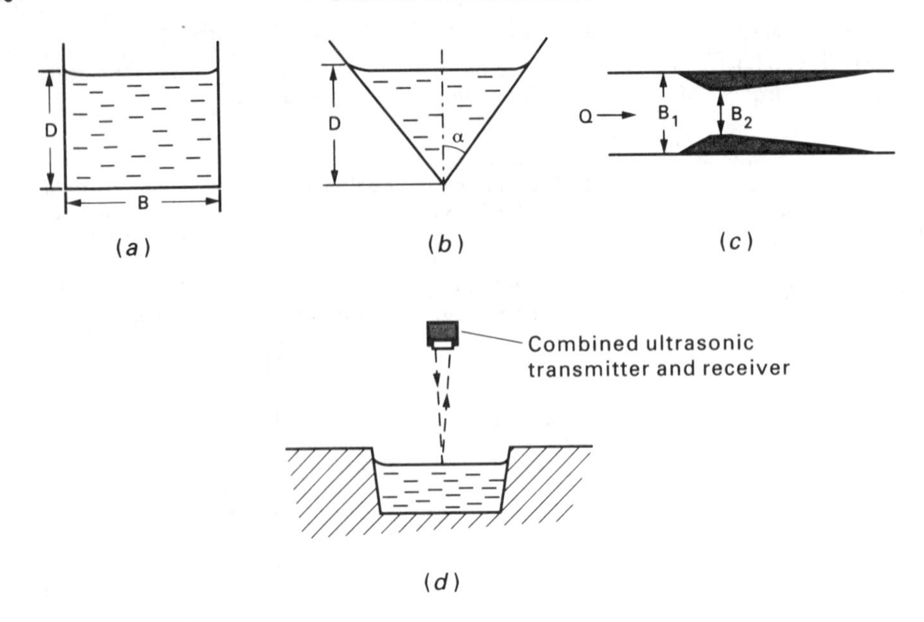

FIG. 6.7. Open channel flow measurement

(c) Symmetric profile with unusually high axial velocity (e.g. due to a large reduction in pipe diameter).
(d) Asymmetric profile (e.g. due to a single elbow).

Two or more of these conditions can occur at the same time, resulting in asymmetric axial, radial and tangential velocity vectors. Some flowmeters are more sensitive than others to particular types of flow distortion, e.g. orifice meters are affected by pure swirl more than venturi meters are; magnetic flowmeters are unaffected by changes in the radial velocity component whereas ultrasonic time-of-flight meters are highly susceptible thereto; swirl and asymmetry have the least effect on positive displacement meters and the greatest effect on variable area meters.

Swirl is produced by the flow being made to change direction twice—each time in a different plane (Fig. 6.8*b*). This can be due to a combination of two successive 90° elbows or of a 90° elbow closely followed by a valve. The strongest swirl occurs when the two elbow planes are at 60° to each other[(2)].

To avoid swirl, elbows should be well separated and have large radii of curvature. If this is not possible then the flowmeter should be sited at least 40 pipe diameters downstream of fittings causing asymmetric flow only and a minimum of 100 pipe diameters downstream when swirl is likely to occur[(15)]. There should also be at least 10 pipe diameters allowed downstream of the meter free of any obstruction or fitting. If the flow is laminar then these distances should be doubled.

Alternatively, a flow conditioning device can be fitted at least 15 pipe diameters upstream of the flowmeter. A large variety of these is available (a few are illustrated in Fig. 6.9) and the right one must be selected for the type of flow distortion occurring. In general the more complex flow conditioners produce the better velocity profiles but they are generally the more expensive and give higher pressure drops.

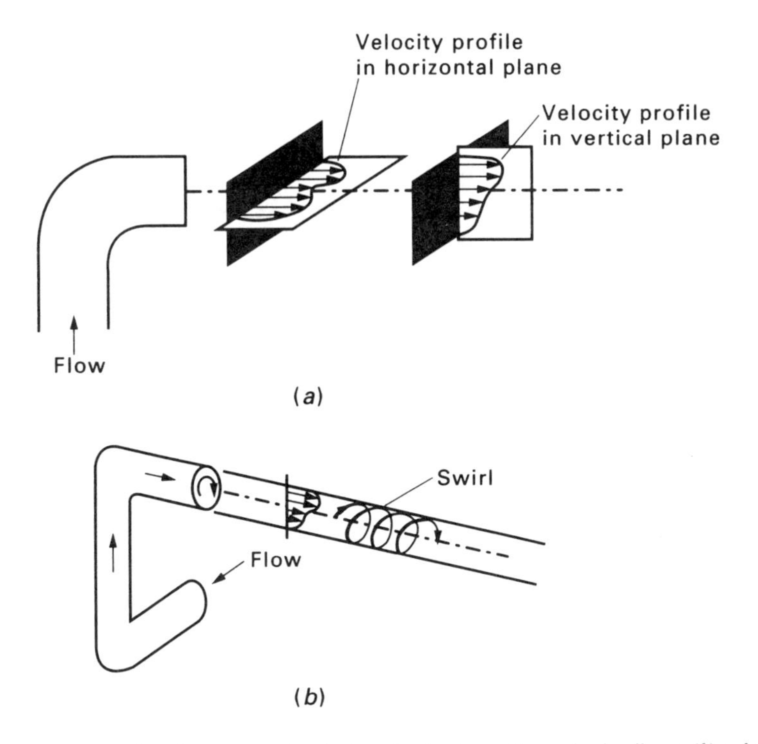

FIG. 6.8. Flow profile distortion: (*a*) velocity distribution after a single elbow; (*b*) velocity distribution and swirl following two 90° elbows in different planes

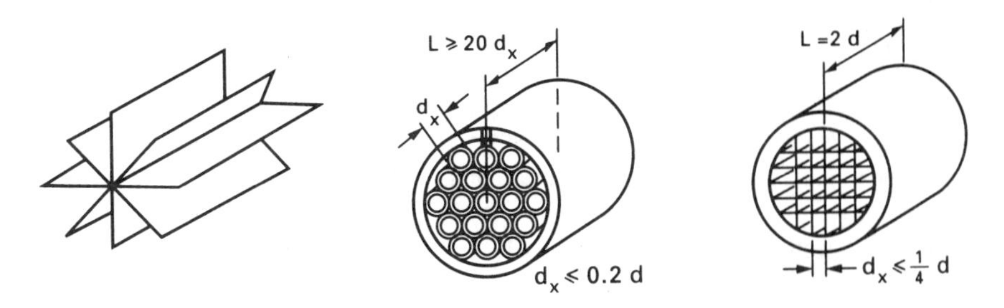

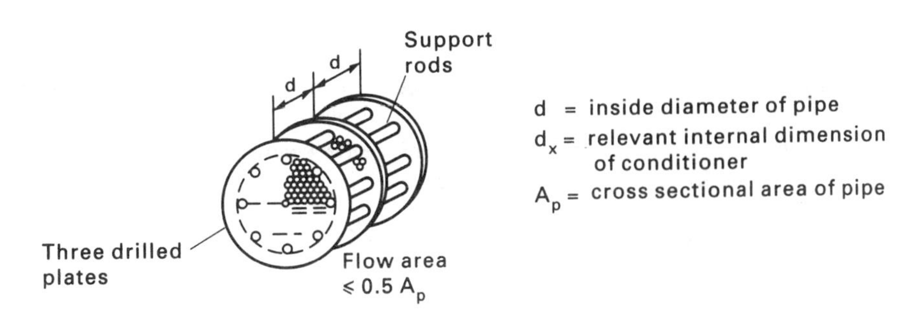

FIG. 6.9. Typical flow conditioners

6.3. THE MEASUREMENT OF PRESSURE

Pressure is measured extensively in the chemical processing industries and a wide variety of pressure measuring methods has been developed. Some of these have already been discussed in Volume 1, Section 6.2.2, viz. the manometer (which is an example of a gravity-balance type of meter), the Bourdon gauge (an example of an elastic transducer); and mention is made of the common first element in most pressure signal transmission systems—the differential pressure (DP) cell (Volume 1, Section 6.2.3). The latter also frequently forms part of a pneumatic transmission system and further discussion of this can be found in Section 6.3.4.

Reasons for pressure measurement in the process industries are three-fold. Firstly, any closed pipe or vessel will have a maximum safe operating pressure. Exceeding this can lead to the destruction of equipment and danger to personnel. In such cases, absolute and highly accurate readings of pressure are often less important than reliability, and it may be sufficient simply to install a high or low pressure limit alarm system. Secondly, the operating pressure of a process often has a significant effect upon other variables within the process and upon the specification of the product and, in such cases, it is usually necessary to have a consistent and accurate measurement including associated automatic control. Finally, pressure measurement can form part of the quality control procedures for a particular process. In this latter context care must be taken in the selection, installation and maintenance of the measuring equipment.

6.3.1. Classification of Pressure Sensors

SAUNDERS[16] has suggested that pressure sensors may be divided into four categories, viz. utility, general industry, process control and precision and test. Utility instruments (gauges) are found on many small pressure-related items (regulators, valve positioners, small pumps, etc.) and should be considered as little more than indicators. General industrial instruments are required to have a reliable design with accuracies of 1 to 2 per cent of full scale deflection (see Section 6.10.1). It is necessary for gauges which are employed as sensors for control systems to be of good quality and reliability, and to have accuracies of the order of 0.5 to 1 per cent of full scale.

HIGHAM[17] has listed three categories of pressure measurement, viz.:

(a) *Absolute pressure* which is the difference between the measured pressure and a perfect vacuum.
(b) *Gauge pressure* which is generally considered to be the difference between the measured pressure and local atmospheric pressure.
(c) *Differential pressure* which is the difference between two measured pressures, neither of which is local atmospheric pressure.

Units of gauge pressure are frequently employed in the process industries and are typically expressed in terms of kN/m^2 *gauge* which is a unit not defined in the SI system. Care must be exercised in interpreting the recorded pressure when there is no indication of the meter reading being gauge or absolute. Some more recently manufactured pressure sensors reported as displaying gauge pressure compare

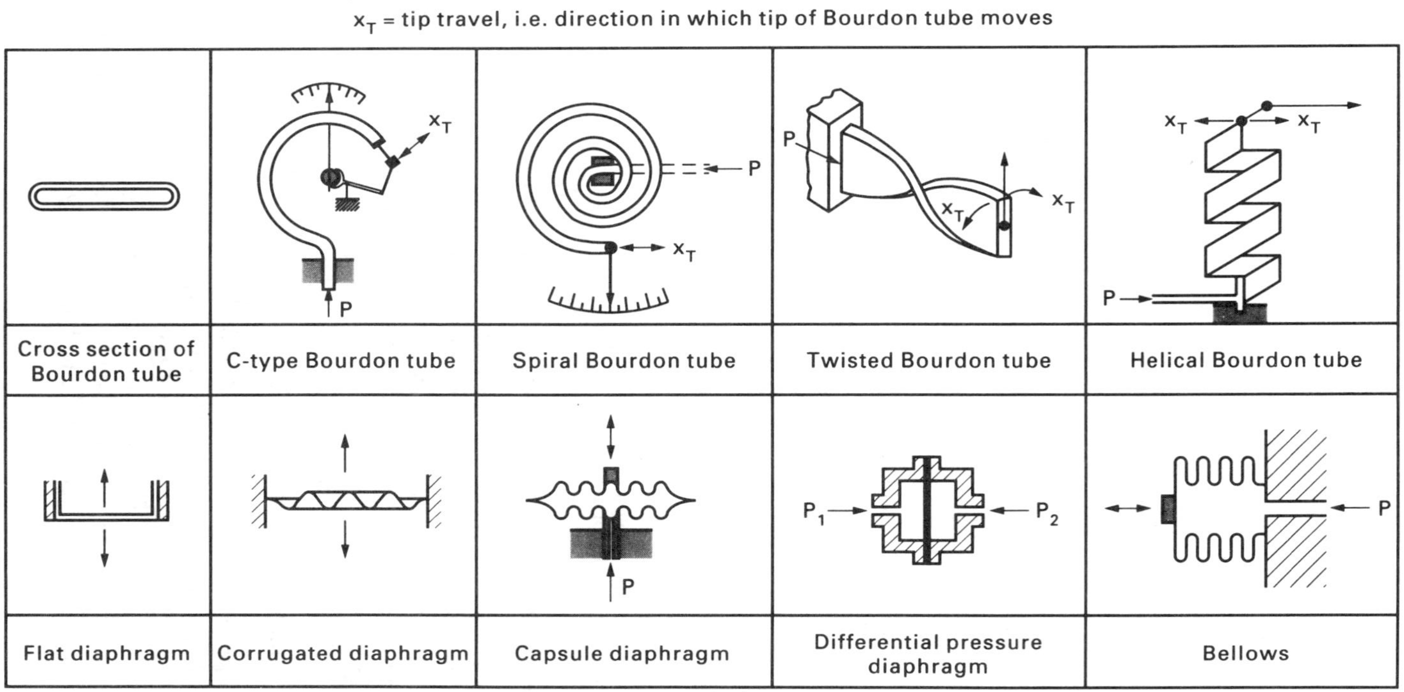

x_T = tip travel, i.e. direction in which tip of Bourdon tube moves

Cross section of Bourdon tube	C-type Bourdon tube	Spiral Bourdon tube	Twisted Bourdon tube	Helical Bourdon tube
Flat diaphragm	Corrugated diaphragm	Capsule diaphragm	Differential pressure diaphragm	Bellows

FIG. 6.10. Some elastic pressure transducers[4, 8, 18]

the measured pressure to a *fixed* pressure—usually the *standard atmosphere* ($101.325\ kN/m^2$).

6.3.2. Elastic Elements

Diaphragms and bellows used in the sensing of variations in pressure are illustrated in Fig. 6.10. Some common Bourdon gauge configurations are included for completeness. In each instance the elastic movement of the element is amplified by systems of linkages and/or gears or by conversion to an electrical signal. The principle of operation of the Bourdon tube lies in a property of its non-circular cross-section. Any increase in the pressure on the inside of the tube in comparison to the outside causes the tube to try to attain a circular cross-section which results in the tube attempting to untwist itself. The twisted and helical tubes are more sensitive than the C-type Bourdon element but are limited in terms of maximum pressure (Table 6.3). Increased sensitivity can also be achieved by linking a number of capsule diaphragms together.

TABLE 6.3. *Ranges of Some Basic Elastic Elements*[18]

Type of Element	Maximum Recommended Pressure (MPa)	Minimum Recommended Span[a] (kPa)
Bourdon C-type tube	700	30[b]
Bourdon twisted tube	7	
Diaphragm	7	0.05[c]
Bellows	6	1.2

[a] See Section 6.10.1.
[b] 100 kPa when measuring vacuum.
[c] For single diaphragm or capsule.

6.3.3. Electric Transducers for Pressure Measurement

Electrical output transducers exist in many forms and consist simply of elastic elements coupled with suitable displacement transducers such as capacitance, inductance and reluctance pick-ups, certain types of strain gauge, piezoelectric elements, potentiometers and eddy current probes. A selection of these is described in the following paragraphs.

Capacitive Pressure Sensors

A typical capacitive sensor is illustrated diagrammatically in Fig. 6.11. The element consists of a gas filled parallel plate capacitor in which one plate is a flat circular diaphragm whilst the other is a fixed metal disc. When the pressure on the outside surface of the flexible diaphragm is increased by an amount δP compared with the pressure within the unit, the diaphragm is deflected by an amount w at a radius r_1 (assuming the distorted section to be smoothly curved) given by (Volume 6, Section 13.3.5):

$$w = \frac{3}{16}\frac{(1-\nu^2)}{\mathscr{E}x_d^3}(r_2^2 - r_1^2)^2\,\delta P \qquad (6.13)$$

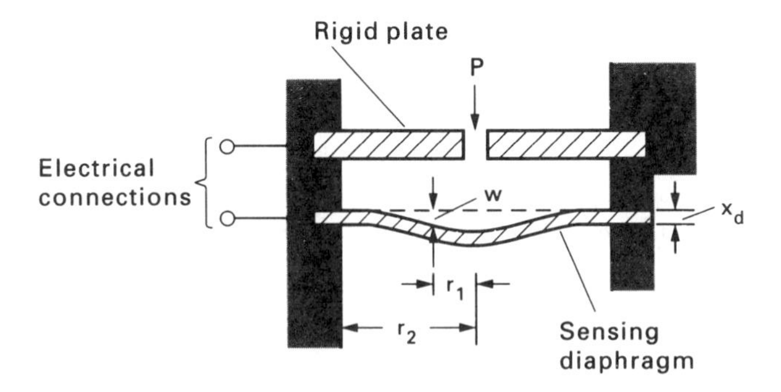

FIG. 6.11. Capacitive pressure sensor with flexible diaphragm

where r_2 is the radius of the diaphragm, $\mathscr{E}$ is Young's modulus, x_d is the thickness of the diaphragm and ν is Poisson's ratio [$\nu = -$ (transverse strain in the diaphragm)/(longitudinal strain in the diaphragm)]. The resulting increase in capacitance is[19]:

$$\delta\mathscr{C} = \frac{\varepsilon_0 \varepsilon_r \pi r_2^6 (1 - \nu^2)}{16 \mathscr{E} x_d^3 \delta_c^2}\, \delta P \tag{6.14}$$

where δ_c is the initial separation of the plates of the capacitor. Usually air is used as the dielectric (i.e. $\varepsilon_r \approx 1$). Capacitive sensing elements are employed in conjunction with either a.c. deflection bridges or with electrical oscillator circuits. Account should be taken of dielectric losses within the capacitor and any capacitance of the cable connecting the sensor to the measurement circuit. Alternative versions of this type of transducer either vary the area of the plates of the capacitor or the displacement of a solid dielectric section situated between the plates.

Capacitive pressure transducers are traditionally used for relatively low pressure ranges (up to about 800 kPa) and for vacuum measurement (Section 6.3.5).

Inductive Pressure Transducers

These employ a magnetically permeable member to increase and decrease respectively the inductances in the coils of a two coil inductance bridge. Changes in inductances are additive due to the way in which the coils are connected within the bridge circuit (Fig. 6.12).

It can be shown that, for such a transducer, the inductance $\mathscr{L}$ of each coil is given by[8]:

$$\mathscr{L} = \frac{\mathscr{L}_0}{1 + k_l x_s} \tag{6.15}$$

where $\mathscr{L}_0$ is the reference inductance (at $x_s = 0$); x_s is a measure of the distance moved by the core of the inductor (the *stroke*) and k_l is a constant determined by the geometry of the transducer.

Such variable inductance transducers are available with strokes of 0.0025 to 0.5 m. The non-linearity of the element can range from 0.02 to 1 per cent with a sensitivity

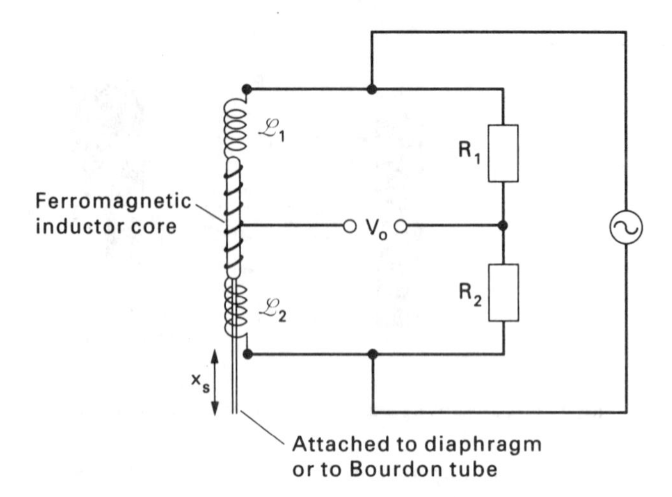

FIG. 6.12. Inductive pressure transducer

of between 200 and 1600 V/m. This type of transducer is normally coupled with a diaphragm or a Bourdon tube as the pressure sensor.

An associated type of transducer is the *Linear Variable Differential Transformer (LVDT)* which is essentially a transformer with a single primary winding and two identical secondary windings wound on a tubular ferromagnetic former. The primary winding is energised by an a.c. source (see Fig. 6.13).

The two secondary windings are connected in series opposition so that the output voltage is the difference of the voltages in the secondary windings (i.e. $V_o = V_1 - V_2$, $V_1 > V_2$). A ferromagnetic core moves inside the primary coil and thus varies the mutual inductance between the primary and secondary coils which in turn varies the secondary voltages. The displacement of the pressure sensor (capsule, bellows or

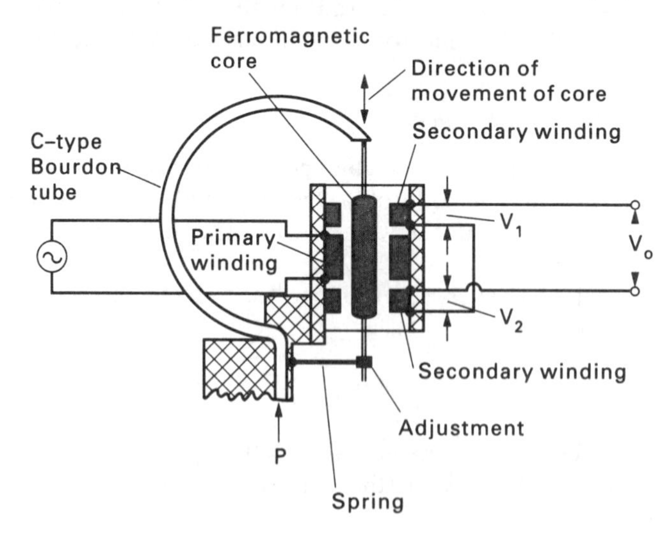

FIG. 6.13. Linear variable differential transformer (LVDT) using C-type Bourdon tube as the pressure sensor

Bourdon tube) is reflected in the position of the ferromagnetic core of the LVDT and is thus a function of the difference between the input and output voltages measured in terms of phase shift and amplitude ratio (see Section 7.8.4).

Commercial LVDTs have a full range stroke from 0.1 to 80 mm with a sensitivity of about 25 to 1250 V/m depending upon the frequency of excitation and size of stroke.

Reluctive Pressure Transducers

These are often associated with inductive transducers. However, in this instance the primary variable is the reluctance of a magnetic circuit. If a magnetic flux is induced in a magnetic circuit (e.g. a loop of ferromagnetic material) then, by analogy with the concept of an electromotive force driving a current through an electrical resistance, we can imagine a magnetomotive force F_{mm} driving a magnetic flux ϕ through a magnetic reluctance $\mathcal{S}$, where:

$$F_{mm} = \phi \mathcal{S} \tag{6.16}$$

If ϕ is induced by passing a current i through N_T turns of a coil wound on the ferromagnetic loop (Fig. 6.14*a*), then:

$$\phi = \frac{F_{mm}}{\mathcal{S}} = \frac{N_T i}{\mathcal{S}} \tag{6.17}$$

This is the flux linked by one turn of the coil. Hence, the total flux ϕ_T linked by N_T turns is:

$$\phi_T = N_T \phi = \frac{N_T^2 i}{\mathcal{S}} = \frac{N_T^2 i \mu_0 \mu_r A}{\ell} \tag{6.18}$$

where ℓ is the total length of the flux path, μ_r is the relative permeability of the loop material, μ_0 is the permeability of free space ($4\pi \times 10^{-7}$ Henry/m), A is the cross-sectional area of the flux path, and $\mathcal{S} = \ell / \mu_0 \mu_r A$[20].

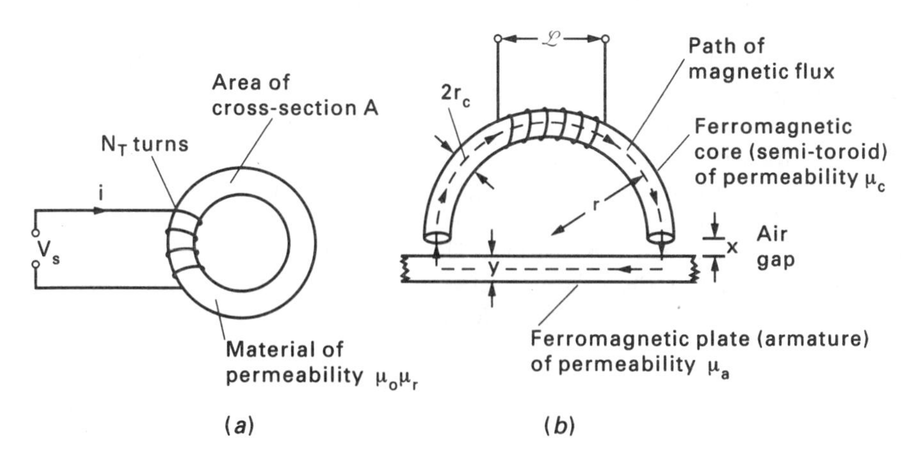

FIG. 6.14. Reluctive transducer: (*a*) basic circuit; (*b*) transducer with air gap and moveable armature

The principle of the reluctive transducer is shown in Fig. 6.14*b*. The total reluctance in the magnetic circuit is:

$$\mathcal{S}_T = \mathcal{S}_{\text{core}} + \mathcal{S}_{\text{air gap}} + \mathcal{S}_{\text{armature}}$$

$$= \frac{\pi r}{\mu_0 \mu_c \pi r_c^2} + \frac{2x}{\mu_0 \pi r_c^2} + \frac{2r}{\mu_0 \mu_a (2ay)} \tag{6.19}$$

where r is the radius of curvature of the semi-toroidal ferromagnetic core, r_c is the radius of the cross-section of the toroid, x is the width of the air gap and y is the thickness of the movable armature.

As $\mathcal{S}_{\text{air gap}} >> \mathcal{S}_{\text{core}}$ or $\mathcal{S}_{\text{armature}}$, then a small change in x will produce a large variation in $\mathcal{S}_T$. The self-inductance $\mathcal{L}$ of the coil in Fig. 6.14*a* is the total flux linked per unit current. Thus, from equation 6.18:

$$\mathcal{L} = \frac{\phi_T}{i} = \frac{N_T^2}{\mathcal{S}_T} \tag{6.20}$$

Consequently, a change in reluctance will produce a variation in $\mathcal{L}$, where, from equation 6.15:

$$\mathcal{L} = \frac{\mathcal{L}_0}{1 + k_I x} = \frac{N_T^2}{\mathcal{S}_T} \tag{6.21}$$

To overcome the non-linearity of equation 6.21, the transducer is generally designed on a push–pull basis and incorporated into an a.c. deflection bridge.

[Note the difference in operation of the inductive transducer (Fig. 6.12) where the core of an inductor is displaced linearly by a distance x_s along the axis of the coil and the reluctive sensor (Fig. 6.14) which responds to a small displacement x of the armature.]

Strain Gauges

The conversion of alterations in pressure into changes in resistance due to strain in two or four arms of a Wheatstone bridge network has been used in commercial pressure transducers for many years. Modern transducers generally employ *deposited metal film* (*thin film*) strain gauges applied either directly to the inside of a diaphragm pressure sensor (Fig. 6.15*a*) or to a secondary sensing member that is moved by a diaphragm via a mechanical linkage.

The principle of the resistance strain gauge(21) is that the electrical resistance of a conductor will change when it is stretched or compressed due to the consequent variation in its physical dimensions. There is an additional effect called the *piezoresistance* which is the relation between the resistivity ρ' of the material and the mechanical strain. The resistance R of a conductor of area of cross-section A and length x is given by:

$$R = \rho' \frac{x}{A} \tag{6.22}$$

When R changes by an amount δR due to equivalent changes in ρ', x and A (Fig. 6.15*b*), then:

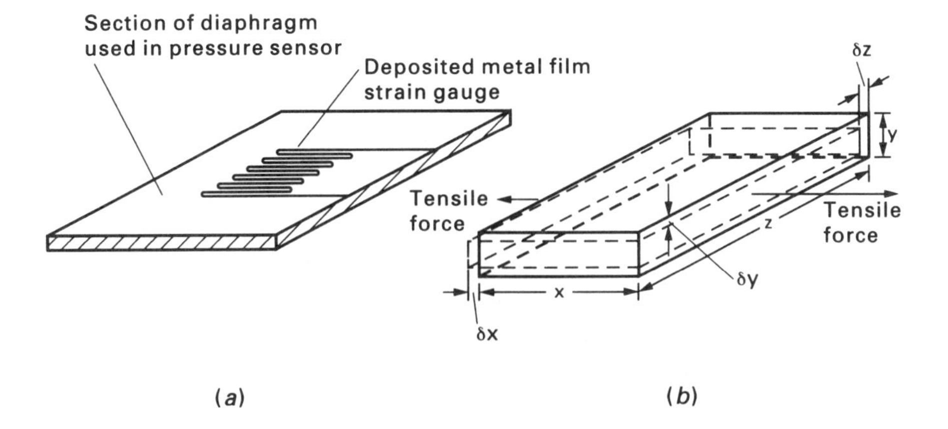

FIG. 6.15. Basis of strain gauge pressure transducer: (*a*) thin-film strain-gauge used as pressure sensor; (*b*) deformation of conductor under tensile stress

$$\delta R = \left(\frac{\partial R}{\partial x}\right)(\delta x) + \left(\frac{\partial R}{\partial A}\right)(\delta A) + \left(\frac{\partial R}{\partial \rho'}\right)(\delta \rho')$$

$$= \frac{\rho'}{A}(\delta x) - \frac{\rho' x}{A^2}(\delta A) + \frac{x}{A}(\delta \rho') \quad (6.23)$$

Dividing through by $R = \rho' x/A$ gives:

$$\frac{\delta R}{R} = \frac{\delta x}{x} - \frac{\delta A}{A} + \frac{\delta \rho'}{\rho'} \quad (6.24)$$

$\frac{\delta x}{x}$ is the longitudinal strain e_x in the conductor and $A = zy$ for a rectangular cross-section (Fig. 6.15*b*).

Hence:

$$\delta A = z\delta y + y\delta z$$

Therefore:

$$\frac{\delta A}{A} = \frac{\delta y}{y} + \frac{\delta z}{z} = 2e_T \quad (6.25)$$

where δy, δz are the equivalent changes in y and z respectively and e_T is the transverse strain in the element.

Thus:

$$\frac{\delta R}{R} = e_x - 2(-\nu e_x) + \frac{\delta \rho'}{\rho'}$$

$$= (1 + 2\nu)e_x + \frac{\delta \rho'}{\rho'}$$

$$= \mathcal{G} e_x \quad (6.26)$$

with the *gauge factor* $\mathcal{G} = \frac{\delta R/R_0}{e_x}$.

i.e.:

$$\mathcal{G} = \frac{\delta R/R_0}{\delta x/x} = \underset{\substack{\text{Resistance}\\\text{change due to}\\\text{length change}}}{1} + \underset{\substack{\text{Resistance}\\\text{change due to}\\\text{area change}}}{2\nu} + \underset{\substack{\text{Resistance}\\\text{change due to}\\\text{piezoresistance}\\\text{effect}}}{\frac{1}{e_x}\frac{\delta\rho'}{\rho'}} \tag{6.27}$$

A typical resistance strain gauge has a gauge factor of about 2, an unstrained resistance of 120 Ω and a maximum change in resistance of approximately 7 Ω between maximum compression and maximum tension. The strain gauge is placed in one arm of a Wheatstone bridge network—usually of the null-balance type so that the gauge carries no current. In order to compensate for temperature effects a second identical gauge in an unstressed diaphragm is attached to the other arm of the bridge. A more effective arrangement is to use all four arms of the Wheatstone bridge by placing a strain gauge in tension in one arm, an equivalent gauge in compression in a second arm, and unstressed gauges in the remaining two arms.

Semiconductor strain gauges have a much larger piezoresistance effect leading to gauge factors of between 100 and 175 for 'P' type material and between −100 and −140 for 'N' type material*. These consequently are much more sensitive to changes in strain than the metal resistance types. On the other hand, they are affected to a greater extent by variations in temperature.

Piezoelectric Sensing Elements

If a force is applied to a crystal, the atoms in the crystal lattice will be displaced by an amount which is proportional to the applied force.

A typical piezoelectric pressure transducer consists of a pressure sensing diaphragm acting against a stack of quartz discs which produces an output signal in the form of an electric charge (Fig. 6.16). This effect is reversible in that, if a charge is applied, then the material will mechanically deform. The discs form part of a capacitor constructed by attaching electrodes to selected faces of the material. Hence, any charge Q created will be reflected in terms of a voltage across the assembly according to $V = Q/\mathcal{C}$ where $\mathcal{C}$ is the capacitance of the transducer. The charge does slowly leak away through the very high leakage resistance of the quartz. Thus the transducer is much better suited to measuring variations in pressure rather than static pressure. However, this type of sensor is gaining in popularity due to its very fast response (typically 10^{-6}s rise time and 15 per cent overshoot in response to a step change in pressure—see Section 7.11).

Potentiometric Transducers

These operate by sliding a movable contact or *wiper* over a resistance element. The contact motion can be translational, rotational or a combination of both depending upon the device providing the displacement (e.g. Bourdon tube, diaphragm, etc.). If no current is drawn from the meter then the output voltage V_0 will

* For an explanation of this difference between semiconductor materials see Ref. 20 or any good elementary text describing the basic principles of electronics.

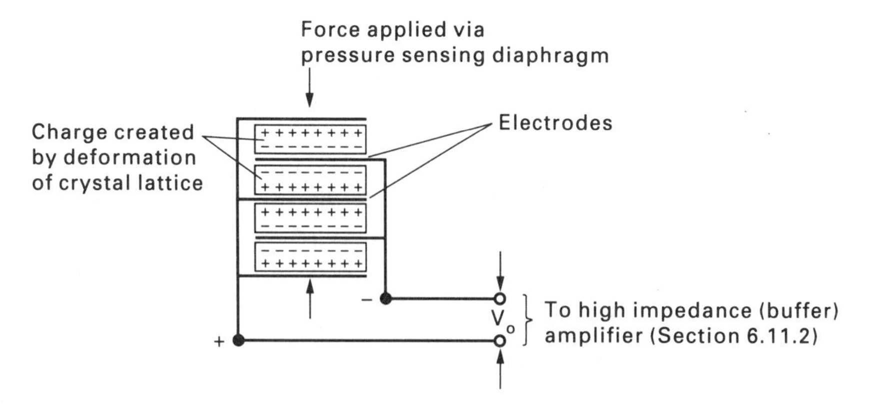

FIG. 6.16. Principle of the piezoelectric transducer

be directly proportional to the motion of the contact moving over the resistance element (provided that the distribution of resistance in the latter is uniform).

The usual situation, however, is that current is drawn from the element to operate a recorder and/or some further element within a control system. Under these conditions the relation between the displacement x of the contact (i.e. the change in process pressure) and V_o will be non-linear due to loading effects (Section 6.11.6) unless the resistance of the load R_L is much greater than that of the potentiometer R_{AB}.

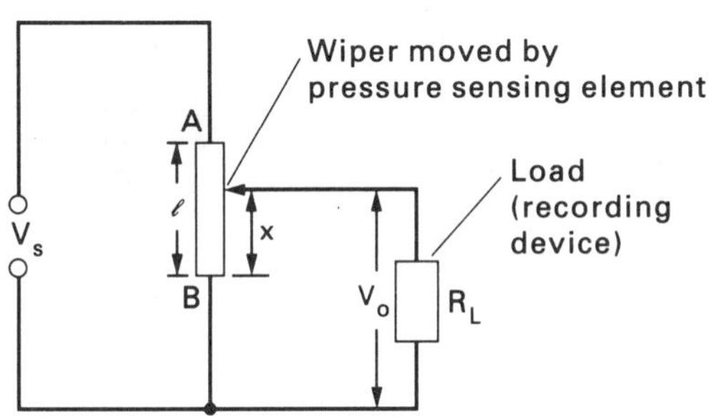

FIG. 6.17. Translational potentiometric transducer

From Fig. 6.17:

$$\frac{V_o}{V_s} = \frac{1}{\dfrac{\ell}{x} + \dfrac{R_{AB}}{R_L}\left(1 - \dfrac{x}{\ell}\right)} \tag{6.28}$$

where ℓ is the length of the potentiometer slide-wire. Hence, if $R_L >> R_{AB}$, we have the linear relationship:

$$\frac{V_o}{V_s} \simeq \frac{x}{\ell} \tag{6.29}$$

although under these conditions the sensitivity of the combination of sensor and recorder will be relatively low.

The resolution of potentiometric transducers is dependent upon the construction of the resistance element. In the case of a *wire-wound* resistance, in order to obtain a high resistance in a small space, the resistance wire is wound on to a *mandrel* or card which is straight or formed into a circle or helix depending upon the motion of the contact. This limits the resolution of the transducer as the wiper moves from one wire to the next on the mandrel. The best resolution that can be obtained is about 0.01 per cent (see Section 6.10.1). Typical wire-wound potentiometers have strokes of between 0.0025 m and 0.5 m and rotational versions from about 10° of arc to 50 turns. An alternative often employed is the *conductive plastic film* element. This provides a continuous resistance element and thus, a zero resolution, but such elements suffer from a higher temperature coefficient of resistance. A more recent development is a combination of earlier types in which a conductive plastic coating is sprayed on to a wire-wound resistor.

Vibrating Element Pressure Transducers

The principle upon which these are based is a change in the resonant frequency of a vibrating section—commonly a wire or a cylinder. A schematic diagram of the wire version is shown in Fig. 6.18.

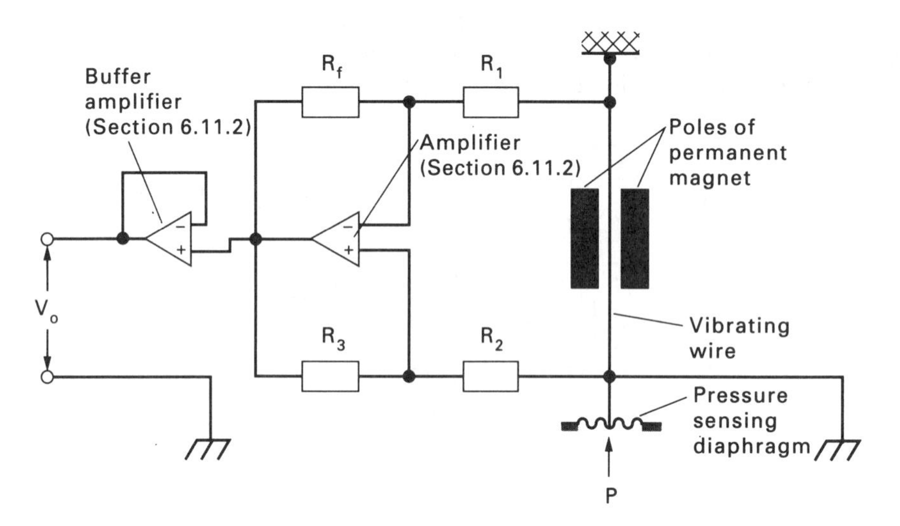

FIG. 6.18. Vibrating wire pressure transducer

A very thin wire (typically made of tungsten) is attached to the centre of a pressure sensing diaphragm. The fixed end of the wire is insulated and the whole is located within a magnetic field provided by a permanent magnet. When a current is passed through the wire it moves in this field sufficiently to induce an electromotive force and an associated current within the wire. This induced emf is amplified and applied back across the wire in order to sustain its oscillation. The design is generally such that an increase of pressure produces a decrease in the tension in the wire and thus a reduction in its angular frequency of oscillation ω according to[22]:

$$\omega = \frac{\pi}{\ell n}\sqrt{\frac{F}{A\rho}} \tag{6.30}$$

where n is the integer mode of frequency, ℓ is the length of the wire, A is the area of cross-section of the wire, ρ is the density of the material of the wire and F is the tensile force acting on the wire.

The vibrating cylinder version of this type of transducer employs a sensing element of the straight tube type and is similar to the vibrating cylinder density meter (Section 6.6.1). The cylinder is maintained in oscillation by a feedback amplifier/limiter combination and its frequency of oscillation varies with the pressure of the fluid within it, which alters the hoop stress on the cylinder. As with the vibrating wire, the frequency of oscillation increases with the fluid pressure. In order to be able to use the sensor for the measurement of absolute pressures, it is surrounded with a cylindrical housing and the annular space between the two is evacuated. Such sensors give excellent repeatability and have a relatively low temperature sensitivity.

6.3.4. Differential Pressure Cells

Differential pressure transmitters (or *DP cells*) are widely used in conjunction with any sensor that produces a measurement in the form of a pressure differential (e.g. orifice plate, venturi meter, flow nozzle, etc.). This pressure differential is converted by the DP cell into a signal suitable for transmission to a local controller and/or to the control room. DP cells are often required to sense small differences between large pressures and to interface with difficult process fluids. Devices are available that provide pneumatic, electrical or mechanical outputs.

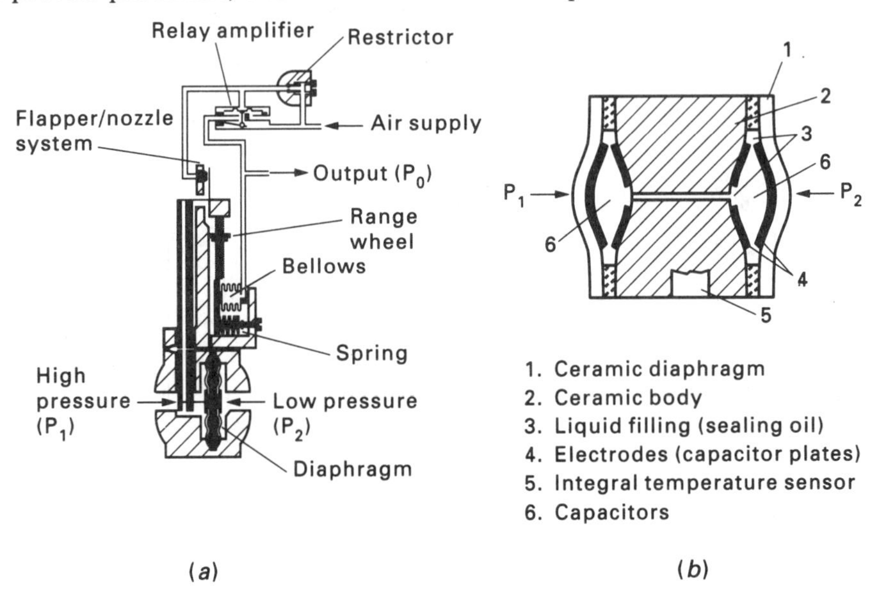

FIG. 6.19. Differential pressure cells (transmitters): (*a*) pneumatic cell; (*b*) capacitive ceramic cell

A typical DP cell is shown in Fig. 6.19a[23]. The associated flapper/nozzle system works on a force-balance principle (see Volume 1, Section 6.2.3). The output pressure P_0 is linearly related to the difference in pressure $(P_1 - P_2)$, thus:

$$P_0 = K(P_1 - P_2) \tag{6.31}$$

K is generally large and such systems exhibit good linearity and accuracy.

Figure 6.19b is an illustration of a DP cell which operates by varying the distance between the plates of two adjacent capacitors (see also Fig. 6.11)[24]. The high and low pressure signals from the sensor are applied to ceramic diaphragms to which one plate of each capacitor is attached. The subsequent change in the separation of the capacitor plates produces a variation in capacitance which is detected by incorporating the cell into a capacitance bridge, as described in Section 6.5.3 (Fig. 6.31b).

The installation and positioning of the DP cell relative to the sensor is often dependent upon the nature of the fluid stream being measured. Some of the more common forms of installation are shown in Fig. 6.20.

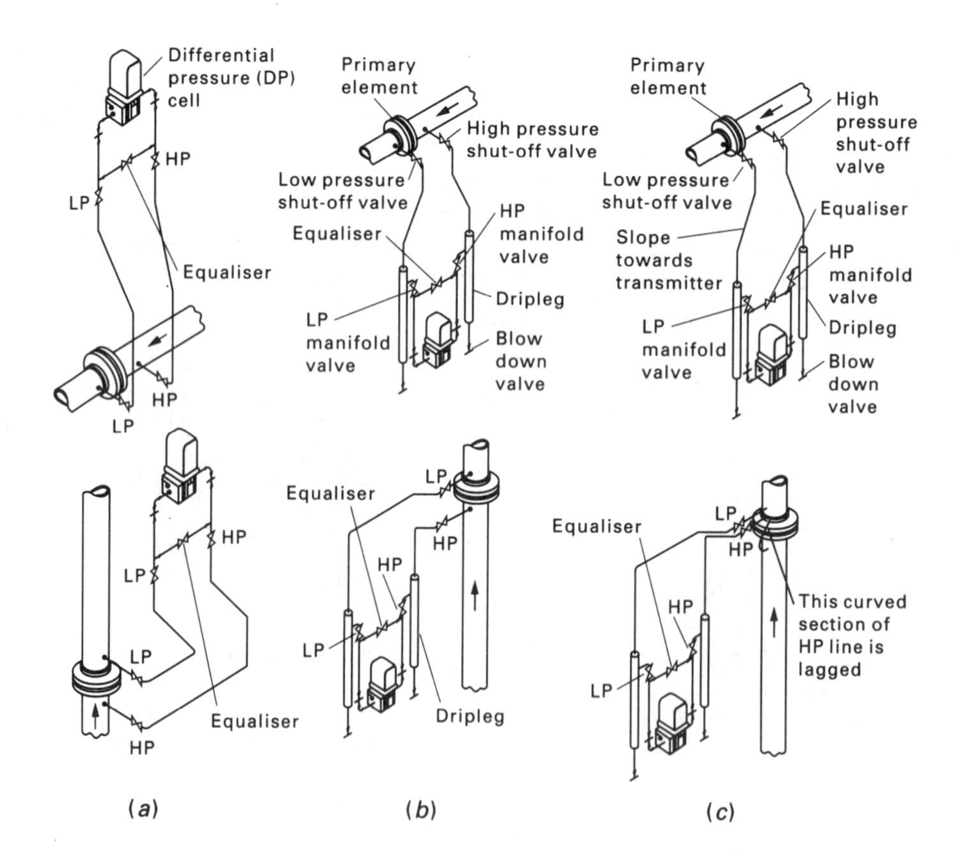

FIG. 6.20. Installation of differential pressure cells (transmitters): (a) arrangement for non-corrosive dry gas and non-corrosive liquid; (b) piping for wet gas (non-condensing) and liquids with solids in suspension; (c) arrangement for steam and condensable vapours

6.3.5. Vacuum Sensing Devices

These are considered separately as vacuum sensing devices differ substantially from other types of pressure sensor.

Two commonly used units of vacuum measurement are the *torr* and the *micrometer*. One torr is 1/760 of one atmosphere (or 1 mm Hg) at standard conditions and, hence, one torr = 133 N/m^2 and one micrometer = 10^{-3} torr = 0.133 N/m^2. Ranges of vacuum are conventionally divided up as follows[25]:

Low vacuum: 760 to 25 torr (100 to 3 kN/m^2)
Medium vacuum: 25 to 10^{-3} torr (3 to 10^{-4} kN/m^2)
High vacuum: 10^{-3} to 10^{-6} torr (10^{-4} to 10^{-7} kN/m^2)
Very high vacuum: 10^{-6} to 10^{-9} torr (10^{-7} to 10^{-10} kN/m^2)
Ultra high vacuum: below 10^{-9} torr (below 10^{-10} kN/m^2).

Elastic elements such as diaphragms and capsules are used as sensing elements in combination with different displacement transducers (e.g. inductive, potentiometric, etc.) for low/medium vacuum measurements down to about 1 torr. Elastic elements have been employed in conjunction with capacitive transducers for the measurement of pressures as low as 10^{-5} torr.

Hot Wire or Thermal Conductivity Gauges

These measure the change in thermal conductivity of a gas due to variations in pressure—usually in the range 0.75 torr (100 N/m^2) to 7.5×10^{-4} torr (0.1 N/m^2). At low pressures the relation between pressure and thermal conductivity of a gas is linear and can be predicted from the kinetic theory of gases. A coiled wire filament is heated by a current and forms one arm of a Wheatstone bridge network (Fig. 6.21). Any increase in vacuum will reduce the conduction of heat away from the filament and thus the temperature of the filament will rise so altering its electrical resistance. Temperature variations in the filament are monitored by means of a thermocouple placed at the centre of the coil. A similar filament which is maintained at standard conditions is inserted in another arm of the bridge as a reference. This type of sensor is often termed a *Pirani gauge*.

Ionisation Gauges

There are two principal types of ionisation gauge, viz. the *hot cathode* type in which electrons are emitted by a heated filament, and the *cold cathode* type in which electrons are released from the cathode by the impact of ions. In both cases the vacuum is measured in terms of the ion current. The electrons are accelerated by a potential difference (usually about 2000 V) across the ionisation tube (see Fig. 6.22). Positively charged ions are formed by the electrons striking gas molecules. The number of positive ions produced is a function of the gas density (i.e. the pressure) and the electron current i_e which is normally held constant. The ions are collected at a negatively charged electrode and the resulting ion current i_i is a direct measure of the gas pressure. The hot cathode version is the most sensitive of the two and can be used to measure vacua down to about 10^{-10} torr ($\approx 10^{-8}$ N/m^2).

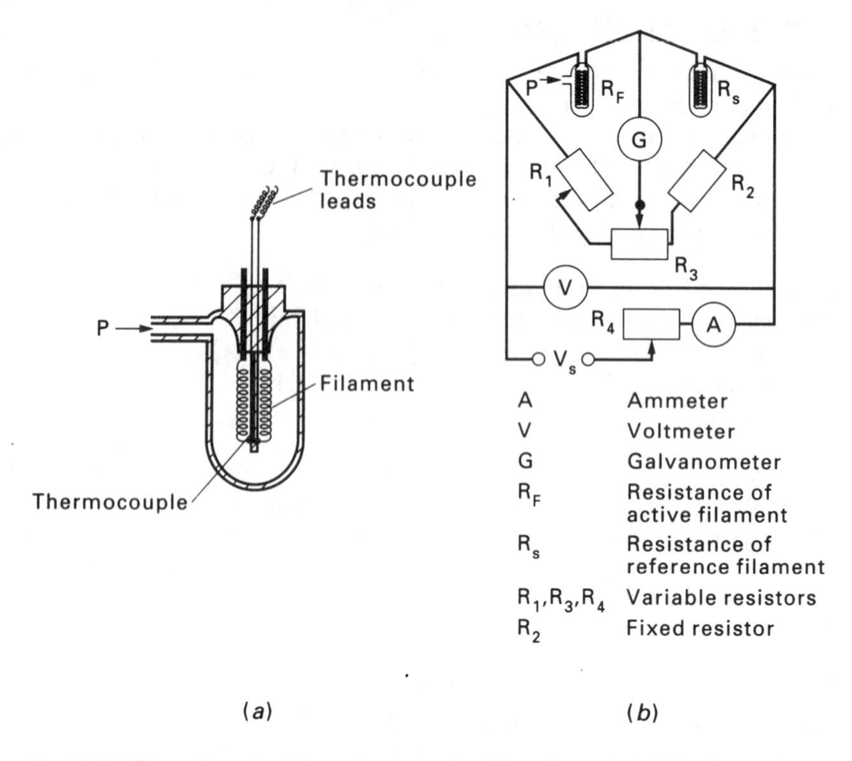

FIG. 6.21. Thermal conductivity vacuum-sensing (Pirani) gauge: (*a*) Pirani transducer; (*b*) typical bridge circuit for Pirani gauge

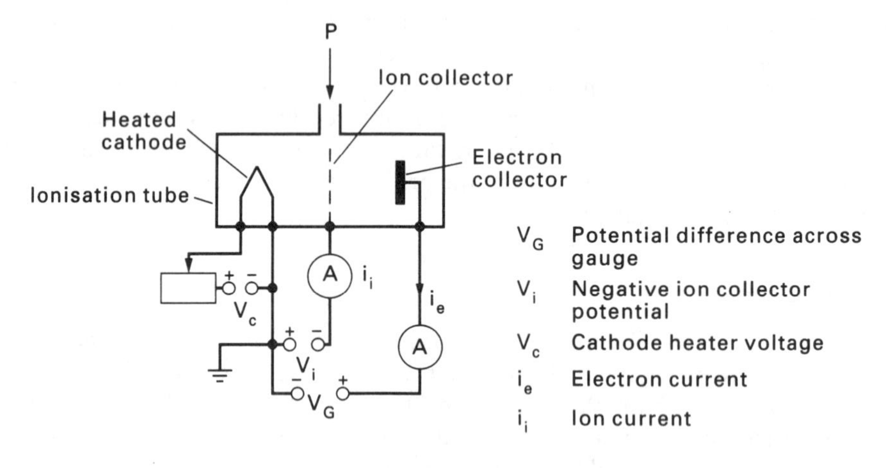

FIG. 6.22. Hot cathode ionisation gauge

6.4. THE MEASUREMENT OF TEMPERATURE

The fundamental meaning of temperature may be described in terms of the zeroth law of thermodynamics. This states that *when two bodies are each in thermal equilibrium with a third body they are then in thermal equilibrium with each other*, i.e.

all three bodies are said to be at the same temperature. Any temperature scale (and the numbers assigned to it) which is based upon one of the physical properties of materials which vary with temperature is, by nature, arbitrary and is, in principle, as good as any other scale based upon another of these properties. In practice, the properties employed will only be those which have easily observable thermometric characteristics, e.g. the pressure of a gas in a closed vessel, the length of a column of mercury in a capillary tube, or the resistance of a platinum wire. It is important that such a temperature scale should be *reproducible* (i.e. will give consistently the same values under the same conditions) and, in order that this should be so, the following must be specified:

(i) the particular thermometric substance and property employed,
(ii) the numbers to be attached to two fixed points, and
(iii) the relation between the temperature and the thermometric property which is to be used for interpolating between the fixed points and extrapolating beyond them.

If the temperature is considered to be a linear function of the thermometric property, then:

$$\theta = a_\theta X + b_\theta \tag{6.32}$$

where θ is the temperature and X is the value of the thermometric property. The constants a_θ and b_θ are determined by the numbers assigned to the fixed points, e.g. on the Celsius scale, a_θ and b_θ can be determined by solving,

$$0 = a_\theta X_1 + b_\theta \quad \text{and} \quad 100 = a_\theta X_2 + b_\theta \tag{6.33}$$

where X_1 and X_2 are the values of the thermometric property at the ice and steam points. Hence from equations 6.32 and 6.33:

$$\theta(^\circ\text{C}) = \frac{100(X - X_1)}{X_2 - X_1}. \tag{6.34}$$

A temperature measured on one scale can always be converted into a temperature measured on another scale (e.g. Fahrenheit) with mathematical *precision* (i.e. to as many significant figures as is required—see Section 6.10.1).

It is important to consider what happens when the substance or property employed is changed. For example, for a liquid in glass thermometer (using the Centigrade scale):

$$\theta(^\circ\text{C}) = \frac{100(\ell - \ell_1)}{\ell_2 - \ell_1} \tag{6.35}$$

where ℓ is the length of the column of liquid at temperature θ°C. For a constant volume gas thermometer:

$$\theta(^\circ\text{C}) = \frac{100(P - P_1)}{P_2 - P_1} \tag{6.36}$$

where P is the pressure at θ°C.

If linear interpolation is applied in this way then two thermometers using different thermometric properties brought into contact with the same heat reservoir will give identical readings only at the fixed points.

e.g. for the same heat source we may obtain:

Constant volume hydrogen thermometer	40°C
Mercury-in-glass thermometer	40.11°C
Platinum resistance thermometer	40.36°C.

An absolute scale of temperature can be designed by reference to the Second Law of Thermodynamics, viz. the *thermodynamic temperature scale*, and is independent of any material property. This is based on the Carnot cycle and defines a temperature ratio as:

$$\frac{T_1}{T_2} = \frac{Q_{H,1}}{Q_{H,2}} \tag{6.37}$$

where $Q_{H,1}$ and $Q_{H,2}$ are the quantities of heat absorbed and rejected respectively by a reversible engine operating between reservoirs at T_1 and T_2. Hence, if a number is selected to describe the temperature of a chosen fixed point, then the temperature scale is completely defined. Currently, this fixed point is the *triple point* of water (273.16 K). The thermodynamic temperature scale is not, in fact, usable in practice as it is based on the ideal Carnot cycle. However, it can be shown that a temperature scale defined by a constant volume or constant pressure gas thermometer employing an ideal gas is identical to the thermodynamic scale. In practice, real gases have to be used in gas thermometers, and the readings must be corrected for deviations from ideal gas behaviour. The determination of a temperature on any thermodynamic scale is a tedious and time-consuming procedure and the International Practical Temperature Scale (IPTS) has been developed to conform as closely as possible to the thermodynamic scale. For this, apart from the triple point of water (at which the two scales are in precise agreement), five other primary fixed points are specified, viz:

Boiling point of liquid oxygen	−182.96°C	(90.17 K)
Boiling point of water	100.00°C	(373.13 K)
Freezing point of zinc	419.58°C	(629.71 K)
Freezing point of silver	961.93°C	(1235.06 K)
Freezing point of gold	1064.43°C	(1337.56 K).

For interpolation between these points, the IPTS specifies the use of specific measuring instruments, equations and procedures[26]. The highest reproducibility of the IPTS is at the triple point of water. The accuracy falls at lower or higher temperatures than the triple point.

6.4.1. Thermoelectric Sensors

The Thermocouple

In 1821 Seebeck discovered that, in an electric circuit consisting of two different materials X and Y in the form of wires, when the two junctions are at different temperatures θ_1 and θ_2 a potential exists at the terminals on open circuit and, if the circuit is closed, a current flows[27]. At each junction there exists a *contact* potential $E_{XY}^{\theta_1}$ and $E_{XY}^{\theta_2}$ respectively which depends on the type of metal employed and the temperature of the junction. When the system is on closed circuit the electromotive force (emf) is given by:

$$E_{XY}^{\theta_1,\theta_2} = E_{XY}^{\theta_1} - E_{XY}^{\theta_2} = \beta_1(\theta_1 - \theta_2) + \beta_2(\theta_1^2 - \theta_2^2) + \ldots\ldots \qquad (6.38)$$

where β_1 and β_2 etc. are coefficients determined by the types of metal employed to construct the junctions.

DOEBELIN[4] lists five laws of thermocouple behaviour:

(i) The emf of a given thermocouple depends only upon the temperatures of the junctions and is independent of the temperatures of the wires connecting the junctions.

(ii) If a third metal Z is inserted between X and Y then, provided that the two new junctions are at the same temperature (θ_3), the emf is unchanged.

(iii) If a third metal Z is inserted between X and Y then, provided that the two new junctions XZ and ZY are both at the same temperature (θ_1 or θ_2), the emf is unchanged.

(iv) If the emf obtained using the metals X and Y is E_{XY} and that using metals Y and Z is E_{YZ} then the emf obtained employing X and Z will be:

$$E_{XZ} = E_{XY} + E_{YZ} \qquad (6.39)$$

(the junction temperature being θ_1 and θ_2 in each case). This is called the *Law of Intermediate Metals.*

(v) If a thermocouple produces an emf $E_{XY}^{\theta_1,\theta_2}$ when its junctions are at θ_1 and θ_2 respectively and $E_{XY}^{\theta_2,\theta_3}$ when its junctions are at θ_2 and θ_3 then it will produce an emf:

$$E_{XY}^{\theta_1,\theta_3} = E_{XY}^{\theta_1,\theta_2} + E_{XY}^{\theta_2,\theta_3} \qquad (6.40)$$

when the junctions are at θ_1 and θ_3. This is termed the *Law of Intermediate Temperatures.*

Law (i) is important in industrial applications where the leads joining measurement and reference junctions may well experience large changes in ambient temperature. Law (ii) means that there will be no change in emf when a voltmeter is placed in the circuit. Law (iii) allows the brazing or soldering of the wires at the measurement junctions without affecting the emf. The use of law (v) is illustrated in the following example.

Example 6.1

The temperature of a gas oil product flowing through a pipe is monitored using a chromel/alumel thermocouple. The measurement junction is inserted into the pipe and the reference junction is placed in the plant control room where the temperature is 20°C. The emf at the thermocouple junction is found to be 6.2 mV by means of a potentiometer connected into the thermocouple circuit adjacent to the reference junction. Find the measured temperature of the gas oil.

Solution

Standard thermocouple tables (BS 4937)[28] can be employed to find the measured temperature, but these require a reference temperature of 0°C. However, using equation 6.40:

$$E^{\theta_1,\theta_3} = E^{\theta_1,\theta_2} + E^{\theta_2,\theta_3}$$

Thus we can write:

$$E^{\theta_1,0} = E^{\theta_1,\theta_2} + E^{\theta_2,0}$$

From tables, for a chromel/alumel junction at 20°C, $E^{\theta_2,0} = 0.8$ mV.

$$E^{\theta_1, 0} = 6.2 + 0.8 = 7.0 \text{ mV}.$$

Hence, from tables: $\theta_1 = \underline{\underline{177°C}}$.

Thermojunctions may be formed by welding, soldering or pressing the materials together. Such junctions give identical emfs (by law (iii)), but may well produce different currents as the contact resistance will differ depending on the joining process utilised. Whilst many materials exhibit thermoelectric effects, only a small number are employed in practice. The characteristics of the more common thermocouple materials are listed in Table 6.4.

A typical thermocouple installation for an industrial application is shown in Fig. 6.23. Instead of placing the reference junction in a temperature controlled environment (which is often inconvenient), an *automatic reference junction compensation circuit* is fitted. This provides a second source of emf $E^{\theta_2, 0}$ in series with the thermocouple emf E^{θ_1, θ_2}. The meter thus measures $E^{\theta_1, \theta_2} + E^{\theta_2, 0} = E^{\theta_1, 0}$ where $E^{\theta_2, 0}$ varies with temperature according to:

$$E^{\theta_2, 0} = \beta_1 \theta_2 + \beta_2 \theta_2^2 + \beta_3 \theta_2^3 + \ldots\ldots$$

$$\approx \beta_1 \theta_2 \quad \text{(as } \theta_2 \text{ is generally small).}$$

In order to protect the thermocouple against chemical or mechanical damage, it is normally enclosed in a sheath of mineral packing or within a *thermowell* (Fig. 6.24). Any material which contains the junction should be a good conductor of heat on the one hand, but an electrical insulator on the other. A potentiometric converter is frequently employed to convert the thermocouple signal to the standard 4–20 mA current range prior to further processing and control room presentation. The *extension wires* which connect the thermocouple element to the control room should have similar thermoelectric properties to those of the thermocouple junction wires.

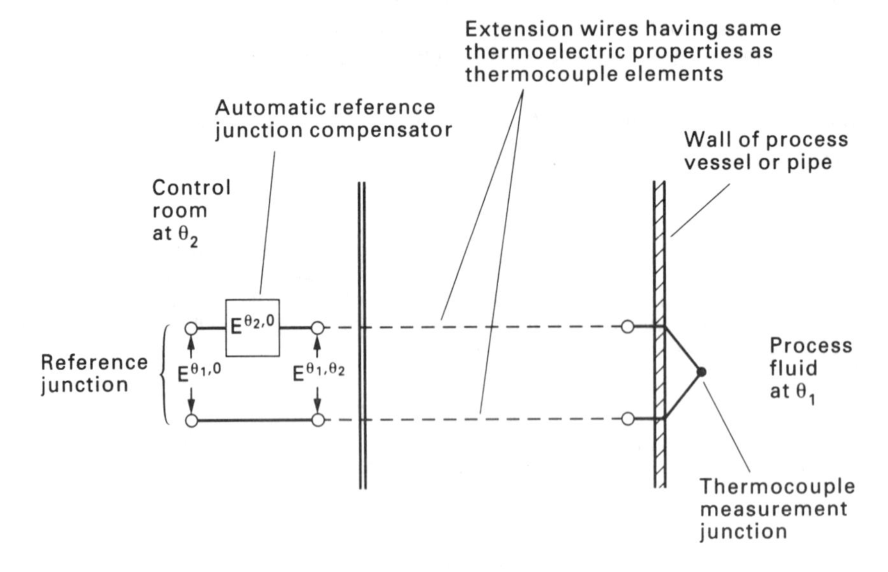

FIG. 6.23. Thermocouple installation with automatic reference junction compensation circuit

TABLE 6.4. *Standard Thermocouple Data and Characteristics*[8, 29, 30, 31]

Thermocouple Type	Operating Range (K)	Best Tolerance	Output Emf (μV)[(a)]			Some Typical Applications	Comments
			600K	800K	1000K		
B Platinum–30%Rhodium/ Platinum–6%Rhodium	273–2000	900–1100K: *tol* = + 4K 1100–2000K: *tol* = ±0.005|*T*–273|K	517	1381	2618	Wherever high stability and long life is required.	Very stable. Poisoned by reducing atmospheres and metal vapours. Use non-metallic sheath.
E Chromel/Constantan	100–1100	230–650K: *tol* = ±1.5K 650–1100K: *tol* = ±0.004|*T*–273|K	23147	39184	55259	Annealing furnaces, acid production (acetic acid, nitric acid, etc.).	High output emf—useful for low temperature sensitivity. Requires good sheath protection in sulphur-bearing or reducing atmospheres.
J Iron/Constantan	100–1100	230–650K: *tol* = ±1.5K 650–1050K: *tol* = ±0.004|*T*–273|K	17818	28906	40823	Polyethylene manufacture, paper making, tar stills, annealing and reheat furnaces, chemical reactors.	Good oxidation resistance. Requires good sheath protection in sulphur-bearing atmospheres above 800K, also in oxygen and in humid surroundings.
K Chromel/Alumel	100–1300	230–650K: *tol* = ±1.5K 650–1300K: *tol* = ±0.004|*T*–273|K	13331	21791	30257	Blast furnace gases, brick kilns, annealing furnaces, boilers, acid production, reactors, superheater tubes, nuclear pile instrumentation, soaking pits, glass tank flues.	Stable and very linear. Recommended for oxidising and neutral conditions. Requires good sheath protection for sulphur-bearing or reducing atmospheres.
N Nickel–Chromium–Silicon *(Nicrosil)*/ Nickel–Silicon *(Nisil)*	100–1300	230–650K: *tol* = ±1.5K 650–1300K: *tol* = ±0.004|*T*–273|K	10303	17780	25587	Blast furnace gases, heat treatment, semiconductor manufacturing, nuclear pile instrumentation, power station boilers, brick kilns.	Very low drift. Particularly good at extreme temperatures and nuclear applications.
R Platinum–13%Rhodium/ Platinum	273–1850	273–1400K: *tol* = ±1K 1400–1900K: *tol* = ±[1 + 0.003(*T*–1400)]K	2666	4766	7063	As for type B.	As for type B.
S Platinum–10%Rhodium/ Platinum	273–1850	273–1400K: *tol* = ±1K 1400–1900K: *tol* = ±[1 + 0.003(*T*–1400)]K	2571	4502	6560	As for type B.	As for type B.
T Copper/Constantan	30–700	230–400K: *tol* = ±0.5K 400–620K: *tol* = ±0.004|*T*–273|K	300K 1073	400K 5566	500K 10741	Sulphuric acid manufacture, food processing, stack gases, plastic moulding, lubricating oils, producer gas plants.	Recommended for low temperatures. Requires protection from acid fumes. Suitable for oxidising and reducing atmospheres < 600K.

(a) Referred to 273.16 K.

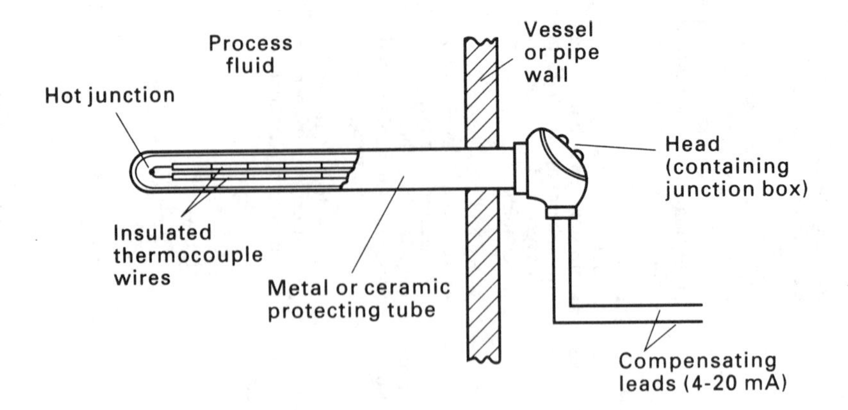

FIG. 6.24. Typical industrial thermocouple arrangement

The dynamics of thermocouple junctions are discussed in Section 7.5.2.

Several thermocouples may be connected in series in order to increase the sensitivity of the instrument. This arrangement is called a *thermopile* and, for n identical thermocouples, gives an output which is n times greater than that from a single thermocouple. Thermopiles are frequently used to detect and measure heat radiation (Section 6.4.2). In this case the surface of the detector is blackened to maximise the absorption of the incoming radiation.

Electrical Resistance Sensors

There are two principal classes of this type of sensor, viz. (i) resistance thermometers (resistance temperature detectors)—which are constructed from normal metallic conducting materials, and (ii) thermistors—which are bulk semiconductor sensors.

The Resistance Temperature Detector (RTD) or Resistance Thermometer

The variation of resistance with temperature θ for metals can be expressed by:

$$R_\theta = R_0(1 + \beta_{\theta,1}\theta + \beta_{\theta,2}\theta^2 + \dots\dots + \beta_{\theta,N}\theta^N) \tag{6.41}$$

where R_0 is the resistance at $\theta = 0°\text{C}$ and $\beta_{\theta,1}$, $\beta_{\theta,2}$, ... are the temperature coefficients of resistance. The RTD is usually constructed from copper, nickel or platinum and, over limited ranges, responds with reasonable linearity (i.e. putting $\beta_{\theta,2} = \beta_{\theta,3} = \dots = \beta_{\theta,N} = 0$ in equation 6.41). More common RTDs are listed in Table 6.5, and a well-type element together with its associated bridge configuration is shown in Fig. 6.25.

A balance over the bridge circuit gives[8]:

$$\theta = \frac{100V_o(R_0 + 2R_{CL})}{R_0 V_s \beta_{\theta,1}} \tag{6.42}$$

Temperature variations in the instrument are a source of error and electrical power dissipation is limited to avoid the effects of self-heating. This is achieved by means of the four lead system shown in Fig 6.25*b*. This minimises any effects of variations in temperature on the resistance R_{CL} of the connections between the RTD and the bridge and is used normally with digital thermometers and data acquisition systems where the sensor non-linearity is corrected within the computer software.

TABLE 6.5. *Typical RTDs and Thermistors*[32-35]

Resistance Temperature Detector (RTD) Type	Normal Operating Range (K)	Accuracy (K)	Comments
Platinum 100Ω	70 to 870	±0.4	Resistance increases with temperature (i.e. temperature coefficient always positive). Good stability. Special types are available which extend the range of temperatures able to be monitored from a minimum of 10K to a maximum of 1120K.
Platinum 200Ω	70 to 870	±0.4	
Platinum 500Ω	70 to 870	±0.2	
Copper 10Ω	170 to 370	±2.4	
Copper 25Ω	170 to 370	±1.0	
Nickel 500Ω	200 to 450	±0.1	
Thermistor Type	**Normal Operating Range (K)**	**Accuracy (K)**	**Comments**
Mixed Metallic Oxides	170 to 470	±0.1 to ±0.2	Temperature coefficient large and negative. Much higher sensitivity and resistance than RTDs so less affected by interconnecting resistances.
Silicon	100 to 430	±0.1 to ±0.2	

The Thermistor

Whereas the RTD exhibits a small positive temperature coefficient, the thermistor has a large negative temperature coefficient and the resistance/temperature relationship is highly non-linear. The latter is typically:

$$R = R_0 \exp\left[\beta_\theta\left(\frac{1}{\theta} - \frac{1}{\theta_0}\right)\right] \tag{6.43}$$

where R and R_0 are the resistances at temperatures θ K and θ_0 K respectively, and β_θ is a constant which is characteristic of the material of the semiconductor. Usually θ_0 is set at 298 K. The maximum range of temperatures over which thermistors can be employed is between 70 and 1300 K (standard ranges are shown in Table 6.5). Commercial thermistor probes are generally more sensitive than RTDs or thermocouples, and the severely non-linear characteristics involved can be accommodated by employing a suitable computerised data monitoring system.

6.4.2. Thermal Radiation Detection

Thermal radiation has a frequency range principally between $7.5 \times 10^{12}\ \text{s}^{-1}$ and $1 \times 10^{15}\ \text{s}^{-1}$ and, as such, covers most of the visible and infra-red sections of the electromagnetic spectrum (EMS). The relation between thermal radiant energy and temperature is discussed in Volume 1 (Section 9.5.3).

The applications of *thermal radiation detectors* (TRDs) now have become more comprehensive in the process industries as they can be employed anywhere it is

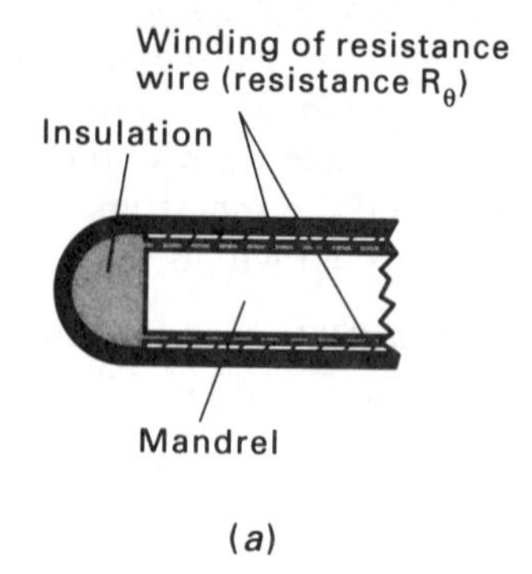

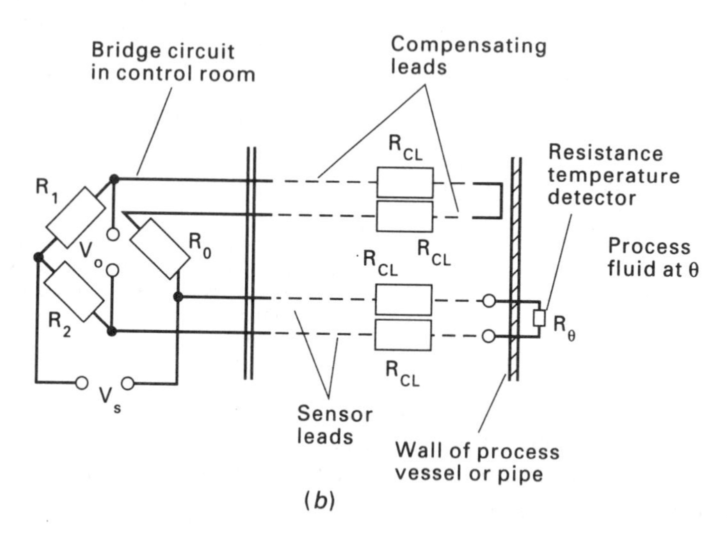

FIG. 6.25. Typical resistance temperature detector element of circular cross-section: (*a*) well-type RTD element of circular cross-section; (*b*) four lead compensation system and associated bridge circuit

desirable that there should be no contact between the process fluid, solid or vessel and the temperature measuring device. Originally this type of sensor was used to measure the heat energy radiated from a *hot* body (>600 K) but modern detectors are able to measure temperatures as low as 220 K[36]. Typical applications for TRDs are[8, 36]:

(i) Measurement of temperature under the following conditions:
  (a) High temperatures at which a normal sensor would melt or decompose (e.g. process heaters, high temperature gas phase reactors).
  (b) Of a moving body.
  (c) Detailed measurement of the temperature distribution over a surface where the number of conventional sensors required would be prohibitive.
  (d) Other applications where contact with the process material is not possible or is undesirable (e.g. in the food industry, frozen foods).
(ii) Measurement of radiant heat flux.
(iii) Measurement of gas composition (Table 6.16).

The total energy emitted per unit area per unit time (i.e. the total power) for a black body at temperature T is given by the *Stefan–Boltzmann law* (see also Volume 1, Section 9.5.3), viz.:

$$\mathbf{E} = \sigma T^4 = \int_0^\infty W_\lambda d\lambda = \int_0^\infty \frac{J_1}{\lambda^5 [\exp(J_2/\lambda T) - 1]} d\lambda \tag{6.44}$$

where W_λ is the power spectral density for a black body radiator (i.e. the quantity of radiation emitted from a flat surface into a hemisphere, per unit wavelength, at the wavelength λ), σ is the *Stefan–Boltzmann* constant, and J_1 and J_2 are constants. (If $\mathbf{E}$ is measured in W/m^2, W_λ in W/m^3, λ in m and T in K, then $\sigma = 5.67 \times 10^{-8}$ W/m^2K^4, $J_1 = 3.74 \times 10^{-16}$ Wm^2 and $J_2 = 0.0144$ mK.) The wavelength λ_{max} at which W_λ has a maximum value decreases as T increases and:

$$\lambda_{max} = \frac{J_3}{T} \tag{6.45}$$

where J_3 is a constant. (If T is measured in K and λ_{max} in m, then $J_3 = 2.89 \times 10^{-3}$ mK.) Hence, at higher temperatures there is more power in the visible light region of the EMS, and at lower temperatures there is more power in the infra-red section.

Equation 6.44 gives the total power emitted by a black body. However, materials and bodies whose temperatures are measured with radiation-type instruments often deviate considerably from ideal black body behaviour. This deviation is expressed generally in terms of the *emissivity e* of the measured body (see also Volume 1, Section 9.5.4), and the energy emitted by the body per unit area per unit time is:

$$\mathbf{E}_g = \sigma e T^4 \tag{6.46}$$

Emissivities are functions of λ, T and properties of the system which are difficult to quantify or measure, such as size, shape, surface roughness, angle of viewing, etc. This often leads to uncertainties in the estimation of the temperature of a body from the measurement of the thermal radiation it emits. The emissivity of a black body is 1, and a body which has an emissivity which is independent of λ and T is termed a *grey body*. A further source of error is the loss of energy in transmitting radiation from the source to the radiation detector. Such losses may occur by absorption in glass lenses and mirrors used to focus the energy and/or in absorptive gases such as H_2O, CO_2 and O_2.

Thermal Radiation Measurement Systems

The Broadband Radiation Thermometer (Total Radiation Pyrometer)

This instrument (Fig. 6.26) is the most common type used in day-to-day industrial applications and is sensitive to all wavelengths present in the incoming radiation[37]. The total power detected can be obtained from equations 6.44 and 6.46,

i.e.:
$$E_g = J_{RT}e \int_0^\infty \frac{J_1}{\lambda^5 [\exp(J_2/\lambda T) - 1]} d\lambda = J_{RT} e \sigma T^4 \quad (6.47)$$

assuming that the emissivity e is independent of λ and T. J_{RT} is a factor dependent upon the source lens/detector geometry.

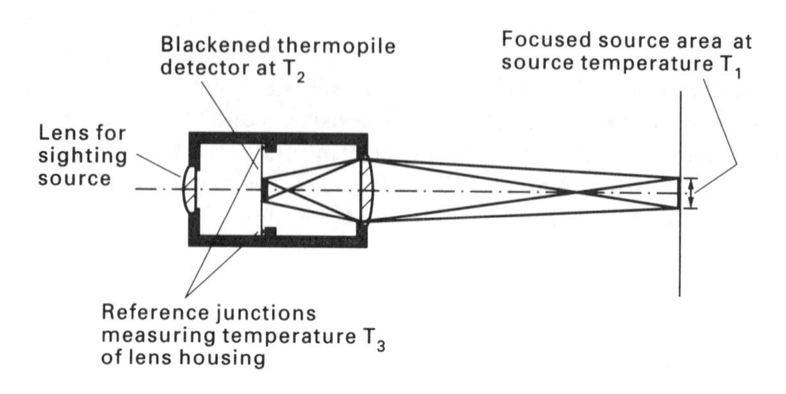

FIG. 6.26. Lens-type broadband radiation thermometer

These instruments employ a blackened thermopile as a detector and focus the radiation using lenses or mirrors. For a given source temperature T_1 the incoming radiation heats the measuring junction until losses due to conduction, convection and radiation just balance the heat input, i.e.

$$\underset{\text{Heat loss}}{h_{RT}(T_2 - T_3)} \approx \underset{\text{Radiant heat input}}{J_{RT} e \sigma T_1^4} \quad (6.48)$$

where h_{RT} may be considered as a heat transfer coefficient for combined conduction, convection and radiation and T_2 and T_3 are the temperatures of the detector and the lens housing respectively (Fig. 6.26). T_1 and T_3 are both influenced by ambient conditions and suitable compensation must be provided for any variations in the latter—particularly when low temperatures are being monitored. The thermopiles can have from 1 to 30 junctions. A smaller number of junctions will give a faster response (due to the smaller mass) but will have a lower sensitivity.

Narrow-Band Radiation Thermometer

These instruments use photon detectors which are semiconductor devices that are responsive only to a narrow band of wavelengths. Many narrow-bond thermometers further restrict the wavelength bandwidth by means of a narrow pass filter. In this case the total power detected is[4]:

$$E_g = J_{RT}e \int_b^a \frac{J_1}{\lambda^5 [\exp(J_2/\lambda T) - 1]} d\lambda$$

(where $a = \lambda_0 + \delta\lambda/2$, $b = \lambda_0 - \delta\lambda/2$ and λ_0 is the mean wavelength detected)

i.e.:

$$E_g \approx J_1 J_{RT} e \frac{\delta\lambda}{\lambda_0^5} \exp\left(-\frac{J_2}{\lambda_0 T}\right) \tag{6.49}$$

where $\delta\lambda << \lambda_0$ and $J_2/\lambda T >> 1$.

Chopped Radiation Thermometers

With several different types of instrument a technique is employed in which the radiation between the source and the detector is chopped (i.e. periodically interrupted) at a controlled frequency. This enables high gain a.c. amplifiers to be employed, leading to greater instrument sensitivity as well as good temperature compensation. Thermal detectors employing this method include the blackened chopper radiometer (Fig. 6.27) in which a mirror focuses the radiation on to a square thermistor detector. The motion of the chopper enables the detector to "see" radiation from the source and radiation reflected from the black surface of the chopper. The instrument then compares the two. A typical temperature range for this instrument is ambient to 1600 K with a time constant of about 0.01 s and a chopping frequency of 200 Hz. If a faster response is required then a photon detector can be substituted for the thermal detector.

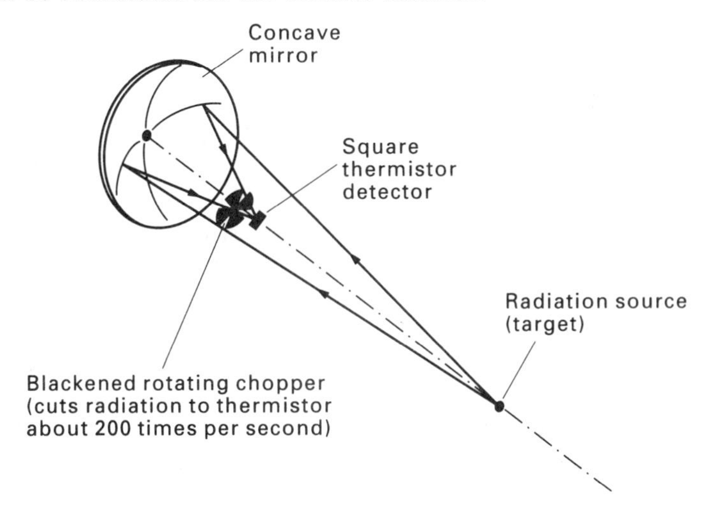

FIG. 6.27. Blackened chopper radiometer

Optical Pyrometers

The most accurate of all radiation thermometers is the disappearing filament optical pyrometer which is employed in setting the IPTS above 1336 K. An image of the source is superimposed on the filament of a previously calibrated tungsten lamp. The calibration enables the brightness temperature of the filament to be related to the current through the filament. A red narrow-band filter is placed between the tungsten lamp (which itself lies between the observer and the source) and the eye of the observer. The current through the lamp filament is varied until the image of the filament disappears in the superimposed image of the source. At this point it is considered that the brightness of the filament (at temperature T_f) and of the source (at temperature T_s) are equal. For these conditions, from equations 6.44 and 6.46:

$$\underset{\text{Source}}{\frac{J_1 e_{\lambda s}}{\lambda^5[\exp(J_2/\lambda T_s) - 1]}} = \underset{\text{Filament}}{\frac{J_1 e_{\lambda f}}{\lambda^5[\exp(J_2/\lambda T_f) - 1]}}$$

where $e_{\lambda s}$ is the emissivity of the source at wavelength λ and the emissivity of the filament $e_{\lambda f}$ can be considered as 1 at the same wavelength. If SI units are employed, then $\exp(J_2/\lambda T) >> 1$ for $T < 4000$ K, and:

$$\frac{e_{\lambda s}}{\exp(J_2/\lambda T_s)} = \frac{1}{\exp(J_2/\lambda T_f)}$$

Hence:

$$\frac{1}{T_s} - \frac{1}{T_f} = \frac{\lambda}{J_2} \ln(e_{\lambda s}) \tag{6.50}$$

For $e_{\lambda s} \neq 1$, equation 6.50 gives the required temperature correction. The system is not very sensitive to errors in $e_{\lambda s}$; e.g. for a source at 1000°C, an error of 10 per cent in $e_{\lambda s}$ results in an error of 0.45 per cent in T_s[4].

Difficult Applications

The use of TRDs has been facilitated by the use of fibre-optic cable assemblies (Section 6.12.4). This permits sensing in confined spaces as well as enabling the sensor to 'see round corners' and into regions containing nuclear or electromagnetic radiation[36]. Low levels of source visibility can be overcome by employing *two-colour* or *ratio* sensors which measure radiation at two different wavelengths simultaneously. Such instruments can be used where dust or dirt obscures the source from the sensor.

Lower temperature applications commonly occur in the food industry (e.g. the monitoring of frozen foods). This is particularly important over a frequency range of $2 \times 10^{13}\ s^{-1}$ to $4 \times 10^{13}\ s^{-1}$ within which most organic materials emit infra-red radiation of significant power. An additional advantage is that the infra-red absorption bands for CO_2 and water vapour (the presence of which can cause errors because they attenuate the radiation emitted from the source material) are outside this range of wavelengths.

6.5. THE MEASUREMENT OF LEVEL

Liquid level measurement is of considerable importance in the process industries and methods range from a simple dip stick or sight glass to sophisticated computerised systems giving the total mass of liquid in a large storage tank to accuracies of ±0.1 per cent. A number of the methods described in this section can be applied also to level measurement of such materials as slurries, powders and granular solids and this is indicated where appropriate. Often a level measuring device is not required to give a continuous measurement but simply an indication of whether the material inside the vessel has risen or fallen to a certain limiting height. Such devices simply initiate *high* and *low level limit alarms* respectively and are termed *float switches.*

6.5.1. Simple Float Systems

Floats are independent of the static pressure and are of either the *buoyant* or the *displacement* type (Fig. 6.28)[38]. The former is less dense than the liquid in which it is partially immersed and the latter is slightly denser than the liquid. Consequently, the buoyant or *moving* float has a constant immersion and will rise and fall precisely the same distance as the liquid level. Its position at any instant is measured via a potentiometric or reluctive transducer (Section 6.3.3) to give an output which is directly proportional to the level. The displacement or *static* type has a variable immersion and a variable upthrust upon it. The level is then measured in terms of the net weight of the float. In this case the buoyant force exerted on the float is converted into an output signal by a strain gauge or force-balance transducer (Section 6.3.3). Both types of float are used to give a continuous measurement of level. In the case of corrosive liquids, the number of parts immersed is reduced to a minimum by coupling the float magnetically with the transducer which is sealed from the contents of the vessel. If the liquid surface is likely to be turbulent (e.g. when the liquid is boiling, as in the base of a distillation column), then the float is placed in a float chamber or side arm where calmer conditions will prevail (Fig. 6.29).

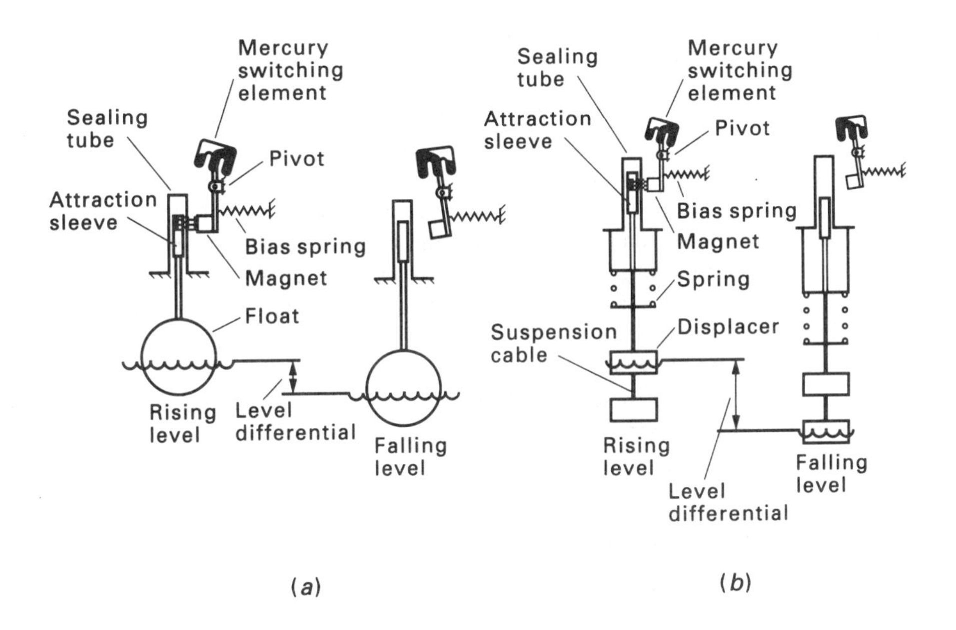

FIG. 6.28. Simple float switches: (*a*) buoyant type; (*b*) displacement type

Floats are simple, reliable and rugged. They are suitable for use at high pressures and temperatures and in the presence of foams. If constructed of suitable materials, they can be employed in aggressive environments and for liquids with densities as low as 300 kg/m^3. When used in the level limit alarm mode, float switches can detect variations in level down to about 2 mm. However, as they are mechanical devices, floats can suffer from jamming, fouling by the material being measured, and general wear and tear.

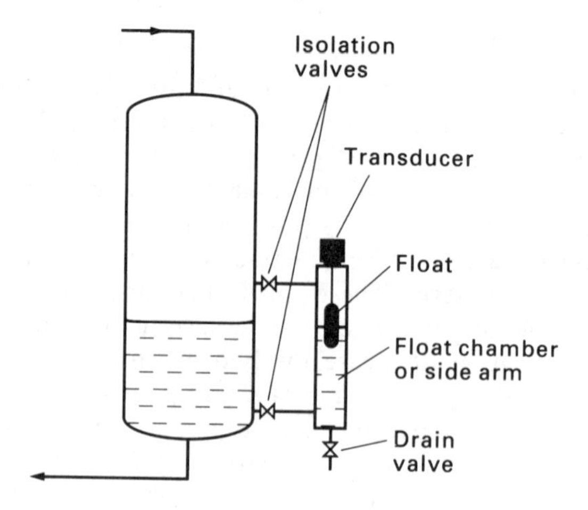

FIG. 6.29. Use of float chamber to measure liquid level

6.5.2. Techniques Using Hydrostatic Head

Several techniques of varying complexity are used for directly measuring the hydrostatic head exerted by a liquid in a vessel. These include:

(i) Flush-mounting a pressure transducer (Section 6.3) close to the base of the vessel. This is appropriate for an open vessel and gives a direct measure of the liquid head. This technique can be employed also for slurries, powders, highly viscous media such as toothpaste, etc. However, it is not suitable for applications where the measured material can solidify or crystallise over the transducer diaphragm. Careful consideration should also be given to the constancy of the specific gravity of the material with variations in temperature or composition.

(ii) If the vessel is closed, then the difference in pressure between that in the vapour space and that near the base is measured using a differential pressure (DP) cell (Section 6.3.4). Generally the latter will not be flush with the vessel wall and a typical arrangement for a volatile liquid is illustrated in Fig. 6.30.

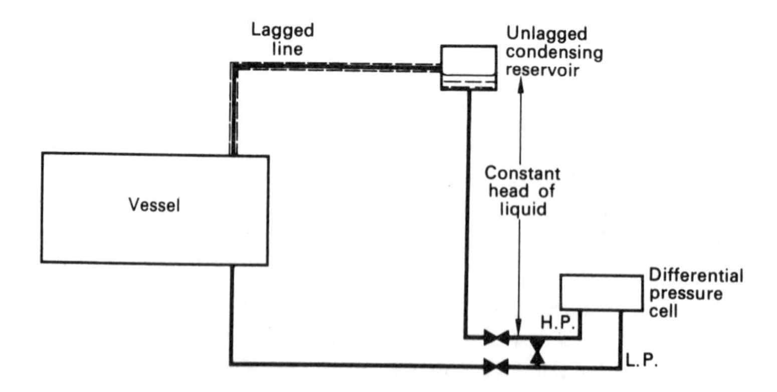

FIG. 6.30. Measurement of the level of a volatile liquid using a differential pressure cell

If the liquid is not volatile, the condensing reservoir shown is not required and the high and low pressure connections on the DP cell are reversed. The DP cell is normally located *below* the lower measuring point (or *tapping*) in the vessel. Any lines connected to the DP cell which are liquid filled should be bled through the DP cell *bleed port* to make sure that there is no trapped vapour present.

Hydrostatic head sensors are easily installed, contain no moving parts, and have good accuracy (approaching ±0.5 per cent of full scale) and repeatability.

6.5.3. Capacitive Sensing Elements

A common method of level measurement is to use a capacitance bridge. A typical arrangement is shown in Fig. 6.31*a* in which the sensor consists of two concentric metal cylinders. In the case of a circular tank, the wall of the tank can be employed as the outer cylinder of the sensor. The capacitance of the sensor is:

$$\mathscr{C} = \frac{2\pi\varepsilon_0[\ell_c + (\varepsilon_r - 1)z]}{\ln\left(\dfrac{r_2}{r_1}\right)} \tag{6.51}$$

where the cylinders are of height ℓ_c and contain liquid up to depth z within the annulus. ε_0 is the *permittivity of free space* (8.85×10^{-12} farads/metre), ε_r is the *relative permittivity* of the liquid involved, and r_2 and r_1 are the radii of the cylinders ($r_2 > r_1$). (Note that frequent reference is made to the *dielectric constant* of a material. This is identical to the material's permittivity ε, where $\varepsilon = \varepsilon_0\varepsilon_r$.)

Capacitive sensing elements are incorporated into either electrical oscillator circuits or into a.c. deflection bridge circuits. An example of the latter is shown in Fig. 6.31*b*. In this the output voltage is:

$$V_o = V_s\left(\frac{1}{1 + \mathscr{C}_0/\mathscr{C}} - \frac{1}{1 + R/R_0}\right) \tag{6.52}$$

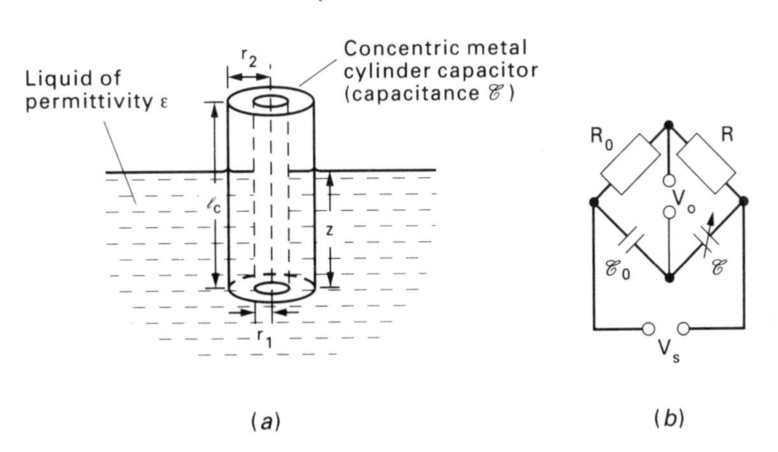

FIG. 6.31. Level-sensing capacitor and associated bridge network: (*a*) concentric cylinder capacitator; (*b*) capacitance bridge

If $\mathscr{C} = \mathscr{C}_{min}$ when $V_o = 0$, then $\mathscr{C}_0/\mathscr{C}_{min} = R/R_0$, i.e. $\mathscr{C}_0 = \mathscr{C}_{min}R/R_0$.

Hence:

$$V_o = V_s\left(\frac{1}{1+\dfrac{\mathscr{C}_{min}R}{\mathscr{C}R_0}} - \frac{1}{1+\dfrac{R}{R_0}}\right)$$

If R/R_0 is made >> 1:

$$V_o \approx V_s\frac{R_0}{R}\left(\frac{\mathscr{C}}{\mathscr{C}_{min}} - 1\right) \tag{6.53}$$

The level in the tank can then be determined from equations 6.51 and 6.53.

If the material is non-conducting then a bare capacitance probe can be employed—otherwise the probe must be insulated. Difficulties can arise due to unexpected variations in the dielectric constant of the material being measured (e.g. in the case of a powder which is prone to absorb water vapour).

6.5.4. Radioactive Methods (Nucleonic Level Sensing)

A γ-ray source (e.g. ^{137}Cs, ^{60}Co, ^{226}Ra) is frequently used for liquid level *measurement* and for liquid and solid level *detection*. The absorption of the γ radiation varies with the thickness x and nature of the absorbing material between the source and the detector according to the Beer–Lambert law, viz:

$$\mathscr{I} = \mathscr{I}_0 \exp(-\xi \rho x) \tag{6.54}$$

where $\mathscr{I}$ and $\mathscr{I}_0$ are the intensities of the radiation received at the detector with and without the presence of the absorbing material respectively, ξ is the mass absorption coefficient (constant for a given source and absorbing material) and ρ is the density of the absorbing material. If two phases are present between the source and the detector (e.g. gas and liquid or gas and solid), then:

$$\mathscr{I} = \mathscr{I}_0 \exp(-\xi_1\rho_1 x_1 - \xi_2\rho_2 x_2) \tag{6.55}$$

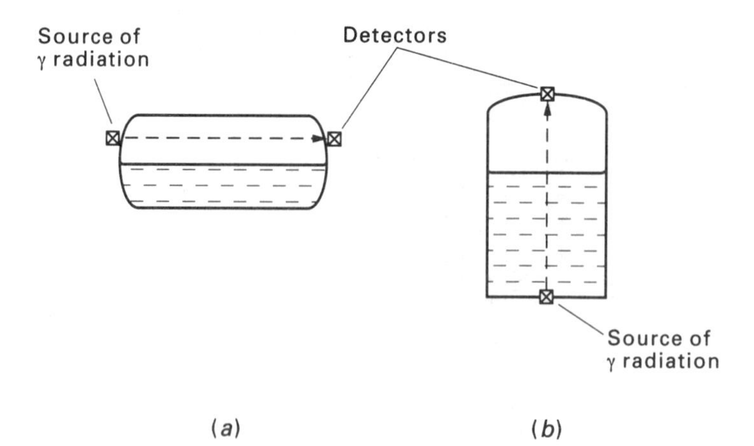

FIG. 6.32. Nucleonic level-sensing

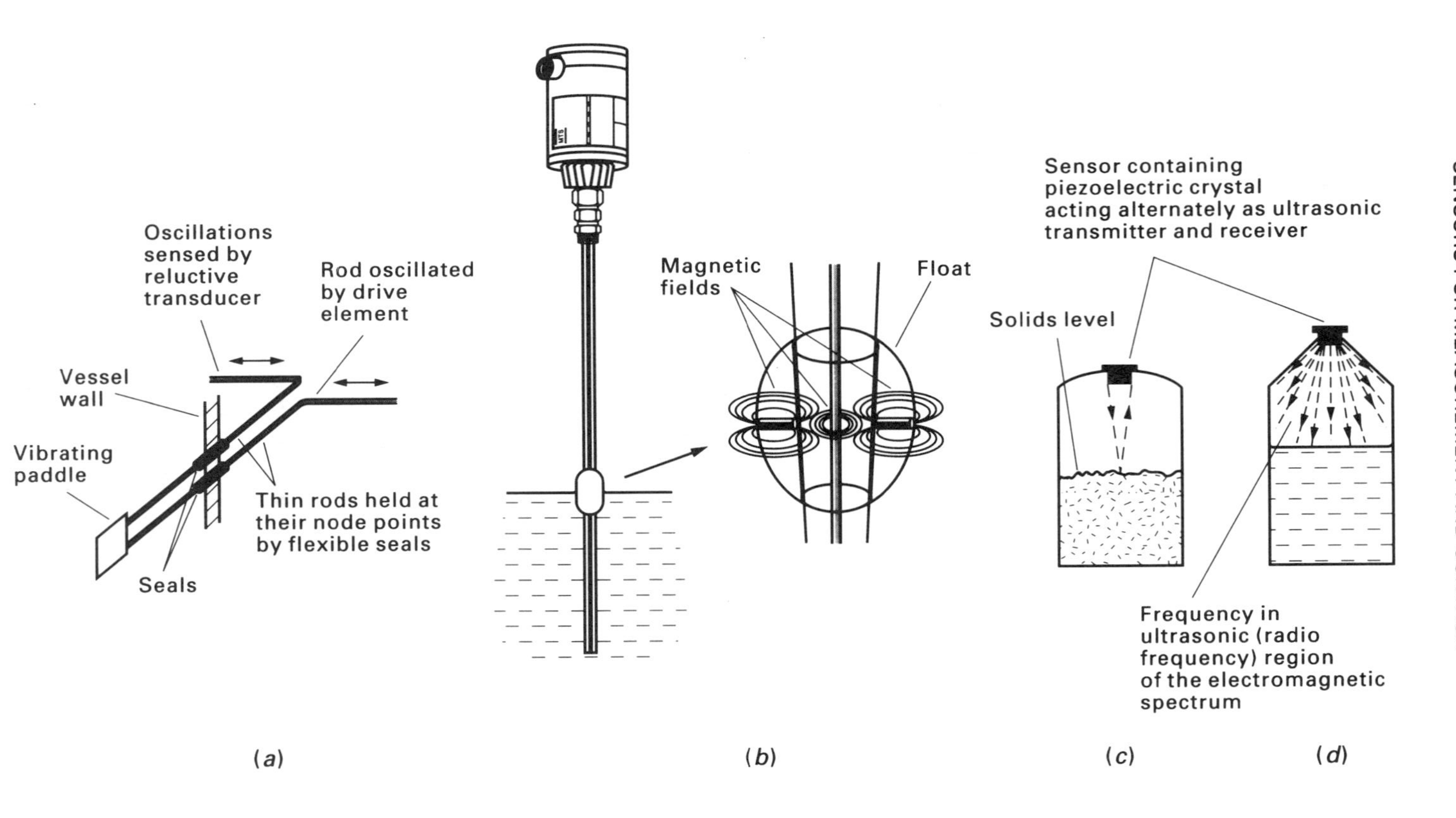

FIG. 6.33. Various methods of level-sensing: (*a*) vibrating-paddle level sensor; (*b*) level measurement using magnetostriction; (*c*) ultrasonic path method; (*d*) ultrasonic cavity-resonance method

Two different arrangements are illustrated in Fig. 6.32. Figure 6.32*a* shows a typical installation where a high level limit alarm is required. In this case the intensity of the radiation $\mathscr{I}$ reaching the detector would be determined by equation 6.54. If a continuous measurement of level is necessary, then the radioactive source and detector would be positioned as in Fig. 6.32*b* and $\mathscr{I}$ is then given by equation 6.55.

6.5.5. Other Methods of Level Measurement

These are summarised in Table 6.6.

6.6. THE MEASUREMENT OF DENSITY (SPECIFIC GRAVITY)

Instruments are available for the continuous measurement of the density and relative density (specific gravity) of both liquids and gases. When specific gravity is a sufficiently sensitive function of composition for a particular process stream, sensors employed to monitor density can also be employed as composition analysers.

6.6.1. Liquids

The Use of Level Measuring Devices

Certain instruments which have been designed to measure liquid level (Section 6.5) can be modified easily to record variations in liquid density. Such instruments are those that measure the pressure P exerted by a given depth of liquid z in a vessel, i.e. $P = \rho g z$. If z is fixed then:

$$\rho = \frac{1}{gz} P \tag{6.56}$$

Typical are the pressure transducer (Section 6.3.2) and the DP cell (Section 6.3.4). In these cases, of course, the level must be maintained constant.

Weighing Meters

These are generally in the form of a U-tube attached by flexible connectors to the inlet and outlet lines of the instrument (Fig. 6.34). The instrument has a substantial volume (ID $\approx$ 0.025 m) and, where flows are small (recommended minimum velocity $\approx$ 1 m/s), it is probable that the whole process fluid stream, rather than simply a sample, will be sent through the instrument.

The operation of the instrument is a typical and useful example of pneumatic feedback[39]. Initially the counterbalance weight is so positioned that balance is achieved. When the weight of the U-tube increases, due to an increase in the density of the process liquid, the weighing beam tends to turn in a clockwise direction, raising the flapper and thus allowing more air to escape from the nozzle. The resulting decrease in pressure on top of the diaphragm in the relay allows the valve spool to rise and this enlarges the pressure in the feedback bellows. The latter increases until sufficient extra force is applied to the U-tube through the pivoted transmitter arm to restore the original balance. The U-tube is thus continuously

TABLE 6.6. *Other Methods of Level Measurement*[25, 40, 41, 42, 43]

Sensing by	Continuous Measurement	Discrete Measurement	Suitable for	Comments
Conductivity	yes	yes	liquids	Tank wall can be used as one electrode if metallic. Liquid must be electrically conductive. Electrodes form part of a conductivity bridge network.
Weight	yes	no	liquids, powders, granular solids.	Generally uses load cells positioned beneath the vessel. Typical accuracy—about 0.1 per cent of FSD[a].
Heat transfer	no	yes	liquids	Uses resistive element in which a current produces a self-heating effect. Element cools when liquid comes into contact with it. The resulting change in resistance is measured.
Photo-electric effect	no	yes	liquids, powders, granular solids.	Electromagnetic signal in visible light or infra-red region of electromagnetic spectrum is sensed by photo-electric cell. Beam is attenuated by liquids and cut off by solids. Liquids can transmit or reflect the beam.
Damped oscillations	no	yes	liquids and some dry products.	Employs oscillating element which is normally a vibrating fork or paddle driven mechanically (Fig. 6.33*a*) or by a piezoelectric crystal vibrating at its resonant frequency. When immersed in the material there is a frequency or amplitude shift due to viscous damping which is sensed usually by a reluctive transducer (Section 6.3.3).
Damped oscillations (magnetostrictive type)	no	yes	liquids, powders, granular solids.	When a ferromagnetic rod is exposed to a longitudinal magnetic field it increases slightly in length. The field is created and oscillated by drive and feedback coils wound on the rod—the end of which is in contact with the inside surface of the tip of a sealed probe sheath. The oscillator circuit is set so as to maintain the oscillation in the rod when the tip of the probe is in contact with any compressible fluid (e.g. gas, froth, foam, etc.). When the end of the probe touches an incompressible fluid (viz. a liquid) the vibration is damped, the oscillation ceases and a signal is generated. This type of instrument is used increasingly—particularly with difficult systems. The probe should not be immersed in any material which may harden on the probe tip (Fig. 6.33*b*).
Ultrasonic type—cavity resonance method	yes	no	liquids, powders, granular solids.	Volume sensing technique in which electromagnetic oscillations at ultrasonic or radio frequencies are projected from a transmitter placed at the top of the vessel into the space above the material (the *ullage*). As the level rises the volume of the ullage decreases and the resonant frequency of the oscillation changes. This method is not recommended where the density of the vapour above the material varies significantly (e.g. when measuring the level of solvents) or in the presence of foams. Typical accuracies are of the order of 1 per cent of FSD[a]. (Fig. 6.33*d*.)
Ultrasonic type—ultrasonic path method	yes	yes	liquids, powders, granular solids.	For continuous level sensing the sensor uses a reflected pulsed ultrasonic signal. The travel time of a pulse reflected back from the surface is measured. For discrete level sensing a directly transmitted continuous signal is employed. The solid or liquid being measured intervenes in the sonic path and either cuts the signal (solids) or attenuates it (liquids). Unsuitable where the density of the vapour varies and in the presence of foams. Suitable for vessels, flumes, weirs and open channels. Typical accuracies are of the order of 1 per cent of FSD[a] (Fig. 6.33*c*.)
Microwaves and Radar	yes	yes	liquids, powders, granular solids.	Action very similar to ultrasonic signals. Has been employed for discrete sensing. Pulsed energy used in transmitted or reflected modes. Transmitter and receiver can be mounted outside the vessel. Very useful for determining mean levels in closed vessels under difficult conditions, e.g. in cases of high turbulence, in the presence of vapour or condensate, or where there is encrustation of surfaces with deposits. Typical accuracies are of the order of 0.2 per cent FSD[a].

(a) FSD: full scale deflection (see Section 6.10.1).

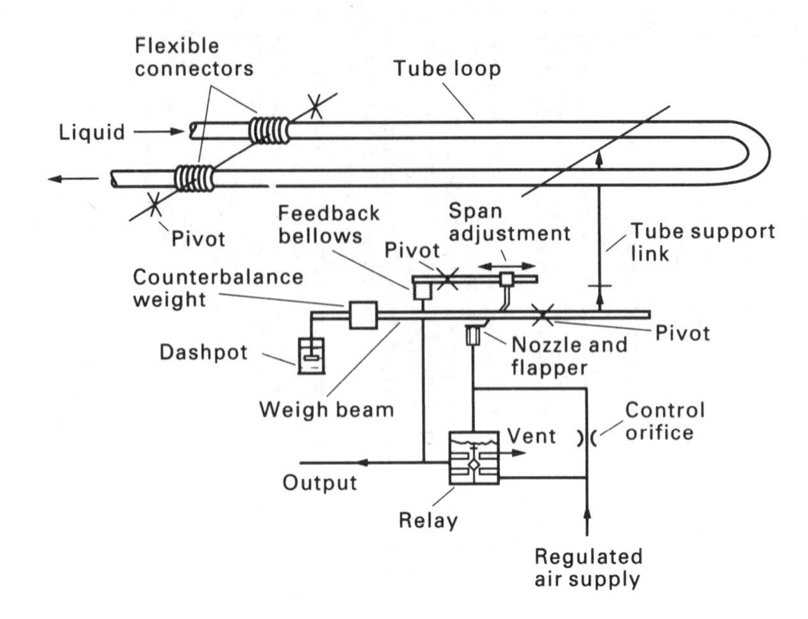

FIG. 6.34. U-Tube specific gravity (density) meter and pneumatic transmitter

balanced and its movement, which is damped, is no greater than a few seconds of arc. Since the area of the feedback bellows and lever lengths are fixed, the increase in pressure required to restore balance is directly proportional to the density of the fluid flowing in the U-tube. The output air pressure (usually in the range 20–100 kN/m^2) is a measure of the density of the fluid.

This type of meter is used generally to measure liquid densities ranging from 500 kg/m^3 to 1600 kg/m^3 with an accuracy of about ± 0.5 per cent of FSD. The effects of variations in ambient temperature are negligible, but mechanical vibration (e.g. from pumps, etc.) can create substantial errors.

Buoyancy Meters

These are typically based on the principle of the hydrometer. For automatic measurement and control an inductance bridge may be employed to detect the position of the hydrometer. The level of the liquid in which the hydrometer is floating must be maintained constant. The float or *displacer* may be partially or totally immersed in the liquid (similar to the float systems used to measure level—Section 6.5.1 and Fig. 6.28). A schematic diagram of an instrument using a totally immersed displacer is shown in Fig. 6.35.

As the density of the liquid increases the float also rises and lifts the chain. The float continues to ascend until the additional weight of the chain raised equals the additional buoyancy due to the increased density. The reverse occurs when the density of the liquid is reduced. The position of the float is detected by a linear variable differential transformer (LVTD) in which the movement of the ferromagnetic core of the displacer changes the inductance between the primary and secondary windings of a differential transformer (see also Fig. 6.13). Such meters

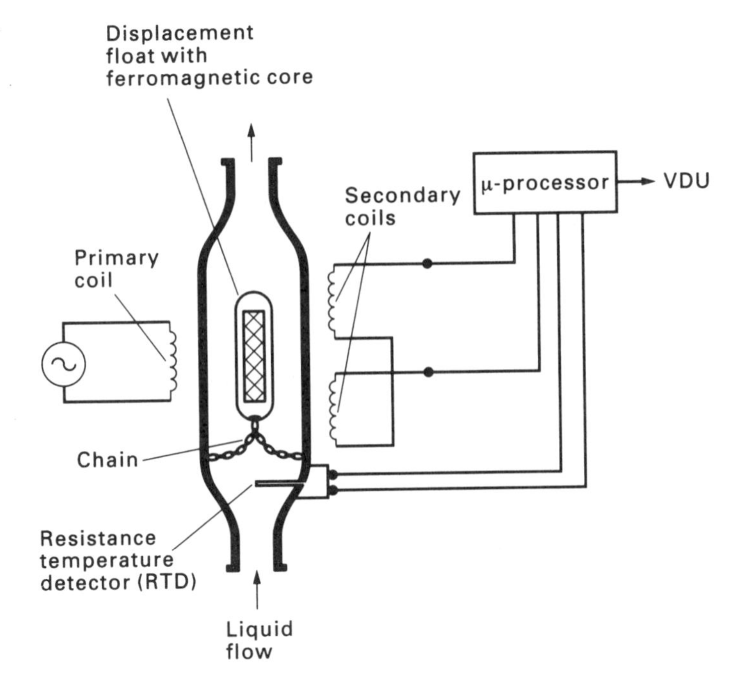

FIG. 6.35. Buoyancy meter with totally immersed float and LVDT sensing-element

are used generally to measure densities in the range 500–3500 kg/m^3 with spans of between 5 kg/m^3 and 200 kg/m^3. Accuracies of ±3 per cent of span or ±0.2 kg/m^3 (whichever is the greater) have been reported[(39)]. A resistance thermometer is provided to record changes in temperature and to modify the output accordingly.

Acoustic Meters

These operate on the basis of the measurement of the resonant frequency of an oscillating system. The oscillation is impressed on the system, either via a cylinder immersed in the liquid, or via a tube through which the liquid flows. A typical arrangement is shown in Fig. 6.36 in which the sensing element consists of a single pipe through which the process fluid flows[(44)]. Heavy masses are fixed to each end of the pipe and the whole is attached to the pipe flanges and outer casing of the instrument by bellows and ligaments which allow the pipe to oscillate freely. Positioned along the pipe is an electromagnetic drive and a pick-up coil assembly. Any variation in the mass per unit length of the oscillating section (due to a change in the density of the fluid flowing through it) produces a change in the resonant angular frequency ω of the system. This is transmitted by magnetically coupling the oscillating system with a suitable amplifier. The latter provides both a frequency modulated output signal and the excitation to produce the required oscillation. The output is automatically compensated for any change in the length of the tube due to variations in temperature and is expressed in terms of the density ρ of the sample liquid by:

$$\rho = K_{LAM}\left[\left(\frac{\omega_0}{\omega}\right)^2 - 1\right] \tag{6.57}$$

where ω_0 is the resonant angular frequency obtained with the pipe empty and K_{LAM} is the transducer constant. Although equation 6.57 is non-linear, linearity can be assumed with little error for changes in density up to about 20 per cent. Meters of this kind are very robust and give excellent results. Any range of densities can be monitored and accuracies better than 0.01 kg/m^3 have been reported[17].

6.6.2. Gases

The principle of the acoustic meter can be equally applied to the monitoring of gas densities. A high sensitivity is required as the variations in density involved are

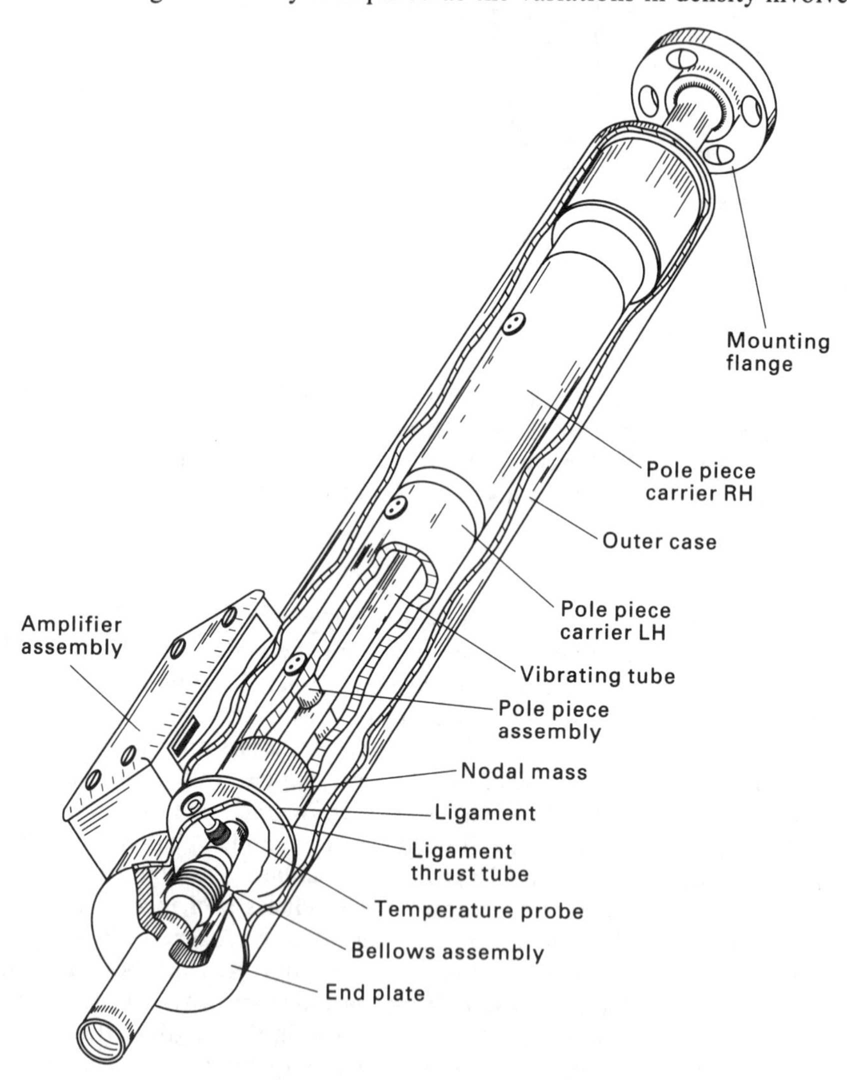

FIG. 6.36. Acoustic liquid-density meter

much smaller than for liquids. One method is to induce a resonant frequency in a thin-walled cylinder immersed in the sample gas. This causes the gas surrounding the cylinder to oscillate at the same frequency. The total mass of the vibrating system is thus increased. This reduces the resonant angular frequency and the change in frequency is related to the density of the gas by:

$$\rho = K_{GAM}\frac{\omega_0 - \omega}{\omega_0^2\omega}\left[1 - K'_{GAM}\left(\frac{\omega_0 - \omega}{\omega_0^2\omega}\right)\right] \tag{6.58}$$

where ω_0 is the resonant angular frequency under vacuum conditions and K_{GAM} and K'_{GAM} are calibration constants. The output of the instrument is automatically compensated for variations in gas temperature. Densities from 0 to 400 kg/m^3 can be monitored with accuracies of 0.1 per cent of the meter reading[17].

There are several instruments which employ highly sensitive weighing techniques. One particular meter is based upon the measurement of the upthrust on a float[45]. Any displacement of the float due to a change in the gas density (and consequently buoyancy) is counterbalanced by a magnetic field produced by an electromagnet. The float is thus maintained in the null position by the force generated by the flow of a *direct* current in the coil of the electromagnet. This current is directly proportional to the density of the gas. A temperature correction is applied to the readout and densities up to 500 kg/m^3 at pressures up to 17 MN/m^2 can be measured.

6.7. THE MEASUREMENT OF VISCOSITY

The concept of viscosity as a physical property of a fluid is considered in Volume 1, first for Newtonian fluids (Section 3.3) and secondly for the non-Newtonian case (Section 3.7). A further discussion of viscosity on the molecular scale in terms of the momentum transfer occurring in a gas flowing over a flat surface appears in Volume 1, Section 12.2.4. Methods for estimating viscosities are described in Volume 6, Section 8.7.

There are a substantial number of ways in which the viscosity of a fluid can be measured. Not all of these can be adapted easily to on-line process measurement. It is convenient to classify viscometers according to the type of physical measurement made.

6.7.1. Off-line Measurement of Viscosity

Measurement of the Rate of Flow of Fluid in a Capillary

The Ostwald U-tube instrument is the most common type of viscometer based upon this principle (Fig. 6.37*a* and Table 6.7). The viscometer is filled with liquid until the liquid level is such that the bottom of the meniscus in the right-hand limb coincides with the mark C. The fluid is drawn up the left-hand limb to a level about 5 mm above A and then released. The time t taken for the bottom of the meniscus to fall from A to B is recorded. The dynamic viscosity μ is determined from Poiseuille's law (Volume 1, equation 3.30 and Section 9.4.3) which under the

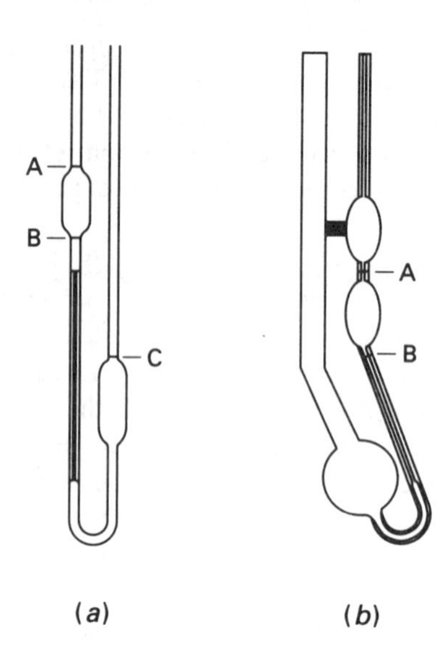

FIG. 6.37. Common capillary viscometers: (a) simple Ostwald type; (b) Cannon–Fenske type

conditions applied (viz. the driving force remaining nearly constant throughout) gives:

$$\mu = K_v \rho t \tag{6.59}$$

or:

$$\nu = K_v t \tag{6.60}$$

where K_v is the viscometer constant (not dimensionless), ρ is the density of the liquid and ν is the kinematic viscosity ($= \mu/\rho$). Capillary viscometers with different dimensions are employed for different viscosity ranges. For fluids having kinematic viscosities below 10^{-5} m^2/s (fast flow, giving small values of t), it may be necessary to apply a correction to equation 6.60 for kinetic energy effects, thus:

$$\nu = K_v t - K_v'/t. \tag{6.61}$$

Many modifications of the basic Ostwald geometry are employed in different situations. One example is the Cannon–Fenske routine viscometer (Fig. 6.37*b*) which is used in the oil industry for measuring kinematic viscosities of 0.02 m^2/s and less[46]. As viscosity is sensitive to variations in temperature, these types of viscometer are always immersed in a constant temperature bath. They are not normally suitable for non-Newtonian fluids although FAROOQI and RICHARDSON[47] have employed a capillary viscometer to characterise a power-law fluid.

Measurement of the Time Taken for a Body to Fall Freely Through a Fluid

The falling sphere instrument is representative of this class of viscometer and is employed with higher viscosity liquids. A stainless steel ball is allowed to sink through a column of the liquid under test, and the time taken to fall from one designated level to another at its terminal falling velocity is recorded. In order that the terminal falling velocity u_0 is attained, the upper timing mark is positioned at a distance of at least six diameters of the ball below the point at which the ball is released. The conditions of the test should be such that Stokes' law can be applied (i.e. at $Re'_0 < ca.\ 0.2$, where $Re'_0 = u_0 d' \rho/\mu$ is the particle Reynolds number and d' is the diameter of the ball (Volume 2, Section 3.2)), and the ratio of the ball to tube diameter should not exceed 1/10 to minimise the drag effects of the tube wall on the terminal falling velocity (Volume 2, Section 3.3.3). British Standard BS 188[48] makes recommendations regarding suitable ball diameters and rates of fall for kinematic viscosities in the range $5 \times 10^{-4} - 0.25\ \text{m}^2/\text{s}$. The viscosity is generally determined from[48]:

$$\nu = \frac{K_v(\rho_s - \rho)}{\rho} t \tag{6.62}$$

where K_v is the viscometer constant (not dimensionless), ρ_s and ρ are the densities of the ball material and the liquid respectively and t is the time taken for the ball to travel between the two timing marks.

Measurement of the Torque Exerted on a Stationary Surface by an Adjacent Moving Surface

Typical of this class of viscometer is the coaxial or Couette type of instrument described in Volume 1, Section 3.7.4. The sample fluid is contained within the annular space between two coaxial cylinders, either of which may be rotated by a motor with the remaining cylinder suspended elastically in such a way that the torsional couple exerted on the latter can be measured. If the outer cylinder of radius r_2 rotates with an angular velocity ω_v and the inner cylinder of radius r_1 is stationary, and the torque (or viscous drag) per unit length of cylinder exerted on the inner cylinder is $\mathbf{T}'$, then, for a Newtonian fluid[49]:

$$\mu = \frac{(r_2^2 - r_1^2)}{4\pi\omega_v r_1^2 r_2^2} \mathbf{T}' \tag{6.63}$$

Hence μ can be determined by measuring $\mathbf{T}'$ at given values of ω_v (Table 6.7).

Major difficulties which arise with such viscometers are end effects due to the drag exerted on the ends of the rotating section, viscous heating due to the work done on the test fluid, and possible misalignment of the axes of the rotating and stationary

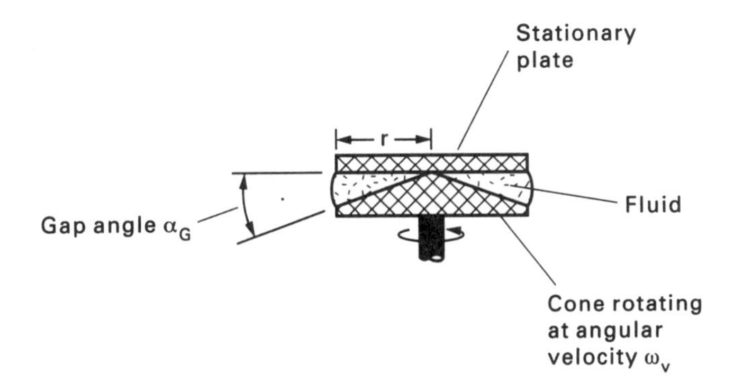

FIG. 6.38. Basis of the cone-and-plate viscometer

sections[50]. Equation 6.63 is established assuming negligible end effects, i.e. that the concentric cylinders are of effectively infinite length. Non-Newtonian behaviour of the test liquid can be taken into account by the use of suitable correction factors[49].

Another frequently employed viscometer of this type involves the cone and plate arrangement shown in Fig. 6.38. If **T** is the torque required to maintain the plate of radius r in a stationary position, then, for Newtonian liquids[51]:

$$\mu = \frac{3}{2\pi\dot{\gamma}r^3}\mathbf{T} \tag{6.64}$$

where edge effects are neglected and $\dot{\gamma}$ is the rate of shear ($\approx \omega_v/\alpha_G$, where ω_v is the angular velocity of the cone and the angle α_G between the cone and the plate is sufficiently small that $\sin \alpha_G \approx \alpha_G$). Equation 6.64 can also be applied in the case of non-Newtonian fluids provided that $\dot{\gamma}$ is reasonably constant throughout the measurement[49].

TABLE 6.7. *Ranges of Common Viscometers*

Viscometer type	Lowest viscosity (Ns/m^2)	Highest viscosity (Ns/m^2)	Shear rate range (s^{-1})
Off-Line (Laboratory) Instruments			
Capillary (Fig. 6.37)	2×10^{-4}	100	$1 - 1.5 \times 10^4$
Falling ball	1	200	indeterminate
Couette	5×10^{-4}	4×10^6	$10^{-2} - 10^4$
Cone and plate (Fig. 6.38)	10^{-4}	10^9	$10^{-3} - 10^4$
Oscillating cylinder (Weissenberg rheogoniometer)	10^{-4} [b]	5×10^6 [b]	$7 \times 10^{-3} - 9 \times 10^3$
On-Line Instruments			Applications
Capillary (Fig. 6.39)	2×10^{-3}	4	Newtonian fluids (e.g. lubricating oils, fuel oils). Repeatability ±0.5 per cent of FSD[a].
Couette type (Fig. 6.40)	10^{-3}	5×10^3	In-line or in-tank. Suitable for Newtonian and non-Newtonian fluids. Repeatability ±0.2 per cent of FSD[a].
Vibrating rod or cylinder (Fig. 6.41)	10^{-4} [b]	2×10^3 [b]	In-line or in-tank. Suitable for Newtonian and non-Newtonian fluids. Can be used for mineral slurries. Repeatability ±0.25 per cent of FSD[a].

[a] Repeatability and FSD (full scale deflection)—see Section 6.10.1.
[b] This is the *dynamic viscosity in small amplitude oscillatory shear* which is the *real* component of the *complex shear viscosity* which is a function of the angular frequency of oscillation.

By the Reaction of a Vibrating Element Immersed in the Liquid

This instrument employs an oscillating sphere, cylinder, paddle or reed. The range of a typical instrument (Weissenberg Rheogoniometer) is given in Table 6.7. The amplitude of the oscillation varies with the viscosity of the liquid under test and this variation in amplitude is detected by a suitable transducer. Such devices are sensitive to changes in density as well as viscosity and, in order to minimise variations in density, the temperature of the measured fluid and the transducer is carefully controlled. This procedure is also suitable for the characterisation of fluids exhibiting non-Newtonian properties[49].

6.7.2. Continuous On-line Measurement of Viscosity

The on-line measurement of viscosity under plant conditions poses particular difficulties. This is due to the wide range of viscosities that can occur within a process plant, to the difficulty of obtaining reliable measurements (particularly for non-Newtonian fluids) and to the accuracy that is often required (e.g. better than within ±1 per cent for lubricating oils). Variables which can affect the measured viscosity are the temperature, pressure and rate of flow of the sampled stream—quite apart from the normal errors that can occur in any similar instrument (e.g. due to variations in supply voltage and frequency, sample contamination, sample not being representative of the bulk fluid, etc.). Automatic temperature compensation is always required and, in the case of multiphase systems, the difficulty of obtaining a representative sample is considerable (see Section 6.9). In this instance

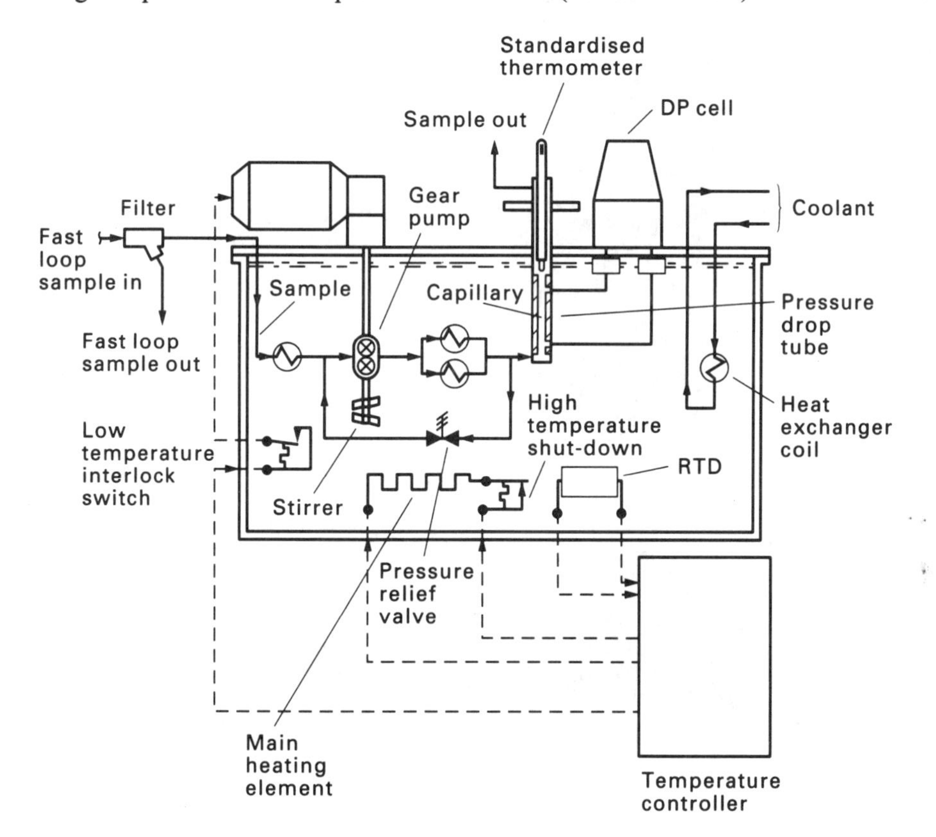

FIG. 6.39. On-line capillary viscometer

it is advisable to pass the whole of the process stream through the instrument where possible.

Some instruments are available for process on-line application (Table 6.7). Capillary types give good results with Newtonian fluids and are based upon Poiseuille's law (Volume 1, equation 3.30). The pressure drop across the capillary

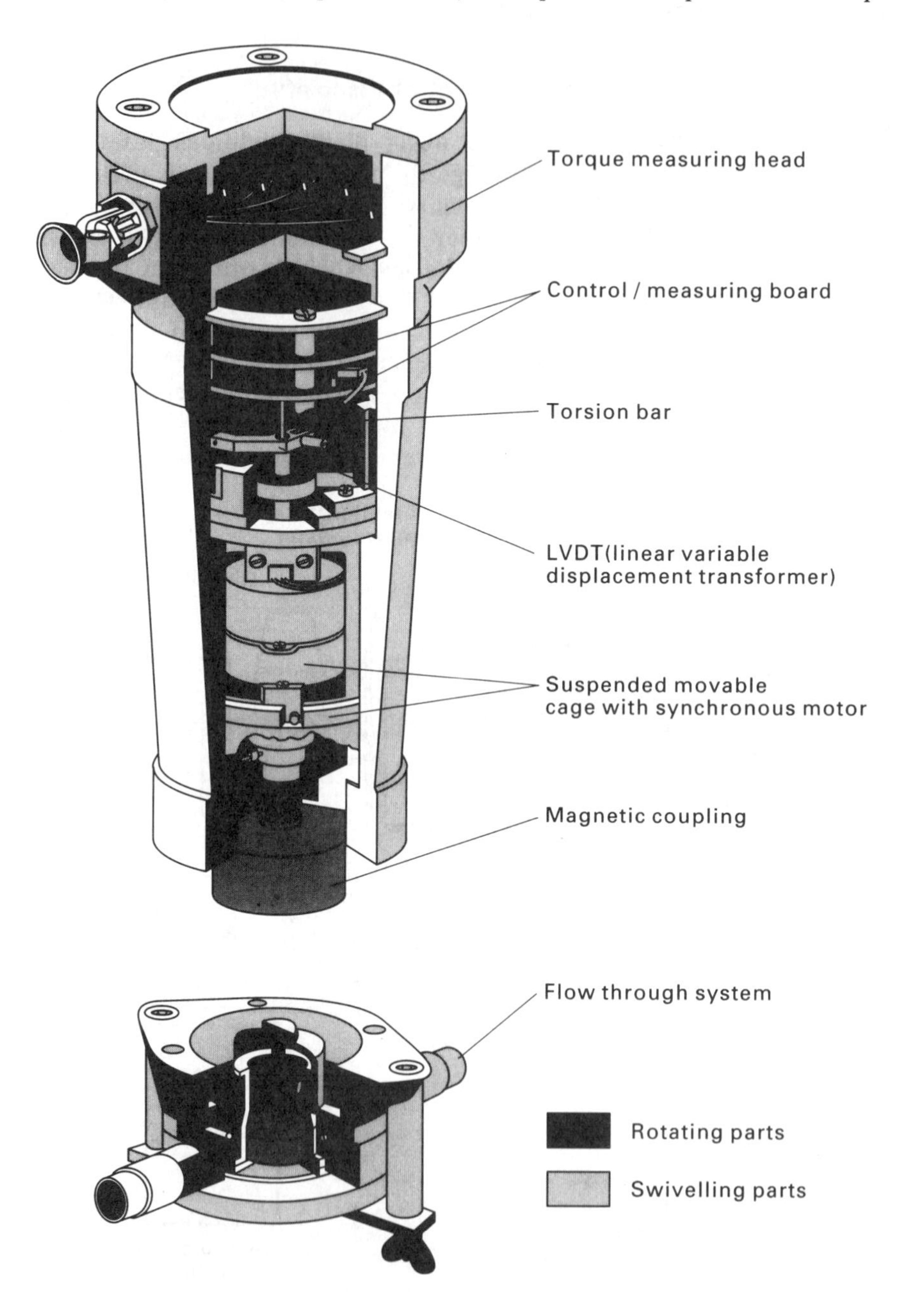

FIG. 6.40. Continuous process viscometer based upon the Couette principle

section and the temperature of the sample stream are carefully controlled (Fig. 6.39)[52]. A fast loop sampling system (Section 6.9) is often provided as a means of reducing the total response time of the instrument (usually about 3 min).

Continuous viscometers based upon the Couette principle are able to measure the viscosity of both Newtonian and non-Newtonian fluids over a wide range (Table 6.7). A typical instrument of this type is illustrated in Fig. 6.40[53].

A cylinder is rotated within the fluid by means of a magnetic coupling with a rotating magnet driven by a synchronous motor. The latter is suspended by a torsion bar and the torque reaction of the motor caused by the viscous drag on the rotating cylinder is balanced by the resistance of the torsion bar. The angle of deflection of the bar (which is a function of the viscosity) is measured and converted into a 4–20 mA output signal. At small rates of flow ($< 7 \times 10^{-4}$ m^3/s) the whole stream can be passed through the instrument—otherwise a sampling system is necessary.

Direct in-line measurement of viscosity can also be achieved by employing an oscillating sensor. The latter may be driven by an electromechanical system (Fig. 6.41)[54] or by a pulse of current through a piece of magnetostrictive alloy[55]. The electromechanical type is the more complex, but also the more rugged, and can cover a much wider range of viscosities (as listed in Table 6.7)—oscillating sphere versions being used over lower viscosity ranges and oscillating rods for higher ranges. A temperature sensor is inserted within the sphere or at the tip of the rod. The less expensive pulsed current viscometer produces ultrasonic oscillations within a thin blade sensor. The blade protrudes through a metal diaphragm and, once energised, oscillates longitudinally at its natural frequency which is determined by its length. The amplitude of the oscillation is less than 5×10^{-7} m and is damped by the flowing liquid. The degree of damping depends upon the viscoelastic properties of the material being tested. When the amplitude has been damped to a predetermined level, a further drive pulse is initiated. Hence, as the viscosity increases, the frequency N_p of the drive pulse also increases such that, for Newtonian materials:

$$\mu = K_v'' \left(\frac{N_p^2}{\rho} \right) \tag{6.65}$$

where K_v'' is the sensor constant (not dimensionless). Temperature compensation is provided and the technique can be applied to non-Newtonian fluids. Care must be taken when positioning oscillation viscometers on a plant as they are sensitive to plant vibration.

6.8. THE MEASUREMENT OF COMPOSITION

A knowledge of the composition of a process stream is often of major importance. Such information may be necessary to determine whether a particular product has the required specification or whether the composition of a particular stream is changing. Frequently, variations in either will require some kind of control action to maintain the planned operational strategy of the plant. Direct measurement of composition is, at first sight, the obvious method of determining the nature of the materials entering, leaving or passing from point to point within a process. In practice, however, there are many difficulties. In the first instance, it is necessary to

(*a*)

Push-pull
driver and
pickup

(*b*)

FIG. 6.41. Electromechanically-driven rod viscometer

decide what property of the material can be employed as a suitable measure of its composition. This property may be physical or chemical and should be a quantity that differs widely between the various components of the material stream being analysed. It is also important to ascertain whether off-line or on-line analysis is required. The latter is necessary if the analyser is to be the measuring element for an automatic control system (see Chapter 7) in which case the response time (i.e. time constant) of the analyser must be substantially smaller than that of the plant it is being used to control. If off-line analysis is sufficient, then such considerations are of less importance.

There are numerous properties of materials which can be used as measures of composition, e.g. preferential adsorption of components (as in chromatography), absorption of electromagnetic waves (infra-red, ultra-violet, etc.), refractive index, pH, density, etc. In many cases, however, the property will not give a unique result if there are more than two components, e.g. there may be a number of different compositions of a particular ternary liquid mixture which will have the same refractive index or will exhibit the same infra-red radiation absorption characteristics. Other difficulties can make a particular physical property unsuitable as a measure of composition for a particular system, e.g. the dielectric constant cannot be used if water is present as the dielectric constant of water is very much greater than that of most other liquids. Instruments containing optical systems (e.g. refractometers) and/or electromechanical feedback systems (e.g. some infra-red analysers) can be sensitive to mechanical vibration. In cases where it is not practicable to measure composition directly, then indirect or *inferential* means of obtaining a measurement which itself is a function of composition may be employed (e.g. the use of boiling temperature in a distillation column as a measure of the liquid composition—see Section 7.3.1).

The composition analysers having the widest application are those which separate the components within a mixture as part of the analysis. The most common examples of this are the chromatograph and the mass spectrometer. Mass spectrometers are expensive and on-line chromatographic analysis can suffer from substantial time lapses (regarded as distance–velocity lags for control purposes—Section 7.6) before the result of the analysis is known. A further limitation of these two methods is that they are suitable generally for volatile liquids only, although liquid phase chromatographs are available (Volume 2, Section 19.4.3).

In the following account, the more common means of composition analysis are presented. The descriptions are not intended to be exhaustive as many methods of analysis are specific to particular processes. Reference should be made to manufacturers' literature for further information.

6.8.1. Photometric Analysers

These include such instruments as opacity monitors, turbidimeters, colorimeters, refractometers and spectrophotometers. A selection of these is described—particularly where the instrument has a more general application as an on-line process analyser and/or to illustrate a general principle of operation. It is likely that development of fibre-optic techniques (Section 6.12.4) will extend the use of this type of sensor in the future[56].

Spectrophotometers

These instruments constitute a class of *spectroradiometric* analysers generally encountered in the chemical process industries and employed to monitor wavelengths between the middle infra-red (MIR) and ultra-violet (UV) regions of the electromagnetic spectrum (Table 6.8).

TABLE 6.8. *Regions of the Electromagnetic Spectrum*

Region	Wavelength range
FIR (far infra-red)	15–800 μm
MIR (middle infra-red)	3–15 μm
NIR (near infra-red)	780–3000 nm
VIS (visible light)	380–780 nm
UV (ultra-violet)	200–380 nm
XUV (extreme ultra-violet)	10–200 nm

The radiation may be due to emissions from a hot source, or to the luminescence, fluorescence or phosphorescence of the sample. An *emission spectrum* consists of a number of generally very narrow peaks (called *spectral lines*) occurring at certain wavelengths which are characteristic of the materials contained within the source. The amplitudes of the peaks are related to the *abundance* or concentration of the materials present. Alternatively, radiation from a source is passed through a sample. In this case the quantity absorbed by the sample at a particular wavelength is again characteristic of the materials present in the sample. This is termed *absorption spectrometry* and produces spectral transmission lines in the form of equally narrow valleys—or peaks (Fig. 6.42) where the information is expressed in terms of *absorbance* ($\mathcal{A}$) rather than *transmittance* ($\mathcal{T}$)[57], and:

$$\mathcal{A} = \log\left(\frac{1}{\mathcal{T}}\right) \tag{6.66}$$

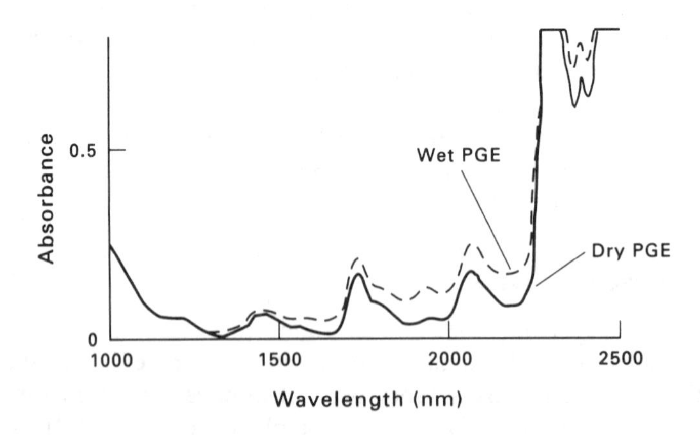

FIG. 6.42. Absorbance spectra of wet and dry polyalkylene glycol ether (brake fluid) in the near infra-red region of the electromagnetic spectrum[57]

Dispersive Photometers

These are distinguished from non-dispersive instruments by the presence of a wavelength dispersing device or *monochromator*. This receives a wide band of wavelengths and transfers a selection of these (i.e. a narrow band of wavelengths) via an exit slit to a suitable detector. Photometers using visible light employ interference filters, prisms or diffraction gratings as monochromators, the latter being the most common. The grating is rotated by a suitable drive and the band of wavelengths of interest is scanned by sweeping the diffracted beam over a very narrow exit slit so that the detector senses one wavelength at a time. The wavelength sensed is dependent upon the characteristics of the grating and its angular displacement. Similar diffraction-grating monochromators are used in photometers operating in the infra-red, visible light and ultra-violet regions and can give resolutions better than 0.01 nm for wavelengths from 30 nm to 50 μm[25].

Non-Dispersive Photometers

These differ from dispersive instruments in that the monochromator is replaced by a *narrow-band-pass* filter which is selected for the wavelength of maximum emission (or maximum absorption) of a specific material whose relative abundance is to be determined. The instrument of this type most commonly employed in the process industries to measure concentrations of gases in flowing gas mixtures is the *non-dispersive infra-red* (*NDIR*) analyser. This instrument works on the basis of the absorption of infra-red radiation at specific wavelengths (generally in the NIR region) peculiar to particular gases. A simplified diagram of a typical arrangement is shown in Fig. 6.43[58].

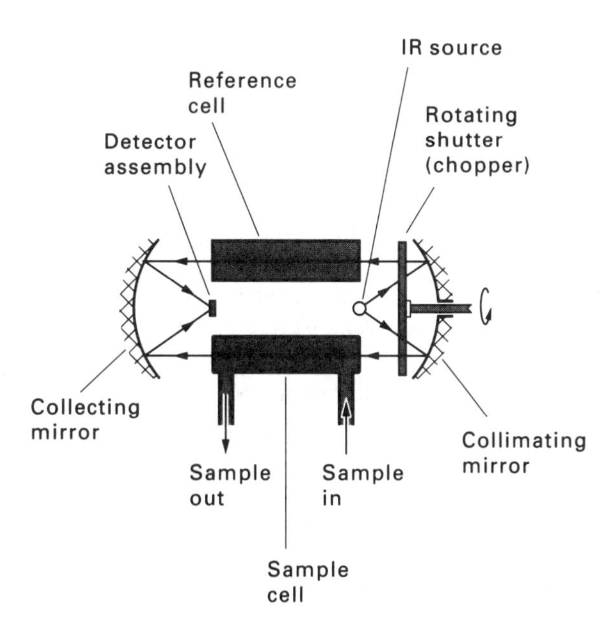

FIG. 6.43. Schematic diagram of a dual-beam NDIR analyser

Two beams of infra-red radiation of equal energy (see also Section 6.4.2) are chopped by a rotating shutter which allows the beams to pass simultaneously but intermittently through a sample cell which contains the flowing sample and a reference cell. The latter is filled with a non-absorbing gas. The detector assembly consists of two sealed absorption chambers separated by a thin metal diaphragm which acts as one plate of a parallel-plate capacitor. The quantity of infra-red radiation at a particular wavelength passing through the sample cell is a direct function of the concentration of absorbing gas in the sample. This is compared with the equivalent radiation passing through the non-absorbent gas in the reference cell.

Absorbent gases which can be monitored using this type of detector include CO, CO_2, N_2O, SO_2, H_2O, CH_4, C_2H_5OH, CH_3COCH_3, C_6H_6 and other hydrocarbons. Common non-absorbing gases are O_2, N_2, Cl_2, A and He. Relative sensitivities to various absorbing gases are given in Table 6.4.

TABLE 6.9. *Sensitivities of a Typical Infra-red Analyser to Various Absorbing Gases Expressed as a Comparison to that for Water Vapour*[58, 59]

Gas	Sensitivity
H_2O	1.0
CO	0.5
CO_2	0.1
CH_4	0.5
C_2H_4	1.0
N_2O	0.1
NO_2	1.0
SO_2	0.2
HCN	1.0
CH_3COCH_3	2.5
C_6H_6	2.5

The length of the absorption tube depends upon the gas being sampled and the concentration range to be covered. The energy absorbed $\mathbf{E}_{abs}$ by a column of gas of path length ℓ_p is:

$$\mathbf{E}_{abs} = \mathbf{E}_{inc}\{1 - \exp(-k_{abs}C\ell_p)\} \tag{6.67}$$

where $\mathbf{E}_{inc}$ is the incident energy, C is the concentration of the gas being measured and k_{abs} is an absorption constant. When $k_{abs}C\ell_p << 1$, i.e. at low concentrations, the relation between $\mathbf{E}_{abs}$ and C is approximately linear. The instrument is relatively insensitive to variations in temperature. For water vapour a figure of 3 per cent variation in sensitivity per 1 K change in temperature has been reported[60].

The conventional dual beam instrument is now being superseded by totally solid-state devices based upon single beam, multi-wavelength technology[61]. Advantages claimed for the latter are improved stability, resistance to shock and vibration, and insensitivity to the effects of sample cell contamination—which is a common problem when using IR absorption techniques to analyse dirty gases (e.g. flue gases).

The Interaction of Light with Materials

Refractometry

This involves the measurement of the *refractive index* of an optically transparent material. When light passes between two optical media of differing density, it will be refracted and Snell's law applies, viz:

$$\frac{\eta_1}{\eta_2} = \frac{\sin \alpha_2}{\sin \alpha_1} \tag{6.68}$$

where α_1 and α_2 are the *angles of incidence* and η_1 and η_2 are the *refractive indices* for the two media respectively.

If the composition of an optically transparent sample changes, then its refractive index will also change. The *continuous process refractometer* generally measures the effect that this change in refractive index has on the *critical angle* occurring at the interface between a glass prism and the sample. When the angle of incidence within the prism α_3 becomes so large that the light cannot emerge, then the ray is totally internally reflected (Fig. 6.44). The minimum value of α_3 at which this occurs is called the critical angle α_c, where:

$$\frac{\eta_2}{\eta_1} = \frac{1}{\sin \alpha_c} \tag{6.69}$$

Thus, if η_2 (the refractive index of the glass of the prism) is known, then η_1 (the refractive index of the sample) can be determined by measuring α_c.

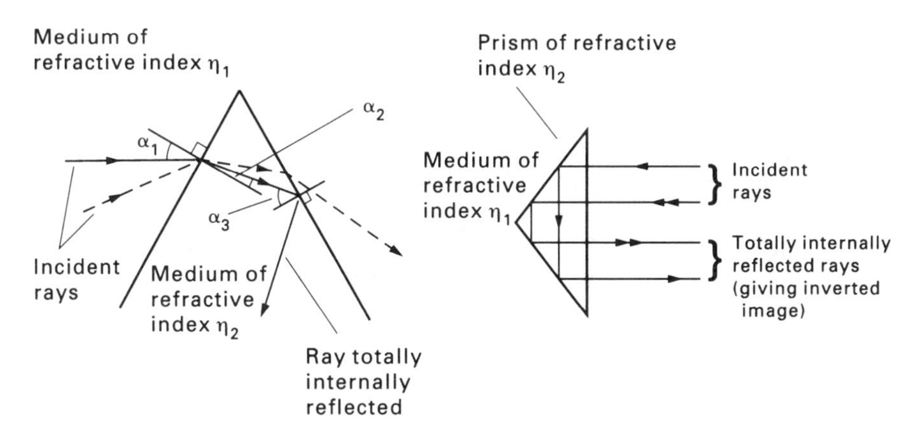

FIG. 6.44. Examples of total internal reflection

A typical arrangement for an on-line process refractometer is shown in Fig. 6.45. Besides the prism, which is in contact with the process liquid, there is commonly a light source, a beam collimator and a photo-conductive sensor. As the nature or concentration of the process liquid varies, the critical angle changes, so moving the internally reflected collimated beam of light across the photo-detector. The latter is divided into two sections, one of which is arranged to receive a constant amount of

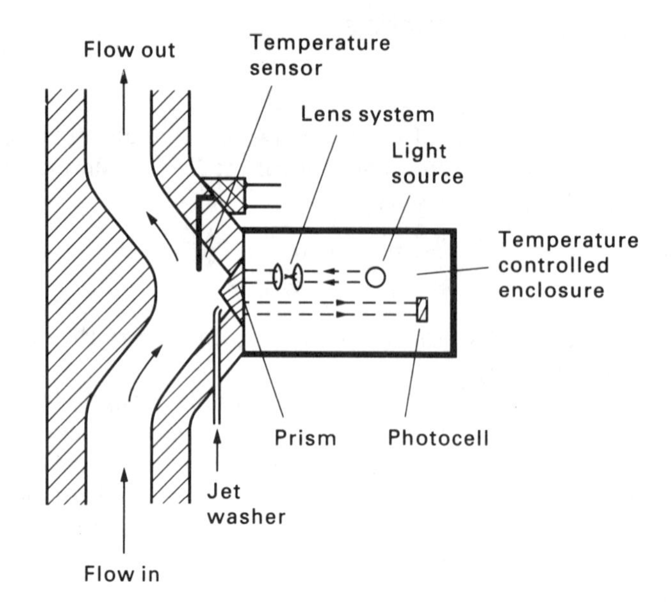

FIG. 6.45. Continuous valve-body type refractometer suitable for measuring concentration of dissolved solids

light (as a reference) and the other which responds to light which varies in intensity as the critical angle changes. Such instruments can be used also for measuring concentrations of solids in liquids but, in this application, fouling of the optical surface can be a problem.

Turbidimeters

These measure *turbidity*, usually by sensing light scattered at 90° to the direction of the incident light beam. This type of turbidimeter is frequently termed a *nephelometer* or *nephelometric turbidimeter*. A typical arrangement is shown in Fig. 6.46. The nephelometer is commonly used for detecting solid particles present in water (e.g. waste water) or in air (e.g. smoke). The instrument requires periodic cleaning due to fouling of the optical system.

Opacity Monitors

These are similar to nephelometers, except that they measure the attenuation of a light beam due to the combined effects of absorption and scattering by the sample. The instrument consists of a light source, a collimator and a photo-detector. The most common application is the measurement of smoke density in chimney stacks. In this case the optical surfaces exposed to the smoke are kept clean by flows of clean air. The density of the smoke is expressed in terms of *per cent opacity, per cent transmittance* or *optical density*, where:

$$\text{per cent opacity} = 100 \times \text{opacity} = 100 - \text{per cent transmittance} \qquad (6.70)$$

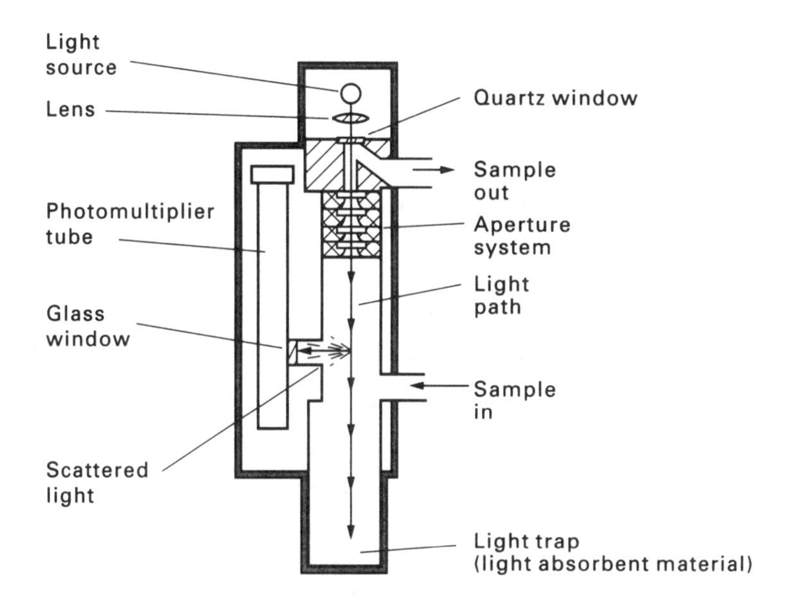

FIG. 6.46. Continuous nephelometric turbidimeter

and: optical density = – log (1 – opacity) (6.71)

More modern instruments employ lasers as the light source.

6.8.2. Electrometric Analysers

These instruments use electrochemical methods to determine composition.

Conductivity Cells

The conductivity of a conducting liquid is very sensitive to the presence of small quantities of electrolytes. Dilution not only increases the proportion of dissolved electrolyte which forms ions in solution, but also reduces generally the number of ions per unit volume. To negate the latter effect, a *molar conductivity* Λ is defined as:

$$\Lambda = \frac{\kappa}{C} \tag{6.72}$$

where κ is the conductivity (i.e. conductance per metre) and C is the concentration. Conductance is measured in *Siemens* (S) and is the reciprocal of the electrical resistance measured in ohms. (Hence, if C is measured in kmol/m^3 and κ in S/m then the units of Λ will be Sm^2/kmol.)

The molar conductivity of an electrolyte at infinite dilution Λ^0 is given by Kohlrausch's law, i.e.:

$$\Lambda^0 = \lambda_a^0 + \lambda_c^0 \tag{6.73}$$

where λ_a^0, λ_c^0 are the ionic conductivities per unit charge of the anions and the cations, respectively, of the dissolved electrolyte at infinite dilution (see Table 6.10).

Kohlrausch's law can be assumed to apply at concentrations up to about 10^{-4} kmol/m^3. For a fully dissociated electrolyte at these concentrations, from equations 6.72 and 6.73:

$$\kappa = (n_a Z_a \lambda_a^0 + n_c Z_c \lambda_c^0) C \tag{6.74}$$

where n_a, n_c represent respectively the numbers of anions and cations produced by the dissociation of one molecule of the electrolyte and Z_a, Z_c are the respective charges on each ion. Values of λ_a^0 and λ_c^0 vary greatly with temperature (e.g. for Na^+, $\lambda_c^0(298K) = 5.01$ Sm^2/kmol and $\lambda_c^0(373K) = 14.50$ Sm^2/kmol) and approximate values can be calculated from:

$$\lambda_T^0 = \lambda_{298}^0[1 + \beta_c(T - 298)] \tag{6.75}$$

where β_c is a temperature coefficient for a specific ion and T is in K.

TABLE 6.10. *Some Ionic Conductivities at Infinite Dilution at 298* K (λ_{298}^0 Sm^2/kmol)[62]

Anion	λ_a^0	Cation	λ_c^0
OH^-	19.91	H^+	34.98
$\frac{1}{2} SO_4^{2-}$	8.00	NH_4^+	7.36
Cl^-	7.64	$\frac{1}{2} Ca^{2+}$	5.95
NO_3^-	7.15	$\frac{1}{2} Cu^{2+}$	5.36
$\frac{1}{2} CO_3^{2-}$	6.93	$\frac{1}{2} Mg^{2+}$	5.31
CH_3COO^-	4.09	Na^+	5.01

Many different types of conductivity cell are available but the principle of operation is the same in the great majority of cases. A small current is passed through the sample liquid between two electrodes using a square wave voltage source (this type of excitation minimises the effects of polarisation)[63]. If the electrodes are of fixed area A_e and constant distance x_e apart, then the conductance is:

$$\kappa' = \frac{\kappa A_e}{x_e} = \frac{1}{R_c} \tag{6.76}$$

assuming that the liquid between the electrodes maintains a uniform cross-section. Thus, if the resistance R_c is measured (usually by means of a Wheatstone bridge network), the concentration of the electrolyte can be determined from equation 6.74. Often equation 6.76 is expressed in terms of a cell constant k_c such that $\kappa = k_c/R_c$. The cell constant employed depends upon the conductivity of the solution being measured. Conductivities of between 5 μS/m and 1000 S/m can be monitored using corresponding cell constants of between 1 m^{-1} and 5000 m^{-1} respectively. Cell constants can be affected by the polarisation of the sample material. The extent of the polarisation depends largely upon the frequency of the excitation current used and the nature of the electrode surfaces. This is particularly so at higher values of conductivity (>0.1 S/m) when it is necessary to treat or coat the electrodes. Increasing the frequency of the current reduces polarisation but increases errors due to capacitance effects.

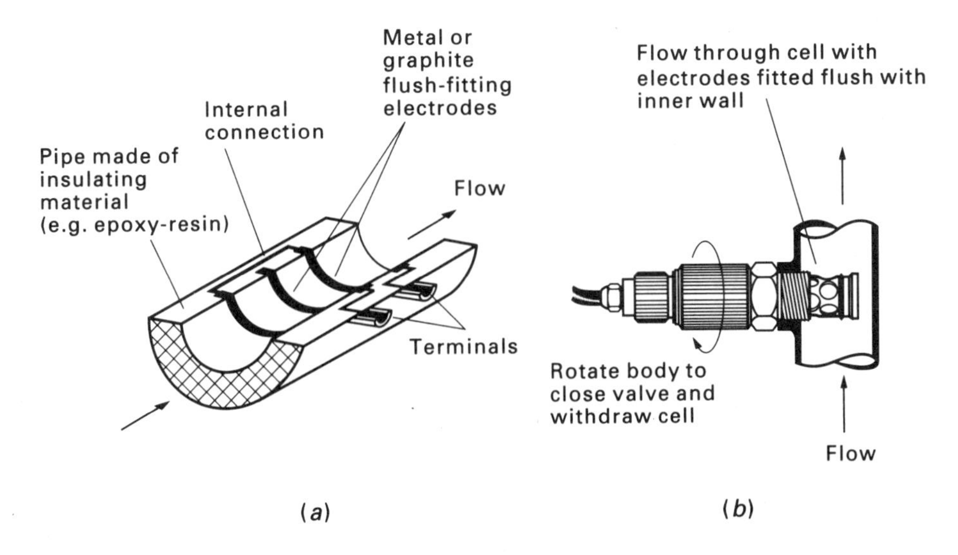

FIG. 6.47. Electrical conductivity cell: (*a*) section through flow-through conductivity cell showing arrangement of electrodes; (*b*) screw-in cell inserted in pipeline

Flow-through conductivity sensors suitable for insertion in pipelines (see Fig. 6.47*a*) are now available for use at temperatures up to 480 K and pressures up to 1700 kN/m^2 [64]. As conductivity is temperature sensitive, a thermistor is usually included in the detector circuit as part of a temperature compensator. Screw-in cells (Fig. 6.47*b*) will withstand higher pressures. More recently, electrodeless methods of measuring conductivity have become available. In this case the solution is placed between two energised toroids. The output voltage of the instrument (from the output toroid circuit) is proportional to the conductivity of the solution provided that the input voltage remains constant. This type of conductivity meter can be used under much more severe conditions, e.g. with highly corrosive or dirty systems[63].

pH Measurement and Other Ion Selective Electrodes

pH is defined as:

$$\text{pH} = \log\left(\frac{1}{[\text{H}^+]}\right) \tag{6.77}$$

where $[\text{H}^+]$ is the concentration of hydrogen ions in solution in appropriate units. Thus pH varies between 0 for solutions having a hydrogen ion concentration of 1 $kmol/m^3$ (strong acid) and 14 when the hydrogen ion concentration is 10^{-14} $kmol/m^3$ (strong alkali) with the neutral point (pure water) having a pH of 7. pH values of some common materials are listed in Table 6.11. A pH sensor enables the degree of acidity or alkalinity of a liquid to be determined and is of particular use in monitoring effluent and discharges from chemical works, refineries, etc.

pH as described by equation 6.77 is difficult to measure and an operational definition of pH (BS 1647[65]) is obtained by specifying the EMFs E_A and E_B of two standard cells, where:

E_A is the EMF of the cell: Pt–H_2/solution **A**/concentrated KCl solution/reference electrode.

E_B is the EMF of the cell: Pt–H_2/solution **B**/concentrated KCl solution/reference electrode.

Both cells are at the same temperature and the reference and bridge solutions are identical in the two cells. Then:

$$pH_A - pH_B = \frac{E_A - E_B}{2.303 \dfrac{\mathbf{R}T}{\mathbf{F}}} \tag{6.78}$$

where **R** is the universal gas constant, T is the temperature in K and **F** is the *Faraday Number*, i.e. 9.6490×10^7 coulombs/kmol (96,490 coulombs/mole).

TABLE 6.11. *pH Values of some Common Materials at 298K*[62]

Material	Concentration (kmol/m^3)	pH
	Acids	
Hydrochloric acid	1.0	0.1
Sulphuric acid	1.0	0.3
Hydrochloric acid	0.1	1.1
Sulphuric acid	0.1	1.2
Sulphurous acid	0.1	1.5
Oxalic acid	0.1	1.6
Acetic acid	1.0	2.4
Benzoic acid	saturated	2.8
Acetic acid	0.1	2.9
Carbonic acid	saturated	3.8
Hydrogen sulphide	0.1	4.1
Ammonium sulphate	0.1	5.5
	Alkalis	
Sodium bicarbonate	0.1	8.4
Calcium carbonate	saturated	9.4
Magnesium hydroxide	saturated	10.5
Ammonia	0.1	11.1
Ammonia	1.0	11.6
Sodium carbonate	0.1	11.6
Calcium hydroxide	saturated	12.4
Sodium hydroxide	0.1	13.0
Sodium hydroxide	1.0	14.0

It is not easy to set up a hydrogen electrode, hence subsidiary reference electrodes are used where the potentials of these relative to the standard hydrogen electrode have been determined previously. Practical considerations limit such electrodes to those consisting of a metal in contact with a solution which is saturated with a sparingly soluble salt of the metal and which contains also an additional salt with a common anion. Examples of commonly used electrodes are the silver/silver chloride electrode, i.e. Ag/AgCl(s)KCl(aq), and the mercury/mercurous chloride electrode (known as the *calomel* electrode), i.e. Hg/Hg_2Cl_2(s)KCl(aq). In each case the potential of the reference electrode is governed by the activity of the anion in solution, which can be shown to be constant at a given temperature.

A typical pH measuring circuit consists of two electrodes, viz. the *pH sensing electrode* and the *reference electrode*, immersed in the fluid to be measured together with a high-impedance voltage measuring system (see Fig. 6.48*a* and Section 6.11.6). The most commonly employed pH electrode is the glass electrode (Fig. 6.48*b*) which consists of a glass body and an electrode wire. The latter connects an external coaxial cable to the internal reference electrode which is immersed in a solution of constant pH. This solution wets a specially constructed glass membrane which is selective to H^+ ions (i.e. to pH). The exterior of this membrane is immersed in the solution to be measured. Figure 6.48*c* is a diagrammatic representation of the reference electrode which is similar to the pH sensing electrode, except that the glass membrane is replaced by a porous plug allowing a liquid/liquid interface to exist between the salt solution and the sample. The Ag/AgCl(s)KCl(aq) reference electrode is a silver wire coated with solid silver chloride and the salt solution is 3 molar KCl ($3 kmol/m^3$).

The measured potential across the system (Fig. 6.48*a*) is:

$$E = E_{ir} + E_m + E_j - E_{er} \qquad (6.79)$$

where E_{ir}, E_m, E_j and E_{er} are the EMFs generated at the internal reference electrode, the membrane, the liquid junction and the external reference electrode, respectively. Under normal conditions, E_j is negligible and E_{ir} and E_{er} are constant. Hence, equation 6.79 becomes:

$$E = E_0 + E_m \qquad (6.80)$$

where E_0 is a constant.

A typical half-cell reaction may be written:

$$\alpha\mathbf{A} + \beta\mathbf{B} + \ldots\ldots \rightarrow \gamma\mathbf{C} + \delta\mathbf{D} + \ldots\ldots$$

where α, β, etc. are the numbers of the respective ions denoted by **A**, **B**, etc. Then the electrode potential can be expressed in terms of the *Nernst Equation*[66].

i.e.:

$$E = E_0 + \frac{\mathbf{R}T}{n\mathbf{F}} \ln\left(\frac{[\mathbf{A}]^\alpha[\mathbf{B}]^\beta \ldots}{[\mathbf{C}]^\gamma[\mathbf{D}]^\delta \ldots}\right) \qquad (6.81)$$

where [**A**] is the ionic concentration of component **A**, etc., and n is the net number of negative charges transferred in the reaction.

For any ion-selective electrode, equation 6.81 reduces to:

$$E = E_0 + \frac{\mathbf{R}T}{n\mathbf{F}} \ln a \qquad (6.82)$$

where a is the activity of the particular ions involved (i.e. of H^+ ions for a pH meter). Thus, it should be noted that pH measuring devices actually measure the effective concentration (or activity) of the H^+ ions and not the actual concentration. Hence, defining pH in terms of the H^+ activity a_{H^+} (i.e. $pH = -\log(a_{H^+})$ where $a_{H^+} = (\gamma_{H^+})[H^+]$ and γ_{H^+} is the activity coefficient), then from equation 6.82 (assuming that the electrode is entirely selective to H^+ ions):

$$pH = \frac{\mathbf{F}(E_0 - E)}{2.303\mathbf{R}T} \qquad (6.83)$$

and for any positive ion carrying one elementary charge (e.g. H^+) at 298 K:

$$E - E_0 = 59.2 \times 10^{-3} \log a \tag{6.84}$$

where E, E_0 and the coefficient (59.2×10^{-3}) are expressed in V. Thus the nominal output of a pH electrode is 59.2 mV/pH at 298K. The actual output can vary from

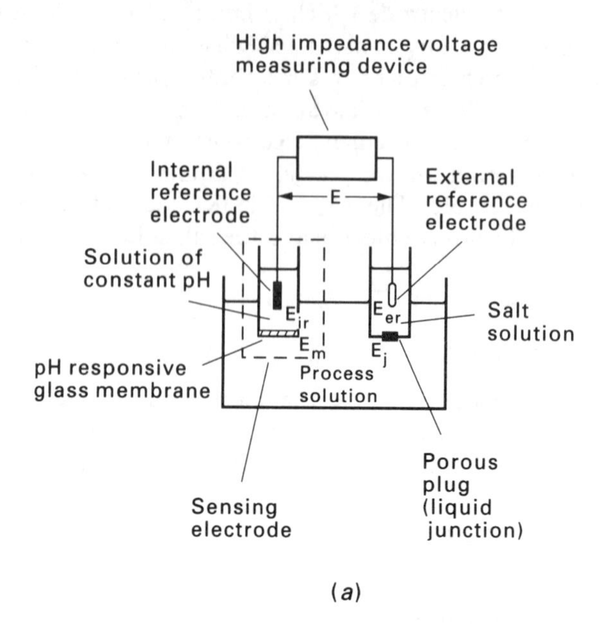

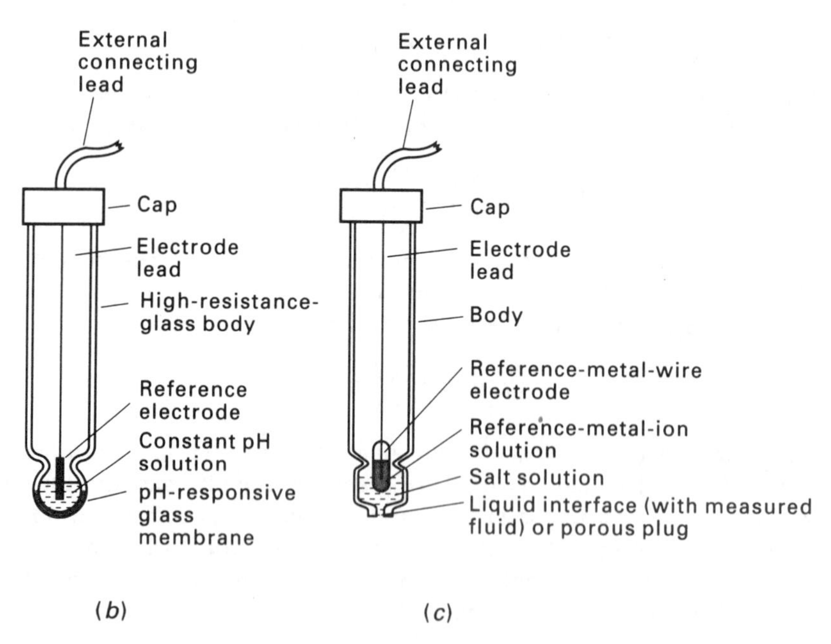

FIG. 6.48. Schematic diagrams of pH sensing electrodes: (*a*) pH measurement system; (*b*) pH electrode; (*c*) reference electrode

this value and a means of adjustment is provided to correct for this against a solution of known pH.

Electrodes which are selective to ions other than H^+ are commonly available. These are called *specific ion* or *pIon* sensors and are listed in Table 6.12. Major difficulties with pH and pIon sensors are non-linearity, noise and sensitivity. Fouling of the probe can also be troublesome and many industrial instruments employing these principles are provided with the means of cleaning electrode surfaces ultrasonically.

TABLE 6.12. *Some Commonly Available Specific Ion (pIon) Electrodes*[58, 67]

Type of Electrode	Type of Construction	Selective to Ions
Glass	As for H^+ and pH.	Na^+, K^+, NH^+, Ag^+ and other univalent cations
Solid state	Membrane consists of compacted disc or single crystal of active material.	F^-, Cl^-, Br^-, I^-, S^{2-}, Ag^+, Cu^{2+}, Pb^{2+}, Cd^{2+}
Heterogeneous membrane	Similar to solid-state type but active material dispersed in inert matrix.	Cl^-, Br^-, I^-, S^{2-}, Ag^+
Liquid ion exchange	Internal reference solution and measured solution are separated by a porous layer holding liquid of low water solubility in which is dissolved large molecules containing the ions of interest.	Ca^{2+}, Cu^{2+}, Pb^{2+}, Cl^-, ClO_4^-, NO_3^-, BF_4^-
Gas sensing membrane	Not true membrane electrodes as no current passes across the membrane. The ion being determined diffuses through the membrane into an electrochemical cell. The consequent change in the chemistry of the cell is monitored by an ion sensitive electrode.	NH_3, SO_2, CO_2

ORP (Redox) Sensors

When a suitable electrode is immersed in a solution containing ions of a material in two different states of oxidation (e.g. Cu^+ and Cu^{2+}, Fe^{2+} and Fe^{3+}) the electrode acquires a potential which depends upon whether the ions tend to move towards the higher or towards the lower oxidation state. If the tendency is towards the higher oxidation state then the solution has reducing properties and the ions will preferentially discharge electrons to the electrode which will become negatively charged relative to the solution. If, on the other hand, the ions in solution tend to move towards the lower oxidation state (i.e. if the solution has oxidising properties) then the ions will take electrons from the electrode which will then become positively charged with respect to the solution. Hence, the electrode potential (*oxidation – reduction* potential, *ORP*, or *redox* potential) is a measure of the oxidising or reducing power of the solution in which it is immersed. The ORP (E_{orp}) can be calculated from the activities of the ion in the oxidation states using equation 6.82:

$$E_{orp} = E_0 + \frac{\mathbf{R}T}{n\mathbf{F}} \ln\left(\frac{\text{activity of ion in higher oxidised state}}{\text{activity of ion in lower oxidised state}}\right) \tag{6.85}$$

The ORP electrode is similar to the pH electrode except that it is made from platinum, gold or silver on a platinum base. A reference electrode is required which may be either the Ag/AgCl or the calomel electrode used in pH determination. Combined pH/ORP instruments are also available and the user can switch to whichever facility is required.

Polarographic Sensors

When two dissimilar metals are immersed in an electrolyte and connected together, a current will flow due to the build-up of electrons on the more electropositive electrode. This current soon stops due to the polarisation of the cell. If a suitable depolarising agent is added, then current will continue to flow. In the polarographic sensor the sample is arranged as the depolarising agent and the magnitude of the current produced is a function of the concentration of the material to be detected in the sample. In some polarographic instruments the reaction and the current in the cell are maintained by applying an external voltage (≈ 0.7 V) across the cell. This type of sensor is used to measure pollutants such as Cl_2, HCl, HBr and HF in air; for determining the concentration of O_2 in flue gas (up to about 25 per cent); and for measuring quantities of inert gases. Accuracies of ±1 per cent of full scale deflection have been reported[63].

High Temperature Ceramic Sensors (Zirconia Cells)

Zirconium oxide (*zirconia*—ZrO_2) doped with small quantities of yttria (Y_2O_3) has a stable lattice structure which permits oxygen ion conduction at high temperatures. The sensor is constructed by placing a heated section of the doped ZrO_2 in contact with the sample gas containing O_2 and a reference gas (usually air). Figure 6.49*a* shows a particular arrangement in which two porous platinum electrodes are attached to the opposite sides of a disc of the zirconia electrolyte which is heated electrically to about 1000 K[68]. A potential difference E_{Zr} is set up between the two electrodes which is a function of the partial pressures of the concentrations of O_2 in the sample and in the reference. From equation 6.82:

$$E_{Zr} = E - E_0 = 0.0496T \log\left(\frac{P'_0}{P'}\right) + k_{Zr} \tag{6.86}$$

where E, P' and E_0, P'_0 are the potentials and partial pressures of the sample and reference respectively and k_{Zr} is the cell constant (V). The constant 0.0496 has the units V/K in the SI system. Equation 6.86 should not be used with any other system of units.

Zirconia cells are specific to O_2 and are employed generally to monitor the O_2 content of flue gases. They can be used for concentrations of O_2 from about 0.1 to 20 mole per cent with accuracies of ±0.1 per cent up to 10 per cent O_2 and of ±2.5 per cent between 10 and 20 per cent O_2[69]. The sensor can be inserted directly into the flue or can employ a sampling system (Fig. 6.49*b*). The response time is short (≈ 7 s for 90 per cent of the response to take place—depending on the application) but because of the high temperature of operation the instrument is not suitable for measuring the O_2 content of flammable gases.

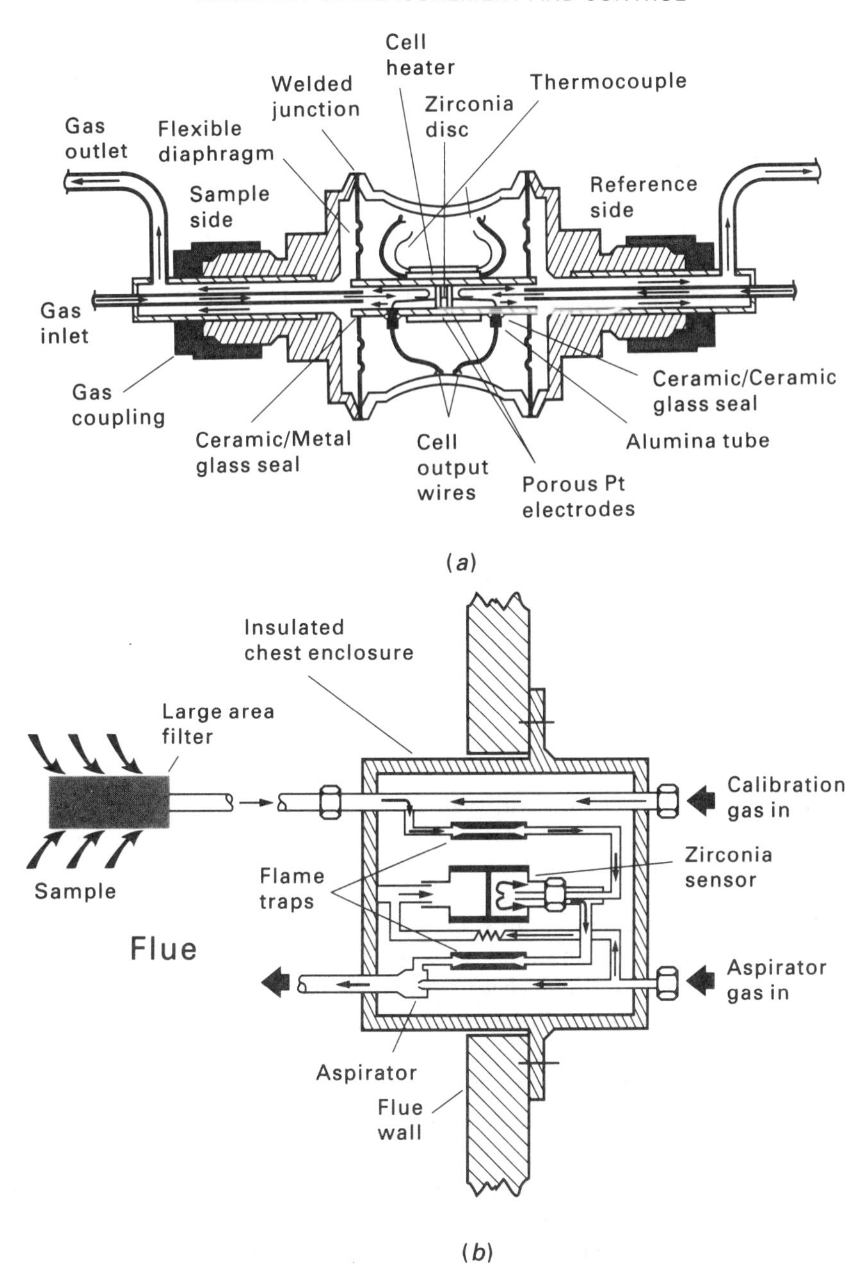

FIG. 6.49. Zirconia oxygen-analyser: (*a*) details of typical analyser; (*b*) analyser used in the extractive mode for measuring the oxygen concentration in flue gas

6.8.3. The Chromatograph as an On-Line Process Analyser

One of the most widely employed instruments for the on-line analysis of samples of multicomponent gases or volatile liquids is the gas–liquid or gas–solid chromatograph (GLC or GSC). This is an instrument which analyses discrete samples of material, and the total time taken between the extraction of the sample from the process stream and the production of a result in the form of a signal to a relevant control system can be considerable. This constitutes one of the major difficulties in using such instruments for on-line control purposes (Section 7.17).

The principles underlying the operation of the chromatograph and its use in a commercial separation process are discussed in Volume 2, Chapter 19. In this section emphasis is placed on its function as an on-line process analyser in which form it consists of three major subdivisions (apart from any electronic readout system), viz. the sampling assembly, the chromatograph column and the detector. All three are generally contained within the same temperature-controlled environment (Fig. 6.50).

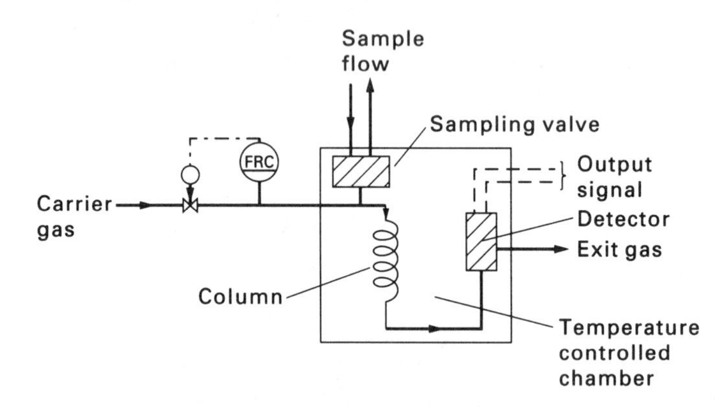

FIG. 6.50. Schematic diagram of gas chromatograph

The Sampling Assembly

This is of major importance and must send to the chromatograph representative and homogeneous samples of the material in vapour or gaseous form such that they can be analysed at the required intervals of time. A typical sampling valve is illustrated in Fig. 6.51. A continuous flow of sample passes through the sampling valve and the *sample loop* to minimise sample dead-time (Fig. 6.51*a*). The sample loop provides the correct volume of sample for analysis. The sample must be vaporised where necessary and filtered to ensure that it is clean and dry. Sample flowrates are usually of the order of 10^{-7}–10^{-6} m^3/s (6–60 ml/min) (see Section 6.2.3). More complex sampling systems are often required either when more than one column is used in the analysis, or where one installation is employed to analyse a number of different streams (Fig. 6.52)[70]. In the latter case, *back-flushing* is necessary to prevent inter-sample contamination.

The Chromatograph Column

This typically consists of a 2–10 mm OD stainless steel tube of 1–2 m length wound into a single or double helix or shaped into a U-tube. Many different packings (*solid supports*) are available (e.g. glass beads, activated charcoal, activated alumina, etc.) which are coated with a selected solvent (*stationary phase*). There are many of the

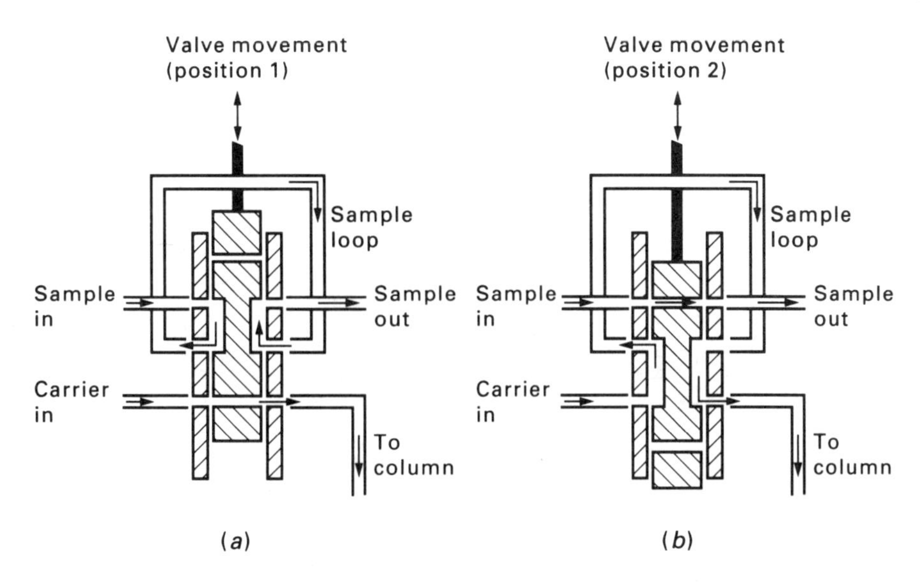

FIG. 6.51. Typical gas chromatograph sampling assembly: (*a*) sample flow by-passing column; (*b*) sample conveyed into column by carrier gas

latter and they are generally identified by trade names. The choice of stationary phase is dependent upon its thermal stability, upon the components present in the sample and upon the degree of separation required within the column. This process is called *gas–liquid* chromatography (GLC) and is now being superseded by *gas–solid* chromatography (GSC) which uses columns packed with molecular sieves and other porous polymeric materials which give the required component separations[71]. No liquid stationary phase is required and, hence, no difficulties can occur due to either its loss or its thermal instability.

The Detector

Ideally, this should exhibit a fast response, a high sensitivity and reproducibility, give a linear output over a wide range of compositions, be suitable for different carrier gases and be sufficiently rugged for use in the field. Two kinds of detector generally meet these requirements, viz. the flame ionisation detector (FID) (Table 6.16) and the thermal conductivity detector (TCD) (Section 6.8.5). Other detectors that can be used in special applications are described in Table 6.16. Generally, the detector and the carrier gas are chosen together such that the components eluted from the column generate large signals, e.g. helium is used with TCDs because of its high thermal conductivity (Table 6.13) and because it is safer to use than hydrogen. TCDs are more suitable for the analysis of permanent gases whereas FIDs are generally employed for the detection of organic compounds.

The Liquid Phase Chromatograph (LC)

Liquid Phase Chromatography is employed to analyse liquids of low volatility (Volume 2, Section 19.4.3). No carrier gas is used and the sampling valve (Fig. 6.51)

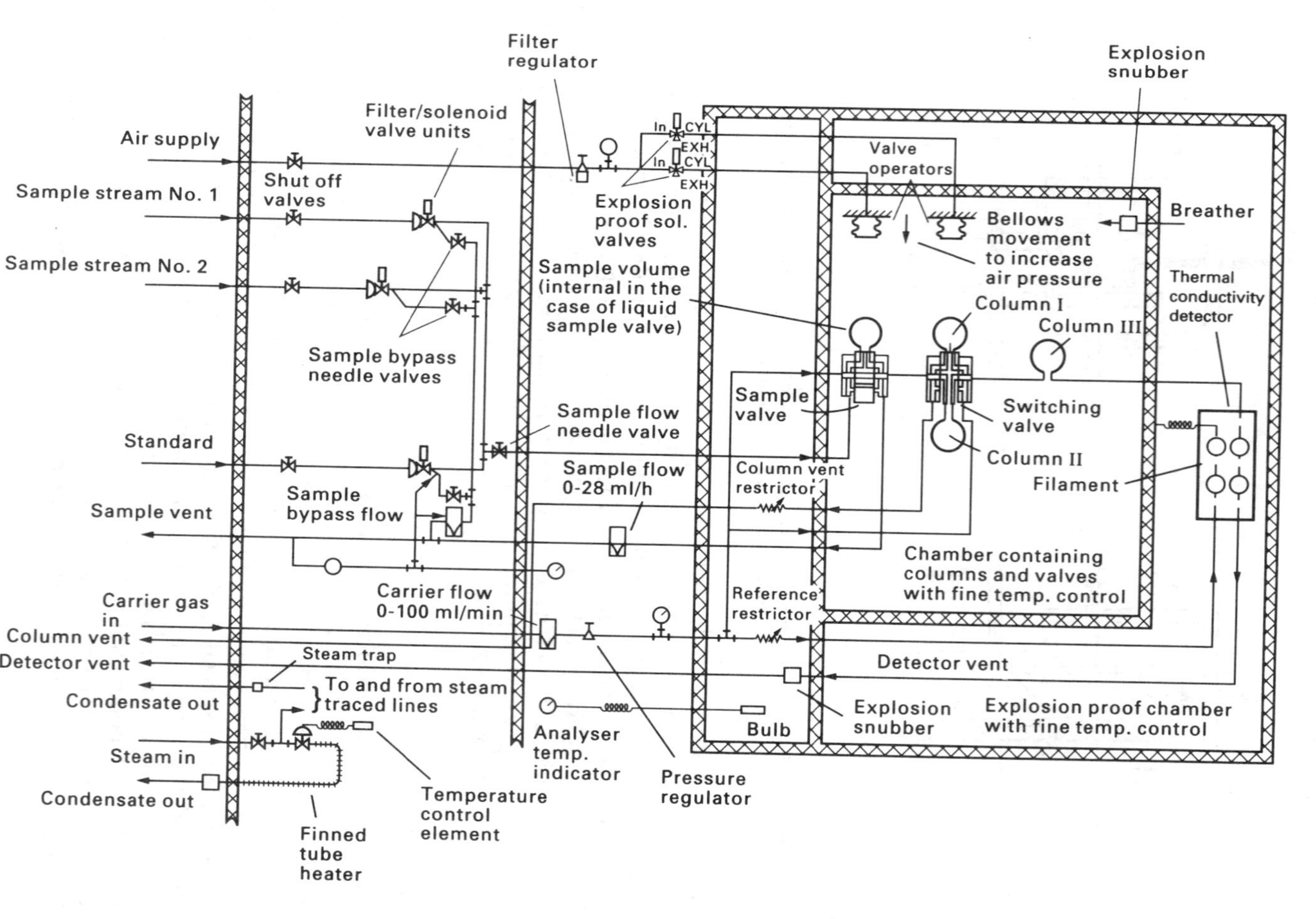

FIG. 6.52. Flow diagram of multi-stream chromatograph with thermal conductivity detector

does not require a sample loop (as a smaller sample suffices—Fig. 6.51). LC columns may simply contain solid particles or use a support and stationary phase as in the GLC. The most widely employed detectors for LCs are UV photometric and refractive index sensors (Section 6.8.1).

6.8.4. The Mass Spectrometer

This instrument is able to make a quick and accurate analysis of a wide range of materials either on or off line. Many different configurations of the instrument are available. Figure 6.53 is a schematic diagram of the *double focusing* type which has the advantage of providing a much more uniform range of ion energies than the *single focusing* version. Analysis times as low as 1 s to 3 s can be obtained although cycle times experienced under on-line process conditions (which include the DV lag in the sampling system) are of the order of 10 s to 15 s[72]. When used in combination with the gas chromatograph (Section 6.8.3) the mass spectrometer is able to identify and quantify very small traces of substances (of the order of 1 in 10^9). Sensitivities of 1 to 5 ppm can be achieved with standard process instruments with repeatabilities of ±1 per cent—although these figures are very much dependent upon the stream compositions and the particular application[73]. It is a highly specific instrument but unfortunately too expensive for many process applications.

The basic principle of the instrument is that positive ions are produced from the various components in the sample to be analysed by subjecting the sample material to a series of electric sparks or by bombarding it with high energy photons. These ions are passed through a magnetic or electric field (or a combination of both) which resolves them into their component ions according to their charge/mass ratio. The deflecting field is arranged to focus ions of a given mass on to an electron multiplier or scintillation counter. By arranging the deflecting field to vary in a predetermined manner, a range of masses can be scanned and the abundance of ions of each particular mass recorded. This *mass spectrum* can be analysed automatically

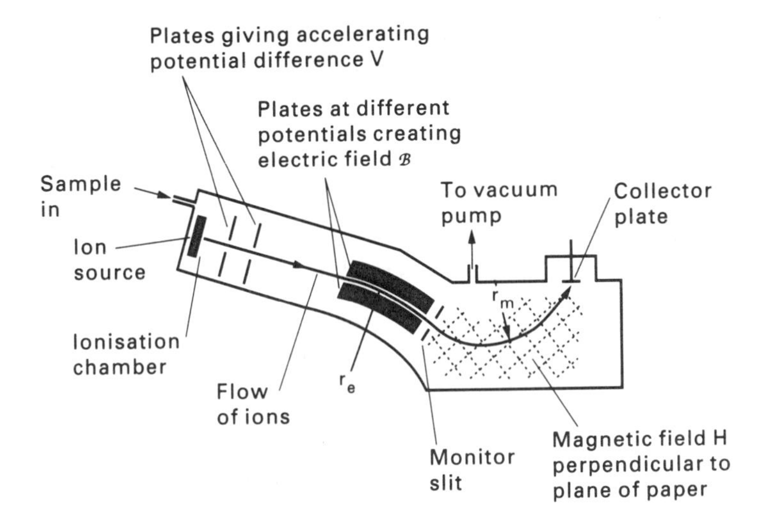

FIG. 6.53. Schematic diagram of double-focusing mass spectrometer

using a computer-based analysis system and compared with mass spectra of known materials held in a data bank. Thus the identity and concentration of the various components within the sample are derived on a continuous basis. A simplified treatment of the underlying principle of the mass spectrometer follows.

Consider an ion of mass m_i and charge e accelerated through a controlled potential difference of V. The work done on the ion as a result of the potential difference V will accelerate it from rest to a velocity u_i, thus:

$$\tfrac{1}{2} m_i u_i^2 = eV \tag{6.87}$$

Suppose the ion now enters a deflecting electric field $\mathcal{F}$ at right angles to the direction of the field (Fig. 6.53). The ion will then move in a circular path of radius r_e and the consequent reaction to the centripetal force will be:

$$\frac{m_i u_i^2}{2r_e} = e\mathcal{F} \tag{6.88}$$

Hence, from equations 6.87 and 6.88:

$$r_e = \frac{m_i u_i^2}{e\mathcal{F}} = \frac{2V}{\mathcal{F}} \tag{6.89}$$

Thus the radius of the path of the ion in the field depends on the accelerating and deflecting fields only and is independent of e/m_i. Consequently, if $\mathcal{F}$ is maintained constant, the electrostatic analyser will focus the ions at the monitor slit in accordance with their translational energies imparted by the potential difference V. The monitor slit can be arranged to intercept a given portion of the beam. This passes to the electromagnetic analyser in which a magnetic field of magnetic flux density $\mathcal{B}$ is applied at right angles to the direction of the beam. The force F_i exerted on the ions is:

$$F_i = \mathcal{B} e u_i \tag{6.90}$$

which is at right angles to their motion (Fleming's left hand rule), and thus the ions once more follow a circular path of radius r_m described by:

$$\frac{m_i u_i^2}{r_m} = \mathcal{B} e u_i \tag{6.91}$$

From equations 6.87 and 6.91:

$$\frac{e}{m_i} = \frac{2V}{\mathcal{B}^2 r_m^2} \tag{6.92}$$

Hence, if the electric and electromagnetic fields are kept constant then all ions having the same e/m_i ratio will have the same value of r_m.

6.8.5. Thermal Conductivity Sensors for Gases

Gas thermal conductivity sensors are used to detect variations in the composition of mixtures of gases by monitoring changes in the thermal conductivity of the mixture. Such instruments are used (a) as detectors for gas chromatographs (Section

6.8.3), and (b) as stand-alone instruments for gas analysis. In some applications the pressure exerted by a pure gas is determined directly from its thermal conductivity (Section 6.3.5).

Several forms of gas sensor based upon thermal conductivity are available. The most common type of detector (the *katharometer*) consists of a number of hot-wire sensors arranged in a Wheatstone Bridge circuit (Fig. 6.54)[8]. A small current i is supplied to heat each arm of the bridge. The heat transfer coefficient h for

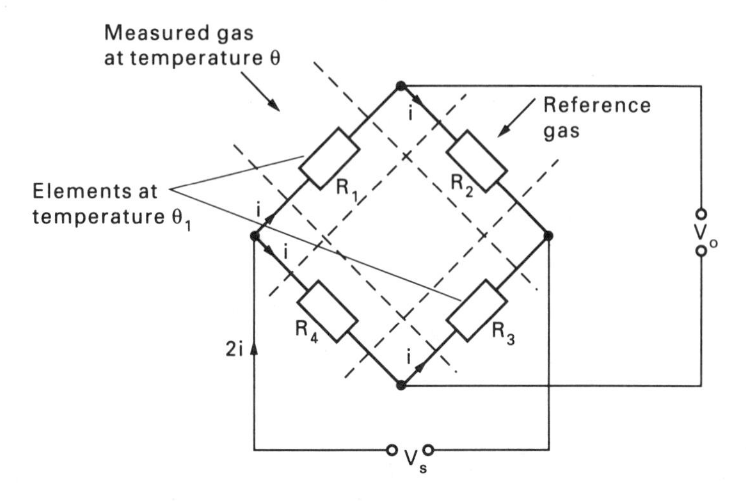

FIG. 6.54. Katharometer deflection bridge

convection from the wire to the gas (under the conditions applied within the instrument) is dependent largely upon the velocity of the gas and its thermal conductivity k (see also Volume 1, Section 9.4). If the gas velocity is maintained constant and small then h will be predominantly a function of k, and from a steady-state heat balance over the resistance R_1:

$$i^2 R_1 = h_{mix} A_R(\theta_1 - \theta) \tag{6.93}$$

where A_R is the surface area of R_1 and θ_1 and θ are the temperatures of R_1 and the gas respectively ($\theta_1 > \theta$).

But:
$$R_1 = R_0(1 + \beta_\theta \theta_1) \quad \text{and} \quad R_\theta = R_0(1 + \beta_\theta \theta) \tag{6.94}$$

where β_θ is a temperature coefficient and R_θ is the resistance of R_1 at the temperature of the gas.

Hence:
$$\frac{R_1}{R_\theta} = \frac{1 + \beta_\theta \theta_1}{1 + \beta_\theta \theta} \approx 1 + \beta_\theta(\theta_1 - \theta) \tag{6.95}$$

assuming that $(\theta_1 - \theta)$ is small. Thus, from equations 6.93 and 6.95:

$$i^2 R_1 = \frac{h_{mix}}{\beta_\theta}\left(\frac{R_1}{R_\theta} - 1\right) \tag{6.96}$$

$$\therefore \quad R_1 = \frac{R_\theta h_{mix}}{h_{mix} - \chi} = R_3 \tag{6.97}$$

as the sample flows over both R_1 and R_3 (Fig. 6.54) and where $\chi = i^2 \beta_\theta R_\theta / A_R$ is a constant if θ is maintained constant. Similarly, for the reference gas:

$$R_2 = R_4 = \frac{R_\theta h_{ref}}{h_{ref} - \chi} \tag{6.98}$$

Hence:

$$V_o = i(R_1 - R_4) = \frac{iR_\theta \chi(1/h_{mix} - 1/h_{ref})}{(1 - \chi/h_{mix})(1 - \chi/h_{ref})}$$

$$= iR_\theta \chi(1/h_{mix} - 1/h_{ref}) \text{ as } \chi/h_{mix} \text{ and } \chi/h_{ref} \text{ are generally} << 1$$

$$= \frac{i^3 R_\theta^2 \beta_\theta}{A_R}\left(\frac{1}{h_{mix}} - \frac{1}{h_{ref}}\right) \tag{6.99}$$

Consequently, the output voltage varies as the cube of the sensor current and is a non-linear function of the heat transfer coefficient, i.e. of the thermal conductivities, provided that the velocities of the gases passing through the detector are small. Equation 6.99 indicates that the katharometer has a maximum sensitivity ($\approx$ 1000 ppm) when the bridge current is high and when it is used to measure composition changes in binary or pseudo-binary mixtures of gases whose components have thermal conductivities which differ widely (Table 6.13).

TABLE 6.13. *Relative Thermal Conductivities of Some Common Gases at 298K (Air = 100)*[74, 75]

Gas	Thermal conductivity	Gas	Thermal conductivity
Cl_2	34	CO	96
SO_2	36	Air	100
C_6H_6	40	N_2	100
CO_2	64	O_2	103
H_2O	68	CH_4	132
C_2H_4	80	He	577
NH_3	94	H_2	700

(Thermal conductivity of air at 298 K is 0.026 W/mK.)

Example 6.2

A katharometer is employed to determine the concentration of H_2 in a H_2/CH_4 mixture. The proportion of H_2 can vary from 0 to 60 mole per cent. The katharometer is constructed as shown in Fig. 6.54 from four identical tungsten hot-wire sensors for which the temperature coefficient of resistance β_θ is 0.005 K^{-1}. The gas mixture is passed over sensors R_1 and R_3 whilst the reference gas (pure CH_4) is passed over sensors R_2 and R_4. The total current supplied to the bridge is 220 mA and it is known that the resistance at 25°C and surface area of each sensor are 8 Ω and 10 mm^2 respectively. Assuming the heat transfer coefficient h between gas and sensor filaments to be a function of gas thermal conductivity k only under the conditions existing in the katharometer and that in this case $h = k \times 10^4$ (h in W/m^2K and k in W/mK), draw a graph of the output voltage V_o of the bridge network as a function of mole per cent H_2.

Solution

From Table 6.13, at 298 K:

$$k_{H_2} = 700 \times 0.026/100 = 0.182\ \text{W/mK}$$
$$k_{CH_4} = 132 \times 0.026/100 = 0.034\ \text{W/mK}.$$

Consider one particular gas composition, e.g. 50 mole per cent H_2. This is equivalent to $(100/(100+800)) \times 100 = 11.1$ mass per cent H_2. From Volume 6, Section 8.8.4, the thermal conductivity of the mixture k_{mix} can be considered as a simple weighted average, i.e:

$$k_{mix} = 0.111 \times 0.182 + 0.889 \times 0.034 \approx 0.050\ \text{W/mK}.$$

Hence:

Heat transfer coefficient for the gas mixture $h_{mix} = k_{mix} \times 10^4 = 0.050 \times 10^4\ \text{W/m}^2\text{K}$

and for the pure reference gas $h_{CH_4} = k_{CH_4} \times 10^4 = 0.034 \times 10^4\ \text{W/m}^2\text{K}$.

From equations 6.96 and 6.97 and noting that the current i in each arm of the bridge is $220/2 = 110\ \text{mA} = 0.11\ \text{A}$, then $\chi/h_{mix} = i^2\beta_\theta R_\theta/(A_R h_{mix}) = 0.11^2 \times 0.005 \times 8/(10^{-5} \times 0.050 \times 10^4) \approx 0.10$ which is sufficiently smaller than unity for equation 6.99 to be applicable. Hence, from equation 6.99:

$$V_o = \frac{i^3 R_\theta^2 \beta_\theta}{A_R}\left(\frac{1}{h_{mix}} - \frac{1}{h_{CH_4}}\right) = \frac{0.11^3 \times 8^2 \times 0.005}{10^{-5}}\left(\frac{1}{0.050 \times 10^4} - \frac{1}{0.034 \times 10^4}\right)$$
$$\approx (-)40\ \text{mV}.$$

(The negative sign is not significant.)
Thus a range of output voltages can be calculated for different gas mixture compositions. These are shown in Fig. 6.55.

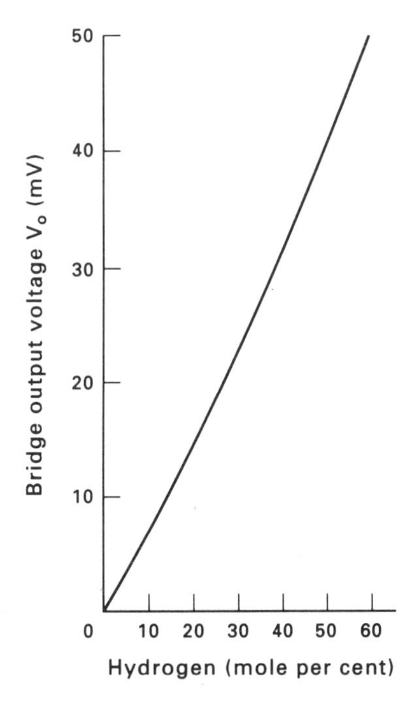

FIG. 6.55. Katharometer output voltage as a function of hydrogen concentration in sample

6.8.6. The Detection of Water

The Measurement of Water in Gases—Humidity Sensors

Methods of determining humidity by physical or chemical absorption or adsorption or using wet and dry bulb thermometry and associated psychrometry can be

TABLE 6.14. *Techniques Suitable for the On-line Monitoring of Moisture*[76–78]

Type of Sensor	Principle of Operation	Accuracy, Range, etc.	Comments
Resistive hygrometer elements	Used in both wafer and cylindrical form. Wire electrodes are inserted into a thin flat section of suitable insulation (e.g. polystyrene) or are wound round a cylinder of the same material. This is covered with a film of hygroscopic salt and thus the resistance of the element changes with the quantity of water vapour absorbed. The latter is a function of the concentration of water vapour in the gas to which the hygrometer surface is exposed.	Reasonably accurate if calibrated regularly.	Other resistive hygrometers employ a humidity sensitive substrate only, e.g. SiO_2, CrO_2, Al_2O_3, and various polymers. These materials often exhibit a change in capacitance as well as resistance and, in such cases, the change in total impedance is measured. They can be used at high moisture concentrations and have a wide response range. They have a rapid response, are simple to use and are cheap. They are prone to contamination. The salt can be etched away and the sensor requires frequent recalibration due to the resulting drift. Very limited on-line use.
Electrolytic (coulometric) hygrometers	The quantity of electricity required to carry out a chemical reaction is measured. The principle is based upon Faraday's law of electrolysis. Water is absorbed on to a thin film of dessicant (e.g. P_2O_5) and electrolysed. The current required for the electrolysis varies according to the amount of water vapour absorbed. The current depends also upon the flowrate.	Capable of high precision. Used in the range 1000 to 3000 ppm of water by volume.	Somewhat complicated procedure. Recombination of products to water is necessary after electrolysis. Density, pressure and flowrates have to be maintained precisely. Contamination can poison the cell. It is ideal for binary mixtures but is of limited range. Suitable for on-line operation.
Aluminium oxide sensor	Consists of a strip of aluminium anodised with a porous oxide layer and coated with a thin layer of gold. The aluminium and gold form two electrodes of what is effectively an Al_2O_3 capacitor. Water vapour rapidly transports through the gold layer and absorbs on to the oxide layer. This equalises in a manner which can be related to the vapour pressure by the change in electrical impedance.	Capable of high precision. Limited operating range—generally between 200 K and 300 K.	Sample has to be clean to avoid poisoning the sensor. Sensor has to be dried if oversaturated, then recalibrated. Has fast response and is ideal for binary mixtures. Relatively cheap. Suitable for on-line use.
Electrolytic type sensors	Uses thick film techniques, e.g. capacitor coated in glass bonded on to a ceramic disc mounted on a thermoelectric (Peltier effect) cooler. Control is by a platinum resistance thermometer which adjusts the temperature of the cooler to regain equilibrium after a change in capacitance due to moisture deposit.	Range depends on technique. Capable of high precision.	Limitations are similar to those for Al_2O_3 sensor. Capable of being direct mounted. Relatively cheap. Suitable for on-line use.
Piezoelectric hygrometers (Fig. 6.56)	These consist of a quartz crystal with a hygroscopic coating (Fig. 6.56*a* and Section 6.3.3). Two crystals are usually employed and the wet gas (sample) and dry gas (reference) are passed over them alternately (normally every 30s—Fig. 6.56*b*). The crystals absorb and desorb. The difference in angular frequency $\Delta\omega$ is proportional to the concentration of water vapour by volume.	Capable of high precision. Can monitor moisture contents of the order of 1 to 3000 ppm by volume. Usually measures up to dewpoint of 310 K.	Fast response but expensive. Sample must be clean to avoid contamination of the crystals. Complex sampling system. Suitable for on-line use.
Optical dewpoint sensors	The humidity of a gas is equal to the humidity of that gas in its saturated state at its dew point. A typical dewpoint sensor uses a plate with a mirror surface in close contact with a thermoelectric (Peltier effect) cooler. A beam of light is directed at the mirror and the reflected ray is detected with a photocell. The amount of condensation occurring on the mirror during cooling affects directly the quantity of light reaching the photocell.	High precision. Maximum range from 220K to 350K. Typical accuracy ± 0.2 K and typical resolution ± 0.01 K.	Can be used as a standard instrument for calibration purposes. Ideal for laboratories. Control has to be very precise. Mirror contamination can occur in dirty gases. Cannot be employed in presence of other gases which condense at similar temperatures to the moisture. Some instruments fitted with heated mirror to boil off contaminants—does not necessarily eliminate the problem. Suitable for on-line use—can have continuous or semi-continuous operation.
Photometric methods (non-dispersive infra-red techniques) (Figs 6.57 and 6.43)	Narrow bands of infra-red energy at both reference and measurement wavelengths (in the 1 to 2 μm range) are passed alternately through a sample cell and a reference cell (Fig. 6.57), the remaining energy is collected by a solid-state detector and suitable hardware is used to calculate the ratio of these energy levels. The latter is a direct measure of the concentration of water vapour in the sample. Simpler dual-beam instruments are also available (Fig. 6.43).	Single-beam instrument maximum range from 240 K dew point to 380 K dew point with concentration ranges of 0 to 5000 ppm by volume to 0 to 30 per cent by volume. Dual-beam type range 220 K to 300 K.	Single-beam version gives good selectivity with complex mixtures—dual-beam type much less selective. Less sensitive than electrolytic hygrometers. Higher cost than conventional hygrometers. Suitable for on-line use. Both resistant to contamination—however, single-beam instruments are less affected by deposits on the cell—give better calibration stability in polluted gases.

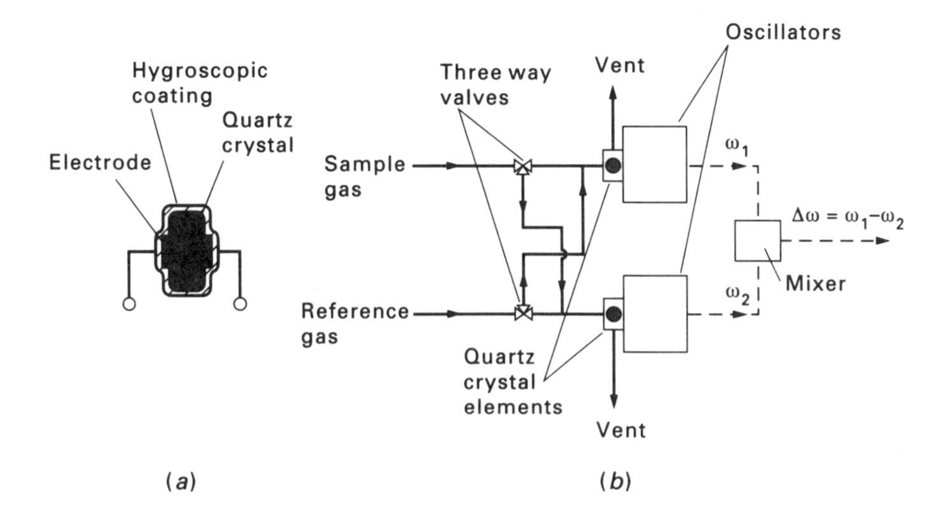

FIG. 6.56. Quartz-oscillator moisture-meter

found in Volume 1, Sections 13.2 and 13.3. These methods of determining the moisture content of process gases are however being superseded by more direct means. A brief resumé of the more common on-line humidity sensors (*hygrometers*) is given in Table 6.14.

Most common moisture probes are prone to contamination. The most serious causes of the latter are exposure to Cl_2, NH_3 or wet acids. Common causes of errors in hygrometer readings are too infrequent calibration and the use of the wrong materials for sampling pipework. The allowable concentration of water vapour is so small in some processes that diffusion of water vapour through the walls of

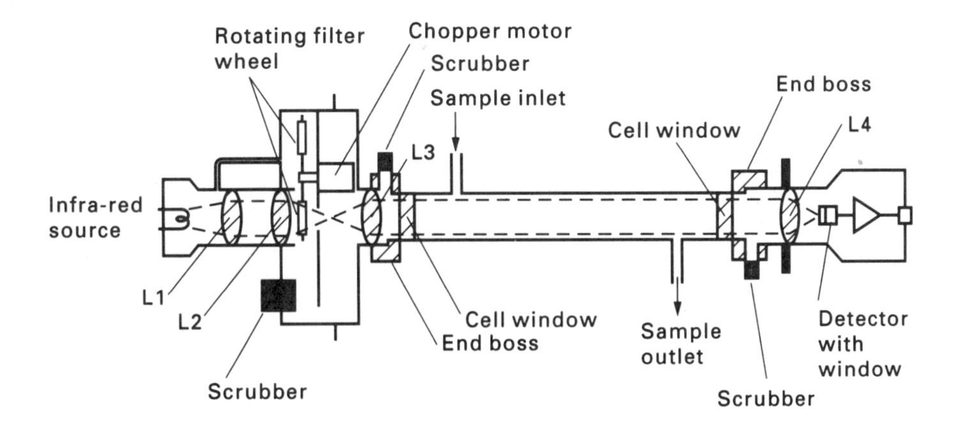

FIG. 6.57. Single-beam infra-red analyser suitable for the detection of water in liquids (L1–L4: lenses)

sampling tubes can cause significant errors. In general, plastics are quite unsuitable as sample pipe material—stainless steel or copper being the best choice[79].

The Detection of Water in Liquids and Solids (The Measurement of Moisture Content)

The detection of water in some organic liquids is of considerable importance, e.g. both 'antifreeze' (ethylene glycol) and aviation fuel (kerosene) should contain as little water as possible. Such materials can be monitored for water content by the use of non-dispersive infra-red techniques or turbidimeters (Section 6.8.1).

TABLE 6.15. *Typical Industrial Methods for the Measurement of Moisture in Solids*[80]

Method	Approximate practical range (per cent by mass)	Limitations
Conductance	0 to 25	Affected by other conductive substances in the sample.
Capacitance	0 to 30	The conductivity of the material and variations in the density will affect the measurement.
Near infra-red	0 to 90	By surface measurement. Highly reflective surfaces, crystalline materials and very dark coloured materials can cause difficulty.
Microwave absorption	*By sample*—0 to 100 *In-line*—3 to 45	Sensitive to changes in the density of the material.
Nuclear magnetic resonance	0 to 100	Relatively small sample. Affected by any other hydrogen atoms present in the sample other than water, e.g. oils and fats.

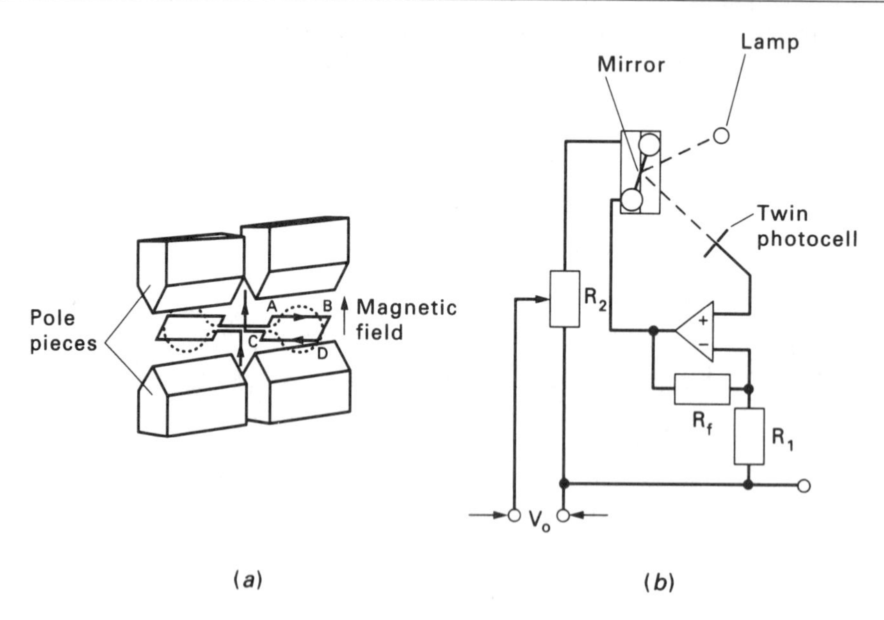

FIG. 6.58. Dumbell-type paramagnetic oxygen analyser: (*a*) arrangement of dumbell and magnetic poles; (*b*) principle of operation

There are many solid materials where the control of moisture content is essential, e.g. in manufactured foodstuffs, animal feed, pharmaceuticals, paper and board, tobacco, sand and cement, and a wide range of chemicals and plastics. It has to be considered whether the total moisture content is required or whether it is sufficient simply to measure surface moisture—although generally a relationship can be established between the two. Table 6.15 lists typical industrial methods for the measurement of the water content of solid materials.

6.8.7. Other Methods of Gas Composition Measurement

There are numerous instruments available for the measurement of the compositions of (or simply for the detection of) gases. These sensors employ a wide variety of physical and chemical methods and many can also be used for analysing liquids by vaporising the liquid sample before it is passed through the instrument. A selection is described in Table 6.16.

6.9. PROCESS SAMPLING SYSTEMS

6.9.1. The Sampling of Single-Phase Systems

The importance of a well designed sampling system cannot be overemphasised and poor sampling techniques will lead to erroneous results even with the most sophisticated instruments. Sampling systems vary widely and are often unique to a particular situation and application. However, certain guidelines are common to many fluid sampling arrangements and the objective in all cases is to obtain a truly representative sample of the material to be analysed[81]. Procedures applied to solids are very specialised and are not considered here.

Fluid sampling systems generally consist of four main sections, viz. the probe, the sample line, sample treatment and sample disposal.

The Probe

The probe is used to remove the sample from the main stream of material and it is important to ensure that the material to be sampled is well mixed prior to the sampling point so that the probe is able to extract a representative sample. Figure 6.59*a* shows a simple probe which is suitable for sampling liquids and gases at low pressures and is generally made from 11.7 mm ID stainless steel tubing. The fluid is sampled at the centre of the main stream unless this requires the probe to be of such a length that the stream velocity conditions create unacceptable vibrational stress within the probe. The maximum permissible length of any probe from its support required to avoid such stresses can be determined from[82]

$$L_{max} = K_x \left(\frac{\mathscr{E} d_0^2 (d_0^2 + d^2)}{u_{max}^2 \rho} \right)^{0.25} \tag{6.100}$$

where $\mathscr{E}$ and ρ are the modulus of elasticity and density of the probe material respectively, d_0 and d are the OD and ID of the probe tube, u_{max} is the maximum

TABLE 6.16. *Further Methods of Gas Measurement and Detection*[74, 83]

Type of Gas Detector	Used to Detect	Bulk Gas (or Carrier Gas where Appropriate)	Principle of Operation	Relative Sensitivity [(a)]	Application
Magnetodynamic (magnetic susceptibility) (Fig. 6.58)	O_2	Various	The magnetic susceptibility of O_2 is very large compared with that of other gases (except NO, NO_2 and ClO_2). Such gases are called *paramagnetic*. Two quartz spheres are attached to the ends of a bar to form a dumb-bell which is suspended in an atmosphere of N_2 between the poles of a powerful magnet. The sample is fed into the chamber containing the dumb-bell. The magnetic field is disturbed and the dumb-bell is deflected. This deflection is detected by an optical system. The dumb-bell is returned to its null position by passing a current through a coil surrounding it. This current is proportional to the concentration of O_2.	0.01	Measurement of O_2 from 0 to 100 per cent in various streams. Can be used with flammable gas mixtures
Ultrasonic	Mainly gases of low molar mass	H_2, N_2, CO_2, He, Ar.	A quartz crystal transducer transmits a sound wave through a sample of gas with a similar crystal used as the receiver. The velocity of the sound wave is proportional to the square root of the molar mass of the sample. The phase shift of the sound signal is measured by comparison with a reference signal. Precise temperature control is required.	1–10	Gas chromatography
Catalytic (Pellistor)	Flammable gases	Air	Measures the heat output due to the catalytic oxidation of flammable gas molecules. A stream of the sample is passed over the sensor which is usually a ceramic bead impregnated with Pt or Pd. The temperature variations in the sensor due to reaction are monitored.	Dependent on individual design.	Flammable gas detector. Usually portable
Semiconductor	Flammable and other gases	Air	Two small coils of platinum wire are fixed side by side within a bead of semiconductor material of diameter 2–3 mm. The sample gas flows over the bead and molecules of gas are adsorbed on the surface of the semiconductor—thereby changing the resistance of the semiconductor as measured between the two coils. One coil is also employed as a heater to ensure that the adsorption is reversible. The device is sensitive but has poor selectivity.	Dependent on individual design.	Flammable gas detector. Inexpensive
Flame photometric (FPD)	Compounds of S, P, and halogens	N_2, He	The sample is mixed with a H_2/O_2 or H_2/air mixture and ignited. The resulting flame produces simple molecular species and excites them to higher electronic states. The excited species subsequently return to their ground states—at the same time producing characteristic molecular band-spectra. These emissions are monitored by a photomultiplier via a suitable filter, thus making the detector highly selective. The instrument is very sensitive but suffers from a highly non-linear response.	100	Sulphur analysers; gas chromatography
Flame ionisation (FID)	Organic compounds	N_2	The sample is fed to a H_2/air flame. The latter contains relatively few ions but does contain atoms of high kinetic energy. The flame produces large numbers of positive ions and secondary electrons from the trace organic materials in the sample gas. A potential is applied across the detector chamber and the ions and electrons move towards the electrodes thus forming an *ionisation current* which is suitably amplified. The FID is mass rather than concentration sensitive and the sensitivity is thus the minimum detectable mass of carbon passing through the flame per second. The FID has a high sensitivity and a linear response.	100[(b)]	Hydrocarbon analysers; gas chromatography
Photo ionisation (PID)	Organic compounds (except low molar mass)	N_2	Uses monochromatic UV radiation to ionise molecules of organic compounds. The ions formed are driven to a collector electrode by applying an electric field across the ionisation chamber. The resulting ion current is measured by means of an electrometer. The instrument can be made selective by using UV sources of different wavelengths. The instrument is highly sensitive and has a wide linear range.	100–1000	Gas chromatography
Helium ionisation	Traces of permanent gases	He	Helium gas containing the trace materials is passed through a small chamber where it is exposed to β radiation from a tritium source. The helium atoms are excited by collision with the high energy β electrons. The excited helium atoms lose their energy by collision with atoms or molecules of the trace gases. The resulting ionisation is measured by application of an electric field so producing an *ionisation current*. The performance of this detector is highly dependent upon its geometry.	100	Gas chromatography
Electron capture	Oxygen and halogen compounds	Ar, N_2	This detector employs a source of β radiation (usually ^{63}Ni—cf. helium ionisation detector). The velocities of the high energy electrons are reduced to thermal velocities by collision with atoms of inert gas with which the detector chamber is purged. When a sample of a gas with a greater electron affinity than the inert gas is introduced into the cell, some of the electrons are captured and form negative ions. Hence the current across the cell (produced by an appropriate electric field) is reduced. This instrument is extremely sensitive to electronegative species, particularly halogenated and oxygenated compounds. Carrier gas, detector current and detector temperature are selected for maximum sensitivity.	1000–10,000	Trace gas and explosive detector; gas chromatography

[(a)] Rough guide only—highly dependent upon design and application—unity represents one part in 10^9, 1000 represents one part in 10^{12} etc.

possible velocity of the process fluid in the pipe and K_x is a dimensionless constant (≈ 3.4).

High Pressure Probes

If the fluid to be sampled is liquid which is to be vaporised, or is a gas at high pressure, then a narrow bore pipe (typically 2 mm ID) is inserted through the original low pressure probe (Fig. 6.59*b*).

Probes for Fluids at High Temperatures

Figure 6.59*c* is an illustration of a typical furnace gas probe suitable for temperatures up to 700 K. For higher temperatures (up to 1900 K) it is necessary to provide water cooling.

Probes for Dirty Gases

To sample dirty gases the sample probe should be turned downstream as solid particles have a much greater inertia than gas molecules and are much less likely to change direction and enter the probe[39]. Alternatively the end of the probe may be covered with an unglazed porcelain filter. Often a water spray is provided at the inlet to the probe in order to wash solids out of the gas as it enters the sampling system (Fig. 6.59*d*). The gas and the water vapour are separated in a water trap and care should be taken to avoid any water entering the measuring instrument. This type of probe is unsuitable for water soluble gases (e.g. CO_2, SO_2, H_2S, etc.).

Probes for Corrosive Gases

If the gas is particularly corrosive the steam ejector system shown in Fig. 6.59*e* may be employed to sample gases at temperatures of up to 450 K (e.g. flue gases). The possibility of corrosion occurring in the sample lines when the steam/gas sample cools to the dew point is much reduced due to the dilution of the corrosive condensate by the condensed steam.

Probes for Large Diameter Pipes

In large gas ducts or large diameter pipes carrying liquids it may be necessary to use multiple probes (Fig. 6.59*f*) or long probes with multiple inlets. ASTM Standard D4177[82] specifies sampling procedures required to obtain representative samples of petroleum and petroleum products and similar methods can be employed in sampling most non-corrosive liquid industrial chemicals.

The Sample Line

The second part of the system consists of the sample line or pipe which transports the sample to the analyser. The distance between the analyser and the sample point

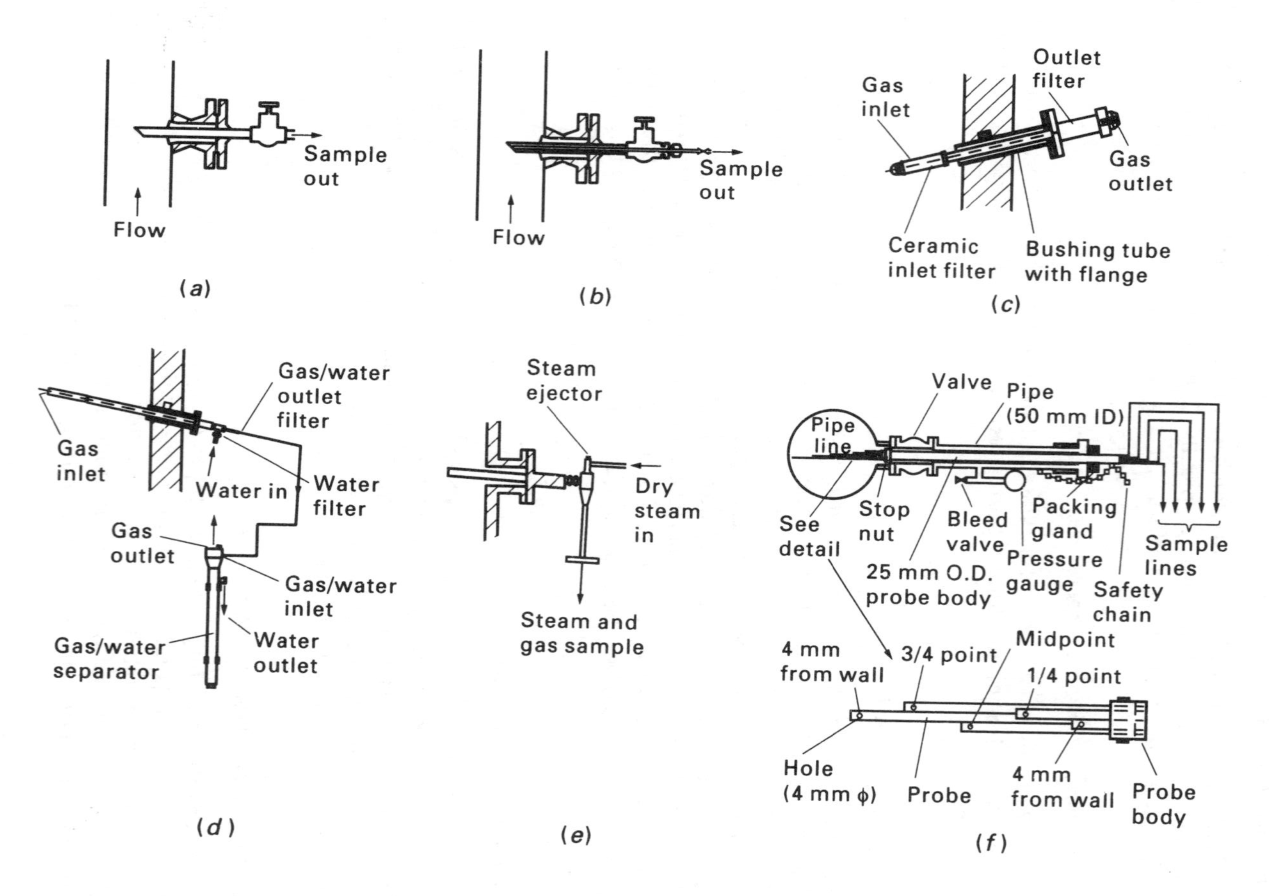

FIG. 6.59. Fluid sampling probes: (*a*) normal sample probe; (*b*) small volume probe; (*c*) gas-sampling probe; (*d*) water-wash probe for furnace gases; (*e*) probe for flue gas, using steam ejector; (*f*) multipoint pipeline probe

should be kept to a minimum so that the dead time in the sample line is as small as possible. In any event the dead time should not exceed 30 s[39]. If a long sample line is unavoidable, then a *fast loop* should be installed (Fig. 6.60) which operates at a substantially higher rate of flow than that required by the analyser.

Other techniques can also be used for reducing the sample line dead time, e.g. allowing volatile liquids under pressure (such as propane or butane) to vaporise within the probe as gases flow at much higher velocities than liquids under the same conditions. The material from which the sample line is constructed should not change the characteristics of the sample. In some instances (e.g. when sampling low concentrations of H_2S or water vapour through copper tubing) processes of adsorption/desorption can affect the response of the measuring instrument quite severely[79]. Errors can also arise due to the permeability of the walls of sample tubing made of various plastics to certain gases (e.g. PTFE tubing is permeable to O_2 and to water vapour (Section 6.8.6) and PVC tubing is permeable to the smaller hydrocarbon molecules such as CH_4).

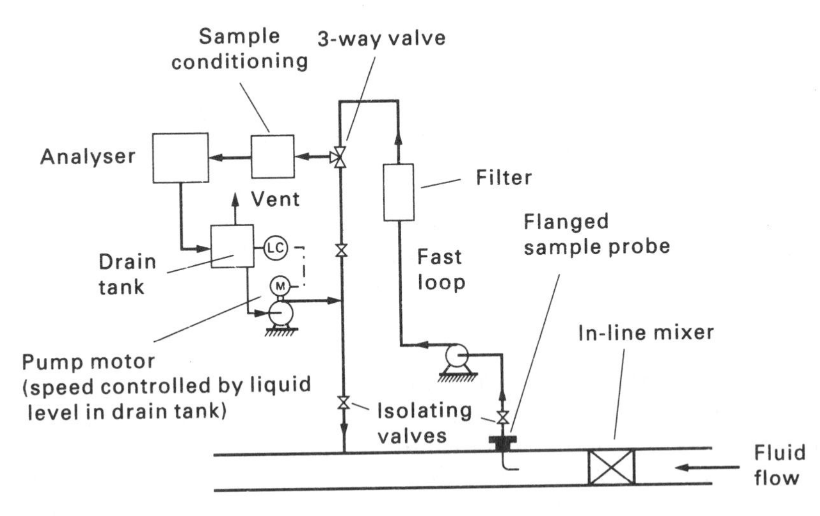

FIG. 6.60. Typical liquid-sampling system with fast loop

Sample Treatment

It is necessary to treat the sample so that it is in a suitable state for analysis. This can involve the adjustment of the temperature, pressure and/or rate of flow of the sample. It is often necessary to monitor carefully the rate of flow of sample through the measuring instrument itself. Such flowrates are generally small and any of the techniques listed in Table 6.2 can be employed. Most commercial analysers are designed to operate at or near atmospheric pressure. All solid material should be filtered from the sample as soon as possible and corrosive and deposit-forming materials should be removed using suitable chemical absorbants. Condensation should be avoided in vapour sample lines by keeping the sample above its dew point. This may require lagging and/or heating of the line. Facilities for washing out should always be provided.

Sample Disposal

Finally, a means must be available for the disposal of the sample after it leaves the analyser. Direct discharge to atmosphere or to drain is possible only where it is safe to do so. In some cases this is done after suitable treatment and in others the sample is pumped back to the process.

6.9.2. The Sampling of Multiphase Systems (Isokinetic Sampling)

Sampling of materials consisting of more than one phase at the sampling point is particularly difficult. Common examples are the sampling of liquid or vapour from a plate absorber or distillation column or from a packed column, or the sampling of a multiphase mixture flowing in a pipe or duct. In such cases it is not always known whether the liquid or vapour has been sampled or some combination of both.

When sampling process gases, either to determine dust concentration or to obtain a representative dust sample, it is necessary to take special precautions to avoid segregation through inertial effects. For this purpose the gas duct is traversed by the nozzle of the sampling tube and samples are taken at predetermined points across the traverse. The nozzle is pointed directly into the gas stream and the gas velocity at the mouth of the nozzle must be the same as the local gas velocity at that point. This procedure is termed *isokinetic sampling*. If the sampling velocity is too high, the dust sample will contain a lower concentration of dust particles than the stream being sampled with a greater proportion of fine particles. If the sampling velocity is too low, then the dust sample may contain a higher concentration of dust with a greater percentage of coarse particles.

6.10. THE STATIC CHARACTERISTICS OF SENSORS

The selection of the most suitable instrument for a required measurement from a range of commercially available instruments necessitates the knowledge of certain important factors. These can be divided into the *static* and *dynamic* characteristics of the instrument. The dynamic properties of instruments are fundamentally no different from those of any other system or process and are described, therefore, by the analysis of system dynamics presented in Chapter 7. Static properties, which are specific to instrumentation, are discussed in this section.

6.10.1. Definitions

There is a wide variety of important static terms and properties related to instrumentation which require careful definition and understanding. Some of these are frequently used in a misleading or even incorrect context. Expressions such as the *accuracy* of an instrument or its *precision* or its *sensitivity* are often employed by manufacturers in a casual or imprecise manner. A brief description of the more common terms follows.

Range

The *range* of an instrument is the region over which a quantity may be measured or received (input range) or transmitted (output range) and is defined by stating the *lower* and *upper range values*, e.g. a thermocouple may be quoted as having an input range of 100–250°C and an output range of 4–10 mV. The lowest and the highest quantities to which a device can be adjusted are termed the *lower* and *upper range limits*, respectively.

Span

This is often termed the *full scale deflection* (FSD) and is the magnitude of the range of the instrument. In the above example, the input span of the thermocouple is 150°C and the output span is 6 mV.

Turndown

The *turndown* is a direct function of the range of an instrument and is the *ratio of the upper range value of the instrument to the lower range value.* For example, if an instrument is required to measure pressures which range from 100 kN/m^2 to 1000 kN/m^2 then it is said to have a turndown of 10:1. Care must be taken in specifying a suitable turndown for an instrument when it is only occasionally exposed to a very high value of the measured variable. Suppose that a mass flowmeter is specified to measure a process stream which has a normal rate of flow of about 1 kg/s but very intermittently produces flows of the order of 50 kg/s. In this case the important characteristic of the instrument is that it should have a good *over-range capability* rather than a turndown of 50 to 100 since the latter may be unattainable or may require a much more expensive transducer (see Table 6.1).

Sensitivity

This may be defined as the ratio of the change in magnitude of the output signal corresponding to the change in the magnitude of the input after a steady state has been reached. In the limit this becomes the rate of change of the output with respect to the input. For example, consider a thermocouple for which the output E in mV is given by:

$$E = -3.275 \times 10^{-6}(T - 273.15)^2 + 0.04153(T - 273.15) - 0.03043$$

where T is the temperature in K and the reference junction is held at 273.15 K. Then the sensitivity (mV/K) will be:

$$\frac{\mathrm{d}E}{\mathrm{d}T} = -6.55 \times 10^{-6}(T - 273.15) + 0.04153.$$

Note that, in this case, the sensitivity is a function of T, e.g. at 300 K, $\frac{\mathrm{d}E}{\mathrm{d}T} = 0.04135$ mV/K and at 500 K, $\frac{\mathrm{d}E}{\mathrm{d}T} = 0.04004$ mV/K.

Resolution

This is defined as the minimum difference in the values of a quantity that can be discriminated by a device. In essence this is the largest change in input that can occur without the output changing. Strictly this is a function of elements that have a discrete output and an analog input, e.g. the analog to digital converter (Section 6.11.5) for which the resolution is the change in input voltage required to make the output change by the least significant bit.

Example 6.3

A stepper motor having a resolution of 1/256 of a revolution is used as direct drive for a gate valve controlling a liquid flow. The mass rate of flow G kg/s is given by:

$$G = \alpha(2.7 + 0.2\alpha)$$

where α is the angular rotation of the shaft of the stepper motor in radians. Calculate the rate of flow of the liquid and its resolution when the stepper motor shaft has rotated 0.15 radians.

Solution

$$G = \alpha(2.7 + 0.2\alpha) = 0.15[2.7 + 0.2(0.15)]$$
$$= \underline{\underline{0.410 \text{ kg/s.}}}$$

$$\frac{dG}{d\alpha} = 2.7 + 0.4\alpha = 2.7 + 0.4(0.15) = 2.76 \text{ kg/s per unit angular rotation of the shaft}$$

$$\text{resolution} = \delta G = 2.76\ \delta\alpha = \frac{2.76 \times 2\pi}{256}$$
$$\approx \underline{\underline{0.068 \text{ kg/s.}}}$$

Repeatability

This may be defined as the closeness of agreement among a number of consecutive outputs from the given device for the same value of the device input under the same operating conditions when approached from the same direction over a full range sweep of readings. It is necessary to specify the direction of approach because of the possibility of *hysteresis* in the instrument where hysteresis is the difference between the output of a device for a given value of the input depending on whether the output is increasing or decreasing.

The most common reason for a lack of repeatability is the existence of random fluctuations in the environment surrounding the instrument x_1, possibly in its power supply x_2, and also in its input signal x_3 due to random variations in the operation of a device upstream of the instrument in question. These random fluctuations frequently display a normal or Gaussian distribution. The output of a device in response to such random fluctuations may then be expressed as:

$$y = k_0 + k_{x_1}x_1 + k_{x_2}x_2 + k_{x_3}x_3 + f(x_1, x_2, x_3) \tag{6.101}$$

where $f(x_1, x_2, x_3)$ contains any non-linear contributions and k_0, k_{x_1}, k_{x_2} and k_{x_3} are constants. If δy is a small variation in y due to small deviations in x_1, x_2 and x_3, (viz. δx_1, δx_2 and δx_3 respectively) then:

$$\delta y = \left(\frac{\delta y}{\delta x_1}\right)\delta x_1 + \left(\frac{\delta y}{\delta x_2}\right)\delta x_2 + \left(\frac{\delta y}{\delta x_3}\right)\delta x_3 \tag{6.102}$$

and $\dfrac{\delta y}{\delta x_1}$, etc. can be determined from equation 6.101.

It can be shown[84] that if y is a linear function of x_1, x_2 and x_3, i.e. $f(x_1, x_2, x_3)$ in equation 6.101 is zero, then:

$$\sigma_y = \sqrt{k_{x_1}^2 \sigma_{x_1}^2 + k_{x_2}^2 \sigma_{x_2}^2 + k_{x_3}^2 \sigma_{x_3}^2} \tag{6.103}$$

where σ_y, σ_{x_1}, etc. are the standard deviations of y, x_1, etc., respectively. Hence the standard deviation of δy, i.e. that of y about the mean of y, is:

$$\sigma_{\delta y} = \left(\left(\frac{\partial y}{\partial x_1} \right)^2 \sigma_{x_1}^2 + \left(\frac{\partial y}{\partial x_2} \right)^2 \sigma_{x_2}^2 + \left(\frac{\partial y}{\partial x_3} \right)^2 \sigma_{x_3}^2 \right)^{1/2} \tag{6.104}$$

The corresponding mean value y' of the element output will be obtained from equation 6.101, viz.

$$y' = k_0 + k_{x_1} x_1' + k_{x_2} x_2' + k_{x_3} x_3' + f(x_1, x_2, x_3) \tag{6.105}$$

Accuracy (Precision), Bias and Measurement Error

There is often confusion concerning the distinction between *accuracy* and *precision*. The term precision is frequently used to describe the *reproducibility* or repeatability of results[85] (see previous section). Accuracy denotes the nearness of a measurement to its accepted value and should be expressed in terms of *error* (see below). Thus, accuracy involves a comparison with respect to a true or accepted value, and precision compares a result with the best value of several measurements made in the same way. An alternative (but less carefully worded) distinction[86] is to describe precision as the number of digits employed in a calculation (which may not all be correct) and accuracy as the number of digits to which the result of that calculation is correct.

The *error* in the reading of an instrument is accepted generally as being the difference between the actual measurement and the corresponding *true* value of the measurement where:

$$\text{positive error} = \text{actual measurement} - \text{true value.} \tag{6.106}$$

The true value or true reading is the value that would be obtained if the quantity were measured by an *exemplar* method, i.e. a method agreed upon by experts as being sufficiently accurate for the purpose to which the data will be put. However, it is important to distinguish between the error in a single measurement as defined in this way and the error intrinsic within the measurement process associated with the relevant instrument. In the latter instance, a single reading will constitute a sample from a statistical population generated by the measurement process. If the characteristics of that measurement process are known, then it is possible to determine the limits of the error in the single measurement, although the error itself is not known as this would presume knowledge of the corresponding true value. In order to achieve this, a statistical testing of the instrument is required in which a large number of readings (outputs) is recorded under the same apparent process conditions and with the same input to the instrument. However, as previously pointed out, these readings will differ due to random fluctuations in the environment in which the instrument is placed. Generally, such fluctuations and the instrument outputs corresponding to them are assumed to exhibit a normal or Gaussian distribution. This may not be so in practice, and the closeness of the data to a Gaussian distribution can be ascertained by application of appropriate statistical tests(87).

When instruments are calibrated, in practice it is usual to vary the input to the instrument incrementally over a particular range of true values and to record the corresponding measured values. Often there is no multiple repetition of the test at any one true value of the input. The resulting set of input/output data provides an *average calibration curve* for that instrument. Frequently (and often simply for convenience), this is assumed to be a straight line and the well-known least squares procedure is employed to fit a straight line to the data. It is then possible to estimate such quantities as the *instrument bias* and the *uncertainty* of the instrument reading, as outlined in the following example.

Example 6.4

A certain mass flowmeter (see Section 6.2.3) was tested (calibrated) by comparing the readings given by the instrument G_R with true (known) values G_T of the flow of a gas as measured by the instrument in a 0.15 m ID pipeline. True and measured values are compared in Fig. 6.61 and Table 6.17. Estimate the errors in the flowmeter due to bias and imprecision. Assume that variations in the input and output of the instrument are normally distributed.

TABLE 6.17. *True and Measured Mass Rate of Flow*

True flowrate G_T (kg/min)	Measured flowrate G_R (kg/min)	True flowrate G_T (kg/min)	Measured flowrate G_R (kg/min)	True flowrate G_T (kg/min)	Measured flowrate G_R (kg/min)
0.00	0.75	3.50	3.98	7.00	7.25
0.50	1.16	4.00	4.40	7.50	7.76
1.00	1.34	4.50	4.73	8.00	7.79
1.50	1.94	5.00	5.28	8.50	8.33
2.00	2.24	5.50	5.67	9.00	8.67
2.50	3.12	6.00	6.16		
3.00	3.26	6.50	6.70		

Solution

A straight line was fitted to these data using the method of least squares(87). The equation obtained was:

$$G_R = mG_T + k$$

where:

$$m = \frac{N\sum_{j=1}^{N}(G_T)_j(G_R)_j - \sum_{j=1}^{N}(G_T)_j\sum_{j=1}^{N}(G_R)_j}{N\sum_{j=1}^{N}(G_T)_j^2 - \left(\sum_{j=1}^{N}(G_T)_j\right)^2}$$

$$= \frac{19 \times 537.9 - 85.5 \times 90.53}{19 \times 527.3 - 7310}$$

$$= \underline{\underline{0.916}}$$

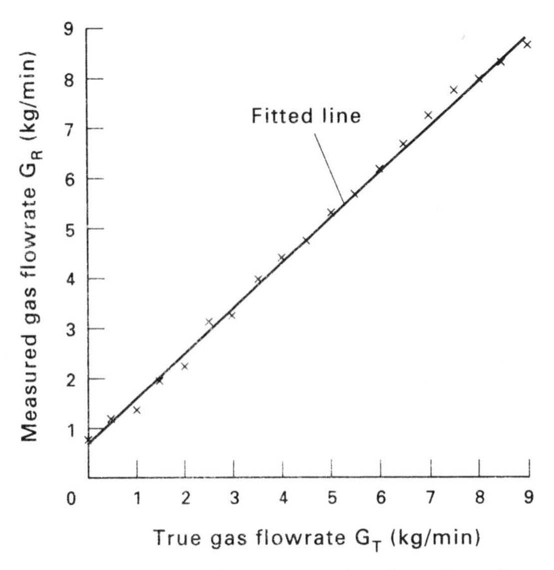

FIG. 6.61. Measured gas flowrate as a function of true flowrate

and:

$$k = \frac{N\sum_{j=1}^{N}(G_R)_j \sum_{j=1}^{N}(G_T)_j^2 - \sum_{j=1}^{N}(G_T)_j(G_R)_j\sum_{j=1}^{N}(G_T)_j}{N\sum_{j=1}^{N}(G_T)_j^2 - \left(\sum_{j=1}^{N}(G_T)_j\right)^2}$$

$$= \frac{90.53 \times 527.3 - 537.9 \times 85.5}{19 \times 527.3 - 7310}$$

$$= \underline{\underline{0.643}}$$

The fitted line is shown in Fig. 6.61. In order to estimate the scatter of the data it is necessary to compute the relevant standard deviations.

The variance of the measured values $= \sigma_{G_R}^2 = \frac{1}{N}\sum_{j=1}^{N}(mG_T + k - G_R)^2.$

$$= 0.0214$$

Thus the standard deviation of the measurements

$$= \sigma_{G_R} = \sqrt{(0.0214)} = \underline{\underline{0.146}}$$

The variance of the gradient $= \sigma_m^2 = \dfrac{N\sigma_{G_R}^2}{N\sum_{j=1}^{N}(G_T)_j^2 - \left(\sum_{j=1}^{N}(G_R)_j\right)^2}$

$$= 1.50 \times 10^{-4}$$

Thus:

$$\sigma_m = \underline{\underline{0.0123}}$$

The variance of the intercept $= \sigma_k^2 = \dfrac{\sigma_{G_R}^2 \sum_{j=1}^{N} (G_T)_j^2}{N \sum_{j=1}^{N} (G_T)_j^2 - \left(\sum_{j=1}^{N} (G_R)_j \right)^2}$

$$= 0.00417$$

Hence: $\sigma_k = \underline{\underline{0.0645}}$

In this calculation it has been assumed that the value of σ_{G_R} is sufficiently nearly the same for all values of G_T so that all the data points in the table can be used, thus making it unnecessary to repeat the experiments many times. Furthermore, if the distribution is considered to be Gaussian, then the *confidence limits* for the estimates of the gradient and intercept of the fitted line can be determined. For this type of distribution it can be shown[87] that 95 per cent of the readings will lie within $\pm 2\sigma$ of the mean. Hence it is possible to say with 95 per cent certainty (or with 95 per cent confidence limits) that the equation of the fitted line is:

$$G_R = mG_T + k$$

where $m = 0.916 \pm 2\sigma_m = \underline{0.916 \pm 0.02}$ and $k = 0.643 \pm 2\sigma_k = \underline{0.643 \pm 0.13.}$

Corresponding values are $\pm 3\sigma$ for 99.7 per cent confidence limits and $\pm \sigma$ for 68 per cent confidence limits for a Gaussian distribution. The true values G_T of the process variables must also contain some positive or negative errors. These can be determined from σ_{G_T} where:

$$\sigma_{G_T}^2 = \frac{1}{N} \sum_{j=1}^{N} \left(\frac{G_R - k}{m} - G_T \right)^2 = \frac{\sigma_{G_R}^2}{m^2}$$

Hence:

$$\sigma_{G_T} = \frac{\sigma_{G_R}}{m}$$

$$\approx \underline{\underline{0.16}}$$

Thus, if the instrument output records a reading of $G_R = 5.50$ kg/min then the estimate of the true value of the rate of flow will be the corresponding value of $G_T \pm 2\sigma_{G_T}$, i.e. 5.68 ± 0.32 kg/min, with 95 per cent confidence limits. The total error obtained in the calibration may be split into two parts, viz. the *bias* (or *systematic error*), i.e. $5.50 - 5.68 = -\underline{0.18}$ kg/min, and the *error due to imprecision (random error or non-repeatability)*, i.e. $\pm\underline{0.32}$ kg/min. The bias is assumed to be the same each time this reading is obtained and consequently can be corrected for, whereas the random error is generally different each time and thus cannot be removed—although the limits of this error can be estimated as above. The *total inaccuracy* of the measuring element is the sum of the bias and the random error. A better method of estimating uncertainty when the number of samples is small (such as in this example) is to use *Student's t-Test*[87].

Care must be taken in evaluating instrument error from manufacturers' literature. Frequently the accuracy of an instrument is specified without any explanation as to what the precise meaning of the number is. The error quoted is often simply the largest horizontal deviation from the fitted line of any data point. In Fig. 6.61, this is 0.25 kg/min and occurs at a number of points. This would therefore be quoted as an error of *2.8 per cent of full scale* (full scale being 9 units). This is reasonable if the bias is known to be zero. The form of the specification for an instrument quoted by manufacturers is generally in terms of a percentage of the full scale reading of that instrument. Thus, if a thermocouple is quoted as having an inaccuracy of ± 0.5 per cent of full scale and covers a range of 100 K to 1000 K, then this can be taken to mean that no error greater than 5 K will occur anywhere in the range. Of course, at 100 K, this would constitute a maximum error of ±5 per cent.

Threshold

This is defined as the minimum value of the input below which no output from the instrument can be detected (*cf.* resolution).

Dead Band, Dead Space, Dead Zone

When applied to the performance of an instrument these terms signify the range over which an input can be varied without the instrument output responding in any way. They are usually expressed in terms of the percentage of span of the instrument and should not be confused with the concept of dead time (Section 7.6).

Scale Readability

This is dependent upon the accuracy to which a human observer can read an analog record of the output of the instrument in question, or how closely an indicator marking a particular position on a scale can be read. For example, can a particular pressure gauge be read to the nearest 0.1 kPa, to the nearest 1 kPa, or to the nearest 10 kPa? It is important to consider this carefully when presented with data from an instrument with a digital output where the readability will be the same over the whole range of the instrument. In many cases the data obtained are only as precise as those indicated in the analog form.

Zero Shift (Zero Error)

It is important to check the zero setting (or the setting of the lower range value) for an instrument as a zero error will cause the whole of the instrument span to be displaced. The zero setting may drift or change over a period of time (*zero shift*). Such drifting is frequently due to variations in ambient conditions—most commonly temperature. In addition to zero shift, point values of the measured variable in different regions of the span may drift by different amounts.

6.11. SIGNAL CONDITIONING

Data produced by any basic measuring device commonly have to pass through two further stages before being presented to the process operator and/or used in a process control system (see Fig. 6.1). First, an individual signal may require some form of conditioning before it can be transmitted to the control room (e.g. amplification, conversion to a variable more suitable for transmission) and, secondly, it has to be transmitted—often simultaneously with other suitably conditioned signals. Further conditioning (e.g. noise reduction, further variable conversion, amplification, etc.) will frequently take place in the control room prior to the quantity being indicated or recorded, and/or employed as the measured variable of a given control scheme.

6.11.1. Bridge Circuits

Primary variables such as pressure, flow, temperature, etc., are frequently converted into resistance, capacitance or inductance by the use of a suitable transducer (Sections 6.2–6.4). These latter quantities are generally measured using a variety of bridge circuits. Both null and deflection bridges are employed with facilities such as the provision of sensitivity adjustment, calibration, and zeroing of the output voltage when the measured physical quantity is itself zero. Literature concerning bridge circuits is commonplace and the reader is referred to NEUBERT[19].

6.11.2. Amplifiers

Most electrical signals produced by transducers via bridges or otherwise are of low voltage (0–100 mV) or low power. Such signal levels are not generally suitable for data transmission, analog or digital processing, energising (deflection of) an indicator or for remote recording; hence they require amplification. The basis of most amplifier circuits, filters, converters and much other data processing equipment is the *operational amplifier (op-amp)*. There are many types and makes, and a typical simplified version of a voltage amplifying op-amp circuit is illustrated in Fig. 6.62*a*.

It is easy to show that this configuration gives:

$$V_o = \left(\frac{R_f}{R} + 1\right) V_i \tag{6.107}$$

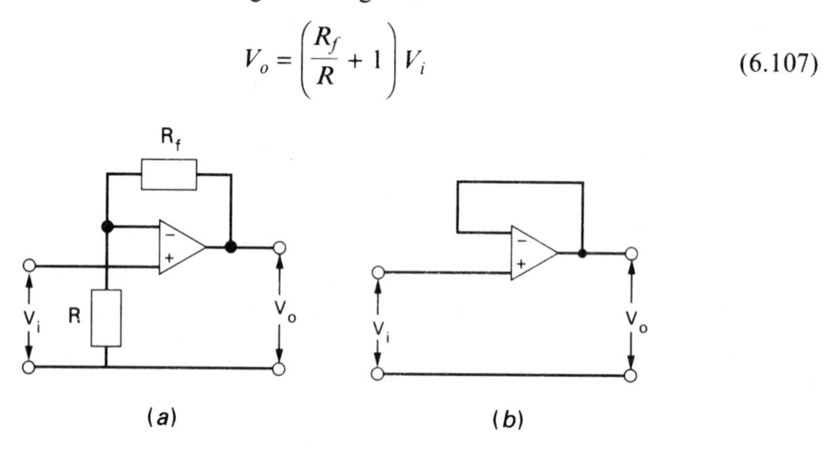

FIG. 6.62. Typical operational amplifiers: (*a*) non-inverting amplifier; (*b*) buffer amplifier

Thus, the *signal gain* depends upon the ratio R_f/R. Figure 6.62*b* represents a *buffer amplifier* or *voltage follower* in which $V_o = V_i$. This has the high input and low output impedances necessary to obviate the kind of inter-element loading problems illustrated in Section 6.11.6. For a more detailed treatment the reader is referred to SMITH[88].

The *instrumentation amplifier* is a high-performance differential amplifier consisting of a number of closed-loop op-amps. An ideal instrumentation amplifier gives an output voltage which is proportional only to the difference between two input voltages V_{i_1} and V_{i_2}, viz.:

$$V_o = K_{\text{amp}}(V_{i_2} - V_{i_1}) \tag{6.108}$$

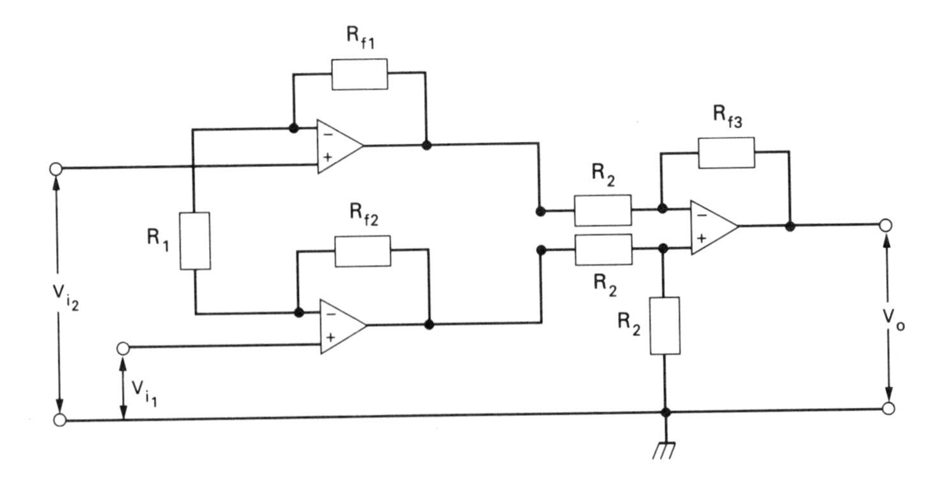

FIG. 6.63. Typical instrumentation amplifer

where K_{amp} is the amplification factor which can be varied over a wide range by manipulating the values of the resistors in the network[88].

The *isolation amplifier* is an instrumentation amplifier which is used in more difficult applications, e.g. where low level signals are superimposed upon high common-mode voltages, where processing circuitry has to be protected from faults and power transients, or where interference from motors etc. is severe. "Ordinary" instrumentation amplifiers require a return path for the bias current. If this is not provided then the bias current will charge stray capacitances which, in turn, produce large drifts in the output. Thus, when "floating" sources such as thermocouples are amplified, a connection to amplifier ground must be provided (Fig. 6.63) and this can produce excessive noise. An isolation amplifier does not need a ground connection of this type as the signal is *isolated* from ground and interfering noise is largely rejected[4].

6.11.3. Signals and Noise

Noise occurs for many reasons in signals that are transmitted between elements of a control loop or in data transmission systems. Sources of noise may be internal or external. Internal sources are those such as random temperature-induced motion of electrons or other charge carriers leading to a corresponding random voltage variation (*Johnson noise*). Transistors produce a similar type of noise (*shot noise*) due to random fluctuations in the way in which carriers diffuse across a junction. External sources of noise are most commonly near a.c. power circuits. This latter form of noise is frequently termed *mains pick-up* or *mains hum.* Heavy generators and turbines can also cause severe disturbances. Other electrical sources of interference are high voltage power lines, fluorescent lighting, and the switching on and off of electric motors. These external sources cause interference by inductive and/or capacitive coupling with the signal (in Fig. 6.64 this coupling effect is indicated in lumped form whereas, in practice, it will be distributed along the entire length of

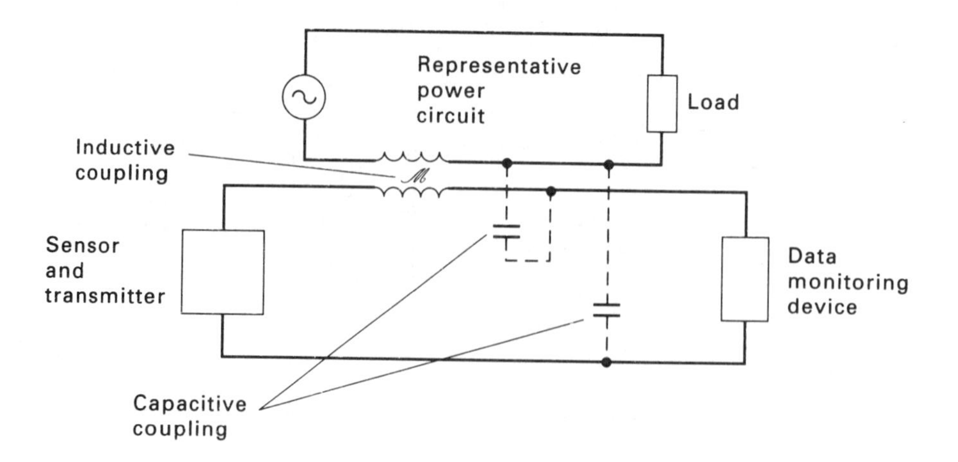

FIG. 6.64. Interference from external sources due to inductive and capacitive coupling

the circuit). For example, if there is inductive coupling present, then an alternating current i in a power circuit will induce an interference signal V_{int} in the measuring element circuit, where:

$$V_{int} = \mathscr{M} \frac{di}{dt} \tag{6.109}$$

where $\mathscr{M}$ is the mutual inductance. Inductive coupling will occur even if the measuring element circuit is isolated from earth.

Heavy electrical equipment can cause interference through the creation of *multiple earths* where there are leakages to earth at different points of the measuring device circuit. These earth points will be at different potentials due to the existence of the ground current which produces *common* and *series mode* interference voltages in the measurement circuit.

Clearly, such interference can be reduced by increasing the distance between the source and the measurement circuit—mutual inductance and capacitance both being inversely proportional to the distance. Inductive coupling can also be much reduced by the use of twisted pair cable (Fig. 6.65). If adjacent loops in the circuit have the same area (e.g. loops 1 and 2) and are coupled with the same magnetic field, then the induced voltages between points A and B and between B and C will cancel each other out, and this will be repeated along the whole section of twisted pairs.

Capacitive coupling can be suppressed by enclosing the entire measurement circuit within an earthed metal screen which provides a low impedance path to earth for the interfering currents. This is called *electrostatic screening*.

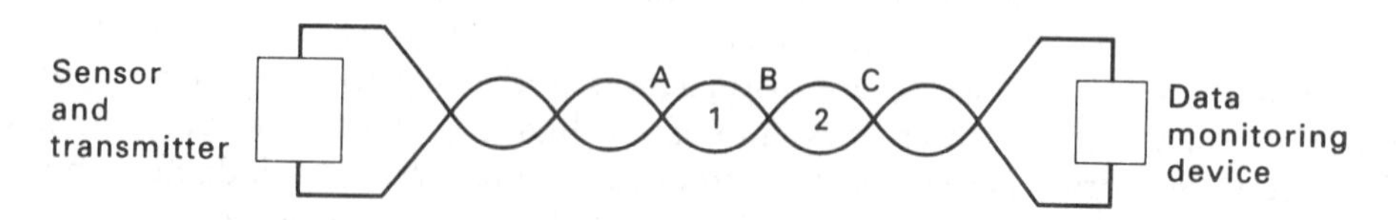

FIG. 6.65. The twisted-pairs technique for reducing inductive coupling

6.11.4. Filters

Filters are devices which transmit a specific range of the frequency spectrum of a signal and which reject the remainder. They can take many physical forms—but the most highly developed and convenient types are electrical. The latter can be divided into *analog filters* and *digital filters*. Analog filters are electrical networks of resistors, capacitors and amplifiers used to condition continuous signals. Digital filters consist of software within a microcomputer which processes sampled data signals. If the power spectrum of a measurement signal occupies a different frequency range from that of any noise or interference, then a filter will substantially improve the signal to noise ratio of that measurement signal. A large variety of such filters is available. The one having the required characteristics should be selected in terms of the ranges of frequencies required to be transmitted and rejected. The effects of employing the four most common types of filter are illustrated in Fig. 6.66. Of course, if the power spectrum of the measurement signal and that of the interfering noise overlap, then filtering will be of limited use and other procedures have to be considered.

6.11.5. Converters[89, 90]

Analog to Digital (A/D) Conversion

A/D conversion is of increasing importance as, in many instances, the analog outputs of various measuring elements are now frequently connected to microcom-

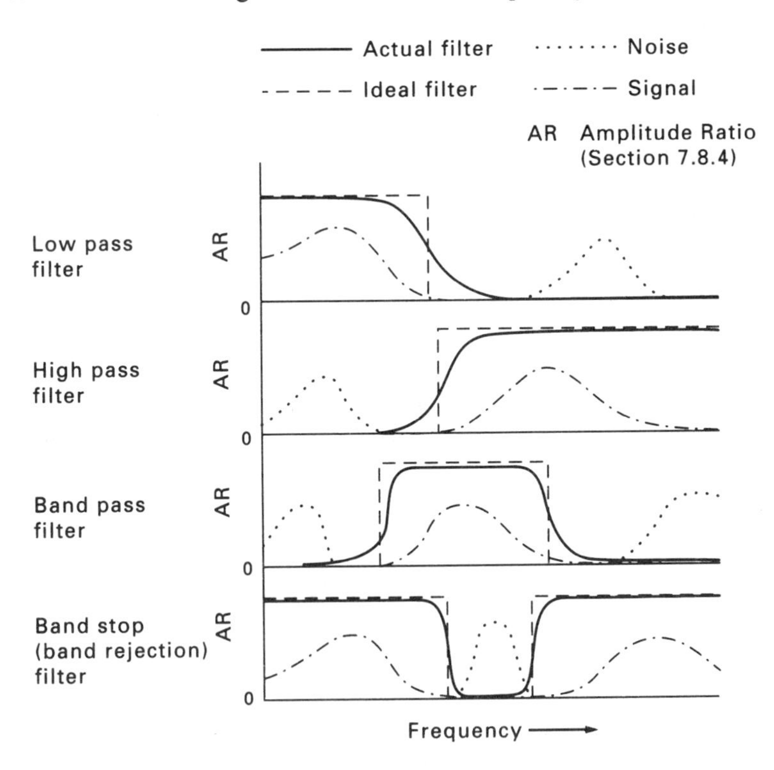

FIG. 6.66. Basic filter characteristics

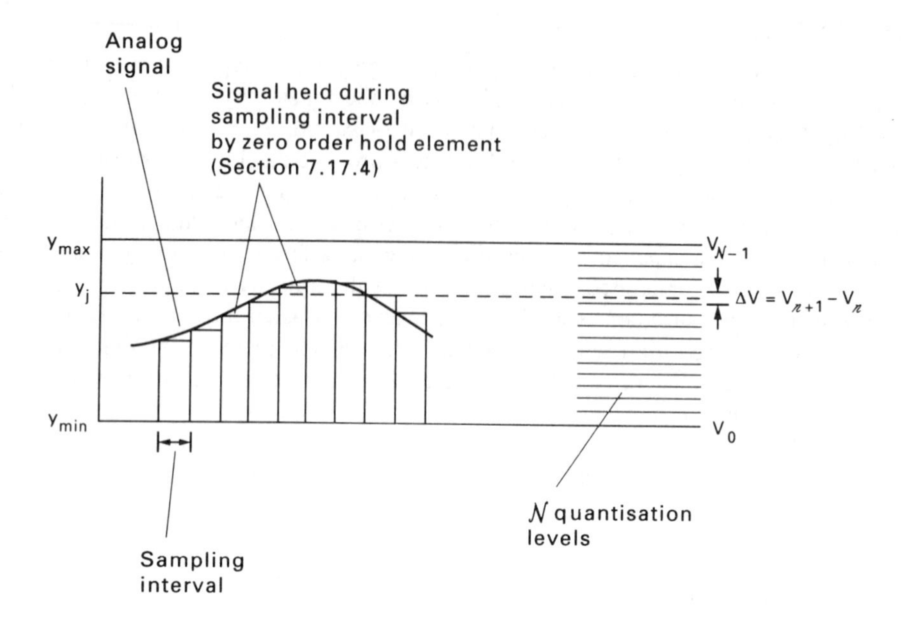

FIG. 6.67. Sample, hold and quantisation of an analog signal

puter installations which operate on the basis of discrete information. The function of an A/D converter can be divided up into three stages, viz. *sampling*, *quantisation* and *encoding*. Sampling (i.e. *sample* and *hold* in this context) is discussed in Section 7.17. Figure 6.67 illustrates the basic operations of the sample and hold process. This element is necessary as the A/D conversion requires a finite time to operate (up to a few ms). The signal from the hold element y_j is rounded to the nearest of $\mathcal{N}$ quantisation levels, i.e. V_n where $n = 0, 1, 2, \ldots\ldots, \mathcal{N} - 1$. The *quantisation interval* can then be written as:

$$\Delta V = \frac{y_{max} - y_{min}}{\mathcal{N} - 1} \tag{6.110}$$

where $y_{max} - y_{min}$ is the span (Section 6.10.1) allowed for y_j with $y_{max} = V_{\mathcal{N}-1}$ and $y_{min} = V_0$.

The maximum *quantisation error* expressed as a percentage of the span is:

$$\Xi_{max} = \pm \frac{100\,\Delta V}{2(y_{max} - y_{min})} \text{ per cent} = \pm \frac{50}{\mathcal{N} - 1} \text{ per cent} \tag{6.111}$$

and the resolution (Section 6.10.1) of the converter due to the quantisation step is:

$$\mathcal{R} = \frac{100\,\Delta V}{V_{\mathcal{N}-1} - V_0} \text{ per cent of the full-scale deflection.} \tag{6.112}$$

The encoder converts V_n into a *parallel digital signal* corresponding to a binary coded version of the denary numbers $0, 1, 2, \ldots\ldots, \mathcal{N} - 1$, where the number of binary digits (*bits*) required to encode a decimal number x is:

$$n \geqslant \frac{\log x}{\log 2} \tag{6.113}$$

and n wires in parallel will be required to produce an n-bit parallel electrical signal corresponding to the n-bit binary number.

TABLE 6.18. *Operation of a Successive Approximation A/D Converter*

Input voltage = 1.843 V

Clock Pulse	D/A Converter Input	D/A Converter Output (V_{dac} Volts)	Comparator Output (V_{comp} Volts)	Bit Value
1 (clear register)	00000000	0	0	—
2 (first guess)	01111111	1.27	0 (LO)	$b_7 = 1$ (MSB)
3 (second guess)	10111111	1.91	1 (HI)	$b_6 = 0$
4	10011111	1.59	0 (LO)	$b_5 = 1$
5	10101111	1.75	0 (LO)	$b_4 = 1$
6	10110111	1.83	0 (LO)	$b_3 = 1$
7	10111011	1.87	1 (HI)	$b_2 = 1$
8	10111001	1.85	1 (HI)	$b_1 = 0$
9 (final guess)	10111000	1.84	0 (LO)	$b_0 = 1$ (LSB)

Output digital signal = 10111001, i.e. 1.85 V

Equivalent serial digital signal: 1 0 0 1 1 1 0 1
(see also Fig. 6.75)

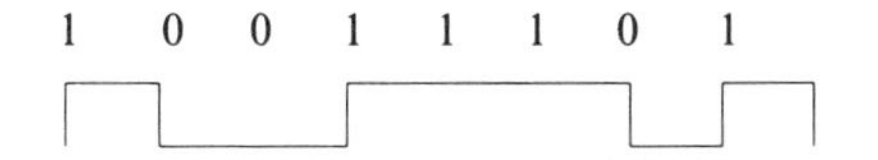

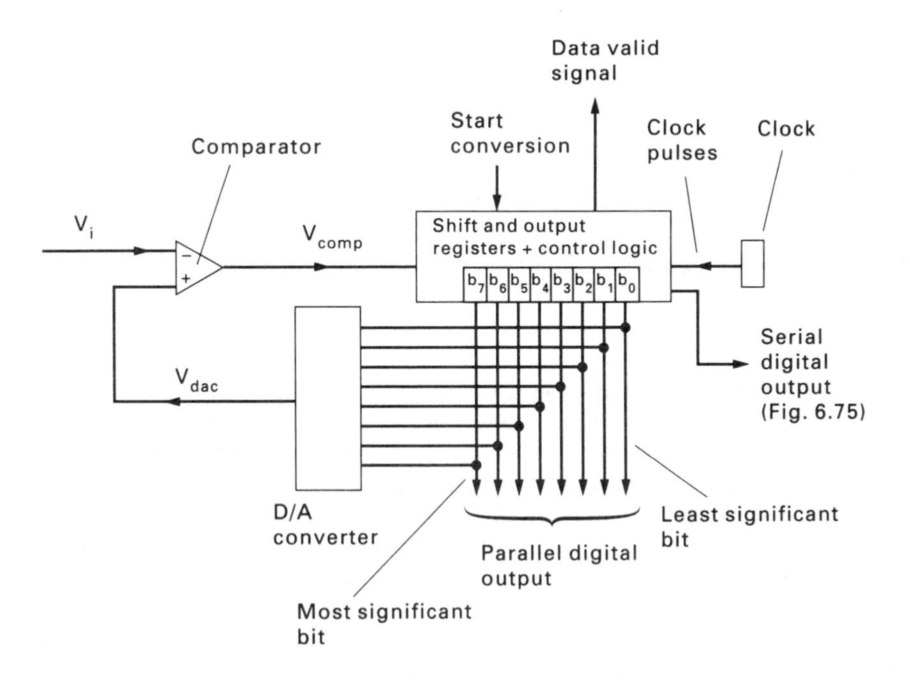

FIG. 6.68. Successive approximation 8-bit binary A/D converter (Table 6.18)

Another commonly employed encoding scheme is that of *binary coded decimal* (*bcd*). In this case each decade of the denary (decimal) number is coded separately into its binary equivalent and each denary digit has a maximum of four bits. The number of bits required in this case to encode a decimal number x is:

$$n_{bcd} \geqslant 4 \log x \tag{6.114}$$

There are many types of commercial A/D converter. The most common is the *successive approximation* type (shown in Fig. 6.68) which employs a D/A converter (see below) in a closed-loop arrangement. Converters of this type are available with resolutions of up to 16 bits and conversion times of 30 μs or less. As an example of the way in which such a converter operates—consider an analog input voltage of $V_i = 1.843$ V supplied to the A/D converter shown in Fig. 6.68. At each successive clock-pulse after the conversion is initiated, one bit of the digitally encoded output will be produced, the most significant bit (MSB) being first, and the least significant bit (LSB) being last. The operation of the converter is represented in Table 6.18.

The *dual slope* converter is superior to the successive approximation type in many respects, e.g. in having an automatic zero capability, good noise suppression, and accuracy which is not affected by capacitor rating or clock frequency. However, the use of the latter type of converter is limited because of its slow conversion time (about 30 ms).

Digital to Analog (D/A) Conversion

This is required when a digital signal (e.g. the output from a microcomputer) is used to drive an analog device (e.g. a control valve). Most D/A converters utilise the so-called *R–2R ladder network*. The accuracy of these devices is dependent upon the stability of the reference voltage and of the resistors employed. Figure 6.69 illustrates a simple 3 bit D/A circuit using an R–2R ladder network. The *current steering* switches are connected to ground whether they are at 0 ("off") or 1 ("on"). This is because the op-amp negative input is a *virtual ground*[90]. The op-amp sums the currents "steered" to it by the switches and produces an output voltage which is proportional to this sum. Devices giving 8–12 bits of resolution and 4–20 mA current output are the most commonly employed in process control applications.

6.11.6. Loading Effects

When devices are connected together, the action of one device may affect the performance of the device immediately preceding it. This is termed *inter-element* loading. If a total measurement system is constructed out of basic building blocks (e.g. sensor, amplifier, transmitter, receiver, recorder), then loading problems must be considered carefully if the measurement system is to work satisfactorily. Not only has inter-element loading to be taken into account, but also the effect of placing the measuring element within the process itself. For example, the insertion of the thermocouple into a flowing fluid will change the flow pattern in the vicinity of the thermocouple and the temperature distribution at the end of the thermocouple will not then be the same as when the thermocouple was not present[91]. This is called *process loading*. On the whole, errors due to the latter are of less importance than those caused by inter-element loading. These effects are additional to any random instrument error as described in Section 6.10.1. The extent of the error that can occur through inter-element loading can easily be illustrated in Example 6.5.

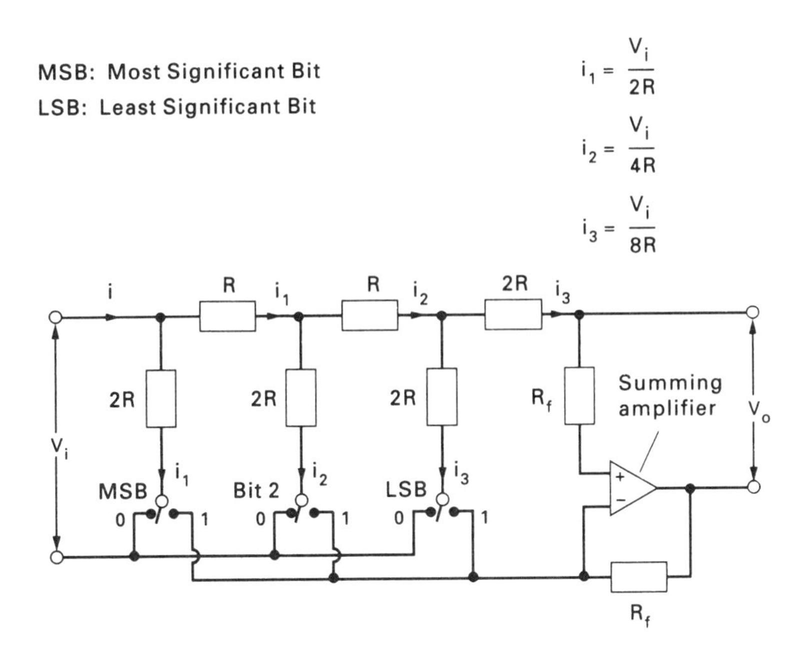

FIG. 6.69. An R–2R ladder-network 3-bit D/A converter

Example 6.5

A pH glass electrode (Section 6.8.2) having a sensitivity of 50 mV per unit pH and a resistance of 250 MΩ is connected directly to a recording device which has a sensitivity of 0.02 pH scale reading per mV and a resistive load of 100 kΩ (Fig. 6.70). If the electrode is immersed in a solution of pH = 5, determine the pH reading displayed by the recorder.

Solution

$$\text{e.m.f. due to pH meter} = 50 \times 5 = 250\ \text{mV}$$

$$\text{Total resistance of circuit} = 250\ \text{M}\Omega + 100\ \text{k}\Omega \approx 250\ \text{M}\Omega$$

$$\text{Current in circuit} = \frac{0.25}{2.5 \times 10^8} = 10^{-9}\text{amps.}$$

$$\text{Hence, potential difference across recorder} = V_L = 10^{-9} \times 10^5 = 10^{-4}\text{V.}$$

$$\text{Meter reading in pH units} = 10^{-4} \times 20 = \underline{\underline{2 \times 10^{-3}}}$$

which is clearly nonsense. This result is due to inter-element loading.

Suppose now that a buffer amplifier (Section 6.11.2) is inserted between the pH sensor and the recorder. A buffer amplifier has a large input impedance (of the order of 10^{12} Ω), a low output impedance (about 10 Ω) and a voltage gain of unity. The circuit is now represented by Fig. 6.71.
Thus:

$$V_i = \frac{0.25 \times 10^{12}}{2.5 \times 10^8 + 10^{12}} = 0.25\ \text{V}$$

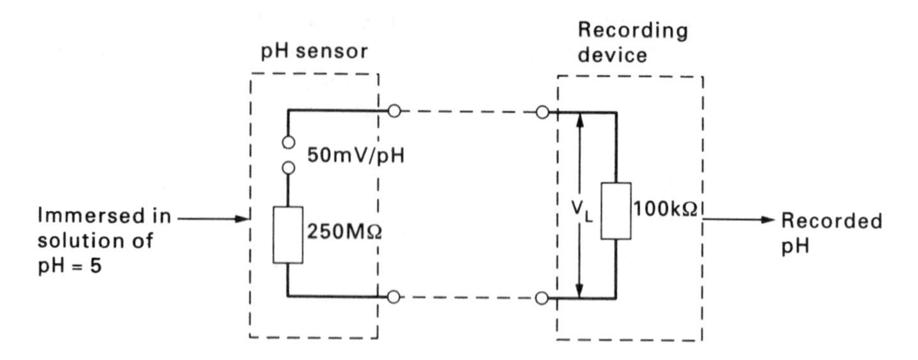

FIG. 6.70. An example of inter-element loading with a pH sensor

For a buffer amplifier:

$$V_o = V_i$$

Thus:

$$V_L = \frac{0.25 \times 10^5}{10 + 10^5} \approx 0.25 \text{ V}$$

Hence the reading of the pH meter will be 0.25 × 20

= 5 pH units.

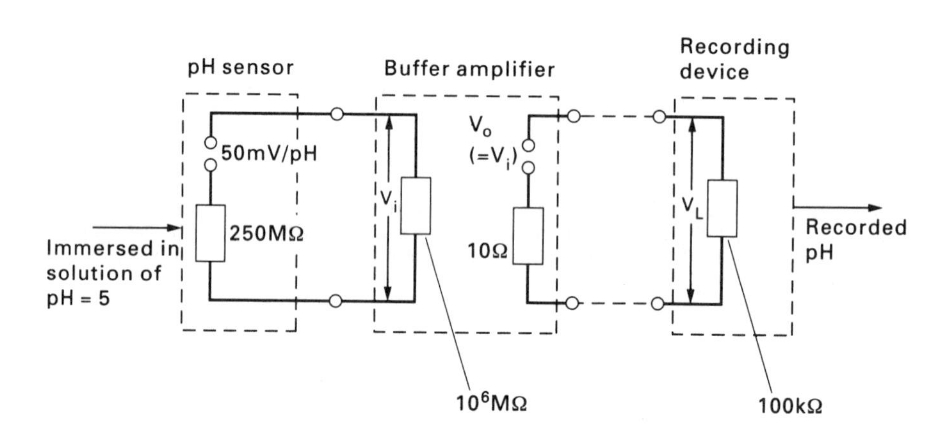

FIG. 6.71. Avoidance of inter-element loading by the use of a buffer amplifier

Example 6.5 shows that the insertion of a buffer amplifier reduces the error due to inter-element loading to negligible proportions (about 0.1 per cent in terms of pH). As a general guide, the impedance of the recording device should be much greater than that of the sensor where there is direct voltage transfer from sensor (source) to recording element (load). The relationship between the load voltage and source in a complex circuit is much simplified by the application of *Thévenin's theorem*[20] which states that any linear network of impedances and voltage sources can be substituted by an equivalent circuit containing a single voltage source E_{Th}

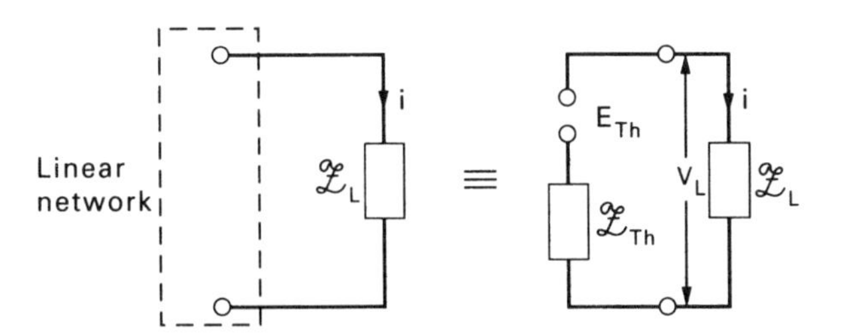

FIG. 6.72. Thévenin equivalent circuit

and an impedance $\mathscr{Z}_{Th}$ in series with it, where E_{Th} is the open-circuit voltage of the network across its output terminals and $\mathscr{Z}_{Th}$ is the network impedance with all source voltages put equal to zero and replaced by their respective internal impedances.

An *equivalent Thévenin circuit* is shown in Fig. 6.72. If V_L is the potential difference across the impedance of the load $\mathscr{Z}_L$ (e.g. the recorder in the above example), then:

$$V_L = i\,\mathscr{Z}_L = \left(\frac{E_{Th}}{\mathscr{Z}_{Th} + \mathscr{Z}_L}\right)\mathscr{Z}_L. \tag{6.115}$$

Hence, to reduce the effect of inter-element loading to a minimum, $\mathscr{Z}_L$ must be much greater than $\mathscr{Z}_{Th}$, i.e. $V_L \approx E_{Th}$ (as in the second part of Example 6.5 where $\mathscr{Z}_{Th}$ is effectively 10 Ω and $\mathscr{Z}_L$ is 100 kΩ).

If the sensor or transmitter is a current source rather than a voltage source, then a converse approach is required. This can be illustrated by a further example.

Example 6.6

An electronic torque balance transmitter incorporating an LVTD (Fig. 6.13) has a sensitivity of 1 mA output current per 1 kN/m^2 change in measured differential pressure, where the current is measured across an output impedance of 100 kΩ. This transmitter is connected to a recorder which has an input impedance of 10 kΩ and a sensitivity of 1 kN/m^2 per mA change in current from the transmitter (Fig. 6.73). If the pressure sensor is measuring a true pressure differential of 5 kN/m^2, what will be the corresponding reading on the recorder?

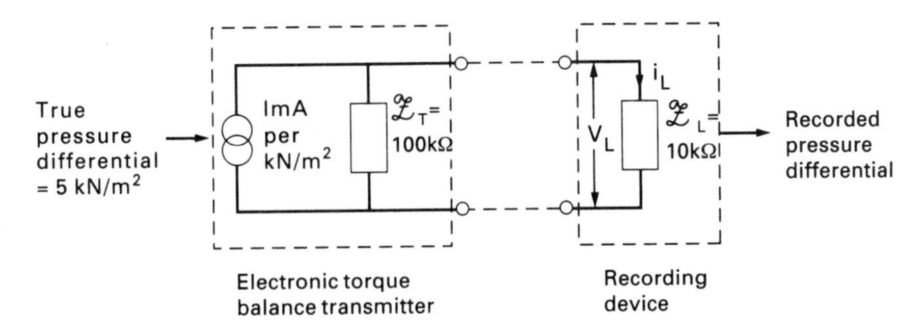

FIG. 6.73. The recording of a transmitter signal when the transmitter is acting as a current source

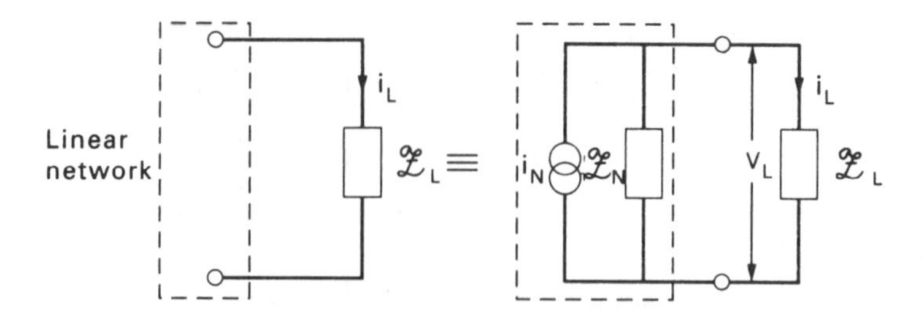

FIG. 6.74. Norton equivalent circuit

Solution

From the transmitter sensitivity, for a true pressure differential of 5 kN/m², the transmitter output current i_T is 5 mA. Neglecting any resistance in the cables connecting the transmitter and the recorder, the total impedance of the circuit is:

$$\mathscr{Z}_X = \frac{\mathscr{Z}_T \mathscr{Z}_L}{\mathscr{Z}_T + \mathscr{Z}_L} = \frac{100 \times 10}{100 + 10} = \frac{100}{11}\ \text{k}\Omega$$

Thus:

$$V_L = i_T \mathscr{Z}_X = \frac{500}{11}\ \text{V}$$

and:

$$i_L = V_L/\mathscr{Z}_L = 4.5\ \text{mA}$$

Hence, the recorded pressure differential = 4.5 kN/m² which represents an error of about 10 per cent.

If, however, $\mathscr{Z}_T$ is increased to 1 MΩ and $\mathscr{Z}_L$ is reduced to 1 kΩ, then, for a transmitter output of 5 mA, $\mathscr{Z}_X = 1000/1001$ kΩ and $V_L = 5 \times 1000/1001$ V.

Hence:

$$i_L = \frac{5 \times 1000 \times 1000}{1001 \times 1000} = 4.995\ \text{mA}.$$

In this case the recorded pressure differential is 4.995 kN/m², i.e. an error of 0.1 per cent.

For more complex current sources, it is necessary to employ *Norton's theorem*[20] which states that any linear network of impedances and voltage sources can be substituted by an equivalent circuit containing a current source i_N in parallel with an impedance $\mathscr{Z}_N$, where i_N is the current which flows when the output terminals of the network are short-circuited and $\mathscr{Z}_N$ is the network impedance with all source voltages put equal to zero and replaced by their internal impedances.

From the *equivalent Norton circuit* (Fig. 6.74):

$$V_L = \frac{i_N \mathscr{Z}_N \mathscr{Z}_L}{\mathscr{Z}_N + \mathscr{Z}_L} \tag{6.116}$$

Hence, it is necessary to make $\mathscr{Z}_L << \mathscr{Z}_N$ in order to minimise the effect of inter-element loading (i.e. to make $V_L \approx i_N \mathscr{Z}_L$) when the sensor or transmitter is a current source.

6.12. SIGNAL TRANSMISSION (TELEMETRY)

Some type of signal transmission is required for all measuring systems that have a display (indicator, recorder) remote from the measuring element and transmitter. In many instances, measurements may be required of temperatures, flows, pressures, compositions, etc. from a plant in a particular preset order, or upon demand from a central point (e.g. in a distributed computer control system (Section 7.20)). In such cases a *multi-input, multi-output (MIMO) data acquisition* system using *time division multiplexing* may be employed. Instruments may be connected together in the form of *local area networks (LANs)* as part of a distributed control system. The way in which the instruments (termed *nodes* in this context) are connected together is known as the *topology* of the system. Various topologies are in common use and the particular topology of the LAN specifies the way in which each instrument or device within the LAN supplies information (i.e. *talks*) to the other instruments or devices in the LAN.

The term *telemetry* is reserved, generally, for multiple data systems which use a modulated high frequency carrier to transmit information concerning required sets of measurements from one point to another. This is necessary where plant items are located in remote areas at considerable distances from each other (e.g. pumping stations for the transfer of fluids through long pipelines in the oil, gas and water industries). A typical telemetry system comprises a *master station* acting as a central control point, with several *out-stations* placed at strategic positions. It must be capable of transmitting large quantities of information in both directions between master station and out-station and, as such, it involves the principles of *serial digital signalling, error detection* and *frequency shift keying*(92).

6.12.1. Multiplexers (Time Division Multiplexing)

A multiplexer is essentially a device which switches between a number of different inputs (*channels*) and is constructed of solid-state switches(93). Channels are addressed usually in a fixed order (*sequential addressing*) with a parallel channel address to specify the order in which the input channels are to be connected to the output channel. *Random addressing* is also possible, in which the operator can select a channel as required. A sample and hold device (Section 7.17.4) is attached which maintains the value of a particular signal until it is sampled again. The final output is generally encoded either as a *serial digital binary* signal or as a bcd signal (Section 6.11.5). Different measured variables often have frequency spectra with different maximum frequencies, i.e. the rates of change of different process variables can vary widely. Thus, it may be necessary to sample a flow measurement much more often than (say) a temperature measurement, or a composition measurement more often (if possible) than a pressure measurement.

6.12.2. Serial Digital Signals

Where large quantities of information have to be transmitted between control room, measuring devices, local controllers, etc., parallel analog signalling systems (in which individual devices are connected separately to the control room) are being replaced rapidly by two wire serial digital connections(94). The latter have a number of advantages, the two major ones being the cost of wiring and the ability to withstand corruption by noise. Generally, the analog signal is changed to digital form prior to transmission using an A/D converter. The latter produces a parallel

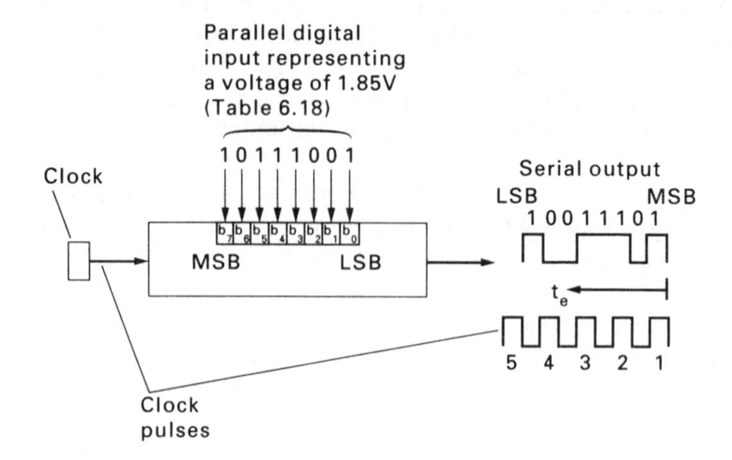

FIG. 6.75. Parallel-to-serial conversion of digital signals (t_e = time elapsed after initial clock pulse)

digital signal which requires conversion to serial form. Figure 6.75 illustrates diagrammatically how this can be achieved.

On receipt of the first clock pulse, the contents of the 8-stage shift register are shifted one place to the right, and hence b_0 (*the least significant bit—LSB*) is transmitted first. The second clock pulse causes the next bit b_1 to be transmitted, and so on until the register is empty. This type of data transmission is frequently termed *pulse code modulation (PCM)*. It is resistant to considerable noise levels as the final receiver has only to decide whether a 1 or a 0 has been transmitted (i.e. whether there is a pulse or not) and the length of the pulse. The transmission link over which the PCM is sent may be a cable, an optical fibre, or a radio link. Such links have a variety of speeds of data transmission which are expressed in terms of *baud* (or *baud rate*) where 1 baud = 1 bit/s. A single signal encoded into an n-bit code and sampled f times per second requires a transmission bit rate of[8]:

$$\mathcal{R}_b = n_f \text{ baud.} \tag{6.117}$$

If there are k_m multiplexed signals in the PCM transmission then:

$$\mathcal{R}_b = n k_m f \text{ baud} \tag{6.118}$$

and the bandwidth of the PCM for a single or multiplexed signal is:

$$\psi = 0 \quad \text{to} \quad \mathcal{R}_b/2 \tag{6.119}$$

It is important that any errors occurring during the decoding of a noise-affected PCM are detected. This is achieved by the addition of extra *check digits* to the *information digits* containing the measurement data. Hence, each complete *code word* transmitted of n digits will consist of k' information digits and $(n - k')$ check digits. This is termed an (n, k') code and is described as having a *redundancy* of $\dfrac{100(n - k')}{n}$ per cent.

In industrial telemetry systems, the quantity of random noise is generally small and, consequently, there is a low probability of errors occurring. Occasionally, large errors extending over short time periods do occur (e.g. due to the switching of pumps, etc.). In such cases no attempt is made to correct the signal and the receiver, having detected an error, simply requests that the information be transmitted again or ignored for that time period.

High transmission rates can be achieved, if necessary, over the relatively short distances required in a process plant. The PCM equivalent of the 4–20 mA analog transmission system shown in Fig. 6.1 can operate at up to 9,600 baud for distances up to 3000 m. The standard RS-232C transmission link is limited to about 15 m at rates up to 20,000 baud. Higher speed interfaces (such as versions of the IEEE-488 connection) used for computer control systems can handle up to 20,000 bytes/s (which for an 8-bit system is about 1.6×10^5 baud). However, in this case, the distance between devices is limited to about 2 m[4]. The more recent RS-422A standard allows the transmission of data rates of 10^7 baud over distances not exceeding 16.4 m and 10^5 baud over distances not exceeding 1220 m[95].

In many telemetry systems the PCM signal is frequency modulated on to a *carrier wave*. This is termed *frequency shift keying (FSK)*. In other arrangements, the output of the transducer is converted into fixed step changes of the phase of the modulating signal. A device for this purpose is termed a modulator/demodulator or *modem*.

6.12.3. The Transmission of Analog Signals

This is the traditional mode of transmitting data between plant and control room and is being steadily superseded by serial digital links (Section 6.12.2). Such hard-wire connections are prone to the effects of noise and exhibit inductance and capacitance effects which may be serious if the frequency of the transmitted signal is high. This has not generally been the case in the process industries and the frequency response of such analog data links is quite adequate. A common hard-wired analog transmission link widely used in the process industries is the 4–20 mA two-wire current loop. The 4–20 mA current is known as an *offset zero* or *live zero* signal. The advantage of employing a live zero is that it allows instrument or line faults to be detected. This system transmits both power and signal. The voltage appearing across the transmitter output terminals (which varies as the output current changes) forms the transmitter power supply, but the transmitter operation is insensitive to changes in this voltage as long as it stays above a given minimum value. In this case the transmitter is a true current source which makes the system relatively insensitive to induced noise voltages and line resistance changes.

6.12.4. Non-electrical Signal Transmission

Optical Fibres[96, 97]

The role of fibre-optics in high speed telemetry systems is increasing rapidly. Necessary components are an electrically controllable light source (e.g. a *light emitting diode* (*LED*)), an optical fibre, and a photo detector as the receiver. Very high data rates with wide bandwidths can be achieved. Optical fibres have the following advantages:

(a) They are immune from common electromagnetic interference sources such as power switching.
(b) They exhibit a high degree of safety in hazardous environments.
(c) They provide considerable data security and minimal fibre-to-fibre leakage (or cross-talk).
(d) Data can be transmitted at much higher frequencies (up to 10 MHz) with much lower losses than in equivalent electrical systems. Hence, many more signals can be multiplexed into one cable and these can travel much greater distances (at present up to 30 km) without further amplification.
(e) The fibres are highly resistant to corrosion.
(f) The fibres are lighter and of smaller diameter than metal conductors of the same information carrying capacity.
(g) Earth loop difficulties are reduced or eliminated.

One area where further development of fibre-optic transmission systems is required is in the design of suitable interfaces or converters to provide sufficient mechanical force to drive control valve actuators (Section 7.22.3)[98].

Pneumatic Transmission

Pneumatic transmission lines have been used in the process industries for many years and, although increasingly being superseded by hard-wired, optical fibre and even radio telemetry systems, they are still employed in flammable environments and where signals are conveyed over distances of up to 100 m—particularly in older plants. The essential components of a pneumatic transmission system are a supply of clean, dry air (usually at a pressure of 125–200 kN/m^2 gauge—see Section 6.3.1); a suitable transmitter which produces an output pressure generally in the range 20–100 kN/m^2 gauge; a receiver; and the necessary pipework. Pneumatic transmitters

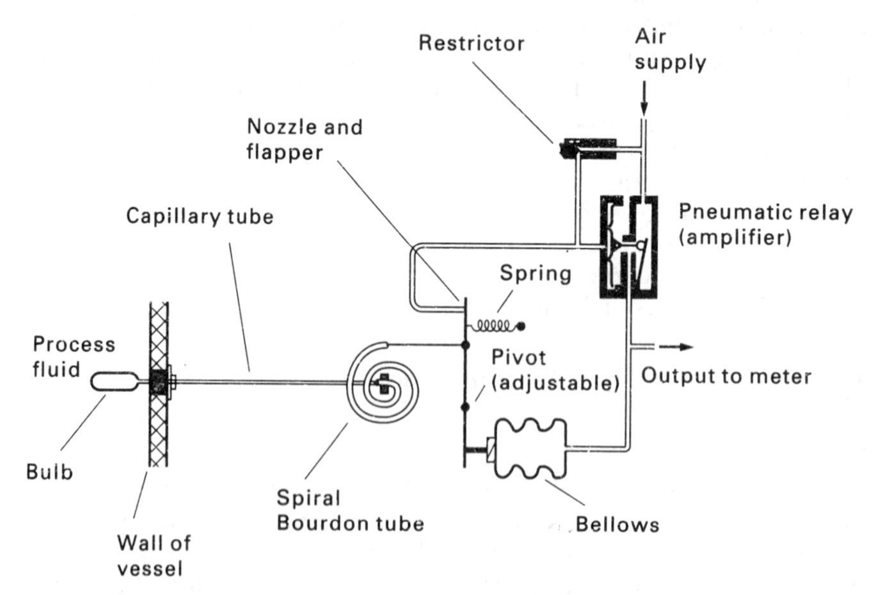

FIG. 6.76. Temperature-sensing bulb and associated pneumatic transmitter

consist of three main elements, viz. a sensing element such as a bellows, diaphragm or Bourdon tube; a sensor such as a flapper/nozzle system; and a feedback element which is usually another bellows. A typical pneumatic transmitter linked to a temperature sensing bulb is shown in Fig. 6.76.

The distance over which pneumatic signals can be transmitted is limited by the volume of the tubing and the resistance to flow. The dynamics of pneumatic systems can generally be approximated by a first order lag plus a dead time (Sections 7.5 and 7.6). Tubing may be made of copper, aluminium or plastic, and is normally of 5 mm ID. Pneumatic receivers can be in the form of indicators, recording devices and/or controllers.

Electropneumatic (E/P) Converters

One important application of pneumatic transmission is in the operation of diaphragm actuators. These are the elements generally employed to drive the spindles of control valves (Section 7.22.3) and, if hard-wired transmission systems are employed, require devices which convert electric current into air pressure or air flowrate, i.e. *electropneumatic (E/P) converters*. The basic construction of a typical E/P converter is illustrated in Fig. 6.77. A coil is suspended in a magnetic field in such a way that when a current is passed through the coil it rotates. This rotation is sensed by a flapper/nozzle system (Section 7.22.1). The nozzle is supplied with air via a restrictor and its back pressure actuates a pneumatic relay. The output from the latter is applied to the feedback bellows and also acts as output from the E/P converter. Electropneumatic valve positioners employ the same principle of operation.

Force–Balance Transducers

These elements also operate on the flapper/nozzle principle and are similar in construction to the pneumatic differential pressure (DP) cell (Section 6.3.4 and Figs 6.19*a* and 6.34) and the pneumatic controller (Section 7.22.1).

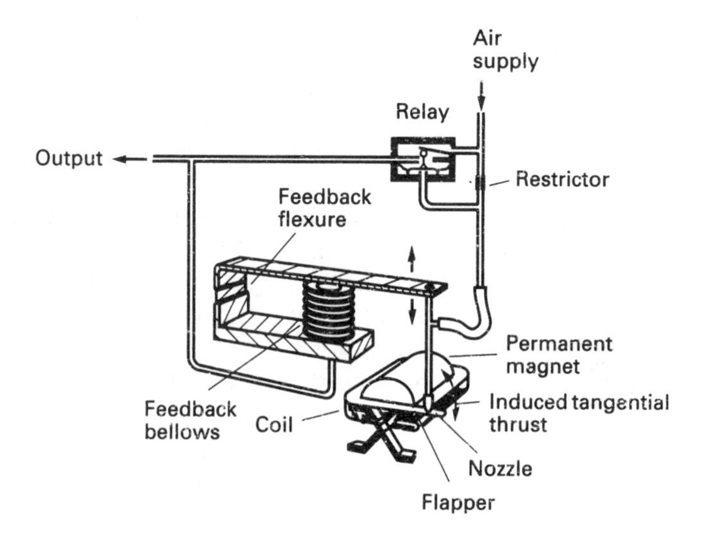

FIG. 6.77. Schematic diagram of electropneumatic (E/P) converter

6.12.5. Smart Transmitters and Associated Protocols—Intelligent Hardware

An instrument which not only measures a variable, but also carries out further processing in order to refine the data obtained before presentation either to an observer or to some other stage of the system, is generally termed *intelligent* or *smart*. In practice, additional functions are usually available as well as the data-processing facility and these are all normally contained within the transmitter of the measuring device. These so-called *smart transmitters* are microprocessor-based and enable the device:

(a) To produce a higher accuracy of measurement through automatic compensation for systematic errors and changes in ambient conditions.
(b) To change measurement ranges automatically as required.
(c) To provide automatic calibration.
(d) To communicate to the operator or maintenance personnel via a computer interface or hand-held communicator.

The cost of an intelligent instrument can be twice that of the equivalent device without the smart facility (the latter is termed a *dumb* instrument)[(99)]. However, the use of a smart transmitter does generally improve the inherent accuracy of the sensor itself.

A common feature of smart devices is the ability, either to transmit the normal 4–20 mA analogue output (which is digitally linearised and compensated where necessary), or to provide digital communication with other devices as desired. Digital communication with a smart transmitter can be implemented from a microprocessor within the control room, or by the use of a hand-held terminal. The latter can be inserted at any point within the 4–20 mA current loop and instructions to change the range, calibration, etc. can be sent to a specific smart device[(100)].

One continuing difficulty with intelligent devices is their compatibility within a given distributed control system or LAN. This problem is particularly acute when individual instruments and other devices within the LAN are purchased from different manufacturers, each of whom employs its own protocol[(94)]. In such cases, the devices will generally be unable to talk to each other without substantial additional hardware. Attempts are being made to overcome this problem by establishing a common industry communication standard termed *Fieldbus*. Currently, the nearest approach to a cross-company or *open standard* protocol is *HART* (the *Hiway Addressable Remote Transducer* protocol)[(99,101)]. This is effectively an interim solution in which existing 4–20 mA two wire systems can be used to transmit digital information. It employs the FSK technique (Section 6.12.2) and permits two modes of operation, viz. a 4–20 mA mode with digital signals imposed upon it and an all-digital mode which allows up to 15 intelligent devices to be connected to one cable. Hence, instruments that employ the HART protocol can be accessed by other HART devices, although they may have been purchased from different sources.

6.13. FURTHER READING

BANNISTER, B. R. and WHITEHEAD, D. G.: *Transducers and Interfacing* (van Nostrand Reinhold, Wokingham, U.K., 1986).
BENTLEY, J. P.: *Principles of Measurement Systems* (Longman, New York, 1983).

BILLING, B. F. and QUINN, T. J.: *Temperature Measurement* (Adam-Hilger, New York, 1975).
CLEVETT, K. J.: *Handbook of Process Stream Analysis* (Wiley, New York, 1986).
CONSIDINE, D. M.: *Process Instruments and Controls Handbook* (McGraw-Hill, New York, 1957).
DOEBELIN, E. O.: *Measurement Systems Application and Design*, 4th edn (McGraw-Hill, New York, 1990).
MILLER, R. W.: *Flow Measurement Engineering Handbook* (McGraw-Hill, New York, 1983).
NORTON, H. N.: *Sensor and Analyser Handbook* (Prentice-Hall, Englewood Cliffs, New Jersey, 1982).
PARADINE, C. G. and RIVETT, B. M. P.: *Statistical Methods for Technologists* (English Universities Press, London, 1966).
WILLARD, H. H., MERRITT, L. L., DEAN, J. A. and SETTLE, F. A.: *Instrumental Methods of Analysis*, 6th edn (Wadsworth, Belmont, 1981).

6.14. REFERENCES

1. STROUHAL, F.: *Ann. Phys. Chem.* **5** (1878), 216. Über eine besondere Art der Tonerregung.
2. MILLER, R. W.: *Flow Measurement Engineering Handbook* (McGraw-Hill, New York, 1983).
3. GIMSON, C.: *Processing* (April, 1993), 14. Guide to techniques for gas flow measurement.
4. DOEBELIN, E. O.: *Measurement Systems Application and Design*, 4th edn (McGraw-Hill, New York, 1990).
5. FOWLES, G., in *Instrumentation Reference Book* (ed. NOLTINGK, B. E.), Chapter 1, Measurement of flow (Butterworths, London, 1988).
6. Endress and Hauser plc: *Technical Product Group Report* PG001D/05/e (1993), 20. Flow measurement.
7. LYNNWORTH, L. C., in *Physical Acoustics*, **14** (eds. MASON, W. P. and THURSTON, R. N.), Chapter 5, Ultrasonic flowmeters (Academic Press, New York, 1979).
8. BENTLEY, J. P.: *Principles of Measurement Systems* (Longman, New York, 1983).
9. WASSON, R.: *Trans. Inst. Meas. Cont.* **5** (12) (1972), 479. Mass flow metering.
10. PLACHE, K. O.: *Mech. Eng.* (March, 1979), 36. Coriolis/gyroscope flow meter.
11. VAN DER BENT, H.: *Processing* (May, 1993), 24. Metering mass flow.
12. GERRARD, D.: *Control and Instrumentation* **24** (2) (1992), 37. Very low flows in sharp focus.
13. HENDERSON, F. M.: *Open Channel Flow* (Macmillan, London, 1966).
14. British Standard 3680: *Methods of Measurement of Liquid Flow in Open Channels* (1969–1983).
15. STOOR, P. G. J., in *Instrumentation and Automation in Process Control* (eds. PITT, M. J. and PREECE, P. E.), Chapter 5, Flow (Ellis Horwood, New York, 1990).
16. SAUNDERS, H.: *Control and Instrumentation* **22** (12) (1990), 27. Pressure gauging.
17. HIGHAM, E. H., in *Instrumentation Reference Book* (ed. NOLTINGK, B. E.), Chapter 8, Measurement of density (Butterworths, London, 1988).
18. CONSIDINE, D. M.: *Process Instruments and Controls Handbook* (McGraw-Hill, New York, 1957).
19. NEUBERT, H. K. P.: *Instrument Transducers: An Introduction to their Performance and Design*, 2nd edn (Oxford University Press, London, 1975).
20. MORLEY, A. and HUGHES, E.: *Principles of Electricity*, 4th edn (Longman, New York, 1986).
21. WINDOW, A. L. and HOLISTER, G. S.: *Strain Gauge Technology* (Applied Science Publ., New Jersey, 1982).
22. SMITH, C. J.: *Intermediate Physics*, 4th edn (Edward Arnold, London, 1965).
23. Fischer and Porter Ltd.: *Instruction Bulletin* 10B1465, *Revision 3* (1986). Differential pressure transmitters.
24. STOKES, D.: *Processing* (May, 1993), 22. Putting the pressure on ceramics.
25. NORTON, H. N.: *Sensor and Analyser Handbook* (Prentice-Hall, Englewood Cliffs, New Jersey, 1982).
26. BILLING, B. F. and QUINN, T. J.: *Temperature Measurement* (Adam Hilger, New York, 1975).
27. BLEANEY, B. I. and BLEANEY, B.: *Electricity and Magnetism*, 3rd edn (Oxford University Press, Oxford, 1976).
28. BS 4937: Part 4:1973. British Standard 4937 (British Standards Institution, London). Nickel-chromium/nickel-aluminium thermocouples. Type K.
29. BS 4937: Parts 1–4:1973. Parts 5–7:1974. Part 8:1986. British Standard 4937 (British Standards Institution, London). International thermocouple reference tables.
30. Thermocouple Instruments Ltd.: *Publication* TIL 714 MI (1987). Mineral insulated thermocouples.
31. BARNEY, G. C.: *Intelligent Instrumentation*, 2nd edn (Prentice-Hall, Englewood Cliffs, New Jersey, 1988).
32. BS 1041: Part 3:1989. British Standard 1041 (British Standards Institution, London). Temperature measurement—guide to selection and use of industrial resistance thermometers.
33. BS 1904:1984. British Standard 1904 (British Standards Institution, London). Industrial platinum resistance thermometers sensors.
34. HOROWITZ, P. and HILL, W.: *The Art of Electronics* (Cambridge University Press, New York, 1988).
35. Honeywell Control Systems Ltd.: *Technical Bulletin* 34–ST–03–28/E (1992), 2.

36. NOAKES, W.: *Control and Instrumentation* **23** (8) (1991), 27. Pyrometers play it cool.
37. HARRISON, T. R.: *Radiation Pyrometry and Its Underlying Principles of Radiant Heat Transfer* (Wiley, New York, 1960).
38. SOR Europe Ltd.: *Technical Bulletin* 345E (1986), 6. Industrial instrumentation.
39. JONES, E. B.: *Instrument Technology*, 2nd edn, **2**. On-line analysis instruments (Butterworth, London, 1976).
40. CHETTLE, T.: *Processing* (Jan/Feb 1993), 14. Choosing good solutions for liquid level control.
41. Vega Controls Ltd.: *Technical Bulletins* 2.14763 and 2.14725 (August 1992). Ultrasonic and radar based level transmitters.
42. GRASSBY, A.: *Control and Instrumentation* **24** (2) (1992), 23. Level instrumentation.
43. FARRANT, D.: *Control and Instrumentation* **24** (9) (1992), 55. Radar measurement is on the level.
44. Schlumberger Electronics (U.K.) Ltd.: *Technical Information on Density Transducers* (1986).
45. Rubotherm Präzisionsmeßtechnik GmbH: *Technical Information on Magnetic Suspension Balances* (1993).
46. American Society for Testing and Materials (ASTM): *Annual book of ASTM Standards*, **05.01**, *Petroleum Products and Lubricants* (I) (1988), 184. D446—Standard specifications and operating instructions for glass capillary kinematic viscometers.
47. FAROOQI, S. I. and RICHARDSON, J. F.: *Trans. I. Chem. E.* **58** (1980), 116. Rheological behaviour of kaolin suspensions in water and water–glycerol mixtures.
48. BS 188:1977. British Standard 188 (British Standards Institution, London). Methods for the determination of the viscosity of liquids.
49. WALTERS, K.: *Rheometry* (Chapman and Hall, London, 1975).
50. WHORLOW, R. W.: *Rheological Techniques* (Wiley, New York, 1980).
51. WALTERS, K. and JONES, W. M., in *Instrumentation Reference Book* (ed. NOLTINGK, B. E.), Chapter 2, Measurement of Viscosity (Butterworths, London, 1988).
52. Precision Scientific Inc.: *Technical Bulletins* 41070 and 44860 (1992). GCA/IEG continuous viscometer.
53. Contraves AG.: *Technical Bulletin* T224e–8801 CZ/IND (1992). Covimat 101/105 process viscometer.
54. FITZGERALD, J. V., MATUSIK, F. J. and WALSH, T. M.: *Measurements and Control* (December, 1987), 24. In-line viscometry.
55. Combustion Engineering Inc.: *Specification* SC4–7–190 (1989). The model 1800 viscometer.
56. DORZLAW, G. and WEISS, M. D.: *Chem. Eng. Prog.* **89** (9) (1993), 42. Improve on-line process control with new infra-red analysis.
57. BRUCE, S. H.: *Proceedings of the First International Near Infrared Spectroscopy Conference*, University of East Anglia, Norwich, U.K. (1987). Process NIR analysis.
58. CLEVETT, K. J.: *Handbook of Process Stream Analysis* (Wiley, New York, 1986).
59. Servomex (U.K.) Ltd.: *Technical Bulletin* 7981–6014 (1993). Servomex 2500 infra-red process analyser.
60. SMITH, A. C., in *Instrumentation Reference Book* (ed. NOLTINGK, B. E.), Chapter 3, Chemical analysis —Spectroscopy (Butterworths, London, 1988).
61. Servomex (U.K.) Ltd.: *Technical Bulletin* 7986–0679 (1991). Process NIR analysis.
62. LIDE, D. R. (ed.): *CRC Handbook of Chemistry and Physics*, 73rd edn (CRC Press, Boca Raton, 1992).
63. CUMMINGS, W. G. and TORRANCE, K. in *Instrumentation Reference Book* (ed. NOLTINGK, B. E.), Chapter 4, Chemical analysis—electrochemical techniques (Butterworths, London, 1988).
64. Foxboro Great Britain Ltd.: *Technical Bulletin* K44B 15M (1989). Electrochemical measurements.
65. BS 1647: Parts 1 and 2:1984. British Standard 1647 (British Standards Institution, London). pH measurement.
66. LEWIS, M. and WALTER, G.: *Advancing Chemistry* (Oxford University Press, Oxford, 1982).
67. BAILEY, P. L.: *Analysis with Ion-Selective Electrodes* (Heyden, New York, 1976).
68. Servomex (U.K.) Ltd.: *Technical Bulletin* 7981–5167 (1990). 700 series Zirconia cell—how it works.
69. Servomex (U.K.) Ltd.: *Technical Bulletin* 7981–3565 (1992). 700 B Zirconia oxygen analysers.
70. Foxboro Great Britain Ltd., *Technical Bulletin* L58X 11P (1993). On-line chromatography.
71. COOPER, C. J. and DE ROSE, A. J.: *Pergamon Series in Analytical Chemistry*, **7**. The analysis of gases by chromatography (Pergamon, Oxford, 1983).
72. WILLARD, H. H., MERRITT, L. L., DEAN, J. A. and SETTLE, F. A.: *Instrumental Methods of Analysis*, 6th edn (Wadsworth, Belmont, 1981).
73. Asea Brown Boveri: *Technical Bulletin* SC7–12–1092 (1993). Fundamentals of mass spectroscopy.
74. TAIRD, C. K., in *Instrumentation Reference Book* (ed. NOLTINGK, B. E.), Chapter 5, Chemical analysis —gas analysis (Butterworths, London, 1988).
75. LILEY, P. E., REID, R. C. and BUCK, E., in *Perry's Chemical Engineers' Handbook*, 6th edn (eds. GREEN, D. W. and MALONEY, J. O.), Section 3, Physical and chemical data (McGraw-Hill, New York, 1984).
76. MEADOWCROFT, D. B., in *Instrumentation Reference Book* (ed. NOLTINGK, B. E.), Chapter 6, Chemical analysis—moisture measurement.
77. Michele Instruments Ltd.: *Technical Bulletin* (1988). Advanced industrial hygrometry.

78. Servomex (U.K.) Ltd.: *Technical Bulletin* 7986–0763 (1992). The measurement of water by infra-red.
79. STOCKWELL, P.: *Process Industry Journal* (February, 1993), 29. Dewpoint measurement—and its pitfalls.
80. SLIGHT, H.: *Control and Instrumentation* **23** (2) (1991), 53. Measurement of moisture content.
81. ARCHER, A.: *Control and Instrumentation* **24** (3) (1992), 45. On-line fluid sampling systems.
82. American Society for Testing and Materials (ASTM): *Annual Book of ASTM Standards*, **05.03**, *Petroleum Products and Lubricants* (III) (1988), 304. D4177—Standard method for automatic sampling of petroleum and petroleum products.
83. Servomex (U.K.) Ltd.: *Technical Bulletin* 7981–3130 (1992). 1100 A/H paramagnetic oxygen analysers.
84. PARADINE, C. G. and RIVETT, B. M. P.: *Statistical Methods for Technologists* (English Universities Press, London, 1966).
85. SKOOG, D. A. and WEST, D. M.: *Fundamentals of Analytical Chemistry* (Holt, Rinehart and Winston, New York, 1966).
86. JEFFREY, A.: *Mathematics for Engineers and Scientists* (Van Nostrand Reinhold, Wokingham, U.K., 1982).
87. MAISEL, L.: *Probability, Statistics, and Random Processes* (Simon and Schuster, New York, 1971).
88. SMITH, J. I.: *Modern Operational Amplifier Design* (Wiley, New York, 1971).
89. LAM, H. Y. F.: *Analogue and Digital Filters: Design and Realisation* (Prentice-Hall, New Jersey, 1979).
90. HOESCHELE, D. F.: *Analog-to-Digital/Digital-to-Analog Conversion Techniques* (Wiley, New York, 1968).
91. MONGKHOUSI, T., LOPEZ-ISUNZA, H. F. and KERSHENBAUM, L. S.: *Chem. Eng. Res. Des.* **70** (1992), 255. The distortion of measured temperature profiles in fixed bed reactors.
92. O'REILLY, J.J.: *Telecommunications Principles* (van Nostrand Reinhold, Wokingham, U.K., 1984).
93. BANNISTER, B. R. and WHITEHEAD, D. G.: *Transducers and Interfacing* (van Nostrand Reinhold, Wokingham, U.K., 1986).
94. JONES, J.: *Control and Intrumentation* **24** (4) (1992), 57. How do you get cheap distributed control?
95. HALL, D. V.: Microprocessors and Interfacing (McGraw-Hill, New York, 1986).
96. WILSON, J. and HAWKES, J. F. B.: *Optoelectronics: An Introduction* (Prentice-Hall, London, 1983).
97. SENIOR, J.: *Optical Fibre Communications, Principles and Practice* (Prentice-Hall, London, 1985).
98. BRAMLEY, C.: *Process Industry Journal* (April, 1993), 33. Optical control of pneumatic actuators.
99. WRIGHT, C.: *Control and Instrumentation* **23** (2) (1991), 55. How to select a temperature transmitter.
100. SHARROCK, P.: *Control and Instrumentation* **23** (2) (1991), 59. Smart transmitters versus not so smart.
101. BOWDEN, R.: *Process Industry Journal* (January, 1993), 34. HART enhancement takes the brakes off 'smart' innovation.

6.15. NOMENCLATURE

		Units in SI System	Dimension in **M,N,L,T,θ,A**
A	Cross-sectional area of flux path	m^2	$\mathbf{L^2}$
A_e	Area of electrodes	m^2	$\mathbf{L^2}$
A_p	Cross-sectional area of pipe	m^2	$\mathbf{L^2}$
A_R	Surface area of resistor	m^2	$\mathbf{L^2}$
$\mathscr{A}$	Absorbance	—	—
a	Activity coefficient	—	—
a_θ	Constants in equations 6.32 and 6.33	—	—
B	Width of notch, channel or venturi throat	m	**L**
B_f	Width of bluff body	m	**L**
$\mathscr{B}$	Magnetic flux density	T (tesla)	$\mathbf{MT^{-2}A^{-1}}$
b_θ	Constants in equations 6.32 and 6.33	—	—
b_0, b_1 etc.	Bits (b_0—least significant bit)	—	—
C	Molar concentration	$kmol/m^3$	$\mathbf{NL^{-3}}$
C_D	Coefficient of discharge	—	—
$\mathscr{C}$	Capacitance	F (farad)	$\mathbf{M^{-1}L^{-2}T^4A^2}$
D	Depth of liquid above bottom of notch or in flume	m	**L**
d	Pipe internal diameter	m	**L**
d'	Particle diameter, ball diameter—falling sphere viscometer	m	**L**

		Units in SI System	Dimension in **M,N,L,T,θ,A**
d_0	Pipe external diameter	m	**L**
d_x	Internal dimension of flow conditioner in Fig. 6.9	m	**L**
E	Electromotive force	V	$\mathbf{ML^2T^{-3}A^{-1}}$
E_{orp}	Oxidation–reduction potential	V	$\mathbf{ML^2T^{-3}A^{-1}}$
E_{Th}	Thévenin emf	V	$\mathbf{ML^2T^{-3}A^{-1}}$
E_{ZA}	Potential difference across Zirconia cell	V	$\mathbf{ML^2T^{-3}A^{-1}}$
$\mathbf{E}$	Total energy emitted by black body per unit area per unit time	W/m^2	$\mathbf{MT^{-3}}$
$\mathbf{E}_{Abs}$	Energy absorbed by gas	J	$\mathbf{ML^2T^{-2}}$
$\mathbf{E}_g$	Total energy emitted by grey body per unit area per unit time	W/m^2	$\mathbf{MT^{-3}}$
$\mathbf{E}_{inc}$	Incident energy	J	$\mathbf{ML^{-2}T^{-2}}$
$\mathscr{E}$	Young's modulus (modulus of elasticity)	N/m^2	$\mathbf{ML^{-1}T^{-2}}$
e	Emissivity	—	—
e_T	Transverse strain	—	—
e_x	Longitudinal strain	—	—
e_λ	Emissivity at wavelength λ	—	—
$\mathscr{e}$	Charge on gaseous ion	C (coulomb)	**TA**
F	Tensile force	N	$\mathbf{MLT^{-2}}$
F_i	Force exerted on ion	N	$\mathbf{MLT^{-2}}$
F_{mm}	Magnetomotive force	AT (ampere-turn)	**A**
$\mathbf{F}$	Faraday number	96493 C	**TA**
$\mathscr{F}$	Electric field strength	V/m	$\mathbf{MLT^{-3}A^{-1}}$
f	Sampling frequency	s^{-1}	$\mathbf{T^{-1}}$
f	Function	—	—
G	Mass rate of flow	kg/s	$\mathbf{MT^{-1}}$
$\mathscr{G}$	Gauge factor (equation 6.26)	—	—
g	Acceleration due to gravity	m/s^2	$\mathbf{LT^{-2}}$
h	Heat transfer coefficient	W/m^{2}K	$\mathbf{MT^{-3}\theta^{-1}}$
h_{mix}	Heat transfer coefficient in equation 6.93	W/m^{2}K	$\mathbf{MT^{-3}\theta^{-1}}$
h_{ref}	Heat transfer coefficient in equation 6.98	W/m^{2}K	$\mathbf{MT^{-3}\theta^{-1}}$
h_{RT}	Heat transfer coefficient in equation 6.48	W/m^{2}K	$\mathbf{MT^{-3}\theta^{-1}}$
I	Moment of inertia	kg m^2	$\mathbf{ML^2}$
$\mathscr{I}$	Intensity of nuclear radiation	W/m^2	$\mathbf{MT^{-3}}$
i	Current	A	**A**
i_e	Electron current	A	**A**
i_i	Ion current	A	**A**
i_L	Current through load	A	**A**
i_{Th}	Thévenin current	A	**A**
J_1	Constant in equations 6.44, 6.47 and 6.49	Wm^2	$\mathbf{ML^4T^{-3}}$
J_2	Constant in equations 6.44, 6.47 and 6.49	mK	$\mathbf{L\theta}$
J_3	Constant in equation 6.45	Wm^2	$\mathbf{L\theta}$
J_{RT}	Sensitivity factor for radiation thermometer	—	—
j	Number of samples or data points	—	—
K_1	Constant in equation 6.5	m^3/s^2	$\mathbf{L^3T^{-2}}$
K_2	Constant in equation 6.6	kg/m^{3}s^2	$\mathbf{ML^{-3}T^{-2}}$
K_3	Constant in equation 6.6	kg/m^{3}s	$\mathbf{ML^{-3}T^{-1}}$
K_{amp}	Amplification factor for instrumentation amplifier	—	—
K_{CM}	Coriolis meter constant	Nm	$\mathbf{ML^2T^{-2}}$
K_D	Constant for Doppler meter	m^3/s^2	$\mathbf{L^3T^{-2}}$
K_{GAM}	Calibration constant for acoustic gas density meter	kg/m^{3}s^2	$\mathbf{ML^{-3}T^{-1}}$
K'_{GAM}	Calibration constant for acoustic gas density meter	s^{-2}	$\mathbf{T^{-2}}$
K_{LAM}	Transducer constant for acoustic liquid density meter	kg/m^3	$\mathbf{ML^{-3}}$
K_{TF}	Constant for time of flight meter	m^3	$\mathbf{L^3}$

		Units in SI System	Dimension in **M,N,L,T,θ,A**
K_v	Capillary and falling ball viscometer constant	m^2/s^2	L^2T^{-2}
K'_v	Capillary viscometer correction factor for fast flow	m^2	L^2
K''_v	Pulsed current viscometer constant	kg^2/m^4s	$M^2L^{-4}T^{-1}$
K_{vp}	Correction for velocity profile	—	—
K_x	Constant in equation 6.97	—	—
k	Thermal conductivity	W/mK	$MLT^{-3}\theta^{-1}$
	or intercept of straight line	—	—
k_{Abs}	Absorption constant	m^2/kmol	$N^{-1}L^2$
k_c	Conductivity cell constant	m^{-1}	L^{-1}
k_I	Constant for inductive pressure transducer	m^{-1}	L^{-1}
k_m	Number of multiplexed signals in equation 6.115	—	—
k_{Zr}	Zirconia cell constant	V	$ML^2T^{-3}A^{-1}$
$\mathscr{k}'$	Number of information digits	—	—
L	Length of tube defined in Figs 6.6 and 6.9	m	L
L_p	Path length	m	L
L_{max}	Maximum permissible length of probe (equation 6.97)	m	L
$\mathscr{L}$	Inductance	H (henry)	$ML^2T^{-2}A^{-2}$
ℓ	Total length of flux path, length of liquid column	m	L
ℓ_c	Height of cylindrical capacitive sensing element	m	L
ℓ_p	Path length of column of gas	m	L
$\mathscr{M}$	Mutual inductance	H (henry)	$ML^2T^{-2}A^{-2}$
m	Mass (of fluid)	kg	M
	or gradient of straight line	—	—
m_i	Mass of ion	kg	M
N	Total number of data points, number of terms of series	—	—
N_p	Number of drive pulses per unit time	s^{-1}	T^{-1}
N_T	Number of turns	—	—
$\mathscr{N}$	Number of quantisation levels	—	—
n	Integer mode of frequency	—	—
n_a	Number of anions	—	—
n_c	Number of cations	—	—
$\mathscr{n}$	Number of binary digits (bits)	—	—
$\mathscr{n}_{bcd}$	Number of binary coded decimal bits	—	—
P	Pressure	N/m^2	$ML^{-1}T^{-2}$
P'	Partial pressure	N/m^2	$ML^{-1}T^{-2}$
Q	Average volumetric flowrate	m^3/s	L^3T^{-1}
	or electric charge	C (coulomb)	TA
Q_H	Quantity of heat	J	ML^2T^{-2}
R	Resistance	Ω	$ML^2T^{-3}A^{-2}$
R_c	Resistance of conductivity cell	Ω	$ML^2T^{-3}A^{-2}$
R_{CL}	Resistance of compensating leads	Ω	$ML^2T^{-3}A^{-2}$
R_F	Resistance Pirani gauge active filament	Ω	$ML^2T^{-3}A^{-2}$
R_f	Resistance of amplifier	Ω	$ML^2T^{-3}A^{-2}$
R_L	Resistance of load	Ω	$ML^2T^{-3}A^{-2}$
R_s	Resistance of Pirani gauge reference filament	Ω	$ML^2T^{-3}A^{-2}$
R_θ	Resistance at temperature θ	Ω	$ML^2T^{-3}A^{-2}$
$\mathbf{R}$	Universal gas constant	J/kmol K	$MN^{-1}L^2T^{-2}\theta^{-1}$
$\mathscr{R}$	Resolution of A/D converter	—	—
$\mathscr{R}_b$	Transmission bit rate	s^{-1}	T^{-1}
r	Radius	m	L
r_e	Radius of path of ion in electric field	m	L
r_g	Radius of gyration	m	L
r_m	Radius of path of ion in magnetic field	m	L
$\mathscr{S}$	Magnetic reluctance	A/Wb (ampere/weber)	$M^{-1}L^{-2}T^2A^2$

		Units in SI System	Dimension in M,N,L,T,θ,A
s	Magnitude of position vector	—	—
T	Absolute temperature	K	$\boldsymbol{\theta}$
$\mathbf{T}$	Torque	Nm	$\mathbf{ML^2T^{-2}}$
$\mathbf{T}'$	Torque per unit length of cylinder	N	$\mathbf{MLT^{-2}}$
$\mathcal{T}$	Transmittance	—	—
t	Time	s	$\mathbf{T}$
t_{down}	Downstream elapsed time	s	$\mathbf{T}$
t_e	Elapsed time (Fig. 6.75)	s	$\mathbf{T}$
t_{up}	Upstream elapsed time	s	$\mathbf{T}$
u	Velocity	m/s	$\mathbf{LT^{-1}}$
u_0	Terminal falling velocity	m/s	$\mathbf{LT^{-1}}$
u_ℓ	Velocity of liquid	m/s	$\mathbf{LT^{-1}}$
u_{free}	Free stream velocity	m/s	$\mathbf{LT^{-1}}$
u_i	Velocity of ion	m/s	$\mathbf{LT^{-1}}$
$u_{\max}$	Maximum velocity of process fluid	m/s	$\mathbf{LT^{-1}}$
u_s	Speed of sound in liquid	m/s	$\mathbf{LT^{-1}}$
$\mathbf{u}$	Fluid velocity vector	m/s	$\mathbf{LT^{-1}}$
V	Voltage	V	$\mathbf{ML^2T^{-3}A^{-1}}$
V_o	Output voltage	V	$\mathbf{ML^2T^{-3}A^{-1}}$
V_i	Input voltage	V	$\mathbf{ML^2T^{-3}A^{-1}}$
V_{int}	Interference voltage due to inductive coupling	V	$\mathbf{ML^2T^{-3}A^{-1}}$
V_L	Potential difference across output load *or* width of air gap in equation 6.19	V	$\mathbf{ML^2T^{-3}A^{-1}}$
V_s	Voltage source	V	$\mathbf{ML^2T^{-3}A^{-1}}$
W_λ	Power spectral density for black body radiator	W/m^3	$\mathbf{ML^{-3}}$
w	Deflection of diaphragm in Fig. 6.11	m	$\mathbf{L}$
X	Thermometric property	—	—
x	Length or distance in x direction *or* width of air gap in equation 6.19	m	$\mathbf{L}$
x_d	Thickness of diaphragm in Fig. 6.11	m	$\mathbf{L}$
x_e	Distance between conductivity cell electrodes	m	$\mathbf{L}$
x_s	Distance travelled by inductor core (stroke)	m	$\mathbf{L}$
x_T	Tip travel of Bourdon tube	m	$\mathbf{L}$
y	Length in y direction	m	$\mathbf{L}$
Z_a	Number of charges carried by anion	—	—
Z_c	Number of charges carried by cation	—	—
$\mathfrak{Z}_L$	Input impedance of load	Ω	$\mathbf{ML^2T^{-3}A^{-2}}$
$\mathfrak{Z}_{Th}$	Thévenin impedance	Ω	$\mathbf{ML^2T^{-3}A^{-2}}$
$\mathfrak{Z}_N$	Norton impedance	Ω	$\mathbf{ML^2T^{-3}A^{-2}}$
$\mathfrak{Z}_T$	Transmitter output impedance	Ω	$\mathbf{ML^2T^{-3}A^{-2}}$
z	Length in z direction *or* liquid level in tank	m	$\mathbf{L}$
α	Half angle of triangular notch *or* angle between pipe wall and ultrasonic signal	—	—
α_c	Critical angle	—	—
α_{CM}	Coriolis angle of twist	—	—
α_G	Gap angle—cone and plate viscometer	—	—
β_1	Thermoelectric coefficient in equation 6.38	V/K	$\mathbf{ML^2T^{-3}\theta^{-1}A^{-1}}$
β_2	Thermoelectric coefficient in equation 6.38	V/K^2	$\mathbf{ML^2T^{-3}\theta^{-2}A^{-1}}$
β_c	Temperature coefficient for ionic conductivity	1/K	$\boldsymbol{\theta}^{-1}$
$\beta_{\theta,1}$	Temperature coefficient of resistance	Ω/K	$\mathbf{ML^2T^{-3}\theta^{-1}A^2}$
$\beta_{\theta,2}$	Temperature coefficient of resistance	Ω/K^2	$\mathbf{ML^2T^{-3}\theta^{-2}A^{-1}}$
$\dot{\gamma}$	Rate of shear	s^{-1}	$\mathbf{T^{-1}}$
δ_c	Vertical separation of capacitor plates (equation 6.14)	m	$\mathbf{L}$
ε	Permittivity	(coulomb)^2/Nm^2	$\mathbf{M^{-2}L^{-5}T^6A^2}$

		Units in SI System	Dimension in M,N,L,T,θ,A
ε_0	Permittivity of free space	$(\text{coulomb})^2/\text{Nm}^2$	$\mathbf{M^{-2}L^{-5}T^6A^2}$
ε_r	Relative permittivity	—	—
η	Refractive index	—	—
θ	Temperature	K	$\boldsymbol{\theta}$
κ	Conductivity (conductance per metre)	S/m	$\mathbf{M^{-1}L^{-3}T^3A^2}$
κ'	Conductance	S (Siemen)	$\mathbf{M^{-1}L^{-2}T^3A^2}$
Λ	Molar conductivity defined by equation 6.72	Sm^2/kmol	$\mathbf{M^{-1}N^{-1}T^3A^2}$
Λ^0	Molar conductivity at infinite dilution	Sm^2/kmol	$\mathbf{M^{-1}N^{-1}T^3A^2}$
λ_a^0	Ionic conductivity per unit charge of anion at infinite dilution	Sm^2/kmol (anions)	$\mathbf{M^{-1}N^{-1}T^3A^2}$
λ_c^0	Ionic conductivity per unit charge of cation at infinite dilution	Sm^2/kmol (cations)	$\mathbf{M^{-1}N^{-1}T^3A^2}$
λ_T^0	Ionic conductivity per unit charge of ion at infinite dilution	Sm^2/kmol	$\mathbf{M^{-1}N^{-1}T^3A^2}$
μ	Dynamic viscosity	Ns/m^2	$\mathbf{ML^{-1}T^{-1}}$
μ_0	Permeability of free space	N/A^2	$\mathbf{MLT^{-2}A^{-2}}$
μ_r	Relative permeability	—	—
ν	Poisson's ratio	—	—
	or kinematic viscosity	m^2/s	$\mathbf{L^2T^{-1}}$
ξ	Mass absorption coefficient	m^2/kg	$\mathbf{M^{-1}L^2}$
ρ	Density	kg/m^3	$\mathbf{ML^{-3}}$
ρ'	Resistivity	Ωm	$\mathbf{M^2L^2T^{-3}A^{-2}}$
σ	Stephan—Boltzmann constant	$\text{W/m}^2\text{K}^4$	$\mathbf{MT^{-3}\boldsymbol{\theta}^{-4}}$
	or standard deviation	—	—
ϕ	Magnetic flux	Wb (Weber)	$\mathbf{ML^2T^{-2}A^{-1}}$
Ξ_{max}	Maximum quantification error	—	—
χ	$i^2\beta_\theta R_\theta/A_R$ (equation 6.97)	$\text{W/m}^2\text{K}$	$\mathbf{MT^{-3}\boldsymbol{\theta}^{-1}}$
ψ	Bandwidth	s^{-1}	$\mathbf{T^{-1}}$
ω	Angular frequency	radians/s	$\mathbf{T^{-1}}$
ω_v	Angular velocity	radians/s	$\mathbf{T^{-1}}$
$\Delta\omega_D$	Doppler frequency shift	s^{-1}	$\mathbf{T^{-1}}$
Re'	Particle Reynolds number at terminal falling velocity $u_0 d'\rho/\mu$	—	—
S	Strouhal number $\omega B_f/u_{free}$	—	—

Prefix

δ represents small finite changes in relevant variable.

Suffix

j refers to jth sampling instant, or to jth data point.

CHAPTER 7

Process Control

7.1. INTRODUCTION

Control in one form or another is an essential part of any chemical engineering operation. In all processes, there arises the necessity of keeping flows, pressures, temperatures, compositions, etc. within certain limits for reasons of safety or specification. It is self-evident that *automatic* control is highly desirable, as manual operation would necessitate continuous monitoring of the controlled variable by a human operator and the efficiency of observation of the operator would inevitably fall off with time. Furthermore, fluctuations in the controlled variable may be too rapid and frequent for manual adjustment to suffice.

In its simplest form, the control of a process is most often accomplished by measuring the variable it is required to control (the *controlled variable*), comparing this measurement with the value at which it is desired to maintain the controlled variable (the *desired value* or *set point*), and adjusting (in a prescribed way until the desired value is attained) some further variable (the *manipulated variable*) which has a direct effect on the controlled variable.

In order to design a control system to operate not only automatically but efficiently, it is frequently necessary to obtain both the steady-state and dynamic (unsteady-state) relationships between the particular variables involved. The manner in which this information is obtained is dependent largely upon the process being controlled and the control strategy to be employed. The latter may consist of one or more types of control, e.g. single or multiple feedback loops, adaptive control, combined predictive and feedback arrangements, distributed architectures, optimal control, etc. The use of the more modern and complex control systems is feasible only because of the development of the microprocessor and, as a result, control has been one of the areas of major growth in the process industries over the last twenty years. This growth is reflected equally in the very substantial changes that have occurred in the plant control room and in the methods of presentation of information to the process operator.

7.2. FEEDBACK CONTROL

A simple feedback control system is illustrated in Fig. 7.1, the function of which is to control at Y the temperature of the stream leaving the heat exchanger. The temperature θ of the stream at Y (the controlled variable—Fig. 7.1*a*—usually denoted by the letter C as in Fig. 7.1*b*) is measured by means of a thermocouple,

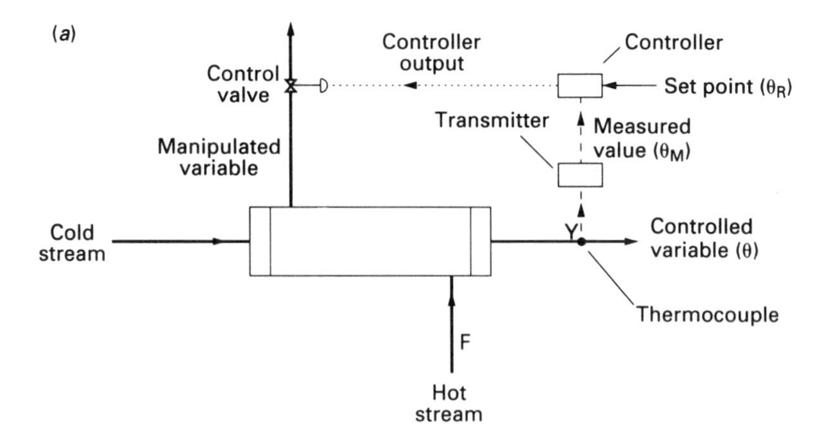

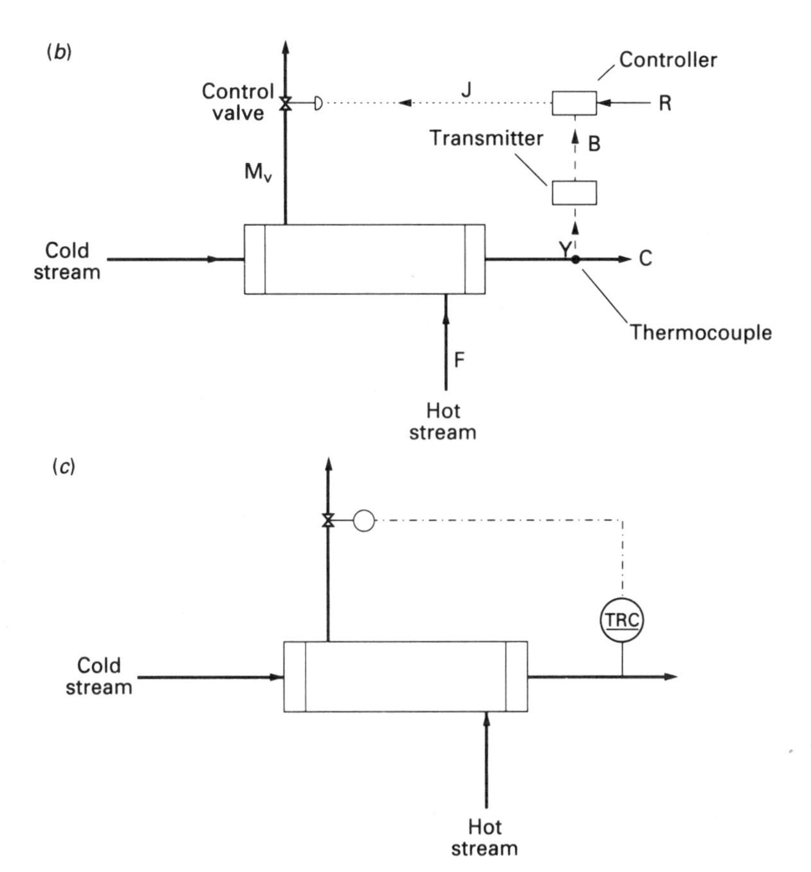

TRC: temperature recorder controller (see Volume 6, Section 5.2.2)

FIG. 7.1. Simple feedback control system: (*a*) illustrating components; (*b*) standard nomenclature; (*c*) representation according to British Standard BS 1646[1]

the output of which is fed via a suitable signal transmission system (see Section 6.12) to a controller. The controller consists of two parts. The first compares the measured temperature θ_M (i.e. the measured value—indicated usually by the letter B—Fig. 7.1*b*) with the desired value or set point θ_R (generally denoted by R). (Note that θ and θ_M will not necessarily be identical at the same instant of time—particularly when θ is varying.) This part of the controller is termed the *comparator* and produces an *error* (ε) such that:

$$\varepsilon = R - B = \theta_R - \theta_M \tag{7.1}$$

The second part of the controller provides the required control action. The most common types of control action (in terms of fixed parameter control) and their effects upon the controlled variable are described in Section 7.2.2—other control strategies are discussed later in this chapter. It will suffice at present to observe that the controller produces an output J which is a function of ε (Figs 7.1*a* and 7.1*b*). The controller output signal may be in the form of an air pressure from a pneumatically operated controller or a current or voltage supplied by an electronic controller or by a microprocessor simulating the appropriate control action. This signal is transmitted to the control valve (which is called the *final control element*) and is connected in such a way that the valve starts to open further when ε becomes positive, i.e. when $\theta_M < \theta_R$ the control system calls for more heat to be supplied by increasing the flowrate F of the hot stream. When ε is negative (i.e. $\theta_M > \theta_R$) then the valve starts to shut. When θ_M is at the set point (i.e. $\theta_M = \theta_R$) then $R = B$ and $\varepsilon = 0$. In this instance there is no control action and the position of the valve stem does not change.

There are two principal functions of a control loop of this kind, i.e. two reasons why a difference might occur between θ_M and θ_R thus producing an error. The first is that changes may occur in such variables as the cold or hot stream inlet temperature and cold stream inlet flow. Even F may vary due to reasons other than the setting of the control valve. All these are termed *load changes*, or are collectively described as the *load*. Control of the controlled variable in the face of variations in load is often termed the *regulator* problem or the *load rejection* case. The second reason is that we may wish to raise or lower θ for various production or operational reasons. This can be achieved by raising or lowering θ_R as desired—so creating a positive or negative error respectively. The control system will seek to minimise ε, i.e. to bring θ_M to the new value of θ_R. This is called the *servo* problem or the *set point following* case. It is not possible (or necessary) for θ_M or, indeed, θ to adjust to θ_R precisely under all forms of control action (see Section 7.2.2).

Figure 7.1*c* shows the same control loop using symbols given in British Standard BS 1646[1].

7.2.1. The Block Diagram

Any control system may be more simply represented in the form of a *block diagram*. This shows how information flows round the *control loop*, the function of each constituent section or *block*, and its relation to adjacent sections or blocks. Each block represents the relation between the signal entering the block and the signal leaving it. Figure 7.2 is the block diagram of the process illustrated in Fig. 7.1.

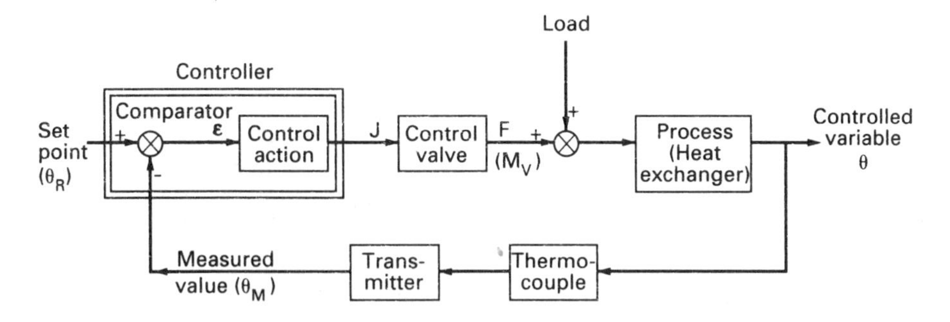

FIG. 7.2. Block diagram of the control system shown in Fig. 7.1

It can be seen that the term *control loop* is appropriate as information passes around a closed loop of components. This form of control is called *closed-loop* or *feedback* (referring to the feedback of information from the controlled variable to the comparator). A simple loop of this kind can be represented in general terms as in Fig. 7.3.

By comparing Figs 7.1, 7.2 and 7.3 it is obvious that M_V (the manipulated variable) represents the flowrate of the hot stream. The load U enters the loop at this point as changes in any of the load variables will affect the heat entering the system. Thus the total heat input to the process will be due to U and to M_V. The reasons when and why the net effect of U and M_V may be represented by a simple

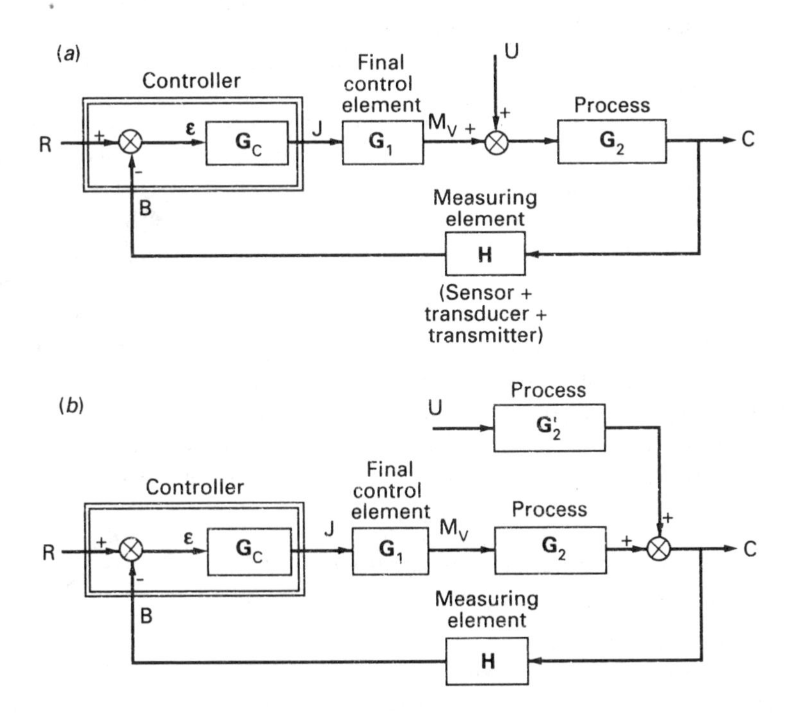

FIG. 7.3. Standard block diagrams

summation are discussed in Section 7.4.1. The process is the heat transfer mechanism which relates changes in U and/or M_V to variations in θ. In general terms, the blocks are lettered by convention as in Fig. 7.3, $\mathbf{G}_C$ representing the control action of the controller, $\mathbf{G}_1$ the final control element, $\mathbf{G}_2$ the process and $\mathbf{H}$ the measuring element. Departures from this lettering do occur, but the convention employed in Fig. 7.3 will be retained throughout this chapter. The number of blocks and their configuration can also vary substantially, depending upon the complexity of the control strategy and the number of control system components involved. Numerous examples of this are presented later in the chapter. An often-preferred alternative to the standard block diagram of Fig. 7.3*a* is Fig. 7.3*b*. This treats the effects of M_V and U on the process separately and facilitates the analysis of the control system in a number of cases.

Apart from the controller itself, the major components of the control system are considered elsewhere. The reader is referred to Chapter 6 for the description of various measuring elements and their associated transmission systems. Final control elements in the form of control valves are discussed in Section 7.22.3, and a generalised approach to the representation of the processes themselves is described in Sections 7.5 and 7.6.

It is useful now to examine some basic control mechanisms.

7.2.2. Fixed Parameter Feedback Control Action

As indicated previously, the control action of the controller seeks to alter the position of the final control element in such a way as to minimise the error in the least possible time with the minimum disturbance to the system. The control action selected depends largely upon the dynamic behaviour of the other components in the control system.

The simplest type of control which is commonly experienced is that of having an *on–off* or *two-position* action (often called *bang-bang* control). A typical example is the thermostatically controlled domestic immersion heater. Depending on the temperature of the water in the tank, the power supply is either connected to, or disconnected from, the heater. The relationship between controller input and output might appear as in Fig. 7.4. Such a system is simple and inexpensive. However, the oscillatory nature of the control makes it suitable only for those purposes where close control is not essential and/or where its non-linear action can be taken into account. This is considered in more detail in Section 7.16.2.

There are three principal types or *modes* of control action which are more generally employed, viz. *proportional* (**P**), *integral* (**I**), and *derivative* (**D**). In the first the controller produces an output signal J which is proportional to the error, i.e.:

$$J = J_0 + K_C\varepsilon \tag{7.2}$$

where K_C is the proportional gain or sensitivity, and
J_0 is the controller output when $\varepsilon = 0$.

Hence, with proportional control, the greater the magnitude of the error the larger is the corrective action applied.

It is generally assumed when considering control system dynamics that at $t < 0$ the control system is at a steady state and that ε is zero. Hence, it is necessary to

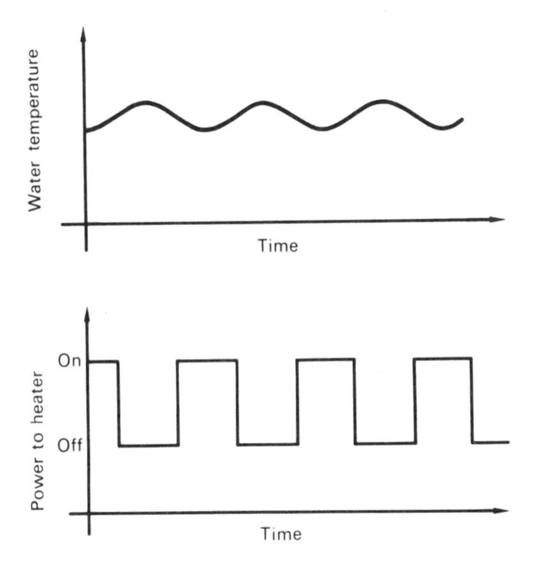

FIG. 7.4. On–off temperature control of a domestic hot water tank

include the term J_0 in the controller output in order to maintain the final control element (almost invariably a control valve) at its steady-state setting when $\varepsilon = 0$. The insertion of J_0 in the control algorithm can be considered as setting the *operating point* for the controller and thus be a possible means of providing so-called *bumpless transfer* from *manual* to *automatic* control. This is discussed further in Section 7.17.6.

The integral and derivative modes are normally used in conjunction with the proportional mode. Integral action (or *automatic reset*) gives an output which is proportional to the time integral of the error. Proportional plus integral (**PI**) action may be represented thus:

$$J = J_0 + K_C\varepsilon + K_I \int_0^t \varepsilon \, dt \tag{7.3}$$

where K_I is a constant.

Derivative action (often termed *rate* control) gives an output which is proportional to the derivative of the error. Hence, for **PD** control:

$$J = J_0 + K_C\varepsilon + K_D \frac{d\varepsilon}{dt} \tag{7.4}$$

where K_D is a constant.

Frequently all three modes are used together as **PID** control, i.e.:

$$J = J_0 + K_C\varepsilon + K_I \int_0^t \varepsilon \, dt + K_D \frac{d\varepsilon}{dt} \tag{7.5}$$

Sampled data, or *discrete* forms of equations 7.2, 7.3, 7.4 and 7.5 are employed to simulate these various control actions in the form of software within a *microprocessor based control* (*MBC*) system and are described in Section 7.17.6. The same relationships (whether in continuous or discrete form) are said to describe *fixed parameter* controllers when the parameters (viz. K_C, K_I and K_D) are left unaltered throughout the entire period of control action. Controllers in which the parameters are continually and automatically adjusted to take account of changing process conditions and dynamics are termed *adaptive* and these are discussed in Section 7.18.

7.2.3. Characteristics of Different Control Modes—Offset

Figure 7.5 shows typical responses of a controlled variable to a step disturbance in load for a simple control loop of the type illustrated in Figs. 7.1 and 7.3. The effects of different control actions are summarised in the following sections.

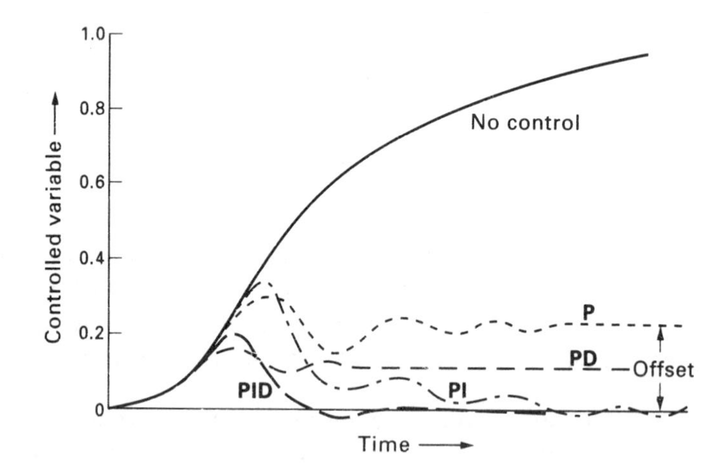

FIG. 7.5. Response of controlled variable to step disturbance in load using different control modes

Proportional Control

The response has a high maximum deviation and there is a significant time of oscillation (*response time*). The period of this oscillation is moderate. For a sustained change in load, the controlled variable is not returned to its original value (the desired value) but attains a new equilibrium value termed the *control point*. This difference between the desired value and the control point is called the *offset* or *droop*. The reason for offset with proportional action can be seen if it is remembered that the control action is proportional to the error.

Consider again the temperature control system fitted to the heat exchanger in Fig. 7.1. Suppose that the temperature of the cold stream decreases. Then, clearly, the temperature at Y, i.e. θ, will also begin to fall. In response to this the controller will open the control valve further in proportion to the error—where the error is given by equation 7.1. In order to maintain this new steady state, i.e. with the increased rate of flow of the hot stream, a constant additional output must be applied to the control valve by the controller. This additional output can exist only if there is an error signal applied to the controller. In order to maintain this error, θ (and θ_M) will rise above the desired value until the system comes to equilibrium with the valve open wide enough to keep θ at the new control point, hence creating an offset. It will be shown later (Section 7.9.3) that the offset is reduced as the gain K_C of the proportional controller is increased. However, K_C cannot be increased indefinitely as this leads to oscillatory behaviour. The usual setting for K_C is a compromise between offset and degree of oscillation. Being a simple form of control, proportional action is frequently employed on its own where offset is not an important consideration and where the system is sufficiently stable to enable a fairly high value of K_C to be tolerated (e.g. for the control of liquid level or for the control of the pressure of gas streams).

A setting knob is provided on the more traditional *stand-alone* controllers for the adjustment of K_C as well as a pointer to set the desired value. The setting knob for K_C is sometimes graduated in terms of *proportional band.* This quantity is defined as *the error required to move the final control element over its whole range* and is expressed as a percentage of the *total* range of the measured variable (see Example 7.1). In the newer MBC installations, the control action is simulated in the form of software and values of K_C and of the desired value are entered via an appropriate keyboard with the relevant values displayed on a *VDU* (*Visual Display Unit*)—see Section 7.19.

Example 7.1

The level of liquid in a tank is controlled using a pneumatic proportional controller as shown in Fig. 7.6. The level sensor is able to measure over the range 1.85 to 2.25 m. It is found that, after adjustment, the controller output pressure changes by 4 kN/m^2 for a 0.01 m variation in level with the desired value held constant. If a variation in output pressure of 80 kN/m^2 moves the control valve from fully open to fully closed, determine the gain and the proportional band.

Solution

From equation 7.2:

$$J_A = J_0 + K_C\varepsilon_A$$

$$J_B = J_0 + K_C\varepsilon_B$$

Subtraction gives:

$$J_A - J_B = K_C(\varepsilon_A - \varepsilon_B)$$

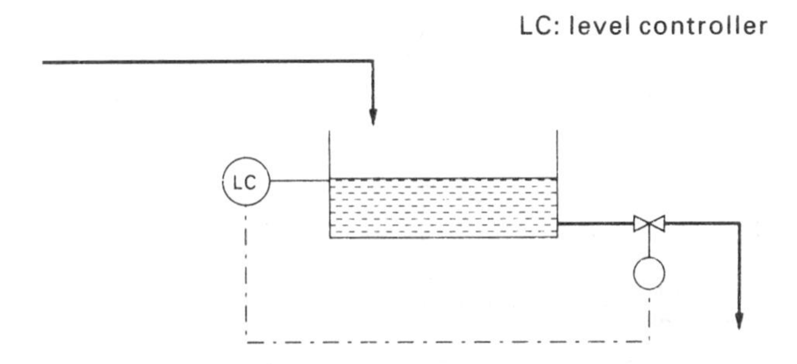

FIG. 7.6. Level control system

As the level changes by 0.01 m whilst the desired value is held constant, the error must change by 0.01 m with a change in output pressure of 4 kN/m^2.

$$\therefore \qquad K_C = \frac{4}{0.01} = 400\ kN/m^2 \text{ per metre change in level}$$

If an error change of 0.01 m causes a change in output pressure of 4 kN/m^2 then the error required to move the valve from fully open to fully shut, assuming valve movement, output pressure and error to be linearly related, will be:

$$\frac{80}{4} \times 0.01 = 0.2\ m$$

therefore, by definition:

$$\text{Proportional band} = \frac{0.2}{2.25 - 1.85} \times 100$$

$$= \underline{\underline{50 \text{ per cent}}}$$

PI Control

It can be seen from equation 7.3 and Fig. 7.7 that the controller output will continue to increase as long as $\varepsilon > 0$. With proportional control an error (offset) had to be maintained so that the controlled variable (i.e. the temperature at Y—Fig. 7.1) could be kept at a new control point after a step change in load, i.e. in the inlet temperature of the cold stream. This error was required in order to produce an additional output from the proportional controller to the control valve. However, with **PI** control, the contribution from the integral action does not return to zero with the error, but remains at the value it has reached at that time. This contribution provides the additional output necessary to open the valve wide enough to keep the level at the desired value. No continuous error (i.e. no offset) is now necessary to maintain the new steady state. A quantitative treatment of this is given later (Section 7.9.3).

The disadvantages of **PI** control are that it gives rise to a higher maximum deviation, a longer response time and a longer period of oscillation than with proportional control alone (see Fig. 7.5). Hence, this type of control action is used

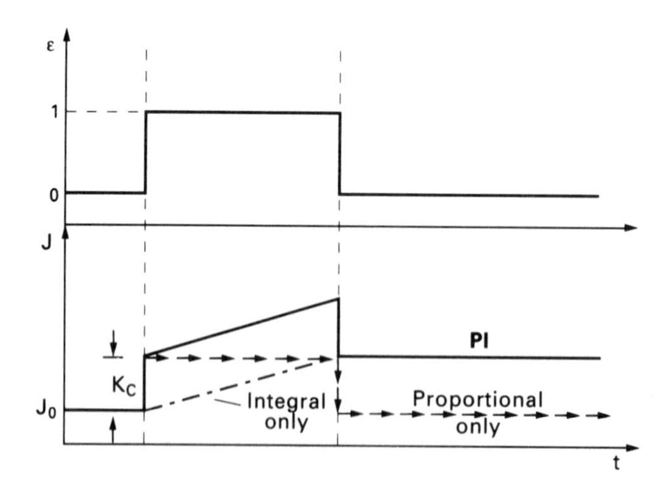

FIG. 7.7. Response of **PI** controller to unit step change in error

where these disadvantages can be tolerated and offset is undesirable. It is a frequently employed combination, especially when the responses of the other components in the control loop are rapid, e.g. the control of liquid (or gas) flowrate where the flow responds more or less instantaneously to a movement in the spindle of the control valve. On the other hand, if **PI** control is employed for a system in which the errors cannot be eliminated quickly, or indeed persist, it is possible for the integral term to produce increasingly larger signals until the system is saturated, i.e. the control valve will remain fully open or fully shut. This is termed *integral windup* and usually occurs when plants are shut down or started up. For systems that have a slow response to changes in load or set point, e.g. heat transfer processes, **PI** action can lead to an even slower response. PETERS[2] has illustrated how increasing the integral time increases the tendency for a control system to cycle.

Equation 7.3 may be rewritten thus:

$$J = J_0 + K_C\left(\varepsilon + \frac{1}{\tau_I}\int_0^t \varepsilon \, dt\right) \tag{7.6}$$

where τ_I is termed the *integral time*.

The maximum deviation etc. of the controlled variable is determined by the settings of both K_C and τ_I which can be keyed into an MBC system as required. A second knob is provided on a stand-alone controller and is calibrated either in terms of τ_I min or $1/\tau_I$ (min)$^{-1}$. The latter is called the *reset rate*.

PD Control

Control action due to the derivative mode occurs only when the error is changing (equation 7.4). The presence of the derivative mode contributes an additional output, $K_D(d\varepsilon/dt)$, to the final control element as soon as there is any change in error. When the error ceases to change, derivative action no longer occurs (Fig. 7.8). The effect of this is similar to having a proportional controller with a high gain

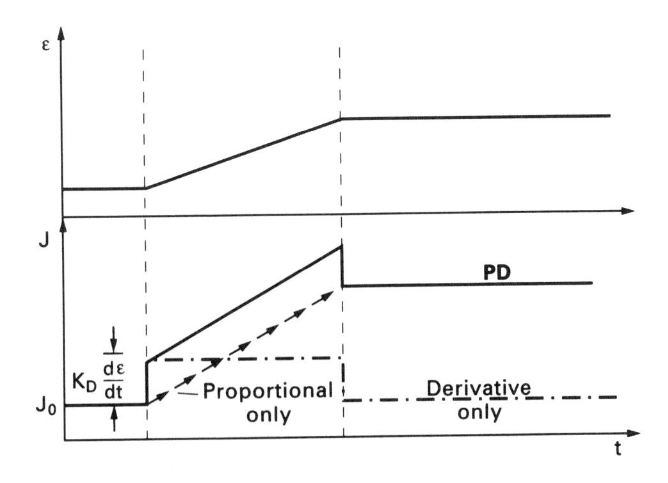

FIG. 7.8. Response of **PD** controller to ramp change in error

when the measured variable is changing most rapidly, and a low gain when the latter is varying slowly. Due to this action the controlled variable exhibits the least oscillation and the lowest maximum deviation (Fig. 7.5). The same amount of offset occurs as with proportional control alone with the same proportional gain. However, the addition of derivative action allows a higher gain to be used before the control system becomes unstable (see Example 7.7 at the end of Section 7.10.4). By this means a smaller offset can be obtained than with proportional action alone. Derivative action is sometimes termed *anticipatory* or *rate control* and is employed where substantial and quick variations in the controlled variable have to be suppressed. However, if the controlled variable has settled at the set point but is subject to low amplitude but rapidly changing noise, then derivative action will exacerbate the situation by magnifying these small rapid changes in noise into large and unnecessary controller output signals.

Equation 7.4 may be rewritten:

$$J = J_0 + K_C\left(\varepsilon + \tau_D \frac{\mathrm{d}\varepsilon}{\mathrm{d}t}\right) \tag{7.7}$$

where τ_D is the *derivative time*, usually measured in minutes. Means of setting both K_C and τ_D on the controller are provided.

PID Control

This is essentially a compromise between the advantages and disadvantages of **PI** and **PD** control. Offset is eliminated by the presence of integral action and the derivative mode reduces the maximum deviation and time of oscillation, although the latter are still greater than with **PD** control alone (Fig. 7.5).

Here we have, from equation 7.5:

$$J = J_0 + K_C\left(\varepsilon + \frac{1}{\tau_I}\int_0^t \varepsilon \mathrm{d}t + \tau_D \frac{\mathrm{d}\varepsilon}{\mathrm{d}t}\right) \tag{7.8}$$

PID action should be employed for systems which respond rather sluggishly. The presence of the derivative term allows a higher proportional gain to be used which will speed up the control action. Such controllers are frequently installed because of their versatility and not because analysis of the system has indicated the need for the presence of all three modes of control.

A description of the control action of a typical industrial three term controller is given in Section 7.7.2.

7.3. QUALITATIVE APPROACHES TO SIMPLE FEEDBACK CONTROL SYSTEM DESIGN

A distillation column provides a good example of *multiple-input/multiple-output* (*MIMO*) control and illustrates well the qualitative methodology involved in determining a suitable control strategy for a process. The first requirement is to decide the primary objective of the process, i.e. what is its principal purpose? Let us suppose that, for the column shown in Fig. 7.9, it is required to produce an overhead product D of a particular specification x_D without attempting to control

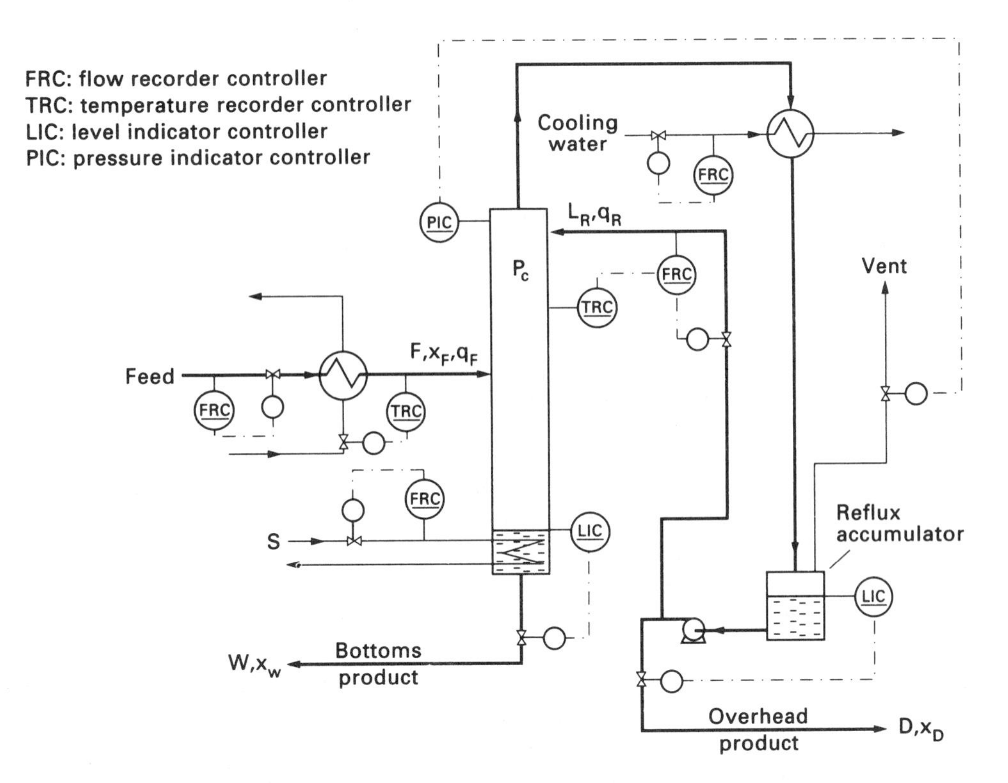

FIG. 7.9. Multiple input/multiple output (MIMO) control of a distillation column

the composition x_W of the bottoms product. The next step is to identify those variables which can affect x_D—albeit to different degrees. These are:

Feed composition	x_F
Feed flowrate	F
Feed quality	q_F
Heat per unit time to reboiler	Q_B
Reflux flowrate	L_R
Reflux quality	q_R
Column pressure	P_c

This is a substantial list and is by no means exhaustive. We can approach the selection of suitable measured and manipulated variables either by using *degrees of freedom* or in a more heuristic manner.

7.3.1. The Heuristic Approach

This, in many instances, is easier and more direct, but does not necessarily produce the best control configurations. There are well-established guidelines that can be followed.

(i) Endeavour to control any load variable prior to it affecting the process.

(ii) As far as is possible, measure the variable that you are controlling, e.g. when controlling the level in a reflux accumulator then use the level as the measured variable.
(iii) Each controlled variable will require one unique manipulated variable.
(iv) Make sure that you do not have more than one control valve manipulating the same stream (a surprisingly easy trap to fall into when designing control systems for a complex plant which requires a large number of control loops).
(v) Avoid, where possible, interaction between control loops (Section 7.15).

Thus, some load variables can be eliminated by the application of guideline (i) to Fig. 7.9, e.g. it is relatively easy to maintain the feed quality by installing a preheater with the outlet temperature controlled as shown, and the feed flow can be fixed by using a simple flow control loop (as long as intermediate storage is provided between the distillation plant and any plant upstream which supplies the feed to the column).

Certain variables are less suitable than others as manipulated variables, e.g. column pressure is unsuitable as the response of x_D to any change in P_c will be slow and possibly insensitive. The column pressure may be controlled itself in a variety of ways, e.g. by manipulating the flow of vapour in the column overheads line, the flow of cooling water to the condenser, or the flow of non-condensables through the vent on the reflux accumulator. The last is often the preferred option and, in this case, the flow of cooling water to the condenser may be manipulated using a simple flow control loop (guideline (i)—Fig. 7.9). The steam flowrate S to the heating coil can be controlled in the same way. Then, provided that the quality of the steam to the plant is maintained and that ambient conditions do not vary significantly, the remaining load variable for the plant in Fig. 7.9 is the composition x_F of the feed.

If the overhead product is sent directly to storage, then its flowrate may be allowed to vary and, consequently, can be used to maintain the level in the reflux accumulator (the latter is necessary to provide a suction supply for the reflux pump). Similarly the liquid level in the base of the column may be controlled by manipulating the bottoms product flowrate, and it is necessary to keep the steam-heated coil covered with liquid. This leaves one suitable manipulated variable which can be used to control x_D, viz. the reflux flowrate L_R. Guideline number (ii) suggests that x_D should be measured directly as it is the controlled variable. This will require a suitable composition analyser. Such instruments can be difficult to use (Section 6.8) and may introduce large dead times into the control system which can degrade the control action (see Section 7.6 and Example 7.7). A commonly employed alternative is to measure the boiling temperature at a suitable point in the column (which is related directly to the composition of the material in the column at that point) and to use L_R to maintain this at a particular desired value. Employing one variable (viz. boiling temperature) to control another (viz. overhead composition) is termed *inferential control.*

A further consideration is the position of the thermocouple measuring the boiling temperature. Clearly, if the sensor is placed near to the point at which the disturbance occurs (i.e. the feed entry point) then the controller will know immediately that action has to be taken and will adjust the reflux according to the change in boiling temperature on the feed tray. However, the change in reflux flow will take some time to propagate itself down the column to the feed tray This time is dependent upon the number of stages, the volume of liquid contained within each stage (viz. the liquid hold-up) and the liquid flowrate down the column. Hence, there may be a substantial delay before any corrective action occurs on the feed tray. In the meantime the controller will continue to provide the change in reflux flow required to correct the initial deviation of the feed tray boiling temperature. The result is that more (or less) reflux is sent to the column than is necessary to bring the feed tray temperature back to its previous value. If, on the other hand, the measuring element is placed on the top tray where the corrective action of the reflux can be sensed immediately then there may be some delay whilst the feed composition disturbance is transmitted up the column. Thus a significant time may elapse before the control system takes any action. Positioning the sensor on the top tray is preferred as the delay is then generally not large. Alternatively, a compromise is made and the sensor is placed midway between the feed point and the entry of the reflux. This is the procedure illustrated in Fig. 7.9. In this case the temperature sensor is not used to control the reflux flow directly but manipulates the desired value of a flow controller which, in turn, controls the reflux flowrate. This is termed *cascade control* (see Section 7.13). An alternative to employing the reflux as the manipulated variable is to keep L_R constant and to adjust the reboil rate, i.e. the steam flowrate S to the reboiler in Fig. 7.9. This control strategy would be more appropriate to the control of x_W than of x_D. A similar argument can be applied to the positioning of the sensor which, in this case, would be located within the stripping section.

Controlling both x_D and x_W simultaneously requires two manipulated variables (guideline (iii)) and L_R and S present themselves as obvious candidates. However, severe problems of interaction between the corrective actions of both loops can result (Section 7.15).

A great many assumptions have been made in this qualitative approach. Little account has been taken of the dynamics of the plant at this stage and the stability of x_D in the face of changes (*perturbations*) in x_F. The plant may be affected by ambient conditions, e.g. a sudden shower of rain will have a cooling effect which will cause vapour rising in the column to condense and increase the column internal reflux; a long reflux line can lead to considerable degradation of the control (see Section 7.6). However, even with this simplistic approach, it is possible to examine a number of the control problems associated with the plant.

Many alternative methods of controlling distillation columns have been suggested and implemented[3,4] and the strategy suggested here is in the nature of an example (see also Volume 6, Section 5.8.7).

7.3.2. The Degrees of Freedom Approach

The *degrees of freedom* of any process constitute the number of independent variables that must be specified in order to define the process completely. Thus it is only possible to achieve the control of a process when all the degrees of freedom have been fixed. By considering the degrees of freedom involved in a process, it is possible to obtain a fuller understanding of the controllability of that process and to design an effective controller. One definition[5] of the number of degrees of freedom n_f that can be assigned to a process is:

$$n_f = n_v - n_e \tag{7.9}$$

where n_e is the number of independent equations containing n_v independent variables that are required to describe the process.

If the process illustrated in Fig. 7.9 is a binary distillation with N stages, then $4N + 5$ relationships can be established as in Table 7.1. The corresponding number of independent variables are listed in Table 7.2. A fuller description of the relationships involved can be found in STEPHANOPOULOS[6], Example 4.13.

Hence, from equation 7.9:

$$n_f = (4N + 13) - (4N + 5)$$

$$= \underline{8} \tag{7.10}$$

We have not included every possible independent variable or relationship (e.g. it is assumed that the vapour holdup on each tray is negligible, that the molar latent heats of vaporisation of both components are approximately equal, that there are no heat losses to the surroundings, etc.). Generally, such relationships are omitted in an attempt to simplify the process model.

TABLE 7.1. *Relationships for an N-Stage Binary Distillation Process*

Type of relationship	Number
Equilibrium relationships	$N + 1$
Liquid flow relationships	N
Mass balances around the ith tray (where $i = 2, \ldots, N - 1, i \neq f$)	
Total mass balances	$N - 3$
Component mass balances	$N - 3$
Balances around feed tray	2
Balances around top tray	2
Balances around bottom tray	2
Balances around column base	2
Balances around reflux accumulator	2
	$4N + 5$

TABLE 7.2. *Independent Variables for an N-Stage Binary Distillation Process*

Independent variables	Number
Vapour compositions	$N + 1$
Liquid compositions	$N + 2$
Liquid hold ups	$N + 2$
Liquid flows	N
Other variables (F, q_F, x_F, D, W, P_c, L_R, S)	8
	$4N + 13$

This is equivalent to an assumption that any variations in these quantities have a negligible effect upon the output variables of interest. Care must be taken when modelling a process that all the relevant variables and equations are included and that, on the other hand, the model does not contain redundant equations or variables. In general, a carefully modelled process will contain one or more degrees of freedom, i.e. $n_f > 0$. This is termed an *under-specified system*. In order to convert this to an *exactly specified* system it is necessary to introduce n_f additional relationships involving some or all of the n_v independent variables. These can be derived by considering:

(a) any systems external to the process (the *external world*) which affect any of the independent variables.
(b) the introduction of suitable control systems for the plant.

Although we do not necessarily know the relationships involved in the external world, they do exist and determine the values of the relevant load disturbances. Consequently, we can say that the external world removes as many degrees of freedom as the number of disturbances. In Fig. 7.9 we are introducing a flow control loop to keep F constant, and a temperature control loop with a preheater to maintain q_F. The feed composition is fixed by a relationship that we do not know—but all the same must exist. The control objectives are achieved by fitting suitable control systems to the plant and we can say that these control systems remove as many degrees of freedom as the number of control objectives in the overall control strategy.

From equation 7.10, $n_f = 8$. The feed flow and feed temperature control loops remove two of these and the unknown relationship specifying x_f (which may be dependent upon the configuration of a plant upstream) reduces n_f again by one—giving $n_f = 5$. Hence, we require five control objectives to specify the system uniquely, i.e. to control it adequately. These will be to maintain x_D (as a market requirement) and to control S, P_c, and the levels in the base of the column and the reflux accumulator (for operational feasibility).

Care must be taken not to specify more control objectives than the available remaining degrees of freedom. If this is done then the system becomes *over specified* and there is no solution to the system, i.e. the system cannot be controlled. For example, it is impossible to design a control system for the distillation column illustrated in Fig. 7.10 which will control not only x_D, S, P_c and the two levels—but also (say) the overhead product flowrate D and the bottoms product flowrate W.

It can be seen that the heuristic and the degrees-of-freedom approaches lead to similar conclusions. Both methods may be usefully employed—the former to give a common-sense preliminary view of a suitable control strategy, and the second to check that the strategy is neither over-controlling nor under-controlling the plant.

7.4. THE TRANSFER FUNCTION

Before an efficient control system can be designed it is necessary to consider how all sections of the control loop will behave under the influence of variations in load (i.e. what are the load rejection properties of the system?) and/or its set point (i.e. what are the set point following characteristics?). This requires experimental investigation, or a time-dependent mathematical analysis (i.e. in the unsteady state), or both. Although each section of the loop will necessitate a separate analysis, there are several basic simple physical systems which can be treated in much the same way. These quantitative procedures can be simplified by the use of the Laplace transform. This operational approach requires that the differential equations describing the behaviour of the sections be linear in form, but this is not very often the case. Fortunately, non-linear relationships can frequently be represented by linear approximations with little error, subject to certain limitations (see Sections 7.5.2 and 7.16.1). Once each section or system has been described in this way it is possible to form its appropriate *transfer function*, where:

$$\text{Transfer function} = \frac{\text{Laplace transform of output}}{\text{Laplace transform of input}} \tag{7.11}$$

(the Laplace transform $\bar{x}$ of any variable x (say) with respect to the independent variable t being defined by the equation:

$$\bar{x} = \int_0^\infty x e^{-st} dt$$

where s is a parameter)*.

Thus the transfer function is basically an input–output mathematical relationship. This is a most appropriate concept to use in conjunction with block diagrams (Section 7.2.1) which are themselves basically input–output schematic diagrams so that each block may be represented by the transfer function describing its behaviour.

The advantage of defining a transfer function in terms of Laplace transforms of input and output is that the differential equations developed to describe the unsteady-state behaviour of the system are reduced to simple algebraic relationships (e.g. cf. equations 7.17 and 7.19). Such relationships are much easier to deal with, and normal algebraic laws can be used to relate the various transfer functions of each component in the control loop (see Section 7.9). Furthermore, the output (or response) of the system to a variety of inputs may be obtained without classical integration.

The transfer function approach will be used where appropriate throughout the remainder of this chapter. Transfer functions of continuous systems will be expressed as functions of s, e.g. as $\mathbf{G}(s)$ or $\mathbf{H}(s)$. In the case of discrete time systems, the transfer function will be written in terms of the z-transform, e.g. as $\mathbf{G}(z)$ or $\mathbf{H}(z)$ (Section 7.17). An elementary knowledge of the Laplace transformation on the part of the reader is assumed and a table of the more useful Laplace transforms and their z-transform equivalents appears in Appendix 7.1.

7.4.1. Linear Systems and the Principle of Superposition

A number of the systems described in the remainder of this chapter are assumed to be linear with respect to time, i.e. their time-dependent properties can be described by linear differential equations. Such systems follow the *principle of superposition*[7,8]. This property is such that *if the individual output of a system is*

* The Laplace transform parameter in this chapter is denoted by s which is now common practice in the control field. Previous editions of this book and other volumes of this edition employ p as the Laplace parameter.

known for each of several different inputs acting separately, then the total output for all the inputs acting together can be obtained by simply summing those individual outputs.

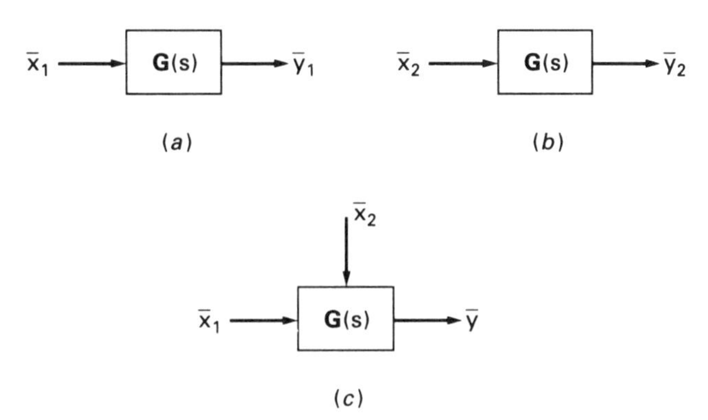

FIG. 7.10. Illustration of the principle of superposition

In Fig. 7.10 **G**(s) is the transfer function of a particular linear system. If, for an input $\bar{x}_1$, an output $\bar{y}_1$ is obtained, both being expressed in terms of the Laplace transform:

$$\mathbf{G}(s) = \frac{\bar{y}_1}{\bar{x}_1} \qquad \text{(Fig. 7.10}a\text{)}$$

For an input $\bar{x}_2$:

$$\mathbf{G}(s) = \frac{\bar{y}_2}{\bar{x}_2} \qquad \text{(Fig. 7.10}b\text{)}$$

For an input $\bar{x}_1 + \bar{x}_2$, by the principle of superposition:

$$\bar{y} = \bar{y}_1 + \bar{y}_2 = \mathbf{G}(s)\bar{x}_1 + \mathbf{G}(s)\bar{x}_2 = \mathbf{G}(s)(\bar{x}_1 + \bar{x}_2) \qquad \text{(Fig. 7.10}c\text{)}$$

Similarly, if the behaviour of the components of the block diagram illustrated in Fig. 7.3 can be assumed to be linear, then the effects of load and manipulated variable on the process can be added using the principle of superposition.

7.4.2. Block Diagram Algebra

Blocks in Series

In Fig. 7.11 $\mathbf{G}_a(s)$, $\mathbf{G}_b(s)$ and $\mathbf{G}_c(s)$ are transfer functions describing the input–output relationships for each block respectively, where:

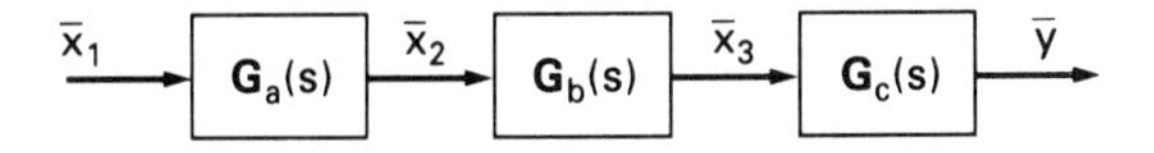

FIG. 7.11. Blocks in series

$$\mathbf{G}_a(s) = \frac{\bar{x}_2}{\bar{x}_1}, \quad \mathbf{G}_b(s) = \frac{\bar{x}_3}{\bar{x}_2}, \quad \mathbf{G}_c(s) = \frac{\bar{y}}{\bar{x}_3}$$

The transfer function of the whole system is given by $\mathbf{G}_x(s) = \bar{y}/\bar{x}_1$. This can also be obtained by multiplying together the three blocks in series, viz.:

$$\frac{\bar{y}}{\bar{x}_1} = \frac{\bar{x}_2}{\bar{x}_1}\frac{\bar{x}_3}{\bar{x}_2}\frac{\bar{y}}{\bar{x}_3}$$

i.e.

$$\mathbf{G}_X(s) = \mathbf{G}_a(s)\mathbf{G}_b(s)\mathbf{G}_c(s) \tag{7.12}$$

Blocks in Parallel

In Fig. 7.12:

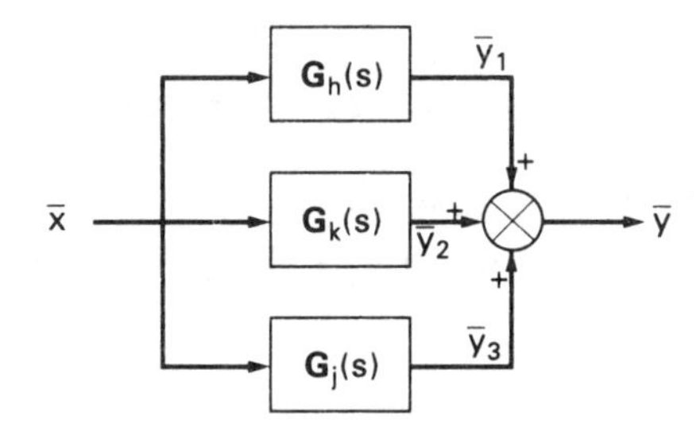

FIG. 7.12. Blocks in parallel

$$\mathbf{G}_h(s) = \frac{\bar{y}_1}{\bar{x}}, \quad \mathbf{G}_j(s) = \frac{\bar{y}_2}{\bar{x}}, \quad \mathbf{G}_k(s) = \frac{\bar{y}_3}{\bar{x}}$$

(signal $\bar{x}$ is applied equally as the input to all three blocks).

The transfer function of the whole system is $\mathbf{G}_Y(s) = \bar{y}/\bar{x}$, and by the principle of superposition $\bar{y}$ is obtained from the additive effects of $\bar{y}_1$, $\bar{y}_2$ and $\bar{y}_3$. Therefore:

$$\frac{\bar{y}}{\bar{x}} = \frac{\bar{y}_1 + \bar{y}_2 + \bar{y}_3}{\bar{x}}$$

and:

$$\mathbf{G}_Y(s) = \mathbf{G}_h(s) + \mathbf{G}_j(s) + \mathbf{G}_k(s) \tag{7.13}$$

Junctions of Signals

The summation of signals is represented on a block diagram as shown in Fig. 7.13. Figure 7.13*a* represents the relationship:

$$\bar{y} = \bar{x}_1 + \bar{x}_2$$

and Fig. 7.13*b*:

$$\bar{y} = \bar{x}_3 - \bar{x}_2 - \bar{x}_1$$

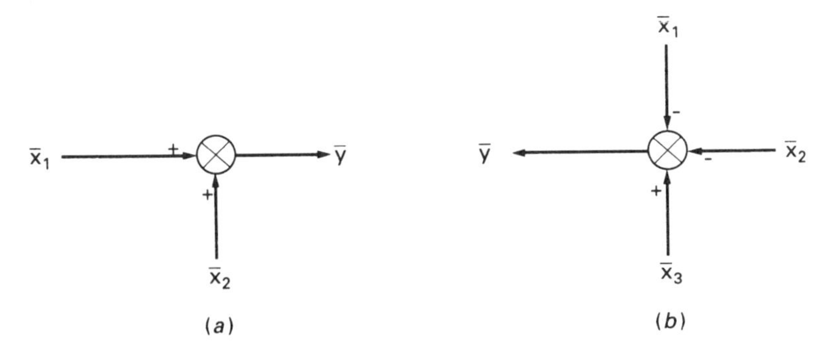

FIG. 7.13. Summation of signals

7.4.3. The Poles and Zeros of a Transfer Function

In general terms, transfer functions arising within a chemical engineering context can almost always be written as a ratio of two polynomials in s. Thus, referring to Fig. 7.10*a*:

$$\mathbf{G}(s) = \frac{\bar{y}_1}{\bar{x}_1} \tag{7.14}$$

where $\bar{x}_1$ and $\bar{y}_1$ are polynomials in s. The only exceptions are systems containing dead-times which introduce exponential terms (Section 7.7). For physically realisable systems (i.e. systems that can exist in practice), $\bar{y}_1$ will be always of lower order than $\bar{x}_1^{(6)}$. The roots of the equation $\bar{x}_1 = 0$ are termed the *poles of the transfer function* or the *poles of the system* represented by $\mathbf{G}(s)$. At the poles of the system (i.e. when s is put equal to a pole or root of $\bar{x}_1$) the transfer function becomes infinite. The roots of the equation $\bar{y}_1 = 0$ are called the *zeros of the transfer function* or the *zeros of the system*. At the zeros of the system, the transfer function becomes zero. (See also Section 7.10.1.)

7.5. TRANSFER FUNCTIONS OF CAPACITY SYSTEMS

7.5.1. Order of a System

If the unsteady-state behaviour of a system is described by a first order differential equation, it is termed a *first-order system*. Other descriptions frequently used are *first-order lag* and *single capacity system*. Similarly, a component described by a second-order differential equation is termed a *second-order system*, and so on.

7.5.2. First-Order Systems

The behaviour of many control loop components can be described by first-order differential equations provided that certain simplifying assumptions are made. Great care should be taken that the assumptions made are reasonable under the conditions to which the component is subjected. Two examples of a first-order system are described—a measuring element and a process. An illustration of a multivariable system which approximates to first order with respect to each input variable can be found in Example 7.11.

A Measuring Element—The Thermocouple

Consider a thermocouple junction immersed in a fluid whose temperature θ_0 varies with time (Figs 7.14 and 6.19). Assume that all resistance to heat transfer resides in the film surrounding the thermocouple wall, that all the thermal capacity lies in the junction, and that for $t < 0$ there is no change of temperature with time, i.e. the system is initially at a steady state.

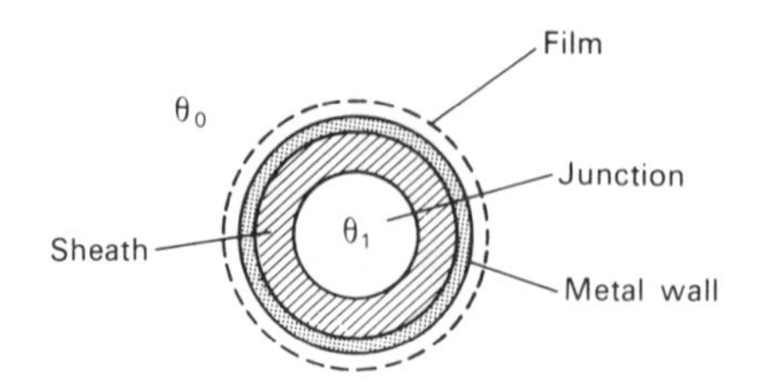

FIG. 7.14. Cross-section of thermocouple junction

The temperature of the thermocouple θ_1 will also vary with time. It is required to determine the relationship between θ_0 and θ_1. Writing an unsteady-state energy balance for the junction:

$$\begin{pmatrix}\text{Energy input}\\ \text{per unit time}\end{pmatrix} - \begin{pmatrix}\text{Energy output}\\ \text{per unit time}\end{pmatrix} = \begin{pmatrix}\text{Rate of accumulation}\\ \text{of energy in junction}\end{pmatrix}$$

i.e.:

$$h_1 A_1(\theta_0 - \theta_1) - 0 = \frac{\mathrm{d}}{\mathrm{d}t}(m_1 C_1 \theta_1)$$

where h_1 is the film heat transfer coefficient,
A_1 is the mean area for heat transfer in the film,
C_1 is the average specific heat of the junction material, and
m_1 is its mass.

These are all assumed constant with respect to time, hence:

$$h_1 A_1(\theta_0 - \theta_1) = m_1 C_1 \frac{\mathrm{d}\theta_1}{\mathrm{d}t} \tag{7.15}$$

In the steady state there is no accumulation; therefore from equation 7.15:

$$h_1 A_1(\theta_0' - \theta_1') = 0 \tag{7.16a}$$

or

$$\theta_0' = \theta_1' \tag{7.16b}$$

where θ_0', θ_1' are the steady-state values of θ_0 and θ_1 respectively. Subtracting equation 7.16*a* from equation 7.15 gives:

$$(\theta_0 - \theta_0') - (\theta_1 - \theta_1') = \frac{m_1 C_1}{h_1 A_1}\frac{\mathrm{d}}{\mathrm{d}t}(\theta_1 - \theta_1')$$

But $\frac{\mathrm{d}}{\mathrm{d}t}(\theta_1 - \theta_1') = \frac{\mathrm{d}\theta_1}{\mathrm{d}t}$ as θ_1' is constant, so:

$$\vartheta_0 - \vartheta_1 = \tau \frac{\mathrm{d}\vartheta_1}{\mathrm{d}t} \tag{7.17}$$

where:
$$\vartheta_0 = \theta_0 - \theta_0',$$
$$\vartheta_1 = \theta_1 - \theta_1', \text{ and}$$
$$\tau = \frac{m_1 C_1}{h_1 A_1} = (m_1 C_1)\left(\frac{1}{h_1 A_1}\right).$$

ϑ_0 and ϑ_1 are termed *deviation variables* which represent the difference between the values at any time and at the steady state. τ has the dimensions of time and is called the *time constant** of the system which, in this case, is the product of the heat capacity of the junction and the resistance to heat transfer of the surrounding film.

The Laplace transform of equation 7.17 is:

$$\bar{\vartheta}_0 - \bar{\vartheta}_1 = \tau[s\bar{\vartheta}_1 - \vartheta_1|_{t=0}]$$

since
$$\frac{d\bar{\vartheta}}{dt} = \int_0^\infty \frac{d\vartheta_1}{dt} e^{-st}\, dt = [e^{-st}\vartheta_1]_0^\infty + s\int_0^\infty e^{-st}\vartheta_1\, dt$$
$$= -\vartheta_1|_{t=0} + s\bar{\vartheta}_1$$

But:
$$\vartheta_1|_{t=0} = \theta_1' - \theta_1' = 0 \tag{7.18}$$

So:
$$\bar{\vartheta}_0 - \bar{\vartheta}_1 = \tau s\bar{\vartheta}_1$$

and:
$$\frac{\bar{\vartheta}_1}{\bar{\vartheta}_0} = \mathbf{G}(s) = \frac{1}{1 + \tau s} \tag{7.19}$$

$\mathbf{G}(s)$ is the transfer function relating θ_0 and θ_1. It can be seen from equation 7.18 that the use of deviation variables is not only physically relevant but also eliminates the necessity of considering initial conditions. Equation 7.19 is typical of transfer functions of first order systems in that the numerator consists of a constant and the denominator a first order polynomial in the Laplace transform parameter s. The numerator represents the steady-state relationship between the input θ_0 and the output θ_1 of the system and is termed the system *steady-state gain*. In this case the steady-state gain is unity as, in the steady state, the input and output are the same both physically and dimensionally (equation 7.16*b*). Note that the constant term in the denominator of $\mathbf{G}(s)$ must be written as unity in order to identify the coefficient of s as the system time constant and the numerator as the system steady-state gain. (e.g. If $\mathbf{G}(s) = \dfrac{4}{s+2}$ then, rewriting $\mathbf{G}(s)$ in the form of equation 7.19, $\mathbf{G}(s) = \dfrac{2}{1 + 0.5s} = \dfrac{K}{1 + \tau s}$. Hence, the system time constant is 0.5 units and the steady-state gain is 2 units). (See also equation 7.26.)

A Process—Liquid Flowing Through a Tank

Consider liquid flowing through a tank of uniform area of cross-section A with variable volumetric inlet rate of flow Q_0 (Fig. 7.15). Let the volumetric rate of flow

* It is the convention in chemical engineering process control to use minutes as the normal unit of time for a time constant. This convention is followed in this text.

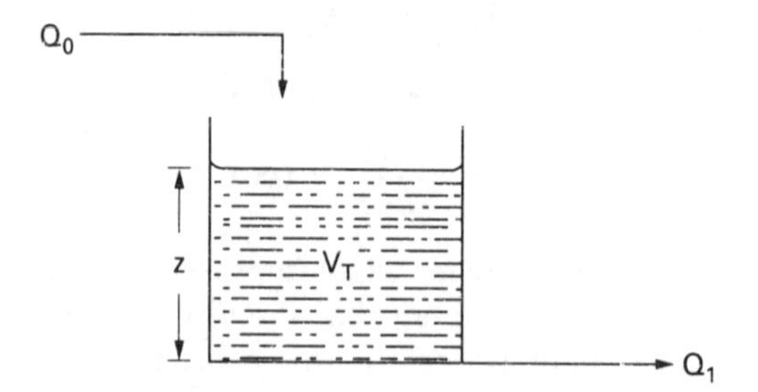

FIG. 7.15. Liquid flowing through a tank (self-regulating system)

of liquid leaving and the volume of liquid in the tank at any time be Q_1 and V_T, respectively. Assume the liquid density ρ to be constant. If the inflow remains constant, the liquid head z will adjust itself until:

$$Q_0' = Q_1' \tag{7.20}$$

where Q_0', Q_1' are the steady-state values of Q_0 and Q_1 respectively. This is termed *self-regulation* and is a phenomenon exhibited by a number of process systems. The outflow is a function of the liquid head and the resistance to flow in the outlet line. If the head lost due to friction is neglected, then from an energy balance for turbulent flow (Volume 1, Section 2.5.1) in the outlet pipe:

$$Q_1 = K_1 z^{1/2} \tag{7.21}$$

where K_1 is a constant for a particular pipe.

A mass balance over the tank in the unsteady state yields:

$$Q_0\rho - Q_1\rho = \frac{\mathrm{d}}{\mathrm{d}t}(\rho V_T) = \rho A \frac{\mathrm{d}z}{\mathrm{d}t} \tag{7.22}$$

Substituting from equation 7.21:

$$Q_0 - K_1 z^{1/2} = A \frac{\mathrm{d}z}{\mathrm{d}t} \tag{7.23}$$

Equation 7.23 cannot be transformed in the usual way as it contains the non-linear term $z^{1/2}$ for which there is no simple transform. This term can be approximated, however, by a linear expression (i.e. *linearised*) in which variations (*perturbations*) in level are considered to occur around the steady-state level z'. To achieve this Q_1 is first expanded as a Taylor's series in terms of these variations[9], thus:

$$Q_1 = Q_1' + \left[\frac{\mathrm{d}Q_1}{\mathrm{d}z}\right]_{z'}(z - z') + \left[\frac{\mathrm{d}^2Q_1}{\mathrm{d}z^2}\right]_{z'}\frac{(z - z')^2}{2} + \text{higher order terms} \tag{7.24a}$$

If the variation in level is small, it is possible to neglect powers greater than unity with little error, i.e.:

$$Q_1 \approx Q_1' + \left[\frac{\mathrm{d}Q_1}{\mathrm{d}z}\right]_{z'}(z - z') \tag{7.24b}$$

Substituting from equation 7.21, we have:

$$Q_1 = Q_1' + K_2(z - z') \tag{7.25}$$

where $K_2 = K_1/(2\sqrt{z'})$.

From equations 7.22 and 7.25, we obtain:

$$(Q_0 - Q_1') - K_2(z - z') = A\frac{dz}{dt}$$

Substituting from equation 7.20 and using deviation variables:

$$\mathcal{Q}_0 - K_2\mathfrak{z} = A\frac{d\mathfrak{z}}{dt}$$

where $\mathcal{Q}_0 = Q_0 - Q_0'$ and $\mathfrak{z} = z - z'$.

Transforming yields the transfer function between inflow and liquid level, viz.:

$$\mathbf{G}(s) = \frac{\bar{\mathfrak{z}}}{\bar{\mathcal{Q}}_0} = \frac{1/K_2}{1 + \tau s} \tag{7.26}$$

where $\tau = A/K_2$ and $1/K_2$ is the steady-state gain (cf. equation 7.19).

Differentiating equation 7.25 gives:

$$\frac{dz}{dt} = \frac{1}{K_2}\frac{dQ_1}{dt}$$

Substituting for dz/dt in equation 7.22 leads to the transfer function between inflow and outflow, i.e.:

$$\mathbf{G}(s) = \frac{\bar{\mathcal{Q}}_1}{\bar{\mathcal{Q}}_0} = \frac{1}{1 + \tau s} \tag{7.27}$$

In this case the steady-state gain is unity (cf. equations 7.19 and 7.26).

The degree of variation around the steady state to which systems can be subjected such that it is possible to approximate them by using linear relationships (e.g. as in equation 7.24) differs from system to system. The dynamics of a highly non-linear reactor might be described satisfactorily by a linear analysis for perturbations of up to ±3 per cent[10]. On the other hand the dynamics of some distillation columns have been shown to remain reasonably linear in the face of variations of ±25 per cent in some process variables[11].

For a further discussion of non-linear systems see Section 7.16.

7.5.3. First-Order Systems in Series

It is frequently required to examine the combined performance of two or more processes in series, e.g. two systems or capacities, each described by a transfer function in the form of equations 7.19 or 7.26. Such *multicapacity* processes do not necessarily have to consist of more than one physical unit. Examples of the latter are a protected thermocouple junction where the time constant for heat transfer across the sheath material surrounding the junction is significant, or a distillation column in which each tray can be assumed to act as a separate capacity with respect to liquid flow and thermal energy.

Systems or processes operating in series may be considered as *non-interacting*, as *interacting*, or as a mixture of both.

Non-Interacting Systems

When two or more systems are operating in series and the behaviour of any system is not affected by any system subsequent to it, then those systems are said to be *non-interacting*.

Two tanks in series. Consider the two tanks shown in Fig. 7.16. In this system neither the rate of flow through tank 1 nor the level in tank 1 is affected by what occurs in tank 2. Thus the two processes (or capacity systems) are non-interacting and we can model their dynamic behaviours individually. To establish the relationship between the level in tank 2 and the volumetric flowrate entering tank 1 at any instant of time, we need only to determine the individual transfer functions between Q_0 and Q_1 and between Q_1 and z_2.

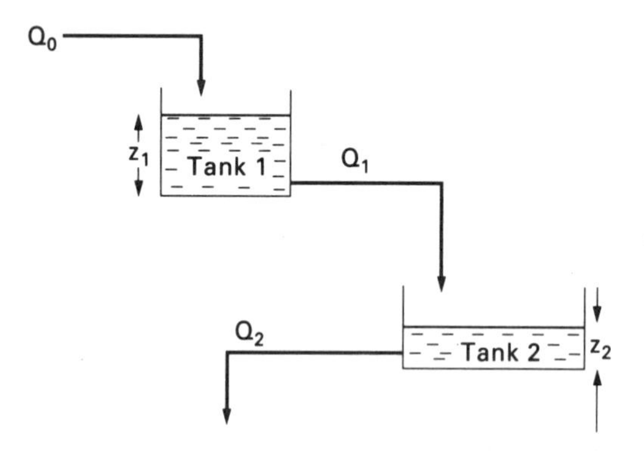

FIG. 7.16. Two non-interacting tanks in series

For tank 1, from equation 7.27:

$$\mathbf{G}_1(s) = \frac{\bar{\mathcal{Q}}_1}{\bar{\mathcal{Q}}_0} = \frac{1}{1 + \tau_1 s} \tag{7.28}$$

where $\tau_1 = A_1/K_2$, $K_2 = K_1/(2\sqrt{z_1'})$ and K_1 is a constant for the outlet pipe from tank 1.

For tank 2, from equation 7.26:

$$\mathbf{G}_2(s) = \frac{\bar{\mathfrak{z}}_2}{\bar{\mathcal{Q}}_1} = \frac{1/K_4}{1 + \tau_2 s} \tag{7.29}$$

where $\tau_2 = A_2/K_4$, $K_4 = K_3/(2\sqrt{z_2'})$ and K_3 is a constant for the pipe leaving tank 2.

By eliminating $\bar{\mathcal{Q}}_1$ between equations 7.28 and 7.29, the transfer function relating Q_0 and z_2 is obtained, viz.:

$$\mathbf{G}_3(s) = \frac{\bar{\mathfrak{z}}_2}{\bar{\mathcal{Q}}_0} = \frac{\bar{\mathfrak{z}}_2}{\bar{\mathcal{Q}}_1}\frac{\bar{\mathcal{Q}}_1}{\bar{\mathcal{Q}}_0} = \frac{1/K_4}{(1 + \tau_1 s)(1 + \tau_2 s)}$$

$$= \frac{1/K_4}{\tau_1 \tau_2 s^2 + (\tau_1 + \tau_2)s + 1} \tag{7.30}$$

The relationship between Q_0 and Q_2 can be obtained similarly.

A distillation process. The behaviour of liquid and vapour streams in any stagewise process can usually be approximated by a number of non-interacting first order systems in series. For example, ROSE and WILLIAMS[12] employed a first order transfer function to represent the dynamics of liquid and vapour flow in a 5-stage continuous distillation column. Thus for stage n in Fig. 7.17:

$$\frac{\bar{\mathscr{V}}_n}{\bar{\mathscr{V}}_{n-1}} = \frac{1}{1 + \tau_1 s} \quad \text{and} \quad \frac{\bar{\mathscr{L}}_n}{\bar{\mathscr{L}}_{n-1}} = \frac{1}{1 + \tau_2 s} \tag{7.31}$$

where τ_1 is the time required for vapour to pass through the liquid on tray n and τ_2 for liquid to pass across the plate (viz. the vapour and liquid *holdup times* respectively).

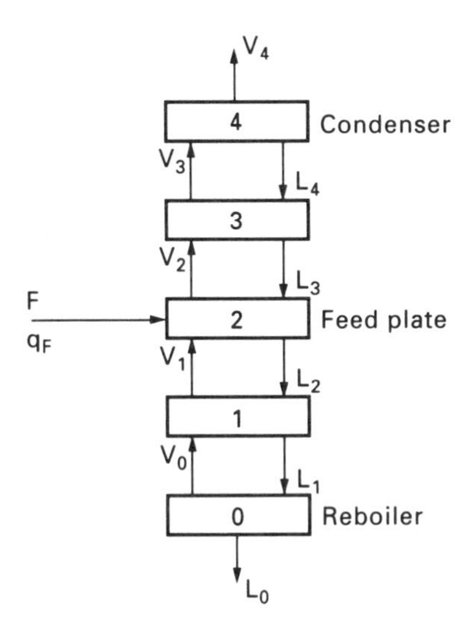

FIG. 7.17. Mass flows in continuous plate distillation column

For the feed plate:

$$\left.\begin{aligned} \bar{\mathscr{V}}_F &= \frac{\bar{\mathscr{V}}_{F-1}}{1 + \tau_1 s} + (1 - q_F)\frac{\bar{\mathscr{F}}}{1 + \tau_3 s} \\ \bar{\mathscr{L}}_F &= \frac{\bar{\mathscr{L}}_{F+1}}{1 + \tau_2 s} + q_F \frac{\bar{\mathscr{F}}}{1 + \tau_4 s} \end{aligned}\right\} \tag{7.32}$$

Equation 7.32 takes into account the variations in feed, vapour and liquid rates on the column vapour and liquid flows. The feed quality q_F is assumed to remain constant (q_F is the fraction of the feed which flows down the column as liquid).

For an N-stage stripping column (Figs. 7.18*a* and 7.18*b*), from equation 7.12:

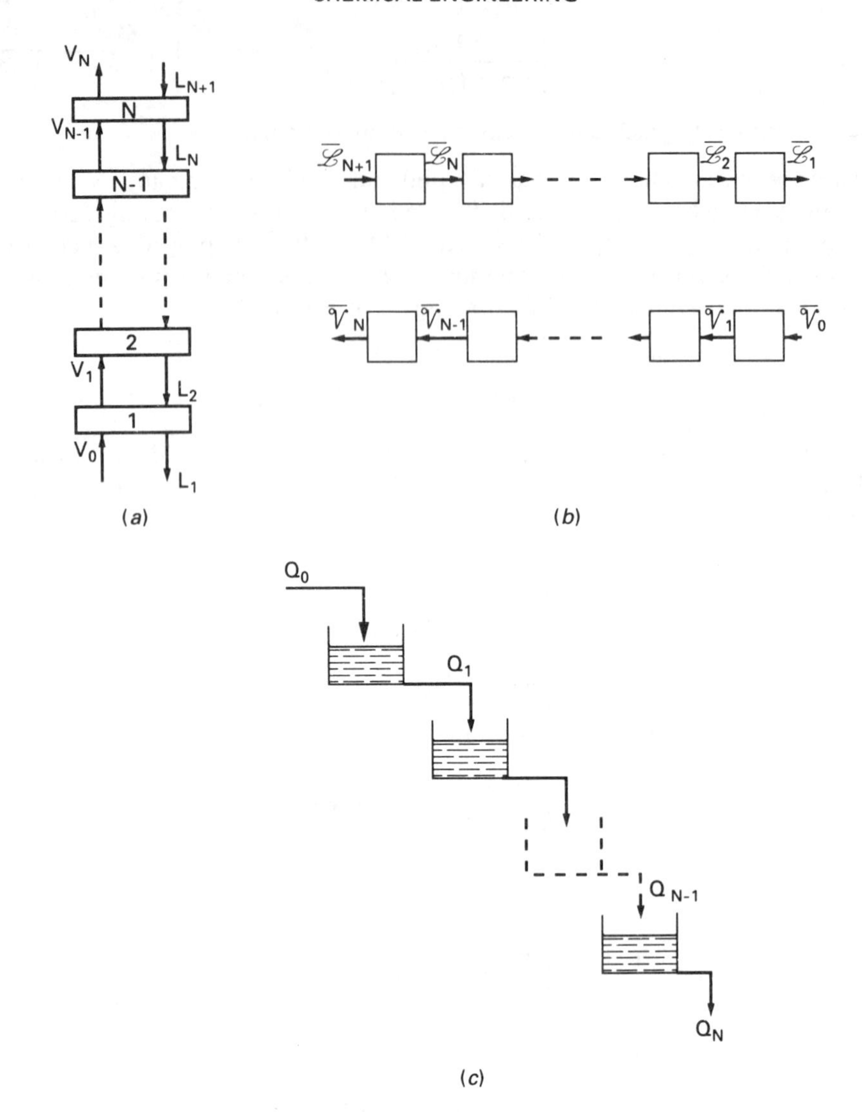

FIG. 7.18. Non-interacting stages in series: (*a*) stripping column; (*b*) block diagram for stripping column; (*c*) tanks in series

$$\frac{\bar{\mathscr{L}}_1}{\bar{\mathscr{L}}_{N+1}} = \frac{\bar{\mathscr{L}}_N}{\bar{\mathscr{L}}_{N+1}} \frac{\bar{\mathscr{L}}_{N-1}}{\bar{\mathscr{L}}_N} \cdots \frac{\bar{\mathscr{L}}_2}{\bar{\mathscr{L}}_3} \frac{\bar{\mathscr{L}}_1}{\bar{\mathscr{L}}_2}$$

If the liquid phase time constants are all the same, from equation 7.31:

$$\frac{\bar{\mathscr{L}}_1}{\bar{\mathscr{L}}_{N+1}} = \frac{1}{(1 + \tau_2 s)^N} \tag{7.33}$$

Similarly:

$$\frac{\bar{\mathscr{V}}_N}{\bar{\mathscr{V}}_0} = \frac{1}{(1 + \tau_1 s)^N} \tag{7.34}$$

Transfer functions analogous to equations 7.33 and 7.34 can be obtained for any number of non-interacting first order systems in series, e.g. for N tanks in series (Fig. 7.18c) having the same time constant τ, from equation 7.27:

$$\frac{\bar{\mathcal{Q}}_N}{\bar{\mathcal{Q}}_0} = \frac{1}{(1 + \tau s)^N}$$

Interacting Systems

When two or more systems are operating in series and the behaviour of any system *is* affected by any system or systems subsequent to it then those systems are said to be *interacting*.

Two interacting tanks in series. In the arrangement shown in Fig. 7.19 variations in the level in tank 2 will affect the flow through tank 1. Consequently, the two tanks are interacting and cannot be treated separately. Equation 7.22 gives the mass balance for each tank, viz., for constant density ρ throughout:

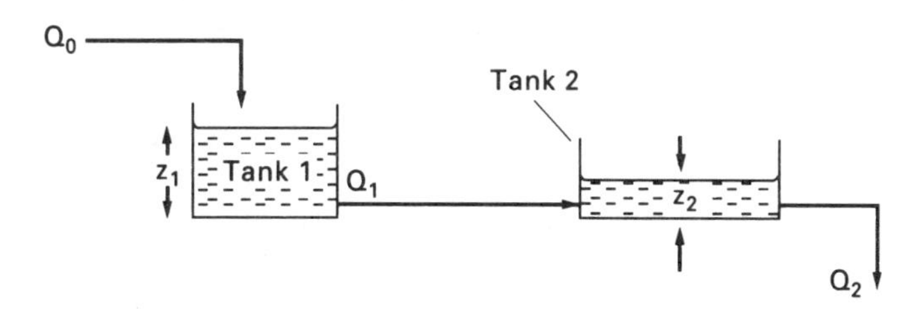

FIG. 7.19. Two interacting tanks in series

$$Q_0 - Q_1 = A_1 \frac{dz_1}{dt} \tag{7.35}$$

and

$$Q_1 - Q_2 = A_2 \frac{dz_2}{dt} \tag{7.36}$$

Expressing equations 7.35 and 7.36 in deviation variables and transforming yields:

$$\bar{\mathcal{Q}}_0 - \bar{\mathcal{Q}}_1 = A_1 s \bar{\mathfrak{z}}_1 \tag{7.37}$$

and

$$\bar{\mathcal{Q}}_1 - \bar{\mathcal{Q}}_2 = A_2 s \bar{\mathfrak{z}}_2 \tag{7.38}$$

The flow-head relationships in this case are:

$$Q_1 = K_1 \sqrt{(z_1 - z_2)} \tag{7.39}$$

and

$$Q_2 = K_3 \sqrt{z_2} \tag{7.40}$$

Linearising equations 7.39 and 7.40 using Taylor's series (Section 7.5.2), expressing them in deviation form and transforming leads to:

$$\bar{\mathcal{Q}}_1 = K_2(\bar{\mathfrak{z}}_1 - \bar{\mathfrak{z}}_2) \tag{7.41}$$

and:
$$\bar{\mathfrak{Q}}_2 = K_4 \bar{\mathfrak{z}}_2 \tag{7.42}$$

where $K_2 = K_1/[2\sqrt{(z_1' - z_2')}]$ and $K_4 = K_3/2\sqrt{z_2'}$.

The transfer function relating Q_0 to z_2 can now be derived by eliminating $\bar{\mathfrak{Q}}_1$, $\bar{\mathfrak{Q}}_2$ and $\bar{\mathfrak{z}}_1$ between equations 7.37, 7.38, 7.41 and 7.42:

$$\mathrm{G}(s) = \frac{\bar{\mathfrak{z}}_2}{\bar{\mathfrak{Q}}_0} = \frac{1/K_4}{\tau_1 \tau_2 s^2 + (\tau_1 + \tau_2 + \kappa)s + 1} \tag{7.43}$$

where $\tau_1 = A_1/K_2$, $\tau_2 = A_2/K_4$ and $\kappa = A_1/K_4$.

The parameter κ is termed the *interaction factor* and the degree of interaction increases with the magnitude of κ. When $\kappa = 0$ equation 7.43 becomes the same as equation 7.30, i.e. the equivalent non-interacting case.

Interacting capacities are always overdamped and the interaction can be considered as modifying the time constants of the system. In this case, any variation in the interaction factor will alter the ratio of the two tank time constants. The tank with the larger time constant is the controlling tank and the total system response becomes more sluggish as the degree of interaction increases. Hence, interacting capacities respond more slowly to disturbances than those which are non-interacting.

Thermocouple junction with protective sheath. Suppose the resistance to heat transfer of the sheath surrounding the thermocouple described in Section 7.5.2 is not negligible. The unsteady-state heat transfer mechanism must then be considered in two stages.

An energy balance over the fluid film surrounding the metal wall (Fig. 7.20), neglecting the heat capacity of the sheath material and the fluid film, gives:

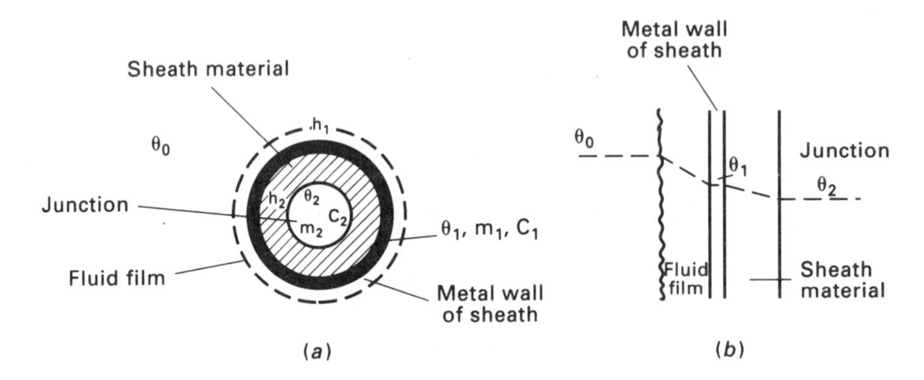

FIG. 7.20. Thermocouple junction including resistance of sheath: (*a*) cross-sectional view; (*b*) temperature profile

$$h_1 A_1(\theta_0 - \theta_1) - h_2 A_2(\theta_1 - \theta_2) = \frac{\mathrm{d}}{\mathrm{d}t}(m_1 C_1 \theta_1) \tag{7.44}$$

and over the sheath material:

$$h_2 A_2(\theta_1 - \theta_2) - 0 = \frac{\mathrm{d}}{\mathrm{d}t}(m_2 C_2 \theta_2) \tag{7.45}$$

where θ_1, m_1, C_1 and θ_2, m_2 and C_2 are the temperature, mass and specific heat of the metal wall and junction respectively, and h_1, A_1 and h_2, A_2 are the heat transfer coefficients and mean areas for heat transfer for the film and sheath respectively.

Introducing deviation variables and transforming, we obtain from equation 7.44:

$$\bar{\vartheta}_0 - \bar{\vartheta}_1 - \frac{h_2 A_2}{h_1 A_1}[\bar{\vartheta}_1 - \bar{\vartheta}_2] = \tau_1 s \bar{\vartheta}_1 \tag{7.46}$$

where $\tau = \dfrac{m_1 C_1}{h_1 A_1}$ and C_1, C_2, h_1 and h_2 are assumed constant with respect to time.

Similarly, from equation 7.45:

$$\bar{\vartheta}_1 - \bar{\vartheta}_2 = \tau_2 s \bar{\vartheta}_2 \tag{7.47}$$

where $\tau_2 = \dfrac{m_2 C_2}{h_2 A_2}$.

Elimination of $\bar{\vartheta}_1$ from equations 7.46 and 7.47 yields:

$$\mathbf{G}(s) = \frac{\bar{\vartheta}_2}{\bar{\vartheta}_0} = \frac{1}{\tau_1 \tau_2 s^2 + [\tau_1 + \tau_2(1 + h_2 A_2/h_1 A_1)]s + 1} \tag{7.48}$$

In this case the interaction factor κ is $h_2 A_2/h_1 A_1$. Interaction occurs because the rate of heat transfer from the first capacity (the metal wall) to the second (the junction) is dependent upon the temperature of the latter. τ_1 is the time constant of the first capacity and τ_2 the time constant of the second.

Interaction between control loops is discussed in Section 7.15.

7.5.4. Second-Order Systems

Consider the U-tube manometer shown in Fig. 7.21. Suppose that the liquid level is initially at **BB**. If the pressure differential ($P_1 - P_2$) increases, the liquid in the

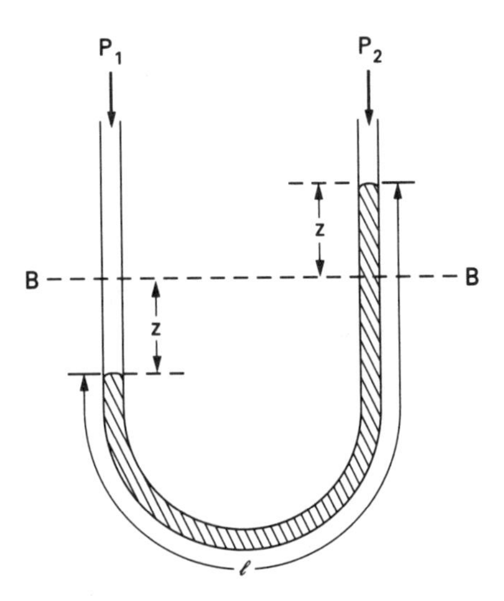

FIG. 7.21. U-Tube manometer

manometer will move as indicated. During this movement the difference in the force exerted on each limb will provide momentum to the column of liquid which will be opposed by inertia, frictional drag and the force due to gravity.

If z is the distance moved by the liquid in the manometer tube in time t, then:

$$\text{Inertial force} = \text{Mass of column of liquid} \times \text{Acceleration}$$

$$= \frac{\pi d^2}{4} \ell \rho \frac{d^2 z}{dt^2}$$

where d is the diameter of the manometer tube and ρ is the density of the liquid.

From Volume 1, Section 3.4.3:

$$\text{Frictional drag per unit cross-sectional area} = -\Delta P_f = 4\left(\frac{R}{\rho u^2}\right)\frac{\ell}{d}\rho u^2$$

If fully-developed laminar flow is assumed, then (from Volume 1, Section 3.4.2):

$$\frac{R}{\rho u^2} = 8Re^{-1} = \frac{8\mu}{\rho u d}$$

Hence:

$$-\Delta P_f = \frac{32\mu\ell}{d^2} u = \frac{32\mu\ell}{d^2}\frac{dz}{dt}$$

Finally:

$$\text{Force due to gravity} = 2\frac{\pi d^2}{4}\rho z g$$

Thus, a force balance gives:

$$(P_1 - P_2)\frac{\pi d^2}{4} = \frac{\pi d^2}{4}\ell\rho\frac{d^2 z}{dt^2} + \frac{32\mu\ell}{d^2}\frac{\pi d^2}{4}\frac{dz}{dt} + 2\frac{\pi d^2}{4}\rho z g$$

Simplification leads to:

$$\frac{d^2 z}{dt^2} + \frac{32\mu}{d^2\rho}\frac{dz}{dt} + \frac{2g}{\ell} z = \frac{-\Delta P}{\ell\rho} \tag{7.49}$$

where $\Delta P = P_2 - P_1$.

In order to introduce deviation variables, we need to write the equivalent steady-state relationship, viz.:

$$\frac{2g}{\ell} z' = \frac{-\Delta P'}{\ell\rho} \tag{7.50}$$

where z', $\Delta P'$ are the steady-state values of the variables.

Subtracting equation 7.50 from equation 7.49 leads to:

$$\frac{d^2 z}{dt^2} + \frac{32\mu}{d^2\rho}\frac{dz}{dt} + \frac{2g}{\ell}(z - z') = \frac{-(\Delta P - \Delta P')}{\ell\rho}$$

Thus:

$$\frac{d^2 \mathfrak{z}}{dt^2} + \frac{32\mu}{d^2\rho}\frac{d\mathfrak{z}}{dt} + \frac{2g}{\ell}\mathfrak{z} = \frac{-\Delta\mathcal{P}}{\ell\rho} \tag{7.51}$$

where:

$$\mathfrak{z} = z - z',\ \Delta\mathcal{P} = \Delta P - \Delta P',\ \frac{d\mathfrak{z}}{dt} = \frac{dz}{dt} \text{ and } \frac{d^2\mathfrak{z}}{dt^2} = \frac{d^2 z}{dt^2}$$

Transforming equation 7.51 gives:

$$s^2\bar{z} + \frac{32\mu}{d^2\rho}s\bar{z} + \frac{2g}{\ell}\bar{z} = \frac{-\Delta\bar{\mathcal{P}}}{\ell}$$

Hence:

$$\frac{\bar{z}}{-\Delta\bar{\mathcal{P}}} = \frac{1/(\rho\ell)}{s^2 + \dfrac{32\mu}{d^2\rho}s + \dfrac{2g}{\ell}}$$

In order to write a transfer function in standard form, the constant in the denominator should always be unity (see Section 7.5.2). Thus the numerator and the denominator must both be multiplied by $\ell/2g$ giving:

$$\mathbf{G}(s) = \frac{\bar{z}}{-\Delta\bar{\mathcal{P}}} = \frac{1/(2\rho g)}{\dfrac{\ell}{2g}s^2 + \dfrac{16\mu\ell}{d^2\rho g}s + 1}$$

$$= \frac{K_{MT}}{\tau^2 s^2 + 2\zeta\tau s + 1} \tag{7.52}$$

where $\tau = \sqrt{\left(\dfrac{\ell}{2g}\right)}$, $\zeta = \dfrac{8\mu}{d^2\rho}\sqrt{\left(\dfrac{2\ell}{g}\right)}$, and K_{MT} (the steady-state gain) $= \dfrac{1}{2\rho g}$.

Equation 7.52 is the standard form of a second-order transfer function arising from the second-order differential equation representing the model of the process. Note that two parameters are now necessary to define the system, viz. τ (the time constant) and ζ (the *damping coefficient*). The steady-state gain K_{MT} represents the steady-state relationship between the input to the system ΔP and the output of the system z (cf. equation 7.50).

There are distinct similarities between second order systems and two first-order systems in series. However, in the latter case, it is possible physically to separate the two lags involved. This is not so with a true second order system and the mathematical representation of the latter always contains an acceleration term (i.e. a second-order differential of displacement with respect to time). A second-order transfer function can be separated theoretically into two first-order lags having the same time constant by factorising the denominator of the transfer function; e.g. from equation 7.52, for a system with unit steady-state gain:

$$\mathbf{G}(s) = \frac{1}{\tau^2 s^2 + 2\zeta\tau s + 1} = \frac{1}{\tau^2(s - \alpha_1)(s - \alpha_2)} \tag{7.53}$$

where the poles of $\mathbf{G}(s)$ are:

$$\alpha_1 = -\frac{\zeta}{\tau} + \frac{\sqrt{(\zeta^2 - 1)}}{\tau} \quad \text{and} \quad \alpha_2 = -\frac{\zeta}{\tau} - \frac{\sqrt{(\zeta^2 - 1)}}{\tau}$$

α_1 and α_2 are real and different, real and repeated, or are complex conjugates depending on the value of ζ. The same factors are obtained by equating the denominator of the transfer function (equation 7.52) to zero.

i.e.:

$$\tau^2 s^2 + 2\zeta\tau s + 1 = 0 \tag{7.54a}$$

or:

$$(s - \alpha_1)(s - \alpha_2) = 0 \tag{7.54b}$$

Equations 7.54*a* and 7.54*b* are two different forms of the *characteristic equation* of the system which is a relationship of considerable importance in determining the system stability (see Section 7.10).

Note that true second-order systems are quite rare in the chemical process industries, although second-order terms may arise when fitting transfer functions to experimental data obtained from complex processes (see also Section 7.8.6).

7.6. DISTANCE–VELOCITY LAG (DEAD TIME)

A typical example of this type of lag is experienced in the continuous sampling of a process stream (Fig. 7.22 and Section 6.9). It can be seen that a change in the composition x_m of the main stream will be detected by the analyser only after a time τ_{DV}, where τ_{DV} is the time required for the change to travel through the sample line. Hence, for plug flow:

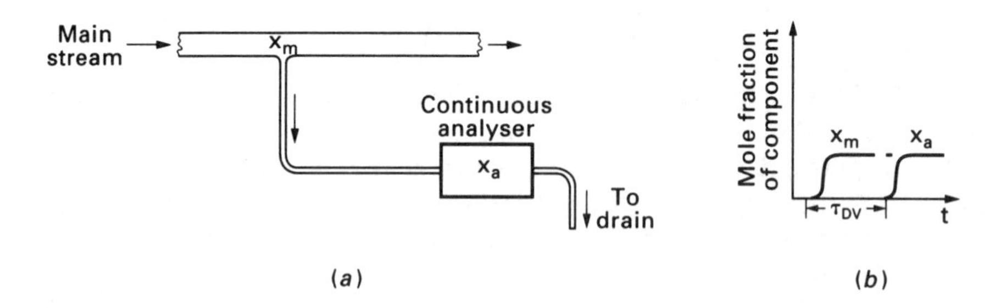

FIG. 7.22. Distance–velocity lag in sample line

$$\tau_{DV} = \frac{\text{Volume of sample line}}{\text{Volumetric flowrate of sample}}$$

and:

$$x_a|_{t=t} = x_m|_{t=t-\tau_{DV}} \tag{7.55}$$

where x_a is the response of the analyser, i.e. the composition of the main stream as recorded by the analyser. At steady state:

$$x_a|_{t=0} = x_m|_{t=-\tau_{DV}} \tag{7.56}$$

Subtracting equation 7.56 from equation 7.55 and introducing deviation variables:

$$\mathscr{x}_a|_{t=t} = \mathscr{x}_m|_{t=t-\tau_{DV}} \tag{7.57}$$

Transforming equation 7.57 gives:

$$\bar{\mathscr{x}}_a|_{t=t} = \bar{\mathscr{x}}_m|_{t=t-\tau_{DV}} \tag{7.58}$$

But, from the translation theorem*:

$$\bar{x}_m|_{t=t-\tau_{DV}} = \exp(-\tau_{DV}s)\bar{x}_m|_{t=t} \tag{7.59}$$

Hence, combining equations 7.58 and 7.59 gives the transfer function for a distance–velocity (DV) lag, i.e.:

$$\mathbf{G}(s) = \frac{\bar{x}_a|_{t=t}}{\bar{x}_m|_{t=t}} = \exp(-\tau_{DV}s) \tag{7.60}$$

x_m can be any type of function and then x_a will be the same function but occurring τ_{DV} units of time later.

This type of lag is encountered frequently in flow systems and may also be termed *dead time* or *transportation lag*. The presence of much DV lag in any control loop or configuration can lead to instability in the control action[13] (see Fig. 7.49).

7.7. TRANSFER FUNCTIONS OF FIXED PARAMETER CONTROLLERS

7.7.1. Ideal Controllers

From equation 7.2, for proportional action:

$$J - J_0 = K_C\varepsilon \tag{7.61}$$

$(J - J_0)$ represents the deviation of the controller output from its steady-state value and may therefore be replaced by a deviation variable $\mathcal{J}$. At time $t = 0$ we assume that there is no error, thus ε is itself a deviation variable. Hence equation 7.61 becomes:

$$\mathcal{J} = K_C\varepsilon$$

Transforming gives:

$$\mathbf{G}_C(s) = \bar{\mathcal{J}}/\bar{\varepsilon} = K_C \tag{7.62}$$

which is the transfer function of an ideal proportional controller.

By similar procedures we obtain from equations 7.3, 7.4 and 7.5, for **PI** control:

* The translation theorem may be established as follows:

$$\bar{x}_{t=t-\tau} = \int_0^\infty x_{t=t-\tau}\exp(-st)\,\mathrm{d}t$$

Putting $t - \tau = t_0$:

$$\bar{x}_{t=t-\tau} = \int_{-\tau}^\infty x_{t=t_0}\exp\{-s(t_0+\tau)\}\,\mathrm{d}t_0$$

But for $t_0 \leqslant 0$, $x_{t=t_0} = 0$, therefore:

$$\bar{x}_{t=t-\tau} = \int_0^\infty x_{t=t_0}\exp\{-s(t_0+\tau)\}\,\mathrm{d}t_0$$

$$= \exp(-s\tau)\int_0^\infty x_{t=t_0}\exp(-st_0)\,\mathrm{d}t_0$$

$$= \exp(-s\tau)\,\bar{x}_{t=t}$$

$$\mathbf{G}_C(s) = K_C\left(1 + \frac{1}{\tau_I s}\right) \tag{7.63}$$

For **PD** control:

$$\mathbf{G}_C(s) = K_C\left(1 + \tau_D s\right) \tag{7.64}$$

and for **PID** control:

$$\mathbf{G}_C(s) = K_C\left(1 + \frac{1}{\tau_I s} + \tau_D s\right) \tag{7.65}$$

7.7.2. Industrial Three Term Controllers

Normal commercial **PID** controllers are generally constructed by adding a lead compensator (Section 7.12.2) as the derivative mode to a **PI** controller. This type of derivative module typically has the transfer function:

$$\mathbf{G}_{C(\text{deriv})} = \frac{1 + \tau_{D1} s}{1 + \tau_{D2} s} \tag{7.66}$$

where τ_{D2} normally has a value which is in the range $\tau_{D1}/20$ to $\tau_{D1}/6$[14].

Such an element provides the *high frequency roll-off* that is necessary with derivative action (i.e. it avoids the tendency of the ideal derivative mode to amplify noise in the error signal). The inclusion of such an element leads to the transfer function of the relevant industrial controller as being:

$$\mathbf{G}_C = \frac{K_C(1 + \tau_I s)(1 + \tau_{D1} s)}{\tau_I s(1 + \tau_{D2} s)} \tag{7.67}$$

7.8. RESPONSE OF CONTROL LOOP COMPONENTS TO FORCING FUNCTIONS

7.8.1. Common Types of Forcing Function

Forcing function is a term given to any disturbance which is externally applied to a system. A number of simple functions are of considerable use in both the theoretical and experimental analysis of control systems and their components. Note that the response to a forcing function of a system or component without feedback is called the *open-loop* response. This should not be confused with the term *open-loop* control which is frequently used to describe feed-forward control. The response of a system incorporating feedback is referred to as the *closed-loop* response. Only three of the more useful forcing functions will be described here.

The Step Function

This is defined (Fig. 7.23) by:

$$\mathrm{f}(t) = \begin{Bmatrix} 0, & t < 0 \\ M, & t \geqslant 0 \end{Bmatrix} \tag{7.68}$$

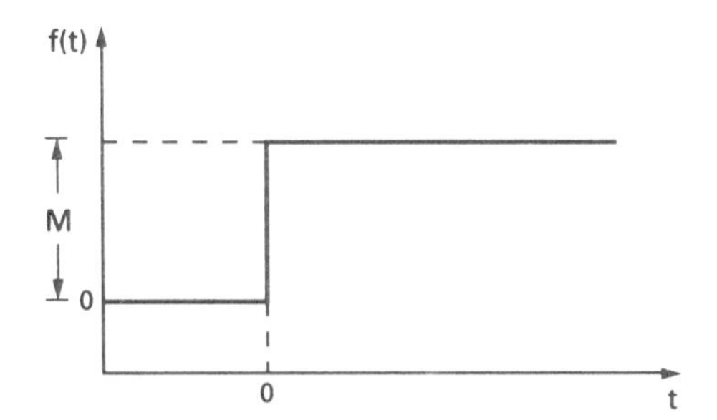

FIG. 7.23. Step function of magnitude M

or:

$$f(t) = M\mathrm{u}(t) \tag{7.69}$$

where $\mathrm{u}(t)$ represents a step function of unit magnitude. The transform of $f(t)$ is given by:

$$\bar{f} = \frac{M}{s} \tag{7.70}$$

The Sinusoidal Function

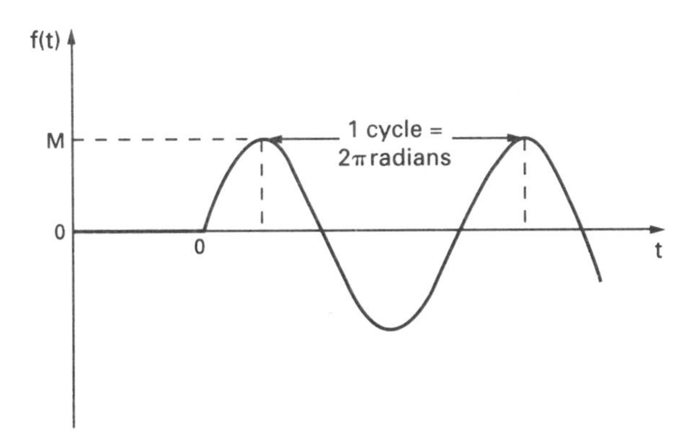

FIG. 7.24. Sinusoidal function of amplitude M

For this function (Fig. 7.24):

$$f(t) = \begin{Bmatrix} 0, & t < 0 \\ M \sin \omega t, & t \geqslant 0 \end{Bmatrix} \tag{7.71}$$

or:

$$f(t) = M\mathrm{u}(t) \sin \omega t \tag{7.72}$$

$\therefore$

$$\bar{f} = \frac{M\omega}{s^2 + \omega^2} \tag{7.73}$$

Here M is termed the *amplitude* of the signal and ω its angular frequency in radians/unit time. The frequency may also be expressed as:

$$f = \omega/2\pi \tag{7.74}$$

The Pulse Function

This is defined (Fig. 7.25) by:

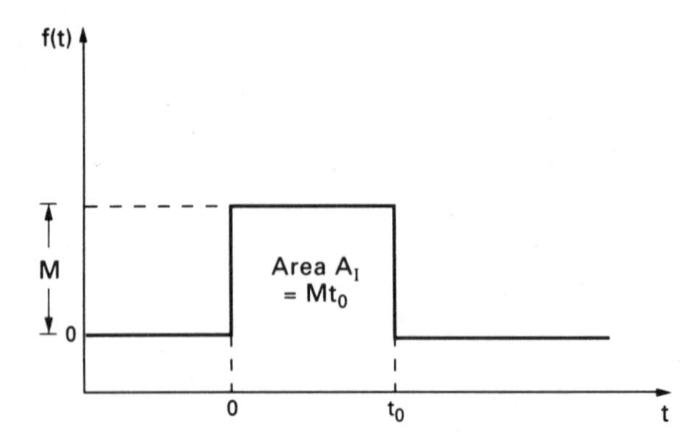

FIG. 7.25. Pulse function of magnitude M

$$f(t) = \begin{Bmatrix} 0, & t < 0 \\ M, & 0 \leqslant t \leqslant t_0 \\ 0, & t < t_0 \end{Bmatrix} \tag{7.75}$$

This may alternatively be represented by two successive step functions of the same magnitude but of opposite sign, i.e.:

$$f(t) = M[u(t) - u(t - t_0)] \tag{7.76}$$

Using the translation theorem:

$$\bar{f} = M\left[\frac{1}{s} - \frac{1}{s}e^{-st_0}\right]$$

$$= \frac{M}{s}[1 - e^{-st_0}] \tag{7.77}$$

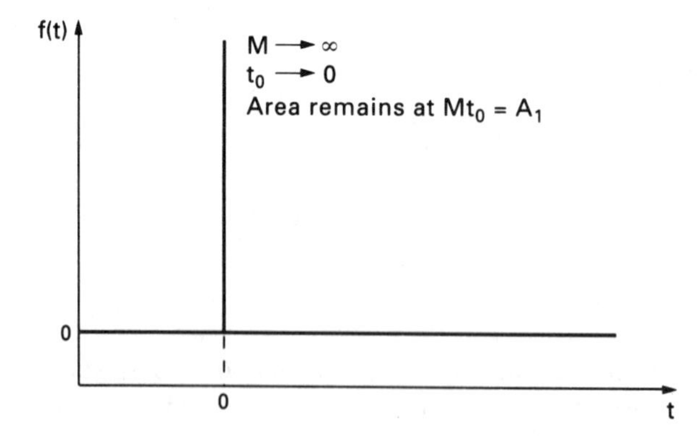

FIG. 7.26. Impulse function of area A_I

The *impulse* is a special case of the pulse function in which $t_0 \to 0$ but the area A_I under the impulse function remains constant and finite (Figure 7.26). Thus, for the impulse function:

$$\bar{f} = \lim_{t_0 \to 0} \left\{ \frac{M}{s}[1 - e^{-st_0}] \right\}$$

$$= \lim_{t_0 \to 0} \left\{ \frac{A_I}{t_0 s}[1 - e^{-st_0}] \right\} = \frac{0}{0}$$

By l'Hôpital's rule[15] the limit of an indeterminate quantity is the same as that of the function obtained by independently differentiating numerator and denominator with respect to the variable which is approaching the limit.

$$\bar{f} = \lim_{t_0 \to 0} \left\{ \frac{\dfrac{d}{dt_0}[A_I(1 - e^{-st_0})]}{\dfrac{d}{dt_0}[t_0 s]} \right\}$$

$$= \lim_{t_0 \to 0} \left\{ \frac{A_I s e^{-st_0}}{s} \right\} = A_I \tag{7.78}$$

When $A_I = 1$ we have the special case of the *unit* impulse, often termed the *Dirac function* $\delta(t)$.

7.8.2. Response to Step Function

First-Order System

From equation 7.19:

$$\mathbf{G}(s) = \frac{\bar{\vartheta}_1}{\bar{\vartheta}_0} = \frac{1}{1 + \tau s} \qquad \text{(equation 7.19)}$$

For a step input (equation 7.70):

$$\bar{\vartheta}_0 = \frac{M}{s}$$

$$\therefore \quad \bar{\vartheta}_1 = \frac{1}{1 + \tau s}\frac{M}{s} \tag{7.79}$$

$$= -\frac{M}{s + 1/\tau} + \frac{M}{s}$$

Inversion of equation 7.79 gives the response of the thermocouple to a step change in temperature of the surrounding fluid, i.e.:

$$\vartheta_1 = \left\{ \begin{matrix} M\left[1 - \exp\left(-\dfrac{t}{\tau}\right)\right], & t \geqslant 0 \\ 0, & t < 0 \end{matrix} \right\} \tag{7.80}$$

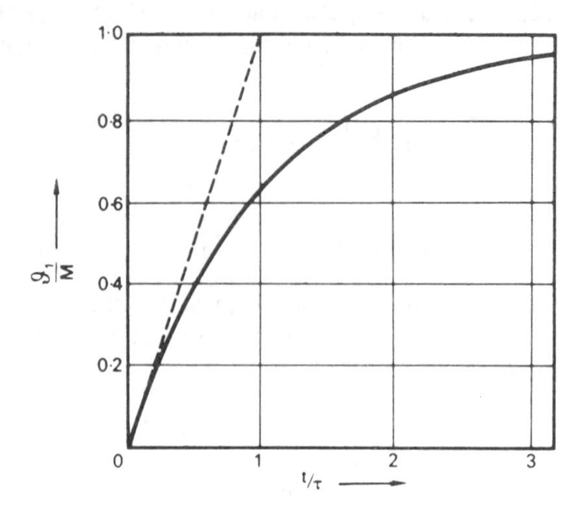

FIG. 7.27. Response of first-order system to step forcing function

Plotting ϑ_1/M vs t/τ illustrates the distinctive characteristics of the step response of a first-order system (Fig. 7.27). The response is immediate and the slope is a maximum at $t = 0$ and decreases with increasing time. If the response is measured experimentally, the time constant may be evaluated thus:

(a) Put $t/\tau = 1$ in equation 7.80. Then:

$$\vartheta_1/M = 1 - \exp(-1) = 0.632$$

Hence the response will have reached 63.2 per cent of its final value when $t = \tau$.

(b) Also from equation 7.69:

$$\frac{d}{dt}(\vartheta_1/M) = \frac{1}{\tau}\exp\left(-\frac{t}{\tau}\right)$$

Therefore at $t = 0$, the slope of the graph is $1/\tau$. Hence, if the initial rate of change were maintained, the response would be completed in a time equivalent to one time constant.

If the value of the time constant obtained by both methods is not the same within the limits of experimental error, the response is not truly first order.

Second-Order System

Consider a step change in pressure differential applied to a manometer (Fig. 7.21). From equation 7.52 the displacement of the column of liquid is given by:

$$\bar{z} = \frac{1/(2\rho g)}{\tau^2 s^2 + 2\zeta\tau s + 1}\,\frac{M}{s}$$

$$= \frac{MK_{MT}}{\tau^2 s(s - \beta_1)(s - \beta_2)} \tag{7.81}$$

where $\beta_{1,2} = -\frac{\zeta}{\tau} \pm \frac{\sqrt{(\zeta^2 - 1)}}{\tau}$ and $K_{MT} = 1/(2\rho g)$.

Three different cases of equation 7.81 must be considered for inversion (see Appendix 7.2):

(a) $\zeta < 1$. The roots of the characteristic equation (equation 7.54) are complex conjugates. Inversion gives[16]:

$$\mathfrak{z} = MK_{MT}\left\{1 - \frac{1}{\phi}\sin\left(\frac{\phi t}{\tau} + \tan^{-1}\frac{\phi}{\zeta}\right)\exp\left(-\frac{\zeta t}{\tau}\right)\right\} \quad (7.82)$$

where $\phi = \sqrt{(1 - \zeta^2)}$.

(b) $\zeta > 1$. The characteristic equation has real roots, and:

$$\mathfrak{z} = MK_{MT}\left\{1 - \left(\cosh\frac{\nu t}{\tau} + \frac{\zeta}{\nu}\sinh\frac{\nu t}{\tau}\right)\exp\left(-\frac{\zeta t}{\tau}\right)\right\} \quad (7.83)$$

where $\nu = \sqrt{(\zeta^2 - 1)} = i\phi$.

(c) $\zeta = 1$. The roots are repeated, and:

$$\mathfrak{z} = MK_{MT}\left\{1 - \left(1 + \frac{t}{\tau}\right)\exp\left(-\frac{t}{\tau}\right)\right\} \quad (7.84)$$

Note that the term MK_{MT} appears in each case. This is simply the product of the steady-state gain and the magnitude M of the forcing function.

The effect of the value of the damping coefficient ζ on the response is shown in Fig. 7.28. For $\zeta < 1$ the response is seen to be oscillatory or *underdamped*; when $\zeta > 1$ it is sluggish or *overdamped*; and when $\zeta = 1$ it is said to be *critically* damped, i.e. the final value is approached with the greatest speed without overshooting the final value. When $\zeta = 0$ there is no damping and the system output oscillates continuously with constant amplitude.

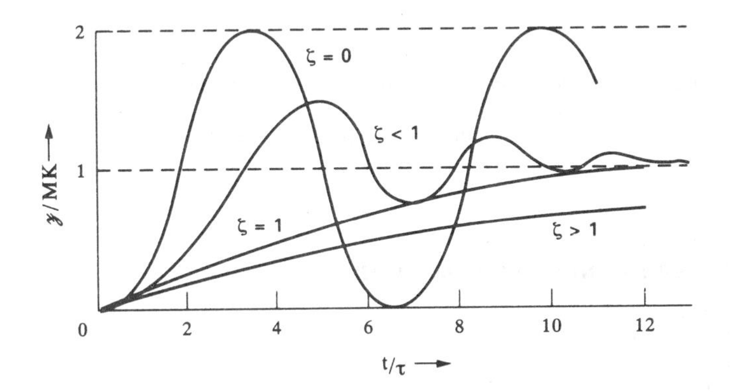

FIG. 7.28. Response of second-order system to step forcing function

Equation 7.83 may be expressed in the form:

$$\mathfrak{z} = MK_{MT}\{1 - A_3\exp(-B_1 t/\tau) - A_4\exp(-B_2 t/\tau)\}$$

A similar expression is obtained for the response of two first-order systems in series (see Section 7.5.3).

7.8.3. Initial and Final Value Theorems[17]

Frequently, it is required to determine the initial or final value of the system response to some forcing function. It is possible to evaluate this information without inverting the appropriate transform into the time domain.

The value of the response at $t = 0$ is given by the *initial value* theorem which states that:

$$\lim_{t \to 0} \{f(t)\} = \lim_{s \to \infty} \{s\bar{f}\} \tag{7.85}$$

where $\bar{f}$ is the transform of $f(t)$.

The *final value* theorem gives the value of the response at $t = \infty$:

$$\lim_{t \to \infty} \{f(t)\} = \lim_{s \to 0} \{s\bar{f}\} \tag{7.86}$$

provided that $s\bar{f}$ does not become infinite for any value of s satisfying $\mathcal{R}e(s) \geqslant 0$, where $\mathcal{R}e$ denotes the real part of a function. If this condition does not hold, $f(t)$ does not approach a limit as $t \to \infty$. This condition does not apply to equation 7.85.

The latter theorem enables the system steady-state gain to be determined as the relation between the input and output as $t \to \infty$ after a step change of unit magnitude has been applied to the system.

Example 7.2

Determine the steady-state gain of the U-tube manometer described in Fig. 7.21.

Solution

From equation 7.81, the transformed form of the response of the level of the column of liquid in the manometer tube to a unit step change in pressure differential is:

$$\bar{z} = \frac{1/(2\rho g)}{\tau^2 s^2 + 2\zeta\tau s + 1} \frac{M}{s}$$

Using the final value theorem (equation 7.86) ($s\bar{f}$ does not become infinite for $Re(s) > 0$):

$$\text{Steady-state gain} = \lim_{t \to \infty} \{f(t)\} = \lim_{s \to 0} \{s\bar{f}\}$$

$$= \underline{\underline{1/(2\rho g)}}$$

7.8.4. Response to Sinusoidal Function

First-Order System

From equations 7.19 and 7.73:

$$\bar{\vartheta}_1 = \frac{1}{(1 + \tau s)} \frac{M\omega}{(s^2 + \omega^2)}$$

Expanding by means of partial fractions gives:

$$\bar{\vartheta}_1 = \frac{M\omega\tau^2}{(1 + \omega^2\tau^2)} \frac{1}{(1 + \tau s)} - \frac{M\omega\tau}{(1 + \omega^2\tau^2)} \frac{s}{(s^2 + \omega^2)} + \frac{M\omega}{(1 + \omega^2\tau^2)} \frac{1}{(s^2 + \omega^2)}$$

The inverse transform follows:

$$\vartheta_1 = \frac{M}{1+\omega^2\tau^2}\{\omega\tau\exp(-t/\tau) - \omega\tau\cos\omega t + \sin\omega t\}$$

A more useful form is obtained by using the identity:

$$\mathbf{x}\sin\alpha + \mathbf{y}\cos\alpha = \mathbf{z}\sin(\alpha+\psi) \tag{7.87}$$

where $\mathbf{z}^2 = \mathbf{x}^2 + \mathbf{y}^2$ and $\tan\psi = \mathbf{y}/\mathbf{x}$.

Hence:

$$\vartheta_1 = \frac{M\omega\tau}{1+\omega^2\tau^2}\exp(-t/\tau) + \frac{M}{\sqrt{(1+\omega^2\tau^2)}}\sin(\omega t+\psi) \tag{7.88}$$

where $\psi = \tan^{-1}(-\omega\tau)$.

When $t \to \infty$ the *ultimate* periodic response is obtained, i.e.:

$$\lim_{t\to\infty}\vartheta_1 = \frac{M}{\sqrt{(1+\omega^2\tau^2)}}\sin(\omega t+\psi) \tag{7.89}$$

This type of function is said to be *stationary*, i.e. its value varies with time, but in a regularly repeating pattern. Stationary conditions should not be confused with the steady state.

Comparison of equations 7.71 and 7.89 shows that when $t \to \infty$:

(a) both input and output are sinusoidal functions of frequency ω.
(b) the amplitude of the output is always less than that of the input (i.e. the signal is *attenuated*), and
(c) the output differs in phase from the input by an angle ψ.

If $\psi < 0$ the output is said to *lag* the input, and if $\psi > 0$ it is said to *lead* the input. The relationship between input and output for a sinusoidal forcing function as $t \to \infty$ constitutes an important tool in the analysis and design of control systems termed *frequency response analysis*. Of particular importance are the *amplitude ratio* (AR) and the *phase shift* ψ. The AR represents the relationship between the output and the input amplitudes as $t \to \infty$, i.e.:

$$\text{AR} = \left(\frac{\text{amplitude of output signal}}{\text{amplitude of input signal}}\right)_{t\to\infty} \tag{7.90}$$

The phase shift is the difference in angle measured in degrees* between input and output as $t \to \infty$ (Fig. 7.29).

For the above first-order system, from equations 7.71 and 7.89:

$$\text{AR} = \frac{1}{\sqrt{(1+\omega^2\tau^2)}} \tag{7.91}$$

and:

$$\psi = \tan^{-1}(-\omega\tau) \tag{7.92}$$

Both AR and ψ are functions of ω and for a first-order system. AR < 1 and $\psi < 0$ for $\omega > 0$. Thus the output is attenuated and lags the input. As $\omega \to \infty$,

* It is the convention in chemical engineering process control to measure phase shift in degrees. This convention is followed in this text.

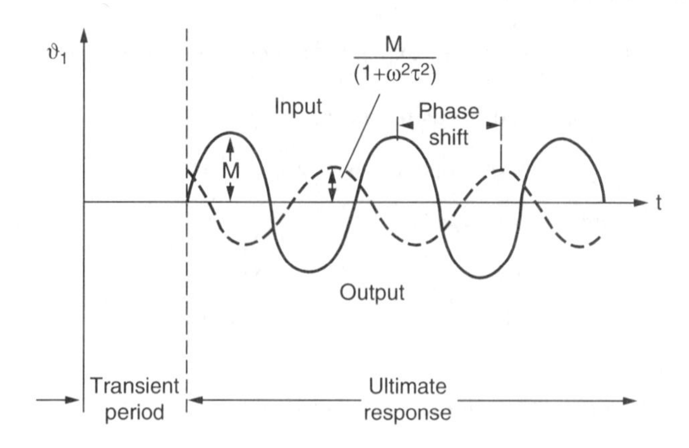

FIG. 7.29. Ultimate response of first-order system to sinusoidal forcing function

AR $\rightarrow$ 0 and $\psi \rightarrow -90^\circ$, thus the maximum lag which can be attained with a first-order system is 90°.

The Substitution Rule

The frequency response of a system may be more easily determined by effecting the substitution $s = \mathrm{i}\omega$ in the appropriate transfer function [where $\mathrm{i} = \sqrt{(-1)}$] (i.e. mapping the function from the s domain onto the frequency domain[17]). Hence substituting into equation 7.19:

$$\mathbf{G}(\mathrm{i}\omega) = \frac{1}{1 + \tau \mathrm{i}\omega}$$

$$= \frac{1}{1 + \tau^2\omega^2} + \mathrm{i}\frac{(-\tau\omega)}{1 + \tau^2\omega^2} \tag{7.93}$$

The AR and phase shift are obtained as the magnitude and argument of equation 7.93. Hence:

$$\mathrm{AR} = \frac{1}{\sqrt{(1 + \omega^2\tau^2)}} \qquad \text{(equation 7.91)}$$

and:

$$\psi = \tan^{-1}(-\omega\tau) \qquad \text{(equation 7.92)}$$

It can be shown[18] that this method may be applied to any system described by a linear differential equation or to a distance–velocity lag in order to obtain the relevant frequency response characteristics.

Second-Order Systems

The frequency response is again of principal interest. Substituting $s = \mathrm{i}\omega$ in equation 7.52:

$$G(i\omega) = \frac{K_{MT}}{-\tau^2\omega^2 + 2\zeta\tau i\omega + 1}$$

Hence: $$AR = \text{magnitude of } G(i\omega)$$

$$= \frac{K_{MT}}{\sqrt{[1 - \omega^2\tau^2)^2 + (2\zeta\omega\tau)^2]}} \tag{7.94a}$$

For a second-order system with a unit steady-state gain, from equation 7.53:

$$AR = \frac{1}{\sqrt{[(1 - \omega^2\tau^2)^2 + (2\zeta\omega\tau)^2]}} \tag{7.94b}$$

$$\psi = \text{argument of } G(i\omega)$$

$$= \tan^{-1}\left(\frac{-2\zeta\omega\tau}{1 - \omega^2\tau^2}\right) \tag{7.95}$$

In this case, for certain values of $\omega\tau$ with $\zeta < 1/\sqrt{2}$, the amplitude ratio is greater than unity (equation 7.94*b* and Fig. 7.46). Also $\psi \to -90°$ as $\omega\tau \to 1$ (equation 7.95). As $\omega\tau$ increases above unity tan ψ approaches zero but is always positive. Thus, as $\omega\tau \to \infty$, $\psi \to -180°$. The maximum phase lag is therefore 180°, whereas it is 90° for a first-order system. For an *n*th-order system the maximum lag obtainable is 90*n* degrees ($n\pi/2$ radians).

Distance–Velocity Lag

In order to determine the frequency response characteristics, substitute $s = i\omega$ in equation 7.60:

$$G(i\omega) = \exp(-i\omega\tau)$$

$$= \cos\omega\tau - i\sin\omega\tau$$

∴ $$AR = \sqrt{(cos^2\omega\tau + \sin^2\omega\tau)} = 1 \tag{7.96}$$

and: $$\psi = \tan^{-1}\left(-\frac{\sin\omega\tau}{\cos\omega\tau}\right) = -\omega\tau \tag{7.97}$$

7.8.5. Response to Pulse Function

Only the special case of the impulse will be considered (Section 7.8.1). This is a particularly useful function for testing system dynamics as it does not introduce any further s terms into the analysis (equation 7.78). The determination of the response of any system in the time domain to an impulse forcing function is facilitated by noting that:

$$\bar{f}_{impulse} = s\bar{f}_{step} \tag{7.98}$$

i.e. the transform of the impulse response is simply the transform of the step response multiplied by the operator s.

Inverting equation 7.98 gives:

$$\{f(t)\}_{impulse} = \frac{d}{dt}\{f(t)\}_{step} \tag{7.99}$$

Thus, if the step response in the time domain is known, the impulse response can be determined by differentiating the former.

First-Order System

For unit step change (putting $M = 1$ in equation 7.80):

$$\{\vartheta_1\}_{\text{step}} = 1 - \exp(-t/\tau) \qquad \text{(equation 7.80)}$$

Therefore, using equation 7.99, the response to a unit impulse is:

$$\{\vartheta_1\}_{\text{impulse}} = \frac{\mathrm{d}}{\mathrm{d}t}(1 - \exp(-t/\tau)) \qquad (7.100)$$

$$= 1/\tau \exp(-t/\tau)$$

Figure 7.30 shows that the theoretical response to an impulse immediately rises to its maximum value and then decays exponentially. The broken line indicates the probable response obtained in practice to an experimentally applied impulse having a finite timescale.

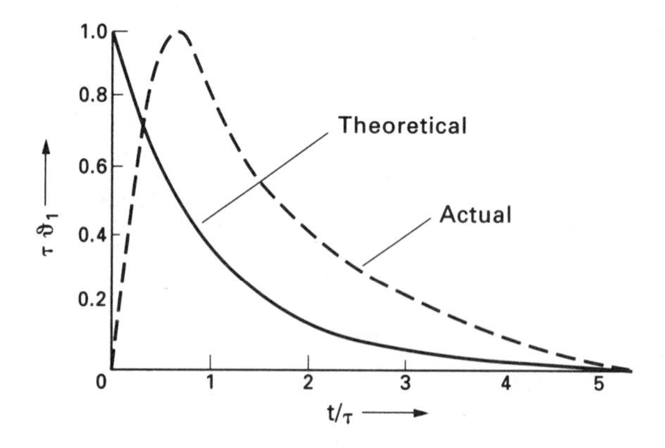

FIG. 7.30. Response of first-order system to unit impulse forcing function

Second-Order System

There are three cases to consider dependent on the form of the roots of the characteristic equation (see Fig. 7.31).

(a) $\zeta < 1$. From equation 7.82 with $M = A_I$:

$$\{\mathfrak{z}\}_{\text{impulse}} = \frac{\mathrm{d}}{\mathrm{d}t}\{\mathfrak{z}\}_{\text{step}} = A_I K_{MT}\left\{\frac{1}{\tau\phi}\exp\left(-\frac{\zeta\tau}{\tau}\right)\sin\frac{\theta t}{\tau}\right\} \qquad (7.101)$$

(b) $\zeta > 1$. From equation 7.83:

$$\{\mathfrak{z}\}_{\text{impulse}} = A_I K_{MT}\left\{\frac{1}{\tau\nu}\exp\left(-\frac{\zeta t}{\tau}\right)\sinh\frac{\nu t}{\tau}\right\} \qquad (7.102)$$

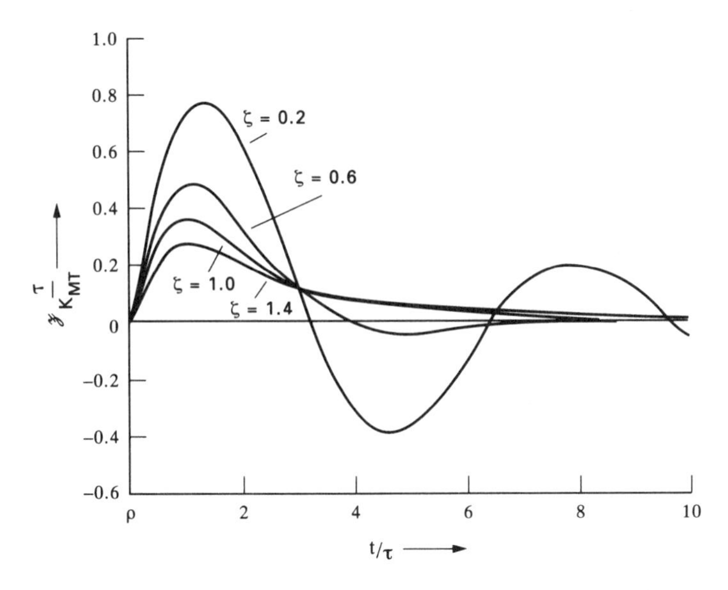

FIG. 7.31. Response of second-order system to unit impulse forcing function

(c) $\zeta = 1$. From equation 7.84:

$$\{\mathfrak{z}\}_{\text{impulse}} = A_I K_{MT} \left\{ \frac{1}{\tau^2} \exp\left(-\frac{t}{\tau} \right) \right\} \tag{7.103}$$

Again put $A_I = 1$ for a unit impulse.

7.8.6. Response of More Complex Systems to Forcing Functions

Transfer functions involving polynomials of higher degree than two and decaying exponentials (distance–velocity lags) may be dealt with in the same manner as above, i.e. by the use of partial fractions and inverse transforms if the step response or the transient part of the sinusoidal response is required, or by the substitution method if the frequency response is desired. For example, a typical fourth-order transfer function:

$$\mathbf{G}(s) = \frac{8}{18s^4 + 36s^3 + 45s^2 + 10s + 1}$$

may be written as:

$$\mathbf{G}(s) = 8\left\{ \frac{1}{(1 + 3s)(1 + 6s)(1 + s + s^2)} \right\}$$
$$= K\{\mathbf{G}_a(s)\,\mathbf{G}_b(s)\,\mathbf{G}_c(s)\}$$

where $K = 8$ is the system steady-state gain.

If is desired to determine the frequency response of this system and if the amplitude ratio and phase shift due to $\mathbf{G}_a(s)$ are $(\text{AR})_a$ and ψ_a respectively (and similarly), then the amplitude ratio and the phase shift of $\mathbf{G}(s)$ are given by:

$$\text{AR} = K(\text{AR})_a(\text{AR})_b(\text{AR})_c \tag{7.104}$$

and:
$$\psi = \psi_a + \psi_b + \psi_c \tag{7.105}$$

(These relationships may be extended to include any number of terms on the right hand side).

Hence, from equations 7.91, 7.92, 7.94 and 7.95:

$$\mathrm{AR} = 8\frac{1}{\sqrt{(1-9\omega^2)}}\frac{1}{\sqrt{(1+36\omega^2)}}\frac{1}{\sqrt{\{(1-\omega^2)^2+\omega^2\}}}$$

$$\psi = \tan^{-1}(-3\omega) + \tan^{-1}(-6\omega) + \tan^{-1}\left\{\frac{-\omega}{1+\omega^2}\right\}$$

(see also Example 7.3).

Note:

Equations 7.104 and 7.105 may be established as follows:

Consider $\mathbf{G}(s) = K\mathbf{G}_1(s)\mathbf{G}_2(s)$. Putting $s = \mathrm{i}\omega$ we obtain a relationship in terms of the complex variable of the form:

$$\frac{\beta + \mathrm{i}\gamma}{K} = (\beta_1 + \mathrm{i}\gamma_1)(\beta_2 + \mathrm{i}\gamma_2)$$
$$= \beta_1\beta_2 - \gamma_1\gamma_2 + \mathrm{i}(\gamma_1\beta_2 + \gamma_2\beta_1)$$

Equating real and imaginary parts:

$$\frac{\beta}{K} = \beta_1\beta_2 - \gamma_1\gamma_2 \quad \text{and} \quad \frac{\gamma}{K} = \gamma_1\beta_2 + \gamma_2\beta_1$$

$$\mathrm{AR} = \text{magnitude of } (\beta + \mathrm{i}\gamma) = \sqrt{(\beta^2 + \gamma^2)}$$
$$= K\sqrt{(\beta_1^2\beta_2^2 + \gamma_2^2\gamma_1^2 + \gamma_1^2\beta_2^2 + \gamma_2^2\beta_1^2}$$
$$= K\sqrt{(\beta_1^2 + \gamma_1^2)}\sqrt{(\beta_2^2 + \gamma_2^2)} = K(\mathrm{AR})_1(\mathrm{AR})_2$$

Also:
$$\psi = \tan^{-1}\frac{\gamma}{\beta} = \tan^{-1}\left\{\frac{K(\gamma_1\beta_2 + \gamma_2\beta_1)}{K(\beta_1\beta_2 - \gamma_1\gamma_2)}\right\}$$

and:
$$\psi_1 + \psi_2 = \tan^{-1}\frac{\gamma_1}{\beta_1} + \tan^{-1}\frac{\gamma_2}{\beta_2}$$

$\therefore$
$$\tan(\psi_1 + \psi_2) = \frac{\gamma_1/\beta_1 + \gamma_2/\beta_2}{1 - (\gamma_1/\beta_1)(\gamma_2/\beta_2)} = \frac{\gamma_1\beta_2 + \gamma_2\beta_1}{\beta_1\beta_2 + \gamma_1\gamma_2}$$

$\therefore$
$$\tan\psi = \tan(\psi_1 + \psi_2)$$

Hence:
$$\psi = \psi_1 + \psi_2$$

This may be extended to any number of functions.

Example 7.3

In the heat exchanger arrangement illustrated in Fig. 7.32 the following are known:

(a) The response of the temperature θ_A of stream 2 leaving exchanger A to a change in the rate of flow F of stream 1 entering is first-order with a time constant of 0.67 min and a steady-state gain of 20.

(b) The response of θ_B to a change in θ_A is underdamped second-order with a time constant of 3.2 min and a damping coefficient of 0.48. The steady-state gain is unity.

Determine (i) the response of θ_B to a step change of unit magnitude in F, and (ii) the frequency response relationships between θ_B and F. Assume the temperatures of all the inlet streams and the rate of flow of stream 2 through A and that of stream 3 through B to remain constant.

Solution

The system block diagram is shown in Fig. 7.33.

(a)
$$\mathbf{G}_A(s) = \frac{\bar{\vartheta}_A}{\bar{\mathscr{F}}} = \frac{20}{1 + 0.67s} \quad \text{(equation 7.26)}$$

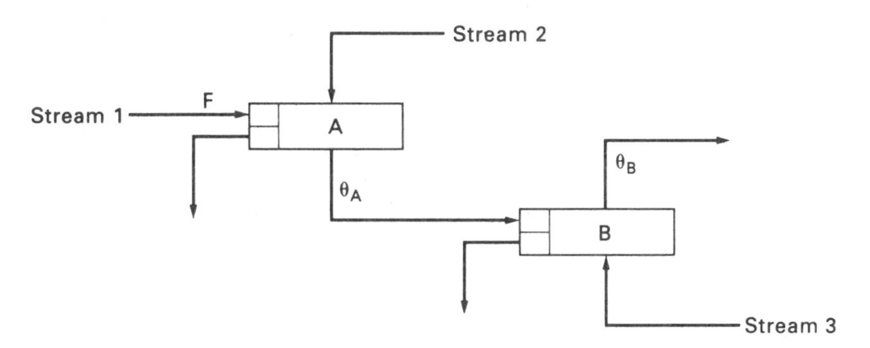

FIG. 7.32. Heat exchanger arrangement

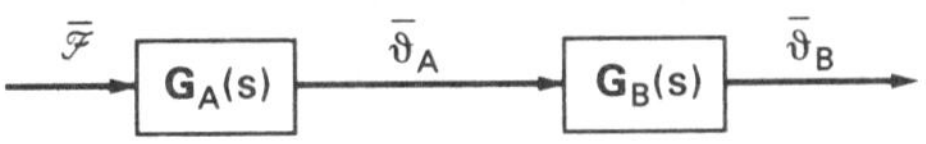

$$\bar{\mathscr{F}} \rightarrow \boxed{\mathbf{G}_A(s)} \xrightarrow{\bar{\vartheta}_A} \boxed{\mathbf{G}_B(s)} \xrightarrow{\bar{\vartheta}_B}$$

FIG. 7.33. Block diagram of heat exchanger system

(b)

$$\mathbf{G}_B(s) = \frac{\bar{\vartheta}_B}{\bar{\vartheta}_A} = \frac{1}{10.2s^2 + 3.1s + 1} \qquad \text{(equation 7.53)}$$

$\therefore$

$$\mathbf{G}_2(s) = \frac{\bar{\vartheta}_B}{\bar{\mathscr{F}}} = \mathbf{G}_A(s)\mathbf{G}_B(s) \qquad \text{(equation 7.12)}$$

$$= \frac{20}{(1 + 0.67s)(10.2s^2 + 3.1s + 1)}$$

(i) For step change of unit magnitude in F:

$$\bar{\vartheta}_B = \frac{20}{s(1 + 0.67s)(10.2s^2 + 3.1s + 1)}$$

$$= \frac{20}{s} - \frac{0.69}{1 + 0.67s} - \frac{194s + 74.7}{10.2s^2 + 3.1s + 1}$$

$$= \frac{20}{s} - \frac{1.03}{s + 1.5} - 19\left\{\frac{s + 0.152}{(s + 0.152)^2 + 0.075} + \frac{0.233}{(s + 0.152)^2 + 0.075}\right\}$$

Hence, from Table of Laplace transforms (Appendix 7.1):

$$\therefore \qquad \vartheta_B = 20 - 1.03\exp(-1.5t) - 19\exp(-0.152t)[\cos(0.274t) + 0.85\sin(0.274t)]$$

(ii) The frequency response may be determined by applying the substitution rule to $\mathbf{G}_2(s)$. It is easier, however, to make use of equations 7.104 and 7.105. If $(AR)_A$ and $(AR)_B$ are the amplitude ratios of $\mathbf{G}_A(s)$ and $\mathbf{G}_B(s)$ respectively, then:

$$(\mathrm{AR})_A = \frac{20}{\sqrt{[1 + (0.67\omega^2)]}} = \frac{20}{\sqrt{(0.45\omega^2 + 1)}} \qquad \text{(equation 7.91)}$$

$$(\mathrm{AR})_B = \frac{1}{\sqrt{\{[1 - (3.2\omega)^2]^2 + [3.1\omega]^2\}}} = \frac{1}{\sqrt{(104\omega^4 - 10.8\omega^2 + 1)}} \qquad \text{(equation 7.94)}$$

Thus, from equation 7.104:

$$\mathrm{AR} = \frac{20}{\sqrt{[(0.45\omega^2 + 1)(104\omega^2 - 10.8\omega^2 + 1)]}}$$

Similarly, using equations 7.92, 7.95 and 7.105:

$$\psi = -\tan^{-1}(0.67\omega) - \tan^{-1}\left\{\frac{3.1\omega}{1 - 10.2\omega^2}\right\}$$

7.9. TRANSFER FUNCTIONS OF FEEDBACK CONTROL SYSTEMS

Once each element in any feedback control loop has been described in terms of its transfer function, the behaviour of the closed-loop can be determined by the formulation of appropriate *closed-loop transfer functions*. Two such are of importance, i.e. those relating the controlled variable C to the set point R and to the load U, respectively.

7.9.1. Closed-Loop Transfer Function Between *C* and *R*

This is obtained by assuming that no changes occur in U, i.e. if $\mathscr{U}$ represents a deviation variable, then $\mathscr{U} = 0$. Hence $\mathscr{U}$ has no effect on the control loop and can be omitted. Thus Fig. 7.3 is replaced by Fig. 7.34.

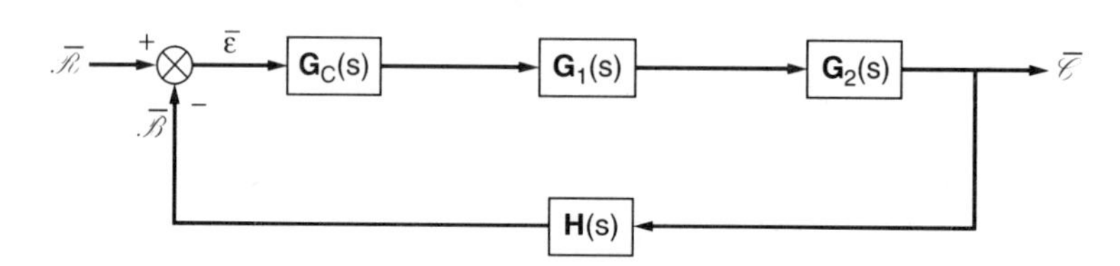

FIG. 7.34. Block diagram for changes in set point only (viz. set point following case)

From Fig. 7.34 and equation 7.12:

$$\bar{\varepsilon} = \bar{\mathscr{R}} - \bar{\mathscr{B}} \tag{7.106}$$

$$\bar{\mathscr{C}} = \mathbf{G}_C(s)\mathbf{G}_1(s)\mathbf{G}_2(s)\bar{\varepsilon} \tag{7.107}$$

$$\bar{\mathscr{B}} = \mathbf{H}(s)\bar{\mathscr{C}} \tag{7.108}$$

The closed-loop transfer function is obtained by eliminating $\bar{\varepsilon}$ and $\bar{\mathscr{B}}$ from equations 7.106, 7.107 and 7.108, i.e.:

$$\frac{\bar{\mathscr{C}}}{\bar{\mathscr{R}}} = \frac{\mathbf{G}_X(s)}{1 + \mathbf{G}_X(s)\mathbf{H}(s)} \tag{7.109}$$

where $\mathbf{G}_X(s) = \mathbf{G}_C(s)\mathbf{G}_1(s)\mathbf{G}_2(s)$.

Thus Fig. 7.34 may be replaced by an equivalent single block, as shown in Fig. 7.35.

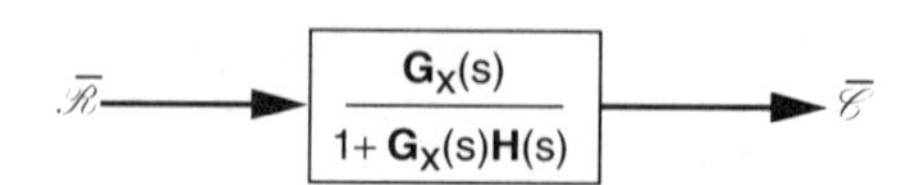

FIG. 7.35. Equivalent single block

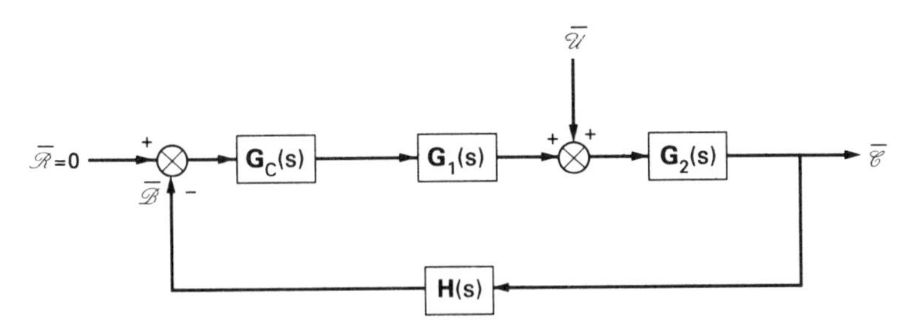

FIG. 7.36. Block diagram for changes in load only (viz. load disturbance rejection case)

7.9.2. Closed-Loop Transfer Function Between *C* and *U*

In this case $\mathcal{R} = 0$ (and $\bar{\mathcal{R}} = 0$), i.e. the set point is held constant.
From Fig. 7.36:

$$\bar{\varepsilon} = -\bar{\mathcal{B}}$$

$$\bar{\mathcal{C}} = \mathbf{G}_2(s)[\bar{\mathcal{U}} + \mathbf{G}_C(s)\mathbf{G}_1(s)\bar{\varepsilon}]$$

$$\bar{\mathcal{B}} = \mathbf{H}(s)\bar{\mathcal{C}}$$

$$\frac{\bar{\mathcal{C}}}{\bar{\mathcal{U}}} = \frac{\mathbf{G}_2(s)}{1 + \mathbf{G}_X(s)\mathbf{H}(s)} \quad (7.110)$$

where $\mathbf{G}_X(s) = \mathbf{G}_C(s)\mathbf{G}_1(s)\mathbf{G}_2(s)$.

In general, for single loops of the type illustrated in Figs 7.3, 7.34 and 7.36, the closed-loop transfer function may be determined from:

$$\frac{\bar{\mathcal{X}}_o}{\bar{\mathcal{X}}_i} = \frac{\mathbf{G}_{Xa}(s)}{1 + \mathbf{G}_{Xb}(s)} \quad (7.111)$$

where $\mathbf{G}_{Xa}(s)$ is the product of the transfer functions of all blocks between $\bar{\mathcal{X}}_i$ (input) and $\bar{\mathcal{X}}_o$ (output), and $\mathbf{G}_{Xb}(s)$ is the product of all blocks in the whole loop—often termed the *open-loop transfer function* of the control system. It is possible to apply the same rule successively to simplify certain multiple loop control schemes (e.g. cascade control—Section 7.13).

The effect of simultaneous changes in load and set point can be determined by the principle of superposition, i.e. by summing the separate variations due to each type of disturbance alone.

7.9.3. Calculation of Offset from the Closed-Loop Transfer Function

Load Change with Proportional Control

Referring to Fig. 7.36, for a proportional controller $\mathbf{G}_C(s) = K_C$ (equation 7.62). Assume for simplicity that the time constants of $\mathbf{G}_1(s)$ (the final control element) and $\mathbf{H}(s)$ (the measuring element) are negligible compared with that of $\mathbf{G}_2(s)$ (the process), i.e. that $\mathbf{G}_1(s)$ and $\mathbf{H}(s)$ are constants. Suppose also that $\mathbf{G}_2(s)$ is first order, then $\mathbf{G}_2(s) = 1/(1 + \tau s)$. Thus, from equation 7.110—for variations in load only:

$$\frac{\bar{\mathscr{C}}}{\bar{\mathscr{U}}} = \frac{1/(1 + \tau s)}{1 + K_C\mathbf{G}_1(s)\mathbf{H}(s)/(1 + \tau s)} = \frac{1}{1 + \tau s + K_C\mathbf{G}_1(s)\mathbf{H}(s)}$$

And for a unit step change in load:

$$\bar{\mathscr{C}} = \frac{1}{s}\left(\frac{1}{1 + \tau s + K_C\mathbf{G}_1(s)\mathbf{H}(s)}\right) \tag{7.112}$$

Offset is defined as the difference between the desired and actual response of the output as $t \to \infty$ (Section 7.2.3), i.e. in this case:

$$\text{Offset} = \lim_{t\to\infty} \mathscr{R}(t) - \lim_{t\to\infty} \mathscr{C}(t) = \mathscr{R}(\infty) - \mathscr{C}(\infty) \tag{7.113}$$

the desired response being the same as the change in set point. But, for a load change, the requirement of the control loop is to keep the output steady, i.e. the set point is fixed and $\mathscr{R}(\infty) = 0$. $\mathscr{C}(\infty)$ is given by the final value theorem (Section 7.8.3), i.e. from equations 7.112 and 7.86:

$$\lim_{t\to\infty} \mathscr{C}(t) = \lim_{s\to 0}\left\{s\frac{1}{s}\left(\frac{1}{1 + \tau s + K_C\mathbf{G}_1(s)\mathbf{H}(s)}\right)\right\}$$

$$= \frac{1}{1 + K_C\mathbf{G}_1(s)\mathbf{H}(s)}$$

Substituting in equation 7.113:

$$\text{Offset} = -\frac{1}{1 + K_C\mathbf{G}_1(s)\mathbf{H}(s)} \tag{7.114}$$

Thus the offset is reduced as the controller gain K_C is increased.

*Load Change with **PI** Control*

From equation 7.63:

$$\mathbf{G}_C(s) = K_C\left(1 + \frac{1}{\tau_I s}\right) \qquad \text{(equation 7.63)}$$

Keeping the remaining transfer functions as before:

$$\frac{\bar{\mathscr{C}}}{\bar{\mathscr{U}}} = \frac{1/(1 + \tau s)}{1 + K_C\left(1 + \dfrac{1}{\tau_I s}\right)\mathbf{G}_1(s)\left(\dfrac{1}{1 + \tau s}\right)\mathbf{H}(s)}$$

$$= \frac{\tau_I s}{\tau_I s(1 + \tau s) + K_C\mathbf{G}_1(s)\mathbf{H}(s)(\tau_I s + 1)}$$

Hence, for a unit step change in load, from equations 7.86 and 7.113:

$$\text{Offset} = \mathscr{R}(\infty) - \mathscr{C}(\infty) = 0 \tag{7.115}$$

Thus the presence of the integral action has removed the offset indicated by equation 7.114. Similar procedures may be employed with other control actions and also for the set point following case.

The response of the controlled variable to different types of perturbation (forcing function) in set point or load can be determined by inverting the appropriate transform (e.g. equation 7.112). This is possible only for simple loops containing low order systems. More complex control systems involving higher order elements require a suitable numerical analysis in order to obtain the time domain response.

7.9.4. The Equivalent Unity Feedback System

The analysis of feedback control systems can often be facilitated by conversion to an equivalent *unity feedback* system, i.e. a feedback loop in which the feedback path is represented by a steady-state gain of unity. There are two principal cases to consider.

Conversion to Unity Feedback when the Transfer Function in the Feedback Part of the Loop is Represented by a Steady-State Gain K

Consider a simple feedback loop (Fig. 7.3*a*) in which the feedback path consists of elements which approximate to a steady-state gain K (Fig. 7.37). In this instance, the equivalent unity feedback loop is determined by placing $1/K$ in the set point input to the main loop and compensating for this by adding an additional factor K in the forward part of the loop prior to the entry of the load disturbance, as in Fig. 7.38. It is easy to confirm that the standard closed loop transfer functions $\bar{\mathscr{C}}/\bar{\mathscr{R}}$ and $\bar{\mathscr{U}}/\bar{\mathscr{R}}$ are the same for the block diagrams in Figs 7.37 and 7.38.

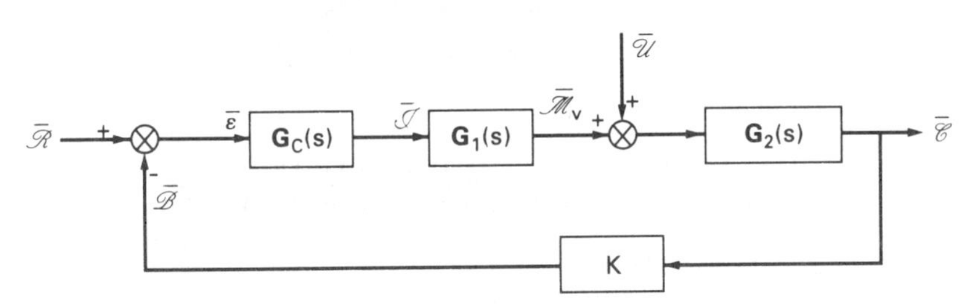

FIG. 7.37. Simple feedback control loop with feedback path represented by a steady-state gain K

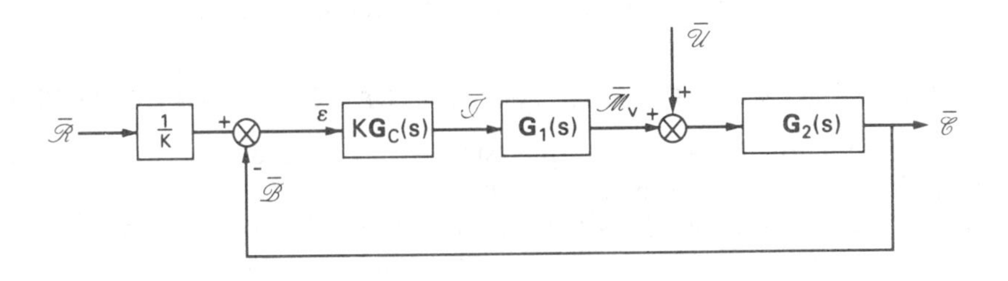

FIG. 7.38. Unity feedback system equivalent to the control loop shown in Fig. 7.37 when the dynamics of the feedback path are not significant

Conversion to Unity Feedback when the Dynamics of the Feedback Part of the Loop are Significant

In this case **H** in Fig. 7.3*a* cannot be treated as a constant and we write:

$$\mathbf{H}(s) = K[1 + \mathbf{H}_1(s)] \tag{7.116}$$

K is inserted in the forward part of the loop as before and the factor $[1 + \mathbf{H}_1(s)]$ is incorporated within the loop in the form of two feedback paths as shown in Fig. 7.39.

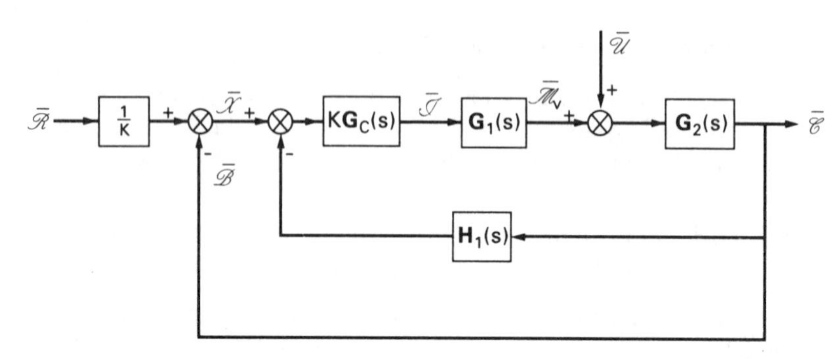

FIG. 7.39. Unity feedback system equivalent to the control loop shown in Fig. 7.37 when the dynamics of the feedback path are significant

From Fig. 7.39 and equation 7.111, the transfer function of the inner loop is:

$$\mathbf{G}_{il}(s) = \frac{\bar{\mathscr{C}}}{\bar{\mathscr{X}}} = \frac{K\mathbf{G}_C(s)\mathbf{G}_1(s)\mathbf{G}_2(s)}{1 + K\mathbf{G}_C(s)\mathbf{G}_1(s)\mathbf{G}_2(s)\mathbf{H}_1(s)}$$

Hence, for the set point following case:

$$\frac{\bar{\mathscr{C}}}{\bar{\mathscr{R}}} = \frac{1}{K}\frac{\mathbf{G}_{il}(s)}{1 + \mathbf{G}_{il}(s)} = \frac{\mathbf{G}_C(s)\mathbf{G}_1(s)\mathbf{G}_2(s)}{1 + K\mathbf{G}_C(s)\mathbf{G}_1(s)\mathbf{G}_2(s)(1 + \mathbf{H}_1(s))}$$

Thus, substituting equation 7.116:

$$\frac{\bar{\mathscr{C}}}{\bar{\mathscr{R}}} = \frac{\mathbf{G}_C(s)\mathbf{G}_1(s)\mathbf{G}_2(s)}{1 + \mathbf{G}_C(s)\mathbf{G}_1(s)\mathbf{G}_2(s)\mathbf{H}(s)} \tag{7.117}$$

which is the closed loop transfer function for the original loop shown in Fig. 7.3*a*.

7.10. SYSTEM STABILITY AND THE CHARACTERISTIC EQUATION

The normal function of any control system is to ensure that the controlled variable attains its desired value as rapidly as possible after a disturbance has occurred, with the minimum of oscillation. Determination of the response of a system to a given forcing function will show what final value the controlled variable will attain and the manner in which it will arrive at that value. This latter is a function of the stability of the response. For example, in considering the response of a second order system to a step change, it can be seen that oscillation increases as ζ decreases (Fig. 7.28). The stability is therefore considered to decrease with ζ. The limiting case occurs when $\zeta = 0$ and the response oscillates with constant amplitude—it is then said to be *conditionally stable*. With more complex systems it is possible to obtain oscillations of increasing amplitude such that the output never attains a new stable (steady-state) value but increases to a level which is only restricted by the physical limits of some part of the system or control loop. This type of response is termed *unstable* (or *unbounded*). The majority of processes involving fluid flow and/or heat transfer give overdamped responses when uncontrolled (i.e. on open-loop). Whenever a closed-loop system is formed incorporating a controller, however, there is always the possibility of an unstable response occurring.

7.10.1. The Characteristic Equation

Consider the transfer function given by equation 7.109. For a unit step change in the set point R, the Laplace transform of the controlled variable C is given by:

$$\bar{\mathscr{C}} = \frac{1}{s}\frac{\mathbf{G}_X(s)}{1 + \mathbf{G}_X(s)\mathbf{H}(s)}$$

$$= \frac{\mathbf{G}_X(s)\mathrm{f}(s)}{s(s-\alpha_1)(s-\alpha_2)\ldots\ldots(s-\alpha_r)} \tag{7.118}$$

where $\alpha_1, \alpha_2, \ldots\ldots, \alpha_r$ are the r roots of the equation:

$$1 + \mathbf{G}_X(s)\mathbf{H}(s) = 0 \tag{7.119}$$

[i.e. the r poles of the transfer function $\mathbf{G}_X(s)/(1 + \mathbf{G}_X(s)\mathbf{H}(s))$] and $\mathrm{f}(s)$ is some function of s.

Equation 7.119 is the *characteristic equation* of the system shown in Fig. 7.34 and is dependent only upon the open-loop transfer function $\mathbf{G}_X(s)\mathbf{H}(s)$ and is therefore the same for both set point and load changes (equations 7.109 and 7.110).

The determination of the nature of the roots of the characteristic equation (or the poles of the corresponding system transfer function) forms the basis of many techniques used to establish the nature of the stability of the system. In order to calculate the step response, equation 7.118 must be split into partial fractions for inversion, thus:

$$\bar{\mathscr{C}} = \frac{B_0}{s} + \frac{B_1}{s-\alpha_1} + \frac{B_2}{s-\alpha_2} + \ldots\ldots + \frac{B_r}{s-\alpha_r}$$

Inversion gives:

$$\mathscr{C} = B_0 + B_1 e^{\alpha_1 t} + B_2 e^{\alpha_2 t} + \ldots\ldots + B_r e^{\alpha_r t} \tag{7.120}$$

The roots of the characteristic equation may be real and/or complex, depending on the form of the open-loop transfer function. Suppose α_1 to be complex, such that:

$$\alpha_1 = \beta_1 + \mathrm{i}\gamma_1$$

where β and γ are real. Then:

$$e^{\alpha_1 t} = e^{\beta_1 t} e^{i\gamma_1 t} = e^{\beta_1 t} (\cos \gamma_1 t + i \sin \gamma_1 t) \tag{7.121}$$

It can be seen from equations 7.118, 7.120 and 7.121 that should any root of the characteristic equation have a positive real part, the resulting step response (equation 7.120) will contain an exponentially increasing term, i.e. it will be unstable. Thus, for stability, the roots of the characteristic equation (i.e. the poles of the system transfer function) must have negative real parts. Some idea can be obtained of the degree of stability of the system from the magnitude of the real parts of the roots. If the latter are negative and large, the transients will decay rapidly and the response will be more stable than for a system having roots with small negative real parts.

A qualitative assessment of the stability of a given system can be made conveniently by considering the positions of the system poles (i.e. the roots of the characteristic equation) on the complex plane. This is illustrated in the following example.

Example 7.4

Map the poles of the heat exchanger system described in Example 7.3 on to the complex plane. Hence make a qualitative assessment of the general stability of the process.

Solution

The poles of the system transfer function $\mathbf{G}(s)$ (Example 7.3) are the roots of the corresponding characteristic equation which is obtained by equating the denominator of $\mathbf{G}(s)$ to zero. Hence:

$$(1 + 0.67s)(10.2s^2 + 3.1s + 1) = 0$$

Thus:

$$s_1 = -1/0.67 = -1.5 + 0.i$$

and:

$$s_{2,3} = \frac{-3.1 \pm \sqrt{(3.1^2 - 4 \times 10.2)}}{20.4}$$

$$= -0.152 \pm 0.274i$$

Hence, all the poles of the system transfer function lie in the left half of the complex plane (all have negative real parts—Fig. 7.40). Thus the system is stable (i.e. it is *bounded*) and θ_B will attain a steady-state after a perturbation in F and after all intermediate transients have decayed. Two of the poles are complex conjugates which signifies that the system response will oscillate as it moves to the new steady-state (Fig. 7.41).

Should any pole of the system lie on the imaginary axis (i.e. when its real part is zero) then the system is *conditionally* stable. Poles lying to the right of the imaginary axis (corresponding to roots of the characteristic equation with positive real parts) indicate *unbounded* behaviour, i.e. that the system is unstable.

7.10.2. The Routh–Hurwitz Criterion

It is often difficult to determine quickly the roots of the characteristic equation. HURWITZ[19] and ROUTH[20] developed an algebraic procedure for finding the number of roots with positive real parts and consequently whether the system is unstable or not.

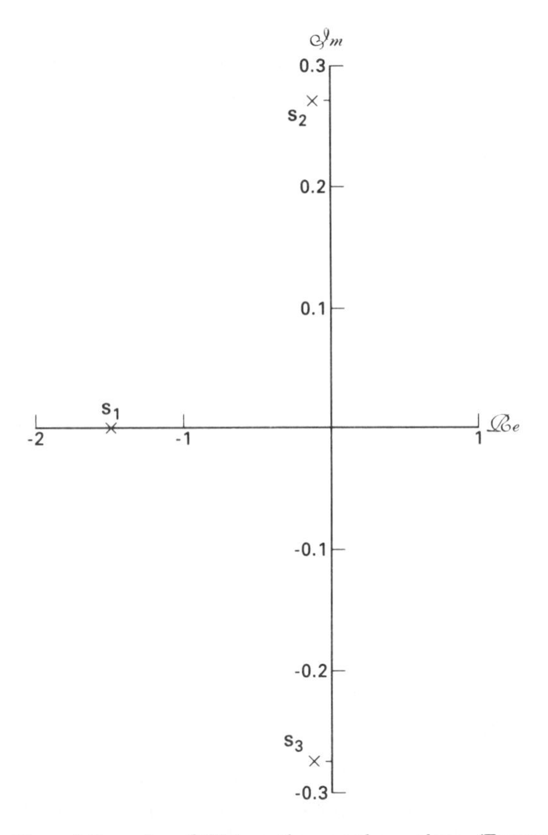

FIG. 7.40. Position of the poles of **G**(*s*) on the complex *s*-plane. (Examples 7.3 and 7.4)

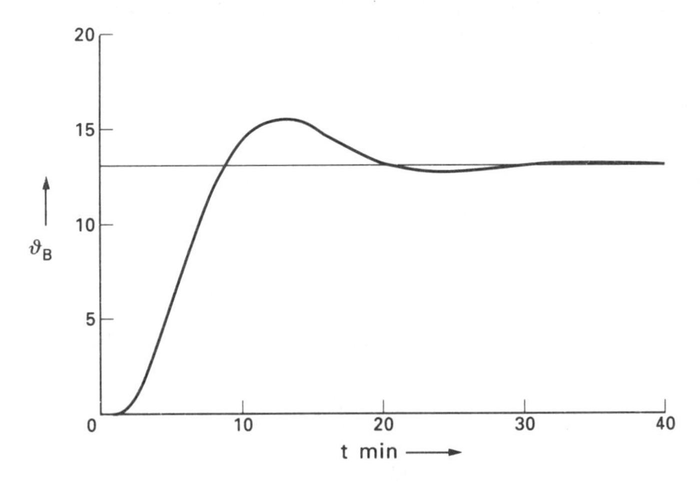

FIG. 7.41. Response of θ_B to a step change of unit magnitude in *F*. (Examples 7.3 and 7.4)

The characteristic equation is first written in the form:

$$c_r s^r + c_{r-1} s^{r-1} + \ldots\ldots + c_1 s + c_0 = 0 \tag{7.122}$$

where c_r is positive. If any of the remaining coefficients is negative, the system is unstable. If, however, all are positive then a further test must be applied using the *Routh array*. This is formed from the coefficients of the characteristic equation thus:

Row 1	c_r	c_{r-2}	c_{r-4}	c_{r-6}	.	.	.	.
Row 2	c_{r-1}	c_{r-3}	c_{r-5}	c_{r-7}	.	.	.	.
Row 3	b_{31}	b_{32}	b_{33}	.	.	.	.	.
Row 4	b_{41}	b_{42}	b_{43}	.	.	.	.	.
Row 5	b_{51}	b_{52}	b_{53}	.	.	.	.	.
...	.	.	.	.	.	.	.	.
...	.	.	.	.	.	.	.	.

where:

$$b_{31} = \frac{c_{r-1}c_{r-2} - c_r c_{r-3}}{c_{r-1}}, \qquad b_{32} = \frac{c_{r-1}c_{r-4} - c_r c_{r-5}}{c_{r-1}}, \qquad \text{etc.}$$

$$b_{41} = \frac{b_{31}c_{r-3} - b_{32}c_{r-1}}{b_{31}}, \qquad b_{42} = \frac{b_{31}c_{r-1} - b_{33}c_{r-5}}{b_{31}}, \qquad \text{etc.}$$

$$b_{51} = \frac{b_{41}b_{32} - b_{42}b_{31}}{b_{41}}, \qquad b_{52} = \frac{b_{41}b_{33} - b_{43}b_{31}}{b_{41}}, \qquad \text{etc.}$$

etc. etc. etc.

From this we can conclude that:

(i) If any of the elements in the first column of the array is negative then there is at least one root of the characteristic equation (i.e. pole of the system) lying in the right half of the complex plane.

(ii) The number of roots in the right half of the complex plane is given by the number of sign changes in the first column of the array.

Example 7.5

A system has the characteristic equation:

$$s^5 + 7s^4 + 18s^3 + 23s^2 + 17s + 6 = 0$$

By using the Routh–Hurwitz criterion determine whether or not the system is stable.

Solution

The characteristic equation is already in the form of equation 7.122 and all the coefficients are positive. Thus, to examine the stability of the system further, the Routh array must be constructed, viz.:

1	18	17
7	23	6
14.7	16.1	0
15.3	6	0
10.3	0	
6	0	

All the elements in the first column of the array are positive and thus the system is bounded and stable.

(A further example of the use of the Routh array in connection with discrete systems appears in Section 7.17.5.)

7.10.3. Destabilising a Stable Process with a Feedback Loop

An inherently stable process can be destabilised by the addition of a feedback control loop—particularly where integral action is included. This is illustrated in the following example using the characteristic equation and the Routh–Hurwitz stability criterion.

Example 7.6

Examine the stability of the heat exchanger system described in Example 7.3 using the Routh–Hurwitz criterion with θ_B controlled by manipulating F using a **PI** controller as in Fig. 7.42. Assume that the dynamics of the measuring element and the control valve are such that their time constants are negligible in comparison to that of the process. Under these circumstances the transfer functions of the measuring element and the control valve can be represented by their steady-state gains. Let these be 5 units and 0.01 units respectively. The assumptions regarding flows and temperatures in Example 7.3 still apply.

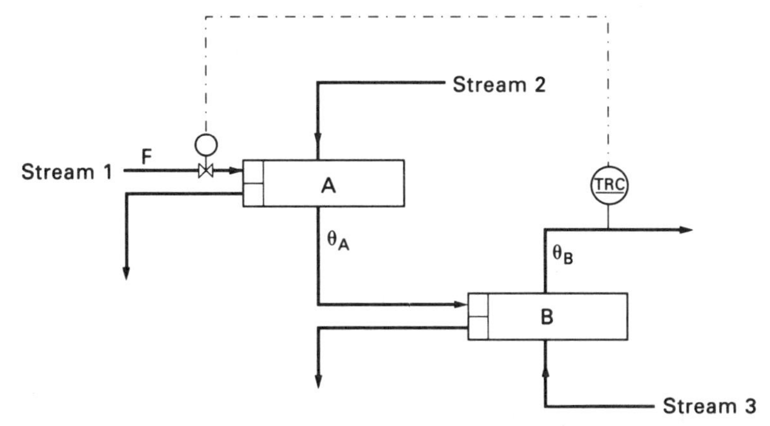

FIG. 7.42. Heat exchanger temperature control system

Solution

The corresponding block diagram is shown in Fig. 7.43 where:

$$\mathbf{G}_C(s) = K_C\left(1 + \frac{1}{\tau_I s}\right) \quad \text{(equation 7.63)}$$

$$\mathbf{G}_1(s) = 0.01$$

$$\mathbf{G}_2(s) = \mathbf{G}_A(s)\mathbf{G}_B(s) = \frac{20}{(1 + 0.67s)(10.2s^2 + 3.1s + 1)} \quad \text{(from Example 7.3)}$$

$$\mathbf{H}(s) = 5$$

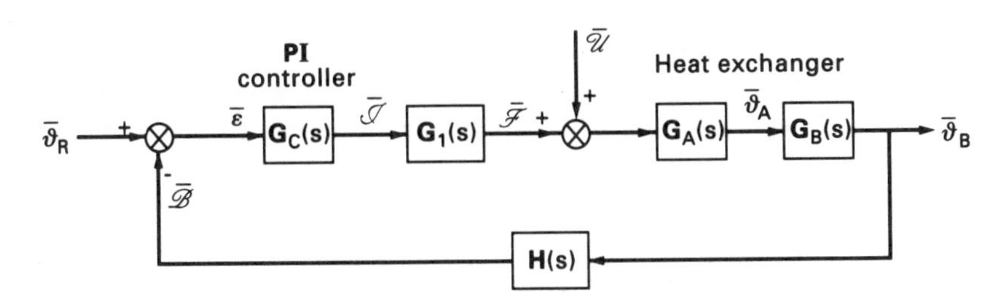

FIG. 7.43. Block diagram of system illustrated in Fig. 7.42

Hence, using equation 7.109, the closed-loop transfer function for the set point following case, i.e. relating C to R, is:

$$\frac{\bar{\mathscr{C}}}{\bar{\mathscr{R}}} = \frac{\mathbf{G}_C(s)\mathbf{G}_1(s)\mathbf{G}_2(s)}{1 + \mathbf{G}_C(s)\mathbf{G}_1(s)\mathbf{G}_2(s)\mathbf{H}(s)}$$

Thus, the characteristic equation for the closed-loop system is:

$$1 + \mathbf{G}_C(s)\mathbf{G}_1(s)\mathbf{G}_2(s)\mathbf{H}(s) = 0$$

i.e.:

$$(1 + 0.67s)(10.2s^2 + 3.1s + 1) + K_C\left(1 + \frac{1}{\tau_I s}\right) = 0$$

$\therefore$

$$\tau_I s(1 + 0.67s)(10.2s^2 + 3.1s + 1) + K_C(\tau_I s + 1) = 0$$

i.e.:

$$6.8\tau_I s^4 + 12.3\tau_I s^3 + 3.8\tau_I s^2 + (1 + K_C)\tau_I s + K_C = 0 \qquad (7.123)$$

All the coefficients in the characteristic equation have the same sign; hence the Routh array must be constructed in order to ascertain the stability of the system. The Routh array corresponding to equation 7.123 is:

$6.8\tau_I$	$3.8\tau_I$	K_C
$12.3\tau_I$	$(1 + K_C)\tau_I$	0
$0.55\tau_I(5.9 - K_C)$	K_C	0
$(1 + K_C)\tau_I - \dfrac{22.4K_C}{5.9 - K_C}$	0	0
K_C	0	0

As K_C and τ_I must be positive, then the conditions for any of the terms in the first column to be negative (i.e. for which the system is unstable) are that:

(i) $K_C > 5.9$

(ii) $\dfrac{22.4K_C}{5.9 - K_C} > (1 + K_C)\tau_I$

Hence, we must choose initially a value of $K_C < 5.9$ for stability. Suppose K_C is put equal to 3, then, for the system to be stable (from (ii)), τ_I must be greater than 5.8 min. If K_C is reduced to 1.8, then the minimum possible value of τ_I to retain stability will be 3.5 min. This is as expected since a reduction in K_C will increase system stability—thus allowing the amount of integral action to be increased before the system becomes unstable (reflected in a smaller value of integral time). (See also Section 7.2.3 and Example 7.8.)

We can similarly stabilise an unstable process using **P** or **PD** action—but not **PI** action. This situation is unlikely, however, as most process systems met by the chemical engineer are operated under conditions where the process is inherently stable.

The disadvantages of the Routh–Hurwitz test are that it is necessary to know the system transfer functions and that it gives no information concerning the degree of stability of the system. For cases where it is desired to determine the latter, the *Nyquist* or *Bode* stability criterion may be used. The Nyquist criterion[21] can be applied to all systems and involves plotting the system frequency response in the complex plane. This procedure is discussed later. A rather simpler approach is supplied by the criterion of BODE[22], which can be shown to be an extension of the Nyquist procedure[23] but applies only to systems for which the amplitude ratio and phase shift vary monotonically with frequency (fortunately most control systems are of this type). Although the Nyquist and Bode criteria find less application in modern process control practice than previously, they are useful in illustrating the relative effects of different lags within a control loop and of varying the parameters of the controller. Hence, methods involving both criteria are presented here in some detail.

7.10.4. The Bode Stability Criterion

Consider the control loop shown in Fig. 7.44. Suppose the loop to be broken after the measuring element, and that a sinusoidal forcing function $M \sin \omega t$ is applied to the set point R. Suppose also that the open-loop gain (or amplitude ratio) of the system is unity and that the phase shift ψ is -180°. Then the output $\mathscr{B}$ from the measuring element (i.e. the system open-loop response) will have the form:

$$\mathscr{B} = M \sin(\omega t - 180^\circ)$$
$$= -M \sin \omega t$$

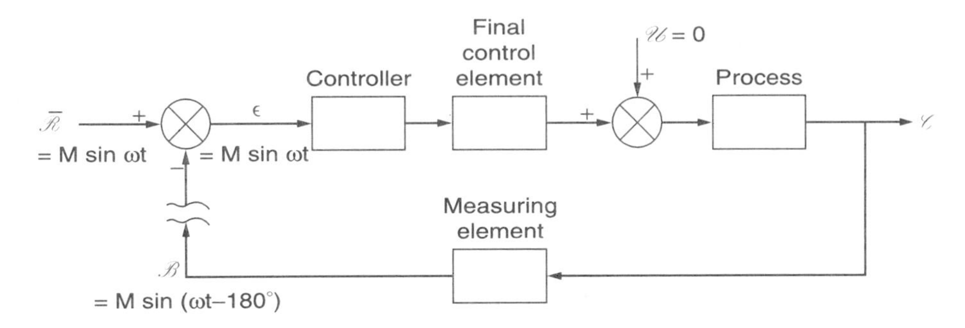

FIG. 7.44. Establishment of Bode stability criterion

Now, if at some instant of time $\mathscr{R}$ is reset to zero and the loop is instantaneously closed, we will have:

$$\varepsilon = \mathscr{R} - \mathscr{B}$$
$$= 0 - (-M \sin \omega t)$$
$$= M \sin \omega t$$

This indicates that the oscillation, once set in motion, will be maintained with constant amplitude around the closed-loop for $\mathscr{U} = \mathscr{R} = 0$. If, however, the open-loop gain or AR of the system is greater than unity, the amplitude of the sinusoidal signal will increase around the control loop, whilst the phase shift will remain unaffected. Thus the amplitude of the signal will grow indefinitely, i.e. the system will be unstable.

This heuristic argument forms the basis of the *Bode stability criterion*[22,24] which states that *a control system is unstable if its open-loop frequency response exhibits an AR greater than unity at the frequency for which the phase shift is* -180°. This frequency is termed the *cross-over frequency* (ω_{co}) for reasons which become evident when using the *Bode diagram* (see Example 7.7). Thus if the open-loop AR is unity when $\psi = -180^\circ$, then the closed-loop control system will oscillate with constant amplitude, i.e. it will be on the verge of instability. The greater the difference between the open-loop AR (< 1) at ω_{co} and $AR = 1$, the more stable the closed-loop system will be. This difference is normally measured in terms of *gain margin* where, if $(AR)_{co}$ is the open-loop AR at ω_{co}:

$$\text{Gain margin} = \frac{1}{(AR)_{co}} \tag{7.124}$$

Hence the gain margin may be used to indicate the degree of the stability of the system and a gain margin < 1 signifies an unstable system.

Another measure of the degree of stability can be expressed in terms of *phase margin* (*PM*), where:

$$\text{PM} = 180^\circ - \text{phase lag in degrees for which the AR is unity} \tag{7.125}$$

A negative phase margin indicates an unstable system. Clearly, first and second order systems are inherently stable as the maximum phase shift of the former is -90° and of the latter -180° (Section 7.8.4). (Note that when such a system is included within a feedback control loop this innate stability may no longer exist—see Section 7.10.3.)

Normally, design specifications require a gain margin > 1.7 and a phase margin $> 30^\circ$. Simultaneous specification of both gain and phase margins may lead to different controller designs, i.e. different sets of controller parameters. In such cases the parameters giving the more stable closed-loop response are usually selected.

The Bode Diagram

In order to facilitate the application of the Bode criterion the system frequency response may be represented graphically in the form of a Bode diagram or plot. This consists of *two* graphs which are normally drawn with the axes:

(a) $\log_{10}$ (amplitude ratio) versus $\log_{10}$ (frequency)
(b) phase shift versus $\log_{10}$ (frequency)

with frequency plotted as the abscissa in both cases.

First-order system. From equations 7.91 and 7.92:

$$AR = \frac{1}{\sqrt{(1 + \omega^2\tau^2)}} \qquad \text{(equation 7.91)}$$

$$\psi = \tan^{-1}(-\omega\tau) \qquad \text{(equation 7.92)}$$

Thus:

$$\log(AR) = -\tfrac{1}{2}\log[1 + (\omega\tau)^2]$$

$\therefore$

$$\lim_{\omega\tau \to \infty} [\log(AR)] = -\log(\omega\tau) \tag{7.126}$$

since $(\omega\tau)^2 >> 1$ as $\omega\tau \to \infty$.

Equation 7.126 is termed the *high frequency asymptote* (HFA) and is a straight line of gradient -1 passing through the point (1, 1) on a plot of log (AR) versus $(\omega\tau)$. Here $\omega\tau$ is employed rather than ω in order that the resulting diagram represents a first-order system with any time constant. The *low frequency asymptote* (LFA) is given by:

$$\lim_{\omega\tau \to 0} [\log (\text{AR})] = 0$$

i.e.:

$$\lim_{\omega\tau \to 0} (\text{AR}) = 1 \qquad (7.127)$$

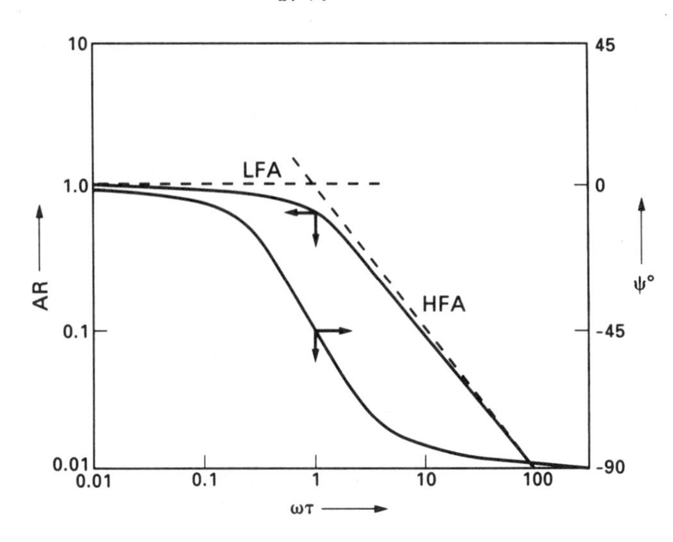

FIG. 7.45. Bode diagram for first-order system

Hence the two asymptotes, as shown in Fig. 7.45, intersect at the point (1,1). The frequency at this point is given by:

$$\omega_c = \frac{1}{\tau} \qquad (7.128)$$

where ω_c is termed the *corner* or *break* frequency. From equation 7.91, when $\omega = \omega_c$:

$$\text{AR} = \frac{1}{\sqrt{(1 + \omega_c^2 \tau^2)}} = \frac{1}{\sqrt{2}} = 0.707$$

An AR plot may be sketched from its asymptotes and the AR at ω_c. This is sufficiently accurate for most purposes.

The phase shift diagram may also be easily sketched, for, from equation 7.92:

$$\lim_{\omega\tau \to 0} (\psi) = 0° \quad \text{and} \quad \lim_{\omega\tau \to \infty} (\psi) = -90°$$

and when $\omega_c = \frac{1}{\tau}$, $\psi = -45°$.

Intermediate points must be calculated and it is worth noting that the plot is symmetrical about ψ at ω_c.

Second-order system. From equations 7.94*b* and 7.95:

$$\text{AR} = \frac{1}{\sqrt{[(1 - \omega^2\tau^2)^2 + (2\omega\tau)^2]}} \qquad \text{(equation 7.94}b\text{)}$$

$$\psi = \tan^{-1}\left(\frac{-2\zeta\omega\tau}{1 - \omega^2\tau^2}\right) \qquad \text{(equation 7.95)}$$

The Bode diagram in this case (Fig. 7.46) is distinguished by the fact that ζ is a parameter which affects both the AR and the ψ plots. However, the asymptotes may be determined in the same manner as for the first-order system. It is found that, for all ζ, the AR high frequency asymptote is a straight line of slope -2 passing through the point (1, 1) and the LFA is represented by the line AR = 1. The ψ plots all tend to zero degrees as $\omega\tau \rightarrow 0$ and to $-180°$ as $\omega\tau \rightarrow \infty$. When $\omega_c = 1/\tau$, $\psi = -90°$ independently of ζ. For $\zeta < 1/\sqrt{2}$ the AR plots exhibit maxima in the neighbourhood of $\omega\tau = 1$ and also give values of AR > 1. The frequency at which the AR is a maximum for any given values of $\zeta(< 1/\sqrt{2})$ is termed the *resonant* frequency.

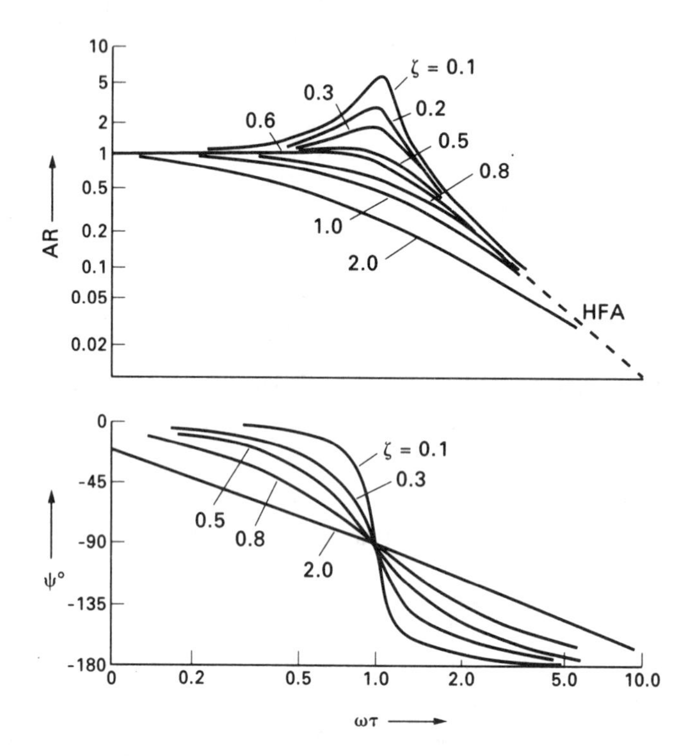

FIG. 7.46. Bode diagram for second-order system

Other functions may be plotted in a similar manner (Fig. 7.47).

Systems in series. The usefulness of the logarithmic plot becomes apparent when it is desired to determine the frequency response of systems in series. The resultant amplitude ratio and phase shift may be obtained using equations 7.104 and 7.105, i.e.:

$$\text{AR} = K(\text{AR})_1(\text{AR})_2(\text{AR})_3 \ldots\ldots \qquad \text{(equation 7.104)}$$

$$\therefore \quad \log(\text{AR}) = \log K + \log(\text{AR})_1 + \log(\text{AR})_2 + \log(\text{AR})_3 + \ldots\ldots \qquad (7.129)$$

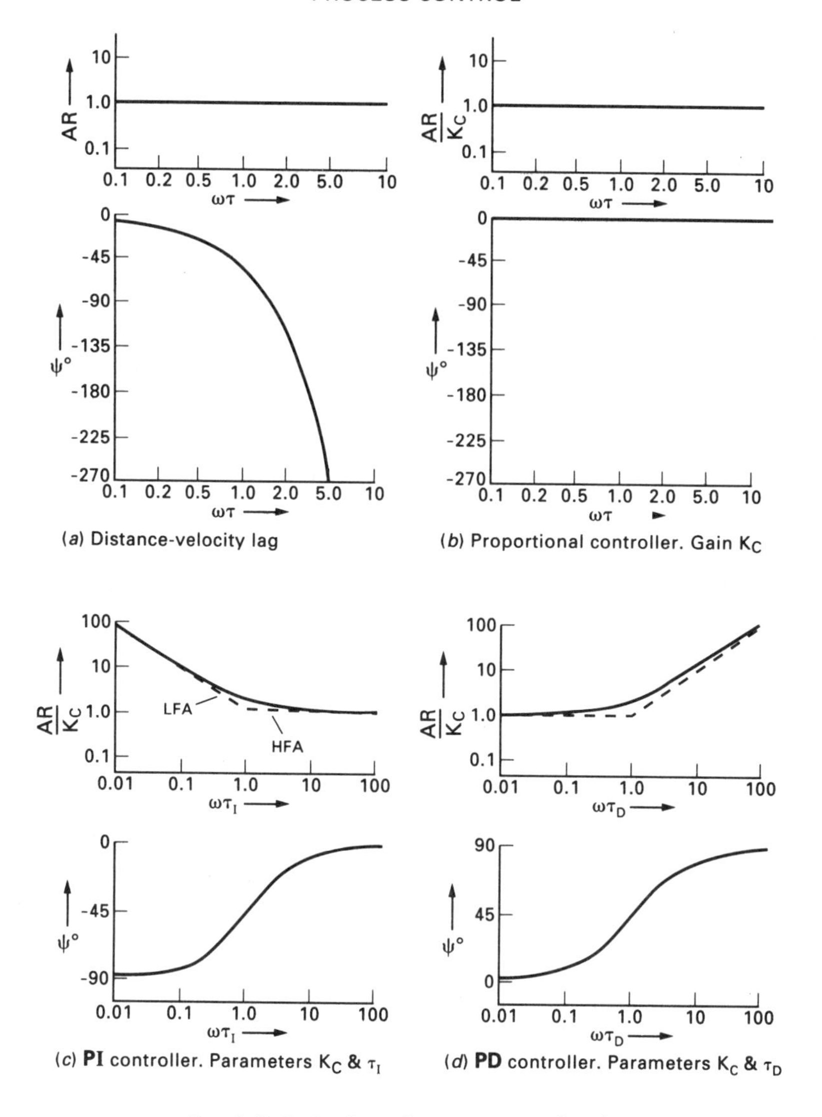

FIG. 7.47. Bode plots of some common functions

and $$\psi = \psi_1 + \psi_2 + \psi_3 + \ldots\ldots \qquad \text{(equation 7.105)}$$

The overall Bode plot of a number of systems in series may be obtained therefore as follows:

(i) Plot the individual ARs and phase shifts as functions of frequency ω and *not* of $\omega\tau$.
(ii) To determine the resultant AR, add the individual ARs on the log/log plot, treating values above unit AR as positive and below unit AR as negative. It is normally sufficiently accurate simply to add the asymptotes.
(iii) Add the individual phase shifts on the linear/log plot to obtain the resultant phase shift.

(iv) If the resultant transfer function is multiplied by some constant, e.g. the steady-state gain, the entire AR plot is moved vertically by an amount equal to that constant (see equation 7.129). The phase shift is unaffected.

In the following example the effect of the various fixed parameter control modes on the stability of a simple feedback control loop are examined using the Bode stability criterion and the concept of gain and phase margins.

Example 7.7

A control system using **PI** control is represented by the block diagram shown in Fig. 7.48. The transfer functions describing the various blocks are as shown with $K_c = 10$, $\tau_I = 1$ min, $K_1 = 0.8$ and $K_v = 0.5$. By determination of the gain and phase margins, show the effect on the stability of the control system of introducing derivative action with $\tau_D = 1$ min.

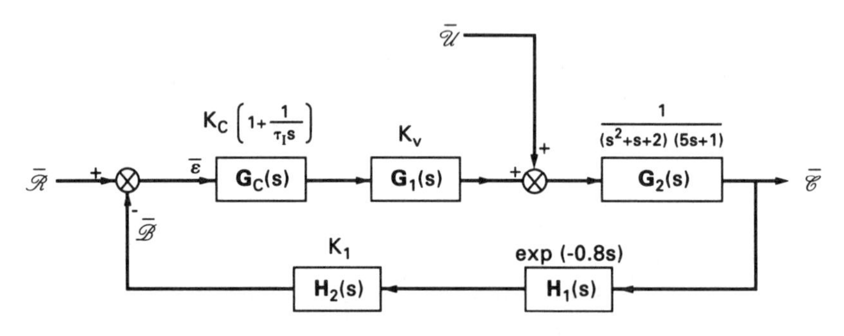

FIG. 7.48. Block diagram of control loop described in Example 7.7

Solution

Using equation 7.12, the open-loop transfer function of the control system without derivative action is written as:

$$\mathbf{G}_X(s) = \frac{10\left(1 + \dfrac{1}{s}\right) \times 0.5 \times \exp(-0.8s) \times 0.8}{(s^2 + s + 2)(5s + 2)}$$

$$= \frac{\left(1 + \dfrac{1}{s}\right) \times \exp(-0.8s)}{(0.5s^2 + 0.5s + 1)(2.5s + 1)}$$

$$= \mathbf{G}_d(s)\mathbf{G}_e(s)\mathbf{G}_f(s)\mathbf{G}_g(s)$$

The Bode diagram of $\mathbf{G}_X(s)$ is obtained by breaking down the transfer function into its constituent parts, plotting each separately and performing a graphical summation.

(a) $\mathbf{G}_d(s) = 1/(2.5s + 1)$ is the transfer function of a first-order system with a time constant of 2.5 min. Hence, from equations 7.91, 7.92 and 7.128:

$$\text{AR} = \frac{1}{\sqrt{(1 + 6.25\omega^2)}}, \quad \psi = \tan^{-1}(-2.5\omega) \quad \text{and} \quad \omega_c = 0.4 \text{ radians/min.}$$

(b) $\mathbf{G}_e(s) = 1/(0.5s^2 + 0.5s + 1)$ is the transfer function of a second-order system with a time constant of $1/\sqrt{2}$ min and a damping coefficient of $1/(2\sqrt{2})$. From equations 7.94*b* and 7.95:

$$\text{AR} = \frac{1}{\sqrt{[(1 - 0.5\omega^2)^2 + (0.5\omega)^2]}} \quad \text{and} \quad \psi = \tan^{-1}\left(-\frac{0.5\omega}{1 - 0.5\omega^2}\right)$$

The asymptotes intersect at the point AR = 1, $\omega = \sqrt{2}$.

(c) $\mathbf{G}_f(s) = \exp(-0.8s)$ represents a distance–velocity lag. From equations 7.96 and 7.97:

$$\text{AR} = 1 \text{ for all } \omega \quad \text{and} \quad \psi = -0.8\omega \text{ (radians)}$$

(d) $\mathbf{G}_g(s) = (1 + 1/s)$. Using the substitution rule (Section 7.8.4) we obtain:

$$\text{AR} = \sqrt{\left(1 + \frac{1}{\omega^2}\right)} \quad \text{and} \quad \psi = \tan^{-1}\left(-\frac{1}{\omega}\right)$$

The AR low frequency asymptote is a straight line of gradient -1. The HFA is AR = 1. These intersect at the point AR = 1. The phase shift plot is the inverse of that for a first-order system, i.e. $\psi \to -90°$ as $\omega \to 0$ and $\psi \to 0°$ as $\omega \to \infty$ (see Fig. 7.47*c*).

The overall Bode diagram (Fig. 7.49) is obtained using the procedures outlined in Section 7.10.4. From Fig. 7.49, $\omega_{co} = 0.77$ radians/min. Hence, from equation 7.124:

$$\text{Gain margin} = \frac{1}{(\text{AR})_{co}} = \frac{1}{0.96} = \underline{\underline{1.04}}$$

Now $\omega = 0.74$ radians/min at AR = 1, thus from equation 7.125:

$$\text{Phase margin} = 180° - 177° = \underline{\underline{3°}}$$

(This phase margin is too small to be shown in Fig. 7.49.)

Although by definition the system is stable, the phase margin is so close to zero and the gain margin so close to unity that any slight variation in any of the control system parameters or, indeed, in the process conditions, could make the system unstable, i.e. could cause a pole or poles of the system closed-loop transfer function to move into the right half of the complex plane (Section 7.10.1).

(e) When derivative action is introduced with $\tau_D = 1$ min the controller transfer function becomes:

$$\mathbf{G}_5(s) = \left(1 + \frac{1}{s} + s\right)$$

giving:

$$\text{AR} = \frac{1}{\omega}\sqrt{(\omega^4 - \omega^2 + 1)}$$

and:

$$\psi = \tan^{-1}\left(\omega - \frac{1}{\omega}\right)$$

The LFA on the amplitude ratio plot is a straight line of gradient -1 passing through (1, 1). The HFA is a straight line of gradient $+1$ passing also through (1, 1). The phase shift approaches $-90°$ as $\omega \to 0$ and tends to $+90°$ as $\omega \to \infty$. At $\omega = 1$ radian/min, $\psi = 0°$. The effect of adding the new controller transfer function is illustrated in Fig. 7.50. From the modified overall Bode diagram $\mathbf{G}_X(s) = \mathbf{G}_d(s)\mathbf{G}_e(s)\mathbf{G}_f(s)\mathbf{G}_g(s)$ and $\omega_{co} = 1.3$ radians/min giving:

$$\text{Gain margin} = \frac{1}{0.48} = \underline{\underline{2.08}}$$

and:

$$\text{Phase margin} = \underline{\underline{29°}}$$

The addition of derivative action has stabilised the control loop substantially, increasing the gain margin by a factor of 2 and the phase margin by 26°. In fact, the gain margin is now such that the response of the control system will tend to be overdamped. Although this appears to support the case for the addition of derivative action, some caution is necessary as this control mode can degrade the control under certain circumstances (see Section 7.2.3).

7.10.5. The Nyquist Stability Criterion

The Polar Plot (Nyquist Diagram)

The polar plot is an alternative to the Bode diagram for representing frequency response data and is the locus of all points occupied by the tip of a vector in the complex plane whose magnitude and direction are determined by the amplitude ratio and phase shift, respectively, as the frequency of the forcing function applied to the system is varied from zero to infinity.

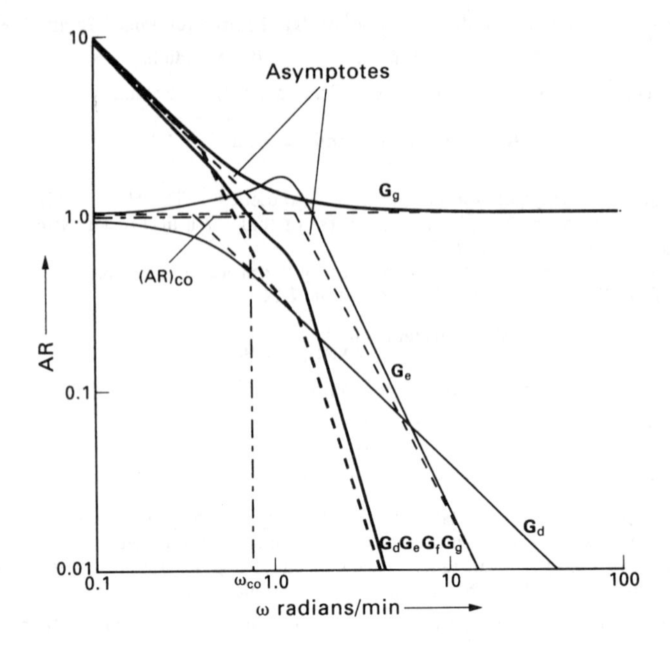

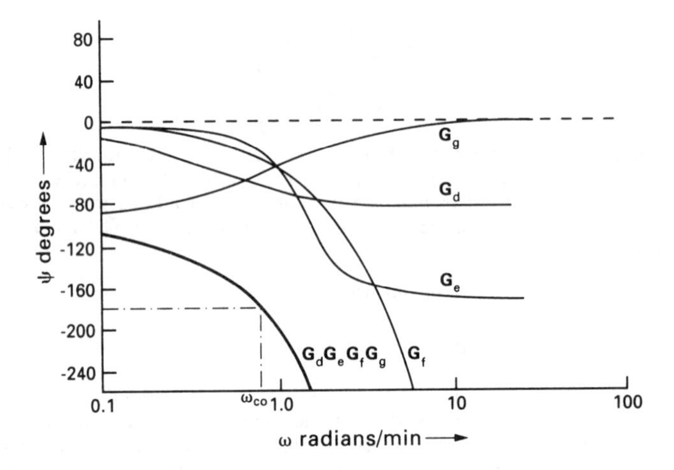

FIG. 7.49. Open-loop Bode diagram for control system shown in Fig. 7.48 (no derivative action)

Thus, for a first-order system with unit steady-state gain:

$$AR = \frac{1}{\sqrt{(1 + \omega^2 \tau^2)}} \qquad \text{(equation 7.91)}$$

$$\psi = \tan^{-1}(-\omega\tau) \qquad \text{(equation 7.92)}$$

If $\mathbf{V}_N$ is the vector concerned, then for $\omega = 0$:

$$\text{magnitude of } \mathbf{V}_N = AR|_{\omega=0} = 1$$

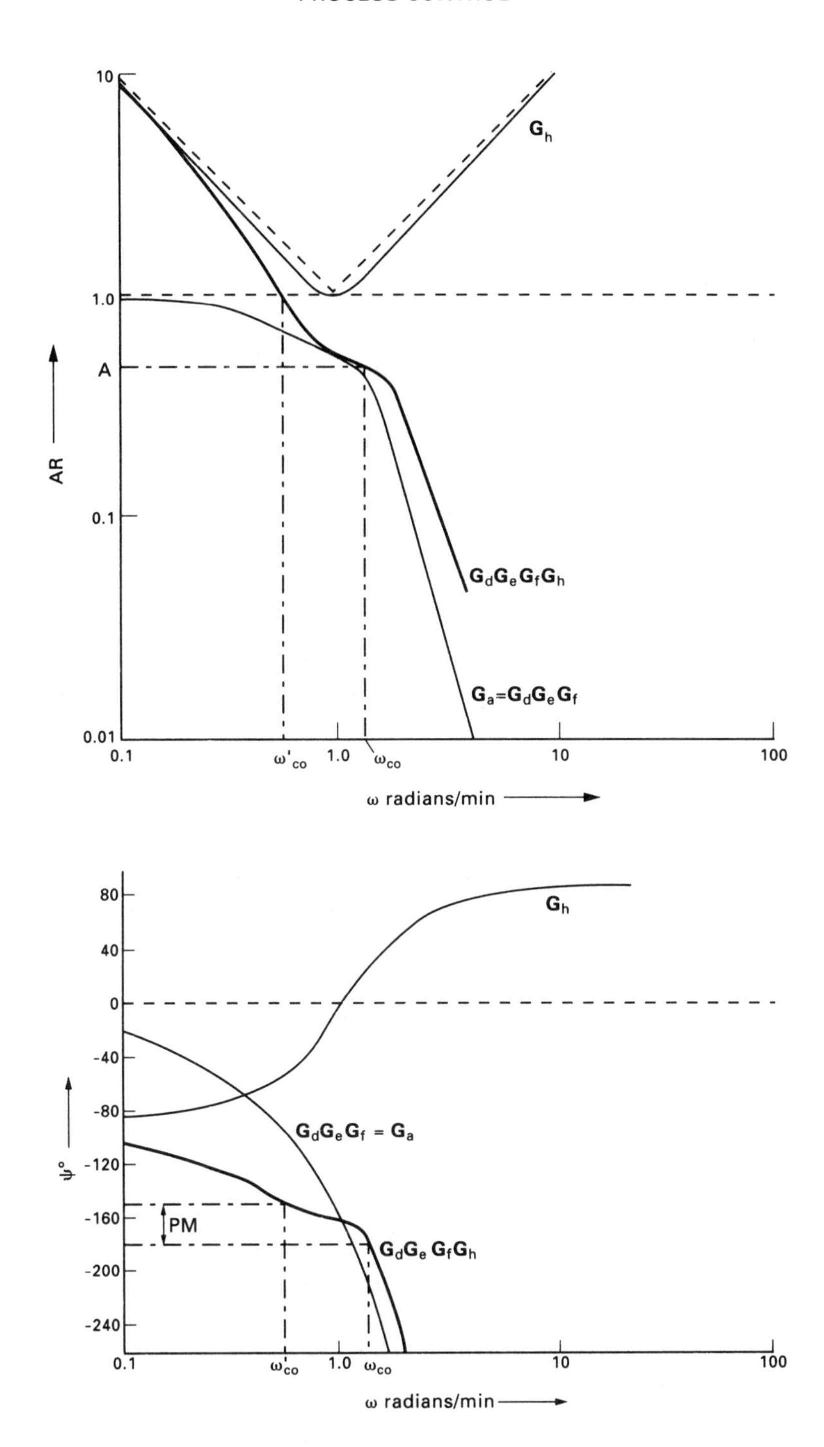

FIG. 7.50. Open-loop Bode diagram for control system shown in Fig. 7.48 with the addition of derivative action

and the direction of $\mathbf{V}_N$ relative to the real axis is given by:

$$\psi|_{\omega=0} = 0°$$

For $\omega = \infty$:

$$\text{magnitude of } \mathbf{V}_N = 0 \quad \text{and} \quad \psi = -90°$$

When $\omega = 1/\tau$:

$$\text{magnitude of } \mathbf{V}_N = 1/\sqrt{2} \quad \text{and} \quad \psi = -45°$$

In this case the entire locus is a smooth semicircle which commences on the positive real axis and approaches the origin from the direction of the negative imaginary axis (Fig. 7.51*a*). An alternative procedure for constructing a polar plot is described in the solution to Example 7.8. Other commonly employed polar plots are illustrated in Fig. 7.51.

Polar plots of type n systems. In general, from Section 7.4.3, we can write:

$$\mathbf{G}(s) = \frac{\bar{y}}{\bar{x}} \qquad \text{(equation 7.14)}$$

where $\bar{x}$ and $\bar{y}$ are polynomials in the Laplace transform parameter s. More specifically, these polynomials can be expressed in terms of first and second-order factors which may be written in the general form:

$$\mathbf{G}(s) = \frac{K(1+\tau_{a1}s)(1+\tau_{a2}s)\ldots\ldots(\tau_{b1}^2 s^2 + 2\zeta_{b1}\tau_{b1}s + 1)\ldots\ldots}{s^n(1+\tau_{c1}s)(1+\tau_{c2}s)\ldots\ldots(\tau_{d1}^2 s^2 + 2\zeta_{d1}\tau_{d1}s + 1)\ldots\ldots} \qquad (7.130)$$

The value of n is significant in the construction of polar (Nyquist) plots and when $n = 0, 1, 2$, etc. the system represented by $\mathbf{G}(s)$ is termed *type 0*, *type 1*, *type 2*, etc. respectively. It is rare to find values of $n > 2$. Figure 7.52 shows polar plots of typical type 0, type 1, type 2 and type 3 systems.

System Stability from the Nyquist Diagram

The Nyquist stability criterion may be employed in cases where the Bode criterion is not applicable (e.g. where the phase shift has a value of −180° for more than one value of frequency). It is usually stated in the form[17]:

$$n_E = n_Z - r \qquad (7.131)$$

where n_E is the net number of encirclements of the point (−1, 0) by the frequency response on the complex plane in the same direction as the path of s values chosen (i.e. in the direction of increasing frequency). n_Z and r are the numbers of zeros and poles of the function respectively.

It can be shown[16] that, if there are any net encirclements of the point (−1, 0) on the Nyquist diagram (i.e. if $n_E > 0$), then the system characteristic equation will have roots lying to the right of the imaginary axis and consequently the system will be unstable (Fig. 7.53).

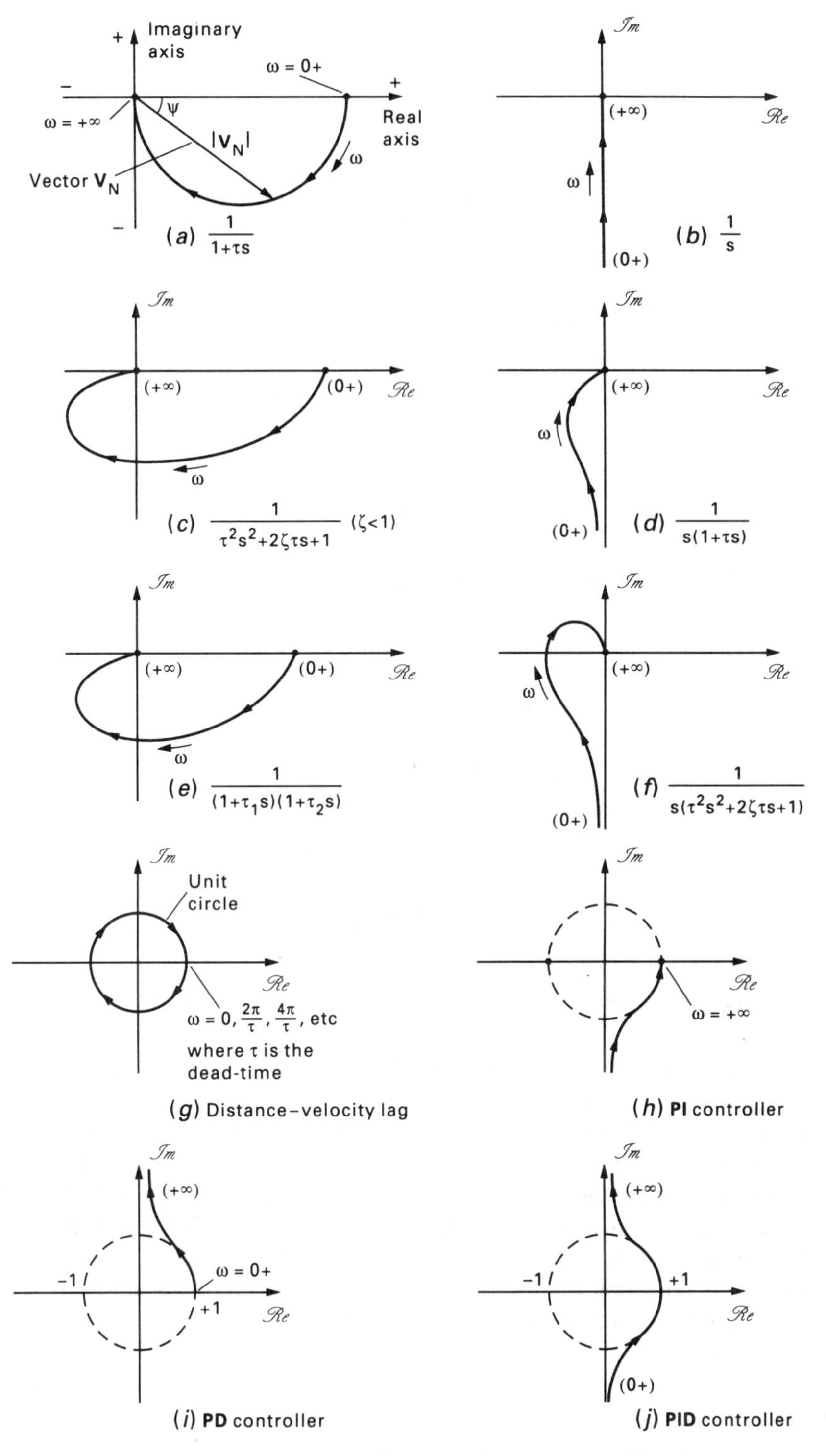

FIG. 7.51. Polar plots (Nyquist diagrams) of some common functions

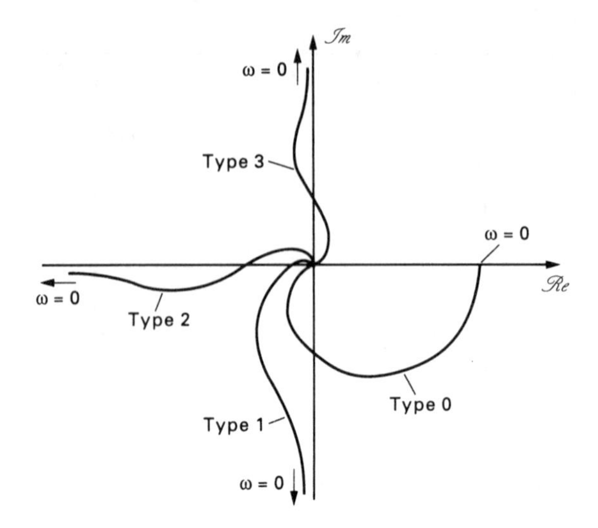

FIG. 7.52. Comparison of polar plots of Type 0, Type 1, Type 2 and Type 3 systems

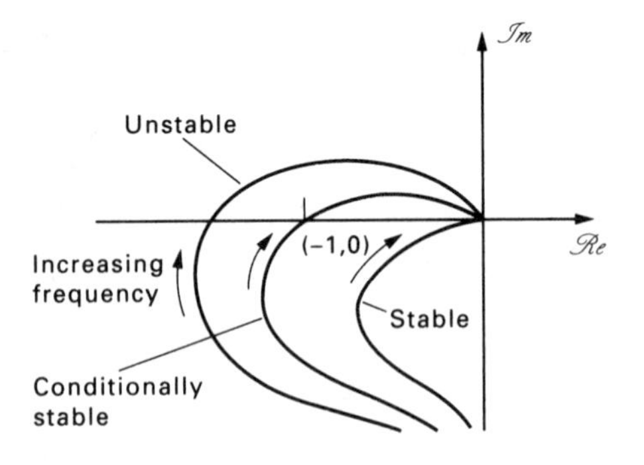

FIG. 7.53. The use of the polar plot in applying the Nyquist stability criterion

Gain and phase margins on the polar plot. These can be determined also from the polar plot as indicated in Fig. 7.54. A circle of unit radius, centre the origin (the *unit circle*) is drawn. The magnitude of the vector from the origin to the intersection of the polar plot and the unit circle is clearly unity. As this is also the amplitude ratio at this frequency then ψ_{PM} must represent the phase margin (Section 7.10.3). If the magnitude of the vector at the frequency where the polar plot cuts the negative real axis is K_M, then $1/K_M$ is the gain margin, i.e. the additional gain that would be required by the system for the plot to pass through the point $(-1, 0)$. An unstable system is indicated when $1/K_M < 1.0$ or $\psi_{PM} < 0$ (as with the Bode criterion).

Example 7.8

Examine the stability of the control system described in Example 7.6 using the Nyquist criterion.

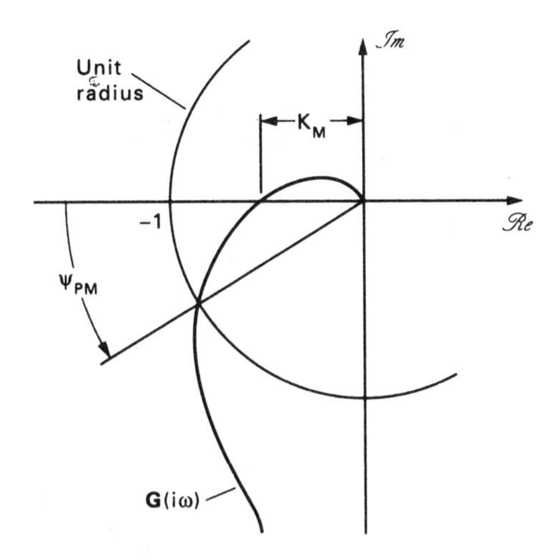

FIG. 7.54. Gain and phase margins on the polar plot

Solution

The system open-loop transfer function for this system is:

$$\mathbf{G}_X(s) = \mathbf{G}_C(s)\mathbf{G}_1(s)\mathbf{G}_2(s)\mathbf{H}(s)$$

$$= K_C\left(1 + \frac{1}{\tau_I s}\right) \times 0.01 \times \frac{20}{(1 + 0.67s)(10.2s^2 + 3.1s + 1)} \times 5$$

Mapping this on to the complex plane (i.e. putting $s = \mathrm{i}\omega$) gives:

$$\mathbf{G}_X(\mathrm{i}\omega) = K_C\left(1 - \frac{\mathrm{i}}{\tau_I\omega}\right) \times \frac{1}{(1 + 0.67\mathrm{i}\omega)(-10.2\omega^2 + 3.1\mathrm{i}\omega + 1)}$$

$$= K_C\left(1 - \mathrm{i}\left(\frac{1}{\omega\tau_I}\right)\right) \times \frac{(1 - 12.3\omega^2) - i(3.77\omega - 6.83\omega^3)}{(1 - 12.3\omega^2)^2 + (3.77\omega - 6.83\omega^3)^2}$$

Hence:

$$\mathscr{R}e\,(\mathbf{G}(\mathrm{i}\,\omega)) = \frac{K_C\,((1 - 12.3\omega^2) - (3.77 - 6.83\omega^2)/\tau_I)}{(1 - 12.3\omega^2)^2 + (3.77\omega - 6.83\omega^3)^2} \quad (7.132a)$$

and

$$\mathscr{I}m\,(\mathbf{G}(\mathrm{i}\,\omega)) = \frac{K_C\,((6.83\omega^3 - 3.77\omega) - (1 - 12.3\omega^2)/(\omega\tau_I))}{(1 - 12.3\omega^2)^2 + (3.77\omega - 6.83\omega^3)^2} \quad (7.132b)$$

From example 7.6 we know that critical stability occurs for $K_C = 1.8$, $\tau_I = 3.5$. Hence, by the Nyquist criterion, when these conditions are applied, the polar plot will pass through the point $(-1, 0)$ on the complex plane, i.e. for these values of the controller parameters, $\mathscr{I}m\,(\mathbf{G}(\mathrm{i}\omega)) = 0$. Thus, from equation 7.132*b*:

$$6.83\omega^3 - 3.77\omega = (1 - 12.3\omega^2)/(3.5\omega)$$

$\therefore$

$$23.9\omega^4 - 0.9\omega^2 - 1 = 0$$

and:

$$\omega^2 = +0.224 \quad \text{or} \quad -0.187$$

Taking positive values only gives $\omega = 0.47$ radians/min. Substituting for ω, K_C and τ_I in equation 7.132*a* leads to:

$$\mathscr{R}e\,((\mathbf{G}(\mathrm{i}\omega)) = \frac{1.8[(1 - 2.76) - (3.77 - 1.53)/3.5]}{(1 - 2.76)^2 + 0.224(3.77 - 1.53)^2}$$

$$= -1.02 \approx \underline{\underline{-1}}$$

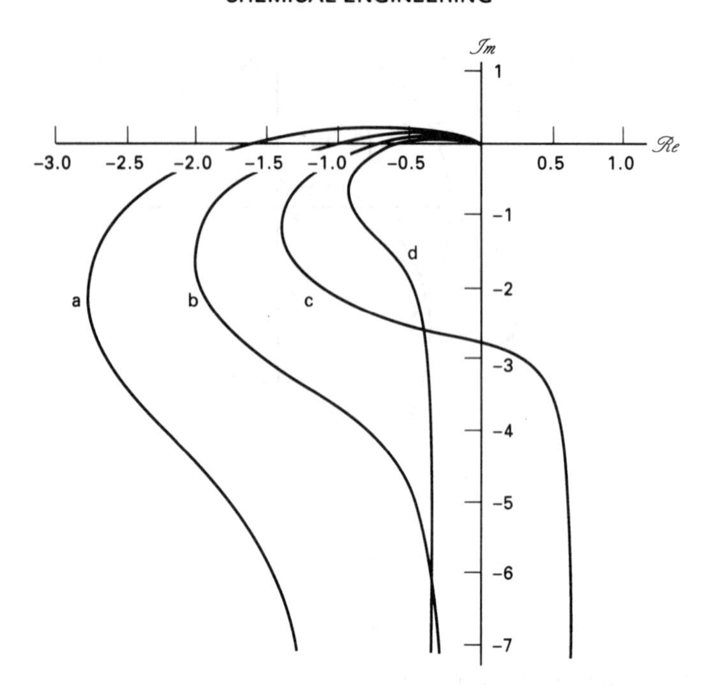

FIG. 7.55. Polar plots of the open-loop transfer function $\mathbf{G}_X(s)$ in Example 7.8: (*a*) $K_C = 1.8$, $\tau_I = 2.5$ (unstable); (*b*) $K_C = 1.8$, $\tau_I = 3.5$ (conditionally stable); (*c*) $K_C = 1.8$, $\tau_I = 6$ (stable); (*d*) $K_C = 0.6$, $\tau_I = 2.5$ (stable)

Hence with $K_C = 1.8$ and $\tau_I = 3.5$, the polar plot of the open-loop transfer function passes through the point $(-1, 0)$. This confirms the result obtained in Example 7.6 using the Routh–Hurwitz criterion, i.e. that with these controller parameters, the response of the controlled variable θ_B is conditionally stable.

Figure 7.55 shows polar plots of the open-loop transfer function $\mathbf{G}_X(s)$ for different values of K_C and τ_I.

7.10.6. The Log Modulus (Nichols) Plot

In addition to the Bode and Nyquist diagrams, frequency response information can be usefully presented in the form of a *Nichols* or *log modulus* plot[25]. This is essentially a combination of the two separate Bode diagrams in that it is a plot of log (amplitude ratio) (i.e. log (AR) = $\log|\mathbf{G}(i\omega)|$) vs phase shift (i.e. $\psi = \arg\mathbf{G}(i\omega)$) with frequency ω as a parameter. Hence both AR and ψ can be obtained for any given value of ω from one diagram. Figure 7.56 illustrates log modulus plots for the transfer function $\mathbf{G}(s)$ described in Example 7.8 with different values of K_C and τ_I (*cf.* Fig. 7.55). In Fig. 7.56 gain and phase margins for curve *d* are given by $10^{-A_{LM}}$ and ψ_{PM} respectively. In this case, the gain margin $\approx 10^{0.28} \approx 1.9$ and the phase margin $\approx 29°$ indicating a stable system (Section 7.10.4). For further use of this type of plot, see Example 7.10.

7.11. COMMON PROCEDURES FOR SETTING FEEDBACK CONTROLLER PARAMETERS

Many procedures exist for estimating optimum settings for controllers and the most common are described here. One of the usual bases employed is that the response of the controlled variable to a change in load or set point should have a

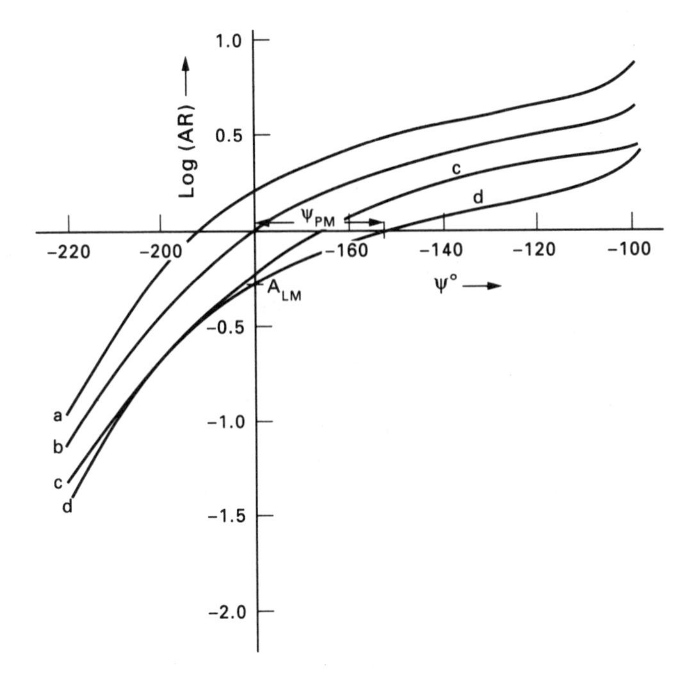

FIG. 7.56. Log modulus plots of the open-loop transfer function $\mathbf{G}_X(s)$ in Example 7.8: (*a*) $K_c = 1.8$, $\tau_I = 2.5$ (unstable); (*b*) $K_c = 1.8$, $\tau_I = 3.5$ (conditionally stable); (*c*) $K_c = 1.8$, $\tau_I = 6$ (stable); (*d*) $K_c = 0.6$, $\tau_I = 2.5$ (stable)

decay ratio of $^1/_4$, i.e. the ratio of the *overshoot* of the first peak of the response curve to the overshoot of the second peak is 4:1 as in Fig. 7.57. There is no direct mathematical justification for this[5] but it is a compromise between a rapid initial response and a short *response time*. [The response time, *settling time* or *line-out time* is

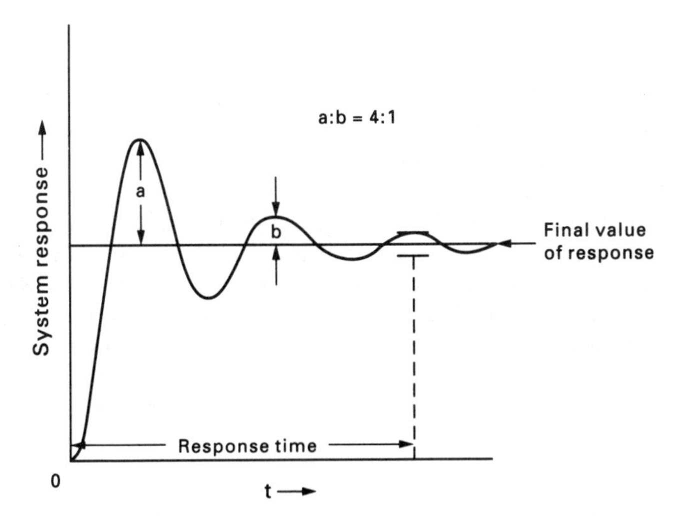

FIG. 7.57. System response with decay ratio of $^1/_4$

the time required for the absolute value of the system response to come within a small specified amount of the final value of the response—say ±5 per cent (Fig. 7.57).]

7.11.1. Frequency Response Methods

Determination of Settings from Frequency Response Data—The Method of Ziegler and Nichols

If the frequency response characteristics of the control system are known then it is possible to estimate values of controller parameters which will give specified gain and phase margins. However, this necessitates trial and error procedures. The semi-empirical method of ZIEGLER and NICHOLS[26] is more easily applied as follows.

The open-loop Bode diagram for all the components in the control loop, excepting the controller, is plotted and the cross-over frequency determined. If the total open-loop amplitude ratio at ω_{co} is $(AR)_{co}$, then, by the Bode criterion, the gain of a proportional controller which would cause the system to be on the verge of instability will be:

$$K_u = \frac{1}{(AR)_{co}} \tag{7.133}$$

where K_u is termed the *ultimate gain*. The *ultimate period* is defined as that period of sustained cycling which would occur if a proportional controller of gain K_C were used, thus:

$$T_u = \frac{2\pi}{\omega_{co}} \tag{7.134}$$

TABLE 7.3. *Ziegler–Nichols Controller Settings*[26]

Control action	Controller settings		
	K_C	τ_I	τ_D
P	0.5 K_u	—	—
PI	0.45 K_u	$T_u/1.2$	—
PID	0.6 K_u	$T_u/2$	$T_u/8$

The Ziegler–Nichols settings are derived from K_u and T_u on the basis of gain and phase margins of 2 and 30°, respectively, for proportional control alone. The addition of integral action introduces more phase lag at all frequencies, and hence a lower value of proportional gain K_C is required to maintain the same phase margin. Adding derivative action introduces phase lead and thus a greater value of K_C can be tolerated. The controller settings recommended are shown in Table 7.3. Note that no settings are given for a **PD** controller, although the values for K_C and τ_D from the **PID** settings may be employed. However, these tend to produce overdamped responses of the controlled variable. A further difficulty occurs if the open-loop phase lag (without the controller) does not exceed 180° at any frequency, in which case no value of ω_{co} can be obtained. In both these instances an alternative method should be employed.

Example 7.9

Determine the settings of the **PI** and **PID** controllers used in Example 7.7 by the method of Ziegler and Nichols.

Solution

The open-loop Bode diagram without the controller is given by $\mathbf{G}_a$ in Fig. 7.50.

Hence: $\omega_{co} = 1.12$ radians/min.

$$\therefore \quad K_u = \frac{1}{(\text{AR})_{co}} = \frac{1}{0.49} = 2.04$$

and:
$$T_u = \frac{2\pi}{1.12} = 5.6 \text{ min}$$

For a **PI** controller, from Table 7.1, the desired settings are:

$$K_C = \underline{\underline{0.92}} \quad \text{and} \quad \tau_I = \underline{\underline{4.7 \text{ min}}}$$

For a **PID** controller, the desired settings are:

$$K_C = \underline{\underline{1.22}}, \quad \tau_I = \underline{\underline{2.8 \text{ min}}} \quad \text{and} \quad \tau_D = \underline{\underline{0.7 \text{ min}}}$$

Loop Tuning

This procedure may be used when the frequency response of the system is unknown. The system is tested as a closed-loop with any integral or derivative action removed (i.e. by making τ_I as large as possible and putting $\tau_D = 0$ (equation 7.8)). It may be necessary to inject small disturbances to initiate the cycling because of process inertia. The gain at which the cycling commences is considered to correspond to the K_u of the Ziegler–Nichols method and the period of oscillation is taken to be T_u. The controller settings are determined then using Table 7.3.

7.11.2. Process Reaction Curve Methods

The Cohen–Coon Procedure

The controller is placed on manual control (i.e. effectively removing it from the control loop) and the response of the measured variable to a small step change in the manipulated variable is recorded as shown in Fig. 7.58*a*[27]. This response is called the *process reaction curve*. A tangent is drawn to this curve at the point of inflexion (Fig. 7.58*b*). The intercept of this tangent on the abscissa is termed the *apparent dead time* (τ_{ad}) of the system. The gradient of the tangent is given by:

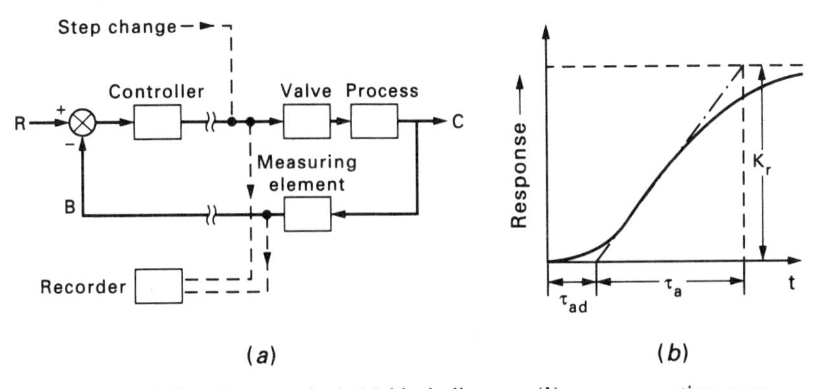

FIG. 7.58. Cohen–Coon method: (*a*) block diagram; (*b*) process reaction curve

$$m = K_r/\tau_a \tag{7.134}$$

where τ_a and K_r are the *apparent time constant* and the steady-state gain of the response, respectively. The relevant controller settings are given in Table 7.4.

TABLE 7.4. *Cohen–Coon Controller Settings*

Control action	Controller settings		
	K_C	τ_I	τ_D
P	$\dfrac{1}{K_r}\dfrac{\tau_a}{\tau_{ad}}\left(1+\dfrac{\tau_{ad}}{3\tau_a}\right)$	—	—
PI	$\dfrac{1}{K_r}\dfrac{\tau_a}{\tau_{ad}}\left(\dfrac{9}{10}+\dfrac{\tau_{ad}}{12\tau_a}\right)$	$\tau_{ad}\left(\dfrac{30+3\tau_{ad}/\tau_a}{9+20\tau_{ad}/\tau_a}\right)$	—
PD	$\dfrac{1}{K_r}\dfrac{\tau_a}{\tau_{ad}}\left(\dfrac{5}{4}+\dfrac{\tau_{ad}}{6\tau_a}\right)$	—	$\tau_{ad}\left(\dfrac{6-2\tau_{ad}/\tau_\alpha}{22+3\tau_{ad}/\tau_a}\right)$
PID	$\dfrac{1}{K_r}\dfrac{\tau_a}{\tau_{ad}}\left(\dfrac{4}{3}+\dfrac{\tau_{ad}}{4\tau_a}\right)$	$\tau_{ad}\left(\dfrac{32+6\tau_{ad}/\tau_a}{13+8\tau_{ad}/\tau_a}\right)$	$\tau_{ad}\left(\dfrac{4}{11+2\tau_{ad}/\tau_a}\right)$

The Cohen–Coon settings are based on the assumption that the open-loop system behaves in the same manner as the transfer function:

$$\mathbf{G}(s) = \frac{K_r \exp(-\tau_{ad}s)}{1+\tau_a s} \tag{7.135}$$

COHEN and COON[27] determined the relationships in Table 7.4 so as to give responses having large decay ratios, minimum offset and minimum area under the closed-loop response curve.

Integral Criteria

MURRILL[5] has pointed out that the $^1/_4$ decay ratio constraint upon which the previous methods are based has disadvantages, e.g. the decay ratio is determined from the first two peaks and a comparison of the second and third peaks may yield a different result. Since a perfect response would correspond to there being no error at any time, some criterion based upon the minimisation of some function of the total error under the response curve would be logical. Three such functions are commonly employed, viz.:

(a) The *integral of the square error (ISE)*, where:

$$\text{ISE} = \int_0^\infty [\varepsilon(t)]^2 \mathrm{d}t \tag{7.136}$$

(b) The *integral of the absolute value of the error (IAE)*, where:

$$\text{IAE} = \int_0^\infty |\varepsilon(t)| \mathrm{d}t \tag{7.137}$$

(c) The *integral of the time-weighted absolute error (ITAE)*, where:

$$\text{ITAE} = \int_0^\infty t|\varepsilon(t)|\,dt \tag{7.138}$$

In cases where the controlled variable does not converge to the set point (e.g. when offset occurs) the error $[R(t) - B(t)]$ may be replaced by $[C(\infty) - C(t)]$ assuming that $C(\infty)$ is known. If $C(\infty)$ is not known, then the infinite limit is replaced by a finite limit which is sufficiently large for all significant transients to have died away. Thus, the performance of the control strategy in terms of the response of the controlled variable is reduced to a single number which is sometimes termed the *Figure of Merit*.

Which of the three criteria is chosen is very much dependent upon the characteristics of the system. If the controlled variable response exhibits large deviations (i.e. there are large errors), then use of the ISE function will lead to the determination of controller parameters which will give the best control characteristics. As the errors are squared, minimisation of the ISE will be particularly effective. Use of the IAE function will suppress small errors better than will the ISE criterion as it treats errors of all magnitudes in a uniform manner. The ITAE function should be employed where there are persistent errors since this will lead to an amplification of the effect (and thus a weighted influence on the consequent estimation of the controller parameters) of the latter at large values of t.

Workers at Louisiana State University[28] have used the minimisation of these error criteria to produce controller settings from the process reaction curve. Values of controller parameters based upon the ITAE criterion are based upon a tuning relation of the form:

$$Y = \sigma_1 \left(\frac{\tau_{ad}}{\tau_a}\right)^{\sigma_2} \tag{7.139}$$

where τ_{ad} and τ_a are obtained from the process reaction curve and values of σ_1 and σ_2 are listed in Table 7.5. The relevant controller settings are calculated from:

$$Y = K_r K_C \text{ for the proportional gain} \tag{7.140}$$

$$Y = \tau_a/\tau_I \text{ for the integral time} \tag{7.141}$$

$$Y = \tau_D/\tau_a \text{ for the derivative time} \tag{7.142}$$

Equivalent values of σ_1 and σ_2 based upon the ISE and IAE criteria are also available[28].

TABLE 7.5. *Constants for Equation 7.139 Based upon the ITAE Criterion*[28]

Control action	Controller mode					
	Proportional		Integral		Derivative	
	σ_1	σ_2	σ_1	σ_2	σ_1	σ_2
P	0.490	−1.084	—	—	—	—
PI	0.859	−0.977	0.674	−0.680	—	—
PID	1.357	−0.947	0.842	−0.738	0.381	0.995

7.11.3. Direct Search Methods

The integral criteria described in equations 7.136, 7.137 and 7.138 form convenient objective functions for any optimisation scheme employed. This approach is best applied to the control of systems with known dynamics so that the controlled response can be simulated. Starting with an initial set of controller parameters (whether for single, two term or three term control), the controlled response can be determined over a suitable period of time (generally the response time) and the appropriate integral relation (IAE, ISE or ITAE) evaluated by summing the values of the function obtained at successive intervals of time. The magnitude of these sampling intervals is chosen according to the amount of damping shown by the response. The sampling interval for a highly oscillatory response should clearly be much shorter than that for a sluggish and overdamped response in order to obtain a suitably accurate summation. By varying the controller parameters according to a predetermined optimisation routine, the parameters required for a minimum value of the integral relation can be found. Any suitable constrained or unconstrained multivariable optimisation scheme can be employed for this purpose[29].

7.12. SYSTEM COMPENSATION

7.12.1. Dead Time Compensation

The presence of significant amounts of dead time in a control loop can cause severe degradation of the control action due to the additional phase lag that it contributes (see Example 7.7). One method for compensating for the effects of dead time in the control loop has been suggested by SMITH[30]. This consists of the insertion of an additional element which is often termed the *Smith predictor* as it attempts to predict the delayed effect that the manipulated variable will have upon the process output.

In the unity feedback system shown in Fig. 7.59 (set-point following case), it is assumed that all the dead time is contained within the process and is represented by the transfer function $\mathbf{G}_3(s)$, where:

$$\mathbf{G}_3(s) = \exp(-\tau_{DV}s) \qquad \text{(equation 7.60)}$$

Consider the open-loop response of B to a change in R. Clearly it is desirable that the measured value should consist of current information and not information which has been delayed by the dead time represented by $\mathbf{G}_3(s)$. Suppose that the measured value without dead time is B_1, i.e. that:

$$\bar{\mathscr{B}}_1 = \mathbf{G}_C(s)\mathbf{G}_1(s)\mathbf{G}_2(s)\bar{\mathscr{R}}$$

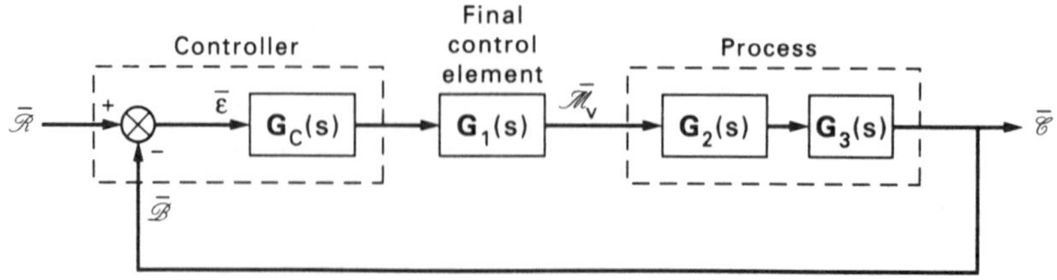

Controller — Final control element — Process

$\bar{\mathscr{R}}$, $\bar{\varepsilon}$, $G_C(s)$, $G_1(s)$, $\bar{\mathscr{M}}_V$, $G_2(s)$, $G_3(s)$, $\bar{\mathscr{C}}$, $\bar{\mathscr{B}}$

FIG. 7.59. Simple feedback control loop with significant dead time in the process

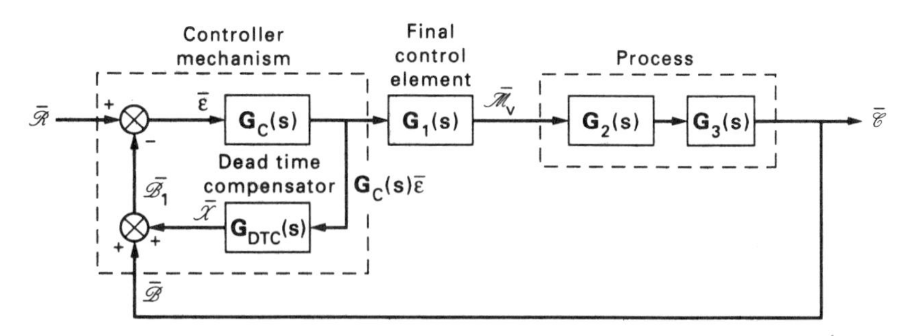

FIG. 7.60. As Fig. 7.59 but with dead time compensation included in the controller mechanism

If B is the open-loop response of the measured value including the dead time, then we can write (Fig. 7.60):

$$\bar{\mathscr{B}}_1 = \bar{\mathscr{B}} + \bar{\mathscr{X}} \tag{7.143}$$

where $\bar{\mathscr{B}} = \mathbf{G}_X(s)\mathbf{G}_3(s)\mathscr{R}$, $\bar{\mathscr{X}} = \mathbf{G}_X(s)[1 - \mathbf{G}_3(s)]\bar{\varepsilon}$ and $\mathbf{G}_X(s) = \mathbf{G}_C(s)\mathbf{G}_1(s)\mathbf{G}_2(s)$.

The signal X can be introduced by means of an internal feedback loop within the controller (Fig. 7.60) which constitutes the dead time compensator $\mathbf{G}_{DTC}(s)$ such that:

$$\mathbf{G}_{DTC}(s) = \mathbf{G}_1(s)\mathbf{G}_2(s)[1 - \mathbf{G}_3(s)] \tag{7.144}$$

From Fig. 7.60 and equations 7.143 and 7.144:

$$\bar{\mathscr{B}}_1 = \bar{\mathscr{B}} + \bar{\mathscr{X}} = \bar{\mathscr{C}} + \mathbf{G}_X(s)[1 - \mathbf{G}_3(s)](\bar{\mathscr{R}} - \bar{\mathscr{B}}_1)$$

Hence:
$$\bar{\mathscr{B}}_1 = \frac{\bar{\mathscr{C}} + \mathbf{G}_X(s)[1 - \mathbf{G}_3(s)]\bar{\mathscr{R}}}{1 + \mathbf{G}_X(s)[1 - \mathbf{G}_3(s)]} \tag{7.145}$$

Also:
$$\bar{\mathscr{C}} = \mathbf{G}_X(s)\mathbf{G}_3(s)(\bar{\mathscr{R}} - \bar{\mathscr{B}}_1) \tag{7.146}$$

Eliminating $\bar{\mathscr{B}}_1$ from equations (7.145) and (7.146) gives:

$$\bar{\mathscr{C}} = \mathbf{G}_X(s)\mathbf{G}_3(s)\bar{\mathscr{R}} - \mathbf{G}_X(s)\mathbf{G}_3(s)\left(\frac{\bar{\mathscr{C}} + \mathbf{G}_X(s)[1 - \mathbf{G}_3(s)]\bar{\mathscr{R}}}{1 + \mathbf{G}_X(s)[1 - \mathbf{G}_3(s)]}\right)$$

$$\bar{\mathscr{C}}\,(1 + \mathbf{G}_X(s)[1 - \mathbf{G}_3(s)] - \mathbf{G}_X(s)\mathbf{G}_3(s)) = \mathbf{G}_X(s)\mathbf{G}_3(s)\bar{\mathscr{R}}$$

$$\therefore \quad \frac{\bar{\mathscr{C}}}{\bar{\mathscr{R}}} = \frac{\mathbf{G}_X(s)}{1 + \mathbf{G}_X(s)}\mathbf{G}_3(s)$$

$$= \frac{\mathbf{G}_X(s)}{1 + \mathbf{G}_X(s)}\exp(-\tau_{DV}s) \tag{7.147}$$

This means that the Smith compensator has moved the effect of the dead time outside the closed-loop as in the equivalent schematic representation in Fig. 7.61.

It can be seen from equation 7.144 that implementation of the Smith predictor assumes knowledge of the transfer functions (i.e. the dynamics) of the process (including the dead time) and the final control element. These are unlikely to be

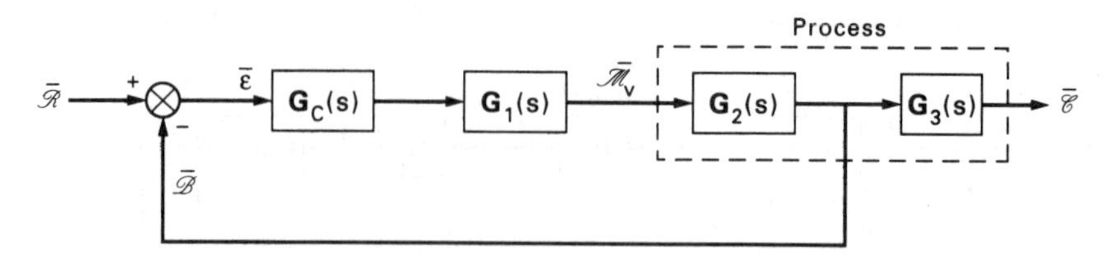

FIG. 7.61. Schematic representation of the effect of the dead time compensation shown in Fig. 7.60. Dead time now effectively outside the control loop

either known or modelled perfectly. Consequently, the effectiveness of the compensator will depend upon how well $\mathbf{G}_1(s)$, $\mathbf{G}_2(s)$ and τ_{DV} are known. It is more important to estimate τ_{DV} accurately than the parameters and form of either $\mathbf{G}_1(s)$ or $\mathbf{G}_2(s)$. Furthermore τ_{DV} could well change with time due to variations in process flowrates. This information would have to be supplied to the compensator in order that $\mathbf{G}_{DTC}(s)$ can be continually updated.

7.12.2. Series Compensation

It can be seen from Example 7.8 and Fig. 7.55 (curves (*a*) and (*d*)) that changing the proportional gain K_C of the controller varies the scale factor of the relevant open-loop polar plot, but not its basic shape. This applies equally to the total open-loop gain. If the system is unstable, i.e. the open-loop polar plot encircles the point $(-1, 0)$ on the complex plane, then clearly stability can be achieved by the requisite reduction in the system open-loop gain. This, however, means that the gain will be reduced equally over the whole frequency range. There are circumstances where this is not desirable. For example, it may be necessary to have a high gain only at low frequencies. Such a system would be one where there is substantial resistance to the movement of a valve due to (say) friction. A signal which has a high gain towards the lower end of the frequency spectrum would overcome this. Various other applications can also make it necessary to change the shape of the polar plot in order to obtain the desired dynamic performance. Series compensation is a method by which this can be achieved and entails the insertion of a suitable compensating element (or *series compensator*) in the feed-forward section of the control loop. The frequency response characteristics of such a component may be designed so that its output can either lead or lag its input. It is often advantageous to use a component for which the output lags the input at certain frequencies and leads at other frequencies.

Lead Compensation

A lead compensator has the transfer function:

$$\mathbf{G}_{\text{comp}}(s) = \frac{\tau_2}{\tau_1}\left(\frac{1 + \tau_1 s}{1 + \tau_2 s}\right) \quad (\tau_1 > \tau_2) \tag{7.148}$$

The corresponding frequency response is obtained by substituting $i\omega$ for s (Section 7.8.4), viz:

$$\mathbf{G}_{\text{comp}}(\mathrm{i}\omega) = \frac{\tau_2}{\tau_1}\left(\frac{1 + \tau_1 \mathrm{i}\omega}{1 + \tau_2 \mathrm{i}\omega}\right) \quad (\tau_1 > \tau_2) \tag{7.149}$$

The lead compensator contributes phase advance to the system and thus increases the overall system stability (Section 7.10.4). The degree of phase advance provided is a function of frequency. At the same time this type of compensator increases the overall system amplitude ratio, which has the effect of reducing the the stability of the system. However, the major contribution of phase advance occurs at those frequencies where the open-loop polar plot is adjacent to the $(-1, 0)$ point on the complex plane. The increase in amplitude ratio takes place at lower frequencies and, consequently, the effect of this is much less significant. As the ratio of τ_1/τ_2 is increased, the maximum phase advance supplied by the lead compensator also increases, i.e. the greater is the stabilising effect of the compensating element[31].

Curve a in Fig. 7.62 shows the open-loop polar plot for the heat exchanger system described in Example 7.6. with $K_C = 1.8$ and $\tau_I = 2.5$ (see also Example 7.8 and Fig. 7.55). Clearly this indicates an unstable system (Section 7.10.5). If a lead compensator with $\tau_1 = 1$ min and $\tau_2 = 0.1$ min ($\tau_1/\tau_2 = 10$) is inserted into the loop, as shown in Fig. 7.63, then the system becomes stable (curve b in Fig. 7.62) due to the additional phase lead supplied by the compensator. (Using these values of τ_1 and τ_2, K_C can now be increased by almost a factor of ten before the system becomes unstable).

Note that it is necessary to include an extra element τ_1/τ_2 in the loop in order to maintain the original open-loop steady-state gain.

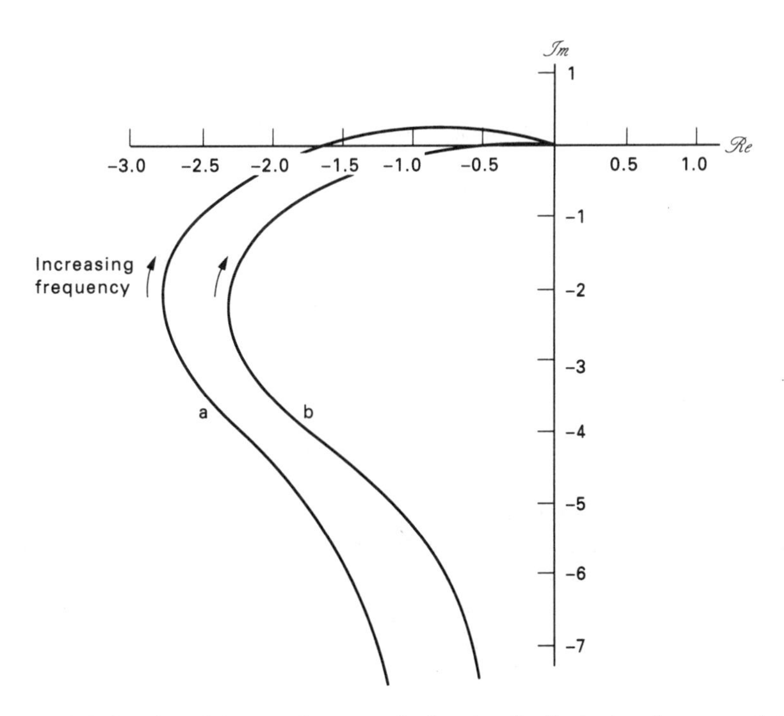

FIG. 7.62. Polar plot of the open-loop transfer function for the heat exchanger control system described in Example 7.6 with $K_c = 1.8$ and $\tau_I = 2.5$: (*a*) uncompensated system; (*b*) compensated system with $\tau_1 = 1$ min, $\tau_2 = 0.1$ min

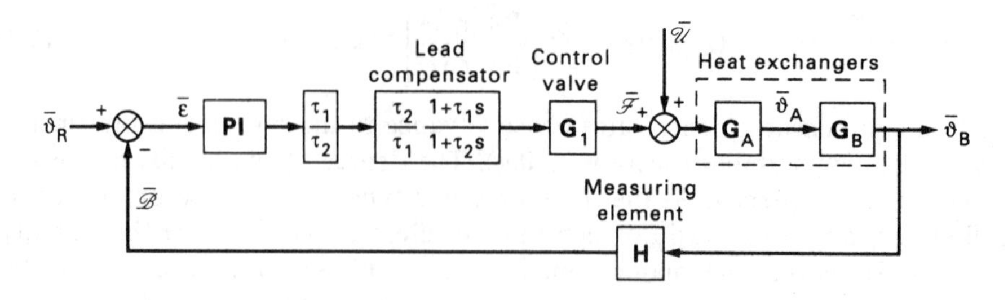

FIG. 7.63. Block diagram of heat exchanger control system illustrated in Fig. 7.42 with lead compensator added

The properties of the lead compensator must be considered in order to select suitable values for τ_1 and τ_2. The AR and phase shift of such a compensator are (from equations 7.104, 7.105 and 7.149):

$$\text{AR}_{\text{comp}} = \sqrt{\left(\frac{1 + \omega^2 \tau_1^2}{1 + \omega^2 \tau_2^2}\right)} \tag{7.150}$$

and:

$$\psi_{\text{comp}} = \mathscr{Arg}\left(\frac{1 + \tau_1 i\omega}{1 + \tau_2 i\omega}\right) = \tan^{-1}(\omega\tau_1) - \tan^{-1}(\omega\tau_2) \tag{7.151}$$

Hence, for maximum phase shift:

$$\frac{d\psi_{\text{comp}}}{d\omega} = \frac{\tau_1}{1 + (\omega\tau_1)^2} - \frac{\tau_2}{1 + (\omega\tau_2)^2} = 0$$

Thus the value of ω for which ψ_{comp} is a maximum is:

$$\omega_{\max} = \frac{1}{\sqrt{(\tau_1 \tau_2)}} \tag{7.152}$$

Substituting equation 7.152 in equations 7.150 and 7.151 gives:

$$(\text{AR}_{\text{comp}})_{\max} = \sqrt{\left(\frac{\tau_1}{\tau_2}\right)} \tag{7.153}$$

and:

$$(\psi_{\text{comp}})_{\max} = \tan^{-1}\left\{0.5\left(\sqrt{\left(\frac{\tau_1}{\tau_2}\right)} - \sqrt{\left(\frac{\tau_2}{\tau_1}\right)}\right)\right\} \tag{7.154}$$

The frequency at which $(\psi_{\text{comp}})_{\max}$ occurs is located at the mid-point between the frequencies $\omega_{c1} = 1/\tau_1$ and $\omega_{c2} = 1/\tau_2$ (i.e. the two corner frequencies in the corresponding Bode diagram—see Section 7.10.4). The determination of the appropriate values of τ_1 and τ_2 to provide a given stability specification requires a trial and error procedure as shown in Example 7.10.

Example 7.10

The log modulus plot of $\mathbf{G}(i\omega)$ for the heat exchanger system illustrated in Figs 7.42 and 7.43 with the controller parameters set at $K_C = 1.8$ and $\tau_I = 6$ is shown in Fig. 7.64 (curve (i)—from curve (c) in

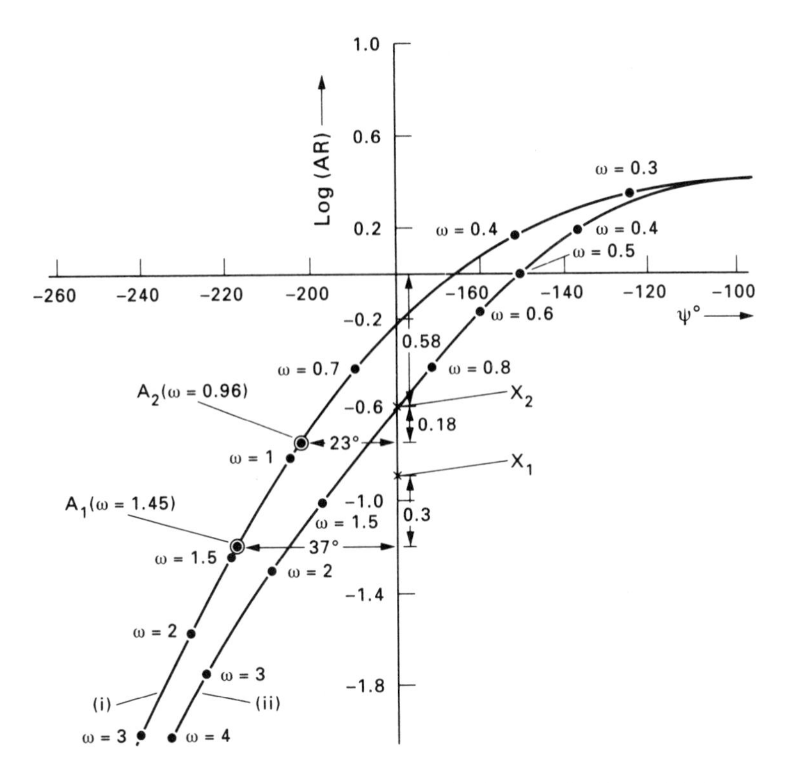

FIG. 7.64. Log modulus plots of open-loop transfer function for Example 7.10: curve (i)—uncompensated system; curve (ii)—compensated system

Fig. 7.56). Determine the values of τ_1 and τ_2 for a series lead compensator such that the compensated system will exhibit a gain margin of 3.8.

Solution

If the gain margin is to be 3.8, then the log modulus plot of the compensated system must pass through the point log(AR) = log 3.8 = 0.58 when $\psi = -180°$ (point X_2 in Fig. 7.64).

As a first trial, put $\tau_1/\tau_2 = 4$. The corresponding maximum AR and phase shift can now be obtained from equations 7.153 and 7.154, viz:

$$(AR_{comp})_{max} = \sqrt{\left(\frac{\tau_1}{\tau_2}\right)} = 2$$

and:
$$(\psi_{comp})_{max} = \tan^{-1}\{0.5(\sqrt{4} - \sqrt{1/4})\} \approx 37°$$

Hence, point A_1 on curve (i) in Fig. 7.64 (plot of uncompensated system) is found by measuring 37° to the left of the vertical −180° axis. The corresponding frequency (which is a parameter on the log modulus plot) is $\omega_{max} \approx 1.45$ radians/min. Thus the effect of including this compensator in the control loop is to move the entire log modulus plot such that point A_1 shifts 37° to the right (corresponding to a decrease in phase lag of 37°) and vertically upwards by an amount log $[(AR_{comp})_{max}]$ = log 2 ≈ 0.3 units (which represents the corresponding increase in AR). This brings point A_1 to point X_1 where log AR = 0.88. Hence the gain margin for this compensated system is $10^{0.88} \approx 7.6$ which does not meet the required stability specification.

As a second trial, put $\tau_1/\tau_2 = 2.3$. This gives $(AR_{comp})_{max} \approx 1.5$ and $(\psi_{comp})_{max} \approx 23°$. Marking off 23° to the left of the vertical −180° axis gives point A_2 on the uncompensated plot (curve (i)—Fig. 7.64). The frequency at this point is $\omega_{max} \approx 0.96$ radians/min. Thus, inclusion of this compensator in the control loop will shift the log modulus plot such that point A_2 moves 23° to the right and vertically upwards by $\log[(AR_{comp})_{max}] = \log 1.5 \approx 0.18$ units. This coincides with point X_2 and thus has the required compensation. Curve (ii) (Fig. 7.64) is the log modulus plot for the system including the compensator. Individual values of τ_1 and τ_2 can be found by combining the estimated ratio $\tau_1/\tau_2 = 2.3$ with equation 7.152, i.e.:

$$\omega_{max} = \frac{1}{\sqrt{(\tau_1 \tau_2)}} = 0.96$$

Hence:

$$\tau_1 \approx 1.6 \text{ min} \quad \text{and} \quad \tau_2 \approx 0.7 \text{ min}$$

Lag Compensation

A lag compensator is described by:

$$\mathbf{G}_{comp}(s) = \frac{1 + \tau_1 s}{1 + \tau_2 s} \qquad (\tau_1 < \tau_2) \tag{7.155}$$

and the frequency response is:

$$\mathbf{G}_{comp}(i\omega) = \frac{1 + \tau_1 i\omega}{1 + \tau_2 i\omega} \qquad (\tau_1 < \tau_2) \tag{7.156}$$

The output of the element represented by equation 7.155 lags the input. However, the destabilising effect of this additional lag is more than offset by an associated decrease in amplitude ratio. This decrease is more pronounced as the difference between τ_1 and τ_2 is increased. Lag compensators can be designed to produce different total open-loop stability specifications (e.g. in terms of allowable gain margin, phase margin, etc.) in a manner similar to that for lead compensators.

Lag–Lead Compensation

The transfer function of a *lag–lead* element is:

$$\mathbf{G}_{comp}(s) = \left(\frac{1 + \nu\tau_1 s}{1 + \nu\tau_2 s}\right)\left(\frac{1 + \tau_2 s}{1 + \tau_1 s}\right) \tag{7.157}$$

where $\tau_1 < \tau_2$ and $\nu > 1$ is a constant.

The corresponding frequency response is:

$$\mathbf{G}_{comp}(i\omega) = \left(\frac{1 + \nu\tau_1 i\omega}{1 + \nu\tau_2 i\omega}\right)\left(\frac{1 + \tau_2 i\omega}{1 + \tau_1 i\omega}\right) \tag{7.158}$$

The Bode diagram of such a function is shown in Fig. 7.65 which is a combination of the corresponding plots for the separate lead and lag elements. Figure 7.65 shows that the magnitude of ν controls the separation between the ranges of frequency over which the individual influences of the lag and lead portions of the lag-lead element are each significant. A typical value of ν is 5[31]. The maximum phase advance contributed by the compensator occurs at $\omega = 1/\sqrt{(\tau_1 \tau_2)}$ and the corresponding amplitude ratio is $\sqrt{\tau_1/\tau_2}$. This is the same as for the lead compensator only, except that, for the lag–lead compensator, the sign of the amplitude ratio is negative. This property makes the lag–lead compensator much more effective than the lead compensator alone as it reduces the AR of the compensated system

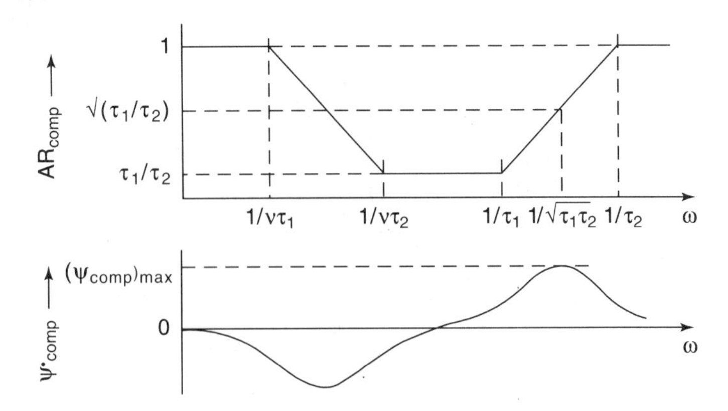

FIG. 7.65. Bode diagram for a lag-lead compensator

at the frequency at which the phase advance of the compensator is a maximum. Design procedures can be employed which are similar to those for the lead compensator.

7.13. CASCADE CONTROL

One example of the use of cascade control has already been demonstrated in Fig. 7.9 (Section 7.3) where the output of the temperature controller is used to adjust the set point of the reflux flow controller. In general, a cascade system consists of a *primary* and a *secondary* control loop. Disturbances occurring in the secondary loop (i.e. in reflux flowrate in this case) are corrected by the secondary (or *slave*) controller before they can affect the controlled variable of the primary loop (i.e. the measured temperature in the column). The latter is controlled by the primary (or *master*) controller. For the system to operate satisfactorily, the response of the secondary control loop must be significantly faster than that of the primary loop. Hence proportional action alone with a high gain (i.e. with a large value of K_C) is often used in the secondary controller. Any offset associated with this proportional control action in the secondary loop is dealt with by including integral action in the primary controller. Hence, the latter is either a **PI** or **PID** controller and the secondary controlled variable is automatically adjusted according to the requirements of the primary control loop.

A further example of cascade control is illustrated in Fig. 7.66 in which the temperature of the outlet stream of a process heater is controlled by regulating the flow of fuel to the burners. The output of the primary controller is again used to adjust the set point of the secondary controller which maintains a steady flow of fuel until the temperature of the heater outlet stream changes and the primary controller calls for a consequent change in the fuel flowrate. Figure 7.67*a* is the block diagram for this cascade control system and the corresponding closed-loop transfer function of the secondary loop for a set point change is (from equation 7.111):

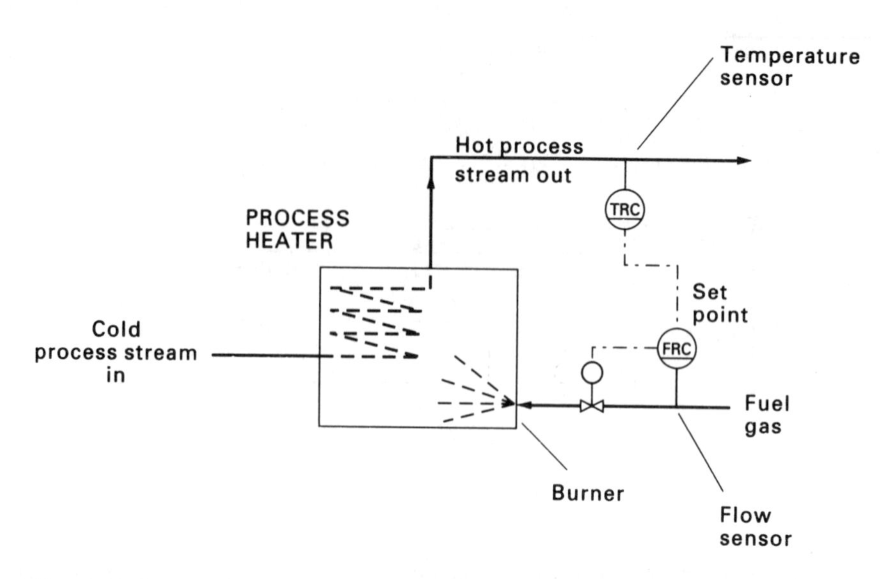

FIG. 7.66. Cascade control of the temperature of the outlet stream from a process heater

$$\mathbf{G}_{scl}(s) = \frac{\bar{\mathscr{C}}_s}{\bar{\mathscr{R}}_s} = \frac{\mathbf{G}_{Cs}(s)\mathbf{G}_{1s}(s)\mathbf{G}_{2s}(s)}{1 + \mathbf{G}_{Cs}(s)\mathbf{G}_{1s}(s)\mathbf{G}_{2s}(s)\mathbf{H}_s(s)} \tag{7.159}$$

Hence, from Fig. 7.67*b*, the primary closed-loop transfer function is:

$$\mathbf{G}_{pcl}(s) = \frac{\bar{\mathscr{C}}_p}{\bar{\mathscr{R}}_p} = \frac{\mathbf{G}_{Cp}(s)\mathbf{G}_{scl}(s)\mathbf{G}_{2p}(s)}{1 + \mathbf{G}_{Cp}(s)\mathbf{G}_{scl}(s)\mathbf{G}_{2p}(s)\mathbf{H}_p(s)} \tag{7.160}$$

Usually the dynamics of the secondary loop are sufficiently faster than those of the primary loop for $\mathbf{G}_{scl}(s)$ to be approximated by its steady-state gain. For the same reason it is possible to tune the cascade system by tuning first the inner loop and then the outer loop.

Cascade control is particularly useful in reducing the effect of a load disturbance that moves through the control system slowly. The inner loop has the effect of reducing the lag in the primary loop such that the multiloop cascade arrangement responds more quickly than an equivalent single loop system and exhibits a higher degree of oscillation(18). A combination of cascade and ratio control is shown in Fig. 7.71.

7.14. FEED-FORWARD AND RATIO CONTROL

7.14.1. Feed-forward Control

This is also termed *predictive control* and can be usefully employed where there are significant time lags between a load change and its subsequent effect upon a system output variable. A typical example is the response of the composition x_D of the overhead product stream of a distillation process to a perturbation in feed

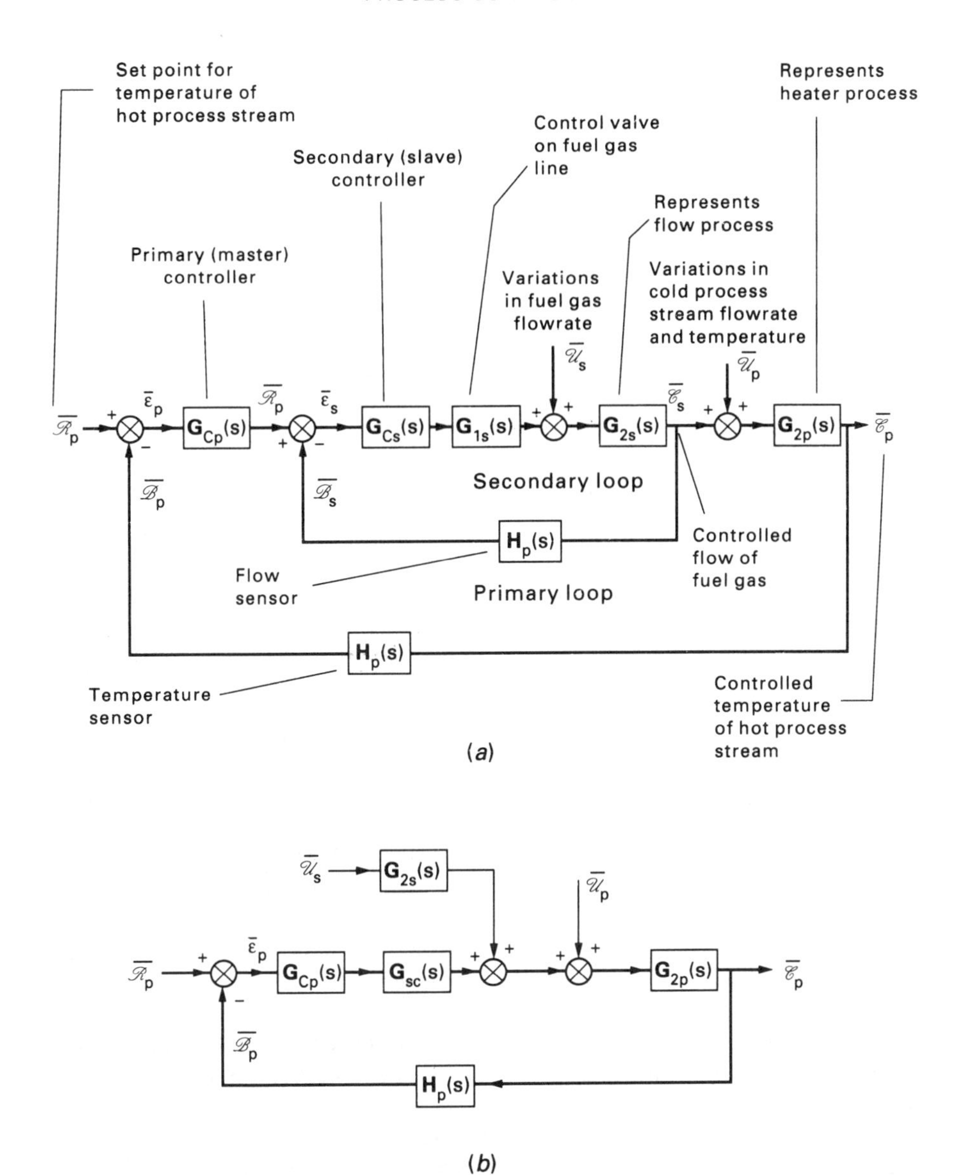

FIG. 7.67. Block diagrams of cascade control system shown in Fig. 7.66: (*a*) original block diagram; (*b*) condensed block diagram

composition. The problems of employing fixed parameter feedback control in such a case are discussed in Section 7.3.1. An important feature of the feedback control system is that it will not take action until the controlled variable moves away from its set point. In the particular example of the feeback control of x_D (using boiling temperature at some appropriate point within the column as the measured variable), the effect of the control action may be significantly delayed due to the time lags in the column. In the meantime further disturbances may have entered the column.

An : composition analyser
FRC : flow recorder controller
TRC : temperature recorder controller
LIC : level indicator controller
PIC : pressure indicator controller

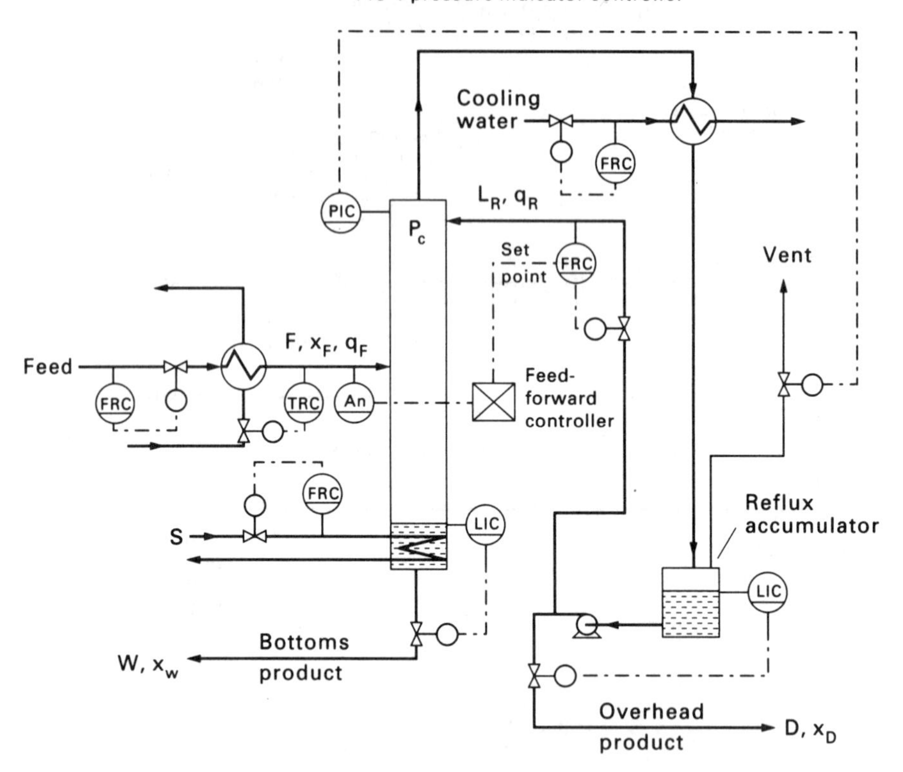

FIG. 7.68. Feed-forward control of the composition of the overhead product stream of a distillation column where feed composition is the only load variable

Feed-forward or predictive control (or compensation) can be applied in such instances in order that control action may be taken before the controlled variable deviates from its specified value.

A typical arrangement is illustrated in Fig. 7.68 where the variations in feed composition are measured by a suitable composition analyser (An) (see Section 6.8). The signal from the analyser is fed directly to the feed-forward controller, the output of which is cascaded on to the set point of the reflux flow controller (see also Section 7.13). If the transfer functions relating feed composition x_F, reflux flowrate R and overhead product composition x_D are known, then we can write:

$$\mathbf{G}_1(s) = \frac{\bar{\mathscr{x}}_D}{\bar{\mathscr{x}}_F} \quad \text{and} \quad \mathbf{G}_2(s) = \frac{\bar{\mathscr{x}}_D}{\bar{\mathscr{R}}}$$

i.e.

$$\bar{\mathscr{x}}_D = \mathbf{G}_1(s)\bar{\mathscr{x}}_F \quad \text{and} \quad \bar{\mathscr{x}}_D = \mathbf{G}_2(s)\bar{\mathscr{R}}$$

If perturbations in x_F and R occur simultaneously, then, by the principle of superposition, the resultant variation in x_D will be:

$$\bar{x}_D = \mathbf{G}_1(s)\bar{x}_F + \mathbf{G}_2(s)\bar{\mathcal{R}} \tag{7.161}$$

The desired control criterion is that there should be no variation in x_D, i.e. that:

$$\bar{x}_D = x_D = 0$$

Substitution of this condition in equation 7.161 yields:

$$0 = \mathbf{G}_1(s)\bar{x}_F + \mathbf{G}_2(s)\bar{\mathcal{R}}$$

$$\therefore \qquad \mathbf{G}_{FF}(s) = \frac{\bar{\mathcal{R}}}{\bar{x}_F} = -\frac{\mathbf{G}_1(s)}{\mathbf{G}_2(s)} \tag{7.162}$$

$\mathbf{G}_{FF}(s)$ is the transfer function of the feed-forward controller which is obtained from the known transfer functions $\mathbf{G}_1(s)$ and $\mathbf{G}_2(s)$ using equation 7.162.

If it is desired to control both overhead and bottoms compositions simultaneously in the face of variations in feed composition then it is necessary to employ two controlling variables as in Fig. 7.69. In this case knowledge is required of the following transfer functions:

$$\mathbf{G}_{11}(s) = \frac{\bar{x}_D}{\bar{x}_F}, \qquad \mathbf{G}_{12}(s) = \frac{\bar{x}_D}{\bar{\mathcal{R}}}, \qquad \mathbf{G}_{13}(s) = \frac{\bar{x}_D}{\bar{y}} \tag{7.163a}$$

$$\mathbf{G}_{21}(s) = \frac{\bar{x}_W}{\bar{x}_F}, \qquad \mathbf{G}_{22}(s) = \frac{\bar{x}_W}{\bar{\mathcal{R}}}, \qquad \mathbf{G}_{23}(s) = \frac{\bar{x}_W}{\bar{y}} \tag{7.163b}$$

Hence, the control criteria for the two feed-forward controllers are:

$$\bar{x}_D = \mathbf{G}_{11}(s)\bar{x}_F + \mathbf{G}_{12}(s)\bar{\mathcal{R}} + \mathbf{G}_{13}(s)\bar{y} = 0 \tag{7.164a}$$

$$\bar{x}_W = \mathbf{G}_{21}(s)\bar{x}_F + \mathbf{G}_{22}(s)\bar{\mathcal{R}} + \mathbf{G}_{23}(s)\bar{y} = 0 \tag{7.164b}$$

These yield the relevant transfer functions, viz.:

$$\mathbf{G}_{FFA}(s) = \frac{\bar{\mathcal{R}}}{\bar{x}_F} = \frac{\mathbf{G}_{11}(s)\mathbf{G}_{23}(s) - \mathbf{G}_{21}(s)\mathbf{G}_{13}(s)}{\mathbf{G}_{22}(s)\mathbf{G}_{13}(s) - \mathbf{G}_{12}(s)\mathbf{G}_{23}(s)} \tag{7.165a}$$

$$\mathbf{G}_{FFB}(s) = \frac{\bar{y}}{\bar{x}_F} = \frac{\mathbf{G}_{11}(s)\mathbf{G}_{22}(s) - \mathbf{G}_{21}(s)\mathbf{G}_{12}(s)}{\mathbf{G}_{23}(s)\mathbf{G}_{12}(s) - \mathbf{G}_{13}(s)\mathbf{G}_{22}(s)} \tag{7.165b}$$

Substantial effort in modelling and/or experimental measurement is required in order to derive $\mathbf{G}_{FFA}(s)$ and $\mathbf{G}_{FFB}(s)$. Due to errors in determining the individual transfer functions ($\mathbf{G}_{11}(s)$, $\mathbf{G}_{12}(s)$, etc.), to errors in measurement, and to load variables which have not been accounted for in the models, feed-forward compensation can never be perfect, and considerable drifting of the controlled variable(s) can occur. On the other hand, the two variable feed-forward control model expressed by equation 7.165 automatically takes into account any interaction between the reflux and steam flow control loops (see also Section 7.15).

In some instances it is found that the derived transfer functions produce a physically unrealistic or *unrealisable* situation. For example, consider the control of x_D only where x_F is the load variable and R is the manipulated variable. Suppose that:

$$\mathbf{G}_1(s) = \frac{\bar{x}_D}{\bar{x}_F} = \frac{2\exp(-0.1s)}{2s+1}$$

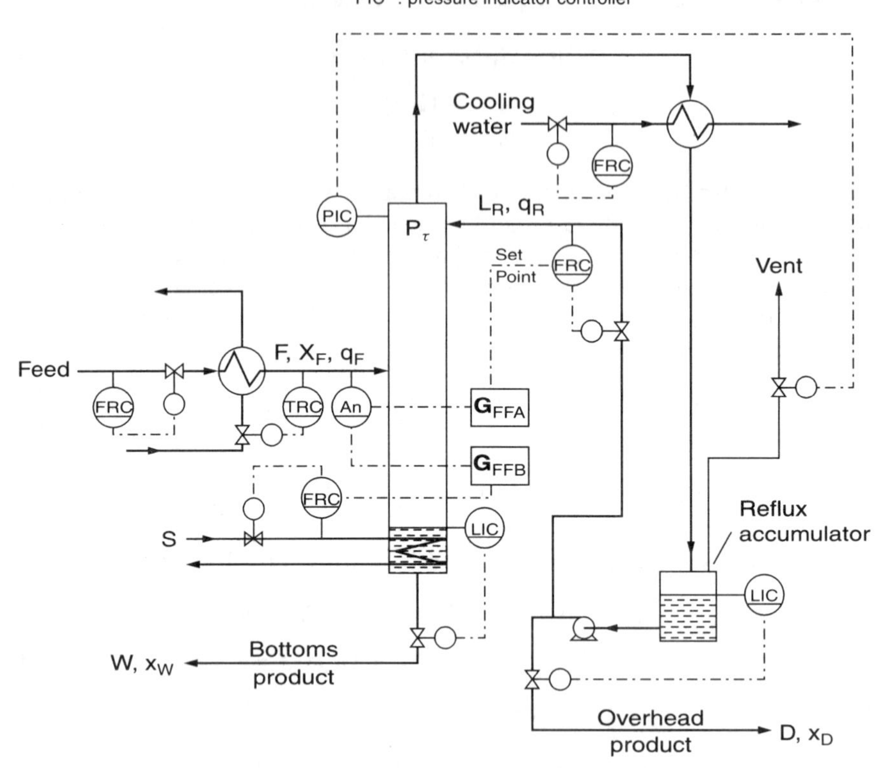

FIG. 7.69. Simultaneous feed-forward control of overhead and bottoms compositions in the face of variations in feed composition

and:

$$\mathbf{G}_2(s) = \frac{\bar{x}_D}{\bar{\mathscr{R}}} = \frac{10\exp(-0.5s)}{2s+1}$$

then, from equation 7.162:

$$\mathbf{G}_{FF}(s) = \frac{\bar{\mathscr{R}}}{\bar{x}_F} = -\frac{\mathbf{G}_1(s)}{\mathbf{G}_2(s)} = -\frac{\exp(0.4s)}{5}$$

The positive power of the exponent in $\mathbf{G}_{FF}(s)$ implies that the feed-forward controller requires values of the load disturbance measured at some future time. This is clearly impossible.

Imperfections in feed-forward control can often be overcome by the addition of suitable feedback action. A typical design is shown in Fig. 7.70 where any variations in x_D which occur bring the feedback control loop into action. The reflux flow is shown on flow control in cascade with the boiling temperature of the liquid at an appropriate point within the column. The inner (or slave) flow controller maintains

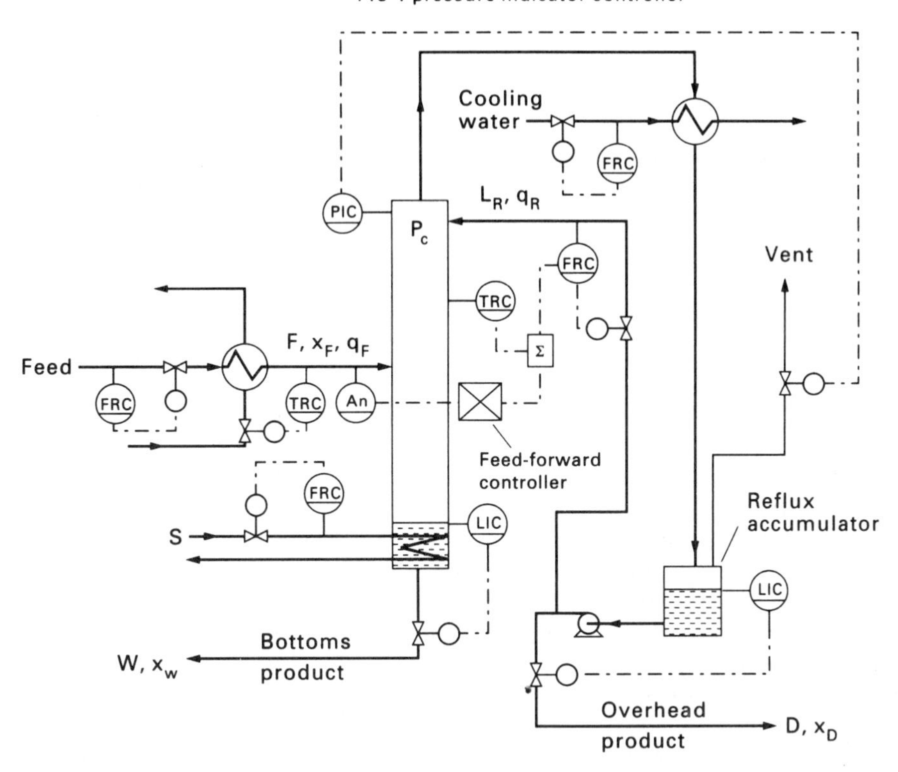

FIG. 7.70. Combined feed-forward/feedback control of the overhead product stream of a distillation column. Feed composition is the only significant load variable

a steady reflux flowrate until a change occurs in boiling temperature or in feed composition—either of which will vary the set point of the flow controller as required. Good results have been obtained using a feed-forward controller designed on a steady-state basis with this control configuration[32].

7.14.2. Ratio Control

Ratio control is generally defined as a means of holding two or more disturbances in a constant ratio to each other[6]. A common example is where the flowrate of one particular stream (sometimes called the *wild* stream) is measured and this measurement is used to adjust the set point of a flow controller which, in turn, regulates the flow of another process stream (the *controlled* stream) such that the flowrates of the two streams maintain a fixed ratio to each other. In this sense, ratio control bears a similarity to cascade control, although, in fact, it is a form of predictive control. Figure 7.71 demonstrates the use of a ratio control strategy for diluting concentrated sodium hydroxide solution, the level in the tank being also

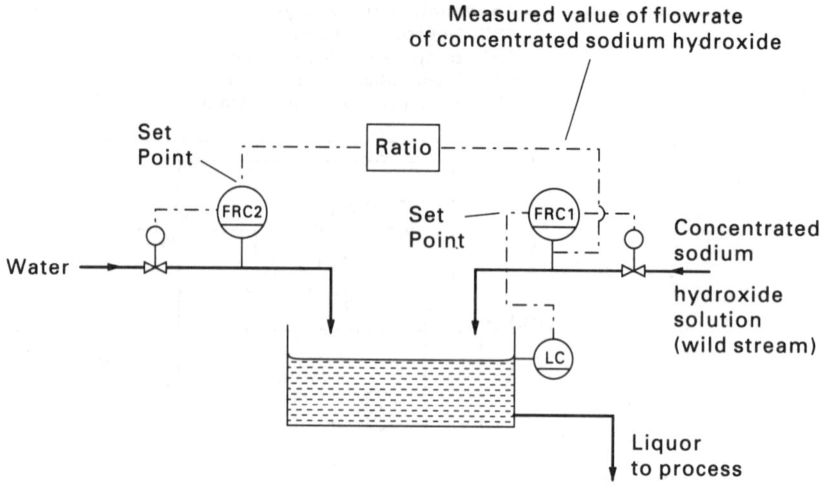

Measured value of flowrate of concentrated sodium hydroxide
Set Point
Ratio
FRC2
Set Point
FRC1
Water
Concentrated sodium hydroxide solution (wild stream)
LC
Liquor to process

FIG. 7.71. Level control cascaded on to a ratio flow control system

controlled by a cascade arrangement with the level controller as the primary (master) controller.

An alternative type of ratio control system is shown in Fig. 7.72 where both flowrates are measured and the ratio between them determined. This measured ratio is compared to the desired ratio (acting as the set point) and the difference is used as the error signal for the controller which adjusts the flowrate of the controlled stream accordingly.

STEPHANOPOULOS[6] lists common uses of ratio control as follows:

(a) To keep a constant ratio between the feed flowrate and the steam flowrate to the reboiler of a distillation column.
(b) To maintain constant a distillation column reflux ratio.

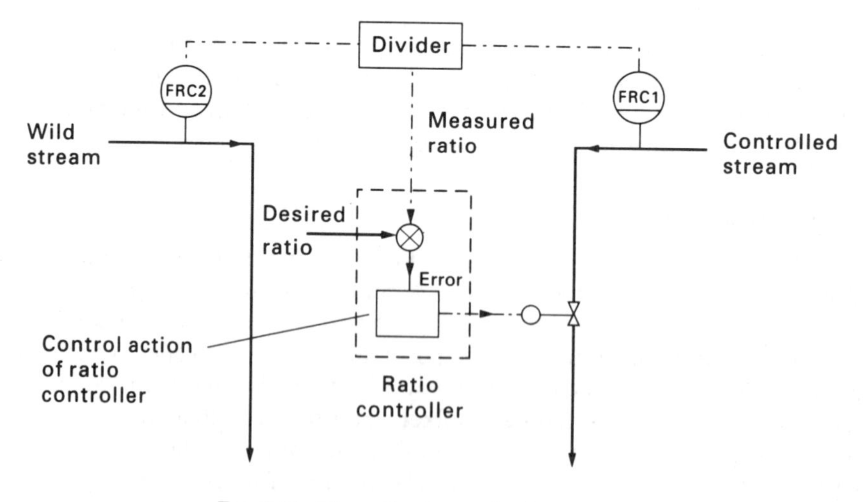

FIG. 7.72. Alternative form of ratio control

(c) To control the ratio of two reactants entering a reactor.
(d) To hold constant the ratio of two feed streams entering a blending operation in order to maintain the composition of the blend.
(e) To maintain the ratio of a purge stream to a recycle stream.
(f) To keep the fuel/air ratio of the feed to a burner at that value which gives the most efficient combustion.
(g) To maintain constant the ratio of the liquid flowrate to gas flowrate entering an absorber in order to achieve the desired composition in the exit stream.

7.15. MIMO SYSTEMS—INTERACTION AND DECOUPLING

7.15.1. Interaction Between Control Loops

Interaction can be between two or more processes or between actions produced by different control loops applied to a single process. The former has already been discussed in Section 7.5.3. Some degree of interaction between control loops will nearly always occur in a multiple-input/multiple-output (MIMO) system. For example, consider the distillation process described in Section 7.3 (Fig. 7.9). Suppose it is desired to control simultaneously the compositions of both the overheads product stream (by manipulating the reflux flowrate) and the bottoms product stream (by regulating the steam flowrate to the reboiler). A typical arrangement is shown in Fig. 7.73.

Adjustment of the reflux in order to maintain a constant overhead composition (generally employing a boiling temperature measured at a selected point within the rectifying section) will inevitably affect the composition of the bottoms product (and, of course, the boiling temperature in the stripping section). The consequent adjustment of the reboil vapour flowrate (via the steam-to-reboiler flowrate) in order to maintain the bottoms composition at its desired value will, in turn, disturb the overheads composition—and so on. Often such control loop interactions lead to *cycling* (continuous oscillations of slightly varying amplitude) in the controlled variables of the interacting loops. This can often be quite severe, i.e. the amplitude of the oscillation can be large. The block diagram (Fig. 7.74) corresponding to Fig. 7.73 illustrates the way in which interaction occurs between the two control loops.

In Fig. 7.74, $\mathbf{G}_{11}(s)$ is the transfer function between the reflux flowrate and the measured variable (the boiling temperature at the selected stage in the rectifying section). $\mathbf{G}_{22}(s)$ is the transfer function between the steam-to-reboiler flowrate and the boiling temperature at the selected stage in the stripping section. Control loop interactions are represented by the transfer functions $\mathbf{G}_{21}(s)$ and $\mathbf{G}_{12}(s)$, where $\mathbf{G}_{21}(s)$ is the relationship between the reflux flowrate and the temperature in the stripping section and $\mathbf{G}_{12}(s)$ represents that between the steam-to-reboiler flowrate and the temperature in the rectifying section of the column.

In general, we can write, from Fig. 7.74, for any pair of interacting control loops:

$$\bar{\mathscr{C}}_1 = \mathbf{G}_{11}(s)\bar{\mathscr{M}}_{v1} + \mathbf{G}_{12}(s)\bar{\mathscr{M}}_{v2} \tag{7.166a}$$

$$\bar{\mathscr{C}}_2 = \mathbf{G}_{21}(s)\bar{\mathscr{M}}_{v1} + \mathbf{G}_{22}(s)\bar{\mathscr{M}}_{v2} \tag{7.166b}$$

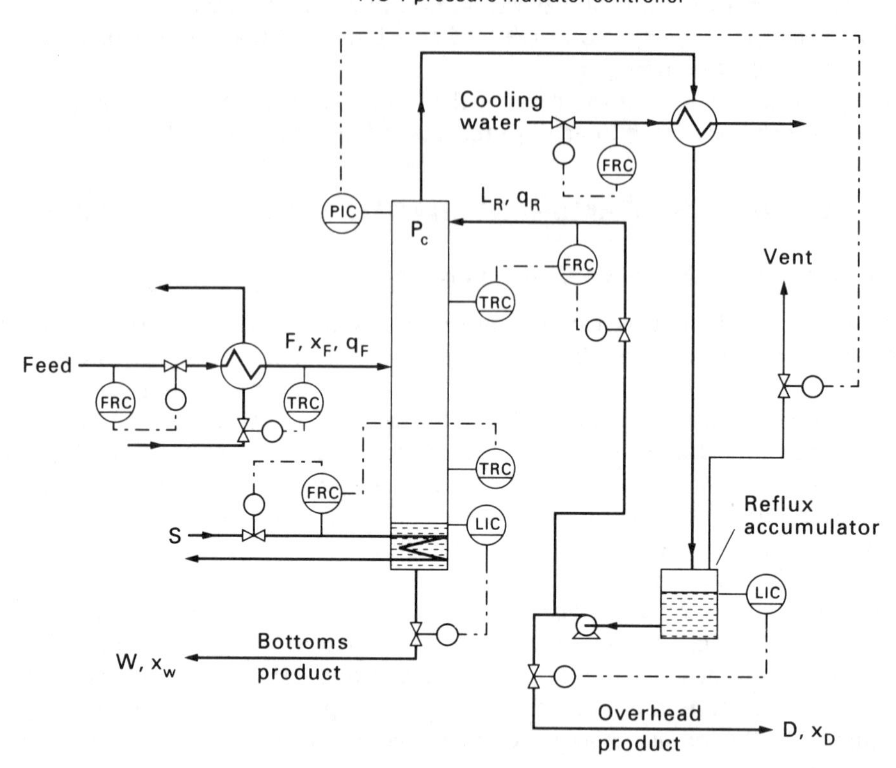

FIG. 7.73. Simultaneous control of the compositions of the overhead and bottom product streams of a distillation process

and:

$$\bar{\mathcal{M}}_{v1} = \mathbf{G}_{C1}(s)(\bar{\mathcal{R}}_1 - \bar{\mathcal{C}}_1) \tag{7.167a}$$

$$\bar{\mathcal{M}}_{v2} = \mathbf{G}_{C2}(s)(\bar{\mathcal{R}}_2 - \bar{\mathcal{C}}_2) \tag{7.167b}$$

Eliminating $\bar{\mathcal{M}}_{v1}$ and $\bar{\mathcal{M}}_{v2}$ from equations 7.166 and 7.167 gives:

$$\bar{\mathcal{C}}_1 = \mathbf{G}_{C1}(s)\mathbf{G}_{11}(s)(\bar{\mathcal{R}}_1 - \bar{\mathcal{C}}_1) + \mathbf{G}_{C2}(s)\mathbf{G}_{12}(s)(\bar{\mathcal{R}}_2 - \bar{\mathcal{C}}_2) \tag{7.168a}$$

$$\bar{\mathcal{C}}_2 = \mathbf{G}_{C2}(s)\mathbf{G}_{22}(s)(\bar{\mathcal{R}}_2 - \bar{\mathcal{C}}_2) + \mathbf{G}_{C1}(s)\mathbf{G}_{21}(s)(\bar{\mathcal{R}}_1 - \bar{\mathcal{C}}_1) \tag{7.168b}$$

Hence the closed-loop transfer function $\bar{\mathcal{C}}_1/\bar{\mathcal{R}}_1$ can be determined for R_2 held constant ($\bar{\mathcal{R}}_2 = 0$) and $\bar{\mathcal{C}}_2/\bar{\mathcal{R}}_2$ when R_1 is held constant ($\bar{\mathcal{R}}_1 = 0$).

7.15.2. Decouplers and their Design

Suppose that, in the system represented by the block diagram shown in Fig. 7.74, both controlled variables C_1 and C_2 are at their respective set point values R_1 and R_2. A perturbation in either R_2 or in the load U_2 will cause a consequent change in

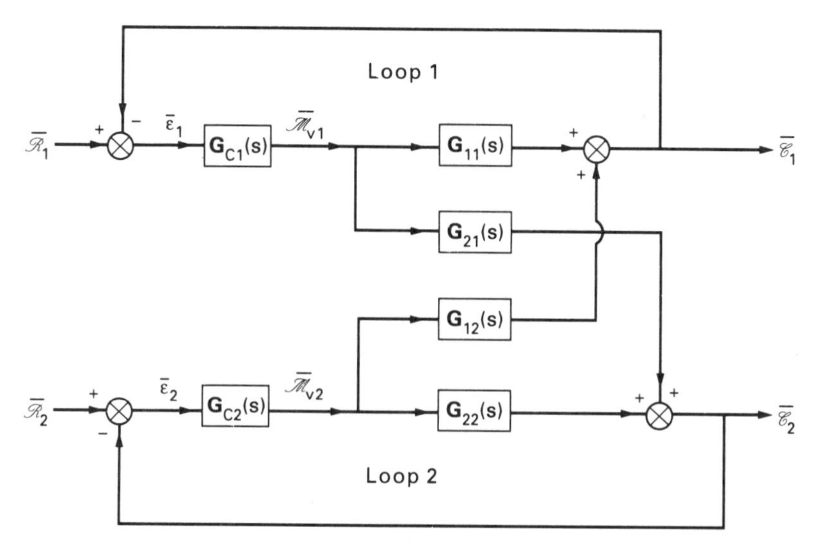

FIG. 7.74. Block diagram illustrating interaction between control loops on distillation column shown in Fig. 7.73

M_{v2}. This will not only act as a suitable control signal for C_2 but will also produce an undesired disturbance in control loop 1 and cause C_1 to deviate from its desired value. However, M_{v1} can be varied by a quantity M_{v1}^+ which compensates for this interaction effect from M_{v2}. If the deviation in C_1 due to interaction is $\mathscr{C}_1^+$, then M_{v1}^+ can be determined from equation 7.166*a*, i.e.:

$$\mathbf{G}_{11}(s)\bar{\mathscr{M}}_{v1}^+ + \mathbf{G}_{12}(s)\bar{\mathscr{M}}_{v2} = \bar{\mathscr{C}}_1^+ \tag{7.169}$$

If there is to be no deviation in C_1, then the appropriate control criterion is that $\bar{\mathscr{C}}_1^+ = \mathscr{C}_1^+ = 0$. Applying this to equation 7.169 gives:

$$\mathbf{G}_{11}(s)\bar{\mathscr{M}}_{v1}^+ + \mathbf{G}_{12}(s)\bar{\mathscr{M}}_{v2} = \bar{\mathscr{C}}_1^+ = 0$$

Hence:

$$\bar{\mathscr{M}}_{v1}^+ = -\frac{\mathbf{G}_{12}(s)}{\mathbf{G}_{11}(s)}\bar{\mathscr{M}}_{v2} = \mathbf{D}_1(s)\bar{\mathscr{M}}_{v2} \tag{7.170}$$

where $\mathbf{D}_1(s)$ is termed a *decoupler*. The action of the decoupler can be represented as in Fig. 7.75*a*.

The same procedure can be used to design a decoupler $\mathbf{D}_2(s)$ to compensate for the interaction effects caused in loop 2 (Fig. 7.74) by disturbances in loop 1. In this case, from equation 7.166*b*:

$$\mathbf{G}_{21}(s)\bar{\mathscr{M}}_{v1} + \mathbf{G}_{22}(s)\bar{\mathscr{M}}_{v2}^+ = \bar{\mathscr{C}}_2^+ = 0$$

Hence:

$$\bar{\mathscr{M}}_{v2} = -\frac{\mathbf{G}_{21}(s)}{\mathbf{G}_{22}(s)}\bar{\mathscr{M}}_{v1} = \mathbf{D}_2(s)\bar{\mathscr{M}}_{v1} \tag{7.171}$$

From Fig. 7.75*b* (which illustrates both decouplers applied at the same time) and equation 7.166*a*:

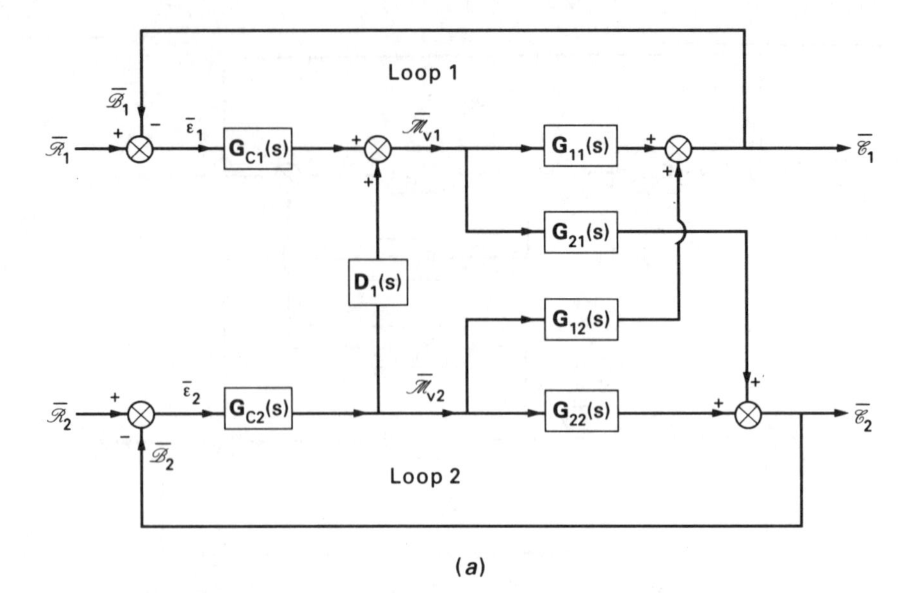

(a)

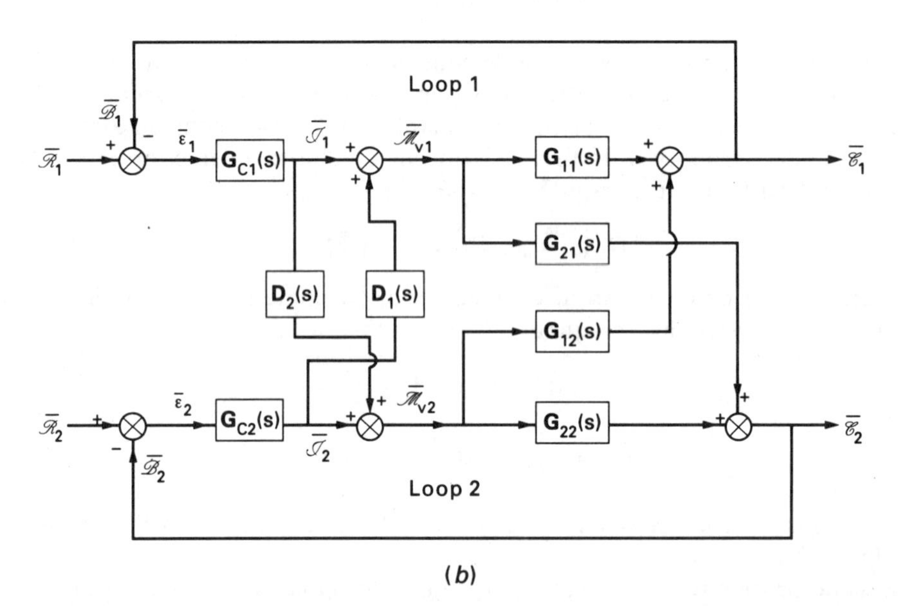

(b)

FIG. 7.75. The use of decouplers to compensate for the interaction shown in Fig. 7.74: (a) the decoupling of loop 1 from the effects of loop 2; (b) the decoupling of both loops

$$\bar{\mathscr{C}}_1 = \mathbf{G}_{11}(s)\bar{\mathscr{M}}_{v1} + \mathbf{G}_{12}(s)\bar{\mathscr{M}}_{v2}$$

$$= \mathbf{G}_{11}(s)[\mathbf{D}_1(s)\bar{\mathscr{T}}_2 + \bar{\mathscr{T}}_1] + \mathbf{G}_{12}(s)\bar{\mathscr{M}}_{v2}$$

Substituting from equation 7.170 for $\mathbf{D}_1(s)$:

$$\bar{\mathscr{C}}_1 = -\mathbf{G}_{12}(s)\bar{\mathscr{T}}_2 + \mathbf{G}_{11}(s)\bar{\mathscr{T}}_1 + \mathbf{G}_{12}(s)\bar{\mathscr{M}}_{v2} \tag{7.172}$$

But, from Fig. 7.75*b* and equation 7.171:

$$\bar{\mathcal{M}}_{v2} = \bar{\mathcal{J}}_2 + \mathbf{D}_2(s)\bar{\mathcal{J}}_1 = \bar{\mathcal{J}}_2 - \frac{\mathbf{G}_{21}(s)}{\mathbf{G}_{22}(s)}\bar{\mathcal{J}}_2 \tag{7.173}$$

Thus, from equations 7.172 and 7.173:

$$\bar{\mathscr{C}}_1 = -\mathbf{G}_{12}(s)\bar{\mathcal{J}}_2 + \mathbf{G}_{11}(s)\bar{\mathcal{J}}_1 + \mathbf{G}_{12}(s)\bar{\mathcal{J}}_2 - \frac{\mathbf{G}_{21}(s)\mathbf{G}_{12}(s)}{\mathbf{G}_{22}(s)}\bar{\mathcal{J}}_1$$

$$= \bar{\mathcal{J}}_1\left(\mathbf{G}_{11}(s) - \frac{\mathbf{G}_{21}(s)\mathbf{G}_{12}(s)}{\mathbf{G}_{22}(s)}\right) \tag{7.174a}$$

Similarly, it can be shown that:

$$\bar{\mathscr{C}}_2 = \bar{\mathcal{J}}_2\left(\mathbf{G}_{22}(s) - \frac{\mathbf{G}_{12}(s)\mathbf{G}_{21}(s)}{\mathbf{G}_{11}(s)}\right) \tag{7.174b}$$

Thus, the equivalent completely decoupled control loops can be represented by the block diagram in Fig. 7.76.

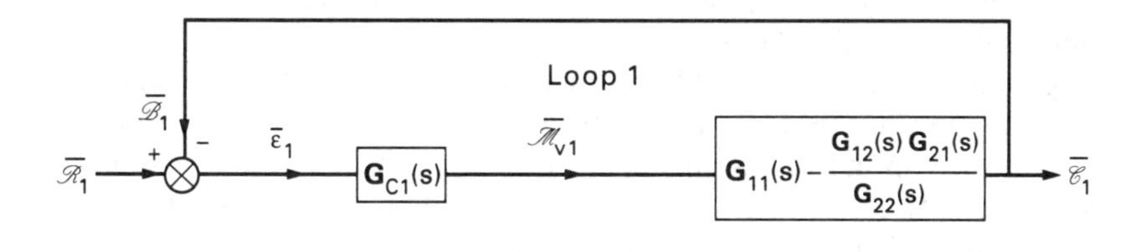

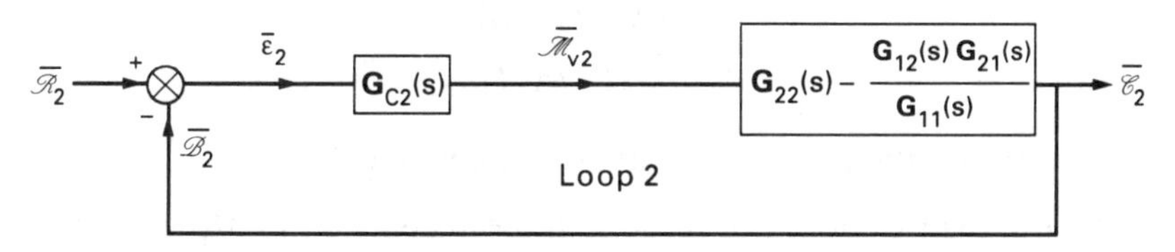

FIG. 7.76. Equivalent decoupled control loops corresponding to the system illustrated in Fig. 7.75*b*

It should be noted that:

(a) A decoupler is essentially a type of feed-forward compensator.

(b) Perfect decoupling requires perfect knowledge of the process in order that the transfer functions $\mathbf{G}_{11}(s)$, $\mathbf{G}_{22}(s)$, $\mathbf{G}_{12}(s)$ and $\mathbf{G}_{21}(s)$ are exact. This is rarely so in practice. Hence, most systems produce only partial decoupling and, in such cases, considerable success can be achieved by designing the decoupler on a steady-state basis. (Compare steady-state feed-forward control—Section 7.14.1.)

(c) If the decoupling is good, then independent tuning of each decoupled loop can be carried out without detriment to the stability of the whole system.

(d) For highly non-linear processes, it may be necessary to employ adaptive decouplers whose characteristics change as the magnitude and level of the disturbances imposed upon the system vary.

SHINSKEY has discussed in detail both interaction and decoupling in general[33] and in respect to the distillation process[34]. ROSENBROCK[35] has presented a more theoretical treatment of the problem.

7.15.3. Estimating the Degree of Interaction Between Control Loops

The degrees of interaction between different sets of control loops controlling a MIMO plant can usually be determined by the use of the *relative gain array* technique due to BRISTOL[36]. This approach can be illustrated by considering the process shown in Fig. 7.74 which has two inputs and two outputs. The procedure is as follows:

(a) Both feedback loops are first imagined to be disconnected by opening the feedback path of each loop. A step change ΔM_{v1} is applied to M_{v1} and the corresponding steady-state change ΔC_1 in C_1 is recorded whilst keeping M_{v2} constant. (For good results ΔM_{v1} should be as small as possible[37].) Thus the open-loop steady-state gain between C_1 and M_{v1} for constant M_{v2} can be determined as $(\Delta C_1/\Delta M_{v1})_{M_{v2}}$.
(b) Loop 2 (containing C_2 and M_{v1}) is then reconnected. The step change in M_{v1} is repeated—but, in this case, M_{v2} is also varied by an amount necessary to control C_2. The corresponding change $\Delta C_1'$ in C_1 is measured. The open-loop steady-state gain between C_1 and M_{v1} is now $(\Delta C_1'/\Delta M_{v1})_{M_{v2}}$. In general, $\Delta C_1'$ will be different to ΔC_1 as $\Delta C_1'$ is the combined result of variations in M_{v1} *and* M_{v2}.
(c) The *relative gain* λ_{11} can now be computed as:

$$\lambda_{11} = \frac{(\Delta C_1/\Delta M_{v1})_{M_{v2}}}{(\Delta C_1'/\Delta M_{v1})_{C_2}} \tag{7.175}$$

Different numerical values of λ_{11} indicate the following:

(i) If $\lambda_{11} = 0$, then C_1 does not respond to perturbations in M_{v1} and, in this case, M_{v1} should not be employed as the manipulated variable for the purpose of controlling C_1.
(ii) If $\lambda_{11} = 1$, then M_{v2} does not affect C_1 and the behaviour of loop 2 does not affect the performance of loop 1. (Note that the converse may not be so.)
(iii) If $0 < \lambda_{11} < 1$, then the behaviour of loop 2 does affect loop 1. The degree of interaction which occurs can be estimated by constructing the relative gain array (see below).
(iv) If $\lambda_{11} < 0$, then C_1 is severely affected by variations in M_{v2} and consequent changes in C_1 will be in the opposite direction to any variations due to perturbations in M_{v1}. This type of interaction can lead to considerable control difficulties[36].

Similarly λ_{12}, λ_{21} and λ_{22} can be defined as the relative gains between C_1 and M_{v2}, C_2 and M_{v1}, and C_2 and M_{v2} respectively, where:

$$\lambda_{12} = \frac{(\Delta C_1/\Delta M_{v2})_{M_{v1}}}{(\Delta C_1'/\Delta M_{v2})_{C_2}}, \quad \lambda_{21} = \frac{(\Delta C_2/\Delta M_{v1})_{M_{v2}}}{(\Delta C_2'/\Delta M_{v1})_{C_1}} \quad \text{and} \quad \lambda_{22} = \frac{(\Delta C_2/\Delta M_{v2})_{M_{v1}}}{(\Delta C_2'/\Delta M_{v2})_{C_1}}$$

Paragraphs (i) to (iv) above apply equally to numerical values of these relative gains.

The Relative Gain Array

The two possible control configurations for a system with two inputs and two outputs are shown in Fig. 7.77. One example of this is illustrated in Fig. 7.73 where the overhead and bottoms product compositions of a distillation process are controlled using the reflux and steam-to-reboiler flowrates respectively as the manipulated variables. Theoretically, we could employ the reflux flowrate to control the bottoms product composition and the steam-to-reboiler flowrate to control the overhead product composition. It is possible to determine which configuration produces the least interaction by forming the system relative gain array Λ, where:

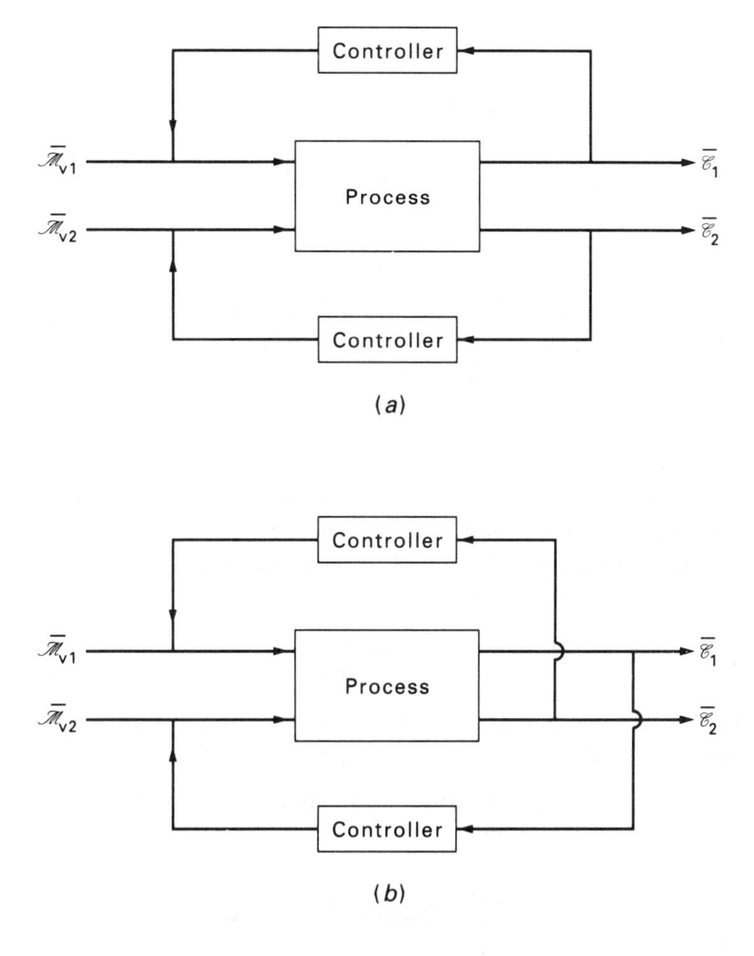

FIG. 7.77. Alternative control loop configurations for a system with two inputs and two outputs

$$\Lambda = \begin{array}{c} (M_{v1}) \quad (M_{v2}) \\ \begin{bmatrix} \lambda_{11} & \lambda_{12} \\ \lambda_{21} & \lambda_{22} \end{bmatrix} \end{array} \begin{array}{c} \\ (C_1) \\ (C_2) \end{array} \tag{7.176}$$

It can be shown that the sum of the relative gains in any row or column of the RGA is equal to one[36]. Hence, it is only necessary to compute one of the relative gains in the array and the other three can then be calculated. There are five forms of the RGA which are worthy of note and which can be expressed in terms of different values or ranges of values of λ_{11}, viz.:

(a) If $\lambda_{11} = 1$, then:

$$\Lambda = \begin{bmatrix} 1 & 0 \\ 0 & 1 \end{bmatrix}$$

This indicates that the two loops are entirely non-interacting, C_1 being coupled only with M_{v1} and C_2 with M_{v2} (Fig. 7.77*a*).

(b) If $\lambda_{11} = 0$, then:

$$\Lambda = \begin{bmatrix} 0 & 1 \\ 1 & 0 \end{bmatrix}$$

The unit elements in the opposite diagonal indicate that there are two non-interacting loops which can be formed—but, in this case, they are constructed by coupling C_1 with M_{v2} and C_2 with M_{v1} (Fig. 7.77*b*).

(c) If $\lambda_{11} = 0.5$, then:

$$\Lambda = \begin{bmatrix} 0.5 & 0.5 \\ 0.5 & 0.5 \end{bmatrix}$$

In this case, the degree of interaction will be the same, whichever of the two configurations is employed—Fig. 7.77 either *a* or *b*.

(d) If $0 < \lambda_{11} < 0.5$, e.g. $\lambda_{11} = 0.25$, then:

$$\Lambda = \begin{bmatrix} 0.25 & 0.75 \\ 0.75 & 0.25 \end{bmatrix}$$

The two larger numbers (i.e. 0.75) indicate the recommended control loop configuration (i.e. that with the smaller amount of interaction) which is constructed by coupling C_2 with M_{v1} and C_1 with M_{v2} (Fig. 7.77*b*).

(e) If $0.5 < \lambda_{11} < 1$, e.g. $\lambda_{11} = 0.8$, then:

$$\Lambda = \begin{bmatrix} 0.8 & 0.2 \\ 0.2 & 0.8 \end{bmatrix}$$

In this case the recommended control loop configuration is constructed by coupling C_1 with M_{v1} and C_2 with M_{v2} (Fig. 7.77*a*).

Configurations which produce relative gains outside the range 0 to 1 indicate situations where control is difficult.

7.16. NON-LINEAR SYSTEMS

Many of the techniques described in this chapter depend upon the assumption (whether explicit or implicit) that the systems being investigated can be defined as linear, i.e. that their time-dependent properties can be expressed in terms of linear differential equations and that the principle of superposition[7,8] can be applied (see Section 7.4.1). However, no real physical system (or process) can be described as being truly linear, especially if its behaviour is considered over a wide range of operating conditions. Consequently, it is often necessary to consider the degree of non-linearity of any given part of a control system. Many processes can be regarded as acting according to a set of linear relationships (which approximate their behaviour) provided that the relevant variables only change over a fairly narrow range (see Section 7.5.2). On the other hand, there are processes which exhibit a high degree of non-linearity (e.g. those involving a highly exothermic chemical reaction) where linear approximations will not be appropriate and the standard transfer function approach, which is so useful in the analysis of linear (or relatively linear) systems, cannot be employed.

Although there is no universal procedure for analysing non-linear systems, there are several methods which can be applied to particular types or classes of system. One of these (viz. the describing function technique) is discussed in this section. First, however, the linearisation procedure employed in Section 7.5.2 is expanded to include relationships containing more than one independent variable.

7.16.1. Linearisation Using Taylor's Series

The linearisation method employed in Section 7.5.2 considers the dependent variable (volumetric flowrate of liquid leaving a tank Q_1) being perturbed about its steady-state value Q_1' in response to changes in the level of liquid z in the tank about its steady-state level z'. Thus, Q_1 can be expressed as a function of z in terms of a Taylor's series, viz.:

$$Q = Q_1' + \left[\frac{dQ_1}{dz}\right]_{z'} (z - z') + \left[\frac{d^2Q_1}{dz^2}\right]_{z'} \frac{(z - z')^2}{2} + \text{higher order terms} \qquad \text{(equation 7.24}a\text{)}$$

Neglecting higher order terms (assuming that $(z - z')$ is small) yields the corresponding linear approximation:

$$Q - Q' = K_2(z - z') \qquad \text{(equation 7.25)}$$

This procedure can be extended to functions of any number of variables[9]. For example, suppose that the mass flowrate w and temperature θ of a particular process stream both vary about their steady-state values w' and θ' respectively. The rate of heat flow in the stream at any instant Q_H (which is assumed also to fluctuate about its steady-state value Q_H' due to the changes in w and θ) can be written in terms of Taylor's series:

$$Q_H = Q_H'(w', \theta') + \left[\frac{\partial Q_H}{\partial w}\right]_{w',\theta'} (w - w') + \left[\frac{\partial Q_H}{\partial \theta}\right]_{w',\theta'} (\theta - \theta') + \frac{1}{2!}\left\{\left[\frac{\partial^2 Q_H}{\partial w^2}\right]_{w',\theta'} (w - w')^2\right.$$

$$\left. + 2\left[\frac{\partial^2 Q_H}{\partial w \partial \theta}\right]_{w',\theta'} (w - w')(\theta - \theta') + \left[\frac{\partial^2 Q_H}{\partial \theta^2}\right]_{w',\theta'} (\theta - \theta')^2\right\} + \text{higher order terms} \qquad (7.177)$$

For small variations in w and θ, terms involving powers higher than unity in equation 7.177 can be neglected, i.e.:

$$Q_H \approx Q'_H(w', \theta') + \left[\frac{\partial Q_H}{\partial w}\right]_{w', \theta'}(w - w') + \left[\frac{\partial Q_H}{\partial \theta}\right]_{w', \theta'}(\theta - \theta') \qquad (7.178)$$

The use of equation 7.178 is illustrated in Example 7.11.

Example 7.11

Water is heated by passing it through a steam-heated kettle (Fig. 7.78) at a mass flowrate w. The inlet and outlet temperatures of the water are θ_i and θ_o respectively. Steam condenses in the jacket of the kettle at a temperature θ_v and a pressure P_v. It is intended to control the temperature of the water by placing a temperature sensor in the water in the kettle and using this measurement to manipulate the flow of steam to the kettle jacket. In order to tune the controller it is necessary to derive the transfer functions relating θ_o to θ_i, θ_v and w.

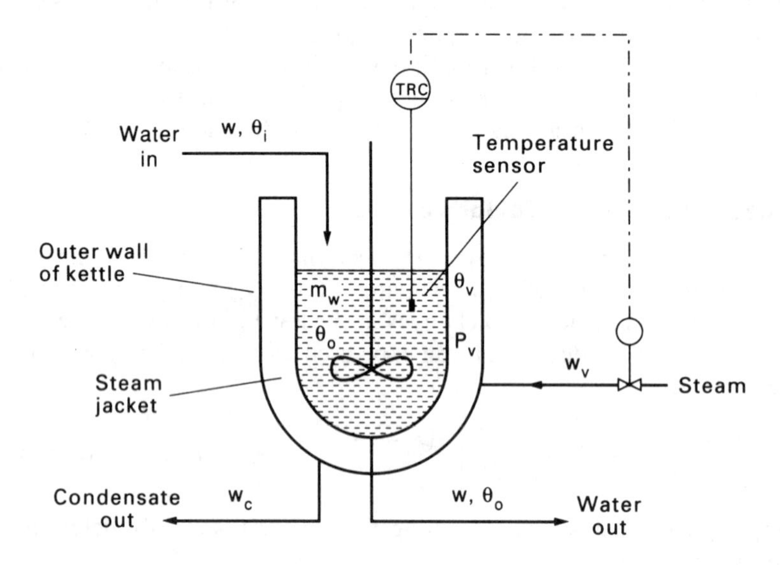

FIG. 7.78. Steam-heated kettle

Solution

This is a complex system and, in order that it can be treated relatively simply, we need to make a number of assumptions, viz.:

(a) Heat losses to the atmosphere are negligible.
(b) The hold-up volume of water in the kettle remains constant.
(c) The thermal capacity of the kettle wall between the steam and the water is negligible.
(d) The thermal capacity C_{ow} of the outer wall of the kettle is finite.
(e) The temperature of the outer wall of the kettle is uniform and equal to the temperature of the steam at any instant.
(f) The temperature of the water in the kettle is uniform and is equal to θ_o at all times.
(g) The rate of heat transfer from the steam to the water is approximated by the expression $Q_s = UA(\theta_v - \theta_o)$ where U is an overall heat transfer coefficient and A is the heat transfer area.
(h) The heat capacity of the metal wall remains constant.
(i) The heat capacity of the water C_w remains constant.
(i) The density of the water ρ_w is the same throughout the system.
(j) The steam in the jacket of the kettle is saturated.

An energy balance on the water side gives:

$$Q_{Hi} - Q_{Ho} + Q_s = \frac{\mathrm{d}}{\mathrm{d}t}(m_w C_w \theta_o) \tag{7.179}$$

where Q_{Hi} and Q_{Ho} are the rates of heat energy entering and leaving the kettle respectively via the water flow and m_w is the mass of the water in the kettle. The corresponding steady-state relationship is:

$$Q'_{Hi} - Q'_{Ho} + Q'_s = 0 \tag{7.180}$$

Subtracting equation 7.180 from equation 7.179 gives the equivalent relationship in terms of deviation variables (Section 7.5.2), viz.:

$$\mathcal{Q}_{Hi} - \mathcal{Q}_{Ho} + \mathcal{Q}_s = \frac{\mathrm{d}}{\mathrm{d}t}(m_w C_w \theta_o) \tag{7.181}$$

From equation 7.179:

$$w C_w(\theta_i - \theta_o) + UA(\theta_v - \theta_o) = m_w C_w \frac{\mathrm{d}\theta_o}{\mathrm{d}t} \tag{7.182}$$

As w, θ_i and θ_o are all variables, the quantities $Q_{Hi} = wC_w\theta_i$ and $Q_{Ho} = wC_w\theta_o$ are non-linear. However, they can be approximated by their linear equivalents using equation 7.178, i.e.:

$$Q_{Hi}(w, \theta_i) \approx Q'_{H_i}(w', \theta'_i) + \left[\frac{\partial Q_{Hi}}{\partial w}\right]_{w', \theta'_i}(w - w') + \left[\frac{\partial Q_{Hi}}{\partial \theta}\right]_{w', \theta'_i}(\theta_i - \theta'_i) \tag{7.183}$$

Thus: $$Q_{Hi} - Q'_{Hi} = C_w \theta'_i (w - w') + C_w w'(\theta_i - \theta'_i)$$

i.e.: $$\mathcal{Q}_{Hi} = C_w(\theta'_i \varpi + w'\vartheta_i) \tag{7.184a}$$

Similarly: $$\mathcal{Q}_{Ho} = C_w(\theta'_o \varpi + w'\vartheta_o) \tag{7.184b}$$

From equations 7.181, 7.182 and 7.184:

$$(\theta'_i \varpi + w'\vartheta_i) - (\theta'_o \varpi + w'\vartheta_o) + \frac{UA}{C_w}(\vartheta_v - \vartheta_o) = m_w \frac{\mathrm{d}\vartheta_o}{\mathrm{d}t}$$

Transforming gives:

$$\bar{\varpi}(\theta'_i - \theta'_o) + w'\bar{\vartheta}_i + \frac{UA}{C_w}\bar{\vartheta}_v = \bar{\vartheta}_o\left(w' + \frac{UA}{C_w} + m_w s\right)$$

Rearranging yields the required transfer functions:

$$\bar{\vartheta}_o = \frac{K_1}{1 + \tau_w s}\bar{\vartheta}_i + \frac{K_2}{1 + \tau_w s}\bar{\vartheta}_v + \frac{K_3}{1 + \tau_w s}\bar{\varpi}_i \tag{7.185}$$

where: $\tau_w = \dfrac{m_w C_w}{UA + w' C_w}$, $K_1 = \dfrac{w' C_w}{UA + w' C_w}$, $K_2 = \dfrac{UA}{UA + w' C_w}$, and $K_3 = \dfrac{C_w(\theta'_o - \theta'_i)}{UA + w' C_w}$.

Hence the response of θ_o to θ_i, θ_v or w is approximately first order with a time constant τ_w.

An energy balance on the steam side yields:

$$H_v w_v - H_c w_c - UA(\theta_v - \theta_o) = \frac{\mathrm{d}}{\mathrm{d}t}(V_v \rho_v Z_v) + \frac{\mathrm{d}}{\mathrm{d}t}(m_{ow} C_{ow} \theta_v)$$

$$= V_v \frac{\mathrm{d}}{\mathrm{d}t}(\rho_v Z_v) + m_{ow} C_{ow} \frac{\mathrm{d}\theta_v}{\mathrm{d}t} \tag{7.186}$$

where V_v is the volume of the jacket steam space and w_v and w_c are the mass flowrates of the steam and condensate respectively. H_v, ρ_v and Z_v are the enthalpy, density and internal energy of the steam respectively. H_c is the enthalpy of the condensate and m_{ow} is the mass of the outer wall of the kettle (see also assumption (e)).

A mass balance on the steam side gives:

$$w_v - w_c = \frac{\mathrm{d}}{\mathrm{d}t}(V_v \rho_v) = V_v \frac{\mathrm{d}\rho_v}{\mathrm{d}t} \tag{7.187}$$

Eliminating w_c from equations 7.186 and 7.187:

$$w_v(H_v - H_c) - UA(\theta_v - \theta_o) = (Z_v - H_c) V_v \frac{\mathrm{d}\rho_v}{\mathrm{d}t} + m_{ow} C_{ow} \frac{\mathrm{d}\theta_v}{\mathrm{d}t} + V_v \rho_v \frac{\mathrm{d}U_v}{\mathrm{d}t} \tag{7.188}$$

ρ_v, U_v, H_v and H_c are functions of the steam and condensate temperatures (θ_v and θ_c) and can be expressed in terms of Taylor's series using equation 7.24*a*. Hence:

$$\left.\begin{aligned}
\rho_v &\approx \rho'_v + \chi_1\vartheta_v \quad \text{where } \chi_1 = \left[\frac{d\rho_v}{d\theta_v}\right]_{\theta'_v} \\
Z_v &\approx Z'_v + \chi_2\vartheta_v \quad \text{where } \chi_2 = \left[\frac{dZ_v}{d\theta_v}\right]_{\theta'_v} \\
H_v &\approx H'_v + \chi_3\vartheta_v \quad \text{where } \chi_3 = \left[\frac{dH_v}{d\theta_v}\right]_{\theta'_v} \\
H_c &\approx H'_c + \chi_4\vartheta_c \quad \text{where } \chi_4 = \left[\frac{dH_c}{d\theta_c}\right]_{\theta'_c}
\end{aligned}\right\} \tag{7.189}$$

χ_1, χ_2, χ_3 and χ_4 can be found from the steam tables in Volume 1 at the particular steady-state applying.

From equations 7.188 and 7.189 and assuming that $\theta_v = \theta_c$:

$$w_v[(H'_v - H'_c) - (\chi_3 - \chi_4)\vartheta_v] - UA(\theta_v - \theta_o) = \left[(Z'_v - H'_c) + (2\chi_2 - \chi_4)\vartheta_v + \frac{\chi_2}{\chi_2}\rho'_v + \frac{m_{ow}C_{ow}}{\chi_1 V_v}\right]\chi_1 V_v \frac{d\theta_v}{dt} \tag{7.190}$$

The terms $(\chi_3 - \chi_4)\vartheta_v$ and $\left[(2\chi_2 - \chi_4)\vartheta_v + \frac{\chi_2}{\chi_1}\rho'_v\right]$ for steam are generally negligible in comparison to $(H'_v - H'_c)$ and $(Z'_v - H'_c)$ respectively. Discarding these terms from equation 7.190 and writing the remainder in terms of deviation variables and transforming leads to:

$$\bar{\vartheta}_v = \frac{1}{1 + \tau_v s}\bar{\vartheta}_o + \frac{K_4}{1 + \tau_v s}\bar{w}_v \tag{7.191}$$

where:
$$\tau_v = \frac{\chi_1 V_v(U'_v - H'_c) + m_{ow}C_{ow}}{UA} \quad \text{and} \quad K_4 = \frac{H'_v - H'_c}{UA}$$

Thus the response of θ_v to variations in θ_o and w_v approximates to first order in each case.

7.16.2. The Describing Function Technique

The describing function method is applicable to any non-linearity which has the characteristic that if the input is a sinusoidal signal then the output is a periodic function[38–40]. Because of its simplicity and wide range of applicability, the describing function technique is one of the most versatile procedures for analysing non-linear effects.

Many non-linearities are such that, for a sinusoidal input (e.g. $x = M\sin\omega t = M\sin(2\pi ft)$), the output will exhibit the same period (where the period $= 1/f = 2\pi/\omega$) as the input signal. Moreover, these outputs satisfy the condition pertaining to *odd periodic functions*, viz. that the signal over the second half of the period is identical to that which would be obtained if the signal over the first half of the period were rotated π radians about the mid-point of the whole period. An example of such a function of period 2π is shown in Fig. 7.79. Clearly, if this function over the period 0 to π is rotated π radians about the point $\omega t = \pi$, it will then coincide with itself over the period π to 2π. Odd functions of period 2π can be represented by a Fourier series of the form[17]:

$$f(t) = \Xi_1 \sin\omega t + \Xi_2 \sin 2\omega t + \Xi_3 \sin 3\omega t + \ldots\ldots \tag{7.192}$$

where:
$$\Xi_n = \frac{1}{\pi}\int_0^{2\pi} f(t)\sin(n\omega t)\,d(\omega t)$$

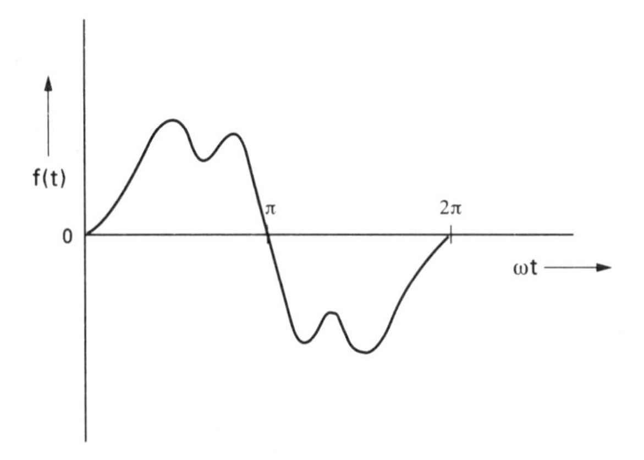

FIG. 7.79. Odd function of period 2π

$$= \frac{2}{\pi}\int_0^{\pi} \mathrm{f}(t)\sin(n\omega t)\,\mathrm{d}(\omega t) \quad \text{for } n = 1, 2, 3, \ldots \tag{7.193}$$

The describing function approach will now be applied to two common non-linear elements which fulfil these conditions

The Describing Function of an On–Off Element

The operating characteristics of an on–off element (Fig. 7.80*b*) can be written as:

$$\mathrm{f}(t) = \begin{Bmatrix} Y_0 & \text{for} & x > 0 \\ -Y_0 & \text{for} & x < 0 \end{Bmatrix} \tag{7.194}$$

The effect of applying a sinusoidal perturbation to this element is also shown in Fig. 7.80. Hence, from equation 7.193:

$$\Xi_n = \frac{2}{\pi}\int_0^{\pi} Y_0 \sin(n\omega t)\,\mathrm{d}(\omega t) = -\frac{2Y_0}{n\pi}\Big[\cos n\omega t\Big]_0^{\pi}$$

$$= \begin{Bmatrix} \dfrac{4Y_0}{n\pi} & \text{for} & n = 1, 3, 5, \ldots \\ 0 & \text{for} & n = 2, 4, 6, \ldots \end{Bmatrix} \tag{7.195}$$

From equation 7.192, the corresponding Fourier series which describes the output y of the on–off element is:

$$y = \frac{4Y_0}{\pi}\left(\sin\omega t + \tfrac{1}{3}\sin 3\omega t + \tfrac{1}{5}\sin 5\omega t + \ldots\ldots.\right) \tag{7.196}$$

[n.b. The even coefficients (Ξ_0, Ξ_2, Ξ_4 etc.) are all zero when the output function has the additional symmetry that, when the portion of the function from 0 to $\pi/2$ is reflected about a vertical axis through $\pi/2$, it coincides with the portion from $\pi/2$ to π. This is the case for this function.]

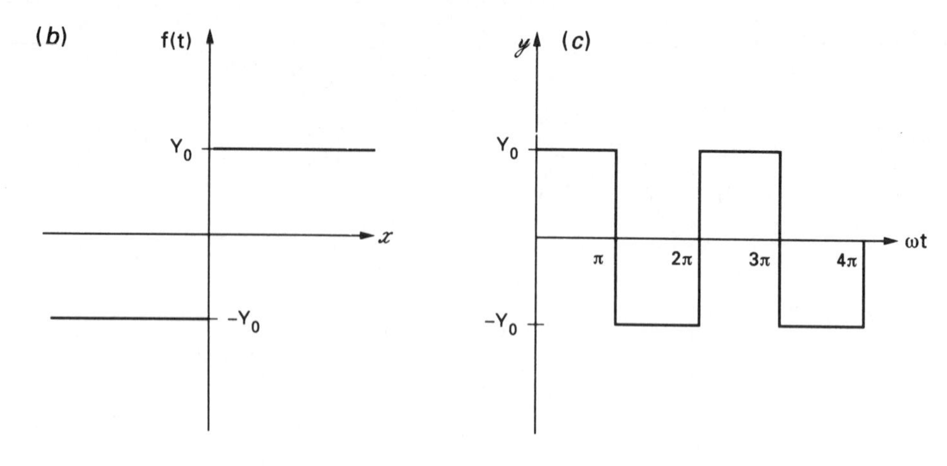

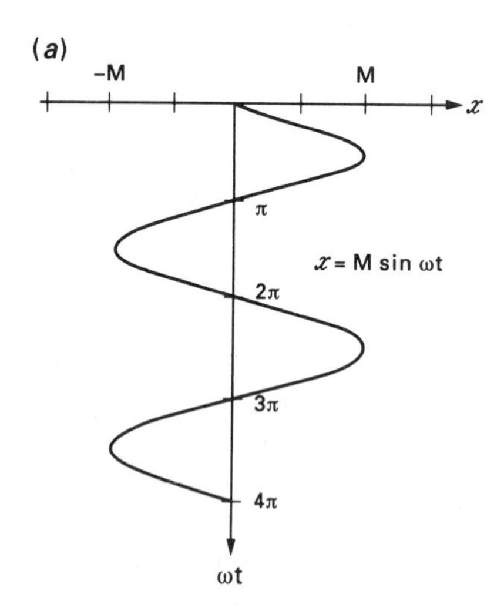

FIG. 7.80. Response of on–off element to sinusoidal perturbation: (*a*) input; (*b*) characteristic of element; (*c*) output

Figure 7.81 compares the output y of the on–off element over the period 0 to 2π with the values of the first two terms of the equivalent Fourier series representation (equation 7.196). Clearly the most significant contribution is given by the fundamental component $\Xi_1 \sin \omega t$. Moreover, the higher frequency contributions of $\Xi_3 \sin \omega t$, $\Xi_5 \sin \omega t$ etc. are progressively attenuated more by other linear components in the system and thus have less effect on the operating characteristics of the system. Hence, the output of the non-linearity is well represented by the fundamental component, i.e.:

$$y \approx \Xi_1 \sin \omega t \tag{7.197}$$

Thus, the behaviour of the non-linear on–off element is approximated by the input/output relationship:

$$\mathrm{N} = \frac{y}{x} \approx \frac{\Xi_1 \sin \omega t}{M \sin \omega t} = \frac{\Xi_1}{M} = \frac{4 Y_0}{\pi M} \tag{7.198}$$

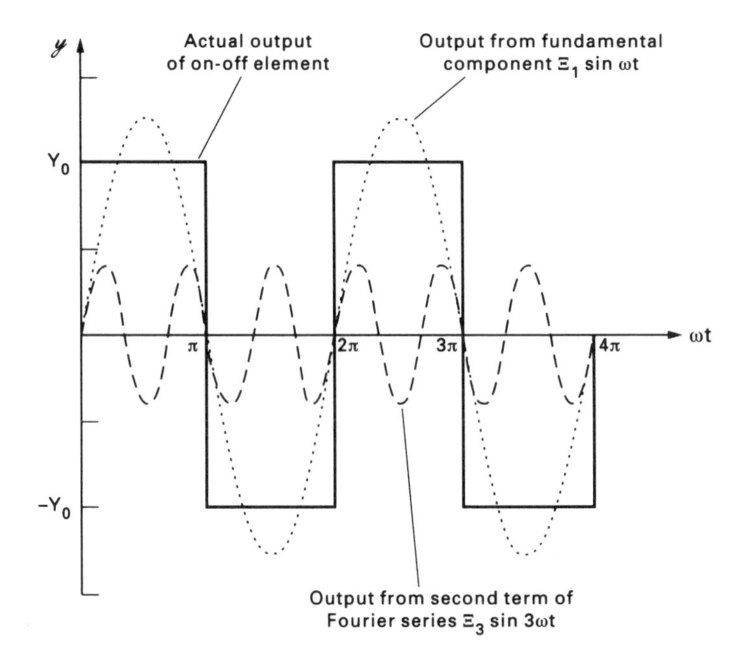

FIG. 7.81. Comparison of output of on–off element with those produced by first and second terms of corresponding Fourier representation

N is called the *describing function* for the non-linear element. **N** can be considered as a non-linear equivalent of the linear system transfer function.

The Describing Function of a Dead-Zone Element

The output of a *dead-zone* or *dead-band* element in response to a sinusoidal input is shown in Fig. 7.82 and is written:

$$\mathrm{f}(t) = \begin{Bmatrix} k(M\sin\omega t - \delta) & \text{for } x > \delta \\ 0 & \text{for } 0 \leqslant x \leqslant \delta \end{Bmatrix} \tag{7.199}$$

It can be seen from Fig. 7.82 that $\delta = M\sin\beta$, i.e. that $\beta = \sin^{-1}(\delta/M)$. For the dead-zone element, when $M > \delta$, the fundamental component is given by equation 7.193 with $n = 1$, i.e.:

$$\begin{aligned}
\Xi_1 &= \frac{2}{\pi}\int_0^{\pi} \mathrm{f}(t)\sin(\omega t)\mathrm{d}(\omega t) \\
&= \frac{2}{\pi}\int_{\beta}^{\pi-\beta} kM(\sin(\omega t) - \sin\beta)\sin(\omega t)\,\mathrm{d}(\omega t) \\
&= \frac{2kM}{\pi}\left[\frac{\omega t}{2} - \frac{\sin 2\omega t}{4} + \cos(\omega t)\sin\beta\right]_{\beta}^{\pi-\beta} \\
&= \frac{2kM}{\pi}\left(\frac{\pi - 2\beta}{2} - \frac{\sin 2\beta}{2}\right)
\end{aligned} \tag{7.200}$$

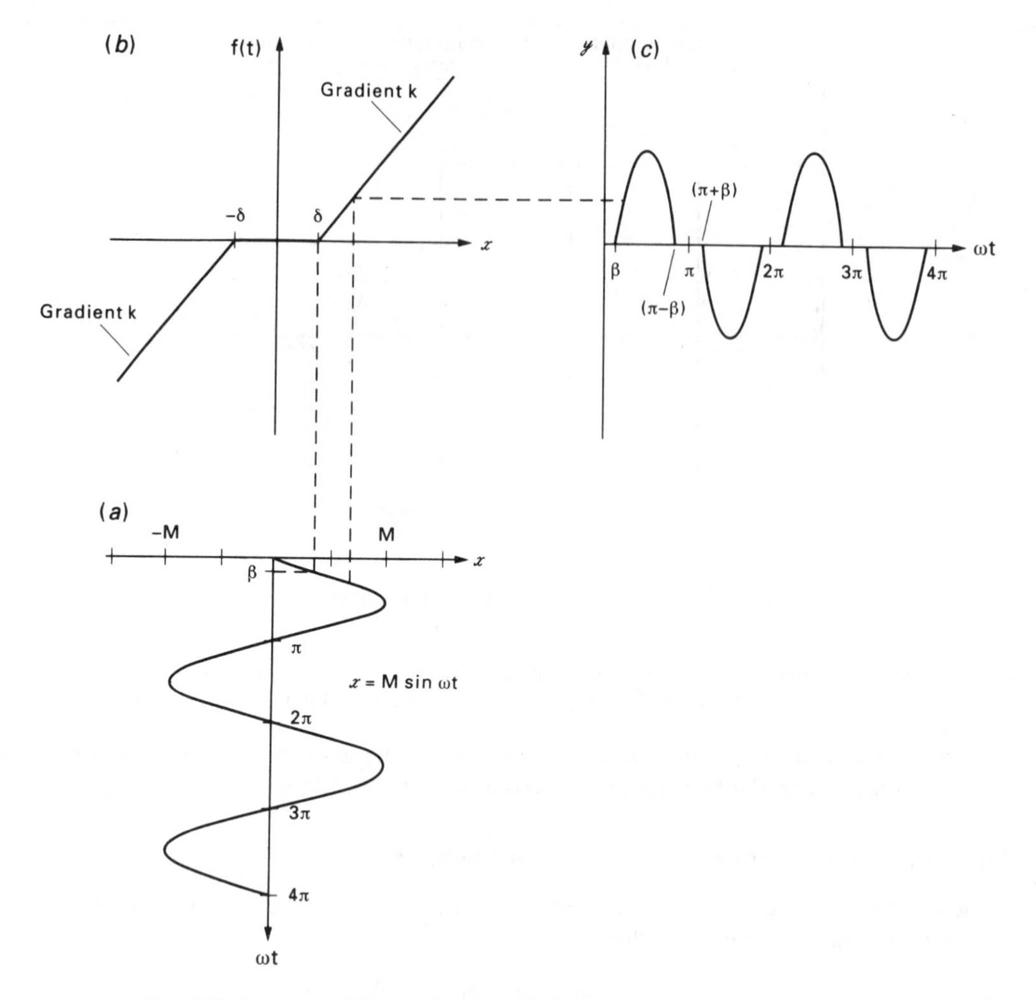

FIG. 7.82. Response of linear element with dead-zone to sinusoidal perturbation: (a) input; (b) characteristic of element; (c) output

Thus, the describing function is given by:

$$\mathbf{N} = \frac{y}{x} \approx \frac{\Xi_1}{M} = \left\{ \begin{array}{ll} \dfrac{k}{\pi}(\pi - 2\beta - \sin 2\beta) & \text{for} \quad M > \delta \\ 0 & \text{for} \quad 0 \leqslant M \leqslant \delta \end{array} \right\} \tag{7.201}$$

The characteristic of this describing function is conveniently plotted as $\mathbf{N}/k$ versus M/δ in Fig. 7.83.

Combining Describing Functions

It is possible to determine the characteristics of more complex non-linear functions and to derive their corresponding describing functions by adding and subtracting the describing functions of simpler functions. For example, the characteristic of

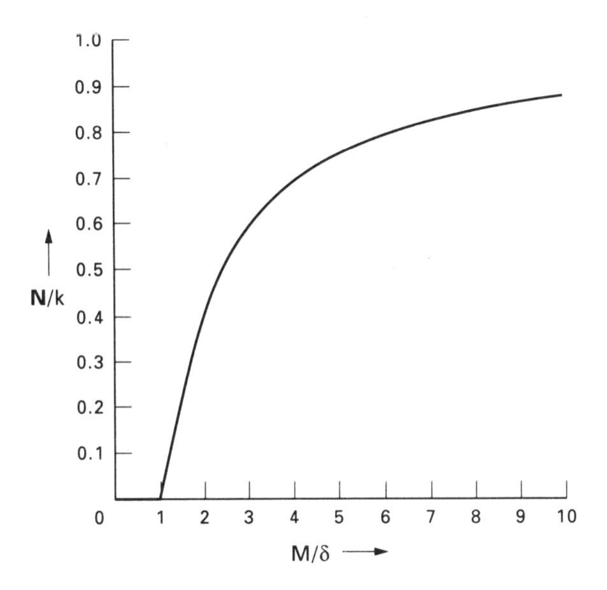

FIG. 7.83. Characteristic of describing function for linear element with dead-zone

an element which behaves linearly in response to an input signal over the range $x = -\sigma$ to $x = +\sigma$ but which saturates outside that range ($F_3(t)$ in Fig. 7.84*a*) can be constructed by subtracting the characteristics of a linear element with a dead zone ($F_2(t)$ in Fig. 7.84*a*) from a linear element with no saturation ($F_1(t)$). The corresponding plots of $\mathbf{N}/k$ as a function of $M/\delta = M/\sigma$ are shown in Fig. 7.84*b*.

The describing function corresponding to $F_1(t)$ is:

$$\mathbf{N}_1 = \frac{y}{x} = k \qquad \text{for all values of } M$$

The describing function for the element with the dead zone $\mathbf{N}_2$ is given by equation 7.201. Hence, the describing function for the saturating element $\mathbf{N}_3$ is (Fig. 7.84*b*):

$$\mathbf{N}_3 = \mathbf{N}_1 - \mathbf{N}_2 = \left\{ \begin{array}{ll} \dfrac{k}{\pi}(2\beta + \sin 2\beta) & \text{for} \quad M > \sigma \\ k & \text{for} \quad 0 \leqslant M \leqslant \sigma \end{array} \right\} \tag{7.202}$$

Stability Analysis Using the Describing Function Approach

The analysis of the stability of a control system which contains a non-linear element that can be characterised by a describing function $\mathbf{N}$ is facilitated by considering the non-linear element as a variable gain for which $\mathbf{N}$ depends upon the amplitude M of the input signal[40]. The behaviour (in terms of frequency response) of linear elements in the control loop will be a function of frequency only.

Consider a control loop containing a non-linear component represented by its describing function $\mathbf{N}$ and a number of linear components which can be combined together to be described by a linear transfer function $\mathbf{G}_X(s)$ (Fig. 7.85). The closed-loop transfer function for this system is (from equation 7.111):

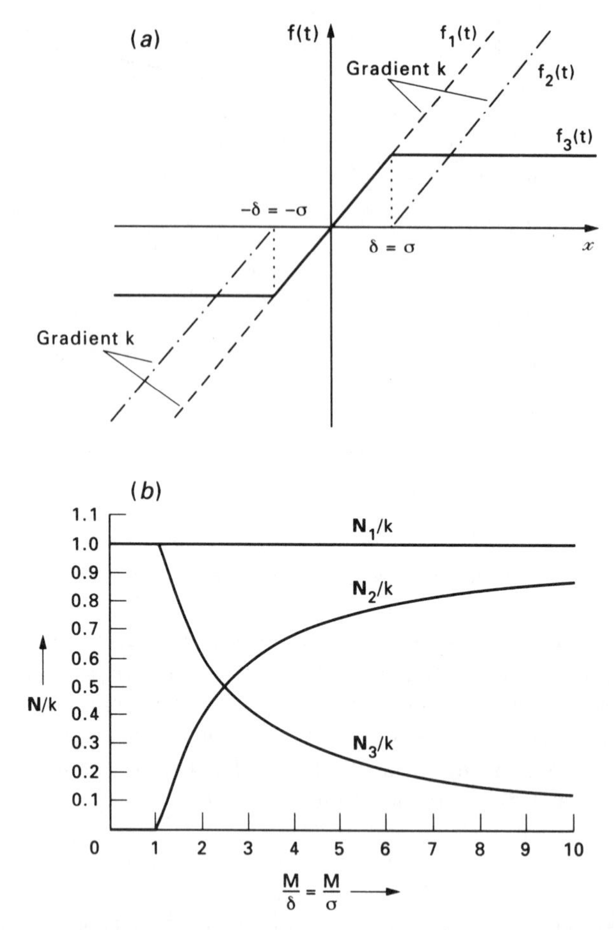

FIG. 7.84. Graphical method of obtaining describing function from known describing functions: (*a*) operating characteristics; (*b*) describing function characteristics

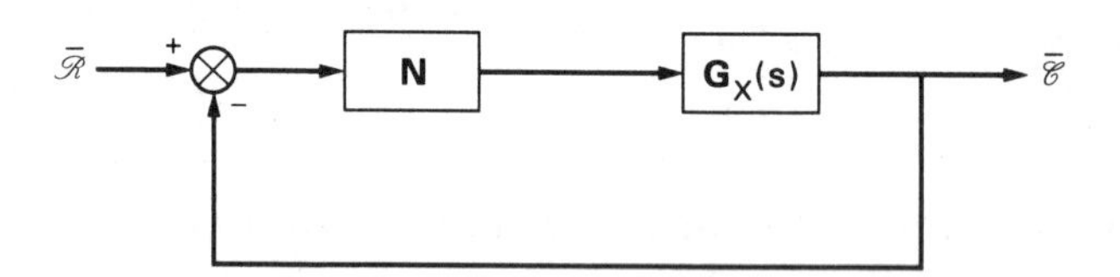

FIG. 7.85. Control system containing non-linear element

$$\frac{\bar{\mathscr{C}}}{\bar{\mathscr{R}}} = \frac{\mathbf{N}\mathbf{G}_X(s)}{1 + \mathbf{N}\mathbf{G}_X(s)} \tag{7.203}$$

Thus, from the Nyquist criterion (Section 7.10.5), the system will be conditionally stable if the polar plot of $\mathbf{NG}(s)$ passes through the point $(-1, 0)$ on the complex plane. Under these circumstances there will be one frequency ω_x at which:

$$\mathbf{N}\mathbf{G}_X(\mathrm{i}\omega_X) = -1$$

i.e.:

$$\mathbf{G}_X(\mathrm{i}\omega_X) = -\frac{1}{\mathbf{N}} \tag{7.204}$$

Hence, if $\mathbf{G}_X(\mathrm{i}\omega)$ and $-1/\mathbf{N}$ are both plotted on the same complex plane, then a conditionally stable system is indicated by the conditions existing at the point of intersection of the two plots. This is illustrated in the following example.

Example 7.12

The controller parameters in the control system described in Example 7.6 are set at $K_C = 0.6$ and $\tau_I = 2.5$ minutes. The linear control valve used in the control loop develops a dead-zone over the mid-section of its range of operation. The dead-zone is variable and its magnitude affects the sensitivity of the valve over the remaining range of operation. Examine the effect of this dead-zone on the stability of the control system.

Solution

The block diagram of the equivalent unity feedback system is shown in Fig. 7.86 (as $\mathbf{H}(s)$ approximates to its steady-state gain) with the non-linearity represented by its describing function $\mathbf{N}$. $\mathbf{N}$ can be considered as being additional to the original valve transfer function (a steady-state gain of 0.01) and has the characteristic given by equation 7.199 (see also Fig. 7.82). The value of k will then be the factor by which the valve sensitivity can be changed by the size of the dead-zone. From Fig. 7.86 and equation 7.203:

$$\frac{\bar{\mathscr{C}}}{\bar{\mathscr{R}}} = \frac{\mathbf{N}\mathbf{G}_C(s)\mathbf{G}_1(s)\mathbf{G}_2(s)}{1 + \mathbf{N}\mathbf{G}_C(s)\mathbf{G}_1(s)\mathbf{G}_2(s)\mathbf{H}(s)}$$

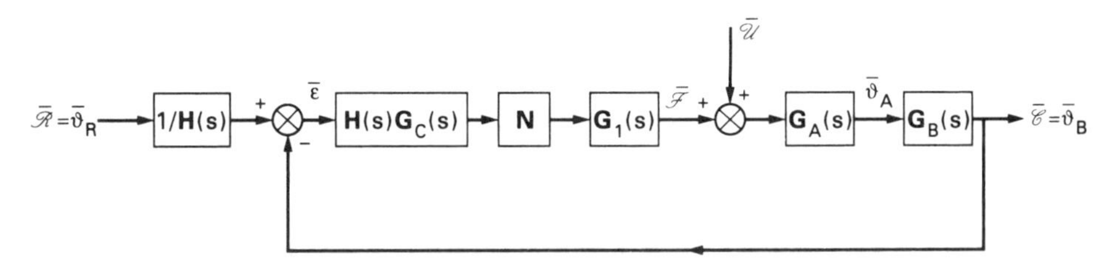

FIG. 7.86. Equivalent unity feedback block diagram of control system illustrated in Fig. 7.42 including non-linear element

The stability of the system can be examined by plotting $-1/\mathbf{N}$ and $\mathbf{G}_X(\mathrm{i}\omega) = \mathbf{G}_C(\mathrm{i}\omega)\mathbf{G}_1(\mathrm{i}\omega)\mathbf{G}_2(\mathrm{i}\omega)\mathbf{H}(\mathrm{i}\omega)$ on the same plot and examining the conditions applying at the point of intersection (equation 7.204). The polar plot of $\mathbf{G}_X(\mathrm{i}\omega)$ is obtained from Fig. 7.55 (curve (d)) and is shown in Fig. 7.87. Clearly the plot of $-1/\mathbf{N}$ will coincide with the negative side of the real axis. At the point of intersection of the two graphs, from equation 7.204, $-1/\mathbf{N} = \mathbf{G}_X(i\omega_X) = -0.53$ (Fig. 7.87). Thus $\mathbf{N} = 1.89$. Hence, from equation 7.201:

$$1.89 = \frac{k}{\pi}(\pi - 2\beta - \sin 2\beta) \qquad \text{for } \frac{M}{\delta} > 1 \text{ and } \beta = \sin^{-1}\delta/M \tag{7.205}$$

Values of M/δ corresponding to this value of $\mathbf{N}$ can be determined for different values of k. For example, putting $k = 4$ in equation 7.205 yields:

$$2\beta + \sin 2\beta = 1.66$$

Hence, by trial:

$$\underline{\underline{\beta \approx 0.44 \text{ radians and } M/\delta = \frac{1}{\sin\beta} \approx 2.3}}$$

Thus, for $k = 4$, if $M/\delta > 2.3$, then the system will be unstable, i.e. if the amplitude of the input disturbance is greater than 2.3 times the size δ of the dead-zone, then instability will result (Fig. 7.87). If k is reduced to 2, then $\beta \approx 0.043$ (by trial) and $M/\delta \approx 23$, i.e. the amplitude of the input must now exceed

23δ for the non-linear element to make the system unstable. In general, the system stability is increased if k is reduced, and, in this case, if $k < 1/\mathbf{G}_X(i\omega_X)$ (i.e. $k < 1.89$), the two graphs will not intersect and, however much the amplitude of the input signal is increased, the system will not become unstable.

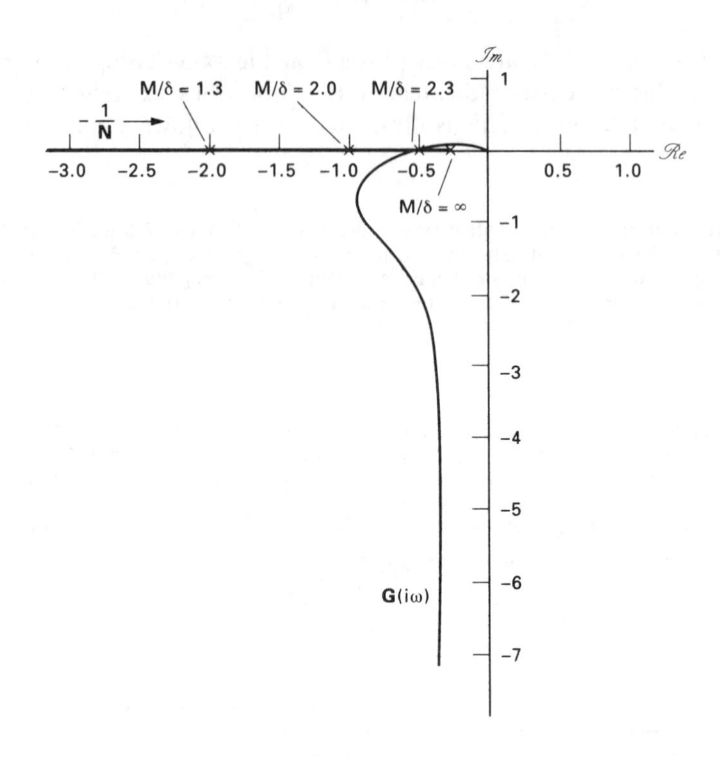

FIG. 7.87. Plots of $\mathbf{G}(i\omega)$ and $-1/\mathbf{N}$ for $k = 4$

7.17. DISCRETE TIME CONTROL SYSTEMS

7.17.1. Sampled Data (Discrete Time) Systems

In many control applications information is received and transmitted in the form of a succession of samples. For example, a process temperature might be sampled via a multiplexer (Section 6.12.1) every few seconds, or information concerning a process flowrate might be transmitted along a distributed computer control system data highway at different intervals of time according to a particular protocol (Section 7.20.4). A control system containing any kind of microprocessor installation will require data in continuous (*analog*) form to be converted by an A/D converter (Section 6.9.5) to a *discrete time* signal before transmission to the microprocessor. In such cases the information received by the controller is classed as *sampled* or *discrete time data* and the system as a *sampled data* or *discrete time system.*

Consider the sampler switch shown in Fig. 7.88*a* where $f(t)$ is the continuous input. Suppose the switch closes every $\mathcal{T}$ units of time (the *sampling time*) and opens again instantaneously. Then the output of the sampler (generally written as $f^*(t)$) will be a sequence of impulses each of magnitude equal to that of $f(t)$ at that particular sampling time (Figs 7.88*b* and 7.88*c*). [In practice this will not be so as a

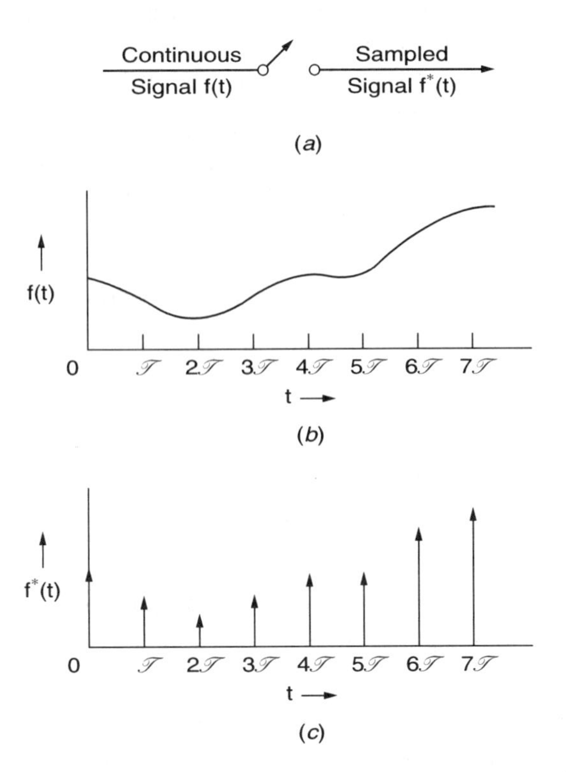

FIG. 7.88. Sampling a continous signal: (*a*) sampler switch; (*b*) continuous signal—input to sampler; (*c*) sampled signal—output from sampler

true impulse is impossible to produce in reality and the sampled output will approximate more to a series of narrow pulses.]

Hence:
$$f^*(t) = f(0)\delta t + f(\mathscr{T})\delta(t - \mathscr{T}) + f(2\mathscr{T})\delta(t - 2\mathscr{T}) + \ldots\ldots$$
$$= \sum_{n=0}^{\infty} f(n\mathscr{T})\delta(t - n\mathscr{T}) \tag{7.206}$$

where $\delta(t)$ is a unit impulse (Dirac function) and $\delta(t - n\mathscr{T})$ is a unit impulse at $t = n\mathscr{T}$. Taking the Laplace transform of equation 7.206 we obtain:

$$\bar{f}^*(s) = f(0) + f(\mathscr{T})\exp(-\mathscr{T}s) + f(2\mathscr{T})\exp(-2\mathscr{T}s) + \ldots\ldots$$
$$= \sum_{n=0}^{\infty} f(n\mathscr{T})\exp(-n\mathscr{T}s) \tag{7.207}$$

Putting $z = \exp(\mathscr{T}s)$ yields:

$$f(z) = \mathscr{Z}\{f^*(t)\} = f(0) + f(\mathscr{T})z^{-1} + f(2\mathscr{T})z^{-2} + \ldots\ldots$$
$$= \sum_{n=0}^{\infty} f(n\mathscr{T})z^{-n} \tag{7.208}$$

where $f(z)$ signifies the z-transform of $f^*(t)$.

It can be seen from the above that the output of a sampler and its transform can both be represented by infinite series. The use of the z-transform simplifies the treatment of such systems, and relationships for sampled data processes can be derived in terms of z-transforms which are similar to those obtained for equivalent continuous systems employing the Laplace transform.

For simple functions the z-transforms can be obtained easily from their equivalent Laplace transforms by the use of partial fractions and Appendix 7.1.

Example 7.13

Determine the z-transform equivalent of the transfer function of the heat exchanger system described in Example 7.3.

Solution

From Example 7.3: $\mathbf{G}_2(s) = \dfrac{\bar{\vartheta}_B}{\bar{\mathscr{F}}}$

$$= \frac{20}{(1 + 0.67s)(10.2s^2 + 3.1s + 1)}$$

$$= 1.55\left(\frac{1}{s + 1.5} - \frac{s - 1.19}{(s + 0.152)^2 + 0.0749}\right)$$

$$= 1.55\left(\frac{1}{s + 1.5} - \frac{s + 0.152}{(s + 0.152)^2 + 0.0749} + \frac{1.34}{(s + 0.152)^2 + 0.0749}\right)$$

$$= 1.55\,(\bar{\mathrm{f}}_a - \bar{\mathrm{f}}_b + \bar{\mathrm{f}}_c)$$

From Appendix 7.1, the z-transform of $\bar{\mathrm{f}}_a = 1/(s + 1.5)$ is:

$$\mathrm{f}_a(z) = z/[z - \exp(-1.5\mathscr{T})].$$

The second term requires the simultaneous use of the z-transform of $\mathrm{f}(s + a)$ (i.e. $\mathrm{f}[z\exp(a\mathscr{T})]$) where $a = 0.152$ and the z-transform of:

$$\bar{\mathrm{f}}_d = \frac{s}{s^2 + 0.274^2}$$

i.e.:

$$\mathrm{f}_d(z) = \frac{z^2 - z\cos(0.274\mathscr{T})}{z^2 - 2z\cos(0.274\mathscr{T}) + 1}$$

Hence, the z-transform of the second term in the bracket is:

$$\mathrm{f}_b(z) = \frac{[z\exp(0.152\mathscr{T})]^2 - z\exp(0.152\mathscr{T})\cos(0.274\mathscr{T})}{[z\exp(0.152\mathscr{T})]^2 - 2z\exp(0.152\mathscr{T})\cos(0.274\mathscr{T}) + 1}$$

The z-transform of $\bar{\mathrm{f}}_c$ can be obtained in a similar manner, viz.:

$$\mathrm{f}_c(z) = \frac{4.93z\exp(0.152\mathscr{T})\sin(0.274\mathscr{T})}{(z\exp(0.152\mathscr{T}))^2 - 2z\exp(0.152\mathscr{T})\cos(0.274\mathscr{T}) + 1}$$

Thus the z-transform of $\mathbf{G}_2(s)$ is:

$$\underline{\underline{\mathbf{G}_2(z) = 0.0256\,[\mathrm{f}_a(z) - \mathrm{f}_b(z) + \mathrm{f}_c(z)]}}$$

A combination of Appendix 7.1 and partial fractions may also be used to derive the inverse of a z-transform (i.e. to determine the corresponding time domain function) in a similar manner to that for Laplace transforms. This is illustrated by the following example.

Example 7.14

Derive the inverse transform of:

$$\mathrm{f}(z) = \frac{z\{\mathscr{T}[\exp(-2\mathscr{T}) - z] + 2(z - 1)^2\}}{(z - 1)^2[z - \exp(-2\mathscr{T})]}$$

Solution

$$\mathrm{f}(z) = \frac{z\{\mathscr{T}[\exp(-2\mathscr{T}) - z] + 2(z-1)^2\}}{(z-1)^2[z - \exp(-2\mathscr{T})]} = \frac{z\{2(z-1)^2 - \mathscr{T}[z - \exp(-2\mathscr{T})]\}}{(z-1)^2[z - \exp(-2\mathscr{T})]}$$

$$= \frac{2z}{z - \exp(-2\mathscr{T})} - \frac{\mathscr{T}z}{(z-1)^2}$$

Hence, from Appendix 7.1, the corresponding time function is:

$$\mathrm{f}(t) = 2\exp(-2t) - t$$

However, the z-transform $\mathrm{f}(z)$ represents a sampled data signal of which $\mathrm{f}(t)$ is the continuous form. Thus the inverse of $\mathrm{f}(z)$ is the sampled data signal $\mathrm{f}^*(t)$ corresponding to $\mathrm{f}(t)$, where, from equation 7.206:

$$\mathrm{f}^*(t) = \sum_{n=0}^{\infty}[2\exp(-2n\mathscr{T}) - n\mathscr{T}]\delta(t - nT)$$

7.17.2. Block Diagram Algebra for Sampled Data Systems

Consider a sampler switch followed by an element for which we know the continuous transfer function $\mathbf{G}(s)$ as in Figure 7.89.

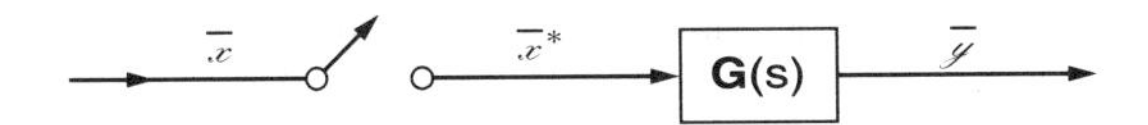

FIG. 7.89. Basic sampler configuration

In this case:

$$\mathbf{G}(s) = \frac{\bar{y}}{\bar{x}^*}$$

where $\bar{x}^*$ is defined by equation 7.207 as $\sum_{n=0}^{\infty} x(n\mathscr{T})\exp(-n\mathscr{T}s)$.

$$\bar{y} = \bar{x}^*\mathbf{G}(s) \tag{7.209}$$

The sampled data equivalent of $\bar{y}$, i.e. $\bar{y}^*$, is then given by[31]:

$$\bar{y}^* = \bar{x}^*\mathbf{G}^*(s) \tag{7.210}$$

where $\mathbf{G}^*(s) = \sum_{n=0}^{\infty} \mathbf{G}(n\mathscr{T})\exp(-n\mathscr{T}s)$ is called the *pulse transfer function* of the system and $\bar{y}^* = \sum_{n=0}^{\infty} \bar{y}(n\mathscr{T})\exp(-n\mathscr{T}s)$. Putting $z = \exp(\mathscr{T}s)$ in equation 7.210 produces the z-transform relation:

$$y(z) = x(z)\mathbf{G}(z)$$

i.e.:

$$\mathbf{G}(z) = \frac{y(z)}{x(z)} \tag{7.211}$$

A comparison of equations 7.209 and 7.210 establishes that, in order to derive the pulse transfer function $\mathbf{G}^*(s)$ from the continuous transfer function $\mathbf{G}(s)$, we must '*star*' both sides of equation 7.209 by writing:

$$[\bar{y}]^* = \bar{y}^* \tag{7.212}$$

and

$$[\bar{x}^*\mathbf{G}(s)]^* = \bar{x}^*\mathbf{G}^*(s) \tag{7.213}$$

i.e. when 'starring' a product, only the last term of the product is starred. Summed quantities which are starred are equal to the sum of the separate individually starred quantities[31], viz.:

$$[\bar{x} + \bar{y}]^* = \bar{x}^* + \bar{y}^* \tag{7.214}$$

Two further common sampled data configurations are illustrated in Fig. 7.90. From Fig. 7.90*a*:

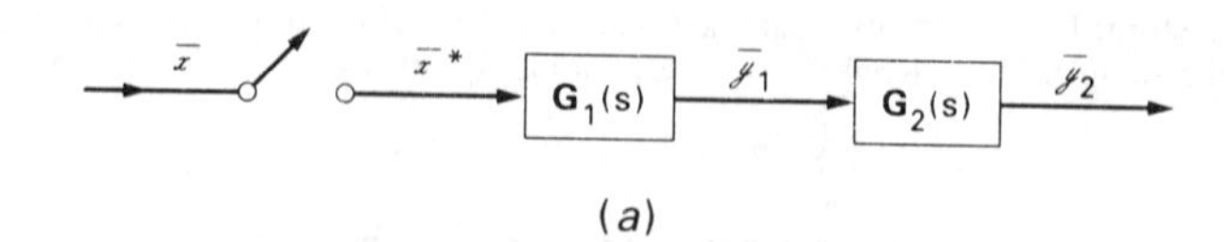

$\bar{x}$ $\bar{x}^*$ $G_1(s)$ $\bar{y}_1$ $\bar{y}_1^*$ $G_2(s)$ $\bar{y}_2$

(*b*)

FIG. 7.90. Common sampler configurations

$$G_1(s) = \frac{\bar{y}_1}{\bar{x}^*} \quad \text{and} \quad G_2(s) = \frac{\bar{y}_2}{\bar{y}_1}$$

$$\therefore \quad \bar{y}_2 = \bar{y}_1 G_2(s) = \bar{x}^* G_1(s) G_2(s) \tag{7.215}$$

Starring both sides of equation 7.215 gives:

$$\bar{y}_2^* = \bar{x}^*[G_1(s)G_2(s)]^* = \bar{x}^* G_1G_2^*(s) \tag{7.216}$$

where $[G_1(s)G_2(s)]^*$ is written conventionally as $G_1G_2^*(s)$. Thus, the equivalent relationship for equation 7.216 in terms of the z-transform is:

$$y_2(z) = x(z)G_1G_2(z)$$

Hence:

$$G(z) = \frac{y_2(z)}{x(z)} = G_1G_2(z) \tag{7.217}$$

where $G(z)$ is the transfer function in terms of the z-transform for the two blocks in series.

From Fig. 7.90*b*:

$$\bar{y}_1 = \bar{x}^* G_1(s) \quad \text{and} \quad \bar{y}_2 = \bar{y}_1^* G_2(s)$$

Starring both equations gives:

$$\bar{y}_1^* = \bar{x}^* G_1^*(s) \quad \text{and} \quad \bar{y}_2^* = \bar{y}_1^* G_2^*(s)$$

Thus:

$$y_1(z) = x(z)G_1(z)$$

and:

$$y_2(z) = y_1(z)G_2(z) = x(z)G_1(z)G_2(z)$$

Hence, in this case, the transfer function of the two blocks in series is:

$$\mathbf{G}(z) = \frac{\mathscr{Y}_2(z)}{\mathscr{X}(z)} = \mathbf{G}_1(z)\mathbf{G}_2(z) \qquad (7.218)$$

The essential difference between equations 7.217 and 7.218 is illustrated in the following example.

Example 7.15

A measuring instrument consists basically of a sensor and a transducer. The sensor transmits a signal x to the transducer every second and the transducer responds as a first-order system with a time constant of 5s and a steady-state gain of 2 units. The output y_1 of the transducer drives a transmitter which also approximates to first-order behaviour with a time constant of 4s and a steady-state gain of 5.

(a) Derive the transfer function between the output y_2 of the transmitter and the sensor signal x.

(b) If several such instruments are joined to the same transmitter via a multiplexer such that each transducer is sampled once every two seconds, what will be the effect on the transfer function relating y_2 and x?

Solution

(a) In this case the sampler configuration is the same as that shown in Fig. 7.90*a* where $\mathbf{G}_1(s)$ and $\mathbf{G}_2(s)$ are the transfer functions of the transducer and transmitter respectively. The transfer function for this configuration is given by equation 7.217, i.e.:

$$\mathbf{G}(z) = \frac{\mathscr{Y}_2(z)}{\mathscr{X}(z)} = \frac{\mathscr{Y}_2(z)}{\mathscr{X}(z)} = \mathbf{G}_1\mathbf{G}_2(z) \qquad \text{(equation 7.217)}$$

Now:

$$\mathbf{G}_1(s) = \frac{2}{1+5s} = \frac{0.4}{s+0.2}$$

and:

$$\mathbf{G}_2(s) = \frac{5}{1+4s} = \frac{1.25}{s+0.25}$$

$\therefore$

$$\mathbf{G}_1(s)\mathbf{G}_2(s) = \frac{0.5}{(s+0.2)(s+0.25)} = 10\left(\frac{1}{s+0.2} - \frac{1}{s+0.25}\right)$$

Hence, the corresponding z-transform is (from Appendix 7.1):

$$\mathbf{G}_1\mathbf{G}_2(z) = 10\left(\frac{z}{z-\exp(-0.2\mathscr{T})} - \frac{z}{z-\exp(-0.25\mathscr{T})}\right)$$

Thus, putting $\mathscr{T}$ = 1s, we obtain the transfer function:

$$\underline{\underline{\mathbf{G}(z) = \mathbf{G}_1\mathbf{G}_2(z) = \frac{0.4z}{z^2 - 1.6z + 0.64}}}$$

(b) This corresponds to the sampler configuration illustrated in Fig. 7.90*b* with the sampling time for the first sampler as 1s and for the second sampler as 2s. Thus, from equation 7.218:

$$\mathbf{G}(z) = \frac{\mathscr{Y}_2(z)}{\mathscr{X}(z)} = \mathbf{G}_1(z)\mathbf{G}_2(z) \qquad \text{(equation 7.218)}$$

In this case we can obtain $\mathbf{G}_1(z)$ and $\mathbf{G}_2(z)$ directly from $\mathbf{G}_1(s)$ and $\mathbf{G}_2(s)$, where:

$$\mathbf{G}_1(z) = \frac{0.4z}{z-\exp(-0.2\mathscr{T}_1)}$$

and:

$$\mathbf{G}_2(z) = \frac{1.25z}{z-\exp(-0.25\mathscr{T}_2)}$$

Putting $\mathscr{T}_1$ = 1s and $\mathscr{T}_2$ = 2s yields the transfer function:

$$\underline{\underline{\mathbf{G}(z) = \mathbf{G}_1(z)\mathbf{G}_2(z) = \frac{0.5z^2}{z^2 - 1.4z + 0.50}}}$$

7.17.3. Sampled Data Feedback Control Systems

Closed-loop transfer functions for feedback loops containing samplers can be derived in terms of the z-transform in a manner analogous to that for completely

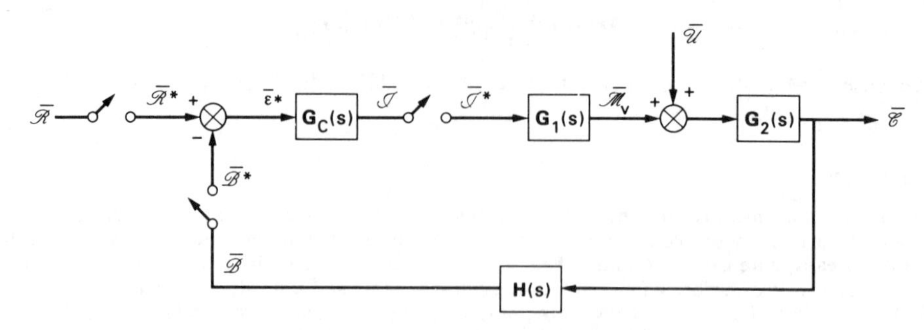

FIG. 7.91. Closed-loop control system with discrete time controller

continuous systems using the Laplace transform (see Section 7.9) except that the simplifying rule suggested by equation 7.111 should not be employed. The exact form of the sampled data closed-loop transfer function depends upon the position of the sampler, or samplers, in the control system. A common situation is where the analog controller in a typical continuous system (e.g. Fig. 7.3) has been replaced by an equivalent discrete time controller. In this case there will be samplers in the form of D/A and A/D converters sampling the measured variable, the controller output and the controller set point as in Fig. 7.91.

The following relationships can be established:

$$\bar{\mathscr{C}} = \bar{\mathscr{M}}_v \mathbf{G}_2(s) = \bar{\mathscr{T}}^* \mathbf{G}_1(s)\mathbf{G}_2(s)$$

Starring (Section 7.17.2) gives:

$$\bar{\mathscr{C}}^* = \bar{\mathscr{T}}^* \mathbf{G}_1\mathbf{G}_2^*(s)$$

Hence:
$$\mathscr{C}(z) = \mathscr{T}(z)\mathbf{G}_1\mathbf{G}_2(z) \tag{7.219}$$

Also:
$$\bar{\mathscr{T}} = \bar{\varepsilon}^* \mathbf{G}_C(s) = (\bar{\mathscr{R}}^* - \bar{\mathscr{B}}^*)\mathbf{G}_C(s)$$

Starring gives:
$$\bar{\mathscr{T}}^* = (\bar{\mathscr{R}}^* - \bar{\mathscr{B}}^*)\mathbf{G}_C^*(s)$$

Thus:
$$\mathscr{T}(z) = (\mathscr{R}(z) - \mathscr{B}(z))\mathbf{G}_C(z)$$

$$= \mathscr{R}(z)\mathbf{G}_C(z) - \mathscr{B}(z)\mathbf{G}_C(z) \tag{7.220}$$

Eliminating $\mathscr{T}(z)$ between equations 7.219 and 7.220 yields:

$$\mathscr{C}(z) = \mathscr{R}(z)\mathbf{G}_C(z)\mathbf{G}_1\mathbf{G}_2(z) - \mathscr{B}(z)\mathbf{G}_C(z)\mathbf{G}_1\mathbf{G}_2(z) \tag{7.221}$$

Also:
$$\bar{\mathscr{B}} = \bar{\mathscr{T}}^* \mathbf{G}_1(s)\mathbf{G}_2(s)\mathbf{H}(s)$$

Starring:
$$\bar{\mathscr{B}}^* = \bar{\mathscr{T}}^* \mathbf{G}_1\mathbf{G}_2\mathbf{H}^*(s)$$

and thus, using equation 7.220:

$$\mathscr{B}(z) = \mathscr{T}(z)\mathbf{G}_1\mathbf{G}_2\mathbf{H}(z) = [\mathscr{R}(z) - \mathscr{B}(z)]\mathbf{G}_C(z)\mathbf{G}_1\mathbf{G}_2\mathbf{H}(z)$$

i.e.:

$$\mathscr{B}(z) = \frac{\mathscr{R}(z)\mathbf{G}_C(z)\mathbf{G}_1\mathbf{G}_2\mathbf{H}(z)}{1 + \mathbf{G}_C(z)\mathbf{G}_1\mathbf{G}_2\mathbf{H}(z)} \tag{7.222}$$

Substituting for $\mathscr{B}(z)$ from equation 7.222 into equation 7.221 gives

$$\mathscr{C}(z) = \mathscr{R}(z)\mathbf{G}_C(z)\mathbf{G}_1\mathbf{G}_2(z) - \frac{\mathscr{R}(z)[\mathbf{G}_C(z)]^2\mathbf{G}_1\mathbf{G}_2\mathbf{H}(z)\mathbf{G}_1\mathbf{G}_2(z)}{1 + \mathbf{G}_C(z)\mathbf{G}_1\mathbf{G}_2\mathbf{H}(z)}$$

Hence:

$$\frac{\mathscr{C}(z)}{\mathscr{R}(z)} = \frac{\mathbf{G}_C(z)\mathbf{G}_1\mathbf{G}_2(z)}{1 + \mathbf{G}_C(z)\mathbf{G}_1\mathbf{G}_2\mathbf{H}(z)} \tag{7.223}$$

Other transfer functions can be similarly derived for other sampler and control loop configurations (see also Example 7.16 and Section 7.17.8).

7.17.4. Hold Elements (Filters)

Theoretically there is no output from a sampler between sampling times (Fig. 7.88*c*). In practice the sampler output is controlled between samples by a *filter* or *hold element*. The most common type of filter is the *zero-order hold element* (*ZOH*) in which the value of the previous sample is retained until the next sample is taken (Fig. 7.92*a*).

Consider a ZOH inserted in a closed-loop system as shown in Fig. 7.92*b*. The combined output of the sampler and the ZOH can be considered as a series of positive and negative step changes of equal magnitude occurring at successive sampling instants (Fig. 7.92*a*), i.e.

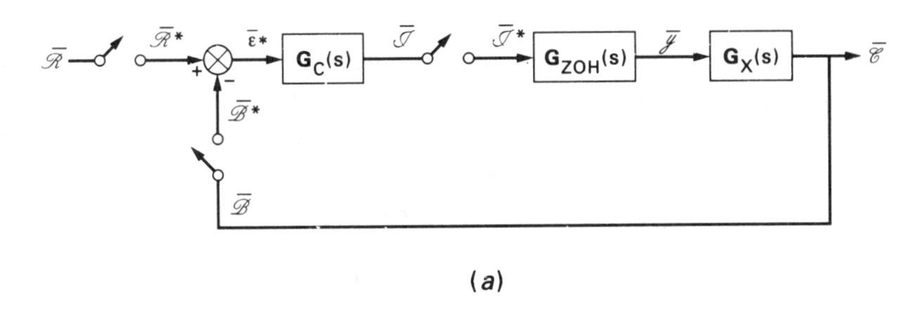

(*a*)

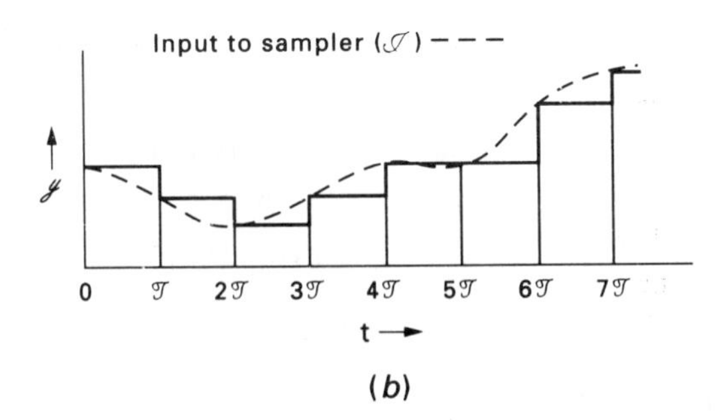

(*b*)

FIG. 7.92. Control loop containing a sampler and a zero-order hold element: (*a*) block diagram; (*b*) output of hold element (filter)

$$y = f(0)[u(t) - u(t - \mathscr{T})] + f(\mathscr{T})[u(t - \mathscr{T}) - u(t - 2\mathscr{T})] + \ldots\ldots \tag{7.224}$$

Transforming equation 7.224 and substituting equation 7.207:

$$\bar{y} = \mathrm{f}(0)\left(\frac{1 - \exp(-\mathscr{T}s)}{s}\right) + \mathrm{f}(\mathscr{T})\left(\frac{\exp(-\mathscr{T}s) - \exp(-2\mathscr{T}s)}{s}\right) + \ldots\ldots$$

$$= \left(\frac{1 - \exp(-\mathscr{T}s)}{s}\right)[\mathrm{f}(0) + \mathrm{f}(\mathscr{T})\exp(-\mathscr{T}s) + \ldots\ldots]$$

$$= \left(\frac{1 - \exp(-\mathscr{T}s)}{s}\right)\bar{\mathrm{f}}^*(s) \tag{7.225}$$

As $\bar{\mathrm{f}}^*(s)$ is the Laplace transform of the output of the sampler without the ZOH, the remaining term must represent the Laplace transform for the zero-order hold element, i.e.:

$$\mathbf{G}_{ZOH}(s) = \frac{1 - \exp(-\mathscr{T}s)}{s} \tag{7.226}$$

Suppose that, in Figure 7.92*a*, $\mathbf{G}_X(s) = \dfrac{2}{1 + 2s}$. Normally $\mathbf{G}_{ZOH}$ is incorporated into the transfer function of the adjacent element as follows.

$$\mathbf{G}(s) = \mathbf{G}_{ZOH}(s)\mathbf{G}_X(s) = \left(\frac{1 - \exp(-\mathscr{T}s)}{s}\right)\left(\frac{2}{1 + 2s}\right)$$

$$= [1 - \exp(-\mathscr{T}s)]\left(\frac{1}{s(s + 0.5)}\right)$$

$$= \mathbf{G}_e(s)\mathbf{G}_f(s)$$

The factor $\mathbf{G}_e(s) = [1 - \exp(-\mathscr{T}s)]$ is the Laplace transform of a function of time $f_e(t)$ consisting of a positive unit impulse occurring at $t = 0$ and a negative unit impulse occurring at $t = \mathscr{T}$. Thus $f_e(t)$ exists only at the sampling instants and the sampled function $f_e^*(t)$ will be the same as $f_e(t)$. Hence $\mathbf{G}_e^*(s) = \mathbf{G}_e(s)$, and:

$$\mathbf{G}(s) = \mathbf{G}_e(s)\mathbf{G}_f(s) = \mathbf{G}_e^*(s)\mathbf{G}_f(s)$$

Starring gives:

$$\mathbf{G}^*(s) = \mathbf{G}_e^*(s)\mathbf{G}_f^*(s)$$

Thus:

$$\mathbf{G}(z) = \mathbf{G}_e(z)\mathbf{G}_f(z) \tag{7.227}$$

where $\mathbf{G}_e(z) = (1 - z^{-1})$ from Appendix 7.1 and $\mathbf{G}_f(z)$ is the z-transform of $\mathbf{G}_f(s)$, i.e.:

$$\mathbf{G}_f(s) = \frac{1}{s(s + 0.5)} = \frac{2}{s} - \frac{2}{(s + 0.5)}$$

$$\therefore \quad \mathbf{G}_f(z) = \frac{2z}{z - 1} - \frac{2z}{z - \exp(-0.5\mathscr{T})}$$

Thus, from equation 7.227:

$$\mathbf{G}(z) = (1 - z^{-1})\left(\frac{2z}{z - 1} - \frac{2z}{z - \exp(-0.5\mathscr{T})}\right) \tag{7.228}$$

7.17.5. The Stability of Sampled Data Systems

It is shown in Section 7.10.1 that a continuous system is unstable if any root of the associated characteristic equation (i.e. any pole of the system transfer function) lies in the right half of the complex s-plane (Fig. 7.93a). If this root is s_1, then s_1 can be expressed in terms of its real and imaginary parts, i.e.:

$$s_1 = \beta_1 + i\gamma_1$$

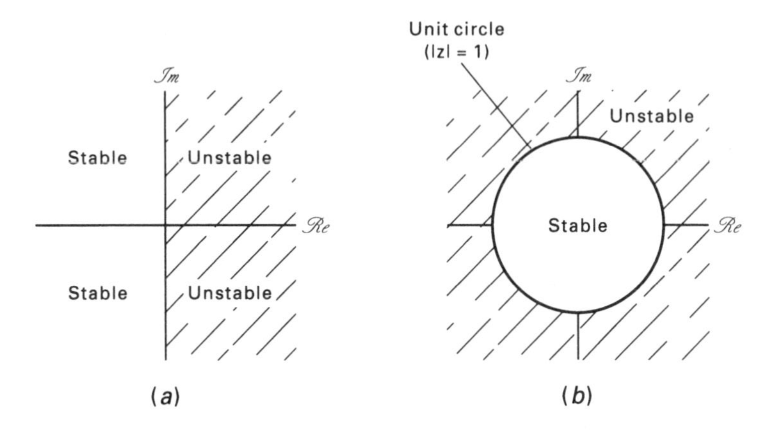

FIG. 7.93. Regions of stability and instability: (a) s-plane; (b) z-plane

Putting $z = \exp(s_1\mathcal{T})$ to obtain the corresponding z-transform (i.e. to map the root on to the complex z-plane) gives:

$$z = \exp(s_1\mathcal{T}) = \exp(\beta_1\mathcal{T})\exp(i\gamma_1\mathcal{T})$$

Thus:

$$|z| = \exp(\beta_1\mathcal{T}) \tag{7.229}$$

If s_1 lies in the right half of the s-plane, then $\beta_1 > 0$ and $|z| > 1$. The latter condition applies to the region outside the unit circle (i.e. outside $|z| = 1$) in the z-plane (Fig. 7.93b). Thus the roots of the z-transformed characteristic equation must all lie within the unit circle in the z-plane for the corresponding system to be stable.

The Use of the Routh–Hurwitz Criterion

In order to determine the number of roots of the z-transformed characteristic equation that lie outside the unit circle, a procedure analogous to the Routh–Hurwitz approach for continuous systems (Section 7.10.2) can be used. The Routh–Hurwitz criterion cannot be applied directly to the characteristic equation $f(z) = 0$. However, by mapping the interior of the unit circle in the z-plane on to the left half of a new complex variable ξ-plane, the Routh–Hurwitz criterion can be applied as for continuous systems to the corresponding characteristic equation in terms of the new variable[41]. This mapping can be achieved using the bilinear transformation[17]:

$$z = \frac{\xi + 1}{\xi - 1} \tag{7.230}$$

This is illustrated in Example 7.16.

Example 7.16

The pressure P in the vessel shown in Fig. 7.94 is controlled by a digital (discrete time) proportional controller of gain K_C. The measurement from the pressure sensor is sampled every minute for transmission via a ZOH element to the controller which is connected directly to a control valve on a line which vents the vessel. The transfer function representing the process (i.e. between P and the flowrate Q through the vent) approximates to $\bar{\mathscr{P}}/\bar{\mathscr{Q}} = 2/(1 + 2s)$. It is known that the steady-state gains of the control valve and measuring element are 2 units and 0.5 units respectively and that their dynamics are negligible in comparison with that of the process. Find the maximum value of K_C for which the system will remain stable and compare this with the stability of the equivalent entirely continuous (analog) control loop.

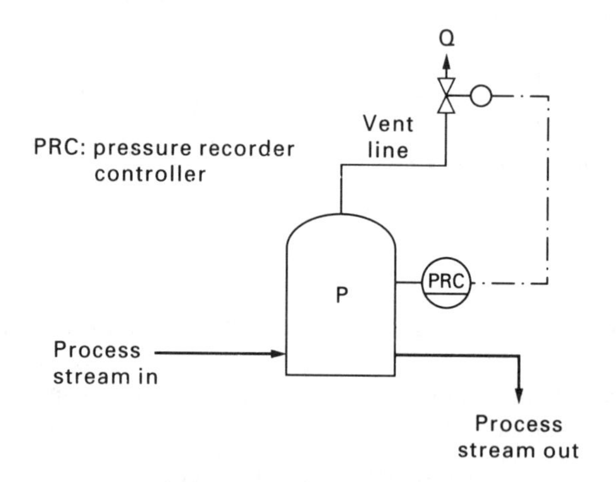

FIG. 7.94. Control of pressure in a vessel

Solution

Figure 7.95 is the block diagram of the sampled data control system. By comparison with Fig. 7.91 and from equation 7.221:

$$\mathscr{C}(z) = \mathscr{R}(z)\mathbf{G}_C(z)\mathbf{G}_1\mathbf{G}_2(z) - \mathscr{B}(z)\mathbf{G}_C(z)\mathbf{G}_1\mathbf{G}_2(z) \quad \text{(equation 7.221)}$$

Now:

$$\bar{\mathscr{B}} = \bar{x}^*\mathbf{G}_{ZOH}(s)$$

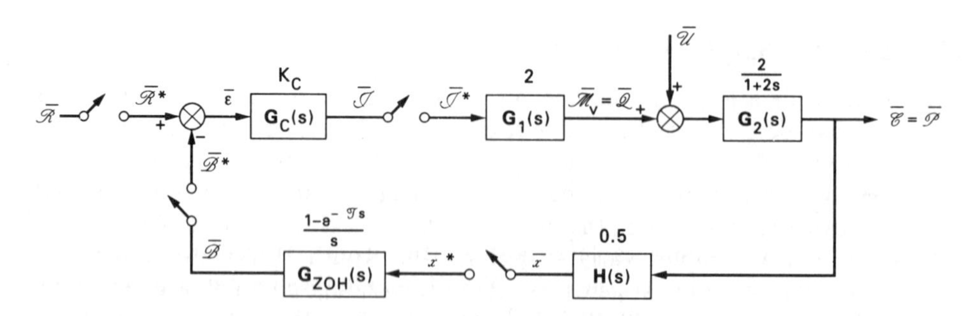

FIG. 7.95. Block diagram of sampled data pressure control loop shown in Fig. 7.94

Starring: $\bar{\mathscr{B}}^* = \bar{x}^* \mathbf{G}_{ZOH}(s)$

Thus: $$\mathscr{B}(z) = x(z)\mathbf{G}_{ZOH}(z) \tag{7.231}$$

Also: $\bar{x} = \bar{\mathscr{T}}^* \mathbf{G}(s)\mathbf{G}_2(s)\mathbf{H}(s)$

Hence: $$x(2) = \mathscr{T}(z)\mathbf{G}_1\mathbf{G}_2\mathbf{H}(z) \tag{7.232}$$

From equations 7.231 and 7.232:

$$\mathscr{B}(z) = \mathscr{T}(z)\mathbf{G}_1\mathbf{G}_2\mathbf{H}(z)\mathbf{G}_{ZOH}(z) \tag{7.233}$$

By comparing equation 7.233 with the relationships derived in Section 7.17.3, it is clear that the corresponding closed loop transfer function is:

$$\frac{\mathscr{C}(z)}{\mathscr{R}(z)} = \frac{\mathbf{G}_C(z)\mathbf{G}_1\mathbf{G}_2(z)}{1 + \mathbf{G}_C(z)\mathbf{G}_1\mathbf{G}_2\mathbf{H}(z)\mathbf{G}_{ZOH}(z)} \tag{7.234}$$

Hence, the system characteristic equation is:

$$1 + \mathbf{G}_C(z)\mathbf{G}_1\mathbf{G}_2\mathbf{H}(z)\mathbf{G}_{ZOH}(z) = 0 \tag{7.235}$$

But: $\mathbf{G}_1(s)\mathbf{G}_2(s)\mathbf{H}(s) = 2 \times 2/(1 + 2s) \times 0.5 = 2/(1 + 2s)$

Thus: $$\mathbf{G}_{ZOH}(s)\mathbf{G}_1(s)\mathbf{G}_2(s)\mathbf{H}(s) = \left(\frac{1 - \exp(-\mathscr{T}s)}{s}\right)\left(\frac{2}{1 + 2s}\right)$$

By comparison with the example in Section 7.17.4, the z-transformed form of the characteristic equation (equation 7.235) is:

$$1 + K_C(1 - z^{-1})\left(\frac{2z}{z - 1} - \frac{2z}{z - \exp(-0.5\mathscr{T})}\right) = 0$$

Putting $\mathscr{T} = 1$ min yields:

$$1 + 2K_C\left(1 - \frac{z - 1}{z - 0.61}\right) = 0 \tag{7.236}$$

In order to apply the Routh–Hurwitz criterion, the transformation given by equation 7.230 must be applied. Hence equation 7.236 becomes:

$$1 + 2K_C\left(1 - \frac{2}{0.39\xi + 1.61}\right) = 0$$

Thus: $$(0.78K_C + 0.39)\xi + (1.61 - 0.78K_C) = 0$$

and, for the system to be stable:

$$1.61 - 0.78K_C > 0$$

i.e.: $$\underline{\underline{K_C < 2.06}}$$

If the control loop is completely continuous, then the closed loop transfer function (from Section 7.9.1) is:

$$\frac{\bar{\mathscr{C}}}{\bar{\mathscr{R}}} = \frac{4K_C}{2s + (1 + 2K_C)}$$

This is first order and, consequently, the continuous system will always be stable, however large is the value of K_C.

Thus the discrete time form of the control system is less stable than the equivalent continuous case.

Polar Plots and the Nyquist Criterion

It can be shown that there is a Nyquist criterion for sampled data systems which is equivalent to that for continuous systems (see Section 7.10.5) and equation 7.131 can be applied in its comparable z-transformed form[42]. In practice it is generally sufficient to ascertain whether the polar plot of $\mathbf{G}(z)$ in the complex z-plane encircles the $(-1, 0)$ point (as with continuous systems in the s-plane) where $1 + \mathbf{G}(z) = 0$ is the system z-transformed characteristic equation. The polar plot is constructed from

G(z) by using the transformation $z = \exp(s\mathcal{T}) = \exp(i\omega\mathcal{T})$ (Section 7.17.1). Thus a family of plots may be obtained, each plot corresponding to a different value of the sampling time $\mathcal{T}$. It should be noted that, as z traverses the unit circle in the z-plane, the real frequency ω varies from zero to the sampling frequency $\omega_s = 2\pi/\mathcal{T}$, and, for sampled data systems, the entire frequency response is contained within the frequency range $0 \leqslant \omega \leqslant \omega_s$.

7.17.6. Discrete Time (Digital) Fixed Parameter Feedback Controllers

Many industrial applications employ discrete time controllers. These operate on the basis of discrete signals rather than on a continuous signal as with analog controllers. They are ideally suited to the digital environment produced by computer control and/or sampled data systems.

Discrete time equivalents of the normal analog **P**, **PI**, **PD** and **PID** controllers are employed most commonly. Sampling rates are generally high and consequently the performance of the discrete-time controller approximates very closely that of the corresponding analog arrangement.

Discrete time controllers will not normally be stand-alone units but will be simulated within the software of a digital computer. The capacity of the computer can be used if necessary to produce more complex forms of feedback control than those provided by the standard algorithms of the classical fixed parameter controller.

The Position Form of the Discrete Time Control Algorithm

The general relationship for the classical **PID** analog controller is given by (Section 7.2.2):

$$J = J_0 + K_C\left\{\varepsilon + \frac{1}{\tau_I}\int_0^{\tau} \varepsilon\,dt + \tau_D\frac{d\varepsilon}{dt}\right\} \qquad \text{(equation 7.8)}$$

The discrete time equivalent to this can be expressed in finite difference form as:

$$J_n = J_0 + K_C\left\{\varepsilon_n + \frac{\mathcal{T}}{\tau_I}\sum_{k=0}^{n}\varepsilon_k + \frac{\tau_D}{\mathcal{T}}(\varepsilon_n - \varepsilon_{n-1})\right\} \qquad (7.237)$$

where $\mathcal{T}$ is the sampling interval and J_n is the value of J at the nth sampling instant.

Equation 7.237 is termed the *position form* of the algorithm as it gives the actual value (or position) of the controller output signal. In this form, at the nth sampling instant, the **PI** algorithm saves only the current value of the error ε_n and the sum of all previous errors, viz. $\mathbf{S}_{n-1} = \sum_{k=0}^{n-1}\varepsilon_k$, i.e.:

$$J = J_0 + K_C\left\{\varepsilon_n + \frac{\mathcal{T}}{\tau_I}(\mathbf{S}_{n-1} + \varepsilon_n)\right\} \qquad (7.238)$$

whilst the **PID** algorithm saves ε_n, ε_{n-1} and the sum of all previous errors, i.e.:

$$J = J_0 + K_C\left\{\varepsilon_n + \frac{\mathcal{T}}{\tau_I}(\mathbf{S}_{n-1} + \varepsilon_n) + \frac{\tau_D}{\mathcal{T}}(\varepsilon_n - \varepsilon_{n-1})\right\} \qquad (7.239)$$

The Velocity Form of the Discrete Time Control Algorithm

In this case, the actual value of the controller output signal at the nth sampling instant is not computed but the change from its value at the preceding sampling time is calculated instead.

At the nth sampling instant, for a **PID** controller:

$$J_n = J_0 + K_C\left\{\varepsilon_n + \frac{\mathcal{T}}{\tau_I}\sum_{k=0}^{n}\varepsilon_k + \frac{\tau_D}{\mathcal{T}}(\varepsilon_n - \varepsilon_{n-1})\right\} \qquad \text{(equation 7.237)}$$

and at the $(n - 1)$th sampling instant:

$$J_{n-1} = J_0 + K_C\left\{\varepsilon_{n-1} + \frac{\mathcal{T}}{\tau_I}\sum_{k=0}^{n-1}\varepsilon_k + \frac{\tau_D}{\mathcal{T}}(\varepsilon_{n-1} - \varepsilon_{n-2})\right\} \qquad (7.240)$$

Subtracting equation 7.240 from equation 7.237 gives:

$$\Delta J_n = J_n - J_{n-1} = K_C\left\{1 + \frac{\mathcal{T}}{\tau_I} + \frac{\tau_D}{\mathcal{T}}\right\}\varepsilon_n - K_C\left\{1 + \frac{2\tau_D}{\mathcal{T}}\right\}\varepsilon_{n-1} + K_C\frac{\tau_D}{\mathcal{T}}\varepsilon_{n-2} \qquad (7.241)$$

Equation 7.241 is the *velocity* form of the **PID** algorithm and the equivalent transfer function is:

$$\frac{\Delta J(z)}{\varepsilon(z)} = K_C\left\{\left(1 + \frac{\mathcal{T}}{\tau_I} + \frac{\tau_D}{\mathcal{T}}\right) - \left(1 + \frac{2\tau_D}{\mathcal{T}}\right)z^{-1} + \frac{\tau_D}{\mathcal{T}}z^{-2}\right\} \qquad (7.242)$$

(It is necessary to express equation 7.242 in terms of z-transforms as it represents a sampled data system. The transform of ε_{n-1} is $z^{-1}\varepsilon(z)$ and the transform of ε_{n-2} is $z^{-2}\varepsilon(z)$, i.e. multiplication of a signal in its z-transformed form by z^{-1} represents the same signal delayed by one sampling period; multiplication by z^{-2} corresponds to a delay by two sampling periods, etc.)

The velocity form of the algorithm has certain advantages over the position form, viz.:

(a) It does not need any initialisation—in other words, it does not require the initial value of the controller output J_0.

(b) Protection against integral windup (Section 7.2.3) is built in. If errors cannot be eliminated quickly then the presence of integral action will cause the controller output to saturate. In the velocity form (equation 7.241) there is no error summation so this does not occur.

(c) Protection can be achieved against computer failure by connecting the output of the controller to an integrating amplifier or to a stepper-motor. Such devices will retain the last calculated position of the control valve in case the computer fails—thus avoiding total loss of control of the process.

(d) Because the output involves only a change in the position of the controller, this algorithm automatically provides bumpless transfer (Section 7.2.2). However, if a substantial error exists when the control is changed from manual to automatic, the controller response may be slow, especially if a large value of τ_I is employed[14].

7.17.7. Tuning Discrete Time Controllers

In general, the usual analog tuning procedures (Section 7.11) will suffice—particularly with small sampling intervals. There is an optimum sampling interval because, as the sampling interval decreases, the load placed upon the computer that is simulating the control action increases. The consequence of sampling is effectively to introduce dead time (DV lag) into the control loop. This can cause the control to be degraded (e.g. Example 7.16) and hence the sampling period should be chosen as the maximum that does not cause the controlled system response to deteriorate (i.e. to become significantly less stable). The optimum sampling interval is directly related therefore to the operating conditions of the controlled process. This sampling time is nearly always found to be between 1/10 and 1/5 of the system dominant time constant or system dead time—whichever is the smaller[6]. For oscillating systems the sampling period should be smaller than half the period of oscillation. In several references[6,43] are quoted typical optimum sampling times for systems controlling different controlled variables, viz. $\mathcal{T} \approx 1$s is for flow control loops, $\mathcal{T} \approx 5$s for level and pressure control loops, and $\mathcal{T} \approx 20$s for temperature control loops. These sampling times reflect the general speed of response of the control loop in question.

7.17.8. Response Specification Algorithms

The computational power and flexibility of the computer is much used now to simulate controllers having characteristics other than the standard **P**, **PI**, etc., modes. Controllers are described in the following for which the design algorithm is derived directly from a specification of the discrete time character of the response of the controlled variable to a given change in set point.

Consider the closed-loop transfer function for the system shown in Fig. 7.92*a*. The closed-loop transfer function $\mathscr{C}(z)/\mathscr{R}(z)$ is derived as in Section 7.17.3, i.e.:

$$\frac{\mathscr{C}(z)}{\mathscr{R}(z)} = \frac{\mathbf{G}_C(z)\mathbf{G}_{ZOH}\mathbf{G}_X(z)}{1 + \mathbf{G}_C(z)\mathbf{G}_{ZOH}\mathbf{G}_X(z)}$$

$$\therefore \qquad \mathbf{G}_C(z) = \frac{1}{\mathbf{G}_{ZOH}\mathbf{G}_X(z)}\left(\frac{\mathscr{C}(z)/\mathscr{R}(z)}{1 - \mathscr{C}(z)/\mathscr{R}(z)}\right) \qquad (7.243)$$

There is a variety of specifications that can be imposed on the system closed-loop response for a given change in set point. These lead to a number of alternative discrete-time control algorithms—the best known of which are the *Deadbeat* and *Dahlin's* algorithms.

Design of a Discrete Time Controller Based Upon a Deadbeat Response

It is required, in this case, that the response of the controlled variable to a step change in set point exhibits zero error at all sampling instants after the first. Such a response would be described theoretically by a step change of the same magnitude but delayed by one sampling instant. Now, for a step change in set point of unit magnitude (from Appendix 7.1):

$$\mathscr{R}(z) = \frac{z}{z-1} = \frac{1}{1-z^{-1}} \tag{7.244}$$

For a similar response in the controlled variable but delayed by one sampling interval:

$$\mathscr{C}(z) = z^{-1}\left(\frac{1}{1-z^{-1}}\right) \tag{7.245}$$

Hence:

$$\frac{\mathscr{C}(z)}{\mathscr{R}(z)} = \frac{1}{z} \tag{7.246}$$

Substituting for $\mathscr{C}(z)/\mathscr{R}(z)$ in equation 7.243 gives:

$$\mathbf{G}_C(z) = \frac{1}{\mathbf{G}_{ZOH}\mathbf{G}_X(z)}\left(\frac{z^{-1}}{1-z^{-1}}\right) \tag{7.247}$$

It should be noted that:

(a) The requirement that the error should be zero at each sampling instant does not preclude large overshoots and/or highly oscillatory behaviour. Deadbeat controllers suffer significantly from such difficulties (Fig. 7.96).
(b) The specification that the controlled variable must reach its final value after one sampling interval (Fig. 7.96) gives a minimum rise time and a very strong control action.
(c) The deadbeat controller algorithm is physically unrealisable (i.e. is physically unrealistic) if any dead time inherent in $\mathbf{G}_{ZOH}\mathbf{G}(z)$ is more than one sampling period.
(d) If a dead time larger than one sampling interval is present in $\mathbf{G}_{ZOH}\mathbf{G}(z)$ then the controlled variable response specification must be modified such that the response is not required to exhibit zero error until after two, three, etc. sampling periods. This is accomplished by multiplying $\mathscr{R}(z)$ by z^{-2}, z^{-3}, etc. respectively as necessary.

Dahlin's Algorithm

DAHLIN[44] suggested that, in order to avoid the large overshoots and oscillatory behaviour which are characteristic of the deadbeat algorithm, the specification of the system closed-loop response to a step change in set point should be the same as that for a first-order system with dead time. The first-order time constant can then be employed as a design parameter which can be adjusted to give the desired closed-loop response. Hence:

$$\overline{\mathscr{C}} = \frac{\exp(-\tau_{DV}s)}{1+\tau s}\frac{1}{s} \tag{7.248}$$

If the dead time is equal to j sampling periods (i.e. $\tau_{DV} = j\mathscr{T}$), then the z-transform of $\overline{\mathscr{C}}$ is:

$$\mathscr{C}(z) = z^{-j}\frac{[1-\exp(-\mathscr{T}/\tau)]\,z^{-1}}{[1-z^{-1}][1-\exp(-\mathscr{T}/\tau)z^{-1}]} \tag{7.249}$$

From equations 7.244 and 7.249:

$$\frac{\mathscr{C}(z)}{\mathscr{R}(z)} = z^{-j} \frac{[1 - \exp(-\mathscr{T}/\tau)]\, z^{-1}}{1 - \exp(-\mathscr{T}/\tau)\, z^{-1}} \tag{7.250}$$

Hence, substitution for $\frac{\mathscr{C}(z)}{\mathscr{R}(z)}$ in equation 7.243 yields the required controller specification algorithm, i.e.:

$$\mathbf{G}_C(z) = \frac{1}{\mathbf{G}_{ZOH}\mathbf{G}_X(z)} \left(\frac{[1 - \exp(-\mathscr{T}/\tau)]z^{-(j+1)}}{1 - \exp(-\mathscr{T}/\tau)z^{-1} - [1 - \exp(-\mathscr{T}/\tau)]z^{-(j+1)}} \right) \tag{7.251}$$

Note that:

(a) The time constant τ is used as a design parameter to give the required closed-loop response (Fig. 7.96). The response gets more sluggish as τ is increased (in this respect τ has an effect similar to that of ζ in the case of second order systems).
(b) Dahlin's algorithm is physically realisable provided that any dead time included in $\mathbf{G}_{ZOH}\mathbf{G}_X(z) \leqslant (j+1)\mathscr{T}$.
(c) Controllers designed using Dahlin's algorithm avoid the excessive control action produced by the deadbeat algorithm (Fig. 7.96).

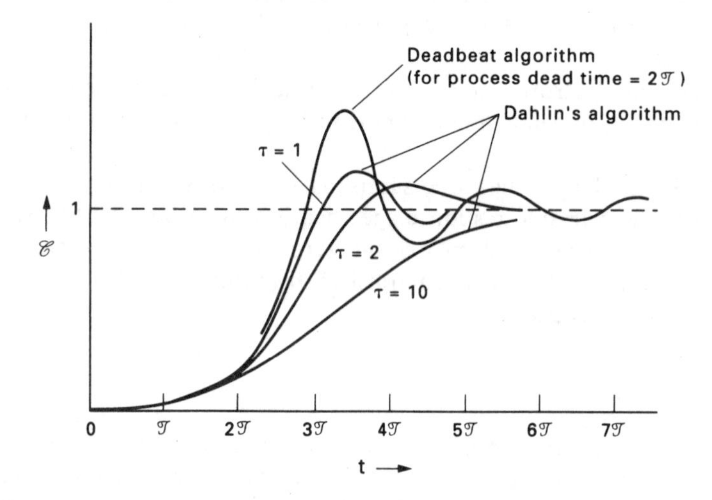

FIG. 7.96. Comparison of response of the controlled variable using deadbeat and Dahlin's response specification algorithm

7.18. ADAPTIVE CONTROL

A system or controller which can adjust its parameters automatically in such a way as to compensate for variations in the characteristics of the process it controls is termed an *adaptive system* or *adaptive controller*.

The concept of adaptive control, in that it can be considered as a kind of non-linear feedback action, is not new. For example, KALMAN[45] described a self-optimising controller in 1958. However, it was impossible to implement Kalman's procedure at that time due to digital computer limitations. The more recent considerable expansion in the application of adaptive strategies is due, in large measure, to the rapid evolution of the microprocessor and thus to the ability to simulate adaptive controllers in software as discrete time systems in computer control applications (Sections 7.19 and 7.20).

Adaptive controllers can be usefully applied because most processes are non-linear (Section 7.16) and common controller design criteria (Section 7.12) are based on linear models. Due to process non-linearities, the controller parameters required to give the desired response of the controlled variable change as the process steady state alters. Furthermore, the characteristics of many processes vary with time, e.g. due to catalyst decay, fouling of heat exchangers, etc. This leads to a deterioration in the performance of controllers designed upon a linear basis.

Generally, an objective function is required for the adaptation strategy which guides the adaptation mechanism to produce the 'best' settings of the controller parameters. For example, the $^1/_4$ decay ratio specification could be employed, or the ISE criterion (Section 7.11). For instance, if the $^1/_4$ decay ratio criterion is used, then, if any change in process parameters leads to decay ratios other than $^1/_4$, the adaptation mechanism adjusts the controller parameters until a $^1/_4$ decay ratio in the controlled response is achieved once again.

ÅSTRÖM and WITTENMARK[46] have listed three basic schemes that have been devised for the adaptive control of processes, viz.:

(i) gain scheduling or feed-forward adaptation,
(ii) model reference control, and
(iii) self-tuning regulators.

7.18.1. Scheduled (Programmed) Adaptive Control

This is employed where a process is well-known and an adequate mathematical model is available. If there is an *auxiliary* process variable which correlates well with any changes occurring in process dynamics, then the 'best' values of the controller parameters can be related ahead of time to the value of the auxiliary variable. Consequently, by measuring the value of the auxiliary variable, the adaptation of the controller parameters can be scheduled (or programmed).

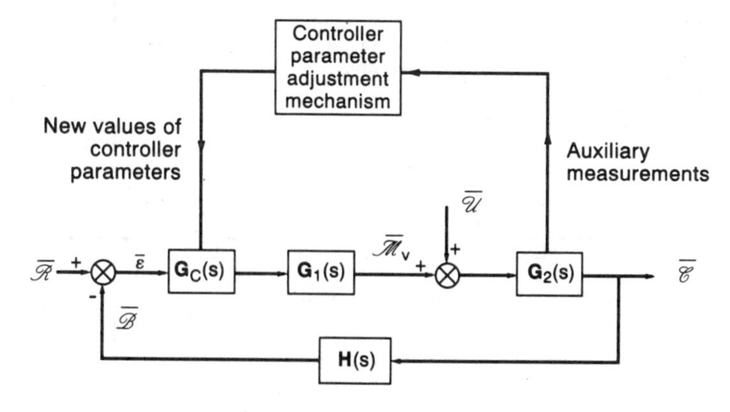

FIG. 7.97. Block diagram of scheduled (programmed) adaptive control system

The relevant block diagram is composed of two loops (Fig. 7.97)—a normal feedback loop and a loop which contains the parameter adjustment mechanism (comparable to feed-forward compensation—Section 7.14.1).

Gain Scheduling Adaptive Control is a special application of this procedure. For example we may have a control valve whose characteristic (input signal/valve stem position relationship) is non-linear. In this case, the valve stem position would be measured in order to obtain the gain of the valve (the appropriate relationship must be known) and the valve gain is used then to adjust the gain of the controller. If the auxiliary variable relationships are more complicated then it may be necessary to employ a *Programmed Adaptive Control* procedure.

Difficulties that can be experienced with this form of control are[47]:

(a) A good theoretical knowledge of the process is required in order to determine the auxiliary variables.
(b) It is necessary to compute the regulator parameters for a large number of operating conditions.
(c) The control strategy is open-loop feed-forward. There is no feedback compensation involved (see Section 7.14.1).

7.18.2. Model Reference Adaptive Control (MRAC)

This is employed when the process is not well-known. The *Model Reference Adaptive Controller* contains a reference model to which the command signal or set point change is applied as well as to the process itself (Fig. 7.98)[48]. The output of the reference model is postulated as the desired controlled process output and this is compared with the actual process output. The difference (or error) ε_m between the two outputs is used to adjust the controller parameters so as to minimise the relevant integral criterion. For example, if the ISE criterion is employed then the quantity

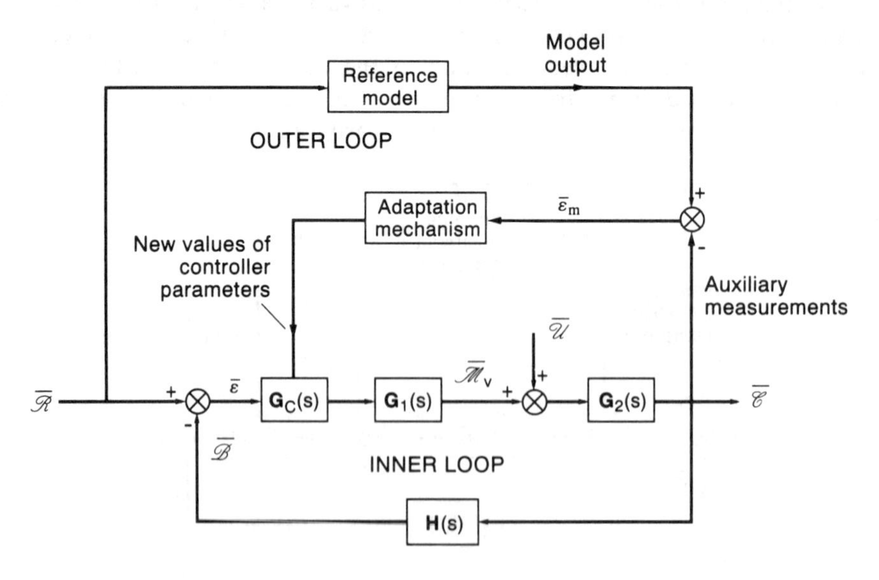

FIG. 7.98. Block diagram of model reference adaptive control system

$\int_0^t [\varepsilon_m(t)]^2 \mathrm{d}t$ (or, for a discrete time system, $\sum_{n=0}^{N} [\varepsilon_m^2]_n$) is minimised. The model chosen as the reference is to a certain extent arbitrary and generally a simple linear model is used. The major difficulty with the MRAC strategy is in designing an adaptation mechanism which provides a stable system, i.e. which brings ε_m to zero as quickly as possible. The application requires mathematical analysis and computing facilities.

7.18.3. The Self-Tuning Regulator (STR)

The architecture of the *self-tuning regulator* is shown in Fig. 7.99. It is similar to that of the Model Reference Adaptive Controller in that it also consists basically of two loops. The inner loop contains the process and a normal linear feedback controller. The outer loop is used to adjust the parameters of the feedback controller and comprises a recursive parameter estimator and an adjustment mechanism.

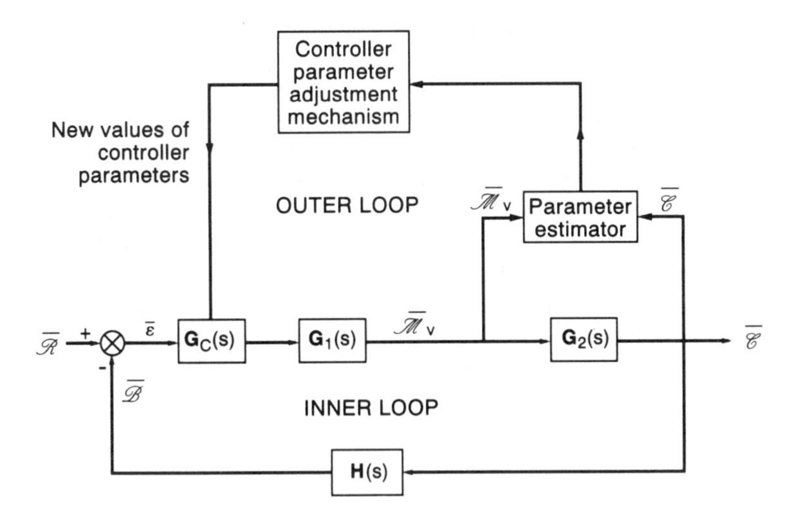

FIG. 7.99. Block diagram of self-tuning regulator

The estimator assumes a simple linear model for the process. For instance, a first-order system with dead time may be employed for which the transfer function may be written as:

$$\mathbf{G}_{\text{est}}(s) = \frac{\overline{\mathscr{C}}}{\overline{\mathscr{M}}_v} = \frac{K \exp(-\tau_{DV} s)}{1 + \tau s} \tag{7.252}$$

(cf. equation 7.135). Occasionally, a second-order linear model is used.

A series of corresponding values of $\mathscr{M}_v$ and $\mathscr{C}$ is measured and these measurements are used to estimate K, τ, and τ_{DV} employing a least squares technique. After the model parameters have been estimated, the parameter adjustment mechanism tunes the controller parameters (i.e. K_C, τ_I, τ_D) using one of the well known design criteria (e.g. minimisation of the ISE). Both the process parameter estimator and the controller parameter adjustment mechanism require programming on a microcomputer. The values of $\mathscr{M}_v$ and $\mathscr{C}$ are sampled at finite intervals of time. Changing this sampling time can affect the control action substantially and the controller can be finely tuned by adjustment of the sampling time.

There are several disadvantages associated with this type of adaptive controller:

(a) It may generate excessive control signals.
(b) Unknown or varying time delays can result in a poor or even unstable performance.
(c) It cannot be applied to all systems (e.g. those exhibiting non-minimum phase characteristics).

Other self-tuning controllers have been designed which overcome these difficulties, e.g. the *Generalised Minimum Variance* self-tuning controller (*GMV*)[49] and the *Generalised Predictive Controller* (*GPC*)[50].

7.19. COMPUTER CONTROL OF A SIMPLE PLANT—THE OPERATOR INTERFACE

7.19.1. Direct Digital Control (DDC) and Supervisory Control

In its simplest form, computer control can be considered as the replacement of a fixed parameter analog controller by a microcomputer with the equivalent digital control algorithm (Section 7.17.6) supplied in the form of appropriate software. In such cases the microcomputer will normally possess an architecture especially designed for the purposes of process control and the software language employed will be of a type which facilitates the construction of suitable control algorithms. If the remaining elements within the control loop are all analog in operation then it will be necessary to convert the analog signal corresponding to the measured value into its equivalent digital form before it can be supplied as the input signal to the microprocessor (which is a discrete time controller). This is achieved by inserting a suitable A/D converter (Fig. 7.97 and Section 6.11.5). Equally a D/A converter will be required to convert the output of the controller into its equivalent analog form.

The elementary configuration shown in Fig. 7.100 is rarely employed in isolation and is generally extended to include the control of a number of loops (i.e. multiloop control) by connecting the input and output of each loop sequentially to the computer via a multiplexer (Section 6.12.1) together with the necessary software. Thus, all the fixed parameter feedback controllers for a given process plant can be substituted by a single microprocessor. Such an arrangement is termed *direct digital control* or *DDC*.

A microcomputer may also be employed to adjust the set points or parameters of the various controllers (which may be analog, or digital, or both) according to some predetermined criterion (Fig. 7.101). This is generally described as *supervisory control* and may, for instance, be employed to maximise plant profitability or to minimise plant energy consumption, etc. The microcomputer involved requires substantially greater power (i.e. faster clock speed, larger RAM, etc.) than that necessary for DDC. Information is fed back to the supervisory computer concerning the status of the various control loops (e.g. values of controlled and manipulated variables, etc.) through a suitable *input/output* (*I/O*) interface. Clearly, it is of advantage to combine supervisory control with DDC and employ the same microprocessor for both operations. In the case of *distributed computer control systems* (*DCCS*) the

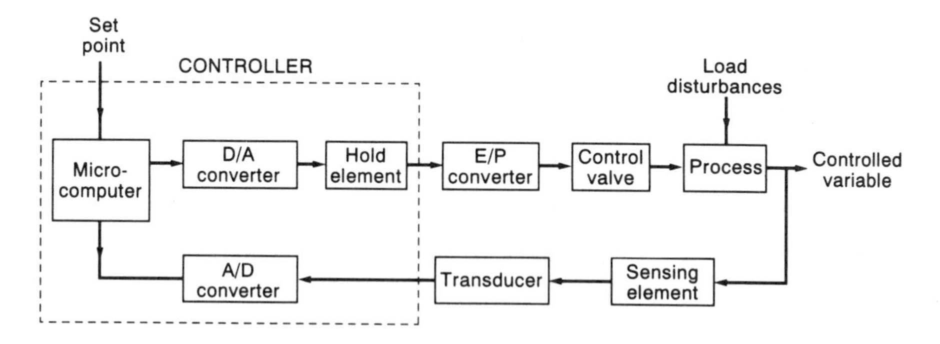

FIG. 7.100. A simple DDC system

supervisory and DDC functions may well be performed by different computers (Section 7.20.2).

The advantages of DDC/supervisory computer control over conventional analog systems have been listed by POPOVIC and BHATKAR[51] as:

(i) easy configuration of control loops,
(ii) greatly facilitated estimation of controller parameters,
(iii) easy self-tuning of controller parameters (see Section 7.18),
(iv) much simplified introduction of new control loops,
(v) enabling the application of more advanced control algorithms,

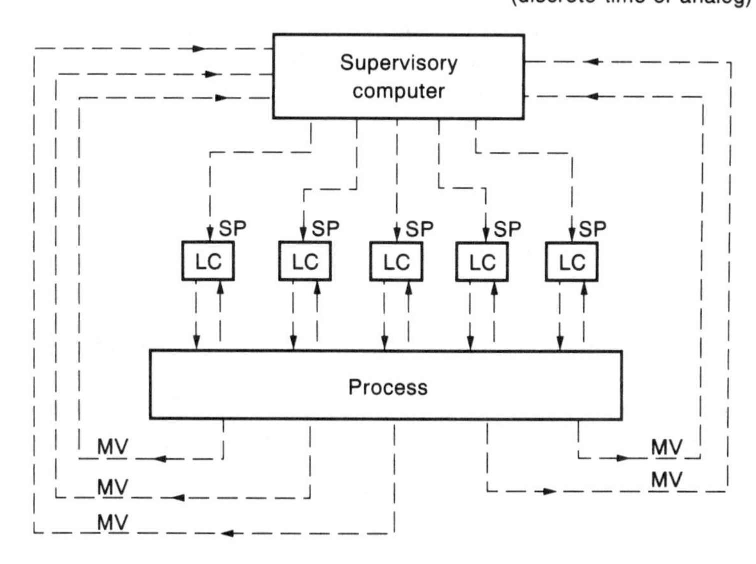

FIG. 7.101. Supervisory control

(vi) provision of the ability to calculate optimal set points for all control loops based on models of the systems involved (i.e. allowing the implementation of *model-based control*), and

(vii) enabling comprehensive data acquisition and presentation to be accomplished.

7.19.2. Real Time Computer Control

The external devices and processes to which the process control computer is connected all behave with respect to their own particular time-scales. When the actions of the computer are related to the time-scales of these external systems (often termed the external world) the computer is said to operate in *real time*[14]. Process control computers are normally concerned with real-time operations and CIVERA *et al.*[52] have divided the latter into:

Type 1. systems which are required to have a mean execution time measured over a defined time interval which is lower than a specified maximum, and

Type 2. systems in which the cycle of computations must be completed within a specified maximum time on each and every occasion.

Type 2 operations are more difficult to achieve than Type 1. An example of a Type 2 procedure is the computer control of a fixed-bed reactor in which a highly exothermic reaction takes place (Fig. 7.102). The reactor is cooled by an internal cooling coil and is controlled by a sampled data system. An important consideration is the choice of sampling interval. Once this has been specified then a Type 2 operation may be necessary in order that all computations required to produce the output signal to the coolant inlet line control valve are performed within that sampling interval. This operation cannot be delayed by other procedures that the computer may be performing. A Type 1 operation in which the mean execution time is less than one sampling interval may not be sufficiently stringent as it allows for the occasional execution time longer than one sampling interval prior to the next adjustment of the flow of coolant. This could lead to a rapid and possibly dangerous increase in reactor temperature before control action is taken.

Real-time operations may be *clock based*, *sensor based*, or *interactive* and may involve combinations of these.

Clock Based Operations

The ability of the system to complete the required cycle of calculations and other activities (interfacing, signal processing, etc.) within each sampling interval is dependent upon the number of operations to be performed and the speeds of the various components involved (particularly the computer itself). Synchronisation (and timing) of these various activities is facilitated by incorporating a *real time clock* which is generally supplied within the computer hardware (although software clocks can be provided) and presents the time in hours, minutes and seconds—often with a back-up battery to avoid the necessity for resetting each time the system is energised.

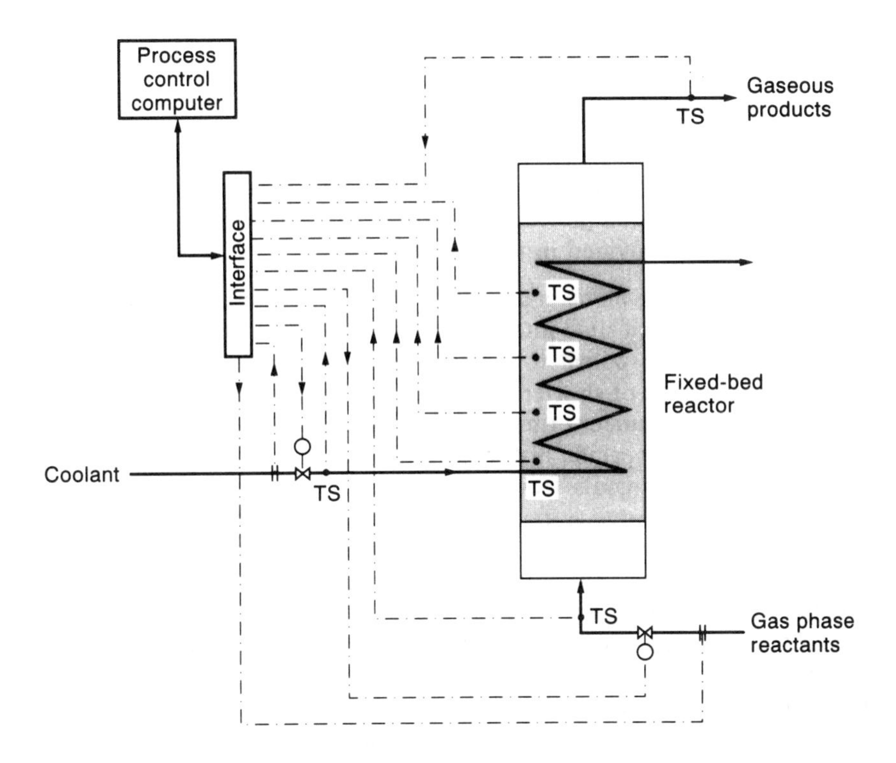

FIG. 7.102. Computer control of a highly exothermic reaction in a fixed-bed reactor

Sensor Based Operations

These are actions of the computer which are initiated by events, e.g. switching off a pump when a tank has been filled to a certain level. They form the basis of the response of computer-controlled plant to alarm conditions, e.g. if the temperature in the reactor described in Fig. 7.102 exceeds a specified value then action must be taken by the computer, such as rapidly increasing the flow of coolant or reducing the feed to the reactor. Generally, sensor-initiated operations must be performed within a specified maximum time interval and are often in the form of *alarm interrupts* (Section 7.18.3). Smaller systems will regularly scan a set of specific process variables via a multiplexer to see if any action is necessary. The latter arrangement is called *polling*.

Interactive Systems

These are generally of Type 1, i.e. the real time requirement is that the mean response time is not greater than some predetermined value. Such systems are similar to the sensor based type, but the response of the computer control system depends upon the number of operations that are being performed at any given time.

This type of architecture is not generally suited to the computer control of processing plant.

7.19.3. System Interrupts

An *interrupt* is a procedure by which the normal sequence of software instructions within the computer is suspended whilst another operation or routine is carried out. When the latter has been completed the suspended software sequence is resumed. Interrupts are normally provided in terms of:

(a) *Real Time Clock*. The clock is used to interrupt the operation of the computer at predetermined fixed time intervals and the different activities of the system are performed in response to the designed interrupt sequence. The clock interrupt is generally of shorter duration than the sampling interval and is the basis of a record of current time and date, and additionally is responsible for the identification of specific historical records.
(b) *Alarm Inputs*. The signals from particular sensors can be used to suspend normal operations and to initiate emergency action if necessary. For example, if the flow of coolant in Fig. 7.102 falls below a certain value then it may be necessary to *enable* a suitable emergency routine such as the temporary reduction of the feed flowrate.
(c) *Manual Override*. This is used to suspend automatic operation for the purpose of repair and/or maintenance.
(d) *Operating System Entry*. This is an interrupt which allows entry to the computer operating system during the normal cycle of operations.
(e) *System Failures*. These can vary in significance. Interrupts resulting from the failure of external hardware or interfacing usually enable software routines which allow the operation of the plant to be continued by alternative means, e.g. by bringing on line a spare D/A converter to replace a failed component or the substitution of the output from a faulty thermocouple with a signal from an alternative sensor. Power failure is more serious but circuits are available which can detect loss of power in the system sufficiently quickly for an emergency shut-down procedure to be performed before the loss of power is such that the system ceases to work altogether.

7.19.4. The Operator/Controller Interface

The operator forms an essential link in any total plant instrumentation and control strategy, and the conveying of information to the process plant operator and the speed at which the operator is able to respond are important considerations. Methods of presentation of information (concerning plant status) to the plant operator, manager or engineer are undergoing a considerable evolution and the effectiveness of the presentation of process data is dependent upon the hardware employed to visualise the data and the protocol used to display it.

Many different methods of information display (of varying degrees of sophistication) can be found on process plants. These can be divided into:

(a) simple indicators,
(b) circular or linear charts,

(c) control panel mimic displays, and
(d) mimic displays on semi-graphic or graphic video terminals.

Often there will be combinations of these and plants will have some means of visually recording data (e.g. by the use of charts) as well as employing indicators to show immediate readings. *Retrofitted* (updated) units may contain all four methods of presentation. Newer plant is more likely to employ the video terminal as the principal source of information, although local indicators may well be fixed to pumps, etc.

The most efficient data visualisations are obtained by using semi-graphic and graphic video terminals (i.e. method (d)) on which process mimic diagrams can be shown and on which up-to-date process data are provided. The *operator station* (*OS*) consists of one or more of these video terminals accompanied by either special purpose or standard QWERTY keyboards (or both) (Fig. 7.103), and is designed to replace the older conventional instrument and control panel. The mimic diagram can increase substantially the speed of decision making by the operator—particularly when the mimic is in colour[53]. The screen displays enable the operator to view the structure of the process and the interactions between specific basic process units, mainly by means of the static part of the mimic diagrams. The operator can also estimate quantitative relationships between process variables from the variable part of the diagram which presents actual and historic process data.

The way in which information is presented on the video display to the operator is of considerable significance. BAILEY[54] recommends that:

(a) Information that is continually being transmitted or received should be grouped sequentially.
(b) Information should be grouped in the order of its frequency of use.
(c) If frequency of use is not a major concern then the information should be grouped in a manner commensurate with the function that it represents.
(d) When some items are critical to the success of the operation of the system, the information should be grouped according to its degree of importance.

FIG. 7.103. Operator station showing video terminals and keyboards (Fisher PROVOX system)

Attention should be paid to the *loading* of the video screen (*display density*) as there is evidence that there is a significant decrease in operator performance for screen loadings greater than 50 per cent (i.e. when more than 50 per cent of the screen is occupied by information). A good design criterion is to keep screen loadings below 25 per cent[55]. Other important considerations in screen design are the employment of a consistent convention in graphics coding (emphasising specific items in a consistent way by using underlines, special fonts, etc.), the use of colour to distinguish between different sets of data, the rate of movement of the display (e.g. the speed of a graphic illustrating the flow of a liquid), and the format in which the information is presented. The interested reader is referred to GILMORE *et al.*[55]

7.20. DISTRIBUTED COMPUTER CONTROL SYSTEMS (DCCS)

7.20.1. Hierarchical Systems

The concentration of all automation functions within a single computer (Section 7.19.1) may be possible for a very simple plant, but this type of configuration is inefficient for more complex processes for which there could be many thousands of connections between plant and computer. Currently, small industrial processes are controlled by a hierarchical architecture consisting of a central computer (usually a minicomputer), which is used to solve central automation problems, together with a series of peripheral computers (generally microprocessors which are called *front-end computers*) which control different sections of the plant (Fig. 7.104*a*). This type of architecture is termed a *decentralised computer system*.

A two-level structure is inadequate, however, for large-scale plants due to overloading of the central computer. In this case, a three-stage hierarchy is employed with intermediate (or *group*) computers running some of the duties which are common to all the local microprocessors (Fig. 7.104*b*).

7.20.2. Design of Distributed Computer Control Systems

The hierarchical arrangements described in Section 7.20.1 form a natural basis for the modern *Distributed Computer Control System* (*DCCS*). It is difficult to classify a standard DCCS as there are currently over 50 different systems available[51] each having a somewhat different architecture. Furthermore, the design of the DCCS depends to a large extent upon the type and size of the process. A typical off shore oil production complex, an oil refinery or a pulp and paper mill, is likely to employ a DCCS with several dozen microcomputers (often called *microcontrollers* in this context) plus a *host*, which is generally a powerful process control computer. The microcontrollers' tasks include control, data acquisition, logic operations and operator interfacing.

If the DCCS is required to replace an existing control system (i.e. if the plant is being retrofitted) then the microcontrollers will normally be located in the *central control room* (*CCR*). In this case, the distribution features of the DCCS will not be fully utilised and the significance of the *digital data highway* (*DH*) will also be reduced.

In an entirely new plant, however, various DCCS components will be housed in *local control rooms* (*LCR*) or in *outstations*. The primary operator stations (OS) and

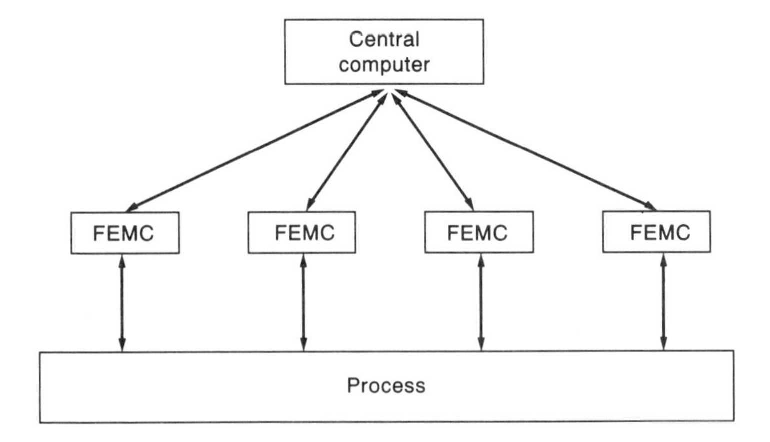

(a)

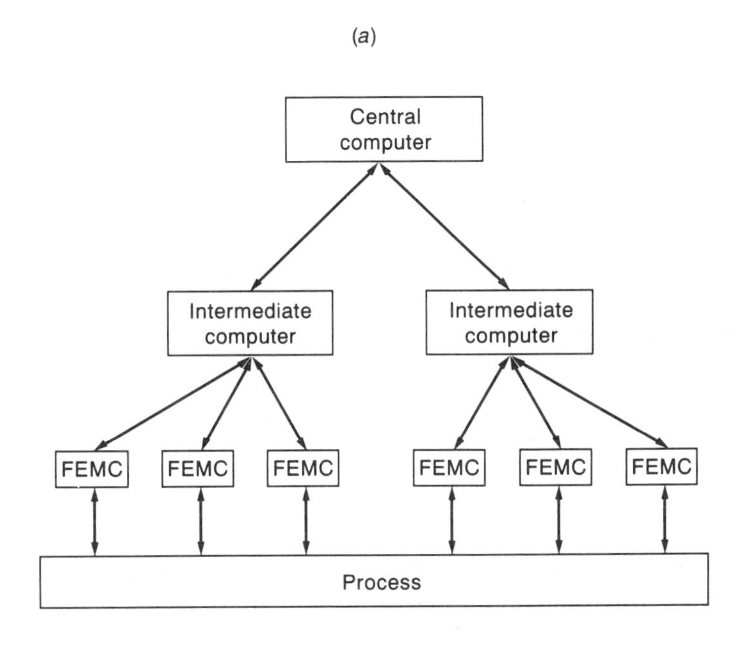

(b)

FIG. 7.104. (a) Two-level hierarchical computer control system (decentralised computer system). (b) Three-stage hierarchical computer control

the host are located in the CCR, and the secondary operator stations together with various microcontrollers are placed in LCRs.

The data highway (which is the main artery of communication) connects the various DCCS components to the host and operator stations. Figure 7.105 demonstrates the distributed architecture of a typical DCCS.

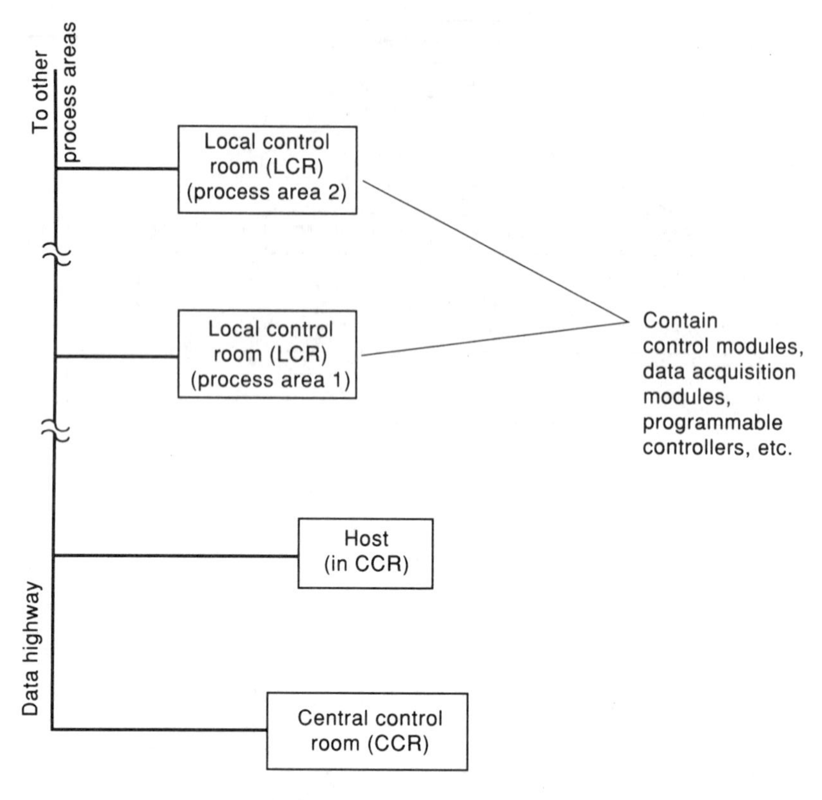

FIG. 7.105. DCCS architecture

7.20.3. DCCS Hierarchy

The main elements of a typical large scale DCCS are shown in Fig. 7.106. There are five principal levels of hierarchy in most large scale distributed computer control systems.

Level One

At the lowest hierarchical level (often called the *loop level*), the functional units of the system are distributed and placed in the vicinity of the plant to which they are linked by direct wiring via the *process interface*. The loop level commonly consists of three components—the *control module* (*CM*), the *data acquisition module* (*DM*), and the *programmable controller* (*PLC*). The characteristics of these elements are a high scanning rate, a high degree of dependability, and direct communication with the host. Control modules are shared multiloop controllers which can perform various advanced control strategies. Data acquisition modules are used primarily for the collection of large quantities of data from the process and some DMs can accommodate over 1000 inputs and outputs. Programmable controllers (PLCs) are discussed in Section 7.21 and form important elements of any DCCS. They are used in such areas as monitoring for fire and gas, to implement emergency shutdowns, and also where batch and semi-batch processes are involved.

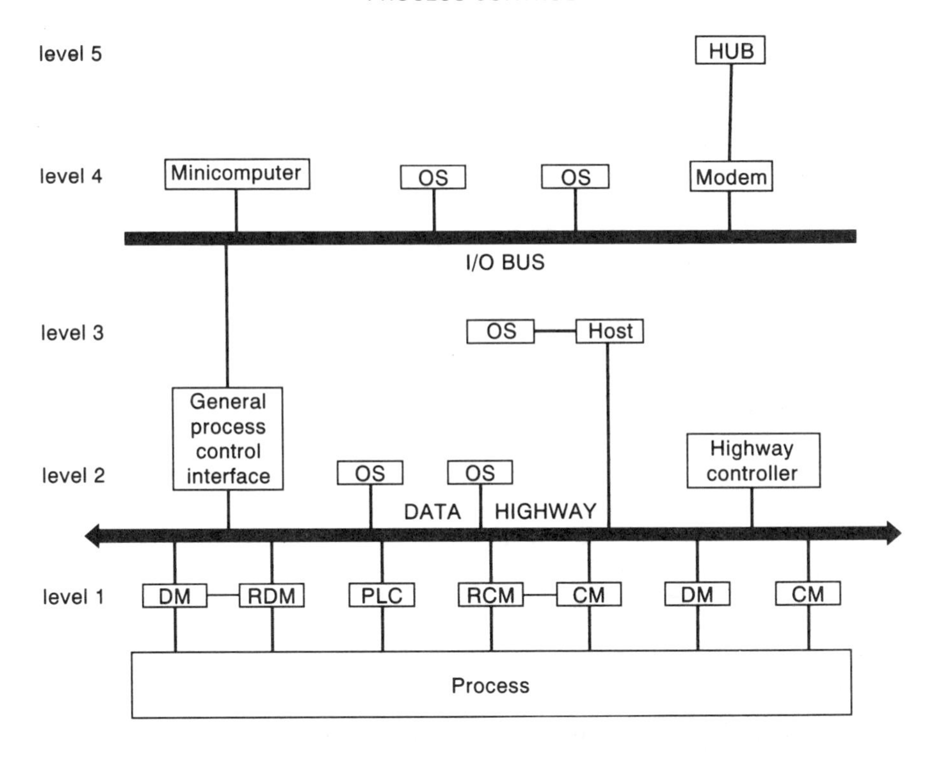

Key: OS: Operator station. DM: Data acquisition module
RDM: Reserve data aquisition module. CM: control module
RCM: Reserve control module. PLC: Programmable controller.

FIG. 7.106. Principal levels of hierarchy in a DCCS

Level Two

This level contains the highway based operator stations which are the most significant elements of the DCCS (see Section 7.20.5). They are the primary man–machine interfaces and are independent of the host. Only if they are designed and configured properly will the DCCS be acceptable to operators. A poor operator station configuration will discourage operators from using the system properly and may lead to costly process shutdowns and production losses.

Level Three

This consists principally of the host which is used to supervise several process areas and may well have six or more data highways linked to it. Hosts are powerful computers with their own operator stations and other peripherals. They are an integral element of modern control systems and only very small processes may be controlled and monitored effectively without a host. However, if not configured properly, a host computer can be overloaded quite easily and will then fail to operate satisfactorily. One serious loading problem for a host is the fast scanning rate required for DDC loops (usually once per second and up to three times per second). As it is a main operator interface, a host must update displays every one to five seconds and, for a large process, may be supervising several thousand points[56]. In addition, hosts are the least reliable of all the elements of a DCCS and, if the DCCS is configured such that host failure can cause the loss of substantial numbers of operator stations and DDC loops, then it is likely that frequent (and costly) shutdowns will result.

Typical tasks for a host are:

(a) supervision and coordination between several units of a plant,
(b) graphic and mimic displays beyond the capability of the stand-alone operator station,
(c) automatic shutdowns and start-ups,
(d) optimisation,
(e) process simulation,
(f) long-term trends, averages and summaries,
(g) postmortem analysis,
(h) interface to foreign computers and data transmission systems,
(i) advanced control strategies beyond the capability of a control module, and
(j) alarm handling beyond the scope of the stand-alone operator station.

Level Four

This is occupied by a minicomputer or a superminicomputer with considerably more power and peripherals than the host. Whilst the host's function is limited to process control and monitoring, the minicomputer at this level may perform such tasks as maintenance scheduling, production control, longterm historical data recording and acquisition, simulation, and optimisation.

Level Five

Level five is generally called the *hub*. This is the highest level in the system hierarchy and is applicable to organisations which coordinate the operations of a number of plants often distributed over a large geographical area. Examples are oil companies, pulp and paper companies, the water and gas industries, etc.

The hub is normally served by a main frame computer and only large corporations employ such a machine to monitor the operation of several plants or works. The hub is located in the company's headquarters and is linked to the various plants by telephone line or by direct cabling (often employing fibre-optic technology).

Possible tasks for a hub are:

(a) production control, scheduling and allocation,
(b) optimisation procedures relevant to general production planning,
(c) coordination between various plants,
(d) stock control,
(e) cost control, and
(f) control and monitoring of the movement of goods.

7.20.4. Data Highway (DH) Configurations

Communication within a DCCS is implemented by means of the *system bus* or *data highway* along which data are exchanged between elements of the system at the same or at different hierarchical levels. The bus represents the basic *system interface* and should provide the following features[51]:

(a) the ability to transfer large quantities of process data (plant status data),
(b) a message-structured transfer of control instructions and plant data,
(c) a flexibly structured data highway,
(d) the ability to transmit data at high transfer rates, and
(e) a relatively high reliability.

The system architecture is a major factor in determining the capacity/constraints of a DCCS and this will be substantially affected by the type of data highway employed and the way in which it is configured. Various DH configurations are illustrated in Fig. 7.107.

Each type of DH possesses particular advantages and disadvantages. The multidrop DH was the first to be used in a distributed control system (i.e. in the Honeywell TDC 2000). It is the simplest and the least expensive, but is less flexible than some of the other types. The most flexible is the clustered or partially clustered/multidrop type. This highway is the least likely to cause data traffic jams, but is also the most expensive and bulky. The ring highway is the most fault-tolerant configuration, but it is the most complex to operate. The star configuration is the least fault-tolerant and is expensive in terms of the amount of cabling required.

A more recent development is to employ Local Area Networks (LANs) where the LAN is a serial communication link used as a data highway in the DCCS for the integration of intelligent terminals at the higher hierarchical levels. There are two principal types of LAN applicable to DCCS, viz.:

(i) the high performance LAN which uses coaxial cable as the data transmission medium and is capable of transmission rates exceeding 50 Mbps (megabits per second) over distances up to 3000m, and
(ii) the low cost LAN which employs shielded twisted pairs (Fig. 6.65). This type of LAN can transfer data over distances up to 300m at rates not exceeding 3 Mbps.

An important aspect of the DH is the *highway protocol* which is a set of rules which define the format, timing, priority, and control of the transmission and exchange of data on the DH. The performance of the DH depends to a very large extent upon the protocol employed. If the protocol is poorly designed then the DH will almost certainly jam with data and fail.

7.20.5. The DCCS Operator Station

A number of the basic characteristics of operator stations (OS) are described in Section 7.19.4. Distributed computer control systems normally offer two types of operator station, a stand-alone OS, and a host-dependent OS. Both provide a logical display hierarchy which, if properly configured, improve significantly the

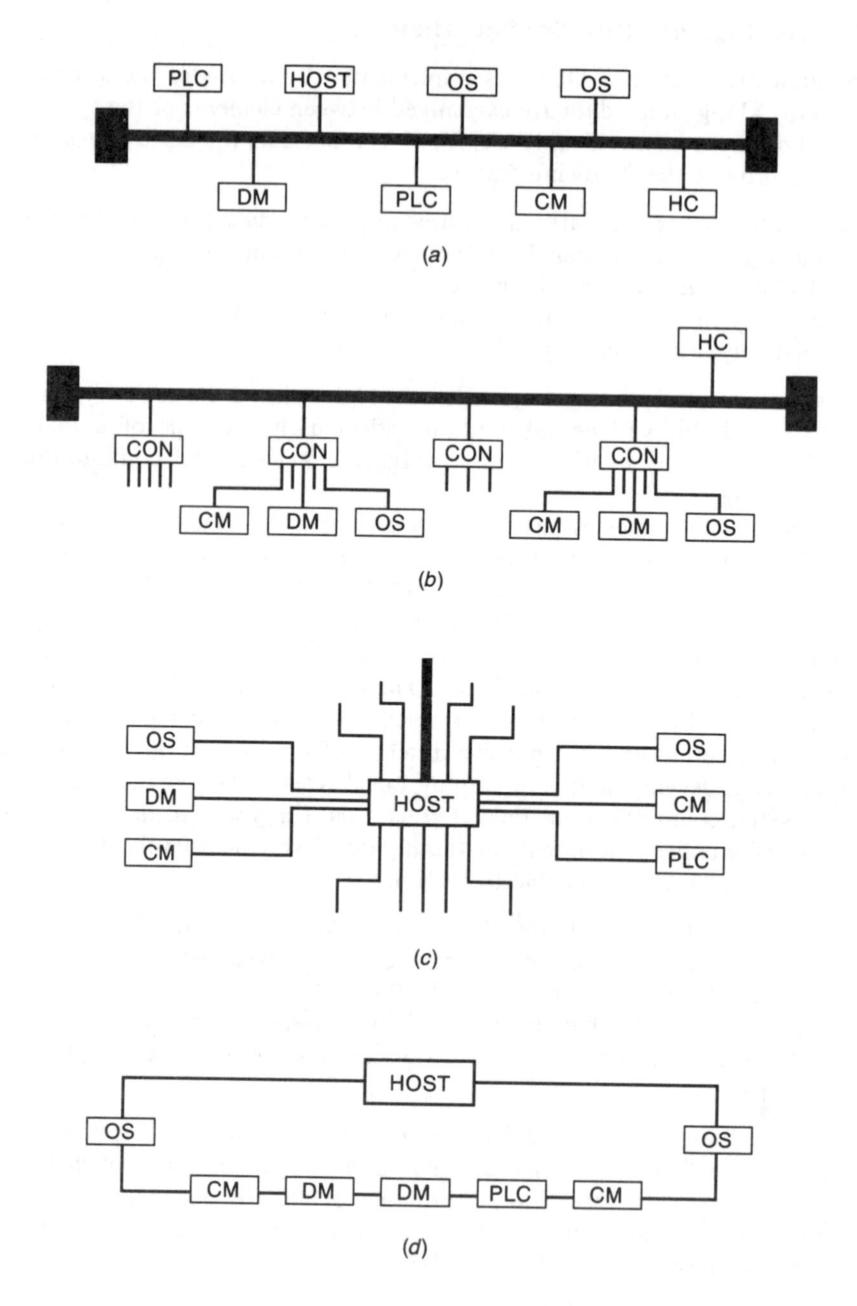

FIG. 7.107. Some typical data highway configurations: (*a*) multidrop DH; (*b*) partially clustered/multidrop DH; (*c*) star configuration; (*d*) ring DH

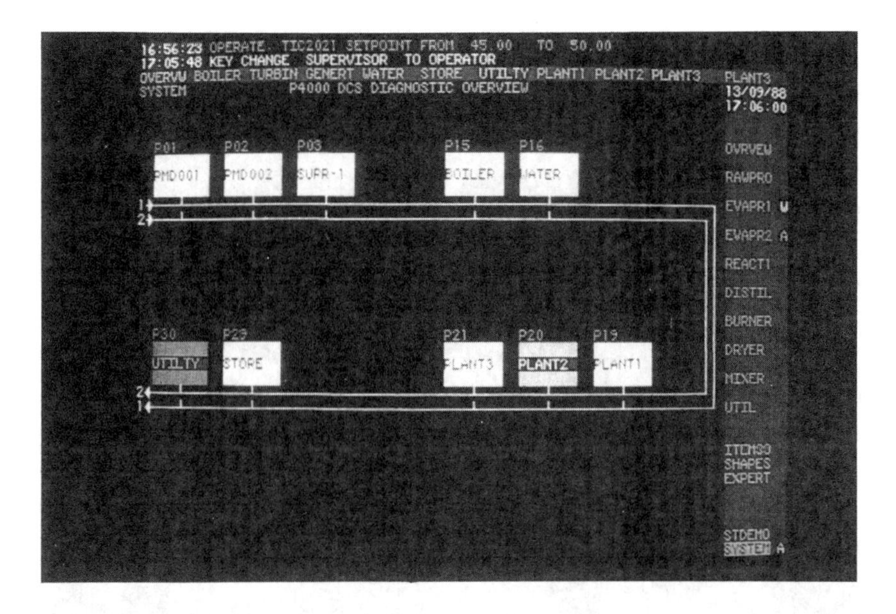

(*a*)

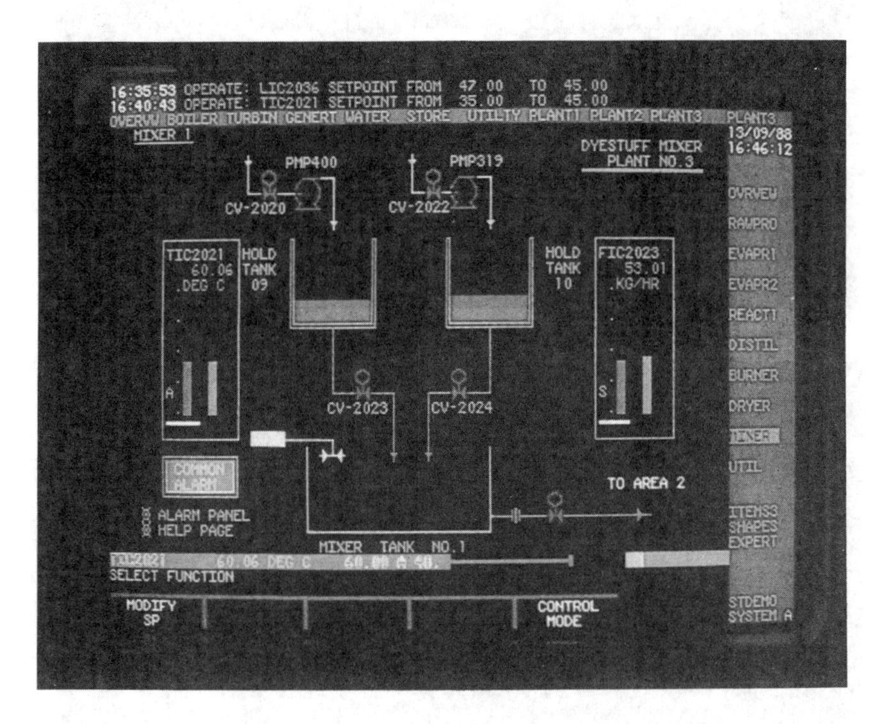

(*b*)

FIG. 7.108. Typical DCCS video displays (Kent P4000 system): (*a*) overview of data highway; (*b*) plant overview as mimic diagram; (*c*) alarm summary; (*d*) unit (group) display; (*e*) historic and current trends

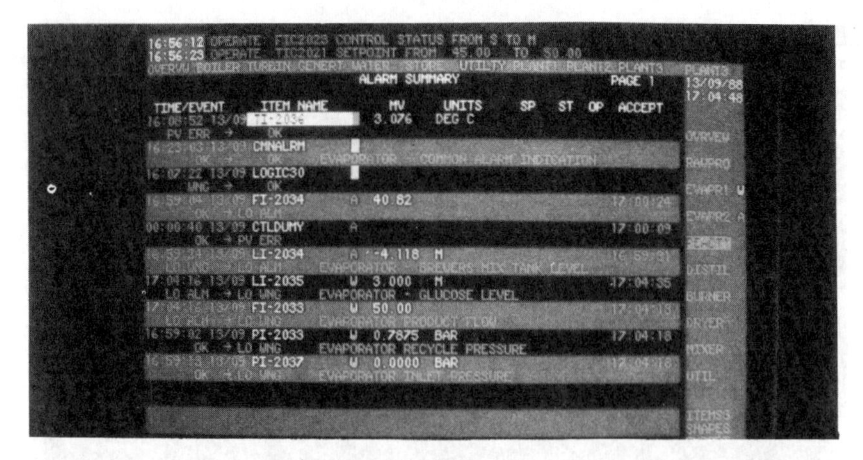

(c)

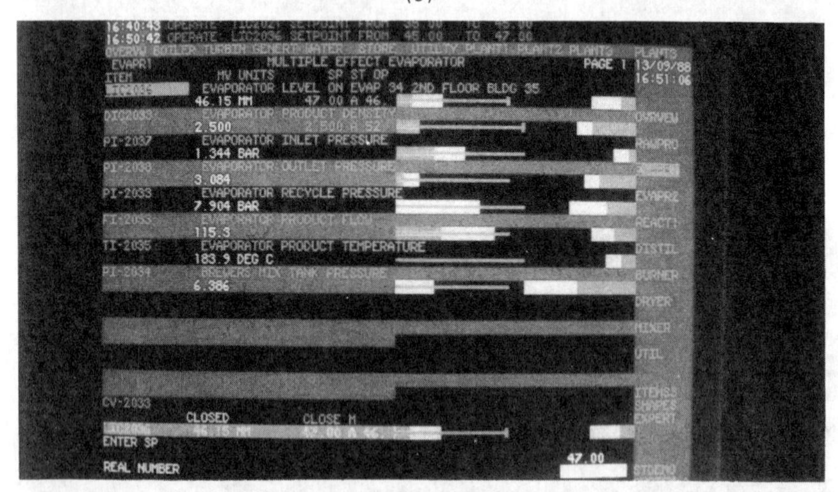

(d)

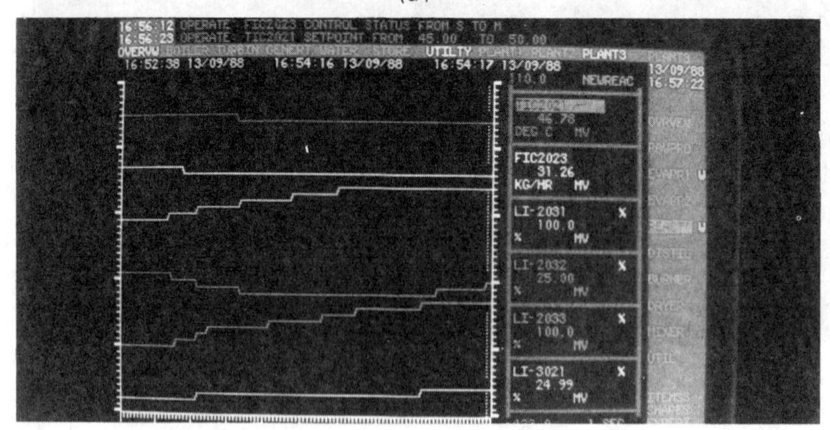

(e)

Fig. 7.108. (*contd*).

operator's vision and ability to control the process. Some typical screen displays are shown in Fig. 7.108.

The more important displays are:

(a) *Overview*. This is regarded as the 'window to the process'. It is an extremely powerful tool and brings all process changes and problems to the attention of the operator. It works on the basis of *operation by exception*. This means that only those process units or parts of units which are not operating satisfactorily will attract the operator's attention. It is recommended that one operator station should be allocated to overview. If any part of the process or control system needs attention, then by looking at the overview the operator will be led to the appropriate display, e.g. unit display, alarm summary/group, or loop display. At these lower level hierarchical displays, the operator will take the necessary action to bring the process under control.

(b) *Alarm Summary*. Because of the importance of alarm handling, a separate OS is normally assigned to the alarm summary. This display may show several important items of information, e.g. point description, process value, type of alarm, level of priority, and to which unit of display the point belongs. It is used also to accept/reset alarms. The time at which a point has gone into the alarm condition and the time at which the operator has acknowledged the alarm are both indicated, and if several points are in alarm, they are displayed in chronological order.

(c) *Unit (Group) Display*. This includes process variables belonging to a particular unit operation. Normally bar graphs are employed to show process variables, controller outputs and set points. Vertical bar graphs are the most suitable for this purpose.

(d) *Alarm Group*. This normally corresponds to the unit display, i.e. for each unit operation one alarm group may be configured. From this display, points in alarm can be acknowledged, disabled, and enabled. A dedicated key (with associated LED) on the OS keyboard is assigned to each of these alarm groups. If a point in a particular alarm group goes into alarm condition, its allocated key starts to flash. By pressing this key the alarm group is displayed and the operator will take the necessary action(s).

(e) *Loop Display*. All the elements of a particular control scheme may be included in one display, viz. the loop display (e.g. where two or three controllers are operating in cascade). The same display may be employed for the control of a complete process or for the tuning of individual controllers.

(f) *Trends and Averages*. The historical data for one or more process variables and/or trends in different variables can be reviewed as one display. Such displays can be used for the study of process upsets, PID algorithm tuning, effects of the changes in one process variable upon other process parameters, etc.

(g) *Block Display*. Each point of the control system has a block display. This is used for the configuration of that point.

In addition to the above, a number of on-line and off-line diagnostic displays are provided in order to locate system failures promptly and to enable faulty components to be repaired. Most operator stations offer facilities which allow graphic and mimic displays to be built.

All the displays described above can be implemented on the host's operator station. However, the latter must not be used as the primary operator interface.

It is important to employ the correct type and quantity of operator stations. For a typical process, with say 2000 I/O, a minimum of three stand-alone operator stations and one host operator station is recommended[56]. The host operator station is used for those tasks which are beyond the capability of the stand-alone operator station (see Section 7.20.3).

7.20.6. System Integrity and Security

Integrity is the confidence that the system is performing correctly. Because of the hierarchical and distributed nature of DCCS, these modern control arrangements can be designed with a high degree of integrity. Graceful degradation (i.e. resistance to catastrophic failure) is attainable by the use of redundancy and correct system configuration. Extensive diagnostic routines incorporated into the various modules of a DCCS ensure that some corrective action will be taken before an error can be transmitted to the rest of the system. The DCCS *watchdog* checks the status of all the tests run within the module. If a test is not completed satisfactorily, the watchdog will generate an interrupt sequence and one or more of the following actions will take place:

(i) The operator will be alerted.
(ii) Process outputs will be frozen.
(iii) A redundant module will take over (either automatically or manually).
(iv) A diagnostic display or report will be generated.

The system must be so designed that it cannot initiate an undesired function due to failure or to operator error. A failure in the higher levels of hierarchy must not degrade the elements in the lower levels. Some of the usual means of maintaining a secure system are by the use of:

(a) keylocks,
(b) security codes and passwords,
(c) configurable keys (viz. disable/enable alarm, parameter change, prohibition of control),
(d) safety caps for some pushbuttons,
(e) grouping of pushbuttons on a keyboard, e.g. diagnostic keys in one group, control parameter keys in another, etc, and
(f) different keyboards for operators and engineers.

7.20.7. SCADA (Supervisory Control and Data Acquisition)

Microprocessor-based control systems are marketed under a variety of acronyms and not all will be true DCCSs. Of particular note are the *Supervisory Control and Data Acquisition* (*SCADA*) systems which form an increasingly central part of the operations of many industries (e.g. offshore oil and gas platforms, petrochemical complexes, pulp and paper works). SCADA describes a communications system whose principal function is to join instrumentation and control into a cohesive package[57] and, particularly, the supervision and management of 'remote' installations. Thus, a SCADA system might consist of process computers, a variety of data highways, operator workstations and remote terminals and will, typically, manage data acquisition, communication and processing, events and alarm reporting and partial process control[51]. Full process control functions may then be provided by special control units connected to the SCADA system.

7.21. THE PROGRAMMABLE CONTROLLER

The *programmable controller* (sometimes called the *programmable logic controller*) is a small microprocessor-based device suitable for the control of a single unit operation, particularly batch processes. It operates by examining a set of input logic signals derived from sensors on the plant, combines them as appropriate with timing signals from its own clock and produces output signals of various kinds. Standard interfaces built into PLCs allow them to be connected directly to process actuators and transducers without the necessity for intermediate circuitry.

Programmable controllers are employed primarily for:

(a) sequential control (which is the most common application),
(b) as fixed parameter PID controllers used to control such process variables as temperature, pressure, flow, level, etc, and
(c) the analysis and manipulation of plant data.

Benefits ascribed to the use of PLCs in comparison to the older stand-alone analog controller are[51]:

(i) flexibility due to the ease with which they can be reprogrammed,
(ii) the facility with which they can be installed on the plant due to their compactness and the ease with which several PLCs can be arranged to form the basis of a distributed control system, and
(iii) the simplicity of their general maintenance as a result of their basic modular structure and the diagnostic provisions built into them.

7.21.1. Programmable Controller Design

Programmable controllers are purpose-built computers which contain three functional areas, viz. those associated with programming, memory and input/output (I/O). Inputs to the PLC are sensed and stored in memory. The action taken by the PLC in response to the resulting input states is dependent entirely upon the control program that has been inserted into the memory. The central processing unit (CPU) of the PLC carries out the programmed instructions held in memory as well as controlling and supervising the performance of all the necessary operations. This culminates in the generation of output signals which are transmitted to the appropriate equipment. Information is transmitted between the CPU, memory and input/output units via a communications highway (or bus). An internal clock controls the operating speed of the PLC and provides the timing and synchronisation of all the functions of the system.

The input/output units of the programmable controller constitute the interface between the PLC and the outside world. Hence, they are provided with the necessary signal conditioning functions which enable the PLC to be connected directly to the process actuators and transducers as desired. For this purpose the

PLC is provided with a number of different types of I/O unit from which those suitable for a particular application can be selected. Smaller PLCs usually have a fixed memory size of capacity between 300 and 1000 instructions, have up to 40 input and 40 output points, and can only be programmed using a ladder diagram language. Medium-sized programmable controllers will contain up to 64 kb (kilobytes) of RAM (read and write memory), be provided with up to 1024 I/O points and have the additional facility of a high level language. Larger programmable controllers with large numbers of I/O points (4048 and more) employ *memory modules* (varying in size between 1 kb and 64 kb) which enable the system to be expanded as required. Table 7.6 lists typical input and output channels for a small PLC.

TABLE 7.6. *Typical PLC I/O Channels*

Inputs		Outputs	
Signal level	Description	Signal level	Description
5V (usually DC TTL[a] level)	Switched[b]	Typically up to 240V AC or DC	Relay output
24V (usually DC but sometimes AC)	Switched	Typically up to 240V AC only	SSR (solid-state relay or triac)
110V (usually AC)	Switched	Typically up to 30V DC only	Transistor output
240V (usually AC)	Switched		

[a] Transistor-transistor logic.
[b] 'Switched' indicates that the signal is either on or off. For example, if the liquid in a tank reaches a particular level, then the input from the level sensor will be energised and the PLC will take the appropriate action.

WARNOCK[58] has compared the capabilities of the modern programmable controller with other controller hardware (Table 7.7). It is clear from this that the PLC has very attractive software and hardware features which make it a natural choice for a wide range of control strategies.

TABLE 7.7. *A Comparison of the Capabilities of Various Controller Hardware*[58]

Characteristic	Relay systems	Digital logic	Computers	PLC systems
Price per function	Fairly low	Low	High	Low
Physical size	Bulky	Very compact	Reasonably compact	Very compact
Operating speed	Slow	Very fast	Reasonably fast	Fast
Immunity from electrical noise	Excellent	Good	Quite good	Good
Ease of Installation	Time-consuming to design and install	Design time-consuming	Programming can be very time-consuming	Simple to program and install
Capable of complicated operations	No	Yes	Yes	Yes
Ease with which function can be changed	Very difficult	Difficult	Quite simple	Very simple
Ease of maintenance	Poor—large number of contacts	Poor if integrated circuits soldered	Poor—several custom boards	Good—few standard cards
Reliability	Fair	Very good	Good	Very good

7.21.2. Programming the PLC

Programmable controllers were designed originally as electronic replacements for hard-wired relay* and timer logic units and provide a means for the programming and executing of simple logic instructions by employing the use of internal functions such as counters, shift registers† and timers. Historically, relay logic circuit operation has been described by the use of *ladder* diagrams and this procedure is now commonly employed as the basis for designing logic control systems. A standard method of programming small PLCs is to draw the required ladder diagram and convert this into a mnemonic instruction set similar in appearance to assembly-type codes but representative of the physical inputs, outputs and functions contained within the PLC. These instructions are keyed into a programming panel which can be programmed as a unit separate from the PLC itself. Once the program has been completed and checked, the programming panel is plugged into the PLC and the program transferred into the CPU of the PLC. Often a *graphic* programmer is employed where the program is displayed on a small screen as a symbolic ladder circuit employing the standard logic symbols provided by the PLC[59]. Small hand-held graphic programmers are available for small to medium-size PLCs. Larger PLCs generally employ larger, static, graphic units. Alternatively, a personal computer can be used to program the PLC.

From its historical connection with relay systems, the ladder diagram frequently contains elements termed contacts and/or coils even though, in most cases, these are notional devices which have no physical presence. Counters and timers are similarly represented. A timer, for example, starts to run when the 'power' is applied to it and becomes *true* when the power has been applied for the length of time specified in the program. If the power is removed before this period has ended, then, as in the case of a relay timer, the time is reset to zero and the timer waits to be turned on again. In the case of a counter, a reset command is issued if the counter is required to be zeroed. Logical 'AND' functions are represented in the ladder diagram by placing contacts in series. Contacts in parallel correspond to logical 'OR' functions.

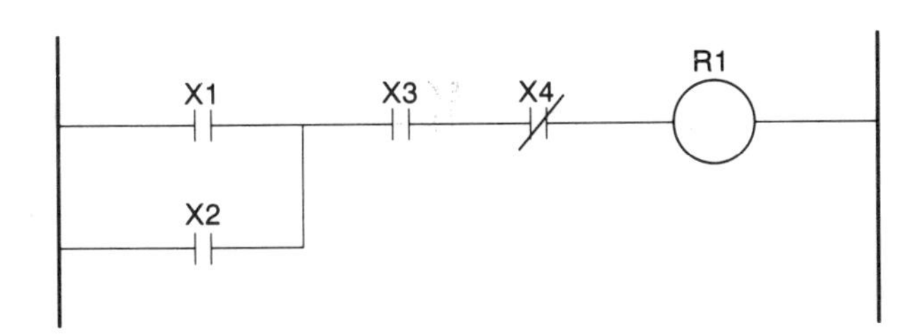

FIG. 7.109. Simple ladder-logic diagram

* A relay is an electrical switch with a high current capacity which is operated indirectly by a relatively low control current.

† In the context of the PLC, a shift register is an address in the CPU of the PLC containing a group of binary digits arranged in such a way that the binary digits can be shifted to the left or to the right as required in order to produce a sequentially controlled chain of events in the process being controlled by the PLC. (See also Section 6.12.2.)

Figure 7.109 shows a simple ladder-logic diagram[60]. This represents the logic statement (or *pseudocode*):

IF (X1 OR X2) AND X3 AND (NOT X4) THEN R1 = TRUE

This statement may be interpreted as—'if X1 or X2 is true and X3 is true, but X4 is not, then output R1 is energised'. This is programmed into the PLC by pressing the appropriate keys on the keypad of the programming panel using the equivalent code (which in this case is for a Mitsubishi Melsec F1 series programmable controller—but is fairly typical):

```
LD   X1
OR   X2
AND  X3
ANI  X4
OUT  R1
```

An important consequence of this method of programming is that the individual commands in the program are tested almost instantaneously and that the program cannot become locked into a small loop and thus cease to service any other control function.

7.22. REGULATORS AND ACTUATORS (CONTROLLERS AND CONTROL VALVES)

7.22.1. Electronic Controllers

Electronic modules are the industry standard for controllers employing a wide range of control strategies. Although, more recently, there has been rapid development of microprocessor-based controllers (see Sections 7.20 and 7.21) where control actions are simulated using software, *hard wired* systems* based upon the integrated circuit (*IC*) and operational amplifier (*op-amp*) are still much in evidence.

The Operational Amplifier

The output voltage V_o of the op-amp shown in Fig. 7.110 can be expressed by equation 6.105, viz.:

$$V_o = K_{amp}(V_{i_1} - V_{i_2}) \quad \text{(equation 6.105)}$$

where K_{amp} is the gain of the amplifier.

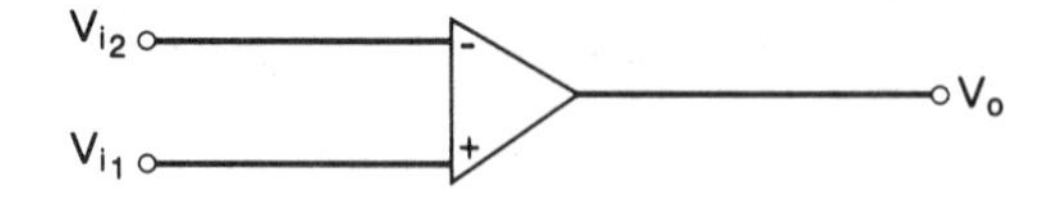

FIG. 7.110. Operational amplifier

* A hard wired system is one made up from a fixed set of electronic components whose characteristics cannot be changed by reprogramming software—if, indeed, software is available within the system.

The properties of an op-amp should approach the following as closely as possible[(61)]:

(a) K_{amp} should be infinite.
(b) $V_o = 0$ when $V_{i_1} = V_{i_2}$ (termed the *common mode rejection ratio* (*CMRR*)).
(c) The input impedance should be infinite.
(d) The output impedance should be zero.
(e) There should be no delay in the response of the output to a perturbation in input.

These properties give the general characteristics of operational amplifiers, viz.:

(i) Since the gain is infinite, then V_o can only be finite if $V_{i_1} = V_{i_2}$.
(ii) As the input impedance is infinite, no current flows to the positive or negative terminals of the amplifier.
(iii) Since the output impedance is zero, an op-amp is able to provide the output voltage V_o whatever might be the impedance of the load connected to the output of the amplifier.

Proportional Action

From Fig. 7.111 and the general characteristics of the op-amp:

Across resistor 1:
$$V_{i_1} - V = i_A R \tag{7.253}$$

Across resistor 3:
$$V = i_A R \tag{7.254}$$

From equations 7.253 and 7.254:
$$2V = V_{i_1} \tag{7.255}$$

Across resistor 2:
$$V_{i_2} - V = i_B R \tag{7.256}$$

Across resistor 4:
$$V - V_3 = i_B R \tag{7.257}$$

From equations 7.256 and 7.257:
$$2V = V_{i_2} + V_3 \tag{7.258}$$

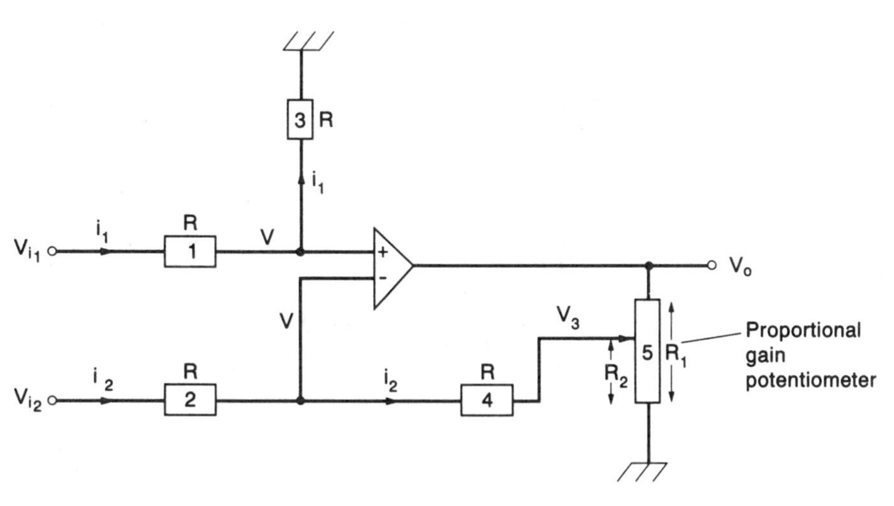

FIG. 7.111. Use of the operational amplifier to generate proportional action

From equations 7.255 and 7.258:

$$V_3 = V_{i_1} - V_{i_2} \tag{7.259}$$

Also, from the proportional potentiometer:

$$\frac{V_o}{V_3} = \frac{R_1}{R_2} \tag{7.260}$$

Hence, eliminating V_3 between equations 7.259 and 7.260:

$$V_o = \frac{R_1}{R_2}(V_{i_1} - V_{i_2}) \tag{7.261}$$

Thus, if V_{i_1} represents the set point of the controller, V_{i_2} the measured value and R_1/R_2 the proportional gain, then equation 7.261 is the equivalent of the relationship for proportional control action given by equation 7.2. The proportional potentiometer enables the proportional gain to be varied.

PI Action

From Fig. 7.112:

$$V_i = iR_I \qquad \text{and} \qquad V_o - V_i = \frac{1}{\mathscr{C}_I}\int_0^t i\,dt$$

where $\mathscr{C}_I$ is the capacitance of the amplifier feedback circuit.

$$\therefore \qquad V_o = V_i + \frac{1}{\mathscr{C}_I}\int_0^t i\,dt = V_i + \frac{1}{R_I\mathscr{C}_I}\int_0^t V_i\,dt \tag{7.262}$$

If V_i corresponds to the error ε, then the **PI** action relationships given by equations 7.6 and 7.262 are equivalent for unit proportional gain. $R_I\mathscr{C}_I$ represents τ_I and the degree of integral action desired is controlled by adjusting R_I, i.e. the resistance of the integral action potentiometer.

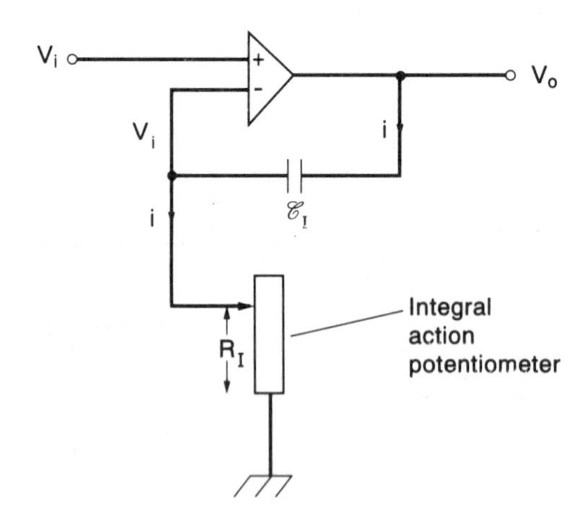

FIG. 7.112. **PI** action using an operational amplifier

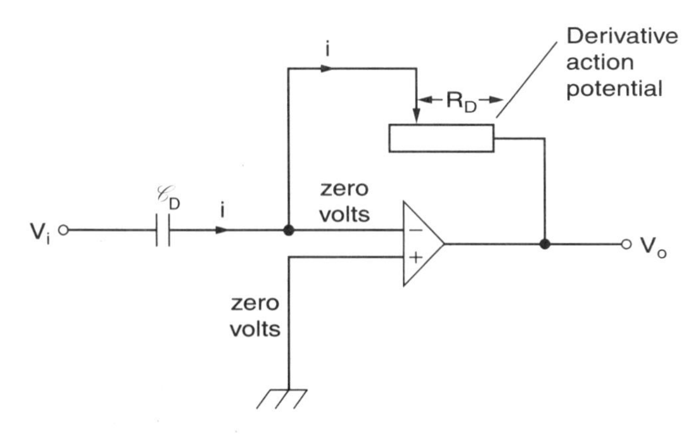

FIG. 7.113. Generation of **PD** action using an operational amplifier

PD Action

From Figure 7.113:

$$V_i = \frac{1}{\mathscr{C}_D}\int_0^t i\mathrm{d}t \quad \text{and} \quad V_0 = -iR_D$$

$$V_0 = -\mathscr{C}_D R_D \frac{\mathrm{d}V_i}{\mathrm{d}t} \tag{7.263}$$

where $\mathscr{C}_D$ is the capacitance of the input circuit.

In this case, if V_i represents the error ε, then equation 7.263 simulates derivative action (Section 7.2.2) with $\mathscr{C}_D R_D$ equivalent to τ_D. Varying R_D regulates the degree of derivative action desired.

***PID** Action*

This is obtained by fitting together the modules illustrated in Figs. 7.111, 7.112 and 7.113.

7.22.2 Pneumatic Controllers

For many years the pneumatic controller was preferred to its electronic counterpart due to its simplicity, its general ruggedness in the process environment, and the fact that its output could be used to operate directly the diaphragm of a pneumatic control valve. Although now largely superseded by software or hard wired electronic equivalents, pneumatic controllers are still employed in special circumstances, e.g. in explosive atmospheres. Furthermore, substantial numbers of pneumatic controllers can be found on older plant and thus an understanding of their principles of operation is necessary.

Proportional Control—Narrow-Band Action

Figure 7.114 shows how a pneumatic controller mechanism may be included in a simple control loop. The pneumatic output from the differential pressure cell

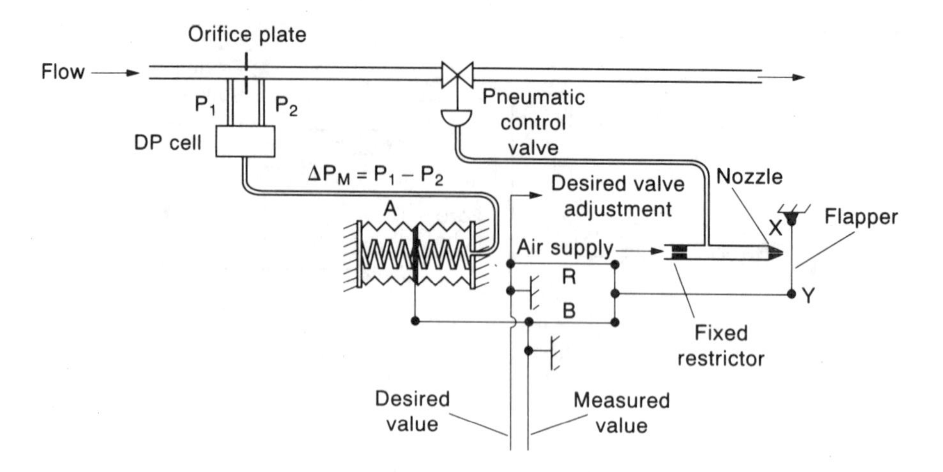

FIG. 7.114. Arrangement of simple pneumatic flow control loop with narrow-band proportional control

($\Delta P_M = -\Delta P$) increases with the flow of fluid through the orifice (see Volume 1, Section 6.3.3). This is transmitted to a recorder and controller, generally located adjacently in the plant control room. The pneumatic signal is converted into a mechanical movement by means of a spring and bellows device A. The bellows movement is linked both to the recorder pen and to the flapper of the controller mechanism which pivots about X. If the fluid flowrate in the line increases, ΔP_M will rise, hence causing the flapper to move closer to the nozzle. This will restrict the flow of air through the nozzle and hence the pressure P upstream from the nozzle will rise. The latter is fed directly to the valve motor which acts in such a way that, as P increases, the valve shuts—thus reducing the flow in the line, and consequently ΔP_M. With this type of controller mechanism the movement of the flapper required for P to vary over its entire range is very small (approx. 2×10^{-5} m). GOULD and SMITH[62] have shown that for such very small movements of the flapper, the change in output pressure ΔP is proportional to the movement of the flapper relative to the nozzle. The controller linkage is arranged so that the displacement of the flapper is proportional to the difference between the set point and the measured value ΔP_M, i.e.:

$$\Delta P \propto R - B = \varepsilon \tag{7.264}$$

(cf. equations 7.1 and 7.2). The extreme sensitivity of this system results in a very high proportional gain and, hence, is termed *narrow-band proportional action*—a high gain corresponding to a small or narrow proportional band (Section 7.2.3).

Proportional Control—Wide-Band Action

The excessive sensitivity of the narrow-band mechanism leads to considerable instability in the control system. This sensitivity is reduced by introducing a feedback bellows as illustrated in Fig. 7.115. In this case, if P increases due to the movement of the flapper towards the nozzle, the bellows will expand. Thus, as pivot Y is moved to the left by a change in measured value or set point, X will move to

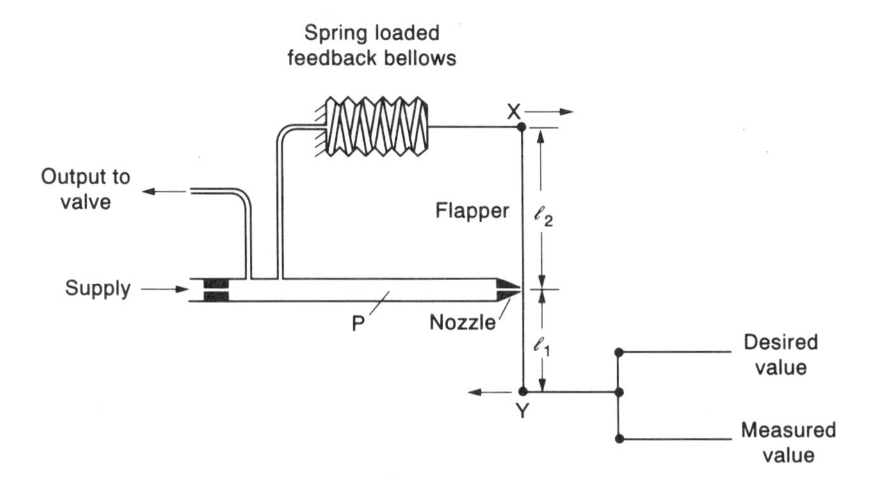

FIG. 7.115. Pneumatic wide-band proportional controller mechanism

the right—so reducing the effective movement of the flapper relative to the nozzle. The resulting decrease in sensitivity is dependent on the extension of the bellows per unit change in P and on the ratio of $\ell_1 : \ell_2$. Such an arrangement can be shown[63] to produce an action which approximates closely to equation 7.2 and proportional bands of up to 600 per cent can be obtained.

PI Action

Integral action is added by the insertion of a restrictor and further bellows (Fig. 7.116). The rate of change of pressure in the integral bellows is proportional to the pressure driving force across the restrictor,

i.e.:
$$\frac{dP_I}{dt} = \mathcal{K}_1(P - P_I) \tag{7.265}$$

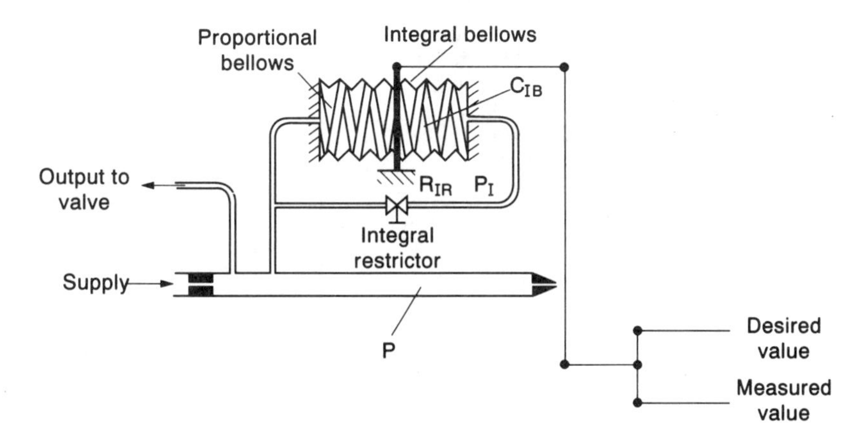

FIG. 7.116. Pneumatic generation of **PI** action

Suppose, for $t < 0$, $P_I = P = P_o$, and that, at $t = 0$, P_o is suddenly increased to $P_o + \Delta P$ by the movement of the flapper, then initially:

$$\frac{dP_I}{dt} = \mathscr{K}_1(P_o + \Delta P - P_I) = \mathscr{K}_1(P_o + \Delta P - P_o) \tag{7.266}$$

$$P_I = \mathscr{K}_1 \int_{t=0}^{t=t} \Delta P \, dt \tag{7.267}$$

But, from equation 7.264, $\Delta P \propto \varepsilon$. Hence:

$$P_I = \mathscr{K}_2 \int_0^t \varepsilon \, dt \tag{7.268}$$

Thus, the pressure of the output to the valve P is the sum of P_I and the pressure produced by the wide-band proportional action contributed by the proportional bellows and the flapper-nozzle system (cf. equation 7.3). Note that for $t > 0$, P_I is no longer equal to P_o and thus equation 7.268 only strictly applies at $t = 0$. The value of $\mathscr{K}_2$ (and consequently τ_I) depends upon the capacity C_{IB} of the integral bellows and the the resistance to flow R_{IR} through the integral restrictor. It is generally assumed that C_{IB} changes little and τ_I is varied by adjusting R_{IR}.

PD Action

By inserting a restrictor in the line to the proportional bellows, any change in P will not be transmitted immediately to the feedback system. Thus, initially, the arrangement shown in Fig. 7.117 will behave as a narrow-band proportional controller changing to wide-band action as the pressure in the feedback bellows P_D approaches P. The rate at which $P_D \rightarrow P$ depends upon the resistance to flow R_{DR} through the derivative restrictor. This mechanism thus simulates derivative action in that it is most sensitive when the error is changing the most rapidly (Section 7.2.3). The derivative time τ_D is varied by adjusting R_{DR}.

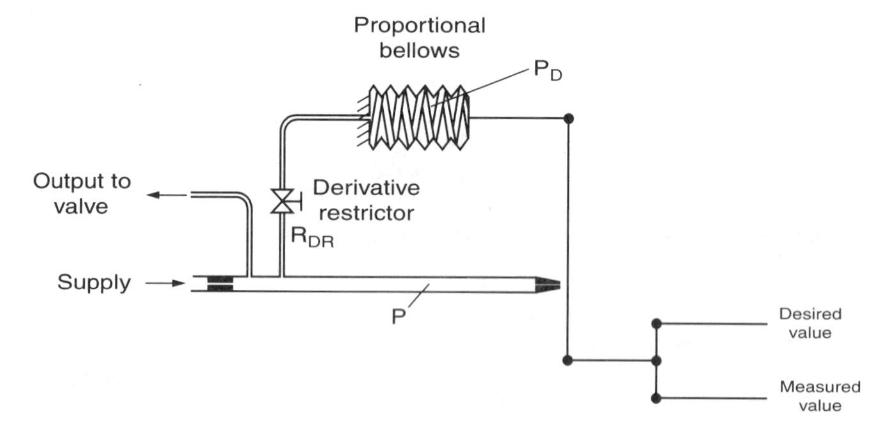

FIG. 7-117. Pneumatic generation of **PD** action

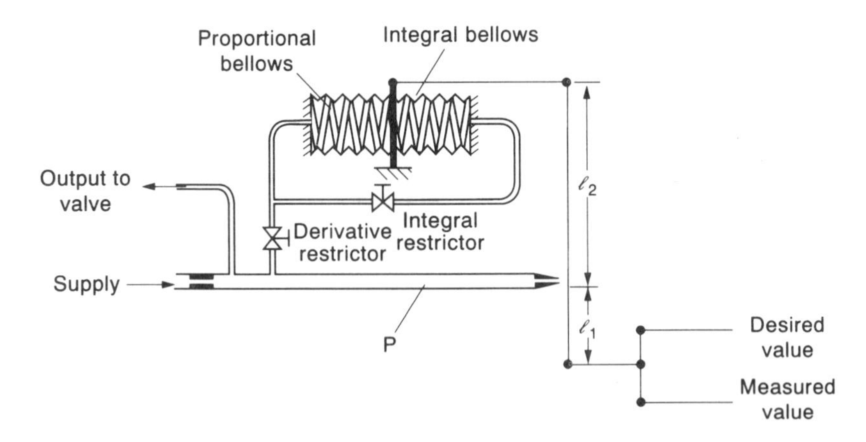

FIG. 7.118. Pneumatic generation of **PID** action

PID Action

A pneumatic signal simulating **PID** action is obtained by a combination of the **PI** and **PD** mechanisms as illustrated in Fig. 7.118.

7.22.3. The Control Valve

The control valve is a variable restriction in a pipeline which receives its position command from a controller—either in the form of a single loop regulator or as part of a more complex control system. As such, the control valve constitutes by far the most common final control element although increasing use is being made of variable speed pumps and fluidics[64] to control the flowrates of process fluids.

A typical control valve is illustrated in Fig. 7.119. It consists of three principal sections, viz.:

(i) The *actuator* or *valve motor*. This is the mechanism which converts the signal from the controller into the motion which positions the *valve plug*.
(ii) The *valve body* which contains the valve seat(s) and is the section of the valve through which the fluid passes.
(iii) The *valve trim* which consists of those sections of the valve in contact with the fluid which are detachable.

The Actuator

Figure 7.120 is a simplified view of a *spring diaphragm* actuator. The actuator receives a pneumatic signal from the controller via a *booster flow enlarger* or a *valve positioner* and can be adapted in the form of an *air-to-open* or an *air-to-close* mechanism.

If the output from the controller is electric (e.g. 4–20 mA), then this can be converted to a pneumatic signal (with pressures up to 600 kN/m^2) using an electropneumatic converter (Section 6.12.4). The resulting pneumatic signal then

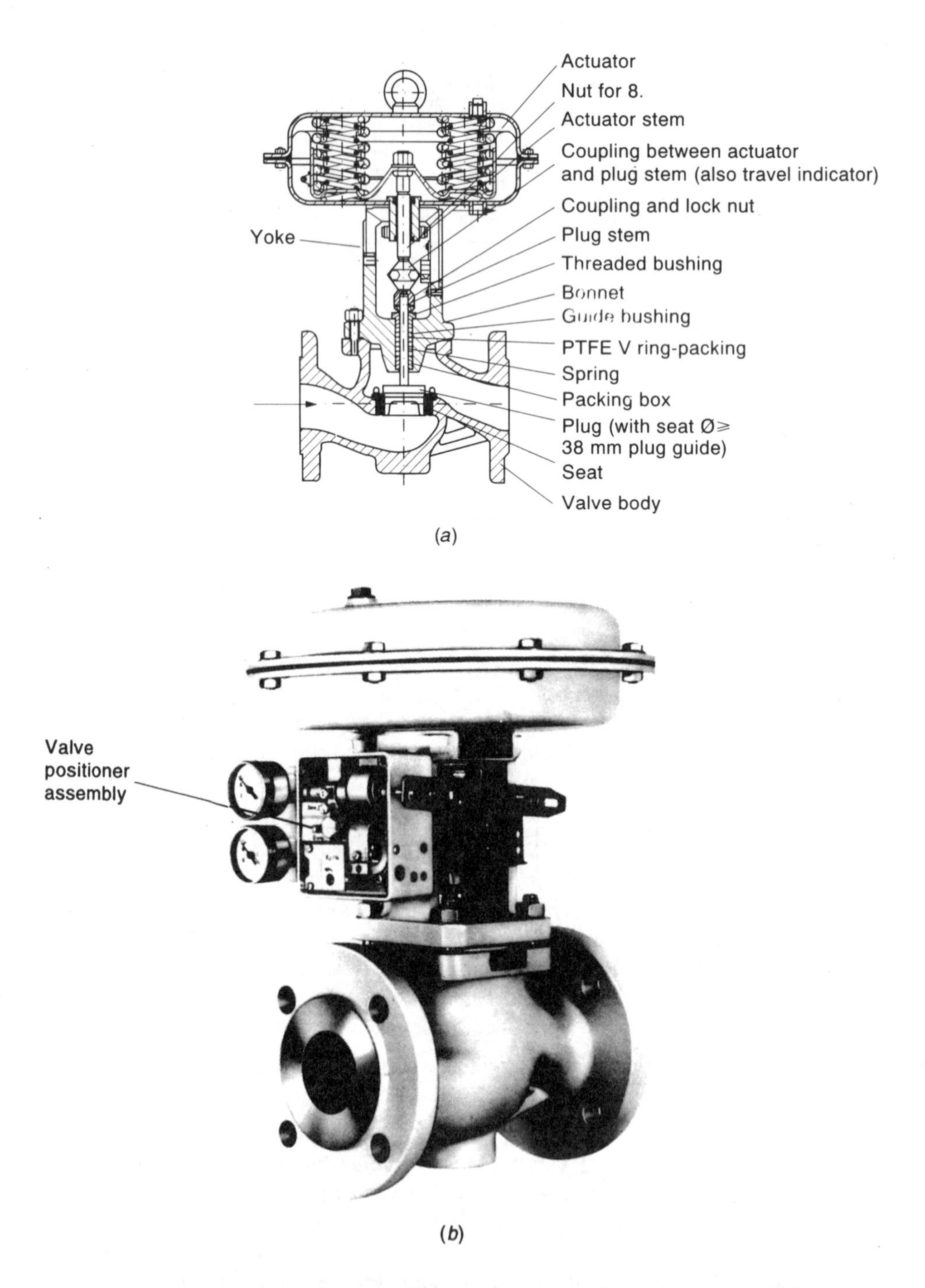

FIG. 7.119. Pneumatically operated control valve: (*a*) double-spring actuator with single-ported globe valve; (*b*) exterior view of double-ported control valve with valve-positioner fitted on the side

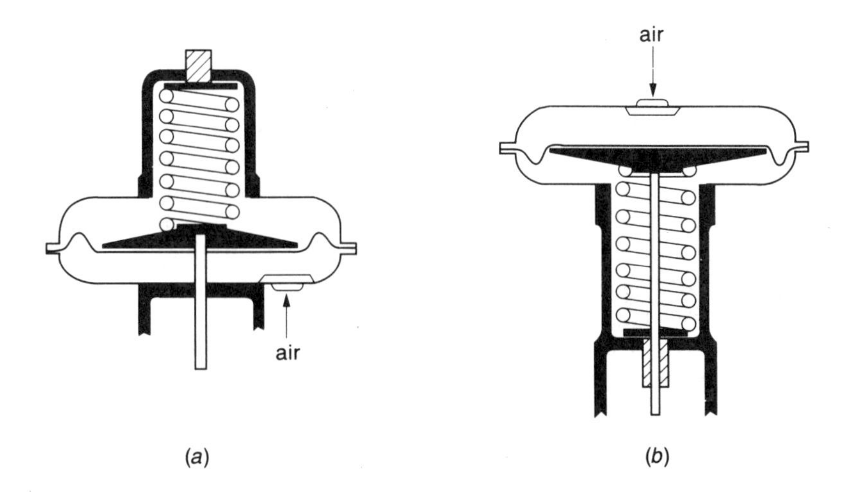

FIG. 7.120. Spring-diaphragm actuator: (*a*) air-to-open (direct) mechanism; (*b*) air-to-close (inverse) mechanism

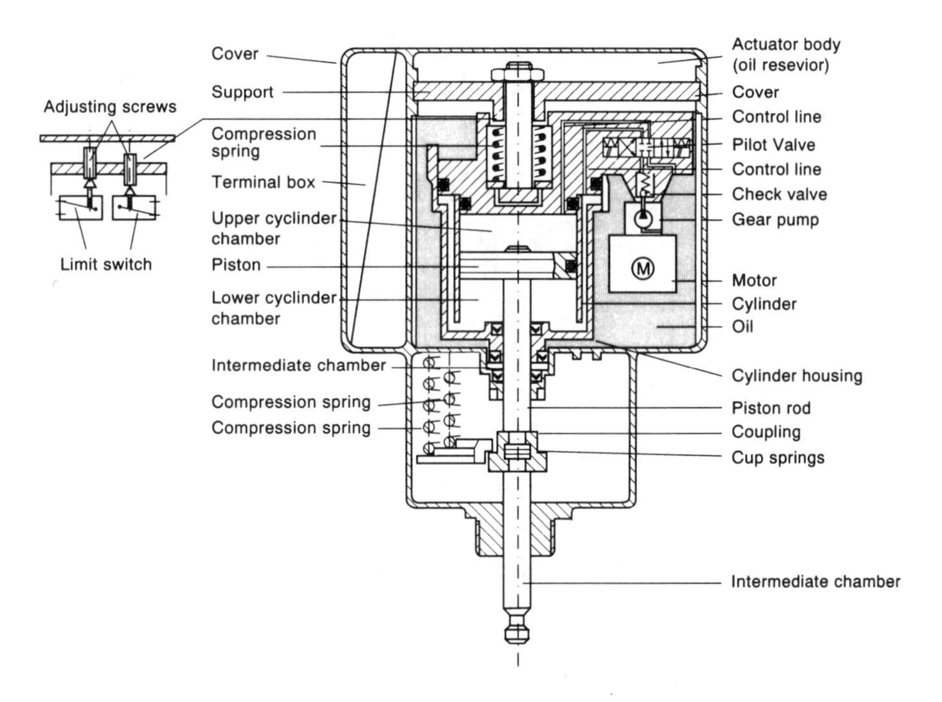

FIG. 7.121. Electrohydraulically operated actuator

operates the actuator in the usual way. Alternatively, an electrohydraulic actuator can be employed (Fig. 7.121)[65] In each case, the actuator is constructed such that the displacement of the stem of the valve is proportional to the force applied by the control signal.

The Valve Positioner

Quite frequently the displacement of the valve stem will not be in proportion to the signal from the controller. This may be due to friction, an imbalance of forces on the valve plug(s) (particularly with single-seated valves) or to hysteresis in the spring/membrane actuator system[66]. This difficulty can be overcome by the fitting of a valve positioner (Figs 7.122 and 7.119*b*). The valve positioner is a high gain, mechanical-pneumatic feedback amplifier with gains of the order of 10 : 1 to 100 : 1. The action of the mechanism is to compare continually the signal from the controller with the position of the valve stem (also represented as an air pressure) and to boost the pressure applied to the actuator membrane until the valve stem is in the correct position. Hence, the positioner also acts as a *power relay* as it amplifies the effect of any changes in pressure supplied from the controller.

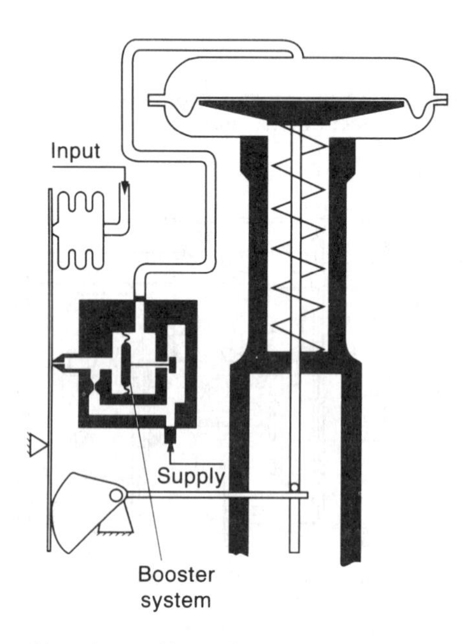

FIG. 7.122. Valve positioner fitted to a diaphragm actuator

The use of a valve positioner is generally beneficial with relatively slow control loops. Fitting a positioner within a fast control loop will decrease the stability of the loop[66].

Valve Body and Valve Trim

There are many different types of valve fitted with actuators to form control valves. Valves may be *single-* or *double-ported* (Fig. 7.123). With single-ported valves the valve plug is subjected to the total differential force across the valve. Such valves are sensitive to pressure fluctuations and powerful actuator elements are required for large pressure drops. Double-ported valves balance out the pressure differential but it is difficult to obtain complete shut-off.

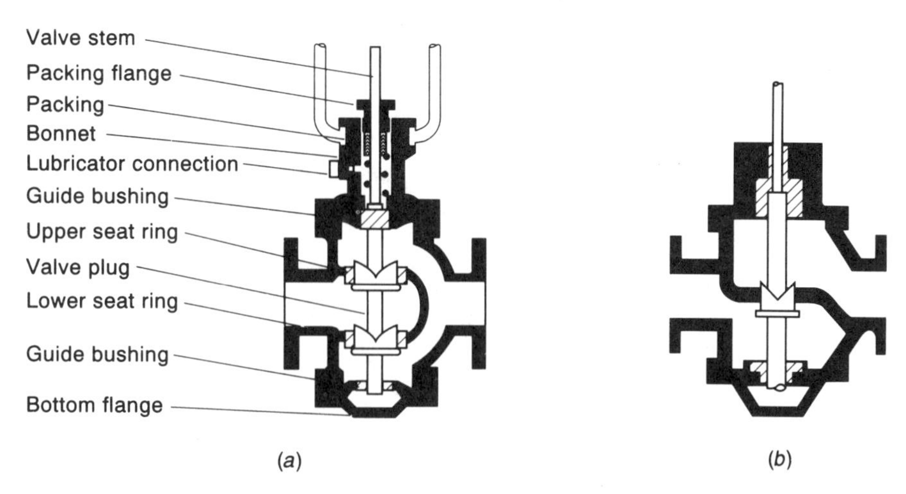

FIG. 7.123. Valve bodies showing typical trims used for control valves: (*a*) double-port globe valve with top and bottom guiding; (*b*) single-port globe valve with top and bottom guiding

Valve bodies other than those illustrated in Fig. 7.123 which are employed for special-purpose control valves are ball valves, butterfly valves and Saunders diaphragm valves. These are described in Volume 1, Section 3.5.4 and Volume 6, Section 5.3.

Control Valve Characteristics

The relation between the valve stem position and the flow through the valve at constant pressure drop is termed the *valve characteristic*. Two characteristics must be evaluated for valve selection, the *inherent* and the *installed* characteristics[67].

The inherent valve characteristic. This is a measure of the theoretical performance of the valve and is divided into:

(i) The decreasing-sensitivity type for which:

$$\left.\frac{\mathrm{d}Q}{\mathrm{d}\ell_s}\right|_{Q=0} > \eta > \left.\frac{\mathrm{d}Q}{\mathrm{d}\ell_s}\right|_{Q=Q_{\max}} \tag{7.269}$$

(ii) The linear (constant) sensitivity type where:

$$\frac{\mathrm{d}Q}{\mathrm{d}\ell_s} = \eta \tag{7.270}$$

(iii) The increasing-sensitivity type for which:

$$\left.\frac{\mathrm{d}Q}{\mathrm{d}\ell_s}\right|_{Q=0} < \eta < \left.\frac{\mathrm{d}Q}{\mathrm{d}\ell_s}\right|_{Q=Q_{\max}} \tag{7.271}$$

where η is the *valve-flow coefficient*, Q is the volumetric flowrate of fluid passing through the valve expressed as a percentage of the maximum flowrate, and ℓ_s is the

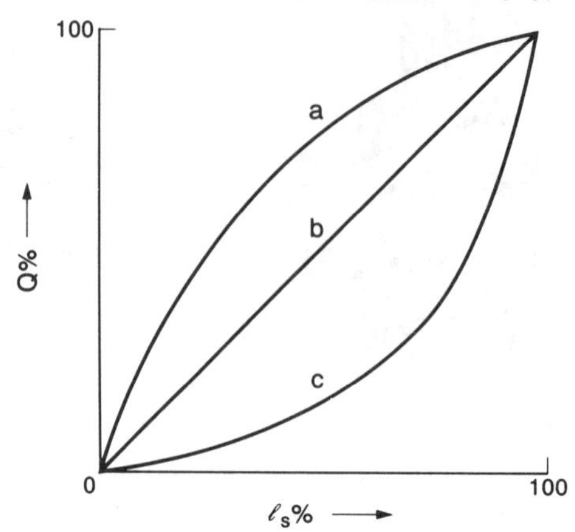

FIG. 7.124. Inherent valve characteristics

displacement of the valve stem expressed as a percentage of the *stroke* or *valve lift* (the maximum valve stem travel).

The different inherent characteristics (Fig. 7.124) are determined by the geometry of the valve orifice and valve plug and describe the volumetric flowrate through the valve as a function of the stem position with constant pressure drop across the valve.

Two other quantities which are important in sizing control valves are the *turndown*, which is the ratio of the normal maximum flowrate through the valve to the minimum controllable flowrate, and the *rangeability*, which is the ratio of the maximum controllable flowrate to the minimum controllable flowrate. Generally, valves are sized such that the turndown is approximately 70 per cent of the rangeability, i.e. that the maximum flowrate through the valve under normal operating conditions is about 70 per cent of the maximum possible flowrate. Typically, increasing-sensitivity valves have higher rangeability values(67).

The installed valve characteristic. This depends upon the ratio of the pressure drop through the valve to the total pressure drop across the whole process line installation including the valve. If a valve of linear sensitivity is handling the entire system pressure drop, then its installed characteristic will also be linear. As the percentage of the pressure drop falls off, the installed characteristic rapidly changes to that of an on–off valve. The increasing-sensitivity valve also loses its characteristic in the same way, but to a far lesser extent. Hence, this type of valve is frequently chosen in preference to the linear type. PETERS(68) has discussed the effects of choosing valves with incorrect characteristics.

7.22.4. Intelligent Control Valves

The *intelligent control valve system* (*ICVS*) is a recent innovation which differs from smart devices (see Section 6.12.5) in that it is a combined module which is able to perform all three major actions of a control loop (viz. measurement, control and process regulation) on a local basis[69]. Thus, an ICVS possesses, at the least, sensors which monitor a number of process variables (e.g. fluid flowrate and temperature, upstream and downstream pressure, valve stem position, valve positioner characteristics, etc.) and a microprocessor of sufficient power to interpret these data, to communicate with a DCCS, and to be capable of acting as a control and self-calibration unit[70]. A distinct advantage of such a system lies in its ability to measure the flowrate of the fluid over the entire controllable range of the valve. This is because the pressure drop across the valve plug remains within its measurable range over the entire stroke of the valve stem due to the accompanying variation in the area for flow within the valve. A further feature of the ICVS is that it can be programmed to bring a process to a safe condition in the case of an emergency (e.g. loss of communication with an associated DCCS).

Intelligent control valves have found increasing application in processes which involve high-value materials and high labour costs such that their higher price can be justified.

7.23. APPENDICES

Appendix 7.1. Table of Laplace and z-Transforms

Description of time function	Time function $f(t)$	Laplace transform $\bar{f} = \int_0^\infty e^{-st} f(t)\,dt$	z-Transform $f(z)$
Unit impulse	$\delta(t)$	1	1
Unit step	$u(t)$	$\dfrac{1}{s}$	$\dfrac{z}{z-1}$
Ramp	t	$\dfrac{1}{s^2}$	$\dfrac{\mathcal{T} z}{(z-1)^2}$
Quadratic	$\dfrac{t^2}{2}$	$\dfrac{1}{s^3}$	$\dfrac{\mathcal{T}^2 z(z+1)}{2(z-1)^3}$
Exponential	e^{-at}	$\dfrac{1}{s+a}$	$\dfrac{z}{z-e^{-a\mathcal{T}}}$
Constant raised to the power t	$a^{t/\mathcal{T}}$	$\dfrac{1}{s-(1/\mathcal{T})\ln a}$	$\dfrac{2}{z-a}\ (a > 0)$
Sinusoidal	$\sin \omega t$	$\dfrac{\omega}{s^2+\omega^2}$	$\dfrac{z \sin \omega\mathcal{T}}{z^2 - 2z\cos\omega\mathcal{T} + 1}$
Cosine	$\cos \omega t$	$\dfrac{s}{s^2+\omega^2}$	$\dfrac{z^2 - z\cos\omega\mathcal{T}}{z^2 - 2z\cos\omega\mathcal{T} + 1}$
Multiplication by e^{-at}	$e^{-at}f(t)$	$f(s+a)$	$f(ze^{a\mathcal{T}})$
Delay by time $n\mathcal{T}$	$f(t-n\mathcal{T})$	$e^{-n\mathcal{T}s}f(s)$	$z^{-n}f(z)$
Zero-order hold element	see Section 7.17.5	$\left\{\begin{array}{l}\dfrac{1-e^{-\mathcal{T}s}}{s}\\ 1-e^{-\mathcal{T}s}\end{array}\right.$	$\dfrac{z-1}{z}$
Exponential hold element		$\dfrac{\mathcal{T}}{1+\mathcal{T}s}$	$\dfrac{z}{z-e^{-1}}$
Bilinear transformation to map the interior of the unit circle in the z-plane onto the left half of the complex variable ξ-plane. (Application of the Routh–Hurwitz stability criterion).			$z = \dfrac{\xi+1}{\xi-1}$

In the above $\mathcal{T}$ indicates the sampling interval.

Appendix 7.2. Determination of the Step Response Function of a Second-Order System from its Transfer Function

From equation 7.81:

$$\bar{\mathfrak{z}} = \frac{MK_{MT}}{\tau^2 s(s-\beta_1)(s-\beta_2)}$$

where:

$$\beta_{1,2} = -\frac{\zeta}{\tau} \pm \frac{\sqrt{(\zeta^2-1)}}{\tau}$$

$$\therefore \quad \frac{\tau^2}{MK_{MT}}\bar{\mathfrak{z}} = \frac{1}{s(s-\beta_1)(s-\beta_2)} = \frac{A}{s} + \frac{B}{s-\beta_1} + \frac{C}{s-\beta_2} \tag{7.A2.1}$$

Inversion of equation 7.A2.1 gives:

$$\frac{\tau^2}{MK_{MT}}\,\mathfrak{z} = A + B\exp(\beta_1 t) + C\exp(\beta_2 t) \tag{7.A2.2}$$

(a) $\zeta < 1$

Now:

$$\beta_1 = -\frac{\zeta}{\tau} + \frac{\sqrt{(\zeta^2 - 1)}}{\tau}$$

Thus, for $\zeta < 1$:

$$\beta_1 = -\frac{\zeta}{\tau} + \mathrm{i}\frac{\sqrt{(1 - \zeta^2)}}{\tau}$$

$$= -\frac{\zeta}{\tau} + \mathrm{i}\frac{\phi}{\tau} \tag{7.A2.3}$$

where $\phi = \sqrt{(1 - \zeta^2)}$ is real.

∴

$$\exp(\beta_1 t) = \exp\left(-\frac{\zeta}{\tau}t + \mathrm{i}\frac{\phi}{\tau}t\right)$$

$$= \exp\left(-\frac{\zeta}{\tau}t\right)\exp\left(\mathrm{i}\frac{\phi}{\tau}t\right)$$

$$= \exp\left(-\frac{\zeta}{\tau}t\right)\left(\cos\frac{\theta}{\tau}t + \mathrm{i}\sin\frac{\theta}{\tau}t\right) \tag{7.A2.4}$$

Similarly:

$$\exp(\beta_2 t) = \exp\left(-\frac{\zeta}{\tau}t\right)\left(\cos\frac{\phi}{\tau}t - \mathrm{i}\sin\frac{\theta}{\tau}t\right) \tag{7.A2.5}$$

From equation 7.A2.1 by partial fractions:

$$A = \frac{1}{\beta_1\beta_2}$$

i.e.:

$$\frac{1}{A} = \frac{\zeta^2}{\tau^2} - \frac{\zeta^2 - 1}{\tau^2} = \frac{1}{\tau^2}$$

∴

$$A = \tau^2$$

Also:

$$B = \frac{1}{\beta_1(\beta_1 - \beta_2)}$$

i.e.:

$$\frac{1}{B} = \beta_1\left(\frac{2\sqrt{(\zeta^2 - 1)}}{\tau}\right) = \beta_1\left(\frac{2\mathrm{i}\phi}{\tau}\right)$$

Substituting from equation 7.A2.3:

$$\frac{1}{B} = \left(-\frac{\zeta}{\tau} + \mathrm{i}\frac{\phi}{\tau}\right)\frac{2\mathrm{i}\phi}{\tau} = -\frac{2\phi}{\tau}\left(\frac{\phi}{\tau} + \mathrm{i}\frac{\zeta}{\tau}\right)$$

∴

$$B = -\frac{\tau^2}{2\phi(\phi + \mathrm{i}\zeta)}$$

Similarly:

$$C = \frac{1}{\beta_2(\beta_2 - \beta_1)}$$

$$= -\frac{\tau^2}{2\phi(\phi - \mathrm{i}\zeta)}$$

Substituting for A, B and C in equation 7.A2.3 and using equations 7.A2.4 and 7.A2.5:

$$\frac{\tau^2}{MK_{MT}}\,\mathfrak{z} = \tau^2 - \frac{\tau^2}{2\phi(\phi + \mathrm{i}\zeta)}\exp\left(-\frac{\zeta}{\tau}t\right)\left(\cos\frac{\phi}{\tau}t + \mathrm{i}\sin\frac{\phi}{\tau}t\right) - \frac{\tau^2}{2\phi(\phi - \mathrm{i}\zeta)}\exp\left(-\frac{\zeta}{\tau}t\right)\left(\cos\frac{\phi}{\tau}t - \mathrm{i}\sin\frac{\phi}{\tau}t\right)$$

$$\frac{1}{MK_{MT}}\,\mathfrak{z} = 1 - \exp\left(-\frac{\zeta}{\tau}t\right)\left(\frac{1}{\phi^2 + \zeta^2}\cos\frac{\phi}{\tau}t + \frac{\phi\zeta}{\phi^2 + \zeta^2}\sin\frac{\phi}{\tau}t\right)$$

But:
$$\frac{1}{\phi^2 + \zeta^2} = \frac{1}{1 - \zeta^2 + \zeta^2} = 1$$

and:
$$\frac{\zeta/\phi}{\phi^2 + \zeta^2} = \frac{\zeta}{\sqrt{(1 - \zeta^2)}}$$

$\therefore$
$$\frac{1}{MK_{MT}}\,\mathfrak{z} = 1 - \exp\left(-\frac{\zeta}{\tau}t\right)\left(\cos\frac{\phi}{\tau}t + \frac{\zeta}{\sqrt{(1 - \zeta^2)}}\sin\frac{\phi}{\tau}t\right) \tag{7.A2.6}$$

Equation 7.A2.6 can be put into a more useful form by using the trigonometric identity:
$$p\cos\alpha + q\sin\alpha = r\sin(\alpha + \rho)\xi$$

where:
$$r = \sqrt{(p^2 + q^2)}$$

and:
$$\tan\xi = \frac{p}{q}$$

Thus:
$$r = \sqrt{\left\{1 + \frac{\zeta^2}{1 - \zeta^2}\right\}} = \frac{1}{\sqrt{(1 - \zeta^2)}} = \frac{1}{\phi}$$

and:
$$\xi = \tan^{-1}\left(\frac{\sqrt{(1 - \zeta^2)}}{\zeta}\right) = \tan^{-1}\frac{\phi}{\zeta}$$

Thus equation 7.A2.6 becomes:
$$\mathfrak{z} = MK\left\{1 - \frac{1}{\phi}\sin\left(\frac{\phi t}{\tau} + \tan^{-1}\frac{\phi}{\zeta}\right)\exp\left(-\frac{\zeta\tau}{\tau}\right)\right\} \qquad \text{(equation 7.82)}$$

(b) $\zeta > 1$

In this case:
$$\beta_{1,2} = -\frac{\zeta}{\tau} \pm \frac{\sqrt{(\zeta^2 - 1)}}{\tau}$$
$$= -\frac{\zeta}{\tau} \pm \frac{\nu}{\tau}$$

where $\nu = \sqrt{(\zeta^2 - 1)}$ and is real.

As for $\zeta < 1$, from equation 7.A2.1 by partial fractions:
$$A = \tau^2$$
$$B = \frac{1}{\beta_1(\beta_1 - \beta_2)}$$

$\therefore$
$$\frac{1}{B} = \left(-\frac{\zeta}{\tau} + \frac{\nu}{\tau}\right)\frac{2\nu}{\tau}$$

and:
$$B = \frac{\tau^2}{2\nu(\nu - \zeta)}$$

Similarly:
$$C = \frac{\tau^2}{2\nu(\nu + \zeta)}$$

Substituting for A, B and C in equation 7.A2.2:
$$\frac{\tau^2}{MK_{MT}}\,\mathfrak{z} = \tau^2 + \frac{\tau^2}{2\nu(\nu - \zeta)}\exp\left(-\frac{\zeta t}{\tau} + \frac{\nu t}{\tau}\right) + \frac{\tau^2}{2\nu(\nu + \zeta)}\exp\left(-\frac{\zeta t}{\tau} - \frac{\nu t}{\tau}\right)$$

Thus:
$$\frac{1}{MK_{MT}}\,\mathfrak{z} = 1 - \exp\left(-\frac{\zeta t}{\tau}\right)\left[-\frac{1}{2\nu(\nu - \zeta)}\exp\left(\frac{\nu t}{\tau}\right) - \frac{1}{2\nu(\nu + \zeta)}\exp\left(-\frac{\nu t}{\tau}\right)\right] \tag{7.A2.7}$$

Putting:
$$\cosh\frac{\nu t}{\tau} = \frac{1}{2}\left[\exp\left(\frac{\nu t}{\tau}\right) + \exp\left(-\frac{\nu t}{\tau}\right)\right]$$

and:
$$\sinh\frac{\nu t}{\tau} = \frac{1}{2}\left[\exp\left(\frac{\nu t}{\tau}\right) - \exp\left(-\frac{\nu t}{\tau}\right)\right]$$

in equation 7.A2.7, leads to:

$$\mathfrak{z} = MK_{MT}\left\{1 - \left(\cosh\frac{\nu t}{\tau} + \frac{\zeta}{\nu}\sinh\frac{\nu t}{\tau}\right)\exp\left(-\frac{\zeta t}{\tau}\right)\right\} \qquad \text{(equation 7.83)}$$

(c) $\zeta = 1$

Under these circumstances:

$$\beta_1 = \beta_2 = -\frac{\zeta}{\tau} = -\frac{1}{\tau}$$

Thus equation 7.A2.1 becomes:

$$\frac{\tau^2}{MK_{MT}}\bar{\mathfrak{z}} = \frac{1}{s(s+1/\tau)^2}$$

$$= \frac{A}{s} + \frac{B}{(s+1/\tau)^2} + \frac{C}{s+1/\tau}$$

Equating coefficients etc. gives:

$$A = \tau, \quad B = -\tau \quad \text{and} \quad C = -\tau^2$$

$$\therefore \qquad \frac{1}{MK_{MT}}\bar{\mathfrak{z}} = \frac{1}{s} - \frac{1/\tau}{(s+1/\tau)^2} - \frac{1}{s+1/\tau} \qquad (7.\text{A}2.8)$$

Inverting equation 7.A2.8:

$$\frac{1}{MK_{MT}}\mathfrak{z} = 1 - \frac{t}{\tau}\exp\left(-\frac{t}{\tau}\right) - \exp\left(-\frac{t}{\tau}\right)$$

$$\mathfrak{z} = MK_{MT}\left\{1 - \left(1 + \frac{t}{\tau}\right)\exp\left(-\frac{t}{\tau}\right)\right\} \qquad \text{(equation 7.84)}$$

7.24. FURTHER READING

ÅSTRÖM, K. J. and WITTENMARK, B.: *Adaptive Control* (Addison-Wesley, Wokingham, U.K., 1989).
BENNETT, S.: *Real-Time Computer Control: An Introduction* (Prentice-Hall, Hemel Hempstead, U.K., 1988).
COUGHANOWR, D. R.: *Process Systems Analysis and Control*, 2nd edn. (McGraw-Hill, New York, 1991).
KUO, B. C.: *Discrete Data Control Systems* (Prentice-Hall, Englewood Cliffs, New Jersey, 1970).
LANDAU, Y. D.: *Adaptive Control—The Model Reference Approach* (Marcel Dekker, New York, 1979).
POPOVIC, D. and BHATKAR, V. P.: *Distributed Computer Control for Industrial Automation* (Marcel Dekker, New York, 1990).
RAVEN, F. H.: *Automatic Control Engineering*, 2nd edn. (McGraw-Hill, New York, 1968).
SHINSKEY, F. G.: *Distillation Control* (McGraw-Hill, New York, 1977).
STEPHANOPOULOS, G.: *Chemical Process Control: An Introduction to Theory and Practice* (Prentice-Hall, Englewood Cliffs, New Jersey, 1984).
WARNOCK, I. G.: *Programmable Controllers* (Prentice-Hall, Hemel Hempstead, U.K., 1988).

7.25. REFERENCES

1. BS 1646: 1979: BRITISH STANDARD 1646 (British Standards Institution, London) Symbolic representation for process measurement control functions and instrumentation.
2. PETERS, J. C.: *Trans. A.S.M.E.* **64** (1942) 247. Experimental studies of automatic control.
3. SHINSKEY, F. G.: *Distillation Control* (McGraw-Hill, New York, 1979).
4. PARKINS, R.: *Chem. Eng. Prog.* **55** (7) (1959) 60. Continuous distillation plant controls.
5. MURRILL, P. W.: *Automatic Control of Processes* (International Textbook Company, Scranton, Pennsylvania, 1967).
6. STEPHANOPOULOS, G.: *Chemical Process Control: An Introduction to Theory and Practice* (Prentice-Hall, Englewood Cliffs, New Jersey, 1984).
7. ACTON, J. R. and SQUIRE, P. T.: *Solving Equations with Physical Understanding* (Adam Hilger Ltd, Bristol, 1985).
8. DEL TORO, V. and PARKER, S. R.: *Principles of Control Systems Engineering* (McGraw-Hill, New York, 1960).

9. BRITTON, J. R., KRIEGH, R. B. and RUTLAND, L. W.: *University Mathematics*, Vol. 2 (Freeman, San Francisco, 1965).
10. CRANDALL, E. D. and STEVENS, W. F.: *A.I.Ch.E.Jl.* **11** (1965) 930. An application of adaptive control to a continuous stirred tank reactor.
11. DOUGLAS, J. M.: *Process Dynamics and Control* (Prentice-Hall, Englewood Cliffs, New Jersey, 1972).
12. ROSE, A. and WILLIAMS, T. J.: *Ind. Eng. Chem.* **47** (1955) 2284. Automatic control in continuous distillation.
13. HARRIOT, P.: *Chem. Eng. Prog.* **60** (8) (1964) 81. Theoretical analysis of components.
14. BENNETT, S.: *Real-Time Computer Control: An Introduction* (Prentice-Hall, Hemel Hempstead, U.K., 1988).
15. MICKLEY, H. S., SHERWOOD, T. K. and REED, C. E.: *Applied Mathematics in Chemical Engineering* (McGraw-Hill, New York, 1957).
16. CEAGLSKE, N. H.: *Automatic Process Control for Chemical Engineers* (Wiley, New York, 1956).
17. WYLIE, C. R.: *Advanced Engineering Mathematics* (McGraw-Hill, New York, 1966).
18. COUGHANOWR, D. R.: *Process Systems Analysis and Control*, 2nd edn. (McGraw-Hill, New York, 1991).
19. HURWITZ, A.: *Math. Annln.* **46** (1895) 273. Über die Bedingungen, unter welchen eine Gleichung nur Wurzeln mit negativen realen Teilen besitzt.
20. ROUTH, E. J.: *Advanced Rigid Dynamics* (Macmillan, London, 1884).
21. NYQUIST, H.: *Bell System Tech. J.* **11** (1932) 126. Regeneration theory.
22. BODE, H. W.: *Bell System Tech. J.* **19** (1940) 421. Relations between attenuation and phase in feedback amplifier design.
23. TAKAHASHI, T.: *Mathematics of Automatic Control* (Holt, Rinehart and Winston, New York, 1966).
24. BODE, H. W.: *Network Analysis and Feedback Amplifier Design* (Van Nostrand Reinhold, New York, 1945).
25. JAMES, H. M., NICHOLS, N. B. and PHILLIPS, R. S.: *Theory of Servomechanisms* (McGraw-Hill, New York, 1947).
26. ZIEGLER, J. G. and NICHOLS, N. G.: *Trans. A.S.M.E.* **64** (1942) 759. Optimum settings for automatic controllers.
27. COHEN, G. H. and COON, G. A.: *Trans. A.S.M.E.* **75** (1953) 827. Theoretical consideration of retarded control.
28. LOPEZ, A. M., SMITH, C. L. and MURRILL, P. W.: *Brit. Chem. Eng.* **14** (1969) 1953. An advanced tuning method.
29. EDGAR, T. F. and HIMMELBLAU, D. M.: *Optimization of Chemical Processes* (McGraw-Hill, New York, 1988).
30. SMITH, O. J. M.: *Chem. Eng. Prog.* **53** (5) (1957) 217. Close control of loops with dead time.
31. RAVEN, F. H.: *Automatic Control Engineering*, 2nd edn. (McGraw-Hill, New York, 1968).
32. WARDLE, A. P. and WOOD, R. M.: *I. Chem. E./European Federation of Chem. Eng. Int. Symp. on Distillation* (1969) *Session* VI, 66. Problems of application of theoretical feed-forward control models to industrial scale fractionating plant.
33. SHINSKEY, F. G.: *Process Control Systems*, 2nd edn. (McGraw-Hill, New York, 1979).
34. SHINSKEY, F. G.: *Distillation Control* (McGraw-Hill, New York, 1977).
35. ROSENBROCK, H. H.: *Computer-Aided Control System Design* (Academic Press, New York, 1974).
36. BRISTOL, E. H.: *IEEE Trans. Autom. Control* **AC-11** (1966) 133. On a new measure of interaction for multivariable process control.
37. STATHAKI, A., MILLICHAMP, D. A. and SEBORG, D. E.: *Can. J. Chem. Eng.* **63** (1985) 510. Dynamic simulation of a multicomponent distillation column with asymmetric dynamics.
38. KOCHENBURGER, R. J.: *Trans. AIEE* **69** (1950) 270. A frequency response method for analysing and synthesising contactor servomechanisms.
39. TRUXAL, J. G.: *Automatic Feedback Control System Synthesis* (McGraw-Hill, New York, 1955).
40. GRAHAM, D. and MCRUER, D.: *Analysis of Non-Linear Control Systems* (Wiley, New York, 1961).
41. KUO, B. C.: *Discrete Data Control Systems* (Prentice-Hall, Englewood Cliffs, New Jersey, 1970).
42. LUYBEN, W. L.: *Process Modeling, Simulation and Control for Chemical Engineers* (McGraw-Hill, New York, 1973).
43. YOUNG, R. E.: *Supervisory Remote Control Systems* (Peter Peregrinus, Stevenage, 1977).
44. DAHLIN, E. B.: *Instrum. Control Syst.* **41** (6) (1968) 77. Designing and tuning digital controllers.
45. KALMAN, R. W.: *Trans. ASME* **80** (Series D) (1958) 468. The design of a self-optimising control system.
46. ÅSTRÖM, K. J. and WITTENMARK, B.: *Adaptive Control* (Addison-Wesley, Wokingham, U.K., 1989).
47. MORANT, F., MARTINEZ, M. and PICÓ, J.: In *Application of Artificial Intelligence in Process Control* by BOULLART, L., KRIJGSMAN, A. and VINGERHOEDS, R. A. eds, Section VI. Supervised adaptive control.
48. LANDAU, Y. D.: *Adaptive Control—The Model Reference Approach* (Marcel Dekker, New York, 1979).

49. CLARKE, D. W. and GAWTHROP, P. J.: *IEE Proc.* **122** (9) (1975) 929. Self-tuning controller.
50. CLARKE, D. W., MOHTADI, C. and TUFFS, P. S.: *Automatica* **23** (1987). Generalised predictive control: parts (i) and (ii).
51. POPOVIC, D. and BHATKAR, V. P.: *Distributed Computer Control for Industrial Automation* (Marcel Dekker, New York, 1990).
52. CIVERA, P., DEL CORSO, D. and GREGORETTI, F.: In *Microprocessors in Signal Processing, Measurement and Control*, TZAFESTAS, S. G. ed. (Reidel, London, 1983). Microcomputer systems in real-time applications.
53. JOVIC, F.: *Expert Systems* (Kogan Page, London, 1987).
54. BAILEY, R. W.: *Human Performance Engineering: A Guide for System Designers* (Prentice-Hall, Englewood Cliffs, New Jersey, 1982).
55. GILMORE, W. E., GERTMAN, D. I. and BLACKMAN, H. S.: *User–Computer Interface in Process Control* (Academic Press, London, 1989).
56. KALANI, G.: *Microprocessor Based Distributed Control Systems* (Prentice-Hall, Englewood Cliffs, New Jersey, 1988).
57. CAUNT, E.: *Process Industry Journal* (January, 1994). It's the application that's important.
58. WARNOCK, I. G.: *Programmable Controllers* (Prentice-Hall, Hemel Hempstead, U.K., 1988).
59. Mitsubishi Electric Corporation: *Instruction Manual H1-1B-041-G* (1992). The Melsec-F programmable controller.
60. JONES, M. G. and WARDLE, A. P.: *Proceedings of the first IASTED International Symposium on Circuits and Systems, Zürich* (1991) 90. Design of a PLC-based control system for a batch reactor.
61. SMITH, J. I.: *Modern Operational Amplifier Design* (Wiley, New York, 1971).
62. GOULD, L. A. and SMITH, P. E.: *Instr.* **26** (1953) 1026. Dynamic behaviour of pneumatically operated control equipment (Part II).
63. COUGHANOWR, D. R. and KOPPEL, L. B.: *Process Systems Analysis and Control* (McGraw-Hill, New York, 1965).
64. SINGH, J.: *Process Industry Journal* (January 1993) 46. Look no moving parts.
65. Samson plc: *Data sheet T8340E* (March, 1992). Electrohydraulic activator for type 240 control valves.
66. SINGH, M. G., ELLOY, J. P., MEZENCEV, R. and MUNRO, N.: *Applied Industrial Control*, Vol. 1 (Pergamon Press, Oxford, 1980).
67. WHERRY, T. C., PEEBLES, J. R., MCNEESE, P. M., WORSHAM, R. E. and YOUNG, R. M.: In *Perry's Chemical Engineers' Handbook*, GREEN, D. W. and MALONEY, J. O. eds., 6th edn. (McGraw-Hill, New York, 1984). Section 22, Process control.
68. PETERS, J. C.: *Ind. Eng. Chem.* **33** (1941) 1095. Getting the most from automatic control.
69. WHITE, B. A.: *Chem. Eng. Prog.* **89** (12) (1993) 31. Rethink the role of control valves.
70. COURT, R.: *Process Industry Journal* (June 1993) 58. Actuators are getting smarter.

7.26. NOMENCLATURE

Symbol	Description	Units in SI System	Dimensions in **M, N, L, T, θ, A**
A	Cross-sectional area	m^2	$\mathbf{L}^2$
A_1, A_2	Mean areas for heat transfer	m^2	$\mathbf{L}^2$
A_3, A_4	Constants	—	—
A_I	Area under pulse function	s	**T**
A_{LM}	Log[1/(gain margin)]	—	—
B	Measured value of a variable	—	—
B_1, B_2, etc.	Constants	—	—
b_{31}, b_{32}, etc.	Elements in Routh–Hurwitz array	—	—
C	Controlled variable	—	—
C_1, C_2	Average specific heats	J/kg K	$\mathbf{L}^2\mathbf{T}^{-2}\boldsymbol{\theta}^{-1}$
C_{IB}	Capacity of integral action bellows	m^3	$\mathbf{L}^3$
C^+	Deviation in controlled variable due to interaction	—	—
$\mathscr{C}_I$	Capacitance of amplifier feedback circuit in hard-wired system generating **PI** action	F (farad)	$\mathbf{M}^{-1}\mathbf{L}^{-2}\mathbf{T}^4\mathbf{A}^2$
$\mathscr{C}_D$	Capacitance of voltage input circuit in hard-wired system generating **PD** action	F (farad)	$\mathbf{M}^{-1}\mathbf{L}^{-2}\mathbf{T}^4\mathbf{A}^2$
c_r, c_{r-1}, etc.	Elements in Routh–Hurwitz array	—	—
D	Overhead product flowrate in mass or moles per unit time	kg/s, kmol/s	$\mathbf{MT}^{-1}, \mathbf{NT}^{-1}$
D	Decoupling element	—	—

		Units in SI System	Dimensions in **M, N, L, T, θ, A**
d	Diameter of manometer tube	m	**L**
F	Feed flowrate in mass or mols per unit time	kg/s, kmol/s	$\mathbf{MT^{-1}}$, $\mathbf{NT^{-1}}$
f	Function	—	—
f(z)	z-transform of function	—	—
f	(Cyclic) frequency	Hz	$\mathbf{T^{-1}}$
G	Transfer function	—	—
$\mathbf{G}_1$	Transfer function of final control element	—	—
$\mathbf{G}_2$, $\mathbf{G}_2'$	Transfer function of process	—	—
$\mathbf{G}_C$	Transfer function of controller	—	—
$\mathbf{G}_{C(deriv)}$	Transfer function of industrial controller derivative module	—	—
$\mathbf{G}_{comp}$	Transfer function of lead, lag or lag–lead compensator	—	—
$\mathbf{G}_{DTC}$	Transfer function of dead time compensator	—	—
$\mathbf{G}_{est}$	Transfer function assumed by estimator (equation 7.252)	—	—
$\mathbf{G}_{FF}$	Feed-forward controller transfer function	—	—
$\mathbf{G}_{il}$	Transfer function of inner loop	—	—
$\mathbf{G}_{pcl}$	Transfer function of primary closed-loop system	—	—
$\mathbf{G}_{scl}$	Transfer function of secondary closed-loop system	—	—
$\mathbf{G}_X$	Transfer function of blocks in series	—	—
$\mathbf{G}_Y$	Transfer function of blocks in parallel	—	—
$\mathbf{G}_{ZOH}$	Transfer function of zero order hold element	—	—
G(z)	Transfer function of discrete time system, i.e. in terms of the z-transform	—	—
$\mathbf{G}^*(s)$	Pulse transfer function (defined in equation 7.210)	—	—
g	Acceleration due to gravity	m/s^2	$\mathbf{LT^{-2}}$
H, $\mathbf{H}_1$	Transfer function of measuring element *or* of elements in feedback path of control loop	—	—
h_1, h_2	Heat transfer coefficients	W/m^2K	$\mathbf{MT^{-3}\theta^{-1}}$
i	$\sqrt{(-1)}$	—	—
i	Current	A	**A**
J	Controller output signal at time t	—	—
J_0	Controller output signal for zero error	—	—
j	Number of sampling periods equivalent to the system dead-time in Dahlin's algorithm	—	—
K	Steady-state gain	—	—
K_1, K_3	Flow-head relationship in equations 7.21 and 7.40 respectively	$m^{2.5}/s$	$\mathbf{L^{2.5}T^{-1}}$
K_2, K_4	Flow-head relationship in equations 7.25 and 7.29 respectively	m^2/s	$\mathbf{L^2T^{-1}}$
K_{amp}	Gain of operational amplifier	—	—
K_C	Proportional gain	—	—
K_D	Constant in equation 7.4	—	—
K_I	Constant in equation 7.3	—	—
K_M	Magnitude of vector at the frequency where polar plot cuts negative real axis	—	—
K_{MT}	Steady-state gain for manometer tube	m^2s^2/kg	$\mathbf{M^{-1}L^2T^2}$
K_r	Steady-state gain of process reaction curve (in Cohen–Coon procedure)	—	—
K_u	Ultimate gain	—	—
$\mathcal{K}_1$, $\mathcal{K}_2$	Constants in equations 7.265–7.268	—	—
k	Input/output relationship for linear section of non-linear characteristic	—	—
L_1, L_2, L_{N-1}, L_F	Liquid flowrates in mass or moles per unit time	kg/s, kmol/s	$\mathbf{MT^{-1}}$, $\mathbf{NT^{-1}}$
L_R	Reflux flowrate in mass or moles per unit time	kg/s, kmol/s	$\mathbf{MT^{-1}}$, $\mathbf{NT^{-1}}$
ℓ	Length of liquid column in manometer tube	m	**L**
ℓ_1, ℓ_2	Distances measured along flapper of pneumatic controller	m	**L**
ℓ_s	Distance of valve stem travel expressed as a percentage of the stroke	—	—
M	Magnitude or amplitude of signal	—	—

		Units in SI System	Dimensions in **M, N, L, T, θ, A**
M_v	Manipulated variable	—	—
M_v^+	Change in manipulated variable required to compensate for interaction	—	—
m_1, m_2	Mass	kg	**M**
N	Number of stages *or* total number of samples	—	—
N	Describing function		
n	Index giving type of polar plot or number of samples or number of terms in Fourier series	—	—
n_E	Net number of encirclements of the point $(-1, 0)$ on the complex plane	—	—
n_e	Number of independent equations	—	—
n_f	Number of degrees of freedom	—	—
n_v	Number of independent variables	—	—
n_Z	Number of zeros of function	—	—
P	Pressure upstream of nozzle in flapper/nozzle system	N/m^2	$\mathbf{ML^{-1}T^{-2}}$
P_1, P_2	Pressures applied to limbs of manometer tube or pressures downstream and upstream of orifice plate	N/m^2	$\mathbf{ML^{-1}T^{-2}}$
P_c	Distillation column pressure	N/m^2	$\mathbf{ML^{-1}T^{-2}}$
P_D	Pressure in feedback bellows of pneumatic controller	N/m^2	$\mathbf{ML^{-1}T^{-2}}$
P_f	Frictional drag per unit cross-sectional area of manometer tube	N/m^2	$\mathbf{ML^{-1}T^{-2}}$
P_I	Pressure in integral bellows of pneumatic controller	N/m^2	$\mathbf{ML^{-1}T^{-2}}$
P_M	Output pressure from differential pressure cell	N/m^2	$\mathbf{ML^{-1}T^{-2}}$
Q	Volumetric flowrate	m^3/s	$\mathbf{L^3T^{-1}}$
	or volumetric flowrate expressed as a percentage of maximum flowrate	—	—
Q_0	Volumetric inlet flowrate	m^3/s	$\mathbf{L^3T^{-1}}$
Q_1, Q_2	Volumetric outlet flowrate	m^3/s	$\mathbf{L^3T^{-1}}$
Q_B	Heat per unit time supplied to boiler	W	$\mathbf{ML^2T^{-3}}$
Q_H	Heat per unit time supplied by flowing stream	W	$\mathbf{ML^2T^{-3}}$
q_F	Heat to vaporise one mole of feed divided by molar latent heat	—	—
q_R	Heat to vaporise one mole of reflux divided by molar latent heat	—	—
R	Set point or desired value of a variable	—	—
	or shear stress on surface	N/m^2	$\mathbf{ML^{-1}T^{-2}}$
	or electrical resistance	Ω	$\mathbf{ML^2T^{-3}A^{-2}}$
R_D	Resistance of derivative action potentiometer	Ω	$\mathbf{ML^2T^{-3}A^{-2}}$
R_{DR}	Resistance of derivative restrictor	s/m^3	$\mathbf{L^{-3}T}$
R_I	Resistance of integral action potentiometer	Ω	$\mathbf{ML^2T^{-3}A^{-2}}$
R_{IR}	Resistance of integral restrictor	s/m^3	$\mathbf{L^{-3}T}$
r	Number of roots of characteristic equation *or* number of poles of closed-loop transfer function	—	—
S	Steam flowrate to reboiler in mass or mole per unit time	kg/s, kmol/s	$\mathbf{MT^{-1}}$, $\mathbf{NT^{-1}}$
S	Sum	—	—
s	Laplace transform parameter	1/s	$\mathbf{T^{-1}}$
T_u	Ultimate period	s	**T**
$\mathcal{T}$	Sampling time	s	**T**
t	Time	s	**T**
t_0	Duration of pulse function	s	**T**
U	Load variable(s)	—	—
$u(t)$	Unit step function	—	—
u	Mean velocity	m/s	$\mathbf{LT^{-1}}$
V	Vapour flowrate in mass or moles per unit time	kg/s, kmol/s	$\mathbf{MT^{-1}}$, $\mathbf{NT^{-1}}$
	or voltage	V	$\mathbf{ML^{-2}T^{-3}A^{-1}}$
V_i	Input voltage for operational amplifier	V	$\mathbf{ML^{-2}T^{-3}A^{-1}}$
V_o	Output voltage for operational amplifiers	V	$\mathbf{ML^{-2}T^{-3}A^{-1}}$
V_T	Volume of liquid in tank	m^3	$\mathbf{L^3}$

		Units in SI System	Dimensions in **M, N, L, T, θ, A**
$\mathbf{V}_N$	Vector in the complex plane	—	—
W	Bottoms product flowrate in mass or moles per unit time	kg/s, kmol/s	$\mathbf{MT}^{-1}$, $\mathbf{NT}^{-1}$
w	Mass flowrate	kg/s	$\mathbf{MT}^{-1}$
X	Output of dead time compensator	—	—
x	Input to control loop component	—	—
x_a	Composition of stream recorded by analyser (mole or mass fraction)	—	—
x_D	Molar mass fraction of component in overhead product	—	—
x_F	Molar mass fraction of component in feed	—	—
x_m	Composition of main stream (mole or mass fraction)	—	—
x_w	Molar mass fraction of component in bottoms product	—	—
$\mathbf{x}$	x dimension	—	—
Y_o	Magnitude of on–off signal (equation 7.194)	—	—
y	Output from control loop component	—	—
$\mathbf{y}$	y dimension	—	—
z	Liquid head	m	**L**
	or z-transform parameter	—	—
$\mathbf{z}$	z dimension	—	—
α	Angle	—	—
α_1, α_2, etc.	Roots of the characteristic equation (poles of the transfer function)	1/s	$\mathbf{T}^{-1}$
β	$\sin^{-1}(\delta/M)$ (Section 7.16.2)	—	—
β_1, β_2	Real parts of the roots of the characteristic equation *or* real parts of the complex variable s	1/s	$\mathbf{T}^{-1}$
γ_1, γ_2	Imaginary parts of the roots of the characteristic equation *or* imaginary parts of the complex variable s	1/s	$\mathbf{T}^{-1}$
δ	Magnitude of dead-zone	—	—
$\delta(t)$	Dirac function	s	**T**
ε	Error	—	—
ε_M	Difference between output of reference model and process output	—	—
ζ	Damping coefficient	—	—
η	Valve-flow coefficient	m^2/s	$\mathbf{L}^2\mathbf{T}^{-1}$
θ	Temperature	K	$\boldsymbol{\theta}$
θ_M	Measured temperature	K	$\boldsymbol{\theta}$
θ_R	Set point value of temperature	K	$\boldsymbol{\theta}$
κ	Interaction factor	s	**T**
Λ	Relative gain array	—	—
λ	Relative gain	—	—
μ	Viscosity of fluid	Ns/m^2	$\mathbf{ML}^{-1}\mathbf{T}^{-1}$
ν	$\surd(\zeta^2 - 1)$	—	—
Ξ	Coefficient in Fourier Series	—	—
ξ	Complex variable defined in equation 7.230	—	—
ρ	Density	kg/m^3	$\mathbf{ML}^{-3}$
σ	Linear range of element with saturation	—	—
σ_1, σ_2	Parameters (constants) in equation 7.139 and Table 7.5	—	—
τ	Time constant	s	**T**
τ_a	Apparent time constant	s	**T**
τ_{ad}	Apparent dead time	s	**T**
τ_{DV}	Distance–velocity lag, dead time	s	**T**
τ_D	Derivative time	s	**T**
τ_I	Integral time	s	**T**
Υ	Tuning parameter in equations 7.139–7.142	—	—
ϕ	$\surd(1 - \zeta^2)$	—	—
ψ	Phase shift	—	—
ψ_{comp}	Phase shift of lead, lag or lag–lead compensator	—	—
ψ_{PM}	Angle representing phase margin on Nyquist diagram	—	—
ω	Angular frequency	1/s	$\mathbf{T}^{-1}$
ω_c	Corner or break frequency	1/s	$\mathbf{T}^{-1}$

		Units in SI System	Dimensions in **M, N, L, T, θ, A**
ω_{co}	Cross-over frequency	1/s	$\mathbf{T}^{-1}$
ω_s	Sampling frequency	1/s	$\mathbf{T}^{-1}$
Re	Reynolds number with respect to pipe diameter	—	—
$\mathcal{R}_e$	Real part of function	—	—
$\mathcal{I}_m$	Imaginary part of function	—	—

Primed symbols represent the steady-state value of the variable (e.g. θ').

Symbols in script (unless listed specifically in the nomenclature) denote the deviation from the steady-state value of the variable (e.g. $\vartheta = \theta - \theta'$).

Symbols with bar represent the Laplace transform of the variable (e.g. $\bar{\theta}$ is the Laplace transform of θ).

Starred symbols represent a sampled output (e.g. $f^*(t)$ is the sampled data equivalent of the continuous signal $f(t)$).

Prefixes: Δ represents a finite change in a variable.

Acronyms

A/D	Analog-to-digital
AR	Amplitude ratio
$(AR)_{co}$	Amplitude ratio at the cross-over frequency
CCR	Central control room
CM	Control module
CMRR	Common mode rejection ratio
CPU	Central processing unit
D/A	Digital-to-analog
DCCS	Distributed computer control system(s)
DDC	Direct digital control
DH	Data highway
DM	Data acquisition module
DV(lag)	Distance–velocity (lag), dead time
E/P	Electropneumatic
FRC	Flow recorder controller
GMV	Generalised minimum variance controller
GPC	Generalised predictive controller
HFA	High frequency asymptote
IAE	Integral of the absolute error
IC	Integrated circuit
ICVS	Intelligent control valve system
I/O	Input/output
ISE	Integral square of the error
ITAE	Integral of the time-weighted absolute error
LAN	Local area network
LC	Level controller
LCR	Local control room
LED	Light emitting diode
LFA	Low frequency asymptote
LIC	Level indicator controller
MBC	Microprocessor-based control
MIMO	Multiple-input/multiple-output
MRAC	Model reference adaptive controller
OS	Operator Station
P	Proportional control action
PD	Proportional plus derivative control action
PI	Proportional plus integral control action
PID	Proportional plus integral plus derivative control action
PLC	Programmable controller (programmable logic controller)
PM	Phase margin
RAM	Read and write memory

RCM	Reserve control module
RDM	Reserve data acquisition module
RGA	Relative gain array
SCADA	Supervisory control and data acquisition
SSR	Solid-state relay
STR	Self-tuning regulator
TRC	Temperature recorder controller
TTL	Transistor–transistor logic
ZOH	Zero order hold element

Problems

1.1. A preliminary assessment of a process for the hydrodealkylation of toluene is to be made:

$$C_6H_5 \cdot CH_3 + H_2 \rightleftharpoons C_6H_6 + CH_4$$

The feed to the reactor will consist of hydrogen and toluene in the ratio $2H_2 : 1C_6H_5 \cdot CH_3$.

(a) Show that with this feed and an outlet temperature of 900 K, the maximum conversion attainable, i.e. the equilibrium conversion, is 0.996 based on toluene. The equilibrium constant of the reaction at 900 K, $K_p = 227$.
(b) Calculate the temperature rise which would occur with this feed if the reactor were operated adiabatically and the products left at equilibrium. For the above reaction at 900 K, $-\Delta H = 50{,}000$ kJ/kmol.
(c) If the maximum permissible temperature rise for this process is 100 degrees K (i.e. just over one-half the answer to (b)), suggest a suitable design for the reactor.

Heat capacities at 900 K (kJ/kmol K): C_6H_6 198; $C_6H_5CH_3$ 240; CH_4 67; H_2 30.

1.2. In a process for producing hydrogen which is required for the manufacture of ammonia, natural gas (methane) is to be reformed with steam according to the equations:

$$CH_4 + H_2O \rightleftharpoons CO + 3H_2; \qquad K_p(\text{at } 1173\text{ K}) = 1.43 \times 10^{13}\text{ N}^2/\text{m}^4$$

$$CO + H_2O \rightleftharpoons CO_2 + H_2; \qquad K_p(\text{at } 1173\text{ K}) = 0.784$$

The natural gas is mixed with steam in the mole ratio $1CH_4 : 5H_2O$ and passed into a catalytic reactor which operates at a pressure of 30 bar. The gases leave the reactor virtually at equilibrium at 1173 K.

(a) Show from the values of the equilibrium constants given above that for every 1 mole of CH_4 entering the reactor, 0.950 mole reacts, and 0.44 mole of CO_2 is formed.
(b) Explain why other reactions such as:

$$CH_4 + 2H_2O \rightleftharpoons CO_2 + 4H_2$$

need not be considered.

(c) By considering the reaction:

$$2CO \rightleftharpoons CO_2 + C$$

for which $K_p = P_{CO_2}/P_{CO}^2 = 2.76 \times 10^{-7}\text{ m}^2/\text{N}$ at 1173 K show that carbon deposition on the catalyst is unlikely to occur under the operating conditions.

(d) Qualitatively, what will be the effect on the composition of the exit gas of increasing the total pressure in the reformer? Can you suggest why for ammonia manufacture the reformer is operated at 30 bar instead of at a considerably lower pressure? The reforming step is followed by a shift conversion:

$$CO + H_2O \rightleftharpoons CO_2 + H_2$$

absorption of the CO_2, and ammonia synthesis:

$$N_2 + 3H_2 \rightleftharpoons 2NH_3$$

1.3. An aromatic hydrocarbon feedstock consisting mainly of *m*-xylene is to be isomerised catalytically in a process for making *p*-xylene. The product from the reactor consists of a mixture of *p*-xylene, *m*-xylene, *o*-xylene and ethylbenzene. As part of a preliminary assessment of the process, calculate the composition of this mixture if equilibrium were established over the catalyst at 730 K.

Equilibrium constants at 730 K are as follows:

$$m\text{-xylene} \rightleftharpoons p\text{-xylene}; \quad K_p = 0.45$$
$$m\text{-xylene} \rightleftharpoons o\text{-xylene}; \quad K_p = 0.48$$
$$m\text{-xylene} \rightleftharpoons \text{ethylbenzene}; \quad K_p = 0.19$$

Why is it unnecessary to consider also reactions such as:

$$o\text{-xylene} \underset{k_r}{\overset{k_f}{\rightleftharpoons}} p\text{-xylene?}$$

1.4. The alkylation of toluene with acetylene in the presence of sulphuric acid is carried out in a batch reactor. 6000 kg of toluene is charged in each batch, together with the required amount of sulphuric acid, and the acetylene is fed continuously to the reactor under pressure. Under circumstances of intense agitation, it may be assumed that the liquid is always saturated with acetylene, and that the toluene is consumed in a simple pseudo first-order reaction with a rate constant of $0.0011\ s^{-1}$.

If the reactor is shut down for a period of 15 min between batches, determine the optimum reaction time for the maximum rate of production of alkylate, and calculate this maximum rate in terms of kg toluene consumed per hour.

1.5. Methyl acetate is hydrolysed by water in accordance with the following equation

$$CH_3 \cdot COOCH_3 + H_2O \underset{k_r}{\overset{k_f}{\rightleftharpoons}} CH_3 \cdot COOH + CH_3OH$$

A rate equation is required for this reaction taking place in dilute solution. It is expected that reaction will be pseudo first order in the forward direction and second order in reverse. The reaction is studied in a laboratory batch reactor starting with a solution of methyl acetate and with no products present. In one experiment, the initial concentration of methyl acetate was 0.05 kmol/m^3 and the fraction hydrolysed at various times subsequently was as follows:

Time (s)	0	1350	3060	5340	7740	∞
Fractional conversion	0	0.21	0.43	0.60	0.73	0.90

(a) Write the rate equation for the reaction and develop its integrated form applicable to a batch reactor.

(b) Plot the data in the manner suggested by the integrated rate equation, confirm the order of reaction and evaluate the forward and reverse rate constants, k_f and k_r.

1.6. Styrene is to be made by the catalytic dehydrogenation of ethylbenzene:

$$C_6H_5 \cdot CH_2 \cdot CH_3 \rightleftharpoons C_6H_5 \cdot CH : CH_2 + H_2$$

The rate equation for this reaction has been reported as follows:

$$\mathcal{R} = k\left(P_{Et} - \frac{1}{K_p} P_{St} P_H\right)$$

where P_{Et}, P_{St} and P_H are partial pressures of ethylbenzene, styrene and hydrogen respectively.

The reactor will consist of a number of tubes each of 80 mm diameter packed with catalyst having a bulk density of 1440 kg/m^3. The ethylbenzene will be diluted with steam, the feed rates per unit cross-sectional area being ethylbenzene 1.6×10^{-3} kmol/m^2 s, steam 29×10^{-3} kmol/m^2 s. The reactor will be operated at an average pressure of 1.2 bar and the temperature will be maintained at 560°C throughout. If the fractional conversion of ethyl benzene is to be 0.45, estimate the length and number of tubes required to produce 20 tonne styrene per day.

At 560°C (833 K) $k = 6.6 \times 10^{-11}$ kmol m^2/N s kg catalyst, $K_p = 1.0 \times 10^4$ N/m^2.

1.7. Ethyl formate is to be produced from ethanol and formic acid in a continuous flow tubular reactor operated at a constant temperature of 30°C. The reactants will be fed to the reactor in the proportions 1 mole HCOOH : 5 moles C_2H_5OH at a combined flow rate of 0.72 m^3/h. The reaction will be catalysed by a small amount of sulphuric acid. At the temperature, mole ratio, and catalyst concentration to be employed, the rate equation determined from small-scale batch experiments has been found to be:

$$\mathcal{R} = k C_F^2$$

where $\mathscr{R}$ is kmol formic acid reacting/(m^3 s)
C_F is concentration of formic acid kmol/m^3, and
$k = 2.8 \times 10^{-4}$ m^3/kmol s.

The density of the mixture is 820 kg/m^3 and may be assumed constant throughout. Estimate the volume of the reactor required to convert 70 per cent of the formic acid to the ester.

If the reactor consists of a pipe of 50 mm i.d. what will be the total length required? Determine also whether the flow will be laminar or turbulent and comment on the significance of your conclusion in relation to your estimate of reactor volume. The viscosity of the solution is 1.4×10^{-3} Ns/m^2.

1.8. Two stirred tanks are available at a chemical works, one of volume 100 m^3, the other 30 m^3. It is suggested that these tanks be used as a two-stage CSTR for carrying out a liquid phase reaction $\mathbf{A} + \mathbf{B} \rightarrow$ product. The two reactants will be present in the feed stream in equimolar proportions, the concentration of each being 1.5 kmol/m^3. The volumetric flowrate of the feed stream will be 0.3×10^{-3} m^3/s. The reaction is irreversible and is of first order with respect to each of the reactants **A** and **B**, i.e. second order overall, with a rate constant 1.8×10^{-4} m^3/kmol s.

(a) Which tank should be used as the first stage of the reactor system, the aim being to effect as high an overall conversion as possible?
(b) With this configuration, calculate the conversion obtained in the product stream leaving the second tank after steady conditions have been reached.

(If in doubt regarding which tank should be used as the first stage, calculate the conversions for both configurations and compare. Note that accurate calculations are required in order to distinguish between the two.)

1.9. The kinetics of a liquid-phase chemical reaction are being investigated in a laboratory-scale continuous stirred tank reactor. The stoichiometric equation for the reaction is $\mathbf{A} \rightarrow 2\mathbf{P}$ and it is irreversible. The reactor is a single vessel which contains 3.25×10^{-3} m^3 of liquid when it is filled just to the level of the outflow. In operation, the contents of the reactor are well stirred and uniform in composition. The concentration of the reactant **A** in the feed stream is 0.5 kmol/m^3. Results of three steady-state runs are as follows:

Feed rate ($m^3/s \times 10^5$)	Temperature (°C)	Concentration of **P** in outflow (kmol **P**/m^3)
0.100	25	0.880
0.800	25	0.698
0.800	60	0.905

Hence determine the constants in the rate equation:

$$\mathscr{R}_A = \mathscr{A} \exp(-E/\mathbf{R}T) C_A^p.$$

1.10. A reaction $\mathbf{A} + \mathbf{B} \rightarrow \mathbf{P}$ which is first order with respect to each of the reactants, with a rate constant of 1.5×10^{-5} m^3/kmol s is carried out in a single continuous flow stirred tank reactor. This reaction is accompanied by a side reaction $2\mathbf{B} \rightarrow \mathbf{Q}$, where **Q** is a waste product, the side reaction being second order with respect to **B**, with a rate constant of 11×10^{-5} m^3/kmol s.

An excess of **A** is used for the reaction, the feed rates to the tank being 0.014 kmol/s of **A** and 0.0014 kmol/s of **B**; ultimately reactant **A** is recycled whereas **B** is not. Under these circumstances the overflow from the tank is at the rate of 1.1×10^{-3} m^3/s, while the capacity of the tank is 10 m^3.

Calculate (a) the fraction of **B** converted into the desired product **P**, and (b) the fraction of **B** converted into **Q**.

If a second tank of equal capacity becomes available, suggest with reasons in what manner it might be incorporated (i) if **A** but not **B** is recycled as above, and (ii) if both **A** and **B** are recycled.

1.11. Consider a model reaction scheme in which a substance **A** reacts with a second substance **B** to give a desired product **P**, but **B** also undergoes a simultaneous side reaction to give an unwanted product **Q**:

$$\mathbf{A} + \mathbf{B} \rightarrow \mathbf{P}; \quad \text{rate} = k_P C_A C_B$$

$$2\mathbf{B} \rightarrow \mathbf{Q}; \quad \text{rate} = k_Q C_B^2$$

The rate equations are given above where C_A and C_B are the concentrations of **A** and **B** respectively.

Let a single continuous stirred tank reactor be used to make these products. **A** and **B** are mixed in equimolar proportions such that each has the concentration C_0 in the combined stream fed at a volumetric flowrate v to the reactor. If the rate constants above are equal $k_P = k_Q = k$ and the total conversion of **B** is 0.95. i.e. the concentration of **B** in the outflow is 0.05 C_0, shows that the volume of the reactor will be $69v/kC_0$ and that the relative yield of **P** will be 0.82 (i.e. case *d*, Fig. 1.24).

Do you consider that a simple tubular reactor would give a larger or smaller yield of **P** than the C.S.T.R. above? What is the essential requirement for a high yield of **P**? Can you suggest any alternative modes of contacting the reactants **A** and **B** which would give better yields than either a single C.S.T.R. or a simple tubular reactor?

2.1. A batch reactor and a single continuous stirred tank reactor are being compared in relation to their performance in carrying out the simple liquid phase reaction **A** + **B** → products. The reaction is first order with respect to each of the reactants, i.e. second order overall. If the initial concentrations of the reactants are equal, show that the volume of the continuous reactor must be $1/(1 - \alpha)$ times the volume of the batch reactor for the same rate of production from each, where α is the fractional conversion. Assume that there is no change in density associated with the reaction and neglect the shutdown period between batches for the batch reactor.

In qualitative terms, what is the advantage of using a series of continuous stirred tanks for such a reaction?

2.2. Explain carefully the *dispersed plug-flow model* for representing departure from ideal plug flow. What are the requirements and limitations of the tracer response technique for determining *Dispersion Number* from measurements of tracer concentration at only one location in the system? Discuss the advantages of using two locations for tracer concentration measurements.

The residence time characteristics of a reaction vessel are investigated under steady-flow conditions. A pulse of tracer is injected upstream and samples are taken at both the inlet and outlet of the vessel with the following results:

Inlet sample point:

Time(s)	30	40	50	60	70	80	90	100	110	120
Tracer conc. (kmol/m^3) × 10^3	<0.05	0.1	3.1	10.0	16.5	12.8	3.7	0.6	0.2	<0.05

Outlet sample point:

Time (s)	110	120	130	140	150	160	170	180	190	200	210	220
Tracer conc. (kmol/m^3) × 10^3	<0.05	0.1	0.9	3.6	7.5	9.9	8.6	5.3	2.5	0.8	0.2	<0.05

Calculate (i) the mean residence time of tracer in the vessel and (ii) the dispersion number. If the reaction vessel is 0.8 m in diameter and 12 m long calculate also the volume flowrate through the vessel and the dispersion coefficient.

3.1. An approximate design procedure for packed tubular reactors entails the assumption of plug flow conditions through the reactor. Discuss critically those effects which would:

(a) invalidate plug flow assumptions, and
(b) enhance plug flow.

3.2. A first-order chemical reaction occurs isothermally in a reactor packed with spherical catalyst pellets of radius r. If there is a resistance to mass transfer from the main fluid stream to the surface of the particle in addition to a resistance within the particle, show that the effectiveness factor for the pellet is given by:

$$\eta = \frac{3}{\lambda r}\left\{\frac{\coth \lambda r - 1/\lambda r}{1 + (2\lambda r/Sh')(\coth \lambda r - 1/\lambda r)}\right\}$$

where $\lambda = (k/D_e)^{1/2}$ and $Sh' = \dfrac{h_D d_p}{D_e}$,

k is the first-order rate constant per unit volume of particle,
D_e is the effective diffusivity, and
h_D is the external mass transfer coefficient.

Discuss the limiting cases pertaining to this effectiveness factor.

3.3. Two consecutive first-order reactions;

$$\mathrm{A} \xrightarrow{k_1} \mathrm{B} \xrightarrow{k_2} \mathrm{C}$$

occur under isothermal conditions in porous catalyst pellets. Show that the rate of formation of **B** with respect to **A** at the exterior surface of the pellet is:

$$\frac{(k_1/k_2)^{1/2}}{1+(k_1/k_2)^{1/2}} - \left(\frac{k_2}{k_1}\right)^{1/2} \frac{C_B}{C_A}$$

when the pellet size is large, and:

$$1 - \frac{k_2}{k_1}\frac{C_B}{C_A}$$

when the pellet size is small. C_A and C_B represent the concentrations of **A** and **B** respectively at the exterior surface of the pellet, and k_1 and k_2 are the specific rate constants of the two reactions.

Comparing these results, what general conclusions can you deduce concerning the selective formation of **B** on large and small catalyst pellets?

3.4. A packed tubular reactor is used to produce a substance **D** at a total pressure of 1 bar utilising the exothermic equilibrium reaction:

$$\mathrm{A} + \mathrm{B} \rightleftharpoons \mathrm{C} + \mathrm{D}$$

36 kmol/h of an equimolar mixture of **A** and **B** is fed to the reactor and plug flow conditions within the reactor may be assumed.

Find the optimal isothermal temperature for operation and the corresponding reactor volume for a fractional conversion z_b of 0.68. Is this the best way of operating the reactor?

The forward and reverse kinetics are second-order with the velocity constants $k_1 = 4.4 \times 10^{13} \exp(-105 \times 10^6/\mathbf{R}T)$ and $k_2 = 7.4 \times 10^{14} \exp(-125 \times 10^6/\mathbf{R}T)$ respectively expressed in m^3/kmols and **R** in J/kmol **K**.

4.1. What is the significance of the parameter $\beta = (k_2 C_{BL} D_A)^{1/2}/k_L$ in the choice and the mechanism of operation of a reactor for carrying out a second-order reaction, rate constant k_2, between a gas **A** and a second reactant **B** of concentration C_{BL} in a liquid? In the above expression D_A is the diffusivity of **A** in the liquid and k_L is the liquid-film mass transfer coefficient. What is the 'reaction factor' and how is it related to β?

Carbon dioxide is to be removed from an air stream by reaction with 0.4 M solution (0.4 kmol/m^3) of NaOH at 25°C at a total pressure of 1.1 bar. A column packed with 25 mm Raschig rings is available for this purpose. The column is 0.8 m internal diameter and the height of the packing is 4 m. Air will enter the column at a rate of 0.015 kmol/s (total) and will contain 0.008 mole fraction CO_2. If the NaOH solution is supplied to the column at such a rate that its concentration is not substantially changed in passing through the column, calculate the mole fraction of CO_2 in the air leaving the column. Say whether you consider a packed column the most suitable reactor for this operation.

Data: Effective interfacial area for 25 mm packing: 280 m^2/m^3
Mass transfer film coefficients:
Liquid—$k_L = 1.3 \times 10^{-4}$ m/s
Gas—$k_G a = 0.052$ kmol/m^3 s bar
For 0.4 M (0.4 kmol/m^3 concentration) NaOH at 25°C:
Solubility of CO_2 i.e. Henry law constant in equation $P_A = \mathscr{H} C_A$
where P_A is partial pressure of CO_2, C_A is equilibrium liquid-phase concentration
$\mathscr{H} = 32$ bar m^3/kmol
Diffusivity of CO_2, $D = 0.19 \times 10^{-8}$ m^2/s

Second-order rate constant for the reaction $CO_2 + OH^- = HCO_3^- = 1.35 \times 10^4$ m^3/kmols
(Under the conditions stated, the reaction may be assumed pseudo first-order with respect to CO_2).

4.2. A pilot scale reactor for the oxidation of o-xylene by air has been constructed and its performance is being tested.

CH$_3$ CH$_3$ + $1.5\ O_2$ = CH$_3$ COOH + H_2O

The reactor, an agitated tank, operates under a pressure of 15 bar (absolute) and at a temperature of 160°C. It is charged with a batch of 60 litres of *o*-xylene and air introduced at the rate of $5.4 \text{ m}^3/\text{h}$ measured at reactor conditions. The air is dispersed into small bubbles whose mean diameter is estimated from a photograph to be 0.8 mm, and from level sensors in the reactor, the volume of the dispersion produced is found to be 88 litres. Soon after the start of the reaction (before any appreciable conversion of the *o*-xylene) the gas leaving the reactor is analysed (after removal of condensibles) and found to consist of 0.045 mole fraction O_2, 0.955 mole fraction N_2.

Assuming that under these conditions, the rate of the above reaction is virtually independent of the *o*-xylene concentration and is thus pseudo first order with respect to the concentration of dissolved O_2 in the liquid, calculate the value of the pseudo first order rate constant.

Further data are as follows:

Estimated liquid-phase mass transfer coefficient $k_L = 4.0 \times 10^{-4}$ m/s.
Henry law coefficient $\mathscr{H}$ for O_2 dissolved in *o*-xylene
$P_A = \mathscr{H}C_A$ where P_A is the partial pressure and C_A the equilibrium concentration in liquid), $\mathscr{H} = 127 \text{ m}^3 \text{ bar/kmol}$
Diffusivity of O_2 in liquid *o*-xylene, $D = 1.4 \times 10^{-9} \text{ m}^2/\text{s}$.
Gas constant, $\mathbf{R} = 8.314$ kJ/kmol K
Composition of air (molar) O_2 : 20.9% N_2 : 79.1%

State clearly any further assumptions that you make and discuss their validity. Do you consider an agitated tank to be the most suitable type of reactor for this process? State your reasons.

4.3. It is proposed to manufacture oxamide by reacting cyanogen with water using a strong solution of hydrogen chloride which acts as a catalyst.

$$(CN)_2 + 2H_2O \rightarrow (CONH_2)_2$$

The reaction is pseudo first order with respect to dissolved cyanogen, the rate constant at the operating temperature of 300 K being $0.19 \times 10^{-3} \text{s}^{-1}$. An agitated tank will be used containing 15 m^3 of liquid with a continuous flow of a cyanogen–air mixture at 3 bar total pressure, composition 0.20 mole fraction cyanogen, and a continuous feed and outflow of the hydrogen chloride solution; the gas feed flowrate will be $0.01 \text{ m}^3/\text{s}$ total and the liquid flowrate $0.0018 \text{ m}^3/\text{s}$. At the chosen conditions of agitation the following estimates have been made:

Liquid phase mass transfer coefficient	$k_L = 1.9 \times 10^{-5}$ m/s
Gas–liquid interfacial area per unit volume of dispersion	$a = 47 \text{ m}^2/\text{m}^3$
Gas volume fraction of dispersion	$\varepsilon_g = 0.031$
Diffusivity of cyanogen in solution	$D = 0.6 \times 10^{-9} \text{ m}^2/\text{s}$
Henry law coefficient ($P_A = \mathscr{H}C_A$) where P_A is partial pressure, C_A is concentration in liquid at equilibrium)	$\mathscr{H} = 1.3 \text{ bar m}^3/\text{kmol}$
Gas constant	$\mathbf{R} = 8314$ J/kmol K

Assuming ideal mixing for both gas and liquid phases, you are required to calculate:

(a) the concentration of oxamide in the liquid outflow (the inflow contains no dissolved oxamide);
(b) the concentration of dissolved but unreacted cyanogen in the liquid outflow, and
(c) the fraction of cyanogen removed from the gas stream. What further treatment would you suggest for the liquid leaving the tank before separation of the oxamide?

Calculate the value of β for this system; suggest another type of reactor that might be considered for this process.

4.4. (a) Consider a gas–liquid–solid hydrogenation such as that described in (b) below in which the reaction takes place within a porous catalyst particle in a trickle bed reactor. Assume that the liquid containing the compound to be hydrogenated reaches a steady state with respect to dissolved hydrogen immediately it enters the reactor and that the liquid is involatile. Show that the rate of reaction per unit volume of reactor space $\mathscr{R}$ (kmol H_2 converted/m^3s for a reaction which is pseudo first order with respect to hydrogen is given by:

$$\mathscr{R} = \frac{P_A}{\mathscr{H}}\left[\frac{1}{k_L a} + \frac{V_p}{k_s S_X(1-e)} + \frac{1}{k\eta(1-e)}\right]^{-1}$$

where P_A = Pressure of hydrogen (bar)
$\mathscr{H}$ = Henry Law coefficient (bar m³/kmol)
$k_L a$ = Gas–liquid volumetric mass transfer coefficient (s^{-1})

k_s = Liquid-solid mass transfer coefficient (m/s)
V_p = Volume of single particle (m^3)
S_X = External surface area of a single particle (m^2)
e = Voidage of the bed
$\boldsymbol{k}$ = First-order rate constant based on volume of catalyst $m^3/(m^3$ catalyst) s
η = Effectiveness factor

(b) Crotonaldehyde is to be selectively hydrogenated to n-butyraldehyde in a process using a palladium catalyst deposited on a porous alumina support in a trickle bed reactor. The particles will be spheres of 5 mm diameter packed into the reactor with a voidage e of 0.4. Estimated values of the parameters listed in (a) above are as follows:

$$k_L a = 0.02 \text{ s}^{-1} : k_s = 2.1 \times 10^{-4} \text{ m/s}$$

$$\boldsymbol{k} = 2.8 \text{ m}^3/(\text{m}^3 \text{ cat})\text{s} : \mathscr{H} = 357 \text{ bar m}^3/\text{kmol}$$

Also for spheres:
$$\eta = \frac{1}{\phi}\left(\coth 3\phi - \frac{1}{3\phi}\right) \quad \text{where } \phi = \frac{V_p}{S_X}\left(\frac{k}{D_e}\right)^{1/2}$$

For the catalyst, the effective diffusivity $D_e = 1.9 \times 10^{-9}$ m^2/s.
If the hydrogen pressure in the reactor is set at 1 bar, calculate $\mathscr{R}$, the rate of reaction per unit volume of reactor, and comment on the relative values of the transfer/reaction resistances involved in the process.

(c) Discuss whether the trickle bed reactor and the conditions described in (b) are the best choices for this process. What alternatives might be considered?

4.5. Aniline present as an impurity in a hydrocarbon stream is to be hydrogenated to cyclohexylamine in a trickle bed catalytic reactor operating at 130°C.

$$C_6H_5 \cdot NH_2 + 3H_2 \rightarrow C_6H_{11} \cdot NH_2$$

The reactor, in which the gas phase will be virtually pure hydrogen, will operate under a pressure of 20 bar absolute. The catalyst will consist of porous spherical particles 3 mm in diameter, and the voidage (i.e. the fraction of bed occupied by gas plus liquid) will be 0.4. The diameter of the bed will be such that the superficial liquid velocity will be 0.002 m/s. The concentration of the aniline in the liquid feed will be 0.055 kmol/m^3.

(a) From the data below, calculate what fraction of the aniline will be hydrogenated in a bed of depth 2 m. Assume that a steady state between the rates of mass transfer and reaction is established immediately the feed enters the reactor.
(b) Describe, stating the basic equations of a more complete model, how you would examine further the validity of the steady state assumption above.
(c) Do you consider a trickle bed to be the most suitable type of reactor for this process; if not, suggest with reasons a possibly better alternative.

Data:
The rate of the above reaction has been found to be first-order with respect to hydrogen but independent of the concentration of aniline. The first-order rate constant $\boldsymbol{k}$ of the reaction on a basis of kmol hydrogen reacting per m^3 of catalyst particles at 130°C is 90 s^{-1}.

Effective diffusivity of hydrogen in the catalyst particles with liquid-filled pores $D_e = 0.84 \times 10^{-9}$ m^2/s.
Effectiveness factor η for spherical particles of diameter d_p.

$$\eta = \frac{1}{\phi}\left(\coth 3\phi - \frac{1}{3\phi}\right) \quad \text{where } \phi = \frac{d_p}{6}\left(\frac{\boldsymbol{k}}{D_e}\right)^{1/2}$$

External surface area of particles per unit volume of reactor: 1200 m^{-1}
Mass transfer coefficient, liquid to particles: 0.10×10^{-3} m/s
Volume mass transfer coefficient, gas to liquid
(basis unit volume of reactor) $(k_L a) = 0.02$ s^{-1}
Henry law coefficient $\mathscr{H}$ for hydrogen dissolved in feed liquid ($P_A = \mathscr{H} C_A$ where P_A is hydrogen pressure and C_A is equilibrium concentration in liquid)
$\mathscr{H} = 2240$ bar m^3/kmol

4.6. Describe the various mass transfer and reaction steps involved in a three-phase gas-liquid-solid reactor. Derive an expression for the overall rate of a catalytic hydrogenation process where the reaction is pseudo first order with respect to the hydrogen with a rate constant k (based on unit volume of catalyst particles).

Aniline is to be hydrogenated to cyclohexylamine in a suspended-particle agitated-tank reactor at 130°C: at this temperature the value of k above is 90 s^{-1}. The diameter d_p of the supported nickel catalyst particles will be 0.1 mm for which the effective diffusivity D_e for hydrogen when the pores of the particle are filled with aniline is 1.9×10^{-9} m^2/s. For spherical particles the effectiveness factor

$$\eta = \frac{1}{\phi}\left(\coth 3\phi - \frac{1}{3\phi}\right) \quad \text{where} \quad \phi = \frac{d_p}{6}\left(\frac{k}{D_e}\right)^{1/2}$$

The proposed catalyst loading, i.e. the ratio by volume of catalyst to aniline, is to be 0.03. Under the conditions of agitation to be used, it is estimated that the gas volume fraction in the three-phase system will be 0.15 and that the volumetric gas–liquid mass transfer coefficient $k_L a$, 0.20 s^{-1} (also with respect to unit volume of the whole three-phase system). The liquid–solid mass transfer coefficient is estimated to be 2.2×10^{-3} m/s and the Henry law coefficient $\mathscr{H}$ for hydrogen in aniline at 130°C, 2240 bar m^3/kmol. ($P_A = \mathscr{H} C_A$ where P_A is the gas-phase partial pressure and C_A is the equilibrium concentration in the liquid).

(i) If the reactor is operated under a hydrogen partial pressure of 10 bar, calculate the rate at which the hydrogenation will proceed per unit volume of the three phase system.
(ii) Consider this overall rate in relation to the operating conditions and the individual transfer resistances. Discuss the question of whether any improvements might be made to the conditions specified for the reactor.

5.1. The hydraulic residence time (based on fresh feed) of an activated-sludge waste-water treatment unit is 5.9 h. The fresh feed has a BOD of 275 mg/l and the settler produces a recycle stream containing 6000 mg/l. Using a sludge age of 6 days, calculate the recycle ratio and the final effluent BOD (assuming that it contains no biomass) given that the yield coefficient Y, is 0.54 and that the specific growth rate μ of the sludge is given by:

$$\mu = \frac{\mu_m S}{K_s + S} - k_d$$

where S is the substrate concentration (BOD), $\mu_m = 0.47\ h^{-1}$, $K_s = 89$ mg (BOD) l^{-1} and the endogenous respiration coefficient $k_d = 0.009\ h^{-1}$.

5.2. A continuous fermenter is operated at a series of dilution rates but at constant (sterile) feed concentration, pH aeration rate and temperature. The data given below were obtained when the limiting substrate concentration was 1200 mg/l and the working volume of the fermenter was 9.8 litres.
Estimate the kinetic constants K_m, μ_m and k_d as used in the modified Monod equation:

$$\mu = \frac{\mu_m S}{K_s + S} - k_d$$

and also the growth yield coefficient Y

Feed flowrate (l/h)	Exit substrate conc. (mg/l)	Dry weight cell density (mg/l)
0.79	36.9	487
1.03	49.1	490
1.31	64.4	489
1.78	93.4	482
2.39	138.8	466
2.68	164.2	465

5.3. When a pilot scale fermenter is run in continuous mode with a fresh feed flowrate of 65 l/h, the effluent from the fermenter contains 12 mg/l of the original substrate. The same fermenter is then connected to a settler–thickener which has the ability to concentrate the biomass in the effluent from the tank by a factor of 3.2, and from this a recycle stream of concentrated biomass is set up. The flowrate of this stream is 40 l/h and the fresh feed flowrate is at the same time increased to 100 l/h. Assuming that the microbial system follows Monod kinetics, calculate the concentration of the final clarified liquid effluent from the system. ($\mu_m = 0.15\ h^{-1}$ and $K_s = 95$ mg/l).

5.4. When a continuous culture is fed with substrate of concentration 1.00 g/l, the critical dilution rate for washout is 0.2857 h^{-1}. This changes to 0.0983 h^{-1} if the same organism is used but the feed concentration is 3.00 g/l. Calculate the effluent substrate concentration when, in each case, the fermenter is operated at its maximum productivity.

5.5. Two continuous stirred-tank fermenters are placed in series such that the effluent of one forms the feed stream of the other. The first fermenter has a working volume of 100 l and the other has a working volume of 50 l. The volumetric flowrate through the fermenters is 18 l/h and the substrate concentration in the fresh feed is 5 g/l. If the microbial growth follows Monod kinetics with $\mu_m = 0.25\ h^{-1}$, $K_s = 0.12$ g/l, and the yield coefficient is 0.42, calculate the substrate and biomass concentrations in the effluent from the second vessel. What would happen if the flow were from the 50 l fermenter to the 100 l fermenter?

$(S = 0.0113\ g/l; \quad X = 2.095\ g/l)$

7.1. After being in use for some time, a pneumatic three-term controller (Fig. 7.118) develops a significant leak in the partition between the integral bellows and the proportional bellows. It is known that the rate of change of pressure in the integral bellows due to the leak is half that due to air flow through the integral restrictor. Show that the leak does not affect the form of the output response of the controller and that the ratio of the gain of the controller with the leak to that of the same controller before the leak developed is given by:

$$\frac{3\tau_2 + \tau_1}{2\tau_2 + \tau_1}$$

where τ_1 and τ_2 are the time constants of the integral and derivative restrictors respectively.

7.2. A mercury thermometer having first-order dynamics with a time constant of 60 s is placed in a bath at 35°C. After the thermometer reaches a steady state it is suddenly placed in a bath at 40°C at $t = 0$ and left there for 60 s, after which it is immediately returned to the bath at 35°C.

(a) Draw a sketch showing the variation of the thermometer reading with time.
(b) Calculate the thermometer reading at $t = 30$ s and at $t = 120$ s.
(c) What would be the reading at $t = 6$ s if the thermometer had only been immersed in the 40°C bath for less than 1 s before being returned to the 35°C bath?

7.3. A tank having a cross-sectional area of $0.2\ m^2$ is operating at a steady state with an inlet flowrate of $10^{-3}\ m^3/s$. Between the liquid heads z of 0.3 m and 0.09 m the flow–head characteristics are given by the equation:

$$Q_2 = 2Z + 0.0006$$

where Q_2 is the outlet flowrate
Determine the transfer functions relating (a) inflow and liquid level, (b) inflow and outflow.
If the inflow increases from 10^{-3} to $1.1 \times 10^{-3}\ m^3/s$ according to a step change, calculate the liquid level 200 s after the change has occurred.

7.4. A continuous stirred tank reactor is fed at a constant rate $F\ m^3/s$. The reaction occurring is:

$$\mathbf{A} \rightarrow \mathbf{B}$$

and proceeds at a rate:

$$\mathcal{R} = \boldsymbol{k} C_0$$

where $\mathcal{R}$ = moles **A** reacting/m^3 (mixture in tank) s,
$\boldsymbol{k}$ = reaction velocity constant, and
C_0 = concentration of **A** in reactor kmol/m^3.

If the density and volume V of the reaction mixture in the tank are assumed to remain constant, derive the transfer function relating the concentration of **A** in the reactor at any instant to that in the feed stream C_i. Sketch the response of C_0 to an impulse in C_i.

7.5. Liquid flows into a tank at the rate of $Q\ m^3/s$. The tank has three vertical walls and one sloping inwards at an angle β to the vertical. The base of the tank is a square with sides of length x m and the average operating level of liquid in the tank is z_0 m. If the relationship between liquid level and flow out of the tank at any instant is linear, develop a formula for determining the time constant of the system.

7.6. Write the transfer function for a mercury manometer consisting of a glass U-tube 0.012 m i.d., with a total mercury-column length of 0.54 m, assuming that the actual frictional damping forces are four times greater than would be estimated from Poiseuille's equation. Sketch the response of this instrument when it is subjected to a step change in an air pressure differential of 14,000 N/m^2 if the original steady differential was 5000 N/m^2. Draw the frequency-response characteristics of this system on a Bode diagram.

7.7. The response of an underdamped second-order system to a unit step change may be shown to be:

$$Y(t) = 1 - \frac{1}{\sqrt{(1-\zeta^2)}} \exp(-\zeta t/\tau) \left\{ \zeta \sin\left[\sqrt{(1-\zeta^2)}\,\frac{t}{\tau}\right] + \sqrt{(1-\zeta^2)} \cos\left[\sqrt{(1-\zeta^2)}\,\frac{t}{\tau}\right] \right\}$$

Prove that the overshoot for such a response is given by:

$$\exp\{-\pi\zeta/\sqrt{(1-\zeta^2)}\}$$

and that the decay ratio is equal to the $(\text{overshoot})^2$.

A forcing function, whose transform is a constant, K, is applied to an under-damped second-order system having a time constant of 0.5 min and a damping coefficient of 0.5. Show that the decay ratio for the resulting response is the same as that due to the application of a unit step function to the same system.

7.8. Air containing ammonia is contacted with fresh water in a two-stage countercurrent bubble-plate absorber.

L_n and V_n are the molar flowrates of liquid and gas respectively leaving the nth plate. x_n and y_n are the mole fractions of NH_3 in liquid and gas respectively leaving the nth plate. H_n is the molar holdup of liquid on the nth plate. Plates are numbered up the column.

(a) Assuming (i) temperature and total pressure throughout the column to be constant, (ii) no change in molar flowrates due to gas absorption, (iii) plate efficiencies to be 100 per cent, (iv) the equilibrium relation to be given by $y_n = mx_n^* + b$, (v) the holdup of liquid on each plate to be constant and equal to H, and (vi) the holdup of gas between plates to be negligible, show that the variations of the liquid compositions on each plate are given by:

$$\frac{dx_1}{dt} = \frac{1}{H}(L_2x_2 - L_1x_1) + \frac{mV}{H}(x_0 - x_1)$$

$$\frac{dx_2}{dt} = \frac{mV}{H}(x_1 - x_2) - \frac{1}{H}L_2x_2$$

where $V = V_1 = V_2$.

(b) If the inlet liquid flowrate remains constant, prove that the open-loop transfer function for the response of y_2 to a change in inlet gas composition is given by:

$$\frac{\bar{\mathscr{Y}}_2}{\bar{\mathscr{Y}}_0} = \frac{c^2/(a^2 - bc)}{\{1/(a^2 - bd)\}s^2 + \{2a/(a^2 - bc)\}s + 1}$$

where $\bar{\mathscr{Y}}_2, \bar{\mathscr{Y}}_0$ are the transforms of the appropriate deviation variables and

$$L = L_1 = L_2, \quad a = \frac{L}{H} + \frac{mV}{H}, \quad b = \frac{L}{H}, \quad c = \frac{mV}{H}$$

Discuss the problems involved in determining the relationship between $\mathscr{Y}$ and changes in inlet liquid flowrate.

7.9. A proportional controller is used to control a process which may be represented as two non-interesting first-order lags each having a time constant of 10 min. The only other lag in the closed loops is the measuring unit which can be approximated by a distance/velocity lag equal to 1 min. Show that, when the gain of proportional controller is set such that the loop is on the limit of stability, the frequency of the oscillation is given by:

$$\tan\omega = \frac{-20\omega}{1 - 100\omega^2}$$

7.10. A control loop consists of a proportional controller, a first-order control valve (time constant τ_v, gain K_v) and a first-order process (time constant τ_1, gain K_1). Show that when the system is critically damped the controller gain is given by:

$$K_C = \frac{(E-1)^2}{4EK_vK_1} \quad \text{where} \quad E = \frac{\tau_v}{\tau_1}$$

If the desired value is suddenly changed by an amount ΔR when the controller is set to give critical damping, show that the error ε will be given by:

$$\frac{\varepsilon}{\Delta R} = \frac{4E}{(1+E)^2} + \left\{\left[\frac{(1-E)^2}{2E(1+E)}\right]\frac{t}{\tau_1} + \frac{(1-E)^2}{(1+E)^2}\right\} \exp\left[-\left(\frac{1+E}{2E}\right)\frac{t}{\tau_1}\right]$$

7.11. A temperature-controlled polymerisation process is estimated to have a transfer function:

$$\mathbf{G}(s) = \frac{K}{(s-40)(s+80)(s+100)}$$

Show by means of the Routh–Hurwitz criterion that two conditions of controller parameters define upper and lower bounds on the stability of the feedback system incorporating this process.

7.12. A process is controlled by an industrial **PI** controller having the transfer function:

$$\mathbf{G}_C = \frac{\tau_I s^2 + (K_C + 2\tau_I)s + 2K_C}{2\tau_I^2 s}$$

The measuring and final control elements in the control loop are described by transfer functions which can be approximated by constants of unit gain, and the process has the transfer function:

$$\mathbf{G}_2 = \frac{1/\tau_2}{\tau_1\tau_2 s^2 - (2\tau_1 - \tau_2)s - 2}$$

If $\tau_1 = \frac{1}{2}$, show that the characteristic equation of the system is given by:

$$(s+2)\{\tau_2^2 s^2 + (1/\tau_I - 2\tau_2)s + K_C/\tau_I^2\} = 0$$

Show also that the condition under which this control loop will be stable:

(a) when

$$K_C \geqslant \left(\frac{1}{2\tau_2} - \tau_I\right)^2$$

(b) when

$$K_C < \left(\frac{1}{2\tau_2} - \tau_I\right)^2$$

is that $1/\tau_I > 2\tau_2$, provided that K_C and $\tau_2 > 0$.

7.13. The following information is known concerning a control loop of the type shown in Fig. 7.3*a*:

(a) the transfer function of the process is given by:

$$\mathbf{G}_2 = \frac{1}{0.1s^2 + 0.3s + 0.2}$$

(b) the steady-state gains of valve and measuring elements are 0.4 and 0.6 units respectively;
(c) the time constants of both valves and measuring element may be considered negligible.

It is proposed to use one of two types of controller in this control loop:
either (i) a **PD** controller whose action approximates to the relationship:

$$J = J_0 + K_C\varepsilon + K_D\frac{d\varepsilon}{dt}$$

where J is the output at time t. J_0 is the output $t = 0$, ε is the error, and K_C (proportional gain) = 2 units and $K_D = 4$ units;

or (ii) an *inverse* rate controller which has the action:

$$J = J_0 + K_C\varepsilon - K_D\frac{d\varepsilon}{dt}$$

(J, J_0, ε, K_C and K_D having the same meaning and values as above.)

Basing your judgement only on the amount of offset obtained when a step change in load is made, which of these two controllers would you recommend? Would you change your mind if you took the system stability into account?

Having regard to the form of the equation describing the control action in each case, under what general circumstances do you think inverse rate control would be better than normal **PD** control?

7.14. A proportional plus integral controller is used to control the level in the reflux accumulator of a distillation column by regulating the top product flowrate. At time $t = 0$ the desired value of the flow controller which is controlling the reflux is increased by 3×10^{-4} m^3/s. If the integral action time of the level controller is half the value which would give a critically damped response and the proportional band is 50 per cent, obtain an expression for the resulting change in level.

The range of head covered by the level controller is 0.3 m, the range of top product flowrate is 10^{-3} m^3/s and the cross-sectional area of the accumulator is 0.4 m^2. It may be assumed that the response of the flow controller is instantaneous and that all other conditions remain the same.

If there had been no integral action, what would have been the offset in the level in the accumulator?

7.15. Draw the Bode diagrams of the following transfer functions:

$$\text{(i)} \quad \mathbf{G}(s) = \frac{1}{s(1 + 6s)}$$

$$\text{(ii)} \quad \mathbf{G}(s) = \frac{(1 + 3s)\exp(-2s)}{s(1 + 2s)(1 + 6s)}$$

$$\text{(iii)} \quad \mathbf{G}(s) = \frac{5(1 + 3s)}{s(s^2 + 0.4s + 1)}$$

Comment on the stability of the closed-loop systems having these transfer functions.

7.16. The transfer function of a process and measuring element connected in series is given by:

$$(2s + 1)^{-2}\exp(-0.4s)$$

(a) Sketch the open-loop Bode diagram of a control loop involving this process and measurement lag (but without the controller).

(b) Specify the maximum gain of a proportional controller to be used in this control system without instability occurring.

7.17. A control system consists of a process having a transfer function $\mathbf{G}_2$, a measuring element $\mathbf{H}$ and a controller $\mathbf{G}_C$.

If $\mathbf{G}_2 = (3s + 1)^{-1}\exp(-0.5s)$ and $\mathbf{H} = 4.8(1.5s + 1)^{-1}$, determine, using the method of Ziegler and Nichols, the controller settings for **P**, **PI**, **PID** controllers.

7.18. Determine the open-loop response of the output of the measuring element in problem 7.17 to a unit step change in input to the process. Hence determine controller settings for the control loop by the Cohen–Coon and ITAE methods for **P**, **PI** and **PID** control actions. Compare the settings obtained with those in problem 7.17.

7.19. A continuous process consists of two sections A and B. Feed of composition X_1 enters section A where it is extracted with a solvent which is pumped at a rate L_1 to A. The raffinate is removed from A at a rate L_2 and the extract is pumped to a cracking section B. Hydrogen is added at the cracking stage at a rate L_3 whilst heat is supplied at a rate Q. Two products are formed having compositions X_3 and X_4. The feed rate to A and L_1 and L_2 can easily be kept constant, but it is known that fluctuations in X_1 can occur. Consequently a feed-forward control system is proposed to keep X_3 and X_4 constant for variations in X_1 using L_3 and Q as controlling variables. Experimental frequency response analysis gave the following transfer functions:

$$\frac{\bar{x}_2}{\bar{x}_1} = \frac{1}{s + 1}, \quad \frac{\bar{x}_3}{\bar{l}_3} = \frac{2}{s + 1}$$

$$\frac{\bar{x}_3}{\bar{q}} = \frac{1}{2s + 1}, \quad \frac{\bar{x}_4}{\bar{l}_3} = \frac{s + 2}{s^2 + 2s + 1}$$

$$\frac{\bar{x}_4}{\bar{q}} = \frac{1}{s + 2}, \quad \frac{\bar{x}_3}{\bar{x}_2} = \frac{2s + 1}{s^2 + 2s + 1}$$

$$\frac{\bar{x}_4}{\bar{x}_2} = \frac{1}{s^2 + 2s + 1}$$

where $\bar{x}_1$, $\bar{x}_2$, $\bar{l}_3$, $\bar{q}$, etc., represent the transforms of small time dependent perturbations in X_1, X_2, L_3, Q, etc.

Determine the transfer functions of the feed-forward control scheme assuming linear operation and negligible distance–velocity lag throughout the process. Comment on the stability of the feed-forward controllers you design.

7.20. The temperature of a gas leaving an electric furnace is measured at X by means of a thermocouple. The output of the thermocouple is sent, via a transmitter, to a two-level solenoid switch which controls the power input to the furnace. When the outlet temperature of the gas falls below 400°C the solenoid switch closes and the power input to the furnace is raised to 20 kW. When the temperature of the gas falls below 400°C the switch opens and only 16 kW is supplied to the furnace. It is known that the power input to the furnace is related to the gas temperature at X by the transfer function:

$$\mathbf{G}(s) = \frac{8}{(1 + s)(1 + s/2)(1 + s/3)}$$

The transmitter and thermocouple have a combined steady-state gain of 0.5 units and negligible time constants. Assuming the solenoid switch to act as a standard on–off element determine the limit of the disturbance in output gas temperature that the system can tolerate.

7.21. A gas-phase exothermic reaction takes place in a tubular fixed-bed catalytic reactor which is cooled by passing water through a coil placed in the bed. The composition of the gaseous product stream is regulated by adjustment of the flow of cooling water and, hence, the temperature in the reactor. The exit gases are sampled at a point X downstream from the reactor and the sample passed to a chromatograph for analysis. The chromatograph produces a measured value signal 10 min after the reactor outlet stream has been sampled at which time a further sample is taken. The measured value signal from the chromatograph is fed via a zero-order hold element to a proportional controller having a proportional gain K_C. The steady-state gain between the product sampling point at X and the output from the hold element is 0.2 units. The output from the controller J is used to adjust the flow Q of cooling water to the reactor. It is known that the time constant of the control valve is negligible and that the steady-state gain between J and Q is 2 units. It has been determined by experimental testing that Q and the gas composition at X are related by the transfer function:

$$\mathbf{G}_{xq}(s) = \frac{0.5}{s(10s + 1)} \quad \text{(where the time constant is in minutes)}$$

What is the maximum value of K_C that you would recommend for this control system?

7.22. A unity feedback control loop consists of a non-linear element **N** and a number of linear elements in series which together approximate to the transfer function:

$$\mathbf{G}(s) = \frac{3}{s(s + 1)(s + 2)}$$

Determine the range of values of the amplitude x_0 of an input disturbance for which the system is stable where (a) **N** represents a dead-zone element for which the gradient k of the linear part is 4; (b) **N** is a saturating element for which $k = 5$; and (c) **N** is an on–off device for which the total change in signal level is 20 units.

Conversion Factors for Some Common SI Units

An asterisk (*) denotes an exact relationship.

Quantity	Unit		SI equivalent
Length	1 in	:	25.4 mm
	1 ft	:	0.304 8 m
	1 yd	:	0.914 4 m
	1 mile	:	1.609 3 km
	*1 Å (angstrom)	:	10^{-10} m
Time	*1 min	:	60 s
	*1 h	:	3.6 ks
	*1 day	:	86.4 ks
	1 year	:	31.5 Ms
Area	1 in.^{2}	:	645.16 mm^2
	1 ft^2	:	0.092 903 m^2
	1 yd^2	:	0.83613 m^2
	1 mile^2	:	2.590 km^2
	1 acre	:	4046.9 m^2
Volume	1 in.^{3}	:	16.387 cm^3
	1 ft^3	:	0.028 32 m^3
	1 yd^3	:	0.764 53 m^3
	1 UK gal	:	4546.1 cm^3
	1 US gal	:	3785.4 cm^3
Mass	1 oz	:	28.352 g
	*1 lb	:	0.453 592 37 kg
	1 cwt	:	50.802 3 kg
	1 ton	:	1016.06 kg
Force	1 pdl	:	0.138 26 N
	1 lbf	:	4.448 2 N
	*1 kgf	:	9.806 65 N
	1 tonf	:	9.9640 kN
	*1 dyn	:	10^{-5} N
Temperature difference	*1 deg F (deg R)	:	5/9 deg C (deg K)
Energy (work, heat)	1 ft lbf	:	1.355 8 J
	1 ft pdl	:	0.042 14 J
	*1 cal (international table)	:	4.1868 J
	*1 erg	:	10^{-7} J
	1 Btu	:	1.055 06 kJ
	1 hp h	:	2.684 5 MJ
	*1 kWh	:	3.6 MJ
	1 therm	:	105.51 MJ
	1 thermie	:	4.185 5 MJ
Calorific value (volumetric)	1 Btu/ft^3	:	37.259 kJ/m^3

Velocity	1 ft/s	: 0.3048 m/s
	1 mile/h	: 0.447 04 m/s
Volumetric flow	1 ft^3/s	: 0.028 316 m^3/s
	1 ft^3/h	: 7.865 8 cm^3/s
	1 UK gal/h	: 1.262 8 cm^3/s
	1 US gal/h	: 1.0515 cm^3/s
Mass flow	1 lb/h	: 0.126 00 g/s
	1 ton/h	: 0.282 24 kg/s
Mass per unit area	1 lb/in.^{2}	: 703.07 kg/m^2
	1 lb/ft^2	: 4.882 4 kg/m^2
	1 ton/sq mile	: 392.30 kg/km^2
Density	1 lb/in^3	: 27.680 g/cm^3
	1 lb/ft^3	: 16.019 kg/m^3
	1 lb/UK gal	: 99.776 kg/m^3
	1 lb/US gal	: 119.83 kg/m^3
Pressure	1 lbf/in.^{2}	: 6.8948 kN/m^2
	1 tonf/in.^{2}	: 15.444 MN/m^2
	1 lbf/ft^2	: 47.880 N/m^2
	*1 standard atm	: 101.325 kN/m^2
	*1 atm (1 kgf/cm^2)	: 98.066 5 kN/m^2
	*1 bar	: 10^5 N/m^2
	1 ft water	: 2.989 1 kN/m^2
	1 in. water	: 249.09 N/m^2
	1 in. Hg	: 3.386 4 kN/m^2
	1 mm Hg (1 torr)	: 133.32 N/m^2
Power (heat flow)	1 hp (British)	: 745.70 W
	1 hp (metric)	: 735.50 W
	*1 erg/s	: 10^{-7} W
	1 ft lbf/s	: 1.355 8 W
	1 Btu/h	: 0.293 07 W
	1 ton of refrigeration	: 3516.9 W
Moment of inertia	1 lb ft^2	: 0.042 140 kg m^2
Momentum	1 lb ft/s	: 0.138 26 kg m/s
Angular momentum	1 lb ft^2/s	: 0.042 140 kg m^2/s
Viscosity, dynamic	*1 P (poise)	: 0.1 N s/m^2
	1 lb/ft h	: 0.413 38 mN s/m^2
	1 lb/ft s	: 1.488 2 N s/m^2
Viscosity, kinematic	*1 S (stokes)	: 10^{-4} m^2/s
	1 ft^2/h	: 0.258 06 cm^2/s
Surface energy (surface tension)	*1 erg/cm^2	: 10^{-3} J/m^2
	*(1 dyn/cm)	: (10^{-3} N/m)
Mass flux density	1 lb/h ft^2	: 1.356 2 g/s m^2
Heat flux density	1 Btu/h ft^2	: 3.154 6 W/m^2
	*1 kcal/h m^2	: 1.163 W/m^2
Heat transfer coefficient	1 Btu/h ft^2 °F	: 5.678 3 W/m^2 K
Specific enthalpy (latent heat, etc.)	1 Btu/lb	: 2.326 kJ/kg
Specific heat capacity	1 Btu/lb °F	: 4.186 8 kJ/kg K
Thermal conductivity	1 Btu/h ft °F	: 1.730 7 W/mK
	*1 kcal/h m °C	: 1.163 W/mK

Taken from MULLIN, J. W.: *The Chemical Engineer* No. 211 (Sept. 1967), 176. SI units in chemical engineering.

Index

This index is intended primarily to direct the reader to entries in this volume, but also contains some cross-references to material in other volumes. Such cross-references will be indicated by the volume number printed in bold Roman numerals. Thus, for example, the entry:

Cascade control 573, 645, 652, **VI**

indicates that reference to that subject will be found on pages 573, 645, and 652 of this volume, and that further such material will be found in Volume 6. Note that the cross-references to other volumes do not include a page number. For this see the index in the volume specified.